ADVANCES IN
MASS SPECTROMETRY

VOLUME 6

535.336.2

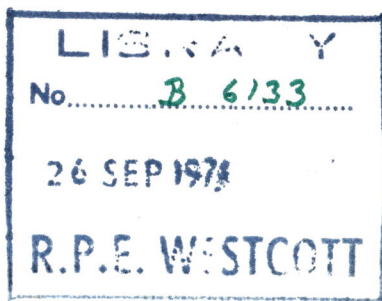
B6133

Rocket Propulsion Establishment
Library

Please return this publication, or request renewal, before
the last date stamped below.

Name	Date 3 days
R G WILLIAMS new	8 . 1 . 75
G J Williams	7 . 2 . 75

CANCELLED

RPE Form 243 (revised 6/71)   739490

ADVANCES IN MASS SPECTROMETRY

VOLUME 6

Proceedings of a Conference held in Edinburgh

Edited by

A. R. WEST

BP Trading Limited, BP Research Centre, Chertsey Road,
Sunbury-on-Thames, Middlesex TW16 7LN, U.K.

APPLIED SCIENCE PUBLISHERS LTD
BARKING, ESSEX

on behalf of

THE INSTITUTE OF PETROLEUM
LONDON

1974

APPLIED SCIENCE PUBLISHERS LTD
RIPPLE ROAD, BARKING, ESSEX, ENGLAND

The symbol IP on this book means that the text has been officially accepted as authoritative by the Institute of Petroleum, Great Britain.

ISBN 0 85334 459 0

WITH 453 ILLUSTRATIONS AND 142 TABLES

© APPLIED SCIENCE PUBLISHERS LTD 1974

Printed in Great Britain by Galliard (Printers) Ltd Great Yarmouth

Contents

ORGANIC FRAGMENTATION MECHANISMS

NEGATIVE IONS

INSTRUMENTS AND TECHNIQUES

INORGANIC APPLICATIONS

METASTABLES

THEORY AND FUNDAMENTALS

COMPUTER SOFTWARE

Organizing Committees

The Conference was sponsored by I.U.P.A.C. and supported by the Royal Society, A.G.M.S./D.P.G. (Germany), A.S.M.S. (U.S.A.), G.A.M.S. (France), H.R.G./I.P. (U.K.), M.S.S.J. (Japan), and other countries as indicated below.

INTERNATIONAL COMMITTEE

A. Quayle (*Chairman*).	U.K.
Dr T. Ast .	Yugoslavia
Prof. J. H. Beynon	U.K.
Dr ir A. J. H. Boerboom	Netherlands
Dr V. Cermak .	Czechoslovakia
Prof. J. Drowart	Belgium
Dr S. Facchetti .	Italy
Dr F. H. Field .	U.S.A.
Prof. Dr H. Fr. Grützmacher	Germany
Prof. J. B. Hasted	U.K.
Prof. Dr J. Kistemaker	Netherlands
C. H. Maynard .	U.K.
Prof. K. Ogata .	Japan
Prof. E. Roth	France
Prof. E. Stenhagen	Sweden
Prof. V. L. Tal'rose .	U.S.S.R.
A. R. West (Honorary Editor)	U.K.

NATIONAL COMMITTEE

A. Quayle (*Chairman*) .	Shell Research Ltd
Prof. J. H. Beynon	Imperial Chemical Industries Ltd
Mrs J. Cummings	Institute of Petroleum
Dr P. A. Cutting	British Gas
Dr R. J. Donovan	Edinburgh University
C. H. Maynard .	Institute of Petroleum
W. L. Mead	British Petroleum Company Ltd
S. F. Noel	Esso Petroleum Company Ltd
Prof. F. M. Page	University of Aston in Birmingham
Dr J. E. Parker .	Heriot-Watt University, Edinburgh
Dr J. M. Price .	Esso Petroleum Company Ltd
Dr R. I. Reed .	Glasgow University
Dr W. Sneddon .	St Bartholomew's Hospital
A. R. West	British Petroleum Company Ltd
Dr D. A. Whan .	Edinburgh University

Chairmen of Plenary Lectures

Professor J. DROWART
Vrije University, Brussels, Belgium

Professor H. D. GRUTZMACHER
University of Hamburg, Germany

Dr J. S. HALLIDAY
A.E.I., Manchester, U.K.

Professor J. B. HASTED
Birkbeck College, London, U.K.

Dr R. E. HONIG
R.C.A. Laboratories, Princeton, U.S.A.

Professor K. R. JENNINGS
University of Warwick, Coventry, U.K.

Dr R. S. LEHRLE
University of Birmingham, U.K.

Professor F. M. PAGE
University of Aston in Birmingham, U.K.

Professor E. ROTH
CEN Saclay, France

Professor H. J. SVEC
Iowa State University, Ames, U.S.A.

Foreword

It was a pleasure to note that the sixth in this series of international mass spectrometry conferences lived up to the standards set previously in London,[1] Oxford,[2] Paris,[3] Berlin[4] and Brussels,[5] and that we were honoured by the presence of about 520 delegates (with a further hundred, or so, ladies) from 30 different countries. Edinburgh was a delightful city in which to hold our conference and we were much indebted to the City Corporation for their help over conference arrangements, as well as the reception they so kindly gave us. The universities of the City—the ancient foundation, the University of Edinburgh, and the new one, Heriot-Watt University—gave us much support and I would particularly thank Professor Ebsworth for his welcoming remarks and Dr Donovan, Dr Parker and Dr Whan for their assistance with the local arrangements.

A conference of this kind depends entirely on the goodwill and energies of many people. We are very grateful to the International Scientific Committee for their support and advice throughout, to the referees of many countries who read all the Abstracts and advised us on the programme and to the Chairmen of the various sessions at the conference itself. I would single out for special mention Mr W. L. Mead who worked so hard and displayed so much ingenuity in getting the exhibition open on time. To the exhibitors go our thanks for what I think was the best exhibition I have seen at any of these conferences. The Appleton Tower proved an excellent location and we were most impressed by the help given so readily by the permanent staff there. As usual, the staff of the Institute of Petroleum worked long and hard over administrative aspects of the conference. We are particularly indebted to Mr C. H. Maynard, Assistant General Secretary (Administration) of the

1. 'Advances in Mass Spectrometry', Vol. 1, Ed. J. D. Waldron, Pergamon Press, London, 1959.
2. 'Advances in Mass Spectrometry', Vol. 2, Ed. R. M. Elliot, Pergamon Press, London, 1963.
3. 'Advances in Mass Spectrometry', Vol. 3, Ed. W. L. Mead, Applied Science Publishers, London.
4. 'Advances in Mass Spectrometry', Vol. 4, Ed. E. Kendrick, Applied Science Publishers, London.
5. 'Advances in Mass Spectrometry', Vol. 5, Ed. A. Quayle, Applied Science Publishers, London.

Institute for his efforts in the three years prior to the conference as well as in Edinburgh itself. To Mr A. R. West, Honorary Editor of these Proceedings, I express my personal thanks, not only for the far from negligible task of editing this volume, but for his work and support in arranging the technical programme for the whole conference.

Finally, I should like to thank the authors of the Review Papers and of the Contributed Papers, without whom there would have been no conference, and the delegates who supported the speakers so strongly. It was indeed a pleasure to note 'capacity houses' on the last afternoon with an attendance of over 300. The technical content of this book will speak for itself, but it is interesting to note that 23 years after the first mass spectrometry conference in Europe (held in Manchester in 1950) at least 26 of the 126 contributed papers were devoted to new instruments and techniques. Analytical or straightforward organic applications proved less popular, but there was a growing interest in biochemical applications and in the study of the decompositions of metastable ions. Revived interest was displayed in ion-molecule reactions and, with the increasing availability and power of computers, many papers described their application as interpretative or as library search instruments, as well as simply data collectors. The organizing committee were pleased to include a somewhat unusual paper, in which a sociologist looked at the early history of mass spectrometry and the effect the personalities of some of the pioneers had on the development of the science: it would have been interesting, if perhaps more dangerous, had he felt able to extend his comments into more recent times. Also included within the programme was a progress report from the Mass Spectrometry Data Centre, Aldermaston (and a demonstration of the 'live' interrogation of a data file in Maryland, U.S.A., via communications satellite) and the opportunity was taken to have an ad hoc international committee meeting on data needs.

To all who contributed to the success of the Edinburgh International Mass Spectrometry Conference, our sincere thanks, and the hope that we may meet again in Italy in 1976.

Shell Research Limited   A. QUAYLE
Chester   (*Chairman, Mass Spectrometry Panel,*
                                                       Hydrocarbon Research Group,
                                                            Institute of Petroleum)

Welcoming Address

By E. A. V. EBSWORTH

(Chemistry Department, Edinburgh University, U.K.)

We are honoured that Edinburgh has been chosen to stand with London, Oxford, Paris, Berlin and Brussels as host city for this International Mass Spectrometry Conference. The importance of mass spectrometry, and the range of topics on which it has had an important influence, grows at a great pace, and international gatherings like this one are invaluable as ganglia from which ideas and collaboration can spread in new and sometimes unexpected ways. We very much hope that you will be able to make full and effective use of your time here both for scientific and for other purposes. To judge from your programme, with its long list of internationally distinguished contributors, the first half of this double hope will certainly be fulfilled. I am not myself a mass spectroscopist, and it would be impertinent for me to try to make any serious scientific comment about the subject, but I am astonished at the range of topics that will be discussed here during the next few days, and delighted to see contributors from all over the world, even from places as far apart as Japan and Brazil.

In the scientific context, too, I owe you all something of an apology. You may well wonder why the conference is housed in this building, which is clearly not a Chemistry Department; you will rightly assume that a University so large by British standards as Edinburgh must have a big Chemistry Department, and you may possibly look around for it. If so, unless you are very energetic you will look in vain, for the Chemistry laboratory is some 5 km away. It is not as new as this building, either. The first Chair of Chemistry was established some 150 years after the foundation of the University, in 1713, and among early Professors of Chemistry was Joseph Black. In terms of such a timescale, our present Chemistry building is modern, but in fact it was completed 50 years ago. You will not be surprised to hear, therefore, that it is in the middle of a process of rejuvenation that is long-drawn-out and expensive. All the ground floor corridors are blocked, and it is in no state to be shown to a conference like yours. None the less, we do extend a very warm welcome to you, and we hope you will understand why we cannot show you more of what we do ourselves.

As for the less formally scientific aspects of your visit, Edinburgh is of course an international city and so well suited for an international conference. I am sure you will have a chance to see the castle and the Palace of Holyroodhouse, and the wide streets and terraces of the New Town; but try to find time as well to see the little courts and wynds off the Royal Mile, and to visit the Museum of Scottish Antiquities. Finally, if you have a clear evening the view from the top of Arthur's Seat is remarkable.

All in all, I am confident that you will take away with you pleasant memories at the end of the week, and perhaps some examples of the distilled essence of Scottish culture as well.

Ionization Processes at Low Energies

By R. STEPHEN BERRY

(*Theoretical Chemistry Department, University of Oxford, U.K.*)†

OUR understanding of how atoms and molecules become ions has deepened a great deal during the past decade. Ten or fifteen years ago we had clear qualitative ideas about some of the mechanisms of ionization, but almost all the quantitative work, especially for molecules, was phenomenological. The quasi-equilibrium theory[1] of fragmentation, a statistical theory, was the frontier method. Now, with the contributions of the intervening years, we look for answers at a different level. Instead of phenomenological correlations, we look for microscopic interpretations of mechanisms. We still use statistical models, and always shall, but these models increasingly use more specific information about the flow of energy among the atoms and electrons of our systems.

Let us look at the picture we now have of ionization at low energies, trying to fit interpretations of many specific processes together into an organized structure. To do this, we must first create a sort of taxonomy. Then we shall go back through this taxonomy and see what new directions it suggests.

I. TAXONOMY

Ionization of atoms and molecules, by which we mean loss of one or more electrons, is one of the three irreversible modes by which an electronically excited system can decay. The other two are dissociation and radiation. Radiation is the slowest of these processes, in the sense that its fastest limit is only about 10^9 sec^{-1}, when we restrict our discussion to energies below about 50 eV. Dissociation is intermediate; its time scale is that of molecular vibration periods, so the upper limit for dissociation rates in individual molecules is about 10^{13} sec^{-1}. Ionization is the fastest; its time scale has, as an upper limit, times of the order of the Bohr period, 10^{-15}–10^{-16} sec. Thus, if ionization is an energetically allowed competitor with radiation or dissociation, ionization usually wins. The time scales of these processes are limited by the characteristic frequencies associated with molecular or atomic spectral line widths, with nuclear vibration, and with the motion of electrons about nuclei.

† Permanent Address: Department of Chemistry and the James Franck Institute, University of Chicago, Chicago, Illinois 60637, U.S.A.

To interpret ionization, it is useful to classify the processes and mechanisms in several different ways. As we go through this classification, we can see how ionization occurs and what factors govern its occurrence.

A. Direct and Indirect Processes

The first distinction we make is between direct and indirect processes. The distinction is somewhat artificial, but extremely useful. It was a distinction much emphasized by my colleague, the late Robert Platzman.[2] Indirect processes, from the most simplistic viewpoint, involve two successive steps, a rapid step of excitation to a state that lives long enough to be characterized, followed by a slower step in which an electron goes into the continuum. Direct processes involve no such intermediate state. A stricter description puts it this way: all ionization processes involve a continuous spectrum of final states, characterized by a cross section that depends on energy. When this cross section exhibits peaks or troughs that are narrow, relative to the energy of the peak or trough itself, then we attribute the peak or trough to a transient compound state whose half-life is comparable to the width—as governed by the uncertainty principle. The compound state can be interpreted as a mixture of bound and continuum states, whenever we are clever enough to think of a way to represent the bound component.

There is a way to think systematically about such transient bound states. The final state, either at times long after excitation or at distances far from the atomic core, is dominated by the ionization continuum: a free electron moving away from a positive core. The active electron has energy sufficient to escape from the Coulomb attraction of the core. By the 'bound component' of the final state we mean the component in which the many-body system contains sufficient energy for ionization but in which insufficient energy is available directly to the active electron. Part of the ionization energy must be stored in some other degree of freedom. The interpretation of indirect ionization processes must begin with identifying the degree (or degrees) of freedom in which the energy is stored. We do this, for example, when we represent the bound state in a particular way. Then, the next step is identifying the coupling mechanism which allows energy to flow from its storage site to the active electron.

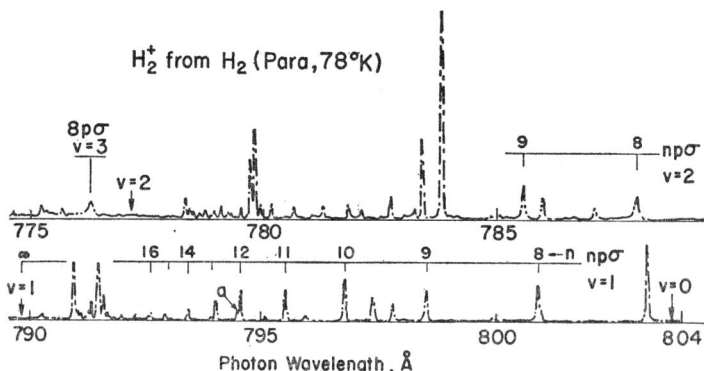

FIG. 1   Intensity of mass-resolved H_2^+ from photoionization of H_2, as a function of the ionizing radiation (taken from ref. 2, with permission).

Indirect processes are sometimes at least as important as direct processes. The photoionization of H_2 illustrates the importance of the indirect processes. Figure 1 is a record of the H_2^+ photoion intensity as a function of the wavelength of ionizing ultraviolet radiation.[3] The area under the peaks is the measure of the contribution of the indirect process of autoionization or preionization: this is clearly at least as great as the contribution from direct photoionization, the part associated with the smooth background between peaks.

B. Mode of Excitation[4]

A second way to classify ionization processes is according to the agent that introduces the required energy. The agents of interest here are electromagnetic radiation, electrons, heavy positive ions and heavy neutrals; the last two are very similar in most of their modes of action. If the process is ordinary ultraviolet photoionization, the coupling is the first-order radiative term, the electric dipole approximation to the operator $\mathbf{p} \cdot \mathbf{A}$ or $\mu \cdot \mathbf{E}$. If the process is high-energy photoionization, we may not be able to use the electric dipole approximation, but have to use the full generalized oscillator form in which we take explicit account of the momentum transferred from photon to electron.

Electron impact ionization can be approximated by the electric dipole approximation in some situations, especially when the Coulombic perturbation by the passing electron has Fourier components in the range of the ionization energy and when the momentum of the passing electron is almost unchanged. When the passing electron transfers a non-negligible amount of momentum to the newly ionized electron (and I am deliberately speaking loosely in classical terms, for the sake of simplicity), the electric dipole approximation does not suffice. This phenomenon is exhibited beautifully by the process of electron impact *excitation*. Forward-inelastically-scattered electrons are associated primarily with excitation of optically allowed—*i.e.* electric-dipole allowed—transitions. Electrons scattered inelastically through larger angles are associated with the excitation of 'forbidden' transitions, either spin-allowed multipole transitions or spin-forbidden transitions, in which electron exchange plays a dominant role. We can look upon the forward-scattering processes as due primarily to distant collisions in which the dipole part of the transient field of the passing electron is the most important part. The large-angle inelastic scattering involves close collisions in which the transient electric field exhibits large gradients over the target.

The same reasoning that applies to electron impact excitation also applies to electron impact ionization, but with a few differences. Electronic excitation to bound states is of course a resonant process; when ionization is allowed energetically, it is allowed to states of all symmetries at all energies. Hence, direct ionization by multipole processes higher than electric dipole is always allowed but the dipole part dominates the other contributions to the total cross section because it provides the strongest interaction for the distant collisions, which are, statistically, the most important. The higher processes occur, and must be included if we want to describe the angular distribution of outgoing electrons, and not just the total probability of ionization.

Impact ionization by heavy ions differs somewhat from ionization by either radiation or electron impact. The positive charge of the ions (the usual case) makes possible a whole set of continuum states that begin with the limit of the Rydberg series of the *projectile*. Hence, there is a whole class of final states of the free electron that move forward with a velocity approximately equal to that of the projectile. These produce a 'forward peak' in the angular distribution of the outgoing electrons that is quite different from anything in the ultraviolet photoeffect or low-energy electron impact ionization.†

The second characteristic of heavy ion collisions is shared by collisions of neutrals. This is the well-known adiabatic nature of the interaction: the relative velocity of the heavy particles is so low that the electronic states adjust almost as adiabatic functions of the interparticle distance. This means that the coupling perturbation in this case should not be thought of as a transient electromagnetic field with Fourier components in some characteristic frequency range of the isolated target. Rather, we must think of such cases in terms of potential curves or surfaces of the compound system, and of the transitions of the compound system among these surfaces. This is the context in which we start to interpret associative ionization, Penning ionization and chemi-ionization.

C. Energy Source Mode

We have still two other useful ways to distinguish among ionization processes, both of which find their application in somewhat limited areas. One applies to indirect processes and to ionization by heavy particles: This is classification according to the degree of freedom from which the requisite energy for ionization is drawn. Is the energy taken from electronic excitation, as in Penning ionization or an Auger process, or it is taken from energy of nuclear motion? If nuclear motion is the source, is it the rotational motion or the radial motion of the system that couples with electronic excitation to generate an electron-ion pair? This method of classification is important because this is where we decide what dynamical picture to use for our theoretical interpretations. Each choice of a degree of freedom implies a particular coupling mechanism, and therefore a particular set of propensity rules (and corresponding hierarchy of transition probabilities), and a particular operator and representation for computations.

D. Interaction Range

The last of the classification schemes applies also to heavy-particle collisions and indirect processes, but, as we have already indicated, is also useful for interpreting some aspects of ionization by electron impact.[5-7] This is interpretation according to the range of interparticle separation most responsible for the ionization process. Classification according to interaction range differs from the other forms of classification insofar as this interpretation is microscopic, rather than observational. It necessarily leans more heavily than the others on a theoretical interpretation. The utility of this interpretation is in its power to rationalize the distribution of energy in the final products and to answer these questions: what causes the observed distribution of

† This was pointed out to me by Yehuda Band.

electron energies and angular distributions, and, if more than one heavy particle occurs in the final state, what determines the relative kinetic energy and angular momentum distributions of the heavy particles?

The classification of ionizing processes according to interaction range appears to have three natural categories, as illustrated in Fig. 2. The first is the 'crossing-point' case, in which transitions between curves or surfaces occur where the surfaces have a common energy. The second is the classical turning point case, in which the transition occurs in the neighbourhood where the relative velocity of colliding particles (or vibrating particles) is almost zero.

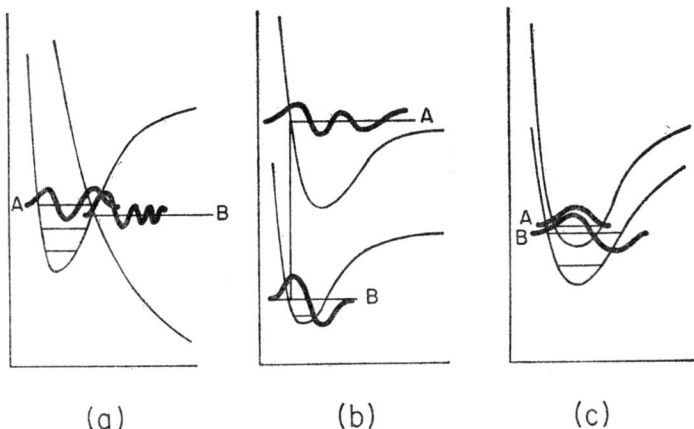

(a)   (b)   (c)

FIG. 2   The three cases for classification according to interaction range: (a) the crossing point case; (b) the classical turning point case; (c) the broad range case. The examples shown here are determined by the overlap of the initial and final vibrational wave functions (heavy lines) but the category could, in principle, be governed by behaviour of the electronic part of the transition amplitude. This occurs in the case of autoionization of the $n\sigma_g$ Rydberg states of H_2, where the Rydberg wave function changes character from s-like to d-like in a narrow interval.

The third case is the broad-range case, in which transition amplitude accumulates over a wide interval, so that evaluation of this amplitude at one internuclear separation would almost certainly give a large error. The first case is illustrated by predissociation and probably the associative ionization of $N + O$.[8] The second case seems to describe some cases of Penning and associative ionization, particularly certain channels with heavy particles such as $Hg + Ar^*$. The vibrationally induced autoionization of some states of H_2 are examples of the third case.[9]

How transition amplitude accumulates is illustrated in Fig. 3. This figure shows the value of the definite integral $\int_0^R \varphi^*_{initial} \mathscr{H}_{coupling} \varphi_{final} \, dR$. When the upper limit is carried to large interparticle separation, this is the transition amplitude, in this case for autoionization of H_2^*. In one case illustrated, the amplitude accumulates near the left turning point, around $R = 2$ a.u. In the other two cases, the amplitude accumulates everywhere, oscillating until the outer turning point of the initial bound state is well passed.

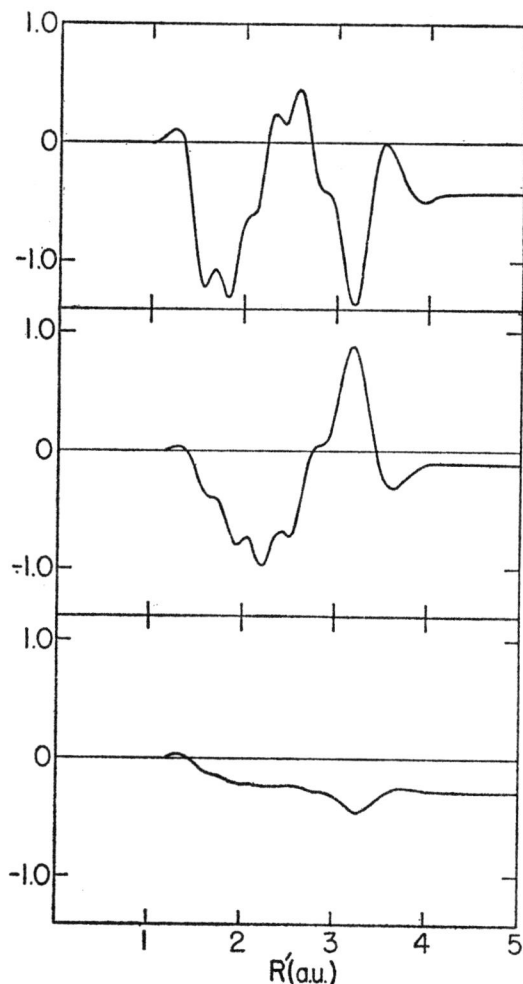

FIG. 3 Examples of the transition amplitude integral $\int_0^R \varphi^*_{\text{initial}} \mathscr{H}_{\text{coupling}} \varphi_{\text{final}} \, dR$, as functions of the upper limit of the integral. The upper two examples are broad-range cases, the third illustrates the classical turning point case.

II. IMPLICATIONS

A. From Direct versus Indirect Process

With this four-variable taxonomy, we can start to get a coherent picture of the processes of ionization at low energies. The indirect processes, we find, are particularly important for energies within the first two or three electron volts of the ionization threshold, especially when the excitation occurs by radiation or electron impact. Indirect processes occur frequently at higher energies—viz. the famous doubly excited autoionizing states of the helium atom—but at energies more than a few electron volts above threshold, direct ionization carries a large density of oscillator strength or generalized oscillator strength, so that the resonances associated with indirect processes may tend to have relatively little effect on the total average probability of ionization. In reality, we do not know the relative importance of direct and indirect ionization processes well above threshold, but we do know examples that appear to involve indirect processes. Near the ionization threshold, however, it is already clear that a large part, sometimes the overwhelming part of the ionization probability is associated with indirect processes. Near the first continuum edge, it is much easier for an electron to find a bound excited level than a continuum level in many atoms and molecules; if the *total* energy in this bound excited state includes enough other internal energy, then the state of the entire molecular system lies in the ionization continuum. If coupling to the ionization continuum is stronger than coupling to the radiation continuum (*i.e.* emission of a photon), then ionization will occur. It should be possible now to predict the average relative importance of direct and indirect ionization processes for many small molecules. Smith[10] has demonstrated how one can do this, incorporating the nature of the line shapes for autoionization. We first estimate the total electronic oscillator strength, per unit energy, for the electronically bound states and the continuum. We then divide the electronic oscillator strength among specific transitions to rotation–vibration states, according to their Franck–Condon and Hönl–London factors. Then, in each energy band, whose width ought to be great enough to include several rovibronic lines if we are really interested in overall average distributions of probability, we simply compare the total integrated oscillator strength for the continuum with that for the bound states. The work is tedious but ideally suited for a computer. (Strictly, one more factor should enter, namely the fraction of decay of each quasibound excited state that leads to ionization; this is actually important to do sometimes when ionization competes with dissociation, but in our own experience, ionization always wins over radiation.)

The reasons for distinguishing direct and indirect processes go beyond estimating the total probability of ionization. Electrons set free by indirect processes generally have very different energy distributions (as shown, for example, by Berkowitz and Chupka[11]) and angular distribution (such as those analyzed by Dill[12]) than directly ionized electrons. Hence, we need to know whether we are dealing with direct or indirect processes when we attempt to collect and analyse the electrons from a low-energy process. This becomes especially important in the high-resolution threshold processes such as those studied by Peatman *et al.*,[13] or by Villarejo *et al.*[14]

B. From Agent of Excitation

We have already pointed out some of the distinctions coming from ionization by different agents. An area of molecular ionization studies that we shall have to explore much further is the comparison of ionization products from different entrance channels that lead to the same or related compound states.[15,16] For example, the ionization products from photo- and electron impact-ionization of formaldehyde could be compared with the products of collisions of H_2 with CO in various electronically excited states, particularly the (presumably) long-lived Rydberg states that, with H_2, can be correlated with excited states of formaldehyde. Triatomic compound systems would be even more instructive in terms of the inferences we could make about the mechanism of ionization.

A favourite of some of us is the HCO system. Chemi-ionization occurs readily from the entrance channel of CH + O in their ground states.[17] Direct observation of the ionization of HCO$^+$ from ionization of HCO could now be accomplished with ion-cyclotron resonance; so, perhaps, could observation of HCO$^+$ produced by collisions of H atoms with excited CO. The reasons for the interest in this system are the competition between the formation of HCO$^+$ and generation of CO + H, from CH + O, the implications of this competition for formation of interstellar molecules, and the capacity of this system to

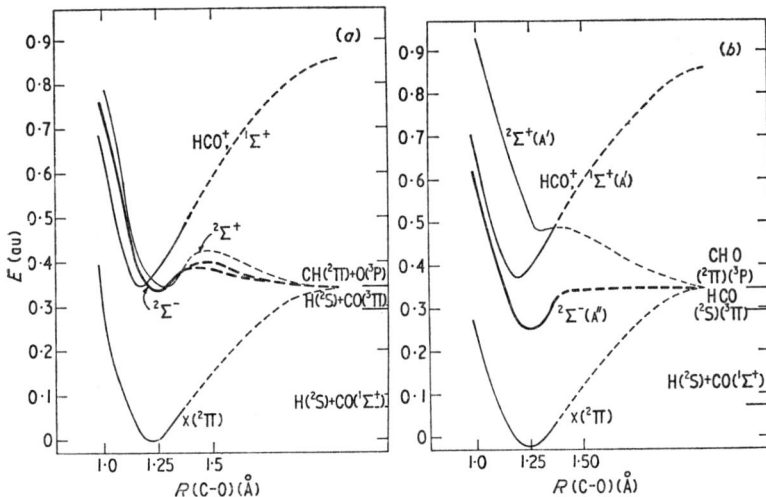

FIG. 4 Some of the potential surfaces, presumably the most important, for HCO and HCO$^+$; the energy is shown as a function of the C–O distance with the H–C–O angle and the C–H distance fixed. The curves correspond to HCO angles of 180° and 140°, as labelled.

illustrate how the shapes and crossings of potential surfaces govern the probabilities of both indirect ionization in polyatomics and collisional ionization by heavy particle collisions. Figures 4 and 5 illustrate two sections of some potential surfaces for HCO and HCO$^+$, showing how the neutral system may cross over to the chemi-ionized system. (From ref. 19.) We shall return to this point when we discuss the microscopic 'interaction range' interpretation. This particular system needs much more study; perhaps, with techniques such as ion-cyclotron resonance, we can learn about its properties.

C. From Energy Storage Mode

The identification of the degree of freedom providing the energy for ionization is at the centre of the problem of connecting observations with any microscopic interpretation. In some cases, the mechanism is obvious or at least unambiguous. The autoionization of the rare gases between their first and second ionization thresholds[20]—between the thresholds giving ions in excited $^2P_{1/2}$ states, and those giving ions in $^2P_{3/2}$ states—has only one interpretation. The

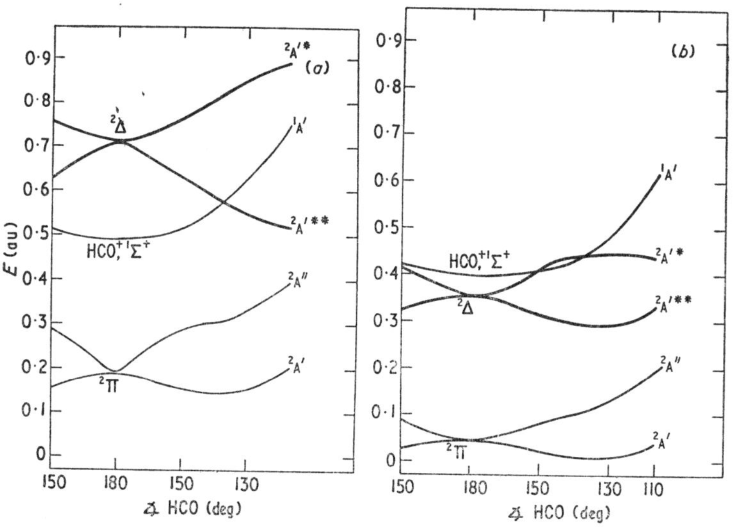

FIG. 5 Some potential curves for the HCO, HCO$^+$ system with E as a function of the HCO angle, with the C–H and C–O distances fixed. This figure is simply a different section through the same surfaces on which Fig. 4 is based.

Rydberg states around $^2P_{1/2}$ cores lie in the continuum of the ions in $^2P_{3/2}$ states, plus free electrons. The coupling mechanism is the spin-orbit interaction that spoils the J-quantum number of the core because of coupling between the excited Rydberg electron and the core. The coupling is quite weak for electrons excited to Rydberg s-orbitals (from the valence p-orbital) so the s-lines are relatively sharp. The coupling between Rydberg d-electrons and the core is stronger, and the corresponding optical absorption exhibits broad lines. For both s- and d-electrons, the coupling decreases as the Rydberg quantum number n increases, so the lines of both s and d series become sharp as n increases.

In molecules, too, the coupling is sometimes unambiguous. Near the ionization threshold, the H_2 molecule has a series of autoionizing states that can only be storing their 'energy deficit' in molecular rotation. These states, studied by Takezawa[21] and by Herzberg and Jungen,[22] and interpreted by Fano[23] and Dill,[12] autoionize because energy is slowly transferred between molecular Rydberg electrons and a rotating molecule–ion core. Each time the Rydberg electron ventures into the core region, it sees the non-spherical core, scatters from this core and thereby spoils its m_l quantum number. Sometimes it also spoils its l quantum number, and when this happens, the opportunity arises for the Rydberg electron to pick up enough energy, from rotation, to ionize. The selection rules for this process restrict the parity to be constant; the propensity rules then say that $\Delta J = \pm 2$ is far more probable than $\Delta J = \pm 4$, and so forth, so we know immediately what transitions are the most likely.

In some cases, molecular vibration is the storage mode.[9,24–26] The H_2 molecule, at energies in which accessible Rydberg states have excited vibrational levels above the $v' = 0$ level of $H_2{}^+$, can pump vibrational energy into Rydberg electrons. Here, the most important rules are the propensity rules that govern the hierarchy of rates. They tell us that $\Delta v = -1$ is much more likely than $\Delta v = -2$, and so forth, up to vibrational quantum changes in which only small non-cancelling residues keep the transition amplitudes from vanishing.

In both rotationally induced and vibrationally induced autoionization, the effective operator is the part of the nuclear kinetic energy associated with breakdown of the Born–Oppenheimer approximation. This tells us quickly how the rates should scale with mass, and therefore what the isotope effect should be. In the case of H_2, HD and D_2, these are borne out so far as we can yet tell, so we have good reason to believe our interpretations.

One other coupling mechanism is also possible; this is electron–electron electrostatic interaction. In mathematical terms this takes the form of configuration interaction between bound and continuum states. In physical terms, it is electron–electron correlation that spoils the pure orbital motion and orbital quantum numbers of the electrons, and sends one electron into the continuum. This is the process responsible for the autoionization of most (electronically) doubly excited states, and was the stimulus for Fano's interpretation of what we now call the Beutler–Fano line shape. Electron

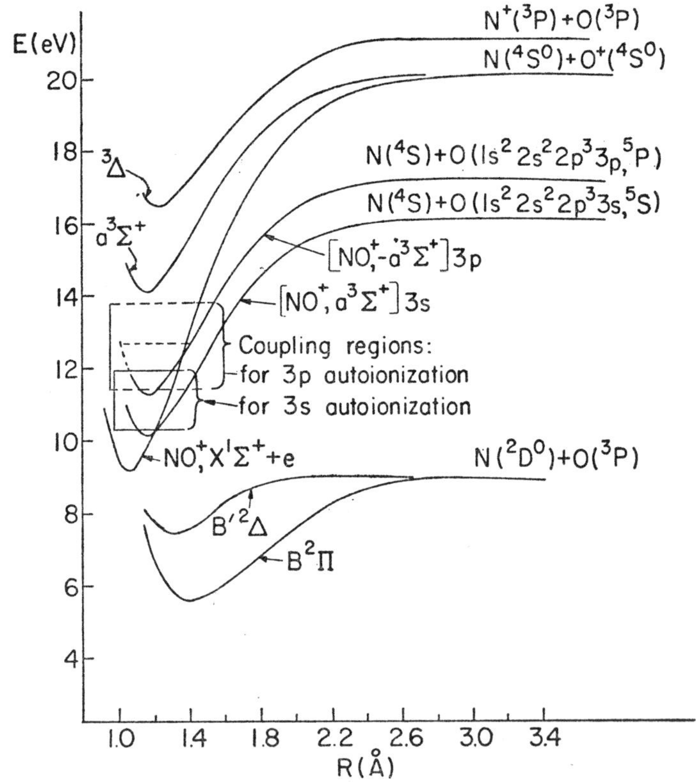

FIG. 6 Potential curves for $N + O \rightarrow NO^+ + e$; the dominant process, according to Bardsley, is the electron-correlation-induced autoionization at the crossing of curves for NO* and $NO^+ + e$ where the B' state curve reaches the NO^+ curve.

correlation is, according to Bardsley,[8] the probable mechanism of associative ionization in $N + O$. The system is isoelectronic with HCO, but the relationships among the states are slightly different, as the curves of Fig. 6 show.

Sometimes we do not know the mechanism that couples bound and continuum states. Berkowitz has pointed out to us that the mechanism of autoionization of F_2 is sometimes electronic and sometimes vibrational.[27] Carolyn Duzy and I are trying now to determine the relative importance of electronic and vibrational contributions to the autoionization of N_2, by carrying out a theoretical analysis of the ionization of this molecule. The energy analysis of autoionized electrons is sometimes a fairly unambiguous experimental means to make this interpretation.

D. Interaction Ranges

In the olden days, we would handle the problem of interaction range effects by looking for the crossing point of two potential curves, one of the final ion and the other of the initial neutral system, and evaluate the transition probability for ionization by applying a Landau–Zener–Stueckelberg treatment. This is a powerful semi-classical approach that holds remarkably well for many systems, as Olsen *et al.* have shown.[28] However, this interpretation is not adequate to give us a full picture of how molecular ionization occurs. First, it is inadequate because it doesn't work well when two crossing potential curves are very close and nearly parallel for a long interval. Second, it is only meaningful for motion in one dimension and is insufficient, by itself, to treat anything more complicated than diatomics. Third, there are many examples in which ionization occurs because of transitions between nearly but *non-crossing* surfaces. The Landau–Zener–Stueckelberg treatment is not adequate for such cases. The state-coupling problem itself is a quantum mechanical exercise that has been treated by many people, and I shall not try to review the methods here. Let me only cite some conclusions. First, it is a very good general rule that the nuclei satisfy the Franck–Condon principle very well in any state-crossing problem: they instantaneously preserve both position and momentum —*locally*—when the electronic state of the system changes. Hence, as Cermak and Herman[6] first pointed out, one can make a connection between the distributions of electronic and nuclear energies, the shapes of potential curves, and the range of internuclear distance over which transitions occur, in a monoenergetic heavy particle collision or even in a thermal system.

This capability, resting on the approximate conservation of nuclear momentum, leads us to return to our three cases, those in which the transitions occur primarily at crossing points, those whose transitions occur in the vicinity of classical turning points and those whose transitions occur over a broad range of distances. The first two give rise to narrow spreads in the kinetic energy of outgoing electrons; the latter may give large spreads. Hotop and Niehaus[7] have been very successful in applying this kind of interpretation to infer the shapes of potential curves from examination of the energy distributions of Penning electrons.

The interaction range approach looks like a useful one for interpreting, at least roughly, the angular distributions of chemi-ionized and Penning electrons, because of the adiabatic character of the heavy particle collision. When two heavy particles approach, the electrons follow a Mulliken-type correlation diagram, in the sense that their orbital energies move according to such pictures. Consider He* on another target M; when the He* and M approach, the half-filled core level of the He* becomes a half-filled inner shell of the combined molecule, and the active electron of He* goes into a highly excited orbital of the compound system. As He* and M approach, at some internuclear separation, the *electronic* energy of (HeM)* exceeds the ionization energy of HeM, and the compound system finds itself in a state capable of undergoing an Auger transition. The active electron goes out and another electron falls into the vacancy in the core molecular orbital—which is, of course, highly localized around the He. The maximum distance at which this transition can occur fixes the maximum distance of closest approach and therefore the maximum impact parameter for the transition, and sets an approximate upper limit on the cross section. It also sets a limit on the possible orientations of the compound system, with respect to the laboratory frame, when the ionization first becomes possible. If we assume that Auger ionization is always fast with respect to molecular rotation or collision, we can predict the angular distribution of outgoing electrons. We can do this because we know the selection and propensity rules for Auger processes, so we know (at least in principle) the possible angular momenta of the outgoing electron with respect to the molecular axis. But we also know the possible orientations of the molecular axis when the transition occurs. Hence we can predict the angular distributions of electrons for Penning and associative ionization, when these processes are fast compared with rotation. We have not yet done this; I put it in here to show that there are soluble, interesting problems still ahead in this field of low-energy ionization.

Finally, I must point out a very important situation that is just beginning to be explored, the problem I mentioned regarding surface crossing, in contrast to curve crossing. Tully and Preston,[29,30] and then MacGregor and I,[19] found that, whereas Landau–Zener–Stueckelberg probabilities could cause transition probabilities to vary by more than an order of magnitude, that in multidimensional problems, the likelihood of reaching a transition region could vary by *many* orders of magnitude. In other words, both for the H_3^+ system and the HCO system, it appears that the factor most important to obtain correctly is the probability that the system reach the reactive part of the potential surface. We are strongly tempted to generalize this, and to say that the interpretation of collision-induced ionization, as well as of reactive processes, in polyatomic systems, must rely most heavily on the knowledge of potential surfaces, and on trajectory calculations or their quantum analogues. For reactive collisions, Polanyi's interpretation[31] in terms of early-downhill, late-downhill and intermediate cases is beginning to give us something better than a naive feeling for generalizations. We do not yet have such bases for interpreting collisional ionization in polyatomic systems, but we can look forward to seeing these interpretations develop as we learn more of the relationships of potential surfaces of ions and neutrals.

REFERENCES

1. Gioumousis, G. *J. chem. Phys.,* 1958, **29**, 294.
2. Platzman, R. L. *Radiation Research,* 1962, **17**, 419; *Int. J. Appl. Radiat. and Isotopes,* 1961, **10**, 116.

3. Berkowitz, J. and Chupka, W. A. *J. chem. Phys.*, 1968, **48**, 5726; Chupka, W. A. and Berkowitz, J. *J. chem. Phys.* 1969, **51**, 2/244.
4. For electron impact ionization, see: Bely, O. and Van Regemorter, H. *Ann. Rev. Astron. Astrophys.*, 1970, **8**, 329; Kieffer, L. J. and Dunn, G. H. *Rev. Mod. Phys.*, 1966, **38**, 1; Rudge, M. R. *Rev. Mod. Phys.*, 1968, **40**, 564.
   For photoionization, see: Fano, U. and Cooper, J. W. *Rev. Mod. Phys.*, 1968, **40**, 441; *ibid.*, 1969, **41**, 724; Marr, G. V. *Photoionization Processes in Gases*, 1967, Pure and Applied Physics, Vol. 28.
   For heavy particle ionization, see: Berry, R. S. Chemiionization, *in* 'Molecular Beams and Reaction Kinetics', Vol. 44, *ed.* C. Schlier (Int. School of Physics 'Enrico Fermi', 1969); Ogurtsov, G. *Rev. Mod. Phys.*, 1972, **44**, 1; Garcia, J. D., Fortner, R. J. and Kavanagh, T. M. *Rev. Mod. Phys.*, 1973, **45**, 111.
5. Berry, R. S. and Nielsen, S. E. *in* 'Recent Developments in Mass Spectroscopy', *ed.* K. Ogata and T. Hayakawa, Proc. Int. Conf. on Mass Spectrometry, Kyoto, 1969, Univ. of Tokyo Press, 1970, p. 811.
6. (a) Herman, Z. and Cermak, V. *Coll. Czech. Chem. Comm.*, 1966, **31**, 649.
   (b) Cermak, V. and Herman, Z. *Chem. Phys. Letters*, 1968, **2**, 359.
7. Hotop, H. and Niehaus, A. *Z. Physik*, 1970, **238**, 452.
8. Bardsley, J. N. *Chem. Phys. Letters*, 1967, **1**, 229;
   Bardsley, J. N. *J. Phys.*, B, 1968, **1**, 365.
9. Berry, R. S. and Nielsen, S. E. *Phys. Rev.*, A, 1970, **1**, 395;
   Berry, R. S. and Nielsen, S. E. *Chem. Phys. Letters*, 1968, **2**, 503.
10. Smith, A. L. *Phil. Trans. Roy. Soc.*, A, 1970, **268**, 169.
11. Berkowitz, J. and Chupka, W. A. *J. chem. Phys.*, 1969, **51**, 2341, and earlier references therein, esp. Doolittle, P. H. and Schoen, R. J. *Phys. Rev. Letters*, 1965, **14**, 348.
12. Dill, D. *Phys. Rev.*, A, 1972, **6**, 160.
13. Peatman, W. B., Borne, T. and Schlag, E. W. *Chem. Phys. Letters*, 1969, **3**, 492;
    Baer, T., Peatman, W. B. and Schlag, E. W. *Chem. Phys. Letters*, 1969, **4**, 243.
14. Villarejo, D., Stockbauer, R. and Inghram, M. A. *J. chem. Phys.* 1968, **48**, 3342; *ibid.*, 1969, **50**, 4599.
15. Hotop, H. and Niehaus, A. *Int. J. Mass Spectrom. Ion Phys.*, 1970, **5**, 415.
16. Klots, C. E. *J. chem. Phys.*, 1972, **56**, 124.
17. cf. Peeters, J. and Van Tiggelen, A. 'XII Int. Symp. on Combustion', Combustion Institute, Pittsburgh, Pa., 1969, p. 969; Fontijn, A., Miller, W. J. and Hogan, J. M. 'X Int. Symp. on Combustion', Combustion Institute, Pittsburgh, Pa., 1965, p. 545.
18. Dalgarno, A., Oppenheimer, M. and Berry, R. S. *Astrophys. J.* 1973, **183**, L21.
19. MacGregor, M. and Berry, R. S. *J. Phys.*, B, 1973, **6**, 181.
20. Beutler, H. *Z. Physik*, 1935, **93**, 177; Fano, U. *Phys. Rev.*, 1961, **124**, 1866.
21. Takezawa, S. *J. chem. Phys.*, 1970, **52**, 5793.
22. Herzberg, G. *Phys. Rev. Letters*, 1969, **23**, 1081;
    Herzberg, G. and Jungen, C. *J. Mol. Spectrosc.*, 1972, **41**, 425.
23. Fano, U. *Phys. Rev.* A, 1970, **2**, 353.
24. Russek, A., Patterson, M. R. and Becker, R. L. *Phys. Rev.*, 1968, **167**, 17.
25. Faisal, F. H. M. *Phys. Rev.*, A, 1971, **4**, 1396.
26. Ritchie, B. *Phys. Rev.*, A, 1971, **3**, 95.
27. Berkowitz, J. (private communication; to be published).
28. Olsen, R. E., Smith, F. T. and Bauer, E. *Appl. Optics*, 1971, **10**, 1848.
29. Preston, R. K. and Tully, J. C. *J. chem. Phys.*, 1971, **54**, 4297.
30. Tully, J. C. and Preston, R. K. *J. chem. Phys.*, 1971, **55**, 562.
31. Kuntz, P. J., Nemeth, E. M., Polanyi, J. C., Rosner, S. D. and Young, C. E. *J. chem. Phys.* 1966, **44**, 1168.

ORGANIC FRAGMENTATION MECHANISMS

Chairmen

R. T. APLIN
University of Oxford, U.K.

F. W. McLAFFERTY
Cornell University, U.S.A.

1
Long-range Intramolecular Interactions in 4-*n*-Alkyl Trimellitic Esters

By SEYMOUR MEYERSON

(*Research Department, Standard Oil Company (Indiana),*
Naperville, Illinois, U.S.A.)

IMRE PUSKAS and ELLIS K. FIELDS

(*Research and Development Department, Amoco Chemicals Corporation,*
Naperville, Illinois, U.S.A.)

IN a preliminary study of the mass spectra of 4-*n*-alkyl esters of trimellitic anhydride (TMA), we found that esters in which the alkyl group contains six or more carbons lose C_nH_{2n-2}, that is, the alkyl group less three hydrogen atoms, in competition with loss of the alkyl group less two hydrogens, generally characteristic of alkyl esters of carboxylic acids.[1] One of the migrating atoms is apparently abstracted by a radical centre at an anhydride oxygen atom. This observation and others involving intramolecular reactions between formally distant parts of a sufficiently long flexible molecule were interpreted in terms of molecular coiling induced by attempts of isolated molecules to solvate themselves.[2] The concept of such coiling and internal solvation and the physico-chemical consequences of this behaviour are a matter of continuing interest both in systems restricted to neutral molecules and in the mass spectrometer.

Our preliminary communication on trimellitic anhydrides prompted a study by Cable and Djerassi,[3] who further explored alkyl hydrogen migration to the second functional group in long-chain *n*-alkyl esters of trimellitic anhydride and trimellitimide and in 4-*n*-alkyl-1,2-dimethyl trimellitates. Similar processes have been found by Winnik and his students,[4] at the University of Toronto, in *m*- and *p*-*n*-alkoxybenzoic acids and the derived methyl benzoates and in *n*-alkyl *p*-benzoylbenzoates, which undergo a similar process under UV irradiation.[5]

In our continuing study of this system, we also included 4-*n*-alkyl trimellitimidates, *n*-alkyl methyl iso- and terephthalates, and 4-*n*-alkyl-1,2-dimethyl trimellitates. To help define the origins of migrating atoms, we prepared and studied the mass spectra of a series of labelled *n*-octadecyl esters—6,7-d_2, 9,10-d_2, and 9,10,12,13-d_4—of TMA and, for comparison, benzoic acid.

Confirming Cable and Djerassi, we find that alkyl hydrogen migrates to the second functional group in the trimellitimidates and in the alkyl dimethyl trimellitates, but not in the iso- and terephthalates. The alkyl dimethyl trimellitates show no measurable triple-hydrogen migration, but their spectra contain other clear evidence that alkyl hydrogens find their way to the carbomethoxy groups.

Two lines of inquiry proved particularly fruitful: first, comparison of the $[M-C_{18}H_{35}]^+$ regions of the spectra of labelled *n*-octadecyl benzoates and TMA esters; second, mechanistic interpretation of the loss of small olefin molecules from 4-*n*-alkyl TMA esters and 4-*n*-alkyl-1,2-dimethyl trimellitates.

TABLE I

$[M-C_{18}H_{35}]^+$ Regions in Spectra of Labelled *n*-Octadecyl Benzoates and Trimellitate Anhydrides

	Benzoates					*TMA esters*			
	Relative intensity[a,b]					*Relative intensity*[a,c]			
Mass	d_0	$6,7\text{-}d_2$	$9,10\text{-}d_2$	$9,10,12,\\13\text{-}d_4$	*Mass*	d_0	$6,7\text{-}d_2$	$9,10\text{-}d_2$	$9,10,12,\\13\text{-}d_4$
122	5·45	5·42	5·45	5·47	192	0·91	1·41	1·98	0·94
123	90·2	83·7	87·0	84·7	193	61·7	57·4	52·9	45·9
124	0·54	6·70	3·42	5·54	194	33·8	33·6	36·2	38·2
125	2·24[d]	0·81	0·57	0·37	195	1·52[d]	5·12	6·29	10·9
126	1·31[d]	1·56	1·56	0·94	196	2·02[d]	2·10	1·98	2·46
127	0·22[d]	1·18	1·44	1·19	197	0·13[d]	0·38	0·66	1·02
128	0·03	0·46	0·50	1·02	198		0·15	0·15	0·42
129		0·11	0·12	0·61	199				0·15
130		0·03	0·02	0·20					
131				0·05					
132				0·01					

[a] Corrected for naturally occurring heavy-isotopic contributions.
[b] Total intensity from mass 122 to 130 (132 in the d_4 spectrum) is set equal to 100·0.
[c] Total intensity from mass 192 to 199 is set equal to 100·0.
[d] Hydrocarbon ions derived from the octadecyl group.

Several observations in our work suggest hydrogen migration to the anhydride—or other second functional—group in paths effecting loss of C_nH_{2n-1} as well as of C_nH_{2n-2}, that is, loss of the alkyl group less 2, as well as 3, hydrogens. With this shift in terminus of the migrating atoms in mind, we focussed attention on the $[M-C_{18}H_{35}]^+$ ions in the spectra of the labelled benzoates and TMA esters. Partial spectra covering the pertinent mass regions are shown in Table I. Intensity attributable to unlabelled $[M-C_{18}H_{36}]^+$ at mass 122 is the same in all the benzoate spectra, implying that the production of this ion involves no measurable hydrogen migration from positions 6, 7, 9, 10, 12 or 13. The intensity variation of the analogous peak at 192 in the TMA ester spectra reflects small contributions from interfering species in the spectra of the labelled esters. In any case, the peaks at

123 and 193 are apparently free of contributions from labelled $[M-C_{18}H_{36}]^{+}$, and can be attributed solely to $[M-C_{18}H_{35}]^{+}$. The percent of labelled (with either one or two deuterium atoms) $[M-C_{18}H_{35}]^{+}$ ions can now be estimated for each labelled species at $(I_0 - I)/I_0$, where I and I_0 are the intensities of the labelled and unlabelled esters, respectively, at mass 123 or 193. The values so arrived at are listed in Table II. The differences between the values for the

TABLE II

Percent Label Retention in $[M-C_{18}H_{35}]^{+}$ from Labelled *n*-Octadecyl Benzoates and Trimellitate Anhydrides

	Benzoates	TMA esters	Difference[a]
6,7-*d*$_2$	7·2	7·0	−0·2
9,10-*d*$_2$	3·5	14·3	10·8
9,10,12,13-*d*$_4$	6·1	25·6	19·5

[a] The TMA ester value minus the benzoate value.

similarly labelled esters measure the preference for migration from the labelled positions in the TMA ester over that in the benzoate; we view them as first approximations to the extent of migration to the anhydride group. Positions 6/7 appear to contribute little or nothing. Positions 9/10 and 12/13 contribute about equally. Positions 9/10 and 12/13 also contribute about equally to migration to the carboxyl group in the benzoate, where migration from positions 6/7 is about twice as probable as from the farther-removed carbons.

Over a wide mass region, the spectra of the longer 4-*n*-alkyl TMA esters are dominated by a homologous series of peaks corresponding to loss of olefin molecules C_jH_{2j} from the molecular ion, where $2 \leq j \leq n - 8$, and $n =$ the number of carbons in the alkyl chain. They intrigued us particularly in view of a second such series corresponding nominally to loss of olefin molecules from $[M-H_2O]^{+}$ and to a similar series seemingly stemming from the $[M-CH_3OH]^{+}$ ions in the spectra of 4-*n*-alkyl-1,2-dimethyl trimellitates.

The $[M-C_jH_{2j}]^{+}$ series can be accounted for plausibly by hydrogen abstraction by an anhydride oxygen from $C-(n - j + 2)$ in the alkyl chain, followed by β cleavage with respect to the resulting free radical.

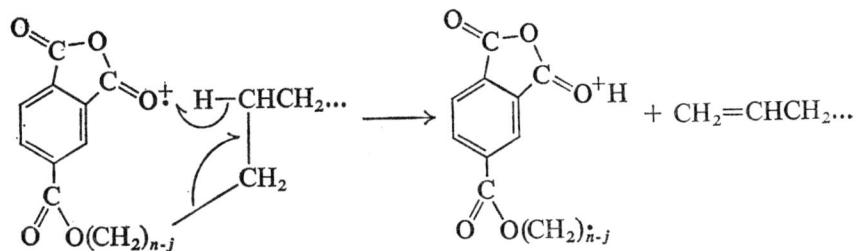

Such a model is particularly attractive if valid, because the intensity distribution of the $[M-C_jH_{2j}]^{+}$ peaks is a direct measure of the probability distribution of abstraction from the various alkyl carbons. The appropriate data, normalized to a total of 100 % for all values of j from 2 to $n - 8$, are listed in Table III. Also listed there for each ester is j', the value of j corresponding to maximum intensity, and n', the position of hydrogen abstraction corresponding to loss of $C_{j'}H_{2j'}$.

The distributions that emerge seem reasonable. Intensity varies smoothly and symmetrically about a maximum which shifts slowly with increasing chain length to positions farther out on the chain. These and other data

TABLE III

Intensity Distributions of $[M-C_jH_{2j}]^+$ Peaks in Spectra of 4-n-Alkyl Trimellitate Anhydrides

j^b	n^a					
	11	12	16	18	20	22
			Relative intensity			
2	40	11	2	2	2	
3	60	63	16	9	5	1
4		26	23	15	9	3
5			25	19	12	6
6			19	21	14	9
7			11	15	16	12
8			4	10	14	14
9				5	12	15
10				4	9	14
11					5	11
12					2	8
13						5
14						2
j'^c	3	3	5	6	7	9
n'^d	10	11	13	14	15	15

[a] Number of carbon atoms in the alkyl group.
[b] Number of carbon atoms in the olefin molecule lost.
[c] The value of j corresponding to maximum intensity.
[d] $n' = n - j' + 2$, the position of hydrogen abstraction corresponding to loss of $C_{j'}H_{2j'}$.

suggest that, for a sufficiently long alkyl group, coiling about the anhydride group brings the hydrogens on carbons 14, 15 and 16 into most intimate contact with the oxygens, and that the density of hydrogen atoms within reactive distance of the oxygens approaches a maximum value, unaffected by further lengthening of the alkyl chain. We have examined similarly the analogous intensity distributions and derived values of j' and n' for the $[M-H_2O-C_jH_{2j}]^{\cdot+}$ peaks in the spectra of 4-n-alkyl TMA esters and for the $[M-CH_3OH-C_jH_{2j}]^{\cdot+}$ peaks in those of 4-n-alkyl-1,2-dimethyl trimellitates, and they resemble closely those found for the $[M-C_jH_{2j}]^{\cdot+}$ peaks as described above.

Despite the attractiveness of the proposed model, some possible complications remained. First, some of the ions in the series under consideration might arise by sequences in which loss of the first olefin molecule, triggered by hydrogen abstraction, is followed by loss of ethylene by β cleavage. Second, hydrogen migration within the alkyl chain might intervene between abstraction and β cleavage, or follow these events and induce loss of a second olefin molecule. In either case, the length of the retained alkyl chain segment would no longer define the position of abstraction. Any such intervening reaction step would presumably entail some additional energy requirement, however, so that the resulting intensity distribution might be energy dependent. We therefore measured the intensity distributions of the $[M-C_jH_{2j}]^{\cdot+}$ peaks in the spectra of n-octadecyl and n-eicosyl TMA esters at 7·5 eV (nominal); they are indistinguishable from the 70-eV data. A more rigorous test was performed by metastable scanning, which showed that 99% of the total metastable-peak intensity leading to the $[M-C_jH_{2j}]^{\cdot+}$ ions corresponds to the molecular ion as the precursor. The remaining 1% consists of small contributions to the yield of $[M-C_6H_{12}]^{\cdot+}$, the most abundant ion in the series, via $[M-C_2H_4]^{\cdot+}$ and $[M-C_4H_8]^{\cdot+}$ intermediates. Metastable scanning of the $[M-H_2O-C_jH_{2j}]^{\cdot+}$ peaks shows that at least 80% of these ions arise by loss of H_2O from the corresponding $[M-C_jH_{2j}]^{\cdot+}$ ions.

The labelling results permit another independent check on the validity of the model, which requires that each $[M-C_jH_{2j}]^{\cdot+}$ ion retain all the hydrogen atoms originating on $C-1$ to $C-(n-j)$ plus one atom from $C-(n-j+2)$. The isotopic compositions so predicted for the isotopically pure $6,7-d_2$, $9,10-d_2$, and $9,10,12,13-d_4$ esters, ignoring possible isotope effects, are listed in Table IV. Because of the presence of isotopic impurities and of interference from ions of neighbouring mass numbers, which may overlap in the labelled esters, precise calculation of the isotopic distributions of the $[M-C_jH_{2j}]^{\cdot+}$ ions derived from the labelled compounds is not feasible. However, in every case where the model calls for a single isotopic species, the peak corresponding to this species is the most intense in the appropriate mass region. Similarly, where the model calls for 50% each of two isotopic species, one of the corresponding peaks is the most intense in the region, and the other is nearly as intense.

In concluding, we would like to offer a semiphilosophical comment. We

TABLE IV

Predicted Isotopic Compositions of $[M-C_jH_{2j}]^+$ Ions in Spectra of n-Octadecyl-6,7-d_2, -9,10-d_2, and -9,10,12,13-d_4 Trimellitate Anhydrides

j	6,7-d_2	9,10-d_2	9,10,12,13-d_4
2	d_2	d_2	d_4
3	d_2	d_2	d_4
4	d_2	d_2	d_4
5	d_2	d_2	d_4
6	d_2	d_2	d_3
7	d_2	d_2	$d_2, d_3{}^a$
8	d_2	d_2	$d_2, d_3{}^a$
9	d_2	d_1	d_1
10	d_2	$d_0, d_1{}^a$	$d_0, d_1{}^a$

^{a} 50% of each species.

like to amuse ourselves at times by looking for parallels between the behaviour of molecules and of people, and the systems we have been discussing comprise extraordinarily apt examples. In initially trying to visualize the molecular behaviour reflected in the TMA ester spectra, we pictured the alkyl group in the gaseous molecule as a long tail waving in the breeze—and it's cold out there! Coiling that tail around the more bulky parts of the molecule would help to keep it warm. The first and perhaps best chemical analogy we identified is with the polymer chemists' model for flexible polymers in dilute solution in poor solvents—random coiling into a ball. The isolated molecules in the rarefied atmosphere of the mass spectrometer may be considered as an infinitely dilute solution in an infinitely poor solvent. The analogy can be extended to the formation of spherical drops of liquid or of gas bubbles in a liquid and of circular islands in monomolecular films, extrapolated again to the limiting case of a single molecule. In each of these instances, a molecule or a collection of molecules, looking out at an environment that it perceives as different from itself and hence unfriendly, tries to attain a configuration that will minimize the surface exposed to the outside world. The parallel with human behaviour requires no elaboration. A non-chemist friend, listening to one of us expound on this model, has suggested that we call it the Conestoga wagon theory.

REFERENCES

1. Meyerson, S., Puskas, I. and Fields, E. K., *Chem. Ind. (London)*, 1968, p. 1845.
2. Meyerson, S. and Leitch, L. C., *J. Amer. Chem. Soc.*, 1971, **93**, 2244.
3. Cable, J. and Djerassi, C., *J. Amer. Chem. Soc.*, 1971, **93**, 3905.
4. Winnik, M. A., Lee, C. K. and Kwong, P. T. Y., personal communication.
5. Breslow, R. and Winnik, M. A., *J. Amer. Chem. Soc.*, 1969, **91**, 3083.

Discussion

H. F. Grützmacher (University of Hamburg, Germany): Would you comment on whether you consider the coiling of the molecules as caused by ionization or as pre-existing in the neutral molecule?

S. Meyerson: We found that the relative probabilities of abstraction from the various positions along the chain are identical as measured by relative intensities of normal peaks at 70 eV and at low voltage and of metastable peaks. Thus the probability distribution of the implied coiled configurations appears constant over the range of energy and reaction time represented by these measurements. Moreover, we have been impressed by the pronounced difference in behaviour between methyl *n*-alkyl iso- and terephthalates and 4-*n*-alkyl-1,2-dimethyl trimellitates. The phthalates, unlike the trimellitates, show no evidence of hydrogen abstraction, hence apparently lack an essential structural feature. All of these observations lead us to believe that coiling takes place in the neutral molecules and that the observed abstractions simply mirror pre-existing orientations. Hydrogen bonding between a methyl or methylene group in an alkyl chain and a heteroatom is known to be weak,[1] but such bonding has been demonstrated in certain biopolymers.[2] Moreover, very recent work from the Polish Academy of Sciences[3] documents the occurrence of 'bifurcate' hydrogen bonding, involving one donor and two acceptors, and the increased strength of the chelated arrangement so attained may well be the factor enabling the trimellitates to attain relatively stable hydrogen-bonded configurations although the phthalates do not.

G. Dijkstra (Utrecht University, The Netherlands): If, as you suppose, the flexible chain is already wound round the polar part of the molecule before ionisation, this should be open to confirmation by infrared spectroscopy. Snyder and Schachtschneider in the U.S. and de Ruig in The Netherlands have assigned the infrared bands of flexible chains to specific *cis–trans* sequences of the single bonds. Provided compounds can be found which have a high enough vapour pressure for infrared absorption spectra to be taken the configuration of the molecules can be evaluated. With the greater sensitivity recently attained by the development of the interferometer this should be possible. What are the smallest molecules to show this curling-up in your investigations?

S. Meyerson: The position from which hydrogen appears to be abstracted closest to the carboxy groups is 5 or 6 in loss of C_nH_{2n-2} and 10 in loss of olefin and of (olefin $+ H_2O$) from TMA esters, and 6 in loss of (olefin $+ CH_3OH$) from 4-*n*-alkyl-1,2-dimethyl trimellitates.

REFERENCES

1. Mark, H. F. and Atlas, S. M. *Bull. Soc. Chim. France,* 1970, p. 3275.
2. Krimm, S., Kuroiwa, K. and Rebane, T. *in* 'Conformation of Biopolymers,' Vol. 2, *ed.* G. N. Ramachandran, Academic Press, New York, 1967, p. 439; Krimm, S. and Kuroiwa, K. *Biopolymers,* 1968, **6**, 401.
3. Dabrowski, J. and Swistun, Z. *J. Chem. Soc.,* B, 1971, 818; Dabrowski, J., Swistun, Z. and Dabrowska, U. *Tetrahedron,* 1973, **29**, 2257; Dabrowski, J. and Swistun, Z. ibid., 1973, **29**, 2261.

2

Stereochemical Approach to the Mechanism of Retro-Diels–Alder Fragmentation Under Electron Impact

By ASHER MANDELBAUM and PETER BEL

(*Department of Chemistry, Technion—Israel Institute of Technology, Haifa, Israel*)

Two suggestions have been made for the possible mechanism of the 'retro-Diels–Alder' (RDA) fragmentation under electron impact: (*a*) the stepwise process (Scheme 1),[1] and (*b*) synchronous cleavage of the two C,C bonds (Scheme 2).[2]

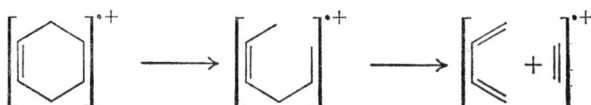

Scheme 1

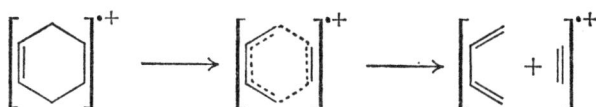

Scheme 2

Thermochemical data for the $C_2H_4^{\cdot+}$ and $C_4H_6^{\cdot+}$ ions obtained from cyclohexene have been cited as favouring the two-step mechanism.[1] On the other hand metastable peak characteristics[3] and charge distribution experiments[4] are in favour of the concerted mechanism in the case of tetralin† and 4-vinylcyclohexene.

We have recently shown that a stereochemical approach to the problem of the mechanism of this fragmentation process is also possible in some specific cases.[5] The RDA is highly stereospecific in the three diketone systems **1–3**, giving rise to very abundant diene ions only for the *cis*-isomers. This

---

† It should be noted that the loss of C_2H_4 from the molecular ion of tetralin has been shown to be not a pure retro-Diels–Alder fragmentation.[5]

behaviour suggests that the two C,C-bonds are cleaved concurrently in the course of the RDA fragmentation in these systems.[6]

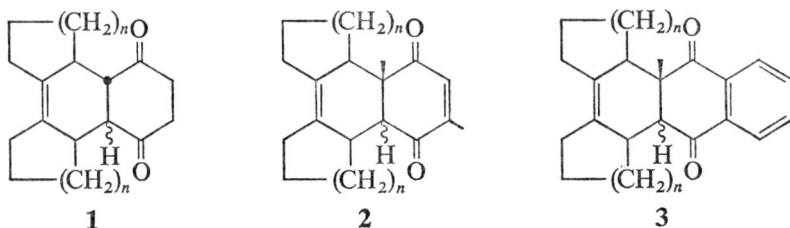

It has been shown on the other hand that in three pairs of stereoisomeric Δ^2-octalins (4–6) the occurrence of abundant ions formed by the RDA reaction is not dependent upon the configuration at the central bond.[7] In various other systems the RDA fragmentation gives rise to abundant ions

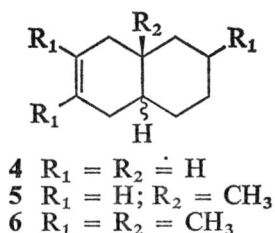

4 $R_1 = R_2 = H$
5 $R_1 = H; R_2 = CH_3$
6 $R_1 = R_2 = CH_3$

despite *trans*-fusion of the relevant rings,[1,8] but no data are available at present for the corresponding *cis*-isomers for comparison.

The different behaviour of various compounds with respect to the RDA fragmentation (and consequently different mechanisms) led us to extend the

TABLE I

Abundance of RDA Ions in Hexahydroanthracene Derivatives

Compound	Ion a/$M^{\cdot+}$	
	cis	*trans*
7	$4\cdot0^a$	$>0\cdot01$
8	$1\cdot6^a$	$0\cdot05$
9	$3\cdot1^a$	$0\cdot16$
10	$8\cdot8^a$	$1\cdot7^b$
11	~200	$>0\cdot1$
12	~200	$>0\cdot1$
13	$4\cdot5^a$	$>0\cdot01$
14	$2\cdot5^{a,c}$	$>0\cdot01$
15	$8\cdot1^a$	$0\cdot29^d$

[a] Most abundant ion.
[b] One of the lowest peaks in the mass spectrum ($1\cdot5\%$ of the most intense one). The ratio ion a/$M^{\cdot+}$ is larger than in other *trans*-isomers because of the very low abundance of $M^{\cdot+}$.
[c] The quinone ion is also present.
[d] Not pure.

7 *cis* R = CH$_3$
8 *cis* R = H

ion a
m/e 110
m/e 82

7 *trans* R = CH$_3$
8 *trans* R = H

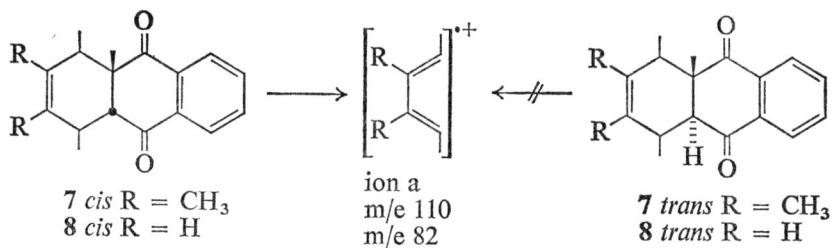

investigation to other systems. One of the points we were interested in was the role of the carbonyl groups in the stereospecific nature of the process.

In the tricyclic diketones (**7** and **8**) the RDA fragmentation is again highly stereospecific, giving rise to very abundant m/e 110 and m/e 82 ions, respectively, in the *cis*-isomers, but not in the *trans*-analogues (see Table I and Fig. 1). The high degree of stereospecificity is not confined to diketones in this tricyclic system. The corresponding diols (**9**) as well as their acetates (**10**), benzoates (**11**) and nitrobenzoates (**12**) also exhibit high stereospecificity in their decomposition by the RDA fragmentation, only the *cis*-isomers yielding abundant m/e 110 ions.

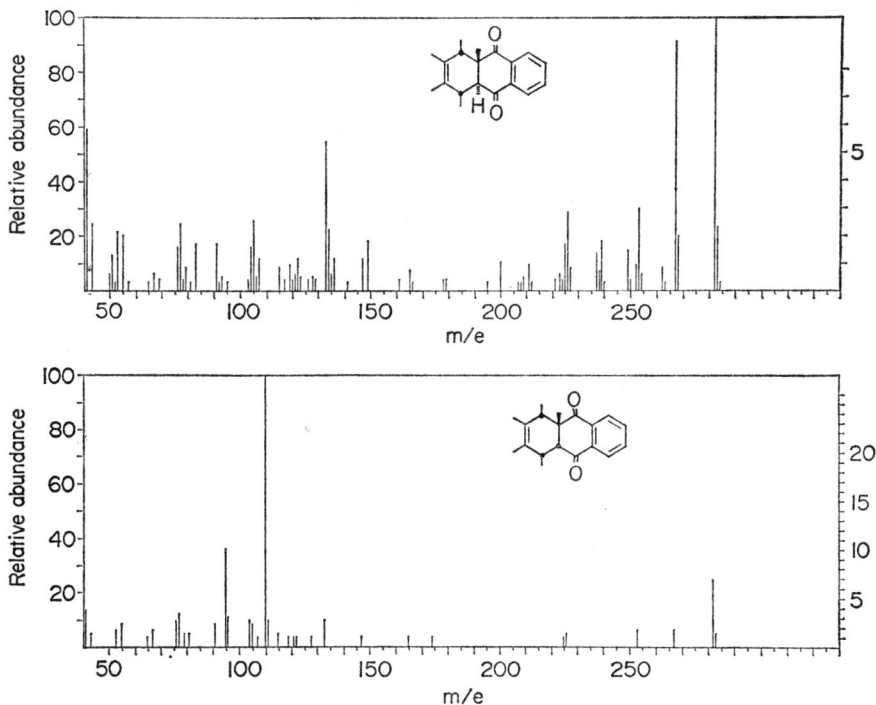

FIG. 1   Mass spectra of *cis*- and *trans*-1,2,3,4,9a-pentamethyl-1,4,4a,9,9a,10-hexahydro-anthracene-9,10-diones (**7** *cis* and **7** *trans*).

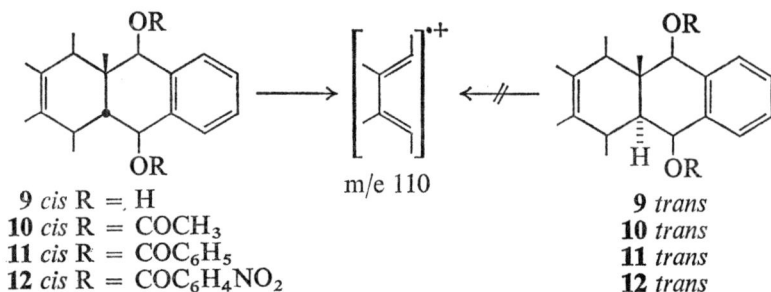

9 *cis* R = H
10 *cis* R = COCH$_3$
11 *cis* R = COC$_6$H$_5$
12 *cis* R = COC$_6$H$_4$NO$_2$

m/e 110

9 *trans*
10 *trans*
11 *trans*
12 *trans*

It was of interest to examine if different charge distribution in the molecular ion has an effect on the stereospecificity of the RDA reaction. For this purpose the substituted nitro- and aminodiketones 13 and 14 were examined. In these two pairs of stereoisomers the RDA process is again highly stereospecific affording abundant ions only in the *cis*-isomers. This implies that in the tricyclic system examined the RDA fragmentation takes place by a concerted mechanism irrespective of charge distribution.

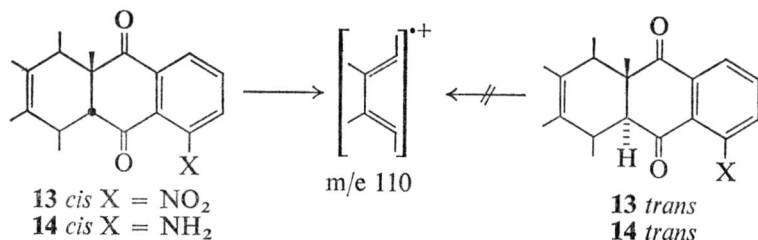

13 *cis* X = NO$_2$
14 *cis* X = NH$_2$

m/e 110

13 *trans*
14 *trans*

At the time of writing this abstract the *trans*-hydrocarbon (15) was not available in the pure state. The preliminary data shown in Table I clearly indicate that the RDA process exhibits stereospecificity in this case too, although it seems that it is to a lower degree.

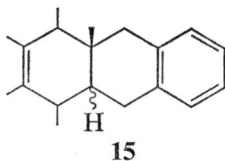

15

The results of this study show that the RDA fragmentation is stereospecific and therefore takes place by a concerted mechanism in the examined tricyclic system irrespective of substituents. It is apparent, however, that stereospecificity is not a general feature of this process. The data reported so far do not seem to allow formulation of a rule in this respect. More work will be required before the effect of configuration on the RDA process can be predicted with any confidence for an unknown system.

REFERENCES

1. Budzikiewicz, H., Brauman, J. I. and Djerassi, C., *Tetrahedron*, 1965, 21, 1855.
2. Dougherty, R. C., *J. Amer. Chem. Soc.*, 1968, 90, 5788.

3. Elwood, T. A., Beynon, J. H. and Rabideau, P. W., 19th Ann. Conf. on Mass Spectrometry and Allied Topics, Atlanta, Georgia, May 1971, p. 173.
4. Smith, E. P. and Thornton, E. R., *J. Am. Chem. Soc.*, 1967, **89**, 5079.
5. Gruetzmacher, H. and Puschmann, M., *Chem. Ber.*, 1971, **104**, 2079.
6. Karpati, A., Rave, A., Deutsch, J. and Mandelbaum, A., *J. Am. Chem. Soc.*, 1973, **95**, 4244.
7. Hammerum, S. and Djerassi, C., in the press; we thank Prof. Djerassi for making his manuscript available to us prior to publication.
8. (a) Budzikiewicz, H., Djerassi, C. and Williams, D. H., 'Structure Elucidation of Natural Products by Mass Spectrometry', Vol. 2, Holden-Day, San Francisco, 1964, and references cited therein. (b) Wünsche, C. and Löw, I., *Tetrahedron*, 1966, **22**, 1893; (c) Brooks, C. J. W. and Draffan, G. H., *Tetrahedron*, 1969, **25**, 2865; (d) Drewes, S. E. and Budzikiewicz, H., *J. Chem. Soc. C.*, 1969, **63**; (e) Berti, G., Bottari, F., Marsili, A., Morelli, I. and Mandelbaum, A., *Chem. Comm.*, 1967, **50**; (f) Berti, G., Marsili, A., Morelli, I. and Mandelbaum, A., *Tetrahedron*, 1971, **27**, 2217; (g) Meister, L., Kaufmann, H., Stocklin, W. and Reichstein, T., *Helv. Chim. Acta*, 1970, **53**, 1659, and many others.

3

On the Electron Impact Induced Rearrangement of 3-Phenyl-1-nitropropane to 3-Phenyl-3-hydroxy-1-nitrosopropane, Revealed by ^{13}C-Labelling; a New Insight into the Loss of Nitric Oxide from the Molecular Ion

By Mrs T. A. MOLENAAR-LANGEVELD
and N. M. M. NIBBERING

(*Laboratory for Organic Chemistry, University of Amsterdam, Amsterdam,*
The Netherlands)

INTRODUCTION

SINCE the early days of organic mass spectrometry it has been known that molecular ions of aromatic nitro compounds expel a molecule of nitric oxide[2-4] by virtue of a nitro–nitrite rearrangement,[5] induced upon electron impact. This rearrangement formally corresponds with a 1,2-aryl shift from nitrogen to oxygen. It is therefore not surprising that such a rearrangement is not observed for saturated aliphatic nitro compounds,[6] as alkyl groups appear to migrate with much more difficulty[7-9] than aryl groups[10] upon electron bombardment.

Yet, in some papers from our laboratory it has been shown that loss of nitric oxide may occur from molecular ions, having the nitro group attached to a saturated carbon atom:

(1) The molecular ion of nitrocyclopropane[6,11] (m/e 87, $C_3H_5NO_2$) appears to lose nitric oxide, as evidenced by the occurrence of a peak at m/e 57 (C_3H_5O) in its mass spectrum. This reaction could be explained in the same way as for aromatic nitro compounds, *i.e.* a 1,2-cyclopropyl shift from nitrogen to oxygen, realizing that a cyclopropane ring has a character intermediate between aromatic and aliphatic.

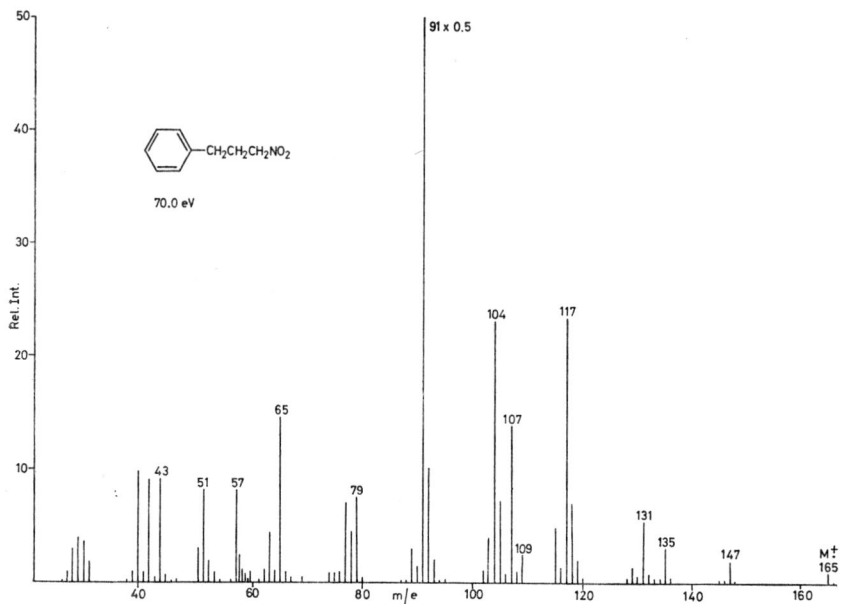

(a)

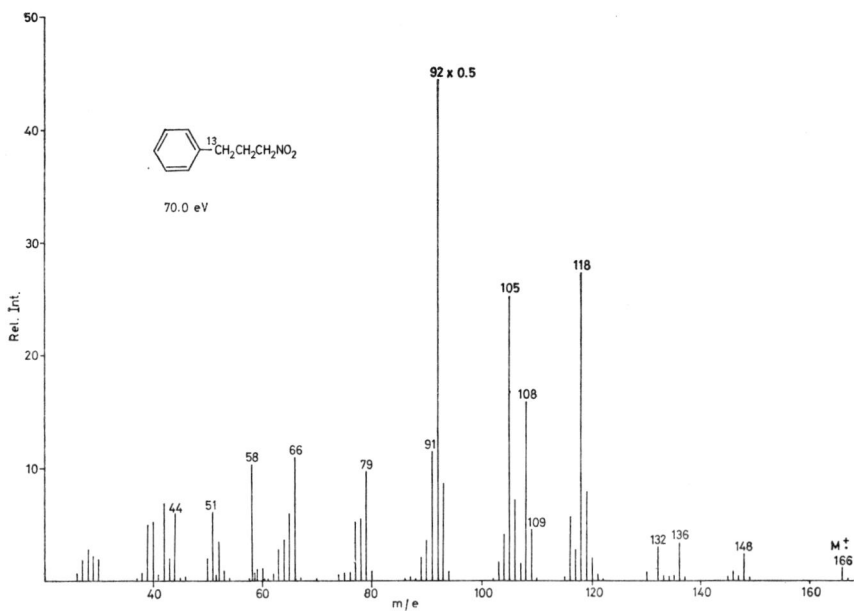

(b)

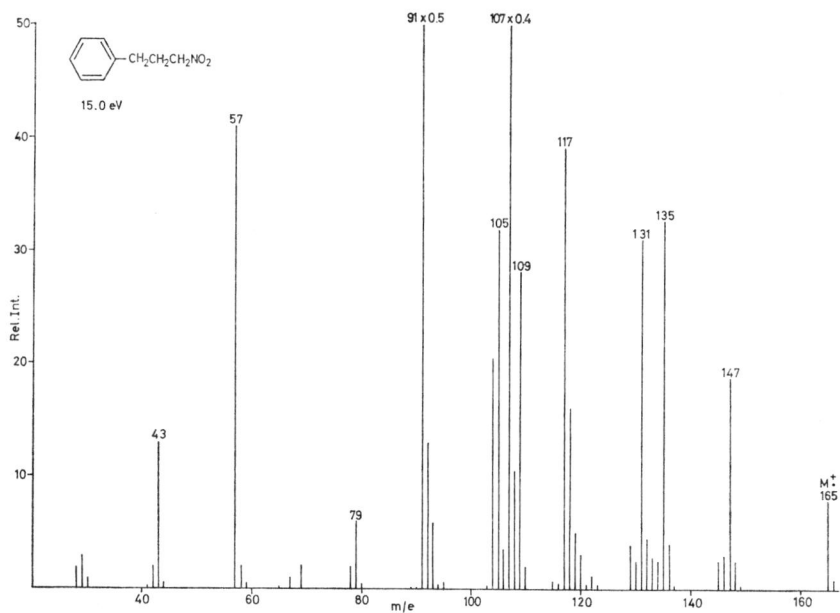

(a)

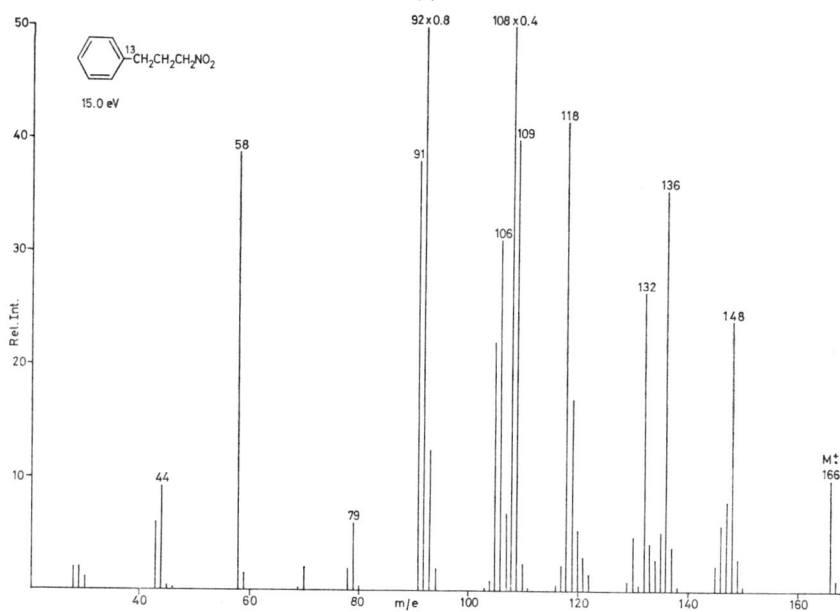

(b)

FIG. 1   Mass spectra of 3-phenyl-1-nitropropane (upper part) and of its analogue, labelled with ^{13}C at the benzylic position (lower part) at 70 eV and at 15 eV. Figures 1(b) have been obtained after correction for unlabelled material (see Experimental); then the sum of intensities of the peaks at m/e 91 and m/e 92 has been taken as 100%. Note also that the actual relative intensities of some peaks have been reduced by the factor given to get them within the limit of 50%.

(2) In the mass spectrum of 3-phenyl-1-nitropropane[1] (m/e 165, $C_9H_{11}NO_2$) a peak at m/e 135 ($C_9H_{11}O$) is observed, possibly arising from an attack of one of the oxygen atoms of the nitro group upon one of the *ortho* carbon atoms of the phenyl ring, followed by expulsion of nitric oxide as suggested by Djerassi.[6]

The explanation presented for the latter case seems very attractive in the light of a recent publication on oxygen migration in primary nitroalkenes under electron impact,[12] showing that such an attack indeed results in the loss of nitric oxide. In a recent paper from our laboratory, however, it has been shown that the $C_9H_{11}O^+$-ion from 2-methyl-2-phenylpropane-1,3-diol has an oxygen protonated phenylacetone structure[9] and exhibits decompositions similar to those of the $(M–NO)^+$-ion of the title compound.[1]

This observation prompted us to consider again the results of D-labelling in 3-phenyl-1-nitropropane, performed earlier,[1] in particular, with regard to the structure of the $(M–NO)^+$-ion; this has led us to the conclusion that ^{13}C-labelling at the benzylic position could provide much more information about its structure, as confirmed in the present paper.

RESULTS AND DISCUSSION

(1) The Successive Loss of Nitric Oxide, Ethylene and Carbon Monoxide from the Molecular Ion of 3-Phenyl-1-Nitropropane

The mass spectrum of 3-phenyl-1-nitropropane is shown in Fig. 1(*a*) and that of its analogue, labelled with ^{13}C at the benzylic position in Fig. 1(*b*).

Comparison of these spectra reveals that the label is completely retained in the $(M–NO)^+$-ion (m/e 135 shifts to m/e 136) and in the $(M–NO–C_2H_4)^+$-ion (m/e 107 shifts to m/e 108), but that it is completely lost in the $(M–NO–C_2H_4–CO)^+$-ion (m/e 79 remains m/e 79).

This observation implies that in some stage of these reactions one of the oxygen atoms of the nitro group is linked to the benzylic carbon atom. The route that is followed to effect this can be deduced from previous D-labelling experiments.[1] These have shown that in the molecular ion one of the benzylic hydrogen atoms migrates to one of the oxygen atoms of the nitro group via a McLafferty rearrangement, resulting in the elimination of nitromethane in the *aci*-form.

Assuming that this rearrangement proceeds in a stepwise manner and indications for it have been presented in the literature[13,14] it is then quite conceivable that the generated hydroxyl group migrates back[15] to the benzylic carbon radical as shown in Scheme 1, sequence $a \rightarrow b \rightarrow c$.

The structure of ion *c*, thus corresponding with ionized 3-phenyl-3-hydroxy-1-nitrosopropane, readily explains the successive expulsion of nitric oxide, ethylene and carbon monoxide as depicted in Scheme 1 through sequence $c \rightarrow d \rightarrow e \rightarrow f$.

The last step is confirmed by the observation of a corresponding diffuse peak; furthermore it may be noted, that ion *e*, produced by loss of the original methyl group from the molecular ion of 1-phenylethanol labelled with ^{13}C at position 1, also expels *exclusively* ^{13}C-labelled carbon monoxide.[16]

The whole sequence *a* to *f* in Scheme 1 fully agrees with previous D-labelling experiments,[1] but invalidates the bond formation between one of the oxygen atoms of the nitro group and one of the *ortho* carbon atoms of the phenyl ring in the molecular ion of 3-phenyl-1-nitropropane, tentatively suggested at that time.[1,6]

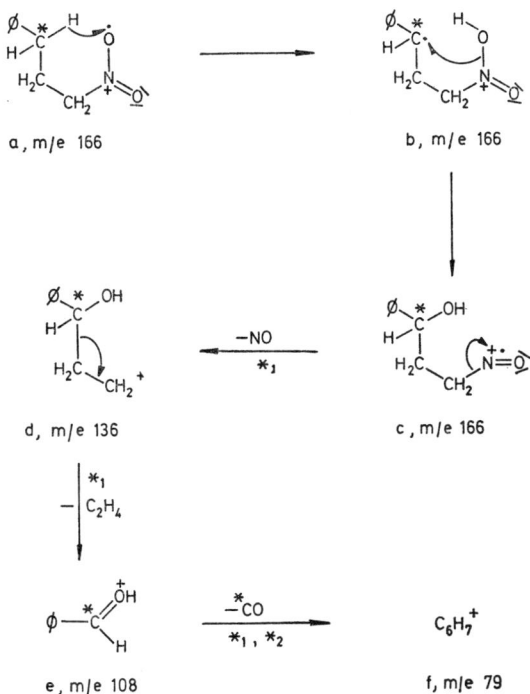

SCHEME 1   Rearrangement of the molecular ion of 3-phenyl-1-nitropropane to that of 3-phenyl-3-hydroxy-1-nitrosopropane and further degradations. *C = ^{13}C; *$_1$ = metastable transition found in the first field free region; *$_2$ = metastable transition found in the second field free region. This notation also holds for the following schemes.

(2) Formation of the $C_6H_7NO^{+\cdot}$-ion from the Molecular Ion

A second argument for rearrangement of the molecular ion of 3-phenyl-1-nitropropane to that of 3-phenyl-3-hydroxy-1-nitrosopropane follows from the peak at m/e 109 in the spectra of the unlabelled- and ^{13}C-labelled compounds (Figs. 1(a) and 1(b)). This observation illustrates that the label is lost during the formation of fragment m/e 109. It is generated in one step from the molecular ion, as shown by the refocussing technique.[17] Inspection of the spectra of the D-labelled analogues, reported earlier,[1] reveals that this fragment contains the original phenyl ring, one hydrogen atom from the 3-position and one hydrogen atom from the 2-position. It must therefore have the composition of C_6H_7NO, which is confirmed by a high-resolution mass measurement.

All these data can be interpreted, starting from the 3-phenyl-3-hydroxy-1-nitrosopropane structure of the molecular ion, as shown in Scheme 2, sequence c to l: step $c \rightarrow g$ corresponds with an attack of a radical upon a phenyl ring, a reaction type extensively described in the literature;[18] the attack of the lone-pair electrons of oxygen upon the primary carbenium ion centre in ion h, generated by a heterolytic cleavage of the C–N bond in ion g, gives ion i and is a reaction type, reported a few years ago by Cooks, Ronayne and Williams;[19] the proton transfer reaction $i \rightarrow j$ seems nowadays to be a common process in gaseous ions;[20,9] the loss of oxetene in step $k \rightarrow l$ involves transfer of a hydrogen atom to a radical centre via a four-membered transition state, a familiar process in mass spectrometry;[21] the 1,2-shift of hydrogen from carbon to nitrogen in reaction $j \rightarrow k$ is assumed because then the aromatic character is restored, although this shift could occur as well after loss of oxetene.

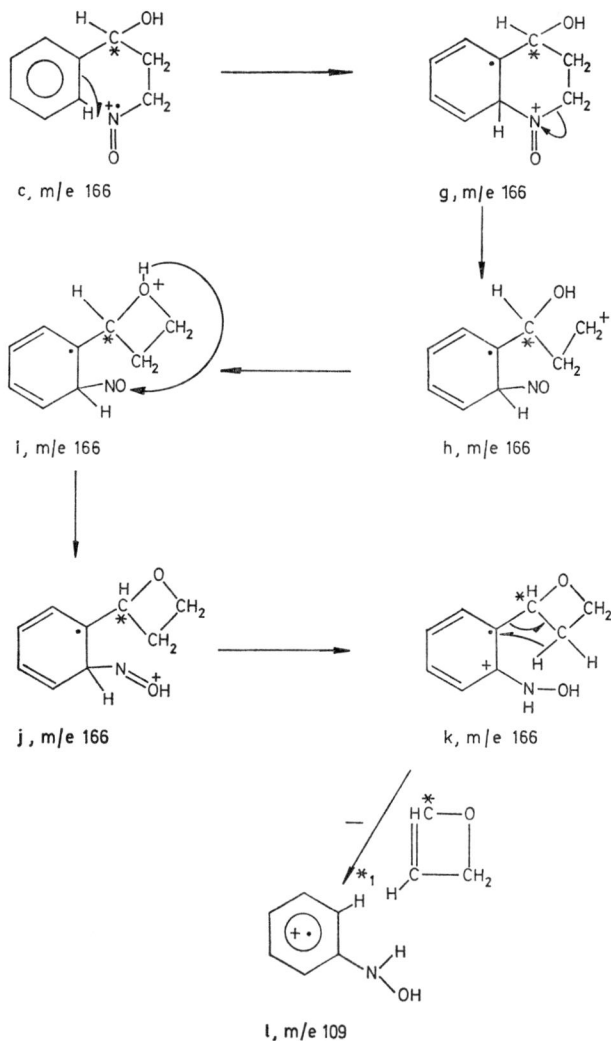

SCHEME 2   Rationalization of the formation of the $C_6H_7NO^{+\cdot}$-ion from the molecular ion of 3-phenyl-1-nitropropane, starting from its rearranged 3-phenyl-3-hydroxy-1-nitroso-propane structure (cf. Scheme 1).

(3) Expulsion of C_6H_6 from the $(M-NO)^+$-ion†

Previous D-labelling experiments and diffuse peaks have shown that the $(M-NO)^+$-ion eliminates a molecule of benzene, containing a benzylic hydrogen atom in addition to the original phenyl ring.[1] This is an interesting reaction, because it formally corresponds with a 1,1-elimination, a rare process in the field of mass spectrometry.[22] The rationalization, given earlier and based upon the assumption of a bond formation between an *ortho* carbon atom and one of the oxygen atoms of the nitro group in the molecular ion,[1,6] must now be abandoned in view of the present ^{13}C-labelling experiment (see above).

It is logical to assume an oxygen-protonated 1-phenyloxetane structure for the present $(M-NO)^+$-ion, either generated by attack of the lone-pair electrons of the oxygen atom upon the primary carbenium ion centre in ion *d* (Scheme 1) or by loss of NO from ion *i* (Scheme 2). Such a structure can easily account for the loss of benzene, as depicted in Scheme 3 through sequence $m \rightarrow n$, and agrees fully with the present ^{13}C-labelling (m/e 57 shifts completely to m/e 58, cf. Figs. 1(*a*) and 1(*b*)).

Ion *n* finally expels carbon monoxide containing the original benzylic carbon atom as evidenced by the ^{13}C-labelling and an appropriate diffuse peak (Scheme 3, sequence $n \rightarrow o$).

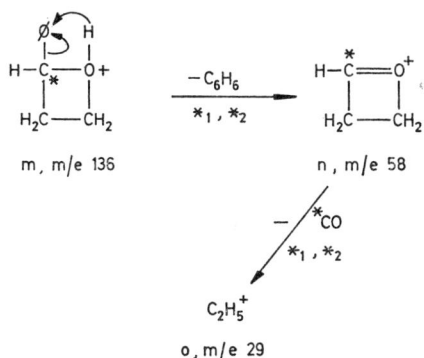

SCHEME 3   Rationalization of the formal 1,1-elimination of benzene from the $(M-NO)^+$-ion of 3-phenyl-1-nitropropane, followed by loss of carbon monoxide (see further text).

CONCLUSION

The principles of the various reactions outlined in this paper appear to be strongly related to those of organic reactions in solution when it is realized that interacting functional groups are now present in one ionic species.

† The $(M-NO)^+$-ion also eliminates C_7H_8 to give fragment m/e 43, as reported earlier.[1] Using the refocussing technique[17] it has now been shown that this fragment is generated not only from the $(M-NO)^+$-ion, but also from the $(M-H_2O)^+$-ion. The corresponding signals, however, are too weak for calculating the relative contributions to its formation, especially in the present case where the ^{13}C-labelling is incomplete. The formation of m/e 43 will therefore not be discussed.

It should further be noted that the rearrangement of the molecular ion described is in favour of a stepwise nature of the well-known McLafferty rearrangement.[23]

EXPERIMENTAL

Mass spectra were obtained with an AEI MS-902 mass spectrometer under the following conditions: Ion source temperature: 150°; inlet temperature of all-glass heated inlet system: 150°; trap current: 98 μA; filament current: 1·1 A; accelerating voltage: 8 kV; repeller potential: $+1·4$ V; pressure in the ion source: $0·6 \times 10^{-6}$ torr; pressure in the analyser region: $0·1 \times 10^{-6}$ torr.

PREPARATION OF 3-^{13}C-3-PHENYL-1-NITROPROPANE

Reaction of benzaldehyde-7-^{13}C with malonic acid in pyridine,[24] followed by reduction with LAH yielded 3-^{13}C-3-phenylpropanol. The corresponding nitro compound was prepared via the bromide as described earlier,[1] and purified by gas chromatography (30% SE30).

3-^{13}C-3-Phenyl-1-nitropropane contained 42·0% unlabelled material.

ACKNOWLEDGEMENT

The authors are greatly indebted to Professor Dr Th. J. de Boer for his interest in this work and to Mr W. J. Rooselaar and Mr F. A. Pinkse for measuring the mass spectra.

REFERENCES

1. Nibbering, N. M. M. and de Boer, Th. J., *Org. Mass Spectrom.*, 1970, **3**, 487 and reference 2 cited therein.
2. Momigny, J., *Bull. Soc. Roy. Sci. Liège*, 1956, **25**, 93.
3. Beynon, J. H., Saunders, R. A. and Williams, A. E., *Ind. Chim. Belge*, 1964, **29**, 311.
4. Budzikiewicz, H., Djerassi, C. and Williams, D. H., 'Mass Spectrometry of Organic Compounds', Holden-Day, San Francisco, 1967, pp. 515–21.
5. Benoit, F. and Holmes, J. L., *Chem. Comm.*, 1970, p. 1031 and references cited therein.
6. Budzikiewicz, H., Djerassi, C. and Williams, D. H., 'Mass Spectrometry of Organic Compounds', Holden-Day, San Francisco, 1967, pp. 512–15.
7. Fischer, M. and Djerassi, C., *Chem. Ber.*, 1966, **99**, 750.
8. Budzikiewicz, H., Djerassi, C. and Williams, D. H., 'Mass Spectrometry of Organic Compounds', Holden-Day, San Francisco, 1967, pp. 153–4.
9. Kerkhoff, M. A. Th. and Nibbering, N. M. M., *Org. Mass Spectrom.*, 1973, **7**, 37.
10. See references 1a–d, quoted in reference 9.
11. Hofman, H. J. and de Boer, Th. J. *in* 'Some Newer Physical Methods in Structural Chemistry', *ed.* R. Bonnett and J. G. Davis, United Trade Press, London, 1967, p. 57.
12. Carney, R. L., *Org. Mass Spectrom.*, 1972, **6**, 1239.
13. Lacey, M. J., MacDonald, C. G. and Shannon, J. S., *Org. Mass Spectrom.*, 1971, **5**, 1391 and references cited therein.
14. Sheehan, M., Spangler, R. J., Ikeda, M. and Djerassi, C., *J. Org. Chem.*, 1971, **36**, 1800.
15. See for other hydroxyl migrations references 1e–h, quoted in reference 9.
16. Unpublished observation from the authors' laboratory.

17. Jennings, K. R. *in* 'Some Newer Physical Methods in Structural Chemistry', *ed.* R. Bonnett and J. G. Davis, United Trade Press, London, 1967, p. 105.
18. Cooks, R. G., *Org. Mass Spectrom.*, 1969, **2,** 481. (Review article.)
19. Cooks, R. G., Ronayne, J. and Williams, D. H., *J. Chem. Soc. (C)*, 1967, p. 2601.
20. a. Molenaar-Langeveld, T. A. and Nibbering, N. M. M., *Tetrahedron*, 1972, **28,** 1043.
    b. Bruins, A. P., Nibbering, N. M. M. and de Boer, Th. J., *Tetrahedron Letters*, 1972, p. 1109.
21. McLafferty, F. W., 'Determination of Organic Structures by Physical Methods', Vol. 2, *ed.* F. C. Nachod and W. D. Phillips, Academic Press, New York, 1962, chap. 2, p. 145.
22. Very recently the three first cases of a specific 1,1-elimination from a *molecular ion* have been found: (a) Kalir, A., Sali, E. and Mandelbaum, A., *J. C. S. Perkin II*, 1972, p. 2262. (b) Meyerson, S. and Karabatsos, G. J., *Org. Mass Spectrom.*, 1973, **7,** 950. (c) Molenaar-Langeveld, T. A. and Nibbering, N. M. M., *Org. Mass Spectrom.*, in press.
23. McLafferty, F. W., *Anal. Chem.*, 1956, **28,** 306.
24. Dutt, S., *Chem. Zentralblatt*, 1925, **II,** 1853.

4

Scope and Limitations in Applying Thermodynamics to the Interpretation of Mass Spectral Data

By P. C. CARDNELL, A. G. LOUDON, R. MAZENGO, I. RUSSELL and K. S. WEBB

(*Christopher Ingold Laboratories, University College, London, U.K.*)

THE object of this paper is to illustrate some of the problems in applying thermodynamics to the interpretation of appearance potential data by discussing three studies involving this undertaken in our laboratory.

The first is the electron-impact induced retro-Diels–Alder type reactions. In the case of the heterocycles Ia–Ic (Ia, X = O; Ib, X = NH and Ic, X = S) it has already been shown[1] that $A[C_8H_8]^{+\cdot}$ fits a concerted mechanism (1)

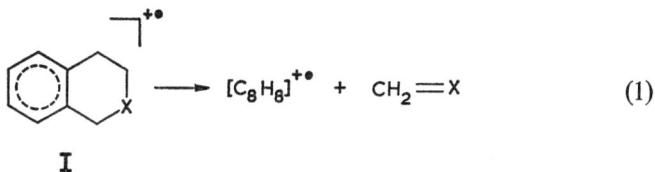

$$[C_8H_8]^{+\bullet} + CH_2 =\!= X \qquad (1)$$

I

for the formation of this ion and gives $\Delta H_f[C_8H_8]^{+\cdot} = 240$ kcal/mole using 2.†

$A[C_8H_8]^{+\cdot}$
$= \Delta H_f[C_8H_8]^{+\cdot} + \Delta H_f \text{ (Neutral product(s))} - \Delta H_f \text{ (Substrate)} \qquad (2)$

Compounds IIa–IIc (IIa, X = O; IIb, X = NH and IIc, X = S) were made in order to investigate if similar reactions occurred. In all cases the $[C_8H_8]^{+\cdot}$ ion was prominent in their mass spectra. In the case of IIa $\Delta H_f[C_8H_8]^{+\cdot}$ calculated from (2) according to the concerted mechanism (3) was 246 kcal/mole, for IIb it was 238 kcal/mole, but for IIc 187 kcal/mole. In

---

† The last two terms in (2) were calculated where necessary by the methods of Benson,[2] otherwise the data were taken from Reference 3. In the case of IIa–IIc no correction was made for the strain in the eight-membered rings which would increase $\Delta H_f[C_8H_8]^{+\cdot}$ slightly. Excess energy terms were ignored, which is reasonable since these are low energy reactions and there are no metastables for $[M]^{+\cdot} \rightarrow [C_8H_8]^{+\cdot}$.

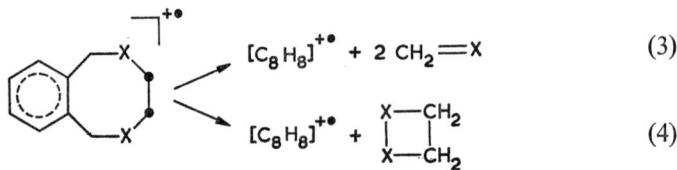

$$[C_8H_8]^{+\bullet} + 2\ CH_2{=}X \tag{3}$$

$$[C_8H_8]^{+\bullet} + \begin{array}{c} X{-}CH_2 \\ | \quad | \\ X{-}CH_2 \end{array} \tag{4}$$

the case of IIc (X = S) this value is below any $\Delta H_f[C_8H_8]^{+\cdot}$ value recorded in the literature.[1,3,4] If a different concerted mechanism (4) is considered then the $\Delta H_f[C_8H_8]^{+\cdot}$ values obtained are 197, 238 and 234 kcal/mole for IIa, IIb and IIc respectively. The $\Delta H_f[C_8H_8]^{+\cdot}$ value for IIa in this case is also below any value recorded in the literature. Thus mechanism (3) can be ruled out for IIc (X = S) and mechanism (4) can be ruled out for IIa (X = O). However, both mechanisms are possible for IIb (X = NH). In this case then the applications of thermodynamics gives a clue to the behaviour of these compounds.

A situation in which the application of thermodynamics to mass spectrometry leads to a more confusing picture is in an attempt to rationalize the $A[R_1R_2N]^+$ and $A[RCO]^+$ values (R_1–R_4 and R = H or Me) determined for compounds of the type $R(X = Y)NR_1R_2$ (Y = O, S), i.e. (methylated) ureas, thioureas formamides, acetamides (X = C) and phosphoramides (X = P NR_3R_4). If the formation of these ions were due to a simple cleavage reaction the appearance potentials should be governed by (5).† For formamide, acetamide and their methylated derivatives $E[C(=O)–NR_1R_2]$ can be calculated from the ΔH_f values

$$A[A]^+_{A-B} = E(A{-}B) + I(A{\cdot}) + E_x \tag{5}$$

of the neutral species[3] to be 83·0 and 90·5 kcal/mole for formamide and acetamide respectively and these values decrease by 1‡ and 3 kcal/mole respectively on mono and dimethylation of the nitrogen. When the problems of competitive shifts are considered, together with the electron impact ionization potentials of HCO· (9·85 eV),[3] CH_3CO· (8·05 eV)[3]†† NH_2·(11·4 eV) and $(CH_3)_2N$· (9·40 eV)[3] it is clear that E_x(5) would be expected to be a minimum for $A[RCO]^+$ when $R_1 = R_2 = H$ and for $A[NR_1R_2]^+$ when $R_1 = R_2 = CH_3$. In the case of formamide, however, $A[M{-}H]^+$ (12·00 eV) is lower than $A[HCO]^+$ (see Table I) and for acetamide $A[M{-}CH_3]^+$ (11·50 eV) is about the same as $A[CH_3CO]^+$ so, particularly for formamide, E_x may not be zero. From the $A[HCO]^+$ and $A[CH_3CO]^+$ values (Table I) for formamide and acetamide respectively the values calculated for $E[C(=O)–NH_2)]$ are 86 ± 6 and 88 ± 6 kcal/mole respectively. These are in good agreement with the values calculated from the thermal data. The agreement is surprisingly good in the case of formamide since $A[NH_2]^+$ is much lower than calculated from (5) and indeed (Table II) lower than

---

† $E(A{-}B)$ is the bond energy, $A[A]^+$ the appearance potential for the ion $[A]^+$ formed from the molecule A–B, and E_x any excess energy.

‡ Value of ΔH_f (monomethylformamide) estimated.[2]

†† The ionization potential calculated for CH_3CO· from appearance potential data[3] is lower than this but this is not the case for HCO.[3]

TABLE I

$A[\text{HCO}]^+$ and $A[\text{CH}_3\text{CO}]^+$ Values^a (eV) for the (Methylated) Formamides and Acetamides of Formulae $\text{RCONHR}_1\text{R}_2$

R/R_1R_2	H, H	H, CH$_3$	CH$_3$	CH$_3$^{b}
H	13·70	13·90	14·50	(14·00)
CH$_3$	11·70	12·40	12·55	(11·35)

All values rounded off to the nearest 0·05 eV; reproducibility about ±0·20 eV.
^{a} Measured under high resolution where necessary.
^{b} The values in brackets are those of Gowenlock et al.[6] The discrepancy in the case of $A[\text{CH}_3\text{CO}]^+$ for N-dimethylacetamide is rather large.

$A[\text{HCO}]^+$. This latter discrepancy is also observed on calculating the $E[\text{C}(=\text{O})-\text{N}(\text{CH}_3)_2]$ from $A[\text{C}_2\text{H}_6\text{N}]^+$ values for dimethyl formamide and dimethyl acetamide (Table II), where values of 51 ± 6 and 63 ± 6 kcal/mole respectively are found.† It seems unlikely that $E[\text{C}(=\text{O})-\text{N}(\text{CH}_3)_2]$ is lower than the E(C–N) bond energy (70 kcal/mole)[7] which would be necessary to make the observed and calculated (5) values agree, since there is some evidence for partial double bond character of the E(C–N) bond in amides.[8a] Thus direct cleavage does not seem to occur here.

TABLE II

$A[\text{NR}_1\text{R}_2]^+$ Values (eV) for Ureas, Thioureas, Acetamides, Formamides and Phosphoramides

Ion^a/Compound	Urea	Thiourea	Phosphor-amide	Acetamide^e	Formamide^e
$[\text{NH}_2]^+$^{b}	12·95	12·65		12·60	13·4
$[\text{CH}_3\text{NH}]^+$^{c}	11·65	11·70	11·50	11·50	11·15
$[(\text{CH}_3)_2\text{N}]^+$^{d}	11·10	10·35	11·20	12·15 (12·40)	11·60 (11·60)

^{a} Nominal structure.
^{b} Non-methylated compound.
^{c} Monomethyl compound except for sym-trimethylphosphoramide.
^{d} For permethyl compound.
^{e} Values in brackets are those determined by Gowenlock et al.[6]

The situation for the amino ions in the case of the ureas and thioureas seems even worse. If we take a value of 86 kcal/mole for $E[\text{C}(=\text{O})-\text{NH}_2]$ in the ureas, which seems reasonable in view of the above arguments and the difference in the 'heat of combustion' of urea calculated using the 'normal' bond energies[8b] and the observed value, then the calculated $A[\text{NH}_2]^+$ value (5) is about 2 eV above the observed value. A similar calculation for thiourea involving the comparison of the 'heat of formation' of thiourea calculated from the 'normal' bond energies[7] and the observed value[3] shows that thiourea is about 28 kcal/mole more stable than would be expected. If this is due only to some double bond character[8a] in the C–N bonds then a reasonable estimate for $E[\text{C}(=\text{S})-\text{NH}_2]$ would be about 84 kcal/mole. Taking this

† The calculations of Gowenlock et al.[6] show similar discrepancies.

value then $A[NH_2]^+$ calculated (5) is about 2·4 eV higher than that observed. In the case of the dimethylamino cation the $A[(CH_3)_2N]^+$ values calculated (5) are about 1·70 (tetramethylurea) and 2·30 (tetramethylthiourea) higher than the observed values. Even in the case of $A[NH_2]^+$ for formamide and acetamide, where any competitive shift would be expected to be a maximum, similar discrepancies (1·40, 2·50 eV respectively) exist. In contrast, using an estimated value for $E(P–N)$ (2·2 eV)[7] gives $A[(CH_3)_2N]^+$ calculated (5) of 11·3 eV. This is in good agreement with the observed value for hexamethyl-phosphoramide, but in the absence of any data for $E[P(=O)–N(CH_3)_2]$ this result must be treated with caution.

TABLE III

$A[M–CH_3]^+$ and $A[M–H]^+$ Values for some n,n'Dimethyl Biphenyls and 1,1-Binaphthyls

Ion/Compound	Binapthyls				Biphenyls		
n	2	3	7	8	2	3	4
$A[M–CH_3]^+$	13·25	12·25	12·75	11·50	11·75	13·50	12·65
$A[M–H]$	—	—	—	—	12·20	13·00	12·85

Several possible explanations for the discrepancy in the case of the $[NH_2]^+$ ion can be ruled out. Pyrolysis† to ammonia followed by electron impact fragmentation seems unlikely since $A[NH_2]^+$ from EI is 16·00 eV[3] and about 15·3 eV from photoionization experiments.[10] $I(NH_2·)$ has been determined both by electron impact (11·40 eV)[3] and indirectly from photo-ionization experiments (11·22 eV)[10] so this value seems reasonably reliable. The $I([CH_3]_2N·)$ value seems also reliable since the value calculated from $A[M–CH_3]^+$ for trimethylamine of 9·3 eV[3] (5) is in good agreement with the value determined directly. However, here for this and the $[CH_4N]^+$ ions the possibility of the formation by rearrangement of the ion corresponding to the $[M–H]^+$ ions of mono and di-methylamine cannot be ruled out since $I(CH_4N·)$ and $I(C_2H_6N:)$ calculated from (5) are 6·45 and 6·25 eV, $A[M–H]^+$ being 11·65 and 10·45 eV respectively.

Calculation for the ureas, thioureas and amides show that the formation of these ions from the $[M–R]^+$ ions is a higher energy process than formation from the molecular ions. In the case of some of the thioureas these ions are formed from the $[M–HS]^+$ ions (m*) but assuming the elements of HCN lost are lost as HCN, the most stable possible species, gives, in the case of thiourea, a higher calculated appearance potential than the direct cleavage value. Unfortunately there are no $I[·NR_1R_2 C = Y]$ values available so no cross-check is available in this case.

This problem of the structure of the ion formed is also very important in the field of correlation of steric strain and appearance potentials,[11] which is the theme of the third piece of work to be discussed. We have reported[12] the

† The intensity ratios $(R_1R_2NH)/R_1R_2N$ are also wrong for a considerable contribution to the intensity of $[R_1R_2N]$ by pyrolysis, but a small contribution could lower the appearance potential.[15]

$A[M-CH_3]^+$ values for some n,n'dimethyl-1,1'-binaphthyls in which there is considerable deformation of the aromatic rings in the 8,8' compound. However, the variation in the $A[M-CH_3]^+$ values (Table III) seems to be much larger than can be explained on the basis of variation in the steric strain.[13] Further from the spectra of the n,n'di-(monodeuteromethyl) compounds it is clear that in the case of 2,2'dimethyl-1,1-binaphthyl† and all the biphenyls the low energy process for the formation of the $[M-CH_3]^+$ ions does not involve simple loss of one of the original methyl groups. The labelling evidence, however, is also not consistent with randomization of the hydrogens and so it seems unlikely that a common ion is formed.[14] Thus the strain energies are not likely to be reflected in the appearance potentials, since this demands common ion formation (see (2)).

In the case of the $[M-H]^+$ ion the labelling evidence suggests that similar if not common ions may be formed. In all the labelled compounds the (M-H)/(M) ratio suggests complete hydrogen scrambling over one or both rings. However, the ion at m/e (M-16) in the unlabelled compounds, which comes from the $[M-H]^+$ ion (m*) moves to m/e (M-17) in the labelled compounds, suggesting the loss of one of the original methyl groups. Assuming that this hydrogen scrambling implies a common structure for one of the original rings then the $[M-H]^+$ ions may have similar heats of formation.‡ If the differences in steric strain were small as is suggested[13] then this would produce similar $[M-H]^+$ values. In the case of the biphenyls this, with the exception of the 2,2' compound perhaps seems to be the case. In the case of the binaphthyls the log ion current versus electron voltage curves tail w.r.t. the standard below 0·5% of the ion current at 50 eV.†† However, at this point the voltage is the same for both the 2,2' and 8,8' compounds.

In this case, although the results are not inconsistent with thermodynamics the $A[M-CH_3]^+$ values cannot be explained on this basis, although the $A[M-H]^+$ values can be to some extent rationalized.

REFERENCES

1. Loudon, A. G., Maccoll, A. and Wong, S. K., *J. Chem. Soc. B.*, 1970, p. 1727.
2. Benson, S. W., *Thermochemical Kinetics*, J. Wiley, New York, 1968.
3. Franklin, J. L., Dillard, J. G., Rosenstock, H. M., Heron, J. T., Draxl, K. and Field, F. H., 'Ionisation Potentials, Appearance Potentials and Heats of Formation of Gaseous Positive Ions', *N.S.R.D.S.–N.B.S.*, 1969, **26**, Washington.
4. Franklin, J. L. and Corral, S. R., *J. Amer. Chem. Soc.*, 1969, **91**, 5940.
5. Henderson, E. and Fisher, I. P., *Trans. Farad. Soc.*, 1967, **63**, 1342.
6. Gowenlock, B. G., Jones, P. and Majer, J. R., *Trans. Farad. Soc.*, 1961, **57**, 23.
7. Cotton, F. A. and Wilkinson, G., p. 100 *in* 'Advanced Inorganic Chemistry', 2nd edn. J. Wiley and Son, London, 1966.

† The 3,3' and 7,7' di-(monodeuteromethyl)-1,1'-binaphthyls were not prepared.

‡ In both cases the near identity of the intensity ratio of two metastable decompositions of the $[M-H]^+$ ion also suggests this.

†† See Reference 16 for a description of the method used to measure ionization and appearance potentials.

8. a. Roberts, J. D. and Caserio, M. C., p. 674, 'Basic Principles of Organic Chemistry', Benjamin, New York, 1965.
   b. Ibid, Idem. p. 245.
9. Isaksson, G. and Sandstrom, J., *Acta Chem. Scand.*, 1970, **24**, 2565 and references therein.
10. Dibeler, V. H., Walker, J. A. and Rosenstock, H. M., *J. Res. Nat. Bur. Stand. A*, 1956, **70**, 459.
11. Jalonen, J. and Pihlaja, K., *Org. Mass. Spectrom.*, **7**. In press.
12. Harris, M. M., Loudon, A. G. and Mazengo, R., *Org. Mass Spectrom.*, 1970, **3**, 1123.
13. Harris, M. M., personal communication.
14. Loudon, A. G. and Mazengo, R. Accepted for publication in *Organic Mass Spectrom.*
15. a. Hvistendahl, G. and Undheim, K., *Org. Mass Spectrom.*, 1973, **6**, 217.
    b. Baldwin, M. A., Maccoll, A. and Miller, S. I., 'Advances in Mass Spectrometry', 1964, **3**, 259.
16. Baldwin, M. A., Kirkien-Konasiewicz, A., Maccoll, A. and Saville, B. *Chem. and Ind.*, 1966, p. 286.

5

Stereoselective Fragmentation Processes in Epimeric Bicyclo(3,3,1)Nonenes

By ISTVÁN LENGYEL

(*Department of Chemistry, St John's University, Jamaica, New York, U.S.A.*)

and USHA R. GHATAK

(*Department of Organic Chemistry, Indian Association for the Cultivation of Science, Calcutta, India*)

THE stereoisomers of most simple organic compounds show identical or nearly identical mass spectra[1] due to the ease of rotation about the bonds attached to the asymmetric centres and/or rearrangements leading to common molecular ions from both stereoisomers. On the other hand, diastereoisomers of alicyclic compounds and epimers of polyfunctional polycyclic compounds with relatively rigid skeletons may exhibit striking mass spectral differences.[2] Recently, Wulfson and co-workers were able to deduce the stereochemistry of epimers of substituted perhydroquinolines by mass spectrometry.[3]

Our investigation of the low and high resolution mass spectra of three pairs of epimeric bicyclo(3,3,1)nonene derivatives (1–6, Scheme 1), accessible

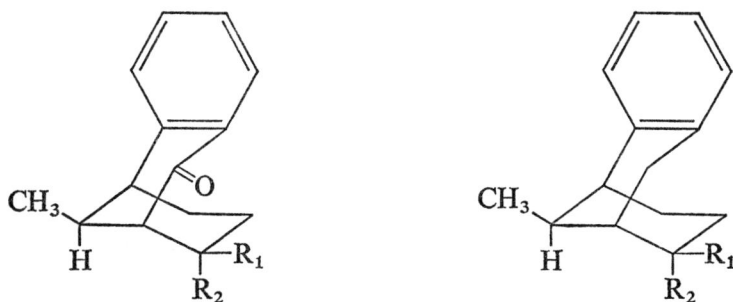

(1) R_1 = COOCH$_3$; R_2 = CH$_3$
(2) R_1 = CH$_3$; R_2 = COOCH$_3$

(3) R_1 = COOCH$_3$; R_2 = CH$_3$
(4) R_1 = CH$_3$; R_2 = COOCH$_3$
(5) R_1 = COOH; R_2 = CH$_3$
(6) R_1 = CH$_3$; R_2 = COOH

SCHEME 1

via the acid-catalysed cyclization of substituted benzylcyclohexanols,[4] established that the large differences observed between the mass spectra of the epimers permit assignment of configuration in this series of compounds.

First, for all three pairs of diastereomeric racemates the more stable (less crowded) *endo* epimer (*axial* COOH or COOCH$_3$ group) exhibits a more abundant molecular ion, *e.g.* 6·63 for *2* versus 3·71 for *1*, expressed in per cent total ionization ($\% \Sigma^{M}_{28}$).

Along with many seemingly non-stereospecific processes in which the original stereochemical difference about C-1 is lost due to C–C bond cleavages in the alicyclic system, some fragmentations are stereoselective and even stereospecific. The stereoselective processes appear to be rearrangements requiring a cyclic transition state. The M-32 ion, m/e 226, is far more abundant

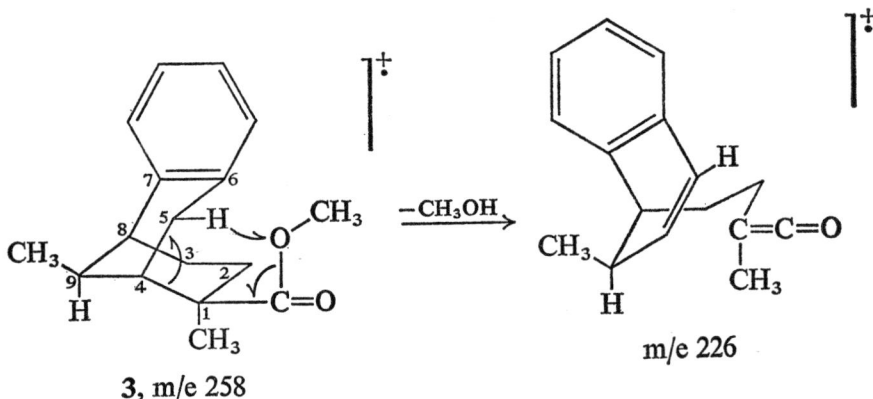

3, m/e 258   m/e 226

SCHEME 2

in the *exo* ester *3*, clearly indicating an orientation-dependent fragmentation process. Elimination of methanol through a six-membered transition state, with participation of a highly activated benzylic hydrogen and formation of a conjugated double bond (Scheme 2) in the product ion is feasible in *3*, not in *4*. It is noteworthy that no ejection of OCH$_3$ is observed from the molecular ions of either ester *3* or *4*. On the other hand, equatorial γ-ketoester *1* in which no hydrogen is available without concomitant C–C bond fission, eliminates OCH$_3$ preferentially, while *axial* γ-ketoester *2* loses methanol (Scheme 3).

An analogous, completely stereospecific process is observed in the carboxylic acids. Equatorial acid *5* eliminates HCOOH (not COOH) to yield ion m/e 198 (M-46), whereas the *endo* isomer *6* loses a COOH radical.

While these fragmentations are very useful for diagnostic purposes, the most abundant ions result from other, more complex processes. The base peak in the spectrum of equatorial ester *3* is at m/e 143, elemental composition exclusively C$_{11}$H$_{11}$. On the other hand, the base peak in the spectrum of the epimeric ester *4* is at m/e 115 (C$_9$H$_7$ and C$_6$H$_{11}$O$_2$), while m/e 143 has a relative abundance of 68%. Thus, initial fission of bonds resulting in free

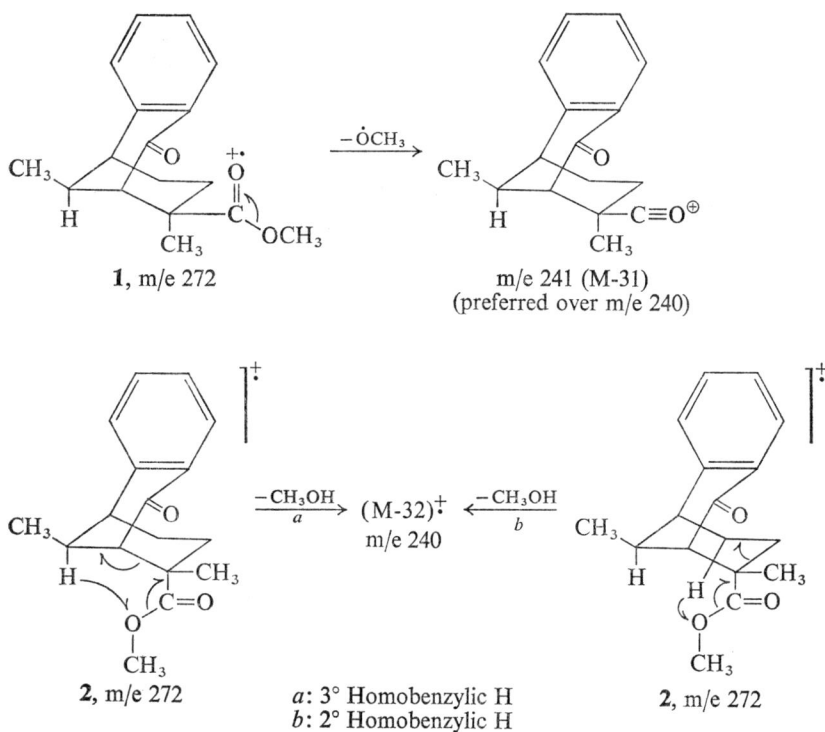

$-\dot{O}CH_3$

1, m/e 272

m/e 241 (M-31)
(preferred over m/e 240)

$\xrightarrow[a]{-CH_3OH}$ (M-32)$^{+\cdot}$ $\xleftarrow[b]{-CH_3OH}$
m/e 240

2, m/e 272

a: 3° Homobenzylic H
b: 2° Homobenzylic H

2, m/e 272

SCHEME 3

rotation about the C_1-C_2 or the C_1-C_4 bond cannot precede fragmentation leading to ion m/e 143, i.e. the original stereochemical difference about C-1 appears to be preserved, at least partially.

According to Dreiding models, equatorial ester 3 has only one γ-hydrogen suitably oriented and at the proper distance for a McLafferty rearrangement, a benzylic one at C-5. Rearrangement of this hydrogen followed by cleavage of the benzylically activated bond a (Scheme 4) leads to m/e 143 ($C_{11}H_{11}$) or alternatively, to m/e 115 ($C_6H_{11}O_2$).

Dreiding models of the axial epimer 4 show that there are two equidistant (but not equivalent) γ-hydrogens suitably oriented for a McLafferty rearrangement, viz., at C-3 and at C-9. The spectrum of 4 indicates a vast preference for the involvement of the C-9 γ-hydrogen, which is tertiary, over the secondary one at C-3: ions m/e 143, 115, 157 and 101, that should result from participation of the former, are all present, while m/e 130 and 128 ($C_7H_{12}O_2$), which would be expected from rearrangement of the latter, are absent.

The most complicated fragmentation processes occur in γ-ketoesters 1 and 2, due to the additional functional group. A non-stereospecific process, leading to m/e 172 (exclusively $C_{12}H_{12}O$, 30% rel. int. in 1, base peak in 2, merits comment. As indicated by a metastable peak at m/e 108·8 for 2, this is a *one-step* transformation ($M^{\cdot +} \rightarrow 172$). Fission of the benzylic bond a

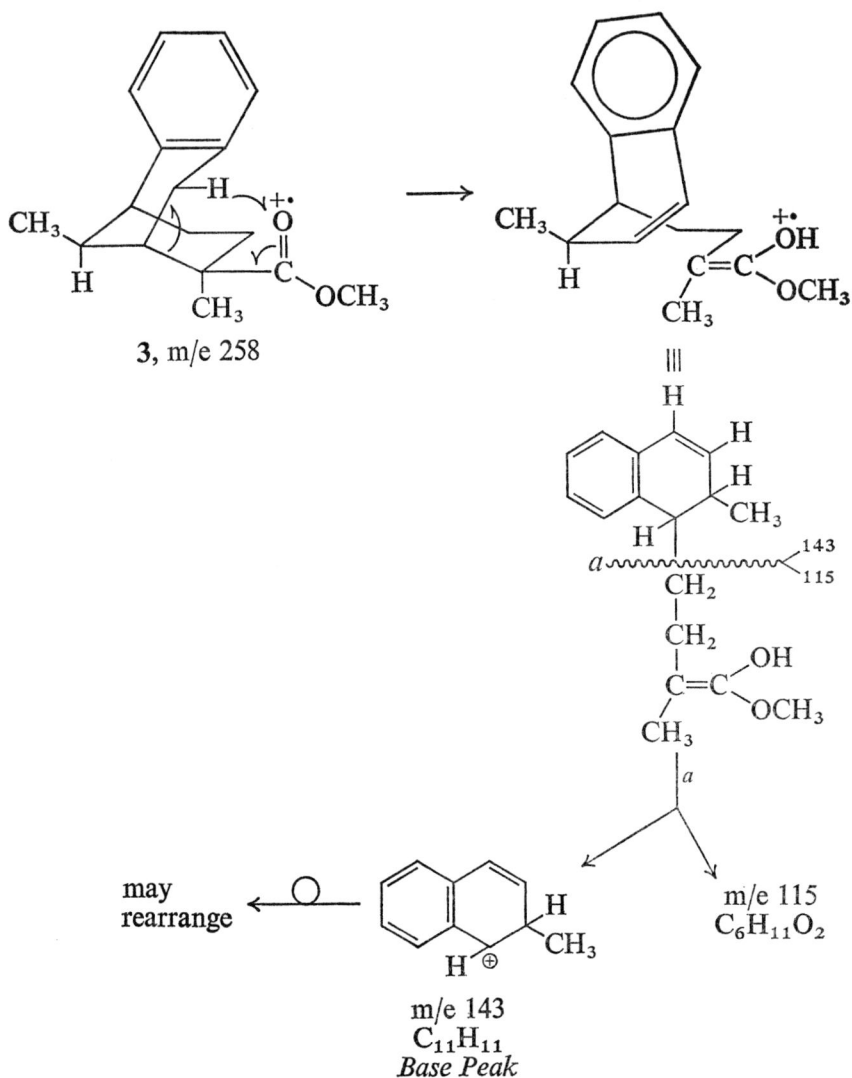

3, m/e 258

m/e 115
$C_6H_{11}O_2$

may
rearrange

m/e 143
$C_{11}H_{11}$
Base Peak

SCHEME 4

(Scheme 5) represents α-cleavage, a typical characteristic of ketones. A 1,5-hydrogen shift with concomitant formation of a conjugated double bond, followed by a retro-Diels–Alder cleavage of the cyclohexene ring with loss of the elements of methyl methacrylate as the neutral fragment leads to the stable through-conjugated ion m/e 172 ($C_{12}H_{12}O$).

The conventional (low resolution) mass spectra were recorded at 70 eV on a CEC 21-103C instrument equipped with a solid probe inlet. For comparison, several pairs were rerecorded on a Hitachi RMU-6D single focussing instrument with little change in the overall pattern. The 70 eV high resolution

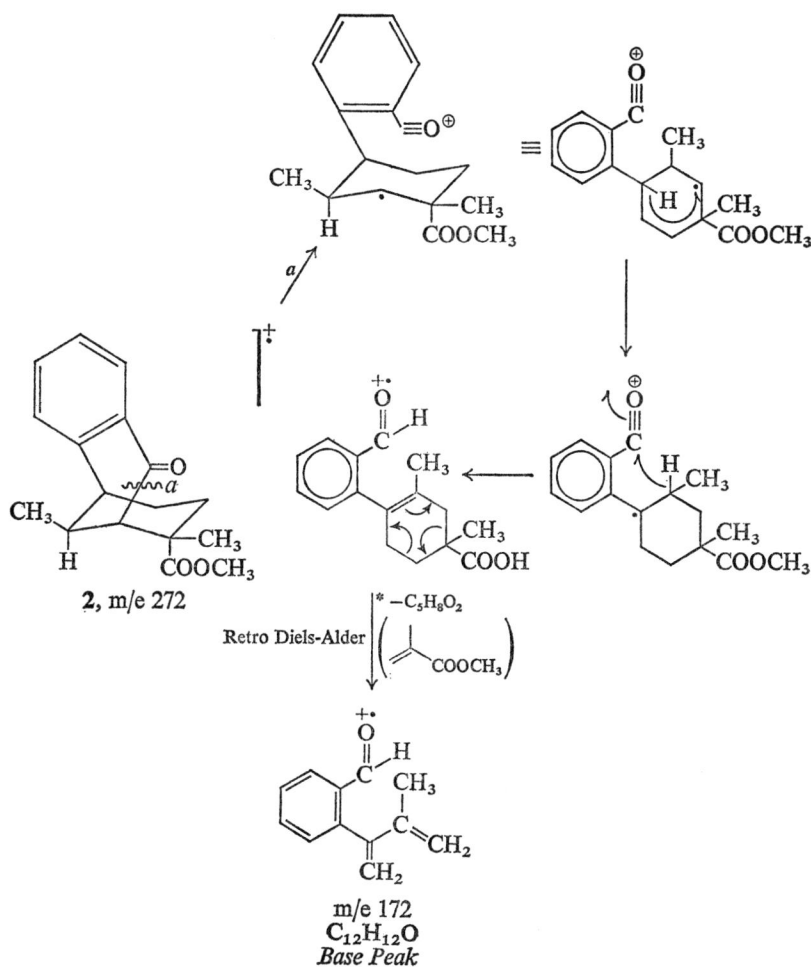

SCHEME 5

spectra were obtained on a CEC 21-110B double focussing mass spectrometer in conjunction with an IBM-1800 computer system.

ACKNOWLEDGMENT

We thank Professor Klaus Biemann, Department of Chemistry, Massachusetts Institute of Technology, for the high-resolution spectra (element maps).

REFERENCES

1. Biemann, K., 'Mass Spectrometry', McGraw-Hill Book Co. Inc., New York, 1962, pp. 144–51.

2. Deutsch, J. and Mandelbaum, A., *J. Amer. Chem. Soc.*, 1970, **92**, 4288.
3. Wulfson, N. S., *et al., Org. Mass Spectrom.*, 1972, **6**, 533.
4. a. Ghatak, U. R., Chakravarty, J. and Banerjee, A. K., *Tetrahedron Letters*, 1965,
   p. 3145.
   b. Ghatak, U. R. and Chakravarty, J., *Tetrahedron Letters*, 1966, p.2449.

6

Kinetic Studies on the Mass Spectral Fragmentation of Some s-Triazolo (4,3-b) and Tetrazolo (1,5-b) Pyridazines

By V. KRAMER, M. MEDVED and J. MARSEL

(*J. Stefan Institute, University of Ljubljana, Ljubljana, Yugoslavia*)

IN a recent paper we presented the mass spectral behaviour of several azolo-pyridazines.[1] The mass spectra of a number of substituted s-triazolo (4,3-b) pyridazines, tetrazolo (1,5-b) pyridazines and some related compounds were reported in that paper. At that time we gave rather a general fragmentation scheme, based solely from inspection of normal 70 eV spectra and accurate mass measurements of a few key fragment ions. To complete these investigations, we have re-examined some selected compounds to obtain more insight into actual fragmentation pathways, as well as the eventual existence of common ion intermediates. For that reason the following compounds were taken:

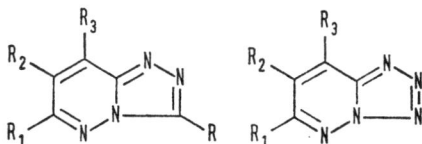

I $R = R_1 = R_2 = R_3 = H$

II $R = R_2 = R_3 = H$, $R_1 = CH_3$

III $R = R_2 = R_3 = H$, $R_1 = Cl$

IV $R = R_2 = CH_3$, $R_1 = Cl$, $R_3 = H$

V $R = R_2 = H$, $R_1 = CH_3$, $R_3 = Cl$

VI $R = R_2 = H$, $R_3 = CH_3$, $R_1 = Cl$

VII $R_1 = R_2 = R_3 = H$

VIII $R_1 = R_3 = H$, $R_2 = CH_3$

IX $R_1 = R_2 = H$, $R_3 = CH_3$

X $R_2 = R_3 = H$, $R_1 = Cl$

XI $R_3 = H$, $R_1 = Cl$, $R_2 = CH_3$

XII $R_2 = H$, $R_1 = Cl$, $R_3 = CH_3$

Their usual 70 eV spectra show that the primary fragmentation event for s-triazolo (4,3-b) pyridazines is the loss of 55 mass units, whereas with tetrazolo (1,5-b) pyridazines it appears to be a loss of 56 mass units. This means the expulsion of a HCN and N_2 from M^+-ions of the former compounds, and two nitrogen molecules from the latter. In this way both types

53

of compounds, if equally substituted, could form at least compositionally equal ions. Thus, *e.g.* the unsubstituted compounds I and VII form ions at m/e 65 (h.r. C_4H_3N), but of very different intensities. These observations alone do not allow any satisfactory conclusions about the fragmentation path. Until the appropriate metastable measurements were made, the correct sequence of HCN and N_2 expulsion could not be given even for *s*-triazolo compounds, because the respective intermediate ion intensities lie in the general background of these spectra. Only for compound I a peak at m/e 92 (M^+–28 mu) of 0·3% rel. int. could give some indication that the decomposition starts with nitrogen expulsion. For tetrazolo-pyridazines no intermediate ions for the reaction M^+–$2N_2$ were found at all. Thus, due to the presence of two isolated N–N linkages in *s*-triazolo-pyridazines and the even more complex tetrazolo–pyridazine system, several interpretations are possible. To clarify the fragmentation mechanisms to some extent we undertook measurements of metastable transitions in the first field free region[2,3] for the major reaction processes. Summarizing the results obtained together with previous observations, many common features of both groups of compounds were registered.

The mass spectra of the *s*-triazolo group display molecular ions as the base peaks, indicating the great stability of this system under electron impact. For all these compounds, except for the 3-methyl derivative, where instead of HCN, CH_3CN is lost, the decomposition sequence M^+–N_2–HCN is found to be very characteristic. Ions formed are of medium intensities for compounds I to III, while they do not exceed 1% rel. int. for chlor-methylated products (IV–VI). Starting from the molecular ion, only for compound I a marked competing process M^+–2HCN appears, whereas isomeric chlor-methyl substituted compounds eject, though to a very low degree, a chlorine radical.

After N_2 and HCN are lost, further fragmentation depends rather on the substituents present in the molecule than their relative positions. Thus for unsubstituted and methyl substituted compounds two competing processes are operative, *i.e.* hydrogen or an additional HCN molecule is lost. In contradistinction, the isomeric chlor-methyl substituted compounds lose a chlorine radical at this step, giving rise to the second most abundant ion at m/e 78, followed by the loss of HCN to afford a peak at m/e 51.

Which of the nitrogens are eliminated in the first step still remains uncertain. Evidently, this problem could only be solved with appropriate ^{15}N labelling studies. But it seems most likely that the nitrogen is expelled first of all from the five-membered triazolo ring (path a, Scheme 1), and that the HCN (3-methyl *s*-triazolo compound IV loses CH_3CN), lost as the second step of this very fast reaction, includes C-3 and the bridgehead nitrogen atom (path b). The following 'paper structures' can be drawn (Scheme 1), *e.g.* for fragmentation of compound I (for substituted *s*-triazolo (4,3-b) pyridazines an appropriate mass shift to higher masses could be imagined).

Equally substituted tetrazolo (1,5-b) pyridazines undergo a similar fragmentation mode; however, the stability of the ions formed differs considerably with those of the former group. The M^+-ions are ranged here from 23% rel. int. at VII down to 3% rel. int. for compound XII, indicating the much lower stability of the tetrazolo system. For the primary two-step transition

M^+-2N_2 we also suggest an analogous fragmentation mechanism shown in the previous Scheme 1. This process as well as further decompositions lead to compositionally and, in some instances, most likely to structurally similar ions as will be discussed later. Other competing fragmentations direct from the molecular ions are also of minor importance here. Further decompositions of M^+-2N_2 ions proceed through an alternative loss of hydrogen or HCN for compounds VII and IX (forming base peaks) and successive loss of chlorine (forming base peaks) and HCN for X to XII.

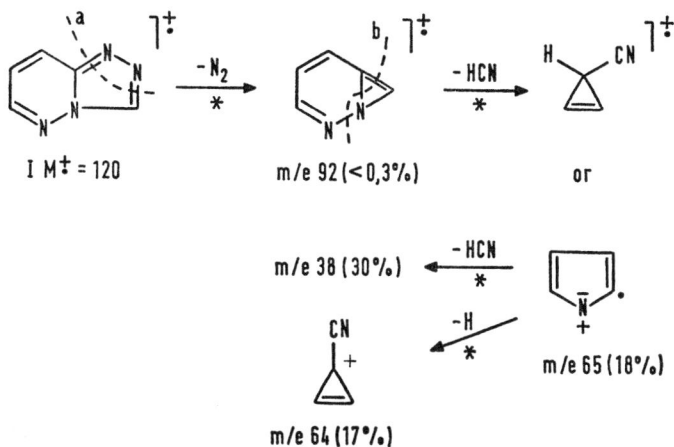

Scheme 1

To find some regularities in these mass spectra with regard to common ionic intermediates, the kinetic approach was applied.[4-8] The method used rests on the characterization of decomposing ions by their metastable transitions. As stated recently, ions identical in energy and structure should behave similarly when decomposing further.[4] For this reason the metastable ion/daughter ion ratios were measured by means of the 'defocussing technique' for several decomposing ionic species, supposed to be identical in structure.

The values given in the Tables are average results of at least five determinations. An error of about 10% at higher electron-volt values should be taken into account. However, it becomes still greater at lower electron energies due to the decreasing peak intensity/noise ratios. Elsewhere in the Tables there are also attached explanatory schemes where the measured entities are depicted.

The results obtained for the competing decomposition of m/e 65 which takes place with unsubstituted *s*-triazolo (4,3-b) pyridazine I and tetrazolo (1,5-b) pyridazine VII are represented in Table I. Considering the errors accompanied by such kind of measurements, reasonable results are found. Constancy of [metastable/daughter] ion ratios for each compound at different electron-volts suggests at least a common structure for m/e 65. The metastable abundance ratios for the competing reactions are very close over a wide range

TABLE I

$$\text{I } (M^+ = 120) \xrightarrow{-N_2,\,-HCN} \quad m/e\ 65 \xrightarrow[m_2^\bullet]{-H \; m_1^\bullet} \begin{matrix} m/e\ 64 \\ \\ m/e\ 38 \end{matrix}$$

$$\text{VII } (M^+ = 121) \xrightarrow{-2N_2} m/e\ 65$$

with $-HCN$ leading to $m/e\ 38$.

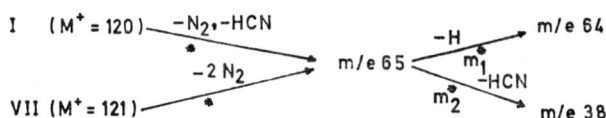

	s-triazolo(4,3-b)pyridazine I			tetrazolo(1,5-b)pyrid. VII		
eV	$Im_1^\bullet/I_{64}$ ·10²	$Im_2^\bullet/I_{38}$ ·10²	$Im_1^\bullet/Im_2^\bullet$	$Im_1^\bullet/I_{64}$ ·10²	$Im_2^\bullet/I_{38}$ ·10²	$Im_1^\bullet/Im_2^\bullet$
70	3,8	2,7	1,4	3,2	2,6	1,2
25	3,7	3,7	1,0	3,6	3,7	1,0
15	9,0	6,8	1,3	13,0	10,0	1,3

of ionizing energies for both I and VII. Thus their decomposition reactions do not occur from isolated electronic states.[7]

Table II represents the relative metastable abundance ratios for a similar competition reaction, but for the methylated products. The values again agree well the supposition that m/e 78 and m/e 52 arise from a common state. Small variations which here and there slightly exceed the estimated error, could be ascribed to the energy and energy distribution differences as well as to experimental uncertainties.[9] Although the substituent positions differ, it seems that after N_2 and HCN for II and $2N_2$ for VIII and IX are lost, a re-arrangement, due to the methyl present, occurs. This observation confirms our foregoing presumptions that eliminated nitrogen and HCN originate mainly from triazolo—and tetrazolo rings including the bridgehead nitrogen atom. In this way a common pyridinium ion m/e 79 can be formed (Scheme 2).

TABLE II

$$\text{II } (M^+ = 134) \xrightarrow{-N_2,\,-HCN} m/e\ 79 \xrightarrow[m_2^\bullet]{-H \; m_1^\bullet} \begin{matrix} m/e\ 78 \\ \\ m/e\ 52 \end{matrix}$$

$$\left. \begin{matrix} \text{VIII} (M^+ = 135) \\ \text{IX } (M^+ = 135) \end{matrix} \right\} \xrightarrow{-2N_2} m/e\ 79$$

with $-HCN$ leading to $m/e\ 52$.

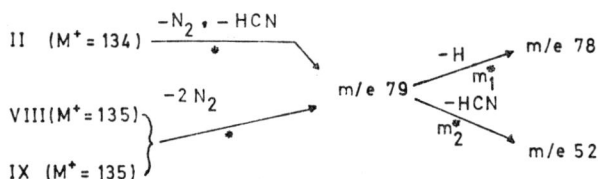

	$Im_2^\bullet / Im_1^\bullet$				
eV	70	30	25	20	18
II	1,4	1,6	0,7	1,2	1,2
VIII	1,3	0,5	0,6	0,8	0,8
IX	1,4	0,5	0,6	0,8	0,8

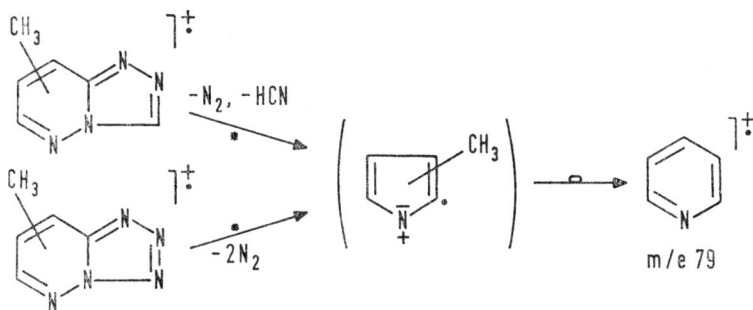

Scheme 2

If this is true, the incorporation of an additional substituent, *e.g.* chlorine, into different positions should cause formation of substituted isomeric pyridinium ions. Table III displays the metastable ion relative abundances of five chloro- and methyl-substituted derivatives for the consecutive decomposition of compositionally common ions m/e 113 and m/e 78.

In fact the metastable relative intensities for this chlorine elimination step markedly differ among this series of compounds. The given values greatly exceed experimental error and therefore it is hardly possible to believe that m/e 78 could arise from a structurally common precursor ion. The results for isomeric 6-chloro methyl substituted tetrazolo (1,5-b) pyridazines (XI and XII) closely relate to that of 6-chloro-8-methyl s-triazolo (4,3-b) pyridazine (VI), which in turn differs from its 8-chloro-6-methyl isomer (V). Thus the following fragmentation pathways, leading to the different m/e 113, could be visualized, *e.g.* for the last two instances (Scheme 3). Their normal mass spectra are similar and could not be differentiated from one another with certainty.

TABLE III

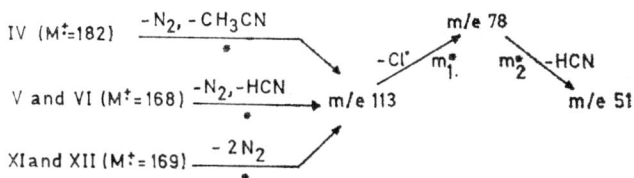

	$I m_1^\bullet / I_{78} \cdot 10^3$					$I m_2^\bullet / I_{51} \cdot 10^2$			
eV	70	25	20	18		70	25	20	18
IV	5,0	16,0	5,3	5,6		4,7	11,6	21,6	24,5
V	5,3	4,9	5,5	6,3		4,2	11,1	20,4	23,7
VI	1,7	1,5	1,9	2,1		4,4	11,7	20,3	23,7
XI	1,3	1,3	1,3	1,3		4,5	8,6	12,5	15,0
XII	1,8	1,8	1,7	1,7		4,5	8,7	10,6	15,6

6-chloro-3,7-dimethyl s-triazolo (4,3-b) pyridazine (IV) offers an unexpected exception which surprisingly correlates to compound V instead of to other 6-chlorinated products. To do so, a further fragmentation mechanism, due to the additional methyl group, must take place. However, up to now we have not found a satisfactory explanation for this behaviour.

The metastable characteristics of the decomposing m/e 78 clearly show that chlorine elimination affords an ion structure which is common to all compounds presented in Table III. Now the compounds IV, V and VI are very

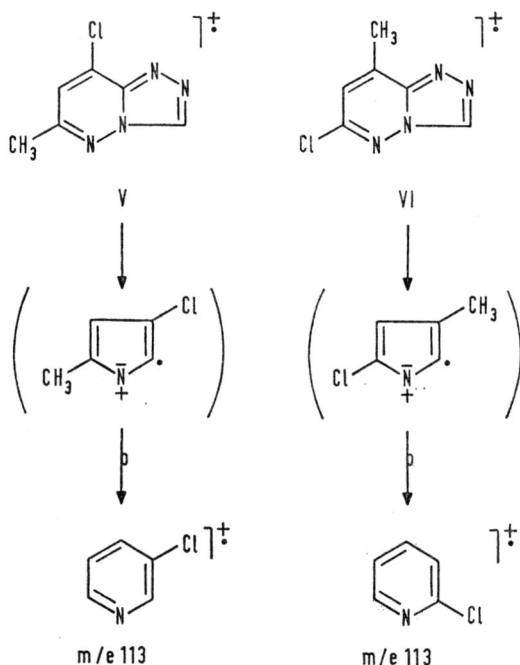

Scheme 3

closely related to one another, but less to XI and XII. Comparing the results of the former compounds to the isomers XI and XII at lower electron energies, a slight difference is found. Having in mind similar investigations appearing in the literature in the last few years for a number of compounds, we assume that these differences can only be due to different internal energies and energy distributions. Thus we conclude that loss of HCN, forming m/e 51, proceeds in all given instances from an identical structure for m/e 78.

Similar investigations were made on two 6-chloro compounds, i.e. 6-chloro-s-triazolo (4,3-b) pyridazine III and 6-chloro-tetrazolo (1,5-b) pyridazine X (Table IV). The results are again consistent with the findings stated above, i.e. abstraction of a chlorine radical takes place from a precursor which is common to both compounds, the somewhat greater discrepancy at 18 eV being largely due to very unfavourable peak intensity to noise ratios.

TABLE IV

III $(M^+ = 154)$ $\xrightarrow{-N_2, -HCN}$

X $(M^+ = 155)$ $\xrightarrow{-2N_2}$ $m/e\ 99 \xrightarrow[m_1^*]{-Cl^*} m/e\ 64 \xrightarrow[m_2^*]{-HCN} m/e\ 37$

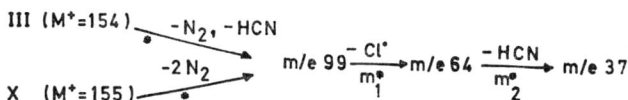

$I m_1^* / I_{64} \cdot 10^2$						
eV	70	50	30	25	20	18
III	0,6	0,8	0,9	1,0	1,3	2,0
X	0,8	1,0	0,9	0,9	1,1	1,2

Summarizing the above findings, we can see that the method used clearly provides more insight into the fragmentation mechanisms, particularly if similarities or dissimilarities in structure and/or energy of decomposing ions need to be identified.

EXPERIMENTAL

All measurements were carried out on a Dupont/CEC-21-110C double-focussing instrument. Normal spectra were recorded at 6 kV, 70 eV and 200 μA ionizing current. Metastable transitions in the first field-free region were obtained by variation of accelerating voltage (starting always at a nominal 4 kV) and constant settings of the electrostatic and magnetic analysers. Care was taken with other operating conditions, *i.e.* the ion repeller voltage, ionizing current control and temperature, which was throughout maintained as low as possible. Samples were introduced directly into the ion source. There were appreciable estimated 'daily' variations (exceeding normal errors) in determinations of metastable abundance ratios. Fortunately, their shifts were always in the same direction for compounds examined on the same day.

ACKNOWLEDGMENTS

We thank the 'Boris Kidric' Fund for financial support. Further we are indebted to B. Stanovnik and M. Tišler for the samples used in this work.

REFERENCES

1. Pirc, V., Stanovnik, B., Tisler, M., Marsel, J. and Paudler, W. W., *J. Heterocyclic Chem.*, 1970, **7**, 639.
2. Barber, M. and Elliott, R. M., *Proc. Seventeenth Amer. Soc. Test. Mat., E* 14 *Conf. Montreal*, 1964.
3. Jennings, K. R., *J. Chem. Phys.*, 1965, **43**, 4176.
4. Shannon, T. W. and McLafferty, F. W., *J. Am. Chem. Soc.*, 1966, **88**, 5021.

5. Bursey, M. M. and McLafferty, F. W., *J. Am. Chem. Soc.*, 1966, **88**, 529.
6. Occolowitz, J. L., *J. Am. Chem. Soc.*, 1969, **91**, 5202.
7. Cooks, R. G., Howe, I. and Williams, D. H., *Org. Mass Spectrom.*, 1969, **2**, 137.
8. Brown, P., *Org. Mass Spectrom.*, 1969, **2**, 1085.
9. Jennings, K. R. and Whiting, A., *Org. Mass Spectrom.*, 1972, **6**, 917.

7

Mass Spectrometry of Dibenzoazepines and 10,11-Dihydro-dibenzoazepines

By D. JOVANOVIC

(*'Roussel Uclaf'*, 93-*Romainville, France*)

THE fragmentation pathways of some benzo- and dibenzodiazepines[1,2] and structurally similar acridones[3] have been examined which permitted an understanding of the principal features of their electron-impact-induced cleavage. No data are available as yet for the fragmentation of dibenzoazepines and 10,11-dihydro-dibenzoazepines. To get a better insight into their fragmentation behaviour a series of dibenzoazepines (*see* Table I) and 10,11-dihydro-dibenzoazepines (*see* Table II) were examined.

TABLE I

Major Peaks and Relative Intensities of Compounds I to VI

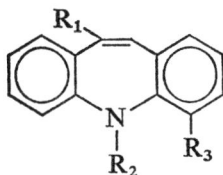

I m/e (I%) $R_1 = H, R_2 = H$ $R_3 = H$	194(16·9), 193(100), 192(13·8), 191(10·7), 190(4·6), 167(8·4), 166(6·3), 165(17·2), 164(3·7), 163(3·9), 152(1·5), 140(4·0), 139(6·2)
II m/e (I%) $R_1 = H, R_2 = H$ $R_3 = COOH$	238(11·0), 237(67·5), 224(6·2), 223(6·2), 220(9·7), 219(45·0), 206(7·1), 191(100), 190(47·5), 180(3·7), 179(4·8), 178(4·5), 177(3·8), 165(11·2), 164(11·0), 163(13·0), 152(3·5), 140(4·4), 139(5·6)
III m/e (I%) $R_1 = H, R_2 = CONH_2$ $R_3 = H$	237(7·4), 236(40·4), 194(15·7), 193(100), 192(31·5), 191(16·3), 190(10·2), 167(4·1), 166(3·8), 165(10·0), 164(3·3), 163(2·9), 152(1·4), 140(1·9), 139(3·1)
IV m/e (I%) $R_1 = OCH_3, R_2 = Me$ $R_3 = H$	238(18·5), 237(100), 223(10·5), 222(74·5), 207(2·7), 206(4·8), 204(4·6), 195(14·3), 194(78·5), 193(6·8), 192(7·2), 191(4·6), 190(4·6), 180(7·1), 179(36·5), 178(10·0), 177(4·3), 166(2·0), 165(8·1), 164(2·2), 163(2·4), 153(3·7), 152(5·5)

Continued overleaf

TABLE I—*continued*

Vm/e (I%) $R_1 = H$, $R_2 = COMe$, $R_3 = H$	236(6·5), 235(32·3), 193(100), 192(63·1), 191(22·4) 190(12·9), 167(2·2), 166(2·8), 165(13·6), 164(4·3), 163(4·2), 152(1·5), 140(2·2), 139(2·7)
VI m/e (I%) $R_1 = (CH_2)_2N(Me)_2$, $R_2 = Me$, $R_3 = H$	279(1·6), 278(8·1), 221(1·0), 220(3·7), 219(1·2), 218(3·5), 217(1·8), 205(1·9), 204(5·7), 194(4·5), 180(1·0), 179(2·2), 178(1·2), 176(1·0), 165(1·1), 152(1·1), 151(1·0), 58(100)

TABLE II

Major Peaks and Relative Intensities of Compounds VII to XII

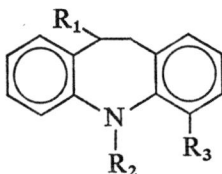

VII m/e (I%) $R_1 = H$, $R_2 = H$, $R_3 = H$	196(14·9), 195(100), 194(62·8), 193(10·4), 192(5·9), 191(4·4), 181(4·6), 180(25·9), 179(5·9), 178(5·3), 168(2·9), 167(7·5), 166(4·0), 165(6·2), 152(5·3), 139(2·2)
VIII m/e (I%) $R_1 = H$, $R_2 = Me$, $R_3 = H$	210(16·3), 209(100), 208(21·8), 195(29·1), 194(98·5), 193(21·8), 192(7·3), 191(3·6), 181(3·6), 180(9·1), 179(10·9), 178(9·1), 167(5·5), 166(3·6), 165(7·2), 152(3·1), 139(1·8)
IX m/e (I%) $R_1 = O$, $R_2 = Me$, $R_3 = H$	224(16·1), 223(100), 222(9·7), 209(4·0), 208(21·8), 197(6·4), 195(9·8), 194(58·1), 193(5·3), 192(3·7), 181(6·2), 180(22·6), 179(14·5), 178(6·6), 177(2·4), 167(3·7), 166(4·0), 165(11·0), 153(3·5), 152(7·1), 151(3·3)
X m/e (I%) $R_1 = H$, $R_2 = H$, $R_3 = COOH$	240(14·1), 239(84·0), 222(17·0), 221(100), 220(11·2), 206(4·7), 194(12·1), 193(34·6), 192(53·3), 191(24·3), 190(11·2), 179(2·8), 178(4·2), 177(4·1), 167(12·2), 166(8·4), 165(22·4), 164(5·5), 163(4·0), 152(8·5), 140(4·0), 139(5·6)
XI m/e (I%) $R_1 = H$, $R_2 = (CH_2)_2OH$, $R_3 = H$	240(4·8), 239(21·8), 209(15·2), 208(100), 195(4·0), 194(11·8), 193(39·0), 192(6·6), 191(5·2), 180(4·7), 179(4·5), 178(5·6), 167(3·7), 166(3·0), 165(5·5)
XII m/e (I%) $R_1 = OH$, $R_2 = Et$, $R_3 = H$	240(15·6), 239(73·8), 224(20·2), 222(9·0), 221(8·9), 220(6·2), 211(14·8), 210(100), 209(10·0), 208(8·9), 207(8·0), 206(50·0), 204(5·9), 196(9·3), 195(9·3), 194(39·2), 193(21·5), 192(19·7), 182(19·5), 181(11·2), 180(53·1), 179(16·4), 178(11·0), 167(12·5), 165(10·9), 152(7·8), 139(3·8).

One of the main characteristics of the examined spectra is a high stability of the molecular ions and primary fragments to further fragmentation. If lateral easily-ejectable chains or other vulnerable groups are absent M^+ usually represents the base peak of the spectrum. Another general feature of these spectra is the appearance of a series of medium-intensity peaks at m/e 180, 179, 167, 165, 152, and 139 whose intensity decreases in the same order. The aforementioned fragments are present in the recorded spectra of

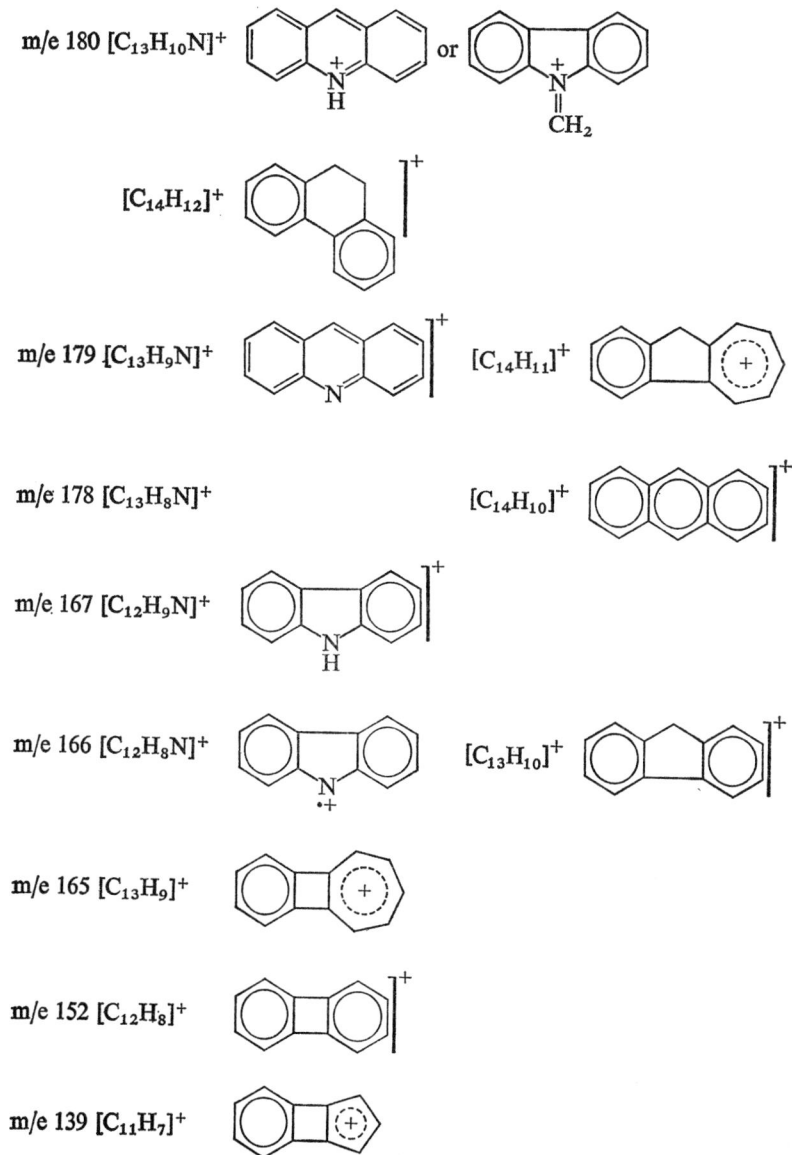

m/e 180 $[C_{13}H_{10}N]^+$ or

$[C_{14}H_{12}]^+$

m/e 179 $[C_{13}H_9N]^+$ $[C_{14}H_{11}]^+$

m/e 178 $[C_{13}H_8N]^+$ $[C_{14}H_{10}]^+$

m/e 167 $[C_{12}H_9N]^+$

m/e 166 $[C_{12}H_8N]^+$ $[C_{13}H_{10}]^+$

m/e 165 $[C_{13}H_9]^+$

m/e 152 $[C_{12}H_8]^+$

m/e 139 $[C_{11}H_7]^+$

SCHEME 1

both dibenzoazepines and 10,11-dihydro-dibenzoazepines. Exact mass measurements pointed invariably to one and the same composition of the fragments: $[C_{14}H_{12}]^+$, $[C_{13}H_{10}N]^+$, $[C_{14}H_{11}]^+$, $[C_{13}H_9N]^+$, $[C_{14}H_{10}]^+$, $[C_{13}H_8N]^+$, $[C_{12}H_9N]^+$, $[C_{12}H_8N]^+$, $[C_{13}H_{10}]^+$, $[C_{13}H_9]^+$, $[C_{12}H_8]^+$, $[C_{11}H_7]^+$. It is highly possible that the structures put forward by Casey et al.[2] and Bowie et al.[3] to explain the fragments m/e 166, 165, 152, 139, 167 and 180 may be considered as valid in this case too (see Scheme 1). The intensity of peaks whose mass (m/e) is less than 139 are in general weak, which makes the spectra of dibenzoazepines and 10,11-dihydro-dibenzoazepines easily recognizable. Moderate-intensity peaks observed at m/e < 139 for some spectra are regularly double-charged ions of the most stable fragments.

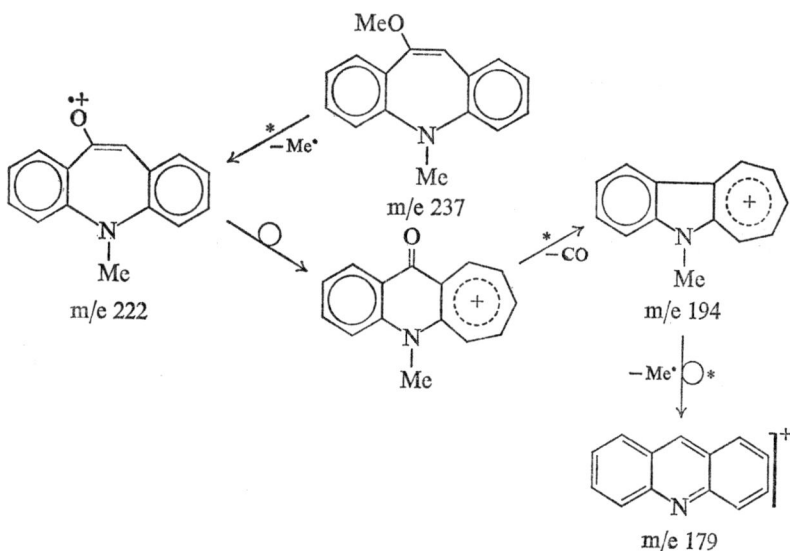

SCHEME 2

The fragmentation of dibenzoazepines is characterized by a pronounced simplicity. So for I the major part of the ionic current is carried by the molecular ion and no significant pathways are observed. The same behaviour is typical for the N-substituted dibenzoazepines III and V which lose predominantly the N-bonded moiety (with or without H transfer) to give I or its radical m/e 192. The subsequent fragmentation pathways are quite analogous to those found for I leading to weak peaks m/e 167, 165, 152, and 139. Of some interest is the fragmentation of IV which according to exact mass measurements and in agreement with the detected metastable peaks can be rationalized by the Scheme 2.

The fragmentation of 10,11-dihydro-dibenzoazepines though made more complex by the presence of the vulnerable allylically-activated C_{10}–C_{11} bond is quite analogous to that of dibenzoazepines especially in the region of lower masses and at m/e < 139 no important peaks are detected as in the preceding case. The allylically-activated rupture of the azepine ring with N-substituted 10,11-dihydro-dibenzoazepines leads to the ejection of the substituent (even when it represents an aliphatic radical) to give

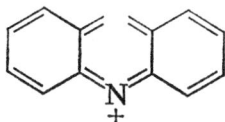

So the $[M-1]^+$ ion of 10,11-dihydro-dibenzoazepine attains a relative abundance of 62·8% and only 21% with 5-methyl-10,11-dihydro-dibenzoazepine thus indicating that the fragmentation pathway

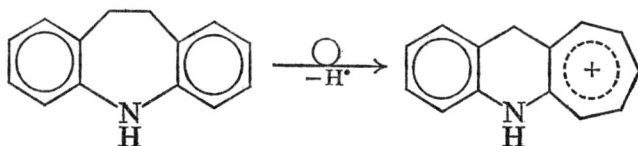

though probably operating is of lesser importance in comparison with the direct ejection of the N-bonded hydrogen. The rupture of the $N–CH_3$ bond with 5-methyl-10,11-dihydro-dibenzoazepine seems not to be the only mechanism leading to the $[M–CH_3]^+$ peak as the analogous fragment m/e 180 appears with a relative abundance of 25·9% in the spectrum of 10,11-dihydro-dibenzoazepine probably due to the azepine-ring contraction.

The direct rupture of the N–alkyl–radical bond is particularly pronounced with 5-ethyl-10,11-dihydro-dibenzoazepine 10-ol, exact mass measurements indicating the next fragmentation scheme (*see* Scheme 3)

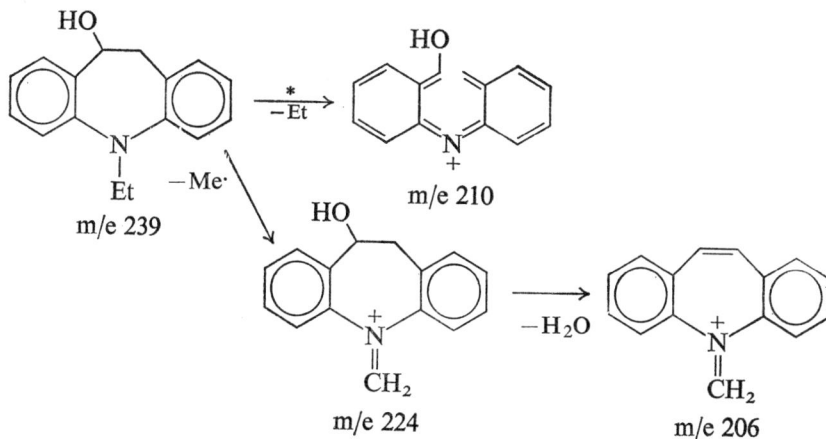

SCHEME 3

The fact that the relative abundance of the $[M–Et]^+$ ion attains 100% should be ascribed to a further weakening of the $C_{10}–C_{11}$ bond in the presence of the OH group. This inhabitual ejection of the N-bonded alkyl radicals may be conditioned by a high stability of the resulting fragment and the potential

resonance between several structures

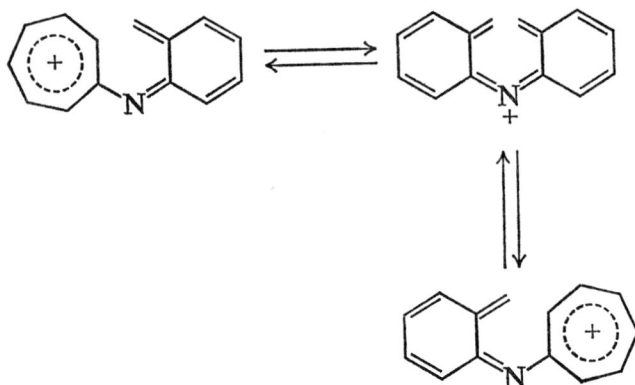

When the ejection of the *N*-bonded alkyl moiety is made more difficult by preceding strongly-directed cleavages as for example with β-(5[10,11-dihydro-dibenzoazepine])-ethanole (in this case the preferential rupture of the chain C–C bond gives rise to m/e 208 with the consolidated *N*-alkyl-moiety bond)

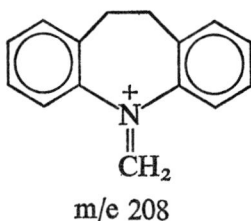

m/e 208

the next step in the fragmentation process is the azepine-ring contraction with the loss of Me˙ as it was observed with 10,11-dihydro-dibenzoazepine.

EXPERIMENTAL

All mass spectra were recorded on a CEC-110C double-focussing mass spectrometer using direct insertion probes at 120–200°. Low-resolution spectra (approximately 2000) were obtained at an electron-beam energy of 70 eV and an accelerating potential of 6 kV. Exact mass values were determined by conventional peak matching using the peaks of PFK as mass standards at a resolution of $\geq 10\,000$.

REFERENCES

1. Sadée, W., *J. Med. Chem.*, 1970, **13**, 475.
2. Casey, A. C., Green, J. H., Lee, A. and Mautner, M., *J. Heterochem.*, 1970, **7** 879.
3. Bowie, J. H., Cooks, R. G., Prager, R. H. and Thredgold, H. M., *Aust. J. Chem.*, 1967, **20**, 1179.

Discussion

A. G. Giumanini (University of Bologna, Italy): I wonder if the author has a mechanism to rationalise the large prevalence of α versus β cleavage?

D. Jovanovic: Yes, in our opinion the driving force for this inhabitual α cleavage may be the presumed azepine-ring opening (the C_{10}–C_{11} bond cleavage) which should lead automatically to the loss of N-bonded substituents.

A. G. Giumanini: An alternative explanation may be a 1,4-hydrogen migration to quaternarize the nitrogen which then would easily lose the substituent. Hydrogen migration to

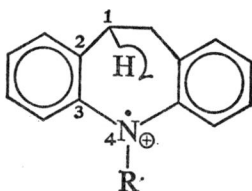

quaternarize the nitrogen may be the driving force for elimination of the benzyl radical, *e.g.* in benzyl amines

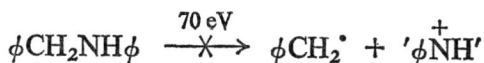

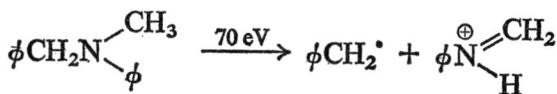

$$\phi CH_2 NH\phi \ \xrightarrow[\text{X}]{70\ eV} \ \phi CH_2^{\cdot} + \ '\phi \overset{+}{N} H'$$

$$\phi CH_2 N \overset{CH_3}{\underset{\phi}{\diagdown}} \ \xrightarrow{70\ eV} \ \phi CH_2^{\cdot} + \phi \overset{\oplus}{N} \overset{CH_2}{\underset{H}{\diagdown}}$$

8
Studies on the Four-Centre Rearrangements of Thioamides

By GYULA HORVÁTH

(*Research Institute for Pharmaceutical Chemistry, Budapest, Hungary*)

INTRODUCTION

SEVERAL papers have reported skeletal and hydrogen rearrangements proceeding via a four-membered ring transition state in order to explain the occurrence of certain ions in the mass spectra of compounds corresponding to the general formula I.[1-4]

$$\begin{array}{cc} D & A \\ | & \| \\ C & -B \end{array} \quad I$$

Based partly on these data, Bentley and Johnstone proposed a formal classification of such rearrangements.[1b,1c] They also discussed various structural factors influencing the abundances of ions arising from these processes.

Concerning thioamides of general formula II, mass spectral investigations of substituted thioacetanilides[5,6] and thioformanilides[7,8] have been reported,

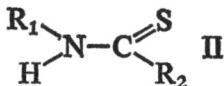

$$\begin{array}{c} R_1 \\ \diagdown \\ H \diagup \end{array} N - C \begin{array}{c} \diagup S \\ \diagdown \\ R_2 \end{array} \quad II$$

but mainly in regard to the loss of *ortho*-substituents from the various aryl groups R_1. In the case of thioformanilide, however, the loss of HCN from the molecular ion has been noted.[7]

RESULTS AND DISCUSSION

Upon investigating the mass spectra of thioamides of general formula II prominent ions were observed, the formation of which can be rationalized by involving four-centre rearrangements of the substituents of nitrogen to the sulphur atom within the thioamide group. These rearrangements are depicted in Scheme 1.

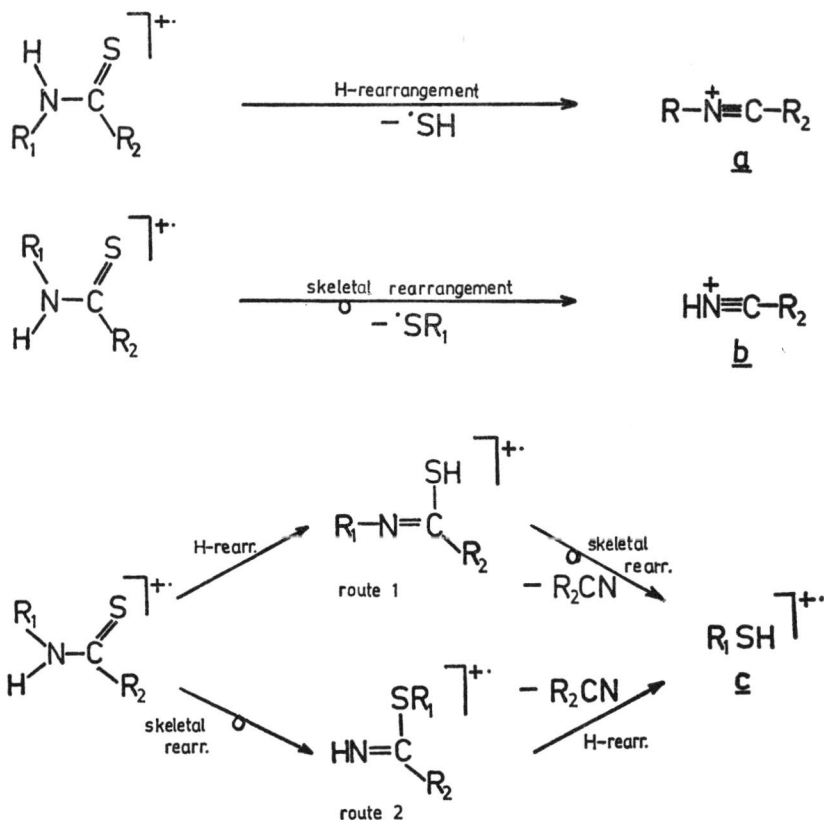

SCHEME 1

The process leading to ion *a* involves the tautomeric thiol-form of thio-amides which is not present, however, in the ground state.[9] Accordingly it can be formulated as being a four-centre hydrogen rearrangement upon electron impact. The loss of ˙SH radical from the intermediate structure formed resembles the mass spectral behaviour of anils.[10] The importance of this cleavage in the case of the thiol-form is well demonstrated by the spectra of the S-methyl derivatives III, in which ion *a* (Scheme 2) yields by far the most abundant fragment-ion (*see* Table I).

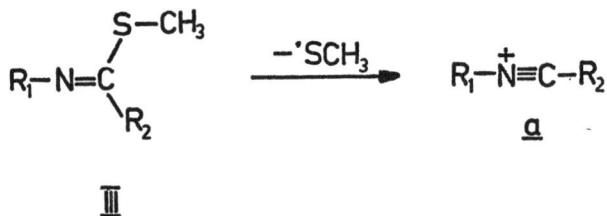

SCHEME 2

Formation of ion *b* requires a four-centre skeletal rearrangement involving the migration of substituent R_1 from the nitrogen to the sulphur atom within the thioamide group. Rearrangements of this type have been reported in the case of various substituted *N*-methyl-benzanilides.[1c] Nevertheless, the corresponding thioamides may represent more suitable models for studying this process, as indicated by the mass spectrum of *N*-methyl-thiobenz-anilide (XI), in which the (M–'S-phenyl)$^+$ ion gives the base peak (*see* Table I).

The most interesting feature of the mass spectra of the thioamides investigated is the formation of ion *c* which requires both hydrogen and skeletal rearrangements to the sulphur atom, followed by elimination of R_2CN. A similar rearrangement has been noted in the case of benzanilide yielding a phenol-ion.[11] In a published spectrum of this compound, however, the corresponding peak is of very low abundance.[12]

Comparison of the abundances of ions *a*, *b* and *c* in the spectra of compounds IV and V or of VII and VIII (Table I) but probably of any other thioamide having structure II and its amide counterpart clearly shows the essential role of the nature of hetero-atom A within general formula I in triggering a four-centre rearrangement.

The abundances of ions *a*, *b* and *c* as well as those of the molecular ions are listed for the compounds investigated in Table I. Elemental compositions of the listed ions of compound IV were corroborated by high-resolution mass measurements, while those of the further compounds were taken by apparent analogy.

The possibility of the formation of ions *a*, *b* or *c* in a thermal rearrangement (*cf.* Reference 4) has been ruled out as none of the spectra showed any

TABLE I

Structures and Selected Mass Spectral Data of the Thioamides Investigated

Compounds		Data	Ions			
Number	Structure		$M^{+\cdot}$	*a*	*b*	*c*
IV		m/e	345	312	104	242
		I (%)	38	19·5	4·8	65
		Σ_{50} %	10·7	5·5	1·35	18·3
V		m/e	329	312	104	226
		I (%)	81	0·25	0·3	1
		—				
VI		m/e	359	312	—	256
		I (%)	69	100	—	0
		—				

GYULA HORVÁTH

TABLE I—*continued*

Compounds		Data	Ions			
Number	Structure		$M^{+\cdot}$	a	b	c
VII	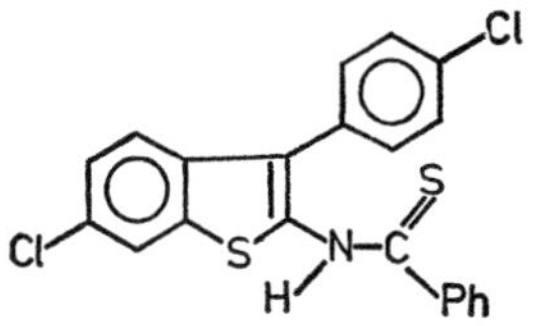	m/e	413	380	104	310
		I (%)	12·5 (11)	4·6 (2·9)	6·25	19·6 (16·6)
		Σ_{50} %^{a}	8·65	2·8	2·3	13·3
VIII	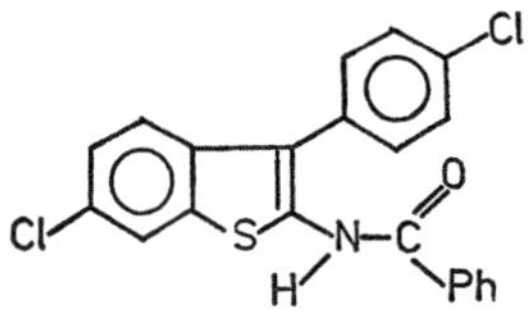	m/e	397	380	104	294
		I (%)	29 (18)	0	0·28	0
		—				
IX	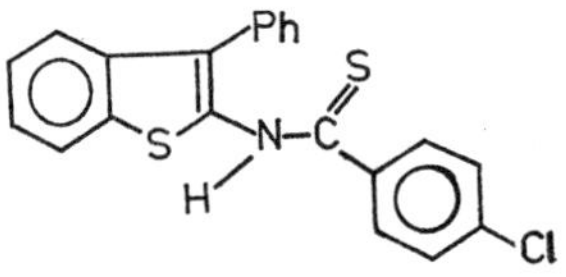	m/e	379	346	138	242
		I (%)	79 (34)	20·5 (8·1)	3·8 (1·2)	100
		Σ_{110} %^{a}	21	5·35	0·9	18·6
X	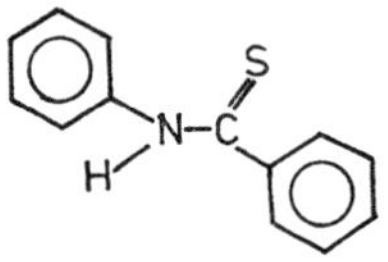	m/e	213	180	104	110
		I (%)	40	15	15	20·5
		Σ_{50} %	11·7	4·4	4·4	6
XI	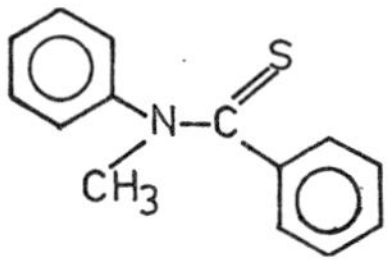	m/e	227	180	118	124
		I (%)	53	2	100	0
		Σ_{50} %	15·1	—	28·5	—
XII	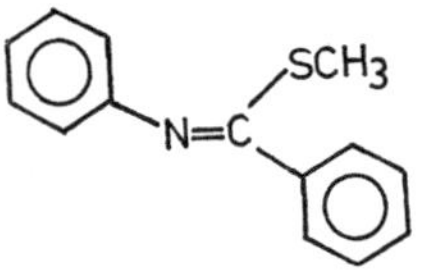	m/e	227	180	—	124
		I (%)	10·2	100	—	0
		Σ_{50} %	5·6	54·3	—	—
XIII	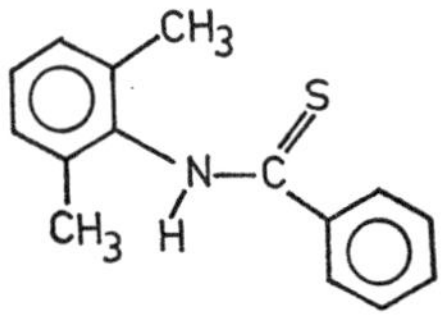	m/e	241	208	104	138
		I (%)	18	14·5	10·6	3·7
		Σ_{77} %	5·7	4·6	3·4	1·2

TABLE I—continued

Compounds		Data	Ions			
Number	Structure		$M^{+\cdot}$	a	b	c
XIV		m/e	214	181	104	111
		I (%)	46	9·2	9·5	0·8
		Σ_{39} %	12·7	2·5	2·6	0·2
XV		m/e	214	181	104	111
		I (%)	100	30	2	2·5
		Σ_{39} %	24·8	7·4	0·5	0·6
XVI		m/e	259	226	149	111
		I (%)	100	27	0	6
		Σ_{39} %	20·5	5·5	—	1·2
XVII		m/e	273	240	149	125
		I (%)	100	34	0	10
		Σ_{39} %	16·5	5·6	—	1·65
XVIII		m/e	151	118	42	110
		I (%)	100	19·6	11·8	39
		Σ_{39} %	17·3	3·4	2·05	6·75

[a] Intensities of the chlorine isotope peaks are summarized.

systematic dependence upon evaporation temperatures. In addition, meta-stable peaks due to the transitions $M^{+\cdot} \to a$ and $M^{+\cdot} \to c$, respectively, have been observed in some spectra† while, if necessary, metastable transitions were measured by using the defocussing technique.‡ Metastable data obtained by this latter method are in agreement with the assumption that ions a, b and c should be formed directly from the (rearranged) molecular ions, as only one metastable ion due to this transition occurred for each of these fragment ions. Data of selected metastable transitions are collected in Table II.

† The instrument used discriminates considerably against metastable ions.
‡ Accelerating voltage defocussing was used.

TABLE II

Characteristic Metastable Transitions Obtained for Representatives of the Various Groups of Compounds Investigated

Compound	Ion symbol	Measured			Observed		
		m_2	Voltage ratio	m_1	m	m_1	m_2
IV	c	242	1·429	345			
VII	c				233·5–7·5	413	310
IX	c				155	379	242
X	b	104	2·046	213			
	c	110	1·934	213			
XIV	b	104	2·055	214			
	a	181	1·183	214			
XV	a				159·5	214	181
XVI	a				197·5	259	226
XVII	a				209·5	273	240

Prompted by the crucial importance of four-centre rearrangements in the fragmentation of certain thioamides (especially compounds IV, VII and IX) it was attempted to get a better insight into the mechanism of these processes† in particular as regards the formation of ion c.

As suggested by the simultaneous occurrence of ions a and b, hydrogen versus skeletal rearrangements can take place as well within the thioamide group upon electron impact. Insofar as being reflected in the abundances of these ions, the migratory aptitudes of the various R_1 aromatic groups might be at least in the same order as the migratory aptitude of hydrogen for the majority of thioamides investigated. Accordingly, in the case of these compounds, four-centre hydrogen rearrangements need not be preferred to the skeletal ones when both types proceed in succession, as upon formation of ion c. Consequently, routes 1 and 2 in Scheme 1 might be, in principle, of similar probability for this process.

Non-interacting molecules of the S-methyl derivatives of thioamides can be considered to be the closest analogues of the hypothetic

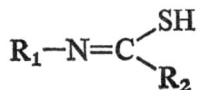

$$R_1-N=C\overset{\displaystyle SH}{\underset{\displaystyle R_2}{}}$$

imido-thiol-form, concerning both electronic and geometric structures in the ground state. There seems to be no reason to assume, that any excited state of the $R_1-N=C-S$ system available for the thiol-form would not be accessible for the S-methyl derivatives, as well. Should the thiol-form be involved in the formation of ion c as intermediate step (Scheme 1, route 1),

† Complications which may arise from such qualitative approaches are admitted by the author.

the corresponding $(R_1SCH_3)^{+\cdot}$ ion should be present in the spectra of the S-methyl derivatives. Spectral data of compounds VI and XII (Table I) show, however, that the $(R_1SCH_3)^{+\cdot}$ ion does not occur in measurable abundance, although ions c give prominent peaks in the spectra of the corresponding thioamides IV and X.

This fact strongly suggests that ion c is formed only if the skeletal rearrangement of the aromatic group R_1 to the sulphur atom *precedes* the hydrogen rearrangement (Scheme 1, route 2).

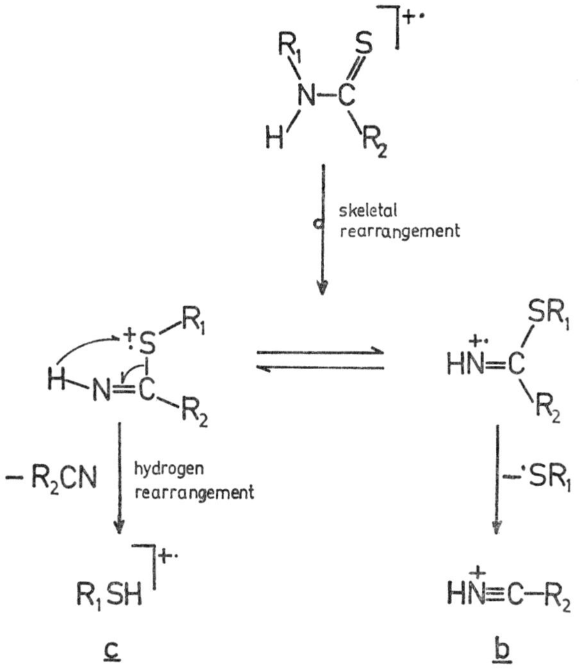

SCHEME 3

By the terms of these considerations, ions b and c seem to arise from the same precursor, which has been formed from the molecular-ion by the migration of substituent R_1 within the thioamide group, as depicted in Scheme 3.

Seeking further evidence of the proposed sequence of rearrangements resulting in ion c, it was attempted to obtain energetic data on the rearrangement processes. For this purpose the ionization potential of compound XVIII as well as appearance potentials of ions a, b and c of the same compound have been determined. On the other hand compound XIX was synthesized[13] in order to obtain its corresponding data for comparison.

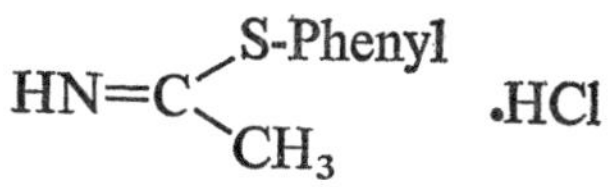

Regrettably, compound XIX decomposed upon evaporation, even at room temperature, yielding thiophenol in high extent, thus crossing this plan.

There are various factors which may affect the abundances of ions b and c relative to the other fragment ions and to each other, as well.

As shown by the data of compounds IV, VII and IX, the steric size of migrating aromatic groups may not be included among these. This also confirms that the new C–S bond formed by the skeletal rearrangement involves the same carbon atom of R_1, which had been bonded originally to the nitrogen.

The influences of some other factors are indicated here, but it is emphasized that further studies will be needed to prove these effects emphatically.

(1) Comparing total intensities of ions b and c for compounds X and XV, respectively, a considerable decrease occurs in the latter. This seems to be due to the electron-withdrawing nature of the pyridyl group, which hinders the skeletal rearrangement by strengthening the R_1–N bond.

(2) Electron-donating substituents on R_1, as in compound XIII, may have a similar effect in the reverse process.

(3) After the skeletal rearrangement had taken place (Scheme 3) the ability of nitrogen and sulphur atoms to stabilize a positive charge is affected by the nature of substituents R_1 and R_2, respectively.

Comparing the intensity ratios of ions b and c for compounds IV and IX or XV and XVI, a decrease in proportion of ion b can be observed in the spectra of the latter compounds. This can be assumed to be the consequence of electron-withdrawal from the nitrogen atom over the –N=C– double bond by the substituents of R_2.

(4) Factors which influence ion abundances in general (e.g. rates of competitive reactions) are acting of course in this instance, too.

CONCLUSION

The rearrangements of thioamides proceeding via four-membered ring transition states have been discussed as being hydrogen and skeletal rearrangements from the nitrogen to the sulphur atom of the thioamide group. A sequence has been proposed for the case when both of these rearrangements occur yielding an aromatic thiol upon electron impact.

Several factors influencing the abundance of the ions discussed were treated briefly.

EXPERIMENTAL

Mass spectra were taken on a Varian MAT SM-1 mass spectrometer. The operating conditions were: resolution 1250; electron energy 70 eV; electron current 300 μA; accelerating voltage 8 kV; source temperature 250°C.

Evaporation temperatures:

compound	IV	V	VI	VII	VIII	IX	X	XI	XII	XIII	XIV	XV	XVI	XVII	XVIII
T_e (°C)	130	150	80	155	190	150	60	30	20	55	95	70	100	110	20

High resolution mass measurements were performed at resolution 10 000 using PFK as the reference standard.

ACKNOWLEDGMENTS

My special thanks are due to Mr E. Koltai for providing me with the samples, except compounds XV–XVII given by Dr L. Farkas, to whom I am indebted, too. The valuable technical assistance of Miss M. Fehér and Mr Gy. Jerkovich is gratefully acknowledged.

REFERENCES

1. a. Cooks, R. G., *Org. Mass Spectrom.*, 1969, **2**, 481.
   b. Bentley, T. W. and Johnstone, R. A. W., *in* 'Advances in Physical Organic Chemistry', *ed.* V. Gold, Vol. 8, Academic Press, London, 1970.
   c. Bentley, T. W. and Johnstone, R. A. W., *J. Chem. Soc.* (*B*), 1971, p. 1804 and the references cited in them.
2. Duffield, A. M., Djerassi, C., Neidlein, R. and Henkelbach, E., *Org. Mass Spectrom.*, 1969, **2**, 641.
3. Ohno, A., Koizumi, T., Ohnishi, Y. and Tsuchihashi, G., *Org. Mass Spectrom.*, 1970, **3**, 261.
4. Tou, J. C. and Rodia, R. M., *Org. Mass Spectrom.*, 1972, **6**, 493.
5. Baldwin, M. A. and Loudon, A. G., *Org. Mass Spectrom.*, 1969, **2**, 549.
6. Baldwin, M. A., Cardnell, P. C., Loudon, A. G., MacColl, A. and Webb, K. S., *in* 'Advances in Mass Spectrometry', *ed.* A. Quayle, Vol. 5, The Institute of Petroleum, London, 1972.
7. Walter, W., Becker, R. F. and Grützmacher, H.-F., *Tetrahedron Letters*, 1968, p. 3515.
8. Grützmacher, H.-F. and Kuschel, H., *Org. Mass Spectrom.*, 1970, **3**, 605.
9. Walter, W. and Voss, J., *in* 'The Chemistry of Amides', *ed.* J. Zabicky, Interscience, London, 1970.
10. Budzikiewicz, H., Djerassi, C. and Williams, D. H., 'Mass Spectrometry of Organic Compounds', Holden-Day, San Francisco, 1967, pp. 392–4.
11. Reference 1(b), p. 219.
12. Goldsmith, D., Becher, D., Sample, S. and Djerassi, C., *Tetrahedron Suppl.*, 1966, **7**, 145.
13. Autenrieth, W. and Brüning, A., *Chem. Ber.*, 1903, **36**, 3464.

Discussion

H. F. Grützmacher (University of Bielefeld, Germany): Did you investigate thioamides of 4-amino-pyridines and did you observe the same effects on the intensities as with 2-amino-pyridine derivatives?

G. Horvath: Yes, the mass spectrum of 4-amino-pyridine-thiobenzoate was also investigated and this compound is included in the paper to appear in the conference proceedings. This compound shows a similar behaviour to 2-amino-pyridine-thiobenzoate, and for that very reason it has been omitted from the presentation.

9

Correlation of Fragmentation Modes of Substituted Stilbenes Under Electron Impact

By H. GÜSTEN

(*Institut für Radiochemie, Kernforschungszentrum, Karlsruhe, Germany*)

L. KLASINC

(*Institut 'Rudjer Boškovic', Zagreb, Yugoslavia*)

V. KRAMER and J. MARSEL

(*J. Stefan Institute, University of Ljubljana, Yugoslavia*)

SUBSTITUENT effects in the fragmentation of aromatic compounds in the mass spectrometer are often explained in terms of changes in electronic structure. It is still not settled whether the effect of the substituents on the formation of particular ionic species could be correlated in a way similar to Hammett's procedure.[1,2,3] Systematic study of substituent effects in the mass spectra of model compounds consisting of two aromatic units connected by a good transmitter of conjugative effects, *e.g.* a –CH=CH– group, should provide information about the transfer of electronic charge or energy between the units, *i.e.* about the electronic structure of the ions. The mass spectra of such substituted diarylethylenes are typical for aromatic compounds, with a dominant molecular ion and relatively few fragmentation modes. Although no transmission effects were observed in monosubstituted stilbenes[4] and monosubstituted styrylquinolines,[5] investigation of the fragmentation modes in stilbene and monosubstituted stilbenes[6] in ortho, meta, para and alpha positions showed that by their influence on the mass spectrum, substituents could be divided into three groups; (A) tightly-bonded, (B) loosely-bonded and (C) rearranging. Compounds of group A and B show a fragmentation similar to that of the unsubstituted stilbene either from their molecular ion or from the [M-substituent] ion, respectively. So, *e.g.* all fluoro, cyano, amino stilbenes, *m*- and *p*-hydroxy-stilbene belong to group A, and the chloro, bromo- and nitro- (with the exception of *o*-nitro) stilbenes to group B. In general the mass spectra are nearly independent of the position of the substituent, but some stabilization of the molecular ion by para-substituents has been observed; the total ion currents for the same substituent tend to be in

79

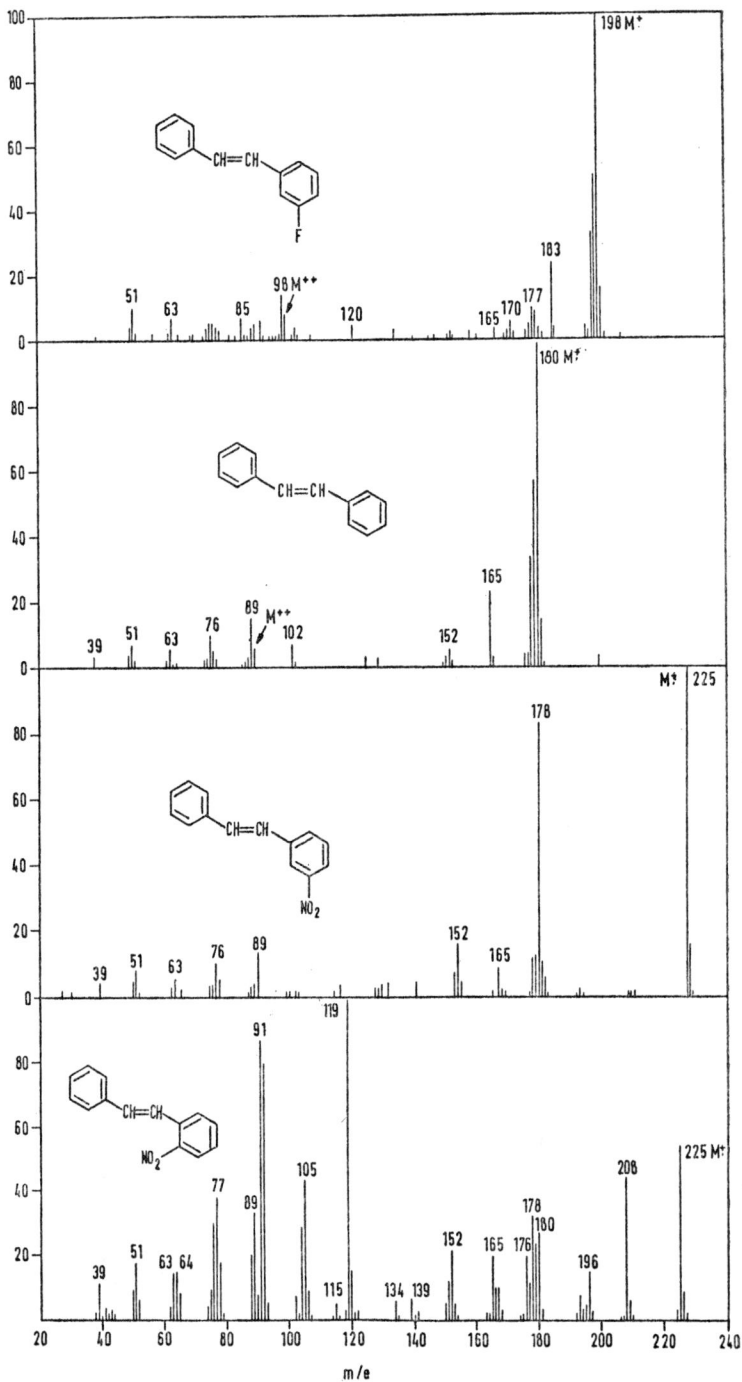

FIG. 1

the order ortho > meta > para. Certain substituents in ortho or alpha-positions cause great changes in the mass spectrum by rearrangement prior to the fragmentation. Such group C substituted stilbenes are, *e.g.* α-methyl-, *o*-methoxy-, α-carboxyl-and *o*-nitrostilbene. In conclusion, with the exception of group C substituted stilbenes (*i.e.* rearrangement reactions), the mass spectra of monosubstituted stilbenes depend on the nature but not on the position of the substituent.

The typical mass spectral characteristics of stilbene and group A, B and C substituted stilbenes are shown in Fig. 1 with *m*-fluorostilbene, *m*-nitrostilbene and *o*-nitrostilbene as representative compounds for the three groups.

Investigations have now been extended to several disubstituted stilbenes. The following 19 disubstituted trans-stilbenes were investigated in the present study (Table I).

TABLE I

4-Nitro-4'-chloro-stilbene,	4-Nitro-4'-amino-stilbene,
4-Nitro-4-dimethylamino-stilbene,	4-Nitro-4'-carboxyl-stilbene,
4-Nitro-4'-methoxy-stilbene,	4-Nitro-4'-hydroxy-stilbene,
4-Nitro-4'-nitro-stilbene,	4-Nitro-3'-hydroxy-stilbene,
4-Nitro-3'-methoxy-stilbene,	2-Nitro-2'-methoxy-stilbene,
3-Nitro-4'-dimethylamino-stilbene,	3-Hydroxy-4'-dimethylamino-stilbene
3-Nitro-3'-methoxy-stilbene,	2-Nitro-4'-amino-stilbene,
4,4'-Difluoro-stilbene,	4,4'-Dimethoxy-stilbene,
2,4'-Dinitro-stilbene,	α-Carboxyl-4'-chloro-stilbene,
	α,α'-Difluoro-stilbene

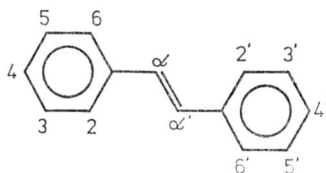

The molecular ion is again the most abundant in the mass spectra of disubstituted stilbenes, with the exception of those having an ortho-nitro substituent. In comparison to the mass spectra of the monosubstituted compounds, a stabilization of the molecular ion by the second substituent can be observed. Regarding the nature of substituent, the presence of a group A substituent is indicated in the mass spectrum by observation of the [M–H] (surprisingly not in substituted methoxystilbenes), and the presence of a group B substituent by abundant ions of m/e 178 and m/e 165. The latter ions are not observed at all in the mass spectra of 4,4'-difluoro and 4,4'-dimethoxy stilbene. Again an ortho–nitro substituent (or other C group substituent) changes the mass spectrum dramatically, but in line with the rearrangements found previously.[6,7]

Thus, one would say that there is no striking difference in comparison to the behaviour of the two corresponding substituted stilbenes in the mass spectrometer. But an inspection of the dependence on the position of the substituents shows that some highly co-operative effects between the two

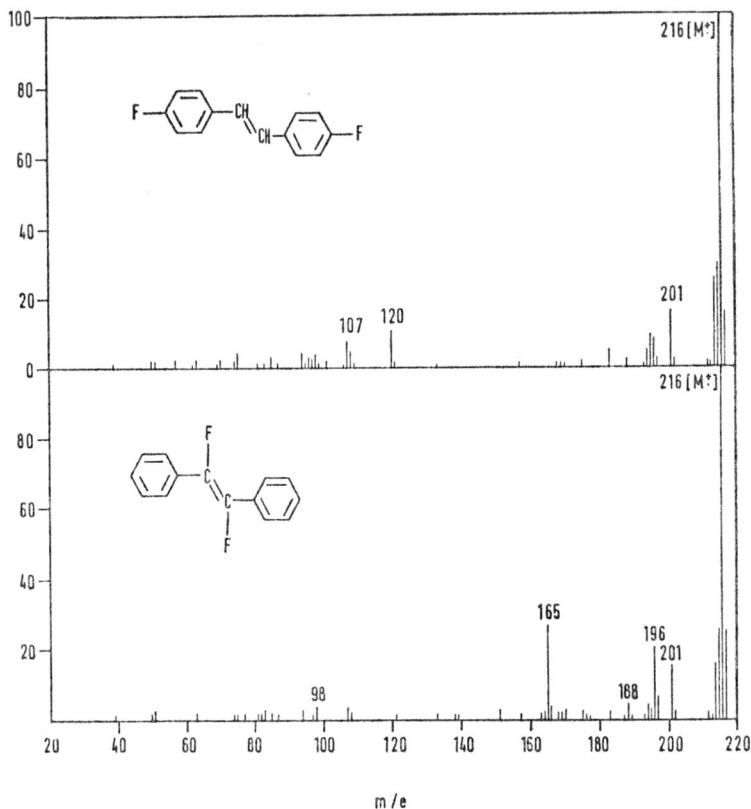

m/e

FIG. 2

substituents must be taking place. Contrary to ortho-, meta-, para- and alpha-fluoro-stilbene, whose mass spectra were found surprisingly similar, 4,4'-difluorostilbene and α,α'-difluorostilbene show great differences in the intensities of ions, especially of m/e 165 (Fig. 2). The same is found for the three isomeric nitromethoxystilbenes (Fig. 3). It is easy to see that stabilization of the molecular ion and the intensities of m/e 178 and m/e 165 proceed in the order $p,p'- \gg m,p'- \gg m,m'$-nitromethoxystilbene. One could argue that the origin of this behaviour lies in conjugative electronic effects which for the ground states of these molecules are in the same order. On the other hand, the operation of such effects was not found although existing in the monosubstituted molecules (meta- and para-nitrostilbene have nearly identical mass spectra). There is also a possibility that substantial changes in the ionization potentials of these compounds are responsible for the observed strong co-operative effects. Measurements of the ionization potentials of all compounds investigated are in progress.

EXPERIMENTAL

All spectra were recorded on a Du Pont double focussing mass spectrometer (CEC 21-110C) by direct introduction.[6] All compounds were purified by

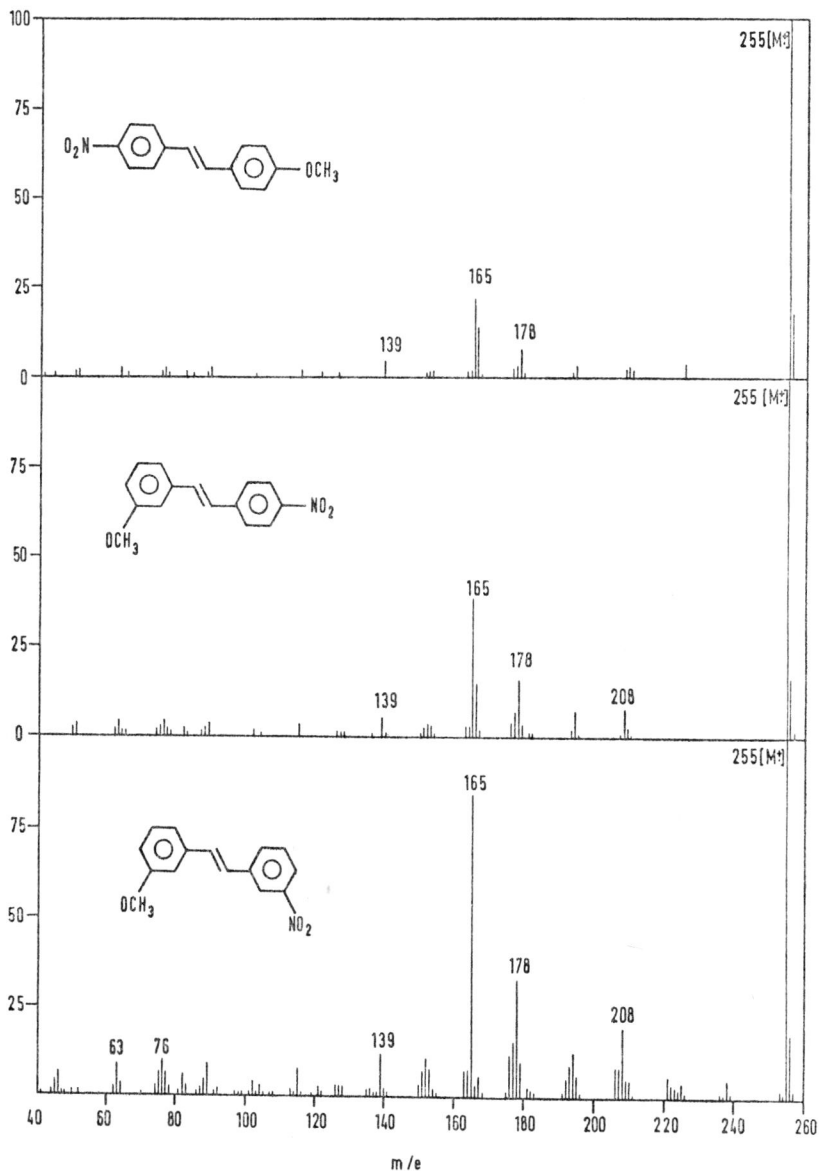

m/e

FIG. 3

column chromatography and final recrystallization. According to spectroscopic data all stilbenes were in the *trans*-form. The spectra reproduced in the figures have been normalized.

ACKNOWLEDGMENTS

This work was performed on the basis of a German–Yugoslav Scientific Co-operation Programme. The financial support of the International Bureau

Jülich, the Republican Council of Scientific Research Work of Croatia and the 'Boris Kidric' Fund of Slovenia is acknowledged.

We thank Prof. E. Fischer of the Weizmann Institute of Science (Rehovoth, Israel) for a sample of α,α'-difluoro-stilbene.

REFERENCES

1. For a review see: Bursey, M. M., *Org. Mass Spectrom.*, 1968, **1**, 31.
2. McLafferty, F. W., *Chem. Commun.*, 1968, p. 956.
3. Bentley, T. W., Johnstone, R. A. and Payling, D. W., *J. Amer. Chem. Soc.*, 1969, **91**, 3978.
4. Güsten, H., Klasinc, L., Marsel, J. and Milivojevic, D., *Euratom Report*, 1972, EUR 4765, 305.
5. Güsten, H., Klasinc, L. and Stefanovic, D., *Org. Mass Spectrom.*, 1973, **7**, 1.
6. Güsten, H., Klasinc, L., Kramer, V. and Marsel, J., *Org. Mass Spectrom.* (in press).
7. Seibl, J. and Völlmin, J., *Org. Mass. Spectrom.*, 1968, **1**, 713.

Discussion

J. Holmes (University of Ottawa, Canada): In *m*- and *p*-nitrostilbenes, the molecular ions appear to lose HNO_2. Is this a one- or two-step process and from where does the H atom come?

L. Klasinc: Measurements of first field free metastables show a two-step loss of OH˙ followed by NO. Additionally, in the normal spectra a metastable corresponding to loss of HNO_2 can be observed. Unfortunately, the origin of the hydrogen could not be established without labelling experiments.

V. Hanus (Czechoslovakian Academy of Science, Prague, Czechoslovakia): What is the behaviour of α-nitrostilbenes?

L. Klasinc: The α-nitrostilbene was in fact synthesized by us after the rearrangement in *o*-nitrostilbene but was not observed in *meta*- and *para*-nitrostilbene. Surprisingly, α-nitrostilbene shows no rearrangement but exhibits a mass spectrum similar to the *meta*- and *para*-isomers.

10

The Mass and Ion Kinetic Energy Spectra of the Chlorinated Anilines†

By S. SAFE, O. HUTZINGER, W. D. JAMIESON and M. COOK

(Atlantic Regional Laboratory, National Research Council of Canada, Halifax, Nova Scotia, Canada)

IONIC decompositions (*e.g.* $m_1^+ \rightarrow m_2^+ + m_3$) which occur in the first field-free region of a double-focussing mass spectrometer do not yield product ions which have the requisite energy to pass through the electrostatic analyser when this has been tuned to transmit the 'normal' (100% energy E) ion beam. However, if the voltage across the plates of the electrostatic analyser is scanned so that the daughter ions having specific non-'normal' $\{[(m_2/e)/(m_1/e)]E\}$ energies are transmitted, then these ions can be duly detected and recorded to give the ion kinetic energy (IKE) spectrum.[1,2]

Electron impact mass spectra of isomeric compounds are often too similar to allow their use to distinguish between compounds.[3,4] However, recent work with isomeric PCBs, DDTs and chlorocyclohexanes[5,6] has shown that the IKE spectra of many isomers are unique. Since isomeric chlorinated anilines are widely used in the pesticide field it was of interest to study and compare their mass and IKE spectra and to determine the possible analytical uses of such data.

The mass spectra of the isomeric chloroanilines (Table I) were virtually indistinguishable. Major fragmentation processes are shown in Scheme 1. Metastable ions were observed at m/e 78·6, 66·7 and 45·9 for the reactions m/e 127 → m/e 100, m/e 127 → m/e 92 and m/e 92 → m/e 65 which occurred

† Issued as NRCC No. 13696.

TABLE I
Mass Spectral Data for the Monochloroanilines

Compound (isomer)	Ion abundances (70 eV)					
	m/e 127	m/e 99	m/e 100	m/e 92	m/e 91	m/e 65
2	100	5	4	13	7	12
3	100	5	5	14	5	13
4	100	5	5	13	4	14

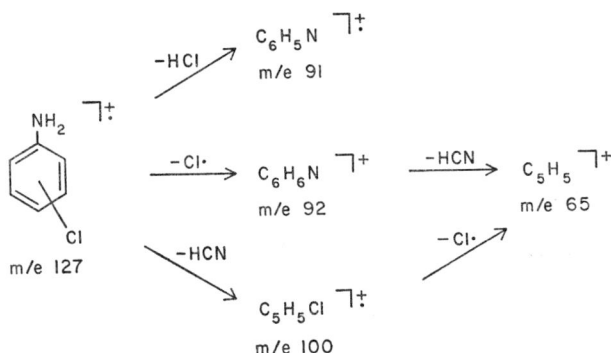

SCHEME 1

in the second field-free region of the mass spectrometer. The ion kinetic energy spectra (Fig. 1) of these isomers gave ionic decomposition peaks at 0·787, 0·718 and 0·701E corresponding to the reactions already noted from the metastable ion data and thus showing the same ionic reactions occurred in the first and second field-free regions. The relative peak intensities (Table II) also indicated significant differences in the IKE spectra of these chloro-aniline isomers. Since the m/e 127 ion decomposes with loss of both Cl· and HCN the ratio of their respective ionic decomposition peaks (*i.e.* [0·718E]/[0·787E]) is a measure of the ratio of the rates for these two reactions. The results obtained were different for each isomer and this indicated incomplete chlorine randomization in the molecular ion and obviates the intermediacy of an azepinium ion intermediate in decomposition of the molecular ion.

The mass spectra of the isomeric dichloroanilines were also similar (Table III) and not useful for distinguishing between the isomers. The fragmentation array (Scheme 2) featured loss of Cl· and HCN from the molecular ion and expulsion of Cl· and HCN from the [M–HCN]$^{·+}$ and [M–Cl]$^{+}$ ions respectively as well as other reactions indicated in the scheme. The IKE

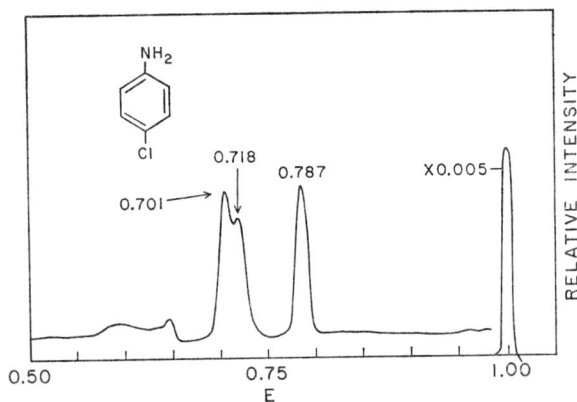

FIG. 1   IKE spectrum of 4-chloroaniline.

TABLE II

Ion Kinetic Energy Data for the Monochloroanilines

Compound (isomer)	Ionic decomposition peak intensities $\times 10^3$			
	0·787E	0·718E	0·701E	0·718E/0·787E
2	10·1	11·7	15·9	1·16
3	10·9	9·20	12·0	0·840
4	7·50	5·90	7·30	0·790

TABLE III

Mass Spectral Data for the Dichloroanilines

Ion (m/e)	Ion abundance data (70 eV) (isomer)					
	2,4	2,3	3,5	2,6	2,5	3,4
161	100	100	100	100	100	100
134	1	1	2	1	1	2
133	4	3	4	3	4	3
126	6	6	6	5	6	6
125	5	4	3	6	4	3
100	2	2	1	2	1	1
99	10	9	9	13	9	12
91	3	3	3	2	3	3
90	14	15·0	9	13	11	11

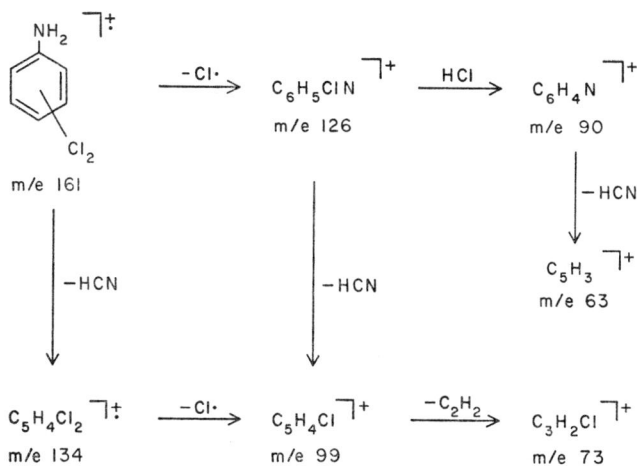

SCHEME 2

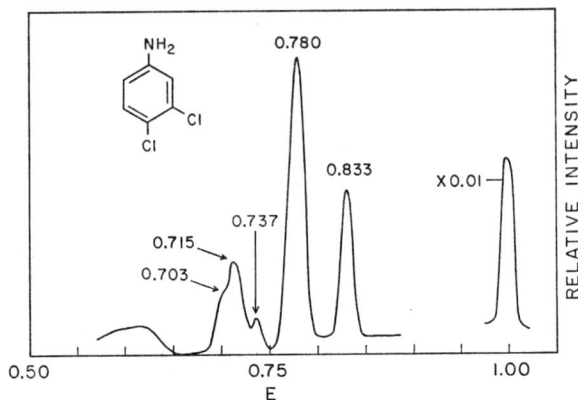

FIG. 2   IKE spectrum of 3,4-dichloroaniline.

spectrum (Fig. 2) of each isomer was recorded. The relative intensities of the ionic decomposition peaks (Table IV) were significantly different for each isomer thus providing useful 'fingerprint' spectra for this series of isomers. The IKE spectra gave peaks at 0·833, 0·780, 0·737, 0·715 and 0·703E corresponding to the reactions m/e 161 → m/e 134, m/e 161 → m/e 126 and m/e 126 → m/e 99, m/e 133 → m/e 99, m/e 126 → m/e 90 and m/e 90 → m/e 63. The two reactions m/e 161 → m/e 126 (calc. 0·783E), m/e 126 → m/e 99 (calc. 0·786E) gave overlapping peaks at 0·780E and the intensity of this peak summed the contribution of both these decomposition reactions. Using photographic plate recording the relative yields of the two daughter ions could be obtained. As with the monochloroanilines the IKE spectra of the isomers are all significantly different.

The mass spectra of the 2,4,6-, 2,3,4- and 2,4,5-trichloroaniline isomers were different and the abundance data for several of the major ions were sufficient to distinguish between these isomers (Table V). The fragmentation scheme (Scheme 3) was complex due to alternate expulsion of HCl, Cl·, HCN and H$_2$CN· moieties. The IKE spectra (Fig. 3) of the trichloroanilines

TABLE IV

Ion Kinetic Energy Data for the Dichloroanilines

Compound (isomer)	Ionic decomposition peak intensities $\times$ 10^3					
	0·833E	0·780E	0·737E	0·715E	0·703E	0·780E^a/0·833E
3,4	9·00	17·3 (0·90)^a	1·80	5·30	3·40	1·73
2,4	5·50	14·1 (0·90)^a	1·00	4·80	3·00	2·31
2,5	5·80	17·9 (0·76)^a	1·30	6·20	4·30	2·34
2,3	5·70	11·5 (0·90)^a	1·10	5·30	4·00	1·82
2,6	4·20	15·9 (0·72)^a	1·00	7·60	5·00	2·71
3,5	9·60	18·5 (0·59)^a	1·70	7·00	4·90	1·14

^{a} Portion of peak intensity due to decomposition of singly or doubly-charged molecular ions as determined from photoplate data.

TABLE V
Mass Spectral Data for the Trichloroanilines

Compound (isomer)	Ion abundances (70 eV)							
	m/e 195	m/e 167	m/e 160	m/e 159	m/e 133	m/e 124	m/e 123	m/e 88
2,4,6	100	2·5	4	6	4	15	3	4·5
2,3,4	100	8	4	3	10	6	15	3
2,4,5	100	3·0	6·0	3·5	12	5	14	3

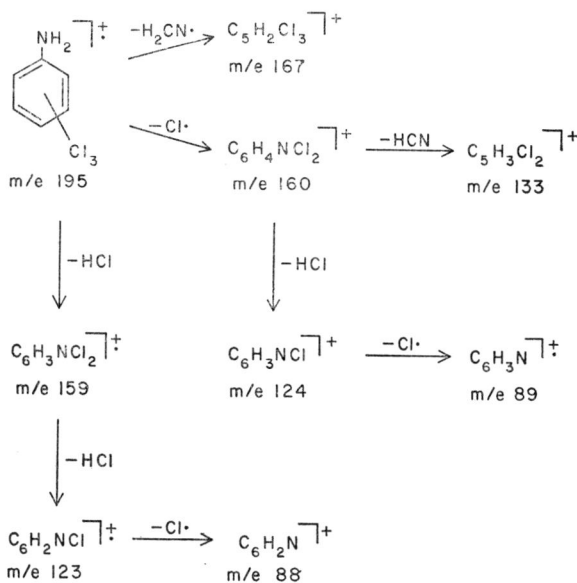

SCHEME 3

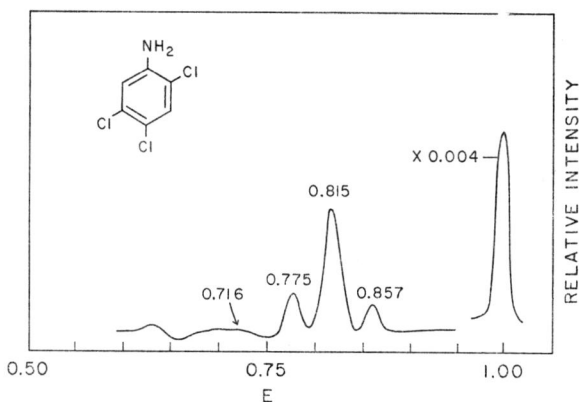

FIG. 3   IKE spectrum of 2,4,5-trichloroaniline.

were also all different and exhibited ionic decomposition peaks at 0·857, 0·815, 0·775, 0·716 corresponding to the reactions m/e 195 → m/e 167, m/e 195 → m/e 159, m/e 159 → m/e 123, and m/e 123 → m/e 88. It was also possible that a shoulder at 0·830 was due to the m/e 160 → m/e 133 reaction. Again in contrast to the mass spectra the IKE spectral data (Table VI) clearly distinguished between the isomers.

TABLE VI

Ion Kinetic Energy Data for the Trichloroanilines

Compound (isomer)	Ionic decomposition peak intensities $\times$ 10^3				
	0·857E	0·815E	0·775E	0·716E	0·815E/0·857E
2,4,6	1·50	7·80	3·10	1·00	5·20
2,3,4	10·1	13·3	5·70	0·90	1·32
2,4,5	3·70	18·7	6·20	1·50	5·05

Examination of the IKE results for all these chlorinated anilines indicates a steric effect in the elimination of HCN (or $H_2CN\cdot$) from the molecular ions decomposing in the first field-free region of the mass spectrometer. Elimination of the amino moiety is inhibited where there is ortho substitution of chlorine.

Thus the IKE technique not only yields more detailed information about fragmentation pathways observed in the primary ion mass spectra of chlorinated anilines but each IKE spectrum clearly distinguished between the isomers thus supporting the analytical utility of this technique.

EXPERIMENTAL

The chlorinated anilines were obtained from Aldrich and Eastman Chemical Co. and crystallized to constant melting point before use. The IKE spectra were recorded on a Dupont CEC 21-110B double-focussing mass spectrometer using a direct introduction wide range probe.[7] The IKE spectra were obtained by scanning the electric sector voltage as previously described[4-6] and the data are the average results obtained for 4 to 5 scans. Relative yields of product ions occurring in the first field-free region could also be recorded photographically using 'Ionomet' vacuum-deposited silver bromide plates and a fixed value for the electrostatic sector voltage.

REFERENCES

1. Beynon, J. H., Baitinger, W. E. and Amy, J. W., *Int. J. Mass Spectrom. Ion Phys.*, 1969, **3**, 47.
2. Beynon, J. H. and Cooks, R. G., *Res/Develop.*, 1971, **22**, 26.
3. Sphon, J. A. and Damico, J. N., *Org. Mass Spectrom.*, 1970, **3**, 51.
4. Safe, S. and Hutzinger, O., *J. Chem. Soc., Perkin Trans.*, 1972, **1**, 686.

5. Safe, S., Hutzinger, O., Jamieson, W. D. and Cook, M., *Org. Mass Spectrom.*, 1973, **7,** 217.
6. Safe, S., Hutzinger, O. and Jamieson, W. D., *Org. Mass Spectrom.*, 1973, **7,** 169.
7. Jamieson, W. D. and Mason, F. G., *Rev. Sci. Instr.*, 1970, **41,** 778.

Discussion

T. Ast (University of Belgrade, Yugoslavia): Have you tried the accelerating voltage scan technique for cases of overlapped energy peaks?

S. Safe: We have tried the technique but in most cases have found that with our setup the electric sector scans give more consistent and reproducible results.

J. Holmes (University of Ottawa, Canada): How do your relative abundances of IKE peaks compare with those of 2nd ffr metastable peaks?

S. Safe: Due to the fact that metastable ion intensities were low for the choroanilines no direct comparison could be made to the IKE peak abundances. It should be noted that for several isomeric PCBs the IKE and metastable results were in contrast.

J. Holmes: Does your energy resolution permit separation of (or allow one to distinguish between) ^{37}Cl and ^{35}Cl losses from molecular ions?

S. Safe: Our energy resolution is not sufficient to distinguish between losses of ^{35}Cl and ^{37}Cl from the molecular ion.

11

Short-Time Pyrolysis as Analogy to Mass Spectral Fragmentation: The Formation of 3H-Azepine

By GERHARD SCHADEN

(*Institut für Organ. Chemie, Technische Hochschule, Darmstadt, Germany*)

THE mass spectrum of 3-(diphenylhydroxymethyl)-3H-azepine[1] (1) shows a strong dependence on the inlet system used. With the direct inlet probe at 100°C (Fig. 1), a weak molecular ion at mass 285 appears.[2] The base peak is at mass 183 according to benzophenone plus one hydrogen. The mass 93

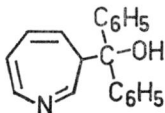

1

corresponds to the azepine part of the molecule plus the hydrogen from the OH-group as could be shown by deuteration. The structure of ions at mass 93 had been elucidated.[3] A chemical problem is to make one of the hitherto unknown azepines by a chemical reaction. Thermal reactions in analogy to mass spectral decompositions have sometimes been successful in preparing

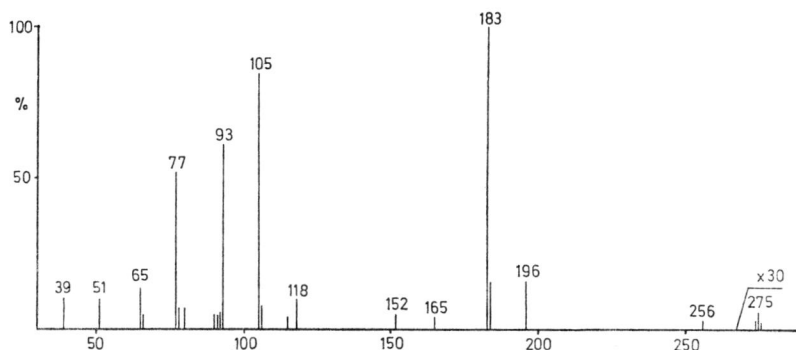

FIG. 1   Mass spectrum of (1), direct inlet, 70 eV.

93

new compounds.[4] Preparative pyrolysis of (1) in the usual manner only gave
benzophenone.[1] In the mass spectrum which was measured through the
heated inlet system at 150°C (Fig. 2) no molecular ion of (1) was present.
The peaks corresponding to benzophenone and again a peak at mass 93
appeared. Addition of D_2O into the heated inlet system effected no exchange.
This showed that the substance with the molecular weight 93 once formed
has no exchangeable hydrogen and could therefore not be aniline or 1H-
azepine. But if first D_2O is introduced into the inlet system and then the

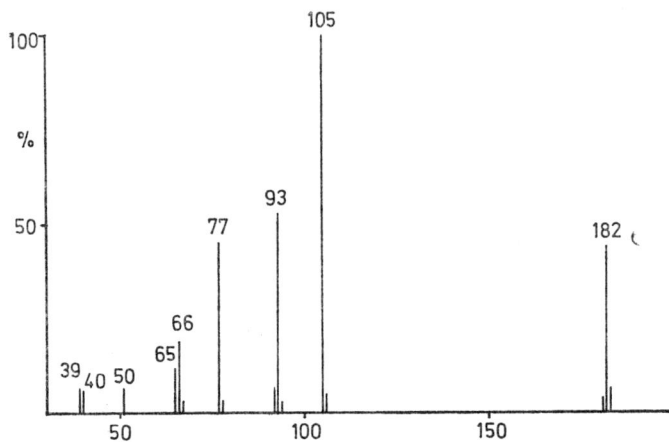

FIG. 2　Mass spectrum of (1), heated inlet 150°C, 70 eV.

substance is added, there is an exchange of one H to mass 94. Therefore the
decomposition of (1) must happen at once after the introduction of the substance into the heated inlet system. Measuring the time dependence of the
spectrum, a decrease of the peak at mass 93 is observed, showing that the
substance with this molecular weight is not very stable. Therefore it cannot
be a picoline which is stable under this condition. So the substance with
the molecular weight 93 seems to be 3H-azepine (2), which is from general
considerations[5] the most stable one of the isomeric azepines. To have
further evidence on the 3H-azepine (2), (1) was pyrolysed with the Curie point

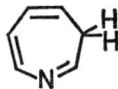

2

pyrolyser.[6] By this method the substance is coated on a ferromagnetic wire
and heated by a high frequency pulse to the final temperature, the Curie
temperature of the wire, without exceeding this temperature. The pyrolysis
products are flushed from the pyrolyser directly into the gas-chromatographic
column. Therefore the thermal reaction time is very short and the pyrolysis
products are at once diluted with the carrier gas. A condensation of the

pyrolysis products is avoided and so unstable substances might be detected. The products which result at different pyrolysis temperatures are examined after gas-chromatographic separation in the mass spectrometer.[7] At 800°C pyrolysis temperature more than 30 substances could be separated; after the mass spectra benzophenone, aniline, cyclopentadiene, benzene, toluene, xylene or ethylbenzene, pyridine, styrene, naphthalene, biphenyl and diphenyl-methane could be identified. The formation of these products showed that the thermolysis conditions were too extreme. Lowering the pyrolysis temperature reduced the number of the resulting substances, also the amount of aniline compared to benzophenone decreased. At 400°C only two substances were formed, one was benzophenone and the other one, which had a retention

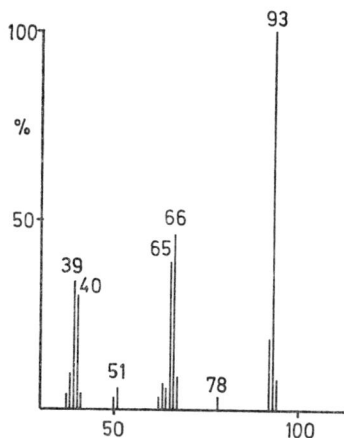

FIG. 3   Mass spectrum of (2), gas chromatographic inlet, 70 eV.

time similar to β- or γ-picoline, showed a molecular weight of 93 and frag-ments similar to a picoline (Fig. 3). D_2O vapour was added after the helium separator but no deuterium exchange was effected. Aniline under these conditions exchanged two hydrogen atoms. As before this also proved that the product formed was not 1H-azepine. α-Picoline had a very different retention time and could be excluded. By comparison of the retention times β- and γ-picoline could not be excluded completely. It was not possible to pyrolyse and to inject at the same moment β- or γ-picoline to compare the retention times. Therefore one half of the pyrolyser wire was coated with (1) and the other part with N-methyl-β-picolinium iodide (3) and N-methyl-γ-picolinium iodide (4). The gas chromatogram from this pyrolysis showed besides benzophenone and methyl iodide three peaks, two being β- and

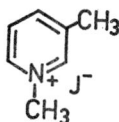

3

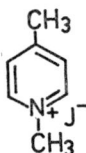

$$\begin{array}{c} CH_3 \\ \\ N^+ \ J^- \\ | \\ CH_3 \end{array}$$

4

γ-picoline formed from (3) and (4), the third the pyrolysis product of (1). Again the substance must be the hitherto unknown 3H-azepine (2). A mixture of 2H-, 3H- and 4H-azepine or of their isomeric compounds could not be present, since they would separate under conditions that separate picolines.

3H-azepine (2) was also formed by the pyrolysis of the potassium salt of azepine-1-carboxylic acid[1] (5) under the same conditions at 400°C and identified in the same way by gas-chromatography mass-spectrometry.

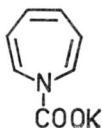

$$\begin{array}{c} \\ N \\ | \\ COOK \end{array}$$

5

EXPERIMENTAL

A Fischer Curie point pyrolyser was used with a Perkin–Elmer F20 gas-chromatograph coupled to a CH4B mass-spectrometer by a two-stage helium separator of the Biemann–Watson type. After the separator D_2O vapour was added till the pressure in the ion source was 10^{-6} torr. This presence of D_2O was sufficient to exchange the two hydrogen atoms of aniline injected into the gas-chromatograph. A low-coated column at low temperature was used to avoid secondary decompositions: 0·1 % Carbowax 20 M or 1500 and 0·002 % KOH on glass beads. The yield of 3H-azepine was 8 %.

REFERENCES

1. Hafner, K., *et al.* to be published; Lindner, H. J., and v. Gross, B., *Chem. Ber.*, 1973, **106**, 1033.
2. Schaden, G., *Chem. Ber.*, 1973, **106**, 1038.
3. Rinehart, Jr, K. L., Buchholz, A. C. and van Lear, G. E., *J. Am. chem. Soc.*, 1968, **90**, 1073; Robertson, A. V. and Djerassi, C., *J. Am. chem. Soc.*, 1968, **90**, 6992.
4. DeJongh, D. C., *in* 'Advances in Mass Spectrometry', Vol. 5, Institute of Petroleum, London, 1971, p. 709; deMayo, P., *Endeavour*, 1972, **31**, 135; Chapman, O. L. and McIntosh, C. L., *Chem. Commun.*, 1971, p. 770; Grützmacher, H. F. and Lohman, J., *Liebigs Ann. Chem.*, 1969, **726**, 47.
5. Huisgen, R., Vossius, D. and Appl, M., *Chem. Ber.*, 1958, **91**, 1; Doering, W. v. E. and Odum, R. A., *Tetrahedron*, 1966, **22**, 81; Maier, G., *Angew. Chem., int.*, 1967, **6**, 402; Carstensen-Oeser, E., *Chem. Ber.*, 1972, **105**, 982.
6. Simon, W. and Giacobbo, H., *Chem.-Ing.-Tech.*, 1965, **37**, 709.
7. Schaden, G., *Chem. Ber.*, 1973, **106**, 2084.

Discussion

H. F. Grützmacher (University of Bielefeld, Germany): Did you determine the ionization potential of the compound of mass 93? This may help to identify the nature of this product.

G. Schaden: I did not determine the ionization potential. It is difficult to obtain a constant pressure of 3H-azepine. In the gas chromatographic method the time is too short and in the heated inlet the substance decomposes further and therefore gives no constant ion current over a long period.

D. C. DeJongh (University of Montreal, Canada): Have you considered the possibility that the compound of mass 93 might be an azabicyclo(2.2.1)heptadiene?

G. Schaden: This structure cannot be excluded but no similar substituted compound is known and therefore I do not believe it to be a bicyclic structure.

A. Prox (K. Thomae GmbH, Germany): How can one exclude other isomeric H-azepines?

G. Schaden: Substituted 2H- and 4H-azepines rearrange at relatively low temperatures to the 3H-isomers.

A. Prox: Can one explain the difference in the mass spectra according to the presence of m/e 65 and m/e 66?

G. Schaden: The stability of the free azepine is lower than the stability of the substituted one. A similar difference is observed between 3H-azepine and the picolines measured under the same conditions.

I. Lengyel (St John's University, New York, U.S.A.): Where does the hydrogen come from in the pyrolysis of

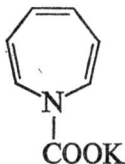

G. Schaden: The hydrogen must come from the traces of water in the carrier gas or adsorbed on the wire.

12

Mass Spectra and Pyrolyses of Azoles

By D. C. DeJONGH, D. C. K. LIN and M. L. THOMSON

(*Département de Chimie, Université de Montréal, Montréal, Québec, Canada*)

THE interpretation of the mass spectra of organic compounds has developed over the years because organic chemists have used their experiences with reactions in the laboratory to explain how and why molecules fragment as they do upon electron impact. The success of this pragmatic approach has been dramatic, and it has been, up to now, more useful than the results of theoretical investigations which try to calculate mass spectra or to explain them. While interpreting mass spectra and postulating structures for ions, interesting reactions and products appear on paper; the analogous reactions in solution or upon pyrolysis or photolysis would often be valuable, or at least interesting.

There are two basic aims of our work in this area. First, we hope to add data to the list of cases in which there are possible correlations between mass-spectral and pyrolytic fragmentation, and to the list of cases in which correlations are missing. In this way, as was the case in the development of guidelines for the interpretation of mass spectra, eventually there will be enough examples in the literature so that empirical guidelines can be established for using the mass spectrum of a compound to predict its pyrolytic reactions. The usefulness of this technique is obvious; not much time or material is required for obtaining a mass spectrum, whereas pyrolysis followed by isolation and identification of products is more complex. The second aim of this work is the development of synthetic routes to organic molecules using pyrolysis.

The apparatus we use is summarized in Fig. 1. Dry nitrogen is passed through a tube (B) fitted with a fritted disc (D) on which solid samples are placed. The flow is monitored and controlled with a rotometer (A). The samples are sublimed into the pyrolysis tube (E) by means of a heating tape wrapped around the outside of the sample holder. Another inlet (C) can be connected to a flask for the introduction of liquid samples or for the introduction of liquid trapping agents. The pyrolysis tube is 24 in. × 1 in. (internal diameter) and is made from quartz. The tube is heated by a 12-in. Hoskin Electric Furnace (F), and the temperature is controlled and read on a Thermolyne Corporation Temcometer (G). Products are collected in a series of traps placed between the furnace and the pump.

In a typical experiment 1·5–2·0 g of sample is introduced, and a nitrogen flow rate of 0·20–0·30 litre/min and a system pressure of 2–3 torr are used.

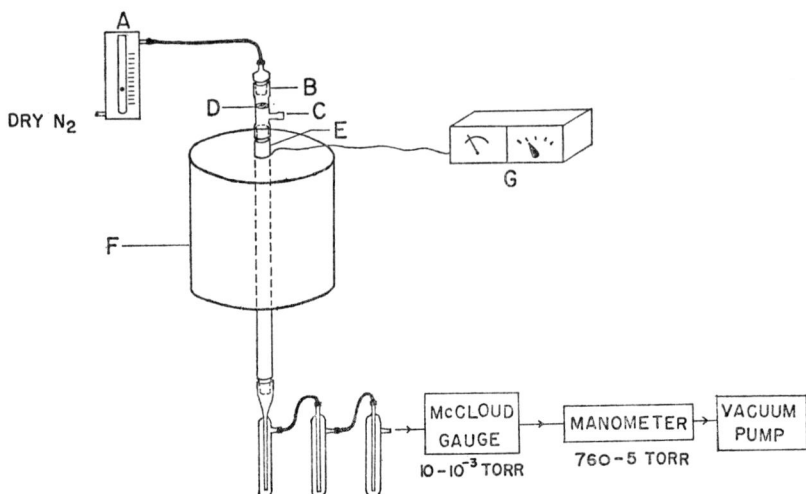

FIG. 1   Diagram of the apparatus used in the pyrolyses.

Products are separated by glpc and/or tlc and identified by standard chemical and spectroscopic methods. Yields are determined by comparison of areas under peaks in the chromatogram of the pyrolysis products, with the areas of peaks of solutions of known concentrations.

The electron-impact and chemical-ionization mass spectra of 2-benzo-thiazolinethione (1), 2-benzimidazolinethione (2) and 2-benzoxazolinethione (3) have been compared with those of their pyrolysis products and parallels

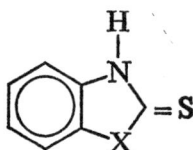

1 X = S
2 X = NH
3 X = O

have been found.[1] In each case, the loss of S from the molecular ions and from the (M + H) ions is the lowest energy-fragmentation. At 800°, 1 gave a 23% yield of benzothiazole (4) and a 13% yield of cyanobenzene (5) (eqn (1)). Loss of S from 1 accounts for the formation of 4, whereas 5 arises from 1 by loss of S_2 and from 4, in a secondary pyrolysis, by loss of S. The one-step loss of S_2 is also observed in the mass spectrum of 1, both in the form of a normal peak and a metastable-ion peak observed while operating in the defocussed mode.

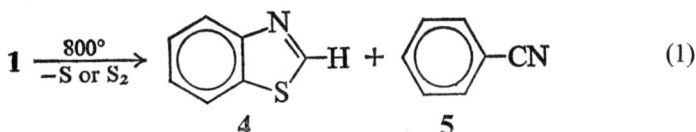

$$1 \xrightarrow[-S \text{ or } S_2]{800°} \quad 4 \quad \text{-H} + \quad 5 \quad \text{-CN} \tag{1}$$

At 950°, **2** gave 62·5% of benzimidazole (**6**) and 7·5% of 2-cyanoaniline (**7**), which are formed by the loss of S from **2** and by the rearrangement of **6** to **7** (eqn (2)).

$$2 \xrightarrow[-\text{S}]{950°}$$

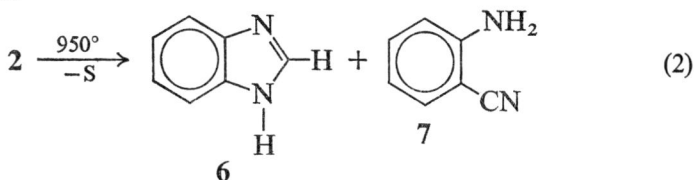

(2)

At 1000°, **3** gave 1·3% of benzoxazole (**8**) and 38% of 2-cyanophenol (**9**) from the loss of S (eqn (3)). At the same temperature, **8** is readily converted to **9** in our pyrolytic system. Compound **3** also gave 12 and 15% of 1- and 2-cyanonaphthalene, respectively, and 7% of naphthalene, presumably by an initial loss of COS from **3**, a low-energy loss also observed in the mass spectrum.

$$3 \xrightarrow[-\text{S}]{1000°}$$

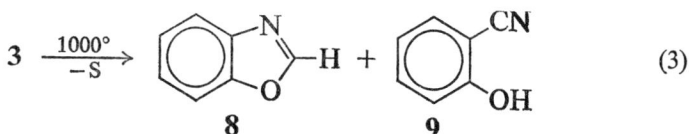

(3)

Thus, the pyrolytic and electron-impact fragmentations of the thiones **1–3** are similar; the lowest-energy paths in the mass spectra can be compared with the lowest-energy pyrolytic paths. Yields are relatively high, ranging from 40–80% under conditions which do not give recovered starting material.

The electron-impact and chemical-ionization mass spectra of 2-benzoxazolinone (**10**), 2-benzimidazolinone (**11**) and 2-benzothiazolinone (**12**) have also been compared with their pyrolysis products.[2] In each case, loss of CO occurs from the molecular ions, followed by CO from **10** and by HCN from **11** and **12**.

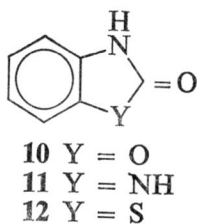

10 Y = O
11 Y = NH
12 Y = S

At 950°, **10** gave 17% of C_5H_5N isomers, related to the loss of 2CO, along with 5·6% of quinoline (eqn (4)).

$$10 \xrightarrow[-2\text{CO}]{950°}$$

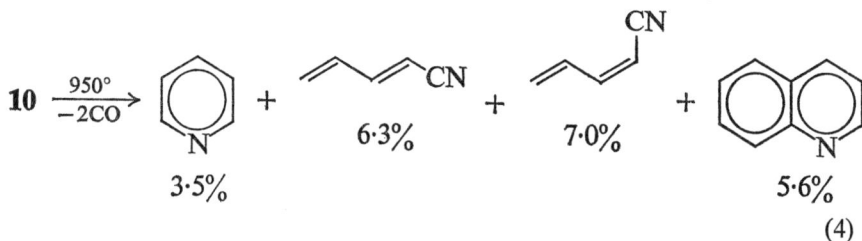

(4)

Compound **11** gave 14% of $C_6H_4N_2$ isomers, related to the loss of CO (eqn (5)).

$$11 \xrightarrow[-CO]{1000°} \left[\text{[structure with two NH groups]} \right] \xrightarrow{-H_2} \text{[CN-substituted ring structure]} + N{\equiv}C-(CH{=}CH)_2-C{\equiv}N \quad (5)$$

10·3%   3·6%

In the mass spectrum of **12**, the major fragmentation is the loss of CO followed by HCN to give a peak at m/e 96 (C_5H_4S). However, no products were isolated from the pyrolyses of **12** which could be related to this fragmentation observed in the mass spectrum. At 950°, 8·7% of starting material was recovered and 14·4% of products were isolated and identified. The major products are 1- and 2-cyanonaphthalene (2·4 and 1·8%, respectively); naphthalene (1·3%), aniline (2·7%), and cyanobenzene (2·6%). The only sulphur-containing product was 2-thiophenecarbonitrile (**13**, 1·1%).

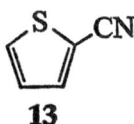

13

The lack of identifiable products related to C_5H_4S is probably due to the fact that C_5H_4S corresponds to cyclopentadienethione which is unstable and polymerizes. So it seems that the major pyrolysis product from **12** does not survive to be isolated. If a large percentage of starting material is not accounted for, this factor must be considered when the comparison is made between mass spectra and pyrolysis.

The overall yields obtained from the pyrolyses of the 2-ones (**10–12**) are substantially lower than the yields obtained from the corresponding 2-thiones (**1–3**). The product from the pyrolyses of the thiones is the azole **14**, formed by the loss of sulphur and rearrangement of a hydrogen. In comparison, the

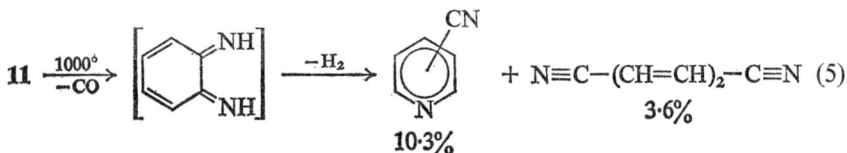 H   Z = S, NH, O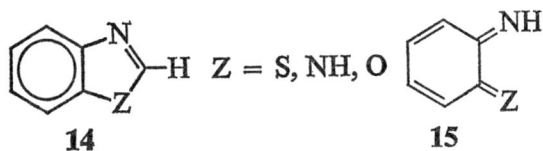

14   **15**

2-ones lose CO and give an unstable intermediate (**15**) which rearranges or fragments further. The mass spectral fragmentations seem to be governed by the elimination of small, neutral species rather than by the stability of the charged species. Also, the pyrolytic fragmentations are driven by the elimination of the same small atoms and molecules such as S, CO and HCN.

We are also studying the mass spectra and pyrolyses of the N-phenyl derivatives of compounds **1–3** and **10–12**. In these cases, intermediates which form initially are trapped internally by the N-phenyl group. Our pyrolysis system has also been used to study o-phenylene sulphite,[3,4,5] o-phenylene carbonate,[6,7] 2-pyrone and 2-pyridone,[8] 2H-naphth[1,8-cd]-isothiazole

1,1-dioxide and its 2-phenyl analogue,[9] naphth[1,8-*cd*]-1,2-oxathiole 2,2-dioxide,[10] and aromatic cyclic diazoketones.[11] We have also used a semi-empirical molecular orbital approach to study *o*-phenylene carbonate and *o*-phenylene sulphite.[12]

REFERENCES

1. DeJongh, D. C. and Thomson, M. L., *J. Org. Chem.*, 1973, **38**, 1356.
2. Thomson, M. L. and DeJongh, D. C., *Can. J. Chem.*, 1973, **51**, 3313.
3. DeJongh, D. C. and Van Fossen, R. Y., *J. Org. Chem.*, 1972, **37**, 1129.
4. DeJongh, D. C., Van Fossen, R. Y. and Dekovich, A., *Tetrahedron Lett.*, 1970, p. 5045.
5. DeJongh, D. C., Van Fossen, R. Y. and Bourgeois, C. F., ibid., 1967, p. 271.
6. DeJongh, D. C. and Brent, D. A., *J. Org. Chem.*, 1970, **35**, 4204.
7. DeJongh, D. C., Brent, D. A. and Van Fossen, R. Y., ibid., 1971, **36**, 1469.
8. Brent, D. A., Hribar, J. D. and DeJongh, D. C., ibid., 1970, **35**, 135.
9. DeJongh, D. C. and Evenson, G. N., ibid., 1972, **37**, 2152.
10. DeJongh, D. C. and Evenson, G. N., *Tetrahedron Lett.*, 1971, p. 4093.
11. DeJongh, D. C. and Van Fossen, R. Y., *Tetrahedron*, 1972, **28**, 3603.
12. DeJongh, D. C. and Thomson, M. L., *J. Org. Chem.*, 1972, **37**, 1135.

Discussion

A. Prox (K. Thomae, GmbH, Germany): How can one exclude the formation of benzoisonitrile in the mass spectrometric breakdown of the thio compounds? The formation of benzonitrile during pyrolysis may occur by thermal rearrangement of benzoisonitrile. This would shed light on the mechanism of ion formation.

D. C. DeJongh: One cannot exclude the formation of benzoisonitrile. However, in our pyrolysis system, benzoisonitrile would rearrange to isomer benzonitrile which is more stable.

P. Bruck (Hungarian Academy of Sciences, Budapest, Hungary): I feel that one of your results gave a warning to mass spectroscopists. I mean the case when *three* products of the thermal loss of 2 CO were identified with a molecular weight of 79. Have you made a structural identification of the m/e 79 ions formed from the same molecule by the loss of 2 CO upon electron impact?

D. C. DeJongh: No, we have not looked at the structure of m/e 79. I think that the formation of isomeric products with a molecular weight of 79 is due to thermal equilibration of the products in the pyrolysis zone. This does not necessarily mean that the ion at m/e 79 has several structures.

H. F. Grützmacher (University of Bielefeld, Germany): Did you try to trap the very interesting intermediate product of the pyrolysis, *i.e.*

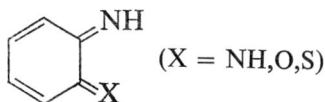

$$(X = NH, O, S)$$

D. C. DeJongh: Yes, we did, but without success. It is often too late if the agent is introduced after the pyrolysis zone. On the other hand, introduction of the agent before the pyrolysis zone often results in the pyrolysis of the agent itself. This limits the number of trapping agents that can be used.

13

Ionization Potentials in Organic Structure Analysis: The Effect of Hydrogen Bonding on the Relative Stabilities of Syn- and Anti-7-Norbornenols

By KALEVI PIHLAJA, JORMA JALONEN and
DAVID M. JORDAN†

(*Department of Chemistry, University of Turku*, 20500 *Turku* 50, *Finland*)

In general, the mass spectra of stereoisomeric compounds closely resemble each other.[1] Accordingly, it has been concluded that mass spectrometry should not be especially well applicable to clarifying stereoisomeric effects. Minor differences in the spectra become, however, more pronounced at lower ion source temperatures and ionizing energies.[1] These differences depend in some way or other on the stereochemistry of the stereoisomers in question and the relative magnitude of these effects can be investigated by measuring the ionization or/and appearance potentials.[2,3]

In recent reviews[2,3] dealing with ionization and appearance potentials in structure analysis it has been shown that the valuable information given by IP and AP measurements can be used in the estimation of various non-bonded interaction energies. With the aid of mass spectrometry both conformational and strain energies of certain stereo- and positional isomers could be determined.[3-6]

In the present paper we report the extension of these studies to the effect of an intramolecular hydrogen bond on the relative stabilities of certain isomeric alcohols. *Syn-* (I) and *anti*-7-norbornenols (II) were originally selected for study since infrared results suggest an intramolecular hydrogen-bonding between the alcoholic OH and the π-electrons of the ethylenic double-bond in the *syn*-7-norbornenols.[7] The oxygen–hydrogen stretching frequencies were 3628 cm^{-1} and 3572 cm^{-1} for the *anti* and *syn* isomers, respectively.[7a] The mass spectra of these isomers have been published by Goto *et al.*[8] and they were very much alike. However, some other norbornane derivatives have mass spectra which exhibit clearer differences. For example, Grützmacher and Fechner[9] observed that the intensities of the molecular ions of some stereoisomeric norbornanediols varied with the tendency of a given isomer to form intramolecular hydrogen bonds.

† Present address: Department of Chemistry, SUCP, Potsdam, N.Y. 13676, U.S.A.

Large differences were also found in the mass spectra of 7-*syn* and 7-*anti* cyclohexyl-2-norbornanols.[10] Kennedy and Kuivila[11] also found some evidence for intramolecular interactions between a double-bond and positively charged tin atom in *syn*-7-trimethyl-stannylnorbornene cations.

Several workers have carried out fragmentation studies on norbornane derivatives, but the IP and AP values of isomer pairs have seldom been measured. A study on *exo*- and *endo*-2-bromonorbornanes revealed identical AP values for the $C_7H_{11}^+$ ion derived from the two isomers.[12] Parallel investigations[13] on the corresponding *exo*- and *endo*-5-bromonorbornenes

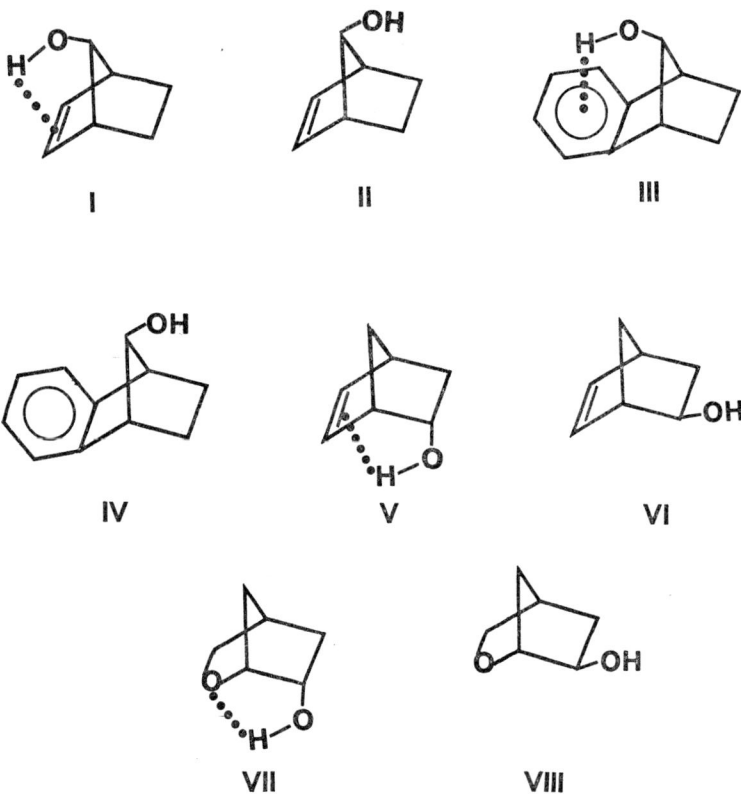

gave equal IP values, but the $AP(C_7H_9^+)$ was 0·1 eV smaller for the *endo* isomer than for the *exo* isomer. This observation was explained as an evidence for anchimeric assistance in the expulsion of a bromine atom.[13] However, we do not exclude the possibility that the lower AP of the *endo* isomer might be due to increased steric interaction as it was the case in isomeric tricyclo [3.2.1.0^{2,4}] octane derivatives.[14] A corresponding study[15] of 5-chloro-norbornenes showed that both IP and $AP(C_7H_9^+)$ values are greater for the *exo* isomer than for the *endo* isomer. In our opinion this difference may again be due to more pronounced steric interactions in the *endo* isomer. Chen *et al.*[10] measured AP's and IP's for a set of isomeric 3- and 7-cyclohexyl-2-norbor-nanols and obtained the same IP values for all isomers and the same AP's for the process M–H$_2$O. However, we have pointed out earlier[2] that the present method is not suitable for processes where both bond-making and bond-breaking processes occur.[3]

The comparison of the AP (or IP) values of stereoisomers rests on the assumption[2,3] that for the 'similar' (stereoisomeric) ions, formed in simple cleavage reactions, the differences in energy terms are likely to be small or at least very similar for both isomers and thus cancelled out. In other words the IP or AP values of ions generated through similar processes differ from the absolute values by the same amount unless extra effects are involved. Generation of a common ion or at least a similar transition state is also necessary for a given isomer pair.[3] In fact, increasing evidence suggests that the ions of many norbornene derivatives and hydrocarbons rearrange to a common intermediate before further decomposition.[10,16] We have pointed out[2,3,5] that on the basis of the foregoing treatment the following equations may be written:

$$AP(A^+) - AP(A_1{}^+) = \Delta H_f(M_1) - \Delta H_f(M) \qquad (1a)$$

$$IP(M^+) - IP(M_1{}^+) = \Delta H_f(M_1) - \Delta H_f(M) \qquad (1b)$$

M and M$_1$ are stereoisomers or positional isomers, $\Delta H_f(M)$ and $\Delta H_f(M_1)$ their enthalpies of formation, and the other quantities the corresponding ionization and appearance potentials.

We have measured the IP's of syn- (I) and anti-7-norbornenols (II) with the method described earlier[4,6] using indene as the reference substance. I is probably stabilized by the intramolecular hydrogen bond and hence it is thermochemically more stable than II.

This stabilisation should be released in the formation of molecular ions and thus eqn (1b) would give a measure for the thermochemical stability difference. Experimentally the value 9 ± 2 kJ mol^{-1} was obtained for $\Delta IP(I-II)$. Recently, Gamba et al.[17] performed molecular orbital calculations for I and II using the CNDO/2 approximation. The formation of the hydrogen bond in the syn isomer (I) was taken into account, and this isomer was found to be 4–7 kJ mol^{-1} thermochemically more stable than the anti form (II). This estimate is very close to our result based on EI ionization potentials. The MO result is also in line with a recent study on corresponding compounds.[18]

For continuation of our study we are preparing syn- (III) and anti-7-benzonorbornanols (IV), endo- (V) and exo-2-norbornenols (VI), and endo- (VII) and exo-2-(6-oxa)norbornanols (VIII). These studies are still underway and additional data are needed before proceeding further. Hence the final discussion will be carried out after completion of the present investigation.

REFERENCES

1. Meyerson, S. and Weitkamp, A. W. Org. Mass Spectrom., 1968, 1, 659.
2. Jalonen, J. and Pihlaja, K. Suomen Kemistilehti, 1972, A45, 116.
3. Jalonen, J. and Pihlaja, K. Org. Mass Spectrom., 1973 7, 1203.
4. Pihlaja, K. and Jalonen, J. Org. Mass Spectrom., 1971 5, 1363.
5. Jalonen, J. and Pihlaja, K. Org. Mass Spectrom., 1972, 6, 1293.
6. Jalonen, J., Pasanen, P. and Pihlaja, K. Org. Mass Spectrom., 1973 7, 949.

7. (a) Bly, R. K. and Bly, B. S. *J. Org. Chem.*, 1963, **28**, 3165; (b) Tichy, M. *Adv. Org. Chem.*, 1965, **5**, 115; (c) Oki, M., Iwamura, H., Onoda, T. and Iwamura, M. *Tetrahedron*, 1968, **24**, 1905; (d) Rochester, C. H. *in* 'The Chemistry of The Hydroxyl Group', Part I, *ed.* S. Patai, Interscience, London, 1971, p. 356.

8. Goto, T., Tatematsu, A., Hata, Y., Muneyuki, R., Tanida, H. and Tori, K., *Tetrahedron*, 1966, **22**, 2213.

9. Grützmacher, H. F. and Fechner, K. H. *Tetrahedron*, 1971, **27**, 5011.

10. Chen, P. H., Kuhn, W. F., Kleinfelter, D. C. and Miller, J. M. Jr, *Org. Mass Spectrom.*, 1972, **6**, 785.

11. Kennedy, J. D. and Kuivila, H. G. *J.C.S. Perkin II*, 1972, p. 1812.

12. DeJongh, D. C. and Shrader, S. R. *J. Amer. Chem. Soc.*, 1966, **88**, 3881.

13. Tomer, K. B., Turk, J. and Shapiro, R. H. *Org. Mass Spectrom.*, 1972, **6**, 235.

14. Brion, C. E., Haywood Farmer, J. S., Pincock, R. E. and Stewart, W. B. *Org. Mass Spectrom.*, 1970, **4**, 587.

15. Steele, W. C., Jennings, B. J., Botyos, G. L. and Dudek, G. O. *J. Org. Chem.*, 1965, **30**, 2886.

16. (a) Shaw, M. A., Westwood, R. and Williams, D. H. *J. Chem. Soc.* (*B*), 1970, p. 1773, (b) Dale, A. J., Weringa, W. D. and Williams, D. H. *Org. Mass Spectrom.*, 1972, **6**, 501; (c) Holmes, J. L. and McGillivray, D., *Org. Mass Spectrom.*, 1971, **5**, 1349; (d) Holmes, J. L. and McGillivray, D. *Org. Mass Spectrom.*, 1973, **7**, 559.

17. Gamba, A., Beltrame, P. and Simonetta, M. *Gazz. Chim. Ital.*, 1971, **101**, 57.

18. Coulombeau, C. and Rassat, A. *Tetrahedron*, 1972, **28**, 4559.

Ions on Surfaces Formed by Field Ionization

By J. H. BLOCK

(Fritz-Haber-Institut der Max Planck-Gesellschaft, Berlin, Germany)

INTRODUCTION

ORIGINALLY this review article was planned to compare field ionization mass spectrometry and secondary ion mass spectrometry, as presently used in surface analysis. Since Dr Honig[1] has discussed problems of surface analysis by secondary ion mass spectrometry in detail, the emphasis of this contribution will now be directed more towards the phenomenon of field ionization at or near surfaces.

During the last three years—*i.e.* since the Fifth International Mass Spectrometry Conference at Brussels—field ion mass spectrometry has come to be characterized by four distinctive areas of development:

1. The identification of single surface atoms or molecules—the atom probe field ion microscope (FIM).
2. The investigation of surface interactions and chemical surface reactions.
3. The observation of field-induced chemical reactions, *e.g.* field-induced adsorption.
4. The analysis of highly unstable ions of macromolecules or biomolecules by the field desorption technique.

PRINCIPLES OF FIELD IONIZATION

The phenomenon of field ionization has been described by different authors.[2-7] The present discussion refers only to the recent findings in field ion formation.

The comparatively least complex mechanism of ion formation is the wave mechanical electron transition of a gas molecule in the homogeneous phase. The superimposed external electric field ($\approx 10^8$ V/cm) diminishes the internal molecular field so that an electron can penetrate the residual potential barrier. The ionization probability D at a certain field strength F is then determined by the ionization potential I_p and the work function Φ of the

neighbouring surface

$$D \propto (I_p - \Phi)(I_p)^{\frac{1}{2}}$$

and ionization occurs only beyond a minimum critical distance x_c from the surface.

This idealized case applies only to rare gases on clean surfaces. If any kind of chemistry affects the interaction between the gas molecule and the surface the mechanism of ion formation is much more complicated and reflects all kinds of chemical interactions, even chemical ionization at the surface.

We can classify these mechanisms as follows:

1. The field ionization of a rare gas atom will be performed at distance x_c solely by the external field F

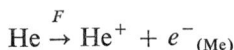

$$He \xrightarrow{F} He^+ + e^-{}_{(Me)}$$

The ionization probability can then be calculated by wave mechanics.

2. In a condensed layer, in a multilayer of gas molecules or during molecular collisions on an adsorbed layer, proton transfer processes are observed very frequently. With water molecules, for instance, molecular associates are normally observed

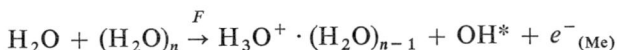

$$H_2O + (H_2O)_n \xrightarrow{F} H_3O^+ \cdot (H_2O)_{n-1} + OH^* + e^-{}_{(Me)}$$

Proton transfer reactions are stimulated by acidic hydrogen atoms and are very common for alcohols, amides, etc. Molecular or dissociatively chemisorbed hydrogen is less qualified for this reaction type. With methane, for instance, CH_5^+ ions are easily observed if traces of water ($p_{H_2O} < 10^{-9}$ torr) are present, while molecular hydrogen does not yield any measurable proton attachment.[8]

3. The ionization process at the surface of a field emitter is also greatly influenced if intermolecular interactions create charge-transfer complexes. Chemical properties of these complexes are explained by a partial electron transfer from a donor to an acceptor molecule, without forming actually separate ions. Onset fields for field ionization are considerably diminished for donor molecules. Benzene, for instance,[9] can be ionized at less than half the usual field strength if an acceptor A, like chloranil, is present at the surface

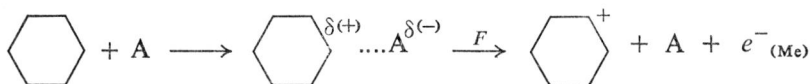

Here an intermediate with partial charges $\delta(+)$ and $\delta(-)$ is formed which facilitates the field ionization process. This kind of chemical interaction is very frequently the reason for the well-known 'promotion' of field ionization, i.e. intensification of ion intensities at reduced field strength. Taking polarization forces into account, charge transfer will also be of importance under field ionization conditions for systems which are not designed for it in the absence of an electric field.

4. The field induced cleavage of surface bonds

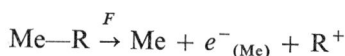

$$\text{Me—R} \xrightarrow{F} \text{Me} + e^-_{\text{(Me)}} + \text{R}^+$$

leads to field desorption of surface molecules or atoms. If R is identical with a bulk metal atom Me, this process is denoted as field evaporation. It is of particular importance for surface analysis and will be discussed later.

These ionization mechanisms at surfaces under extreme electrical fields involve a variety of intricate processes. Consequently, fundamental molecular data, like the ionization potential of an isolated molecule, have much less importance in defining onset fields in field ionization than in comparing the threshold in photo or electron impact ionization. In addition, one further intriguing behaviour in field ion intensities is the large dependence on traces of additive compounds. There are occasional observations that a single water molecule chemisorbed in the ionization zone of a cleaned metal surface will increase ion intensities of field-ionized CO by more than half an order of magnitude. Large fluctuations in ion intensities are the consequence if point emitters are used under such circumstances.

INSTRUMENTAL DEVELOPMENTS

According to the different areas of application which field ionization has found so far, several new technical developments have to be mentioned.

(a) Energy Analysis

The atom probe FIM[10,11]—as is well known—is capable of analyzing mass spectrometrically a single surface atom which is crystallographically identified by a field ion image. The combination of a field desorption pulse with a time-of-flight measurement revealed, however, a rather poor mass resolution ($m/\Delta m \approx 300$) and a frequent occurrence of unidentified masses. The reason for this lies in the effect of time-dependent potentials on the ion trajectory and time-focussing effects which have been discussed in detail.[12-14] Recently, Müller and Krishnaswamy[15] modified the atom probe FIM by

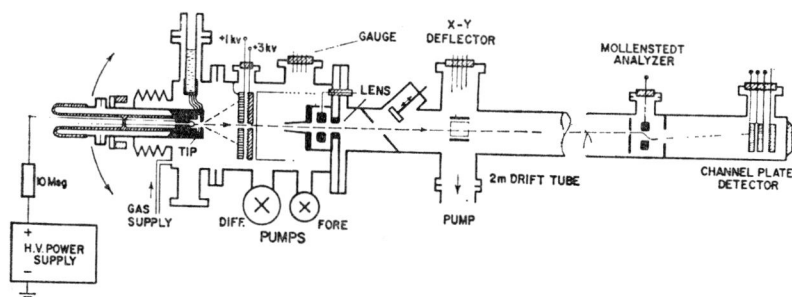

FIG. 1   The atom probe FIM modified with the additional Möllenstedt energy analyser according to Müller and Krishnaswamy.[15]

adding a Möllenstedt energy analyzer (Fig. 1). This technique uses the strong chromatic aberration which a slightly out-of-axis beam suffers in an electro-static saddle-field lens. Different ion energies are displayed at different geometrical positions by an image intensifier. Since a small beam aperture ($\approx 10^{-4}$ rad) and narrow slit ($\approx 7\mu$m) had to be used, an image gas ion current of less than 100 ions/sec originating from a single surface atom had to be measured. The geometrical display of the ion beam was observed and photographed by means of a channeltron array. This device according to

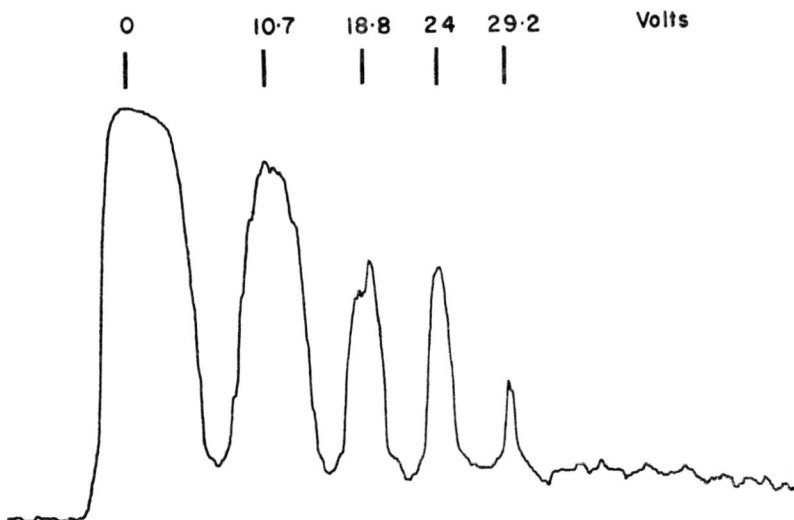

FIG. 2   Energy losses of He$^+$ ions due to resonance tunnelling; centre of the (111) plane of iridium at 5 V/Å, 21°K observed with an atom probe FIM and Möllenstedt filter. Low-order peaks are widened because of overexposure, recorded by a micro-densitometer.[15]

Müller and Krishnaswamy[15] offers an energy resolution of 5×10^{-5}, *i.e.* less than an electron volt for 10 kV field ions. This improvement was applied to confirm resonance tunnelling with a characteristic periodic energy distri-bution of field ions as earlier observed by Jason.[26] As an example of the exceptional energy resolution of a time-of-flight instrument with FI source, the spectrum of He$^+$ ions from a (111) plane of Ir is demonstrated in Fig. 2. The periodic structure of intensities originates from the electron resonance within the potential between the He atom and the surface.

(b) Data Sampling
The small surface area (several Å^2), the low intensities (in the particle/second range) and the resulting high statistical fluctuation of the mass signals demanded new ways for time measurement and signal sampling techniques. The initial use of fast oscilloscopes, where the timescale was triggered by the desorption pulse, has been substituted by two new and interesting methods. The problem to be solved is how to measure flight times within 10 nsec and to register several ions with different masses from a single desorption pulse. Commercial counting devices are not adapted to register these extremely fast

processes. Turner et al.[17] used a timer with a register for the first two arriving particles of a desorption pulse by installing an 82 MHz clock. The counts recorded from this clock were a measure of the time of arrival of an ion. Thimm[18] and Block[19] adapted time-to-pulse-height converters with the ability to register up to eight ions from a single desorption pulse and to discriminate against possible coincidence of simultaneous ionization processes. Small time-windows for higher resolution or large mass scales for rough information about desorbed products can be selected. The evaluation of mass spectra is computerized and the display of data is performed by a plotter. Mass spectrometers may detect a single ion desorbed from a defined site in the atomic scale of a surface. However, laws in physics are determined by statistics. Statistical data sampling as described before is therefore an important requirement for utilizing the sensitivity of the atom probe in evaluating general rules.

Mass spectrometers which have been used so far in combination with a field ion source are of the magnetic sector type or time-of-flight instruments. For many analytical purposes the less expensive quadrupole mass filter would be appropriate. This development is documented in three contributions. Utsumi and Nishikawa[20] used a quadrupole field for the energy analysis of field-desorbed particles. Martin and Block[21] systematically studied the behaviour of field ions in a quadrupole mass spectrometer and Heinen et al.[22] improved the equipment by installing an out-of-axis field ion source for diminishing the troublesome background noise. For many analytical problems the quadrupole mass analyzer has the following particular advantages in combination with a field ion source.

1. Quick scanning of the mass scale to help avoid time-dependent intensity fluctuations in field ion currents.
2. Mass analysis independent of initial ion energies or metastable transitions.
3. A rather large aperture for the ion source which is convenient for a point emitter with approximately a 120° solid angle of ion emission.

The delicate point in field ionization is still the ion source. For surface analysis the point emitter makes extreme demands on vacuum techniques, on care in emitter preparation and on sensitivity in ionic product detection. For analytical purposes, razor blades and needle emitters have been successfully used from which much more intense ion currents are produced but at a less defined field strength, and often from an unknown surface structure. A very important application for analytical purposes is the 'dipping technique' developed by Beckey.[23] Using the advantage of minimal fragmentation in field ionization, low volatile molecules can be impregnated onto a needle emitter and analysed by field desorption. This technique offers the opportunity of analysing macromolecules and substances of biochemical importance which have not been accessible to mass spectrometric techniques so far.

THE ANALYSIS OF INDIVIDUAL SURFACE PARTICLES

In the atom probe FIM[24] single surface atoms or molecules can be identified. The crystallographic position of a surface particle is first determined by imaging the surface with an image gas like He, H_2, Ar or Ne. The emitter is then tilted until the image spot of the required particle is located within the hole of the atom probe. Sometimes a parallax error in ion trajectories at different pulse heights has to be taken into account.[11] A field pulse, sufficient for field desorbing the particle, is then applied, normally after the image gas is removed. The time-of-flight measurement of the desorbed particle yields the mass to charge ratio and a subsequent FIM control ensures that the desired particle has been removed.

With this technique Müller et al.[10,24,25] have analysed several emitter materials. With pure emitter metals the charge on field evaporated ions could be determined. One further unexpected observation was related to surface complexes of inert gas atoms. With tungsten, W^{3+}, WHe^{3+}, WHe_2^{3+} could be detected besides W^{2+} and W^{4+} if He served as an imaging gas. Accordingly, $TaNe^{3+}$, TaN^{3+} were observed as well as Ta^{3+} and Ta^{2+} when a Ne–N_2 mixture was used to image a Ta emitter. Hydride ions TaH^{3+}, however, are unstable and dissociate during the ion trajectory which results in a measurable virtual mass shift of the Ta^{3+} ion. Brenner and McKinney[26] made a statistical evaluation of individual image spots of a tungsten surface, surrounded either by the high vacuum residual gas, by nitrogen or by carbon monoxide. In this case most of the image spots which were formed after chemisorption were desorbed in the form of individual tungsten ions, sometimes clusters of two or three atoms within one ion and only occasionally associated with chemisorbed gas molecules.

There are few examples in the literature for the analysis of composite surfaces. Brenner and McKinney[11] studied an Fe–Mo alloy and could attribute nearly all of the 700 mass signals to Fe^{2+}, Mo^{2+}, Ne^+ and no compounds were observed. Turner et al.[17] have studied a steel specimen containing Mo_2C precipitates, which are characterized by a bright emission of the Ne imaging gas (Fig. 3). The analysis of this region yields C^{2+}, C^+, Mo^{3+}, Mo^{2+} and possibly a trace of MoC^{3+}. No Mo_2C^{3+}, however, has been reported. In a Ni–13 % Ti alloy experimental data are close to the known solid composition.

The analysis of an unknown surface by the atom probe FIM is still in an early stage of development. There are three main factors which prevent a straightforward surface analysis by field desorption or field evaporation:

1. Evaporation probabilities of field ions as a function of the field strength are generally not known, may differ considerably for different parts of a composite surface and are usually drastically altered by gas adsorption. A recent amelioration of the field evaporation theory by McKinstry[27] does not change this situation.
2. Field-induced chemisorption of residual gas components[28] may change the surface composition and influence the desorption probability. Ionic complexes may be formed which are unknown in the chemistry of neutrals.
3. Ionized surface compounds may be unstable and decompose before arriving at the detector of the mass spectrometer.

These complications will be illustrated by a few examples. It is well known in FIM that the appearance of field ion images of one and the same surface

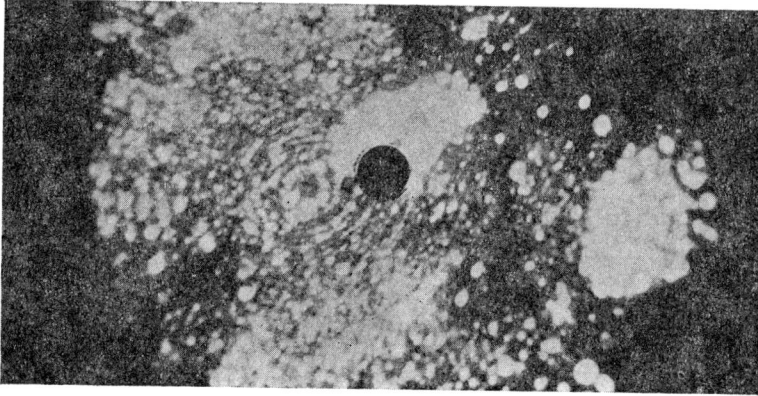

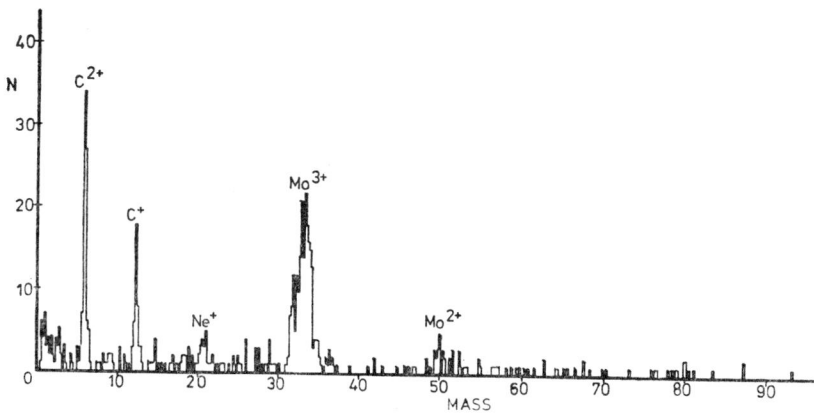

FIG. 3   The analysis of a steel specimen containing Mo_2C precipitates according to Turner *et al.*[17] The Ne^+ FIM image (top) displays a Mo_2C precipitate onto the hole of the atom probe FIM. The mass analysis of this region is shown below. (The aperture of the atom probe is much less than the central spot of the channel plate.)

depends very much on the kind of imaging gas.[29] Lewis and Gomer[30] have now proven that certain chemisorbed gases are not visible in FIM while others are. For instance chemisorbed CO, H and probably O cannot be observed if a W-surface is imaged by Ar.

Analogous properties have recently been observed in FIMS on silver surfaces.[31] Different experimental techniques[32–34] agree fully in concluding that oxygen is chemisorbed on silver. The first part of a monolayer is adsorbed in a dissociative form, the other part as molecular oxygen. Under comparable conditions, but near the field strength of Ag^+ evaporation, no silver–oxygen compounds could be detected mass spectrometrically. Only O_2^+ and Ag^+ are observed at temperatures between 90°K and 425°K. Even after field-free oxygen chemisorption (6 . 10^5 Langmuirs at 10^{-3} torr) at 323°K and 473°K and subsequent field desorption (FD), no silver–oxygen compounds desorb in ionic form. If oxygen is supplied as atoms prepared by the field reaction of H_2O, very small quantities of AgO^+ and AgO_2^+ are formed. More surprisingly, gases which are known not to be adsorbed on silver, like CO,

TABLE I

Field Ions Desorbed From Silver Surfaces After Interaction With Different Gases, According to Reference 31

Gas supplied	Adsorption temp. (°K)	FD temp. (°K)	Observed ions
None		90–375	Ag^+
O_2 1.10^{-5} torr	$90-425^a$ $325-473^b$	90–425 90	Ag^+, O_2^+ (no promotion of Ag^+) Ag^+, O_2^+
H_2O 1.10^{-5} torr	295^a	295	$AgH^+, AgH_2^+, AgH_3^+,$ AgH_4^+ $AgO^+, AgO_2^+,$ $AgO \cdot H_2O^+, AgO_2 \cdot H_2O^+$ $Ag^+, [Ag \cdot (H_2O)_n]^+ \cdot mH_2O$ $H_3O^+ \cdot (H_2O)_n$ $n = 1 \ldots 4, m = 1 \ldots 8$
H_2 5.10^{-6} torr	90^a	90	Ag^+, AgH_2^+, AgH_4^+
CO 5.10^{-6} torr	210^a	210^a	$Ag^+, AgCO^+, Ag(CO)_2^+$ (promotion of Ag^+)

^{a} Field applied during adsorption.
^{b} Adsorption with field off.

Intensities detected $\begin{cases} \text{minimum: } 2.10^{-3} \text{ ions/sec} \\ \text{maximum: } \approx 10^2 \text{ ions/sec} \end{cases}$

yield stable field ions $AgCO^+$, etc. (Table I). With water and other polar compounds, complex molecular ions are formed.

This example clearly demonstrates that observed field ions are not always representative of the actual composition of a metal surface or chemisorption structure. The experimental data, on the contrary, represent a variety of field-induced processes. Reliable analytical values for surface compositions will only be obtained if these reaction types are understood in a quantitative manner.

One is tempted to compare the qualifications, which FIMS and SIMS presently offer for the analysis of an unknown surface. As outlined in Dr Honig's contribution,[1] SIMS may have a sensitivity of $<10^{-6}$ monolayers of a surface area of 0.1 cm^2. This technique has been introduced already to solve various analytical problems in surfaces.

Ionization steps in SIMS are at least as intricate as in FIMS, especially since the high energies of primary ions will cause surface reconstruction. The most common complication, however, is the ignorance of 'ionization cross sections' for compound systems in FIMS and SIMS. The well-known 'promotion effect' of FIMS appears again in SIMS. There are cases known where positive secondary ion currents of a metal increase by orders of magnitudes if oxygen is adsorbed at the metal surface. A comparison of SIMS and FIMS is given in Table II.

TABLE II

Capabilities of Secondary Ion and Field Ion Mass Spectrometry in Surface Analysis

	SIMS	FIMS
Sensitivity	$<10^{-6}$ monolayer $<10^{-14}$ g	1 site 10^{-23} g
Area of analysis	$0 \cdot 1$ cm^2	10^{-16} cm^2
Minimum surface temperature	$\approx 10^3$°K	4°K
Mass resolution $m/\Delta m$	>1000	>300
$\dfrac{\text{[neutrals]}}{\text{[ions]}}$ desorbed	$<10^3$	$\ll 1$
Detection of negative ions	easy	difficult
Ionization cross section, dependence on adsorption	yes	yes
Required vacuum	UHV	extreme UHV
Single crystal faces	possible	only
Material requirement	sufficient conductivity	emitter shape

An advantageous situation in SIMS exists for negative ions which furnish additional information. In FIMS early observations[35] of negative field ions have found some criticism.[4] Recent attempts to detect negative field ions of SF$_6$, HCOOH and other substances from platinum and oxide emitter surfaces have not been successful so far in our laboratory. The desorption of negative field ions faces the following difficulties:

1. Only a small negative field strength can be applied beyond a noticeable emission of field electrons ($<10^{-12}$ Å). Otherwise gas phase ionization at a pressure of $>10^{-6}$ torr will produce positive ions, which destroy the emitter.
2. The discrete energy level of the attached electron is associated with a restricted discrete distance from the surface where only tunnelling can occur.
3. Electron affinities of most substances are smaller than work functions of solid surfaces, so that the back-donation of electrons to the metal surface will compete with the desorption of negative ions.

REACTIONS IN ADSORPTION LAYERS ON SURFACES

For analytical purposes, the minor fragmentation of field ions is generally accepted as a superior property in interpreting mass spectra. Occasionally, however, field ion mass spectra are complicated by mass signals which are usually not observed with electron impact or photo-ionization. Especially at low external fields, gas molecules impinging onto the emitter surface will not be ionized directly, but take part in surface reactions. Several substances may polymerize and form needle structures which are perfect emitter systems for analytical purposes.

The reactions of acetone, which is one of these substances, have been thoroughly studied by Röllgen and Beckey.[37] Such a polymerization reaction may be initiated by surface radical sites which tightly bind acetone molecules, even in an ionic form.

(a) Chemisorption

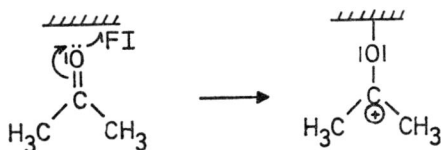

A proton transfer with impinging gas molecules is observed.

(b) Proton Transfer

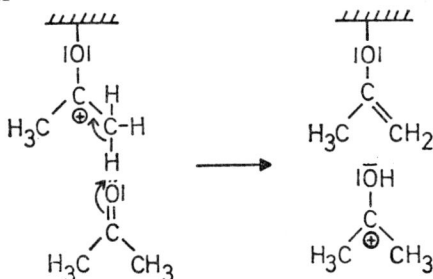

M 59

and the unsaturated surface compound may dimerize.

(c) Dimerization

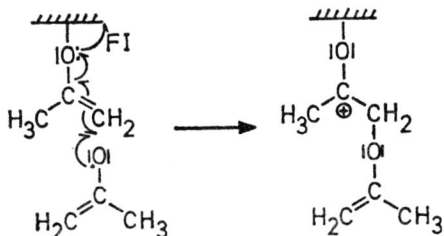

M 116

The loss of oxygen at the surface leads to *M* 99.

(d) Loss of Oxygen

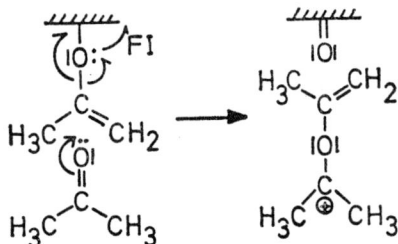

M 99

There are other masses at $M\,115$, $M\,113$, $M\,71$, $M\,50$, $M\,31$ and $M\,29$ (M stands for the mass number of a single charged ion). Reactions forming these products can be described accordingly. In the higher mass region, mass lines extend to values $> M\,200$ as the polymerization proceeds. A notable point in this interpretation is that masses which are smaller than the parent molecular ion are also products of a surface reaction and not fragments due to a monomolecular decay process of parent molecular ions. Arguments which support this interpretation are based on field pulse experiments and will be discussed later.

In a recent publication[36] about the field ionization of hydrazine on Pt-surfaces the ions detected could again be explained by bi-molecular reaction steps in the adsorption layer.

These reaction types indicate that chemisorbed layers play an important part in mechanisms of ion formation. In many cases where surface reactions are observed in field ion mass spectrometry it is not the metal surface of the emitter but a tightly bound surface molecule which interacts with the ionized particle.

This can be demonstrated by another reaction, the surface interaction of water molecules, which has been studied by different authors.[38-40] Proton transfer and molecular association frequently form H_3O^+ and $H_3O^+\cdot(H_2O)_n$ molecular ions. Röllgen and Beckey[41] could prove that these reactions depend on organic chemisorption layers. In mixtures with acetone, benzene or other hydrocarbons reactions, as demonstrated for acetone, will lead to proton-isation:

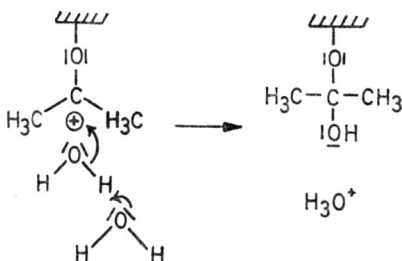

This type of protonisation reaction leads to onset potentials of the H_3O^+ ions which are smaller compared with the ionization potentials of water. These results also indicate that a self-dissociation of water in a H_2O-multilayer according to the Onsager theory

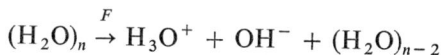

$$(H_2O)_n \xrightarrow{F} H_3O^+ + OH^- + (H_2O)_{n-2}$$

is less probable.

CHEMISORBED IONIC STRUCTURES

The polarization forces of the external field stabilize adsorbed and chemi-sorbed surface structures. In addition, the charge distribution within these molecules will be highly unsymmetric. Therefore, one has to question under

120   J. H. BLOCK

which conditions ionic structures may be chemisorbed in a stabilized form on
field emitter surfaces with high electric fields.

With clean metal surfaces and rare gases—which desorb ionic complexes
like $W \cdot He^{3+}$ [24]—it is difficult to understand that He^+-ions are chemisorbed
on the metal. With the large ionization potential of He (24·5 eV) and the
relatively low work function of W (4·5 eV) ionic surface structures will
scarcely be stable even at a field strength of several V/Å. The situation will be
different if more reactive molecules are chemisorbed on needle emitters (of
high polymer organic material) or on oxide surfaces.

There are three arguments which demonstrate that benzene will be chemisorbed as an ion on the surface of an organic needle emitter: [42,43]

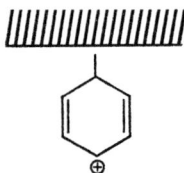

1. The field ionization of molecules like Ar, N_2 and also hydrocarbons is
   promoted by benzene, due to the additional field of the ion charge.
2. Field desorbing benzene, doubly charged molecular ions $C_6H_6^{++}$ are
   observed. For energetic reasons this can be explained only if a charged
   surface molecule is field desorbed

$$S\text{-}C_6H_6^+ \xrightarrow{F} C_6H_6^{++} + e^-_{(solid)}$$

(S = surface site)
3. If field pulses are applied $C_6H_6^{2+}$ ions are initially desorbed and
   reaction times are subsequently not sufficient for a stable chemisorption.
   Only singly charged ions are observed with extended field impulse
   desorption and the promotion effect then disappears.

In the cases discussed so far the chemisorption of ionic structures is field
induced. There are, however, exceptional cases, where surface ions are formed
exclusively by chemical interactions.

Block and Zei[44] investigated the chemisorption of triphenylmethyl (Ph_3C)
compounds on zeolite surfaces. These zeolite structures were deposited onto
Pt-field emitters and contained acidic centres. As also known from IR-
spectroscopy,[45] ionic structures Ph_3C^+ (carbonium ions) are formed on
these surfaces by chemical interaction without the necessity of having an
electric field present. As demonstrated in Fig. 4, the emission of these surface
ions has a completely different behaviour to that of field ionizing Ph_3C
compounds on platinum emitters:

1. On zeolites, a thermal activation but no appreciable electric field is
   necessary for ion desorption; on platinum, as usual in field ionization,
   ion intensities i decrease with increasing temperature and have the
   characteristic onset field.

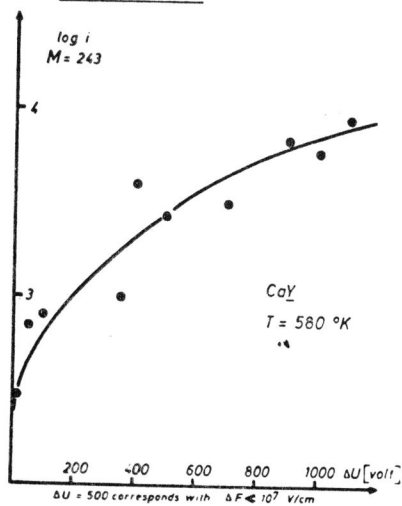

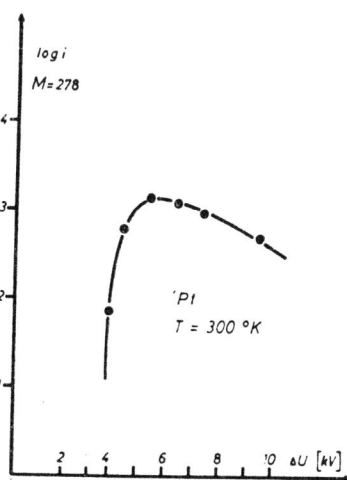

FIG. 4 Ion emission of Ph₃CCl from Pt surfaces (right) and CaY-zeolite surfaces (left). On the zeolite carbonium ions Ph₃C (M 243) are desorbed exclusively, on Pt the parent molecular ion (M 278). The temperature dependence (a) and the field dependence (b) display characteristic differences.

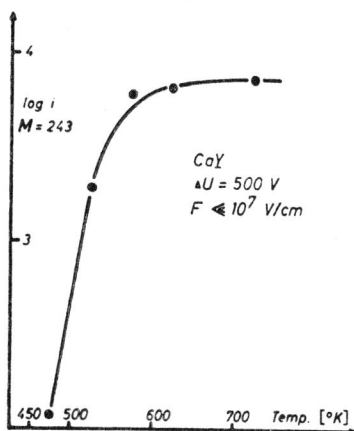

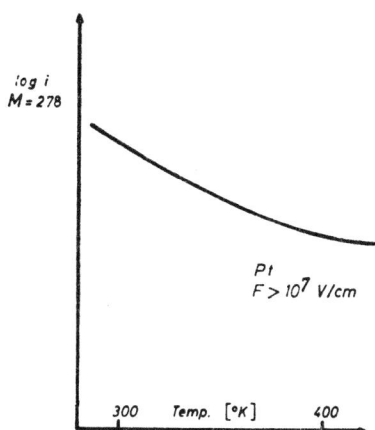

2. On zeolites, carbonium ions Ph_3C^+ (M 243) are observed exclusively; on platinum, however, the parent molecular ion Ph_3Cx^+ and fragments thereof appear.
3. On zeolites, ion intensities increase in accordance with a Schottky barrier assumption $i \sim \exp(eF^{1/2}/kT)$.

These findings and arguments described in detail[46] lead to the conclusion that ions which are observed mass spectrometrically very often are products of specific surface interactions.

FIELD IMPULSE METHODS

An excellent experimental tool for distinguishing between ions formed at surfaces and fragments which originate from the homogeneous decomposition of parent molecular ions is field impulse desorption.[47-49] Ions are desorbed by field pulses ($> 10^{-8}$ sec pulse length) with variable repetition rate (< 70 kc). The kinetics of surface reactions can be investigated by altering the reaction time ($> 10^{-5}$ sec) between two desorption pulses. The field dependence of surface processes can be investigated by applying different reaction fields during the reaction time. Fragments which are formed by an homogeneous decay of parent molecular ions usually have (normalized) intensities which are independent of reaction times. Ions from direct surface interactions will display a reaction time-dependent intensity according to the kinetic laws of their formation.

There is a point of general importance concerning the molecular behaviour in high electric fields. Like temperature or pressure the electric field is a variable of state which alters thermodynamic correlations as well as laws of kinetics. At a field strength of 10^8 V/cm polarization forces acting on molecules reach energetic values which are comparable with chemical bond energies. This interesting area of 'high field chemistry' is most conveniently investigated by field impulse desorption.

A few examples may illustrate the capability of this FID-technique. Earlier it was stated that the protonization, dimerization and polymerization of acetone are field-induced surface processes. Field impulse experiments[37] confirm this mechanism. In Fig. 5 intensities of different ions are given as a function of the reaction (or repetition) time. The $(P + 1)^+$ ion current (P = parent molecule) increases much steeper with the repetition time than P^+, while $(2P - 1)^+$ ions have the steepest dependence. This is in accordance with a three-molecular reaction for $(2P - 1)^+$ and a two-molecular reaction for $(P + 1)^+$ ions. This is further confirmed by the pressure dependence of these molecular ions in FID mass spectra.

Field-induced surface reactions of different hydrocarbons with platinum and tungsten surfaces have been studied by Thimm,[18] Abend et al.[52] and Block et al.[50,53] using a time-of-flight mass spectrometer. A very small surface area (≈ 150 Å^2) of an identified (by FEM) single crystal face was analyzed by FID. For platinum and n-butane, an irreversible contamination of an initially cleaned surface was observed. The FEM pattern after adsorption of n-butane molecules and desorption of ions from platinum is quite different from that of a cleaned surface. The amount of contamination depends on the residual field during two desorption pulses and on the reaction time between these pulses. A low residual field and long reaction time favours contamination. The TOF mass spectrum of the (211) plane of Pt indicates that, except for the fragment $C_2H_5^+$ (M 29), the ion $C_4H_9^+$ (M 57) is the most abundant one and may even exceed the intensity of the parent molecular ion. On other crystal planes the hydrocarbon is dehydrogenated to an even further extent. On the (001) plane, ion intensities of $C_4H_9^+$ (M 57) down to $C_4H_4^+$ (M 52)

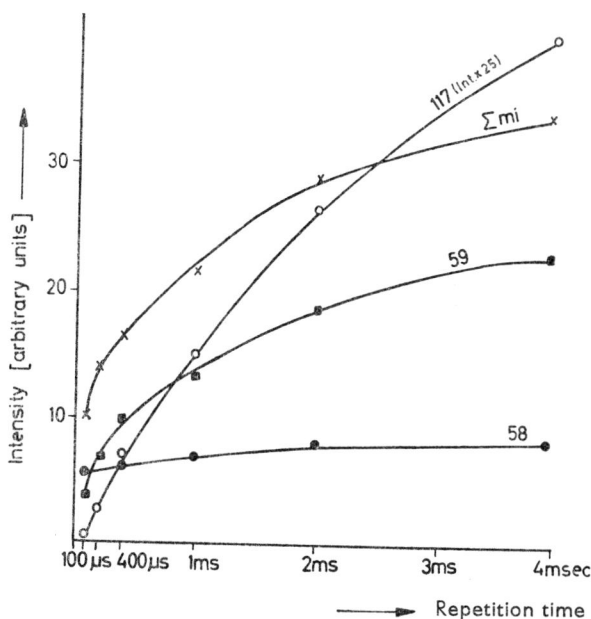

Fig. 5   FID mass spectra of acetone, according to Ref. 37. Dependence of ion intensities M 58, M 59 and M 117 on the pulse repetition time at an acetone gas pressure of 10^{-3} torr.

could be observed. This hydrogenation process is a field-induced reaction in the chemisorption layer and shows a characteristic dependence of ion intensities with increasing residual fields.

Radicals which are formed during these reactions may polymerize and form the observed surface deposit of high molecular weight compounds. These cannot desorb under normal field ionization conditions unless the field reaches values near the desorption field of the bulk material. High molecular weight compounds consisting of Pt, C and H can then be observed. The phenomenon of surface contamination is even more pronounced with W, where high index surface planes are favoured for this reaction type.

A similar polymerization has been observed earlier[51] during the reaction between ethylene and hydrogen on certain platinum surfaces. The time-dependence of the ethylene hydrogenation could only be studied with an excess of hydrogen yielding $C_2H_5^+$ intermediate and $C_2H_6^+$ product ions.

RECENT ANALYTICAL APPLICATIONS

As demonstrated by Beckey,[23] field desorption mass spectrometry can be applied to study thermally unstable substances of low volatility. After initial experiences with amino acids,[53] pesticides[54] and other substances, now pyrolysis products of bacteria and other complex biological materials are being investigated. This method looks very promising for biological and medical applications.

In inorganic chemistry, the structure of sulphur, deposited on metals like W and Pt, is of considerable interest. Recent investigations by Davis[55] which had been performed in combination with FEM studies by Bechtold[56] indicate three different molecular structures of sulphur on a tungsten surface. Even at low temperature ($\approx 78°K$) an irreversible chemisorption occurs if a S_2 molecular beam is directed to a tungsten surface. On top of this 'surface sulphide' a second physisorbed sulphur layer is formed and at even higher doses multilayers of sulphur are observed at room temperature. These adsorption processes are combined with a chemical reaction of S_2 molecules, which either dissociate (within the first monolayer) or recombine in the direction of the stable S_8 structure of condensed sulphur. By mass spectrometric analysis the 'surface sulphide' has not been identified so far. The other structures consist, however, of S_4 to S_8 molecules, with different concentrations. Only if several monolayers are formed on the surface will S_8 be observed in considerable concentration. In all cases, only singly charged S_n^+ ions could be desorbed. This behaviour is particularly interesting in comparison with the properties of Se multilayers. For the semiconducting Se multiply charged ions (Se_n^{2+} and Se_n^{4+}, $n < 33$) could be detected[57] which indicate that a heterolytic bond cleavage is involved in the ion desorption process.

Our knowledge about ion formation and desorption at surfaces and under the influence of extremely high electric fields is still very limited. It is to be regretted that only few scientists have so far encountered this area of research, which faces an encouraging development.

REFERENCES

1. Honig, R. E. 'Analysis of surface and thin films by mass spectrometry', this volume.
2. Gomer, R. 'Field Emission and Field Ionisation', Harvard University Press, Cambridge, Mass., 1961.
3. Müller, E. W. and Tsong, T. T. 'Field Ion Microscopy', Elsevier, New York, 1969.
4. Beckey, H. D. 'Field Ionization Mass Spectrometry', Pergamon Press, Oxford, 1971.
5. Ehrlich, G. Advances in Catalysis, 14, 1963, 255.
6. Beckey, H. D., Knöppel, H., Metzinger, G. and Schulze, P. in 'Advances in Mass Spectrometry', Vol. 3, p. 35, Applied Science Publishers, London, 1966.
7. Block, J. H. in 'Advances in Mass Spectrometry, Vol. 4, p. 791, Applied Science Publishers, London, 1968.
8. Bätjer, K. Dissertation, Freie Universität, Berlin, 1972.
9. Block, J. H. Z. physik. Chem. N. F., 1969, 64, 199.
10. Müller, E. W., Panitz, J. A. and McLane, S. B. Rev. Sci. Instr., 1968, 39, 83.
11. Brenner, S. S. and McKinney, J. T. Surface Sci., 1971, 23, 88.
12. Panitz, J. A., McLane, S. B. and Müller, E. W. Rev. Sci. Instr., 1969, 40, 1321.
13. Thimm, H. Diplomarbeit, Freie Universität, Berlin, 1969.
14. Röllgen, F. W. and Beckey, H. D. Messtechnik, 1972, 115.

15. Müller, E. W. and Krishnaswamy, S. V. *Surface Sci.*, 1973, **36**, 29.
16. Jason, A. J. *Phys. Rev.*, 1967, **156**, 266.
17. Turner, P. J., Regan, B. J. and Southon, M. J. *Surface Sci.*, 1973, **35**, 336.
18. Thimm, H. Dissertation, Freie Universität, Berlin, 1973.
19. Block, J. H. *in* 'Modern Methods of Surface Analysis', DECHEMA—International Symposium, 1973, in press.
20. Utsumi, T. and Nishikawa, O. *J. Vac. Sci. and Technol.*, 1972, **9**, 163.
21. Martin, A. and Block, J. H. Messtechnik, 1973, 149.
22. Heinen, H. J., Hotzel, Ch. and Beckey, H. D. *Int. J. Mass Spectr. and Ion Phys.*, in press.
23. Beckey, H. D. *Int. J. Mass Spectr. and Ion Phys.*, 1969, **2**, 500.
24. Müller, E. W. *Naturwiss.*, 1970, **57**, 222.
25. Müller, E. W. *in* 'Advances in Mass Spectrometry' Vol. 5, p. 427, Applied Science Publishers, London, 1971.
26. Brenner, S. S. and McKinney, J. T. *Surface Sci.*, 1970, **20**, 411.
27. McKinstry, D. *Surface Sci.*, 1972, **29**, 37.
28. Tsong, T. T. and Müller, E. W. *J. Chem. Phys.*, 1971, **55**, 2884.
29. Knor, Z. and Müller, E. W. *Surface Sci.*, 1968, **10**, 21.
30. Lewis, R. L. and Gomer, R. *Surface Sci.*, 1971, **26**, 197.
31. Schmidt, W. A., Frank, O. and Block, J. H. *in* '20th Field Emission Symposium' Penn. State Univ., August 1973.
32. Kilty, P. A., Rol, N. C. and Sachtler, W. M. H. *in* 'Proceedings of the Vth International Congress of Catalysis', Vol. 2, 64–929, ed. J. Hightower, 1972.
33. Czanderna, A. W. *J. Phys. Chem.*, 1964, **68**, 2765; 1966, **70**, 2120.
34. Czanderna, A. W., Frank, O. and Schmidt, W. A. *Surface Sci.*, 1973, **38**, 129.
35. Robertson, A. J. B. and Williams, P. *in* 'Advances in Mass Spectrometry', Vol. 4, p. 847, Applied Science Publishers, London, 1968.
36. Block, J. H. *Z. physik. Chem. N. F.*, 1972, **82**, 1.
37. Röllgen, F. W. and Beckey, H. D. *Surface Sci.*, 1970, **23**, 69.
38. Beckey, H. D. *Z. Naturf.*, 1960, **15a**, 822.
39. Schmidt, W. A. *Z. Naturf.*, 1964, **19a**, 318.
40. Anway, A. R. *J. Chem. Phys.*, 1969, **50**, 2012.
41. Röllgen, F. W. and Beckey, H. D. *Surface Sci.*, 1971, **27**, 321.
42. Röllgen, F. W. and Beckey, H. D. *Berichte der Bunsen-Gesellschaft*, 1971, **75**, 988.
43. Röllgen, F. W. and Beckey, H. D. *Surface Sci.*, 1971, **26**, 100.
44. Block, J. H. and Zei, M. S. *Surface Sci.*, 1971, **27**, 419.
45. Karge, H. *Surface Sci.*, 1973, in press.
46. Block, J. H. *in* 'Proceedings of the Vth International Congress on Catalysis', Vol. 1, E–91, ed. J. Hightower, 1972.
47. Inghram, M. G. and Gomer, R. *Z. Naturf.*, 1955, **10a**, 863.
48. Block, J. H. *Z. physik. Chem. N. F.*, 1963, **39**, 169.
49. Beckey, H. D. and Röllgen, F. W. *Z. Instrumentenk.*, 1966, **74**, 47.
50. Block, J. H., Thimm, H. and Zei, M. S. *Ind. Chim. Belg.*, 1973, **38**, 392.
51. Block, J. H., Thimm, H. and Zühlke, K. *J. Vac. Soc. and Technol.*, 1970, **7**, 63.
52. Abend, G., Thimm, H. and Block, J. H. *in* Ion–Molecule Reactions, International Symposium, Berlin, 1973.
53. Winkler, H. U. and Beckey, H. D. *Organic Mass Spectr.*, 1972, **6**, 655.
54. Schulten, H. R. and Beckey, H. D. 'Field desorption spectrometry of pesticides and their metabolites, ' to be published.
55. Davis, P., Bechtold, E. and Block, J. H. 'Field ion mass spectrometry of sulfur adsorbed on tungsten', to be published.
56. Bechtold, E., Wiesberg, L. and Block, J. H. 'Field emission studies of H_2S and S_2 on tungsten', to be published.
57. Saure, H. and Block, J. H. *Int. J. Mass Spectr. and Ion Phys.*, 1971, **7**, 157.

ORGANIC APPLICATIONS

Chairman

D. H. WILLIAMS
(University of Cambridge, U.K.)

14

The Mass, Metastable and Ion Kinetic Energy Spectra of Some Polycyclic Hydrocarbons Found in Environmental Samples

By R. C. LAO, R. S. THOMAS, J. L. MONKMAN

(Chemistry Division, Technology Development Branch,
Air Pollution Control Directorate, Department of the Environment,
Ottawa, Canada)

and R. F. POTTIE

(Chemistry Division, National Research Council, Ottawa, Canada)

INTRODUCTION

OVER the last decade, a considerable volume of information has been published throughout the world on the polycyclic aromatic hydrocarbon (PAH) content of the airborne particulate portion of air pollution. Attention has been focussed on those compounds whose biological activity has been assayed and found to produce cancerous tumours on the skin of experimental animals.[1] The interest in airborne aromatic hydrocarbons stems from the fact that many of these compounds are present in the organic extract of particulate matter from urban air.[2]

Despite the importance of PAH, which contain five or more rings, as air pollutants, little attention has been given to the mass spectra of individual compounds. For the ten or more isomeric compounds of the formulae $C_{20}H_{12}$ or $_{22}CH_{12}$ which are likely to occur in the 'Benzo[a]pyrene fraction' of air pollutants, no mass spectra data are available in the literature with the exception of Benzo[a]pyrene itself and perylene.[3,4] We have, therefore, initiated a study to obtain mass spectra for these compounds and to evaluate the mass spectrometric techniques as an aid to the analysis of PAH in air. The model systems reported previously include mixtures of Benzo[a]pyrene (BaP), Benzo[e]pyrene (BeP) and Benzo[k]fluoranthene (BkF).[5] These compounds are among the PAH which have been selected by the World Health Organization for intensive investigation. It is usually of most interest to determine BaP since it is carcinogenic. The chromatographic separation of these three isomers is often incomplete using either gas/liquid or liquid/solid phase

129

systems. In addition, mutual interferences preclude UV absorption or fluorescence spectroscopy as analytical techniques for these compounds.[6]

The similarities of E.I. spectra for BaP, BeP and BkF were described by Lao *et al.*[7] Although model systems were developed for simple PAH mixtures, it is difficult to apply this simple approach to an environmental sample containing a broad spectrum of PAH.[6] The additional information concerning the metastable transitions, fragmentation modes and the release kinetic energy for the decomposition process is required to permit the unambiguous identification and accurate measurement of ambient quantities of PAH.

EXPERIMENTAL

Two double-focussing mass spectrometers were used in this study; a Hitachi Perkin-Elmer RMU7L and a unit constructed 'inhouse'.[8]

The RMU7L was modified to permit the ESA voltage to be scanned linearly.[9] Using this technique, the intensities of the apparent ion formed from the metastable transition in the first field-free region to be recorded and

TABLE I

	RMU7L (MS No. 1)	'Inhouse' built (MS No. 2)
Electrostatic analyser voltage	200–250 V	150–250 V
Electron energy	70 eV	70 eV
Probe temperature	100–200°C	100–140°C
Source pressure	5–8 $\times$ 10^{-7} torr	1 $\times$ 10^{-6} torr
Resolution	10 000	1 000

the ion kinetic energy spectrum of the compound were plotted.[10] Using the 'inhouse' mass spectrometer, the ions transmitted through the β slit were mass analysed either by increment ESA voltage decrease (0·1–0·3 V) at constant accelerating potential, or by increment accelerating voltage decrease at constant ESA voltage. The operating conditions for the two instruments are described in Table I.

Using MS No. 1, each apparent ion Ma is focussed magnetically and the ESA voltage E is determined at the centre of the peak. This voltage is then scanned from E to zero and peaks appearing at E_1, E_2, etc. are recorded. The areas of the peaks are measured using a Hewlett–Packard No. 3373B integrator. The values given in this work are the average of three to five scans with a standard deviation of about 5%.

RESULTS AND DISCUSSION

The electron impact mass spectra of some of the BaP isomers are listed in Table II. The main fragmentation patterns of these three PAH are also similar. The spectra suggest the mechanism of fragmentation to be:

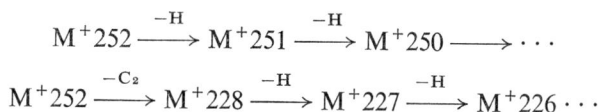

$$M^+252 \xrightarrow{-H} M^+251 \xrightarrow{-H} M^+250 \longrightarrow \cdots$$

$$M^+252 \xrightarrow{-C_2} M^+228 \xrightarrow{-H} M^+227 \xrightarrow{-H} M^+226 \cdots$$

At present there is no direct evidence as to which benzene ring is broken from the compound.

TABLE II

Mass Spectra of PAH

m/e	BaP	BeP	BkF
253	21·49	21·14	21·54
252	100·00	100·00	100·00
251	5·87	7·15	5·17
250	19·34	23·95	18·69
249	2·76	3·68	2·93
248	4·19	5·16	4·46
226	2·33	2·17	1·90
225	2·27	1·65	1·99
224	3·06	2·94	3·21
126	16·39	14·47	20·56
125·5	2·67	4·72	3·27
125	11·46	15·53	12·65
124·5	2·69	3·58	2·93
124	5·31	6·95	5·74
123·5	0·85	0·95	0·92
123	0·92	1·18	0·90
113	8·26	7·76	7·76
112·5	1·79	1·63	2·11
112	5·15	4·72	6·62
111·5	1·42	1·44	1·79
111	1·67	1·75	2·16

The relatively intense ions in the region m/e 113–126 attest to the high degree of multiple ionization in these compounds. These steps are most likely to be:

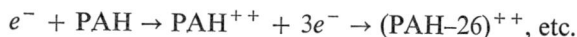

$$e^- + PAH \rightarrow PAH^{++} + 3e^- \rightarrow (PAH\text{--}26)^{++}, \text{etc.}$$

There is no mass spectrometric evidence that those ions are single-charged and are formed by fragmentations. The identical isotopic ratios measured for BaP in the following regions:

$$m/e\ 252:253:254 = 100\cdot00:21\cdot71:2\cdot39$$
$$m/e\ 126:126\cdot5:127 = 100\cdot00:21\cdot63:2\cdot47$$

support the conclusion that almost all the fragmentations are double-charged. This is confirmed by the metastable spectrum of the individual PAH. Although it may be possible that other mechanisms, such as autoionization and secondary ions may be responsible for the formation of ions, they are not important routes to these ions.

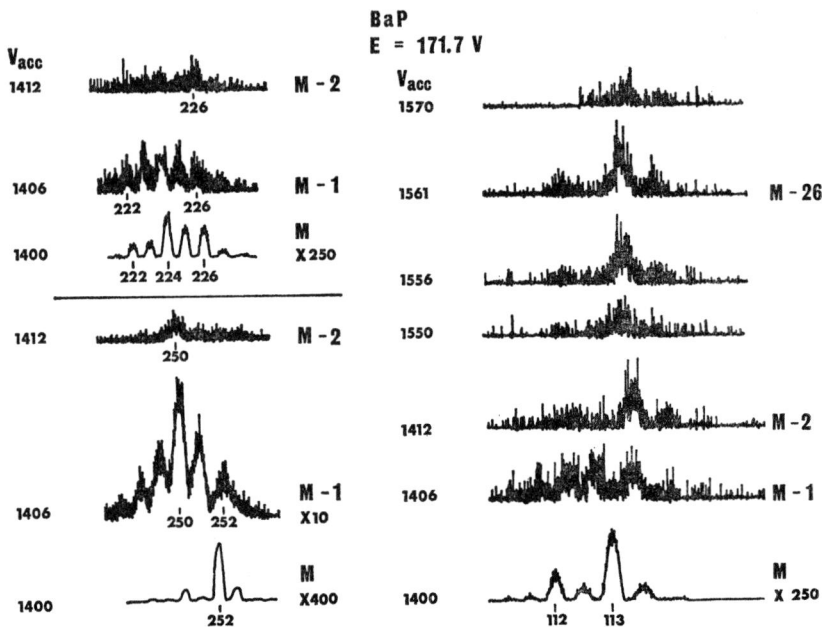

FIG. 1   Metastable spectrum obtained by acceleration potential variation.

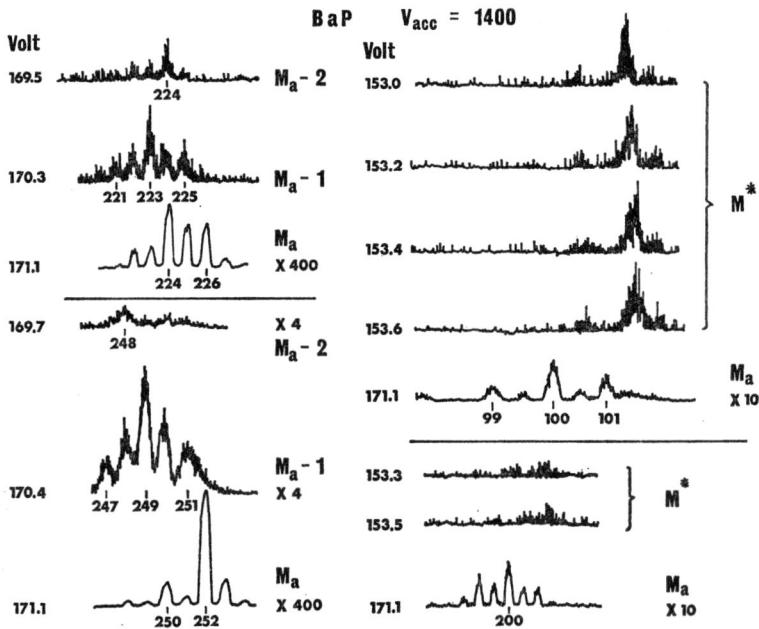

FIG. 2   Metastable spectrum obtained by electric sector voltage decay.

The metastable spectra of BaP and BkF were investigated using MS No. 2. A part of these spectra for BaP are illustrated in Figs. 1 and 2. A comparative group of spectra for BkF are shown in Figs. 3 and 4. Although the similarities of the transitions are evident from these spectra, the relative intensity differences of the major peaks, though small, are measurable. The metastables, M^*, in Figs. 2 to 4 confirm the decomposition processes mentioned previously.

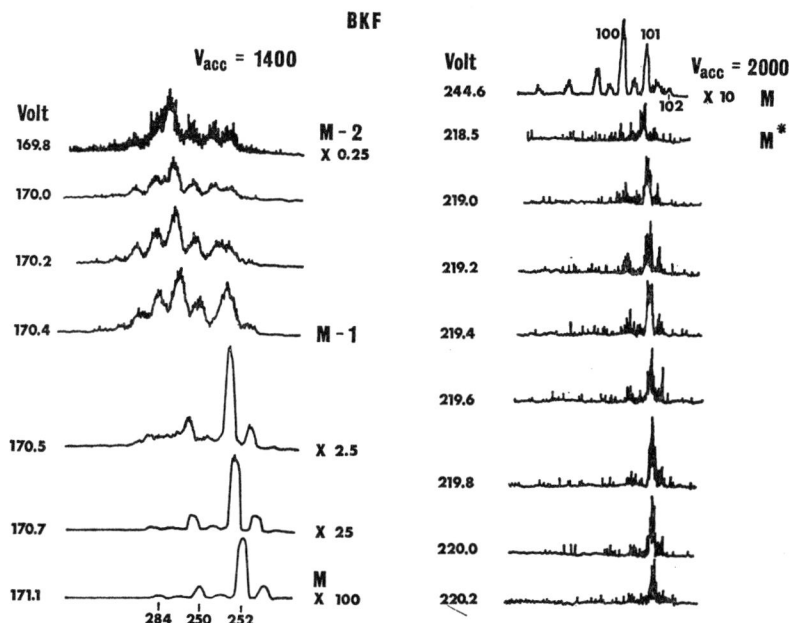

FIG. 3   Metastable spectrum obtained by electric sector voltage decay.

Since our primary interest is the application of mass spectrometric techniques to identify and measure PAH found in environment samples, we concluded that prior to any further detailed study of the metastable transitions and ion kinetic energy spectra, the relative intensities of each apparent ion formed in the decomposition must be examined to evaluate the possibility of using a direct comparison of the differences for PAH analysis.

The results for BkF were first applied to this technique. For the decomposition:

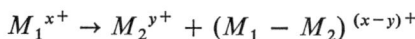

$$M_1{}^{x+} \rightarrow M_2{}^{y+} + (M_1 - M_2)^{(x-y)+}$$

where M_2 is the daughter ion to be calculated from:

$$\frac{M_2}{y} = Ma\frac{E}{E_1}$$

and, M_1 is the parent ion from

$$\frac{M_1}{x} = M_2\frac{E}{E_1}$$

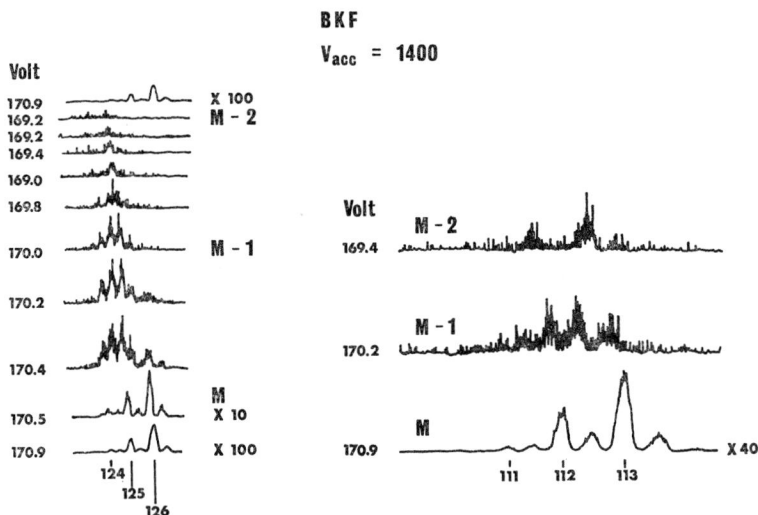

FIG. 4   Metastable spectrum obtained by electric sector voltage decay.

E and E_1 are ESA voltages described in a previous section and M^* is the metastable ion calculated from $(M_2/y)^2/(M_1/x)$. The areas of the metastable decomposition peaks were measured from the spectra relative to the base peak with ESA voltage adjusted to E. No direct comparison of the areas among peaks with difference Ma can be made due to the variation in pressure. No data were taken for peaks with m/e ratio > 98 since the intensities of these peaks was too small to produce meaningful data because of spectral contribution from background.

Work is proceeding to measure the relative peak areas for BaP and BeP. In addition, detailed studies of the metastable transitions and IKE spectra of BaP are being determined using a Spiraltron electron multiplier inserted after the β slit on instrument No. 2.

TABLE III
Data for BkF

Ma	E^a	$E_1{}^a$	M_2	M_1	$M_1 - M_2$	Relative[b] peak area	M^*
253	240·65	239·37	254	255	1	0·000 6	253·105
252	240·45	239·65	253	254	1	0·046 0	252·098
251	240·65	239·70	252	253	1	4·533 4	251·090
250	240·55	239·50	251	252	1	0·041 5	250·082
		238·00	252	254	2	0·001 0	250·094
249	240·45	239·55	250	251	1	0·908 8	249·074
		238·50	251	253	1	0·132 3	249·086
248	240·50	239·40	249	250	1	0·194 2	248·067
		238·50	250	252	2	0·043 3	248·079
		237·60	251	254	3	0·024 5	248·098

TABLE III—*continued*

Ma	E^a	$E_1{}^a$	M_2	M_1	$M_1 - M_2$	$Relative^b$ peak area	$M*$
224	240·45	239·25	225	226	1	0·014 4	224·067
126·5	240·45	239·61	127^c	127·5^c	1	0·003 8	126·553
126	240·50	239·70	126·5	127	1	0·068 2	126·049
125·5	240·35	239·35	126	126·5	1	0·007 6	126·049
125	240·15	239·40	125·5	126	1	0·010 5	125·041
		238·40	126	127	2	0·000 5	125·047
124·5	240·15	239·05	125	125·5	1	0·146 7	145·537
		238·10	125·5	126·5	2	0·004 5	145·543
124	240·05	239·06	124·5	125	1	0·069 4	124·033
		238·10	125	126	2	0·016 8	124·039
123·5	240·50	239·50	124	124·5	1	0·771 1	123·529
		238·60	124·5	125·5	2	0·090 7	123·531
		237·60	125	126·5	3	0·036 0	123·545
123	240·50	239·50	123·5	124	1	0·093 1	123·026
		238·71	124	125	2	0·019 3	123·032
113	240·35	239·10	113·5	114	1	0·000 6	113·041
112·5	240·25	239·10	113	113·5	1	0·100 1	112·537
112	240·35	239·15	112·5	113	1	0·036 9	112·033
111·5	240·55	239·50	112	112·5	1	0·117 7	111·530
		238·30	112·5	113·5	2	0·002 5	111·536
		237·30	113	114·5	3	0·002 0	111·547
111	240·50	239·35	111·5	112	1	0·016 5	111·026
101	240·50	239·50	101·5	102	1	0·006 4	101·042
		216·80	112	124	24	0·007 9	101·192
		216·00	112·5	125	25	0·095 5	101·288
		215·05	113	126	26	0·006 4	101·373
100·5	240·50	239·31	101	101·5	1	0·297 5	100·538
		215·85	112	124·5	25	0·963 6	100·783
		214·95	112·5	126	27	0·244 3	100·472
100	240·45	239·28	100·5	101	1	0·167 3	100·034
		215·00	112	125	26	0·352 6	100·377
99·5	240·45	239·84	100	100·5	1	0·032 0	99·530
		215·85	111	123·5	25	0·071 0	99·785
		214·85	111·5	124·5	26	0·016 3	99·878
		213·95	112	115·5	27	0·011 2	99·973
99	240·45	239·15	99·5	100	1	0·003 8	99·026
		214·65	111	124	26	0·000 3	99·380
98·5	240·45	239·30	99	99·5	1	0·033 2	98·552
		215·45	110	122·5	25	0·017 9	98·788
		214·35	110·5	124	27	0·007 2	98·480
98	240·45	239·20	98·5	99	1	0·008 2	98·018
		214·35	110	123	26	0·010 4	98·383

^{a} In volts.
^{b} Total ion beam at 1·0 E taken as base peak.
^{c} Denotes m/e ratio. All M_1 and M_2 from Ma = 126·5 to 98 are doubly charged.

CONCLUSIONS

The ion kinetic energy spectra of PAH isomers can provide a new approach to the analysis of environmental pollution samples. The ability of IKE spectra to provide additional information about the fragmentation modes and molecular structure will result in a new understanding of the formation and reaction chemistry of these compounds in the gas phase of carbonaceous fuel combustion.

REFERENCES

1. Leiter, J., Shimkin, M. B. and Shear, M. S. *National Cancer Institute,* 1944, **3,** 155–65.
2. Hoffmann, D. and Wynder, E. L. *Cancer,* 1962, **15,** 103–8.
3. Mass Spectra Data, Serial No. 1020, American Petroleum Institute Research Project 44, American Chemical Society, 1954.
4. Stenhage, R., Abrahamsson, S. and McLafferty, F. W. 'Atlas of Mass Spectra Data', Interscience Publishers, New York, 1969, p. 1755.
5. Lao, R. C. *et al.* 'Proceedings of International Symposium on Identification and Measurement of Environmental Pollutants', National Research Council, Canada, Ottawa, 1971.
6. Lao, R. C. *et al.* 'Analytical Chemistry', 1973, Vol. 45, p. 908.
7. Lao, R. C. *et al. International Journal of Environmental Analytical Chemistry,* 1972, **1,** 187.
8. Knewstubb, P. G. and Tickner, A. W. *J. Chem. Phys.,* 1962, **36,** 674.
9. Dowd, G., Thomas, R. S. and Lao, R. C. Manuscript to be submitted to the International Journal of Mass Spectrometry and Ion Physics.
10. Beynon, J. H. *et al. Org. Mass Spectrom.,* 1970, **3,** 455.

15

Field Ionization with Gas Chromatography/Mass Spectrometry in Structural Studies of Natural Products

By D. E. GAMES, A. H. JACKSON, D. S. MILLINGTON and M. ROSSITER

(*Department of Chemistry, University College, Cardiff, U.K.*)

THE advantages of using a combination of field ionization (FI) and electron impact ionization (EI) in the mass spectral study of organic compounds are now well established.[1] Recent reports have described the combinations of FI-MS with on-line data acquisition[2] and with gas chromatography[3] and high resolution FI-MS has been used in the study of complex mixtures.[4] We have applied all these techniques in our studies of mixtures of pyrrole pigments, alkaloids, coumarins and terpenes, and have found that when supplemented with EI-MS they provide an extremely powerful method for the study of such mixtures. Preliminary studies with the Varian combined EI-FI-FD source indicate that un-derivatized pyrrole pigments are amenable to studies using field desorption (FD) and that increased sensitivity should be obtainable in the FI mode since benzonitrile activated emitters may be used with the source, rather than the acetone activated emitters which we have used to date in our field ion studies.[5]

Use of an on-line computer for data acquisition greatly facilitated determination of the m/e values of ions in the FI spectra since the computer's mass-calibrated time scale obtained with the source operating in the EI mode enabled mass measurement to an accuracy of within ± 0.2 amu. The calibration remained constant for at least one day, provided there were no temperature changes in the room. Furthermore, computer processing of the FI spectra was assisted by the configuration of our Varian CH 5D mass spectrometer in which the magnet precedes the electrostatic analyser in the ion path. All the spectral peaks are symmetrical in shape, since metastable ions are not detected under normal operating conditions, and there is no 'tailing' of fragment ions due to fast metastable transitions.

A problem in the use of combined GC-MS is that thermal decomposition of the material under investigation may take place before it reaches the ion source of the mass spectrometer. Examination of the low resolution spectrum obtained by direct introduction of the materials under investigation into the mass spectrometer, provides a partial check for decomposition. The EI spectra obtained are usually extremely complex, and it is difficult to distinguish the molecular species present. Since the molecular ion, or the M + 1 ion, account for the major proportion of the total ionization in FI spectra, the FI spectra of mixtures are much simpler and enable ready identification of the molecular species present, but of course do not enable differentiation between isomeric

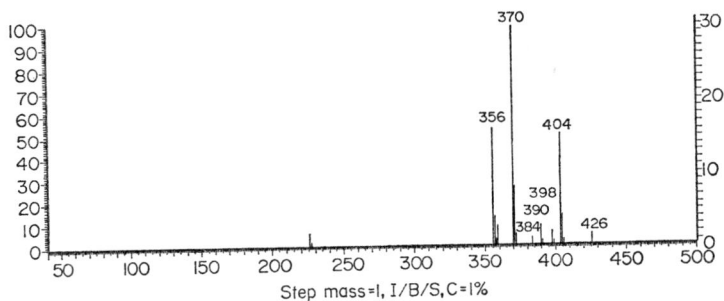

(a)

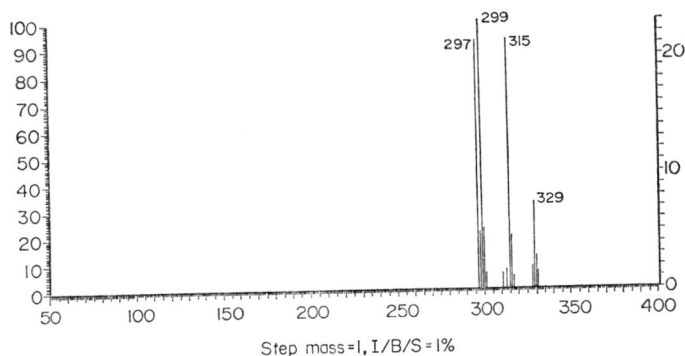

(b)

FIG. 1   (a) FI spectrum of a mixture of coumarins from *Mammea americana*. (b) FI spectrum of a mixture of alkaloids from *Erythrina princeps*.

species. We have used this technique in our studies of porphyrin, alkaloid and coumarin mixtures, and Fig. 1(a) and 1(b) show the FI spectra obtained from a mixture of coumarins isolated from the seeds of *Mammea americana* and a crude alkaloid mixture from the seeds of *Erythrina princeps* respectively. High resolution mass measurement in the FI mode using the peak matching technique with standard compound(s) having molecular weight(s) in the region of interest enables the molecular formulae of the components of the mixture to be established. Table I shows the results obtained using this technique on the alkaloid mixture described earlier.[3]

Large numbers of biologically important compounds are not amenable to EI or FI study due to their lack of volatility. In this context two of our

TABLE I

High Resolution Mass Measurements (R.P. 6000) on Alkaloid Mixture from
Erythrina princeps **Using m/e 299, $C_{18}H_{21}NO_3$ as a Standard**

Found	Calculated	Molecular formula
297·140	297·136	$C_{18}H_{19}NO_3$
315·145	315·147	$C_{18}H_{21}NO_4$
329·157	329·163	$C_{19}H_{23}NO_4$

interests are natural porphyrins and bile pigments which are not normally amenable to mass spectral study without prior derivatization as their methyl esters.[6] FD offers a solution to this problem[1] and satisfactory FD spectra of un-derivatized mesoporphyrin IX (1a), haematoporphyrin (1b) and bilirubin (2) have been obtained. The FD spectrum of mesoporphyrin-IX (MW 566) showed only ions at m/e 566 and 567, while haematoporphyrin (MW 598) exhibited an ion at m/e 562 presumably due to loss of two molecules of water from the hydroxyethyl side-chains, and bilirubin showed only an ion at m/e 584 due to its molecular ion. Clearly this technique will be of great value in

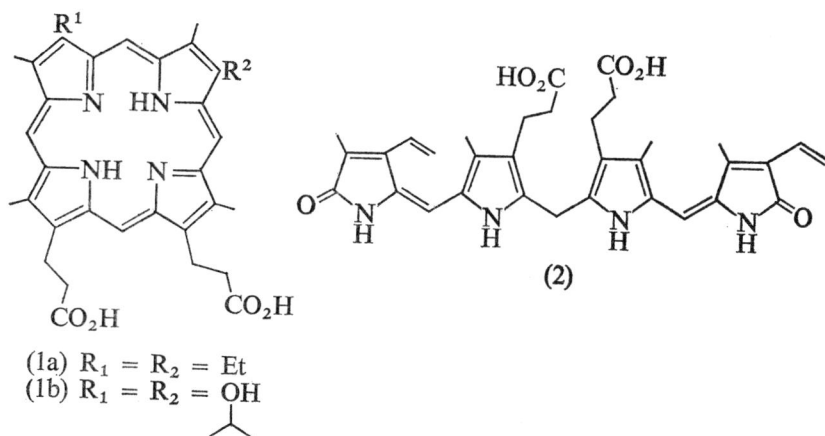

(1a) $R_1 = R_2 = Et$
(1b) $R_1 = R_2 = OH$

the study of crude mixtures of pyrrole pigments and other biologically important compounds which are thermally unstable. In addition to the problem of possible thermal decomposition mentioned earlier, identification of the molecular ions and the detection and interpretation of multi-component peaks can cause serious difficulties in GC-MS. Chemical ionization is one solution to these problems, and the second difficulty may be solved using mass chromatography, if a computer is available. FI offers an alternative solution, which has received little attention as yet.[3] We have investigated the utility of FI in both these aspects when applied to mixtures of alkaloids, coumarins and terpenes and have found that combined GC-FI-MS is feasible at higher sample concentration, but with little loss of resolution compared with GC-EI-MS. Figure 2 shows the total ion current monitor traces obtained

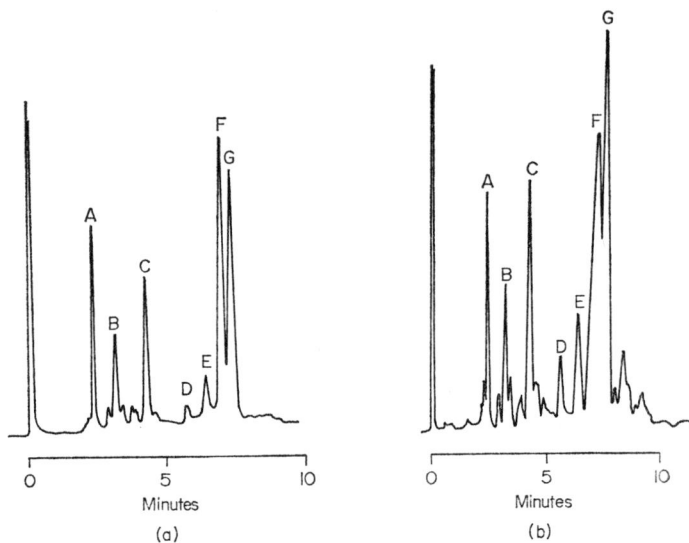

FIG. 2 Total ion current monitor traces from GC-MS of some terpene degradation products: (a) Electron impact. (b) Field ionization 4 m × 2 mm i.d. glass column, 3 % OV1 on gas chrom. Q (100–120 mesh); temperature, programmed at 2°C/min over the range 130–150°; helium flow 30 ml/min.

from a mixture of terpene chemical degradation products[7] using EI and FI sources during GC-MS, and Fig. 3 shows the EI and FI spectra obtained for component G of this mixture. The EI spectrum was typical of all the components, showing extensive fragmentation and low molecular ion intensity, whereas FI in all cases gave intense molecular ion peaks, thus enabling unambiguous molecular weight determination. Identification of multi-component GC peaks is illustrated in Fig. 4, which shows the FI spectrum obtained from one peak on GC-MS of the mixture of coumarins illustrated in Fig. 1. The presence of components having molecular weights of 356 and 370 is indicated since the pure coumarins in this series do not exhibit any other electron fragments apart from the molecular ion.

    Mass spectra were obtained in line diagram and mass list form (with background subtracted) using a Varian 620i computer on line to a Varian CH 5D mass spectrometer. FI measurements were carried out with a Varian combined EI-FI source using acetone activated emitters. Glass columns (3 % OV1 or 3 % OV17 on gas chrom. Q) were used in the GC and coupling to the mass spectrometer was via a two-stage Watson–Biemann separator.

ACKNOWLEDGMENTS

We thank the S.R.C. for financial assistance towards the purchase of the mass spectrometer and for financial support for two of us (D.S.M. and M.R.). We are grateful to Drs K. H. Maurer and P. Schulze (Varian MAT Bremen) for FD spectra, Dr P. V. R. Shannon (University College, Cardiff) for the terpene

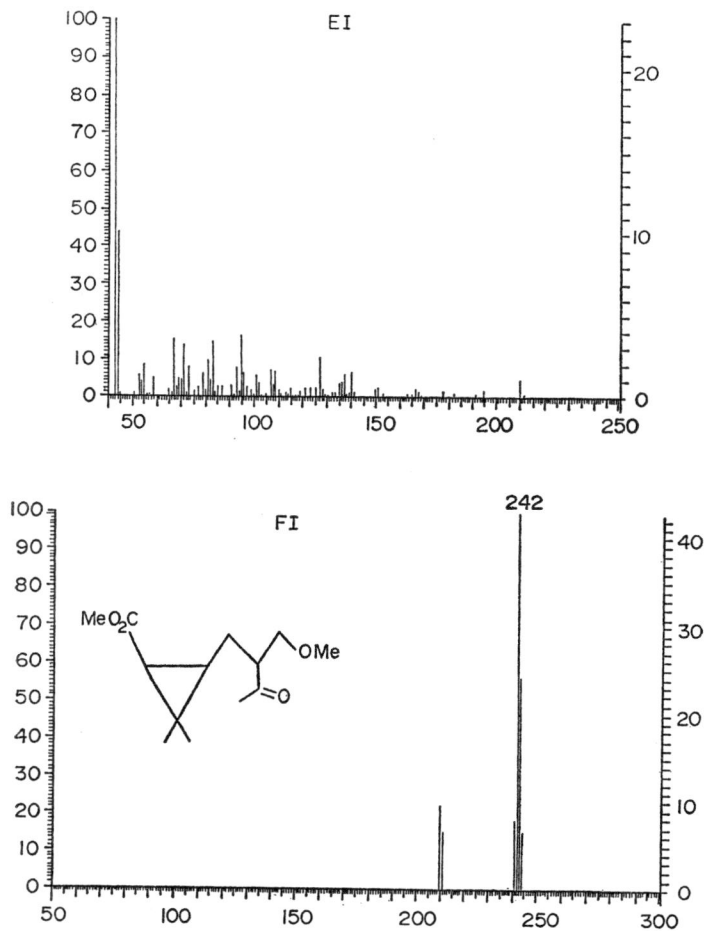

FIG. 3   EI and FI spectra of component G (cf. Fig. 2).

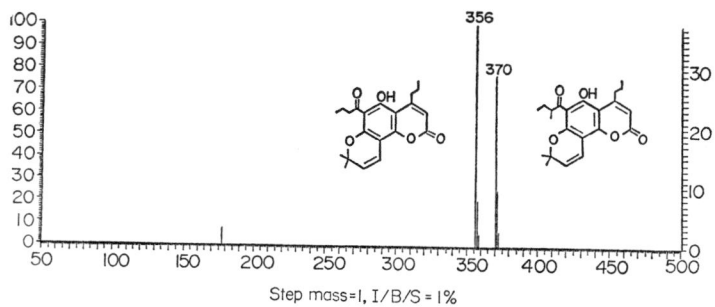

Step mass=1, I/B/S = 1%

FIG. 4   FI spectrum of one peak from GC-MS of coumarin mixture from *Mammea americana* (cf. Fig. 1a).

sample, Dr B. A. Krukoff (Merck, Sharp and Dohm, New York) and Mr A. G. Kenyon (Tropical Products Institute) for the supplies of *Erythrina* and *Mammea* seeds respectively.

REFERENCES

1. Beckey, H. D. *in* 'Biochemical Application of Mass Spectrometry', *ed.* G. R. Waller, Wiley-Interscience, 1972, p. 795.
2. Schulze, P., Simoneit, B. R. and Burlingame, A. L. *J. Mass Spectrometry and Ion Physics,* 1969, **2**, 183.
3. Damico, J. N. and Barron, R. P. *Analyt. Chem.,* 1971, **43**, 17.
4. Forehand, J. B. and Kuhn, W. F. *Analyt. Chem.,* 1970, **42**, 1839; Schulten, H. R., Beckey, H. D., Meuzelaar, H. L. C. and Boerboom, A. J. H. ibid., 1973, **45**, 191.
5. Beckey, H. D., Hilt, E., Maas, A., Migahed, M. D. and Ochterbeck, E. *J. Mass Spectrometry and Ion Physics,* 1969, **3**, 161.
6. Jackson, A. H., Kenner, G. W., Smith, K. M., Aplin, R. J., Budzikiewicz, H. and Djerassi, C. *Tetrahedron,* 1965, **21**, 2913.
7. Cocker, W., Lauder, H. St. J. and Shannon, P. V. R. Unpublished work.

16

GC-MS in the Analysis of Organic Compounds in Meteorites

By JAMES G. LAWLESS and MICHAEL P. ROMIEZ

(Planetary Biology Division, Ames Research Center, NASA, Moffett Field, California, U.S.A.)

THE application of gas chromatography combined with mass spectrometry (GC-MS) to the solution of problems involving the separation and identification of individual organic components in a complex mixture has gained wide acceptance.[1] In recent years this technique has been applied to the analysis of organic compounds in meteorites,[2-15] with probably the most outstanding success coming from its use in the analysis of amino acids. Although many earlier workers found amino acids in carbonaceous chondrite extracts, their results could be accounted for by contamination from the terrestrial biosphere.[16] Since amino acids associated with terrestrial life are generally of the L configuration and chemically synthesized amino acids are racemic mixtures, a procedure to affect resolution and identification of optical isomers of amino acids in carbonaceous chondrite extracts represents a method of distinguishing indigenous chemically synthesized amino acids from terrestrial contaminants. Two gas chromatographic methods have been developed for the resolution of amino acids enantiomers [one using diastereoisomeric derivatives[17] and the other using an optically active stationary phase,[18]] and both have been successfully used with mass spectrometry for the analysis of amino acids in carbonaceous chondrite extracts.[4-10,15] Using the N-trifluoroacetyl (-TFA)-D-2-butyl esters of the amino acids, 18 have been identified in extracts of the Murchison meteorite[4,5] and the same 18 were found in extracts of the Murray carbonaceous chondrite.[8] The suite of amino acids identified included both protein and non-protein types; those amino acids which have an optically active centre and whose diastereoisomeric derivatives could be separated by the gas chromatographic technique employed, showed nearly equal concentrations of the D and L isomers. Further studies show that at least 35 amino acids are present in Murchison meteorite extracts.[6] Of the 103 theoretically possible acyclic monoamino monocarboxylic acids with from 2 to 6 carbon atoms, at least 23 are present in extracts of the Murchison meteorite while commonly only 5 (glycine, alanine, valine, leucine and isoleucine) are found in terrestrial life, for example, *Escherichia coli*.[19] Of the amino acids found in the meteorite, all 4 of the amino acids with 2 and 3 carbon atoms (glycine, α-alanine, β-alanine, and sarcosine) and 7 of the 9 amino acids with 4 carbon atoms (α-aminoisobutyric acid, α-amino-n-butyric acid, β-aminoisobutyric acid, β-amino-n-butyric acid, γ-aminobutyric acid, N-methylalanine and N-ethylglycine) have been identified. Nine isomers of the amino acids with 5 carbon atoms (valine, isovaline, norvaline, plus 6 other isomers) and 3 isomers of the amino acids with 6 carbon atoms are also present. There are at least 5 isomers of the cyclic monoamino monocarboxylic acids present (proline, pipecolic acid, and 3 other isomers). Finally, in addition to aspartic acid and glutamic acid, at least 6 other isomers of the monoamino dicarboxylic acids have been characterized. These results show that the suite of amino acids found in a meteorite extract is indeed complex, and it seems reasonable to assume that given enough sample and a gas chromatographic column with sufficient resolution, mass spectrometry would reveal the presence of many more amino acids.

In order to evaluate rapidly and routinely all the data obtained from these investigations and because of the potential complexity of future investigations of amino acids in extraterrestrial samples, biological samples, and chemical evolution experiments, we have undertaken a computer-based approach to the interpretation of the low resolution mass spectra of N-TFA-2-butyl esters of amino acids. Since the wet chemical procedures preceding the GC-MS analysis eliminate many classes of compounds, the computer programs developed have

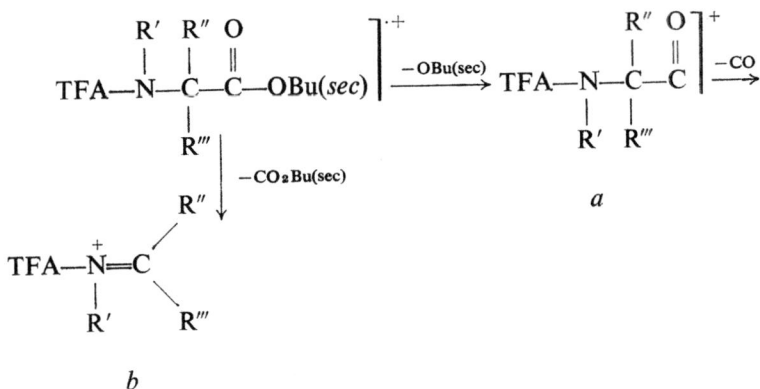

SCHEME 1

been concerned solely with the mass spectra of amino acids rather than taking the general approach of searching a large library.[20-21] Furthermore, because a single class of compounds is being investigated, more detailed programs can be written for structure elucidation. Our technique thus seeks to develop an approach for a limited library of a restricted class of compounds and has been developed along two lines: First, to 'interpret' the mass spectrum of an unknown based upon the mass spectral characteristics of the class; and second, to identify the unknown by the computer matching of the mass spectrum of an unknown with the mass spectra of a library of known compounds.

The aim of the interpretation portion of the program is to elucidate, to at least the point of detailed subclassification, the structure of an unknown amino acid whose mass spectrum is not readily available. It is generally acknowledged that the mass spectra of members of a class of compounds have similar characteristics, and the interpretation of the mass spectrum of an unknown amino acid proceeds on this basis. In the case of the N-TFA-alkyl esters of amino acids, their mass spectra are characterized by the very low abundance (or absence) of molecular ions, with most structural information derived from the formation of the trifluoroacetyl immonium ions produced by fragmentations from the molecular ion.[22-24] Thus, for example, the N-TFA-α-monoamino monocarboxylic acid-2-butyl esters are characterized by ions of types *a* and *b* (Scheme I).

In addition to recognizing the above mass spectral characteristics of N-TFA amino acid esters, the computerized 'interpretation' of the mass spectra of amino acids must fulfill three additional requirements. First, it must not be influenced by variations in ion abundance due to changes in sample flow into the source of the mass spectrometer while operating in the GC-MS mode. Second, it must be amenable to the 'interpretation' of mass spectra acquired in other laboratories on different types of instruments and under various experimental conditions. Finally, when operating in the GC-MS mode, it must not be significantly influenced by low intensity mass spectral peaks associated with column bleed and traces of other amino acids present as the result of unresolved gas chromatographic peaks. An approach which seems potentially useful, given the above considerations, involves the expression of a mass spectrum as an ion series summation.[25] Using this approach, the low resolution mass spectrum of each N-TFA-amino acid-2-butyl ester is represented by the 14 ion series produced using the equation,

$$S_m = \frac{\sum\limits_{n=0}^{\infty} I(98 + m + 14n)}{\sum\limits_{j=99}^{\infty} I(j)} \qquad (1)$$

where I is the relative intensity of the mass spectral peak, $m = 1,...,14$, and S_m is the fractional contribution of the ion series m to the total ion current. Since the significant structural information in the mass spectrum of an N-TFA-amino acid-2-butyl ester results from the trifluoroacetyl immonium ions, it was decided to begin the summation of each mass spectrum from a mass above that of the trifluoroacetyl group (m/e 97). Ions below m/e 97 are fragments produced primarily from the trifluoroacetyl and butyl moieties introduced during the derivatization process. The summation specified in eqn (1) was selected such that it began above m/e 97 and each value of m (from 1 to 14) defines homologous series which would correspond to those starting at masses 1 through 14 respectively. The mass spectrum of each of the 32 amino acid N-TFA-2-butyl esters in our library[26] was subjected to the ion series summation specified above, and the results obtained showed that the amino acids could be subclassified according to their ion series summation. The amino acids evaluated (*see* Table I) were thus classified as acyclic monoamino monocarboxylic acids, cyclic monoamino monocarboxylic acids, and acyclic monoamino dicarboxylic acids, and preliminary indications are that the aromatic monoamino monocarboxylic acids might be similarly classified after a sufficient number of mass spectra of this compound type has been evaluated. The ion series summation data for the three most intense values of m for the

TABLE I

Amino Acid Standards in Library

Acyclic Monoamino Monocarboxylic Acids
Glycine
N-Methylglycine
α-Alanine
β-Alanine
N-Ethylglycine
N-Methylalanine
α-Amino-n-butyric acid
α-Aminoisobutyric acid
β-Amino-n-butyric acid
β-Aminoisobutyric acid
N-Methyl-β-alanine
γ-Aminobutyric acid
Isovaline
Norvaline
Valine
δ-Aminovaleric acid
Leucine
Isoleucine

Cyclic Monoamino Monocarboxylic Acids
4-Piperidine carboxylic acid
3-Piperidine carboxylic acid
2-Piperidine carboxylic acid
Proline
1-Amino-1-cyclopentane carboxylic acid
trans-3-Methylproline
cis-3-Methylproline
cis-4-Methylproline
trans-4-Methylproline

Acyclic Monoamino Dicarboxylic Acids
Aspartic acid
Glutamic acid
Iminodiacetic acid
Iminoacetic-β-propionic acid
Iminoacetic-α-propionic acid

acyclic monoamino monocarboxylic acid derivatives is shown in Fig. 1 as an example. Similar data were obtained from the ion series summation of the cyclic monoamino monocarboxylic acids and the acyclic monoamino dicarboxylic acids. Each of the individual 14 ion series summations for each compound within the above three subclasses was averaged. The resulting

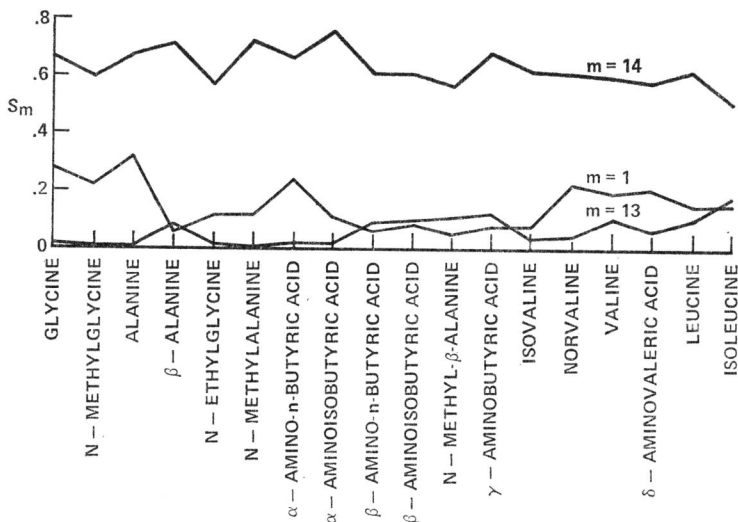

FIG. 1   The variation of S_m for $m = 1, 13, 14$ for the 18 acyclic monoamino monocarboxylic acids used in this study.

average can be expressed as a bar graph called a 'class ion series spectrum'.[25] The 'class ion series spectra' obtained for the acyclic monoamino monocarboxylic acids, cyclic monoamino monocarboxylic acids, and acyclic monoamino dicarboxylic acids are presented in Fig. 2.†

Other specialized programs have been written to determine the number of carbon atoms in an amino acid which is a member of any of the above three subclasses. This process is accomplished by identifying the mass spectral peak corresponding to the loss of the butoxy moiety from the molecular ion. Thus, for example, the last intense ion found in the ion series $m = 14$ of an acyclic monoamino monocarboxylic acid identifies the carbonyl group and therefore the number of carbon atoms in the amino acid; similarly, the last intense ion in the $m = 12$ ion series for cyclic monoamino monocarboxylic acids and in the $m = 2$ ion series for the acyclic monoamino dicarboxylic acids will define the number of carbon atoms for each compound in the subclass. Furthermore, specialized programs have been written for the acyclic monoamino monocarboxylic acids to differentiate α-amino acids from β- and γ-amino acids. It was observed[23] that for α-amino acids, the ion a in Scheme I (loss of a butoxy group from the molecular ion) was of much lower intensity than the ion corresponding to b in Scheme I, while in β- and γ-amino acids, the ion corresponding to the loss of the butoxy group was particularly intense and in some cases, even the base peak. It was also observed that for β-amino acids,

† Thus far only one amino acid (α-amino-3,3,-dimethyl-n-butyric acid) has been found which has an 'individual ion series spectrum' significantly different from the 'class ion series spectrum'. The largest peak above m/e 100 in the mass spectrum of the N-TFA-2-butyl ester of this compound is one produced by the cleavage of the carbon–nitrogen bond α to the nitrogen, with charge retention on the ester portion of the molecule.

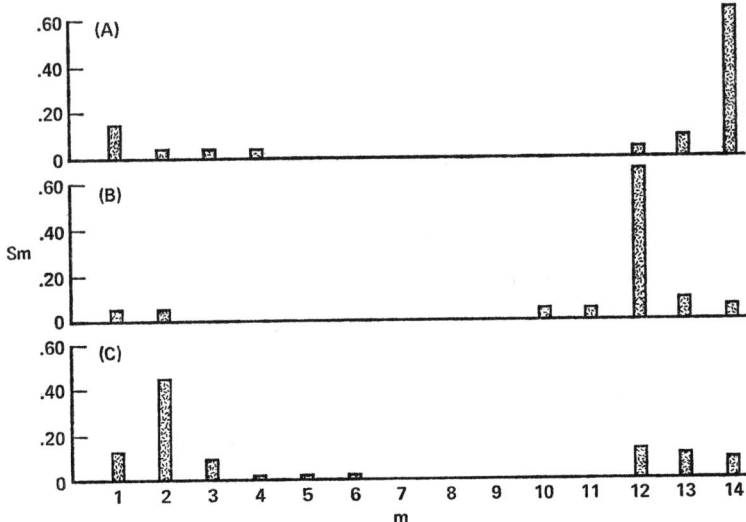

FIG. 2 The 'class ion series spectra' obtained for the (A) acyclic monoamino monocarboxylic acids, (B) cyclic monoamino monocarboxylic acids, and acyclic monoamino (C) dicarboxylic acids.

the ion analogous to b in Scheme I (loss of the butoxycarbonyl moiety from the molecular ion) was, for the compounds studied, less intense than an ion corresponding to the total loss of butoxycarbonyl plus hydrogen from the molecular ion.[23] These two facts have been incorporated into the 'interpretation' program to differentiate α-amino acids from β- and γ-amino acids and to corroborate further the 'identification' of a β-amino acid. Thus, when the mass spectrum of a member of the library is 'interpreted' (Fig. 3) and compared with the spectra in that library, the printout shown in Fig. 4(A) is the result. The first line in Fig. 4(A) is a label entered by the operator to identify the mass spectrum; the second line is the result of the 'interpretation', and finally, the compounds producing the three best similarity indices from the library comparison routine are identified. The library comparison routine is similar to that of Crawford and Morrison[27] and is initiated by making the sum of the observed k peaks, P_n, in each mass spectrum equal unity.

$$\sum_{n=40}^{k} P_n = 1 \tag{2}$$

The spectrum of the unknown is then compared to each reference mass spectrum in the library and a similarity index is produced.

$$\text{Similarity Index} = 2 \cdot 0 - \sum_{n=40}^{k} \mid P_{n(\text{ref})} - P_{n(\text{unk})} \mid \tag{3}$$

A perfect match would produce a similarity index of $2 \cdot 0$, while a complete mismatch would produce a similarity index of 0.

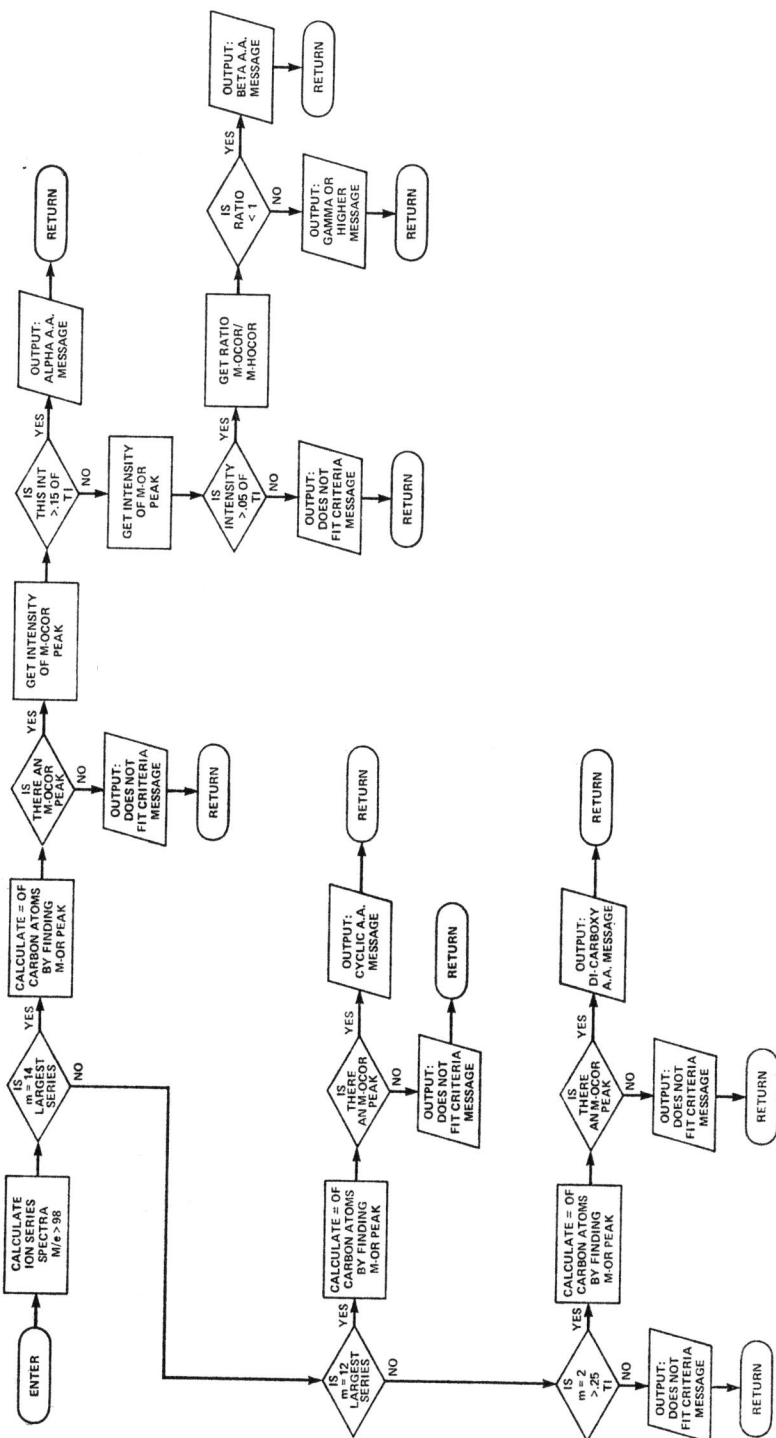

FIG. 3   The computer program subroutines are written in Fortran IV and run on an IBM 360 model 67 operating under an IBM supplied time-sharing system. These are prototype programs; the final version will be written in Data General assembly language for use on a NOVA 1200 with a Decision, Inc., Disk Operating System.

(A) N-ETHYLGLYCINE N-TFA-SEC-BUTYL ESTER

THIS COMPOUND APPEARS TO BE AN ALIPHATIC ALPHA AMINO ACID WITH 4 CARBON ATOMS

SPECTRUM NO.   5   N-ETHYLGLYCINE N-TFA-SEC-BUTYL ESTER   SIMILARITY INDEX 2.00000
SPECTRUM NO.   7   ALPHA-AMINOBUTYRIC ACID N-TFA-SE-BUT EST   SIMILARITY INDEX 1.32982
SPECTRUM NO.   8   ALPHA-AMINOISOBUTYRIC N-TFA-SEC-BUT EST   SIMILARITY INDEX 1.24434

(B) MURCHISON UNK#3

THIS COMPOUND APPEARS TO BE AN ALIPHATIC ALPHA AMINO ACID WITH 5 CARBON ATOMS

SPECTRUM NO. 52   ISOVALINE N-TFA-SEC-BUTYL ESTER   SIMILARITY INDEX 1.66837
SPECTRUM NO. 33   VALINE N-TFA-SEC-BUTYL ESTER   SIMILARITY INDEX 1.38861
SPECTRUM NO. 16   NORVALINE N-TFA-SEC-BUTYL ESTER   SIMILARITY INDEX 1.27389

MURCHISON UNK#5

THIS COMPOUND APPEARS TO BE AN ALIPHATIC ALPHA AMINO ACID WITH 4 CARBON ATOMS

SPECTRUM NO.   8   ALPHA-AMINOISOBUTYRIC N-TFA-SEC-BUT EST   SIMILARITY INDEX 1.71592
SPECTRUM NO.   6   N-METHYLALANINE N-TFA-SEC-BUTYL ESTER   SIMILARITY INDEX 1.62419
SPECTRUM NO.   7   ALPHA-AMINOBUTYRIC ACID N-TFA-SE-BUT EST   SIMILARITY INDEX 1.33899

MURCHISON UNK#36

THIS COMPOUND APPEARS TO BE A GAMMA OR HIGHER ALIPHATIC AMINO ACID WITH 5 CARBON ATOMS

SPECTRUM NO. 45   DELTA AMINOVALERIC ACID N-TFA-SEC-BU EST   SIMILARITY INDEX 0.90910
SPECTRUM NO.   3   ALANINE N-TFA-SEC-BUTYL ESTER   SIMILARITY INDEX 0.77735
SPECTRUM NO. 11   BETA-AMINOBUTYRIC N-TFA-SEC-BUT ESTER   SIMILARITY INDEX 0.61402

FIG. 4   (A) The results of the mass spectrum of the N-TFA-2-butyl ester of *N*-ethylglycine, 'interpreted' and compared with itself and the rest of the library. (B) The result from the 'interpretation' and comparison of the mass spectra of three derivatized amino acids found in extracts of the Murchison meteorite.

Application of this computer-based approach to the mass spectra of compounds obtained from an aqueous extract of the Murchison meteorite is shown in Fig. 4(B) with unknown 3 and unknown 5 being properly identified as isovaline and α-aminoisobutyric acid, respectively. While unknown 36 has not been positively identified, its mass spectrum has been analysed as that of a γ- or higher acyclic monoamino monocarboxylic acid with 5 carbon atoms. This information will be considered for the synthesis of standards necessary to identify the compound.

While the success of these programs to date has been good, further specialized programs are anticipated to make fuller use of the data available. It is anticipated that programs will be added to those outlined in Fig. 3 to search for the presence of a molecular ion; fragment ions from the trifluoroacetyl and butyl portions of the molecule will be sought to verify their incorporation during the derivatization procedure. Furthermore, the search for peaks 16, 17, and/or 18 mass units higher than the m/e value which locates the M-OR ion would provide additional information to verify that the mass spectrum being analysed was that of the amino acids.

The results of our present studies indicate that this technique should provide a rapid evaluation of the GC-MS data obtained from a complex amino acid mixture. Studies of the N-TFA-n-butyl esters of amino acids by this computational procedure has been successful, and the preliminary results from other amino acid N-TFA-alkyl esters has been equally encouraging.

The computer-based approach adopted for this study would appear to be applicable to many classes of compounds, and the GC-MS data obtained from complex mixtures of these classes of compounds could be similarly analysed when a suitable library is available.

REFERENCES

1. Junk, G. A. *Int. J. Mass Spectrom. Ion Phys.,* 1972, **8**, 1.
2. Gelpi, E. and Oro, J. *Geochim. Cosmochim. Acta,* 1970, **34**, 981.
3. Gelpi, E. and Oro, J. *Geochim. Cosmochim. Acta,* 1970, **34**, 995.
4. Kvenvolden, K. A., Lawless, J. G., Pering, K., Peterson, E., Flores, J., Ponnamperuma, C., Kaplan, I. R. and Moore, C. *Nature,* 1970, **228**, 923.
5. Kvenvolden, K. A., Lawless, J. G. and Ponnamperuma, C. *Proc. Nat. Acad. Sci.,* 1971, **68**, 486.
6. Lawless, J. G. *Geochim. Cosmochim. Acta* (in press).
7. Lawless, J. G., Kvenvolden, K. A., Peterson, E., Ponnamperuma, C. and Jarosewich, E. *Nature,* 1972, **236**, 66.
8. Lawless, J. G., Kvenvolden, K. A., Peterson, E., Ponnamperuma, C. and Moore, C. *Science,* 1971, **173**, 626.
9. Oro, J., Gibert, J., Lichtenstein, H., Wikstrom, S. and Flory, D. A. *Nature,* 1971, **230**, 105.
10. Oro, J., Nakaparksin, S., Lichtenstein, H. and Gil-Av, E. *Nature,* 1971, **230**, 107.
11. Folsome, C. E., Lawless, J. G., Romiez, M. and Ponnamperuma, C. *Nature,* 1972, **332**, 108.
12. Folsome, C. E., Lawless, J. G., Romiez, M. and Ponnamperuma, C. *Geochim. Cosmochim. Acta,* 1973, **37**, 455.
13. Pering, K. L. and Ponnamperuma, C. *Science,* 1972, **173**, 237.
14. Studier, M. H., Hayatsu, R. and Anders, E. *Geochim. Cosmochim. Acta,* 1972, **36**, 189.
15. Lawless, J. G. 'Proceedings of the 4th Intl. Mtg. on the Origin of Life' (in press).
16. Hayes, J. M. *Geochim. Cosmochim. Acta,* 1967, **31**, 1395.
17. Nakaparksin, S., Birrell, P., Gil-Av, E. and Oro, J. *J. Chromatogr. Sci.,* 1970, **8**, 177.

18. Pollock, G. E., *Anal. Chem.*, 1972, **44**, 2368.
19. Luria, S. E. *in 'The Bacteria—A Treatise on Structure and Functions', ed.* I. C. Gunsalus and R. Y. Stanier, Academic Press, New York, Vol. 1, 1960, pp. 1–34.
20. Ridley, R. G. *in 'Biochemical Applications of Mass Spectrometry', ed.* G. R. Waller, Wiley-Interscience, New York, 1972, pp. 177–91, and references cited therein.
21. Hertz, H. S., Hites, R. A. and Biemann, K. *Anal. Chem.*, 1971, **43**, 681.
22. Gelpi, E., Koenig, W. A., Gibert, J. and Oro, J. *J. Chromatogr. Sci.*, 1969, **1**, 604.
23. Lawless, J. G. and Chadha, M. S. *Anal. Biochem.*, 1971, **44**, 473.
24. Vetter, W. *in 'Biochemical Applications of Mass Spectrometry', ed.* G. R. Waller, Wiley-Interscience, New York, 1972, pp. 387–404.
25. Smith, D. H., *Anal. Chem.*, 1972, **44**, 536 and references cited therein.
26. Available from the authors.
27. Crawford, I. R. and Morrison, J. D. *Anal. Chem.* 1968, **40**, 1464.

17
Vapour Evolution from Polymers

By L. J. RIGBY

(*STL, Harlow, Essex, U.K.*)

INTRODUCTION

THERMAL volatilisation analysis[1] (TVA) usually describes the differential pressure curve which is obtained when an organic sample is uniformly heated in a continuously pumped vacuum system (Fig. 1). The pressure increase at high temperatures may be attributed to decomposition of the polymer and is therefore a measure of its thermal stability as well as a method of preliminary identification. The low temperature peak is indicative of unpolymerised material and may result from a combination of such ingredients as absorbed gases, plasticiser, unpolymerised resin, crosslinking agents, antioxidants and pigments. When the analysis is carried out in the ion source of a mass spectrometer, it then becomes possible to identify the constituents evolved and to measure their separate evolution rates.

INSTRUMENTATION

Previous work on our Bendix Model 12 Time of Flight Mass Spectrometer (TOFMS) used the standard Bendix furnace which allows polymer samples of

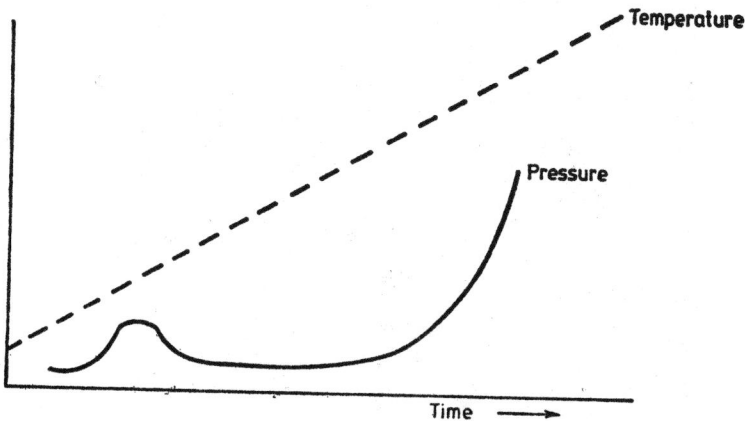

FIG. 1   A typical vapour evaluation curve from a polymer subjected to uniform heating.

153

FIG. 2   The large furnace probe.

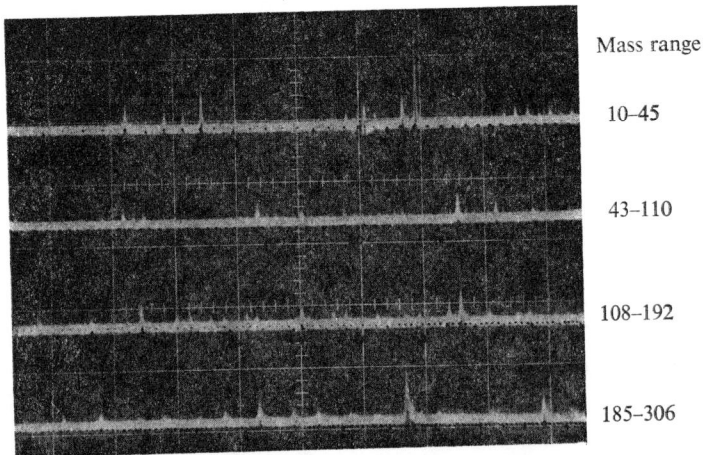

Mass range

10–45

43–110

108–192

185–306

FIG. 3   Four-line mass spectrum of perfluorodiphenyl $C_{12}F_{10}$, mol wt 334.

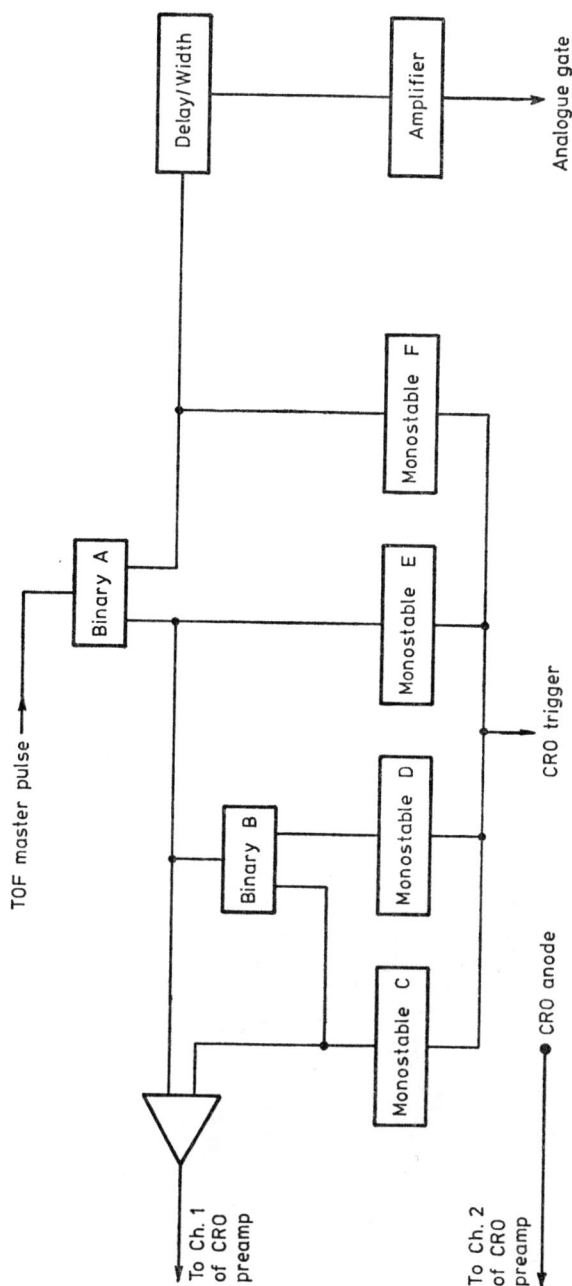

FIG. 4   Schematic diagram of pulse circuit.

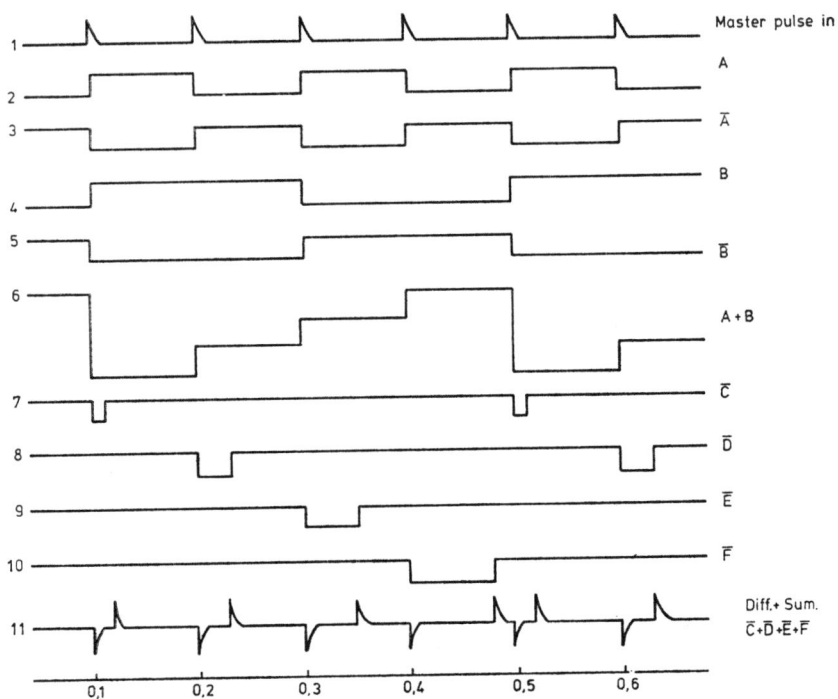

FIG. 5   Timing pulses for 10 kHz operation of TOFMS.

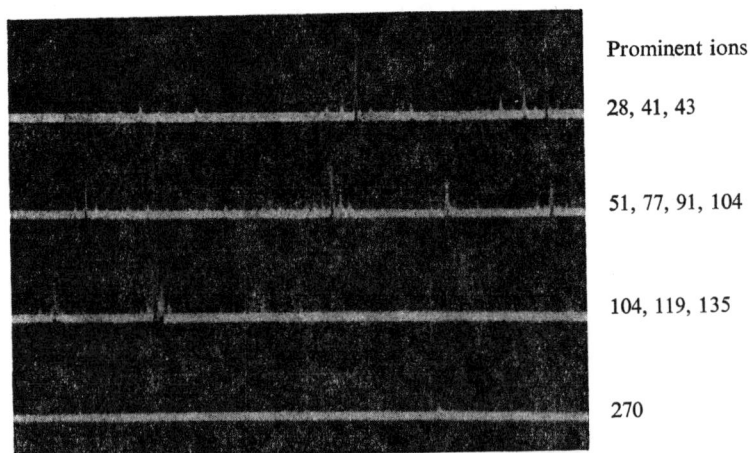

Prominent ions

28, 41, 43

51, 77, 91, 104

104, 119, 135

270

FIG. 6   Mass spectrum of dicumyl peroxide $(CH_3)_2CHC_6H_4OOC_6H_4CH(CH_3)_2$, mol wt 270.

up to 0·2 mg to be inserted through a ball valve into the TOFMS ion source. This amount of sample is ideal for studying the decomposition of the plastic,[2] but in order to obtain adequate sensitivity for low concentrations of additives, a larger furnace has been constructed of thin molybdenum sheet and Thermocoax heating wire† (Fig. 2). A chromel alumel thermocouple sheathed in 1 mm stainless steel‡ was used to measure the temperature and provide feedback to a Eurotherm temperature controller. The furnace capacity is $10 \times 10 \times 3 = 300 \text{ mm}^3$ and it may be heated above $500°C$ by a 21V 5A a.c. supply.

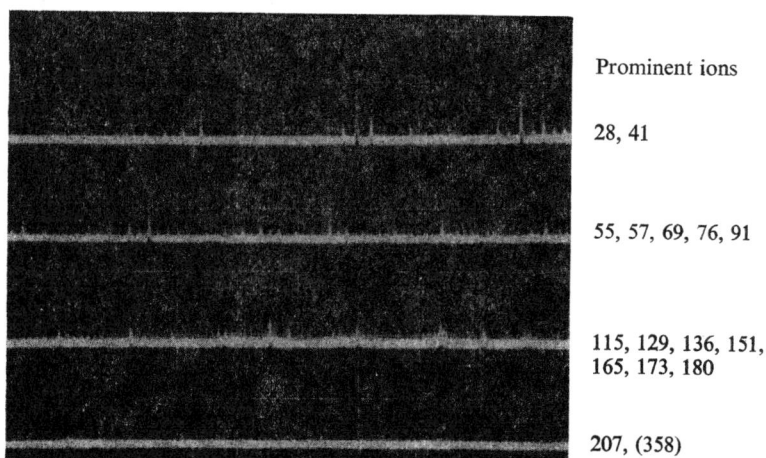

	Prominent ions
	28, 41
	55, 57, 69, 76, 91
	115, 129, 136, 151, 165, 173, 180
	207, (358)

FIG. 7 Mass spectrum of Santanox $((CH_3)_3 C_6H_3 (OH)(CH_3))_2$ S, mol wt 358.

We have found that photography of the oscilloscope display is most convenient for complete mass spectra of volatile species, and we use analogue gates to monitor the partial pressures of the constituents of interest. The TOFMS has been modified to produce a simultaneous four-line mass spectrum (Fig. 3). This allows a mass range of 12–300 to be obtained for time intervals exceeding 1 msec and provides analogue gates without interference with the oscilloscope display. Figure 4 is a schematic diagram of the modified circuit and Fig. 5 indicates the pulses obtained. (The circuit diagrams of the pulse splitting and amplification for analogue and pre-dynode gating are to be published separately.[3])

RESULTS

Typical results which have been obtained for low temperature evolution are illustrated by recent work on single pellets (20–60 mg) of low density polythene. Figures 6 and 7 show respectively the mass spectrum, molecular

† Obtainable from Pye Unicam Ltd, Manor Royal, Crawley, U.K., or D'Etudes et Realisations Nucleaires, 10, Rue de la Passerelle, Suresnes (Seine), France.
‡ Obtainable from Spembley Technical Products, Trinity Estate, Sittingbourne, Kent, U.K.

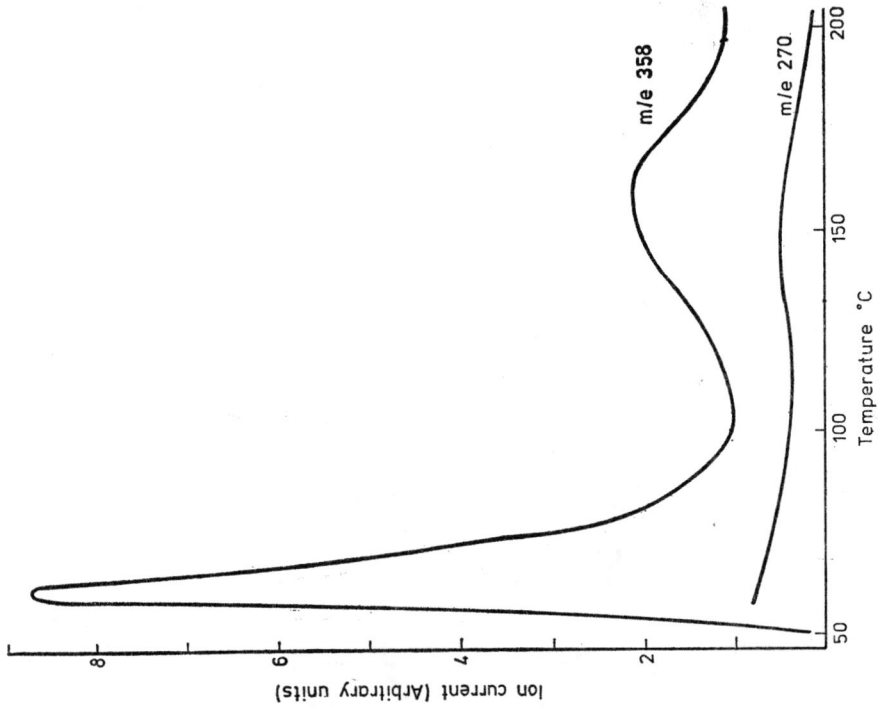

Fig. 9 TVA of sample B showing the evaluation of Santanox with traces of dicumyl peroxide.

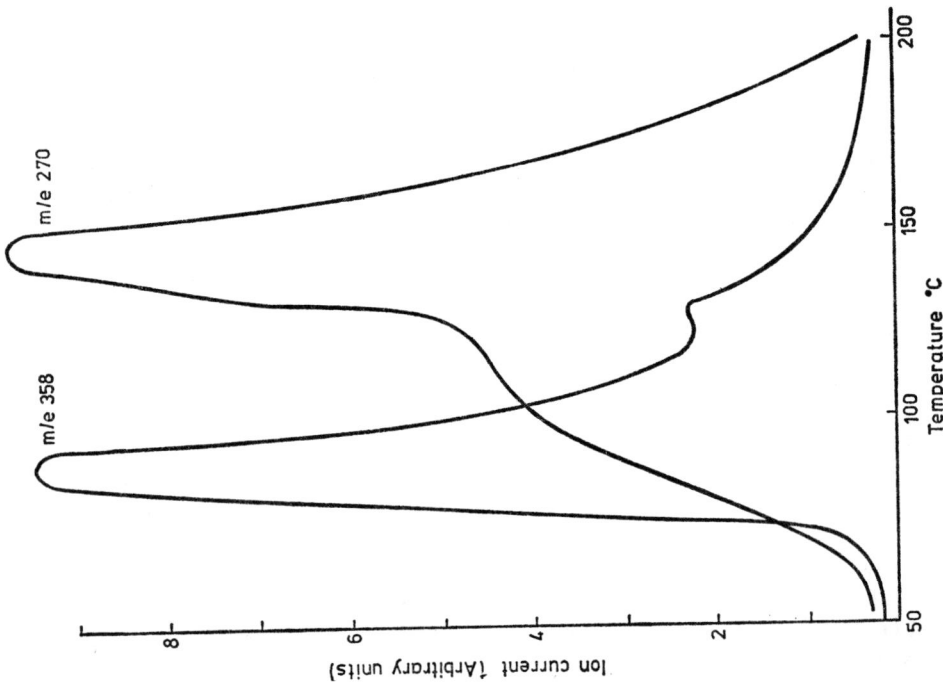

Fig. 8 TVA of sample A.

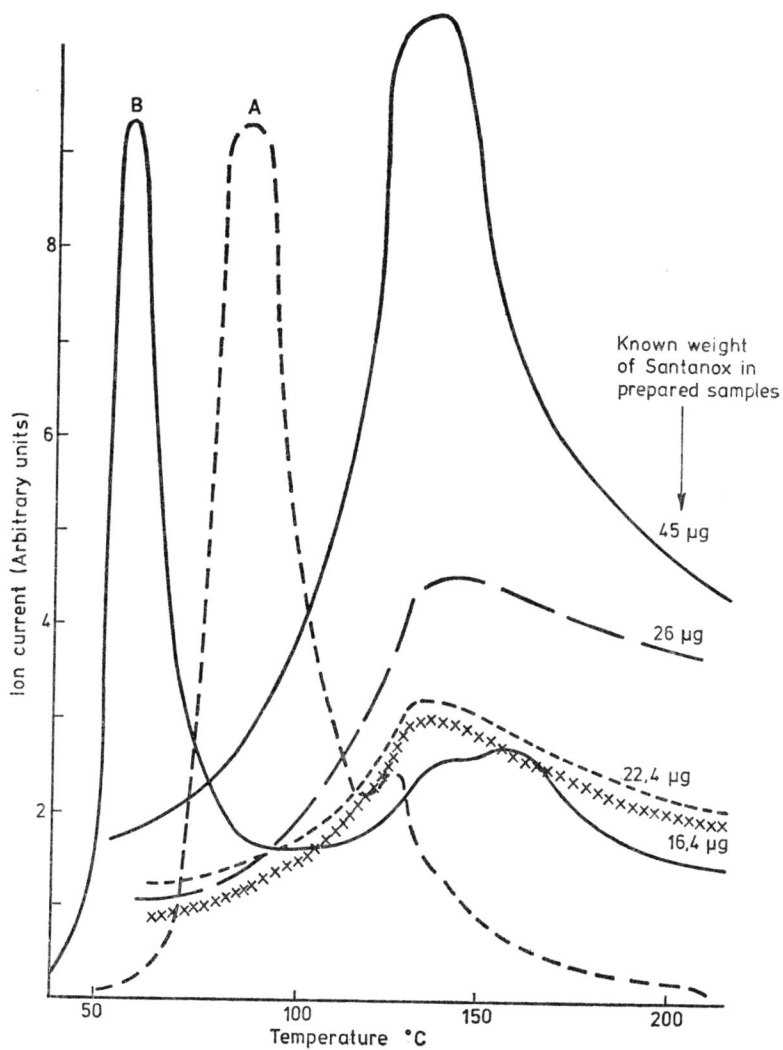

FIG. 10 Replotted TVA curves for Santanox evolved from polythene.

L. J. RIGBY

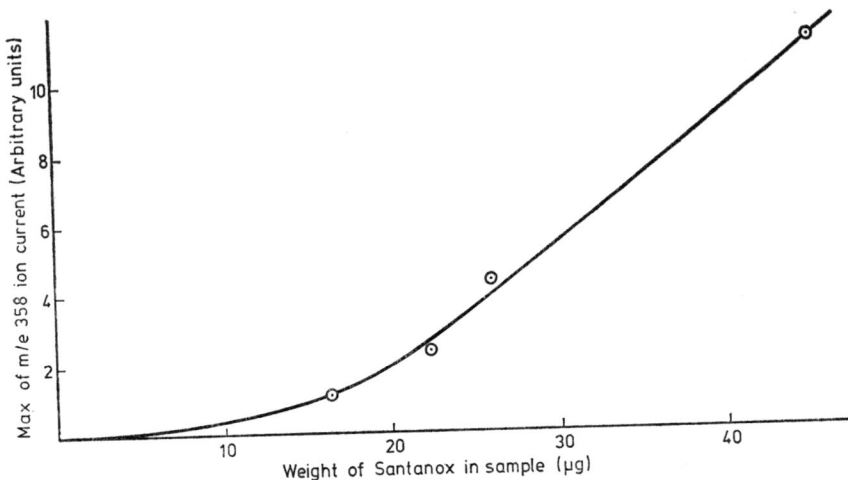

FIG. 11   Maxima of parent ion current for Santanox evolved from polythene.

formula and weight of the crosslinking agent dicumyl peroxide and the antioxidant Santanox which are both present in commercial grades of the polymer.

To avoid confusion in the initial work, it was decided to monitor the evolution of each component by gating each parent ion current even though the TOFMS sensitivity to both of these was very low. Figures 8 and 9 illustrate the TVA curves produced for two commercial samples, both containing Santanox but only the former containing measurable quantities of dicumyl peroxide. Figure 10 shows the replotted TVA curves for both previous samples and laboratory prepared samples. The onset of evolution and the shape of the evolution curve is indicative of the distribution of the additive in the polymer and the area under the curve is a measure of the quantity evolved. The curves in Fig. 10 suggest that the Santanox in the commercial samples is distributed closer to the surface of the sample than in the freshly prepared samples. Figure 11 attempts to relate the evolution maximum to the weight of Santanox in the original sample. This TVA/TOFMS sensitivity curve was found to be non-linear and may result from the diffusion of the additive out of the bulk of the sample.

DISCUSSION

These initial results indicate that a semi-quantitative method is available for measuring the surface and bulk concentrations of specific additives in polymers.

If the compounds investigated are uniformly swept through the mass spectrometer ion source, the area under the evolution curve is linearly proportional to the amount of material evolved:

$$\frac{dW}{dt} = Ksp + V\frac{dp}{dt} \qquad \text{molecules per second}$$

where $K = 3\cdot3 \times 10^{19}$ molecules/torr litre at 25°C, s the pumping speed in litres/second, p the partial pressure in torr of the component of interest and V the ion source volume.

For *complete* evolution

$$\int_0^n dN = Ksp \int_0^t dt + V \int_{p_0}^{p_0} dp$$

or
$$N = Kspt = KS \,(\text{AREA})$$

Figure 10 shows that there is a correlation between the amount of Santanox evolved and the amount originally in the sample. However, accurate measurement is not possible because of the long tail on the evolution curve resulting from diffusion-limited migration from the interior of the sample. If the high temperature peak results only from bulk diffusion from a uniform distribution in the sample, then the height of this peak should be related to the quantity originally present and comparison between similarly crosslinked polymers should be valid.

If the temperature dependence of the diffusion coefficient is of the form

$$D = D_0 \exp\left(-\frac{E}{R}\right) (T_0 + \beta t)$$

where T_0 is the initial temperature, β the heating rate and t the time, and a uniform diffusion gradient exists in the sample,

$$dN/dt = -DA \,(N_0 - N)/x$$

$$= -\frac{A}{x}(N_0 - N)\, D_0 \exp\left(-\frac{E}{R}\right) (T_0 + \beta t)$$

where N_0 is the initial bulk concentration in a sample of thickness x and N is the concentration after time t.

At the maximum of the evolution curve, the evolution rate is balanced by the pumping speed, *i.e.*

$$-\frac{dN}{dt} \max = Ksp_{\max} \qquad \text{molecules per second}$$

where p is the ion source pressure in torr, s the pumping speed in torr litres/ second and $K = 3\cdot3 \times 10^{19}$ molecules/torr litre at 25°C. Therefore the maximum partial pressure of the evolved material is related to the initial bulk composition by

$$P_{\max} = \frac{A}{Ksx}(N_0 - N)\, D_0 \exp\left(-\frac{E}{R}\right) (T_0 + \beta t)$$

which is linear with N_0 at similar conditions of temperature, heating rate and sample matrix and geometry provided that N is small compared with N_0. Since N, the total number of molecules evolved after time t, is probably significant at the peak maximum (particularly for low values of N_0), some curvature in the evolution plot may be expected (Fig. 11). This sets the detection limit for Santanox bulk distributed in polythene to about 0·01 % with the present methods unless the sample surface area and/or the heating rate is increased.

REFERENCES

1. Grassie, N. and McNeill, I. C. *J. Appl. Poly. Sci.,* 1968, **12**(4), 831.
2. Murdoch, I. A. and Rigby, L. J. *in 'Dynamic Mass Spectrometry',* Vol. 3, *ed.* D. Price, Heyden and Son Ltd., London, p. 255, 1972.
3. Rigby, L. J. *J. Phys.,* E, submitted for publication, Aug. 1973.

Biochemical Applications of Mass Spectrometry

By H. BUDZIKIEWICZ

(*Institut für Organische Chemie der Universität, Köln, Germany*)

THE applications of mass spectrometry in biochemistry are so numerous that there are but two ways to treat a topic like this: either to give a general survey highlighting the most recent advances in the various areas or to select one or two special examples and to use them in order to show success and difficulties, limitations and possible paths of further development. Since general reviews of the first type are available, *e.g.* through the Specialist Periodical Reports of the Chemical Society, the second alternative has been chosen here and two classes of compounds will be presented which are still objects of active current research, viz. carotenoids and nucleotides.

CAROTENOIDS

The main areas of the mass spectrometric investigations of carotenoids are (1) identification of known compounds, (2) structure elucidation of new representatives, and (3) metabolic studies.

Identification by mass spectrometry is a welcome method because frequently only small quantities are available allowing additionally only the use of UV and visible spectra, and R_f-values.[1] To avoid wrong conclusions it is therefore necessary to know how reproducible the mass spectra of carotenoids are and whether they can be considered characteristic for certain structures or, at least, partial structures. The main problem of the mass spectrometric investigation of carotenoids is their low volatility and their thermal instability. A series of pyrolytic degradations and isomerisations occurring at the usual sample insertion temperatures (200–300°C) has been described and it is thus remarkable that useful mass spectra can be obtained. This is apparently due to the good reproducibility of these reactions and to the fact that certain isomerisation products still give information in terms of the original structure. From this it is, however, obvious that identical mass spectra will be obtained only under carefully controlled conditions, although the general features of the pattern of a given compound seem to be fairly constant. It has to be realised too that, due to the extensive secondary fragmentation occurring at 70–100 eV, low voltage spectra are usually more characteristic.[2] While certain types of thermal degradation (as, for example, the loss of toluene and xylene, v. infra) may even be of diagnostic value, three types of transformation are rather annoying: the formation of 'M + 2', 'M − 2' (hydrogenation and dehydrogenation) and in the case of epoxides 'M − 16' (loss of oxygen) ions most likely due to catalytic processes in the source.[3] They not only interfere with the purity control but also with isotope analyses.

Mass spectra of carotenoids usually give molecular ions of reasonable abundance and hence molecular weight and elemental composition can be determined. The second question of importance is whether isomeric structures can be distinguished. This is clearly possible where end groups inducing typical fragmentation processes are present[4] as, for example, allylic cleavage (lycopin, 1, Fig. 1), retro-Diels–Alder-decomposition (4,5-double bonds, ε-carotin, 2, Fig. 1) or epoxide fragmentation which, however, has been observed equally for 5,6-, 3 (Fig. 1) and 5,8-epoxides, 4 (Fig. 1), the spectra of this pair of isomers being hardly distinguishable.

These typical fragmentation reactions are of value also for the structure elucidation of new representatives as demonstrated, for example, for the location of additional isoprenoid units in C_{50}-carotenoids (2, R = C_5H_9O).[5]

In some cases relative intensities of certain fragments have to be used for the differentiation of certain isomers[6] as, for example, it has been shown by

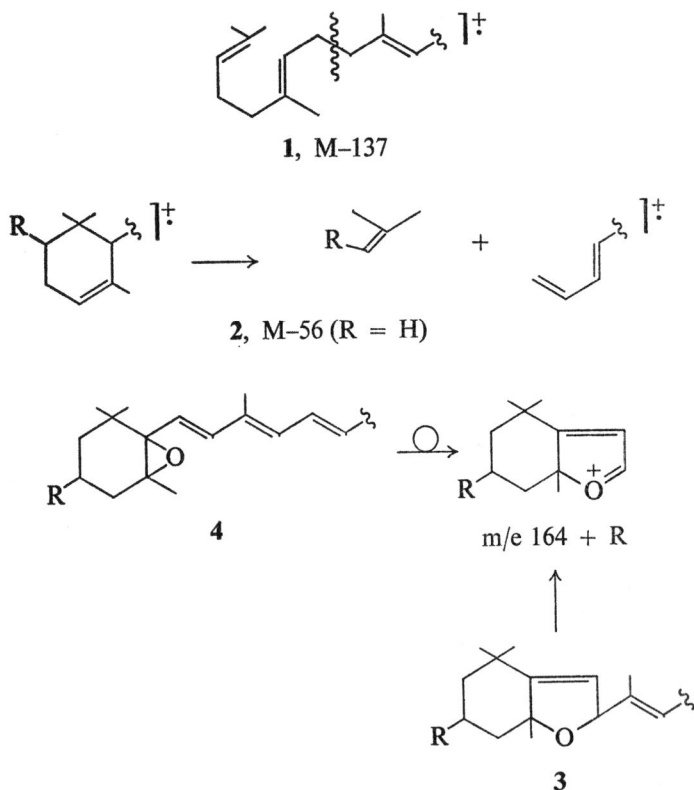

1, M–137

2, M–56 (R = H)

4

m/e 164 + R

3

FIGURE 1

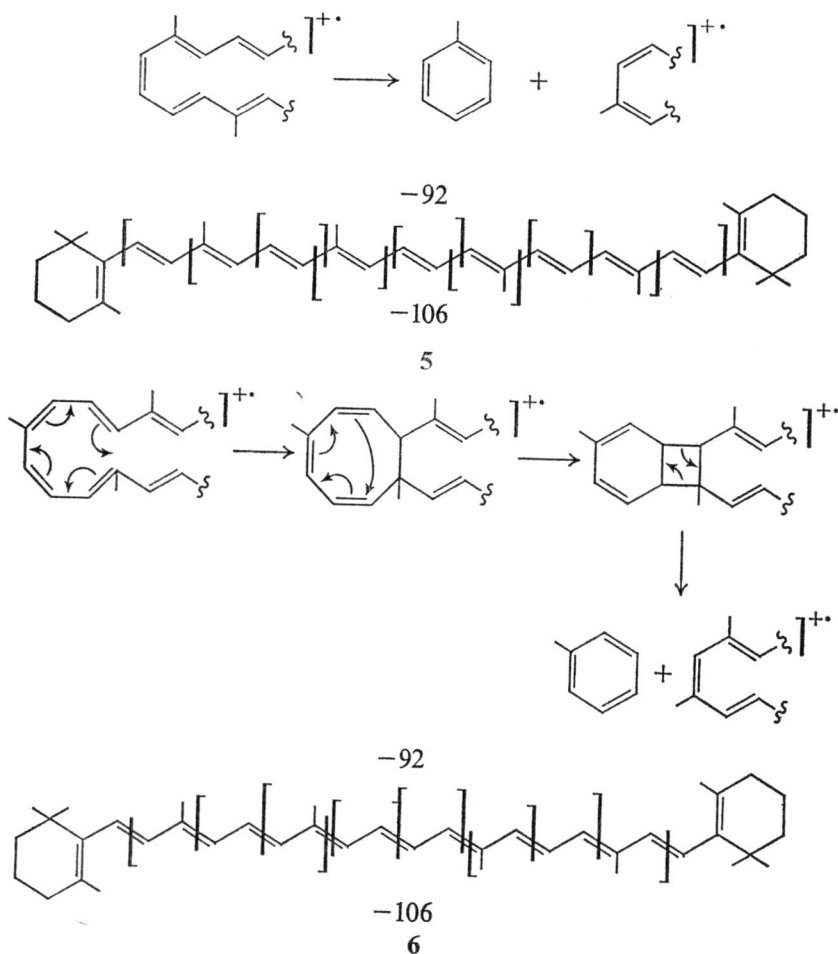

FIGURE 2

Enzell *et al.* that the intensity ratio $[M-92]^+/[M-106]^+$ depends upon the number of conjugated double bonds.[7]

The ions at M-92 and M-106 (loss of toluene and of xylene) are among the most characteristic in the spectra of carotenoids although their genesis is rather complicated, since they are apparently due both to thermal and to electron impact degradation. There is, however, good evidence that both processes obey the same mechanism. Various mechanisms have been discussed,[6,8] two of which are depicted above for the M-106 ion (formal cleavage of single bonds, **5**, Fig. 2, or of double bonds, **6**, Fig. 2). In both cases toluene and xylene do come from specific parts of the molecule. As can be seen from Table I, extensive deuterium labelling clearly rules out mechanism **5**. Analogous results have been obtained for $[M-92]^+$.

The knowledge of the origin of the M-toluene (92) and M-xylene (106) fragments can be used for structure studies of new compounds. Thus the

TABLE I

d-label	Intensity ratio		
	exp.	calculated for mech 5	calculated for mech. 6
$4,4,4'4'd_4$	M-106	M-106	M-106
$7,7'd_2$	M-106	M-106	M-106
$8d$	M-106 / M-107 (1:1)	M-106	M-106 / M-107 (1:1)
$7,7',8,8'd_4$	M-107	M-106	M-107
$8,8',15,15'd_4$	M-107	M-106 / M-108 (2:1)	M-107
$4,4,4',4',8,8',15,15'$ $19,19,19,19',19',19'd_{14}$	M-110	M-108 / M-109 (1:2)	M-110
$10,10'd_2$	M-107	M-106 / M-107 (1:2)	M-107
$11,11',12,12'd_4$	M-108	M-106 / M-108 (1:2)	M-108
$12,20,20,20d_4$	M-106 / M-110 (1:1)	M-106 / M-109 / M-110 (1:1:1)	M-106 / M-110 (1:1)
$14d_1$	M-106	M-106 / M-107 (1:2)	M-106
$15,15'd_2$	M-106	M-106 / M-108 (2:1)	M-106

localisation of the OH– group in loroxanthin[9] can be confirmed since the mass spectrum shows fragments at M-92 (toluene), M-106 (small, xylene) and M-122 (xylenol), but not at M-108 (cresol). Since, as can be seen from 7, Fig. 3, toluene is lost only from the part of the chain which comprises C-13 and C-13' and no loss of cresol is observed, the additional hydroxyl group has to be at the C-9 methyl group:

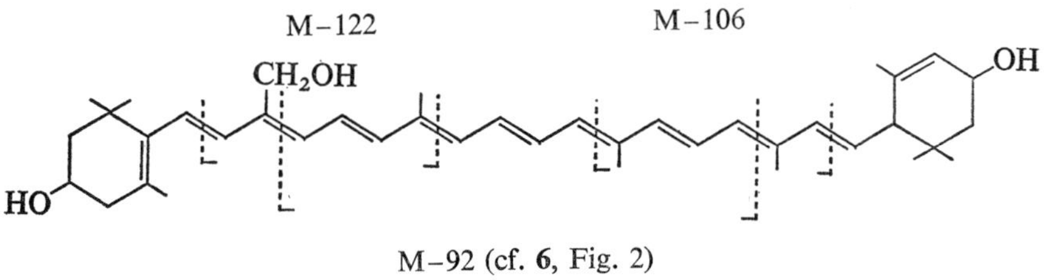

M–92 (cf. 6, Fig. 2)

7

FIGURE 3

Similarly, it is understandable that the presence of a 15,15'- triple bond inhibits the loss of toluene while a small ion of M-90 is observed,[10] but that a triple bond near the end of the chain (e.g. diatoxanthin,[11] heteroxanthin[12]) does not interfere with this fragmentation.

The labelling studies of the polyene chain finally established that fragments containing a terminal epoxide grouping can be used to localize a deuterium

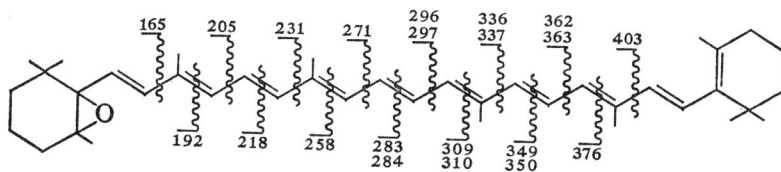

FIGURE 4

label since its formation is accompanied only to a very minor extent by hydrogen randomization (Fig. 4).

Labelling studies were of interest in context with the photosynthesis of green plants where carotenoid epoxides had been discussed as intermediates in the formation of molecular O_2, e.g.

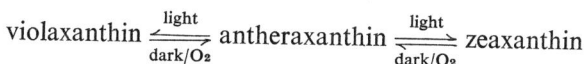

$$\text{violaxanthin} \underset{\text{dark/O}_2}{\overset{\text{light}}{\rightleftarrows} } \text{antheraxanthin} \underset{\text{dark/O}_2}{\overset{\text{light}}{\rightleftarrows}} \text{zeaxanthin}$$

A convenient label would be ^{18}O and there are essentially three methods to determine its incorporation:[13]

(a) Via the nuclear reaction ^{18}O (p,n) ^{18}F which allows determination of the ^{18}O content without interference of ^{16}O from the air, but gives an average for the entire molecule. There are experimental difficulties with small quantities in which case only relative concentrations can be obtained.

(b) Via pyrolysis and analysis of the CO_2 thus obtained by mass spectrometry. In this way the ratio $^{16}O/^{18}O$ is obtained, again an average value for the entire molecule. Erroneous results can be obtained especially with small quantities due to the presence of traces of H_2O and to the lability of carotenoids to oxygen probably resulting in chain epoxidation.

(c) Via direct mass spectrometric investigation. For this purpose neither M^+ nor $[M-92]^+$ or $[M-80]^+$ can be used due to the interference of 'M + 2' and 'M − 2' ions discussed above. However, the typical epoxide fragments m/e 164 + R (v. supra) can be used in the case where the enrichment of ^{18}O is at least 10 % (accuracy of about $\pm 2\%$ abs.). Only the ring oxygen is measured in this way, hence oxidation of the chain during the work-up is of no importance. It is obvious that small samples can be used.

This short survey demonstrates that highly labile compounds may well be subjected to mass spectrometric investigation, but that detailed studies of the behaviour of these compounds are necessary in order to obtain well founded information.

NUCLEOTIDES

Sequential analyses of peptides are amongst the most striking examples of the applicability of mass spectrometry to biochemical problems and several reviews are available on this topic. There is, however, another class of compounds where sequential analysis is of great interest, viz. nucleotides. The advantage with respect to peptides is that there are essentially only four building blocks to be considered, the main difficulty resting in the many polar groups. The following review should give an idea of the state of the art today.

The fragmentation behaviour of pyrimidine and purine bases has been investigated *in extenso*,[14] but it is of minor importance in this context since due to the stability of the aromatic nucleus fragment formation in nucleotides and nucleosides occurs preferentially in the sugar moiety. Identification of the base has therefore to come from mass and elemental composition of the pertinent fragments. That nucleosides (*i.e.* base + sugar) are amenable to mass spectrometry was shown first by Biemann and McCloskey[15] in 1962. They exhibit a rather straightforward fragmentation pattern though the abundance of the typical fragments varies with the compound[14] (Fig. 5).

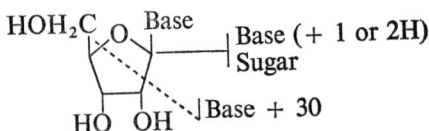

FIGURE 5

Even more clear-cut are the field ionization mass spectra of nucleosides showing M^+ and $[M + 1]^+$ (in cases $[M-H_2O]^+$) ions and the sugar and base fragments.[22,23]

The spectra of nucleosides are especially important for the detection of unusual bases[14] and for biogenetic studies (incorporation of isotopc labels).[16]

The presence of the phosphate ester grouping in nucleotides renders these compounds too involatile for direct electron impact mass spectrometry. Various types of derivatives (TMS, methyl ethers, acetates, acetonates, phenyl boronic acid esters) have been used to enhance volatility of mono-nucleotides, the simplest derivatization being esterification of the phosphoric acid. The usefulness of the spectra thus obtained—as far as structural information is concerned—varies with the newly introduced protecting group, the main disadvantage being the increase in mass especially if one wants to go to higher nucleotides and the amount of material needed for the chemical transformations. Field desorption mass spectrometry allows the measurement of free nucleotides giving $[M + 1]^+$ and ions characteristic of the base and sugar moiety.[23] On the whole, mass spectra of mononucleotides are not, however, of much practical use other than serving as model compounds for higher nucleotides, since structural information is extracted much more readily from the corresponding nucleosides which are also easier to handle.

The applicability of mass spectrometric investigation to dinucleotides has again been demonstrated by Biemann *et al.*[17] who obtained mass spectra of a series of per-trimethylsilyl ethers. The sequence can be deduced from characteristic fragments although one has to resort to differences in abundance. This possible source of ambiguity can be circumvented by asymmetric derivatization,[18] Fig. 6, which is, however, applicable only to ribonucleotides. The use of per-trifluoroacetate derivatives has been suggested, too, but the characteristic ions seem to be of rather low abundance.[19]

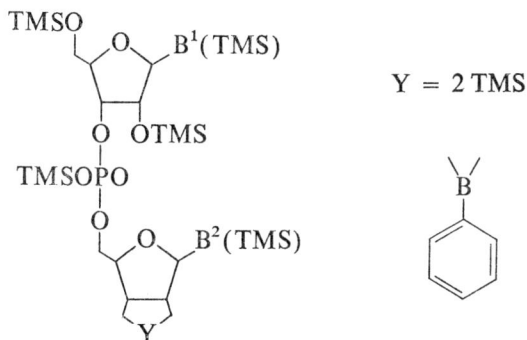

FIGURE 6

There are only a few attempts reported in the literature at obtaining mass spectrometric information from higher nucleotides. Heating of DNA on a direct probe of a mass spectrometer results in ions identifying the four common bases but no sequential information is obtained.[19,20] A combined chemical degradation and mass spectrometric analysis has been suggested for the sequencing of trinucleotides where the right-hand terminus is obtained as an acetylated base, the centre unit as a free base and the left-hand terminus as a nucleoside.[21] The degradation method used restricts the analysis to ribonucleotides. In addition, for higher nucleotides it becomes rather complex.

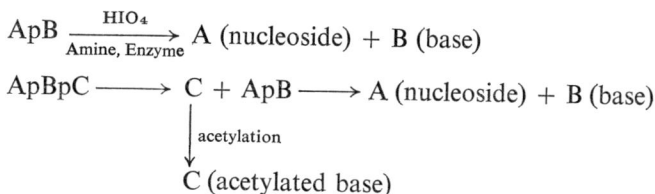

$$\text{ApB} \xrightarrow[\text{Amine, Enzyme}]{\text{HIO}_4} \text{A (nucleoside)} + \text{B (base)}$$

$$\text{ApBpC} \longrightarrow \text{C} + \text{ApB} \longrightarrow \text{A (nucleoside)} + \text{B (base)}$$
$$\downarrow \text{acetylation}$$
$$\text{C (acetylated base)}$$

It has been demonstrated[19] that the mass spectra of the (probably thermal degradation products of) per-trifluoroacetylated oligo-desoxyribonucleotides will identify the two termini and give information about the bases present in the central units, although without indication of their sequence. This may be circumvented by limited enzymatic cleavage yielding a mixture of products of decreasing chain length which can be separated by chromatography, per-trifluoroacetylated and analysed as indicated above to establish the 5'-terminus and sequentially the 3'-termini from which the original chain can be reconstructed.[19]

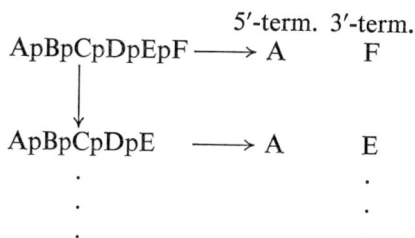

	5'-term.	3'-term.
ApBpCpDpEpF $\longrightarrow$	A	F
$\downarrow$		
ApBpCpDpE $\longrightarrow$	A	E
.	.	.
.	.	.
.	.	.

To summarize, the electron impact analysis of per-derivatized higher nucleotides will apparently soon reach its limit because of the high molecular weights and the increasingly complex fragmentation patterns which make highly pure material indispensable. More complex derivatisation which may facilitate a sequence analysis causes an additional problem if only minute amounts of material are available. Field desorption techniques possibly combined with minor chemical modifications (*e.g.* esterification of the free phosphate hydroxyls only) may lead further. The other promising way seems to be a combined degradation–mass spectrometric procedure.

ACKNOWLEDGMENTS

I wish to thank my co-workers Dr H. Brzezinka and Dr B. Johannes who provided the unpublished material in the carotenoid section, and Dr H. Schiebel, Braunschweig, for unpublished spectra of nucleotide derivatives.

REFERENCES

1. For the use of physical methods in the carotenoid field see: Liaaen-Jensen, S., *Pure Appl. Chem.*, 1973, **35**, 81, and Ref. 4.
2. Budzikiewicz, H., Brzezinka, H. and Johannes, B., *Monatsh. Chem.*, 1970, **101**, 579.
3. Budzikiewicz, H. and Drewes, S. E., *Liebigs Ann. Chem.*, 1968, **716**, 222, and Budzikiewicz, H., *Croat. Chem. Acta*, 1970, **42**, 567.
4. For a complete listing see: Vetter, W., Englert, G., Rigassi, N. and Schwieter, U., *in* 'Carotenoids', Chapter 4, *ed.* O. Isler, Birkhäuser Verlag, Bascl, 1971.
5. Liaaen-Jensen, S., *Acta Chem. Scand.*, 1967, **21**, 1972.
6. Schwieter, U. *et al.*, *Chimia*, 1965, **19**, 294.
7. Enzell, C. R., Francis, G. W. and Liaaen-Jensen, S., *Acta Chem. Scand.*, 1968, **22**, 1054.
8. Schwieter, U., Englert, G., Rigassi, N. and Vetter, W., *Pure Appl. Chem.*, 1969, **20**, 365.
9. Aitzetmüller, K., Strain, H. H., Svec, W. A., Gandolfi, M. and Katz, J. J., *Phytochem.*, 1969, **8**, 1761.
10. Baldas, J., Porter, O. N., Leftwick, A. P., Weedon, B. C. L. and Szaboles, J., *Chem. Comm.*, 1969, 415.
11. Johannes, B., Brzezinka, H. and Budzikiewicz, H., *Z. Naturf.*, 1971, **26b**, 377.
12. Strain, H. H., Aitzetmüller, K., Svec, W. A. and Katz, J. J., *Chem. Comm.*, 1970, 876.
13. Johannes, B., Dissertation TU Braunschweig, 1971.
14. For a compilation see: Hignite, C., *in* 'Biochemical Applications of Mass Spectrometry', *ed.* G. R. Waller, Wiley-Interscience, New York, p. 429 ff, 1972.
15. Biemann, K. and McCloskey, J. A. Jr., *J. Amer. Chem. Soc.*, 1962, **84**, 2005.
16. Capriolo, R. and Rittenberg, D., *Proc. Natl. Acad. Sci.*, 1968, **60**, 1379.
17. Hunt, D. F., Hignite, C. E. and Biemann, K., *Biochem. Biophys. Res. Comm.*, 1968, **33**, 378.
18. Dolhun, J. J. and Wiebers, J. L., *J. Amer. Chem. Soc.*, 1969, **91**, 7755.
19. Wiebers, J. L., *Anal. Biochem.*, 1973, **51**, 542.
20. Chamock, G. A. and Loo, J. L., *Anal. Biochem.*, 1970, **37**, 81.
21. Wiebers, J. L., *Adv. in Mass Spectrom.*, 1971, **5**, 757.
22. Brown, P., Petit, G. R. and Robins, R. K., *Org. Mass Spectrom.*, 1969, **2**, 521.
23. Schulten, H. R. and Beckey, H. D., *Org. Mass Spectrom.*, 1973, **7**, 861.

BIOCHEMICALS

Chairmen

A. FRIGERIO
(*Institute 'Mario Negri', Milan, Italy*)

D. C. DE JONGH
(*University of Montreal, Canada*)

18

The Use of Mass Spectrometry in the Structural Elucidation of Prostaglandins†

By BRIAN S. MIDDLEDITCH and DOMINIC M. DESIDERIO‡

(*Institute for Lipid Research and Department of Biochemistry, Baylor College of Medicine, Houston, Texas, U.S.A.*)

INTRODUCTION

A comparison of the general mass spectral characteristics is made of representative members (PGA_1, PGB_1, PGE_1 and $PGF_{1\alpha}$) of the four classes of prostaglandins. This comparison is being made after studying most of the available prostaglandins with both low and high resolution mass spectrometry, and GC-MS of the trimethylsilyl (TMS) ethers and esters and of the free carbonyl and alkyloxime derivatives.[1-4] In addition, specific d_9-TMS labelling was employed to elucidate various fragmentation pathways.

This detailed study of the prostaglandins has been done for two reasons. On one hand, despite the enormous amount of work done up to now on these biologically potent molecules,[5-7] surprisingly few basic mass spectrometric studies have been done. On the other hand, prostaglandins are being implicated in manifold biological processes.[8-10] As the endogenous concentrations are quite low (on the order of ng/ml fluid, or ng/g tissue), single ion monitoring (SIM) techniques have been developed to quantify PG concentrations at that level.[11-13] Thus it is necessary to understand the genesis of those ions used to identify and quantify a particular PG.

FRAGMENTATIONS COMMON TO ALL SPECTRA

In the spectra discussed below, there are a few general fragmentation processes that occur with all of the compounds. For example, a methyl group, the alkyloxime alkoxy group, the C_5H_{11} chain and trimethylsilanol are readily lost. Thus, the high mass range of these compounds contain ions corresponding to the following list: $M^{+\cdot}$, $[M-15]^+$, $[M-31]^+$, $[M-71]^+$, $[M-90]^{+\cdot}$, $[M-90-31]^+$, $[M-90-71]^+$, $[M-90-15]^+$, and $[M-2 \times 90]^{+\cdot}$.

---

† Last minute circumstances prevented Dr D. M. Desiderio from presenting this paper.

‡ Fellow of the Intra-Science Research Foundation (1971-5).

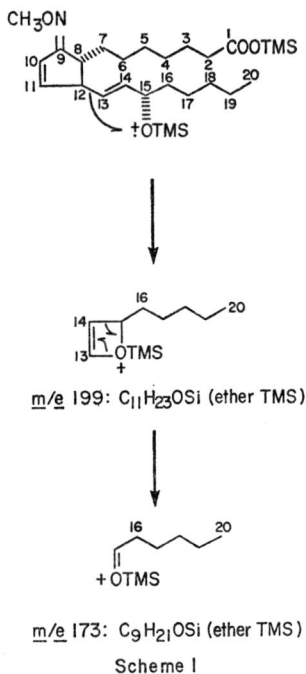

m/e 199: $C_{11}H_{23}OSi$ (ether TMS)

m/e 173: $C_9H_{21}OSi$ (ether TMS)

Scheme 1

In addition, in the low mass region m/e 199 and 173 are found in all spectra. Specific d_9-TMS labelling shows that the TMS group in these two ions originates from an ether TMS. Scheme 1 is a rationalization of the genesis of these two ions.

FRAGMENTATIONS UNIQUE TO EACH COMPOUND

PGA_1—The mass spectrum of the methyl oxime–TMS ester–TMS ether derivative of PGA_1 is given in Fig. 1.[1] On GC, two peaks are found for this derivative. These peaks are presumably *syn-anti* isomers. As we have no method at the moment of distinguishing these two isomers, the mass spectrum of peak 1 is presented. There are only slight qualitative differences between the mass spectra of the two GC peaks.

All but one of the diagnostic ions in this mass spectrum have been listed in the previous section: m/e 509, $M^{+\cdot}$; m/e 494, $[M-15]^+$; m/e 478, $[M-31]^+$; m/e 438, $[M-71]^+$; m/e 419, $[M-90]^{+\cdot}$; m/e 388, $[M-90-31]^+$; m/e 199; m/e 173. Selective d_9-TMS labelling shows that the methyl group lost in the $[M-15]^+$ ion originated from both the ester and ether TMS groups. Similarly, it was found that the loss of trimethylsilanol in the $[M-90]^{+\cdot}$ ion originated from the ether TMS group.

The abundant ion at m/e 148 may be formed as depicted in Scheme 2. The abundance of this ion may be due to the delocalization of the positive charge over nine atoms.

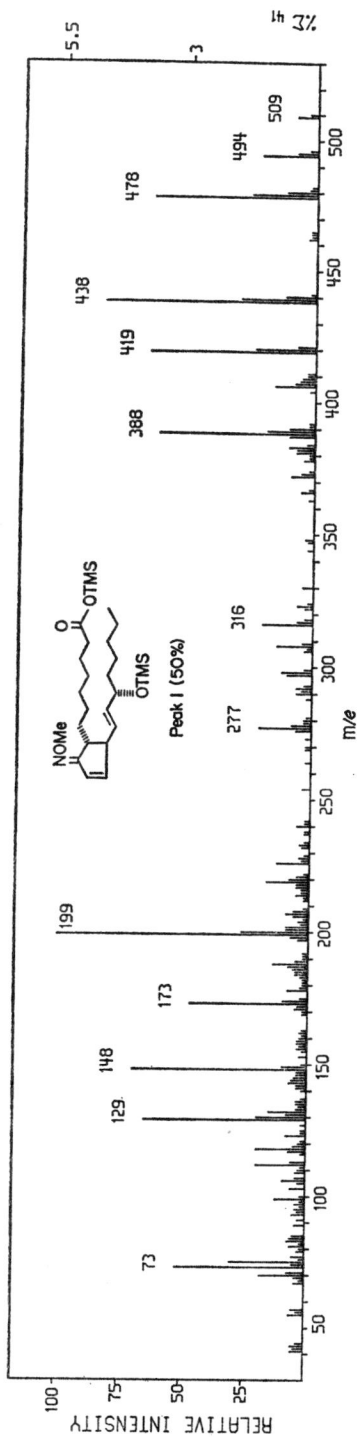

FIG. 1   Mass spectrum of MO–TMS derivative of PGA$_1$, first peak. 6 ft × 0·25 inch silanized glass column, 1% SE-30 on Gas Chrom. Q (100–120 mesh). LKB 9000. Column, 200°; separator, 250°; Ion Source, 270°; accelerating voltage, −3·5 kV, electron energy, 22·5 eV.

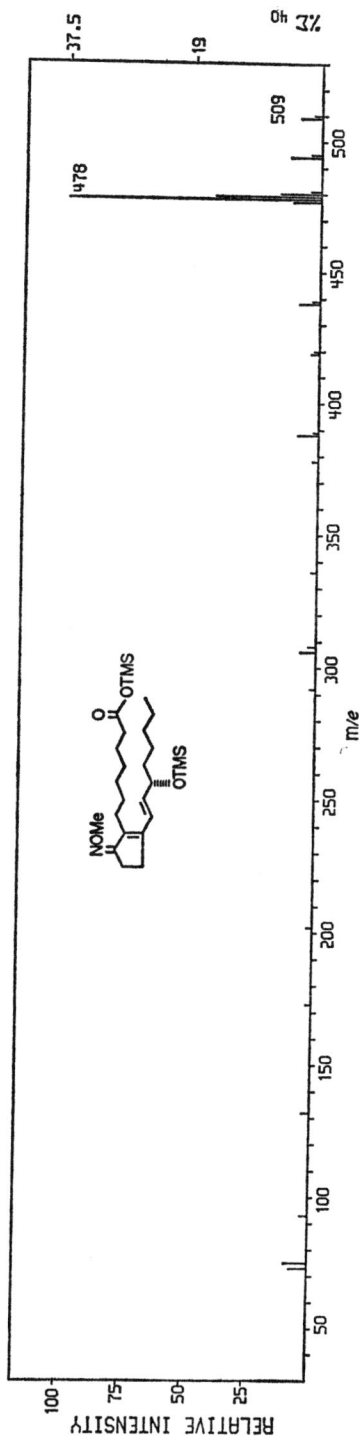

FIG. 2   Mass spectrum of MO–TMS derivative of PGB$_1$. (Same conditions as stated in legend of Fig. 1.)

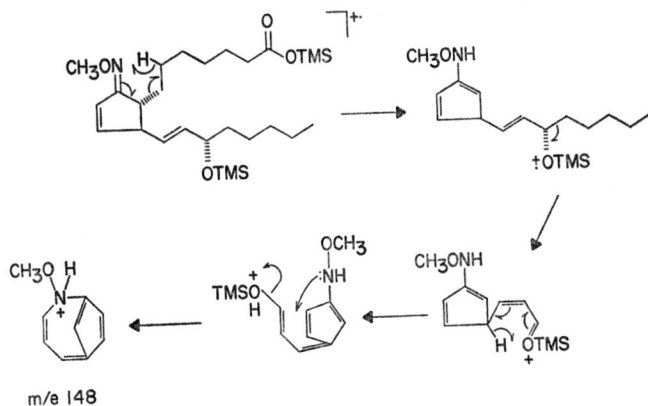

Scheme 2

PGB₁—The mass spectrum of the methyl oxime–TMS ester–TMS ether derivative of PGB$_1$ is given in Fig. 2.[2] On GC, only one peak is found. All of the ions found in this spectrum correspond to the common ions above: m/e 509, $M^{+\cdot}$; m/e 494, $[M-15]^+$; m/e 478, $[M-31]^+$; m/e 438, $[M-71]^+$; m/e 419, $[M-90]^{+\cdot}$; m/e 388, $[M-90-31]^+$.

The most striking feature in this spectrum is that the $[M-31]^+$ ion dominates this spectrum. This type of ion is a very good candidate for SIM due to the sensitivity obtained by having 37·5% of the total ion current concentrated in one ion. The only drawback to this fact is that the PGE compounds must first be treated with base to convert them to the PGB compound.[13] In principle, it would be better to analyse the PGE compounds intact.

PGE₁—The mass spectrum of the methyl oxime–TMS–ester–TMS ether derivative of PGE$_1$ is given in Fig. 3.[3] Again, two peaks were found on the GC, and the first peak is given here, as greater qualitative differences were found between the two spectra and peak 1 had more abundant ions. The ions shared in common with the other spectra are: m/e 599, $M^{+\cdot}$; m/e 584, $[M-15]^+$; m/e 568, $[M-31]^+$; m/e 528, $[M-71]^+$; m/e 509, $[M-90]^+$; m/e 478, $[M-90-31]^+$; m/e 438, $[M-71-90]^+$.

The ion at m/e 225 is formed after cleavage of the ring and structure *a* is compatible with accurate mass and labelling data.

The ion at m/e 133 has an interesting mode of formation. The elemental composition ($C_5H_{13}O_2Si$) of this ion and the fact that this ion shifts to 147 in the ethoxime derivative suggest that this ion forms by migration of the oxime alkoxy group to C-11 to form an ion such as *b*.

The two remaining ions of interest in this mass spectrum are m/e 199 and

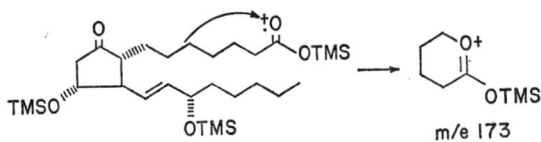

Scheme 3

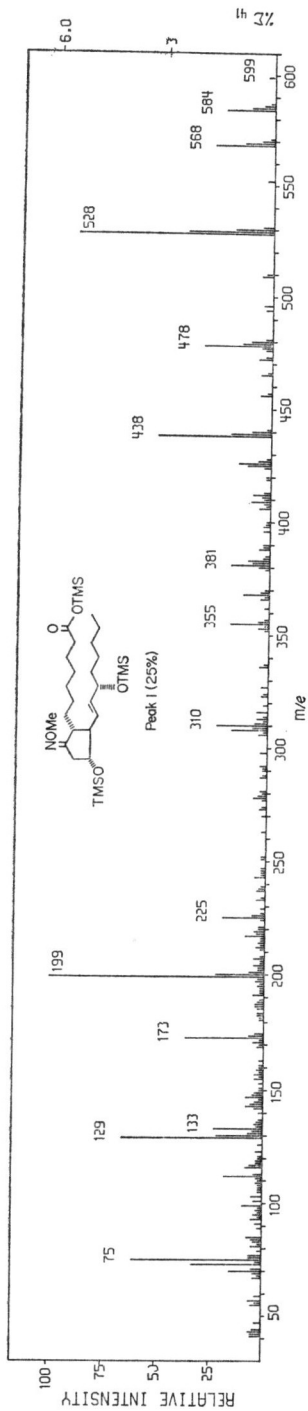

FIG. 3 Mass spectrum of MO–TMS derivative of PGE$_1$, first peak. (Same conditions as stated in legend of Fig. 1.)

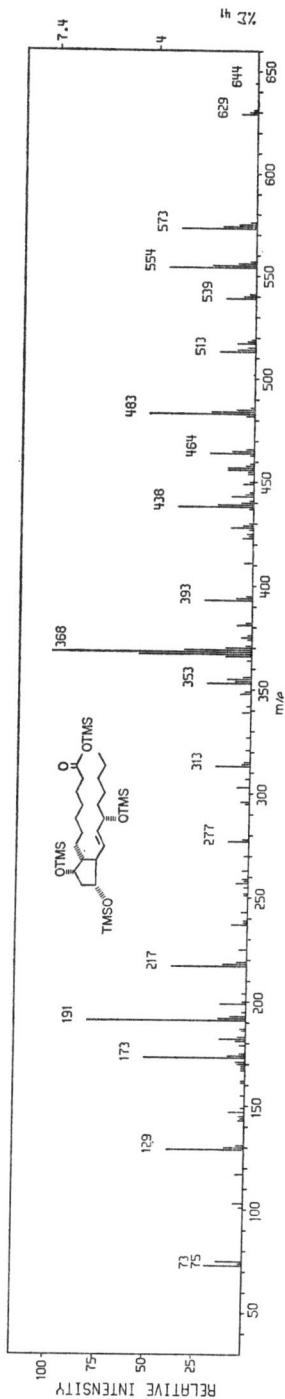

FIG. 4 Mass spectrum of TMS derivative of PGF$_{1\alpha}$. (Same conditions as stated in legend of Fig. 1.)

173, formed as shown in Fig. 1. However, another component of the ion at m/e 173 was found to contain an ester TMS group and to have an elemental composition $C_8H_{17}O_2Si$. This component may form as in Scheme 3.

$PGF_{1\alpha}$—The mass spectrum of the TMS ether–TMS ester derivative of $PGF_{1\alpha}$ is given in Fig. 4.[4] The following ions are observed: m/e 644, $M^{+\cdot}$; m/e 629, $[M-15]^+$; m/e 573, $[M-71]^+$; m/e 554, $[M-90]^{+\cdot}$; m/e 539, $[M-15-90]^+$; m/e 483, $[M-71-90]^+$; m/e 464, $[M-2 \times 90]^{+\cdot}$; m/e 199; m/e 173.

The $[M-15]^+$ ion is formed solely from the loss of a methyl radical from an ether TMS group. The $[M-90]^{+\cdot}$ ion is formed by losing TMSOH from an ether TMS.

The presence of two TMS groups in the ring causes extensive fragmentation to occur as witnessed by the occurrence of ions c–e.

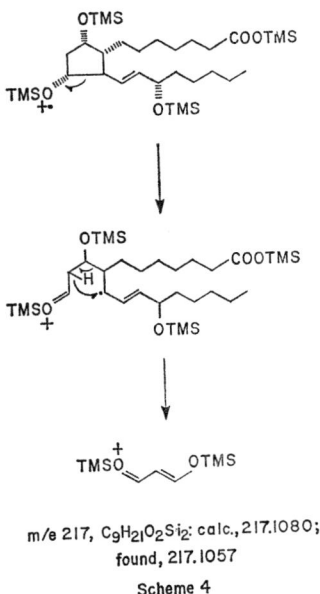

m/e 217, $C_9H_{21}O_2Si_2$: calc., 217.1080; found, 217.1057

Scheme 4

The abundant ion at m/e 217 apparently contains both of the ring TMS groups. It shifts to m/e 235 in the spectra of the d_9-TMS ester d_9-TMS ether and the TMS ester d_9-TMS ether. The genesis of this ion is rationalized in Scheme 4.

The base peak of the spectrum, m/e 368, is formed by a variety of routes. The results of a metastable defocussing study and labelling results are presented in Scheme 5.

CONCLUSIONS

The mass spectra of the alkyloxime–TMS ester–TMS ether derivatives of the prostaglandins A, B, E, F have been studied. It has been found that certain fragmentation processes are common to all spectra. The loss of trimethylsilanol, the C_5H_{11} chain, a methyl radical, and the alkoxyl group all occur,

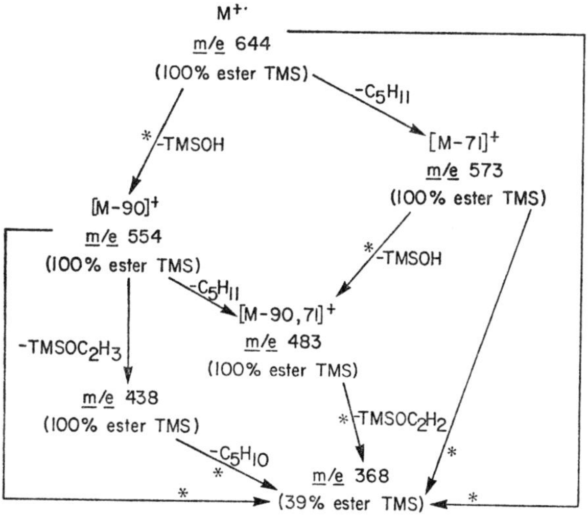

Scheme 5

individually, and in combination with each other. In addition, each type of prostaglandin has a fragmentation unique to itself. For PGA_1, the production of a bicyclic amine ion structure is a unique feature. The PGB_1 derivative is quite unique in that the $[M-31]^+$ ion carries the majority of the total ion current, with very little other fragmentation. The PGE_1 derivative produces some long chain TMS fragments, an alkoxy group migration, and a second component to the common m/e 173 ion. Finally, the $PGF_{1\alpha}$ derivative undergoes extensive fragmentation of the cyclopentane ring.

This detailed analysis of the available prostaglandins has enabled this laboratory to utilize mass spectrometry in both the analysis of endogenous prostaglandins in biological materials and in the quantifying of the compound levels.

ACKNOWLEDGMENTS

The prostaglandins were a generous gift from J. E. Pike and U. Axen of the Upjohn Company, Kalamazoo, Michigan and from K. Sano of the Ono Pharmaceutical Co., Osaka, Japan. This work was supported by grants from the NIH (GM 13901) and the Robert A. Welch Foundation (Q 125). Technical assistance with the high resolution mass spectra was provided by Pam Crain.

REFERENCES

1. Middleditch, B. S. and Desiderio, D. M., Prostaglandins, in press.
2. Middleditch, B. S. and Desiderio, D. M., *Lipids*, 1973, **8**, 267.
3. Middleditch, B. S. and Desiderio, D. M., *J. Org. Chem.*, in press.
4. Middleditch, B. S. and Desiderio, D. M., *Anal. Biochem.*, in press.
5. For a review, see Crain, P. F., Desiderio, D. M. and McCloskey, J. A., *in* 'Methods in Enzymology', *Lipids*, Vol. XIV, Part B, *ed.* J. M. Lowenstein, Academic Press, New York, in press.

6. Gréen, K., *Chem. Phys. Lipids*, 1969, **3**, 254.
7. Vane, F. and Horning, M. G., *Anal. Lett.*, 1969, **2**, 357.
8. Oesterling, T. O., Morozowich, W. and Roseman, T. J., *J. Pharm. Sci.*, 1972, **61**, 1861.
9. Vane, J. R., *Nature*, 1971, **231**, 232.
10. Johnson, M., Rabinowitz, I., Willis, A. L. and Wolf, P. L., *Clin. Chem.*, 1973, **19**, 23.
11. Samuelsson, B., Hamberg, M. and Sweeley, C. C., *Anal. Biochem.*, 1970, **38**, 301.
12. Axen, U., Gréen, K., Hörlin, D. and Samuelsson, B., *Biochem. Biophys. Res. Commun.*, 1971, **45**, 519.
13. Sweetman, B. J., Frölich, J. C. and Watson, J. T., *Prostaglandins*, 1973, **3**, 75.

19

Qualitative and Quantitative Investigation of Barbiturate Metabolites by GC-MS and High Resolution Mass Spectrometry

By N. J. HASKINS, M. J. LEE and B. J. MILLARD

(The School of Pharmacy, London, U.K.)

In view of the long established use of barbiturates as hypnotics and sedatives, it is not surprising that considerable work has been published concerning their metabolism.[1] However, in the case of the medium acting drug allobarbitone (1) a previous study[2] concluded that 30% of the ingested drug was excreted unchanged over a period of three days in humans, and no metabolites were found.

In the present work, urine was collected over a period of 72 hr from a human volunteer after a single dose of 200 mg of allobarbitone. After adjustment to pH 5–6 the urine was extracted with ethyl acetate. The evaporated extract was taken up in methanol and methylated with diazomethane. Half of the methylated extract was then silylated using hexamethyldisilazane and trimethylsilyl chloride in pyridine.

Examination of these two extracts by GC-MS showed both of them to contain the same three components that were not present in a blank sample of urine taken through the same extraction and methylation procedure. The three components, A, B and C had retention times of 5, 7·5 and 12 min. The mass spectra of these are given in Fig. 1. The mass spectrum of A and its retention time were identical to those of an authentic sample of allobarbitone which had been methylated with diazomethane, so that component A is unchanged allobarbitone.

Component B had important ions at m/e 210 and 195. The latter ion is also present in the spectrum of methylated allobarbitone, where it corresponds to the loss of one allyl substituent. M/e 210 could therefore be interpreted as being allobarbitone (methylated) in which a methyl group has replaced one allyl group. Since deallylation has been observed in quinalbarbitone,[3,4] it is conceivable that deallylallobarbitone (2) is a metabolite of allobarbitone, and component B is the result of the introduction of three methyl groups during the methylation step.

In order to test this point, 5-allylbarbituric acid (3) was synthesized and a small amount added to blank urine. The same extraction and methylation procedure was used as before. When examined by GC-MS, two peaks were

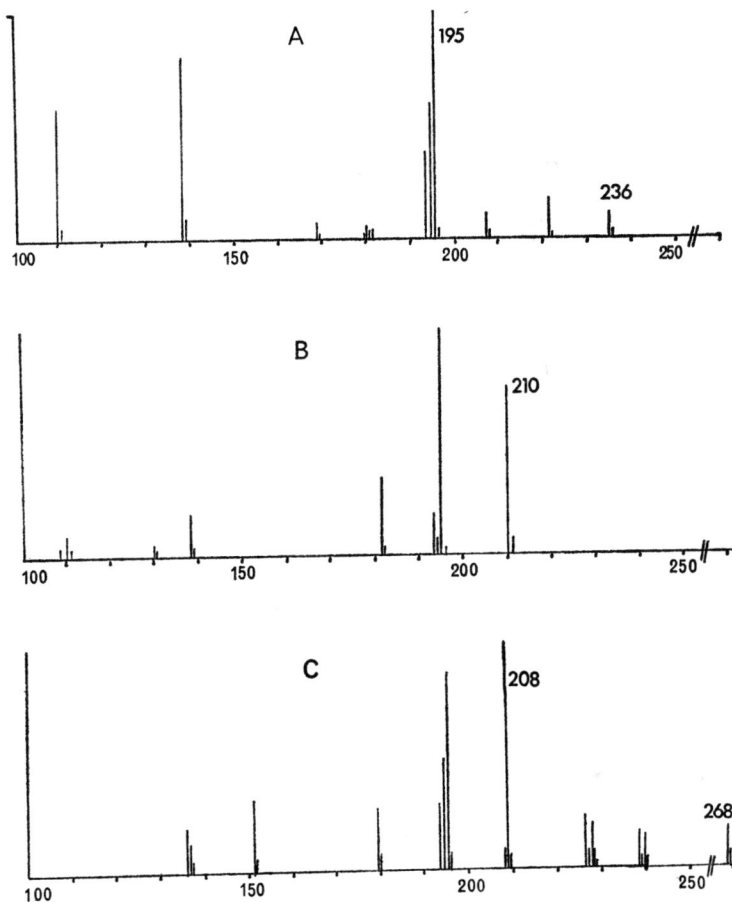

FIG. 1 Mass spectra of components A, B and C in the urine of a volunteer after a dose of 400 mg of allobarbitone.

evident, of retention times 2 and 7·5 min. The mass spectra of both peaks were similar, with that of the latter being identical to the mass spectrum of component B.

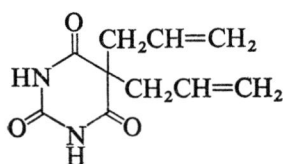

(1)

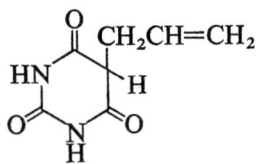

(2)

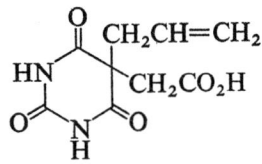

(3)

The formation of two products on methylation can be explained perhaps by formation of an *O*-methyl derivative and by an insertion reaction on the allylic double bond as follows:

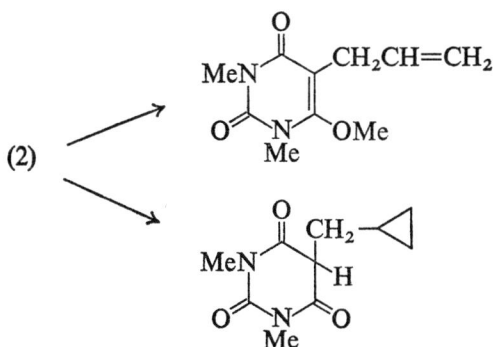

(2)

On re-examination of the urine extracts by monitoring the ions at m/e 210 and 195, responses for these two ions were indeed obtained at 2 and 7·5 min. When methylated allobarbitone was monitored in this way, no such responses were obtained, thus ruling out the possibility that the deallyl compound was present as an impurity in the allobarbitone taken by the volunteer.

Component C had its highest ion at m/e 268 (Fig. 1) with ions at m/e 227, 226, 208 and 195. A structure which would give rise to all these ions would be the ester (4) derived from the acid (3):

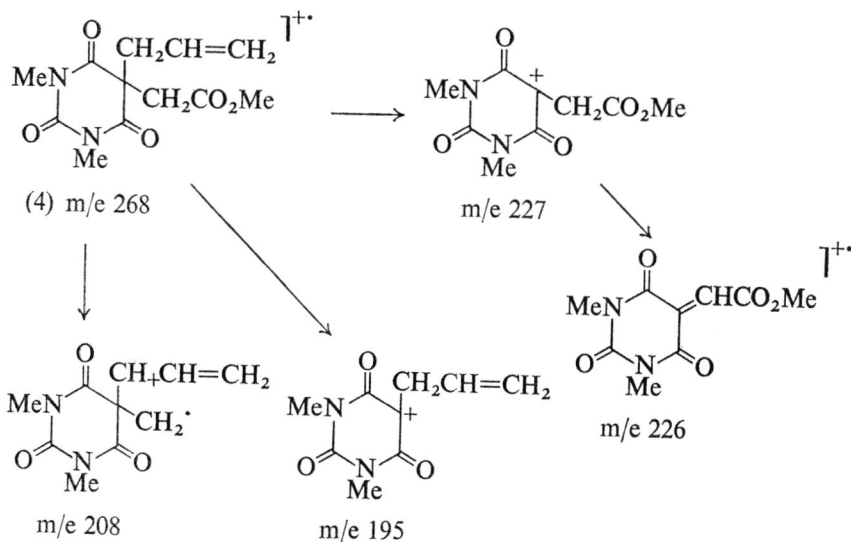

(4) m/e 268

m/e 227

m/e 226

m/e 208

m/e 195

Work is currently in progress to synthesize an authentic sample of this ester.

The transformation of a 5-allyl substituent in barbiturates to a carboxy-methyl group is somewhat unusual. It does not take place for example in nealbarbitone[5] where diol formation is the preferred metabolic pathway. This is also true of quinalbarbitone.[4] Such a diol, at least in unconjugated form, has not been detected in the present work on allobarbitone.

QUANTITATIVE MASS SPECTROMETRY

Because of the non-availability of a synthetic sample of the acid (3), another barbiturate drug heptabarbitone (5) whose major metabolites (6), (7) and (8) have been identified[6] was chosen.

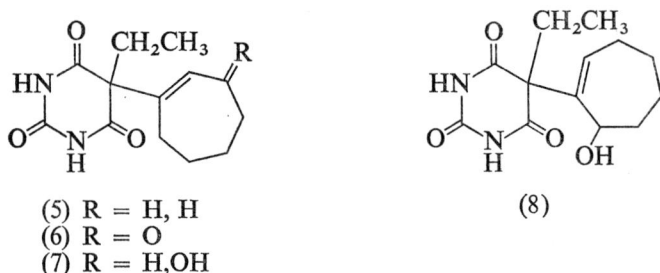

(5) R = H, H
(6) R = O
(7) R = H,OH

(8)

The gas chromatographic procedure[6] for quantitation of heptabarbitone and its metabolites is lengthy, since urine samples have to be extracted and methylated prior to injection on the column. It was decided therefore to investigate the possibility of quantifying these compounds by monitoring selected ions in urine samples directly at high resolution to separate them from the many background ions at the same m/e values. The general procedure was to add 50 mg of a reference compound to the urine and place 1 μlitre samples on the direct inlet probe of the mass spectrometer. An ion from the reference compound and an ion diagnostic of the metabolite were then monitored during the period of evaporation of the sample. The intensities of these two ions were then separately summed. Since in all cases the molecular ions were of low intensity, it was decided to monitor the $(M-C_2H_5)^+$ ions. These had accurate masses of 221·0926 ($C_{11}H_{13}N_2O_3$) for (5), 235·0719 ($C_{11}H_{11}N_2O_4$) for (6), and 237·0875 ($C_{11}H_{13}N_2O_4$) for (7) and (8). It was not possible to distinguish between metabolites (7) and (8) by this method since they are isomeric.

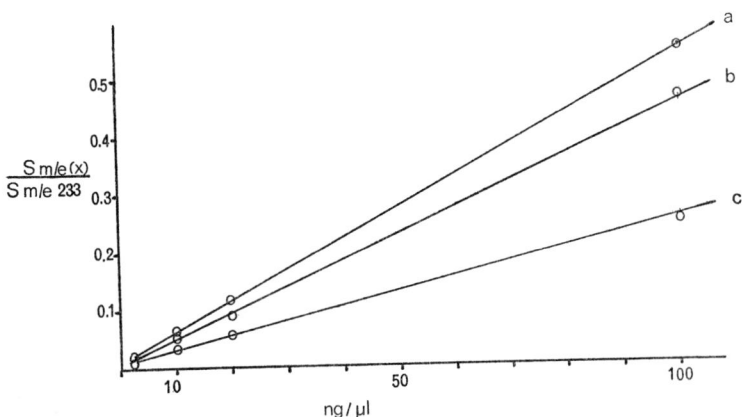

FIG. 2   Calibration lines for (a) heptabarbitone (b) 3-keto-heptabarbitone (c) 3- plus 7-hydroxyheptabarbitone.

The requirements for the internal standard are threefold. Firstly it should be soluble in urine at natural pH, secondly it should have an intense ion near in m/e value to the ion to be monitored and thirdly it should have an unusual composition so that no contribution occurs from compounds already present in urine. The compound 2-chlorophenothiazine, as its hydrochloride was therefore chosen. The molecular ion of this at m/e 233 has composition $C_{12}H_8NSCl$.

Calibration lines were established for heptabarbitone and its metabolites by taking solutions of 500 ml containing 50 mg of 2-chlorophenothiazine as its hydrochloride and (a) 1 mg each (b) 5 mg each (c) 10 mg each and (d) 50 mg each of (4), (5) and (6). Each determination, using 1 μlitre samples on the probe, was repeated four times. The lowest concentration of these

TABLE I

Excretion of Heptabarbitone and its Metabolites

Urine sample	Volume (1)	Nanogrammes per microlitre by mass spec.[a]			Total amount excreted (mg)			GC method
		(5)	(6)	(7 + 8)	(5)	(6)	(7 + 8)	(5)
blank	0·5	0·3	0·4	0·3	—	—	—	—
0–8 hr	0·51	27·4	15·9	68·5	14	8·1	35	12
8–16 hr	0·40	70·0	12·7	92·5	28	5·1	37	30
16–24 hr	0·65	38·5	4·9	47·8	25	3·2	31	25
24–32 hr	0·55	43·6	1·1	20·0	24	0·6	11	22
32–40 hr	0·35	6·0	0·4	17·4	2·1	—	6·1	2
40–48 hr	0·50	0·4	0·4	9·2	0·2	—	4·5	0·4
48–56 hr	0·41	0·3	0·3	2·2	—	—	0·9	—
56–64 hr	0·45	0·4	0·4	0·3	—	—	—	—

[a] Mean of two determinations.

compounds represents 2 ng on the probe. The coefficients of variation at this level were 4·6% for (5), 5·4% for (6) and 4·8% for (7). The calibration lines are shown in Fig. 2.

The same procedure was carried out on the various urine fractions and the quantities of heptabarbitone and metabolites obtained by interpolation on the standard curves. The results and a comparison with a gas chromatographic determination of heptabarbitone are given in Table I.

EXPERIMENTAL

GC-MS determinations were carried out using a Finnigan 1015 Mass Spectrometer. The column was a 9-ft glass packed with 1% QF 1 on gas-chrom. Q (100/120 mesh). The column was programmed from 150 to 200° at 6° per minute.

The high resolution work was carried out on an AEI MS902 mass spectrometer operating at 10 000 resolving power (10% valley).

The synthesis of the metabolites of allobarbitone will be published elsewhere.

REFERENCES

1. *Foreign Compound Metabolism in Mammals*, Vol. 1, pp. 152–5.
2. Reiche, F. and Halberkaun, J., *Munch. med. Wschr.*, 1929, **76**, 235.
3. Cochin, J. and Daly, J. W., *J. Pharm. Exp. Ther.*, 1963, **139**, 154.
4. Waddell, W. J., *J. Pharm. Exp. Ther.*, 1965, **149**, 23.
5. Gilbert, J. N. T., Millard, B. J., Powell, J. W., Whalley, W. B. and Wilkins, B. J., to be published.
6. Gilbert, J. N. T., Millard, B. J., Powell, J. W. and Whalley, W. B., to be published.

20
Gas Chromatography–Mass Spectrometry Applied to Metabolic Profile Analysis

By S. P. MARKEY, W. G. URBAN, A. J. KEYSER and S. I. GOODMAN

(*Departments of Pediatrics and Pharmacology, University of Colorado Medical Center, Denver, Colorado, U.S.A.*)

OVER the past five years, we have designed, constructed and applied a gas chromatograph–mass spectrometer (GC–MS)–computer system for the analysis of inborn errors of metabolism. This paper summarizes the design and construction of this system and illustrates its utility in the diagnosis of two patients with unknown neurological disorders.

In a clinical research environment, it is essential to minimize dependence upon experienced mass spectroscopists, either for operation of the system or interpretation of mass spectral data. Samples often have to be analysed with some urgency, particularly those from newborn infants, and the resulting data interpreted without assistance from an organic mass spectroscopist. Key metabolites may not be the most intense peaks in a complex chromatogram. A completely automated GC–MS computer system coupled with an adequate mass spectral library is a necessity for clinical research.

Our programme began with the design of a GC–MS separator interface useful for a wide range of organic compounds.[1] The published design has been further improved by the use of a precision bore ceramic frit which can be manufactured with controlled porosity.[2] Variation in performance among separators is therefore minimized.

A minicomputer (PDP8i) was then interfaced to the mass spectrometer and programmed for GC–MS.[3] M/e assignment is based upon a repetitive magnetic scan,[4] and the accumulated spectra are saved on disk or DEC tape for a variety of processing options. The entire system is operated by a single technician and does not require computer resources outside our laboratory.

GC–MS metabolic profile analysis of urine acid extracts is most useful due to the wide variety of compounds which can be analysed in a single experiment. In an examination of the normal components identified in a pooled urine extract from ten newborn infants, products from the citric acid cycle (such as fumarate, aconitate and citrate), amino acid catabolism (β-hydroxyisovalerate, β-hydroxy-β-methyl glutarate), sugar metabolism (lactate, tartrate), lipid metabolism (lactate, β-hydroxybutyrate, adipate, suberate), bacterial metabolism (hippurate, hydroxybenzoate), catecholamine metabolism (HVA and VMA), drug metabolism, and artifacts of extraction and derivatization could all be identified by GC–MS.[5]

The identification of mass spectra may be done by one of three options— manual look-up in a catalogue of data; off-line computer search (either batch processing or time-sharing); or on-line, automated library searching using the dedicated data acquisition minicomputer. We have compiled a collection of nearly 300 mass spectra of trimethylsilyl derivatives of compounds of endogenous physiological origin. The compilation (cross-indexed by molecular weight and peak intensity) was printed,[6] and was found to permit rapid identification of frequently encountered compounds by personnel without training in organic mass spectrometry. However, a considerable amount of laboratory effort devoted to plotting and examining mass spectral data may be avoided by the use of a versatile computer library system to match spectra of unidentified compounds which have previously been encountered in normal or abnormal urine extracts. Consequently a series of programs has been written for library searching using the PDP8i.

Previously, the DEC Disk Monitor was used to execute the data acquisition and processing routines. However, it cannot be used to control both DEC tape and disk interchangeably as system devices. All of the data handling routines are being converted to work within OS-8, a relatively new monitor system. A program to generate and edit library data (GLIB) has been written in assembly language. It is capable of accessing n records (where n is not limited) of variable length. Each library sub-set of spectral records is considered a file. For example, TMS-organic acids may form a single file. Accessing of variable length records is possible with the use of MAP blocks into which the starting point of each record is noted. MAP blocks are allocated as needed, and linked by forward pointers. Using GLIB, approximately 730 complete mass spectra can fit onto a single DEC tape, assuming 124 pairs of masses and intensities per spectrum. Aldermaston format D is used to input data from paper tape. An identification number is assigned to each spectrum, the number of masses and intensity pairs are determined, and the minimum m/e is noted with each record.

From the complete mass spectra, a compressed library similar to that described by Grotch[7] is prepared. The compression routine (CLIB) has been written so that any number of compression algorithms may be employed. Currently, the highest peak in a 14 amu window is encoded to one of four intensity levels. Two peaks are stored in a 12-bit word. Compressed spectra are written as one per block or 1460 spectra per tape. Thus a single tape can be used for all routine work for metabolic profile analysis. Because TMS derivatives are employed in these studies, other compression algorithms may eventually be more useful for matching or suggesting structurally related compounds, such as the loss algorithm used by McLafferty.[8]

Mass spectra stored on the disk from a GC–MS can be compared to the library using a search program (SLIB). Any or all of the scans in a run can be specified. Each scan is read from the disk, compressed to the format used for the library, and compared with the library file. The best five fits along with a similarity index factor are printed for each scan.

Applying this system to TMS-urine acid extracts from patients with abnormal chromatograms as detected by our screening program has yielded clinically relevant information. We have encountered a number of patients with metabolic disorders previously described in the literature (methylmalonic acidemia, maple syrup urine disease, phenylketonuria, etc.). Two patients recently screened are discussed below to indicate the use of these methods to metabolic problems not previously reported.

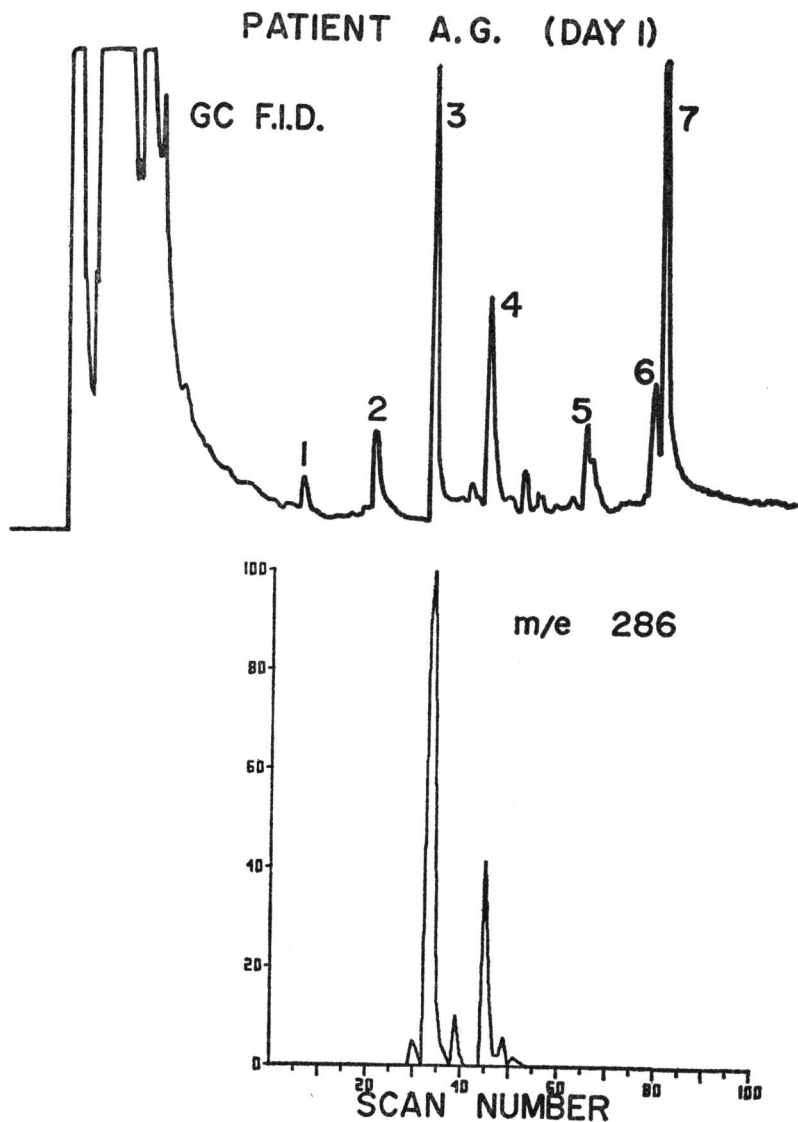

FIG. 1 Upper trace—flame ionization detector recording of a chromatogram of silylated urinary organic acids run on a 6 ft (2 mm i.d.) glass column packed with 5% OV-22 on 80–100 mesh Gas Chrom. Q. Programmed from 50 at 5/min. Peak 1—urea; 2—mol wt 288; 3—5-hydroxymethyl-2-furoic acid; 4—2,5-furan dicarboxylic acid; 5—mol wt 398; 6—HVA; 7—hippuric acid. Lower trace—mass chromatogram of m/e 286 taken from the stored mass spectra recorded while the above chromatogram was made.

Patient A.G. is a 12-month-old male being treated for seizures of unknown etiology. Figure 1 (upper trace) is a chromatogram of his silylated urinary organic acid extract produced on routine screening. Abnormal amounts of 5-hydroxymethyl-2-furoic acid (peak 3) and 2,5 dicarboxylic acid (peak 4) were detected and identified by GC–MS. The only previous report of increased excretion of these furans (not associated with parenteral nutrition) was in the case of two patients with acute lymphocytic leukemia.[9] Because A.G. showed showed no signs of leukemia, further investigation of these metabolites was of interest. A simple assay for these furans is a mass chromatogram of m/e 271 (M-15 of 5-hydroxymethyl-2-furoic acid) and of m/e 285 (M-15 of 2,5 furandicarboxylic acid), or as shown in Fig. 1 (lower trace) of m/e 286 (peak 3, M^+; peak 4, isotope of M-15). The variation in furan excretion with respect to anti-convulsant and steroid medications being administered to A.G. was observed over a period of five days with no correlation apparent. The child was then placed upon a standardized diet of known, but varied, carbohydrate composition for a period of 14 days. No furans were detected throughout the study, but a severe intermittent ketosis developed (Fig. 2). Pettersen and Jellum have suggested that the origin of most urinary furans is dietary.[10] Our findings support this hypothesis, but do not explain why some individuals lack the ability to metabolize ingested furans.

Patient S.W. is a one-year-old female who was investigated because of rapidly progressive neurological deterioration. Routine urinary organic acid

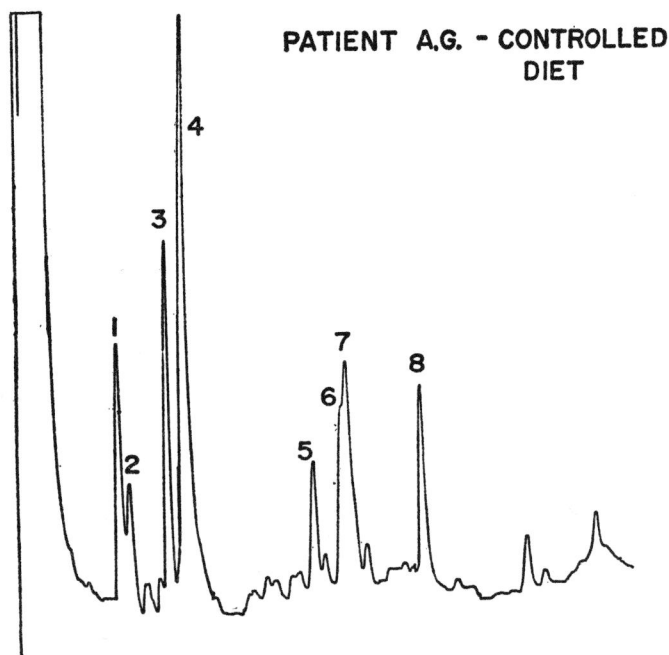

FIG. 2 Chromatogram of the silylated urinary organic acids obtained when patient A.G. was restricted to glucose for carbohydrate intake. Peak 1—β-hydroxybutric; 2—unknown; 3—acetoacetic; 4—urea; 5—p-hydroxyphenylacetic; 6—aconitic; 7—citric; 8—hippuric.

screening revealed a markedly abnormal metabolic profile. A total ion current plot of this chromatogram is shown in Fig. 3. The major peak was rapidly identified as glutaric acid. Screening the family of S.W. revealed that both parents and one normal sibling had normal profiles, but that an older brother with a similar, albeit somewhat milder disorder, also excreted very large quantities of glutaric acid. Enzyme assays are now being performed to determine if this finding represents a hereditary block in the normal metabolism of lysine and tryptophan to glutaconic acid.

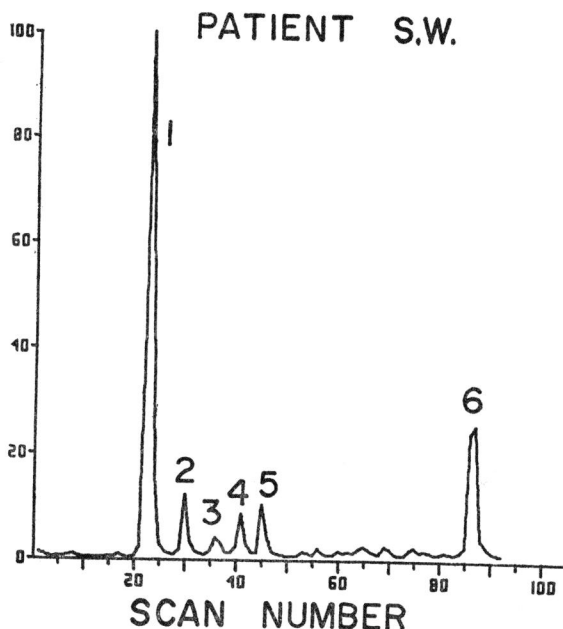

FIG. 3 Total ion current chromatogram of the silylated organic acid extract obtained from patient S.W. Peak 1—glutaric; 2—mol wt 288; 3—unknown; 4—mol wt 364; 5—α-keto-glutaric; 6—hippuric.

CONCLUSION

GC–MS-computer systems are a significant aid in the investigation of inborn errors of metabolism. The time required to identify an abnormal metabolite in a physiological fluid has been dramatically reduced with the use of this technology, and clinical research efforts can now be rapidly focussed upon the biochemistry of the metabolic block and possible dietotherapy to circumvent metabolic crises.

REFERENCES

1. Markey, S. P. *Anal. Chem.*, 1970, **42**, 306.
2. Allen, R. H. Co., Boulder, Colorado.
3. Plattner, J. R. and Markey, S. P. *Org. Mass Spectrom.*, 1971, **5**, 463.

4. Hites, R. A. and Biemann, K., *Adv. Mass Spectrometry*, 1968, **4**, 37.
5. Markey, S. P., Urban, W. G. and Keyser, A. J. *Proc. 21st Ann. Conf. on Mass Spectrometry and Allied Topics*, San Francisco, California, May 1973.
6. Markey, S. P., Thobhani, H. A. and Hammond, K. B. Identification of Endogenous Urinary Metabolites by Gas Chromatography—Mass Spectrometry: A Collection of Mass Spectral Data. Dept. of Pediatrics, University of Colorado Medical Center, Denver, Colorado, 1972.
7. Grotch, S. L., *Anal. Chem.*, 1973, **45**, 2.
8. Venkataraghavan, R., Kwok, K. S., Pesyna, G. and McLafferty, F. W. *Proc. 21st Ann. Conf. on Mass Spectrometry and Allied Topics*, San Francisco, California, May 1973.
9. Mrochek, J. E. and Rainey, W. T. Jr, *Clin. Chem.*, 1972, **18**, 821.
10. Pettersen, J. E. and Jellum, B., *Clin. Chem. Acta*, 1972, **41**, 199.

21

The Measurement of Prostaglandins F by Combined Gas Chromatography–Mass Spectrometry

By R. W. KELLY

(*M.R.C. Unit of Reproductive Biology,*
Department of Obstetrics and Gynaecology, Edinburgh, U.K.)

ONE feature of the application of combined gas chromatography–mass spectrometry (GC–MS) in the biological sciences is the growing use of the mass spectrometer as a detector for the gas chromatograph. These applications have been described by such names as Mass Fragmentography and Multiple Ion Monitoring, but the principle is the same. The accelerating voltage is stepped to bring selected ions into focus sequentially.[1] The gas chromatograms which are produced from such a device are selective and each can be used to monitor independently for chosen components of a mixture. The potential of this type of approach has only been partly realized and the technique is already being applied to many different biochemical analyses.

One field in which the mass spectrometer detector has been used for several years is that of the prostaglandins. The problems of quantitation in analysis of prostaglandin E's by GC–MS were solved by Samuelsson, Hamberg and Sweeley[2] by the use of deuterated methyl oximes as the internal standard. E Prostaglandins (Fig. 1) have a ketone in the cyclopentane ring but this functional group is not shared with the F prostaglandins and consequently analysis of the F series required a different approach. To meet the demand for internal standards a series of nearly perfect compounds (the 3,3,4,4-tetradeutero prostaglandins) was synthesised by the Upjohn Company[3] and these have been invaluable in prostaglandin measurements by GC–MS. The position of substitution of the deuterium atoms in these compounds is such (Fig. 1) that exchange of the deuterium atoms with solvents is minimized. These standards have not only made a good physico-chemical analysis scheme possible but have also provided a good model to study the limits and possibilities of this type of analysis.

The particular aspect of these analyses on which we are concentrating in Edinburgh is the measurement of prostaglandins in crude extracts which have received no purification other than extraction. The main requirement in such measurements is the use of a derivative with a strong fragment ion at a mass as high as possible. In general the higher the mass of the ion chosen the less

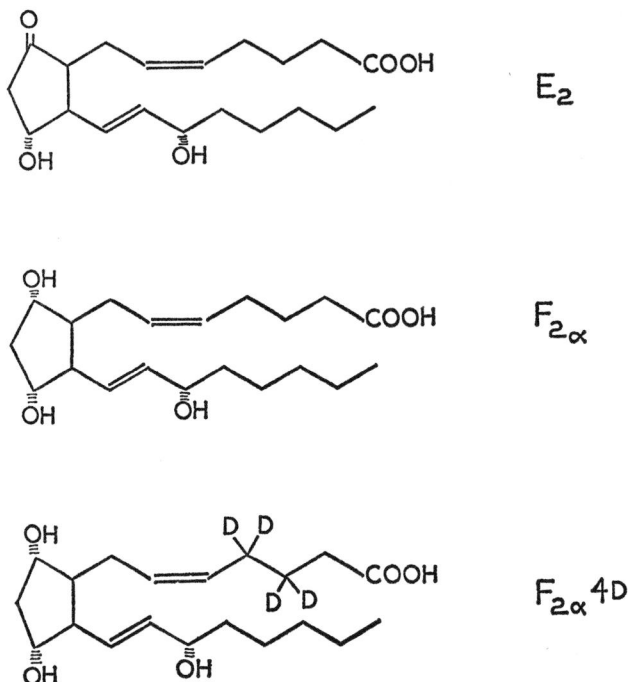

FIG. 1   Prostaglandins E_2, $F_{2\alpha}$ and 3,3,4,4-tetradeutero $F_{2\alpha}$.

the chance of other compounds interfering. The derivative of prostaglandin $F_{2\alpha}$ we have chosen is the methyl ester-9,11-butylboronate-15-trimethyl silyl ether (Fig. 2). Two mass spectrometer–gas chromatograph combinations have been used in these studies, first was an A.E.I. MS12 spectrometer coupled through a Watson Bieman separator to a Hewlett Packard 402 gas chromatograph. The second combination was a Dupont 490B spectrometer coupled through a single-stage all-glass jet separator to a Varian 1400 gas chromatograph. The 490B spectrometer was fitted with extra stability power supplies for the magnet and accelerating voltage. Both instruments were equipped with devices for switching the accelerating voltage between preset

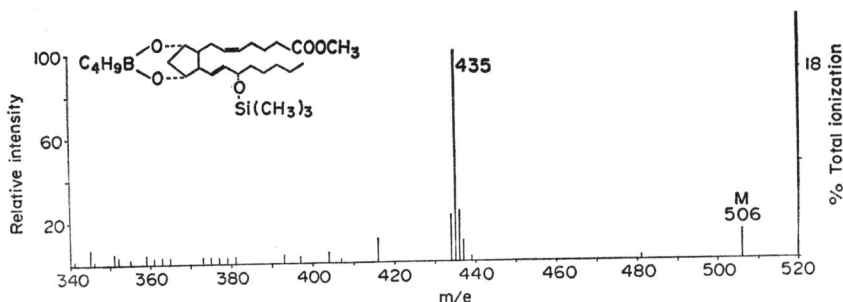

FIG. 2   Upper position of the mass spectrum of methyl ester-9,11-butylboronate-15-trimethyl silyl ether of prostaglandin $F_{2\alpha}$.

values and multiplexing the output from the current amplifier attached to the electron multiplier to give a smooth analogue signal for each ion followed.[4] The multiplexed signals could be heavily smoothed to give a trace of the same quality as that derived from a conventional ionization detector. The current amplifier which was used between the electron multiplier and the multiplexing device was a varactor bridge amplifier (Analogue devices 310 J) with a 1000 MΩ feedback resistor. This combination gave high sensitivity with acceptable bandwidth.

In a typical analysis of a plasma sample, a known amount of deuterated prostaglandin $F_2\alpha$ was added to the plasma and thoroughly mixed with it. The plasma is then extracted with ether-ethylacetate and the extract dried. The

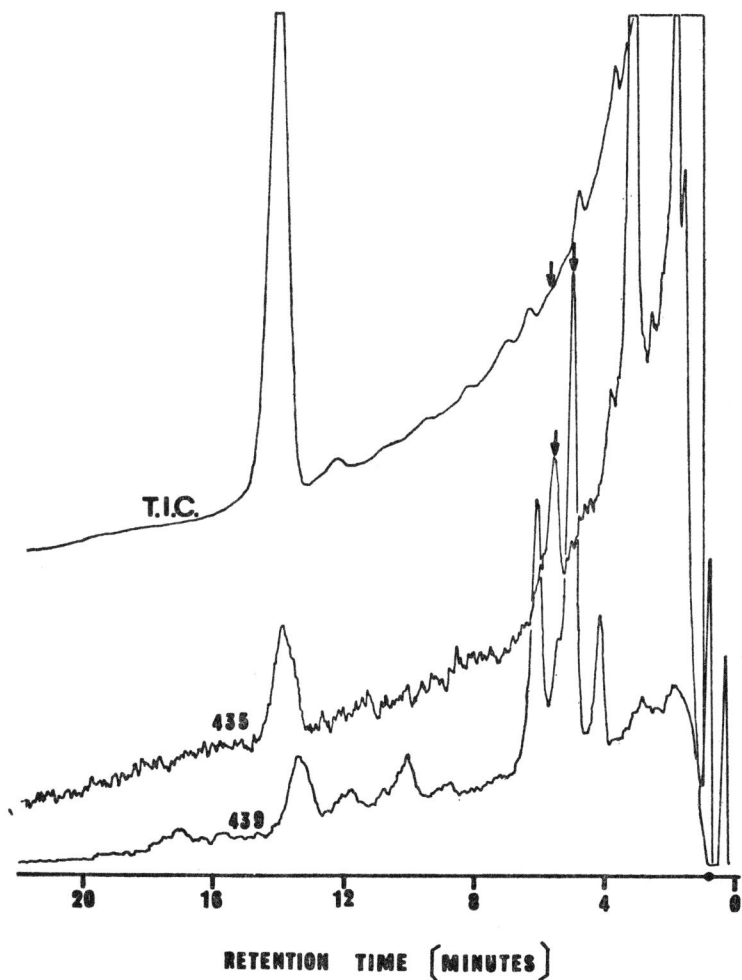

FIG. 3   Typical analysis of a sheep plasma containing 1 ng/ml of prostaglandin $F_2\alpha$. $F_2\alpha$ is followed at m/e 435 while the tetradeutero standard is followed at m/e 439. Arrows show the single point in time at which the $F_2\alpha$ and the tetradeutero $F_2\alpha$ emerge. The peaks do not appear coincident because of the offset of the recorder pens.

extract is derivatized first by treatment with diazo methane to form the methyl ester then by the addition of butyl boronic acid solution followed by a slow distillation of the solvent to form the butyl boronate. Finally *bis* (trimethyl silyl) acetamide (BSA) is added to form the 15 silyl ether. The derivatized mixture in 5–20% BSA is injected directly into the GC–MS combination and the ions at m/e 435 (characteristic of $F_2\alpha$) and m/e 439 (characteristic of tetradeutero $F_2\alpha$) are followed. Quantitation is achieved by measuring the heights of the gas chromatography peaks and taking the ratio. The amount of prostaglandin in the original mixture can be calculated from the measured ratio and the known amount of deuterated material added. The trace from the analysis of a plasma sample with 1 ng/ml prostaglandin $F_2\alpha$ is shown in Fig. 3. A measure of the specificity is seen from an analysis of sheep uterine fluid in Fig. 4. The Total Ion Current (TIC) in this trace shows a large solvent tail and many other peaks beside the $F_2\alpha$ peak, whereas the $F_2\alpha$ selective trace shows one clean peak.

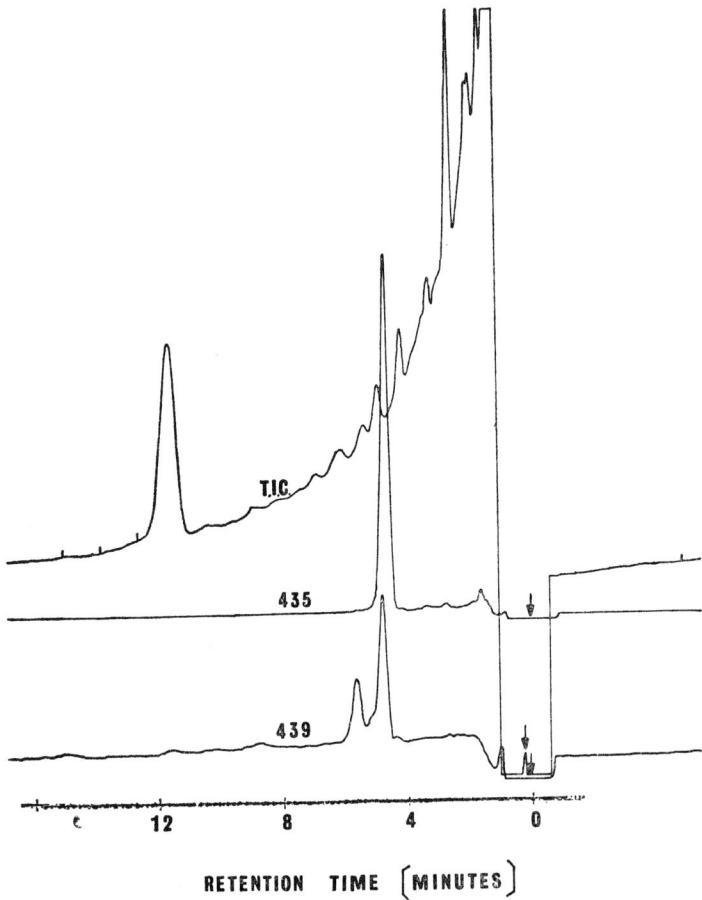

RETENTION TIME [MINUTES]

FIG. 4 An analysis of sheep uterine fluid, arrows denote injection point. The ions monitored are the same as in Fig. 3.

It is important to determine the linearity of the mass spectrometer detector in this type of analysis and to check the response, mixtures were made of 40-ng aliquots of tetradeutero $F_2\alpha$ and varying amounts (from 0·2 to 200 ng) of natural prostaglandin $F_2\alpha$. The mixtures were derivatised and analysed and the results are shown in Fig. 5. The solid triangles denote deviation from linearity when no correction is made for the contribution of natural $F_2\alpha$ to the m/e 439 channel where the tetradeutero material registers. This contribution is about 8 % and if this correction is not made low values are recorded because the tetradeutero response is apparently increased. However, when

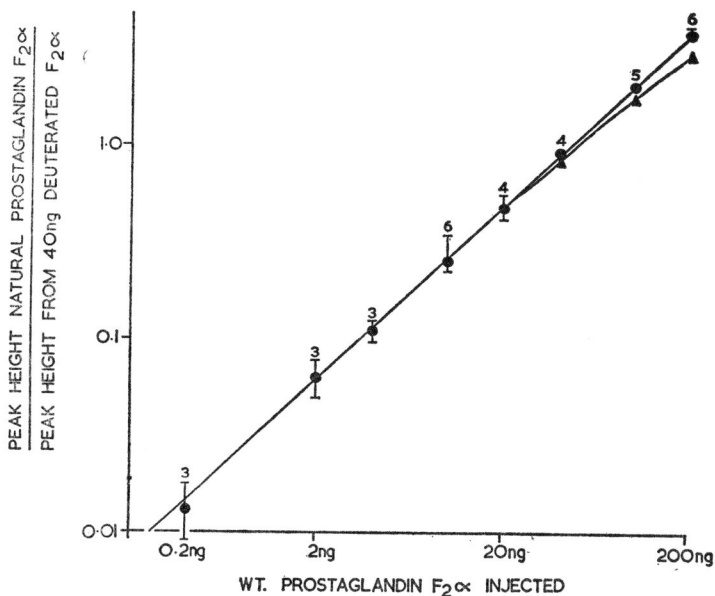

FIG. 5   Linearity of the response of the mass spectrometer detector to prostaglandin $F_2\alpha$. The triangular points are explained in the text.

this correction is made (solid circles) linearity is restored over the full 1000 to 1 dynamic range. The precision in this type of analysis was not as good as might have been expected from an isotope ratio measurement; with samples of less than 10 ng the precision was rarely above 7 % (relative standard deviation) and this value applied equally to the two instrumental combinations. The precision improved with increasing concentration of prostaglandin; for a sheep uterine fluid with 200 ng/ml the precision was 1·3 %.

Two approaches to an increase in precision are possible, one would be to measure peak areas instead of peak heights and the other would be to increase the speed of switching of the accelerating voltage to ensure the maximum sampling of the apex of the GC peak. However, Sweeley et al.[5] have found that the use of peak area measurements actually gives lower precision than peak height measurements. Increasing the switching speed of the accelerating voltage from 0·5 sec per ion to 0·2 sec per ion did not increase the precision.

The imprecision at low nanogram levels is most probably due to the background due to the column bleed; in an experiment to determine the precision of measurement of a series of 5 ng samples the signal-to-noise ratio (peak to peak) was 20:1. The precision in these measurements was 5·3% ($n = 10$) and the noise is sufficient to account for a large proportion if not all of the variation in results. It is the imprecision of the measurements at low prostaglandin levels which places the lower limit on the sensitivity, and it is the problem of precision which has to be tackled to bring these analyses into the low picogram ranges.

ACKNOWLEDGMENTS

The prostaglandins used in this study were the gift of Drs Pike and Axen of the Upjohn Company.

REFERENCES

1. Sweeley, C. C., Elliot, W. H., Fries, I. and Ryhage, R. *Anal. Chem.*, 1966, **38**, 1549.
2. Samuelsson, B., Hamberg, B. M. and Sweeley, C. C. *Anal. Biochem.*, 1970, **38**, 301.
3. Axen, U., Green, K., Horlin, D. and Samuelsson, B. *Biochem. Biophys. Res. Commun.*, 1971, **45**, 519.
4. Kelly, R. W. *J. Chromatog.*, 1972, **71**, 337.
5. Holland, J. F., Sweeley, C. C., Thrush, R. E., Teets, R. E. and Bieber, M. A. *Anal. Chem.*, 1973, **45**, 308.

22

Mass Spectrometry of Peptide Mixtures

By P. ROEPSTORFF and K. BRUNFELDT

(The Danish Institute of Protein Chemistry, Venlighedsvej 4, Hørsholm, Denmark)

INTRODUCTION

Mass spectrometry has for the last decade been recognized as a potential method for the sequence determination of proteins. But the number of reported applications for protein sequencing has been rather limited. This may in part be due to lack of familiarity with or availability of the necessary equipment. Furthermore, when dealing with isolated peptides, the method does not offer any clear-cut advantages such as greater speed or sensitivity as compared with the dansyl-Edman technique.

Mass spectrometry, however, does offer the possibility of carrying out the simultaneous sequence determination of single peptides in a mixture. This was first described by McLafferty and co-workers[1,2] and has since been investigated by our own group,[3,4] by Morris *et al.*,[5,6] and by Stehelin.[7] The advantages of peptide mixture analysis are obvious, first of all, because the time-consuming fractionation and purification procedures prior to sequence determination can be considerably reduced and thereby the loss of sample minimized. Furthermore, the number of derivation procedures needed for analysing an enzymatic hydrolysate of a protein can be reduced.

GENERAL CONSIDERATIONS

The choice for the method for derivation of the peptide mixture was directed by the necessity of obtaining all the sequence-determining ions in a reasonable abundance and with a minimum of secondary fragmentation of the peptides in the mass spectrometer. Acetylation and permethylation were found to be the most advantageous in this respect.

When interpreting the mass spectra of peptide mixtures several possibilities for ambiguities can be foreseen, even when considering only the sequence peaks. Some examples are presented in Table I. The alternative sequences in groups A and B in Table I are identical in elementary composition with the correct sequences and thus can not be eliminated by high resolution mass spectrometry, whereas the alternative sequences in groups C and D can be. The ambiguities in groups A, C and D depend on the derivative used and can be eliminated by comparing the results obtained with two different derivatives of the same sample.[4] In certain cases knowledge of the amino acid composition of the mixture may even be sufficient to eliminate such ambiguities. The alternative sequences in group B are independent of the derivative used, and generally, the amino acid analysis data are not sufficient for a correct interpretation. These alternatives are always due to a cross-over from one peptide chain to another during the interpretation. This may take place when two peptides with different sequences contain the same amino acid residues or the same amino acid residues except for one up to a certain distance from the N-terminal. Certain combinations of homologous amino

TABLE I

Examples of Ambiguities in Mass Spectrometry of Permethylated Peptide Mixtures. When the Sequences in the First Column are Present, the Sequences in the Second Column are also Deduced

	Actual sequence		*Alternative sequence*
(A)	—MeGly—MeLeu	→	—ε—Ac—Me$_2$Lys—
	—MeAla—MeVal—	→	—ε—Ac—Me$_2$Lys—
	—MeGly—MeVal—	→	—δ—Ac—Me$_2$Orn—
	—MeAla—MeAla—	→	—Me$_3$Gln—
	—MeAla—MeGly—	→	—Me$_3$Asn—
(B)	Ac—MeLeu—MeAla—	→	Ac—MeAla—MeLeu—
	Ac—MeAla—		
	Ac—MeGly—Me$_2$Glu—	→	Ac—MeAla—Me$_2$Asp—
	Ac—MeAla—		
(C)	—MeAla—Me$_2$Thr—	→	—Me$_2$Try—
(D)	Ac—MeLeu—MeMet—	→	Ac—Me$_2$Asp—Me$_2$Thr—
	Ac—Me$_2$Asp—		

acid residues may also cause type B alternatives. The performance of one cycle of Edman degradation before the acetylation and permethylation abolishes this identity, except in the case where the N terminals of the two peptides are identical. In certain cases partial vaporization from the solids inlet probe or analysis of metastable ions also allows elimination of type B alternatives.

Most of the erroneous sequences obtained by interpretation of the spectra of peptide mixtures are, however, due to non-sequence peaks or impurity peaks, and not to the above-mentioned combinations of sequence peaks. These erroneous sequences may often be eliminated by mass spectrometric considerations. Also a comparison of the results obtained with two different derivatives of the sample normally allows elimination of these sequences.[4] In order to reduce the number of erroneous sequences of this type, a threshold level must be introduced so that peaks of very low intensity are eliminated. The threshold level cannot be the same throughout the spectrum as the intensity of the sequence determining peaks decreases by a factor of 2–10 from one sequence peak to the next at higher mass.[8] Therefore, a stepwise decreasing threshold level has been used. This is illustrated in Fig. 1, where

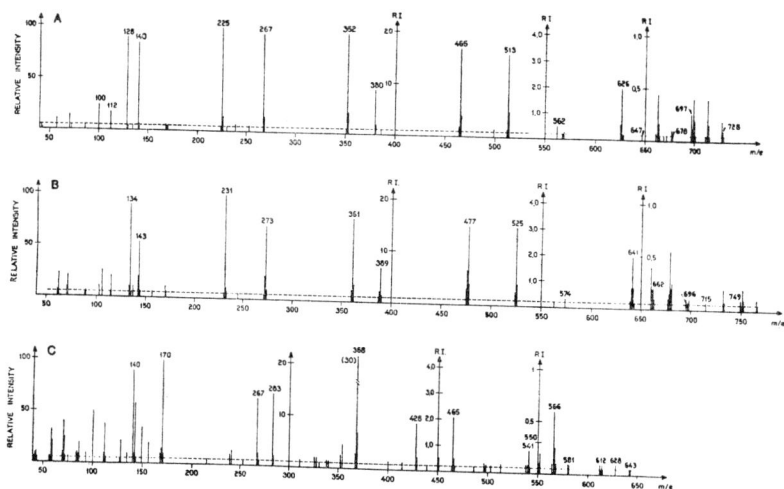

FIG. 1 Mass spectra of a mixture of H-Ala-Pro-Leu-Phe-Val-Gly-OH and H-Pro-Leu-Val-Ala-Pro-Ala-OH. A, acetylated and permethylated, B, deuteroacetylated and deuteropermethylated and C, Edman degraded, acetylated, and permethylated. The masses indicated correspond to sequence ions and molecular ions. The dotted line indicates the threshold level. The mass spectra are drawn relative to the threshold.

the spectra are drawn relative to the threshold level resulting in an intensity-to-threshold ratio of between 20 and 10 except for one sequence peak and the molecular ion region. Another possibility for eliminating some of the smaller peaks is to correct for the ^{13}C-contribution; an example of this will be given below.

EXPERIMENTAL EXAMPLE

In the following is illustrated the effect of introducing the parameters described above in the interpretation of the spectra.

An approximately equimolar mixture of two synthetic peptides, H-Ala-Pro-Leu-Phe-Val-Gly-OH and H-Pro-Leu-Val-Ala-Pro-Ala-OH, was used. The mixture was divided into three parts, part A was acetylated and permethylated as described by Thomas,[9] part B deuteroacetylated and deuteropermethylated, and part C submitted to one cycle of Edman degradation according to Gray[10] and then acetylated and permethylated. The mass spectra obtained are shown in Fig. 1A, B and C respectively.

The interpretation of the mass spectra was carried out by locating peaks possibly corresponding to N-terminal acyl-amino acid residues and from these peaks deducing the possible sequences by trial and error. We have developed a computer program for the interpretation of low resolution mass spectra of peptide mixtures very similar to that described by Kiryushkin et al.[11] The program is used off-line to print out the possible sequences from the different mass spectra. It was found necessary to use a computer, because a manual interpretation is too complicated to carry out when peptide mixtures are analysed. A manual interpretation of the spectrum shown in Fig. 1A accomplished by one of our post-graduate students without prior knowledge of the content of the mixture resulted in an incomplete list of possible peptides and missed one of the correct sequences.

Due to the alternatives listed in Table I the acetylated and permethylated mixture should be expected to give rise to the possible sequences listed in Table II. The table is a computer print-out obtained with an input consisting of a complete list of the permethylated amino acid residues and a reduced mass spectrum containing only the sequence peaks and the molecular peaks. The deuteroacetylated and the deuteropermethylated sample gives rise to a list of similar length and the sample submitted to Edman degradation to a somewhat shorter list. As seen from Table II several alternatives of the type listed in Table I (A and B) are found and a few of the type listed in D, whereas no alternatives of type C are present in this example.

In spite of the rather simple mass spectra the interpretation of the spectra as shown in Fig. 1 resulted in considerably longer lists of possible peptides. The upper curve in Fig. 2 illustrates the number of possible peptides obtained by the interpretation of the acetylated and permethylated sample. The interpretation was based upon the presence of a peak, without considering intensities, possible chain terminations, etc. The curve demonstrates an exponential growth of the number of possibilities as a function of the peptide chain length up to the pentapeptide stage. At the hexapeptide stage the number of possibilities decreases because a continuation of many of the pentapeptide sequences would require peaks beyond the present mass spectrum.

Figure 2 also shows the reduction in the number of possibilities obtained by introduction of different parameters in the interpretation. Reduction of the intensities of the peaks with the ^{13}C-contribution from fragments corresponding to preceding peaks resulted in the elimination of several small peaks, and thus, in a certain, but not sufficient reduction in the number of

TABLE II

Number of Possible Peptides Expected Due to Sequence Peaks and Molecular Peaks in the Mass Spectrum of the Acetylated and Permethylated Mixture. The Intensities and the m/e Values of the Corresponding Sequence-peaks are Indicated Above and Below the Possible Sequences Respectively. * Signifies the Correct Sequences.

50	900	1000	970	170	49	2	1
AC	ALA	PRO	LEU	VAL	PRO	ALA	OME
43	128	225	352	465	562	647	678
50	900	1000	970	170	200	3	1
AC	ALA	PRO	LEU	VAL	PHE	GLY	OME
43	128	225	352	465	626	697	728
50	900	1000	970	160	200	3	1
*AC	ALA	PRO	LEU	PHE	VAL	GLY	OME
43	128	225	352	513	626	697	728
50	900	1000	970	160	1		
AC	ALA	PRO	LEU	PHE	HIS		
43	128	225	352	513	678		

Continued on following page

50 AC 43	900 ALA 128	1000 PRO 225	970 LEU 352	160 PHE 513	3 ORN 697	1 OME 728	
50 AC 43	850 PRO 140	1000 ALA 225	970 LEU 352	170 VAL 465	49 PRO 562	2 ALA 647	1 OME 678
50 AC 43	850 PRO 140	1000 ALA 225	970 LEU 352	170 VAL 465	200 PHE 626	3 GLY 697	1 OME 728
50 AC 43	850 PRO 140	1000 ALA 225	970 LEU 352	160 PHE 513	200 VAL 626	3 GLY 697	1 OME 728
50 AC 43	850 PRO 140	1000 ALA 225	970 LEU 352	160 PHE 513	1 HIS 678		
50 AC 43	850 PRO 140	1000 ALA 225	970 LEU 352	160 PHE 513	3 ORN 697	1 OME 728	
50 AC 43	850 PRO 140	950 LEU 267	970 ALA 352	170 VAL 465	49 PRO 562	2 ALA 647	1 OME 678
50 AC 43	850 PRO 140	950 LEU 267	970 ALA 352	170 VAL 465	200 PHE 626	3 GLY 697	1 OME 728
50 AC 43	850 PRO 140	950 LEU 267	970 ALA 352	160 PHE 513	200 VAL 626	3 GLY 697	1 OME 728
50 AC 43	850 PRO 140	950 LEU 267	970 ALA 352	160 PHE 513	1 HIS 678		
50 AC 43	850 PRO 140	950 LEU 267	970 ALA 352	160 PHE 513	3 ORN 697	1 OME 728	
50 *AC 43	850 PRO 140	950 LEU 267	430 VAL 380	170 ALA 465	49 PRO 562	2 ALA 647	1 OME 678
50 AC 43	850 PRO 140	950 LEU 267	430 VAL 380	170 ALA 465	200 PHE 626	3 GLY 697	1 OME 728
50 AC 43	850 PRO 140	950 LEU 267	170 LYS 465	49 PRO 562	2 ALA 647	1 OME 678	
50 AC 43	850 PRO 140	950 LEU 267	170 LYS 465	200 PHE 626	3 GLY 697	1 OME 728	

P. ROEPSTORFF AND K. BRUNFELDT

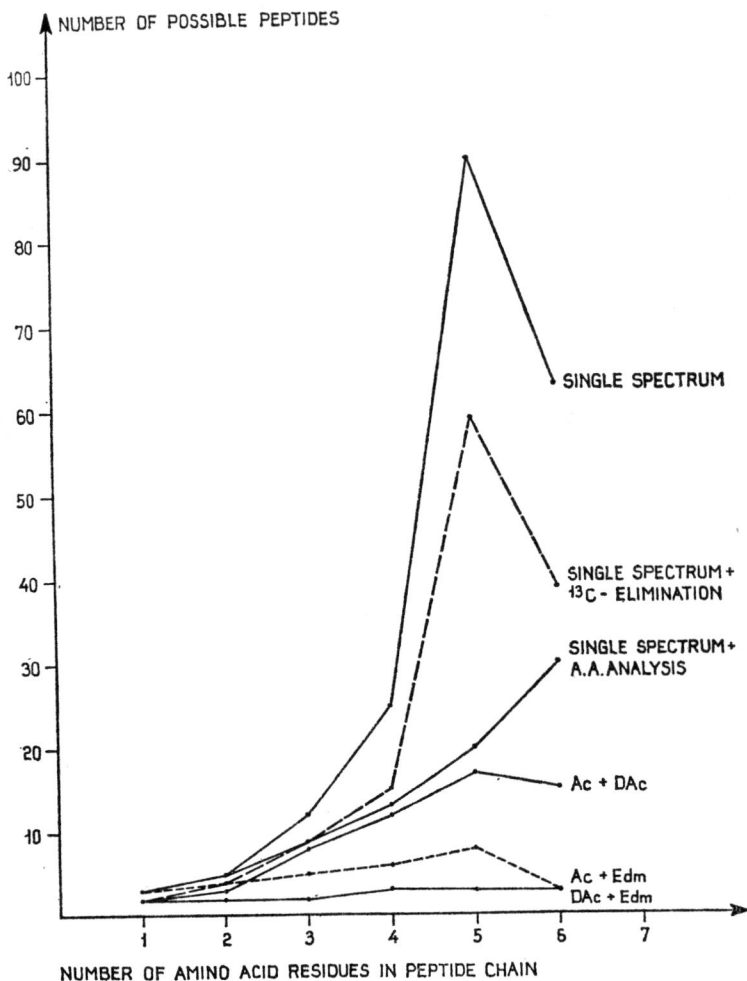

FIG. 2   Number of possible peptides obtained by interpretation of the mass spectra shown in Fig. 1. Ac indicates the acetylated and permethylated mixture, DAc the deuteroacetylated and deuteropermethylated mixture and Edm the Edman degraded, acetylated and permethylated mixture.

possibilities. To a certain extent introduction of the amino acid analysis data also reduces the number of possibilities.

A combination of the results obtained from the acetylated and permethylated sample with those from the deuteroacetylated and deuteropermethylated sample reduced the number of possibilities to the two correct sequences, a series of type B alternatives and three more sequences due to non-sequence peaks, which were multiplied due to type B alternatives. These latter erroneous sequences, which were not terminated, could be eliminated by knowledge of the amino acid analysis or by mass spectrometric considerations. The two peptides in the mixture could not be distinguished from one another by partial vaporization, thus this did not allow an elimination of the type B alternatives. However, examination of metastable ions did allow an elimination of some of them, especially those where the cross-over took place close to the N-terminal; but still seven alternative sequences were possible.

A combination of the results obtained with two acetylated and permethylated samples, one being submitted to one cycle of Edman degradation before the derivation, resulted in elimination of the type B alternatives but not the type A. Furthermore, eight sequences were possible due to non-sequence peaks. None of these were terminated, and all of them could be eliminated by mass spectrometric considerations, as well as by knowledge of the amino acid composition, which in this case also was sufficient to eliminate the type A alternatives.

Finally, a combination of the results obtained with the Edman degraded, acetylated, and permethylated samples and the deuteroacetylated and deuteropermethylated sample resulted in four possible peptides, the two correct ones, a non-terminated pentapeptide, and a non-terminated hexapeptide. The two latter ones could easily be eliminated by either using amino acid analysis data, or by correction for the ^{13}C-contribution, or by mass spectrometric considerations.

DISCUSSION

The previous example demonstrated the reduction of the number of possible sequences by the introduction of various parameters in the interpretation. It was seen that the use of Edman degradation was necessary in order to allow a correct interpretation because type B alternatives were present. A combination of Edman degradation and deuterated derivatives further reduced the number of possibilities. The use of two derivatives combined with either amino acid analysis data or mass spectrometric considerations has been sufficient to ensure a correct sequence determination of the peptides in the mixtures examined so far in our laboratory. Thus in these cases a minor amount of the information already available or obtainable has been found sufficient for a correct interpretation.

When analysing complex mixtures obtained by enzymatic cleavage of proteins one might expect that much more information would be needed in order to ensure a correct interpretation. First of all a knowledge of how the mixture is produced yields some information. Furthermore, since the length of peptides presently amenable to mass spectrometry is about ten residues, a certain fractionation according to mass is necessary. Large peptides can be separated from the rest and digested by another enzyme. Gel filtration has been suggested for this separation,[5] and hence the mixture may also conveniently be fractionated into a number of less complicated mixtures. These samples may be fractionated further by partial vaporization from the solids inlet probe of the mass spectrometer.[3,12] As a result the mixtures dealt with during the interpretation of the mass spectra will contain a rather limited number of peptides. It therefore seems possible to apply mass spectrometric peptide mixture analysis as described above directly to the sequence determination of proteins without the necessity of introducing further mass spectrometric or chemical parameters. A useful supplement, however, might be information about the number of components in the mixture and their molecular weight. This information seems readily obtainable by field desorption mass spectrometry, as the application of this ionization method to mass spectrometry of acetylated and permethylated peptides has been reported to result only in the formation of molecular ions.[13]

REFERENCES

1. Van Lear, G. E. and McLafferty, F. W. *Ann. Rev. Biochem.*, 1969, **38**, 289.
2. McLafferty, F. W., Venkataraghavan, R. and Irving, P. *Biochem. Biophys. Res. Commun.*, 1970, **39**, 274.
3. Roepstorff, P., Spear, R. K. and Brunfeldt, K. *FEBS Letters*, 1971, **15**, 237.
4. Roepstorff, P. and Brunfeldt, K. *FEBS Letters*, 1972, **21**, 320.
5. Morris, H. R., Williams, D. H. and Ampler, R. P. *Biochem. J.*, 1971, **125**, 189.
6. Morris, H. R. *FEBS Letters*, 1972, **22**, 257.
7. Stehelin, D. Thesis, Université L. Pasteur, Strasbourg, 1972.
8. Lederer, E. *Pure Appl. Chem.*, 1968, **17**, 489.
9. Thomas, D. W. *Biochem. Biophys. Res. Commun.*, 1968, **33**, 483.
10. Gray, W. R. *Methods Enzymol.*, 1972, **25**, 333.
11. Kiryushkin, A. A., Fales, H. M., Axenrod, T., Gilbert, E. J. and Milne, G. W. A. *Org. Mass Spectrometry*, 1971, **5**, 19.
12. Brunfeldt, K., Christensen, T. and Roepstorff, P. *FEBS Letters*, 1972, **25**, 184.
13. Winkler, H. U. and Beckey, H. D. *Biochem. Biophys. Res. Commun.*, 1972, **46**, 391.

23

Mass Spectral Studies of Insect Secretions

By D. E. GAMES, A. H. JACKSON, D. S. MILLINGTON
and B. W. STADDON

(*Departments of Chemistry and Zoology, University College, Cardiff, U.K.*)

COMBINED gas chromatography/mass spectrometry provides an excellent method for the study of the secretions of insects, and we have now utilized this technique in our studies of the secretions of the milkweed bug, *Oncopeltus fasciatus*.[1]

The larva of the milkweed bug releases secretions through dorso-abdominal glands and the secretions from the anterior and posterior glands were analysed by GC-MS. Figure 1 shows the total ion current monitor trace obtained for the whole secretion and the GC-MS and GC results obtained from the anterior and posterior gland secretions are summarized in Table I. Components A, C, D and F were readily identified from their mass spectra and the spectra of components of D and F were in agreement with those reported previously for these compounds.[2,3] Poor electron-impact mass spectra were obtained for components B and E, resulting in a tentative identification of B only. The identities of A and C were confirmed by comparison (GC retention time, and mass spectra) with authentic specimens and that of D by comparison with a sample of *trans*-4-oxohex-2-en-1-al obtained from *Sigara falleni*.[2b]

In the adult milkweed bugs the scent gland complex is situated ventrally, in the hind part of the thorax. The secretions stored in the tubular glands and median reservoirs of male bugs and the whole secretions of male and female bugs were studied by GC and GC-MS. The total ion current monitor traces obtained for the female secretion and the male tubular gland secretion are shown in Fig. 2, and the results obtained from GC-MS examination of these and other secretions are given in Table II. Examination of the mass spectra of the components enabled ready identification of most of the more volatile constituents; however, a number of less volatile compounds remain to be identified and work is in hand to complete these identifications. The identities of components B, C, D, E, F and H were confirmed by comparisons with authentic specimens (GC retention times and MS). Marked differences in the compositions of the secretions were observed, the female secretion being composed mainly of unsaturated aldehydes with only traces of unsaturated acetates, whereas the total male secretion contained the same aldehydes together with a corresponding series of acetates. Examination of the secretions of the individual male glands showed that the tubular gland contained almost

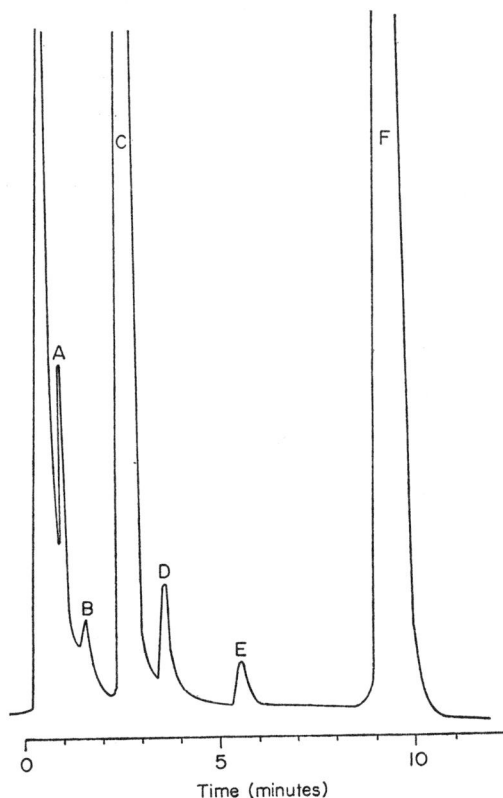

Fɪɢ. 1   Total ion current monitor trace of the whole secretion of *Oncopeltus fasciatus* larva (glass column, 2 m × 2 mm i.d. packed with 3% OV225 on gas chrom.Q 60°C).

TABLE I

Composition of the Secretions from the Larva of *Oncopeltus fasciatus*

Peak	Compound	Mass spectrum (m/e % abundance)[a]	Per cent composition	
			Anterior gland	Posterior gland
A	*trans*-hex-2-en-1-al	41(100), 42(58), 43(39), 55(50) 57(61), 69(67), 83(92), 98(57)	5	0·5
B	hept-2-en-1-al?		1	0·5
C	*trans*-oct-2-en-1-al	41(100), 42(42), 55(93), 57(58) 69(32), 70(75), 82(43), 83(67)	87	15
D	*trans*-4-oxohex-2-en-1-al	41(20), 55(51), 56(7), 57(21) 83(100), 84(15), 112(25)	—	2
E	Unidentified		—	1
F	4-oxo-oct-2-en-1-al	41(47), 55(76), 57(35), 70(45) 83(67), 84(27), 98(100), 111(91)	7	81

[a] Eight largest peaks above m/e 40, 70 eV, ion source 180°C.

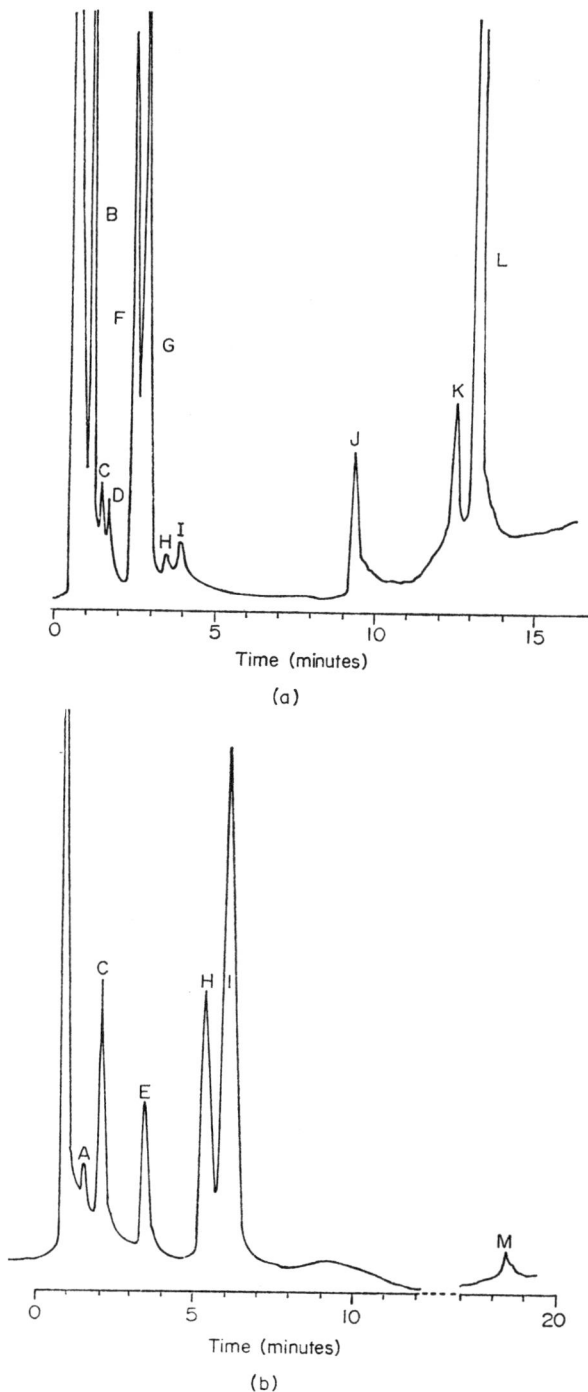

FIG. 2 (a) Total ion current monitor trace of the whole secretion of female adult *Oncopeltus fasciatus* (glass column, 2 m × 2 mm i.d. packed with 3% OV225 on gas chrom. Q 75°C for 5 min, then temperature programmed to 210°C at 14°/min). (b) Total ion current monitor trace of the tubular gland secretion of male adult *Oncopeltus fasciatus* (glass column, 2 m × 2 mm i.d. packed with 3% OV225 on gas chrom. Q, 70° for 8 min, then temperature programmed to 210° at 10°/min).

TABLE II

Composition of the Secretions from Adult Male and Female *Oncopeltus Fasciatus*

Peak	Compounds	Mass spectrum (m/e % abundance)[b]	Male		Female
			Tubular gland	Whole secretion	Whole secretion
A	Unidentified	See Table I	a	a	a
B	trans-hex-2-en-l-al	41(18), 43(100), 55(18), 57(16), 67(44), 71(8), 82(36), 100(18)	a	a	a
C	trans-hex-2-enyl acetate	41(47), 53(35), 65(19), 66(8), 67(30), 81(100), 95(15), 96(45)			
D	hexa-2,4-dienal	41(20), 43(93), 77(34), 79(100), 80(48), 81(22), 98(12), 140(32)		a	a
E	hexa-2,4-dienyl acetate	See Table I	a	a	
F	trans-oct-2-en-l-al	41(64), 54(38), 55(100), 67(50), 70(42), 80(52), 81(72), 95(42)		a	a
G	octa-2,4-dienal	43(100), 54(47), 55(24), 67(54), 68(56), 81(64), 82(41), 110(40)		a	a
H	trans-oct-2-enyl acetate	41(15), 43(100), 54(33), 67(35), 79(20), 80(12), 82(12), 93(18)	a	a	a
I	octa-2,4-dienyl acetate	50(16), 51(27), 53(91), 54(17), 69(11), 81(93), 82(24), 110(100)	a	a	a
J	Unidentified	41(100), 53(49), 70(64), 77(47), 79(64), 81(65), 91(61), 94(48)			
K	Unidentified	41(100), 53(28), 68(42), 69(72), 70(95), 79(42), 81(31), 91(28)			a
L	Unidentified	41(8), 43(16), 55(24), 73(3), 97(100), 98(9), 115(4), 157(5)			a
M	Unidentified		a		

[a] Compound present in secretion.

[b] Eight largest peaks above m/e 40, 70 eV, ion source 180°C.

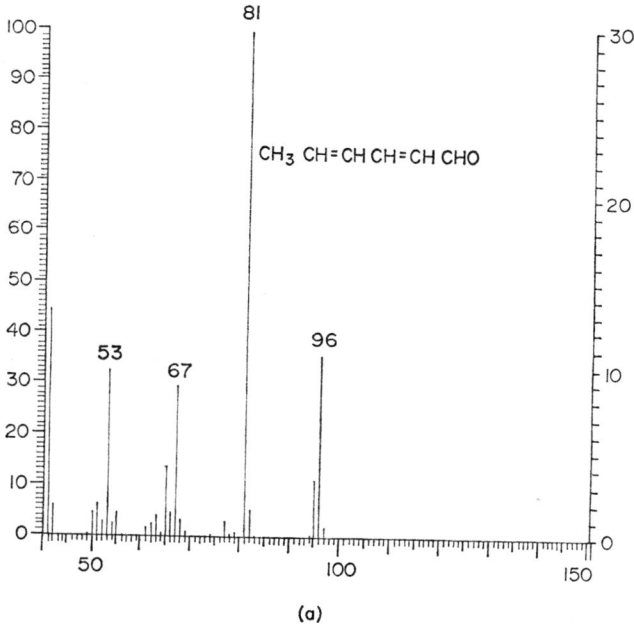

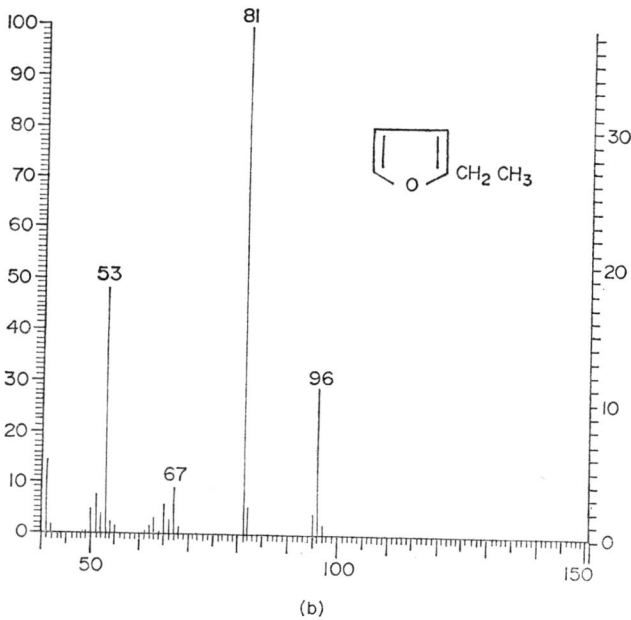

FIG. 3 EI spectra of (a) hexa-2,4-dienal (70 eV, Source 200°). (b) 2-ethylfuran (70 eV, Source 200°).

exclusively unsaturated acetates, and the median reservoir contained unsaturated aldehydes, with traces of unsaturated acetates, the two secretions combining to form the total male secretion.

Mass spectral identification of some of the compounds present in the secretions is complicated by the similarity of their mass spectra to those of their corresponding cyclic isomers. McFadden and Buttery[4] have discussed the difficulties encountered in the mass spectral identification of flavour and aroma components due to similarities in the spectra of alkadienals and alkyl furans. Our studies confirm this in the case of 2-ethylfuran and hexa-2,4-dienal which have very similar spectra (see Fig. 3); however, less similarity was apparent in the cases of 2-n-butylfuran and octa-2,4-dienal. The low resolution mass spectra of hexa-2,4-dienyl and trans-hex-2-enyl acetates and their corresponding alcohols also show similarities with those of cyclohex-2-enyl and cyclohexanyl acetates and their corresponding alcohol respectively. However, in all cases comparison of the eight major peaks above m/e 40 in the mass spectra of the compounds enable distinction to be made between them.

We have commenced a study of the high resolution and metastable (DADI)[5] spectra of these compounds to see if the similarities observed in low resolution spectra are extended to these spectra. Initially the spectra of hexa-2,4-dienal and 2-ethylfuran have been compared and Table III summarizes the high resolution and metastable data obtained for these compounds. The measurements confirm McFadden and Buttery's suggestion[4] that the mass spectral fragmentations of compounds of these types proceed through common ions.

A further problem in the mass spectral identification of the components of the secretions is that in a number of cases the molecular ions are absent or are of low intensity. We have commenced a study of the FI spectra of compounds of the type present in the secretions with a view to using the

TABLE III

Metastable and High Resolution Data for 2-Ethylfuran and Hexa-2,4-dienal

Ion (m/e)	Composition	Metastable	Hexa-2,4-dienal (rel. intensity %)	2-Ethylfuran (rel. intensity %)
		$96 \rightarrow 95$	0·006 3	0·006 5
		$96 \rightarrow 81$	0·025	0·025
95	C_6H_7O	$95 \rightarrow 80$		0·004 0
		$95 \rightarrow 77$	0·010	0·004 0
		$95 \rightarrow 67$	0·010	0·009 0
81	C_5H_5O	$81 \rightarrow 53$	0·000 9	0·001 5
67	C_5H_7	$67 \rightarrow 66$	0·020	0·020
		$67 \rightarrow 65$	0·054	0·060
65	C_5H_5	$65 \rightarrow 64$	0·021	0·024
		$65 \rightarrow 63$	0·046	0·044
53	C_4H_5	$53 \rightarrow 51$	0·002 4	0·004 1
		$41 \rightarrow 40$	0·006 6	0·010
		$41 \rightarrow 39$	0·002 9	0·003 4

technique, either directly on the insect secretions or in combination with GC. FI spectra of hexa-2,4-dienal, *trans*-hex-2-enyl acetate, hexa-2,4-diene-1-ol and *trans*-hex-2-en-1-ol have been obtained; all had their molecular ions as their base peaks and few or no other ions. Other model compounds are currently being studied and it is hoped to apply the techniques to the study of insect secretions in the near future.

Mass spectra were obtained in line diagram and mass list form with background subtracted using a Varian 620i computer on-line to a Varian CH5D mass spectrometer.

We thank the S.R.C. for financial assistance towards the purchase of the mass spectrometer and for financial support for one of us (D.S.M.). We are indebted to Proprietary Perfumes Limited (Ashford, Kent) for samples of some of the chemicals, to Mrs M. J. Evans for the preparation of 2-ethyl- and 2-butylfuran, and to Mr M. Rossiter for running the DADI and high resolution spectra.

REFERENCES

1. Games, D. E. and Staddon, B. W., *Experientia*, 1973, **29**, 532.
2. (*a*) Gilby, A. R. and Waterhouse, D. F., *Proc. Roy. Soc., B*, 1965, **162**, 105.
   (*b*) Pinder, A. R. and Staddon, B. W., *J. Chem. Soc.*, 1965, p. 2955.
3. Calam, D. H. and Youdeowei, A., *J. Insect Physiol.*, 1968, p. 1147.
4. McFadden, W. H. and Buttery, R. G., *in* 'Topics in Organic Mass Spectrometry', *ed.* A. L. Burlingame, Wiley-Interscience, New York, 1970, p. 327.
5. Richter, W. J., Liehr, J. G. and Schulze, P., *Tetrahedron Letters*, 1972, p. 4503.

24

Gas Chromatography–Mass Spectrometry in Degradation and Structure Determination of Porphyrins and Bile Pigments

By A. H. JACKSON, D. S. MILLINGTON and
D. E. GAMES

(Department of Chemistry, University College, Cardiff, U.K.)

MODERN spectroscopic and analytical methods enable most of the structural features of porphyrins and related compounds to be determined provided sufficient material is available. However, many of the intermediates in normal, or abnormal metabolism are present at very low concentrations and it may only be possible to isolate sub-milligram quantities of material for characterization. We have been concerned with the development of methods of structure determination which can be applied to such small amounts of material.

Low and high resolution mass spectra of porphyrins normally enable identification of peripheral substituents to be made; however, their order on the porphyrin nucleus is not delineated. Confirmation of the presence of particular substituents and some information about their order is obtainable by oxidative degradation of the porphyrins to maleimides using chromic acid[1] or to pyrrole-2,5-dicarboxylic acids using permanganate.[2] Mild chromic acid degradation and degradation with dichromate/hydrogen sulphate have also had considerable application in bile pigment structure determination.[3] Reductive degradation of porphyrins to pyrroles[4] and their reductive alkylation to alkyl pyrroles[5] provides similar information. Identification of the degradation products produced has been affected by a wide variety of methods, many of which involve prior separation of the compounds, followed by identification by gas chromatographic or mass spectral methods. Relatively large quantities of material have been used in these investigations and our studies have been directed to exploring the scope of these techniques by the use of combined gas chromatography/mass spectrometry (GC-MS).

Initially our studies centred on the reductive degradation of porphyrins to pyrroles using HI and formaldehyde in acetic acid.[5] The resulting tetra substituted pyrroles were treated with diazomethane and then subjected to GC-MS analysis. Using this technique we were able to differentiate between the two possible formulations (1) and (2) for the new porphyrin 'isocoproporphyrin'.[6] Reductive alkylation of (1) would be expected to give the pyrroles

(3a), (3b) and (4a) whereas (2) would only give the pyrroles (3b) and (4b). Application of the technique resulted in the unambiguous identification of the pyrroles (3a), (3b) and (4a) in approximate proportions 1:2:1 indicating structure (1) for isocoproporphyrin. We have used the technique with 100 μg of isocoproporphyrin and we are currently investigating its lower limit of applicability. Unfortunately, the pyrroles produced by this method, particularly before methylation, are unstable to light and air and this places restrictions on the ultimate sensitivity of the technique.

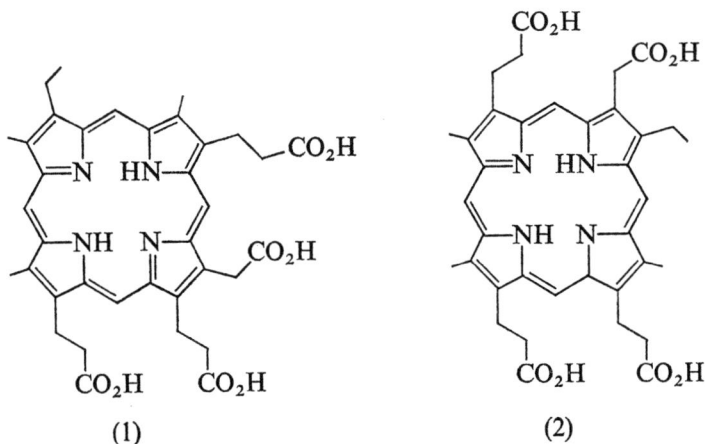

(1)   (2)

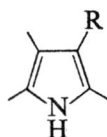   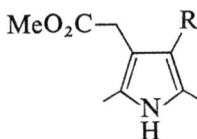

(3a) R = CH_2CH_3 (4a) R = $CH_2CH_2CO_2Me$
(3b) R = $CH_2CH_2CO_2Me$ (4b) R = CH_2CH_3

Recently our studies have been directed to the utility of the various oxidative degradation methods available for porphyrin and bile pigment structure elucidation. Preliminary studies using permanganate as oxidant[2] indicated that this method is not likely to be applicable below the mg level of porphyrin and bile pigment, since the pyrrole–2,5-dicarboxylic acids formed were partially absorbed onto the manganese dioxide formed, and in view of these findings we have restricted our studies to the chromic acid degradation methods.[1,3]

Oxidative degradation of porphyrins and bile pigments using chromic acid yields maleimide derivatives of the pyrrolic units, however nuclei with reactive β substituents, e.g. vinyl, formyl and hydroxyethyl are lost under normal reaction conditions.[7] Rüdiger has found that if milder reaction conditions are used, i.e. chromium trioxide in $2N$–H_2SO_4, it is possible to retain vinyl groups present in the original molecule, in the degradation products.[3] Identification of the maleimides formed in these degradations has usually been by paper or thin layer chromatography;[3] preparative layer chromato-graphy, followed by mass spectral identification[8] or GC comparison[9] have also been used. The technique is applicable with 5 μg of substrate[3] and application of combined GC-MS should enable even smaller quantities of material to be used and give additional certainty in product identification.

Oxidation of mesoporphyrin-IX (5) with chromium trioxide in 50% H_2SO_4, followed by methylation of the reaction products with diazomethane gave two maleimides in approximately equal quantities. A 2 m × 2 mm i.d. glass column packed with 3% OV-17 on gas chrom. Q temperature pro-grammed at 8°/min from 80° gave a good separation of the maleimides and showed their presence in approximately equal amounts. GC-MS showed that the first GC peak corresponded to methylethylmaleimide (6a), and the mass

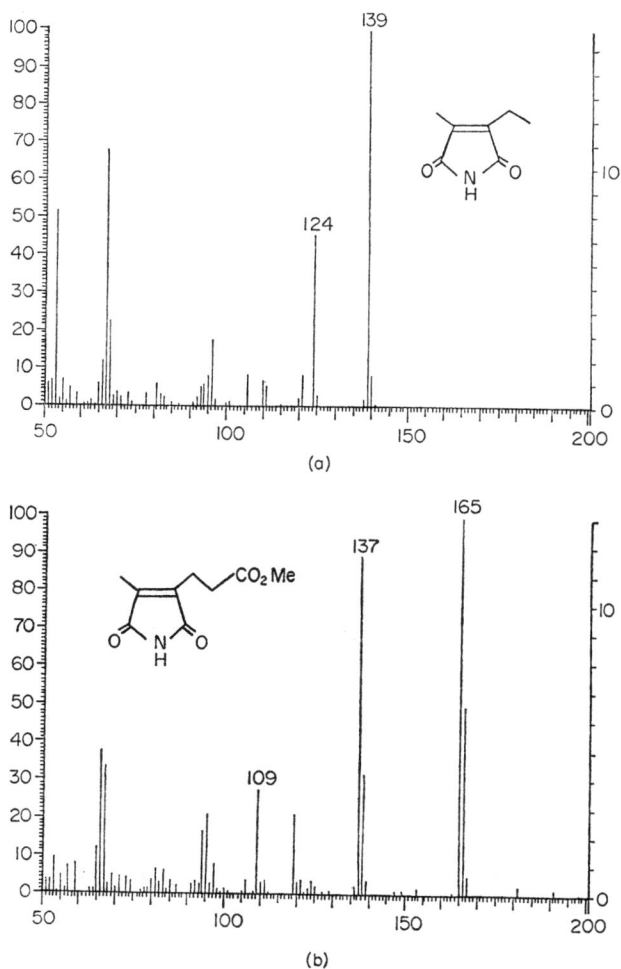

Fig. 1 (a) Electron impact spectrum of methylethylmaleimide obtained by GC-MS (source temperature, 220°). (b) Electron impact spectrum of the methyl ester of hematinimide obtained by GC-MS (source temperature, 220°).

spectrum (Fig. 1a) agrees with those previously reported for this compound.[8,10] The second component, the methyl ester of hematinimide (6b) did not show the expected molecular ion at m/e 197 in its mass spectrum (Fig. 1b) and we believe that the ester (6b) undergoes thermal decomposition by loss of methanol to give the lactone (7) which fragments by successive losses of carbon monoxide to give the ions at m/e 137 and 109 respectively.

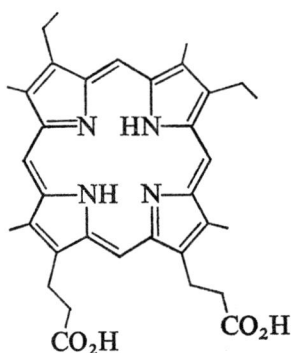

(5)

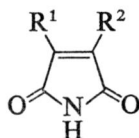

(6a) $R^1 = CH_3$, $R^2 = CH_2CH_3$
(6b) $R^1 = CH_3$, $R^2 = CH_2CH_2CO_2Me$
(6c) $R^1 = CH_3$, $R^2 = CH_2CO_2Me$
(6d) $R^1 = CH_2CO_2Me$, $R^2 = CH_2CH_2CO_2Me$
(6e) $R^1 = R^2 = Et$
(6f) $R^1 = CH_3$, $R^2 = CH=CH_2$

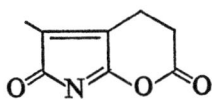

(7)

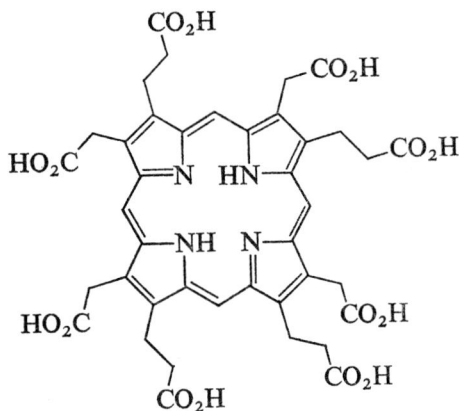

(8)

Studies are underway to establish the site of thermal decomposition and we have found that samples of the ester (6b) run using the direct insertion probe show a molecular ion at m/e 197 with an intensity of 3% and 8% of the base peak at source temperature of 210° and 130° respectively. The methyl esters of tetramethylporphyrin tetra-acetic acid and uroporphyrin-III (8) on similar degradation gave the maleimides (6c) and (6d) respectively as the sole products; the mass spectrum of (6d) recorded during GC-MS at a source temperature of 220° is shown in Fig. 2. Thermal decomposition to the lactone (9) appears to occur but a weak molecular ion (0·5%) is present at m/e 255. Application of the technique to isocoproporphyrin gave the maleimides (6a), (6b) and (6d) in the appropriate ratios 1:2:1, consistent with

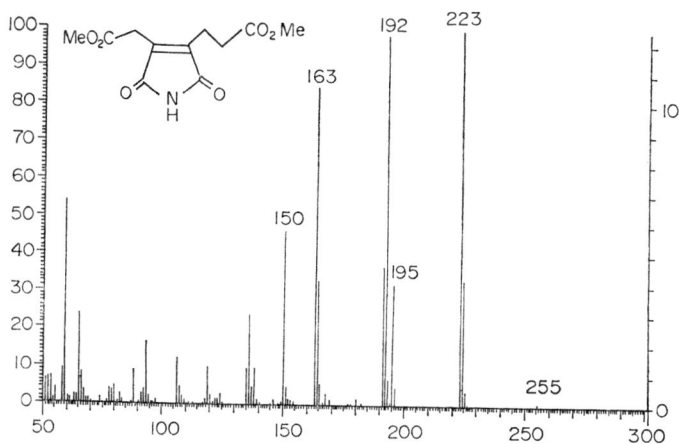

FIG. 2 Electron impact spectrum of the maleimide ester from uroporphyrin-III obtained by GC-MS (source temperature, 220°).

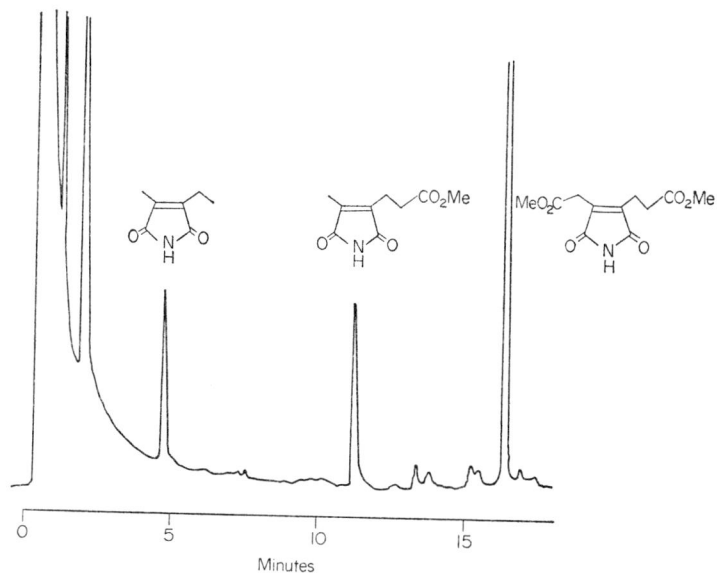

FIG. 3 GC trace of the maleimides from the degradation of mesoporphyrin-IX and uroporphyrin-III with chromium trioxide in 50% H_2SO_4 (glass column 2 m × 2 mm i.d. packed with 3% OV-17 on gas chrom. Q programmed from 80–260° at 8°/min).

structure (1) assigned to isocoproporphyrin on the basis of reductive degrada-tion.[6] Figure 3 shows the GC trace for the mixture of these maleimides obtained by oxidation of a mixture of the esters of mesoporphyrin-IX and uroporphyrin-III.

We have also studied the milder oxidative conditions of chromium trioxide in $2N-H_2SO_4$ described by Rüdiger.[3] Treatment of the β-oxophlorin (10) (related to mesoporphyrin-IX dimethylester) under these conditions, followed

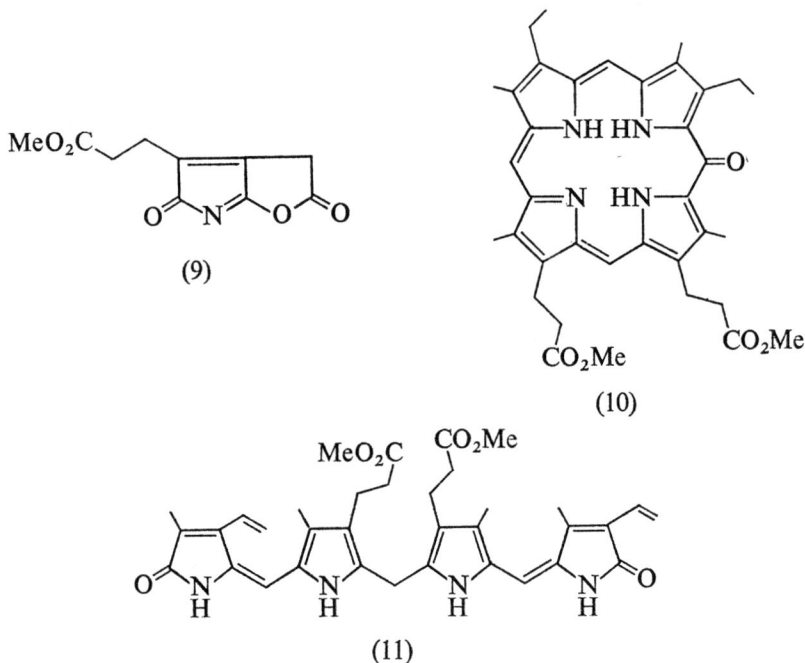

(9)

(10)

(11)

by GC-MS gave the expected maleimides (6a) and (6b) in approximately equal proportions. Similar treatment of octaethyloxophlorin gave the maleimide (6e) as sole product. Bilirubin dimethyl ester (11) under similar conditions gave the maleimides (6b) and (6f) and the total ion current monitor trace obtained from electron impact ionization is shown in Fig. 4. The proportions of the two maleimides are not equal, as would be expected, and we believe that some decomposition of the vinyl maleimide (6f) has occurred either on GC or during oxidation. Previous investigations have suggested that the vinylmaleimide is not amenable to GC study,[11] but our studies have shown that GC-MS is suitable for qualitative studies of bile pigments with vinyl substituents. Quantitative work may require suitable derivatisation before oxidative degradation of tetrapyrroles with vinyl or similar substituents. The mass spectrum of the vinylmaleimide (6f) is shown in Fig. 5 and is in agreement with the spectrum previously reported for this compound.[10]

Our present studies indicate that oxidative degradation using chromium trioxide and $2N-H_2SO_4$ followed by GC-MS is the method of choice for

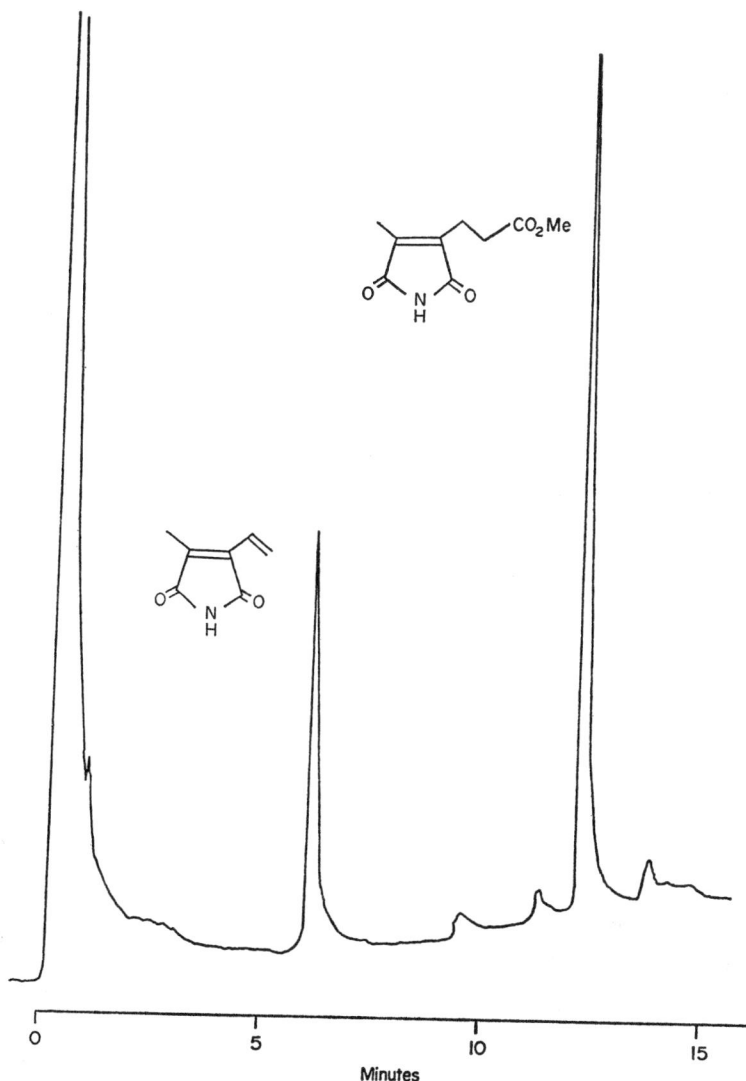

FIG. 4 Total ion current monitor trace of the maleimides from the degradation of bilirubin with chromium trioxide in $2N$-H_2SO_4 (glass column 2 m $\times$ 2 mm i.d. packed with 3% OV-17 on gas chrom. Q, programmed from 80–220° at 8°/min).

the identification of peripheral substituents on porphyrins and bile pigments, but the reductive alkylation technique[5] remains the method of choice for the study of meso-substituted porphyrins and chlorophylls. We have obtained satisfactory results using the oxidative technique with 5 μg of substrate, which is a much lower level than we have been able to achieve with the reductive method and studies are in hand which we hope will enable the identification of hydroxyethyl and formyl groups using this technique.

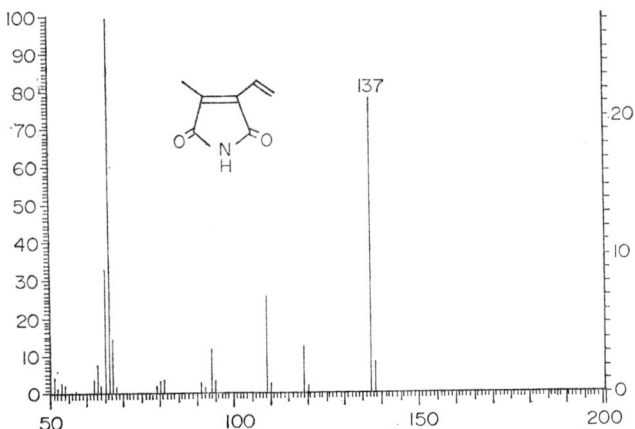

FIG. 5   Electron impact spectrum of methylvinylmaleimide obtained by GC-MS (source temperature 220°).

Mass spectra were obtained in line diagram and mass list form, with background subtracted using a Varian 620i computer on-line to a Varian CH 5D mass spectrometer.

We thank the S.R.C. for financial assistance towards the purchase of the mass spectrometer and for financial support for one of us (D.S.M.).

REFERENCES

1. Morley, H. V. and Holt, A. S., *Canad. J. Chem.*, 1961, **39**, 755; Grassl, M., Coy, U., Seyffert, R. and Lynen, F., *Biochem. Z.*, 1963, **338**, 771; Ellsworth, R. K. and Aronoff, S., *Arch. Biochem. Biophys.*, 1968, **124**, 358.
2. Nicolaus, R. A., Mangoni, L. and Caglioti, L., *Ann. Chim.* (Italy), 1956, **46**, 793; Nicolaus, R. A., Mangoni, L. and Nicoletti, R., *Ann. Chim.* (Italy), 1957, **47**, 178; Tipton, G. and Gray, C. H., *J. Chromatog.*, 1971, **59**, 29.
3. Rüdiger, W., *Angew. Chem. Internat. Edn.*, 1970, **9**, 473.
4. Fischer, H. and Orth, H., 'Die chemie des pyrrols', Vols. 1 and 2, Akademische Verlag, Leipzig, 1937 and 1941.
5. Chapman, R. A., Roomi, M. W., Morton, T. C., Krajcarski, D. T. and MacDonald, S. F., *Canad. J. Chem.*, 1971, **49**, 3544.
6. Stoll, M. S., Elder, G. H., Games, D. E., O'Hanlon, P., Millington, D. S. and Jackson, A. H., *Biochem. J.*, 1973, **131**, 429; Games, D. E., Jackson, A. H. and Millington, D. S., 'Proceedings of the International Symposium on Mass Spectrometry in Biochemistry and Medicine', Milan, 1973. Tamburini, Editore, Milan, in press.
7. Fischer, H. and Wenderoth, H., *Annalen*, 1939, **537**, 170.
8. Ellsworth, R. K. and Aronoff, S., *Arch. Biochem. Biophys.*, 1968, **124**, 358.
9. Morley, H. V. and Holt, A. S., *Canad. J. Chem.*, 1961, **39**, 755.
10. Lightner, D. A. and Quistad, G. B., *Nature New Biology*, 1972, **236**, 203.
11. Rüdiger, W., 'Porphyrins and Related Compounds', *ed.* T. W. Goodwin, Academic Press, London, 1968, p. 121.

25

Structure Elucidation of New Amino Acid Antagonists by Mass Spectrometry

By WILFRIED A. KOENIG

(*Chemisches Institüt der Universität, Tübingen, Auf der Morgenstelle, Germany*)

AMINO acid antagonists are in most cases compounds of low or medium molecular weight, structurally related to amino acids or peptides. These compounds show inhibitory effects on bacteria or fungi, when these microorganisms are grown on chemically defined media. The growths inhibition can be compensated by the presence of one or sometimes more than one amino acid.

The application of gas phase methods to structural investigation of highly polar natural products is limited by the low volatility and thermal lability of these compounds. The most common approach to solving this problem is to increase volatility by removal of the polar sites of the molecules by formation of derivatives.

From a dermatophyte a ninhydrin positive substance was isolated, which showed inhibitory effects on a number of fungi. The activity was compensated by L-histidine and L-threonine.

An amino acid analysis of the antibiotic showed threonine, phenylalanine and arginine in equal molar amounts.

For the structure elucidation of peptides it is necessary to determine the structure and configuration of the optically active amino acids and the sequence of amino acids in the peptide chain.

The GC-MS investigation of the peptide hydrolysate after esterification and trifluoroacetylation yielded three peaks. The mass spectra were identical with the corresponding derivatives of threonine, phenylalanine and arginine. However, the retention time of the standard threonine derivative was substantially shorter than the retention time of the peak from the antibiotic hydrolyzate, which turned out to be allo-threonine.

For the determination of the configuration of the amino acids the gas chromatographic method, first developed by Gil-Av,[1] was used. The separation of D- and L-enantiomers is achieved on capillary columns coated with an optically active dipeptide derivative.[2] The trifluoroacetyl amino acid isopropyl esters are used for this separation. The investigation of the antibiotic showed that arginine and phenylalanine have L-configuration, allo-threonine has D-configuration.

As to the sequence analysis of the peptide antibiotic it is known from arginine peptides to be difficult to derivatize because of the highly polar guanidino
group. In the first step the antibiotic was treated with hydrazine to transform
the arginine residue into an ornithine residue. After esterification and trifluoroacetylation of the product a mass spectrum was obtained which clearly
revealed the structure of a tripeptide ornithyl-threonyl-phenylalanine (Fig. 1).

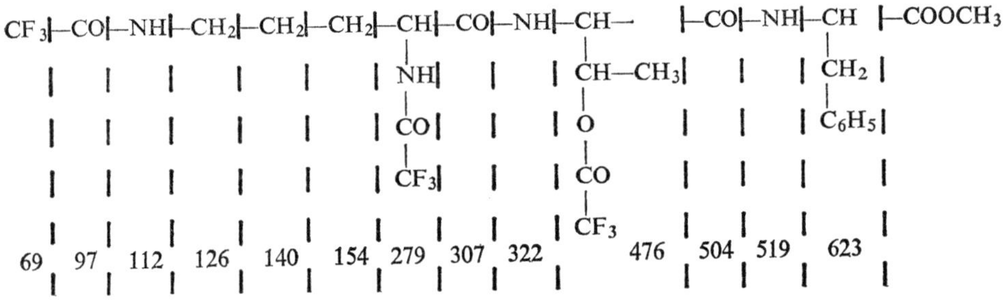

CF_3|–CO|–NH|–CH_2|–CH_2|–CH_2|–CH|–CO|–NH|–CH— |–CO|–NH|–CH |–COOCH$_3$

| | | | | | | NH| | | CH–CH$_3$| | | CH$_2$ |

| | | | | | CO| | | O | | | C$_6$H$_5$|

| | | | | | CF$_3$| | | CO | | | |

| | | | | | | | | CF$_3$ | | | |

69 97 112 126 140 154 279 307 322   476 504 519 623

FIGURE 1

The N-terminal position of ornithine is indicated by fragments containing
the N-TFA groups (m/e 279, m/e 307). The second amino acid in the chain
is the allo-threonine residue, which is proved by sequence peaks at m/e 362
and m/e 390 after elimination of a trifluoroacetic acid molecule, which is
common to threonine and serine peptides. A fragment ion of mass 162
indicates the C-terminal phenylalanine, since it includes the methyl ester
group. The structure assignment is also supported by a molecular ion m/e 682
and a pseudo molecular ion M^+–CF_3COOH at mass 568. The final proof
was obtained by chemical synthesis of an active L-arg-D-allo-thr-L-phe
tripeptide.

From another dermatophyte, Nannizzia gypsea, a ninhydrin positive
substance was isolated, which is active against several bacteria and fungi.
The activity is inhibited by L-arginine and L-citrulline.

Arginine antagonism has been observed in several cases from compounds
structurally related to arginine (Table I).

As a simple method trimethylsilylation is frequently used for formation of
volatile derivatives in gas chromatography and mass spectrometry.

The fragmentation pathway of the TMS derivative of arginine was studied
in detail. The mass spectrum was compared with the spectra of the TMS-
d_9 derivative, of the TMS derivative of the methylester of arginine and of the
TMS derivative of homoarginine.

This technique can be considered as a combination of substituent and
isotope labelling and proved to be very helpful for the interpretation. It
makes high resolution mass spectrometry unnecessary in many cases.

In the mass spectrum of the TMS derivative of arginine an intense molecular
ion can be observed at m/e 534. A fragment ion of structural significance is
formed by loss of the carbodiimide moiety (m/e 348). The peaks at m/e 187
and m/e 171 (187–CH_4) are characteristic for the guanidino group.

The corresponding mass spectrum of the TMS derivative of the antibiotic
shows a molecular ion at m/e 550, 16 mass units above that of arginine. It

TABLE I

$H_2N-\underset{\underset{NH}{\|}}{C}-NH-CH_2-CH_2-CH_2-CH_2-\underset{\underset{NH_2}{\|}}{CH}-COOH$	Homoarginine[3]
$H_2N-\underset{\underset{NH}{\|}}{C}-NH-O-CH_2-CH_2-\underset{\underset{NH_2}{\|}}{CH}-COOH$	Canavanine[4]
$H_2N-\underset{\underset{NH}{\|}}{C}-CH_2-CH_2-CH_2-CH_2-\underset{\underset{NH_2}{\|}}{CH}-COOH$	Indospicine[5]
$H_3C-\underset{\underset{NH}{\|}}{C}-NH-CH_2-CH_2-CH_2-\underset{\underset{NH_2}{\|}}{CH}-COOH$	Iminoethyl-ornithine[6]
$H_2N-\underset{\underset{NH}{\|}}{C}-O-CH_2-CH_2-CH_2-\underset{\underset{NH_2}{\|}}{CH}-COOH$	O-[L-Norvalyl-5]-isourea[7]

can be anticipated that the molecule has an extra oxygen. The fragment ions at m/e 187 and m/e 171 show that the guanidino end group is not modified. The ion, which results from the loss of carbodiimide is now shifted by 16 mass units to m/e 364.

The location of the oxygen inside the carbon chain is very unlikely. The fragment ions m/e 258 and m/e 186 (Fig. 2) probably result from ring formation with loss of the end group, which must include the extra oxygen.

For the binding site of the extra oxygen the α- or δ-nitrogen atoms of arginine are both possible.

R=TMS (m/e 258)   R=TMS (m/e 186)
R=CH$_3$ (m/e 200)   R=CH$_3$ (m/e 128)

FIG. 2

Most important for the localization of the oxygen atom in the molecule was the investigation of a hydrochloric acid hydrolysate of the antibiotic. After silylation the resulting mass spectrum showed a molecular ion at m/e 364. This corresponds to the fragment ion of the same mass in the mass spectrum of the antibiotic, resulting from loss of carbodiimide. The structure of the hydrolysis product could still be α- or δ-N-hydroxy-ornithine. The comparison with authentic δ-N-hydroxy-ornithine finally confirmed the structure of the antibiotic as δ-N-hydroxy-arginine.

ACKNOWLEDGMENT

The financial support of the Deutsche Forschungsgemeinschaft is gratefully acknowledged.

REFERENCES

1. Gil-Av, E. and Feibush, B., *Tetrahedron Lett.*, 1967, p. 3345.
2. König, W. A., Parr, W., Lichtenstein, H. A., Bayer, E. and Oro, J., *J. Chromatogr. Sci.*, 1970, **8**, 183.
3. Bell, E. A., *Biochem. J.*, 1962, **85**, 91.
4. Kitagawa, M. and Tomiyama, T., *J. Biochem. (Japan)*, 1930, **11**, 265.
5. Hegarty, M. P. and Pound, A. W., *Nature*, 1968, **217**, 354.
6. Scannell, J. P., Ax, H. A., Pruess, D. L., Williams, T., Demny, T. C. and Stempel, A., *J. Antibiotics*, 1972, **24**, 179.
7. König, W. A., Kneifel, H., Bayer, E., Müller, G. and Zähner, H., *J. Antibiotics*, 1973, **26**, 44.

Discussion

H. R. Morris (Medical Research Council, Cambridge, U.K.): I would like to make a statement which may be helpful to Dr Koenig. In his formation of methyl esters he carried out *two* separate experiments using CH_3OH/HCl and CD_3OH/HCl to determine the number of carbonyl groups present in the 'unknown'. In our work on structure elucidation we have used the following procedure which has several advantages,[1] not least of which is a saving of valuable sample. By using a 1:1 mixture of $CH_3OH:CD_3OH$ one can easily recognise the number of carbonyl groups in the one spectrum by the pattern in the molecular ion region. For example, a recent examination of a microquantity of a porphyrin metabolite using this technique showed a 1:7:21:35:35:21:7:1 pattern showing the presence of seven carboxyl groups in the unknown. This should also be useful in Professor Jackson's studies described in the previous paper.

1. Hunt, E. and Morris, H. R. A mass spectrometric, ^{13}C and ^1H NMR study on a collagen crosslink, *Biochem. J.*, 1973, in press.

W. A. Koenig: Thank you; yes this should be quite useful in our studies.

26

Biomedical Applications of Chemical Ionization Mass Spectrometry

By RODGER L. FOLTZ

(*Battelle Columbus Laboratories, Columbus, Ohio, U.S.A.*)

DURING the past year Battelle has collaborated with three Columbus, Ohio, hospitals in the rapid identification of drugs in body fluids from persons suspected of ingesting overdoses of drugs. Over 300 emergency samples have been analysed using primarily a computerized gas chromatograph chemical ionization mass spectrometer system.

Chemical ionization (CI) mass spectrometry is now widely recognized as an extremely useful tool for the detection and structural identification of a variety of organic compounds.[1,2,3,4] The technique is particularly well suited to the detection and identification of drugs and drug metabolites for the following reasons:

(1) Sensitivity is very good, in most cases at least equal to that of electron impact (EI) mass spectrometry.
(2) The ion source can be coupled directly to a gas chromatograph without use of a separator.[5,6,7] Either CI or EI mass spectra can be obtained by merely changing the carrier gas, which also serves as the reactant gas.
(3) Methane or isobutane CI mass spectra of drugs characteristically show intense protonated molecule ion peaks which not only facilitate identification, even when reference spectra are not available, but are ideally suited for analyses using selected ion monitoring techniques.[8]
(4) Although methane CI mass spectra show relatively little fragmentation, the fragment ions that are observed are usually adequate for conclusive identification and are often particularly diagnostic of metabolic alterations to the parent drug.

These advantages can be illustrated by describing some recent analyses of drug overdose samples.

We prefer to obtain, whenever possible, samples of the gastric content, blood, and urine from each suspected overdose patient. Although drug levels in the blood are the most significant to the physician, for many drugs the concentration in the blood is too low to permit detection by drug screening methods. The gastric content often contains high concentrations of the ingested drugs; consequently, it is the first sample analysed. After adjustment

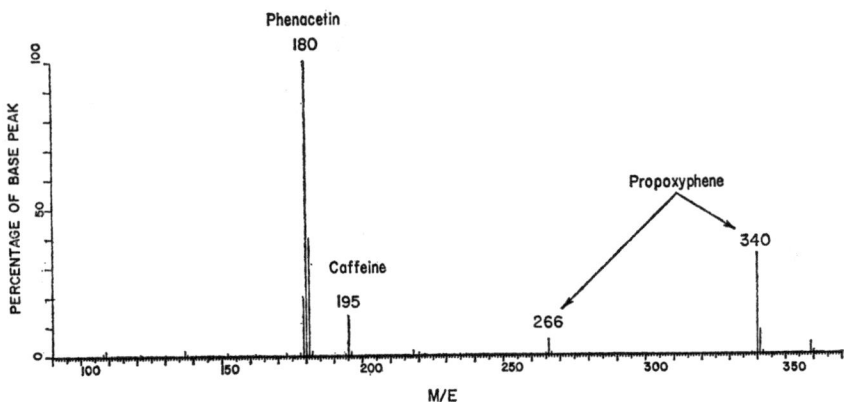

FIG. 1   CI-MS (I-butane) of the gastric content extract from a drug overdose case.

of its pH to 9·3, it is extracted with methylene chloride, and the concentrated extract introduced into the mass spectrometer using the direct probe.[9] Figure 1 shows an isobutane CI mass spectrum of the gastric content extract of a recent drug overdose case. Three drug components of a formulation of Darvon[R] are clearly evident. GC-MS analysis of a chloroform extract of the same patient's urine, after adjusting the pH to 9·3, resulted in the chromatogram shown in Fig. 2. Methane was used as the carrier and reactant gas. Initial identification of all of the labelled peaks was based solely on their methane CI mass spectra. Furthermore, we were able to assign likely structures for each of the propoxyphene metabolites. The structural assignments were

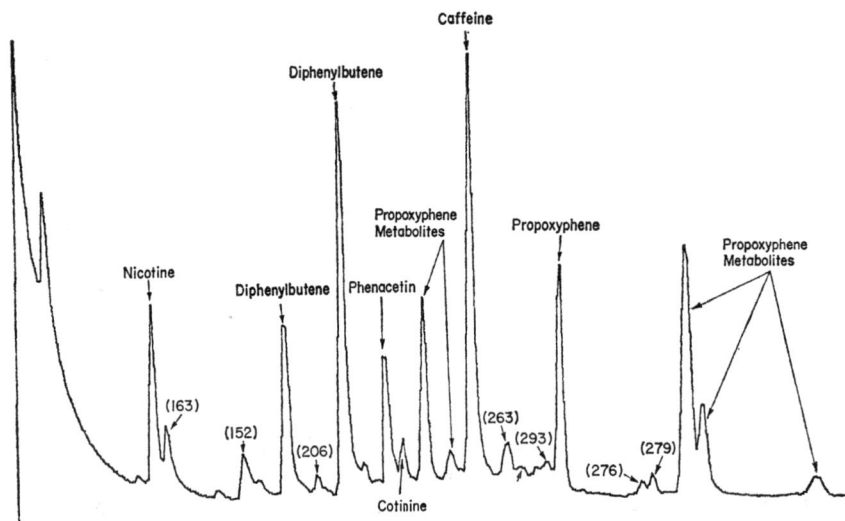

FIG. 2   Total ion current recording of the GC-MS analysis of a urine extract from a propoxyphene overdose case. The numbers in parentheses indicate the molecular weights of otherwise unidentified components.

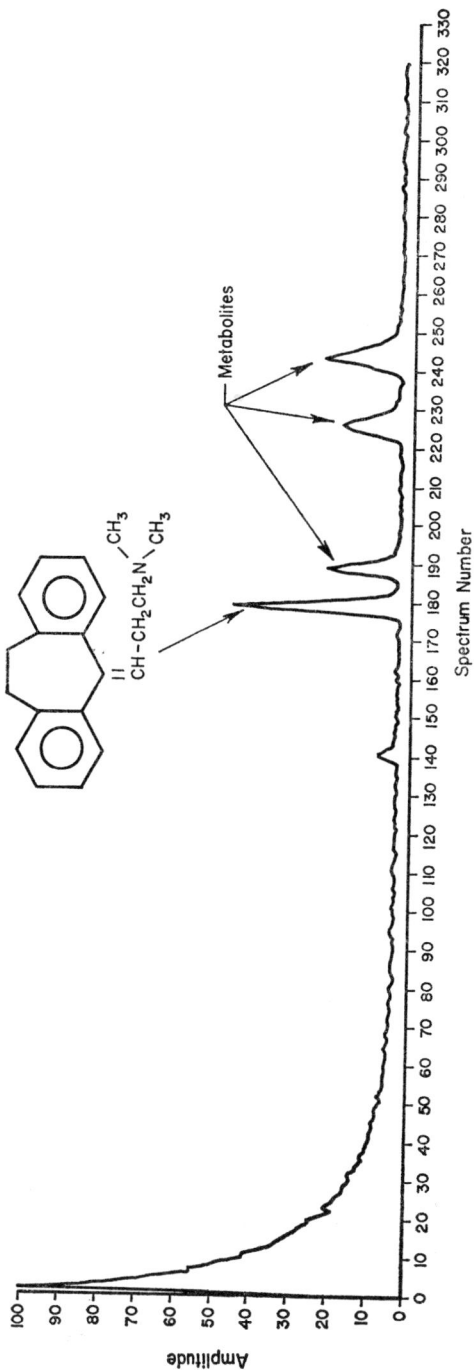

Fig. 3   Total ion current plot of the GC-MS analysis of a urine extract from an amitriptyline overdose case.

RODGER L. FOLTZ

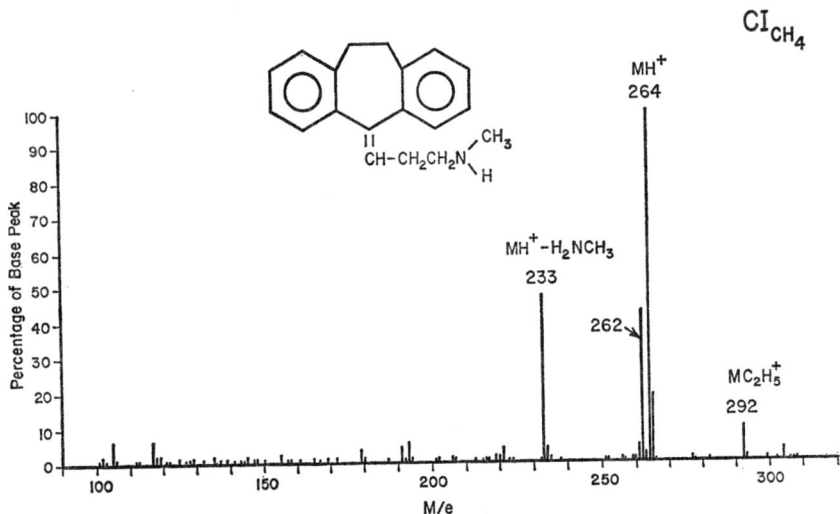

FIG. 4 The methane CI-MS of a metabolite of amitriptyline.

subsequently confirmed by a comparison of their EI mass spectra with those reported by Biemann's group for a similar propoxyphene overdose.[10]

The total ion current recording of the urine extract from an amitriptyline overdose case is shown in Fig. 3. The three metabolite peaks in order of increasing retention time correspond to noramitriptyline, hydroxylated amitriptyline, and hydroxylated noramitriptyline. The methane CI-MS of two of the metabolites shown in Figs. 4 and 5 indicate the ease with which these identifications can be made. In each case a prominent protonated molecule ion is observed ($M \cdot H^+$), as well as an abundant M–H ion which is

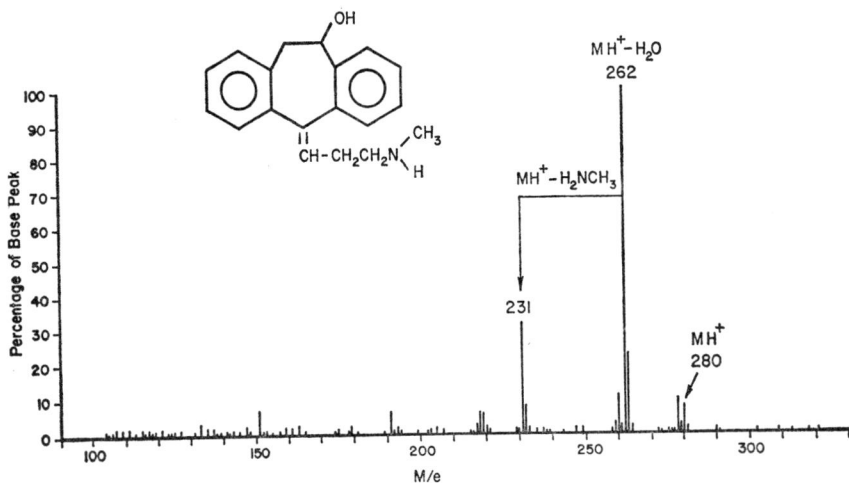

FIG. 5 The methane CI-MS of a second metabolite of amitriptyline.

characteristic of molecules containing an aliphatic amine. The major fragment ions correspond to loss of one or both of the heteroelement functions. The abundance of the $M \cdot H^+ - H_2O$ ion in Fig. 5 strongly suggests the hydroxyl is attached to a saturated carbon, most likely at the location shown. In contrast, the EI mass spectra of the parent drug and its metabolites are dominated by the iminium ions at m/e 58 ($CH_2 = N^+ Me_2$) or 44 ($CH_2 = N^+ HMe$), while the molecular ions are either absent or too weak to be easily detected. Interestingly, the EI mass spectra of amitriptyline and propoxyphene are remarkably similar in spite of their grossly different structures. In this regard it should be kept in mind that although strong EI mass spectra obtained on pure drug samples may be easily distinguished, the spectra obtained from actual body fluid samples are often very weak and require interpretation based on only the most prominent ions.

A third drug overdose example illustrates the advantageous combination of GC-chemical ionization mass spectrometry and computer data processing. The technique of plotting the ion current intensity for selected ions is now commonly used as an effective method of uncovering chromatographic peaks which are obscured by other components. Figure 6 consists of superimposed plots of the total ion current (solid line), the m/e 239 ion current (dashed line), and the m/e 233 ion current (dotted line), resulting from the GC-MS analysis of a blood serum extract from a suspected drug overdose patient. The major peak in the total ion current plot was readily identified as pentobarbital upon examination of spectrum number 37. The presence of smaller amounts of secobarbital and phenobarbital was indicated by the plots of their protonated molecule ion intensities (m/e 239 and 233, respectively). The major problem

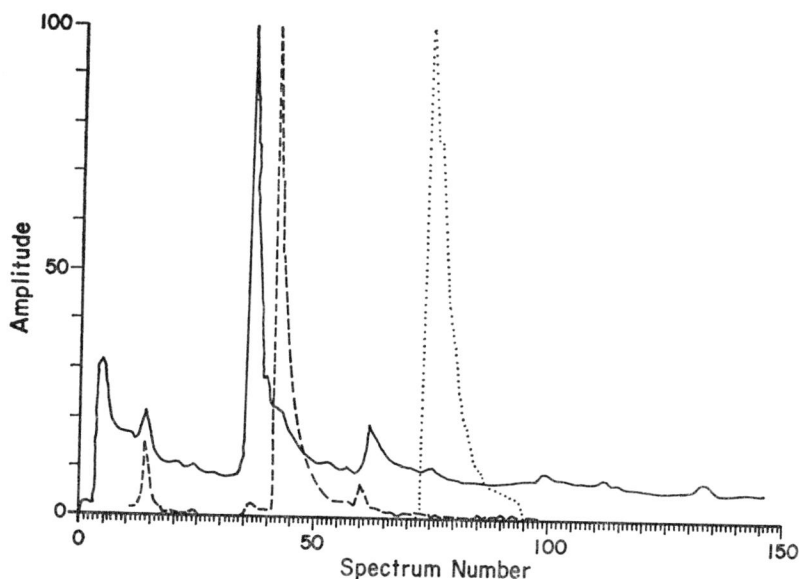

FIG. 6   Computer plots of ion currents from a GC-MS analysis of the extract of blood from a drug overdose patient. Total ion current, solid line; m/e 239 ion current, dashed line; and m/e 233 ion current, dotted line.

with this technique, however, is deciding which m/e ion currents to plot. In order to detect unsuspected drug components, we utilize a computer program which rapidly lists the mass and intensity of the N most intense ions in each spectrum, where N can be any number from 1 to 10. With a cathode ray tube output device a complete listing of these data can be presented and examined in a few minutes. A portion of the listing obtained in this analysis is given in Table I. Phenobarbital is easily detected by the ion intensity at m/e 233 $(M \cdot H^+)$ even though its presence is hardly detectable in the total ion current plot.

TABLE I

M/E and Intensities for the Four Most Abundant Ions in Spectra 70 to 80 of Fig. 6

SPN	m/e	Intensity	m/e	Intensity	m/e	Intensity	m/e	Intensity
70	253	133	251	87	157	57	129	37
71	253	134	251	87	157	55	195	40
72	253	132	251	91	157	71	149	33
73	253	126	251	87	157	69	111	31
74	253	135	233	95	251	84	157	67
75	233	149	253	135	251	86	157	61
76	253	129	233	114	251	82	157	52
77	253	124	233	114	251	85	157	49
78	253	126	233	89	251	88	157	46
79	253	128	251	82	233	63	157	45
80	253	129	251	83	233	54	157	45

A computer plot of spectrum 75 after subtraction of spectrum 72 gave an excellent methane CI-MS of phenobarbital. Although the same computer techniques can be used with EI mass spectra, the abundant protonated molecule ions obtained with chemical ionization are easier to detect and to identify with specific drugs.

REFERENCES

1. Field, F. H., *Accts. Chem. Res.*, 1968, **1**, 42.
2. Munson, B., *Anal. Chem.*, 1971, **43**, 28A.
3. Arsenault, G. P., *in* 'Biochemical Applications of Mass Spectrometry', *ed.* G. R. Waller, Wiley-Interscience, New York, 1972.
4. Foltz, R. L., *Lloydia*, 1972, **35**, 344.
5. Arsenault, G. P., Dolhun, J. J. and Biemann, K., *Anal. Chem.*, 1971, **43**, 1720.
6. Schoengold, D. M. and Munson, B., ibid., 1970, **42**, 1811.
7. Fentiman, A. F., Jr, Foltz, R. L. and Kinzer, G. W., ibid., 1973, **45**, 580.
8. Clarke, P. A. and Foltz, R. L., submitted to *Clinical Chemistry*.
9. Milne, G. W. A., Fales, H. M. and Axenrod, T., *Anal. Chem.*, 1971, **43**, 1815.
10. Althaus, J. R., Biemann, K., Biller, J., Donaghue, P. F., Evans, D. A., Forster, H.-J. Hertz, H. S., Hignite, C. E., Murphy, R. C., Preti, G. and Reinhold, V., *Experientia* 1970, **26**, 714.

Discussion

A. L. Burlingame (University of California, Berkeley, U.S.A.): What mass spectrometer are you using and why?

R. L. Foltz: Most of this work was done using a Finnigan model 1015 quadrupole mass spectrometer, primarily because its data system permits very rapid display of the MS data. Consequently, identification of the drugs can be accomplished very soon after the GC-MS run is completed.

F. W. McLafferty (Cornell University, New York, U.S.A.): How quantitative is your comparison of the sensitivities of chemical and electron ionization?

R. L. Foltz: My statement regarding the sensitivity of chemical ionization compared to electron ionization is based on a general impression resulting from our experience with two different mass spectrometers in our laboratory that have been converted from EI instruments to dual EI/CI instruments, as well as discussions with others who are working with chemical ionization. A specific and really meaningful comparison is very difficult to carry out.

I. Lengyel (St Johns University, New York, U.S.A.): How did you identify the drugs?

R. L. Foltz: We use a simple card file system, currently containing information on about 400 drugs and drug metabolites. The information recorded on the cards includes the m/e and intensities of the most abundant ions in the drug's methane and isobutane chemical ionization mass spectra and its electron impact mass spectrum as well as its gas chromatographic retention time relative to the internal standard, tetraphenylethylene. Since the CI mass spectrum clearly identifies the compound's molecular weight, cards for those drugs having the indicated molecular weight are extracted from the file. Final identification includes consideration of the CI fragment ions and the relative retention time. If ambiguities still exist, the sample is rerun under electron impact ionization conditions.

In this connection I should mention that Dr Henry Fales at N.I.H. has generated a computer listing of all of the compounds in the Merck Index arranged according to molecular weight. This listing has been very useful to us on several occasions and should be of interest to anyone using chemical ionization for drug analysis.

A. Frigerio (Istituto 'Mario Negri', Milan, Italy): Did you find any epoxide as a metabolite of amitriptyline?

R. L. Foltz: No. Since the primary objective of this work was rapid identification of ingested drugs, we did not attempt to detect and identify all of the drug metabolites.

S. Markey (University of Colorado Medical Centre, Denver, U.S.A.): In view of the popular press's attention to the use of mass spectrometry for rapid identification of drugs from overdosed patients by several laboratories in the U.S., it is worthwhile to note that skilled toxicologists are rarely able to make use of this information in designing patient therapy. Experienced toxicologists do not wait for or rely upon positive identification of ingested materials before beginning therapy. Most successful therapy is supportive care and active removal of drugs from the system by gastric lavage. Antidotal therapy is pharmacologically not possible for many compounds, and for the few cases in which it can be used, an astute clinician will outdo a mass spectrometer in speed of diagnosis. Dr Henry Matthew of the Royal Infirmary of Edinburgh, one of the world's leading toxicologists, believes that the current state of generalized therapy for unknown drug overdoses is adequate if the patient is seen by a competent toxicologist soon enough. The impressive low mortality rate in his clinic supports this contention.

In the U.S., new analytical methods for the clinician are given so much attention that these expensive alternatives to good medical training are frequently adopted and passed onto the patient's bill. Dr Foltz has done an elegant job of demonstrating a workable system for analytical toxicology, but I doubt that the information he has supplied to clinicians has resulted in the saving of a human life, a claim being made by other spectro-scopists in this field as well as by salesmen for mass spectrometers.

R. L. Foltz: One of the objectives of this programme is to evaluate the usefulness and need for this type of service. We have found that the physicians are delighted to have this service available. I agree that probably in very few cases have the results we provided significantly altered the treatment prescribed by the attending physician. Nevertheless, the situation may change as physicians gain confidence in the technique and more information is accumulated regarding optimum therapy for specific drug ingestions.

27

The Application of GC-MS to the Study of Urinary Acidic and Neutral Metabolites in Normal Subjects and in Patients with some Metabolic Disorders

By A. M. LAWSON,‡ R. A. CHALMERS,† P. PURKISS,†
F. L. MITCHELL‡ and R. W. E. WATTS†

(*Divisions of Inherited Metabolic Diseases,*† *and Clinical Chemistry,*‡ *M.R.C. Clinical Research Centre, Harrow, Middlesex, U.K.*)

SEVERAL diseases are now known in which one or more carboxylic acids accumulate in the blood and/or are excreted in the urine because of an inherited enzyme deficiency.[1] The methods which have been previously used in this field have only allowed a relatively narrow spectrum of acids to be studied simultaneously. Previous workers[2,3,4] have used solvent extraction or steam distillation as a primary isolation procedure. Neither method is ideal as a screening procedure designed to cover a range of compounds with possibly widely different physical properties. The former method gives poor recoveries of the more polar compounds and the latter may produce artefacts.[5,6] Chalmers and Watts[7,8,9,10] have described quantitative methods for the isolation and gas chromatographic analysis of a very wide range of acidic metabolites in urine. The acids are eluted from a DEAE-Sephadex column into a vessel containing the reagent which forms the derivative of the oxo groups. The preparation of the methoximes, ethoximes or *O*-benzyl oximes is followed by freeze drying and trimethylsilylation to yield the acids as trimethylsilyl (TMS) esters with hydroxyl groups converted to TMS ethers.

Enzyme deficiencies which cause excretion of large amounts of certain carbohydrates in urine are also known. Therefore, similar methods have now been developed to enable a range of the neutral compounds (carbohydrates and the related polyhydric alcohols) which occur in urine to be studied simultaneously. This paper reports some preliminary results obtained by combining these methods with mass spectrometry. Selected ion monitoring and cyclic scanning have been used in some cases to help in the identification of the components of overlapping GC peaks, to determine the presence of other components at very low levels, and to obtain chromatograms of particular groups of metabolites with a common characteristic fragment ion.

235

URINE (≡3 mg creatinine)

+ acidic internal standard (Undecandioic acid)
A-25 DEAE Sephadex anion exchange gel extraction
Water wash (2 × 10 ml)
1·5 M pyridinium acetate eluant (15 ml)

WASHINGS
Neutral and basic fractions

ELUATE
Acidic Fraction

Acidify to pH 2 with 6N HCl
Pass down Dowex 50 × 8 H+ column
Make column eluate and washings up to 50 ml
(Neutral fraction)

Derivatize oxo acids as O-substituted oximes
Freeze dry under controlled conditions of −10°C and 0·5 torr
Trimethylsilylate residue
+ g.c. internal standards (tetracosane and hexacosane)

GAS–LIQUID CHROMATOGRAPHY

Basic Fraction
(On column)

Neutral Fraction

Elute with 4 M NH4OH
Rotary evaporate
Trimethylsilylate residue

Freeze dry aliquot
Trimethylsilylate residue + internal standard

GAS–LIQUID CHROMATOGRAPHY

GAS–LIQUID CHROMATOGRAPHY

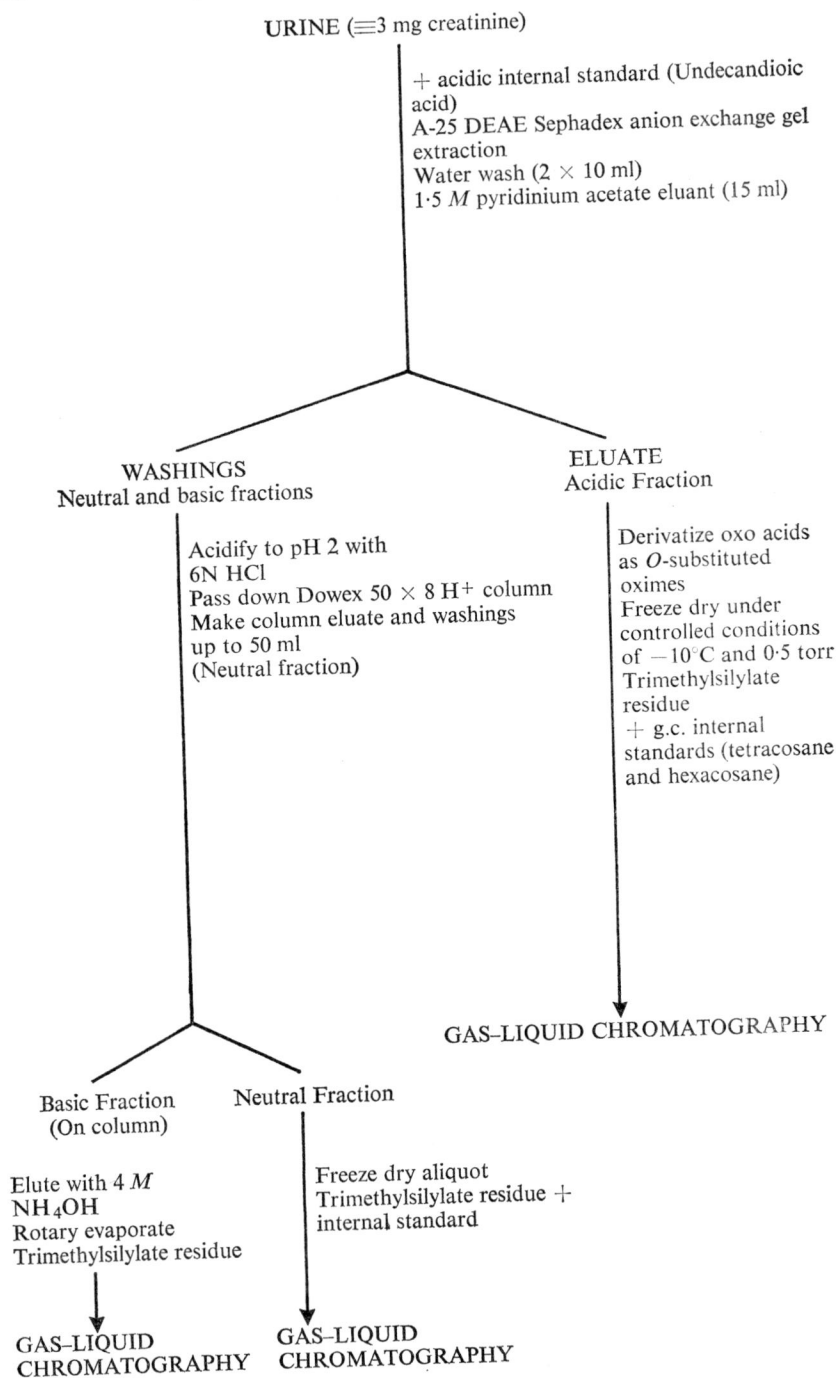

FIG. 1   Flow diagram illustrating the extraction, separation, and derivative preparation of urinary components.

METHODS

Isolation and Derivatization

The methods used for the isolation and derivatization of the acidic and neutral metabolites from urine are summarized in Fig. 1. The full experimental details for the acidic metabolites have been published elsewhere.[10]

Gas–Liquid Chromatography

The TMS and TMS-ethoxime derivatives of the acids are separated by temperature programming,[10] using a Hewlett-Packard (HP) 5750G gas chromatograph equipped with automatic liquid sampler, digital electronic integrator and a teletype output. Data are collected on punched paper tape and processed by an off-line computer.[11]

The neutral fraction, isolated as shown in Fig. 1 is treated with N,O-bis-TMS acetamide in pyridine at 40° until the residue has completely dissolved. Hexacosane dissolved in heptane is added as the internal standard. Gas chromatography is carried out under similar conditions to the above, with temperature programming from 100° to 285°C.

Gas Chromatography–Mass Spectrometry

GC-MS studies were carried out on a Varian Aerograph 2700 gas chromatograph coupled to a Varian MAT 731 double focussing mass spectrometer. The latter was operated at 8 kV and 70 eV. The GC conditions were similar to those used with the Hewlett Packard, with the carrier gas separator and transfer lines maintained at 250°C. Data were processed by the Varian 100 MS SpectroSystem. The identification of the components of the extracts were initially made from compilations of reference spectra and more recently with the help of library search programmes and a library file. In analyses where the mass spectrometer was used in the cyclic scanning mode, 6-sec scans over the mass range 40–1000 amu were taken every 10 sec. Selected ion monitoring of up to two ions was achieved with the standard peak matching device and the output together with the total ionization signal were recorded on a potentiometric recorder.

RESULTS AND DISCUSSION

GC-MS of Substituted Oxime–TMS Derivatives

The methoxime-trimethylsilyl (MO-TMS), the ethoxime-trimethylsilyl (EO-TMS), the benzyloxime-trimethylsilyl (BO-TMS) and the trimethylsilyl (TMS) derivatives of nineteen oxoacids, which included most of the probable urinary metabolites, were prepared and submitted to GC-MS in order to determine their utility both for gas chromatography and for their identification by mass spectrometry.

The TMS derivatives gave M and M-15 ions in most cases. Many of the compounds, particularly the aliphatic oxoacids, have comparatively few other ions with the exception of the common m/e 147, 75 and 73. The latter account for much of the ion current. These spectral features are not ideal either for the computer matching of spectra or for the direct identification of a compound from structurally significant ions. A more serious disadvantage is the appearance of multiple peaks from incomplete derivative formation due to keto-enol tautomerism. This not only increases the complexity of the chromatogram but is unsatisfactory for quantitation.

The MO-TMS and EO-TMS derivatives fragment in similar modes to each other. Although they give small molecular ions, the (M-15) fragments are sufficiently intense for characterization. Other useful ions are (M-OCH$_3$) from methoximes and (M-OC$_2$H$_5$) from the ethoximes. The oxime group tends to direct the fragmentation of the molecule leading to ions such as (M-CO$_2$TMS), and in cases where there is a γ-hydrogen to the oxime available, McLafferty type rearrangements give ions of diagnostic value.

The BO-TMS derivatives are less favourable and suffer from the facile cleavage of the benzylic bond in the oxime moiety to produce a dominant ion at m/e 91 in almost all spectra. This results in the other ions being of low intensity. Although the molecular ion is often small the (M-15) and (M-17) are normally present. The MO-TMS and EO-TMS derivatives are equally good from a mass spectrometric identification standpoint, but the EO-TMS derivative has the advantage of increasing the retention times of low molecular

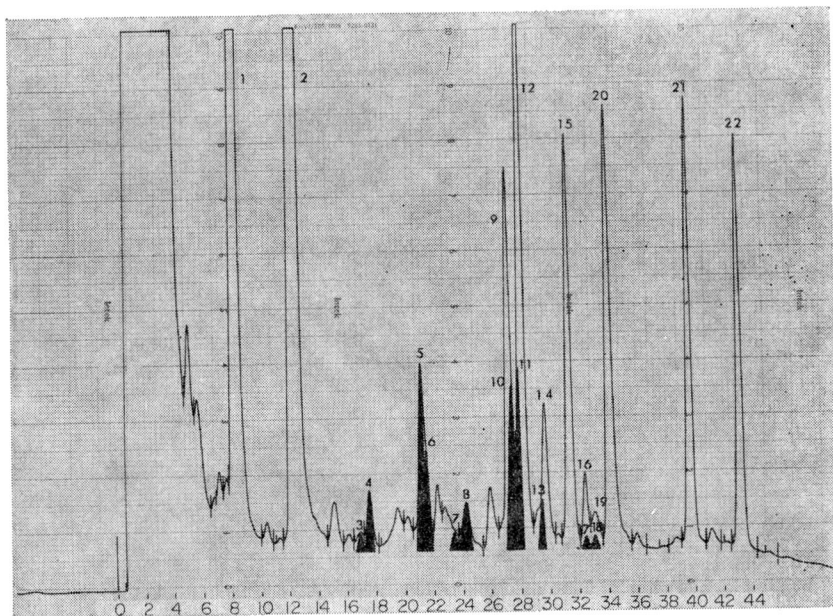

FIG. 2(a) Gas chromatographic profiles of the acidic components extracted from urine from a normal subject, obtained by programmed-temperature gas chromatography of trimethylsilyl derivatives on 10% OV101. The horizontal axis represents time in minutes elapsed from injection in each case. Peak identifications are as follows: 1. sulphate; 2. phosphate; 3. 3-deoxytetronic acid; 4. 2-deoxytetronic acid; 5. erythronic acid; 6. threonic acid; 7. a deoxypentonic acid; 8. a deoxypentonic acid; 9. hippuric acid; 10. a pentonic acid; 11. arabinoic acid; 12. citric acid; 13. a deoxyhexonic acid; 14. glucono-1,5-lactone; 15. undecandioic acid (internal standard); 16. glucuronic acid; 17. a hexonic acid; 18. gluconic acid; 19. saccharic acid; 20. uric acid; 21. tetracosane (internal standard); 22. hexacosane (internal standard). Shaded areas indicate 'sugar' acid positions.

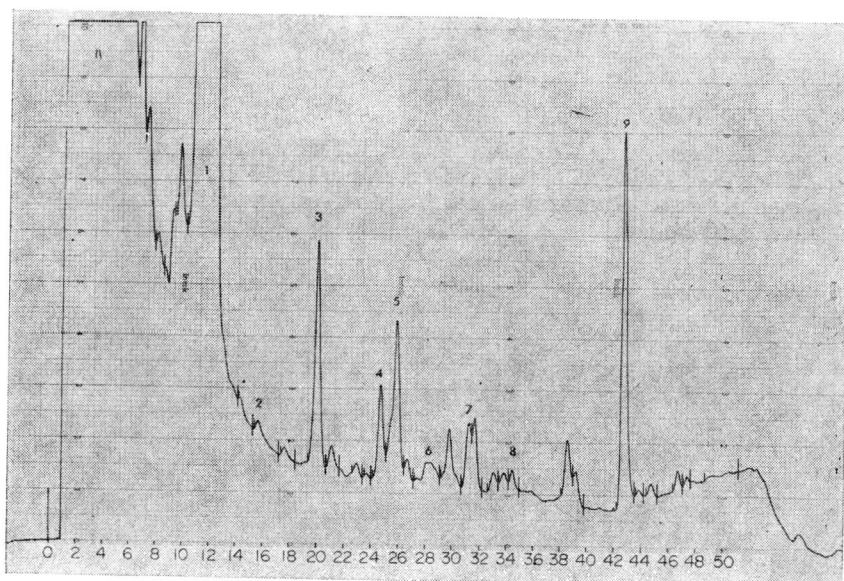

FIG. 2(b)   Gas chromatographic profiles of the neutral components extracted from urine from a normal subject, obtained by programmed-temperature gas chromatography of trimethylsilyl derivatives on 10% OV101. The horizontal axis represents time in minutes elapsed from injection in each case. Peak identifications are as follows: 1. urea; 2. erythrono-1,4-lactone; 3. erythritol; 4. arabitol; 5. a pentose; 6. fructose; 7. sorbitol; 8. inositol; 9. hydrocarbon internal standard.

weight compounds (*e.g.* glyoxylic and pyruvic acids) sufficiently to resolve them from the solvent front.[9] The BO-TMS derivative would only be used in cases where an oxoacid peak is required to be shifted in order to clarify a particular region of the chromatogram.

The Normal Pattern of Organic Acid Excretion

An example of the pattern of organic acids found in the first specimen of urine passed by normal subjects on rising is shown in Fig. 2. Urine collected under these conditions was analysed in order to reduce the possible effect of dietary fluctuation. The shaded areas show the position of the sugar acids. This *qualitative* pattern has so far proved to be consistent in 65 normal subjects of all age groups. Phosphate, sulphate and citrate are present in the largest amounts. The consistent excretion of tetronic, pentonic and hexonic sugar and deoxy sugar acids is striking, and it reflects the efficiency of the method for the extraction of hydrophilic acidic compounds. These substances appear to be mainly of endogenous origin because their quantitative excretion is little affected by grossly changing the amounts of carbohydrate, protein and fat in the diet. There is also little diurnal variation and the day to day excretion of some measured metabolites was consistent in a single subject over an 8-day period. Although we have not quantitated all of the sugar and deoxy sugar acid compounds because their gas chromatographic response factors have not yet been determined, their excretion appears to be between about 5 mg/24 hours (near the detection limit of the method) and 100 mg/24 hours, assuming that the relative response factors are unity.

Selected Ion Monitoring (SIM) and Cyclic Scanning

Gas chromatographic resolution is frequently insufficient to determine if a particular compound is present. This difficulty can often be overcome by monitoring selected ions, characteristic of the compound of interest, in the ion current from the GC effluent. Two examples of this are shown in Fig. 3.

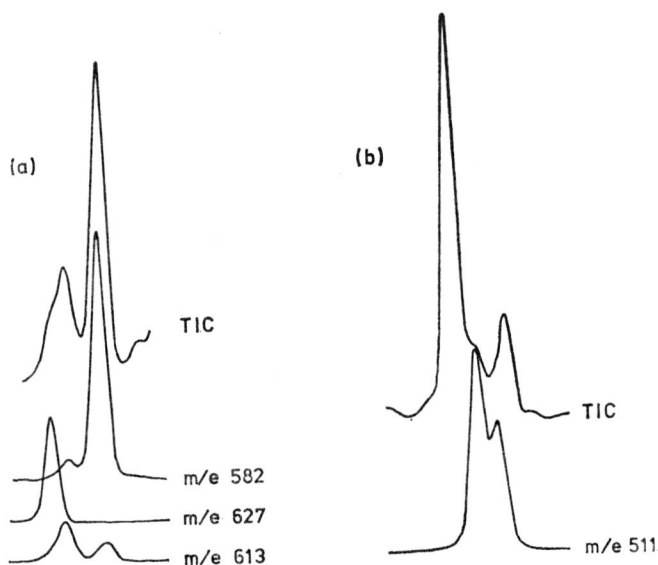

(a)   SIM at m/e 582 (M-15 of EO-TMS glucuronic acid), m/e 627 (M-15 of TMS glucaric (saccharic) acid) and m/e 613 (M-15 of TMS hexonic acids) enabled the presence of these acids to be confirmed.

(b)   The appearance of the hippuric (small peak) and citric acid (large peak) region of the TIC chromatogram indicated the presence of low levels of underlying components, which were suspected to be pentonic acids. SIM m/e 511 (M-15 for TMS pentonic acids) confirmed the presence of these compounds.

FIG. 3   Selected ion monitoring (SIM) in the region of closely overlapping peaks in the total ion current (TIC) chromatogram of components in the acidic fraction of urinary extracts from a normal subject.

In (b) the TIC response in the hippuric (small peak) and citric (large peak) acid region of the chromatogram has been recorded simultaneously with the (M-15) ion at m/e 511 for a pentonic acid. The presence of two isomeric pentonic acids is clearly evident in this multiplet. A more complex situation is represented in (a) where the retention values of glucuronic acid (M-15 of EO-TMS, m/e 582), glucaric acid (M-15 of TMS, m/e 627) and hexonic acids (M-15 of TMS, m/e 613) are too close to allow identification of all the components present.

The procedure is limited to following in a single run a number of ions equivalent to the channels in the peak selection device, in the present case to two ions, and to a maximum mass switching range of 10% of the lowest mass value. This makes SIM a time-consuming process when a number of ions need to be studied.

A useful alternative approach is to make a series of regular scans throughout the gas chromatographic run and then obtain a computer print-out of the occurrence of selected mass values against scan number. A disadvantage of this method is that scans need to be taken close together to give adequate resolution in the mass chromatogram in cases such as those illustrated in Fig. 3. In practice a combination of SIM and cyclic scanning is the most useful approach, although it should be emphasized that prior knowledge of the compounds and their possible presence are necessary if these techniques are to be usefully employed.

The Normal Pattern of Neutral Metabolites
The pattern of neutral compounds observed is illustrated in Fig. 2. Erythritol and arabitol are regularly observed, they appear to be unrelated to the composition of the diet, and therefore to be of endogenous origin. The

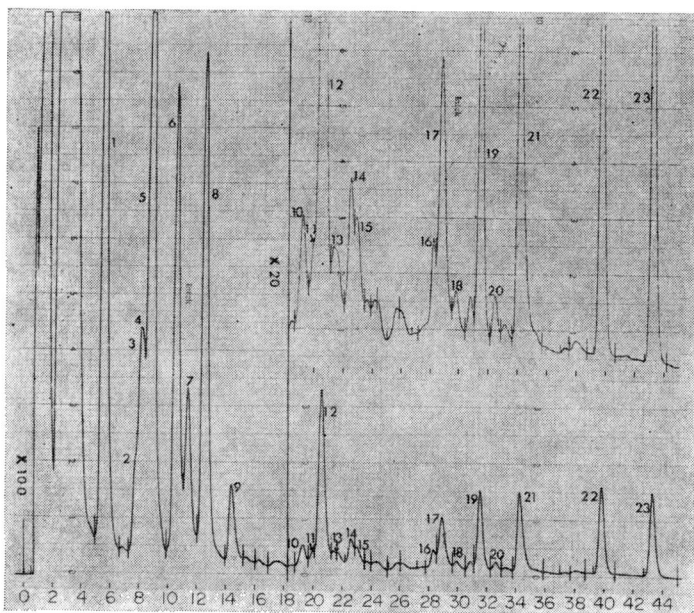

FIG. 4(a) Gas chromatographic profiles of acidic components extracted from the urine of a patient with β-methylcrotonylglycinuria, obtained by programmed-temperature gas chromatography of trimethylsilyl derivatives on 10% OV101. The horizontal axis represents time in minutes elapsed from injection in each case. Peak identifications are: 1. lactic acid; 2. ? β-hydroxyisobutyric acid; 3. 2-deoxyglyceric (β-hydroxypropionic) acid; 4. sulphate; 5. β-hydroxybutyric acid; 6. β-hydroxyisovaleric acid; 7. unidentified; 8. phosphate; 9. unidentified; 10. adipic and malic acids; 11. unidentified; 12. β-methylcrotonylglycine; 13. tetronic acids; 14. β-hydroxy β-methylglutaric acid; 15. 4-hydroxyphenylacetic acid; 16. citric acid; 17. methylcitric acid; 18. 4-hydroxyphenyllactic acid; 19. undecandioic acid (internal standard); 20. glucuronic acid; 21. uric acid; 22. tetracosane (internal standard); 23. hexacosane (internal standard).

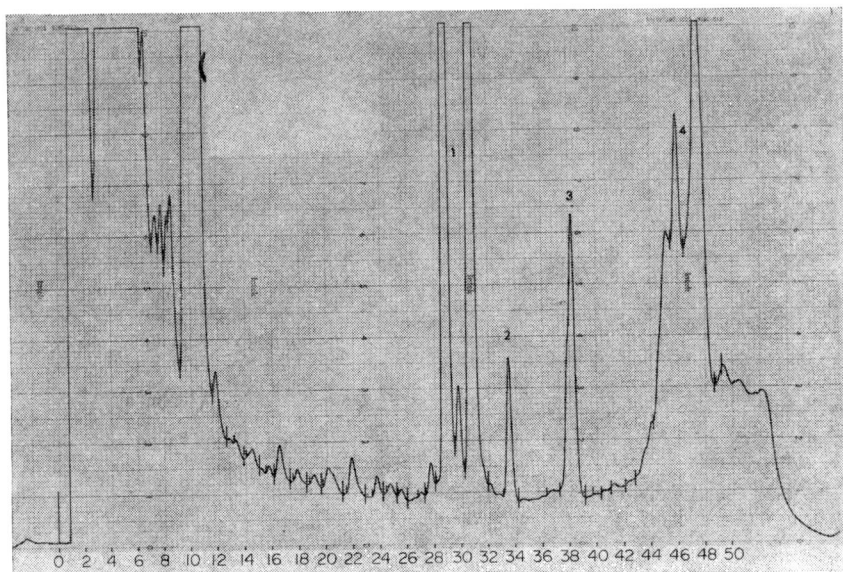

FIG. 4(b)   Gas chromatographic profiles of neutral components extracted from the urine of a patient with diabetes mellitus, obtained by programmed-temperature gas chromatography of trimethylsilyl derivatives on 10% OV101. The horizontal axis represents time in minutes elapsed from injection in each case. Peak identifications are: 1. large glucose peaks; 2. inositol; 3. hydrocarbon internal standard; 4. disaccharide peaks.

following have been identified in some specimens: sorbitol, inositol, fructose, glucose and lactose. The likely range of excretion is between about 5 mg/24 hours and 150 mg/24 hours. Quantitation of these metabolites is particularly difficult because multiple derivatives are formed on trimethylsilylation. The mass spectra of the TMS derivatives of isomeric sugars are very similar, and structural assignments have to be made on the basis of a comparison of the relative intensities of selected ions and on GC retention data.

Studies on Patients

Urine specimens from patients with the following known metabolic disorders have been examined: hyperoxaluria, phenylketonuria, tyrosinaemia, homocystinuria, β-methyl-crotonyl-glycinuria, propionic acidaemia, methylmalonic aciduria, galactosaemia and diabetes mellitus. The results of studies on these patients illustrate the utility of these methods and are exemplified by the chromatograms from the urine of patients with β-methyl-crotonyl-glycinuria and diabetes mellitus (Fig. 4). The presence of β-hydroxy-propionic acid, β-hydroxy-β-methyl-glutaric acid and methyl citric acid in the urine of the patient with β-methyl-crotonyl-glycinuria is of particular interest.

ACKNOWLEDGMENTS

We are pleased to acknowledge our indebtedness to Dr D. Gompertz for the gift of a urine specimen from the patients with β-methyl-crotonyl-glycinuria, propionic acidaemia and methylmalonic aciduria and to Drs K. Bartlett and D. Gompertz for samples of β-methyl-crotonyl-glycine and tiglylglycine. We are also greatly indebted to Miss Sheila Bickel and Mr P. Fiveash for their most skilful assistance. Some of these results will be submitted by R.A.C. in partial fulfillment of the requirements for the Ph.D degree of the Council for National Academic Awards.

REFERENCES

1. Gompertz, D. 'Organicacidaemias', *in* 'Eighth Symposium on Advanced Medicine', *ed*. G. Neale, Pitman Medical London in association with the *Journal of The Royal College of Physicians of London*, 1972.
2. Jellum, E., Stokke, O. and Eldjarn, L. *Scand. J. Clin. Lab. Invest*, 1971, **27**, 273.
3. Crawhall, J. C., Mamer, O., Tjoa, S. and Claveau, J. C. *Clin. Chim. Acta*, 1971, **34**, 47.
4. Gompertz, D. and Draffan, G. H. *Clin. Chim. Acta*, 1972, **37**, 405.
5. Gompertz, D. *Clin. Chim. Acta*, 1971, **33**, 457.
6. Stokke, O. *in* 'Organic acidurias. Proceedings of the Ninth Symposium of the Society for the Study of Inborn Errors of Metabolism. Churchill Livingstone, London, 1972, p. 38.
7. Chalmers, R. A. and Watts, R. W. E. *Analyst*, 1972a, **97**, 224.
8. Chalmers, R. A. and Watts, R. W. E. *Scand. J. Clin. Lab. Invest.*, 1972b, **29**, Suppl. 126, Abstract 30.11.
9. Chalmers, R. A. and Watts, R. W. E. *Analyst*, 1972c, **97**, 951.
10. Chalmers, R. A. and Watts, R. W. E. *Analyst*, 1972d, **97**, 958.
11. Healy, M. J. R., Chalmers, R. A. and Watts, R. W. E. *J. Chromatog.*, 1973, **87**, 365.

Discussion

A. G. Giumanini (University of Bologna, Italy): A large number of laboratories seem to be involved in the interesting problems of the chemical mapping of metabolic diseases and other diseases, as well as the study of drug metabolites with the special aim that the GC-MS methods may become useful diagnostic routines. It is proposed that scattered efforts should be coordinated on all international basis to avoid useless duplications and generate quicker results, simultaneously saving time and money. Method standardization may also be a side-product.

A. M. Lawson: I agree that more effort should be afforded to the interchange of information, data and ideas in the areas you have mentioned. The wider availability of mass spectral files of compounds of pharmacological and biological interest is a good starting point for this process.

International coordination of individual efforts would be useful where the objectives of these individuals are closely defined. If it is envisaged that direct GC-MS methods will be both economic and useful for clinical diagnosis in a particular area then cooperation between groups would be beneficial. On the other hand, I tend to the view that although GC-MS may be a powerful technique to investigate certain biochemical problems it may not often be the direct method of choice for rapid, cost-effective clinical diagnosis. In this situation large-scale cooperation would be less fruitful.

28

Stable Isotope Labelling in Quantitative Drug and Metabolite Measurement by Gas Chromatography–Mass Spectrometry

By G. H. DRAFFAN and R. A. CLARE

(*Department of Clinical Pharmacology, Royal Postgraduate Medical School, London, U.K.*)

B. L. GOODWIN, C. R. J. RUTHVEN and M. SANDLER

(*Queen Charlotte's Maternity Hospital, London, U.K.*)

INTRODUCTION

THE rapid growth of interest in gas chromatography-mass spectrometry (GC-MS) in clinical and biochemical pharmacology is in part attributable to its potential as a means of quantitative assay. The combination of the exceptional sensitivity and specificity inherent in the techniques described as accelerating voltage alternation (AVA)[1] mass fragmentography,[2] or multiple ion detection with the principle of stable isotope dilution,[3] offers unique advantages in the precise determination of drugs and metabolites in biological material (*e.g.* Ref. 4). The labelled compound, in providing the reference in quantification, may also function as a carrier in the isolation procedure and in minimizing adsorptive or degradative loss of sample in gas chromatography.

Two applications in which stable isotope dilution has been required in GC-MS assay are described in this paper. The first concerns a requirement for measurement of the tetrahydroisoquinoline alkaloid salsolinol (Fig. 1), a potentially significant minor metabolite of dopamine, identified in the urine of parkinsonian patients during L-dopa therapy.[5] Salsolinol, as the pentafluoropropionyl (PFP) derivative (I, R = H) has excellent gas chromatographic properties, but its instability and variable recovery from urine preclude accuracy in an assay based on electron capture detection. Standardization was effected in GC-MS with the deuterium labelled material (I, R = ²H) as the added reference.

A contrasting problem was encountered in the specific plasma assay of the new hypotensive drug indoramin (II, R = H). This substance cannot be reproducibly chromatographed at low concentrations carrier free. The

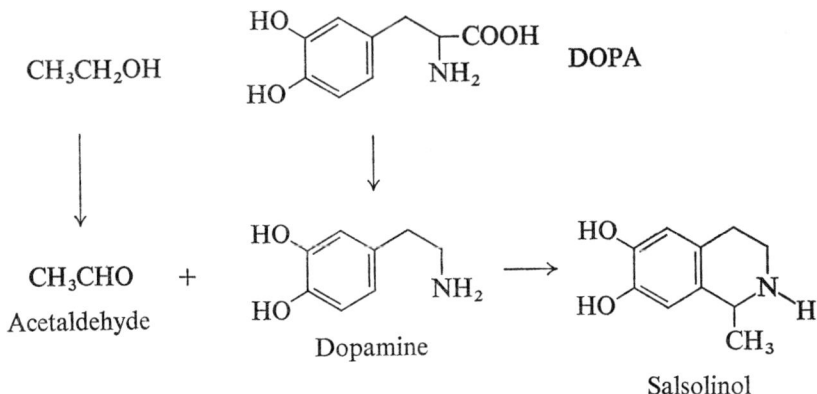

FIG. 1   Postulated mechanism of formation of salsolinol *in vivo*.

expedient of co-injection with highly deuterium-enriched material (II, $R = {}^2H$) provided the basis of a plasma assay to a level of 5 ng/ml.

A single-focussing, magnetic sector instrument (AEI MS-12) was used in multiple ion detection (MID). Basic AVA circuitry has been modified to incorporate sample and hold amplifiers with the provision of independent signal output, back-off and filtering.

METHODS

Instrumentation

A Varian 1400 gas chromatograph coupled via a silicone membrane separator to an A.E.I. MS-12 was employed in GC-MS. The basic accelerating voltage was 8 kV; trap currents of 100 and 250 μA were used at ionizing voltages of 12–18 and 20–24 eV respectively. Sample and hold circuitry for MID is essentially as previously described.[6,7] Most of this work has involved the use of three channels. Typically (Fig. 2, mass range 2·5%) channels are switched every 0·16 sec, with a 0·05-sec sampling time (five times the time constant used in signal averaging). The disconnection of each sample-and-hold circuit and the switching of the high voltage are simultaneous, but a delay is introduced before connection of the next sample-and-hold circuit. Faster switching rates have been briefly investigated (0·05 sec sampling time and 0·05 sec delay). Output is taken via low-pass filters (0·5 or 0·2 sec time constant) to a multichannel pen recorder.

Salsolinol

Salsolinol was isolated from urine (5–10 ml) either by direct adsorption on alumina or in some instances this was preceded by an initial purification on Dowex 50WX8. The sample at pH 5 was passed through a bed of resin (1 g dry wt) and after washing, the amines were eluted with 3N HCl. Catecholamines were protected by addition of 0·4 ml 10% EDTA and 0·4 ml of 20% NaH_2PO_2 and the pH was adjusted to 8·5 with 50% K_2CO_3 prior to adsorption on alumina. The amines were eluted with 3 ml 0·5 M acetic acid.

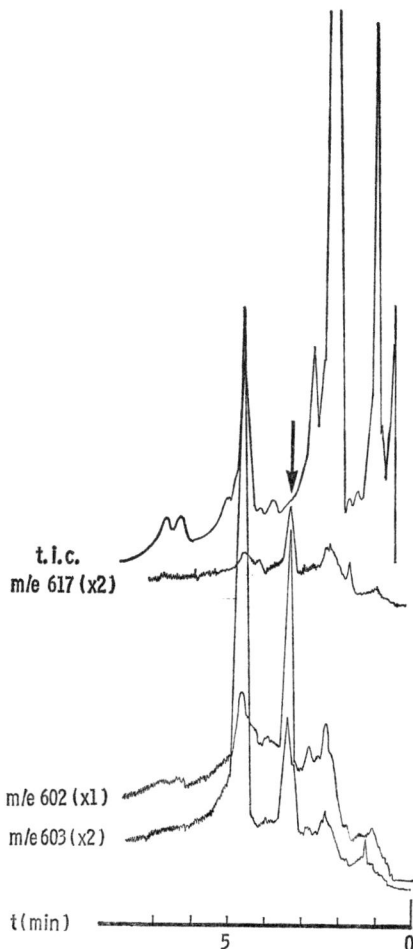

FIG. 2   Parkinsonian patient urine after L-dopa and ethanol. The urine was prefiltered on cation exchange and alumina columns and the extract chromatographed as the PFP derivatives monitoring total ion current, m/e 617, 603 and 602 (for conditions, see text). The indicated point is the elution time of authentic salsolinol-PFP.

Derivatisation with PFP anhydride was carried out on the residue of this eluate after removing the acetic acid under vacuum. Aliquots ($\frac{1}{20}-\frac{1}{50}$) were assayed by GC-MS via a 6 ft × $\frac{1}{8}$ inch o.d. 3% OV-17 column at 150° (helium flow rate, 30 ml/min). Deuterium labelled salsolinol was prepared by condensation of dopamine and acetaldehyde-d$_4$.

Indoramin

Deuterium labelled indoramin (II, R = ^{2}H) was prepared by amide formation using benzoyl chloride-d$_5$. Plasma extraction was effected by addition of indoramin-d$_5$ (30 μg) to plasma (3 ml) made basic with NaOH, extraction into heptane/isoamyl-alcohol, back extraction into HCl and after pH adjustment, recovery into ether. Recovery checked with ^{14}C labelled drug was 33–40% and was effective in the separation of indoramin from its metabolites. Aliquots ($\frac{1}{8}$ to $\frac{1}{5}$) of the final concentrates were assayed by GC-MS via a 3 ft × $\frac{1}{8}$ inch 3% OV-1 column at 260° (helium flow rate, 30 ml/min) monitoring m/e 217 (base peak of II, R = H) and m/e 223 and 224 as the reference ions in indoramin-d_5.

RESULTS AND DISCUSSION

Salsolinol

Reaction of salsolinol, [1,2,3,4-tetrahydro-6,7-dihydroxy-1-methylisoquinoline] with PFP anhydride gave a derivative (I, R = H) with good gas chromatographic properties and a simple mass spectrum in which methyl group loss from the molecular ion provided the base peak (m/e 602, 47% Σ_{40} at 22 eV). In the qualitative identification of this derivative, the ions monitored were m/e 602, 603 and 617 (M). Response ratios in chromatograms obtained from parkinsonian patient urine, following one and two stage purification of the catecholamine fraction, were in good agreement with values for authentic salsolinol-PFP recorded under identical conditions (e.g. channels ratios: m/e 617/602 and 603/602: authentic sample, 0·14 and 0·21 respectively; subject No. 1, 0·13 and 0·18; subject No. 2, 0·15 and 0·20) indicating appreciable concentrations of this metabolite in urine. (The possible presence of related metabolites, suggested by the complexity of chromatograms obtained from some of the patients studied (e.g. Fig. 2), is being further investigated.)

The instability of salsolinol in solution, apparently random losses in the concentration stages and the sensitivity of the derivative to hydrolysis, contributed to a ten-fold variation in the absolute response to recovered standards. Standardization was achieved by addition of 1 μg/ml of salsolinol-d_4 (I, R = ^2H) to urine. Estimation was by ratio measurement, monitoring m/e 617 (molecular ion of the unlabelled compound) and m/e 621 (M) and 623 as the reference ions in the d_4-derivative. Calibration was linear (R = 0·995) in the range 5 ng to 1 μg/ml. The uncertainty of the method was reflected in a standard deviation of 7·3% (n = 25) in the measurement of standard, 1 in 10, dilutions of the unlabelled salsolinol in the deuterated form.

Trace concentrations (2·5 to 10 ng/ml) in the urine of normal subjects were tentatively attributed to salsolinol formed in the absence of exogenous L-dopa. In a parkinsonian patient on L-dopa treatment (4 g daily, by mouth) concentrations of 35 and 220 ng/ml were detected in urine collected 0–3 and 6–9 hr following the ingestion of ethanol (75 ml).

The prediction[8] that tetrahydroisoquinolines should be formed in vivo is supported by the finding of salsolinol and the related tetrahydropapaveroline in human urine[5] and is confirmed for salsolinol in the present study. The raised excretion of salsolinol after administration of L-dopa and ethanol indicates its formation by the Pictet–Spengler condensation between dopamine and acetaldehyde (Fig. 1). This, and analogous tetrahydroisoquinolines may have an important pharmacological role. They are structurally similar to morphine-related alkaloids[8] and can, it is claimed, act as false neurotransmitters.[9] Their role in parkinsonism and alcoholism remains to be defined.

Indoramin (II, R = H)

In the preceding example, isotope dilution was required as a means of standardization for random sample loss in recovery from biological fluid. Favourable gas chromatographic behaviour and the selectivity of MID then allowed trace determination in relatively complex chromatograms. Contrasting methodological problems were encountered in the specific plasma assay of the drug indoramin (II, R = H), 3-[2-(4-benzamido-piperid-l-yl)-ethyl indole]. This substance is chemically stable and lipid soluble and may be concentrated free from its metabolites by organic solvent extraction. However, the detection limit in GC-MS, carrier-free, was of the order of 100 ng with

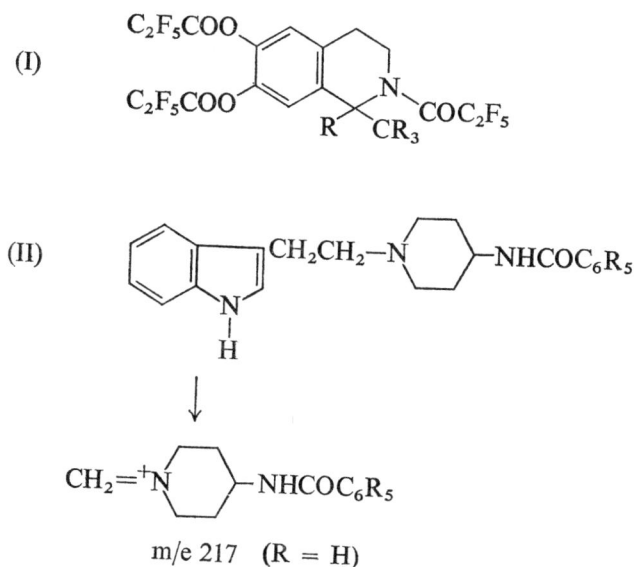

FIG. 3   (I) PFP (COC_2F_5) derivative of salsolinol (R = H or ^{2}H). (II) Indoramin (R = H or ^{2}H) and principal fragment ion.

poor resolution and peak distortion attributable chiefly to variable inlet system (column and separator) adsorptive effects. The lack of a simple alternative assay prompted an investigation of stable isotope dilution and the effect of labelled carrier in compensation for gross chromatographic problems.

The base peak in the spectrum of II (R = H) was m/e 217, $C_{13}H_{17}N_2O$, due to the benzamidopiperidine group (Fig. 3) (60% Σ_{40} at 12 eV). The residual contribution at this mass in the spectrum of indoramin-d_5 (II, R = ^{2}H) was <0·1% of the base peak, m/e 222, allowing the use of high dilutions of sample in labelled carrier. Marked inlet system memory effects were initially encountered; thus a single injection on the column of a 1 μg sample of the hydrogen form required six to eight injections of the deuterium-labelled compound for the complete elimination of residual signal at m/e 217. Such effects were substantially abolished by restricting the dynamic range of measurement to dilutions of 1 in 2000 to 1 in 100 and by ensuring carrier injection at the 1·5 to 3 μg level. These requirements could be met in plasma assay. Recovery of indoramin from plasma (3 ml) proved to be highly reproducible in the presence of 10 μg/ml of deuterated carrier and one-third to one-fifth of the final, comparatively clean extract, could be assayed by GC-MS. Linear calibration was obtained from 5 to 100 ng/ml with a standard deviation of 5–14% throughout this range, providing a method applicable to single and multiple dose pharmacokinetic studies in man.

REFERENCES

1. Sweeley, C. C., Elliot, W. H., Fries, I. and Ryhage, R. *Anal. Chem.*, 1966, **38**, 1549.
2. Hammar, C.-G., Holmstedt, B. and Ryhage, R. *Anal. Biochem.*, 1968, **25**, 532.
3. Rittenberg, D. and Foster, G. L. *J. Biol. Chem.*, 1940, **133**, 737.
4. Gaffney, T. E., Hammar, C.-G., Holmstedt, B. and McMahon, R. E. *Anal. Chem.*, 1971, **38**, 301.
5. Sandler, M., Bonham Carter, S., Hunter, K. R. and Stern, G. M. *Nature*, 1973, **241**, 439.
6. Kelly, R. W. *J. Chromatogr.*, 1972, **71**, 337.
7. Draffan, G. H., Clare, R. A., Williams, F. M., Emons, E. and Jackson, J. L. 'Int. Symposium on Mass Spectrometry in Biochemistry and Medicine', Milan, Italy, May 1973.
8. Sourkes, T. L. *Nature*, 1971, **229**, 413.
9. Cohen, G. *in:* 'Frontiers in catecholamine research', Abstracts of III International Catecholamine Symposium, Strasbourg, France, May 1973.

29

Residue Determination of an Insect Growth Regulator by Mass Fragmentography

By LOREN L. DUNHAM

(*Zoecon Corporation, Palo Alto, U.S.A.*)

ROGER J. LEIBRAND

(*Hewlett–Packard, Palo Alto, U.S.A.*)

THE isolation and structure elucidation of three natural juvenile hormones (JH) in recent years[1-4] has led to increased interest in the possibility of insect control through the use of insect growth regulators.

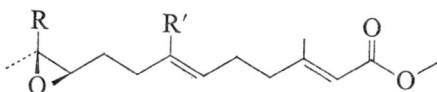

$$I \quad R = R' = Et$$
$$II \quad R = Et, R' = Me$$
$$III \quad R = R' = Me$$

These analogues are compounds which mimic the natural JH's (above) in biological activity. The ideal compound should be one which is stable in the environment, as opposed to the labile epoxide moiety of the known JH, but not persistent and should possess biological activity equal to or greater than the natural JH. The compound should also have low toxicity to non-target organisms.[5]

A highly potent series of Insect Growth Regulators (IGR's) of the alkyl 3,7,11-trimethyl-2,4-dodecadienoate class has been synthesized[6] and tested for biological activity.[7-13]

One of these compounds, isopropyl (2E,4E)-11-methoxy-3,7,11-trimethyl-2,4-dodecadienoate (Altosid™) shows high biological activity by the prevention of adult emergence of certain dipteran insects.[5,6,8,11,12,15]

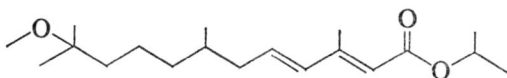

Evaluation of the persistence of compounds of this type in the environment is necessary for correlations of concentrations with biological response[14] as

251

well as to satisfy the requirements of regulatory agencies. Since the compounds are tested at concentrations as low or lower than the ones used for common pesticides, the analysis of residue is difficult. Inasmuch as the IGR's are analysed by gas chromatography and possess only carbon, hydrogen, and oxygen atoms, no highly sensitive or specific GLC detector is normally available. Also since the test compound exhibits many of the chemical characteristics commonly found in the environment, interfering impurities frequently are encountered during analysis.

To increase the sensitivity and specificity of detection of compounds of interest in residue analysis the technique of mass fragmentography[16,18] has been investigated.

The degree of sensitivity attainable in mass fragmentography is dependent upon the relative abundance of the ions to be monitored (Fig. 1) compared

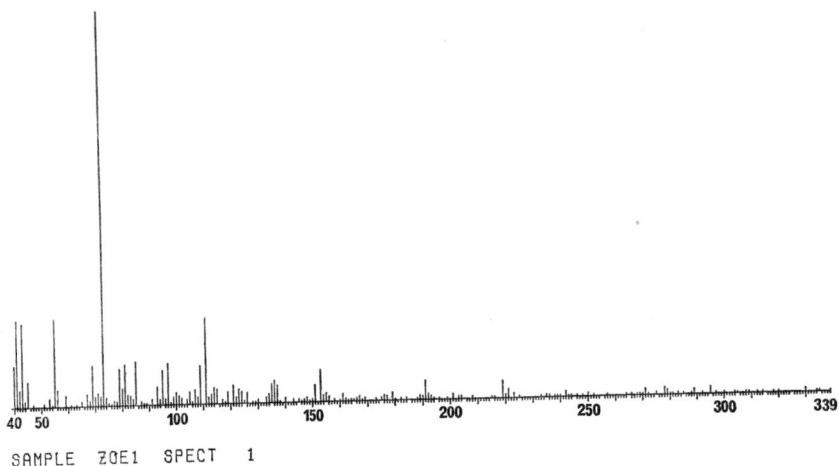

SAMPLE ZOE1   SPECT   1

FIG. 1   Computer constructed diagram representing the 70 eV spectrum of Altosid™.

to those present as background. The specificity of the method is dependent upon the degree of diagnosticity which may be placed upon the monitored ions. As a proof of the genesis of the ions to be monitored by this technique, selective deuteration was performed.[17] Comparison of the fragmentation characteristics of the deuterated samples with those of the corresponding non-deuterated analogues showed that ions occurring at m/e 55, 73, 111 and 153 (Fig. 2) were indeed diagnostic to the presence of the sought compound when encountered in environmental samples.

Although any one of the ions might derive from other sources than the fragmentation of the molecule of interest, the simultaneous monitoring of several diagnostic ions considerably increases the specificity of the method (Fig. 2).

Of the several possible applications for Altosid™, one has been a study of the control of the common housefly (*Musca domestica*) in commercial poultry operation. Since pupation occurs in manure, control of the pest in large egg-producing operations becomes a significant problem. One approach

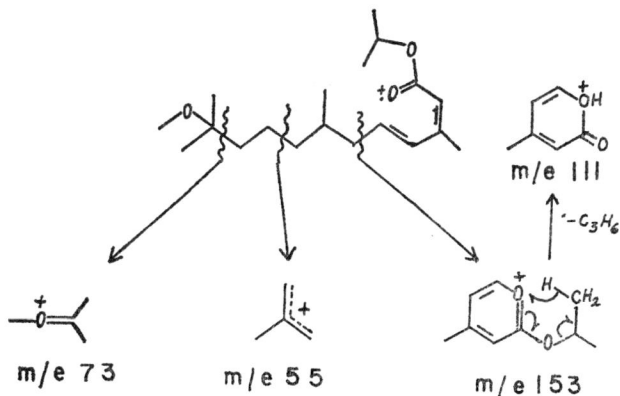

FIG. 2   Electron impact fragmentation routes leading to diagnostic ions of Altosid™.

to administration of the compound has been as a feed additive to be consumed by the poultry. In addition to evaluation of efficacy it becomes necessary to determine the amount of compound present as residues in manure. Also, since poultry meat and eggs are destined for human consumption it is necessary to determine residues therein. Due to the difficulties encountered in clean-up of samples for analysis mass fragmentography was chosen as a confirmation of the presence of Altosid™ residues.

INSTRUMENTAL

Mass spectral data were obtained using a Hewlett–Packard 5930A mass spectrometer operated at an ionizing voltage of 70 eV with an ion source temperature of 190°C with dodecapole at 150°C. The mass spectrometer was coupled to a Hewlett–Packard 5780A gas chromatograph via a silicone membrane molecular separator operated at 200°C with a glass transfer line at 250°C. The chromatography was accomplished utilizing a 0·8 m × 2·0 mm (i.d.) glass column filled with 3% (w/w) methylsilicone (OV-101) coated on 100/120 mesh Chromosorb W AW DWCS.[19] The column was operated at 170°C with helium carrier gas at 40 ml/min.

MULTIPLE ION DETECTION PROGRAM

The mass fragmentography was performed using the multiple ion detection (MID) program in a Hewlett–Packard 5932A data system which employs a Hewlett–Packard 2100A minicomputer with 8K of core memory.

This program monitors the height of up to four mass peaks during a gas chromatograph run. It samples the ion current at four specified masses, stores the samples, and plots them after the run has been completed. It also samples the total ion current, plots it during the run, and stores it for plotting again later. It is useful in the detection of very small quantities of material.

In the first part of the program the dodecapole mass filter is sequentially stepped through the specified masses once per second. While at each mass the ion current is sampled 412 times and averaged. Since the multiplier gain is degraded by long exposure to large currents, only ten samples of total ion current are taken. When the run is completed the total ion current and the currents at the specified masses may be plotted one or more times with different gains and zero offsets.

RESULTS

A chromatogram utilizing multiple monitoring at m/e 55, 73, 111 and 153 of Altosid™ is shown in Fig. 3.

A multiple ion chromatogram of 100 ng of Altosid™ in an extract of poultry tissue is shown in Fig. 4. Multiple ion monitoring of the sample shows that retention time coincidence of only one peak occurs for all four monitored ions. This peak may then be identified as the compound of interest. Further analysis of a 1:10 dilution of the 100 ng sample illustrates that ample material is still present at the 10 ng level to allow confirmation of the residue.

Multiple ion monitoring of a sample derived from poultry manure (Fig. 5) indicates coincidence of two major components from an injection containing 150 ng of Altosid™. The later eluting peak is found to be the sought compound, while the earlier component was identified as an isomer arising from equilibration of the 2-ene double bond.

ALTOSID IGR 100NG STD

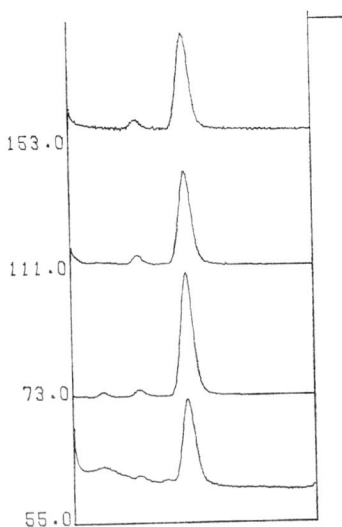

Fig. 3   Multiple ion program representation of a 100 ng standard of Altosid™ monitoring m/e 55, 73, 111 and 153.

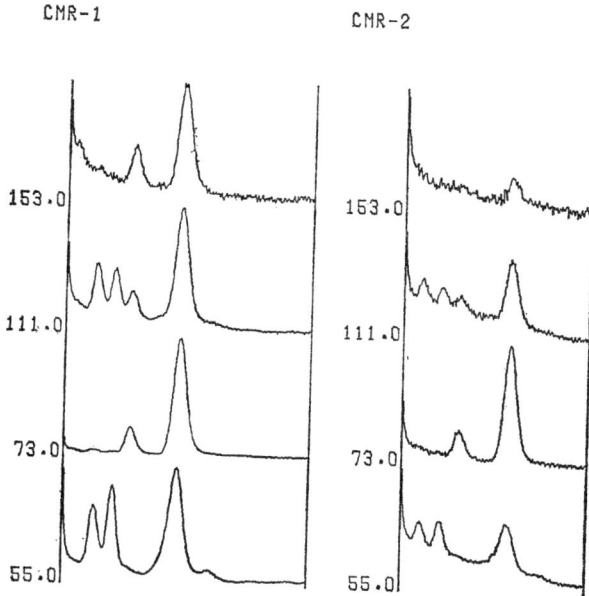

FIG. 4   MID program representation of 100 ng and 10 ng levels of Altosid™ in poultry tissues.

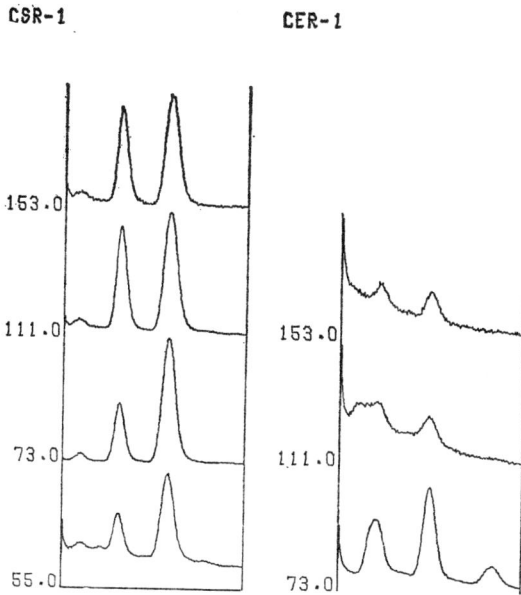

FIG. 5   MID program representation of 150 ng of Altosid™ in poultry manure and 100 ng in poultry eggs.

Figure 5 represents a 100 ng injection of the compound derived from poultry eggs. Due to the extensive amount of lipoid materials present in this type of sample the ions exhibited at m/e 55 become ambiguous for purposes of confirmation. However, ions at m/e 73, 111 and 153 still yield sufficient data to verify the presence of Altosid™.

DISCUSSION

Multiple ion detection is extremely sensitive and specific, as is shown by our studies as well as those of other investigators.

The ultimate minimum detectable limit is usually determined by the relative abundance of the ions in question as opposed to the background interferences in the sample matrix. The signal at 3:1 signal-to-noise ratio at m/e 153 resulting from 10 ng of Altosid™ in the poultry tissue sample (CMR-2) has clearly reached this limit.

In other cases, wherein ion abundances differ from this class of compounds, 10 pg of p-p'-DDT has been detected using the M.I.D. program.

ACKNOWLEDGMENTS

The authors are deeply indebted to Wayne Miller whose untiring efforts have resulted in the gas-chromographic methods necessary for analysis of test compounds in environmental samples and to Norton Bell who developed the Multiple Ion Detection software.

REFERENCES

1. Roller, H., Dahm, K. H., Sweeley, C. C. and Thrust, B. M. *Angew. Chem. Int. Ed. Engl.*, 1967, **6**, 179.
2. Meyer, A. S., Schneiderman, H. A., Hanzmann, E. and Ko, J. H. *Proc. Natl. Acad. Sci. U.S.*, 1968, **60**, 853.
3. Meyer, A. S., Hanzmann, E., Schneiderman, H. A., Gilbert, L. I. and Boyette, M. *Arch. Biophys.*, 1970, **137**, 190.
4. Judy, K. J., Schooley, D. A., Dunham, L. L., Hall, M. S., Bergot, J. B. and Siddall, J. B. 'Isolation, Structure and Absolute Configuration of a New Natural Insect Juvenile Hormone from *Manduca sexta*', *Proc. Nat. Acad. Sci. U.S.*, May 1973, Vol. 70, No. 5, p. 1509–13.
5. Diekman, J. D. 'Use of Insect Hormones in Pest Control', Proc. of a National Extension Insect-Pest Management Workshop, Purdue University, 14–16 March 1972, pp. 69–73.
6. Henrick, C. A., Staal, G. B. and Siddall, J. B. 'Alkyl 3,7,11-Trimethyl-2,4-dodecadienoates, A New Class of Potent Insect Growth Regulators with Juvenile Hormone Activity', *J. Agr. Food Chem.*, May–June 1973, Vol. 21, p. 354–59.
7. Dunn, R. L. and Strong, F. E. 'Control of catch-basin mosquitoes using Zoecon ZR 515 formulated in a slow release polymer—a preliminary report', *MOSQ News*, 1973, 33(1), 110–11.
8. Cupp, E. W. and O'Neal, J. 'Morphogenetic Effects of Two Juvenile Hormone Analogues on Larvae of Imported Fire Ants', *Environ. Entomol.*, 1973, 2(2), 191–4.
9. Cumming, J. E. and McKague, B. 'Preliminary Studies of Effects of Juvenile Hormone Analogues on Adult Emergence of Black Flies (Diptera: Simulidae)', *Can. Entomol.*, 1973, **105**(3), 509–11.

10. Strong, F. E. 'Juvenile Hormone Analogs—Third Generation Pesticides?'; 'Biting Fly Control & Environmental Quality', Proc. of Symposium at Univ. of Alberta, 1972, pp. 35–7; 1973.
11. Schaefer, C. H. and Wilder, W. H. 'Insect Developmental Inhibitors: Practical Evaluation as Mosquito Control Agents', *J. Econ. Ent.*, 1972, **65**(4), 1066–71.
12. Jakob, W. L. 'Additional Studies with Juvenile Hormone-type Compounds Against Mosquito Larvae', *Mosquito News*, 1972, **32**(4), 592–5.
13. Jakob, W. L. 'Insect Development Inhibitors: Tests with House Fly Larvae', *J. Econ. Ent.*, 1973, **66**(3), 819–20.
14. Schaefer, C. H., and Dupras, E. F. Jr, 'Insect Developmental Inhibitors. 4. Persistence of ZR-515 in Water', *Jour. Econ. Ent.*, Vol. 66, No. 4, p. 923.
15. Schaefer, C. H. and Wilder, W. H. 'Insect Developmental Inhibitors. 2. Effects on Target Mosquito Species', *J. Econ. Ent.*, Vol. 66, No. 4, p. 913.
16. Henneberg, D. Z. *Anal. Chem.*, 1961, **183**, 12.
17. Dunham, L. L., Henrick, C. A. and Young, J. W. 'Electron Induced Fragmentation of Conjugated Dienoate Esters', presented at 165th Nat. ACS Mtg., Dallas, Texas, 8–13 April, 1973.
18. Gordon, A. E. and Frigeno, A. *J. Chromatogr.*, 1972, **73**, 401.
19. Leibrand, R. and Dunham, L. *Res./Develop.*, 1973, **24**, 9, 32–8.

Mass Spectrometer Studies with Negative Ions

By J. D. CRAGGS

(*Department of Electrical Engineering and Electronics,
University of Liverpool, U.K.*)

INTRODUCTION

IT is only in the last 20 years or so that research on phenomena involving negative ions has become intensive, despite earlier contributions from Tate and Lozier on single collisions, Massey and D. R. Bates on their theoretical work on the upper atmosphere and others. The extensive development of mass spectrometric techniques in more recent times has been a factor contributing to the growth of research on negative ions, although of course a stimulus has also been provided by the continuing and increasing interest shown in the upper atmosphere and by the relatively new subject of ion-molecule reactions. The latter subject is of concern for example to those working with ionization and electrical discharges in gases and combustion processes and related phenomena.

MASS DISCRIMINATION EFFECTS

The importance of mass discrimination effects in any part of a mass spectrometer/ion collector system needs no emphasis, although work is still appearing, for example on studies of ion-molecule reactions, in which relative collector currents of ions of widely differing mass have to be measured, where little or no attention is paid to these difficult effects. We shall only consider work on negative ions here.

The orifice sampling effect (differential flow) has been fully discussed by Cuthbert[1] who clearly distinguishes not only the forms of molecular and isentropic flow but also the transitional region for the conditions where the particle mean free path is of the order of the orifice diameter. In practice, however, the situation is far more complicated because of the difficulty of using the neutral gas flow formulae in cases of partially ionized gases, where particle collection by the wall of the orifice may well be important. Thus, in practice it seems desirable to test any system in which discrimination must be checked ('quantitative cracking patterns', etc.) by a method which, if possible, measures the overall sensitivity of the complete apparatus, from source to detector (see Kebarle, Haynes and Searles[2]).

Such a method was developed by Parkes[3] in connection with his drift tube/MS work on negative ion reactions in oxygen and is probably the most thorough investigation of its kind to date. This involves, following the work of Kebarle et al. (loc. cit.), a study of the factors governing the sampling of ions by an orifice in a collector plate. Parkes, however, worked at values of reduced field in his source, of up to about 20 V cm^{-1} torr^{-1} (E/p) and at pressures of only a few torr. Some studies of ion-molecule reactions and electrical discharge phenomena involve pressures (20°C) of up to, say, 60 torr, above which little quantitative mass spectrometry has so far been carried out, and E/p values of 40 V cm^{-1} torr^{-1} and upwards. The latter quantity is important because the drift velocity of ions, in the source, to the collector plate is clearly relevant (Parkes, loc. cit.). Parkes, then, gave valuable information on computed values of sampled/collector plate current ratios, and developed an interesting method for controlling the nature of the ion species in his source (for example O_3^- and O^-) by use of the following fast reaction:

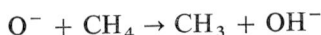

$$O^- + CH_4 \rightarrow CH_3 + OH^-$$

which, even with $<1\%$ CH_4, effectively changes O_3^- to OH^-, the former being formed by the well known dissociative attachment and ion conversion processes:

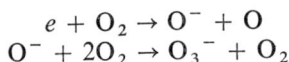

$$e + O_2 \rightarrow O^- + O$$
$$O^- + 2O_2 \rightarrow O_3^- + O_2$$

Removal of O^- by CH_4 thus affects the O_3^- formation. Parkes also used N_2O which acts in the following way (Moruzzi and Dakin).[4]

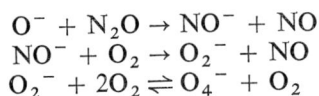

$$O^- + N_2O \rightarrow NO^- + NO$$
$$NO^- + O_2 \rightarrow O_2^- + NO$$
$$O_2^- + 2O_2 \rightleftharpoons O_4^- + O_2$$

Here O^- and O_3^- are being replaced by O_2^- and O_4^-. For 'high' reduced fields in the source i.e. $\sim E/p$ of 10 V cm^{-1} torr^{-1} the relative sensitivities were (quadrupole MS system) for $O^-:O_2^-:O_3^-:O_4^-$ as 1:1·2:2·0:2·75, to be combined with a variable factor lying between 1 and 2 between O^- and the other negative ions which depends on the form of focussing used. This work, therefore, indicates an important amount of overall mass discrimination, for the particular conditions used.

Unpublished work has recently been carried out by McGeehan in this laboratory in which he made careful checks on discrimination in negative ion sampling in SF_6. In the first place he studied the overall linearity of response of the complete system, which comprised a uniform-field Townsend source, sampling orifice (0·076 mm and 0·127 mm diameter), electrostatic collecting lens, GE monopole mass spectrometer and detector (10-stage electron multiplier having Ag–Mg dynodes). Linearity was satisfactory for discharge currents varying from about 5×10^{-10} to 10^{-7} A for pressures in the range 1–37 torr using SF_6^- and SF_5^- separately. The absence of serious discrimination of the orifice, or of ion-molecule reactions taking place

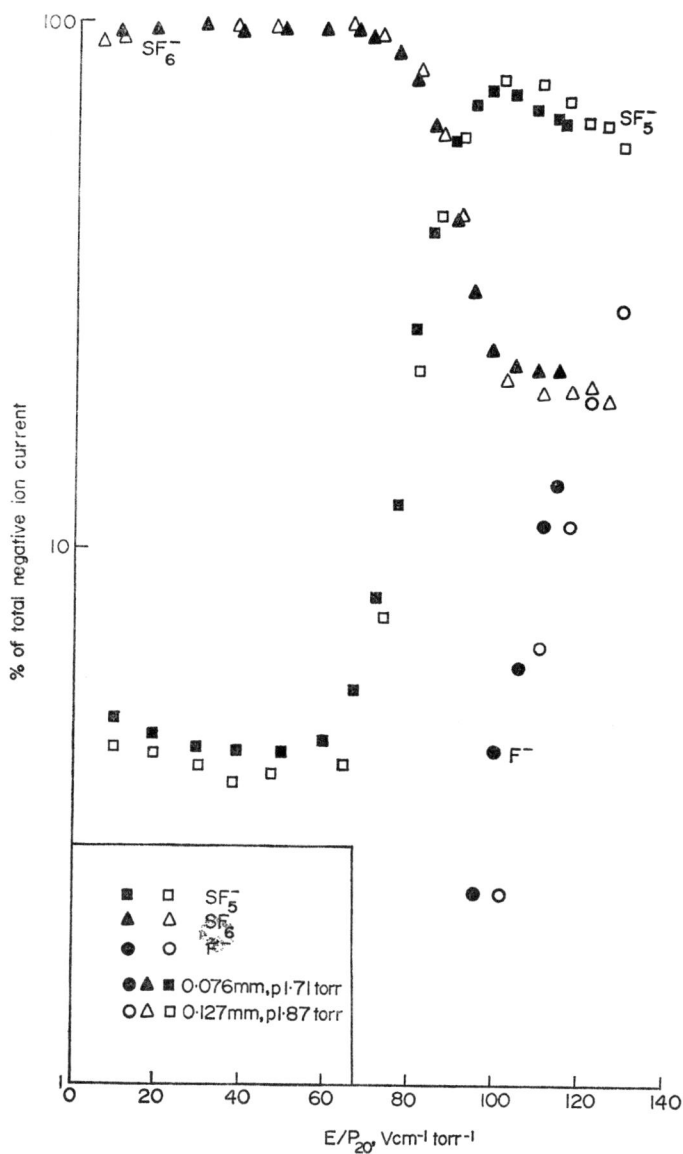

FIG. 1   Sampling of negative ions through apertures 0·076 mm and 0·127 mm in diameter.

downstream of the orifice was confirmed with SF_6^-, SF_5^- and F^- over about the pressure range mentioned, by the use of the two orifice diameters, other parameters being held constant. Some results are shown in Fig. 1 (Kinsman et al.)[5] and which indicate a tolerable consistency ($\pm 10\%$ overall on observed relative ion currents would be highly acceptable). The skimmer electrode extraction field (Fig. 2) is clearly a possible source of error and its effect on observed ion current ratios (SF_6^-/SF_5^-, SF_6^-/F^- and SF_5^-/F^-)

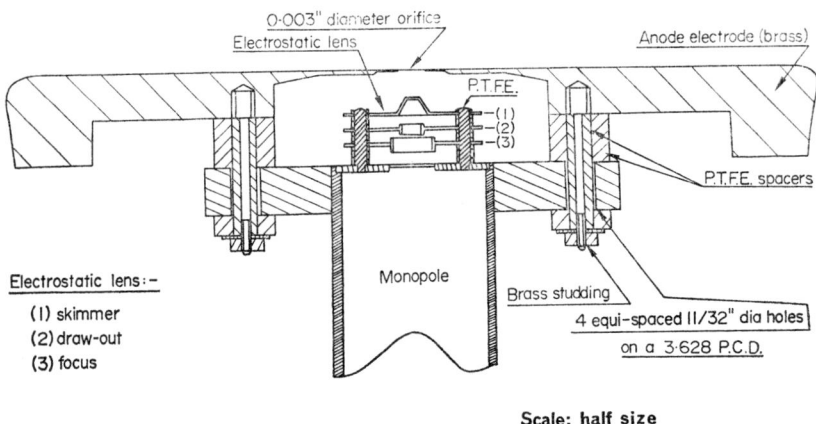

0·003" diameter orifice
Electrostatic lens
P.T.F.E.
Anode electrode (brass)
—(1)
—(2)
—(3)
P.T.F.E. spacers

Electrostatic lens:—

(1) skimmer
(2) draw-out
(3) focus

Monopole
Brass studding
4 equi-spaced 11/32" dia holes
on a 3·628 P.C.D.

Scale: half size

FIG. 2   Negative ion focussing system.

was studied. It was found that providing the anode-skimmer voltage exceeded 70 V the observed ratios did not change appreciably, at least up to 120 V. Mass discrimination in the monopole MS was tested and found to be negligible following Ong and Hasted[6] who varied the resolution of the MS and took successive mass scans to detect discrimination (which would depend on the resolution). Finally, the electron multiplier used as detector was tested by varying the accelerating potential between the MS exit slit and the first dynode of the multiplier. No evidence of substantial discrimination was found for SF_6^-, SF_5^- and F^-.

Kinsman and Rees,[7] working with oxygen and an MS2 mass spectrometer, used a method for measuring discrimination effects beyond those occurring in the sampling orifice jet, in which the total discharge (source) current I_t was maintained constant whilst the relative concentrations of O_2^- and O_3^- (O^- was negligible in these particular experiments) were changed by alteration of the electric field in the source. We have

$$I_t = AI_2 + BI_3$$

where $1/A$ and $1/B$ and I_2 and I_3 are the overall sampling efficiencies and the separate ion currents for O_2^- and O_3^- respectively. By varying E/N in the source, with N constant, and measuring I_2 and I_3 as functions of I_t, the constants A and B could be found. In a particular case at 11 torr source pressure $A \simeq B$, whilst at 20 torr, A was approximately 5% higher than B, although the possible errors in the A and B values were about $\pm 10\%$. In this case the sampling orifice was about 0·12 mm in diameter, so that the above data were taken for a flow regime transitional between molecular and viscous flow. The MS2 with high pressure source has, of course, a carefully designed differential pumping system to take care of gas flow through the orifice (Kinsman and Rees,[8] following Knewstubb and Sugden,[9] who also refer to sampling problems). The results of Kinsman and Rees have been recently confirmed for a quite different experimental system, incorporating a quadrupole mass spectrometer, over an oxygen pressure range of 20–66 torr (20°C) by O'Neill and Craggs[10] and using essentially the same technique.

An interesting investigation into ion-molecule reactions occurring in and near the extraction or sampling orifice has recently been carried out in Innsbruck by Helm et al.[11] They studied in particular, for example, the reaction:

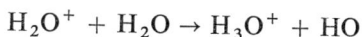

$$H_2O^+ + H_2O \rightarrow H_3O^+ + HO$$

using a hollow cathode discharge tube as source, but their work is of more general interest for positive or negative ion reactions. These authors used an expression due to Walcher[12] for molecular flow conditions through an ideal orifice, namely:

$$n_x = \frac{n_0}{2} \left(1 - \frac{x}{(R^2 - x^2)^{\frac{1}{2}}} \right)$$

where n_0 and n_x are the gas densities in the source and in the escape jet along

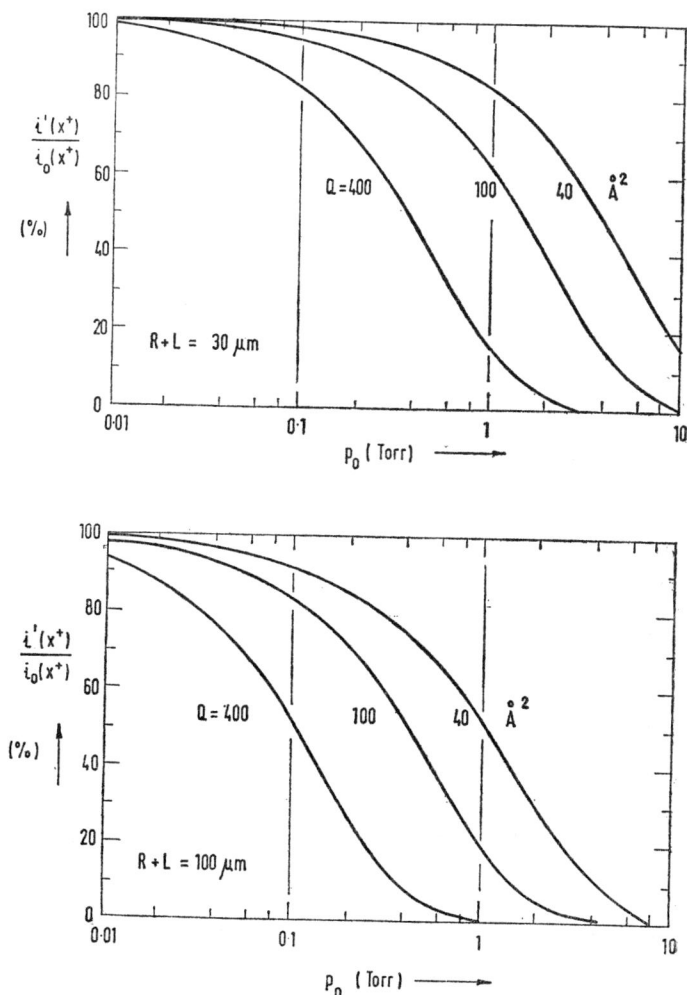

FIGURE 3

the x axis and R is the orifice radius. Thus, on integration:

$$\int_0^\infty n_x d_x = \frac{n_0 R}{2}$$

to which a term $n_0 L/2$ is added to allow for the gas in an orifice of finite length L. The limitation to molecular flow should not be overlooked.

If we now consider a reaction, with cross section Q, in which an ion X^+ is converted into another ion, we can put

$$i'(X^+) = i_0(X^+) \exp\left(-Q(L + R)\frac{n_0}{2}\right)$$

Figure 3 shows the calculations for a range of pressures and various values of Q. It should be pointed out that even for $Q = 40$ Å^2 an equivalent rate coefficient R would be $\sim 10^{-10}$ cm^3/sec for reactions involving typical negative ions and which is considerably faster than many negative ion reactions of interest, with notable exceptions such as the associative detachment reaction O^-/H_2 as described by Moruzzi and Phelps[13] and by Price et al.[14] Reference to the detailed work by Parkes on ion-molecule reactions in sampling jets is made elsewhere in this paper.

SINGLE COLLISIONS

The accurate measurement of threshold energies (critical potentials) and peak heights and widths for electron attachment processes still preoccupies many workers on negative ions, and an outstanding example of modern technique is due to Thynne and his collaborators who have recently developed deconvolution methods, following earlier work by Morrison[15] and others, to eliminate the effect of finite energy spread in the primary electron source.

They used a Bendix 3015 time-of-flight mass spectrometer with refined recording methods (MacNeil and Thynne[16]), and measured the electron energy distribution by the SF_6^- probing method (Hickam and Fox[17]) and, essentially, solved the equation for the probability of negative ion formation for electrons of energy V, i.e.

$$i(V) \propto \sum_{U=0}^\infty I(V + U)m(U)\,\Delta U$$

where $m(U)$ is the fraction of electrons having energy between U and $U + \Delta U$, by computing methods. They also studied the effect of deliberately varying the electron energy distribution from the observed, true, value. An example is shown in Fig. 4, for O^- from SO_2, from the paper by MacNeil and Thynne cited above, and as further examples data on SiF_4 and CF_4 have also been published by these authors.[18]

The cross section $\sigma_{da}(\varepsilon)$ for dissociative attachment may be put approximately (after O'Malley) as

$$\sigma_{da}(\varepsilon) = \sigma_0 \exp\left(-\frac{\tau_s}{\tau_a}\right)$$

where σ_0 is the cross section for the formation of AX^{-*} and τ_s is the time of separation of A and X^- to the crossing point between the negative ion and neutral molecule potential energy curves. τ_a is the mean autoionization lifetime (Christophorou[19]). It is readily seen that the reduced mass of the fragments enters into σ_0 and τ_s so that isotope effects are to be expected, and have been observed by several workers (Schulz, Rapp and others).

Time-of-flight mass spectrometers have been used by various workers since the time of Edelson, Griffiths and McAfee[20] to study mean autoionization (autodetachment) lifetimes $> 10^{-6}$ sec and a good account of this work is due to Christophorou.[19] Essentially the method consists of measuring the number of negative ions decaying into neutrals after a given flight time down

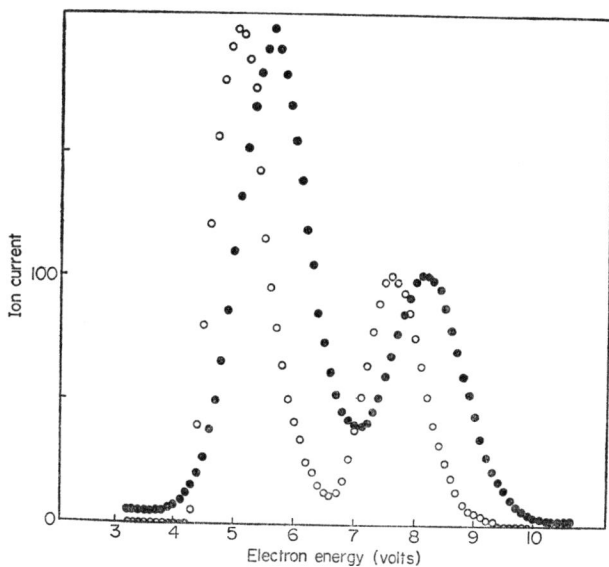

FIG. 4   Ionization efficiency curves for O^- ion formation by sulphur dioxide. Full circles: original experimental data; open circles: deconvoluted results. 15 smoothing and 16 unfolding iterations.

a drift tube; of course various precautions have to be taken, the details of which may be found in the publications exhaustively referred to by Christophorou (loc. cit.). These measurements have relevance to the theory of non-dissociative electron capture, although many of the data, for various reasons, have been taken with complicated molecules which are not readily amenable to precise analytical study. The general physical picture of the autodetachment process involves essentially the distribution of the excess energy available at a collision among the various vibrational degrees of freedom of the molecule concerned. From such work, measurements of mean lifetime and capture cross sections enable estimates of electron affinities to be made. The work of Compton et al.[21] established a model for this kind of study. The dependence of life-times on the number of vibrational degrees of freedom is shown in Fig. 5, taken from the work of Naff, Compton and Cooper.[22]

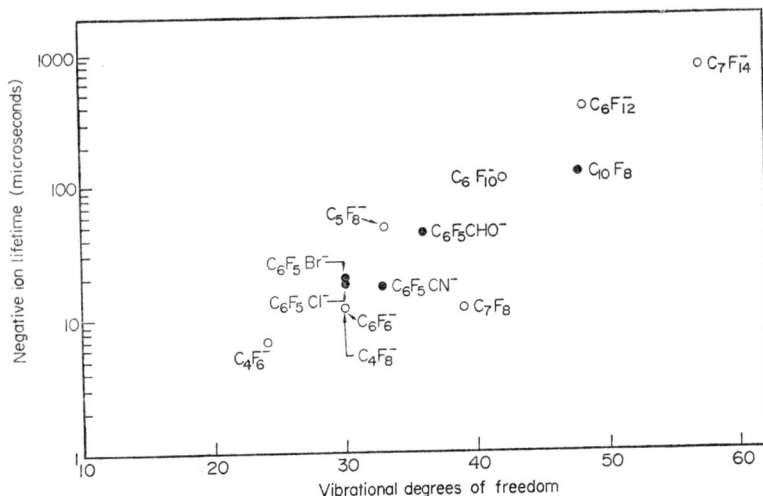

FIG. 5   Dependence of negative ion lifetimes of highly fluorinated parent ions on the number of degrees of vibrational freedom.

Representative data for lifetimes in microseconds taken from Christophorou's book are as follows: SF_6 25–32, C_4F_8 (perfluorocyclobutene) 6·9, and C_7F_{14} (perfluoromethylcyclohexane) 793. These gases are of interest as electrical insulants.

Further work on this subject is due, for example, to Naff et al.[22] working with substituted benzenes. They also used a Bendix t.o.f. mass spectrometer and an r.p.d. source with an energy resolution of about 0·1 eV.

An interesting case of t.o.f. spectrometry is the study by Edelson et al.[20] on SF_6, when the lifetime of SF_6^{-*} was found to be about 10 μsec. The authors point out that the determination of lifetime depends here on the relative sensitivity, of the secondary electron multiplier used as detector, to neutral molecules and ions (accelerating voltage up to 4·5 kV) and they refer to unpublished work where the relative multiplier sensitivities to A, A^+ and A^{++} in presumably similar conditions are within ±20%.

Using a Varian MAT CH4 MS with an r.p.d. source, Lifshitz et al.[23] studied certain polyatomic negative ions, e.g. $C_7F_{14}^-$, SF_6^-, $C_6F_{12}^-$ in relation to their stabilization or break-up mechanisms. An example of the work is shown in Fig. 6 where data on parent, fragment and 'metastable' ions are given. The purpose of this work was to assess the quasi-equilibrium theory (q.e.t.) of mass spectra as applied to negative ions. The fragmentation reactions for Fig. 6a and b are (ms ≡ metastable transition observed)

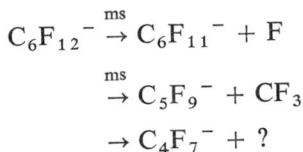

$$C_6F_{12}^- \overset{\text{ms}}{\rightarrow} C_6F_{11}^- + F$$

$$\overset{\text{ms}}{\rightarrow} C_5F_9^- + CF_3$$

$$\rightarrow C_4F_7^- + ?$$

and, as would be expected, the relative yields vary quite critically with

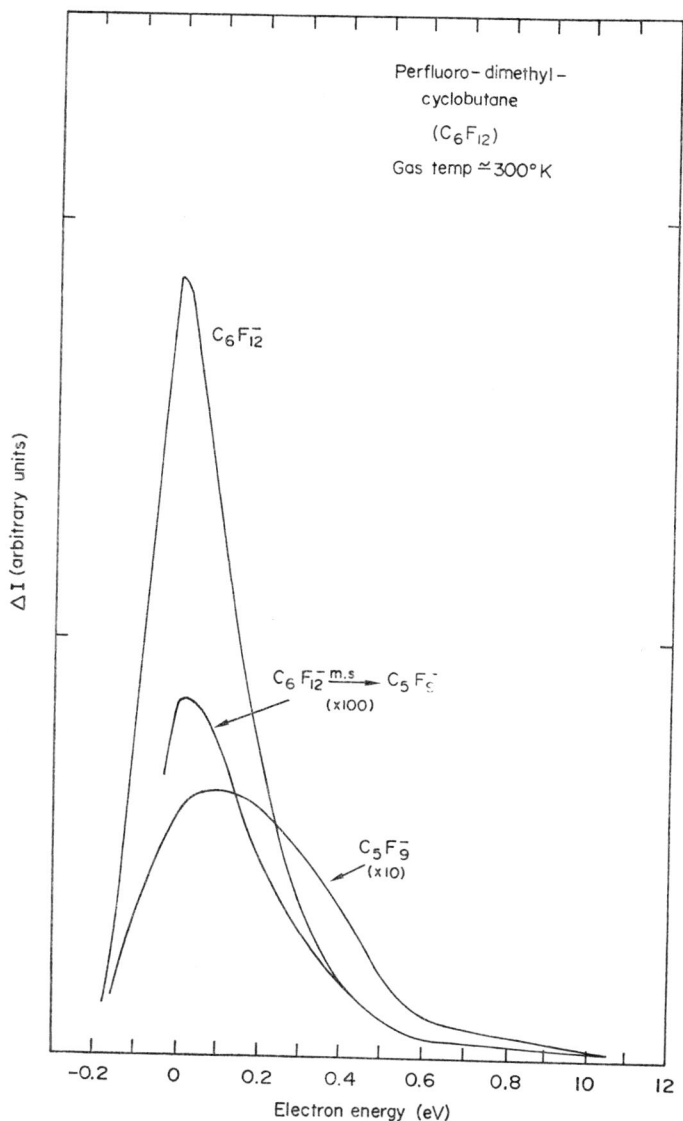

FIG. 6(a) Electron-capture curves for parent, fragment and 'metastable' ions for the reaction: $C_6F_{12}^- \rightarrow C_5F_9^- + CF_3$ in perfluoro-1,2-dimethylcyclobutane. The gas temperature is $\sim$300°K. Notice the scale factors indicated.

temperature. These data enable the fractions of parent, metastable and fragment ions to be determined as a function of electron energy and the resulting functions to be compared with the predictions of q.e.t.

The efficiency with which SF_6 captures low energy electrons has been exploited in various ways to study inelastic electron collision processes in gases, following work of Curran[24] and Jacobs and Henglein.[25] Although

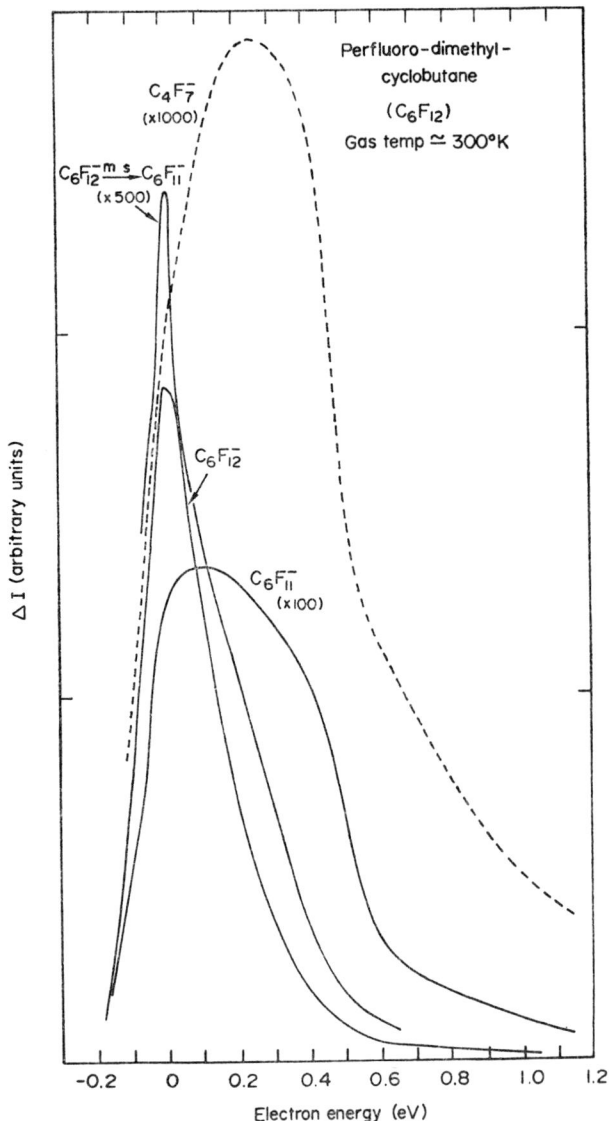

FIG. 6(b) Electron-capture curves for parent, fragment and 'metastable' ions for the reaction: $C_6F_{12}^- \rightarrow C_6F_{11}^- + F$ and for $C_4F_7^-$ in perfluoro-1,2-dimethylcyclobutane. The gas temperature is $\sim 300°K$. ΔI is in arbitrary units but to the same scale as in Fig. 6a. Notice the scale factors.

the capture peak width for SF_6^- has, apparently, not been accurately measured (the best available results due to Schulz apparently indicate a width of some 15 mV) because its observed width is 'instrumental', its extreme sharpness enables SF_6 to be used as a very effective probe for low energy electrons, such as those produced just above a critical potential. The

reactions are

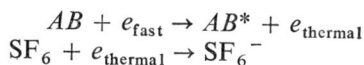

$$AB + e_{\text{fast}} \rightarrow AB^* + e_{\text{thermal}}$$
$$SF_6 + e_{\text{thermal}} \rightarrow SF_6^-$$

The wider SF_5^- peak may also be used in a similar way.

Compton *et al.*[26] have, for example, used a Bendix 14-206 t.o.f. mass spectrometer with an r.p.d. source to study certain excitation processes in He, N_2, HCl, H_2O, D_2O and 8 aromatic molecules; in the latter these authors detected dissociative attachment resonances. Their resolving power was such that the energy of the scattered electrons had to be <30 mV to be detected. The SF_6 partial pressure in the source was typically $<10^{-4}$ torr, with the test gas admitted at $\sim5 \times 10^{-4}$ torr. A typical set of data is shown in Fig. 7 for He.

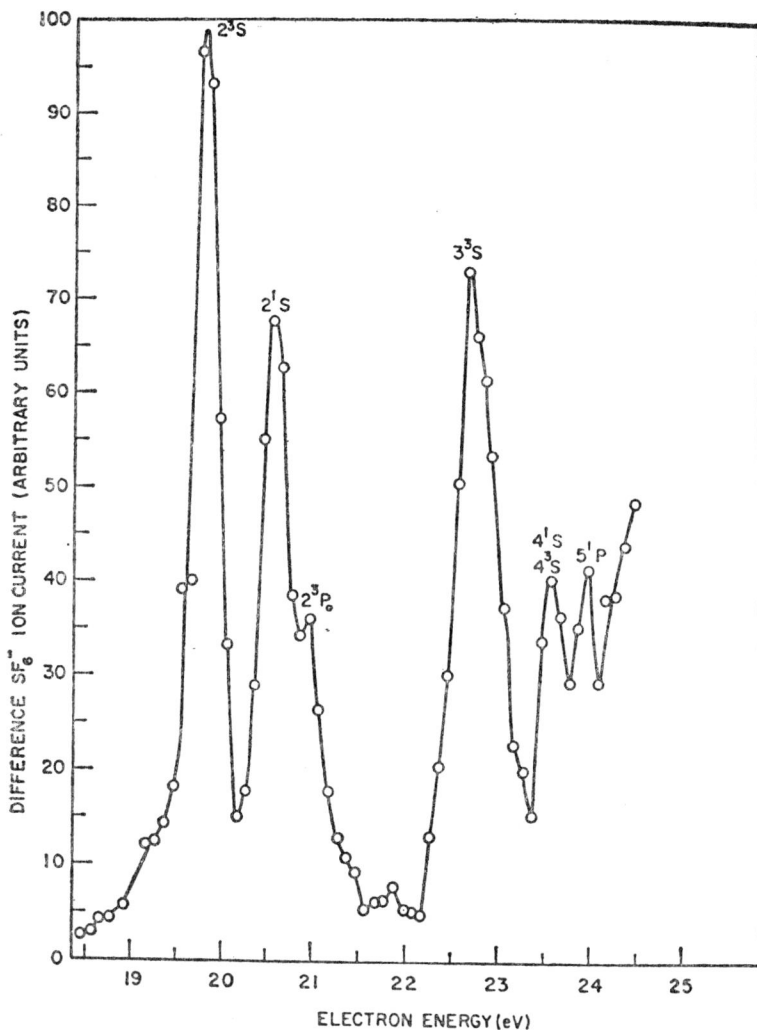

FIG. 7   Electron-impact threshold excitation of He.

SWARM PHENOMENA

A recently developed technique for studying ion-molecule reactions, for thermal energies, is that involving flowing afterglows. It was developed by Fehsenfeld, Ferguson and Schmeltekopf and their colleagues from the early 1960s onwards and has been very productive of new results. Mass spectrometry is an essential part of the experimental method, and a general account of the work has been given by McDaniel et al.[27] Essentially the method consists of introducing into a primary ionized stream of gas, for example He, another gas, such as N_2, and following downstream of the point of injection the products of reaction. In some cases two inlet ports are used, but a typical basic example is the experiment in which N_2 was injected into a stream of He. The loss of He^+ and the gain of N^+ is due to the reaction

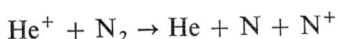

$$He^+ + N_2 \rightarrow He + N + N^+$$

whilst the N_2^+ increase is due to

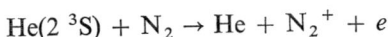

$$He(2\ ^3S) + N_2 \rightarrow He + N_2^+ + e$$

Reaction rate constants can thus be determined at least in simple cases from relative intensities of ion currents, so that mass discrimination effects are not then important. There are now many examples of the valuable results obtained by this method which has certain advantages, for reactions at thermal energies, over time-varying methods. McDaniel et al. (loc. cit.) discuss these advantages. Measured rate constants for a wide variety of negative ion-molecule reactions are given for example by Fehsenfeld, Ferguson and Schmeltekopf.[28] The fast reaction $H_2 + O^- \rightarrow H_2O + e$ (associative detachment) is of interest in discharge studies (Moruzzi and collaborators).

An interesting example of negative ion work with the flowing after-glow technique is that on SF_6 described by Fehsenfeld.[29] Here the helium buffer gas was injected with argon, oxygen or nitrogen to remove the metastable atoms which would otherwise have ionized SF_6 by the Penning effect. The SF_6 of course was injected downstream of the first injection port, and because of its strongly attaching nature it was diluted and injected with argon or helium, e.g. as 0.0495% SF_6 in argon. The attachment rate constants were calculated from the measured production of negative ions as a function of the rate of addition of SF_6. A quadrupole MS was used to identify the ion species, i.e. SF_6^-. Typically the attachment rate was found to be, for thermal electrons at $289°K$, 2.2×10^{-7} cm^3/sec. With rising temperature, 0–$250°C$, the SF_5^-/SF_6^- ratio was found to increase from about 2×10^{-5}:1 to about 7×10^{-2}:1 according, it was suggested, to the following mechanism (Fehsenfeld, loc. cit.).

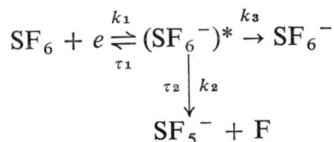

$$SF_6 + e \underset{\tau_1}{\overset{k_1}{\rightleftharpoons}} (SF_6^-)^* \overset{k_3}{\rightarrow} SF_6^-$$
$$\tau_2 \downarrow k_2$$
$$SF_5^- + F$$

τ_1 is here the lifetime of the excited SF_6^- against autodetachment. Only binary kinetics were observed in the present experiment where the pressure could not be reduced below ~ 0.1 torr. τ_1 was indeed measured as 10 μsec by Edelson et al.[20] and as 25 μsec by Compton et al.,[21] whilst the mean time between collisions in Fehsenfeld's experiment was typically 0.5 μsec, k_2 was found to vary rapidly with $1/T$ and the Arrhenius plot indicated an activation energy of 0.43 eV.

As an example of the use of mass-spectrometric techniques in the study of negative ion-molecule reactions we may refer to the research on oxygen, which is a complicated but clearly important case. Burch and Geballe[30] published remarkable work carried out before mass analysis was applied to such a complex system; subsequent experiments have verified their conclusions to a surprising degree. Several authors have, in recent years, studied this case, namely Eiber[31] whose simple spectrometer, based on the Bradbury filter, gave results in excellent agreement with later work, Moruzzi and Phelps,[13] McKnight,[32] Kinsman and Rees[33] and Snuggs et al.,[34,35] but brief reference may perhaps be made to the most recent work of Harrison and Moruzzi[36] whose drift tube technique represents the most modern practice in experiments of this kind. The reaction system is

$$O_2 + e \rightarrow O^- + O$$

followed by, to a relative extent depending on the experimental parameters,

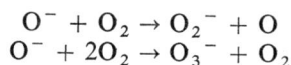

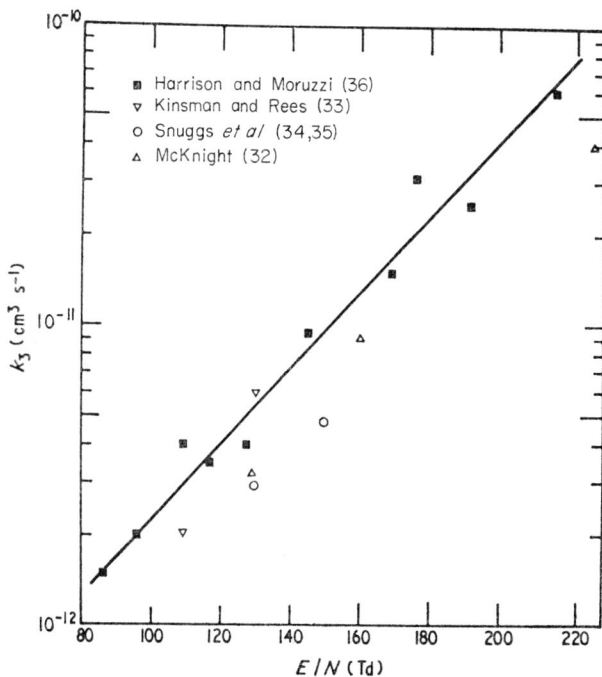

$$O^- + O_2 \rightarrow O_2^- + O$$
$$O^- + 2O_2 \rightarrow O_3^- + O_2$$

FIG. 8(a)   The reaction rate k_3 for the reaction $O^- + O_2 \rightarrow O_2^- + O$ as a function of E/N.

This is of interest in electrical breakdown studies because of the different detachment coefficients to be ascribed to O^-, O_2^- and O_3^-. In certain cases also O_4^- and even O_5^- may be important. Mass spectrometric sampling is clearly essential in this case, because the above three reactions occur to extents depending quite critically upon the reduced electric field and the gas density. The apparatus comprised a uniform field source, with back-illuminated cathode, and a quadrupole MS sampling system, run at low resolution to reduce mass discrimination defects after Hayhurst, Mitchell and Telford.[37]

Data for the reaction rate for the O^- to O_2^- conversion are given in Fig. 8a where the spread of data for given E/N is, where the results of different

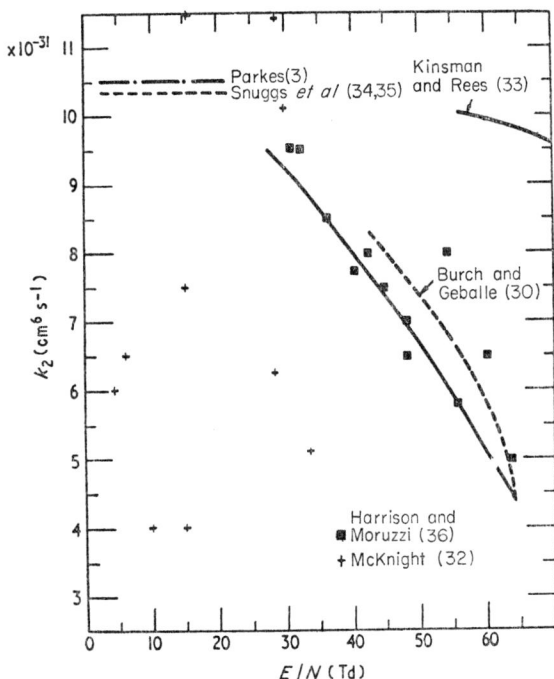

FIG. 8(b)   The reaction rate k_2 for the reaction $O^- + 2O_2 \rightarrow O_3^- + O_2$ as a function of E/N.

workers can be compared, almost 2:1. The reasons for these serious discrepancies are not clear, although mass discrimination, not always allowed for, could contribute. Even so the internal scatter for a given set of observations is sometimes large. There is a clear need for accurate values of rate coefficients of this kind; it is also apparent that cases other than those involving negative ions in oxygen are even more unsatisfactory. Figure 8b shows results for the three-body conversion $O^- \rightarrow O_3^-$. The data at high E/N were taken at different gas densities. Kinsman and Rees worked at pressures of 20 torr or more, and Harrison and Moruzzi at about 1–2 torr. Burch and Geballe did not use mass analysis. The back reaction $O_3^- \rightarrow O^-$ may have been operative in the high pressure case (as indicated by Kinsman and Rees[33]).

The interesting reaction $O_2^- + 2O_2 \rightleftharpoons O_4^- + O_2$ has been studied by Conway and Nesbitt,[38] Parkes[3] and McKnight and Sawina.[39] Parkes used N_2O ($<1\%$) in oxygen to produce O_2^- at higher fields, a faster reaction than the primary process

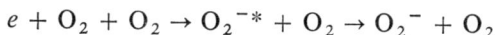

$$e + O_2 + O_2 \rightarrow O_2^{-*} + O_2 \rightarrow O_2^- + O_2$$

and, making full allowance for mass discrimination, measured the equilibrium constant as a function of temperature. E/p was $\sim 0.2\text{--}2 \text{ V cm}^{-1} \text{ torr}^{-1}$ ($E/N \sim 0.6\text{--}5 \times 10^{-16} \text{ V cm}^2$) and a typical forward rate at about $310°\text{K}$ and with $E/N \sim 1 \times 10^{-16} \text{ V cm}^2$ was $k = 3 \times 10^{-31} \text{ cm}^6/\text{sec}$, in good agreement with the data of McKnight and Sawina (loc. cit). who did not correct for mass discrimination effects.

ACKNOWLEDGMENT

I am most grateful to Dr H. Helm and his colleagues in Innsbruck for permission to refer to their work and to reproduce Fig. 3 in advance of publication.

REFERENCES

1. Cuthbert, J. *in* 'Advances in Mass Spectrometry', Applied Science, London, 1966, Vol. 3, p. 821.
2. Kebarle, P., Haynes, R. M. and Searles, S. *Adv. Chem.*, 1966, **58**, 210.
3. Parkes, D. A., *Trans. Far. Soc.*, 1971, **67**, 711.
4. Moruzzi, J. L. and Dakin, J. T. *J. Chem. Phys.*, 1968, **49**, 5000.
5. Kinsman, P. R., McGeehan, J., Prasad, A. N. and Rees, J. A. *in* 'Advances in Mass Spectrometry', Applied Science, London, 1971, Vol. 5, p. 489.
6. Ong, P. P and Hasted, J. B. *J. Phys. B* (*Atom. Molec. Phys.*), 1969, **2**, 91.
7. Kinsman, P. R. and Rees, J. A. *J. Phys. E* (*J. Sci. Instrum.*), 1970, **3**, 444.
8. Kinsman, P. R. and Rees, J. A. *Int. J. Mass Spectrometry & Ion Physics*, 1970, **4**, 393.
9. Knewstubb, P. F. and Sugden, T. M. *Proc. Roy. Soc A*, 1960, **255**, 520.
10. O'Neill, B. C. and Craggs, J. D. *J. Phys. B* (*Atom. Molec. Phys.*), 1973, **6**, 2625.
11. Helm, H., Howorka, F., Handle, F., Egger, F. and Lindinger, W. In course of publication.
12. Walcher, W. *Z. Phys.*, 1944, **122**, 62.
13 Moruzzi, J. L. and Phelps, A. V. *J. Chem. Phys.*, 1966, **45**, 4617.
14. Price, D. A., Lucas, J. and Moruzzi, J. L. *J. Phys. D* (*Appl. Phys.*), 1972, **5**, 1249.
15. Morrison, J. D. *J. Chem. Phys.*, 1963, **39**, 200.
16. MacNeil, K. A. G. and Thynne, J. C. J. *Int. J. Mass Spectrometry & Ion Physics*, 1969, **3**, 35.
17. Hickam, W. M. and Fox, R. E. *J. Chem. Phys.*, 1956, **25**, 642.
18. MacNeil, K. A. G. and Thynne, J. C. J. *Int. J. Mass Spectrometry & Ion Physics*, 1969, **3**, 455.
19. Christophorou, L. G. 'Atomic & Molecular Radiation Physics', Wiley, London, 1971.
20. Edelson, D., Griffiths, J. E. and McAfee, K. B. *J. Chem. Phys.*, 1962, **37**, 917.
21. Compton, R. N., Christophorou, L. G., Hurst, G. S. and Reinhardt, P. W. *J. Chem. Phys.*, 1966, **45**, 4634.
22. Naff, W. T., Compton, R. N. and Cooper, C. D. *J. Chem. Phys.*, 1971, **54**, 212.
23. Lifshitz, C., Peers, A. M., Grajower, R. and Weiss, M. *J. Chem. Phys.*, 1970, **53**, 4605.
24. Curran, R. K. *J. Chem. Phys.*, 1963, **38**, 780.
25. Jacobs, G. and Henglein, A. *Z. für Naturforsch*, 1964, **19a**, 906.
26. Compton, R. N., Huebner, R. H., Reinhardt, P. W. and Christophorou, L. G. *J. Chem. Phys.*, 1968, **48**, 901.

27. McDaniel, E. W., Čermák, V., Dalgarno, A., Ferguson, E. E. and Friedman, L. 'Ion-Molecule Reactions', Wiley, London, 1970.
28. Fehsenfeld, F. C., Ferguson, E. E. and Schmeltekopf, A. L. *J. Chem. Phys.*, 1966, **45**, 1844.
29. Fehsenfeld, F. C. *J. Chem. Phys.*, 1970, **53**, 2000.
30. Burch, D. S. and Geballe, R. *Phys. Rev.*, 1957, **106**, 183, 188.
31. Eiber, H. *Zeits. Angew. Phys.*, 1963, **15**, 103.
32. McKnight, L. G. *Phys. Rev.*, 1970, **2**, 762.
33. Kinsman, P. R. and Rees, J. A. *Int. J. Mass Spectrometry & Ion Physics*, 1970, **7**, 177.
34. Snuggs, R. M., Volz, D. J., Schummers, J. H., Martin, D. W. and McDaniel, E. W. *Phys. Rev.*, 1971, **3**, 477.
35. Snuggs, R. M., Volz, D. J., Schummers, J. H., Martin, D. W. and McDaniel, E. W. *Phys. Rev.*, 1971, **3**, 487.
36. Harrison, L. and Moruzzi, J. L. *J. Phys. D (Appl. Phys.)*, 1972, **5**, 1239.
37. Hayhurst, A. N., Mitchell, F. R. G. and Telford, N. R. *Int. J. Mass Spectrometry & Ion Physics*, 1971, **7**, 177.
38. Conway, D. C. and Nesbitt, L. E. *J. Chem. Phys.*, 1968, **48**, 509.
39. McKnight, L. G. and Sawina, J. M. *Phys. Rev. A*, 1971, **4**, 1043.

NEGATIVE IONS

Chairmen

F. M. PAGE
(*University of Aston in Birmingham, U.K.*)

A. QUAYLE
(*Shell Research Ltd, Chester, U.K.*)

30

Formation and Relative Stability of Negative Clustered Ions by Ion Cyclotron Resonance Spectroscopy

By JOSÉ M. RIVEROS

Instituto de Quimica, University of São Paulo, São Paulo, Brazil

---

THE study of clustered ions in the gas phase has been an active field in the last few years with the advent of high pressure mass spectrometric techniques. Since the direct ion-neutral association reactions are typically thermolecular processes, these techniques have been very successful in detecting successive stages of neutral association around a central ion.

Clustered ions in the gas phase are of primary interest to the chemist because of their obvious relationship with solvation of ions in solution. It is recognized that an understanding of the forces responsible for the stability of these species constitutes an essential step towards unravelling the fashion in which a given solvent affects the chemical reactivity and the chemical properties of these same ions in solution. These complex ions are also important because of their role in ion-molecule reactions, and with regard to nucleation processes in the upper atmosphere.

For the last two years, we have concentrated our efforts on using ion cyclotron resonance techniques to study bimolecular reactions which lead to the formation of negative clustered ions. Such reactions represent an indirect method because they require a neutral fragment to be expelled from the collision complex. A first study of this kind was reported earlier this year based on the reaction of alkoxide ions with alkyl formates (Blair, Isolani and Riveros, *J. Amer. Chem. Soc.*, 1973, **95**, 1057). Reaction (1) is a typical example of this process,

$$CH_3O^- + HCOOCH_3 \rightarrow CH_3O^-(CH_3OH) + CO \qquad (1)$$

Relative solvating ability of the neutrals was determined in these cases by allowing the product ions of reactions similar to (1) to undergo reaction (2).

$$CH_3O^-(CH_3OH) + ROH \overset{?}{\rightarrow} CH_3O^-(ROH) + CH_3OH \qquad (2)$$

Observation of selective solvent displacement reactions was used as a criterion of the relative stability of the complex ions. Such experiments, and some quantitative measurements of the equilibrium constants for these systems, show the following to be the order of solvating ability by one neutral molecule:

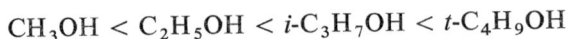

$$CH_3OH < C_2H_5OH < i\text{-}C_3H_7OH < t\text{-}C_4H_9OH$$

A more general method, with apparently wide applications, has been recently found in our laboratories for the case of chloride ions. This method is based on the reactions of a negative ion obtained from phosgene. At 30 eV, the negative ICR spectrum of phosgene at $2\cdot5 \times 10^{-6}$ torr shows the following processes to be important,

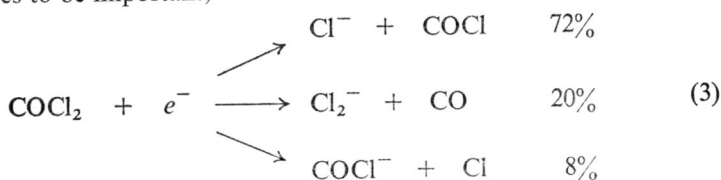

$$
COCl_2 + e^- \longrightarrow
\begin{cases}
Cl^- + COCl & 72\% \\
Cl_2^- + CO & 20\% \\
COCl^- + Cl & 8\%
\end{cases}
\tag{3}
$$

These fragment ions are primarily produced by thermal electrons trapped in a conventional ICR cell after inelastic collisions with neutrals. Addition of a second gas, like water or methanol, gives rise first to a dramatic increase of the negative ions due to the increased number of thermal electrons. Formation of clustered ions is observed to take place under these conditions beginning at pressures of 5×10^{-6} torr. Pulsed double resonance experiments show that the clustered ions are produced exclusively by reactions similar to (4), but not by reactions of Cl^- or Cl_2^-.

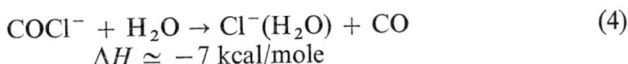

$$COCl^- + H_2O \rightarrow Cl^-(H_2O) + CO \tag{4}$$
$$\Delta H \simeq -7 \text{ kcal/mole}$$

The enhancement of the primary negative ion current depends on the initial energy of the electrons and the cross-section of the neutral for a totally inelastic process, or emission of secondary electrons. This feature is actually a critical consideration of the experiment because the ion $COCl^-$ is in very low abundance.

A general survey of the above reaction shows that a large variety of neutrals like alcohols, acetonitrile, nitromethane, alkyl halides, ketones, aromatic systems, amines, and even xenon are capable of yielding stable clustered chloride ions that can be detected at low pressures in the gas phase by cyclotron resonance techniques. Saturated hydrocarbons and ammonia have been the cases for which reaction (4) has not been observed to take place appreciably.

By maintaining the partial pressure of phosgene at 1×10^{-6} torr, the precursor ion $COCl^-$ is usually observed to react completely at pressures of the neutral above 1×10^{-5} torr. Thus, binary mixtures of selected compounds were used to study relative clustering ability by pulsed double resonance of reaction (5) in both directions, and by pressure dependence of the relative intensities of the solvated ions.

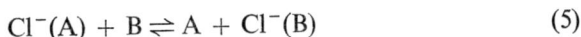

$$Cl^-(A) + B \rightleftharpoons A + Cl^-(B) \tag{5}$$

A few examples of interest show the qualitative trends observed for solvating ability:

(a) $H_2O < CH_3OH < C_2H_5OH$
(b) $CH_3Cl < CH_3NO_2 \sim CH_3CN; CH_3Cl < CH_3CF_3$
(c) $CH_3F < CH_3Cl < C_2H_5Cl < (CH_3)_2CHCl$

The first series is essentially that determined from the experiments with clustered alkoxide ions. In both cases, the complex ions are presumably held together by hydrogen bonding ($Cl^- \cdots HOCH_3$). The results represent the trend of hydrogen bond strength in 'mono-solvated' ions in the gas phase. This trend follows the increasing polarizability of the neutral, and correlates with the gas phase acidity of these compounds as pointed out by Kebarle.

In the second series, the structure of the ion-neutral complex is very likely different from the case of protic 'solvents', and suggests a structure where the dipole moment of the neutral is lined up with the negative ion, ($Cl^- \cdots H_3CCN$). The observed trend seems to favour the notion that ion-permanent dipole forces are important in these situations.

The third series is similar to the second one in the sense that they are aprotic solvents. On the other hand, molecules in this category have similar dipole moments, and not surprisingly the trend of stability follows again the polarizability trend as in the first case.

The discussion for the three cases under consideration is obviously a superficial and partial interpretation of the phenomena associated with ion-neutral bonding in the gas phase. However, a more general description of these clustered species requires quantitative data. In this respect, pulsed ICR techniques coupled with a cell capable of trapping ions for over a tenth of a second have been used to obtain equilibrium constants of ionic gas phase processes with good accuracy. If equilibrium is rapidly achieved, 10 to 50 collisions, Bowers et al. (J. Amer. Chem. Soc., 1971, **93**, 4314) have shown that the high pressure behaviour of the relative intensities yields an alternate method for the determination of equilibrium constants.

The idea of obtaining quantitative data from our experiments has been explored for some cases. Figure 1 shows the behaviour of a 1:1 mixture of CH_3CN and C_2H_5OH in the presence of a small amount of phosgene. At 1×10^{-5} torr, the spectrum clearly shows the preference for formation of $Cl^-(CH_3CN)$ indicating that kinetically the reaction of $COCl^-$ with aceto-nitrile is considerably more favourable. At 1×10^{-4} torr, the species $Cl^-(C_2H_5OH)$ is practically twice as abundant as $Cl^-(CH_3CN)$. At higher pressures, the relative intensities of these two species remain constant within experimental error. Thus, it appears that in this case the chloride transfer reaction approaches equilibrium rapidly.

A particularly well-behaved case with pressure is shown in Fig. 2 where the solvating ability of methanol and ethanol are being compared. The ratio of intensities of the clustered ions is shown to remain practically constant above 5×10^{-5} torr.

If constancy of relative intensities at high pressures (above 1×10^{-4} torr) under conditions where the precursor ion $COCl^-$ has vanished is taken as a sign of achieving equilibrium, the following free energy changes can be quoted for the systems more thoroughly investigated to date.

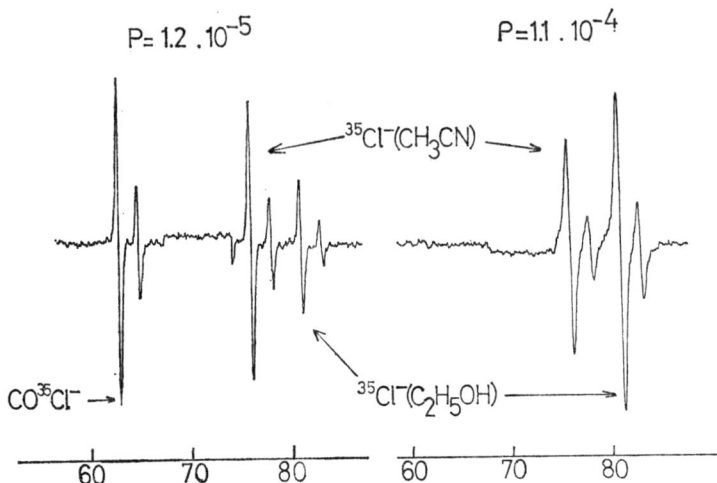

FIG. 1   Negative ICR spectrum at 30 eV of an equimolar mixture of ethanol and aceto-
nitrile in the presence of phosgene. The partial pressure of COCl₂ was 0.26×10^{-5} torr
in both cases. Modulation was turned off between m/e 70 and 74 to avoid the strong peaks
due to Cl_2^-.

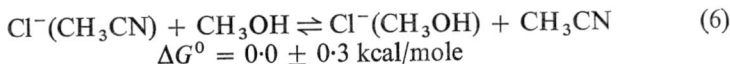

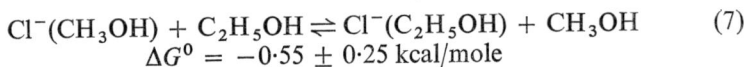

$$Cl^-(CH_3CN) + CH_3OH \rightleftharpoons Cl^-(CH_3OH) + CH_3CN \qquad (6)$$
$$\Delta G^0 = 0.0 \pm 0.3 \text{ kcal/mole}$$

$$Cl^-(CH_3OH) + C_2H_5OH \rightleftharpoons Cl^-(C_2H_5OH) + CH_3OH \qquad (7)$$
$$\Delta G^0 = -0.55 \pm 0.25 \text{ kcal/mole}$$

These experiments were carried out at 295°K, and several different partial
gaseous compositions were used to arrive at the above results. It is important
to notice that the value obtained for reaction (6) is in good agreement with
the data obtained by Kebarle using high pressure mass spectrometry.

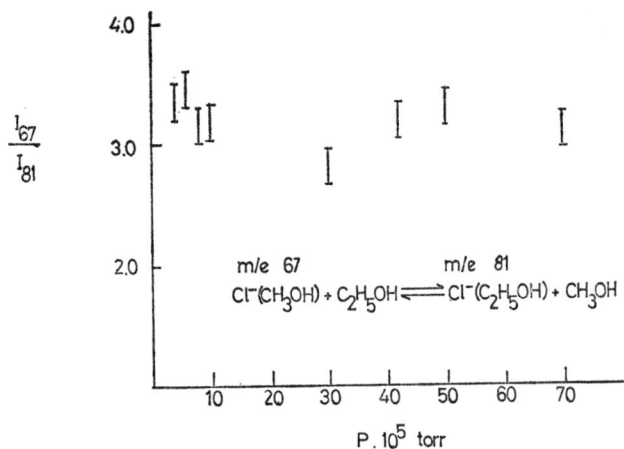

FIG. 2   Intensity ratio of $Cl^-(CH_3OH)$ to $Cl^-(C_2H_5OH)$ plotted as a function of pressure
in a 6:1 methanol–ethanol mixture. Phosgene was kept at 2×10^{-6} torr. Pressures were
read from an ionization gauge calibrated against an MKS Baratron.

As a final remark, other systems presently under study which lead to the formation of chemically interesting 'solvated' ions should be mentioned at this point. Preliminary observations show that reaction (8) is an adequate source of clustered fluoride ions (*see* Fig. 3).

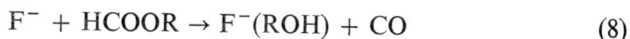

$$F^- + HCOOR \rightarrow F^-(ROH) + CO \qquad (8)$$

Likewise, solvated hydroxide ions have been obtained from reaction (9), although in this case other reactive channels may predominate depending on the size of the alkyl group.

$$OH^- + HCOOR \rightarrow OH^-(ROH) + CO \qquad (9)$$

It is clear that the same techniques applied for the case of the alkoxide and chloride clustered ions can be used to study solvating ability of neutrals.

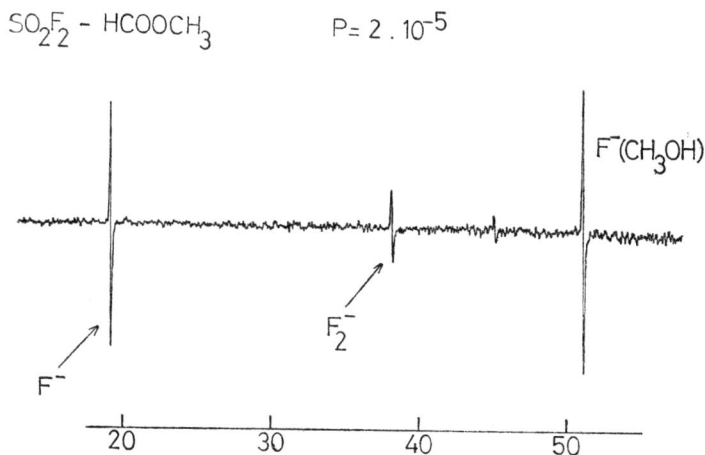

SO_2F_2 – $HCOOCH_3$ $P = 2 . 10^{-5}$

$F^-(CH_3OH)$

F_2^-

F^-

20   30   40   50

FIG. 3   Negative ICR spectrum at 3·7 eV of an approximately equimolar mixture of SO_2F_2 and $HCOOCH_3$. A minor quantity of $HCOO^-$ (m/e 45) is also obtained as a reaction product.

The systems discussed in the present paper suggest that there are many potentially interesting solvated ions that can be studied by ICR, and that valuable quantitative data can be obtained from these experiments.

ACKNOWLEDGMENTS

The work reported in this paper has had major contributions from Antonio Celso Breda, Paulo Celso Isolani and Larry K. Blair. The research was made possible through the generous support of the Conselho Nacional de Pesquisas of Brazil and the Fundação de Amparo à Pesquisa do Estado de São Paulo.

JOSÉ M. RIVEROS

Discussion

R. S. Berry (University of Chicago, U.S.A.): Would you care to comment on multiply solvated ions in your system?

J. M. Riveros: In several of the systems reported here we searched for doubly clustered ions at pressures of 5×10^{-4} torr without success. However, it may be possible to observe such species with larger neutral molecules where the excess energy may be more easily accommodated.

31
Negative Ion High Resolution Mass Spectrometry†

By C. CONE

(*Department of Chemistry, The University of Texas at Austin, Austin, Texas, U.S.A.*)

INTRODUCTION

WITH the rapidly developing interest in alternative methods of studying the formation and fragmentation of ions in the gas phase, such as chemical ionization, field ionization and related techniques, and photoionization, there has also recently been a renewal of interest in the oldest alternative[1] to electron impact positive ionization, namely the observation of the negative ions that are formed on electron bombardment. As with other alternative techniques, compounds which do not yield molecular ions in their positive ion mass spectra do sometimes yield either molecular anions or M-1 peaks in their negative ion mass spectra. A simple example is *t*-butyl alcohol which Melton[2] observed to give a reasonably intense M-1 ion in its negative ion mass spectrum. These results have more recently been extended by Rankin[3] to longer-chain alcohols and aldehydes.

The fragmentation observed in the negative ion spectra of more complex organic molecules is sometimes nicely complementary to that observed in the positive ion spectra of the same compounds. A good example is that of Fontaphillin.[4] While the negative ion spectrum in this example was obtained with a special ion source similar results would be expected with an electron impact ion source. Negative ion spectra are often simpler than positive ion spectra but a note of caution must be sounded as some cases of rearrangements of negative ions have been discovered.[5,6]

From these and other examples that could be cited, it is now apparent that negative ion mass spectrometry is a useful additional method in structure determination. It, therefore, seemed worthwhile to explore the possibility of obtaining high resolution negative ion mass spectra. There were a number of apparent difficulties at the outset of this work. The most important was that of sensitivity. Low resolution negative ion studies have only become reasonably straightforward with modern instruments that have much more sensitivity than is needed for routine positive ion work, as yields of negative ions are typically several orders of magnitude lower. Another difficulty is the much

† Paper II in the series 'High Resolution Negative Ion Mass Spectrometry'.

greater dependence of the processes leading to the formation of negative ions on ionizing voltage than is the case for the formation of positive ions. This makes the recording of a composite spectrum of a mass reference compound, and of the compound of interest, more difficult than is the case in high resolution positive ion mass spectrometry. The processes leading to the formation of negative ions have been reviewed by Melton.[7]

In our initial report[8] of our attempts to obtain high resolution negative ion mass spectra we showed that it was possible to simultaneously record the spectra of high boiling perfluoroalkane with that of a nitroaromatic and of cholesterol dinitrobenzoate. Perfluoroalkane serves as a useful marker compound in negative ion work from m/e 219 to above m/e 700. However, in contrast to the previously published negative ion spectrum of perfluorokerosene[9] we find m/e 331 to be the base peak in the perfluoroalkane spectrum. Large gaps at the low mass end of the spectrum make it an unsuitable mass reference compound over this range. These spectra were recorded using the pressure enhancement technique of Dougherty[10] which gave useful but modest enhancements of the perfluoroalkane peak intensities. The pressure enhancement technique did not change the general appearance of the perfluoroalkane spectrum, and a similar spectrum has also been obtained on a single focussing magnetic deflection instrument. Mass measurement accuracy was comparable to that obtained for routine positive ion spectra and the resolution was about 10 000. Regular reversal of the magnetic field has not led to any loss in mass measurement accuracy for positive ions. These measurements were made by taking advantage of the high sensitivity of photoplate recording of high resolution mass spectra and of the additional sensitivity of the recently introduced evaporated silver bromide plate.[11]

EXPERIMENTAL

Mass spectra were recorded on Ionomet evaporated silver bromide plates using a DuPont (CEC) 21-110-C Mattauch Herzog high resolution mass spectrometer. Marker compounds were introduced through the heated inlet system in amounts 2–3 times those usually used for positive ion spectra. Solids were introduced by the direct introduction probe. Argon was used as pressure enhancing gas and was admitted until the ion source housing gauge read from 5×10^{-5} to 1×10^{-4} mm Hg. The instrument pumping system was not modified; the narrow differential pumping slit had been installed. The standard accelerating voltage supply was disconnected from its output voltage divider and a Fluke 0–10 kV supply (model 410B) connected across that divider. The standard electric sector and magnet supplies were used with the necessary polarity reversals. Because negative arcs feeding back into the ion source electronics were a problem, in more recent experiments we have used batteries to supply the filament current and anode and electron accelerating potentials. The beam monitor was 'zeroed' on the maximum setting of a sensitive scale and the instrument initially tuned for maximum negative deflection of the beam monitor. Generally, good correlation was found between maximum negative deflection of the beam monitor and maximum peak heights of negative ion peaks. Ion source focus conditions for maximum perfluoroalkane peak intensities were found to be quite variable, but in a typical experiment the anode potential was 0 V and the electron accelerating potential was 75 to 100 V. Under these conditions, ionization is presumably due in part to secondary electrons. The repeller voltages were adjusted over their full range for maximum beam intensity.

RESULTS

Mass Reference Compounds

As previously reported, perfluoroalkane is a suitable reference compound at higher masses. We have obtained the negative ion spectra of a number of other perfluorinated compounds as either alternative or supplementary mass standards. Of those commonly used in positive ion work, we find both perfluorotributylamine and perfluorotriheptyl-s-triazine to be unsuitable for general use in negative ion mass spectrometry, because they give negative ion spectra with relatively few peaks. However, we have found that several higher molecular weight perfluorinated, or almost perfluorinated, acids and anhydrides do give useful negative ion spectra for mass reference purposes, although the anhydrides we have studied so far do not yield peaks of useful intensity above the mass of the carboxylate anion which is, of course, roughly only half the molecular weight of the parent molecule. Compounds giving useful spectra include perfluorooctanoic anhydride and 11H-eicosafluoro-undecanoic acid.

We have also been exploring an alternative approach to filling in the gaps in the low mass region of the perfluoroalkane spectrum. This is the use of ions produced by negative ion molecule reactions. As an example, in a mixture of CS_2 and perfluoroalkane in the range of 3:1 to 5:1 by volume in our inlet system, we find, besides a strong molecular anion for CS_2, a series of peaks for S^-, S_2^-, S_3^- up to S_6^- with quite high intensities.

Anthraquinones

One good recent example of the usefulness of negative ion mass spectrometry for structure determination has been the work of Bowie on quinones.[12,13] We, therefore, decided to begin our study of the high resolution negative ion spectra of quinones with some synthetic acetoxyanthraquinones, the low

TABLE I

High Resolution Negative Ion Spectrum of 1,4-diacetoxyanthraquinone

Intensity	Found mass	Calculated mass	Elemental composition
*****	240·0396	240·0423	$C_{14}H_8O_4$
*	241·0430	241·0457	$C_{13}{}^{13}CH_8O_4$
*	281·0450	281·0450	$C_{16}H_9O_5$
**	282·0540	282·0528	$C_{16}H_{10}O_5$
***	324·0631	324·0634	$C_{18}H_{12}O_6$
*	325·0647	325·0668	$C_{17}{}^{13}CH_{12}O_6$

resolution spectra of which are quite well understood through the use of labelled compounds and refocussed metastable studies.

As an example of our initial results in this area, we give in Table I the results obtained for 1,4-diacetoxyanthraquinone.

It is interesting to note that while the general conclusions of Bowie[12] are confirmed by the exact mass measurements, in this compound with both acetoxy groups in peri positions we do observe a small peak for the loss of acetyl radical from the molecular anion as well as the stronger peaks for the sequential loss of two ketene molecules.

Encouraged by the observation that we can obtain suitable composite spectra of mass reference compounds and acetoxyanthraquinones, we are currently extending this study to other synthetic quinones and also to some of the biologically important naturally occurring quinones.

ACKNOWLEDGMENTS

The author is indebted to Mr F. C. Maseles for his skilful assistance with the instrumentation used in this work and to the Robert A. Welch Foundation for their financial support (grant F-493). The high resolution mass spectrometer was purchased with a grant from the National Science Foundation (GP-8509).

REFERENCES

1. Kiser, R. W. 'Introduction to Mass Spectrometry and Its Applications', Chap. 2 and references therein, Prentice-Hall, Englewood Cliffs, N.J., 1965.
2. Melton, C. E. and Rudolph, P. S. J. Chem. Phys., 1959, 31, 1485.
3. Rankin, P. C. Lipids, 1971, 5, 825.
4. Budzikiewicz, H., Herstmann, C., Pufahl, K. and Schreiber, K. Chem. Ber., 1967, 100, 2798.
5. Blumenthal, T. and Bowie, J. H. Aust. J. Chem., 1971, 24, 1853.
6. Alexander, R. G., Bigley, D. B. and Todd, J. F. J. Chem. Comm., 1972, p. 553.
7. Melton, C. E. In 'Mass Spectrometry of Organic Ions', ed. F. W. McLafferty, Chap. 4, Academic Press, New York, 1963.
8. Bouldin, W. J., Cone, C. and Maseles, F. C. 20th Annual Conference on Mass Spectrometry and Allied Topics, Dallas, June 1972; to be published.
9. Gohlke, R. S. and Thompson, L. H. Analyt. Chem., 1968, 40, 1004.
10. Dougherty, R. C. and Weisenberger, C. R. J. Amer. Chem. Soc., 1968, 90, 6570.
11. Masters, J. I. Nature (London), 1969, 223, 611.
12. Ho, A. C., Bowie, J. H. and Fry, A. J. Chem. Soc. (B), 1971, p. 530.
13. Ho, A. C. and Bowie, J. H. Aust. J. Chem., 1971, 24, 1093.

32
High Resolution Photodetachment Studies of Negative Ions†

By H. HOTOP

(*Joint Institute for Laboratory Astrophysics, Boulder, Colorado, U.S.A. and Fakultät für Physik, Universität Freiburg, Germany*)

T. A. PATTERSON

(*Joint Institute for Laboratory Astrophysics, Boulder, Colorado, U.S.A.*)

W. C. LINEBERGER‡

(*Joint Institute for Laboratory Astrophysics, Boulder, Colorado, U.S.A. and Department of Chemistry, University of Colorado, Boulder, Colorado, U.S.A.*)

THE energy resolution in photodetachment experiments has recently been considerably improved to better than 1 meV by the use of continuously tunable laser light sources.[1-4] The determination of electron affinities with an accuracy of some 10^{-4} eV and a thorough test of the theoretical threshold law have thus become possible, as was first demonstrated by Lineberger and Woodward[1] in a study of S⁻-photodetachment. In this paper we give a survey of high resolution ($\Delta\lambda \approx 1$ Å) dye-laser photodetachment studies of atomic (Se⁻, Au⁻, Pt⁻, Na⁻, K⁻) and molecular (OH⁻, OD⁻) negative ions near threshold.

The theoretically predicted behaviour of the cross-section σ_L for photodetachment of atomic negative ions

$$A^- + h\nu \xrightarrow{\sigma_L} A + e^-(k, L) \tag{1}$$

($h\nu$: photon energy; k, L: electron linear and orbital angular momentum) near threshold $h\nu_{thr}$ is given by [5,6]

$$\sigma_L = \text{const } h\nu(h\nu - h\nu_{thr})^{L+1/2L} \propto k^{2+1} \tag{2}$$

It reflects the influence of the dominant long-range interaction between the

† Support of this research by the Advanced Research Projects Agency of the Department of Defense and by the Deutsche Forschungsgemeinschaft is gratefully acknowledged.

‡ Alfred P. Sloan Foundation Fellow, 1972–4.

final state atom and the leaving electron, namely the centrifugal potential. It provides a basis for an extrapolation of experimental cross-sections to threshold and thereby for an accurate determination of the energy levels in negative ions. For a meaningful application, however, it is necessary to know the range of validity of (2); it depends on the details of the $(A + e^-)$ interaction. The polarization potential $-\alpha e^2/2r^4$ (α: polarizability of A) produces a $k^2 \ln k$-correction term[6] in addition to k^2-correction terms of various origins:

$$\sigma_L = \text{const } k^{2L+1} \left(1 - \frac{4\alpha k^2 \ln(\alpha/a_0)^{\frac{1}{2}} k}{a_0(2L + 3)(2L + 1)(2L - 1)} + 0(k^2) \right) \qquad (3)$$

An estimate of its magnitude for $L = 0$ indicates that for atoms of medium polarizability ($\alpha(\text{Se}) \approx 30 \, a_0{}^3$) significant corrections to (2) of order 10% may arise at only 10–20 meV above threshold. For an experimental study of the photodetachment threshold law a photon source of high resolution ($\Delta E \lesssim 1$ meV) should therefore be used.

In the present work a 2 keV negative ion beam is crossed with a flashlamp-pumped, grating-tuned dye-laser of about 1 Å bandwidth ($\Delta E \approx 0.4$ meV), and the fast neutral atoms (molecules) produced in the photodetachment process are detected with an open multiplier. The details of the experimental setup are described elsewhere.[1-4,7]

In the photodetachment of Se$^-$ ions a bound p-electron is set free; one therefore may have a continuum s- or d-electron wave in the final state. Near threshold the d-wave contribution can be neglected, so that one expects $\sigma \propto k$. In all, six fine-structure transitions $\text{Se}^-(^2P_{3/2,1/2}) \rightarrow \text{Se}(^3P_{2,1,0})$ are possible. The relative intensity of the $\frac{3}{2} \rightarrow (2, 1, 0)$ and $\frac{1}{2} \rightarrow (2, 1, 0)$ groups is mainly determined by the relative population of the $\frac{3}{2}$- and $\frac{1}{2}$-levels in the

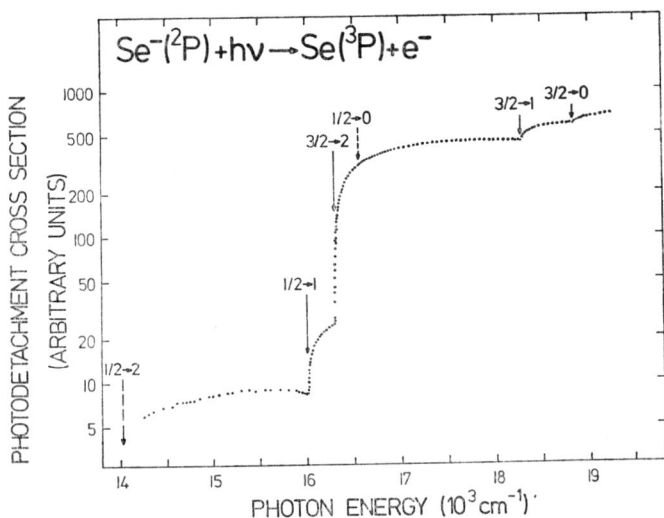

FIG. 1   Se$^-$ photodetachment cross-section in the energy range 14 200–19 200 cm^{-1}. Solid arrows correspond to observed fine-structure onsets, dashed arrows point at position of threshold for unobserved onsets. Note logarithmic cross-section scale.

negative ion beam; about 7% of our Se⁻-beam was found to be in the excited $\frac{1}{2}$-state. The relative transition strengths into the different final states (2, 1, 0) are of considerable interest in connection with the theoretical description of multichannel photodetachment near threshold.[1,7,8]

Our results for Se⁻-photodetachment are summarized in Fig. 1. Four of the six fine-structure onsets can be clearly seen and unambiguously identified; the $\frac{1}{2} \to 2$ onset was just outside the range of our laser, the $\frac{1}{2} \to 0$ onset too weak to be detectable in the rising part of the most intense transition $\frac{3}{2} \to 2$. Threshold plots $(Q/h\nu)^2 = f(h\nu)$ and $(Q/h\nu) = f(k)$ (Q: experimental partial cross-section) were found to be straight lines (deviations $\lesssim 5\%$) over

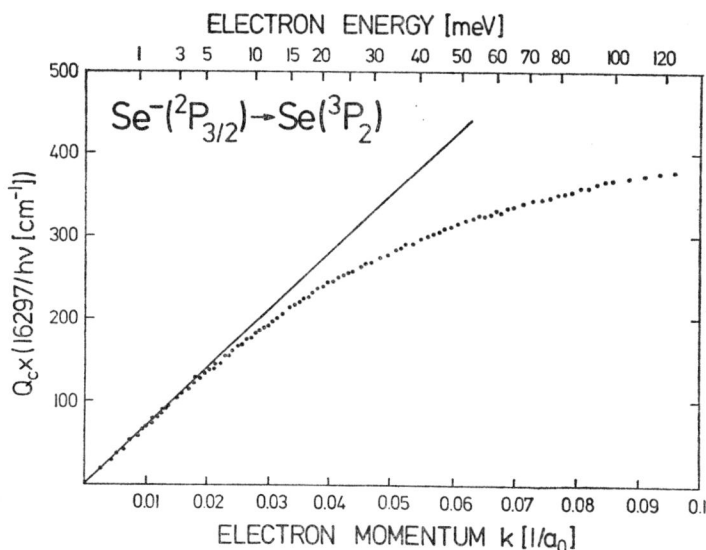

Fig. 2   Threshold plot of Se⁻($^2P_{3/2}$) → Se($^3P_2$) partial photodetachment cross-section (divided by photon energy in units of threshold energy), $Q_c \times 16\,297/h\nu$, versus electron momentum $k[a_0{}^{-1}]$. Electron energy scale is given on top. The straight line corresponds to the theoretical threshold behaviour for an outgoing s-electron-wave, which is seen to present a good fit to the data up to about 5 meV above threshold.

about 5 meV threshold, indicating that the theoretical threshold law (2) is a good description in this range for Se⁻-photodetachment. As an example, Fig. 2 shows the behaviour of the cross section for the $\frac{3}{2} \to 2$ onset. From the thresholds we determined the electron affinity of Se as $EA(\text{Se}) = (16\,297 \pm 2)$ cm⁻¹ and the spin-orbit splitting in the Se⁻ ion as (2279 ± 2) cm⁻¹ (1 eV $\triangleq$ 8065·465 cm⁻¹). The relative strengths for the transitions into the different final channels, which are given by the relative slopes of plots such as the one in Fig. 2, were found to deviate substantially from simple statistical expectations (similar to the findings for S⁻-photodetachment),[1] but to agree well with predictions by Rau and Fano.[7,8]

In order to test the threshold law for the case of an outgoing p-wave ($L = 1$), we studied photodetachment of Au⁻ and Pt⁻ ions, where bound s-electrons are ejected. For both ions we found that the data are well fitted by the threshold law (2) over as much as about 50 meV contrast to the results for Se^-. This result can be explained by an essential cancellation of the $k^2 \ln k$- and the k^2-correction terms in (3) due to a difference in sign. Since the $k^2 \ln k$-term is positive for $L \geq 1$ (which is qualitatively understood by the fact that the attractive polarization potential decreases the centrifugal barrier seen by the escaping electron), the k^2-term is then required to have negative sign in the case of Au^- and Pt^-. For Se^- ($L = 0$) the $k^2 \ln k$-term is negative, and from a fit of the threshold expansion (3) to the Se^--data of Fig. 2 the k^2-term was found to be negative too, corresponding to the observed deviations from (2) close to threshold. A detailed analysis of these results will be presented elsewhere.[3,7] The electron affinities of Au and Pt were determined as $EA(Au) = (18\ 620 \pm 5)$ cm^{-1} and $EA(Pt) = (17\ 160 \pm 16)$ cm^{-1}.

An interesting case in the threshold behaviour of cross-sections may arise in the energy region where new final channels become accessible in addition to already open channels. As a result of flux conservation discontinuities in the cross section slope ('Wigner cusps' or 'Rounded-off steps')[5] for the already open channel (which is coupled to the new one) as well as in the total cross-section slope are expected to occur at precisely the threshold energy $h\nu_{thr}$ for the new channel, if the latter sets in with infinite slope ($\sigma_{new} \propto (h\nu - h\nu_{thr})^{\frac{1}{2}}$, $L = 0$).

Cusps have recently been observed[9] at the first inelastic threshold in the elastic scattering of electrons from alkali atoms, in which the ground and first excited state are very strongly coupled. Norcross and Moores[10] have

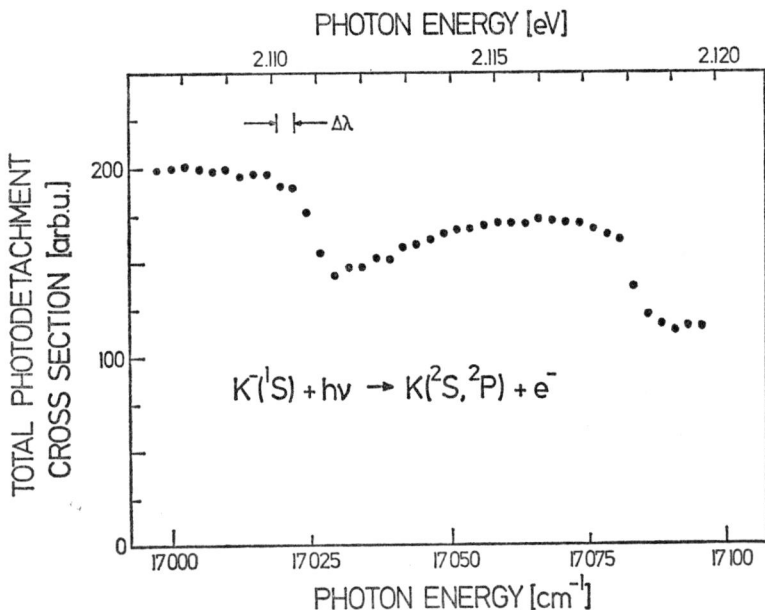

FIG. 3   Total photodetachment cross-section of $K^-(^1S_0)$ in the threshold region for formation of excited $K(4p\ ^2P)$ atoms besides the production of $K(4s\ ^2S)$ ground-state atoms. The sharp structures are interpreted as downward Wigner cusps; the energy separation between the two steep drops in cross-section is consistent with the spectroscopic splitting of $K(4p\ ^2P_{1/2,3/2})$ of 57·7 cm^{-1} within 3 cm^{-1}.

predicted such features to occur in the photodetachment of alkali negative ions at the onset for the production of excited alkali atoms. In an experimental study of Na$^-$ and K$^-$ photodetachment we find an upward cusp in the total cross section for Na$^-$, whereas for K$^-$ a downward cusp is observed (Fig. 3), which appears doubled by the $K(4p\ ^2P_{1/2,\ 3/2})$ spin-orbit interaction (spectroscopic splitting 57·7 cm^{-1}). Our results are in very good qualitative agreement with the calculations.[10] The detailed shape of the observed features (including the fine-structure) is being analysed and presents a sensitive test of theoretical work. From the location of the cusps (we ignore the possibility that a resonance near threshold might contribute to the observed structure) we conclude $EA(\text{Na}) = (0·543 \pm 0·004)$ eV, $EA(\text{K}) = (0·501 \pm 0·001)$ eV. A study of the other alkali negative ions is in progress.

The photodetachment of molecular negative ions near threshold is much more complicated than that of atomic ions as a result of the manifold of

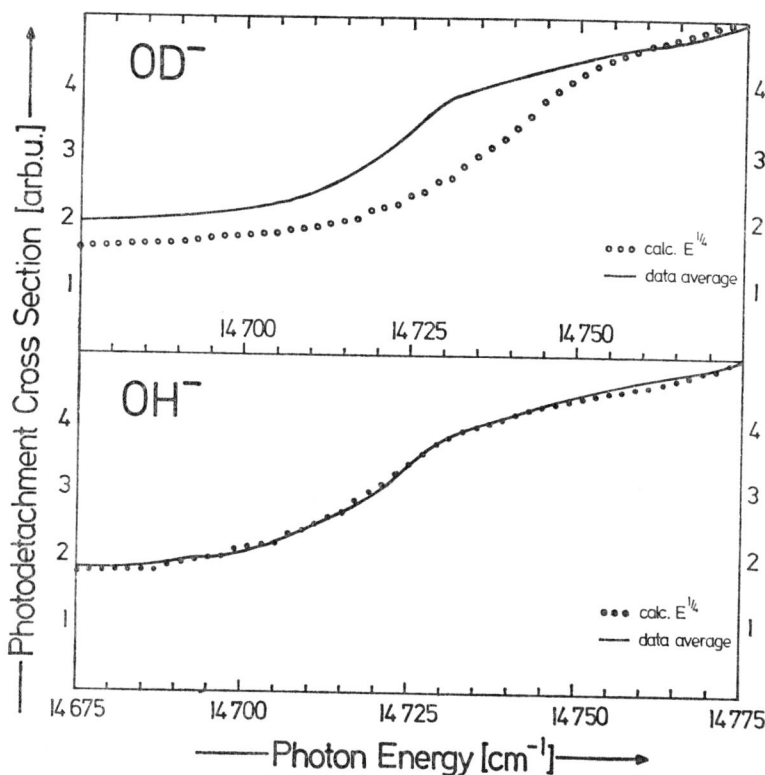

FIG. 4   OH$^-$ and OD$^-$ photodetachment cross section in the range 14 675–14 775 cm^{-1} (lines correspond to average of experimental data) and comparison with calculated 'synthetic' cross-section curves, based on the known rotational constants of the molecular ions and neutral molecules, on an (a priori assumed) electron affinity of OH of 14 723 cm^{-1} (which was also used for the generation of the OD$^-$ curve), and on a threshold law $\sigma \propto E^{1/4}$, assumed to be independent of J' and valid over the energy range of the figure. For OD$^-$ the calculated curve is shifted to higher energies by an amount which corresponds to the electron affinity difference of OH and OD; the one for OD is found to be 20 cm^{-1} smaller than that for OH.

transitions involved and as a consequence of our lack of knowledge of the threshold law and of the precise internal energy state distribution of the negative ions. A rather favourable case is photodetachment of OH^-/OD^- ions, for which it has been established that r_e is the same (within 10^{-3} Å) for the negative ion and the neutral molecule.[11-13] Even though no theoretical prediction exists for the form of the threshold law in this case, where the leaving electron sees the field of a (rotating) permanent dipole, one expects that the individual thresholds will be at least as sharp as $\sigma \propto k \propto E^{\frac{1}{2}}$ (E: electron energy).[6,14]

The general shape of the cross section observed for OH^-/OD^- photo-detachment in the range 7000–6450 Å (14 275–15 475 cm^{-1}) is similar to the one reported by Branscomb,[11] except for a sharp onset at about 14 700 cm^{-1}, which was not resolved in the earlier work.[11] A detailed view of the cross-section in this region is shown in Fig. 4. This onset, which covers essentially 50 cm^{-1}, is due to the opening of what we call the $Q_{3/2}$ rotational branch:[15]

$$OH^-(OD^-)\ {}^1\Sigma^+(v''=0;J'') \to OH(OD)\ {}^2\pi_{3/2}(v'=0;J'=J''+\tfrac{1}{2})\ J'' \geq 1$$

A detailed inspection of this portion of the cross-section along with extensive fitting procedures (described in detail elsewhere)[15] led to the following numbers: $EA(OH) = (14\ 723 \pm 15)\ cm^{-1}$ and $EA(OD) = 14\ 703 \pm 15)$ cm^{-1}. The best fits to the data were obtained by synthetic cross-sections based on the (purely empirical) threshold law $\sigma \propto E^{1/4}$ (for all J'), as shown in Fig. 4 for OH^-. This may be taken as evidence for the influence of the electron-permanent dipole interaction in the final channel. We are aware that, indeed, the threshold law will be dependent on J'. The observed isotope effect for the electron affinity can be almost entirely ascribed to differences in the position of the ground rotational state in $OH({}^2\pi_{3/2})$ and $OD({}^2\pi_{3/2})$ (13 cm^{-1}), which in turn means that the vibrational frequencies in the neutral molecule and in the ion are the same to within 100 cm^{-1}.

In conclusion it should be mentioned that a further improvement in energy resolution by another factor of a hundred can be achieved, so that the experimental determination of electron affinities with an accuracy of the order microelectron volts appears possible in the near future.

REFERENCES

1. Lineberger, W. C. and Woodward, B. W. *Phys. Rev. Lett.*, 1970, **25**, 424.
2. Lineberger, W. C. and Patterson, T. A., *Chem. Phys. Lett.*, 1972, **13**, 40.
3. Hotop, H. and Lineberger, W. C., *J. Chem. Phys.*, 1973, **58**, 2379.
4. Lineberger, W. C. *In* 'Energy, Structure and Reactivity', Proceedings of the Summer Research Conference on Theoretical Chemistry, Boulder, June 1972, Wiley, New York, in press.
5. Wigner, E. P., *Phys. Rev.*, 1948, **73**, 1002.
6. O'Malley, T. F., *Phys. Rev.*, 1965, **137**, A1668; Hinckelmann, O. and Spruch, L., *Phys. Rev.*, 1971, **A3**, 642.
7. Hotop, H., Patterson, T. A. and Lineberger, W. C., *Phys. Rev A*, August 1973.
8. Rau, A. R. P. and Fano, U., *Phys. Rev.*, 1971, **A4**, 1751.
9. Andrick, D., Eyb, M. and Hofmann, H., *J. Phys.*, 1972, **B5**, L15.
10. Norcross, D. W. and Moores, D. L., *in* 'Atomic Physics 3', *ed.* G. K. Walters and S. J. Smith, Plenum, New York, 1973. Moores, D. L. and Norcross, D. W., *Abstract VIII ICPEAC*, Belgrade, July 1973.

11. Branscomb, L. M., *Phys. Rev.*, 1966, **148**, 11.
12. Cade, P. E., *J. Chem. Phys.*, 1967, **47**, 2390.
13. Celotta, R. J., Bennett, R. A. and Hall, J. L. (to be published).
14. Geltman, S., *Phys. Rev.*, 1958, **112**, 176.
15. Hotop, H., Patterson, T. A. and Lineberger, W. C. (to be published).

Discussion

N. R. Daly (AWRE, Aldermaston, U.K.): What contribution to the fast neutral atom beam arises because of collisions between the negative ions and residual gas molecules, and what is the pressure in this region?

H. Hotop: Gated detection of the fast neutral atoms is used, with a window of 1–2 μsec open at the proper time delay after the laser pulse, thereby one discriminates so strongly against the neutrals formed in collisions of the negative ions with the residual gas (pressure about 10^{-8} torr) that signal to noise ratios around 50–100 are obtained for a single shot of the laser (except in the energy region very close to the threshold).

J. Eland (Oxford University, U.K.): In testing Wigner's law how far are you justified in assuming pure s or p outgoing waves?

H. Hotop: Assuming that only electric dipole transitions occur one expects a pure p-wave for the outgoing electron in the photodetachment of $\text{Au}^-(^1S_0)$ to $\text{Au}(^2S_{1/2})$. For the case of Se^-, one may have d-waves besides s-waves, but close to threshold ($K \lesssim 0.1a_0^{-1}$) one can neglect the contribution of the d-wave to the cross-section. The influence of the d-wave is more readily seen in the angular distribution of the photoelectrons, which, incidentally, can also present a good probe of spin–orbit effects (for instance with respect to differences in the $P_{1/2}$ and $P_{3/2}$ continua for outgoing p-waves).

R. S. Berry (University of Chicago, U.S.A.): Your observation of the p-state threshold of the alkalis is perhaps the most direct measurement of electron correlation, in the sense that you measure the amount that the p^2-configuration contributes to the ^1S ground state wave function of the alkali negative ions. Is it practical to go higher in excitation energy to look for contributions from other configurations, such as the d^2?

H. Hotop: Yes, I think it is possible, either with the flashlamp-pumped dye-laser used here or, in the wavelength range below 4500 Å, more readily with an N_2 laser-pumped dye-laser. One could also do laser photodetachment electron spectrometry and thereby directly determine the relative cross-sections for formation of different final states of the atom.

33

Negative Ion-Molecule Reactions Involving Nitrous Oxide and Carbon Dioxide. The Electron Affinities of N_2O and CO_2

By THOMAS O. TIERNAN and ROGER P. CLOW

(Aerospace Research Laboratories, Chemistry Research Laboratory, Wright-Patterson Air Force Base, Ohio, U.S.A.)

NEGATIVE ion-molecule processes involving nitrous oxide and carbon dioxide were investigated using an in-line double mass spectrometer described previously.[1,2] Cross-sections for these reactions have been determined as a function of translational energy. Isotopically labelled reagents were used to obtain information on reaction mechanisms. Thresholds for endothermic charge transfer reactions with N_2O and CO_2 and for collision-induced dissociation reactions of the molecular negative ions with various targets were measured in an effort to derive electron affinity values for these molecules, in the manner which we have previously reported.[3,4] It was anticipated that the latter experiments would also yield an indication of the magnitude of the activation energy which is necessary for formation of N_2O^- and CO_2^- from the corresponding neutral molecules. Such a barrier is to be expected because the neutral species have a linear configuration, while theory predicts that the molecular negative ions are bent.[5]

NITROUS OXIDE

Negative ion reactions with nitrous oxide which were examined in the present study are listed in Table I. Also reported are the maximum cross-sections determined for these reactions and the energies at which the maxima are observed. Absolute cross-sections were estimated by normalizing relative cross-sections for the various reactions to that for the O^-/NO_2 charge transfer reaction, for which an absolute value of $63 \times 10^{-16} \, cm^2$ has been reported at 0·3 eV ion energy.[6] Those reactions which are exothermic at the lowest attainable energy (0·3 eV, lab), (1, 2, 4, 5 and 6), exhibit their maximum cross-sections at this energy, as expected, and decrease sharply with increasing translational energy. This behaviour is illustrated in Fig. 1

295

THOMAS O. TIERNAN AND ROGER P. CLOW

TABLE I
Negative Ion Reactions Involving N_2O

Reaction	Maximum cross-section ($\times 10^{16}$ cm^2)	Energy at maximum cross-section (eV, CM)	Threshold energy (eV, CM)
1. $O^- + N_2O \rightarrow NO^- + NO$	6·0	0·2	a
2. $NO^- + N_2O \rightarrow N_2O^- + NO$	1·4	0·2	a
3. $NO^- + N_2O \rightarrow O^- + N_2 + NO$	3·0	3·8	1·0
4. $N_2O^- + NO_2 \rightarrow NO_2^- + N_2O$	31	0·2	a
5. $N_2O^- + O_2 \rightarrow O_2^- + N_2O$	35	0·1	a
6. $N_2O^- + NO \rightarrow NO^- + N_2O$	6·8	0·1	a
7. $N_2O^- + Ar \rightarrow O^- + N_2 + Ar$	23	7·7	0·8
8. $N_2O^- + Kr \rightarrow O^- + N_2 + Kr$	33	11·5	1·0
9. $N_2O^- + N_2 \rightarrow O^- + 2N_2$	39	6·6	0·9

a Reaction is apparently exothermic at lowest energy attainable in these experiments.

for reactions 1 and 2. At higher translational energies, NO^- undergoes a dissociative electron transfer reaction with N_2O (reaction 3), and this process is endothermic as indicated by the excitation function shown in Fig. 1. Excitation functions are also shown in Fig. 1 for the endothermic collision-induced dissociation reactions of N_2O^- which are listed in Table I (7, 8

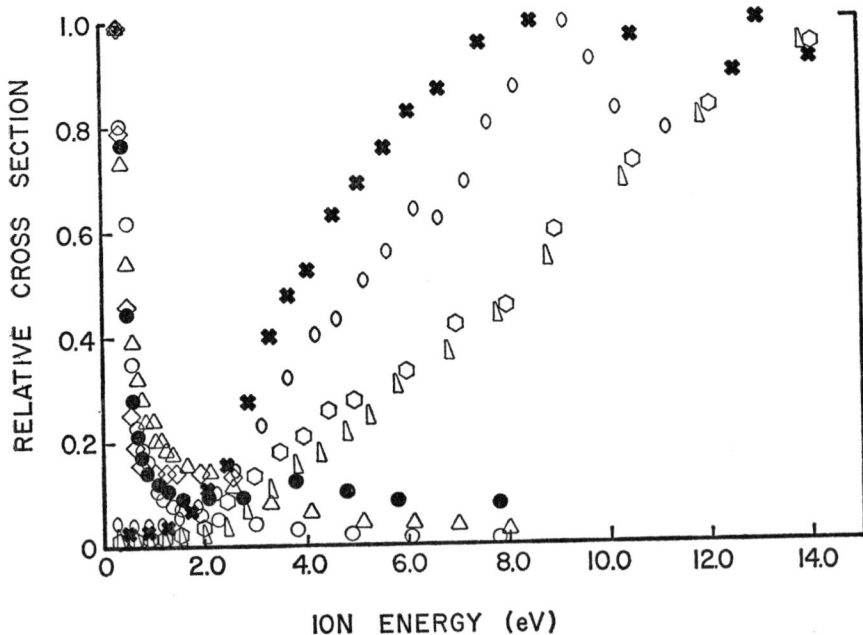

Fig. 1 Cross-sections as a function of translation energy for negative ion-molecule reactions involving nitrous oxide. Reactions are numbered as in the tables. ●, 10a; ◇, 11a; ○, 1; △, 2; 0, 12a; X, 7; ○, 8; �◁, 9.

and 9). As can be seen, these reactions were all observed to exhibit the same energy threshold (CM), within the accuracy of the experimental measurement.

The fact that fast exothermic electron transfer reactions from N_2O^- to NO_2, O_2 and NO are observed at near-thermal energies is an indication that the electron affinity of N_2O is less than that of these three molecules. Recent studies in our own laboratory[3,4] and elsewhere[7] have established that, of these molecules, NO has the lowest electron affinity, which is on the order of 0·02 eV. As indicated in Table I, NO^- is also observed to charge transfer efficiently with N_2O at low energies. It appears from these observations, therefore, that the electron affinities of NO and N_2O are comparable—that is, very nearly zero. However, this value can represent only a lower limit to the adiabatic electron affinity because the N_2O neutral product from the electron transfer reactions is undoubtedly formed in an excited state, owing to the geometry factor.

An independent estimate of the electron affinity of N_2O can be obtained from the thresholds measured for collision-induced dissociation of N_2O^- since these thresholds presumably correspond to the binding energy of the molecular negative ion. If it is assumed that the reactant ions are in the ground state, and that the products are formed in their ground states with zero translational energy at threshold, then it is easily shown that the electron affinity is given by the expression:

$$EA(N_2O) = E_{threshold} - \Delta H_f(O) + EA(O) + \Delta H_f(N_2O)$$

Using the average of the measured collision-induced dissociation thresholds, 0·9 eV, tabulated thermochemical values for the appropriate heats of formation,[8] and the known electron affinity of the oxygen atom,[9] an electron affinity of $EA(N_2O) = 0·6$ eV is calculated from this relation. This value should more nearly represent the adiabatic electron affinity than that obtained from charge transfer experiments, for reasons already discussed. The difference between the values obtained from the two types of experiment ($\sim 0·6$ eV), may indeed be the activation energy required to bend the linear N_2O molecule

TABLE II
Reactions Observed With $N_2{}^{18}O$

Reaction	Notes
10a. $^{16}O^- + N_2{}^{18}O \rightarrow {}^{18}O^- + N_2{}^{16}O$	$\sigma = 5·4 \times 10^{-16}$ cm² at
10b. $^{16}O^- + N_2{}^{18}O \rightarrow N^{16}O^- + N^{18}O$	$E_0{}^- = 0·3$ eV
10c. $^{16}O^- + N_2{}^{18}O \rightarrow N^{18}O^- + N^{16}O$	$\sigma_{10b}/\sigma_{10c} = 1$ for $E_0{}^- < 2$ eV
11a. $N^{16}O^- + N_2{}^{18}O \rightarrow N^{18}O^- + N_2{}^{16}O$	$\sigma = 0·06 \times 10^{-16}$ cm² at $E_{NO^-} = 0·3$ eV
11b. $N^{16}O^- + N_2{}^{18}O \rightarrow N_2{}^{18}O^- + N^{16}O$	
11c. $N^{16}O^- + N_2{}^{18}O \rightarrow {}^{18}O^- + N_2 + N^{16}O$	$\sigma_{11c}/\sigma_{11d} \sim 4$–$10$
11d. $N^{16}O^- + N_2{}^{18}O \rightarrow {}^{16}O^- + N + N_2{}^{18}O$	
12a. $N_2{}^{16}O^- + N_2{}^{18}O \rightarrow {}^{16}O^- + N_2 + N_2{}^{18}O$	$\sigma = 19 \times 10^{-16}$ cm² at $E_{N_2O^-} = 9$ eV
12b. $N_2{}^{16}O^- + N_2{}^{18}O \rightarrow N^{18}O^- + N_2{}^{16}O$	$E_{threshold} = 0·9$ eV

to the angular configuration in which the molecular negative ion is most stable. The N_2O^- ions which are formed in the experiments described here have lifetimes of at least 50 μsec and are probably not metastable. Theoretical considerations by Bardsley,[10] Chantry[11] and Wentworth et al.[12] predict that N_2O^- is bound and should therefore be stable, although the potential well is relatively shallow. This is substantiated by the binding energy measured in the present study, which is about 0·9 eV.

Table II lists the reactions observed in this investigation when isotopically labelled $N_2{}^{18}O$ is used as the neutral reactant. These observations verify the mechanisms implied by the reactions as written in Table I. Reaction 10a shown in Table II is a switching reaction which can, of course, only be observed by using the labelled reactant. This reaction competes quite favourably with the O^-/N_2O reaction which produces NO^-. As can be seen from Table II, reactions 10b and 10c have comparable cross-sections at energies below 0·7 eV (CM), indicating that there is equal probability of incorporating in the NO^- product either the oxygen species from the incident negative ion or from the neutral target. These results suggest that this inter-action proceeds via a symmetric activated complex having a structure such as:

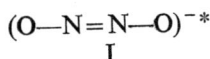

$$(O{-}N{=}N{-}O)^{-*}$$
$$\text{I}$$

In studies of N_2O negative ion reactions under higher pressure conditions in other laboratories (e.g. drift tube, flowing afterglow), this intermediate species has in fact been observed.

At low energies, the long-range charge transfer reaction of NO^- with N_2O is apparently the dominant reaction channel (reactions 11a and 11b), and at higher energies dissociative charge transfer becomes important. The fact that only a small amount of isotopic mixing is evident from the products of the NO^-/N_2O interaction (reactions 11a, 11b, 11c), indicates that an intermediate complex apparently does not play a significant role in this case. Possibly this is due to unfavourable steric factors.

It is interesting that no symmetric charge transfer reaction between N_2O^- and N_2O is observed (reactions 12a and 12b are the only processes detected). This is understandable if the process is not resonant, as would be anticipated from the geometry considerations noted earlier.

On the basis of the results described herein, it is possible to establish a generalized reaction scheme for negative ion processes in nitrous oxide, as indicated below. The extent to which the autodetachment and collisional stabilization processes are involved will, of course, depend upon the pressure and residence time which are characteristic of a particular experiment.

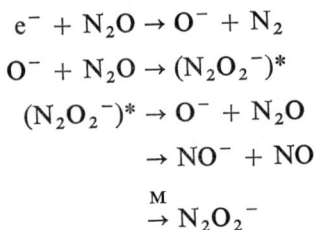

$$e^- + N_2O \rightarrow O^- + N_2$$
$$O^- + N_2O \rightarrow (N_2O_2{}^-)^*$$
$$(N_2O_2{}^-)^* \rightarrow O^- + N_2O$$
$$\rightarrow NO^- + NO$$
$$\overset{M}{\rightarrow} N_2O_2{}^-$$

$$NO^- + N_2O \xrightarrow[\text{range}]{\text{long}} (N_2O^-)^* + NO$$

$$(N_2O^-)^* \xrightarrow{\tau > 50 \ \mu sec} N_2O + e^-$$

$$NO^- + N_2O \rightarrow (N_3O_2^-)^*$$

$$(N_3O_2^-)^* \rightarrow NO^- + N_2O$$

$$\xrightarrow{M} N_3O_2^-$$

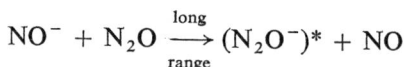

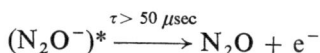

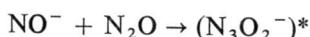

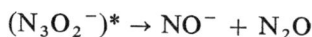

CARBON DIOXIDE

Table III shows the negative ion reactions with CO_2 which were observed in the present study. Corresponding experiments in which ^{18}O-labelled CO_2 was used and the various isotopic products were monitored are reported in Table IV. All the reactions observed in this system with the exception of the

TABLE III
Negative Ion Reactions Involving CO_2

Reaction	Maximum cross-section ($\times 10^{16}$ cm2)	Energy at maximum cross-section (eV, CM)	Threshold energy (eV, CM)
13a. $O^- + CO_2^- \rightarrow CO_2^- + O$	0·2	11·0	2·0, 5·2
13b. $O^- + CO_2 \rightarrow O_2^- + CO$	0·6	7·0	3·2
14a. $O_2^- + CO_2 \rightarrow CO_2^- + O_2$	0·3	9·8	1·2, 4·3
14b. $O_2^- + CO_2 \rightarrow CO_3^- + O$	0·2	4·3	2·5
14c. $O_2^- + CO_2 \rightarrow O^- + CO + O_2$	3·8	8·8	4·6
15a. $NO^- + CO_2 \rightarrow CO_2^- + NO$	0·6	8·9	3·6
15b. $NO^- + CO_2 \rightarrow O^- + CO + NO$	1·5	13·7	6·6

atom exchange process (reaction 16a) are endothermic. Table III also gives the threshold energies determined for these processes, the maximum cross-sections measured, and the energies at which the maxima occur. Two reactions of O^- with CO_2 are observed (13a and 13b), charge transfer and atom abstraction. The excitation functions for these reactions are plotted in Fig. 2. The cross-section for CO_2^- production has a threshold at 2·0 eV, CM, rises rapidly to a maximum at 3·7 eV, CM and rises again to a much broader maximum at about 12 eV, CM. The cross-section for O_2^- formation has a much sharper energy dependence. From the onset at 3·2 eV, CM, this cross-section increases to a maximum at about 9 eV, CM, then drops to zero at higher energies. Distinct structure is evident in both of the excitation functions and is quite reproducible. It should be noted that the O_2^- threshold occurs at the energy where there is a sharp break in the CO_2^-

TABLE IV

Reactions Observed With $C^{18}O_2$

Reaction	Notes
16a. $^{16}O^- + C^{18}O_2 \rightarrow {}^{18}O^- + C^{16}O^{18}O$	$\sigma = 54 \times 10^{-16}$ cm^2, at $E_0{}^- = 0.3$ eV, lab
16b. $^{16}O^- + C^{18}O_2 \rightarrow C^{18}O_2{}^- + {}^{16}O$	$\sigma_{16b}/\sigma_{16c} = 1$, at $E_0{}^- = 5.0$ eV, lab
16c. $\rightarrow C^{16}O^{18}O^- + {}^{18}O$	
16d. $^{16}O^- + C^{18}O_2 \rightarrow {}^{16}O^{18}O^- + C^{18}O$	
17a. $^{16}O_2{}^- + C^{18}O_2 \rightarrow C^{18}O_2{}^- + {}^{16}O_2$	
17b. $^{16}O_2{}^- + C^{18}O_2 \rightarrow C^{18}O_2{}^{16}O^- + {}^{16}O$	
17c. $^{16}O_2{}^- + C^{18}O_2 \rightarrow {}^{18}O^- + C^{18}O + {}^{16}O_2$	$\sigma_{17b}/\sigma_{17c} > 3$
17d. $\rightarrow {}^{16}O^- + {}^{16}O + C^{18}O_2$	
18a. $N^{16}O^- + C^{18}O_2 \rightarrow C^{18}O_2{}^- + N^{16}O$	
18b. $N^{16}O^- + C^{18}O_2 \rightarrow {}^{18}O^- + C^{18}O + N^{16}O$	$\sigma_{18c}/\sigma_{18d} = 1$, at $E_0{}_2{}^-$ $\simeq 45$ eV, lab
18c. $\rightarrow {}^{16}O^- + N + C^{18}O_2$	

excitation function. Also, the structure in the $O_2{}^-$ curve at 3.7 eV, CM, is coincident with a dip in the $CO_2{}^-$ cross-section.

The experiments with $C^{18}O_2$ are particularly informative in interpreting the structure just described. As indicated by the reactions listed in Table IV, when the isotopically labelled neutral is used, an atom switching reaction, (16a), can be detected which is exothermic or thermoneutral, since its cross-section is largest at the lowest energy and decreases sharply as translational

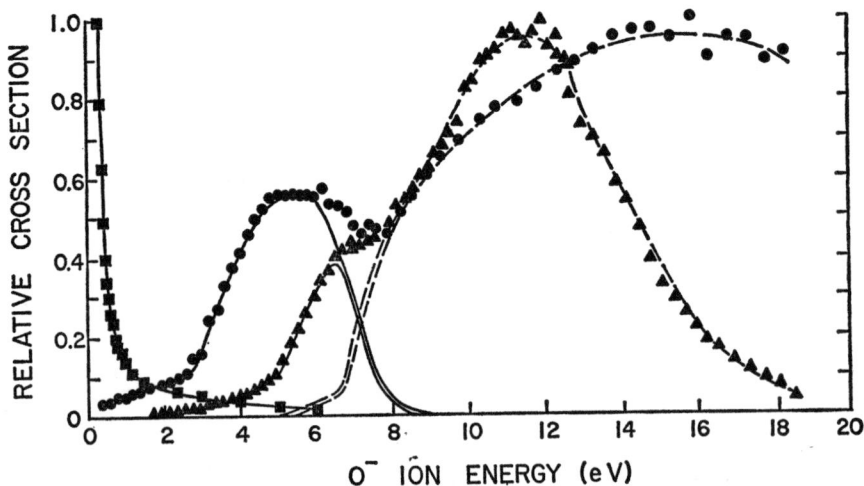

FIG. 2   Cross-sections as a function of translational energy for products from the reaction of O^- with CO_2. Reactions are numbered as in the tables. See text for discussion of extrapolated curves. ■, 16a; ●, 13a; ▲, 13b.

energy is increased (*see* Fig. 2). It can also be seen that the charge transfer reaction of O^- with CO_2 apparently proceeds via both long range and complex mechanisms, since both reactions 16b and 16c are observed. A study of the energy dependence of these products indicates that the ratio of the cross-section for $C^{16}O^{18}O^-$ to that for $C^{18}O_2^-$ increases from ~ 0.1 at threshold to a maximum approaching 1 at 3·2 eV, CM, and then decreases to ~ 0.1 at about 5·2 eV, CM. The $C^{16}O^{18}O$ impurity in the $C^{18}O_2$ used in these experiments is sufficient to account for this ratio at threshold and at higher energy. It is noteworthy that the energy at which the $C^{16}O^{18}O^-$ product attains a maximum is precisely the region in which the structure in the excitation functions of Fig. 2 occurs. It should also be noted that the $^{16}O^-/C^{18}O_2$ reaction yields only isotopically mixed O_2^- products (reaction 16d). Reactions 16a, 16c and 16d, then, give evidence for the formation of a $(CO_3^-)^*$ intermediate which has a sufficient lifetime to permit some energy redistribution and isotopic scrambling to occur. From the energy dependence described, it is concluded that this mechanism is operative only at energies below about 5·2 eV, CM, while at energies above this, long-range charge transfer is the dominant mechanism. In Fig. 2, the solid lines through the data points have been extrapolated to indicate the energy regime in which complex formation occurs, while the dashed lines, similarly extrapolated, indicate the energy regime in which a long-range mechanism is dominant. The stabilized CO_3^- intermediate has actually been detected in experiments at higher pressures using other techniques.[13-16]

As shown in Table III, CO_3^- is also formed at higher energies from the reaction of O_2^- with CO_2 (reaction 14b). In addition, this interaction yields CO_2^- and O^- products. The excitation function for reaction 14a reveals two energy thresholds for this process. Experiments in which $^{16}O_2^-$ was reacted with $C^{18}O_2$ are also reported in Table IV and suggest that isotopic scrambling does not occur in this case.

Reactions of NO^- and S^- with CO_2 were also studied in an effort to derive electron affinity data from the measured thresholds for CO_2^- production (*see* Tables III and IV for a summary of the NO reactions). As in the case of the O_2^- reaction discussed above, charge transfer from S^- to CO_2 is endothermic and the excitation function indicates two energy thresholds. From the lower energy thresholds measured in these several reactions, an electron affinity value of -0.8 ± 0.3 eV is calculated. The higher threshold yields an electron affinity of -3.7 ± 0.2 eV. The observation of two thresholds for the charge transfer reactions producing CO_2^- is apparently not entirely related to the occurrence of two different charge transfer mechanisms, because in the case of the O_2^-/CO_2 reaction, there appears to be only one mechanism operating—the long-range mechanism. A more likely explanation is that the two thresholds are indicative of two different states of CO_2^-. Krauss and Newmann[17] have calculated the energies of two electronic states of CO_2^- and Cooper and Compton[18] have reported lifetimes for two CO_2^- metastables, which could well be the states detected in our experiments. The bound metastable state of CO_2^- and the repulsive states are separated by 3·2 eV according to the theoretical estimates,[17] and this is in reasonably good agreement with the 2·9 eV separation which we observe between the two CO_2^- energy thresholds in our experiments.

ACKNOWLEDGMENT

R. P. Clow was a National Research Council–Air Force Systems Command Postdoctoral Research Associate during the period when this research was accomplished.

REFERENCES

1. Tiernan, T. O. and Marcotte, R. E., *J. Chem. Phys.*, 1970, **53**, 2107.
2. Hughes, B. M. and Tiernan, T. O., *J. Chem. Phys.* 1971, **55**, 3419.
3. Tiernan, T. O., Lifshitz, C. and Hughes, B. M., *J. Chem. Phys.*, 1971, **55**, 5692.
4. Hughes, B. M., Lifshitz, C. and Tiernan, T. O., *J. Chem. Phys.*, 1973, (in press).
5. Herzberg, G. 'Electronic Spectra of Polyatomic Molecules', D. Van Nostrand, Princeton, N.J., 1966, p. 319.
6. Paulson, J. F., *Advances in Chemistry,* 1966, **58**, 28.
7. Celotta, R., Bennett, R., Hall, J., Siegel, M. W. and Levine, J., *Bull. Am. Phys. Soc.*, 1970, **15**, 1515.
8. Franklin, J. L., Dillard, J. G., Rosenstock, H. M., Herron, J. T. and Draxl, K., NSRDS-NBS 26, U.S. Department of Commerce, Washington, D.C., 1969.
9. R. S. Berry, *Chem. Rev.* 1969, **69**, 533.
10. Bardsley, J. N., *J. Chem. Phys.,* 1969, **51**, 3384.
11. Chantry, P. J., *J. Chem. Phys.,* 1969, **51**, 3369.
12. Wentworth, W. E., Chen, E. and Freeman, R., *J. Chem. Phys.,* 1971, **55**, 2075.
13. Paulson, J. F., *J. Chem. Phys.,* 1970, **52**, 963.
14. Moruzzi, J. L. and Phelps, A. V., *J. Chem. Phys.,* 1966, **45**, 4617.
15. Fehsenfeld, F. C., Schmeltekopf, A. L., Schiff, H. I. and Ferguson, E. E. *Planet. Space Sci.,* 1967, **14**, 373.
16. Parker, D. A., *Faraday Transactions I.,* 1972, **68**, 627.
17. Krauss, M. and Newmann, D., *Chem. Phys. Letters,* 1972, **14**, 26.
18. Cooper, C. D. and Compton, R. N., *Chem. Phys. Letters,* 1972, **14**, 29.

34
Van der Waals Complexes by Electron Attachment Mass Spectrometry

By H. KNOF

(B.P. Institut für Forschung und Entwicklung, Wedel, Germany)

and D. KRAFFT

(Institut für Experimentalphysik der Universität, Hamburg, Germany)

THE formation of van der Waals complexes which are agglomerates consisting of atoms or molecules held together by van der Waals forces and which may act as condensation nuclei is well known for several atoms and small molecules.[1-5] Using an electron attachment mass spectrometer it is possible to investigate van der Waals complexes consisting of organic molecules.[6,7]

In addition it is possible to detect free primary radicals as well as complexes between these radicals and organic molecules.

INSTRUMENTATION

The electron attachment ion source of such a mass spectrometer (Fig. 1) consists mainly of an ionization chamber and an electron source which are well separated to prevent pyrolysis of the substance by the hot cathode. The electrons emitted from the cathode are focussed across a distance of 22 cm onto the entrance hole of the ionization chamber. The cathode can be adjusted mechanically to optimize the electron current. The electrons enter the ionization chamber with an energy of about 200 eV, producing positive ions and thereby secondary electrons with low energy. These secondary electrons may be attached to sample molecules and thus give rise to negative ions.

The ionization process reads

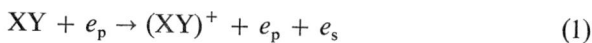

$$XY + e_p \rightarrow (XY)^+ + e_p + e_s \qquad (1)$$
$$XY + e_s \rightarrow (XY^*)^- \qquad (2)$$
$$(XY^*)^- \rightarrow X^- + Y + E_{kin} \qquad (3)$$

Since the yield of negative ions depends on the square of the pressure in the ionization chamber the highest negative ion current is obtained just below a gas pressure at which charge transfer and ion molecule reactions may occur.

The best electron attachment mass spectra were produced at a pressure of
0·01 torr in the ionization chamber.

In order to prevent sparking and gas discharge in the region around the
ionization chamber a metal grid was placed around the ion beam below the
ion source. By this means the migration of charge carriers from the ion beam

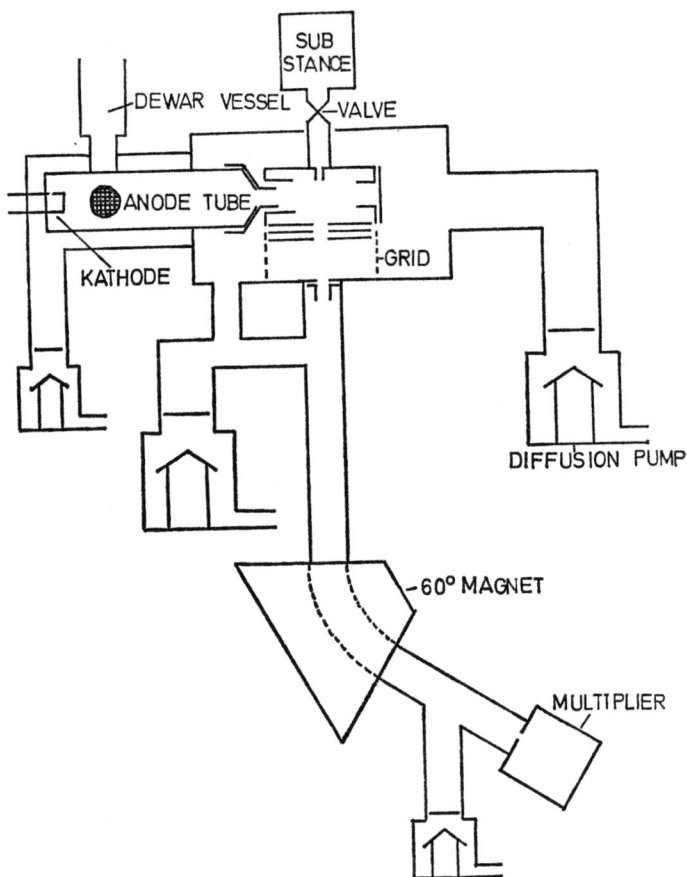

Fig. 1   Schematic view of an electron attachment mass spectrometer.

into the region around the ionization chamber is prevented and measurements
with a pressure up to 10^{-3} torr in this region can be made. Thus excessive
pumping facilities are avoided.

In contrast to the electron attachment ion source of von Ardenne[8] this
ion source makes it possible to investigate pure substances without the need
for an additional gas in which a gas discharge is produced and to measure
spectra electrically which permits quantitative measurements instead of semi-
quantitative measurements with a photoplate. In addition this new electron
attachment ion source can be used with commercial mass spectrometers.

MEASUREMENTS OF VAN DER WAALS COMPLEXES

Various organic substances have been investigated for the existence of van der Waals complexes. Complexes with up to four molecules could be detected.

Alcohols did show very intense complex peaks. Figure 2 gives the mass spectra of ethanol. The identity of the ions is checked by measurements with deuterated ethanol. In addition, by comparing measurements of

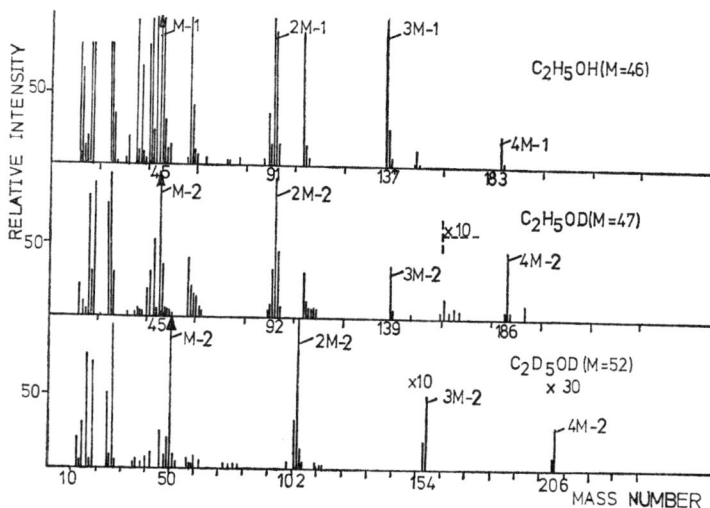

FIG. 2   Mass spectra of ethanol.

deuterated samples with those of non-deuterated alcohols it can be seen that ionization occurs by splitting off the hydrogen from the hydroxy group.

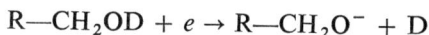

$$R\text{—}CH_2OD + e \rightarrow R\text{—}CH_2O^- + D$$

This is not only true for monomers but for polymers as well. The relative intensities of the first alcohols are given in Table I. They are normalized to the highest polymer which could be measured.

TABLE I

Relative Intensities of Alcohol Complexes

Substance	Masses	Intensities
CH_3OH	31: 63: 95	400: 2: 1
CH_3OD	31: 64: 97	320: 2: 1
CD_3OD	34: 70:106	510: 3: 1
C_2H_5OH	45: 91:137:183	1750: 70:14:1
C_2H_5OD	45: 92:139:186	13600:136: 8:1
C_2D_5OD	50:102:154:206	5000:200:25:1
C_3H_7OH	59:119:179:239	900: 60: 6:1
C_3H_7OD	59:120:181:242	1960:280:20:1
C_3D_7OD	66:134:202:270	300: 10: 1:–

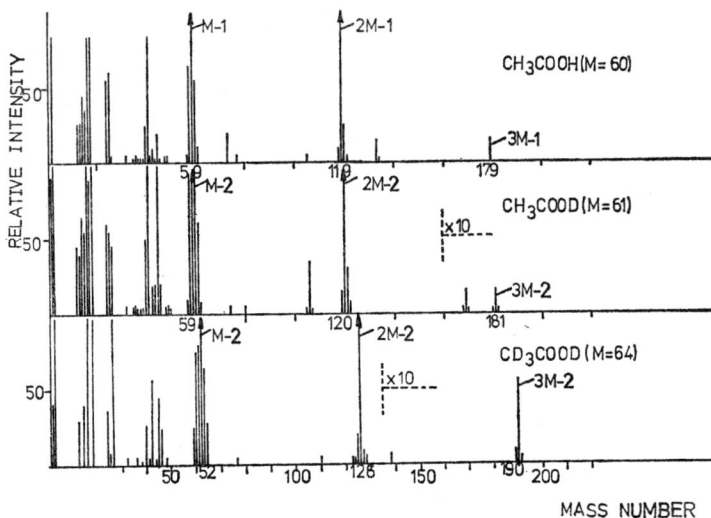

FIG. 3   Mass spectra of acetic acid.

TABLE II

Relative Intensities of Fatty Acid Complexes

Substance	Masses	Intensities
HCOOH	45: 91:137	9100: 65:1
HCOOD	45: 92:139	1200: 30:1
DCOOD	46: 94:142	4800: 40:1
CH_3COOH	59:119:179	4500: 30:1
CH_3COOD	59:120:181	6000: 60:1
CD_3COOD	62:126:190	15400:140:1

TABLE III

Relative Intensities of Various Oxygen and Nitrogen Compounds

Substance	Masses	Intensities
$(CH_3)_2CO$	57:115	50: 1
C_2H_5CHO	57:115	16: 1
CH_3NO_2	60:121:182	1050: 7:1
CD_3NO_2	62:126:190	550: 5:1
$C_3H_7NH_2$	58:117	60: 1
CH_3CN	40: 81:122	2100: 30:1
CD_3CN	42: 86:130	10800:180:1
C_2H_5CN	54:109:164	18000:300:1

Similar spectra are obtained for fatty acids. The mass spectra for acetic acid and for the deuterated samples are given in Fig. 3. The fatty acids show the same ionization behaviour as the alcohols

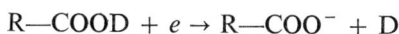

$$R\text{—}COOD + e \rightarrow R\text{—}COO^- + D$$

The relative intensities which are given in Table II drop faster than those of the alcohols and no tetramers could be detected.

Polymers could be identified from other oxygen compounds such as ketones and aldehydes. Similar relations occur for nitrogen compounds. Figure 4 gives the mass spectrum for propion nitrile and Table III shows the relative intensities of various oxygen and nitrogen compounds.

No polymers could be detected from the nonpolar hydrocarbons benzene, cyclopentane and cyclohexane.

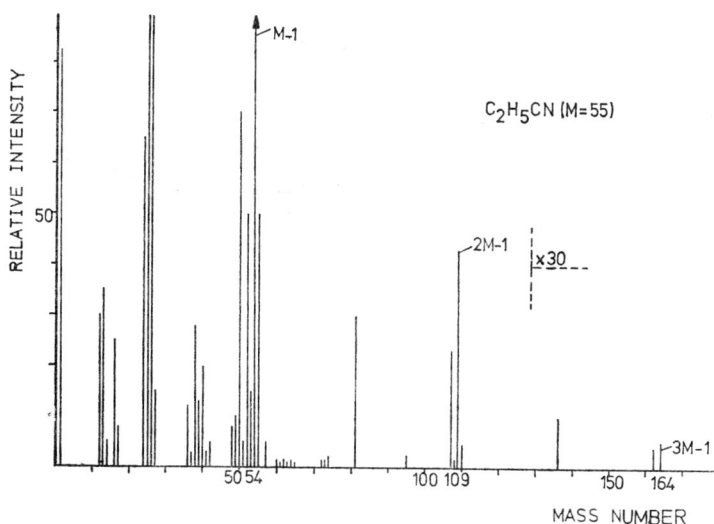

FIG. 4   Mass spectrum of propion nitrile.

PRESSURE DEPENDENCE OF ION INTENSITIES

The dependence of ion intensities on the gas pressure in the ionization chamber is given in Fig. 5 for complex ions as well as for some fragment ions. The pressure dependence is twice as large for the molecular and complex ions than for the fragment ions. This fact is reasonable, since the formation of negative ions by electron attachment is a two step process compared with the ion pair production which is a direct process.

The pressure dependence is the same for monomer, dimer and trimer complex ions. This fact indicates that the neutral complexes do not originate from the ionization chamber but are probably those complexes which exist in equilibrium with the liquid phase of the substance contained in the reservoir.

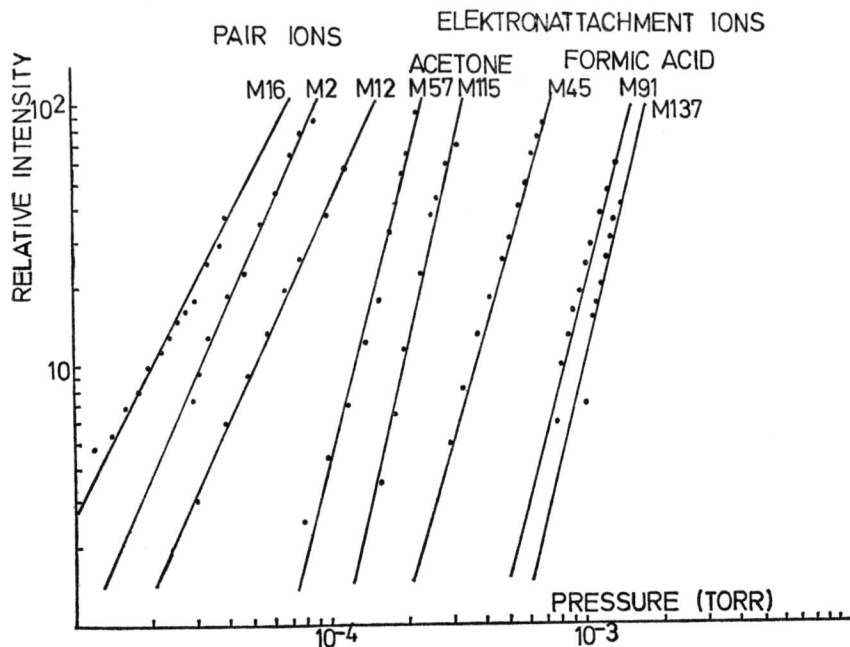

FIG. 5   Pressure dependence of ion intensities for various ions.

MEASUREMENTS OF FREE RADICALS

Electron attachment mass spectrometry is not only good for the measurements of weakly bound van der Waals complexes but also proves to be a useful tool for detecting reactive organic molecules such as free radicals. It is possible to obtain negative ions of primary free radicals produced by ultraviolet light.

For this purpose, the sample gas in the gas inlet tube above the ionization chamber was irradiated with ultraviolet light. The gas pressure there was about 0·1 torr. The light source used was a high pressure xenon arc.

TABLE IV

Free Radicals of Acetone Produced by UV-Light

Mass	Radical
15	CH_3
43	CH_3CO
73	$(CH_3)_2COCH_3$
101	$(CH_3)_2CO\ CH_3CO$
131	$[(CH_3)_2CO]_2CH_3$
159	$[(CH_3)_2CO]_2\ CH_3CO$
33	HO_2

Using a difference method with acetone as sample gas the following radicals (Table IV) could be measured.

These results show that the primary free radicals CH_3 and CH_3CO being produced by UV-irradiation of acetone can be ionized without fragmentation. In addition complexes between these radicals and acetone molecules, *i.e.* $(CH_3)_2COCH_3$, $(CH_3)_2COCH_3CO$ and complexes between radicals and two acetone molecules could be detected. Furthermore there appeared radicals such as HO_2 which probably result from wall reactions.

The primary radicals disappear by undergoing wall reactions. This was shown by forcing all free radicals to interact with a surface within the gas inlet tube.

REFERENCES

1. Becker, E. W., Bier, K. and Henkes, W. *Z. Phys.*, 1956, **146**, 333.
2. Henkes, W. *Z. Naturforsch.*, 1961, **16a**, 842; 1962, **17a**, 186.
3. Greene, F. T. and Milne, F. A. *J. Chem. Phys.*, 1963, **39**, 3150.
4. Leckenby, R. E., Robbins, E. J. and Trevalion, P. A. *Proc. Roy. Soc. London*, 1964, **1280**, 409.
5. Bauchert, J. and Hagena, O. F. *Z. Naturforsch.*, 1965, **20a**, 1135.
6. Knof, H. and Krafft, D. *Z. Naturforsch.*, 1970, **25a**, 849.
7. Knof, H., Hausen, V. and Krafft, D. *Z. Naturforsch.*, 1972, **27a**, 162.
8. von Ardenne, M., Steinfelder, K. and Tümmler, R., 'Elektronenanlagerungs-Massen-spektrographie organischer Substanzen', Springer-Verlag 1971.

Discussion

I. Opanszky (Central Research Institute for Physics, Budapest, Hungary): Am I right that you have seen in your spectra solvated electrons? If so, have you tried to look for $(H_2O)_n{}^-$? This formation in the liquid phase is known to be stable, but in the gas phase—as far as I know—is not known.

H. Knof: So far we have looked for electrons attached to organic complexes only. However the method should be capable of answering your question on the existence of solvated electrons in the water gas phase.

35

A Study of the Negative Ions Produced During Surface Ionization on Hot Metal Filaments

By A. T. CHAMBERLAIN, F. M. PAGE and
M. R. PAINTER

(*University of Aston in Birmingham, Gosta Green, Birmingham, U.K.*)

INTRODUCTION

THE majority of studies of negative surface ionization have been directed to determining the value of the electron affinity of an atom or radical[1,2] and for many years were the only technique for such determination, though now alternative and specific methods[3,4] have been developed. The use of surface ionization studies to determine electron affinities depends upon certain mechanistic assumptions and upon assumptions about the identity of the ion. The former include the assumption that particles incident upon the surface from the gas phase are adsorbed and reach electrical and thermal equilibrium with the surface and may be justified indirectly by the correct values for the electron affinities which are found,[1] and directly by demonstration that the initial energy distribution of emitted particle (ions and electrons) is Maxwellian and corresponds to the filament temperature.[2,5,6,7] The latter assumption of identity may, in certain cases, be justified by mass spectrometry and is the subject of the present paper.

The rate of emission of negative ions from unit area of a surface may be calculated[1] to be

$$ji = (\text{const})\, p \exp\left(-\frac{\chi - E}{kT}\right) \qquad (1)$$

where ji is the current density, p the pressure of the gaseous ion precursor, χ the work function of the surface, T its temperature and E the apparent electron affinity of the ion precursor. This may be compared with Richardson's equation for electron emission

$$je = 120 T^2 \bar{d} \exp\left(\frac{-\chi}{kT}\right) \qquad (2)$$

and the ratio of the current densities written as

$$\frac{je}{ji} = \left(\frac{T^2}{p}\right)(\text{const}) \exp\left(\frac{-E'}{kT}\right) \qquad (3)$$

Alternatively, if the ion precursor yields two negative ions X^- and Y^-, then

$$\frac{jx}{jy} = \exp\left(\frac{E'_x - E'_y}{kT}\right) \tag{4}$$

The constants shown are functions of fundamental physical quantities, but as experimental relations, eqns (1) to (4) may also include non-unitary transmission, or instrument calibration coefficients.

When the mechanism by which the negative ions are formed is unique the results are easy to obtain and analyse, but if several ions are formed by multi-step processes the apparent electron affinity becomes complex and the bond dissociation energy (D_{M-X}) and heat of adsorption (Q_M) of the residue M may be involved in the energetics of formation of X^- for MX so that

$$E' = E_x + Q_M - D_{M-X} \tag{5}$$

EXPERIMENTAL

Two systems were used, a conventional magnetron triode, and an AMP 3 quadrupole mass spectrometer with a planar diode surface ionization source; both systems were diffusion pumped, the magnetron to a background pressure 1×10^{-6} mm Hg ($0{\cdot}13$ mN m^{-2}) and the quadrupole, which was differentially pumped, to a pressure of 5×10^{-7} mm Hg. The thermionic currents were measured in both cases by an AVO 1388B d.c. amplifier with a lower limit of 1×10^{-14} A, or a current density of 10^{-9} A m^{-2}.

RESULTS

Introduction

The magnetron technique is non-specific and deductions drawn about the identity of the charge carriers on energetic grounds may be in error. Direct evidence of the ions formed is therefore essential. The compounds studied were CH_3I, I_2, CF_3I, CH_3Br, CH_3CN, ICl, IBr and ICN and the surface ionization spectra on a tantalum filament in the quadrupole mass spectrometer revealed that the only important ions were the halogens I^-, Cl^- and Br^-, and the cyanide ion CN^-. The ions CH_3^- and CF_3^-, if they were formed, were below the detection limit of the mass spectrometer. The ion currents measured in the magnetron arise therefore from a single ion only.

The results obtained with the compounds mentioned above may be analysed from two viewpoints. Utilizing eqns (3) or (4) and, if necessary, eqn (5) the electron affinity, or other unknown terms, may be evaluated. It was, however, found that the particular compounds being studied could alter the emission properties of some metal filaments. Equation (2) may therefore be utilized and modified to understand these data.

Surface Ionisation of Methyl Compounds

This was initially studied using a tantalum filament. The work function of a clean filament was found to be 461 kJ mol^{-1}. The introduction of CH_3I,

CH_3Br and CH_3CN had the effect of increasing slightly the electron emission (Fig. 1) but only a small reduction in the work function was observed (Table I).

X-ray analysis of the filament used showed that no detectable formation of tantalum carbide had occurred.

The accepted values for the electron affinities of iodine and bromine are 3.06 eV (295 kJ mol^{-1}) and 3.36 eV (325 kJ mol^{-1}) respectively, and as Table I shows, only the surface ionization of molecular iodine (310 kJ mol^{-1})

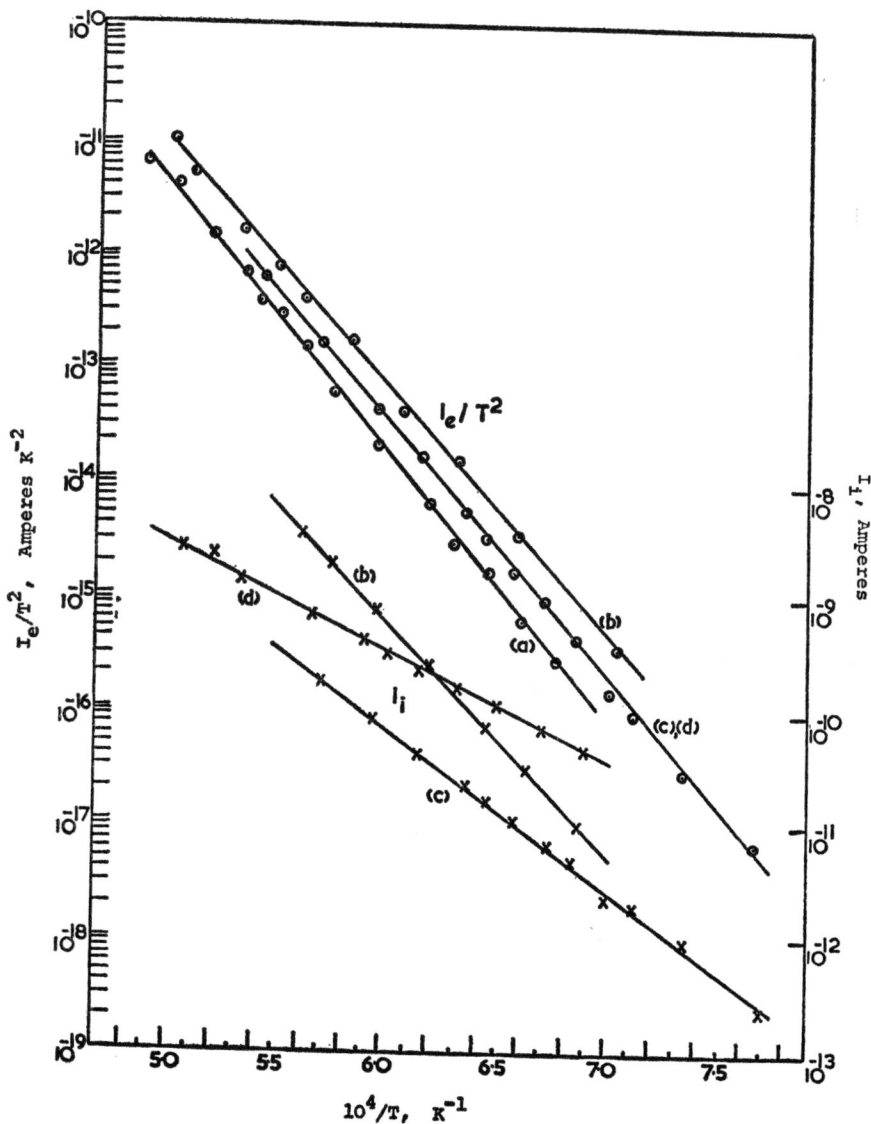

FIGURE 1

TABLE I

Substrate	Ion	χ (kJ mol^{-1})	E' (kJ mol^{-1})
clean (a)	—	459	—
I_2 (d)	I$^-$	445	310
MeI (b)	I$^-$	418	73·8
MeBr (c)	Br$^-$	437	169

yields the correct electron affinity. In the other cases, the rate determining step may be the dissociation of the molecule, so that the apparent electron affinity would be given by

$$E' = E - D$$

or alternatively, the rate determining step may be a concerted action involving the simultaneous rupture of the CH_3–χ bond, the formation of a bond between the methyl and the surface, and the formation of X$^-$, so that

$$E' = E_x + Q_{me} - D_{me-x}$$

The results for various substrate/filament combinations are collected in Table II. It will be seen that there is poor agreement if it is assumed that dissociation is the rate determining step. The hypothesis of a concerted reaction leads to consistent values for the heat of absorption of a methyl group on the different surfaces, values which vary slightly with the heat of sublimation of the filament as expected.[8]

There is some evidence for the existence of a surface methyl group. In the surface ionization[9] of γ picoline (4-methyl pyridine) a peak was observed which could only be attributed to the CH_3O^- ion (m/e = 31). This peak gradually disappeared after prolonged heating of the filament, suggesting that the ion was formed by reaction of a surface methyl group with oxygen or oxide impurities which are gradually removed.

TABLE II

Substrate	Filament	E'_T (kJ mol^{-1})	$E - D$ (kJ mol^{-1})	$E' - (E - D)$ $Q_T CH_3$ (kJ mol^{-1})	Q_0 CH_3 (kJ mol^{-1})
CH_3CN	Ta	11·5	−121·5	133·0	117·5
	W	20·9	−121·5	142·7	130·5
	Pt	2·1	−121·5	123·6	107·5
CH_3Br	Ta	174·0	46·0	128·0	117·5
	W	65·3	46·0	—	—
	Pt	50.7	46·0	—	—
CH_3I	Ta	73·8	69·0	—	—
	W < 1460°K	204·0	69·0	135·0	122·0
	W > 1460°K	75·3	69·0	—	—
	Mo	187·5	69·0	118·5	109·0

It has also been shown[10] that when methane is adsorbed onto metal filaments the initial reaction to form $CH_{3(ads)}$ and $H_{(ads)}$ is rapid but the further decomposition of $CH_{3(ads)}$ is slow.

Electron Emission from Metal and Metal Carbide Filaments

Figure 1 shows that the adsorption of the gases studied on hot tantalum filaments produced relatively minor changes in the emission properties of the filament. X-ray examination showed that the filament remained as pure tantalum (A.S.T.M. Powder Index file No. 4-0788) and that the average crystallite size had grown markedly until it was approximately equal to the diameter of the filament.

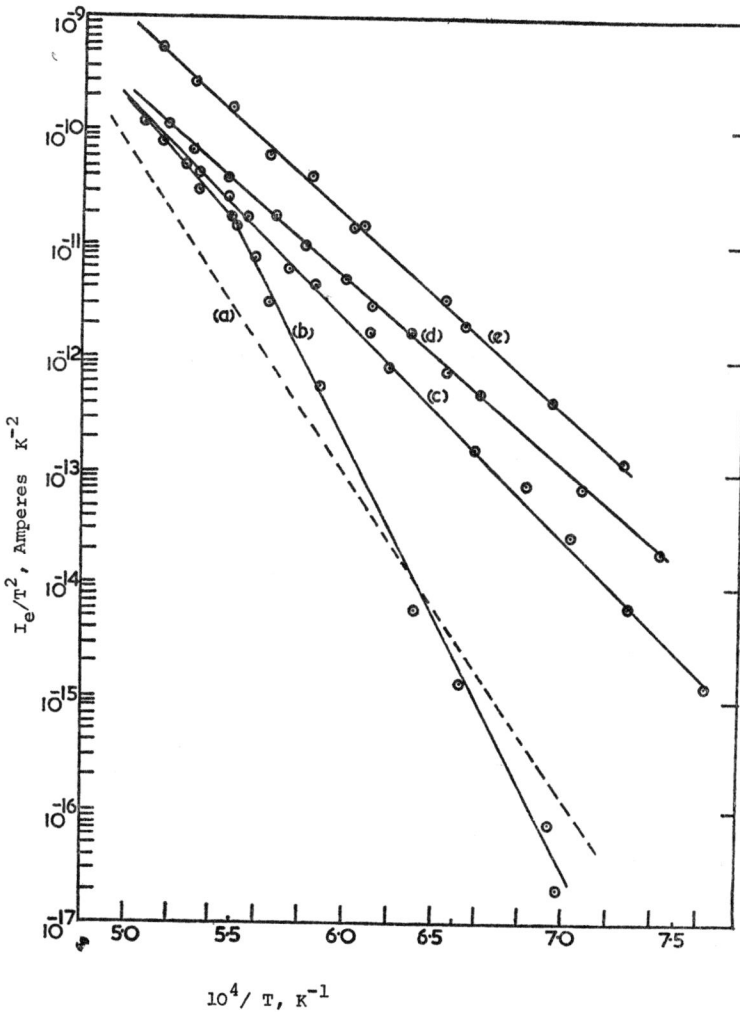

FIGURE 2

However, if a new filament with a preadsorbed film of methyl-containing compound was heated in this same gas a drastic change in the emission properties was observed. Visual, X-ray and scanning electron microscope examination of these filaments showed the formation of a golden yellow carbide with a stoichiometry $TaC_{0.98}$ (A.S.T.M. Powder Index file No. 19-1292). The crystallite size was very small and photomicrographs (up to a magnification of $\times 6500$) showed the filament to appear porous and granular. This carbide is very stable, evaporation of carbon from the surface is reported[11] to be negligible below $2000°K$ and at $3320°K$ the stoichiometry is $TaC_{0.91}$.[12]

The thermionic emission from tantalum carbide filaments depends markedly on the presence of adsorbed gases. This is shown in Fig. 2 where plots of $\log I_e/T^2$ against T^{-1} are shown for clean Ta, TaC and the TaC- MeCN, MBr and MeI systems. The appropriate work functions are shown in Table III.

TABLE III
Reduced Work Function of TaC

Substrate	$-\Delta\chi$ (kJ mol^{-1})	D (Debye)	$\frac{1}{\text{(nm)}}$	$D/1$
MeI	23	0·27	4·0	0·68
MeCN	61	0·44	6·3	0·69
MeBr	85	0·53	6·5	0·78

The work function change of TaC at $1770°K$ is of interest as several workers[10,13] studying the adsorption–desorption reaction of methane on metals found that carbon desorbs with a maximum rate at $1700-1750°K$. It appears that a tenacious carbon film present below $1700°K$ produces a high work function surface.

Experiments in which filaments were saturated with CH_3X at some temperature $<1700°K$ showed that when the gas was pumped away both the ion and electron currents decayed by a first order reaction, but the relative rates differed by a factor of $\sim 10^2$. The decay of the ion current, due to desorption of the ion precursors from the surface was very rapid. The electron current was found to decay with an activation energy of 163 ± 10 kJ mol^{-1} ($\Delta H_0 = 140 \pm 10$ kJ mol^{-1}) to the appropriate point on the TaC emission curve. This value is approximately that expected for diffusion of carbon from the filament interior along grain boundaries.[14,15,16]

The modification of the TaC surface work function by an adsorbed $CH_n (n = 3, 2, 1)$ layer can be considered as a modification of the transmission coefficient in eqn (2).

If the barrier for electron emission is lowered, as it must be if the work function change $\Delta\chi$ is negative, then the modified transmission coefficient is

$$\bar{d} = 8 \frac{[\pi kT(\chi_0 + \Delta\chi)]^{\frac{1}{4}}}{\chi_0} \cdot \exp\left\{-2\left[\left(\frac{8\pi^2 m}{h^2}\right)^{\frac{1}{4}}\right] \cdot l \cdot \Delta\chi^{\frac{1}{2}}\right\} \qquad (6)$$

whilst $\bar{d}_0$ is given by

$$\bar{d}_0 = 2\left(\frac{\pi kT}{\chi_0}\right)^{\frac{1}{4}} \qquad (7)$$

In eqn (6) the thickness of the barrier l can be estimated by assuming it extends over the adsorbed layer and the first layer of surface atoms ($l = 0.60$ nm). Calculations using eqns (6) and (7) give values of l from 0·4 nm to 0·65 nm, in agreement with the estimated figure.

Results and calculations suggest θ is constant and hence at its maximum value which is estimated to be 0·5 of a monolayer.

Using the assumption and the measured values of $\Delta\chi$ the dipole moment of the adsorbed species–surface bond may be estimated. These are given in Table III together with calculated values of l.

It can be seen that the final column, which is a measure of the charge on the adsorbed layer, is reasonably constant.

The Estimation of Electron Affinities

A large number of measurements of the relative negative ion intensities over a tantalum filament have been made in the temperature range 1100–2300°K in the presence of pressure between 10^{-4} and 10^{-6} mm Hg of the interhalogen

TABLE IV

Pseudohalogen (X)	Electron affinity (E_X) (kJ mol^{-1})	$E_I - E_X$ (kJ mol^{-1})	ΔE obs (kJ mol^{-1})
Cl	349	52	64
Br	325	28	19
I	297	—	—
CN	305	8	13

compounds IBr, ICl and ICN. In each case the quadrupole mass spectrometer showed the presence of both ions, and logarithmic plots of the relative ion currents, following eqn (4) gave slopes for IBr and ICN corresponding to differences in the electron affinities of 19·2 and 12·7 kJ mol^{-1} respectively (Table IV).

In the case of ICl, there was a sharp break in the curve at a temperature of 1500°K, and logarithmic plots of the individual ion currents against reciprocal temperature indicated that the iodine electron affinity was 295 kJ mol^{-1} acceptably close to the literature value, over the whole temperature range studied, but that only below 1500 did the chloride ion current behave as expected, leading to the electron affinity of 358 kJ mol^{-1} used in constructing Table IV. Above 1500°K, the chloride ion appears to be formed by a process

$$ICl + e \rightarrow Cl^- + I_{ads}$$

There is no justification for this assertion other than energetic agreement. The heat of adsorption of I on tantalum may be estimated by the method due to Eley to be 149 kJ mol^{-1}. The bond dissociation energy of ICl is 209 kJ mol^{-1}. The observed activation energy for the formation of Cl$^-$ (289 kJ mol^{-1}) therefore leads to an electron affinity of

$$289 + 209 - 149 = 349 \text{ kJ mol}^{-1}$$

for chlorine, which is in accord with the value below 1500°K and with the literature value.

No explanation is offered at this stage for the adsorption process occurring at the higher temperature other than to comment that such inversion of expectation has been postulated before as arising from the inversion of energetics since a high apparent electron affinity implies a low activation energy for the ion current.

REFERENCES

1. Page, F. M. and Goode, G. C., 'Negative Ions and the Magnetron', Wiley Interscience, 1969.
2. Ionov, N. I. Surface Ionisation and its Applications, 'Progress in Surface Science', Vol. 1 (3).
3. Branscomb, L. M. and Smith, S. I., *J. Chem. Phys.*, 1956, **25**, 587.
4. Berry, R. S. and Reimann, G. W., *J. Chem. Phys.*, 1963, **28**, 1540.
5. Müeller, G. *J. Phys. Rev.*, 1934, **45**, 314.
6. Ionov, N. I., *Sov. Phys. J.E.T.P.*, 1948, **18**, 96.
7. Shelton, H., *Phys. Rev.*, 1957, **107**, 1553.
8. Hayward, D. and Trapnell, B., 'Chemisorption'.
9. Chamberlain, A. T., Unpublished work.
10. Rye, R. R. and Hansen, R. S., *J. Chem. Phys.*, 1969, **50**, (8) 3585.
11. Huch, M., *J. Phys. Chem.*, 1955, **59**, 97.
12. Kempler, C. P. and Nadler, M. R., *J. Phys. Chem.*, 1960, **32**, 1477.
13. Hopkins, B. and Shah, *Vacuum*, 1972, **22**, 267.
14. Stean, A. E. and Eyring, H., *J. Phys. Chem.*, 1940, **44**, 955.
15. Andrevskii, R. A. *et al., Fiz. Metal. Metalloved*, 1969, **28**, (2), 298.
16. Suziki, *et al., Bull. Tokyo Inst. Technol.*, 1969, **90**, 105.

Discussion

F. E. Saalfeld (U.S. Naval Research Laboratory, Washington, U.S.A.): Have you tried to detect positive ions with your apparatus?

F. M. Page: A preliminary examination by Dr Farragher showed that when the potentials in the magnetron were reversed, a positive ion current was observed in azulene vapour. No attempt has been made to follow up this observation.

R. S. Berry (University of Chicago, U.S.A.): Have you seen any oxygen ions, either O^- or O_2^-?

F. M. Page: In general, we attempt to remove all traces of oxygen before carrying out our experiments. We did not observe O^- or O_2^-.

D. H. Smith (IAEA/ORNL, Vienna, Austria): Have any studies been made of the effect of oxygen on your measurements? What would such an effect be?

F. M. Page: The effect of oxygen on the ion and electron emission from filaments is complex, owing to the effect upon the work function and the high chemisorption energies involved. In the present series of experiments, the presence of oxygen was considered to be undesirable. However, where an increase in the work function *was* observed and attributed to traces of oxygen, since it disappeared after thermal cleaning of the filament, there was no effect on the measured electron affinity (of Iodine).

36
Studies of the Energies of Negative Ions at High Temperatures

By J. L. FRANKLIN, JOHN LING-FAI WANG,
S. L. BENNETT, P. W. HARLAND and
J. L. MARGRAVE

(*Department of Chemistry, Rice University, Houston, Texas, U.S.A.*)

INTRODUCTION

NEGATIVE ions formed in the gas phase from inorganic compounds have been studied relatively little and those from materials that vaporize only at elevated temperatures have been almost completely neglected. However, it would be expected that various of the metallic halides, oxides, sulphides, nitrates, etc. would form negative ions under the appropriate conditions and that considerable information of interest might be obtained from them. This study has been undertaken to separate negative ions from inorganic materials at high temperatures by dissociative resonance capture processes and to determine their thermochemistry.

When an electron of the proper energy contacts a molecule it may form an unstable ion that will break up in one or more ways in a very short time, thus

$$e + ABC \rightarrow ABC^{-*} \rightarrow A^- + BC$$

If the dissociation asymptote for A^- falls opposite the intersection of the potential energy curve with the Franck–Condon region, it is possible to detect the decomposition products in their ground states. The appearance potential can then be taken as the heat of reaction. If the dissociation asymptote falls below the intersection of the potential energy curve with the Franck–Condon region the appearance potential will include some excess energy which must be detected and deducted in order to obtain the heat of reaction. A portion of any excess energy that is involved will appear as translational energy of the products and this is readily measured. It has been shown by Haney and Franklin[1] with positive ions and by DeCorpo, Bafus and Franklin[2] with negative ions, that the average total translational energy at onset, $\bar{\varepsilon}_t$, is related to the total excess energy, E^*, by the equation

$$E^* = \alpha N \bar{\varepsilon}_t \qquad (1)$$

where N is the number of vibrational modes and α is an empirically determined constant having an average value of 0·43. $\bar{\varepsilon}_t$ is obtained from the measured value $\bar{\varepsilon}_i$, the translational energy of the ion by means of the equation

$$\bar{\varepsilon}_i = \frac{m_i}{m_i + m_n} \times 3/2 \, kT + \frac{m_n}{m_i + m_n} \, \bar{\varepsilon}_t \tag{2}$$

It has subsequently been shown[3] that α in eqn (1) may vary considerably from compound to compound. However, by measuring the translational energy in a dissociative resonance capture process, as a function of the electron energy (and hence of E^*) α can be determined for each process, and considerable improvement in accuracy achieved.

EXPERIMENTAL

A Bendix time-of-flight mass spectrometer model 14-107, modified with a Bendix model 3015 output scanner was used in these studies. Neutral molecules were generated in a Bendix Knudsen Cell Sample Inset System with a tantalum cell/Lucalox liner holding the solid sample for heating by a tungsten wire filament through either radiation or electron bombardment. The temperature of the cell was determined with a tungsten/tungsten-26% rhenium thermocouple inserted in the base of the cell. Gaseous samples were introduced through a conventional handling system.

Appearance potentials were determined by deconvolution of the ion intensities with the MacNeil–Thynne modification[4] of Morrison's[5] method. The electron energy scale was calibrated with O^-/CO, O^-/CO_2, O^-/SO_2 and F^-/SF_6 and the energy distribution of the ionizing electrons was determined from $SF_6{}^-/SF_6$ to be approximately 0·8 eV full width at half maximum. Translational energies were determined from peak shape analysis by the method of Franklin, Hierl and Whan.[6]

For each of the solids investigated the sample was thoroughly degassed and then the temperature of the Knudsen cell increased slowly until shutterable negative ions were observed. The temperature was then further increased until the intensity of the ions was sufficient to permit measurement of the appearance potential and, where possible, the translational energies of the ions.

RESULTS AND DISCUSSION

Figure 1 gives a typical dissociative resonance capture curve (for F^- from GeF_4) along with a plot of the average translational energy of F^- as a function of electron energy. The vertical arrow shows the appearance potential obtained after deconvolution. The translational energy of F^- at onset is 3·5 kcal/mole. The marked flattening of the translational energy at about 6 kcal/mole is due to the fact that at this energy, the more energetic ions begin to strike the wall and are not collected. Over a range of nearly one volt of electron energy the translational energy increases linearly with electron energy. This is characteristic of the behaviour of all of the systems studied. From the slope of the line we compute α for this process to be 0·56 and with

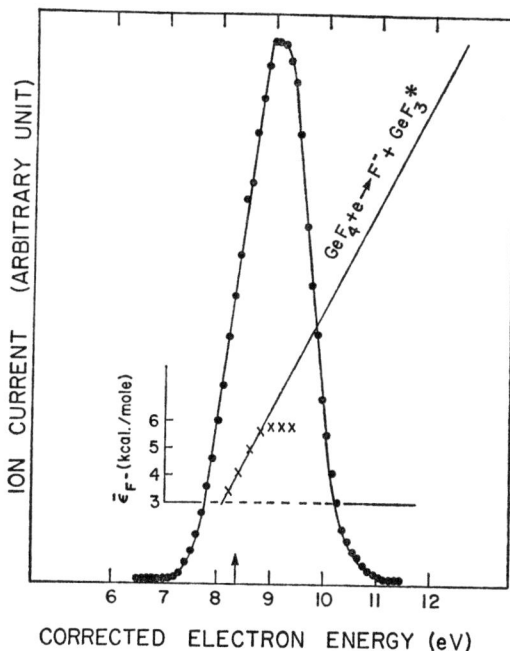

FIG. 1 Resonance and translational energy of F^- ion from GeF_4.

this and the translational energy at onset the excess energy is computed to be 18 kcal/mole.

Group IV Fluorides

The electron affinities of the MF_3 radicals from the Group IV fluorides, except SnF_4, and the heats of formation of the corresponding MF_3^- ions have been similarly determined and the results given in Table I.

Since CF_4, SiF_4 and GeF_4 are gases the measurements were made on them at room temperature. SnF_4 and PbF_4 are solids and had to be vaporized from the Knudsen cell at a temperature of about 520°C.

TABLE I
Thermochemistry of MF_3^- from Group IV Fluorides (kcal/mole).

	Ap	$\bar{\varepsilon}_t$	α	E^*	$\Delta H_f(MF_3^-)$	$EA(MF_3)$
$CF_4 + e \rightarrow CF_3^- + F$	125	15·3	0·33[a]	46	−163·3	50·7
$SiF_4 + e \rightarrow SiF_3^- + F$	247	33	0·43[b]	123	−281	47
$GeF_4 + e \rightarrow GeF_3^- + F$	187	23	0·43[b]	88	−205	37
$SnF_4 + e \rightarrow SnF_3^- + F$	115	—	—	(?)	−156[c]	5[c]
$PbF_4 + e \rightarrow PbF_3^- + F$	5	Thermal	—	0	−212	100

[a] Determined from the slope of ε_i versus electronvolt line.
[b] Taken as the average value at onset.
[c] Experimental values. See text for more probable values.

It is noteworthy that the carbon, silicon and germanium compounds all involved large amounts of excess energy at onset. PbF_3^-, which was formed at a very low appearance potential, appeared to be formed with thermal energy. SnF_3^- was formed with very low intensity and this combined with the multiplicity of interfering isotopic species made it impossible to measure the translational energy of the ion. It is evident from Table I that $\Delta H_f(SnF_3^-)$ should be about -212 kcal/mole and, when this is combined with $\Delta H_f(SnF_3)$ of -151 kcal/mole (see below), we estimate the electron affinity to be approximately 61 kcal/mole. This value is, of course, speculative, but appears reasonable in the light of the other values.

TABLE II

Processes and Energies (kcal/mole) for the Formation of F^- from the Group IV Fluorides.

	Ap	$\bar{\varepsilon}_t$	α	$E^*_{v,t}$	$\Delta H_f(MF_3)$	E^*_{el}
$CF_4 + e \rightarrow F^- + CF_3$	111	19·6	0·43^a	74·1	-120	—
$SiF_4 + e \rightarrow F^- + SiF_3^*$	246	8·4	0·43^a	32	-235	128
$GeF_4 + e \rightarrow F^- + GeF_3^*$	194	3·5	0·56^b	18	-168	123
$SnF_4 + e \rightarrow F^- + SnF_3^*$	124	Thermal	—	0	$(-63)^c$	88
$PbF_4 + e \rightarrow F^- + PbF_3$	21	Thermal	—	0	-112	0

^{a} Average values at onset.
^{b} From the slope of the curve of $\bar{\varepsilon}_i$ against eV.
^{c} Experimental value which includes E^*_{el}.

The most intense ion from all of the Group IV fluorides is F^- and our measurements on it are given in Table II. Considerable amounts of translational energy were found with the first three members of the group and so the total excess translational and vibrational energy, $E^*_{v,t}$, was large with each of these. The tin and lead compounds gave F^- with thermal energy at onset. The heat of formation of CF_3 is in fair agreement with the accepted value of $-112\cdot4$ kcal/mole.[7] This is not true of the silicon, germanium and tin compounds. Thus, in the case of SiF_4 the thermochemical equation for the reaction fails to balance unless an allowance is made for electronic excitation of the SiF_3 product, i.e.

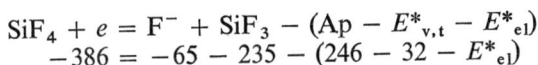

$$SiF_4 + e = F^- + SiF_3 - (Ap - E^*_{v,t} - E^*_{el})$$
$$-386 = -65 - 235 - (246 - 32 - E^*_{el})$$

from which E^*_{el} is found to be 128 kcal/mole or about 5·5 eV. Wang, Krishnan and Margrave[8] have reported a value of 5·47 eV for the energy of the electronic transition $^2B \rightarrow X^2A$ in SiF_3, in excellent agreement with our determination. It should be mentioned that there is strong evidence for a similar excited electronic state of CF_3 which falls in the same resonance region as the one reported above. Similarly, one would expect GeF_3 to be formed in an electronically excited state. When we combine our appearance potential corrected for $E^*_{v,t}$ and apply Wang, Krishnan and Margrave's[9] energy of 5·37 eV for the $^2B - ^2A$ transition of GeF_3, we compute $\Delta H_f(GeF_3)$ to be -168 kcal/mole.

TABLE III

Thermochemical Values (kcal/mole) of Negative Ions from Arsenic.

	Ap	α (slope)	$\bar{\varepsilon}_t$	E^*	ΔH_f (ion)	EA
$As_4 + e \rightarrow As_2^- + As_2$	$70 \pm 1\cdot2$	$0\cdot43$	$6\cdot9 \pm 0\cdot7$	$17\cdot8 \pm 0\cdot7$	$43\cdot8 \pm 4\cdot1$	$2\cdot3 \pm 2\cdot3$
$As_4 + e \rightarrow As_3^- + As$	$81 \pm 2\cdot3$	$0\cdot43$	$4\cdot1 \pm 0\cdot9$	$10\cdot6 \pm 0\cdot9$	$38\cdot1 \pm 5\cdot2$	$18\cdot8 \pm 7$
$As_4 + e \rightarrow As^- + As_3$	$83 \pm 2\cdot3$	$0\cdot43$	$4\cdot5 \pm 0\cdot7$	$11\cdot5 \pm 0\cdot7$	$56\cdot9 \pm 4\cdot9^a$	—

^{a} Neutral As_3 combined with $\Delta H_f(As_3^-)$ gives $EA(As_3)$ to be $18\cdot8 \pm 7$ kcal/mole.

The F^- ion from PbF_4 is formed with thermal energy at an appearance potential of only 0·9 eV. It is apparent that this process yields PbF_3 in the ground electronic state and the computed heat of formation is -112 kcal/mole. From the appearance potential of F^- from SnF_4, it seems probable that the process results in an electronically excited SnF_3. Unfortunately, there is no reliable value of $\Delta H_f(SnF_3)$ which we can use to compute E^*_{el}. From the sequence $D(F-MF_3)$ of 130, 170, 135 and 105 for C, Si, Ge and Pb, a reasonable estimate of $D(F-SnF_3)$ would be about 120 kcal/mole which would give $\Delta H_f(SnF_3)$ to be -151 kcal/mole. Since Ap F^-/SnF_4 gives $\Delta H_f(SnF_3)$ to be -63 kcal/mole, we would then estimate the energy of the $^2B - {}^2A$ transition to be about 88 kcal/mole. The result, while speculative, is not unreasonable.

In further support of the idea that the neutral fragment in certain dissociative resonance capture processes may be electronically excited, we cite other observations of similar behaviour. Thus, Harland, Carter and Franklin[10] attributed the second resonance of O^-/SO_2 to the $^1\Delta$ state of SO, and Petty et al.[11] found that the MF_2 neutrals formed along with F^- ion from the Group III fluorides were electronically excited. Indeed, it now appears that this is a phenomenon that will frequently be experienced.

Arsenic
Arsenic vapour at about 520°C is made up principally of As_4 with small amounts of As_2 also present. As will be seen in Table III, there are three

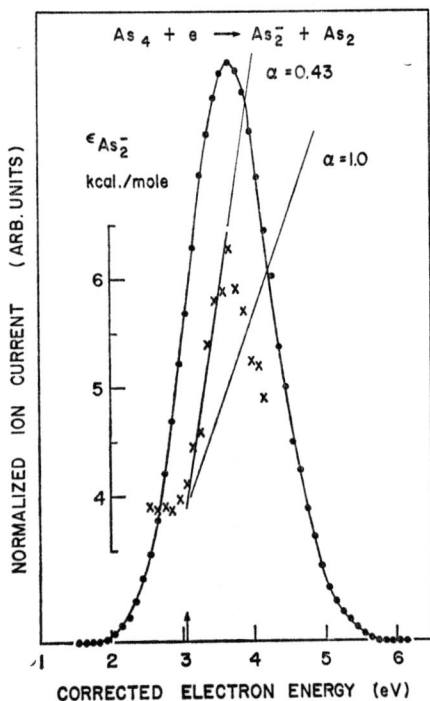

FIG. 2   Resonance and translational energy of As_2^- from As_4.

negative ions formed by dissociative resonance capture, all in our experiment from As_4.

As will be seen in Fig. 2, the translational energy of $As_2{}^-$ rises linearly over the greater part of the resonance and then breaks downward sharply. The downward break is attributable to an instrumental artifact. The slope of the upward portion of the line gives a value of α in eqn (1) of 0·43. Similar behaviour was observed with As^- and $As_3{}^-$. From the computed E^* and Murray, Pupp and Pottie's[12] value of 54·26 kcal/mole for $D(As_4 - As_2)$, we compute the electron affinity of As_2 to be 0·1 eV.

$As_3{}^-$ can come only from As_4 and its heat of formation is $-38·1 \pm 5·2$ kcal/mole. $\Delta H_f(As_3)$ has not previously been determined. Combining our appearance potential and excess energy measurement with Feldman's[13] recent value of $0·75 \pm 0·10$ eV for the electron affinity of As yields $\Delta H_f(As_3)$ to be $56·9 \pm 4·9$ kcal/mole and $EA(As_3) = 18·7 \pm 7$ kcal/mole (0·8 eV). Thus, the electron affinities of As and As_3 are very nearly the same and are considerably greater than that of As_2.

It will be evident from Table III that all three of the processes studied involved considerable excess energy and that serious errors would have resulted had it not been possible to determine and correct for it.

ACKNOWLEDGMENT

The authors are pleased to acknowledge the support of this work by the Office of Naval Research, U.S. Navy, under Contract No. 67-A-0145-0002.

REFERENCES

1. Haney, M. A. and Franklin, J. L., *J. Chem. Phys.*, 1968, **48**, 4093.
2. DeCorpo, J. J., Bafus, D. A. and Franklin, J. L., *J. Chem. Phys.*, 1971, **54**, 1592.
3. Harland, P. W. and Franklin, J. L., to be published.
4. MacNeil, K. A. G. and Thynne, J. C. J., *Int. J. Mass Spectrom. Ion Phys.*, 1969, **3**, 35.
5. Morrison, J. D., *J. Chem. Phys.* 1963, **39**, 200.
6. Franklin, J. L., Hierl, P. M. and Whan, D. A., *J. Chem. Phys.*, 1967, **47**, 3148.
7. JANAF Tables, D. R. Stull, Ed., 1970.
8. Ling-Fai Wang, J., Krishnan, C. N. and Margrave, J. L., *J. Mol. Spectros.*, to be published.
9. Ling-Fai Wang, J., Krishnan, C. N. and Margrave, J. L., to be published.
10. Harland, P. W., Franklin, J. L. and Carter, D. E., *J. Chem. Phys.*, 1973, **58**, 1430.
11. Petty, F., Ling-Fai Wang, J., Steiger, R. P., Harland, P. W., Franklin, J. L. and Margrave, J. L., *High Temp. Science*, 1973, **5**, 25.
12. Murray, J. J., Pupp, C. and Pottie, R. F., *J. Chem. Phys.*, 1973, **58**, 2569.
13. Feldman, D., Reported by W. C. Lineberger at Conference of American Society for Mass Spectrometry, San Francisco, Calif., May, 1973.

37

The Primary Fragmentation Reactions of Organic Negative Molecular Ions: A General Qualitative Rationalisation

By R. G. ALEXANDER,† D. B. BIGLEY,
R. B. TURNER and J. F. J. TODD

(*University Chemical Laboratory, University of Kent, Canterbury, U.K.*)

NEGATIVE ion studies on complex organic compounds must represent one of the most neglected areas of mass spectrometry. The main deterrents appear to be the lower sensitivities encountered in negative ion work and the apparent complexity of the processes by which anions are formed under electron impact. As a result no attempt to rationalise negative ion fragmentation processes in a manner analogous to the positive ion approach had been reported until our recent communication on the subject.[1]

From Melton's treatment[2] it is argued that three processes are of importance for anion formation in a mass spectrometer ion source

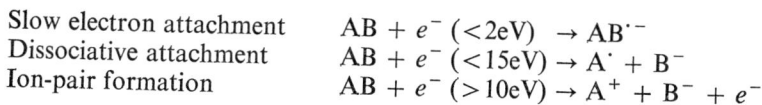

Slow electron attachment	$AB + e^- \ (<2eV)$	$\rightarrow AB^{\cdot -}$
Dissociative attachment	$AB + e^- \ (<15eV)$	$\rightarrow A^{\cdot} + B^-$
Ion-pair formation	$AB + e^- \ (>10eV)$	$\rightarrow A^+ + B^- + e^-$

and it is suggested[2] that with electron bombardment energies in excess of 12–15 eV no parent anions should be observable in the mass spectrum. In practice, however, certain compounds show quite significant yields of parent anions when the bombardment energy is 70 eV, *e.g.* in 3- and 4-nitroacenaphthene the $M^{\cdot -}$ ions form the base peaks of the spectra.[3] In addition a common feature found for all the nitroacenaphthenes is the major yield of NO_2^-, which is not accompanied by the appearance of metastable ion peaks.

To be consistent with Melton's scheme one must assume that even at 70 eV there is an abundance of slow electrons in the ionization region of the source. Several workers have proposed this and the origin of the slow electrons appears to be mainly secondary emission from surfaces[4] and the ejection from molecules in the positive ionization process.[5] In any event it is considered likely that the presence of positive ions trapped in the negative space charge

---

† Present address: Beechams Research Laboratories, Spectroscopy Lab, Brockham Park, Betchworth, Surrey, U.K.

of the 70 eV magnetically collimated electron beam will form an effective trap for slow electrons, in the manner envisaged by Hasted.[6,7]

To determine the possible role which slow electrons may have on the production of negatively charged species in an ion source we have examined the effect which the addition of sulphur hexafluoride, a thermal electron scavenger, has on the ionization efficiency curves for the formation of the $M^{\cdot -}$ and NO_2^- ions from 3-nitrotoluene. Figure 1 shows the curves which are direct recordings obtained from an AEI MS902 instrument operating with a

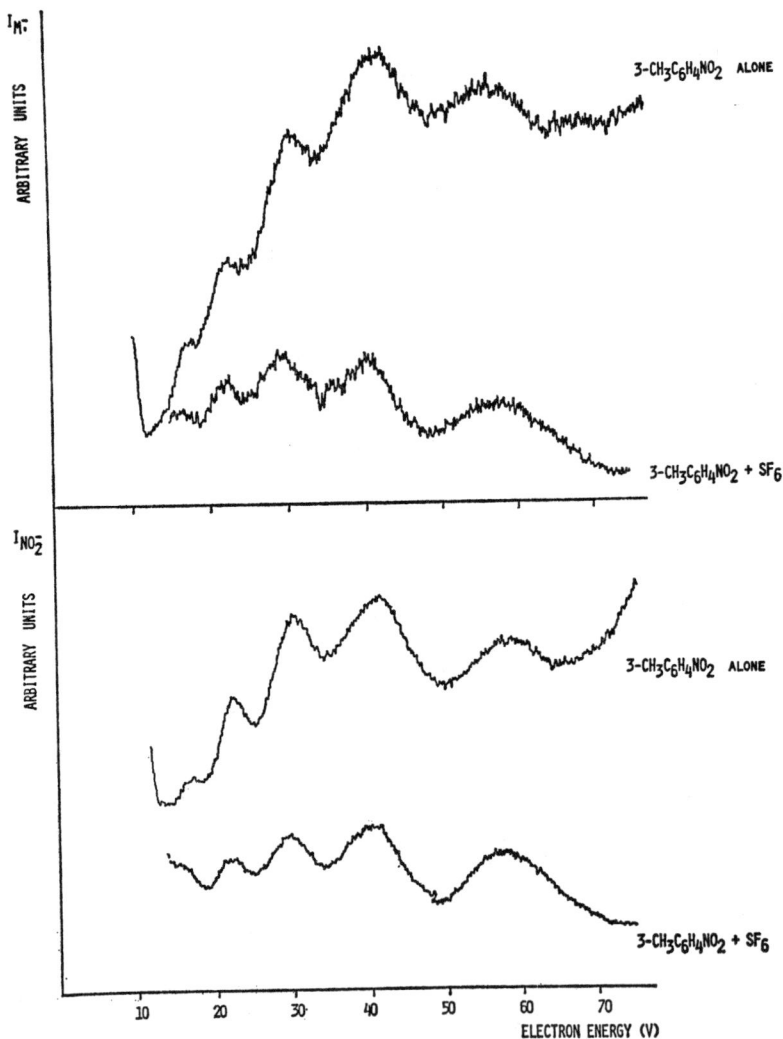

FIG. 1   Ionization efficiency curves for the formation of the molecular ion, $M^{\cdot -}$, and NO_2^- ions from 3-nitrotoluene. For each ion the lower trace illustrates the effect of adding *ca.* 10% sulphur hexafluoride. For these curves and those in Figure 2 the electron energy values are nominal.

trap current of 10 μA and zero repeller voltage, using a digital data transmission interface developed in these Laboratories;[8] for each ion the lower
curve represents the effect of adding *ca.* 10% SF_6 at a total pressure of *ca.*
5×10^{-6} torr. In both cases it is seen that the general shapes of the curves are
maintained, but that the intensities are reduced. Little significance should be
attached to the undulations in the curves which are probably artefacts arising
from the electron optics of the source. Figure 2 shows the complementary

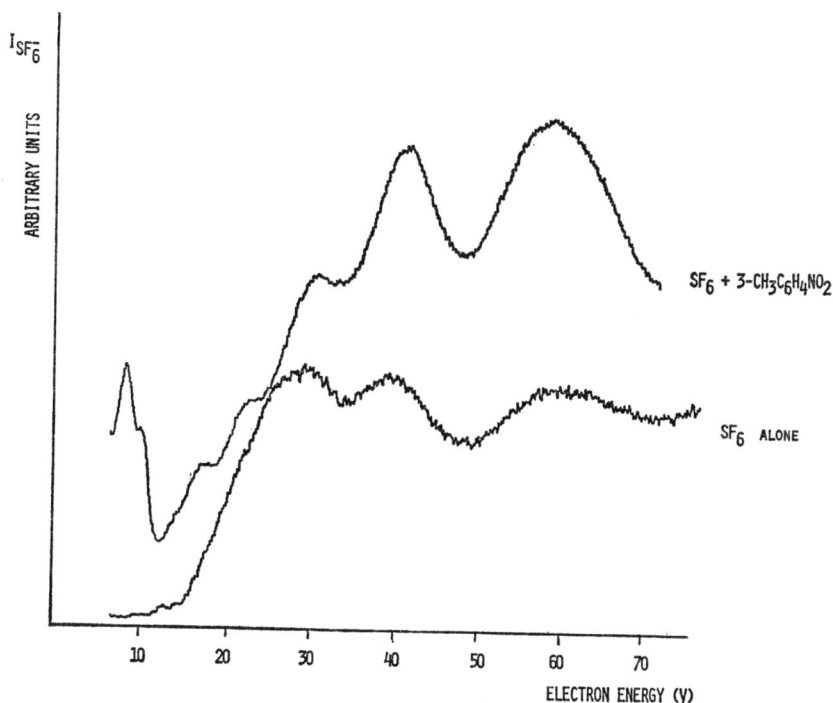

FIG. 2   Ionization efficiency curves for the SF_6^- ion from sulphur hexafluoride, showing
the effect of adding 3-nitrotoluene. For the upper trace the sensitivity has been reduced by a
factor of *ca.* 10.

curve for SF_6^-, illustrating how the addition of the 3-nitrotoluene markedly
enhances the yield of this ion. It is therefore concluded that slow electrons
make a significant contribution to the negative ionization processes, at least
for the compound under investigation.

We must now attempt to explain the apparently conflicting observation,
that certain compounds exhibit very low parent anion abundances in an ion
source operating with 70 eV electrons, since there is no reason *a priori* to
assume that slow electrons will be present during the mass spectral analysis of
certain compounds and not of others. A group of compounds which typify
this behaviour is the phosphoranes[9] shown in Table I, where it is seen that
the base peak in each case corresponds to the M-Ph]$^-$ ion, with very low
yields of M$^{\cdot-}$.

TABLE I

Relative Abundances (Percentages) of Negative Ions Observed in Phosphoranes of the Type Ph$_2$P = CR'COR'' at 70 eV

R'	R''	M$^{-\cdot}$ (c)	[M-Ph]$^-$ (d)	[Ph$_2$PC$_2$O]$^-$ (g)	[Ph$_2$PO]$^-$ (f)	[M-PH$_3$P]$^{-\cdot}$ (e)	[CR'CO]$^-$ (h)
H	Me	1·0	100	1·2	1·3	36·7	33·3
H	Ph	1·3	100	7·1	3·2	1·1	0·9
H	OEt	0·6	100	53·3	1·0	—^{a}	6·0
COMe	Me	0·1	100	0·2	3·0	2·4	14·0
CO$_2$Et	OEt	0·9	100	7·8	0·2	—$^{(a')}$	6·2
COMe	OEt	0·4	100	15·4	3·6	—$^{(a'')}$	15·4 [3·8]^b

Notes (a), (a') and (a'') show [CR'COR'' + H·]$^-$ (1·0%), (0·7%) and (0·4%) respectively.
(b) [3·8] refers to reversal of R' and R'', *i.e.* loss of Me.

These results can be explained if it is assumed that the attachment of slow electrons to a phosphorane molecule results in immediate dissociation. Since the chemical behaviour of the phosphoranes suggests a contribution to the ground state molecular structure from the dipolar form (b) we can write

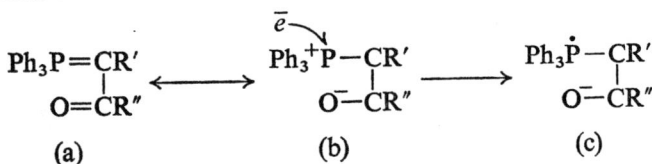

The electron added to the phosphorus atom to yield (c) must be in a high energy non-bonding or an antibonding orbital, with the result that stabilization

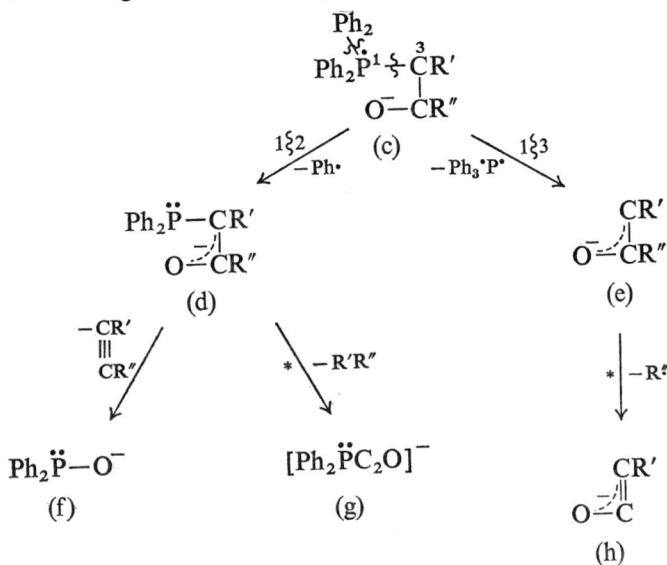

SCHEME 1

will occur either by autodetachment or through rapid 1,2-bond fission, and evidence from the data in Table I together with metastable transitions (*) for secondary decomposition processes suggests that the latter occurs through homolysis and electron coupling at the site of attachment (Scheme 1). We see that the overall effect of scission of either the 1–2 or 1–3 bond is the reduction of phosphorus to the P(III) state. The absence of appropriate metastable peaks in the mass spectrum is consistent with the fragmentation of (c) in a repulsive state.

For such a scheme involving localised electron attachment to be universally applicable it is now necessary to re-consider the case of the nitroaromatics in order to understand why a stable parent anion may be formed. Applying the previous model we have:

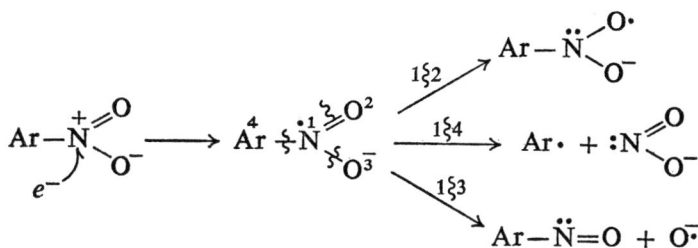

SCHEME 2

and see that when a π-bond is situated at the site of electron attachment homolysis may occur without immediate fragmentation.

From an examination of the primary and secondary fragmentation modes of organic negative ions formed in a number of systems, we conclude that product stability is important in determining the relative significance of competing reactions. However, for the possible routes to be delineated, all those structures in which an electron may reasonably be sited at a positive centre must be considered, and contributions from conjugation taken into account. An example of this may be seen in the predicted fragmentation of a series of phosphinates and thionphosphinates (Scheme 3).

Thus we see that for $Y = O$, a major directing influence is the setting up of the P–O dipole to form (k) and conjugation to give (l). Electron attachment then gives rise to two phosphorus-containing daughter ions (an alternative cleavage of (k) should be to form M-Ph]$^-$ ion: in the spectra of these compounds this is found to be only of minor importance). An alternative dipolar effect may arise from conjugation of the aryl oxygen atom with the ring to form (j) and this is likely to be of particular importance when X is electron withdrawing, e.g. a nitro group. Electron attachment to (j) should result in the formation of a phenoxide ion.

When $Y = S$ the dipolar canonical form (b) is of less importance and (j) should make a greater overall contribution to the ground state configuration. This should result in the yield of the phenoxide anion increasing in importance relative to the M-C$_6$H$_4$X]$^-$ ion in the mass spectra of the thionphosphinates. Reference to Table II shows that this is indeed the case, and that whereas for $Y = O$ the ion (o) is always the base peak, if $Y = S$, the ion M–C$_6$H$_4$X]$^-$

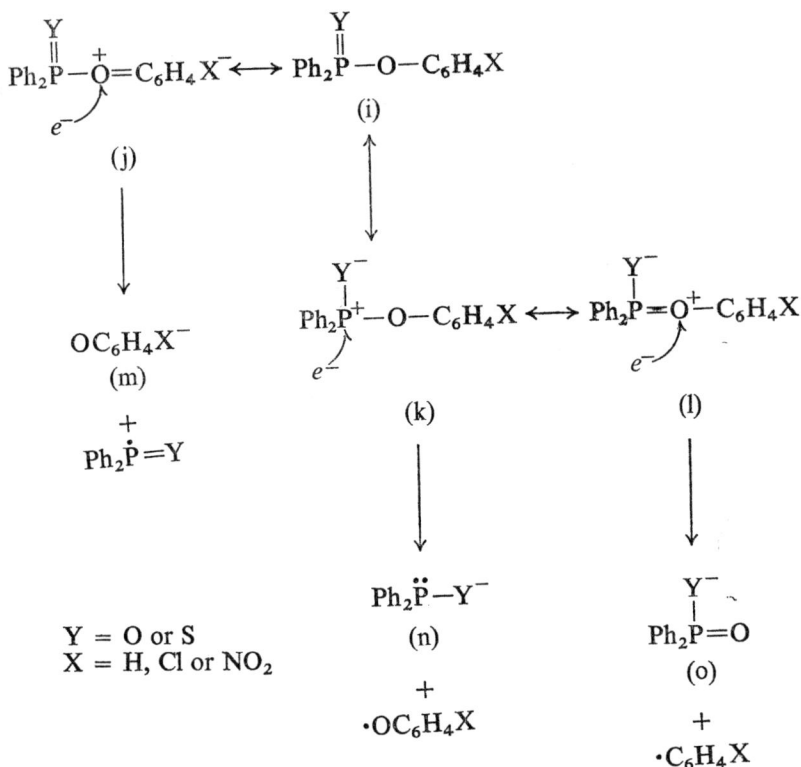

$$\underset{e^-}{\overset{\overset{\displaystyle Y}{\|}}{Ph_2P}}-\overset{+}{O}=C_6H_4X^- \longleftrightarrow \overset{\overset{\displaystyle Y}{\|}}{Ph_2P}-O-C_6H_4X$$

(i)

(j)

↓

$$OC_6H_4X^-$$

(m)

+

$$Ph_2\overset{\cdot}{P}=Y$$

$$\underset{e^-}{\overset{\overset{\displaystyle Y^-}{|}}{Ph_2\overset{+}{P}}}-O-C_6H_4X \longleftrightarrow \underset{e^-}{\overset{\overset{\displaystyle Y^-}{|}}{Ph_2P}}=\overset{+}{O}-C_6H_4X$$

(k)   (l)

↓   ↓

$$Ph_2\overset{\cdot\cdot}{P}-Y^-$$ $$\overset{\overset{\displaystyle Y^-}{|}}{Ph_2P}=O$$

(n)   (o)

+   +

$$\cdot OC_6H_4X$$ $$\cdot C_6H_4X$$

Y = O or S
X = H, Cl or NO$_2$

SCHEME 3

assumes only minor significance. As noted earlier, this importance of conjugation with the aryl nucleus is emphasized for the compounds where X = NO$_2$ since in both instances the relative intensity of the phenoxide ion (m) is increased. For X = H or Cl, replacement of O by S does not appear to have any effect on the (m)/(n) ratio.

The data in Table II emphasize a further point which is common to the negative ion mass spectra of organic compounds, namely that the spectra are extremely simple and consist of a few peaks carrying the bulk of the total ion current, in excess of 74% for the compounds cited. This is consistent with the idea that ionization occurs by means of low energy processes.

SUMMARY AND CONCLUSIONS

(i) There are indications that at 70 eV in a standard mass spectrometer source slow electron attachment is a dominant process in negative ion formation.

(ii) The extent of excitation is small and the negative ion mass spectra are simple.

TABLE II

Primary Negative Ion Yields (% Total Ionization at 70eV) from *para*-Substituted Aryl Diphenylphosphinates and Diphenylthionphosphinates (Ph$_2$—P—OC$_6$H$_4$X)

Y	X	M·⁻	M-C$_6$H$_4$X·]⁻ (o)	M-OC$_6$H$_4$X·]⁻ (n)	OC$_6$H$_4$X⁻ (m)	OC$_6$H$_4$X⁻/[M-OC$_6$H$_4$X·]⁻ (m)/(n)	Σ
O	H	0·5	33·7	9·1	31·3	3·4	74·6
O	Cl	0·7	44·1	4·7	30·5	6·5	79·3
O	NO$_2$	1·6	47·6	—	38·5	—	87·7
S	H	<0·1	3·4	19·8	59·4	3·0	82·7
S	Cl	<0·1	6·1	11·4	75·5	6·6	93·1
S	NO$_2$	<0·1	1·6	8·0	88·1	11·0	97·8

(iii) Molecules appear to dissociate by specific routes which can be related to known chemical properties. Conjugation effects and relative product stabilities are important in determining the most favoured decomposition routes.

(iv) The fragmentation patterns can be correlated with the concept of electron attachment at positive centres followed by adjacent bond homolysis and electron coupling.

(v) The negative ion mass spectra are simple to interpret and using the above approach may even be predicted. Greater consideration should be given to the application of negative ion mass spectrometry as one of the 'gentle' ionization techniques, alongside field and chemical ionization.

ACKNOWLEDGMENT

This research was supported in part by a grant from the SRC to whom one of us (R.G.A.) is also indebted for an 'Instant Award'.

REFERENCES

1. Alexander, R. G., Bigley, D. B. and Todd, J. F. J., *Org. Mass Spectrom.,* 1973, **7**, 643.
2. Melton, R. E., 'Principles of Mass Spectrometry and Negative Ions', Marcel Dekker, New York, 1970, p. 192.
3. Todd, J. F. J., Turner, R. B., Webb, B. C. and Wells, C. H. J., *J. C. S. Perkin II,* 1973, 1167.
4. Fox, R. E., *J. Chem. Phys.,* 1957, **26**, 1281.
5. Thynne, J. C. J., *Chem. Comm.,* 1968, 1075.
6. Baker, F. A. and Hasted, J. B., *Phil. Trans. Roy. Soc.,* 1966, **A261**, 33.
7. Hasted, J. B., 'Sequential Mass Spectrometry', in 'Some Newer Physical Methods in Structural Chemistry', ed. Bonnett, R. and Davies, J. G., United Trade Press, London, 1967, p. 98.
8. Todd, J. F. J., Turner, R. B. and Norris, M. O., *J. Phys. E: Scientific Instruments,* in the press.
9. Alexander, R. G., Bigley, D. B. and Todd, J. F. J., *Org. Mass Spectrom.,* 1973, **7**, 963.

AUTHORS' NOTE ADDED IN PROOF

As the study of negative ion mass spectrometry expands a case can be made for rationalising nomenclature to avoid confusion. Thus conventional positive ion investigations should become known as *Cation Mass Spectrometry (CMS)* and their counterpart as *Anion Mass Spectrometry (AMS)*.

Discussion

H. J. Svec (Iowa State University, Ames, U.S.A.): One source of secondary slow electrons is the bombardment of source electrodes by the primary electrons of 70 eV. It is perhaps the most important source of low energy electrons in the ion source you are using. My own preference would be not to use 70 eV electrons and be forced to take pot-luck on what low energy electrons result. I believe much more definitive information and reproducible spectra would result if several specific low energy electrons could be used to produce dissociative attachment processes. The negative ion currents would be as intense as those obtained in conventional positive ion mass spectra and the possible interferences from positive ions would be absent. Collision induced stabilization of the negative ions would result because of the sample gas pressure rather than an extraneous gas.

J. F. J. Todd: I am not too happy about the idea of using beams of electrons at specific low energies since it must be expected that the particular dissociation reactions consequent upon attachment will therefore be specific to the compound being studied. Our argument is that in our ion source operating with 70 eV electrons there is an abundance of essentially thermal electrons, and the consequences of attachment of these to substrate molecules can be correlated with chemical processes. There is thus a case for exploring the analytical applications of negative ion mass spectrometry on a wider scale. It is not clear to me in what ways positive ions would 'interfere'. Arguments have been published by Professor Hasted which suggest that slow electrons can be trapped within the space charge arising from positive ions trapped within a magnetically collimated higher energy electron beam. In this instance positive ions would have the beneficial effect of concentrating slow electrons in the ionization region, and evidence for some sort of trapping mechanism is apparent from our observation of the sharp dependence of ion intensity upon the repeller potential.

J. B. Hasted (Birkbeck College, London, U.K.): I believe that whilst there may be contributions from surfaces, as Dr Svec has proposed, the bulk of the slow electrons arise from ionization. There may well be failure to reproduce spectra on other instruments, so that a thermal electron source is mandatory for future work. This need not invalidate the theoretical conclusions of Dr Todd.

F. E. Saalfeld (U.S. Naval Research Laboratory, Washington, U.S.A.): What was the ion source pressure in your experiment?

J. F. J. Todd: The pressure of *ca.* 5×10^{-6} torr we employed was measured on an ion gauge situated over the cold trap of the source pump, and thus the source pressure actually employed must be higher than this.

J. Yinon (Weizmann Institute of Science, Israel): I would like to add to the comments made by Professor Svec and Professor Hasted, that we have done an experimental comparison between an open ion source and a closed source in a negative-ion mass spectrometer. We have observed that in the open-structure ion source the amount of secondary electrons is much less than in the closed source. However we could never reduce these secondary electrons, even in the open ion sources, below a certain minimum.

J. F. J. Todd: This confirms our own view that the importance of secondary electrons must not be underestimated.

Analysis of Surfaces and Thin Films by Mass Spectrometry

By RICHARD E. HONIG

(R.C.A. Laboratories, Princeton, New Jersey, U.S.A.)

INTRODUCTION

WHEN a keV ion beam collides with a solid surface, this interaction leads to several processes, as schematically indicated in Fig. 1. A fraction of the ions is *backscattered* by the surface atoms, mostly in elastic or inelastic binary collisions. The remaining ions penetrate the solid and transfer their energy in a series of collision cascades to the lattice. The energetic recoil atoms initiate secondary and tertiary collision cascades, some of which produce *sputtering*, *i.e.* the emission into the vacuum of surface atoms, in either neutral or charged form. In the case of rare gas ions, elastic *scattering* provides energy spectra characteristic of the mass of the scattering centres, thus ion scattering spectrometry (ISS) can be used to determine the composition of the outermost atomic layer of a solid. Since the *sputtering* process expels lattice particles, mostly as neutrals, but some as ions, from the solid into the vacuum, it can be employed in conjunction with a mass spectrometer to characterize the composition of the solid near the surface. Since sputtering continuously uncovers a fresh surface, in-depth concentration profiles of major and minor constituents and trace impurities can be obtained down to a depth of several μm.

While ion scattering resulting from an elastic binary collision is a classical case of particle mechanics,[1,2] it is only quite recently that this process has been utilized as an analytical tool. In an early exploratory study on the reflection of alkali ions from a clean molybdenum surface, Brunnée[3] pointed out that the high-energy edge of the scattering spectra represented elastic binary collisions. It was a decade later that Smith[4] utilized elastic scattering of keV rare gas ions from solid surfaces to determine the mass of the scattering centres. Since then, ion scattering has been applied to the analysis of surfaces in a growing number of laboratories, as evidenced by a recent bibliography on this subject compiled by Honig and Harrington.[5]

The application of sputtered particles to the study of solids has a much longer history. Early investigations by Sloane and colleagues[6,7] and Herzog and Vieböck[8] were followed by studies that utilized secondary ion emission for the analysis of solids.[9-14] These and additional studies have been reviewed by Behrisch,[15] Kaminsky,[1] and Carter and Colligon.[2] Subsequent refinements in the design of the ion beam optics led to the development of the ion microprobe by Castaing and Slodzian,[16] Liebl and Herzog,[17,18] and Drummond and Long.[19] An up-to-date review and evaluation of different types of recently developed ion microprobes and secondary ion mass spectrometers, including Benninghoven's design,[20,21] has been published by Evans.[22,23]

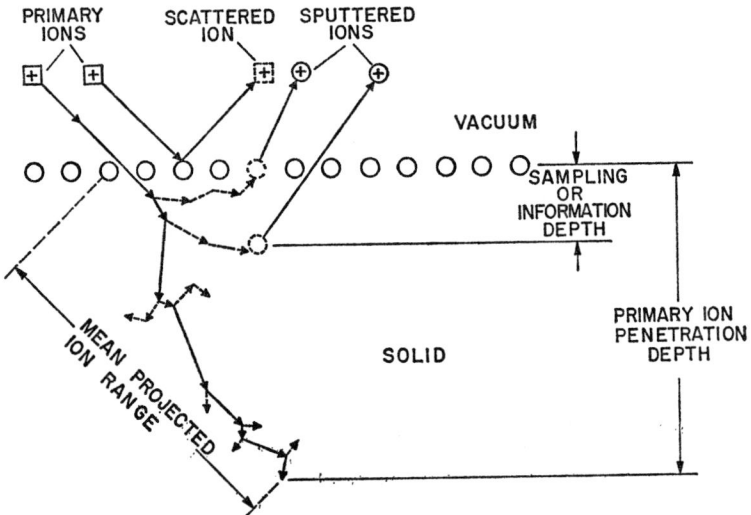

FIG. 1   Schematic of various processes resulting from the collision of a keV ion beam with a solid surface.

The present paper will review the use of ion scattering and sputtering phenomena for the analysis of solid surfaces and thin films, with emphasis on the physical processes involved. The pertinent, interrelated major parameters and considerations that will be taken up in detail below include: size and shape of the sputtered crater and of the detected area; lateral and in-depth concentration profiles and resolution; sampling or information depth, primary ion penetration depth producing lattice damage, and sample consumed in analysis; differential sputtering and sputtering rates; elemental identification, mass resolution, and detection sensitivity; matrix, geometric, and neutralization effects; accuracy and precision; conductivity of surface; and elemental coverage (detectability of impurities in a given matrix).

BASIC CONSIDERATIONS

The Scattering Process

Figure 2 is a schematic representation of the elastic binary collision process, together with scattering formulae, both for the general case and for the special situation when lab angle $\theta = 90°$. In the latter case it is clear that scattering occurs only if $M_2 > M_1$. The intensity of the scattered beam may

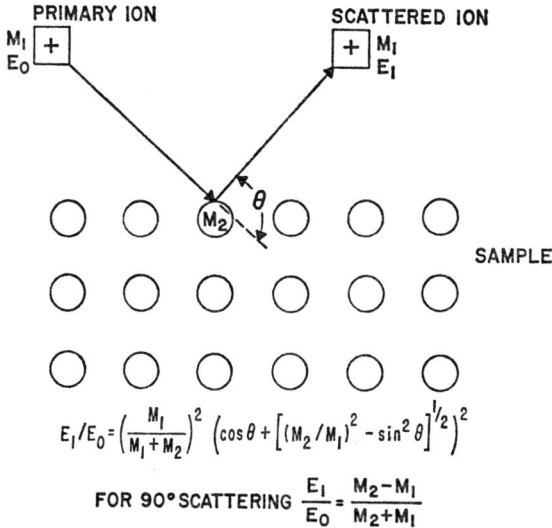

$$E_1/E_0 = \left(\frac{M_1}{M_1+M_2}\right)^2 \left(\cos\theta + \left[(M_2/M_1)^2 - \sin^2\theta\right]^{1/2}\right)^2$$

FOR 90° SCATTERING $\quad \dfrac{E_1}{E_0} = \dfrac{M_2 - M_1}{M_2 + M_1}$

Fig. 2   Schematic and formulae of the ion scattering process.

be written as

$$I_{sc}^+ = k\left(\frac{d\sigma}{d\Omega}\right) [M_1, M_2, Z_1, Z_2, E_0, \theta] \times [1 - P_n(v_1)] \times G(E_0) \qquad (1)$$

where k = a proportionality constant; $d\sigma/d\Omega$ = differential scattering cross section per unit solid angle, in square centimetres (Bingham[24]); P_n = neutralization probability; G = a geometric factor; M_1, Z_1 = atomic mass and number of primary ion, respectively; M_2, Z_2 = atomic mass and number of scattering centre, respectively; E_0 = initial energy of primary ion; θ = scattering angle, in lab co-ordinates; and v_1 = scattered ion velocity. The Bingham report[24] lists cross-section $d\sigma/d\Omega$ in terms of centre-of-mass angle ψ, which is related to lab angle θ by:

$$\sin(\psi - \theta) = \left(\frac{M_1}{M_2}\right)\sin\theta \qquad (2)$$

Figure 3 is a typical plot of scattering cross-section $d\sigma/d\Omega$ shown as a function of atomic number Z_2 of the scattering centre. This curve applies to 1·5 keV $^4He^+$ ions scattered through 90°. Not much is known at this time about neutralization factor P_n, except that it appears to be[25] a function of scattered ion velocity v_1; the quantity $(1 - P_n)$, the probability that the ion is not neutralized, is of the order of 10^{-3}. Geometric factor G indicates that there may be partial or even complete masking at the surface of atom species A by species B which will affect scattered intensity I_A^+.

The Sputtering Process

Sputtering, the emission of neutral and charged particles from the surface region of a solid caused by ion impact, has been reviewed in detail by many writers (*e.g.* Behrisch;[15] Kaminsky;[1] Carter and Colligon[2]). Therefore the

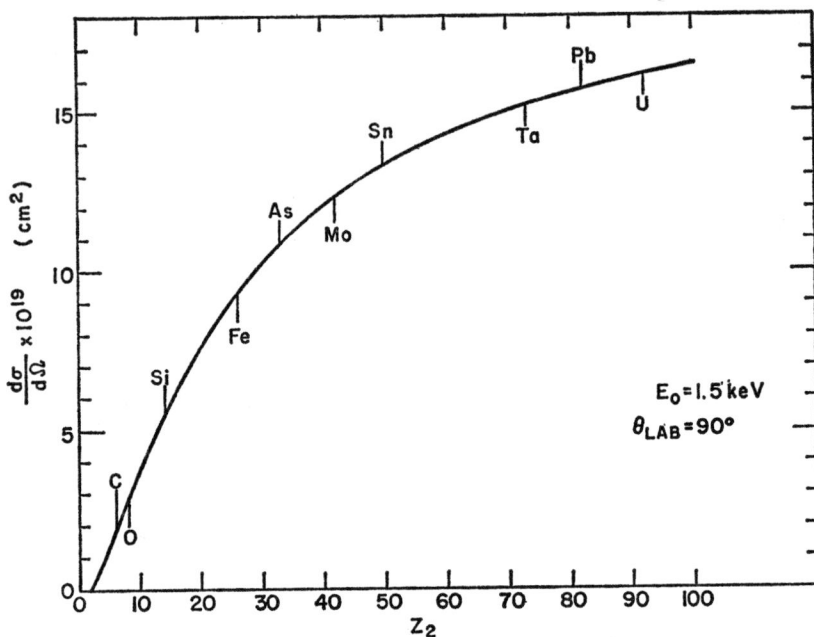

FIG. 3 Differential scattering cross-section $d\sigma/d\Omega$ vs. atomic number Z_2 for 1·5 keV $^4\text{He}^+$ ions scattered through $\theta = 90°$. These data are based on Bingham's[24] calculations which employ a screened Coulomb potential.

present discussion will be limited to those parameters that are of specific interest to the analysis of surfaces and thin films.

Sputtering yield and sputtering rate—The two most important parameters are sputtering yield Y and sputtering rate $\dot{S}$ since they enter into many considerations, in particular the depth scale of an in-depth concentration profile. Yield Y is defined as the total number of secondary particles (mainly neutrals) sputtered per primary ion colliding with the solid surface:

$$Y \equiv \frac{N_s/A}{N_p{}^+/A} = \frac{\dot{N}_s/A}{\dot{N}_p{}^+/A} \qquad (3)$$

where N = number of particles; $\dot{N}$ = their arrival or departure rate, in sec^{-1}; A = target area, in cm^2; and subscripts p and s refer to primary and secondary, respectively. Sputtering yields are known to be a function of many parameters, in particular: atomic mass M_1 and number Z_1, initial energy E_0, and angle of incidence φ of the primary ions; atomic mass M_2 and number Z_2, and binding energy (heat of sublimation) of the surface atoms; crystal structure and orientation of the lattice; and surface roughness of the sample. Values of Y for various 500 eV ions are known, in many instances, from Wehner's work[26] and from Carter and Colligon's tabulations.[2] These sputtering yield data have been collected and are presented in graphical form in Fig. 4 for various 500 eV ions. Values for higher energies can be estimated with the help of a scale factor.

Sputtering rate $\dot{S}$, in particles per second per unit area, is derived from the definition of sputtering yield Y as

$$\dot{S} \equiv \dot{N}_s/A = Y\dot{N}_p{}^+/A = YI_p{}^+/1.60 \times 10^{-19}A \,(\sec^{-1} \mathrm{cm}^{-2}) \qquad (4)$$

where $I_p{}^+ = $ primary ion current, in amperes. Using this definition and assuming that the ion beam has a gaussian density distribution, a sputtering rate in practical units, in $\text{Å}/h$, is derived as

$$\dot{S} = 0.207 \, Y\overline{V}I_p{}^+/d^2 \,(\text{Å}/h) \qquad (5)$$

where $\overline{V} = $ average atomic volume, in Å^3; $I_p{}^+ = $ primary ion current, in nA; and $d = $ ion beam width at half maximum intensity (FWHM), in millimetres. $I_p{}^+$ and d are experimentally measured quantities, and $\overline{V}$ is readily computed for elements and compounds from their respective densities and Avogadro's number.

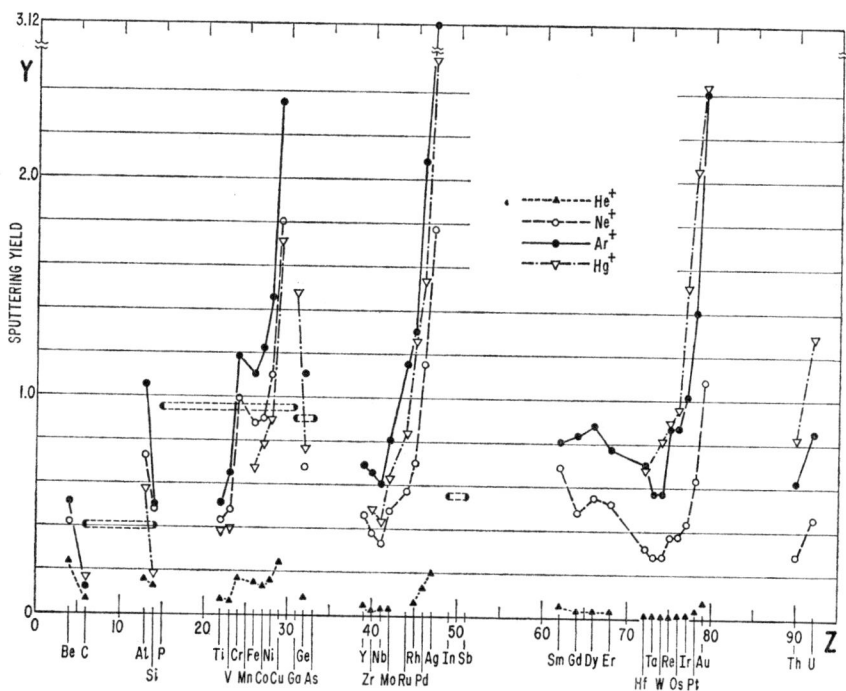

FIG. 4   Sputtering yields Y for 500 eV ions, versus atomic number Z_2 of the surface atoms. The data are based on Wehner's results[26] and the tabulations by Carter and Colligon.[2]

Differential sputtering—Sputtering yield Y and sputtering rate $\dot{S}$ are well-defined quantities provided the solid is made up exclusively of one kind of atom, which is rarely a real-life situation. Thus it is necessary to discuss 'differential sputtering', *i.e.* the effects of individual sputtering rates of a multi-component system on the composition of the surface and on the overall sputtering rate. Recently, Tarng and Wehner[27] used Auger Electron Spectroscopy (AES), and Dahlgren and McClanahan[28] employed a sputtering technique, to demonstrate the following facts which apply under equilibrium conditions to a two-component system (AB) for which $Y_A < Y_B$:

(a) the concentration ratios of surface particles $(C_A/C_B)_{surf} = (Y_B/Y_A) \times (C_A/C_B)_{bulk}$, *i.e.* at the *surface,* the concentration ratio is increased by the ratio of individual sputtering yields, as compared to the concentration ratio existing in the *bulk;*

(b) the overall sputtering rate $\dot{S}_{AB} = \dot{S}_A$, *i.e.* it is limited by and equal to the slower rate; and

(c) the composition of the vapour phase above the *surface* accurately represents the *bulk* composition, *i.e.* $(C_A/C_B)_{vapour} = (C_A/C_B)_{bulk}$.

In the framework of the present review, these findings lead to the following two important conclusions. Surface analytical methods, such as ISS and AES, will yield results that reflect the actual concentration of components at or near the *surface* at a given moment; these concentrations may differ significantly from the underlying *bulk* values. It should be noted here that for ISS the sampling depth is limited to the outermost atomic layer or two, while in the case of AES the 'escape depth' of Auger electrons ranges from about two to six atomic layers.[29] On the other hand, analytical methods based on sputtered particles in the vapour phase, such as Secondary Ion Mass Spectrometry (SIMS), will lead to concentrations typical of the underlying atomic layers contained within the sampling depth.

In this context, it is appropriate to discuss in more detail the different depth concepts already indicated in Fig. 1. The bombarding ion has a mean projected range R in the solid which is a function of primary energy E_0 and of many other variables. While ranges of high keV ions have been studied

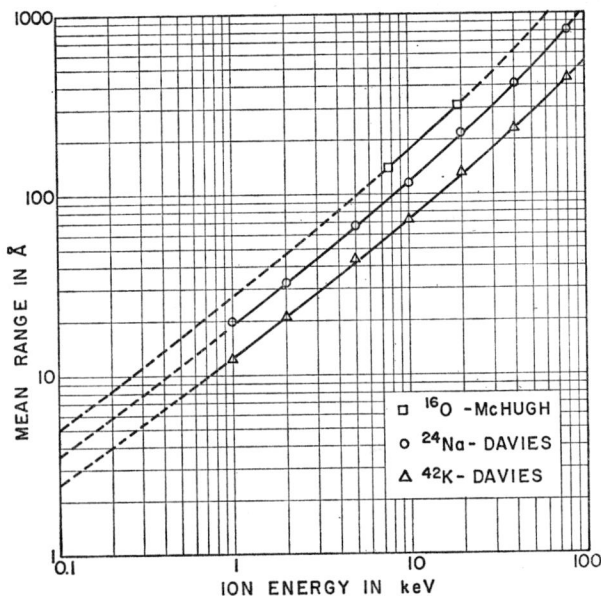

FIG. 5   Mean ranges R of O, Na and K in Ta_2O_5, based on data by Davies[30] McHugh.[31]

extensively, there is scant information available in the range of interest, *i.e.* between one and 20 keV. Figure 5 shows estimates by Davies[30] and McHugh[31] for three different primary particles in a Ta_2O_5 matrix. From the mean projected range and the angle of incidence, a primary ion *penetration* depth is readily derived which represents the region of lattice damage. For a 20 keV ion, this quantity is typically of the order of 250 Å. For surface and thin-film analyses, the *sampling* or *information* depth is of importance as already indicated above. For the sputtering case, a rough estimate indicates that its sampling depth amounts to at most one quarter of the penetration depth, *i.e.* about 60 Å for a 20 keV, and 10 Å for a 3 keV primary ion.

Production of secondary ions—Since the sputtering methods to be described below mostly utilize secondary ions, either positive or negative, rather than sputtered neutrals, it is appropriate to discuss the process of secondary ion production. Present ideas concerning *positive* ions, proposed by Slodzian and Hennequin[32] have been reviewed by Castaing and Hennequin,[33] and more recently by Evans,[22,23] who consider two basic situations: the 'kinetic' process and the 'chemical' process. The following is a brief description of their views. The 'kinetic' process occurs in clean metal and semiconductor samples bombarded by rare gas ions. As a result of the collision cascades mentioned in the Introduction, lattice particles are emitted into the vacuum, some of them in a metastable excited state from which they can, by Auger de-excitation, be transformed into positive ions near the solid surface. Schroeer,[34,35] considering the kinetic case from a somewhat different angle, has developed an expression for the ionization of emitted neutrals by quantum-mechanical transitions of the atoms' valence electrons to the top of the conduction band of the solid. On the other hand, 'chemical' ionization is the term applied to the case where chemically reactive species, such as oxygen, are present in the sample and, because of their high electron affinity, reduce the number of free conduction band electrons. This lowers the neutralization probability for secondary ions formed in the solid and permits them to be emitted as positive ions. The reactive species may be already present in the solid, *e.g.* as an oxide,[16] or they may be introduced into the system, either as a low-pressure gas,[32] or as the primary bombarding ions. Both methods of oxygen introduction were explored by Benninghoven,[36] while the latter method has since been described and used extensively by Andersen.[37,38,39]

Andersen[39] has developed a model which postulates that the sputtering region resembles a dense plasma in local thermal equilibrium and can be characterized by an electron temperature and density (typically, $T_e \simeq 10^{4\circ}K$, and $n_e \simeq 10^{19} cm^{-3}$). The positive ion intensity of a given species can be computed in terms of its neutral concentration and ionization potential, and T_e and n_e.

Since chemical ionization produces positive ion yields that may lie several orders of magnitude above the level associated with the kinetic case, it is obviously important to arrive at a full understanding of the processes involved. In a recent publication, Lewis *et al.*[40] studied in detail two anomalous effects which affect in-depth concentration profiles. They showed that: (1) the surface oxide on a single-crystal Si matrix bombarded by 14·5 keV $^{16}O^-$ ions significantly enhances secondary positive ion emission, in this case roughly within a

10 Å thickness; (2) after surface oxide removal, the secondary Si^+ ion yield within the next few hundred ångstroms reflects the actual concentration of oxygen atoms implanted in the matrix by the primary oxygen ion beam; and (3) the yield depends on the presence of oxygen atoms on the surface and in the matrix, rather than on the species of bombarding ions.

In view of the complexity of secondary ion production discussed above, it is not too surprising that there is only limited information to be found in the literature on absolute positive ion yields, *i.e.* the ratio of secondary ion to primary ion, as a function of matrix elements. Beske[41] has published results for 27 elements bombarded by 12 keV Ar^+, and Jurela[42] for 15 elements sputtered by 40 keV Ar^+. But since these measurements were made at residual gas pressures between 10^{-5} and 10^{-6} torr with primary current densities hardly adequate to keep the sample free of surface impurities, the results are likely to be representative of a combination of the kinetic and chemical processes. Apparently, the only studies made under carefully controlled conditions are those published by Benninghoven and co-workers.[43,44] The first study[43] presents absolute secondary ion yields (secondary ion/primary ion) for eight clean and oxidised metal surfaces. In the oxidised cases (chemical ionization), the yield reaches unity for Al, V, and Cr, and is typically 1000 times larger than for the clean cases (kinetic ionization). In the second study, Benninghoven *et al.*[44] measured positive ion yields from a carefully oxidized tungsten surface sputtered by 3 keV rare gas ions as a function of primary mass and energy in an ultrahigh vacuum. The authors found that under these conditions which favored chemical ionization the secondary ion intensity $I_s{}^+ = k E_0{}^x$, with $2 < x < 3$. This strong dependence on primary energy has yet to be explained and stands in marked contrast to the linear dependence of total yield on E_0, as reported by Wehner.[26]

Secondary *negative* ions have been used for some time to analyse solid surfaces and thin films for elements with high electron affinities. Honig[9,10,11] used rare gas ions to bombard a variety of metals and semiconductors and showed the method to be feasible. This was followed by a study by Krohn[45] who substantially increased the negative ion yields for various metals by bombarding the samples with Cs^+ ions. The yield increased further when an auxiliary neutral Cs beam was directed at the target. This indicates that it is the presence of cesium, regardless of charge, at the sample surface which enhances negative ion emission, presumably by lowering the work function of the target. The Cs^+ bombardment method was later applied by Andersen[38] to a study of negative ion yields from some twelve elements, and found to be effective in producing large negative ion currents, *e.g.* from gold. In a later paper, Andersen[39] applied his thermionic emission model to the negative ion case and presented an equation that predicts ion intensities in terms of electron affinity of the surface, neutral concentration, and electron temperature and density.

In-Depth Concentration Profiles and Resolution

While the analysis of surfaces is of great practical importance, the question more frequently asked is: how does the concentration of a given component in a thin layer vary with depth. This variation, generally called the 'in-depth profile', is an elusive quantity which is difficult to establish for a number of reasons. The fidelity with which a measured profile depicts an actual concentration variation depends on the 'depth resolution' of the method, a term which has been frequently used, but not adequately defined in the literature. Figure 6 serves to illustrate the major parameters that underly the definition of this term. If two layers of width w and concentration C of a given minor constituent are built into a lattice at depths d_1 and d_2, the measured profiles will broaden into the shapes shown at w_1 and w_2. It is evident that the area under the curves must remain constant, i.e. $C \times w = A_1 = A_2$. The deformation of these originally rectangular profiles into the shapes shown is due to several causes, in particular the gaussian distribution of the sputtering beam density. This will produce craters with a bell-shaped cross section,[46] resulting in

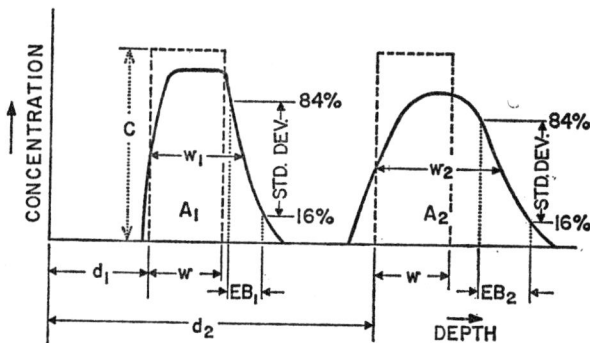

LAYER LOCATIONS: d_1, d_2
LAYER WIDTH: AS DEPOSITED - w
MEASURED (FWHM) - $w_1 < w_2$
AREAS: $A_1 = A_2 = Cw$
EDGE BROADENING: $EB_1 < EB_2$

Fig. 6   Parameters for in-depth concentration profiles.

secondary particles being emitted from different depths. This effect can be largely eliminated by accepting secondary particles only from the central, flat portion of a larger sputtered area created by rastering or by defocusing the primary beam. A second factor is the contribution made by the sampling or information depth of the method employed. A third factor contributing to the deformation is the lattice damage caused by the primary particles and the ensuing collision cascades. Some of the displaced lattice particles can be driven deeper into the lattice in fresh encounters with new primary particles, thus this effect is cumulative. Both sampling depth and lattice damage can be minimized by reducing primary energy E_0, but cannot be eliminated completely since they are characteristic of the basic sputtering process. Referring again to Fig. 6, it seems logical to define 'depth resolution' in terms of the quality of an in-depth profile of an originally rectangular signal, specifically as the broadening of the trailing edges (EB_1 and EB_2). The abscissae associated with 84% and 16% of the signal ($\pm$ one standard deviation from the half maximum value) are chosen to define 'edge broadening'. This is a convention frequently used, but not explained in the literature.

As a rule of thumb, it may be stated that under optimum conditions this edge broadening can be limited to about 10% of the associated depth.

Sample Consumption

In the analysis of surfaces and thin films, sample consumption obviously plays a major role because it limits the resolution (either lateral or in-depth) and detection sensitivity attainable. Morabito and Lewis[47] point out that a minimum sample volume is required to determine the level of a given impurity to a desired precision. This volume is a function of impurity concentration, precision, instrumental efficiency, the secondary ion/sputtered particle ratio, atomic density, and isotopic abundance. As an example, Morabito and Lewis state that a $100 \mu m^3$ sample volume is required to detect 10 parts per million atomic (ppma) of monoisotopic Al with 3% precision, assuming a secondary ion/sputtered particle ratio of 10^{-3}. For a $100 \mu m$ primary ion beam diameter, this amounts to a 130-Å-thick layer removed from the sample during the analysis. McHugh[48] has presented in graphical form detection sensitivity as a function of primary beam diameter. Assuming a primary beam density of 5 mA/cm^2 and a 100-sec collection time, corresponding to the removal of a 10^4-Å layer, he finds a 0·1-ppm detection limit, equivalent to the estimate by Morabito and Lewis.[47]

INSTRUMENTATION

Ion Scattering Spectrometers (ISS)

An ion scattering spectrometer of recent design, commercially available from the 3M Company[49,50] is schematically shown in Fig. 7. Its major components include: a primary ion source with small energy spread, but without (M/q) selector; a target holder and manipulator; a charge neutralization filament; a 127° electrostatic energy analyzer (ESA); and a channel electron multiplier. All of these components are mounted in an ultra-high vacuum system. The primary beam of rare gas ions (initial energy range: 1–3 keV) is focussed into a 1 mm diameter spot on the target, giving rise to a current density of about 10 μA/cm^2. Since the sample surface makes an angle of 45° with the nearly circular primary beam, the sputtered hole will be elliptical in shape, as shown in Fig. 8. The ESA used accepts a band (shown cross-hatched) of ions scattered through 90°. This band includes regions near the crater wall, a fact which limits depth resolution. To eliminate this problem, it is necessary to make the sputtered crater large with respect to the detected area. This could be accomplished by rastering the primary beam, at the expense of the sputtering rate, or else by shortening the entrance slit of the ESA, which lowers the scattered intensity, and thereby detection sensitivity.

A number of sophisticated, laboratory-built ISS instruments have been described in the recent literature by Suurmeijer and Boers,[51] Andrew et al.,[52] Eckstein and Verbeek,[53] Heiland and Taglauer,[54] Brongersma and Mul[55] and Wheatley and Caldwell.[56] All of these instruments are highly versatile: they utilize mass-analysed primary beams; their scattering angle is adjustable; and their target orientation and temperature can be varied. Furthermore, the Suurmeijer and Boers[51] instrument is of special interest since it includes a

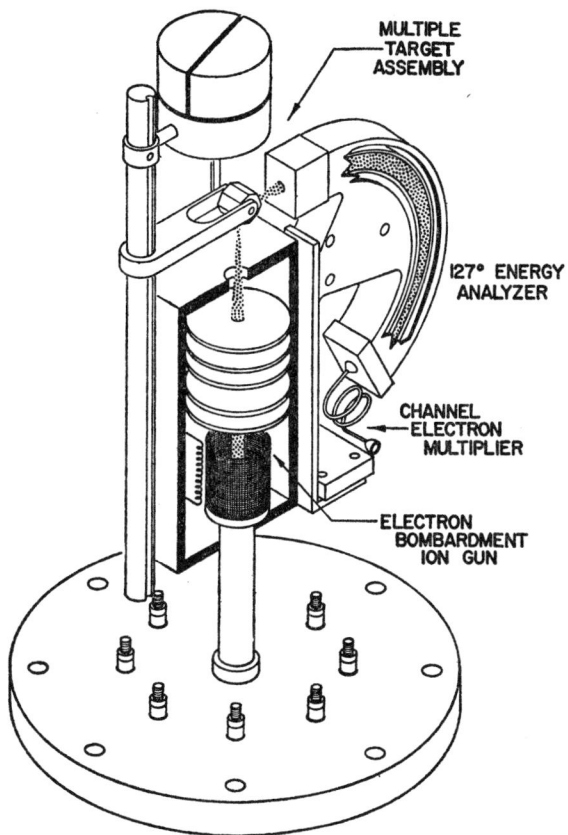

FIG. 7   Schematic of 3M ion-scattering spectrometer.

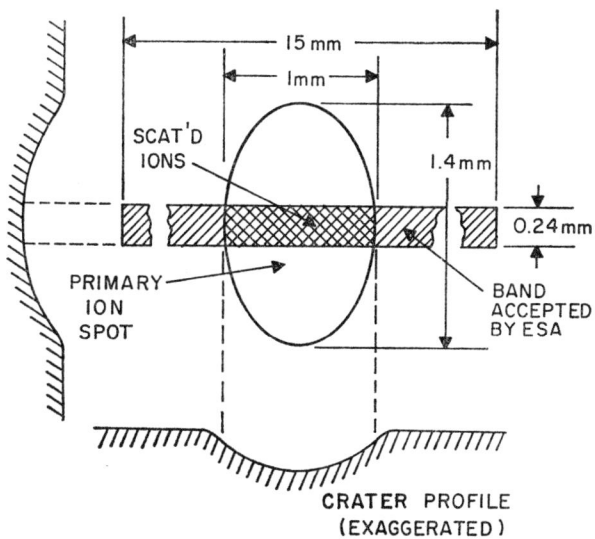

FIG. 8   ISS geometry of sputtered crater and area accepted by the electrostatic analyser.

TABLE I

Intercomparison of Operating Parameters for Ion Scattering, Secondary Ion, and Ionized Neutral Spectrometers

Parameters		Method			
		Ion scattering spectrometer	Secondary ion Microanalyser	Secondary ion Surface analyser	Ionized neutral MS
Primary ion current density (A/cm^2)		10^{-5}	10^{-4}–10^{-2}	10^{-9}–10^{-5}	10^{-2}
Mass resolution $M/\Delta M$	Optimum	30	1000–10,000	300	300
	Typical		300		
In-depth resolution	Optimum (Å)	30	30		
	Depth fraction	$0\cdot3d$	$0\cdot1d$	$0\cdot13d$	$0\cdot13d$
Lateral resolution (μm)	Optimum	500	1	3000	10,000
	Typical		100		
Sampling depth (Monolayers)		1	(20 keV) 20 (5 keV) 5	3	1
Penetration depth (Monolayers)		6	(20 keV) 80 (5 keV) 20	12	3
Sample consumed, complete spectrum (monolayers)		1	300–3000	$0\cdot01$	?
Sputtering rate $\dot{S}$ (Å/sec)		$0\cdot0003$–1	1–100	$0\cdot0001$–$0\cdot1$	$0\cdot05$–10
Detection sensitivity (ppma)		1000	1	100	100
Range factor of elemental sensitivities		5	10^4	10^4	5

second mass analyser in series with the ESA which permits secondary sputtered ions to be identified.

Secondary Ion Mass Spectrometers (SIMS)

In his two comprehensive, up-to-date reviews, Evans[22,23] has discussed the major instrumental features of secondary ion mass spectrometers, and grouped them into three major classes: (1) the direct-imaging analyser developed by Castaing and Slodzian;[16] (2) the scanning ion microprobe designed by Liebl;[18] and (3) the secondary ion analysers developed by Herzog and collaborators[57,58] and more recently by Benninghoven and co-workers.[20,21] Since Evans has discussed in considerable detail the outstanding characteristics of these different types of instruments, the reader is referred in particular to Table I in his 1972 Review[22] for a summary of their essential features. In the present paper, the discussion will be limited to a brief outline of basic operating principles and major components.

Direct-imaging mass analyser—The original Castaing–Slodzian design,[16] modified by the inclusion of a double magnetic analyser and electrostatic mirror, is now commercially available from CAMECA.[47] As indicated in Fig. 9, the primary positive or negative ions are produced in a duoplasmatron and are focused, without mass analysis, into a target spot typically 25–200 μm in diameter. Flat-bottomed craters are obtained by rastering the primary beam. Secondary particles are extracted from an area approximately 200 × 200 μm^2, imaged by an immersion lens, and mass- and energy-analysed in the prism–mirror–prism system. The mass-resolved image is accelerated, converted into electrons, and detected by various means. For quantitative electrical measurements, and to ensure adequate depth resolution, a mechanical

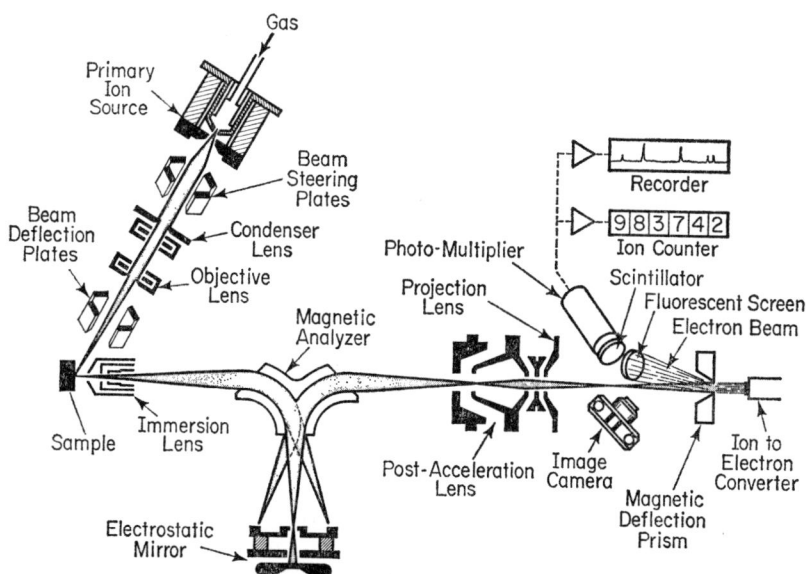

FIG. 9   Schematic of the CAMECA direct-imaging mass analyser.

aperture is placed in the secondary image plane. The analyser has a mass resolution of up to 1000, and an optimum lateral imaging resolution of about 0·8 μm. Since the direct-imaging analyser records all image information *simultaneously,* it yields desired information in much less time than the *sequential* system of the ion microprobe to be discussed below.

Scanning ion microprobe—The first commercial microprobe, available from ARL[59] and shown in Fig. 10, is based on Liebl's[18] design. It utilizes a

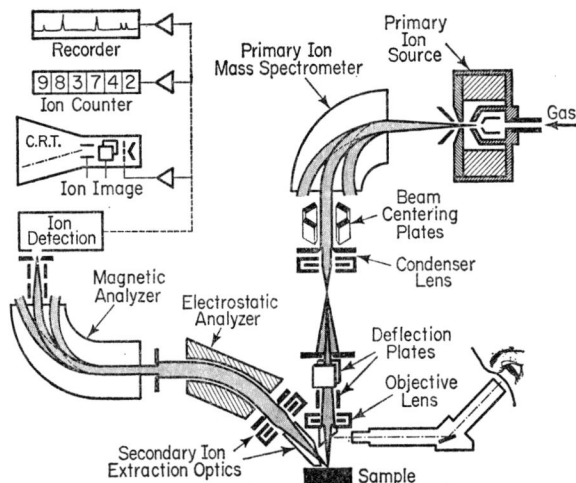

FIG. 10   Schematic of the ARL ion microprobe.

duoplasmatron ion source in combination with a mass selector and focuses the primary beam into a 2 μm spot which can be rastered over the target. The secondary ions are extracted, analysed in a double-focussing mass spectrometer, and detected and imaged by various means. Image information, recorded by rastering, is sequential and requires substantially more time than the simultaneous method. The microprobe has a mass resolution of about 1000, and a lateral imaging resolution of 2 μm. This instrument has provisions for viewing the sample during bombardment.

Recently, AEI has developed a primary ion-generating column, based on a design by Drummond and Long,[19] which in conjunction with the AEI MS 702 constitutes the microprobe shown in Fig. 11. Its technical specifications are equivalent to those of the ARL instrument, but it has a substantially higher mass resolution: 3000 for electrical, and 10 000 for photographic detection.

The latest development among microprobes is a sophisticated design by Liebl[60] which is a combination ion-electron microprobe. The instrument, dubbed 'UMPA' (Universal Microprobe Analyser) by its creator,[61] permits ion or electron bombardment, either separately or simultaneously, by virtue of a novel lens design. Secondary particles are energy-analysed in an ESA and either detected at that point, or else pass into a 180° magnet for mass analysis.

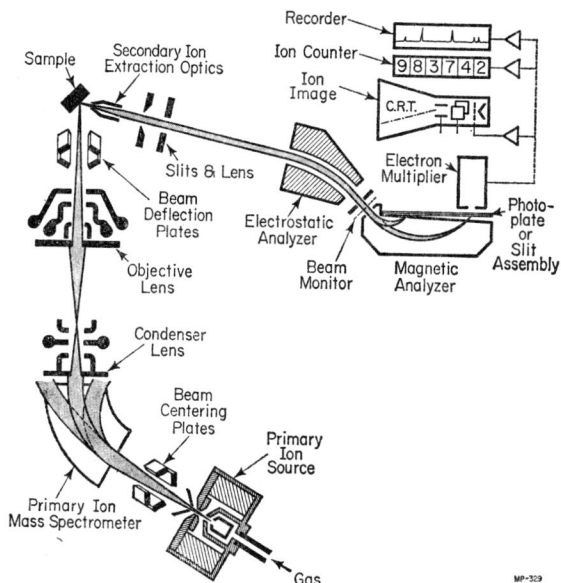

FIG. 11 Schematic of the AEI ion microprobe.

The analyser is housed in an ultra-high vacuum system to facilitate surface studies. With the help of a novel lens design, Liebl expects this complex new instrument to have better resolution and sensitivity than that achieved with the older instrument.

Secondary ion surface analysers—Provided that high lateral resolution is not required, surface analyses and in-depth profiling of thin films can be achieved satisfactorily by combining, in an ultra-high vacuum system, a primary ion gun of simple design with a quadrupole mass spectrometer. The major advantages of this system are: (1) its relatively low cost ($40 000–80 000) as compared to either direct-imaging analysers or microprobes (approximate price: $250 000); and (2) its ultra-high vacuum capability, allowing it to use primary ion beams of low current density, so that only a small fraction of a monolayer is consumed in the study of a surface. The first successful instrument of this type was constructed by Benninghoven and Loebach.[20] In it, primary ions are produced in an axial ion gun, mass-analysed magnetically, and focused into a 3 mm spot on the target. For analyses, primary beam densities ranging from 1–10 nA/cm^2 are used, at energies between 1–3 keV. This density removes only about 1 % of a monolayer during a 'static' analysis lasting 1000 seconds while the system is valved off the pumps. The secondary ions, positive and negative, emitted by the target are analysed in a quadrupole mass spectrometer. Based on this design, a commercial system has been developed by Huber et al.[21] which utilizes ultra-high vacuum techniques, including a Ti sublimation pump, to reduce residual gas pressures to less than 10^{-10} torr. It uses a duoplasmatron source to produce primary ions which are focused directly, without mass analysis, into a 3 mm spot on the target.

Thus current densities can be raised to about $10 \ \mu A/cm^2$, which permits faster sputtering rates for in-depth profiling.

There are two other recent publications (Witmaack et al.;[62] Schubert and Tracy[63]) concerning secondary ion analysers, based essentially on the Benninghoven design. Both groups found, however, that the simple ion gun–quadrupole combination produced a high continuous background which limited the dynamic detection range to at best four decades. For this reason, Witmaack et al.[62] interposed a simple energy analyser between target and quadrupole to prevent sputtered neutrals, light quanta, and energetic scattered primary particles from entering the mass analyser. By this means, a signal-to-noise ratio of 10^8 was achieved in a favourable case. In their parallel study, Schubert and Tracy[63] employed one-half of a cylindrical mirror analyser to improve their signal-to-noise ratio to an estimated value of 10^6 or better.

Ionized Neutral Mass Spectrometers (INMS)

The strong dependence of secondary ion emission from a given matrix on the presence of other constituents, in particular oxygen, makes it difficult to deduce neutral concentrations from positive secondary ion intensities, and suggests the alternative use of sputtered, ionized neutrals. Early exploratory experiments were made by Honig[10] who utilized the same low-pressure ($p \simeq 10^{-4}$ torr), magnetically confined d.c. discharge first to sputter a target and then to ionize the sputtered neutrals. This method was feasible, but the large intensities of the sputtering gas peaks masked wide spectral regions, thus precluding the practical use of that source under those conditions.

Recently, Coburn and Kay[64] employed successfully a rare gas, r.f. glow discharge ($f = 13 \cdot 56 \ \text{MH}, p \simeq 0 \cdot 1$ torr, $E_p \simeq 200$ eV) to bombard targets and ionize the sputtered secondary neutrals. The ionized secondaries are energy-selected by an ESA, and mass-analysed in a quadrupole. Coburn and Kay postulate that ionization of neutral species X proceeds via the Penning mechanism: $(RG)^m + X \rightarrow RG + X^+ + e^-$, where $(RG)^m$ is the rare gas metastable produced in the discharge. The process is operative for all species X with ionization potentials smaller than the rare gas metastable energy. Typical operating conditions produce sputtering rates that range from 1 to 200 monolayers/min and yield ionized particle intensities representative of the matrix composition. The discharge gas ions are about one order of magnitude more intense than the sample ions which come from a target 5 cm in diameter.

A similar r.f. discharge system has been described by Oechsner and Gerhard[65] which operates at a substantially lower Ar pressure (about 10^{-3} torr). Since under these conditions the discharge gas ions are about 4000 times more intense than the target ions, it may be assumed that in this case ionization of neutrals is due to electron impact, rather than the Penning mechanism.

RESULTS

In this section, we shall discuss achievements and limitations of scattering and sputtering methods as applied to surface analyses, lateral imaging, and in-depth profiling, making a special attempt to present an objective picture of real-life situations. Quantitative aspects of the various methods will be summarized at the end in a general comparison.

Surface Analyses

For a true surface analysis, all information should come exclusively from the outermost atomic layer. Thus, ion scattering is strictly speaking the only method capable of performing such an analysis, whereas sputtering methods typically employ primary ion energies between 3 and 20 keV, corresponding to information depths between 10 and 60 Å (4 to 25 atomic layers). However, sputtering methods can supply information essentially limited to the surface layer provided the primary ion energy remains below 1 keV.

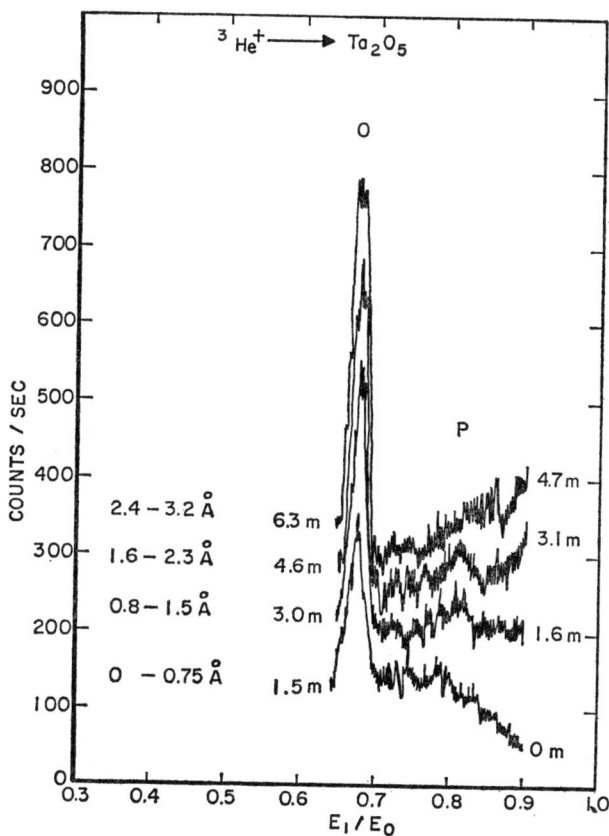

FIG. 12   Phosphorus contamination of a Ta_2O_5 sample revealed by ISS surface analysis.

As an example of a surface analysis by the ion scattering technique,[5] Fig. 12 presents a series of four partial spectra, taken in rapid sequence, of a Ta_2O_5 sample presumed to be partially covered with a phosphorus layer. From run 1 to run 4, the lattice oxygen peak is growing rapidly as gases and radicals, initially adsorbed on the sample surface and shielding the lattice scattering sites, are gradually sputtered away. The small phosphorus peak at $E_1/E_0 = 0.82$

is not visible in run 1 because of initial shielding by surface contaminants; it is observed in runs 2 and 3, but in run 4 it has already disappeared, having been sputtered away. From these and other data it is estimated that phosphorus exists on this sample as a partial monolayer with a surface coverage of a few percent. Sample consumption for each partial spectrum amounted to roughly one quarter of a monolayer.

In a recent review paper, Benninghoven[66] has summarized the considerable volume of surface analyses carried out with his static SIMS method which uses positive as well as negative secondary ions. He studied metals and semiconductors in different states of surface oxidation, and with adsorbed layers of oxygen, various acids, and organic compounds. 3 keV Ar^+ ions, usually at a current density of 10^{-9} A/cm^2, were used for the analyses which consumed about 1 % of a surface layer under these conditions. The instrument appears to have a dynamic detection range of about four decades, and Benninghoven[67] states that by raising the primary current density to 10^{-6} A/cm^2 a detection sensitivity of 10^{-6} monolayer is obtainable. The present author estimates that for the conditions quoted the information depth amounts to about 10 Å, which is comparable to Auger Electron Spectroscopy (AES).

Lateral Concentration Profiles

As mentioned above, the direct-imaging mass analyser and the ion microprobe can provide, respectively, simultaneous or sequential information on lateral elemental distribution through secondary ion images. In either case, a lateral resolution of about 1–2 μm has been achieved. Typical examples of directly-imaged distributions have recently been published by Morabito and Lewis,[47] and of sequentially-imaged distributions by Andersen and Hinthorne.[59]

In-depth Concentration Profiles

As discussed under 'Basic Considerations', there are many requirements that must be met before profiles can be taken that accurately represent concentration as a function of depth. They are summarized in the following list:

(1) The particles used for profiling should come from a well-defined depth, *i.e.* from the central portion of a much larger flat-bottomed crater.

(2) The sampling or information depth should be limited to the outermost layer or two. This presents no problem for ISS and INMS, but requires that SIMS be performed at the minimum primary ion energy compatible with an adequate sputtering rate.

(3) Lattice damage produced by primary ions which causes the mixing of atomic layers should be minimized by employing the lowest practicable primary energy. This requirement applies equally to all methods.

(4) There should be no 'matrix effect', *i.e.* the signal intensity of one component should not be affected by the presence of another.

It will be shown that none of the methods employed for profiling satisfy all of the above requirements.

An example of a depth profile obtained by ISS[5] for a 400 Å-thick Ta_2O_5 film on Ta is shown in Fig. 13. In this case, sputtering was done with a broad ($d = 1 \cdot 8$ mm) beam of $^{20}Ne^+$ ions, while a narrow ($d = 0 \cdot 9$ mm) $^3He^+$ beam was used to permit the detection of O as well as of Ta. Since the

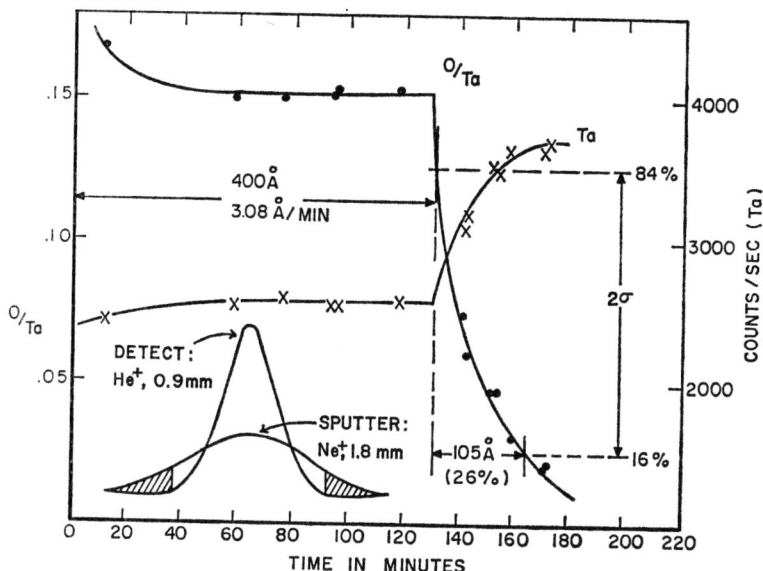

FIG. 13   ISS in-depth concentration profile of 400 Å Ta_2O_5 on Ta.

sputtering rate for Ta_2O_5 was not known *a priori*, the abscissa is shown in terms of sputtering time. However, from the known film thickness a sputtering rate of about 3·1 Å/min could be deduced. The edge broadening shown in Fig. 13 for the O/Ta ratio is approximately 25% of the film thickness, which is substantially worse than the ideal 10% figure quoted above. It reflects the fact that the sputtered crater was not large enough to ensure that the detected signal came exclusively from the central, flat-bottomed crater region. It is also noteworthy that the Ta intensity rise taking place at the Ta_2O_5–Ta interface accurately reflects the difference in Ta concentrations for these two bulk regions. Thus it is seen that in the scattering case the intensity of a given constituent is independent of the presence of oxygen, *i.e.* there is no 'matrix effect'. This situation is very different from that described below for secondary ion emission.

The method of obtaining in-depth profiles with a GCA Ion Microprobe has been discussed in detail by Evans and Pemsler[68] and Pawel *et al.*[69] These studies were carried out with 14 keV Ar^+ ions, under carefully adjusted conditions to ensure a uniform crater profile and constant sputtering rate. In the first study,[68] the interface profiles of a series of $Ta_2{}^{18}O_5/Ta_2{}^{16}O_5$ samples were determined and compared to activation analyses. The results indicate that, regardless of the location of the interface within the sample, the ion microprobe produced depth resolutions that were roughly 40 Å larger than the activation analysis values. The second study[69] included concentration profiles obtained for ^{31}P layers (25 to 100 Å thick) embedded between two 500 Å Ta_2O_5 layers. In this case, depth resolutions of about 70 Å were obtained.

McHugh[70] has employed an ARL ion microprobe to study in detail the effect of primary ion energy on depth resolution using $^{16}O_2{}^+$ and $^{16}O^-$ ions with energies ranging from 1·75 to 18·5 keV. The Ta_2O_5 samples contained a 50 Å P-rich zone located 230 Å below the surface of a 1050 Å film. From the impressive results presented by McHugh, the present author derives depth resolution values of 32Å at 1·75 keV, 45 Å at 4·25 keV, and about 105 Å at 7·75 keV. When these data are plotted against energy and extrapolated, a depth resolution of 25 Å at zero energy is found. The apparent thickness of the P-rich zone, measured at half maximum intensity, ranges from about 80 Å at 1·75 and 4·25 keV to 160 Å at 7·75 keV in a somewhat erratic fashion.

The recent study by Lewis et al.[40] on the effect of oxygen on in-depth profiles has already been mentioned in connection with the mechanism of secondary ion emission. The results obtained by these authors clearly point out that the presence of oxygen on the surface or in the bulk of a sample will enormously enhance the secondary ion emission from any constituent, even if primary oxygen ions are employed. For this reason, it is imperative always to obtain an oxygen profile together with the desired profile, in order to decide whether the shape of the latter truly represents the concentration of the constituent, or simply reflects the presence of oxygen. It is this secondary ion yield enhancement by oxygen and also other electronegative species which frequently limits the usefulness of SIMS information. Thus it is difficult to establish the true concentration profile for a diffused impurity near an oxide–metal interface.

Coburn and Kay[64] as well as Oechsner and Gerhard[65] have explored the feasibility of obtaining in-depth profiles by ionized neutral mass spectrometry. The method fulfils three of the four requirements listed above. The primary ions in the discharge typically have an energy of 200 eV, thus sampling depth is limited to the outermost layer, and lattice damage is minimized. Furthermore, the use of secondary neutrals essentially eliminates matrix effects. On the other hand, there are problems associated with sputtering from the sample edges which may affect in-depth resolution. Preliminary results obtained by Coburn and Kay indicate that a 100 Å Fe layer sandwiched between two 1600 Å EuO layers can be detected, but that beam intensities are subject to substantial fluctuations. Oechsner and Gerhard have demonstrated that the presence of O and N at the surface of a Ta film can be profiled, but their data do not permit an evaluation of the depth resolution achieved.

Insulator Problems

The surface and thin film analysis of insulators presents a special problem because the primary ion beams employed in the scattering and sputtering methods cause the sample surface to charge up. This can produce substantial changes in the target potential and thereby seriously affect the scattered or secondary beams in terms of energy, location, focusing, or intensity. The actual target potential is determined by the rates of charges arriving and leaking off. In other words, it depends on the current density of the primary beam, and the conductivity, area, and thickness of the insulating layer.

Various attempts have been made to resolve this problem. They include the deposition of a thin conducting layer or grid structure, e.g. gold, on the surface, and charge neutralization by flooding with low energy electrons from a filament. The first of these methods has met with only partial success since at least some charge-up reoccurs as soon as the conducting layer or grid has been sputtered away. Charge neutralization by electron flooding works well in many cases, but care must be taken that the hot filament cannot see the target surface in order to avoid sample contamination. In the case of photoconductors, it is often possible to increase conductivity to an adequate level by shining light onto the surface.

Quantitative Aspects

Ion scattering—Equation (1) presented above for the intensity of scattered particles indicated its dependence on scattering cross-section, neutralization probability, and a geometric factor. Only the first of these terms can be evaluated analytically. Therefore, known standards are required in order to determine quantitatively the composition of a surface layer from its scattering spectrum. In the case of some binary compounds, such as Al_2O_3 and SiO_2, the intensity ratios, measured with a precision of about $\pm 4\%$ for the two components, agree quite well with expected values when 1·5 keV $^4He^+$

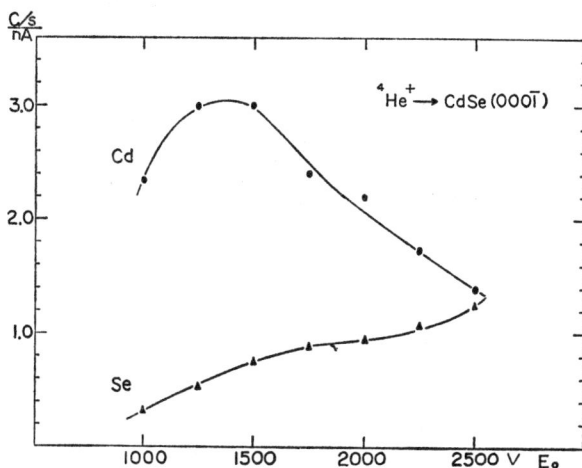

FIG. 14 Scattered intensities for two-component system CdSe versus primary energy E_0.

ions are employed for analysis. However, this agreement may well be fortuitous, as illustrated[5] by Fig. 14 which presents scattered intensities of another two-component system, the 000$\bar{1}$ Se face of CdSe, as a function of primary energy E_0. It appears that, at least in this case, the dependence on E_0 differs widely for the two components, producing intensity ratios I_{Cd}^+/I_{Se}^+ that range from about 7 at $E_0 = 1\cdot1$ keV to unity near $E_0 = 2\cdot6$ keV. It is anticipated that these results eventually can be interpreted in terms of neutralization and geometric effects.

The effect of crystal geometry[5] is clearly demonstrated in Fig. 15 for three different orientations of the polar molecule CdSe: the 0001 Cd face, the neutral 11$\bar{2}$0 face, and the 000$\bar{1}$ Se face. Data were taken with 1·5 keV $^4He^+$ ions. It is interesting to note that Cd intensities change widely while Se intensities stay essentially constant. Figure 15 clearly demonstrates the effects of geometric shielding, and suggests that ISS can be employed to identify crystal orientations, confirming the earlier study by Strehlow and Smith.[71]

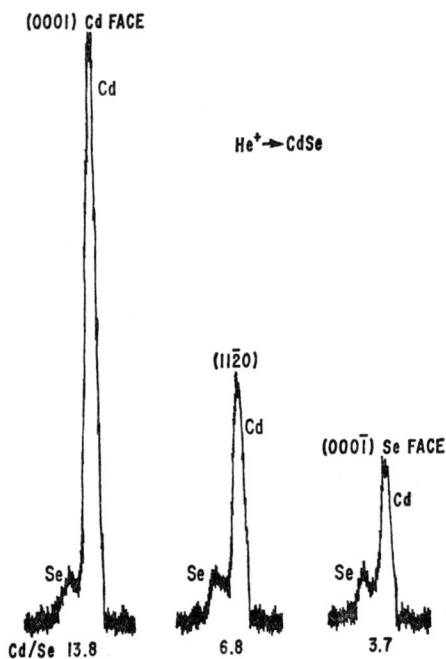

FIG. 15   Effect of CdSe crystal orientation on scattered intensities.

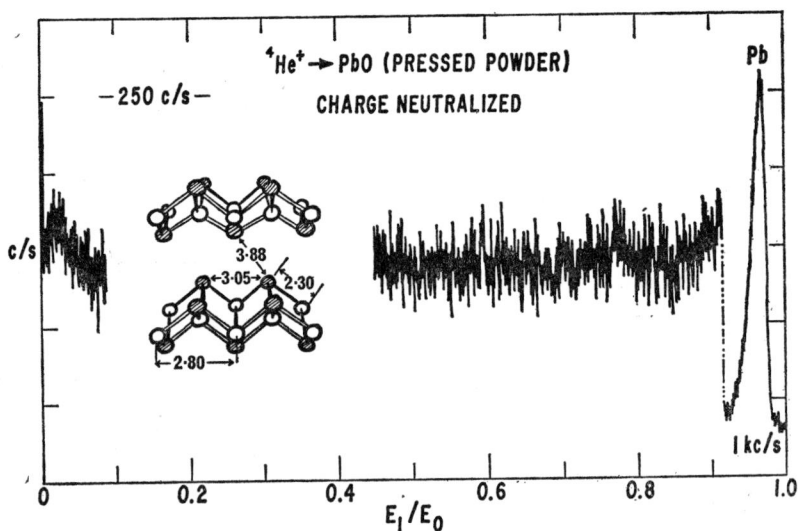

FIG. 16   ISS spectrum of red tetragonal PbO powder, demonstrating geometric shielding and possibly differential sputtering.

An extreme example of geometric shielding is presented[5] in Fig. 16 which shows the spectrum of $^4He^+$ ions scattered from red (tetragonal) pressed PbO powder. While there is a fair-sized Pb peak at the appropriate energy ($E_1/E_0 = 0.96$), no peak could be detected at $E_1/E_0 = 0.6$ which represents scattering from O. This can be explained in terms of the layered, graphite-like structure shown in the insert of Fig. 16. Within a given layer, there is strong bonding between Pb (cross-hatched) and O atoms (open circles), with O being shielded in every conceivable direction. It may be that in this instance sputtering proceeds in a layer-by-layer fashion, thus never exposing the O atoms to the primary beam, or else this represents an extreme case of differential sputtering that leaves the outermost layer bare of O atoms.

Secondary ion mass spectrometry—According to Andersen,[39] the equilibrium equations developed for secondary ion emission (*see above*) can be used to compute the concentration of a constituent in a matrix from the observed ion intensity. The ionization potential must be known for positive ions, or the electron affinity for negative ions, and electron temperature T_e and density n_e are established by internal standards (from two known major component concentrations in the same sample), or else in separate calibration runs. This method has been applied to the microanalysis of metals and alloys, and of various mineral samples, with a stated 10% relative error for major constituents, and a minimum absolute accuracy of a factor of two at the ppma concentration level. Sample areas of about 5000 μm^2 ($d \simeq 80\ \mu m$) were used for the analysis, but there is no statement concerning typical sample depths removed. However, it may be inferred from minimum volume considerations that sample thicknesses of the order of one μm must have been consumed to make analyses at the lower concentration levels. Further work will have to show whether this approach can be applied to in-depth profiling and can, in particular, predict ion intensities of given constituents as a function of the surface and bulk concentration of oxygen present in a sample, a problem which at this point severely limits the general usefulness of the ion microprobe.

From the analytical standpoint, it is important to evaluate objectively the capabilities and limitations of secondary ion mass spectrometry in terms of the parameters listed in the Introduction. A casual reader might gain the impression from some of the literature[72] that under ideal conditions detection sensitivities of 10^{-3} to 10^{-6} parts per billion atomic (ppba) can be achieved; that in-depth profiles of ppba impurities can be measured with a sample consumption of a single monolayer; and that the detection limits for a 1 μm diameter sample are about one ppma. On the other hand, the practical aspects of the method have been considered in a realistic fashion by Morabito and Lewis,[47] by McHugh,[48] and by Werner[73] and DeGrefte.[74] These authors investigated detection limits for typical (rather than favoured) cases in terms of the major interrelated parameters, in particular, primary beam size or sample area used, sputtering rate or sample consumption, depth resolution, and precision. Here is the general consensus abstracted from these studies, quoting order of magnitude values. Primary beam or sample diameters of about 100 μm are required to determine, at a mass resolution near 300, trace concentrations at the ppma level to a precision of a few percent. To obtain complete mass spectra, sample depths ranging from 0·1 to 1 μm may be consumed in the process, at sputtering rates that range from about 1 to 100 Å/sec. When monitoring a single peak under these conditions, a depth resolution of 100 Å or better is attainable, at the expense of lateral resolution. On the other hand, for 'bulk' analyses of micrometer thickness, lateral resolutions of between 1 and 20 μm are possible, depending on the type of instrument employed.

Intercomparison and Evaluation of Methods

The capabilities of the four major methods discussed above—ion scattering spectrometry, secondary ion microanalysis and surface analysis, and ionized neutral spectrometry—are compared in terms of some major parameters in Table I. Such a tabulated intercomparison is, of necessity, incomplete and difficult to achieve because the different methods are forced into a common mould that is only partially applicable to each. To complement Table I, the major capabilities and limitations for each method are listed below in capsule form.

Ion scattering spectrometry determines the composition of the outermost one or two *surface* layers which may differ substantially from the underlying layers because of sputtering. Major and minor constituents are identified with modest mass and lateral resolution, while in-depth resolution will be improved in the near future. Sample consumption and sputtering rate are both low, making this an attractive method for films less than 1000 Å thick. The orientation of polar crystals can be established.

Secondary ion microanalysers, both direct-imaging and scanning, have excellent mass, lateral, and in-depth resolution capabilities and high sensitivity, at the expense of sample consumption. The analytical information comes from a few layers below the sample surface and represents *bulk* concentrations at that level. Their major limitations are the strong effect of surface and bulk oxygen on ion intensities, and the wide range of secondary positive ion yields which differ by as much as four orders of magnitude from element to element. Thus, in-depth profiles are frequently hard to evaluate quantitatively.

Secondary ion surface analysers have good mass and in-depth resolution, but poor lateral resolution. Their information comes from a modest depth, and their outstanding feature is the potentially low sample consumption, frequently only 1 per cent of a monolayer. The problems of secondary ion yields and the effect of oxygen on ion intensities apply equally to this method.

Ionized neutral mass spectrometry has yet to be more fully developed. It has good mass and in-depth resolution, but no lateral resolving capability. Because of the low primary energies employed it presumably samples only the top atomic layers. Its major advantage lies in the fact that it shows no matrix effect for oxygen, and that detection sensitivity varies little from element to element.

CONCLUSIONS

The three-dimensional analysis of solids with the help of scattered and sputtered particles has grown in recent years into a new branch of science and technology. Four major methods—ion scattering spectrometry, secondary ion microanalysis and surface analysis, and ionized neutral mass spectrometry—allow surfaces and thin films to be analysed, laterally and in-depth. High sensitivity, low sample consumption, adequate precision, high resolution, and complete elemental coverage can be achieved, but not necessarily all at once. Further improvements may be expected in the near future.

ACKNOWLEDGMENTS

The author acknowledges with pleasure many helpful discussions with a number of colleagues, in particular with W. L. Harrington and C. W. Magee, RCA Laboratories; F. W. Anderson, IBM, Hopewell Junction, N.Y.; C. A. Evans, Jr, University of Illinois; R. K. Lewis, CAMECA, Elmsford, N.Y.; and J. A. McHugh, KAPL, Schenectady, N.Y. Messrs Evans, Lewis, and McHugh kindly made available prepublication copies of their most recent papers, and the author is especially grateful to C. A. Evans, Jr, for permission to use three of his original drawings as Figs. 9, 10 and 11.

REFERENCES

1. Kaminsky, M., 'Atomic and Ionic Impact Phenomena on Metal Surfaces', Academic Press, New York, 1965.
2. Carter, G. and Colligon, J., 'Ion Bombardment of Solids', McGraw-Hill, London, 1968.
3. Brunnée, C., Z. Physik, 1957, **147**, 161.
4. Smith, D. P., J. Appl. Phys., 1967, **38**, 340.
5. Honig, R. E. and Harrington, W. L., 'Thin Solid Films', 1973, in press.
6. Sloane, R. H. and Press, R., Proc. Roy. Soc. (London), 1938, **A168**, 283.
7. Sloane, R. H. and Watt, B., Proc. Phys. Soc. (London), 1948, **61**, 217.
8. Herzog, R. F. K. and Vieböck, R. P., Phys. Rev., 1949, **76**, 855.
9. Honig, R. E., J. Appl. Phys., 1958, **29**, 549.
10. Honig, R. E., in 'Advances in Mass Spectrometry', ed. J. D. Waldron, Vol. I, Pergamon Press, London, 1959, pp. 162–71.
11. Honig, R. E., in 'Advances in Mass Spectrometry', ed. R. M. Elliott, Vol. II, Pergamon Press, Oxford, 1961, pp. 25–37.
12. Bradley, R. C., J. Appl. Phys., 1959, **30**, 1.
13. Stanton, H. E., J. Appl. Phys., 1960, **31**, 678.
14. Veksler, V. I., Sov. Phys.—J.E.T.P., 1960, **11**, 235.
15. Behrisch, R., 'Festkörperzerstäubung durch Ionenbeschuss', Ergeb, exakt. Naturwiss., Vol. 35, Springer, Berlin, 1964.
16. Castaing, R. and Slodzian, G., J. Microscopie, 1962, **1**, 395.
17. Liebl, H. J. and Herzog, R. F. K., Presented at 11th Ann. Conf. Mass Spectrom., San Francisco, 1963.
18. Liebl, H. J. Appl. Phys., 1967, **38**, 5277.
19. Drummond, I. W. and Long, J. V. P., Nature (London), 1967, **215**, 950.
20. Benninghoven, A. and Loebach, E., Rev. Sci. Instr., 1971, **42**, 49.
21. Huber, W. K., Selhofer, H. and Benninghoven, A., J. Vac. Sci. Technol., 1972, **9**, 482.
22. Evans, C. A., Jr, Anal. Chem., Nov. 1972, **44**, 67A.
23. Evans, C. A., Jr, Proc. Eighth Nat. Conf. Electron Probe Analysis Soc. America, New Orleans, 1973.
24. Bingham, F. W., 'Tabulation of Atomic Scattering Parameters Calculated Classically from a Screened Coulomb Potential', Sandia Research Report SC-RR-66-506, TID-4500 Physics, Aug. 1966.
25. Smith, D. P., Surf. Sci., 1971, **25**, 171.

26. Wehner, G. K., 'Sputtering Yield Data in the 100–600 eV Energy Range', General Mills Report No. 2309, July 1962.
27. Tarng, M. L. and Wehner, G. K., *J. Appl. Phys.*, 1971, **42**, 2449.
28. Dahlgren, S. D. and McClanahan, E. D., *J. Appl. Phys.*, 1972, **43**, 1514.
29. Palmberg, P. W., *Anal. Chem.*, May 1973, **45**, 549A.
30. Davies, J. A., Presented at 15th Ann. Conf. Mass Spectrom., Denver, 1967; see also: Atomic Energy of Canada Ltd. Report AECL-2757, Chalk River, Ont., 1967.
31. McHugh, J. A., Private Communication, 1973.
32. Slodzian, G. and Hennequin, J. F., *Compt. Rend.*, 1966, **263B**, 1246.
33. Castaing, R. and Hennequin, J. F., *in* 'Advances in Mass Spectrometry', *ed.* A. Quayle, Vol. v, Institute of Petroleum, London, 1972, pp. 419–24.
34. Schroeer, J. M., *Vacuum*, 1972, **22**, 603.
35. Schroeer, J. M., Rhodin, T. N. and Bradley, R. C., *Surf. Sci.*, 1973, **34**, 571.
36. Benninghoven, A. *Z. Naturforsch.*, 1967, **22A**, 841.
37. Andersen, C. A., *Int. J. Mass Spectrom. Ion Phys.*, 1969, **2**, 61.
38. Andersen, C. A., *Int. J. Mass Spectrom. Ion Phys.*, 1970, **3**, 413.
39. Andersen, C. A., *Anal. Chem.*, 1973, **45**, 1421.
40. Lewis, R. K., Morabito, J. M. and Tsai, J. C. C., *Appl. Phys. Lett.*, 1973, **23**, in press.
41. Beske, H. E., *Z. Naturforsch.*, 1967, **22A**, 459.
42. Jurela, Z., *in* 'Atomic Collision Phenomena in Solids', *ed.* D. W. Palmer, M. W. Thompson and P. D. Townsend, North-Holland Publishing Co., 1970, pp. 339–49.
43. Benninghoven, A. and Mueller, A., *Phys. Letters*, 1972, **40A**, 169.
44. Benninghoven, A., Plog, C. and Treitz, N., Presented at Int. Conf. on Ion-Surface Interaction, Garching, 1972.
45. Krohn, V. E., Jr, *J. Appl. Phys.*, 1962, **33**, 3523.
46. Socha, A., *Surf. Sci.*, 1971, **25**, 147.
47. Morabito, J. M. and Lewis, R. K., *Anal. Chem.*, 1973, **45**, 869.
48. McHugh, J. A., Paper 98, Pittsburgh Conference, Cleveland, Ohio, 1973.
49. Goff, R. F. and Smith, D. P., *J. Vac. Sci. Technol.*, 1970, **7**, 1.
50. Goff, R. F., *J. Vac. Sci. Technol.*, 1973, **10**, 355.
51. Suurmeijer, E. P. Th. M. and Boers, A. L., *J. Phys. E*, 1971, **4**, 663.
52. Andrew, R., Riley, M., Armour, D. G. and Carter, G., *Vacuum*, 1972, **22**, 587.
53. Eckstein, W. and Verbeek, H., Max-Planck-Institut für Plasmaphysik Report *IPP* 9/7, Garching, 1972.
54. Heiland, W. and Taglauer, E. *J. Vac. Sci. Technol.*, 1972, **9**, 620.
55. Brongersma, H. H. and Mul, P. M., *Surf. Sci.*, 1973, **35**, 393.
56. Wheatley, G. H. and Caldwell, C. W., Jr, *Rev. Sci. Instr.*, 1973, **44**, 744.
57. Herzog, R. F. K., Liebl, H. J., Poschenrieder, W. P. and Barrington, A. E., Geophysics Corp. of American Report No. 65-7-N, 1965.
58. Herzog, R. F. K., Poschenrieder, W. P., Ruedenauer, F. G. and Satkiewicz, F. G., Presented at 15th Ann. Conf. Mass Spectrom., Denver, 1967.
59. Andersen, C. A. and Hinthorne, J. R., *Science*, 1972, **175**, 853.
60. Liebl, H., *Int. J. Mass Spectrom. Ion Phys.*, 1971, **6**, 401.
61. Liebl, H., *Messtechnik*, 1972, p. 358.
62. Witmaack, K., Maul, J. and Schulz, F., *Int. J. Mass Spectrom. Ion Phys.*, 1973, **11**, 23.
63. Schubert, R. and Tracy, J. C., *Rev. Sci. Inst.*, 1973, **44**, 487.
64. Coburn, J. W. and Kay, E., *Appl. Phys. Letters*, 1971, **18**, 435.
65. Oechsner, H. and Gerhard, W., *Phys. Letters*, 1972, **40A**, 211.
66. Benninghoven, A., *Surf. Sci.*, 1973, **35**, 427.
67. Benninghoven, A., *Z. Phys.*, 1970, **230**, 403.
68. Evans, C. A., Jr, and Pemsler, J. P., *Anal. Chem.*, 1970, **42**, 1060.
69. Pawel, R. E., Pemsler, J. P. and Evans, C. A., Jr, *J. Electrochem. Soc.*, 1972, **119**, 24.
70. McHugh, J. A., *Rad. Eff.*, 1973, submitted.
71. Strehlow, W. H. and Smith, D. P., *Appl. Phys. Lett.*, 1968, **13**, 34.
72. Ruedenauer, F. G., *Int. J. Mass Spectrom. Ion Phys.*, 1971, **6**, 309.
73. Werner, H. W., *Vacuum*, 1972, **22**, 613.
74. Werner, H. W. and DeGrefte, H. A. M., *Surf. Sci.*, 1973, **35**, 458.

INSTRUMENTS AND TECHNIQUES

Chairmen
J. H. BLOCK
(Fritz-Haber-Institüt der Max Planck-Gesellschaft, Berlin, Germany)

C. BRUNNEE
(Varian M.A.T. GmbH, Bremen, Germany)

N. R. DALY
(Atomic Weapons Research Establishment, Aldermaston, U.K.)

D. HENNEBERG
(Max-Planck-Institüt für Kohlenforschung, Mülheim/Ruhr, Germany)

R. S. LEHRLE
(University of Birmingham, U.K.)

38

New Mass Spectrometer for Isotopic Analysis of Small Gas Samples

By N. I. BRIDGER, R. D. CRAIG and
J. S. F. SERCOMBE

(*VG-Micromass Ltd., Winsford, Cheshire, U.K.*)

THE new mass spectrometer, called the Isotope Micromass 602C, is a double collector of 6·2 cm radius, 90° deflection, permanent magnet instrument for the analysis of H/D, C, N, O and S isotopes, using hydrogen, nitrogen, carbon dioxide (for C, O) or sulphur dioxide gases.

Design objectives included compatibility with computer coupling (digital ratio readout), the analysis of very small samples (30 μg of carbonate), complete automation after the sample admission and negligible corrections for H_3^+, peak tails, cross mixing, residual background, memory (zero enrichment), sample pressure.

For most applications an absolute accuracy, to 95% confidence, in enrichment values (δ) of $1\cdot0^0/_{00}$ for H_2/HD and $0\cdot1^0/_{00}$ for the other gases is acceptable. The instrument was therefore designed to give a reading reproducibility (σ_{10}) of 20% of the above values while each of the six prime correction terms were to be less than 10% of the same values.

MASS ANALYSER AND VACUUM SYSTEM

The mass spectrometer has two separate analyser heads, number one for H/D and number two for the other gases (*see* Fig. 1). Changeover from H/D analysis to any of the other gases, or vice versa, takes only a few minutes and involves changing the chopper amplifier connections and operating two valves on the inlet line.

The H/D ion source is designed to give a low basic H_3^+ contribution, typically 12 ppm of the H_2^+ ion beam when at 3×10^{-9} A. This 12 ppm is subsequently electronically reduced by a correction signal so that for a 2:1 change in hydrogen pressure the effective H_3^+ correction factor lies below 1·001. The basic H_3^+ contribution is sufficiently constant for the electronic correction control to require setting only once per day.

The 'other gases' ion source is designed for high sensitivity so that the

FIG 1   Twin analyser heads.

analyser tube operates at low pressure, typically $2 \cdot 10^{-7}$ torr for an m/e 44 ion beam from CO_2 of $4 \cdot 10^{-9}$ A. This low pressure gives small scattering of the 3000 V ions, thus the peak tail correction factor for all the gases is less than 1·001, see Fig. 2.

Peaks are flat-topped so that retuning is only required once per day. The collector comprises two deep Faraday buckets of gold so that charging effects are negligible.

The analyser vacuum system comprises a 170 litre/sec polyphenyl ether oil diffusion pump with baffle but no liquid nitrogen trapping, a room temperature molecular sieve foreline trap, automatic magnetic isolation valves for protection with a two-stage rotary pump. Bakeout is to 250°C by oven. Typical residual vacuum, is 10^{-9} torr and the residual spectrum indicates a background correction factor for each of the gases of less than 1·001.

The vacuum system for the sample inlet is identical to the above.

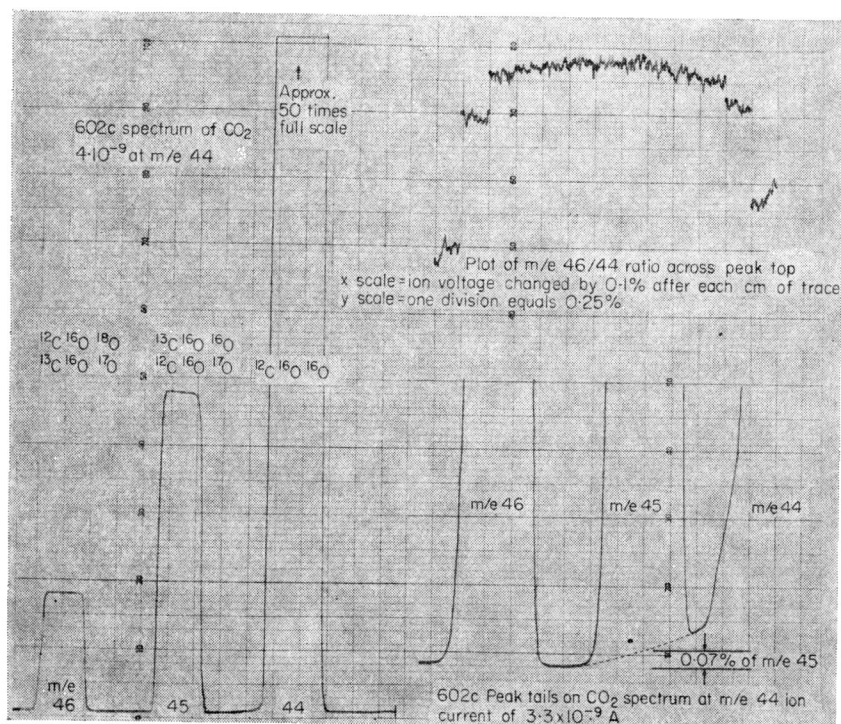

FIG. 2   Peak shapes and tail contribution.

INLET SYSTEM

The inlet system (*see* Fig. 3) is of the twin type, one half being for the sample and the other for the reference gas. Leaks are of the viscous type comprising 800 mm of 0·15 mm bore stainless steel capillary with adjustable crimp at the mass spectrometer ends.

The two leaks feed into the automatically operated, stainless steel change-over valve. This valve uses polished, spherical, tungsten carbide shut-off balls, it is solenoid operated and programmed from the mass spectrometer console. While one gas flows to the mass spectrometer the other gas flows to waste, and the pipework is carefully balanced so that on changeover the flow rate to the analyser remains constant. Thus there is minimal disturbance on changeovers. Cross-mixing correction factor for the valve is of the order of 1·001.

The changeover valve, leaks and inlet metering system are all bakable to 250°C.

The low volume inlet system valves are entirely of metal and are toggle operated to minimize fractionation. A particular feature of the inlet system is the cascade of selectable volumes from 0·5 cm^3 to 70 cm^3 to provide flexibility in handling samples from less than 0·01 atm cm^3 up to 4 atm cm^3.

For small sample work the leaks may be valved off while making the introduction and transferring the sample using the cold finger.

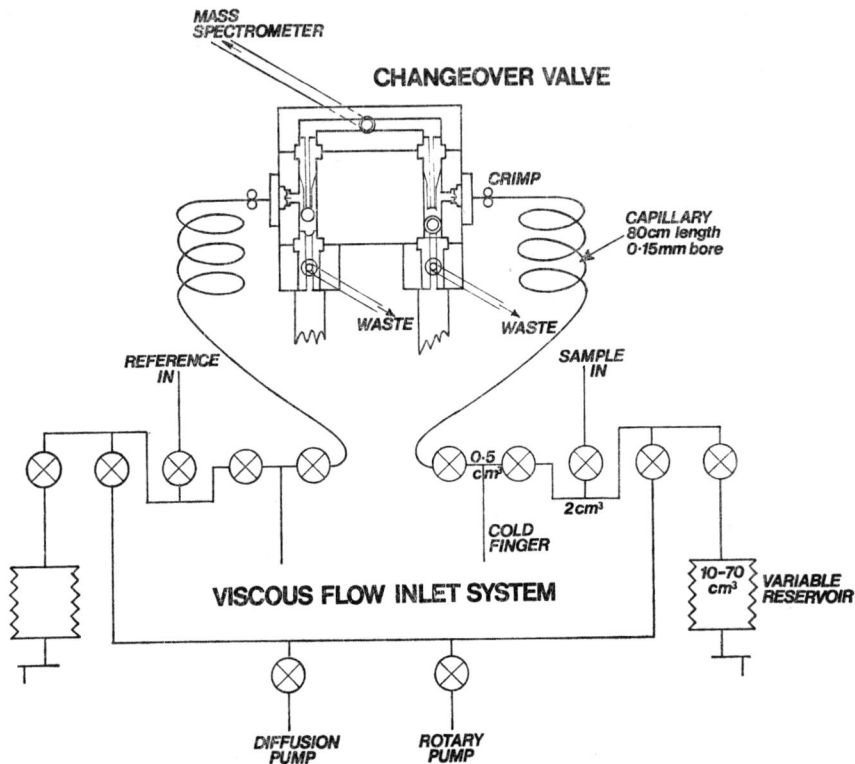

FIG. 3   Sample handling system.

Another facility is the provision to equilibrate the same sample between the two reservoir volumes by closing the common pump valve while opening the two individual pump-out valves. This facility is useful in balancing the leaks by means of the crimp adjustments.

RATIO MEASUREMENT SYSTEM

A true ratio measurement system is used, this means that enrichment readings are independent of absolute signal level. The principle is as follows.

The signal from the major ion beam chopper amplifier is used to determine the sensitivity of the Leeds and Northrup 250 mm potentiometric recorder thus the signal from the minor ion beam chopper amplifier can be fed so that the pen position records the true ratio position, invariant from the major ion beam intensity.

In practice the potentiometer slidewire forms part of a resistor network so that the first two digits of the ratio are read from decade switches.

The recorder serves a quadruple purpose. One, it provides by means of a second slidewire the signal for the digital integrator. This integrator averages the trace over a fixed integration time of 30 sec. Readings are to

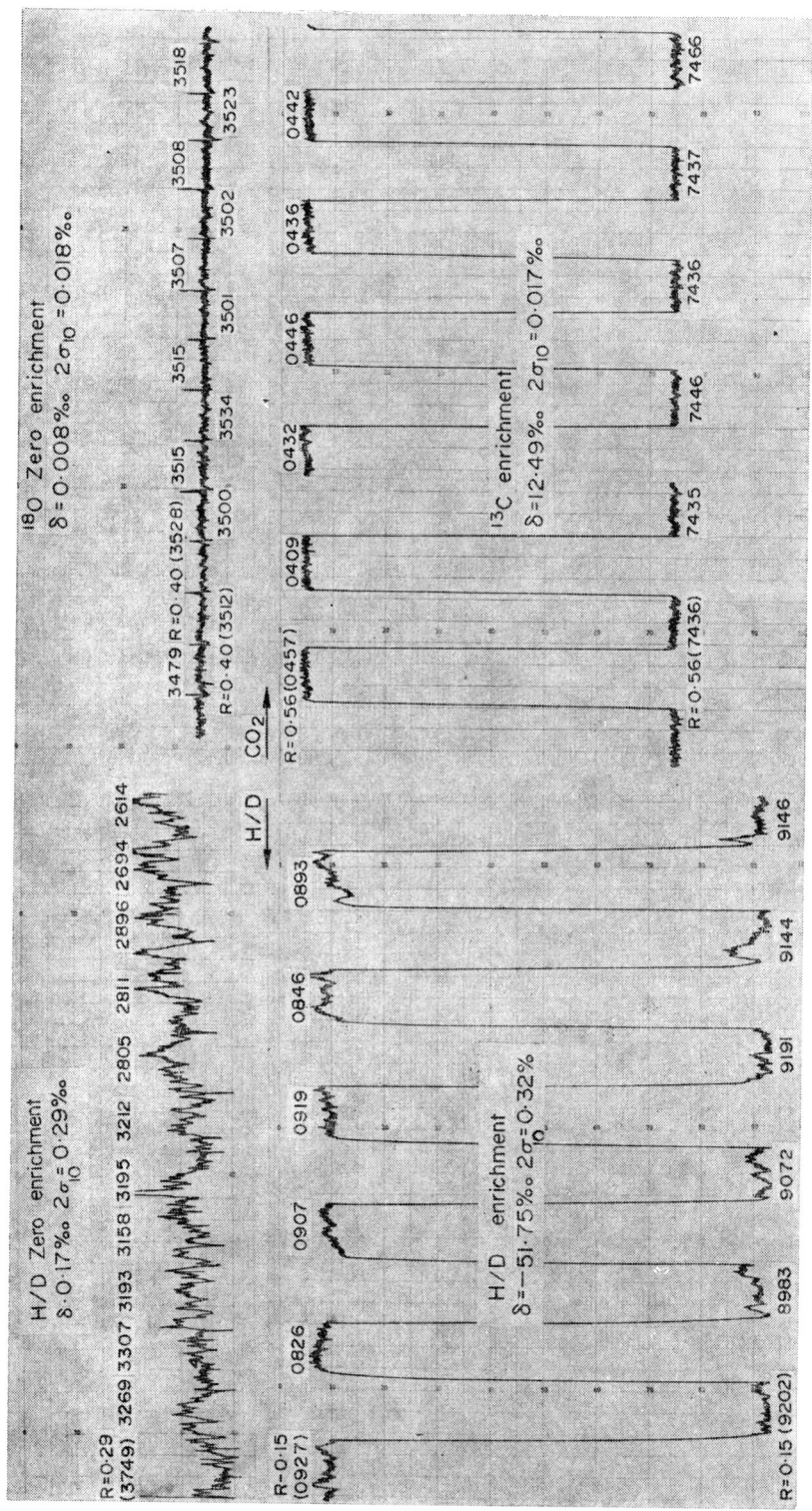

H/D Zero enrichment
δ = 0.17‰ 2σ₁₀ = 0.29‰

¹⁸O Zero enrichment
δ = 0.008‰ 2σ₁₀ = 0.018‰

¹³C enrichment
δ = 12.49‰ 2σ₁₀ = 0.017‰

H/D enrichment
δ = −51.75‰ 2σ₁₀ = 0.32%

FIG. 4 Typical ratio traces.

four figures, 9999 corresponding to 100% chart width. Integration commences at a preset time, usually 15 sec for clean samples, after sample changeover. However, this delay time is adjustable from 15 to 120 sec. to allow for full equilibration after changeover.

Two, the recorder provides square wave traces (*see* Fig. 4) showing the ratio change between sample and reference. This trace is not normally used for measurements but is a valuable check on sample contamination. For example a 'wet' CO_2 sample may show a slow climb to equilibrium which will require setting a longer than normal delay time before commencement of integration.

Three, the recorder provides more sensitive readings than the meters for balancing amplifier zeroes and adjusting peak heights.

Four, the recorder can be used in single beam mode to check peak shapes, peak tails, residual gas background and sample purity.

TABLE I

$m/e \dfrac{46}{44+45}$			$m/e \dfrac{3}{2}$		
Reference	*Sample*	*Diff.*	*Reference*	*Sample*	*Diff.*
400457			150927		
		6979			8275
	407436			159202	
		7027			8376
0409			0826		
		7023			8157
	7435			8983	
		7003			8076
0432			0907		
		7014			8165
	7446			9072	
		7000			8153
0446			0919		
		6990			8272
	7436			9191	
		7000			8345
0436			0846		
		7001			8298
	7437			9144	
		6995			8251
0442			0893		
		7024			8253
	7466			9146	
400441·6	mean	7005·0	150886·3	mean	8238·2

$$\sigma_1\ 0.038^0/_{00} \quad 2\sigma_{10} = 0.024^0/_{00}$$
$$\delta = 17.49^0/_{00}$$

$$\sigma_1 = 0.55^0/_{00} \quad 2\sigma_{10} = 0.35^0/_{00}$$
$$\delta = 54.6^0/_{00}$$

The quoted σ_1 values are the standard deviation for the enrichment (δ) as calculated from a single pair of ratio readings, while the σ_{10} values are the coefficient of variation associated with a group of ten differences.

Other features of the ratio system are the position switch to provide central location of the trace on the chart and the off-set decade to permit automatic measurement of enrichments up to $250^{\circ}/_{00}$ maintaining $0.05^{\circ}/_{00}$ accuracy.

The combined accuracy of the decade network, recorder slidewire and digital integrator system is better than $0.025^{\circ}/_{00}$ on enrichments up to $25^{\circ}/_{00}$.

Readout from the digital integrator is either by printer or by paper-tape punch for use with an off-line computer.

REPRODUCIBILITY

Typical sets of twelve readings for carbon dioxide $\{m/e\ [46/(44 + 45)]\}$ and hydrogen (m/e 3/2) are shown in Table I.

TRIANGULATION TESTS

Analysis of three carbon dioxide samples gave the following results:

^{18}O ratio 46/(44 + 45)

Reference	Sample	Enrichment ($^{\circ}/_{00}$)	$2\sigma_{10}$ $^{\circ}/_{00}$
A	B	−0·324	±0·016
A	C	−9·102	±0·017
B	C	−8·775	±0·014

and for ^{13}C ratio 45/44

Reference	Sample	Enrichment ($^{\circ}/_{00}$)	$2\sigma_{10}$ $^{\circ}/_{00}$
A	B	−2·650	±0·014
A	C	−4·129	±0·015
B	C	−1·471	±0·016

Triangulation test carried out to establish instrument linearity gave:

^{18}O

A–C as read .   .   .   . $-9.102^{\circ}/_{00}$
A–C obtained from A–B and B–C
$[\delta_A (C) = \delta_A (B) + \delta_B (C) + \delta_A (B) \cdot \delta_B (C)\ (\text{Craig 1957})]$
$-9.096^{\circ}/_{00}$

difference $0.006^{\circ}/_{00}$

^{13}C

A–C as read .   .   .   . $-4.129^{\circ}/_{00}$
A–C obtained from A–B and B–C .   .   .   . $-4.117^{\circ}/_{00}$

difference $0.012^{\circ}/_{00}$

Tests agree to well inside the instruments coefficient of variation.

RESULTS AND CONCLUSIONS

Preliminary results show the attainment of better than $1^0/_{00}$ overall to 95% confidence on H/D enrichments and to better than $0.1^0/_{00}$ on ^{13}C, ^{18}O and ^{15}N as measured on a routine basis. Results for ^{34}S have yet to be obtained.

Reading reproducibility is about a factor of five better than the overall precision. However, the difference may be attributed to external sources of error, such as those associated with sample preparation and fractionation during admission.

Correction factors are generally of the order 1·001, or less, and thus may be ignored for δ values less than $50^0/_{00}$ for H/D or less than $5^0/_{00}$ for the other gases.

The smallest samples analysed to date are of carbon dioxide (for ^{18}O) from 30 μg of carbonate.

APPENDIX

Results From Other Laboratories

TABLE IIa

Measurements of Internal Standard versus Working Standard, Showing Reproducibility and Re-introduction of H_2 Samples.

Analysis no.	Date	$\delta D \,{}^0/_{00}$
137	4.4.73	−152·6
158	13.4.	−152·2
167	17.4.	−151.5
180	18.4.	−151·7
192	18.4.	−152·4
194	19.4.	−152·4
203	24.4.	−152·4
213	24.4.	−153·4
223	25.4.	−151·9
228	26.4.	−151·2
235	26.4.	−151.1
258	30.4.	−151·3
261	30.4.	−151·2
272	3.5.	−151·6
281	4.5.	−151·5
309	16.5.	−152·4
321	17.5.	−153·1
329	22.5.	−150·5
367	30.5.	−152·2
368	31.5.	−151·9
402	6.6.	−151·8
403	7.6.	−152·6
440	12.6.	−151·8
467	15.6.	−152·5
503	19.6.	−151·8
526	20.6.	−152·6
527	21.6.	−152·5
550	25.6.	−152·6
551	28.6.	−150·4
572	3.7.	−152·0
573	5.7.	−151·7
600	6.7.	−150·8
614	9.7.	−149·5[a]
626	10.7.	−151·3

Mean value: $\delta D = -151·91 \,{}^0/_{00}$.

$$\sigma_1 = \left(\frac{\Sigma(x_1 - \bar{x})^2}{n-1}\right)^{\frac{1}{2}} = 0·71$$

$$\sigma_m = \frac{\sigma}{n^{\frac{1}{2}}} = 0·13$$

[a] Excluded from the average.
Results contributed by Dr. R. Gonfiantini, I.A.E.A., Vienna.

TABLE IIb

Repeated Measurements on the Same Sample (Analysis Made on Two Successive Days)—Values in $\delta D^0/_{00}$.

	1st Analysis	2nd Analysis	Difference
1	$-4\cdot7$	$-4\cdot5$	$-0\cdot2$
2	$-4\cdot6$	$-4\cdot7$	$0\cdot1$
3	$-5\cdot9$	$-5\cdot5$	$-0\cdot4$
4	$-5\cdot3$	$-4\cdot8$	$-0\cdot5$
5	$-5\cdot5$	$-5\cdot2$	$-0\cdot3$
6	$-6\cdot7$	$-6\cdot5$	$-0\cdot2$
7	$-50\cdot4$	$-49\cdot9$	$-0\cdot5$
8	$-48\cdot8$	$-48\cdot4$	$-0\cdot4$
9	$-50\cdot6$	$-50\cdot4$	$-0\cdot2$
10	$-51\cdot9$	$-51\cdot4$	$-0\cdot5$
11	$-179\cdot8$	$-179\cdot2$	$-0\cdot7$
12	$-183\cdot7$	$-183\cdot8$	$0\cdot1$
13	$-182\cdot7$	$-181\cdot8$	$-0\cdot9$
14	$-422\cdot7$	$-421\cdot3$	$-1\cdot4$
15	$-421\cdot4$	$-420\cdot5$	$-0\cdot9$
16	$-419\cdot3$	$-418\cdot4$	$-0\cdot9$
17	$+6\cdot5$	$+5\cdot8$	$0\cdot7$
18	$-0\cdot8$	$-1\cdot5$	$0\cdot7$
19	$+2\cdot9$	$+2\cdot1$	$0\cdot7$
20	$+2\cdot9$	$+2\cdot0$	$0\cdot9$
21	$+1\cdot6$	$+0\cdot9$	$0\cdot7$
22	$+2\cdot2$	$+2\cdot6$	$-0\cdot4$
23	$-26\cdot3$	$-27\cdot0$	$0\cdot7$
24	$+1\cdot7$	$+1\cdot8$	$-0\cdot1$
25	$-16\cdot0$	$-16\cdot6$	$0\cdot6$
26	$-8\cdot3$	$-8\cdot6$	$0\cdot3$
27	$-16\cdot0$	$-16\cdot3$	$0\cdot3$

$\sigma_1 = 0\cdot61\%$

Dates: 1–16, 21–22/3/1973
17–22, 28–29/3/1973
23–27, 17–18/4/1973

Results contributed by Dr. R. Gonfiantini, I.A.E.A., Vienna.

TABLE III

Reproducibility on Re-introduction of CO_2 Samples Values in $^0/_{00}$.

Aliquot	$\delta\ ^{13}C$
1	−9·79
2	−9·80
3	−9·81
4	−9·74
5	−9·75
6	−9·69
7	−9·73

$\sigma_1 = 0{\cdot}04^0/_{00}$.　　Mean $= -9{\cdot}76^0/_{00}$.

Date	$\delta\ ^{13}C$	$\delta\ ^{18}O$
9.1.73	−4·66	−7·13
10.1.73	−4·52	−7·18
11.1.73	−4·62	−7·02
17.1.73	−4·53	−7·14
18.1.73	−4·40	−7·01
29.1.73	−4·54	−7·19

Mean $= -4{\cdot}55^0/_{00}$.　　Mean $= -7{\cdot}11^0/_{00}$.
$\sigma_1 = 0{\cdot}09^0/_{00}$.　　　$\sigma_1 = 0{\cdot}08^0/_{00}$.

Results contributed by Dr. M. Baxter, Mr. T. D. B. Lyon, Mr. A. E. Fallick, University of Glasgow.

39

Ion Storage Mass Spectrometry: Applications in the Study of Ionic Processes and Chemical Ionization Reactions

By R. F. BONNER, G. LAWSON and
J. F. J. TODD†

(*University Chemical Laboratory, University of Kent, Canterbury, Kent, U.K.*)

R. E. MARCH‡

(*Chemistry Department, Trent University, Peterborough, Ontario, Canada*)

THE results described in this paper were obtained with a novel system in which ions may be created and stored within a three-dimensional quadrupole ion trap (the Quistor)[1] and mass analysed by subsequent ejection and passage through a conventional quadrupole mass filter.

EXPERIMENTAL SYSTEM

The Quistor has the geometry of a hyperboloid of two sheets combined with a hyperboloid of one sheet which form end-cap and ring electrodes respectively. A detailed description of the construction and operation of an early model has been given[2] in which the ion source of a EAI (Process Analysers Inc.) Quad 250A was replaced by a Quistor fabricated from wire mesh and mounted on an adjustable bellows assembly. In the present work, a stainless steel Quistor has been directly attached to the mass filter (Fig. 1) and mounted vertically in the vacuum system. Ions are created by bombardment with a beam of electrons which enter through a hole in the ring electrode, and under appropriate conditions are maintained in stable orbits by a radiofrequency oscillating field developed between the ring and end-cap electrodes. From the theory of operation of the Quistor[2,3] it can be shown that in the absence of d.c. potentials applied between the electrodes the range of m/e-values stable

† Author to whom correspondence should be addressed.
‡ On leave at the University of Kent April–June 1973.

within the trap is given by

$$\frac{m}{e} = \frac{2V_0}{4\pi^2 q_z z_0{}^2 \Omega^2}$$

where V_0 is the amplitude of the oscillating potential (frequency $2\pi\Omega$ radians sec^{-1}) applied between the end-cap electrodes and the ring, z_0 is the closest distance between the end-caps and q_z has values between 0 and 0·9.

FIG. 1   Photograph of the Quistor mounted in place of the ion source of an EAI (Process Analysers Inc.) Quad 250A mass filter. Ions are created by bombardment with a beam of electrons passing through holes in the ring electrode from the filament mounted at the right-hand side. The end-cap nearest the filter is pierced with a series of holes to allow the ions to be ejected for mass analysis.

A block diagram of the system is shown in Fig. 2 together with a representation of the pulse train employed to create, store and eject the ions. Typically, electrons with a nominal energy of 70 eV are admitted to the Quistor in pulses 20 μsec wide, and the ions ejected after a variable storage time by applying a pulse of $+80$ to $+100$ V magnitude to the left-hand end-cap. With a repetition frequency of 200 Hz storage times ranging up to *ca.* 4 ms are possible. The value of Ω is fixed at 2·63 MHz and V_0 is in the range 300–900 V. Other modes of operation are possible, such as continuous ionization with and without ejection, in the presence or absence of r.f. fields; these have been described elsewhere.[2]

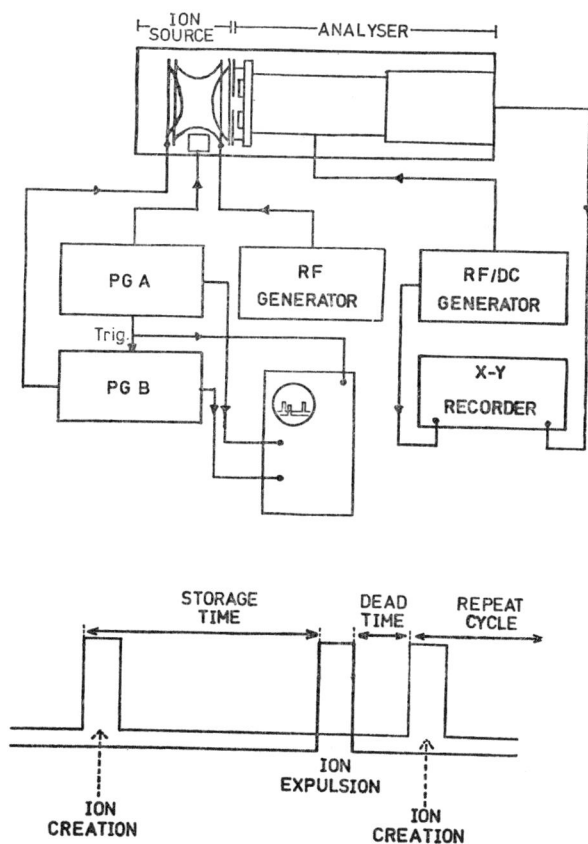

FIG. 2 Schematic layout of the system together with the pulse train for ion creation, storage and ejection. Pulse generator PG A opens the gate for the electron beam to traverse the Quistor, and triggers PG B whose delay determines the storage time; ions are ejected by the positive pulse from PG B being applied to the left-hand end-cap. The radio-frequency oscillating potential is fed to the ring electrode only.

OPERATIONAL CHARACTERISTICS

At first sight it might appear that ionic processes occurring within the Quistor could hardly resemble those within a conventional ion source on account of the high value of r.f. potential applied to the device. Thus it might be expected that the mean ion kinetic energy would be several hundreds of electron-volts. This is, however, not the case. The dynamics of the motion of trapped ions have been extensively studied by Dehmelt and co-workers,[4] whence it has been shown that under the influence of the inhomogeneous oscillatory electric field there is a net restoring force on the ions towards the centre such that they are constrained within pseudo-potential wells of depths $\bar{D}_r$ and $\bar{D}_z$ in the r- and z-directions respectively (Fig. 3). The result is that the influence of the applied r.f. field averages to zero and the ions oscillate with secular frequencies

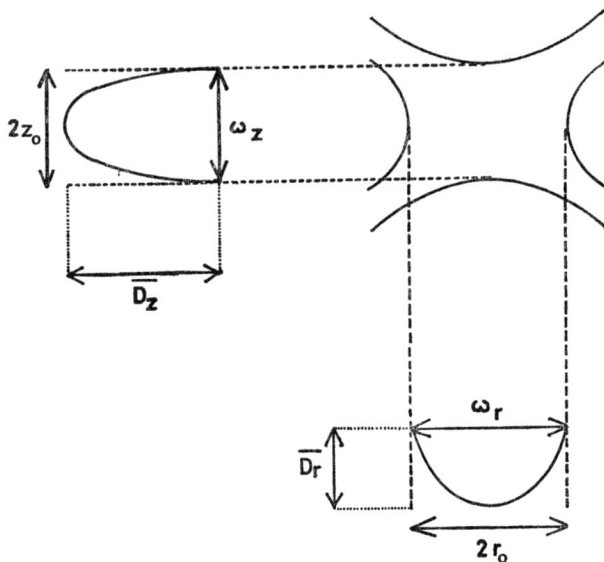

FIG. 3   Representation of the pseudo-potential wells, depths $\bar{D}_z$ and $\bar{D}_r$, retaining ions within the Quistor (after Dehmelt)[4]. In the absence of applied d.c. fields $\bar{D}_z = 2\bar{D}_r \times \omega_z$ and ω_r are the respective secular frequencies of the ion motion within the wells.

ω_r and ω_z, imparting a Lissajous-type motion to the ions.[5] This is analogous to the long period beat motion of ions within quadrupole mass filters.[6]

The value of $\bar{D}_z$ is given by

$$\bar{D}_z = \frac{eV_0{}^2}{16\pi^2 m z_0{}^2 \Omega^2}$$

so that for $V_0 = 360$ V, $z_0 = 7\cdot05 \times 10^{-3}$m and $\Omega = 2\pi \times 2\cdot63 \times 10^6$ radians $\sec^{-1}$ a $CH_4{}^+$ ion is trapped within a well of depth 14·4 V, giving a mean kinetic energy of motion of the ions equal to *ca.* 3·5 eV.[7] The relative yields of products from ion molecule reactions occurring in methane at *ca.* 2×10^{-4} torr suggest that the mean kinetic energy is indeed of this order.[8] From the depth of the pseudo-potential well it is possible to estimate the space-charge-limited number of ions which can be contained within the trap and the value obtained for this quantity is comparable both with the observed number of ions stored and with the results on ion traps reported by others.[9,10]

RESULTS AND DISCUSSION

The general method of approach in all the work described here has been to optimize the signal level and then to examine the spectra recorded at different storage times. In analysing the raw data it is important to verify that (i) the ions observed were indeed subjected to storage and that the gating of the electron beam was effective, and (ii) that there is a 'charge-balance' between the species detected, indicating that all the ions have been ejected from the trap. If this latter point is ignored it is possible for reactant ions to remain in the Quistor for several duty cycles and hence make a 'spurious' contribution to the yield of a product ion.

Metastable Ion Decay Processes

Some preliminary data obtained for the CF^+ and CF_2^+ ions observed in the mass spectrum of C_2F_6 are shown in Fig. 4. The suggestion is that the CF_3^+ ion is decaying to form CF_2^+ and CF^+, although at this stage it is not clear whether the CF_3^+ ion is in a long-lived metastable state or whether the decomposition is collisionally induced. The technique may be seen as being analogous to that of Tatarczyk and von Zahn[11] when the decomposition of mass-selected ions in a quadrupole filter 314 cm long was studied, except that in the present case longer decay times are possible. In principle the Quistor–Quadrupole-combination should permit the determination of the limiting values of kinetic shifts, by recording appearance potential curves for ions formed at longer and longer storage times at pressures low enough to minimise second-order processes.

Charge-transfer Reactions in Argon–Methane Mixtures

In order to examine the effects of ionization within the Quistor by charge-transfer, the argon–methane system was chosen, since this permits comparison with the results of other workers. Thus, Field, Franklin and Head[12] observed

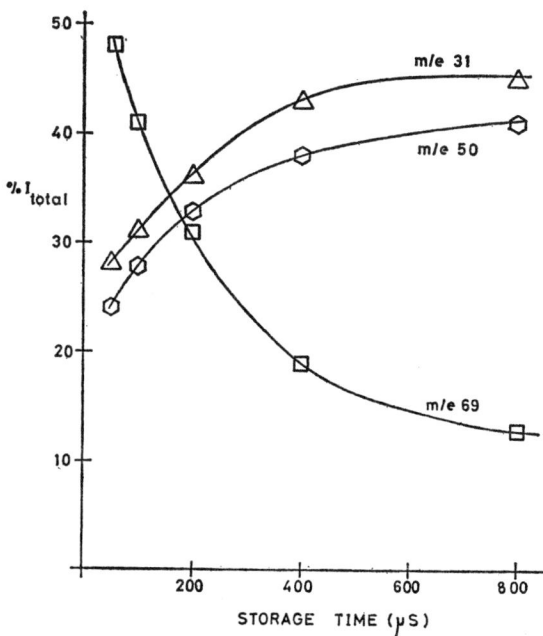

FIG. 4   Plots of percentage total ionization *versus* storage time for CF^+, CF_2^+ and CF_3^+ ions formed from C_2F_6.

charge exchange as the major process and the formation of

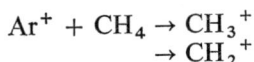

$$Ar^+ + CH_4 \rightarrow CH_3^+$$
$$\rightarrow CH_2^+$$

the ion-molecule reaction products ArH^+, ArC^+, $ArCH_2^+$ and $ArCH_3^+$. We failed to observe any such products and found that the formation of CH_3^+ and CH_2^+, together with ion-molecule reaction products therefrom, accounted entirely for the decreasing intensity of Ar^+ at longer storage times. Thus it was shown that any CH_4^+ (and resultant CH_5^+) ions formed arose only from direct electron-impact ionization of the methane, as noted by Field *et al.*[12] and von Koch.[13] The resultant kinetic plots are shown in Fig. 5 and the rate constants k observed by us for the loss of Ar^+ are as follows:

Argon:Methane ratio	$k/10^{-10}$ cm^3 molec^{-1} sec^{-1}
1:0·11	0·12 ± 0·04
1:0·48	1·8 ± 0·4
1:0·85	2·4 ± 0·6
1:3·44	10·0 ± 1·6

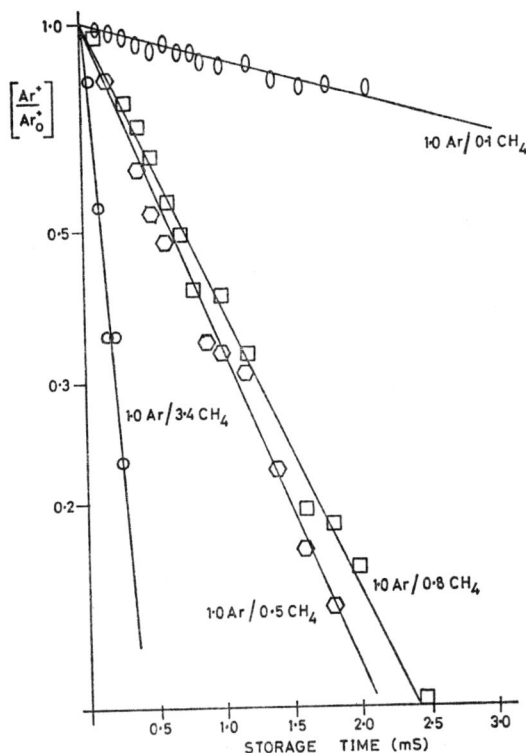

FIG. 5 Semi-logarithmic plot of (argon ion intensity–initial argon ion intensity) *versus* storage times for different argon–methane mixtures as indicated. For these studies the total pressure was in the range 1 to 2 × 10^{-4} torr, V_0 was 300 V and Ω was 2·75 MHz. The line through the hexagonal points is apparently out of order because the pressure for this run was at the top end of the range.

which are in good agreement with the range of values reported by Field et al.[12]

Chemical Ionization Reactions

From the pattern of yields obtained by us[8] for the ions formed from methane at pressures of ca. 10^{-4} torr, it was apparent that chemical ionization should be observable at storage times in excess of 2 ms. Figure 6 shows a series of mass

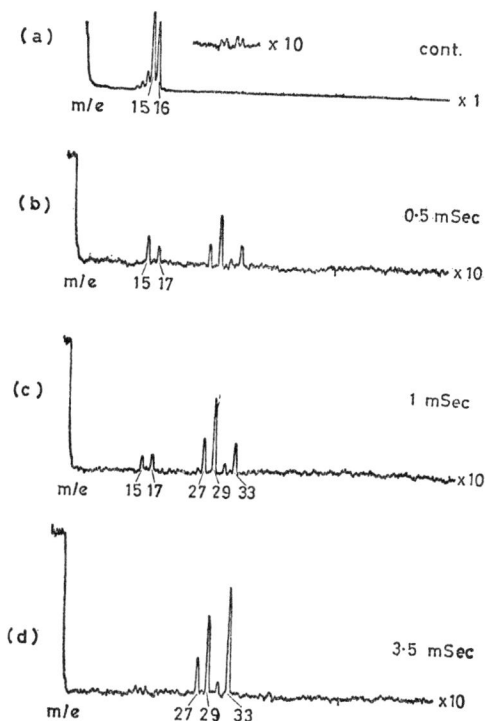

FIG. 6   A series of mass spectra of a methane–methanol (400:1) mixture at a total pressure of 10^{-4} torr run with continuous ionization and no storage (a) and pulsed ionization with increasing storage times (b)–(d).

spectra recorded for a methane–methanol (400:1) mixture at a pressure of 10^{-4} torr. Trace (a) was obtained under continuous ionization conditions and shows minor peaks due to the methanol. With pulsed ionization and increasing storage time (traces (b)–(d)) the growth in secondary product ions is clearly visible, eventually dominated by the species $CH_3OH_2^+$ (m/e 33). The indications are that the Quistor provides a viable means of producing chemical ionization at a total pressure of some 10^4 times lower than that normally employed.

Other Types of Investigation

A number of other kinds of experiment can be carried out with the Quistor–Quadrupole combination. For example, we have observed the process of 'spectrum simplification' when consecutive ion-molecule reactions are allowed to occur in a pure sample at 10^{-4} torr and storage times of 3·5 ms. An entirely different investigation has been the use of a double ionization pulse technique to cause the sequential ionization of trapped neon ions, in a manner analogous to the electron beam trapping method employed by Baker and Hasted[14] and by Redhead.[15]

ACKNOWLEDGMENT

The early stages of this work were carried out with the aid of grants from the Science Research Council to whom we are grateful.

REFERENCES

1. Lawson, G. and Todd, J. F. J., Mass Spectroscopy Group Meeting, Bristol, U.K., 1971, Abstr. No. 44.
2. Lawson, G., Bonner, R. F. and Todd, J. F. J., *J. Phys. E: Scientific Instruments,* 1973, **6**, 357.
3. Dawson, P. H. and Whetten, N. R., *Adv. Electronics Electron Phys.,* 1969, **27**, 59.
4. Dehmelt, H. G., *Adv. Atom. Molec. Phys.,* 1967, **3**, 53 and references cited therein.
5. Wuerker, R. F., Shelton, H. and Langmuir, R. V., *J. Appl. Phys.,* 1959, **30**, 342.
6. Lawson, G. and Todd, J. F. J., *Chem. Brit.,* 1972, **8**, 373.
7. Details of the calculation of mean ion kinetic energies are to be published elsewhere.
8. Bonner, R. F., Lawson, G. and Todd, J. F. J., *Int. J. Mass Spectrom. Ion Phys.,* 1972/3, **10**, 197.
9. Dawson, P. H., Hedman, J. W. and Whetten, N. R., *Rev. Sci. Instrum.,* 1969, **40**, 1444.
10. Harden, C. S. and Wagner, R. E., *Edgewood Arsenal Reports* 1971, EASP 100–93 and EATR 4545.
11. Tatarczyk, H. and von Zahn, U., *Z. Naturforsch.,* 1965, **20a**, 1708.
12. Field, F. H., Franklin, J. L. and Head, H. N., *J. Amer. Chem. Soc.,* 1962, **84**, 1118.
13. von Koch, H., *Arkiv Fysik,* 1965, **28**, 529.
14. Baker, F. A. and Hasted, J. B., *Phil. Trans. Roy. Soc.,* 1966, **A261**, 33.
15. Redhead, P. A., *Can. J. Phys.,* 1971, **45**, 1791.

40
A Programmable Multiple Ion Monitor for Mass Fragmentography with a Quadrupole Mass Spectrometer

By WILLIAM F. FIES and MICHAEL S. STORY

(Finnigan Corporation, Sunnyvale, California, U.S.A.)

A multiple ion monitor has been designed for use in mass fragmentography monitoring up to eight mass fragments, on a time sharing basis. Each mass fragment is monitored by a separate channel module, each channel module is independent of other channel modules except for trigger input and trigger output pulses. During the channel sample time, a signal integration is performed. A sample and hold circuit maintains the final value of integration until this channel is again selected. Integration times 1, 10, or 100 msec are available. Mass is set by means of a Kelvin–Varley control giving a resolution of 3·6 degrees per amu if a 0–1000 mass range is being used. Active low pass filtering and recorder bucking are provided in each channel. This allows optimum signal-to-noise ratio even for mass peaks of widely different amplitudes. Any combination of channel units may be run at any time. Channels may be placed in hold, one at a time while provision is made for injecting sweep voltage during set up. Electrical stability and gas chromatographic data are presented.

The quadrupole mass spectrometer, when used as a detector for a gas chromatograph is particularly suited to mass fragmentography because of its ability to switch in submillisecond times over a mass range in excess of 50 to 1. Mass fragmentography is carried out by substituting a multiple ion monitor for the normal sweep generating circuitry in the mass spectrometer system. The multiple ion monitor supplies to the mass spectrometer a series of voltage steps that set the mass spectrometer, on a time sharing basis, to each of the desired masses. In addition, the multiple ion monitor 'sorts out' the signals from the mass spectrometer's ion detector for each of these masses and routes them to separate recording devices. In order to take best advantage of the available information in the output signal of the mass spectrometer, it is also necessary to provide for some signal processing in the multiple ion monitor.

Since it was desired to give the operator the maximum amount of flexibility in setting the operating conditions, the concept of a front panel programmable multiple ion monitor (PROMIM) was developed.

To obtain the most flexibility from this technique, the instrument should have the following design criteria:

(1) Separate plug-in channel modules for each of the masses to be monitored.
(2) Independent sample time for each channel.
(3) Independent run, skip, hold switch for each channel.
(4) Mass setting voltage stable to better than 666 μV in an 8-hour period. This is equivalent to 0·05 amu in a 0–750 amu range.
(5) Sampling rates and response times to be fast enough for applications such as process control or respiratory analysis.
(6) Signal integration during sample time.
(7) Signal gain to be equal to 1 for all integration times.
(8) Active low pass filter in each channel.
(9) Recorder zero controls sufficient to cover the ± 10-volt output range.
(10) Output noise to the recorder due to sample and hold circuit switching less than 5 mV in a 500 Hz bandwidth at the recorder output.
(11) Electrometer range switching under independent control for each channel.

Item 11 required the design of a new FET electrometer system which could be switched at sufficiently high speeds to minimize the dead time required for settling. This electrometer has five ranges from 10^{-5} A/V to 10^{-9} A/V in 10:1 steps. Switching from any range to the 10^{-5}, 10^{-6}, or 10^{-7} can be carried out

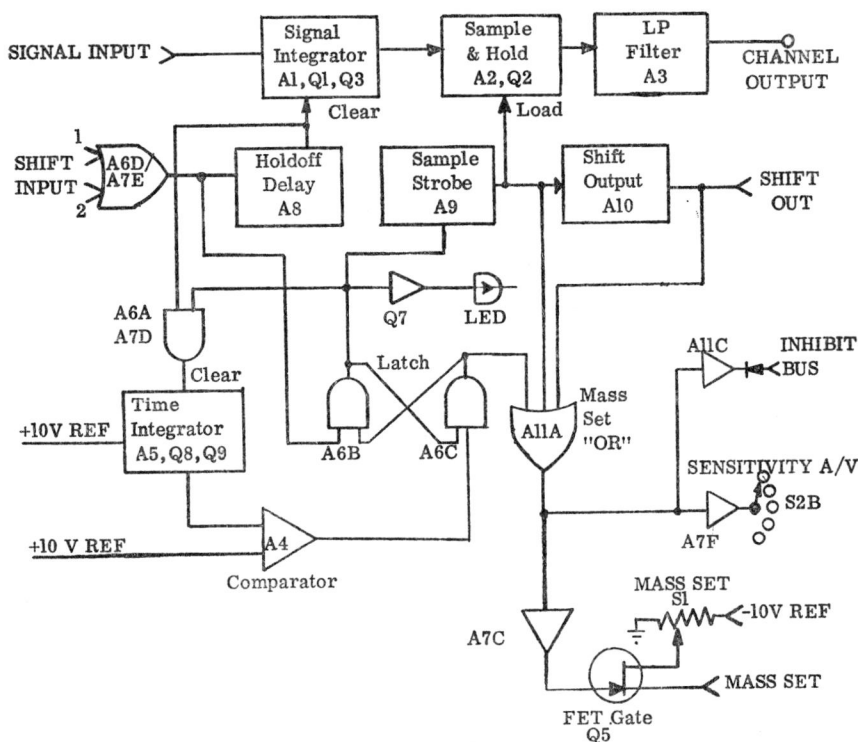

FIGURE 1

in 2 msec for 1 % settling. Switching to the 10^{-8} range requires 6 msec and switching to the 10^{-9} range requires 12 msec. These times are longer than the time required for the mass spectrometer (less than 0·05 msec) and have therefore established the duration of the settling time allowed by PROMIM when switching from one mass to the next.

Figure 1 is a block diagram of one channel of PROMIM. In brief, the sequence for operation for each channel is as follows:

A shift input pulse is received at one of the shift input terminals and triggers a latch and also the hold-off delay. The hold-off delay clears the signal integrators, turns on the mass set voltage, activates the electrometer range selection circuitry and deactivates the time integrator. The hold-off delay is 2 to 12 msec depending on the electrometer range settings as mentioned above. At the end of the hold-off delay, the signal integrator and the time integrator are simultaneously activated. The signal integrator integrates the signal while the time integrator times the signal integration. Integration times are 1, 10 or 100 msec, chosen by a front panel switch.

Signal integration time constants are chosen so that the PROMIM has a gain of 1 regardless of the integration time being used. At the end of the integration time, the signal integrator has a voltage at its output which is equal to the time average of the voltage present at the signal input. When integration is complete, the latch is re-set. This triggers a 25 msec strobe generator which

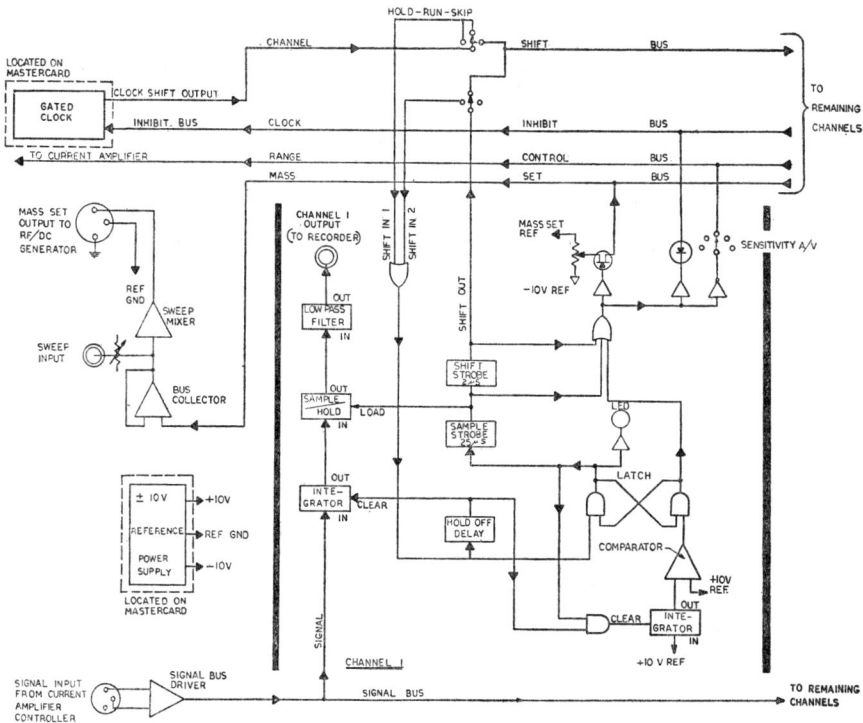

FIGURE 2

loads the integrator output into a sample and hold memory. At the end of the sample strobe a 2 msec shift pulse is generated which is sent out to the next channel. The output of the sample and hold circuit is applied to a low pass filter. The background bucking voltage is added to the signal at this point. From the low pass filter the signal goes to the recorder.

Figure 2 is a block diagram of the complete instrument. The run, skip, hold switch for each channel is located on the main frame of the instrument. By means of this switch the channel may be skipped or it may be placed in 'run', the normal mode, or it may be placed in 'hold', in which the channel retriggers itself instead of the next channel. This function is required to allow the operator to set the channel to the proper mass. Since the channel retriggers itself, the signal processing functions continue to operate in that channel. The inhibit buss and gated clock serve to ensure startup at turnon and also to prevent more than one channel from operating at a time. A sweep control is provided on the main frame of the front panel so that sweep may be mixed with mass set voltage for adjustment purposes.

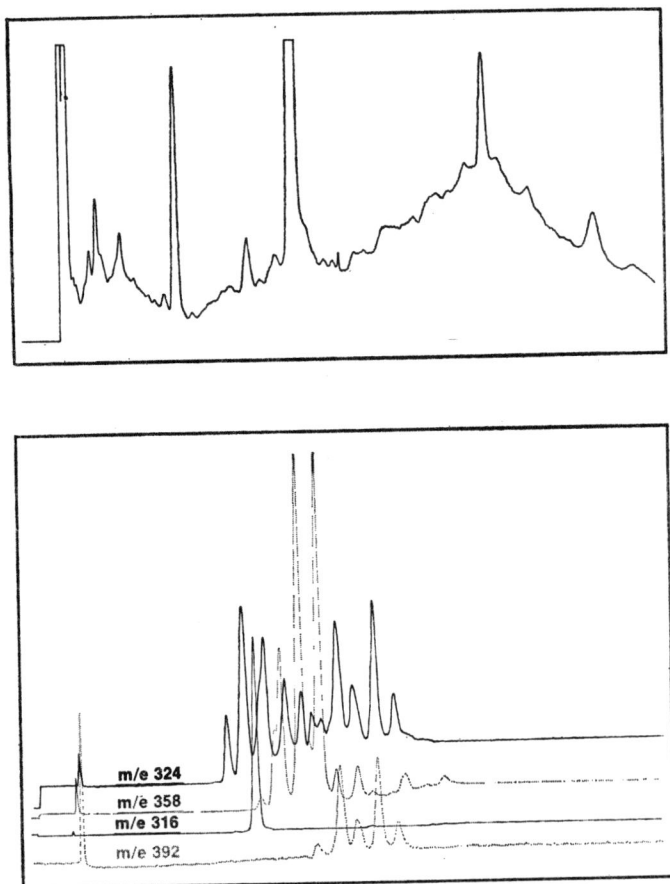

FIG. 3   Top: total ion monitor chromatogram of sturgeon ovary extract. Bottom: multiple ion chromatogram of sturgeon ovary extract.

In operation, one expects the optimum performance will be obtained if the weak mass peaks are monitored at high electrometer sensitivities and long integration times, while the strong mass peaks are monitored with shorter integration times and lower electrometer sensitivities between 0·5 and 5 V, thus allowing a good margin over the switching noise from the sampler hold circuit. These expectations have been supported by both electrical tests and experimental data taken under actual operating conditions.

Drift data on the mass set voltage over a six-day period has shown the worst case drift was 404 μV, equivalent to about 0·03 of an amu in a 0 to 750 amu range.

Figure 3 (lower) is a fragmentogram of an extract of sturgeon ovary, m/e 324, 358 and 392 are characteristic of polychlorinated biphenyls while m/e 316 is a major peak in the spectrum of DDE. For comparison, Fig. 3 (upper) is a total ion monitor chromatogram of the same sample. Figure 4 is a fragmentogram of two prostaglandins. This represents 1 ng of each sample on column.

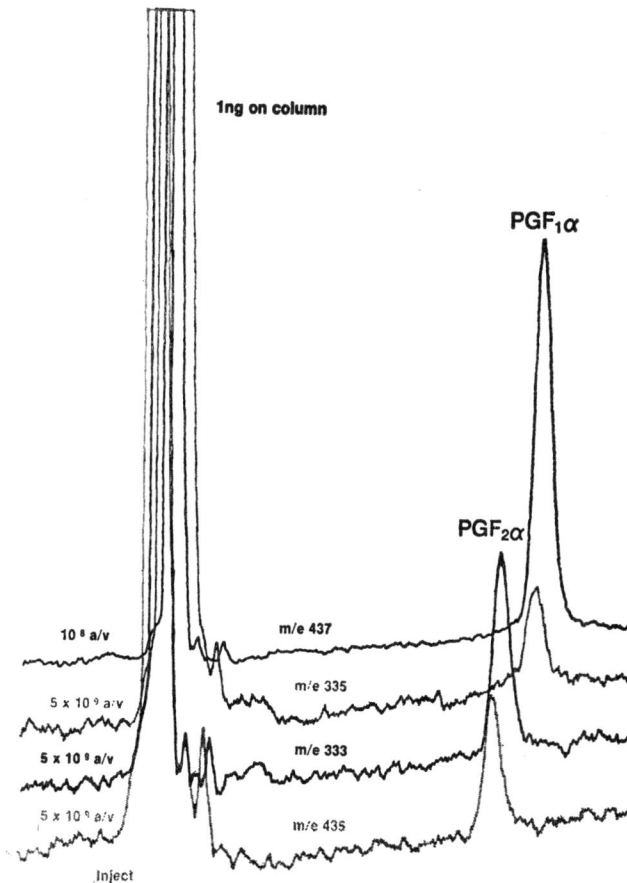

FIG. 4   Multiple ion chromatogram of PGF$_{1\alpha}$ and PGF$_{2\alpha}$ derivatives. (Samples courtesy of Dr. R. W. Kelly, Medical Research Council, Edinburgh, Scotland.)

The solvent peak appears in these fragmatograms because the pressure in the mass spectrometer went so high during the solvent peak that the base line rose over the whole spectral range.

With this instrumentation, the fragmentography technique will allow the quantitative analysis of trace amounts of compounds in complex mixtures even in the presence of interfering substances.

41
The Use of a Tandem Electron Impact–Chemical Ionization Source on AEI MS9 and MS12 Mass Spectrometers

By A. M. HOGG

(*Chemistry Department, University of Alberta,
Edmonton, Alberta, Canada*)

INTRODUCTION

THE use of chemical ionization (CI) as an adjunct to normal electron impact (EI) mass spectrometry is now well established.[1] However, the modifications required to permit a source to operate at the high pressures (0·1–1 torr) needed for CI render it unsuitable for EI work. The electron entrance aperture is usually reduced to a small hole of the order of 0·3 mm diameter which severely attenuates the intensity of the electron beam crossing the source, and the ion exit slit is also reduced in area so that the efficiency of ion extraction is much lower than in a conventional EI source. If the flow of reactant gas is simply cut off and only sample allowed to evaporate into the source, EI operation is possible and sensitivity per microgramme of sample is not greatly reduced because a higher pressure of the sample is generated in the tightly sealed source for the same rate of sample evaporation. This, however, can give rise to unwanted ion-molecule reactions (which have been referred to as 'self chemical ionization'), one example being the ready protonation of amines. Other drawbacks to the use of a CI source for EI are the difficulty in obtaining a stabilized beam of electrons through a small aperture at energies of 70 eV and lower; where EI spectra are usually recorded; and the necessity of having all sample inlet systems tightly sealed to the source block, making source removal considerably more difficult.

Over the last two years CI work in this laboratory has been done on a modified MS12[2] where the design allows the CI source and the original EI source to be quickly interchanged, and recently a viton O-ring has been used in place of the normal gold gasket on the source flange, which further reduces the time required for changeover. When this source is used on a similarly modified MS9 it is necessary to admit a suitable reference compound if exact mass measurement is to be accomplished, and it has been found that no suitable general purpose reference compound is available for use with reactant gases such as ammonia, which can often give virtually one peak

$(M + NH_4^+)$ spectra.[3] There is also no simple means of admitting reference compound to the source.

A possible solution to these difficulties is to combine two ionization chambers, one operating at high pressure for CI and one at low pressure for EI, in the same housing; each provided with its own filament and inlet system. It should then be possible to optimize the conditions in each chamber, and there will be the added advantage that conventional reference compounds (e.g. perfluorokerosene and perfluorotributylamine) can be admitted to the EI chamber in order to measure the mass of sample ions from the CI chamber. A similar tandem ion source for use with a quadrupole GC/CIMS system in order to obtain alternate or simultaneous CI and EI spectra of GC effluent has been reported,[4] but the problems involved are considerably different from those encountered with a high voltage, high resolution instrument.

A further requirement for an acceptable source is that it should have the same overall dimensions as those of the standard A.E.I. MS9/MS12 source and that all existing inlet systems should mate with the EI chamber so that the original source could be replaced at any time.

EXPERIMENTAL

The tandem ion source constructed for use on the MS9/MS12 mass spectrometers is shown in the cut-away sketch (Fig. 1). It is virtually identical in overall dimensions to a standard A.E.I. EI source and uses basically the same support system and beam centering plates.

The source block is constructed in three parts:

(a) A single plate on top, in which is machined a somewhat narrowed EI chamber with an ion exit slit and four sample ports.

(b) A thin plate, in the centre of which is mounted the narrow CI chamber exit slit.

(c) A massive main block which contains the CI chamber with its electron entrance hole, sample entrance port, and guide; on which are mounted the two filament assemblies.

Electrical insulation between (a) and (b) is maintained by means of plates of mica, so that the complete CI source acts as a repeller for the EI source. (a), (b) and (c) are held together by two bolts, the heads of which are insulated from (a) by Rulon washers, and which are threaded into (c).

The filament for the EI chamber is a standard A.E.I. tungsten type while that for the CI chamber differs only in its slightly modified mounting plate. A trap assembly for the EI chamber is mounted behind the CI filament and heating is by means of two Hotwatt $\frac{1}{8}$ in $\times$ 2 in stainless steel encapsulated 35-W heaters.

Critical dimensions of the source are given in Table I.

A source flange with an integral insertion lock and eight dual feedthroughs supports the source and the CI sample, and reactant gas is admitted through this additional lock via a specially constructed probe which has been described

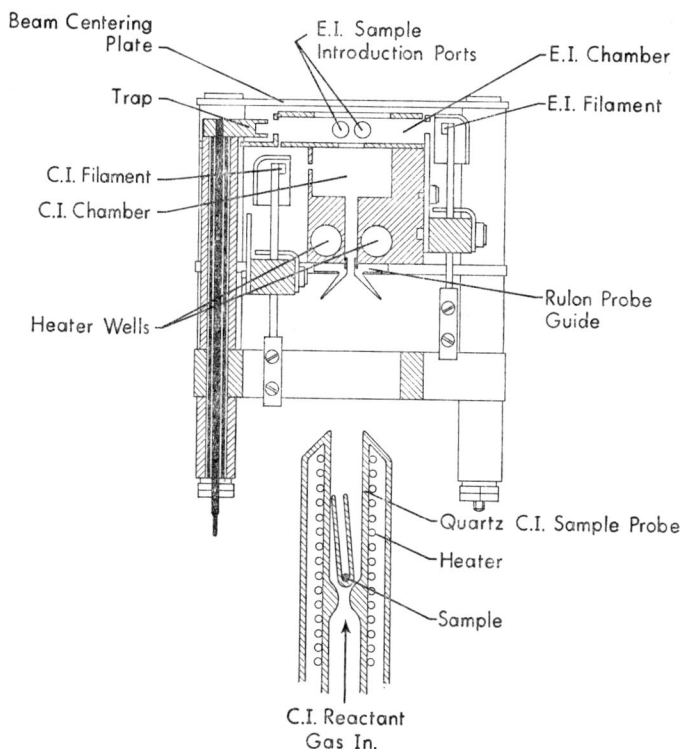

FIGURE 1

previously.[2] Two of the ports on the EI chamber connect with the existing insertion lock and the heated glass re-entrant leading to the several reservoir inlet systems used on the instrument.

An additional floating power supply provides stable current to the CI chamber filament and also up to 500 V electron energy. A minor modification to the existing source supply chassis permits a repeller voltage to be maintained with zero emission from the EI chamber filament.

TABLE I

	EI chamber	CI chamber
Chamber dimensions	19 mm long × 6·5 mm wide × 3·5 mm deep	10 mm diam. × 5·4 mm deep
Electron entrance aperture	3·1 mm × 0·6 mm	0·3 mm diam.
Ion exit slit	10 mm × 1·5 mm	5 mm × 0·05 mm
Distance from electron beam to ion exit slit	0·9 mm	2 mm

RESULTS

The performance of the tandem source has been studied in the three possible modes of operation; electron impact only, chemical ionization only, and simultaneous EI/CI.

Electron Impact—The source has been used in this mode in place of the standard source for running routine samples and despite the large ion exit slit it has been found to have a sensitivity about 20% higher than that of the standard source. This is somewhat surprising but is probably not significant because changes in sensitivity of this magnitude between different EI sources or before and after cleaning of the same source have been experienced in the past and small differences in filament alignment relative to the source collimating magnets are thought to be responsible.

If reactant gas is passed through the CI probe at a rate which will give rise to about 0·3 torr in the CI chambers while at the same time reference compound is flowing into the EI chamber via the heated inlet, the sensitivity of the EI source for the reference ions drops to 35%. No significant evidence of ion-molecule reactions occurring in the EI chamber can be detected. Sample admitted via the CI probe together with reactant gas gives a normal EI spectrum superimposed upon the EI spectrum of the reactant gas.

Chemical Ionization—In this mode the source works well. However, there is no suitable standard source with which to compare it. The electrode potentials for repeller and Y and Z deflection required for maximum sensitivity are virtually identical to those determined for the EI chamber but the beam centering differs slightly due, no doubt, to some misalignment between the CI chamber ion exit slit and the EI chamber electron beam.

Simultaneous EI/CI—When both electron beams are operated simultaneously and reagent gas is flowing the EI chamber operates as it does with reagent gas flowing and the CI filament off. However, no ions can be detected coming from the CI chamber until the repeller voltage is increased to +8V or more. This is now no longer optimum for the EI chamber and so a decrease in EI sensitivity occurs of about a factor of four. The transmission of ions from the CI chamber is very sensitive to repeller voltage and rather abrupt changes in the optimum voltage can occur for no readily apparent reason. It is, however, quite feasible to produce a spectrum where the CI spectrum of a sample is overlaid with its EI spectrum, the EI spectrum of the reactant gas, and the EI spectrum of a reference compound.

When mass measurement by peak matching is attempted using an EI reference ion and a CI sample ion it is found that the measured mass of the sample is dependent upon the repeller voltage used to extract it. It is possible to move a peak, as seen on the oscilloscope display, by as much as 20 ppm; increasing the repeller field leading to a lower apparent mass. At minimum repeller field just above the point at which the CI ions suddenly disappear the measured mass of a known CI ion is within the normal error in measurement.

DISCUSSION

The tandem source described here successfully overcomes the need for changing sources when going from EI to CI operation with no penalty in the form of lower sensitivity or loss of versatility of the EI source. In the CI mode it appears to have good sensitivity and further optimization is possible as more is learned about CI sources in general.

The goal of adequate simultaneous EI and CI operation has yet to be achieved, but it is likely that this can be done through relatively minor changes. Decreased length of the CI chamber exit slit, decreased depth of the EI chamber and a fine mesh over the EI chamber exit slit are among changes which are soon to be evaluated.

An error in determining the mass of CI ions when using EI reference ions was anticipated because of the likelihood that they might have significantly different energies, however, it was expected that the CI ions would have greater energy being formed at higher positive potential which should result in too high a measured mass. The opposite is apparently the case, but until the effects of such things as space charges in the two ionizing regions and field penetration into the EI chamber by the beam centering plates are further investigated, it would be unwise to offer any explanation for this.

As a last resort, it should be possible to switch some source electrode voltages in synchronization with the peak matching system in order to suppress the EI chamber beam while the CI peak is being scanned. This would, of course, preclude the use of a data system for mass measurement.

REFERENCES

1. Munson, B., *Anal. Chem.*, 1971, **43**, 28A.
2. Hogg, A. M., *Anal. Chem.*, 1972, **44**, 227.
3. Hogg, A. M. and Nagabhushan, T. L., *Tetrahedron Lett.*, 1972, **47**, 4827.
4. Arsenault, G. P., Dolhun, J. J. and Biemann, K., *Anal. Chem.*, 1971, **43**, 1720.

Discussion

J. Yinon (Weizmann Institute of Science, Israel): Is it possible that part of the error in mass determination is due to the fact that the ions formed in the chemical ionization source have to pass all the way through the magnetic field of the source, while the ions formed in the electron impact source have a shorter path to travel through that magnetic field.

A. M. Hogg: Ions from the CI chamber do indeed have to travel farther through the field of the source collimating magnets; however, they do not receive the full 8 kV acceleration until the same point as the ions from the EI chamber. Their relative deflection should thus be slight and is unlikely to have much effect on their apparent mass.

42
Analysis of Biopolymers by Curie-Point Pyrolysis in Direct Combination with Low Voltage Electron Impact Ionization Mass Spectrometry

By M. A. POSTHUMUS, A. J. H. BOERBOOM and
H. L. C. MEUZELAAR

(*F.O.M.-Instituut voor Atoom en Molecuulfysica, Amsterdam,
The Netherlands*)

INTRODUCTION

THE direct analysis of biopolymers is limited by the low volatility of these compounds. One existing technique used to overcome this difficulty is the degradation of the macromolecules; for example by hydrolysis and a consecutive derivatization to more volatile compounds. Another technique was reported as early as 1952 by Zemany.[1] He pyrolysed complex organic materials, including biopolymers such as albumin and pepsin, and subsequently analysed the pyrolysis products by mass spectrometry. This pyrolysis–mass spectrometry (Py–MS) technique developed only slowly thereafter.

A fast and reproducible pyrolysis technique (Curie–point pyrolysis[2]) applied directly in the vacuum of a quadrupole mass spectrometer greatly improved the speed, sensitivity and reproducibility of Py–MS.[3,4]

APPARATUS AND EXPERIMENTAL

In our apparatus (Fig. 1) a sample of about 20 μg of biological material is deposited on a 0·5-mm diameter wire of a ferro-magnetic alloy. In a high-frequency magnetic field this wire is heated up to the Curie temperature at which point the temperature stabilizes. Figure 2 shows some typical heating curves. An advantage of this technique is the short rise time and the reproducible and constant temperature. This is of crucial importance for the reproducibility of the pyrolysis. The influence of the temperature on the pyrolysis of insulin, for example, is illustrated in Fig. 3.

The pyrolysis products flow to the ion source of a mass spectrometer. After having passed the ionization region the remaining pyrolysis products are

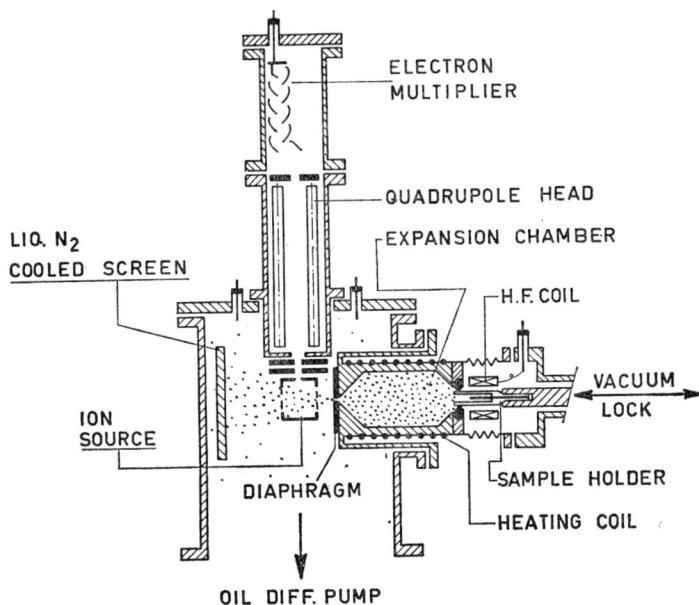

FIG. 1   Curie–point pyrolysis–mass spectrometer.

trapped on a liquid, N_2-cooled screen. The fast pyrolysis reaction causes a
rapid change in the gas pressure and composition, therefore a fast scanning
type of mass spectrometer is necessary. In our system we use a Riber QMM 17
quadrupole mass spectrometer. We investigated the mass range from 30
through 130 m.u. at a repetition rate of 10 spectra per second. The spectra
are summed and stored by a 1024 channel signal averager. The resulting
final mass spectrum is representative of the overall composition of the
pyrolyzate.

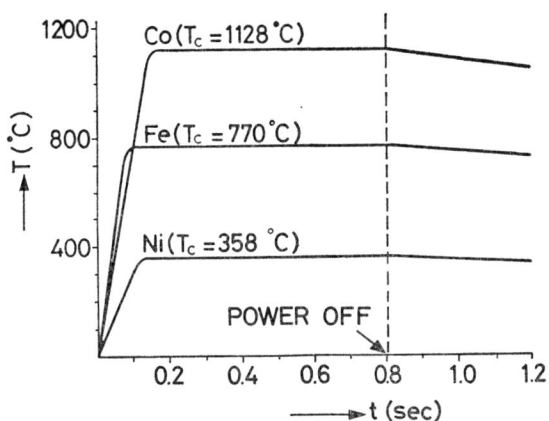

FIG. 2   Typical heating curves.

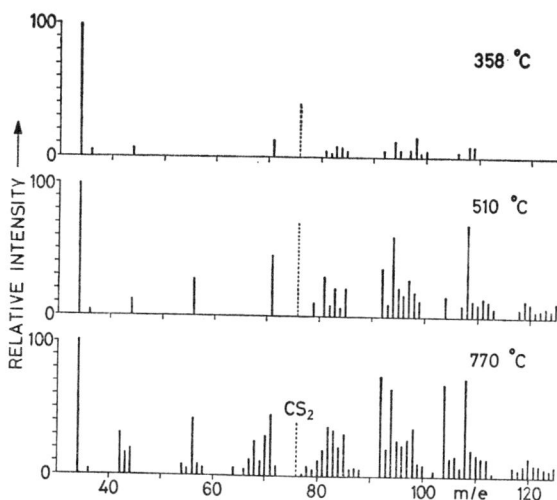

FIG. 3   Influence of pyrolysis temperature on product composition. Sample bovine insulin, suspended in CS_2. The spectra are normalized to 100 for m/e 34.

To obtain a more characteristic finger-print of the complex mixture of pyrolysis products, we applied the well-known, low voltage, electron impact ionization technique. This avoids the accumulation of non-characteristic fragments in the lower mass range, thus enhancing the detection of typical peaks in this range. *See* Fig. 4.

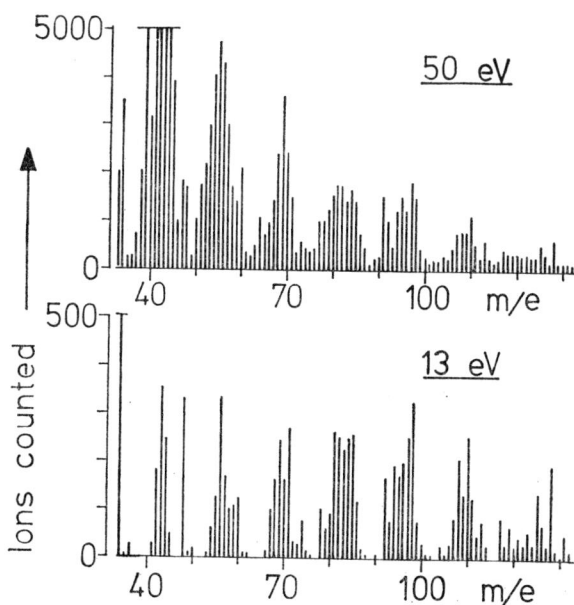

FIG. 4   Pyrolysis–mass spectra of streptococcus F_0 bacteria at 13 and 50 eV electron ionization energy respectively.

Trials with compounds commonly encountered in our pyrolyzates[5] such as acetamide, propionamide, furfuryl alcohol, benzonitrile and indole, indicate that ionization with electrons of 13–15 eV is a good compromise between minimum fragmentation and maximum yield of parent ions.

RESULTS AND DISCUSSION

In Fig. 5 the pyrolysis mass spectra of four bovine proteins are shown. Clearly these spectra show sufficient mutual differences to allow an unambiguous distinction. As these spectra show a large degree of reproducibility, they can be used for identification purposes.

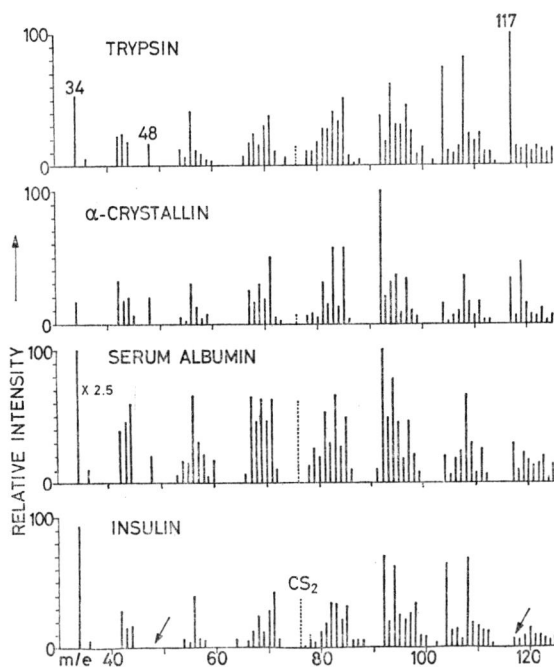

FIG. 5   Pyrolysis–mass spectra of bovine proteins.

Moreover, a definite amount of chemical interpretation of the spectra is possible. We confine our attention to the peaks 34 (H_2S), 48 (CH_3SH) and 117 (C_8H_7N).

Peaks 34 and 48 correspond with cystine, $(-SCH_2CHNH_2COOH)_2$, and methionine, $CH_3SCH_2CH_2CHNH_2COOH$, the only sulphur containing amino acids. Insulin lacks methionine, and in its spectrum the 48 peak is missing. This implies that CH_3SH is exclusively formed as a pyrolysis product of methionine. The low cystine content of α-crystallin in conjunction with the low m/e = 34 peak indicates that H_2S is mainly formed from cystine. Both facts are readily understood from the structural formulae of these compounds.

The large difference in the intensity of the peaks at m/e 117 in Fig. 5 is explained by the fact that insulin also lacks tryptophan, of which indole is a characteristic pyrolysis product. Nevertheless, a small peak at m/e 117 is still present, which is not unexpected as phenylacetonitrile has the same molecular weight as indole and is a pyrolysis product of phenylalanine.

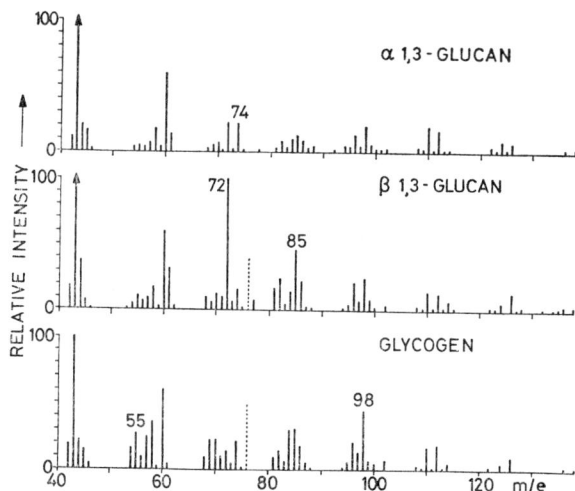

FIG. 6   Pyrolysis–mass spectra of glucose polymers.

As a second example in Fig. 6 the spectra are shown of α- resp β-1,3-glucan, and glycogen, which all are glucose polymers differing only in the type of linkage and branching. Though the chemical structure of these compounds is very similar, the differences in the spectra are obvious.

As to the sensitivity of the method, Table I shows the results of two insulin pyrolyses. We took 50×10^{-9} g of sample and confined our measurements to mass 34. The integrated signal of 50 ng insulin is about ten times higher than the background signal in the same time. The signal-to-noise ratio, however, is about 50. These facts indicate that the present system allows us to monitor a few large peaks at the nanogram level using the step scan mode of the mass spectrometer.

In considering Table I it should be kept in mind that these results were

TABLE I

Total Number of Ion Counts of Mass 34 (H_2S)[a]

	a	b
Permanent background (2×10^{-7} torr)	275	
Clean wire	275	330
Coated wire (solvent only)	330	300
Coated wire (ca. 50 ng insulin)	3 300	2 950

[a] During 10 sec after start of the pyrolysis. Ionization conditions: electron energy 13 eV, ionization current 1 mA. Pyrolysis conditions: 510°C, 1 sec.

obtained with an open-type ion source and 13 eV energy for the ionizing electrons. Moreover, at present quadrupoles having a more than ten times higher transmissivity are commercially available. Disregarding the difficulty inherent in handling nanogram samples it appears possible to improve even further the lower limit of detection.

ACKNOWLEDGMENTS

This research is sponsored by the Organization for Fundamental Research on Matter (F.O.M.) and the Dutch Ministry of Health.

REFERENCES

1. Zemany, P. D., *Anal. Chem.*, 1952, **24**, 1709.
2. Simon, W. and Giacobbo, H., *Chem. Ing. Techn.*, 1965, **37**, 709.
3. Meuzelaar, H. L. C. and Kistemaker, P. G., *Anal. Chem.*, 1973, **45**, 587.
4. Meuzelaar, H. L. C., Posthumus, M. A., Kistemaker, P. G. and Kistemaker, J., *Anal. Chem.*, 1973, **45**, 1546.
5. Schulten, H. R., Beckey, H. D., Meuzelaar, H. L. C. and Boerboom, A. J. H., *Anal. Chem.*, 1973, **45**, 191.

43

The Use of Synchrotron Radiation in Photoionization Mass Spectrometry and Photoelectron Spectroscopy: The Lure Project

By PAUL M. GUYON

(*LURE Molecular Physics Group ER 57, Orsay, France*)

WHEN a molecular system is excited above its ionization potential into a super-excited state, it might decay into ion + electron, ion fragments + electron, neutral fragments, or re-emit light. In order to observe the detailed mechanisms, one has to prepare the system into selected known states and observe the decay into the final states.

The synchrotron radiation from ACO (Orsay Electron Storage Ring) dispersed by a normal incidence monochromator at $\lambda > 350$ Å and a grazing incidence monochromator at $\lambda > 50$ Å gives a source of light of continuously variable energy and bandwidth which is used to selectively excite vibronic levels of molecules. The photoelectrons ejected in the process carry the energy $E_K(e^-) = h\nu - E_{n,v}$ (ion) where $h\nu$ is the absorbed photon energy and $\dot{E}_{n,v}$ (ion) the ion energy. The photoion thus formed might decay into fragments or re-emit light. Hence, detecting by a delayed coincidence technique the photoelectron energy analysed, and the corresponding ion mass analysed, one can observe the individual decay of vibrational levels of molecular ions.[1,2]

THE LURE PROJECT

In 1971, we submitted to CNRS a project of photoelectron photoion coincidence experiments using the synchrotron radiation from ACO. In 1972, this project developed into the LURE synchrotron radiation facility with the collaboration of atomic, molecular and solid state physicists. In 1975, this project will also encompass X-ray radiocrystallography and biology with the construction of a new 1·5 GeV positron electron storage ring: DCI.

ACO Characteristics

ACO is a 22 m long storage ring with eight bending magnets with 1·1 m radius of curvature. A packet of electrons 30 cm long 1·5 mm dia (FWHM) rotates at 13·6 MHz which gives light pulses of 1 nsec (FWHM) duration every 73 nsec. The light emitted has the well-known synchrotron radiation continuous distribution. It is usable from 5 Å to 10 000 Å and peaks at 20 Å when the beam is run at 540 MeV maximum energy. The maximum beam current is 100 mA. A lifetime of 30 hr was measured for a 30 mA beam. The light source is highly stable and the noise level $\Delta I/I$ is less than 10^{-5}. No radiation protection is necessary except during the injection period once or twice a day.

Light Pipes

A drawing of the pipe system is given in Fig. 1. A three-stage differential ion-pumped system isolates the ultra-high vacuum region of the storage ring 10^{-10} torr from the line 10^{-6}–10^{-7} torr. The ring is protected against any

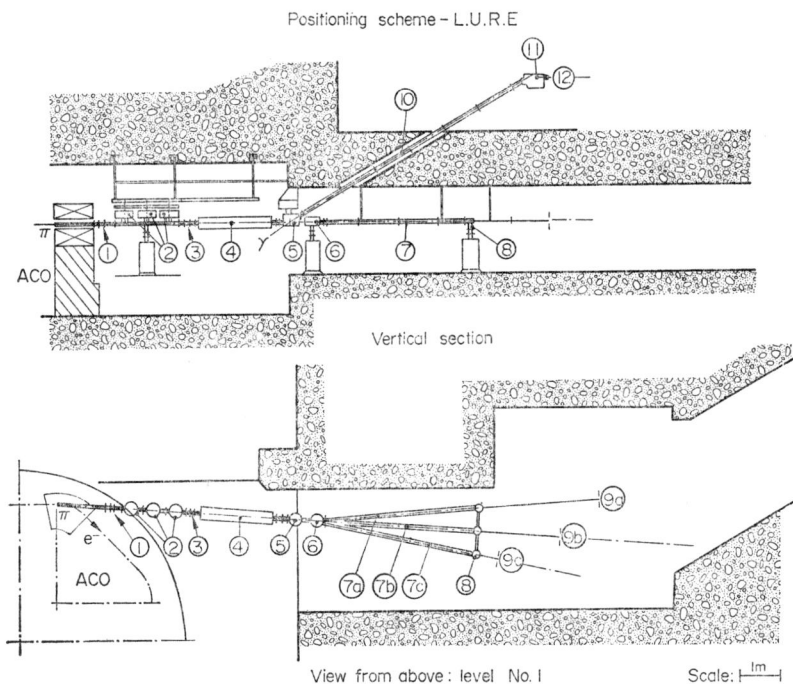

FIG. 1   Diagram of the Lure light pipes system.

1. Ultra-high VG vacuum valve.
2. Three stages differentially pumped.
3. VAT Fast acting valve.
4. Acoustical delay line.
5. First separation chamber.
6. Second separation chamber.
7. 3 horizontal beam lines.
8. Turbomolecular pump.
9. 3 lower level light ports.
10. Higher level beam line.
11. Cylindrical refocussing mirror
12. 2 upper level exit ports for normal incidence monochromators.

accidental opening to the atmosphere by the combination of an acoustical delay line and an ultra-fast acting valve (30 ms closing time).

The light beam has a rectangular shape with a 3 mrad vertical and 10 mrad horizontal aperture. It is separated into two parts by a toroidal mirror which refocusses 70% of it at the upper level, whereas, the remaining 30% can be directed simultaneously into two of the three lower level lines by two vertical mirrors. At the upper level, light can be alternatively directed into two lines and refocussed on the entrance slits of normal incidence monochromators by a cylindrical mirror.

Monochromators

For $\lambda > 350$ Å a 50-cm Seya Namioka has been used by Y. Petroff and R. Pinchaux between 300 and 3000 Å. A 1 m normal incidence McPherson 225 is already installed at the upper level for the photoion photoelectron experiment and a 1·5 m Pouey type monochromator with three exit ports and 0·1 Å resolution will be installed in 1974.

For $\lambda > 50$ Å a double grazing incidence monochromator has been designed by P. Dhez et al.[3] and will be installed in December 1973.

For the X-ray region, a crystal monochromator designed by C. Bonnelle and co-workers has been used on pipe No. 2.

ACKNOWLEDGMENTS

We are indebted to Professor Chabbal, Director of the CNRS who sponsored the project and to its principal contributors, namely: J. Appell, P. Dhez, C. Depautex, J. Durup, Y. Farge, P. Jaéglé, S. Leach, R. Lopez-Delgado, P. Marin, G. Morel, C. Vermeil and F. Wuilleumier.

REFERENCES

1. Brehm, B. and Puttkamer, E. Von., *Adv. Mass. Spectr.*, 1967, **4**, 591.
2. Eland, J., *Int. J. Mass. Spect. and Ion Physics*, 1972, **8**, 143; 1972, **8**, 153; 1972, **9**, 397.
3. International Symposium for Synchrotron Radiation Users; Jan. 1973.

Discussion

J. Eland (Oxford University, U.K.): Can you give an estimate of the usable flux at the exit of the monochromator?

P. M. Guyon: From the actual aperture of the light beam, 2×10^{-5} steradians, one calculates a flux of 10^{13} photons/Å × sec at 1000 Å for 100 mA circulating electron current. After the reflections at 75° incidence necessary to refocus the light on the entrance slit of the monochromator (400 microns image width), one loses about a factor of 2. To get 1 Å band width with the MP225 monochromator one has to use 100-micron slits and thus only one-fourth of the light will enter into the monochromator. Finally, the efficiency of the grating can be estimated to be about one per cent. Hence one expects to get around 10^{10} photons/Å × sec in the range 500–1000 Å. The loss in efficiency at shorter wavelengths is partly compensated for by the increase of the light output of the synchrotron radiation.

R. S. Berry (University of Chicago, U.S.A.): Are you planning to construct apparatus to study angular distributions of photoelectrons with the synchrotron source?

P. M. Guyon: Yes. The polarized character of synchrotron radiation is one of its most interesting properties. Since reflection polarizers are inefficient in the vacuum UV, F. Wuilleumier from Orsay and Professor Melhorn and Dr Schmidt from Freiburg are presently constructing an apparatus to study the angular distribution of photoelectrons from Auger transitions. For this experiment they will use a grazing incidence monochromator designed by P. Jaegle, P. Dhez *et al.* to disperse the light from the storage ring.

44
Advanced Virtual Image Double Focussing Mass Spectrometer

By H. MATUDA

(*Osaka University, Osaka, Japan*)

M. NAITO and M. TAKEUCHI

(*JEOL Ltd., Tokyo, Japan*)

INTRODUCTION

THE virtual image type mass spectrometer has two excellent features which enables it to obtain high sensitivity and resolution.

First, a much wider object slit can be used because of the very small image magnification of the apparatus.

Second, the total ion path can be considerably shortened. In order to obtain good focussing, however, it is quite important to reduce the second-order image aberrations. In the JMS-06 (90°–60° model) reported at the fifth Brussels conference,[1] the focussing properties of second order were not sufficient.

To improve focussing of second order, two different kinds of double focussing mass spectrometers of the virtual image type were designed, constructed and examined. A brief description of these instruments is given in the following.

The influence of fringing fields are taken into consideration in the calculation of their ion optics.

THEORETICAL CALCULATION

The position of an ion at the detector is given by:

$$y_B = r_m(B_1\alpha + B_2\beta + B_{11}\alpha^2 + B_{12}\alpha\beta + B_{22}\beta^2 + B_{33}\alpha_z^2 + B_{35}\alpha_z\zeta + B_{55}\zeta^2)$$

The first-order double focussing requires that $B_1 = B_2 = 0$, which can be satisfied by choosing suitable d and l''_m values. Besides, any two second-order coefficients can be eliminated by selecting suitable values of R'_e and R'_m.

In this calculation we eliminated B_{11} and B_{12}, which are coefficients concerning the radial aperture angle α.

TABLE I

The Field Parameters and Coefficient of Image Aberrations of Virtual Image Double Focussing Mass Spectrometers

	φ_e	φ_m	r_m/r_e	l_e'/r_e	d/r_e	l_m''/r_e	X_e	X_m	R_e'	EK_1	c
(a) 90°–60° model	90°	60°	0·8	1·052	0·7901	1·011 3	0·303	0·59	1·117 5	0·0	2·65
(b) modified 90°–60° model	90°	60°	1·0	1·052	0·648 5	1·628 2	0·303	0·92	−0·790 7	1·327 9	2·65
(c) 118°–60° model	118°	60°	1·0	0·4	0·5311	1·882 3	0·224	1·13	−0·890 7	1·514 4	2·8

	α^2	$\alpha\beta$	β^2	ζ^2	$\alpha_z\zeta$	α_z^2	YY_9	YZ_9
(a) 90°–60° model	0	35	−26·0	−6·0	−22·0	−21·0	−3·8	−5·9
(b) modified 90°–60° model	0	0	−2·43	−5·07	−15·53	−11·58	−3·82	−6·45
(c) 118°–60° model	0	0	−3·44	0·57	−0·48	−2·87	0·71	−3·36

$c = \dfrac{r_e}{R_e}$, $R_e' = \left(\dfrac{dR_e}{dr}\right)_{r=r_e}$ $EK_1 = \dfrac{r_m}{R_m'}$, R_m': The radius of curvature at the boundary of magnetic poles.

$r_e\zeta_0$ and α_z: Axial displacement and inclination at object slit of an ion respectively.

Y_z: Vertical position at final collector slit of an ion.

$Y_z = (YY_9\zeta_0 + YZ_9\alpha_z)r_m$.

Computer calculation was done on many virtual image, double-focussing optical systems, taking into consideration the influence of the fringing field, and changing the values of parameters, φ_e, r_m/r_e, φ_m, R'_e, EK, etc. The results of computer calculation are listed in Table I. The first row is for the 90°–60° model developed previously. The second row is for the model 90°–60° modified from the previous 90°–60° model. It was given the same parameters for the electric field as the former except for R'_e. Most of the parameters for

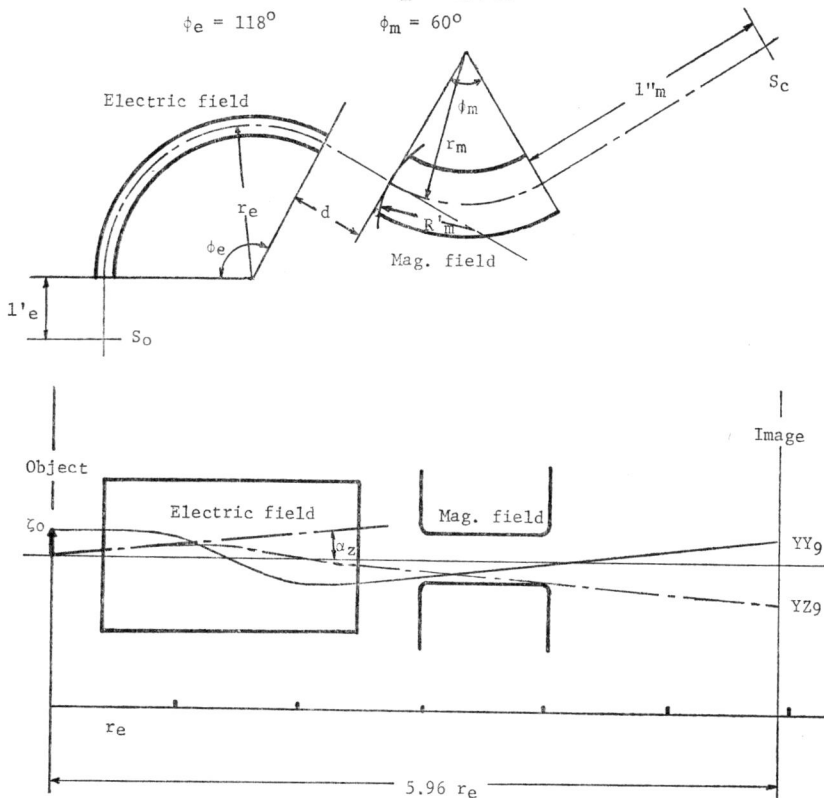

FIG. 1   Schematic diagram of the ion path of new virtual image DFMS.

the magnetic analyser were slightly changed so that the condition $B_{11} = B_{12} = 0$ might be fulfilled and at the same time the axial aberration coefficients B_{33}, B_{35}, B_{55} might be reduced by selecting a combination of suitable values for R'_e, R'_m and r_m/r_e. The third row gives the field parameters and aberration coefficients of the new 118°–60° model which has been constructed very recently for trial purposes.

The schematic diagram of this system is shown in Fig. 1, horizontal ion beam path in (a) and vertical ion beam path in (b). The features of this new

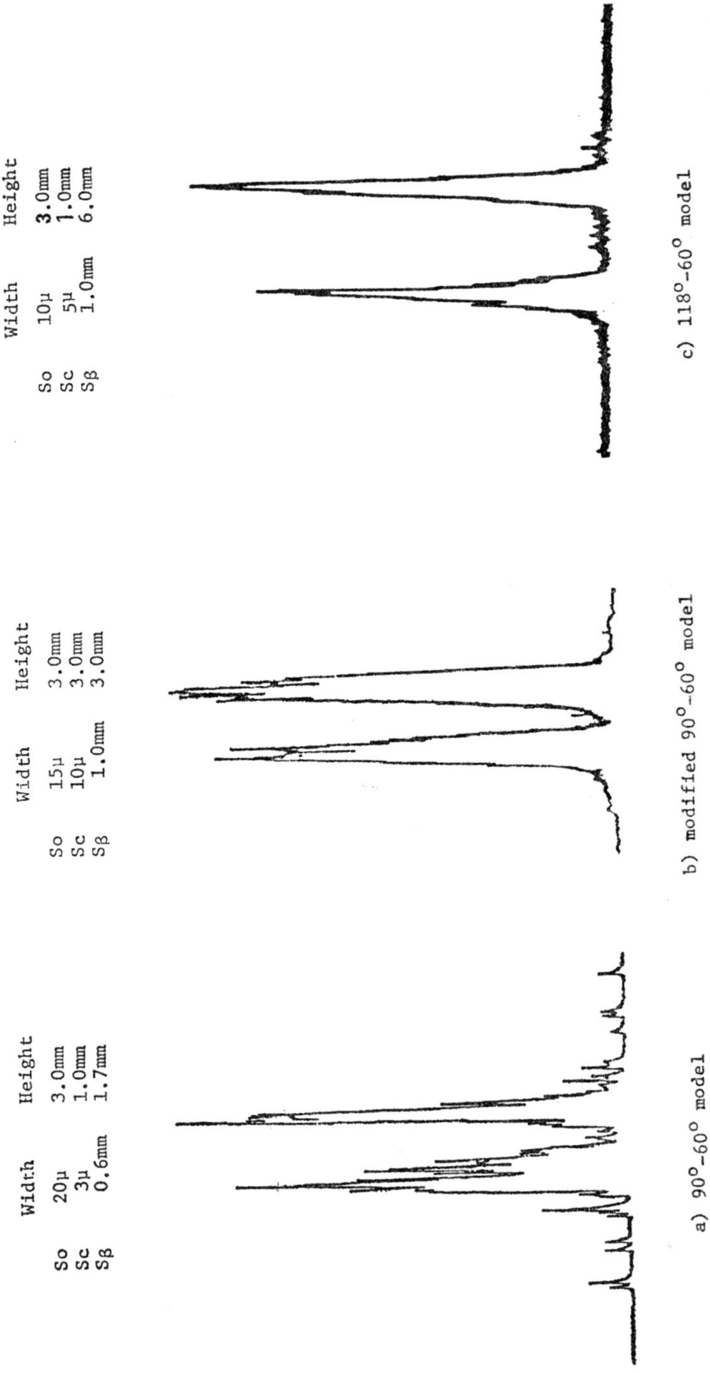

FIG. 2   Resolution measured by $C_5H_5N-^{13}CC_5H_6$ doublet. $M/\Delta M = 9700$. Sample benzene–pyridine mixture. S_o, object slit; S_c, collector slit; S_β, β-slit.

system are as follows:

(1) Improved vertical focussing property: The ion beam coming from the electrostatic field has a cross-point between the magnetic field and the collector, because of using a greater deflection angle of the electrostatic field, $\varphi_e = 118°$. Therefore the intensity enhancement of ion beam at the collector may be expected to be 6 or 7 times that of the 90°–60° models, which can be estimated from the value of YY_9 and YZ_9 listed in Table I, by using the equation for Y_z given in the footnote of Table I.

(2) The image aberration coefficients for off median plane are reduced to about one tenth those of the 90°–60° models. These small aberration coefficients for off median plane allow the use of a larger-height aperture in this instrument than in the others, without reducing the high resolution performance. This leads to its good sensitivity and high resolution.

APPARATUS AND EXPERIMENTAL RESULTS

In Fig. 1, the schematic diagram of the 118°–60° model is shown. The geometrical arrangement of the modified 90°–60° model is similar to (or almost the same as) the previous 90°–60° model.

In the modified 90°–60° model and 118°–60° model, the maximum magnetic

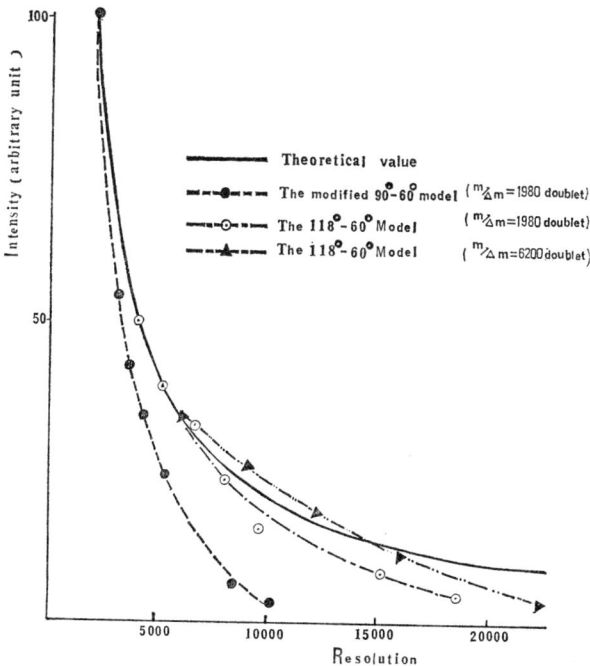

FIG. 3   Characteristic intensity versus resolution.

field intensity was increased to 16 500 Gauss, so that the mass range m/e 800 can be covered at an accelerating voltage of 3 kV.

The magnetic pole gap is 8·5 mm in both instruments and the analyser tube inner height is 6·8 mm. In the case of the 118°–60° model, the 10 mm high parallel beam of ions coming from the object slit can be collected 100% at the collector without being lost even at the narrowest portion of the analyser tube, *i.e.* at the magnetic gap.

Figure 2 shows results of resolving power tests of the three instruments. (The results of the 118°–60° model are preliminary ones.) Figure 3 shows the relationship between the resolution and sensitivity of the instruments. The excellent resolution capability of the 118°–60° model can be seen from these data. In Fig. 3, the experimental values for both the modified 90°–60° model and the 118°–60° model are simultaneously given for the purpose of comparison. The data in Fig. 3 were taken in the following manner. First, with the object slit width set at a certain value, the collector slit is widened enough to give a flat top peak on CRT, and then the slit is narrowed to a point where the flat top disappears without losing the peak intensity. The resolution is now measured from the peak width D of the 5% peak height and from the peak separation L of the doublet used.

From Fig. 3, one can see the reciprocal relationship of resolution/sensitivity is mostly satisfied up to $R = 6000$ for the modified 90°–60° model and up to $R = 15\,000 \sim 20\,000$ or more for the 118°–60° model.

CONCLUSION

Three models of virtual image mass spectrometers have been constructed and examined and it was confirmed experimentally that to consider the influence of the fringing field in the calculation and design of the virtual image DFMS is essentially important.[2] The preliminary experiment of the 118°–60° model has shown that the resolution of about 30 000 can be successfully obtained in such a compact size mass spectrometer.

REFERENCES

1. Ishibashi, M., Nojiri, M. and Watanabe, E., *in* 'Advances in Mass Spectrometry', Vol. 5, *ed.* A. Quayle, Institute of Petroleum, London, 1971, p. 286.
2. Matuda, H. and Matsuo, T., *Int. J. Mass Spectrom. Ion Phys.*, 1971, **6**, 385.

45

A New Digitally Controlled, Computer Compatible Quadrupole Mass Spectrometer

By H. EGLI, W. K. HUBER, H. SELHOFER
and R. VOGELSANG

(*Balzers Aktiengesellschaft für Hochvakuumtechnik und Dünne Schichten, Balzers, F. Liechtenstein*)

INTRODUCTION

THE evolution of quadrupole mass spectrometry over the past twenty years,[1] has shown that this principle is characterized by a number of features (linear mass scale, constant peak width, no magnets) and is worthwhile being considered for use not only in the field of measuring partial pressures in vacuum systems but also in the analytical mass spectrometry field. We may divide the application ranges into two major parts:

(1) Sophisticated vacuum measuring devices (process control, vacuum systems of large accelerators, etc.), and
(2) low resolution analytical mass spectrometry up to 500 to 1000 amu.

The main problems connected with mass spectrometric analysis are the time needed to get relevant information during fast processes and the tremendous amount of information delivered by the mass spectrometer. These problems are associated with all types of mass spectrometers.

If one needs as much information as possible on a specific species within a short time, this can be accomplished with a programmed mass scan, so that only a limited number of peaks are taken into account.[2] In this case the information on other ion species is lost.

The more sophisticated way, especially for the reduction of the normally large amount of data, but also for running the mass spectrometer, is the combination of the mass spectrometer with a computer. Because of the organization normally used in the control of mass spectrometers it is not a trivial task to connect the mass spectrometer by an adequate number of interface elements to a processor. In principle it is possible to do this, and it has been realized by different laboratories.[3]

Starting from this state of the art we followed a concept that makes it possible to extend the range of operation of the basic instrument in such a

413

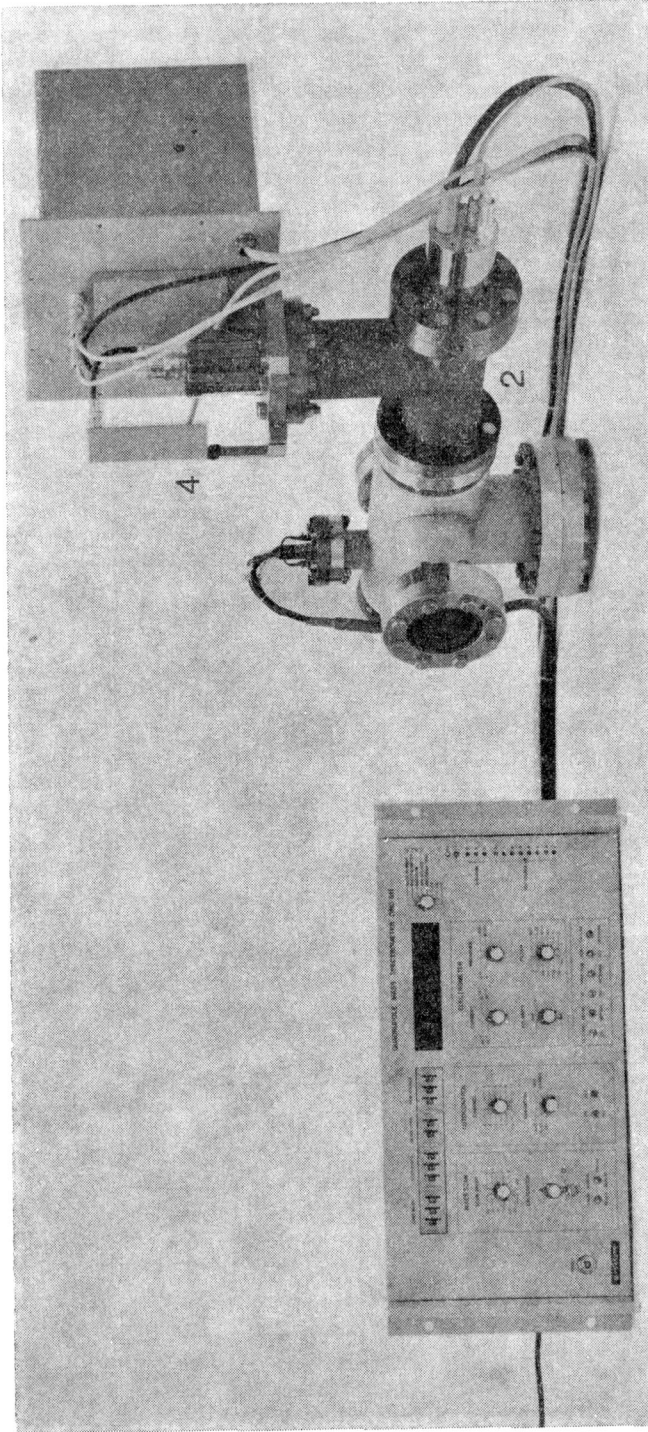

FIG. 1 View of the quadrupole mass spectrometer QMG 511. Basic unit (extreme left); r.f./d.c.-generator (top right). (2) Analyser system; 4) electrometer head.

way that stages of expansion to be expected in the future can be added without difficulty.

DESCRIPTION OF THE MASS SPECTROMETER STRUCTURE

The instrument consists, both mechanically and electrically, of units of modular design, thus allowing high flexibility.

Mass Spectrometer Probe

All parts of the mass spectrometer head are built according to the rules of ultra-high vacuum techniques. Bakeability up to at least 400°C for extended periods guarantees, together with an adequate vacuum system (*e.g.* turbomolecular pump), extremely good vacuum conditions and correspondingly excellent background states.

The quadrupole rod system (8 mm rod diameter, 200 mm length) is machined with high precision and consists of high alumina ceramics and circular molybdenum rods. The overall accuracy is in the low micrometer range.

Different types of ion sources can be combined with the analyser to assure the fit with the problem to be solved. From partial pressure measurements in the extreme ultra-high vacuum (down to the 10^{-16} torr range) up to mass analysis of molecular beams out of evaporation sources ($>10^{-4}$ torr) there is always a specific ion source available.

For the detection of the resolved ions either a Faraday cup or a secondary electron multiplier can be used. For low background or large dynamic range measurements the multiplier is arranged in an off-axis version with a 90° electrostatic deflector between the analyser and the multiplier. Negative as well as positive ions can be analysed.

Electronic Unit

The electronics consists of a basic unit in a single 19-in. rack ($8\frac{3}{4}$ in. high), the r.f./d.c. stage and the electrometer probe, both mounted close to the analyser. The distance between the rack and the analyser can be normally up to 10 m.

This compact arrangement, with spare space for additional options, was only made possible by consequent application of modern techniques; details will be given later. An impression of the dimensions of the complete system can be seen in Fig. 1.

Within the limits of this paper only a few words can be said in relation to the different modules. In the block diagram of Fig. 2 the basic concept as well as some of the possible options are presented.

The Spectrobus consists of a number of connectors mounted on a printed circuit board. Each pin number on any connector is associated with a specific signal line. Electrically the Spectrobus is a high speed, digital and analog information channel. All digital data are transferred in binary code two's complement arithmetic (commonly used in modern digital computers).

The Console is the man–machine interface of the mass spectrometer, containing select switches, BCD–Binary converters, Binary–BCD converters and display.

The Ion Source Supply delivers the voltages, stabilizes the electron emission

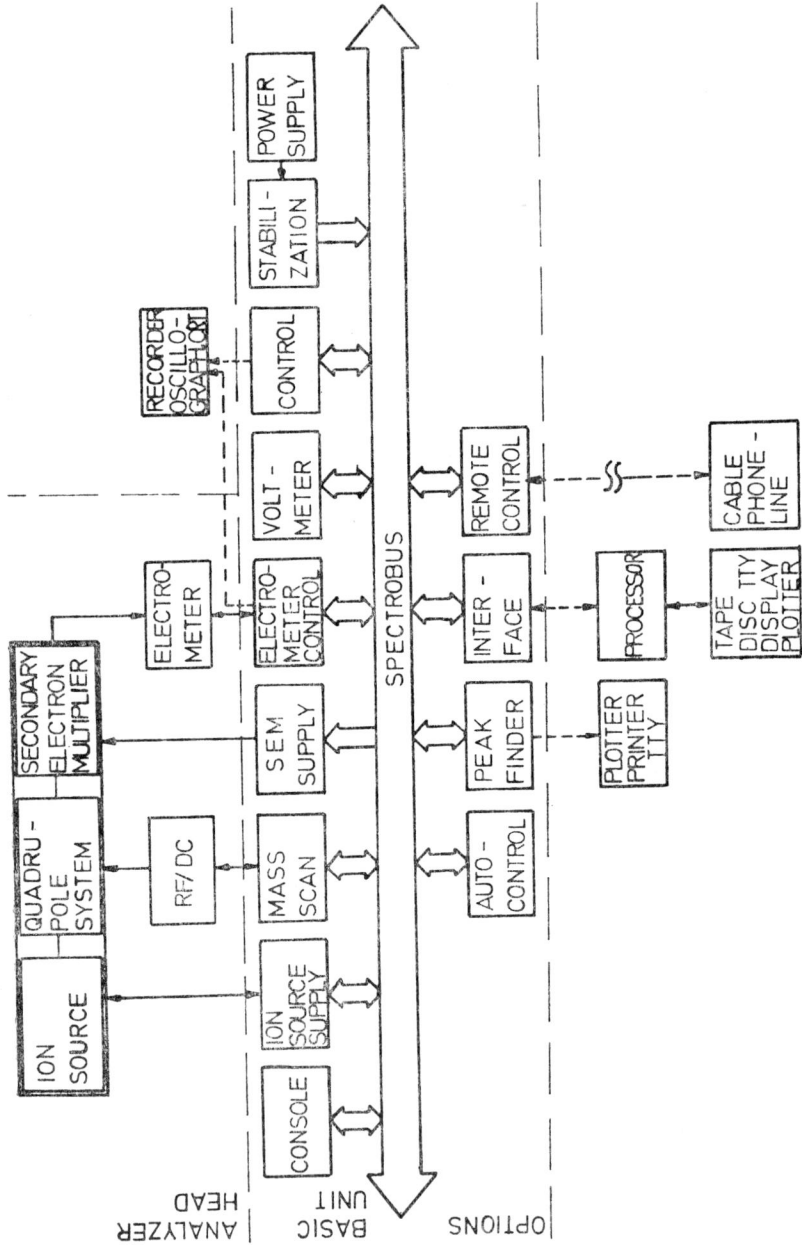

Fig. 2 Schematic presentation of the instruments functions.

and gives the protection necessary for the operation of the different types of ion sources.

The Mass Scan controls the r.f.-amplitude of the r.f./d.c. generator by means of a binary counter and a DAC of high resolution. It also contains the necessary circuitry for the control of the mass resolution by means of the corresponding d.c. component.

The r.f./d.c. stage consists of an oscillator (quartz stabilized), an r.f. amplifier, two rectifiers (one for the regulation of the r.f. amplitude, the other for the d.c. component) an amplifier for the regulation of the r.f. amplitude and an automatic current limiting circuit. The r.f. output circuit is tuned with the capacitance of the quadrupole.

The SEM supply delivers and stabilizes the high voltage for the operation of the secondary electron multiplier in steps of 10 V up to 3550 V.

The Electrometer and Electrometer Control form an extremely fast and sensitive current–voltage converter of high stability. The four ranges (10^{-5}, 10^{-7}, 10^{-9} and 10^{-11} A f.s.) combined with an automatic magnifier ($\times 1$, $\times 10$, $\times 100$) allow measuring and recording of intensities in a wide dynamic range. With the automatic magnifier in operation the analog signal is delivered in a logarithmic–linear mode (3·33 V per decade, linear within each decade, defined zero).

The electrometer can be replaced by a fast, single-ion counting device of high dynamic range, followed by a rate meter (QMR 101). By these means it is possible to measure extremely low ion intensities, especially with the off-axis multiplier. The advantages of this method cannot be discussed in detail here. The Voltmeter is basically an ADC of extremely high conversion rate and wide dynamic range (conversion time 10 μsec, resolution 4×10^{-5}). The digital output is in floating point notation.

The Controller module controls and monitors the data flow on the Spectrobus. Its efficient priority system combined with a short cycle time (800 ns) allows optimal data throughput in connection with processors, process controls, or any external data handler.

The Power Supply and Stabilization deliver the different regulated voltages. A special protection circuit is responsible for a regular switching of the electronics in case of power failure.

The option Autocontrol will be able to store up to twelve complete sets of operation data (for each of the twelve sets: First Mass, Scan Width, Resolution, SEM Voltage, Scan Speed, Calibration, Electrometer) to be programmed from the Console. In the mode Autocontrol the quadrupole is operated according to the program stored.

The Peak Finder allows the feeding of mass numbers and the corresponding values for the intensity of the ions to a printer, plotter or TTY.

The Interface module to the processor will be a single card in the mass spectrometer which is responsible for the dialogue with the processor.

The basic differences in operating the mass spectrometer in combination with a processor for example are clearly shown in the block diagram of Fig. 3. In the case (A) many elements are needed for the interface with the processor, while we (case B) shall need no more than an interface card on the Spectrobus. All the other functions necessary for the combination are included in the sophisticated circuitry of the quadrupole mass spectrometer.

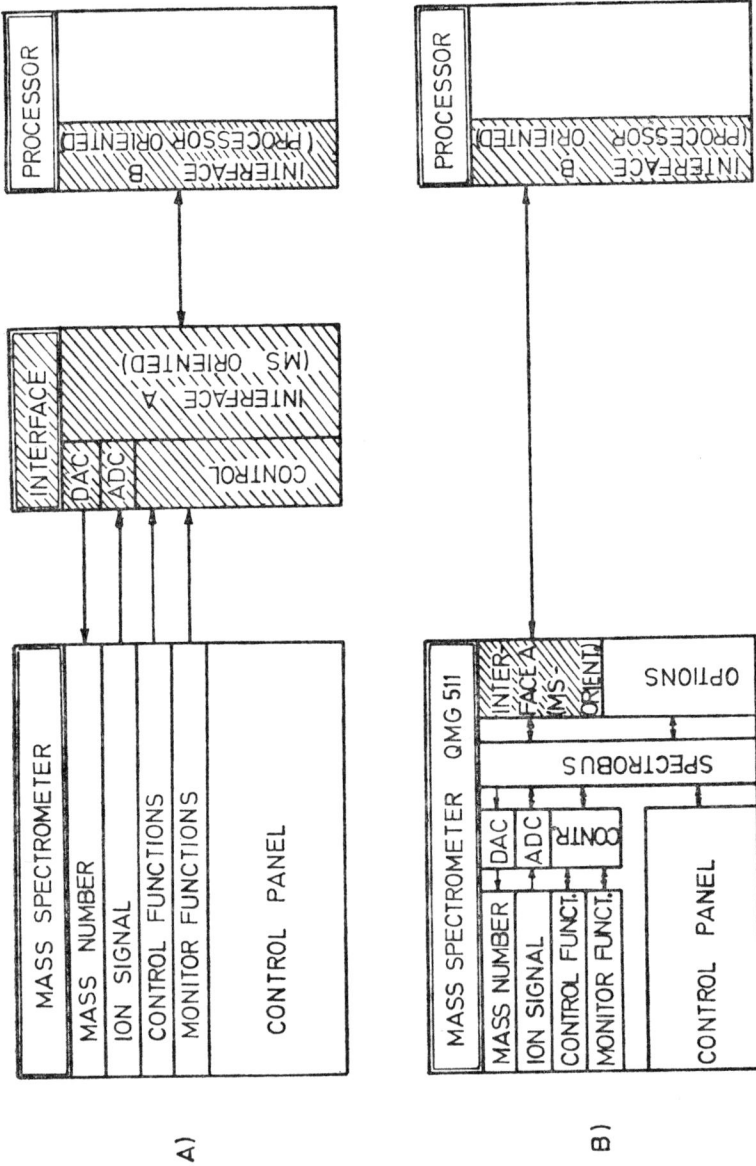

Fig. 3 Schematics of the possible combination of the mass spectrometer and processor. (A) Conventional method. (B) Solution with the QMG 511.

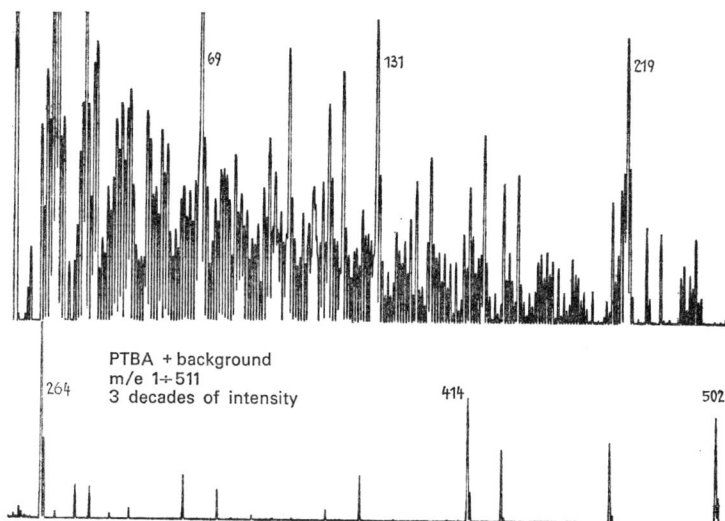

FIG. 4   Mass spectrum of perfluorotributylamine.

PERFORMANCE DATA

The combination described in this paper allows operation in a single mass range from m/e = 1 up to m/e = 511. The linear mass range can be scanned at speeds of 10^{-4} s per mass up to 10 s per mass. The resolution can be adjusted in such a way that resolution is proportional to the mass number (a special feature of the quadrupole principle) or in other words the peak width is constant over the whole mass range. As an example, the well known typical spectrum of perfluorotributylamine is shown in Fig. 4. It can be seen that the peak width throughout the mass range is less than one ($\Delta m \leq 1$, 10% definition). The FWHM definition gives a resolving power of about 2000 at mass 500.

The sensitivity for argon at a constant peak width of $\Delta m = 1$ (10%) and an electron current of 1 mA is with the standard ion source $>2 \times 10^{-4}$ A torr^{-1}.

CONCLUSION

The quadrupole mass spectrometer has reached its well-defined domain in mass spectrometry. It has been our intention to develop a modular quadrupole system incorporating the advantages of modern digital technologies, that not only satisfies the immediate need but can also be modified and extended to include future developments.

REFERENCES

1. Paul, W. and Steinwedel, H., *Z. Naturforsch.*, 1953, **8a**, 448. Paul, W., Reinhard, H. P. and von Zahn, U., *Z. Phys.*, 1958, **152**, 143.
2. Huber, W. K. and Rettinghaus, G., *Messtechnik.*, 1970, **78**, 65–70.
3. Reynolds, W. E. *et al.*, *Analytical Chemistry*, 1970, **42**, 1122–1129. Houseman, J. and Hafner, F. W., *Journal of Physics E: Scientific Instruments*, 1971, **4**, 46–50.

46

Discharge Current and Associated Ions in Radiofrequency Spark Source Mass Spectrometry

By J. BERTHOD

(*Centre d'Etudes Nucleaires de Grenoble, Service de Chimie Analytique, Section d'Etudes et d'Analyses Physico-Chimiques, B.P. 85, Grenoble, France*)

To characterize the ionization processes taking place during the individual discharges of a radio-frequency spark source, our first task was the design of a convenient detector. We used a broad bandpass current transformer which develops a voltage proportional to the current flowing in the secondary spark circuit.

This paper describes the exploratory experiments, the modifications of the spark circuit and the significant improvements observed on the photoplate recorded mass spectra.

EXPLORATORY EXPERIMENTS

The spark unit of the MS 7 mass spectrometer delivers to the sample electrodes a pulsed 600 kHz voltage of 5 kV to 80 kV (peak to peak). The pulses are variable in length between 25 and 200 μs and the repetition rate may be adjusted between 10 Hz and 30 000 Hz.

When the electrodes are too wide to allow a spark, any voltage or current probe gives the signal of Fig. 1(a)1[1] and, by closing the gap to allow sparking, the signal of Fig. 1(a)2 is observed.

Now with our detector, it is possible to analyse what happens during breakdown. The discharge current (Fig. 1(a)3) appears to be a 20 MHz damped sinewave which occurs and lasts up to a half period of the spark voltage. This 20 MHz sinewave depends only on the electrode circuit, *i.e.* resistance, self-inductance and capacitance,[2] and the first peak current may attain a value of 10 A.

In addition to these observations, we measure the r.f. residual voltage in the accelerating voltage line, particularly on the extraction slit, Fig. 1(b)1.

Without a spark, the peak to peak 600 kHz residual voltage is approximately 900 volts, in spite of the 1000 pF filter capacitance, and we reduce this voltage down to 600 V using a 2000 pF capacitance.

1a Oscillating mode

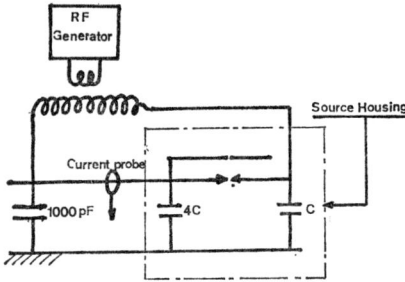

1b₁ Damping mode

R = 330 Ω

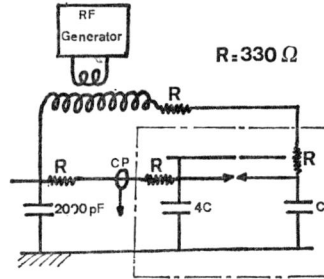

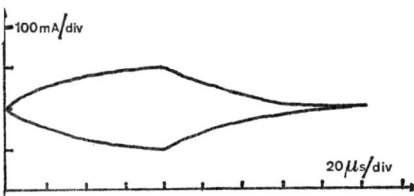

1a₁ Output of C.P. open gap

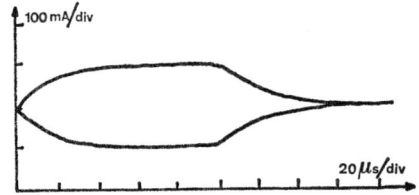

1b₂ Output of C.P. open gap

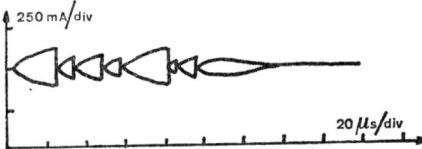

1a₂ Output of C.P. with breakdowns

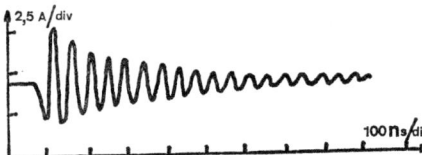

1a₃ Oscillating discharge

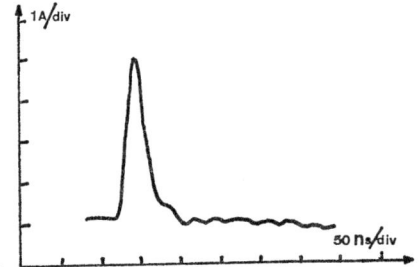

1b₃ Damping discharge

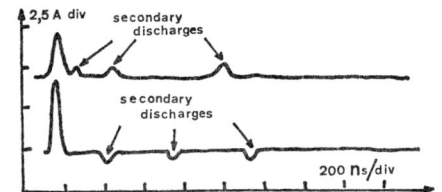

1b₄ Damping discharges

FIGURE 1

During the breakdown we think that the peak current generates a very high voltage on the extraction slit, due to the impedance of the line between the electrode and the filter capacitance. Such a voltage may modify the energy distribution of ions accelerated to the object slit.

MODIFICATIONS OF THE SPARK CIRCUIT

Critical Damped Discharge

We introduced in the secondary spark circuit (Fig. 1(b)1) resistances R of experimentally adjusted value to obtain the critical damp of the discharge. Figure 1(b)3 shows a signal of the damped current discharge, the length of which is close to 100 ns, and we see in Fig. 1(b)2 that the spark voltage rise time is reduced. The loading effect of the resistances reduces the r.f. voltage signal between electrodes and, to compensate for that, it is necessary to increase the primary voltage developed across the primary of the tesla transformer.

As a first result, this pure signal allows us to observe sometimes identical signals of lower amplitude called, in this paper, 'secondary discharges', which occur between 100 and 800 ns after the 'principal discharge'—this without increase of the r.f. voltage (Fig. 1(b)4).

Such a signal is also very convenient to study the number of breakdowns for a 200 μs pulse. In agreement with literature this number, which varies between 0 and up to 70, strongly depends on the spark gap.

Moreover, for a given spark gap, the first discharge of each pulse alone occurs after a long dead time. The relaxation time of the r.f. voltage for this first breakdown is approximately constant, whereas the relaxation time varies statistically from breakdown to breakdown during a pulse.

The Self-Triggered High Voltage Discharge

As a last improvement, we modified the spark circuit oscillator to use only discharges of the first type.

After the trigger is disconnected, the oscillator is monitored with a start signal delivered by a Rutherford B14 generator and with a stop signal taken from the current probe (Fig. 2(a)).

These signals are correctly shaped and their voltage adjusted through a specific electronic device. Without a discharge, the oscillator continuously runs; when a discharge occurs, the discharge signal holds the oscillator cut off until it receives a new 'start signal' from the generator. The frequency of these pulses can be set to any desired value in the range 5 Hz to 10 000 Hz.

Now, the relaxation time of the r.f. voltage for all the breakdowns (i.e. the breakdown voltage) and the dead time between two discharges are approximately constant. However, only a precise value of the spark gap allows this very stable operating mode; any variation of the spark gap destroys these stable conditions. Such a design allows us to control accurately the breakdown (Fig. 2(b-c)).

In addition, this operating mode is convenient for getting a selection of ion swarms according to the breakdown voltage or to the amplitude of the discharge current.

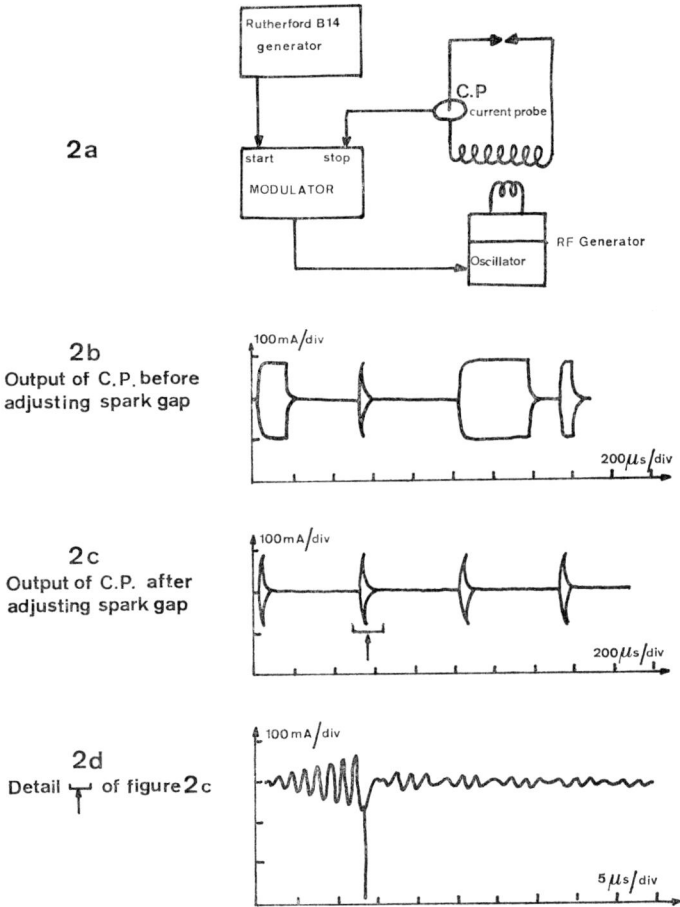

2a

2b
Output of C.P. before
adjusting spark gap

2c
Output of C.P. after
adjusting spark gap

2d
Detail of figure 2c

FIG. 2   Self-triggered damping discharge mode.

RESULTS

This type of discharge affects the vaporization process in the source and different properties of the ion beam.

Source Processes

In contrast to normal oscillating discharges, damped discharges produce tiny sparks (without halo), well localized on the electrodes; therefore the positioning of the electrodes on the optical axis is very critical. Any variation of electrode location involves dramatic change of the monitor current. We observe also for long operating time a noticeable reduction of the sample deposit on the source housing which can be attributed to a less important vaporization process; the monitor current remaining at the same value, these effects show a better ionization efficiency. As to this efficiency, the single shot operation of the self-triggered discharge allows us to analyse the effects of single discharges, and the fact that we observe approximately a ratio ten to one between the number of current probe signals and the number of ion swarms at the monitor shows the relative importance of ion extraction among the factors contributing to the source efficiency.

Photoplate Recorded Mass Spectra

Photoplate detection is convenient for these studies since it records the entire spectrum at the same time and gives correct ion images.

Resolving power—In a short communication[2] we reported improvement of the resolving power on NBS steel spectra using damped discharges. Since these experiments, the α-slit of our mass spectrograph has been reduced from 2 mm to 1 mm, and the resulting increase of the resolving power prevents us comparing the absolute values of these reported experiments with those of the new ones.

TABLE I

Mode	Mass							
	208	123	109	71	55	29	14	12
Damped discharge	5 500	5 700	5 900	3 700	2 800	2 700	3 000	2 800
Oscillating discharge	4 300	4 100	3 900	2 400	2 100	2 000	2 600	—

Table I gives the data obtained on mass spectra recorded from Johnson Matthey copper CA2 with a 30 μ object slit. Line width is measured at half peak height on the lines recorded with an optical density 0·5 to 1·5.

The improvement in resolving power obtained is 40 to 50% between masses 71 to 123 and 30% at masses 29, 55 and 208. We can observe the unusually high values of 3000 and 2800 at masses 14 and 12, due to a curve-shaped plate at the low mass end, determined experimentally for different positions of the plate with regard to the front of the magnetic analyser. Figure 3 shows the profiles of the carbon line recorded both on a flat plate and a curved plate.

Doubly charged ions—Table II gives the relative ion intensities for the doubly charged ions, corrected for line width but not for plate response, measured in three different operating modes.

TABLE II

M^{2+}/M^{1+}	Pb (%)	Sb (%)	Ag (%)	Ga (%)	Mn (%)
Oscillating discharge	39	35	21	19	20
Damped discharge	25	26	13	7	5
Self-triggered damped discharge	10	15	8	3·5	2·5

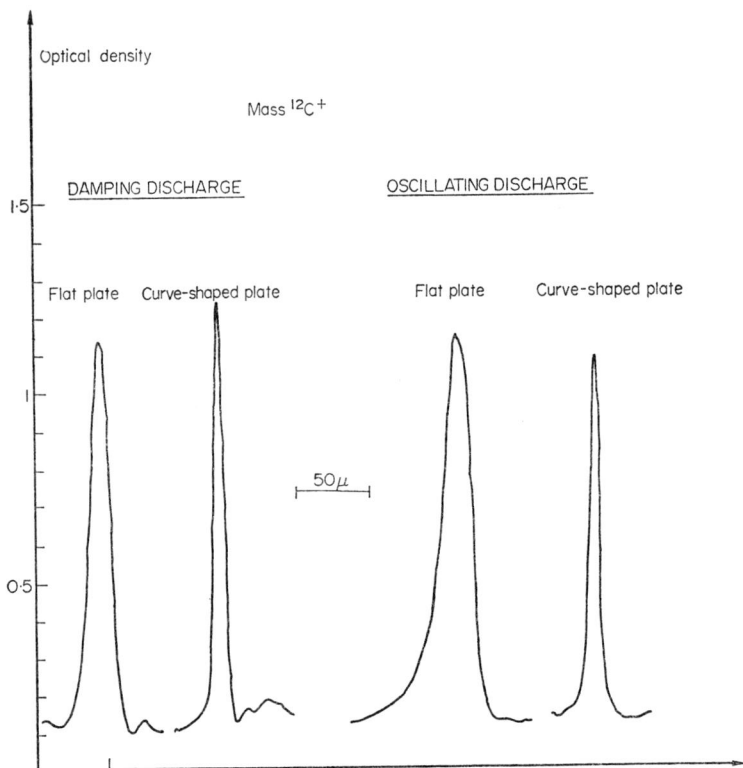

Optical density

Mass $^{12}C^+$

DAMPING DISCHARGE   OSCILLATING DISCHARGE

1·5

Flat plate   Curve-shaped plate   Flat plate   Curve-shaped plate

1

50μ

0·5

FIGURE 3

CONCLUSIONS

The broad bandpass current probe is a fruitful tool for the studies of spark discharges and associated processes such as vaporization and ion beam formation. The damping discharges routinely used in our laboratory for one year gave the discussed improvement of resolving power and we observe some interesting features in the analytical results. For instance, in geological standards pelletized with gold the values of the relative sensitivities of Rb and Cs, normalized at Fe, experimentally obtained were 5·5 and 5 with the oscillating discharge; they are now 1·8 and 1·5. Our experiments are progressing with the self-triggered spark discharge and we hope, in the near future, to compare our results with those of other laboratories.

ACKNOWLEDGMENTS

It is a pleasure to acknowledge several stimulating discussions with R. Stefani and A. M. Andreani and to thank B. Alexandre and C. Riou for their competent assistance in the design and the construction of electronic circuits.

REFERENCES

1. Franzen, J., *in* 'Trace analysis by mass spectrometry', *ed.* Arthur J. Ahearn, Academic Press, 1972.
2. Berthod, J., Alexandre, B., Stefani, R., *International Journal of Mass Spectrometry and Ion Physics*, 1972–73, **10**, 478–80.

47

The Design and Performance of a Versatile, New, Ultra-High Resolution Mass Spectrometer

By S. EVANS and R. GRAHAM

(*AEI Scientific Apparatus Ltd, Manchester, U.K.*)

INTRODUCTION

THE techniques of analytical mass spectrometry have developed rapidly and in many directions over the past few years. For many applications a dynamic resolution of 10 000 is still adequate, but the quantity of sample available may now be only a few nanograms, and the time for which the sample is available may be only 20 to 30 seconds. Consequently, sensitivity at 10 000 resolution is extremely important. In addition, there is an increasing demand for resolution in excess of 100 000. Sulphur containing compounds found in petroleum fractions can give rise to C_3–SH_4 doublets which require a resolution of 70 000 at m/e 200 to resolve; similarly ^{13}C–^{12}CH and H_2–D doublets at m/e 200 require resolutions of 45 000 and 130 000 respectively.

Improvements in ion-optics and electronics over the past decade have now made it possible to design a new instrument, the A.E.I. MS5074, to fulfil these requirements.

BASIC DESIGN AND PERFORMANCE

Modified Nier-Johnson geometry (Fig. 1) similar to that of the A.E.I. MS9, but with an improved electrostatic analyser, has been used for the new instrument. Two hexapole lenses,[1] one situated between the e.s.a. and magnet, and the other between the magnet and collector, are used to correct for image curvature effects and for rotational misalignment of the image and the collector defining slit. These lenses improve sensitivity, particularly at high resolution, thereby making it possible to carry out magnetic scans at resolutions up to 40 000. An electrostatic lens situated between the source and electrostatic analyser provides fine electrical control of the final image position. By varying the lens voltage as the magnet is scanned, optimum conditions can be maintained over the mass range covered.

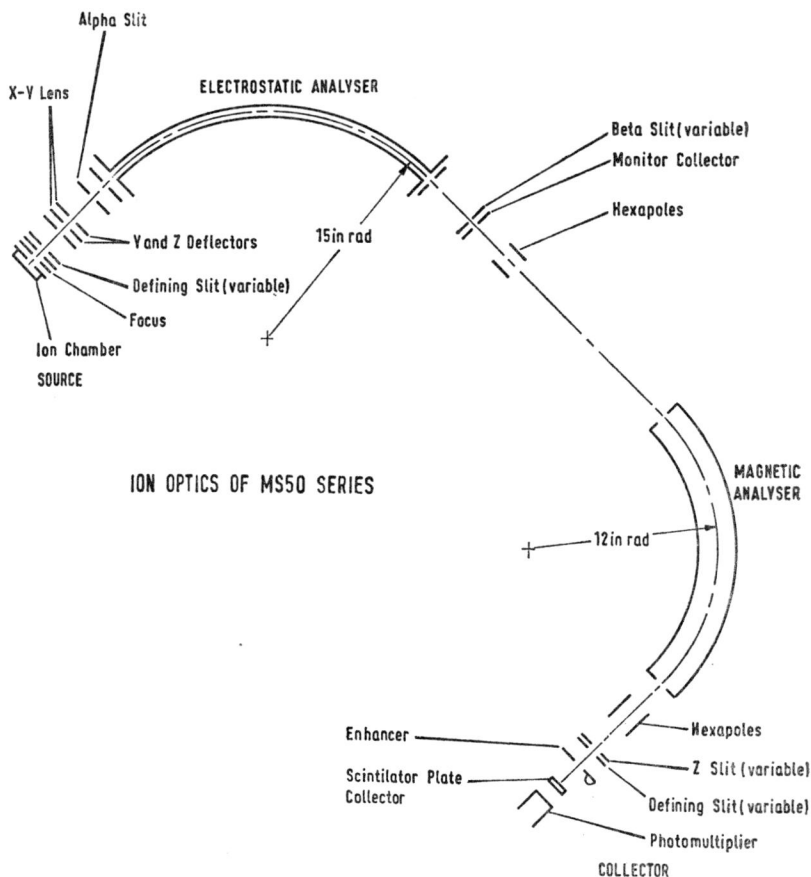

FIGURE 1

The performance specification of the instrument is summarised in Table I.

The sample flow rates quoted are based on obtaining a 20 to 1 intensity range with a signal to noise ratio of 10 to 1 on the smallest peak, over a mass range from m/e 600 to m/e 60.

The mass measurement accuracy using peak matching, specified as 0·3 ppm r.m.s. for 5% mass differences, results from the improved ion optical design of the instrument together with highly stable electronics and the use of a 7 digit, 0·1 ppm voltage divider. To obtain this accuracy using visual matching, it is necessary to operate at a resolution of 50 000, which is now practical due to the increased sensitivity at this resolution resulting from the use of hexapole lenses. The results of a typical experiment, using C_4Cl_6 as the 'unknown' sample, are given in Table II. The four runs were carried out over a total period of 2 days, and it is apparent that the r.m.s. accuracy is well within the 0·3 ppm specified.

To achieve resolution and mass measurement accuracies of this order it has been necessary to design a suitably rigid structure for the basic frame. As a

TABLE I

Static Resolution	150 000 on 10% valley definition	
Resolution	Dynamic Performance Scan Rate (s/decade)	Sample Flow Rate (ng/s)
1 000	3†	0·3
3 000	3	1
10 000	10	1
40 000	100	10
Mass Difference	Mass Measurement (Peak Matching) Mass Measurement Accuracy (ppm R.M.S.)	
5%	0·3	
10%	1·0	
100%	5·0	

† Maximum scan rate 2s/decade.

result of trial experiments, the basic frame is constructed from 12″ × 8″ box section girders, free from resonances in the critical range for floor-borne vibrations of 10–15 Hz, which can be supported directly on isolating mounts where floor vibration levels are high. Rotary pumps are mounted remote from this main frame to eliminate induced vibrations.

The ultimate resolution is also critically dependent on the precision and control of the collector defining slit. Here, this slit uses parallel action swinging arms which give increasingly fine control as the slit width is reduced.

In addition to diffusion pumps using polyphenyl ether, together with Peltier baffles, a 25 l/s ion pump is positioned near the collector to reduce collision scattering of the ion beam and hence improve peak profiles.

A Daly metastable detector[2] is used as the standard detector, giving very stable high gain and low noise in addition to its facilities for examining metastable peaks.

The layout of the instrument has been designed for maximum operator convenience by arranging all the major functional tube unit controls, with one or two minor exceptions, within easy reach of an operator seated at the control console. A desk top is positioned under the source, on which a variety of inlet systems can be mounted, grouped conveniently around the source and fully integrated with the instrument. A gas chromatograph interface, consisting of a separator and its associated valves and pumping lines within a simple oven, mounts directly to one of the source inlet ports. A gas chromatograph mounted on the desk top mates directly with this interface.

SCANNING AND DISPLAY FACILITIES

Display

The built-in oscilloscope display unit can be used for peak matching, spectrum display, setting resolution both statically and dynamically, and checking ion beam stability.

TABLE II

Peak Matching Accuracy. The Conditions Were: Sample C_4Cl_6, Resolution 50 000, Reference Compound Perfluorotributylamine

Composition	True Mass	Corrected Errors (ppm)				Mean Error† (ppm)	Mean Corrected Mass
		Run 1	Run 2	Run 3	Run 4		
$C_4\ ^{35}Cl_5$	222·844 265	0·21	0·27	0·45	−0·1	0·21	222·844 22
$C_4\ ^{35}Cl_4\ ^{37}Cl$	224·841 316	0·54	−0·24	−0·11	−0·06	0·03	224·841 31
$C_4\ ^{35}Cl_3\ ^{37}Cl_2$	226·838 366 1	0·07	0·20	0·46	0·5	0·31	226·838 3
$C_4\ ^{35}Cl_2\ ^{37}Cl_3$	228·835 416 4	0·27	0·19	1·00	−0·04	0·36	228·835 33
$C_4\ ^{35}Cl\ ^{37}Cl_4$	230·832 466 7	−0·33		0·39	0·08	0·05	230·832 46

† RMS of mean 0·23 ppm.

When displaying magnet scans the oscilloscope time base is proportional to the square of the magnet current, thus providing a linear mass display across the oscilloscope screen. The same signal can be used to drive the UV recorder if linear mass chart display is required. Limited portions of the magnet scan can be displayed on the screen, minimum mass display being 20 amu per sweep. This facility enables the presence of a sample in the source to be observed before recording the spectra. For example if it is intended to introduce cholesterol, the magnet could be set to scan repetitively between mass 500 and mass 50, with the spectrum display starting at mass 400 and ending at mass 380. Observation of the intensity of the parent peak of cholesterol at mass 386 is then quite simple and when the desired intensity is reached, the entire spectrum can be readily recorded using any of the output facilities. Automatic single peak triggering is also provided to enable dynamic resolution to be checked at several points in a scan.

The maximum scan in the peak switching mode corresponds to a mass range of 40 000 ppm. This corresponds to a 10 mass unit display at mass 250 which, in conjunction with a reference compound such as p.f.k. enables unknown peaks to be identified very rapidly. Once an unknown peak has been identified, it can be moved to the centre of the screen using the magnet controls and then expanded using the peak switch controls to display a single peak only. The instrument can also be tuned in this way to an expected mass, for example when looking for the characteristic peak of a particular drug metabolite. As the oscilloscope sweep is calibrated directly in ppm, it can be used for the measurement of resolution or for estimating mass.

Metastable Scanning

In addition to all the conventional features of a high resolution instrument, comprehensive metastable scanning facilities are included. Metastable

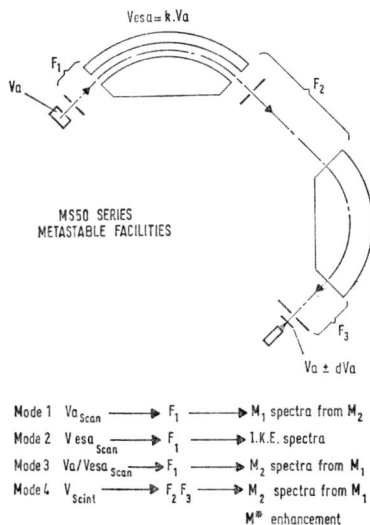

FIGURE 2

transitions can be observed in all three field-free regions F1, F2 and F3 (Fig. 2) using 5 distinct forms of scan. Two of these use facilities of the Daly metastable detector[2] to examine transitions occurring in regions F2 and F3.

Reactions in region F1 can be studied in three ways:

(i) The electrostatic analyser voltage can be scanned, with the accelerating voltage V_a constant, to produce an Initial Kinetic Energy (IKE) spectrum.

(ii) The accelerating voltage V_a can be scanned with the esa voltage constant to produce a Barber-Elliott precursor spectrum for a daughter ion.

(iii) The accelerating voltage V_a can be scanned at the same time as the esa voltage V_e such that V_a is proportional to $(V_e)^2$.

This is a new form of scan which produces a daughter ion spectrum from a precursor ion, over a mass range of $2:1$.

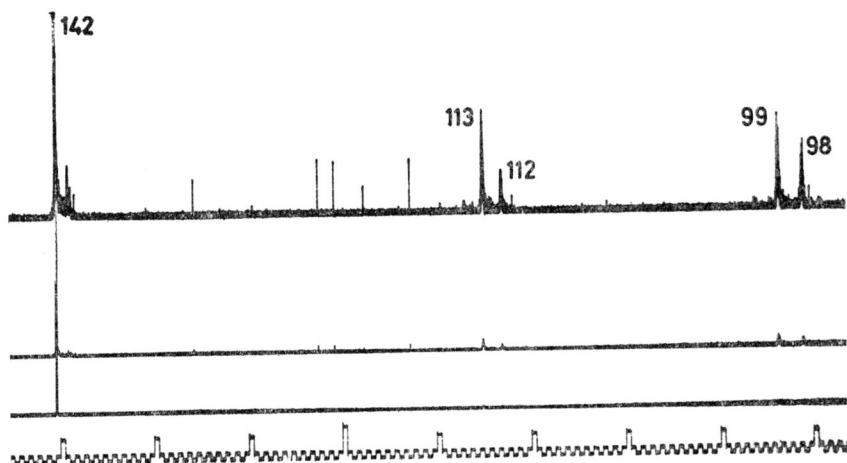

Metastable Combined Scan from Mass 142 of n-decane

FIGURE 3

As an example of the quality of information which can be obtained, Fig. 3 shows part of a combined esa/eht scan from the parent peak of n-decane at m/e 142. The instrument resolution was approximately 1000, and it is apparent that the resolution of the metastable peaks, at m/e 113, 112, 99 and 98 is similar.

Figure 4 summarises the fragmentation information obtained from a series of combined and conventional accelerating voltage metastable scans of n-decane. The circled numbers show the peak in the normal spectrum to which the instrument was tuned. The lower left segment shows the daughter ions produced from these, using the combined scan, whilst the upper right segment shows the precursor ions of these peaks detected using an accelerating voltage scan.

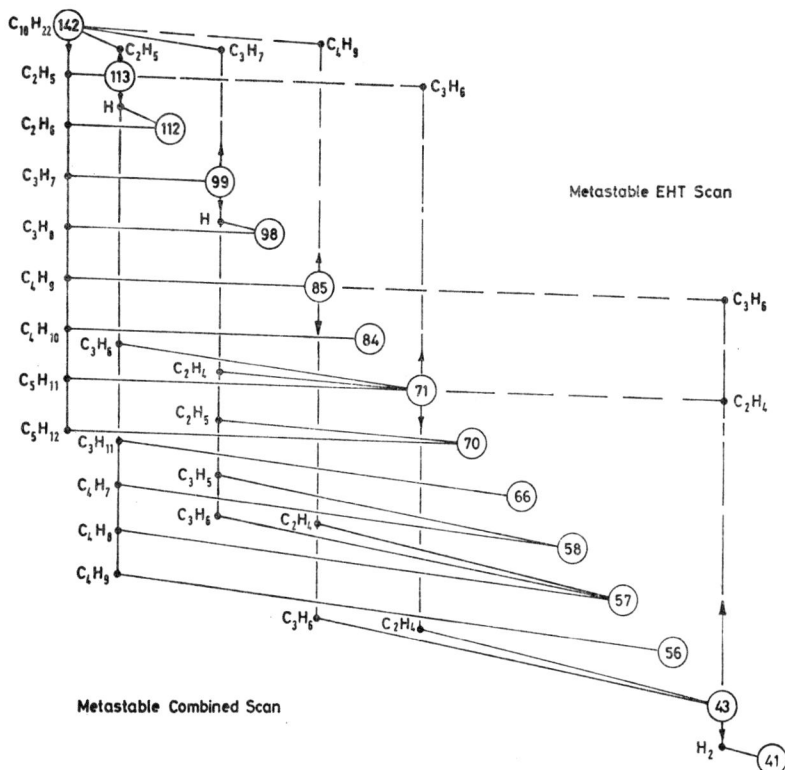

FIGURE 4

Matrix Control

The number of parameters to be set up on an instrument of this type is necessarily large, and for research applications, it is desirable that all its operating parameters should be under manual control. Conversely, once a particular set of parameters has been determined for a specific application, it is equally desirable that the number of manual adjustments be reduced to a minimum. To satisfy these conflicting requirements, a programme switch has been included. With this switch in position 0, all parameters are under manual control. When in one of the remaining 10 positions, the manual control of 14 parameters is overriden by a matrix, which preselects these functions. Thus most of the conditions required for a particular mode of operation can be re-obtained at any time by operation of a single switch.

REFERENCES

1. Boerboom, A. J. H., *Int. J. Mass Spectrom. Ion Phys.*, 1972, **8**, 475–92.
2. Daly, N. R., McCormick, A., Powell, R. E. and Hayes, R., *Int. J. Mass Spectrom. Ion Phys.*, 1973, **11**, 255–76.

48
Design and Operation of a Multipoint Field Ionization Mass Spectrometer

By WILLIAM H. ABERTH, CHARLES A. SPINDT,
MARTIN E. SCOLNICK, RUSSELL R. SPERRY
and MICHAEL ANBAR

(*Stanford Research Institute, Menlo Park, California, U.S.A.*)

A new type of mass spectrometer has been constructed for the purpose of identifying and comparing mass fingerprints of a broad variety of samples including those with very low volatility. The mass spectrometer consists of a multipoint type field ionization source coupled with a modified Wien E × B velocity filter for mass separation. By varying the E field, mass scanning can be obtained at a very fast rate of more than 1000 mass units in 1 sec. Scanning is performed repetitively and the beam detector output is coupled with a 4096-channel multichannel analyser operating in the multiscaling mode. The mass scan rate is synchronized with the multiscaling rate and mass spectra, integrated over many scans, are obtained.

The sample is field ionized by a multipoint structure[1] (*see* Fig. 1). This structure consists of a uniform array of points that are spaced 25 μ apart, are 40 μ high and cover an area of about 1 mm^2. These points are formed on a grid which is supported on a hollow tube and held in place by a press fitted ring. A grid is placed 125 μ above the multipoint array and a potential difference of about 3000 V between points and grid is sufficient to induce field ionization. Sample material enters the field ionizing source from behind the multipoint grid support. This physical arrangement ensures efficient introduction of the sample material into the ionizing region and yields high ionization efficiencies of about 0·1 %. Micro-needle growth can be induced by simply introducing toluene or other organic compounds into the multipoint region. The growth of micro-needles on the points is not essential to the source operation but has the effect of lowering the grid-point voltage required to achieve a given efficiency. The overall source structure is shown in Fig. 2. Both gases and solid samples of low volatility can be effectively ionized. A solid sample contained in a sample holder (shown at the bottom of Fig. 2) can be inserted from the outside through a vacuum lock into the back of the source. The sample can then be heated by means of two heating cartridges inserted into the base of the source structure. A temperature of 400°C can be reached in about 20 min and another 20 min is needed to cool back down to 100°C. The source can be used to ionize an externally located gas sample

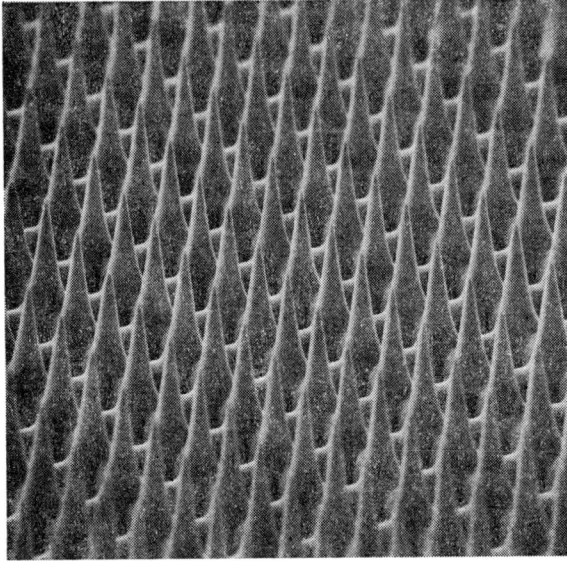

FIG. 1   Scanning electron micrograph of new multipoint structure.

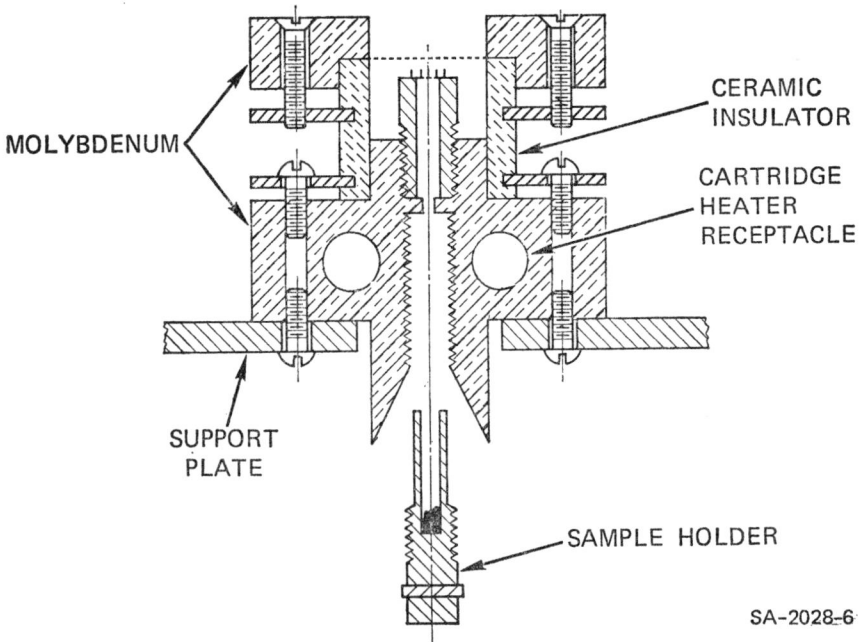

SA-2028-6

FIG. 2   Schematic cross section of a multipoint source showing method of introducing a solid sample and heating technique.

by replacing the sample holder with a ceramic tube. The changing of solid samples or converting from solid sampling to gas sampling mode is accomplished in only a few minutes while maintaining the source at a high vacuum.

The sample ions are extracted from the source and focussed by means of a set of electrodes attached to the source and an einzel lens positioned about 60 cm from the source (*see* Fig. 3). The ion beam then undergoes mass separation by means of a Colutron[2] E × B ion velocity filter.[3] This filter has a 60-cm long mass separating region. Its design is based on that of the Wien filter but with electrostatic and magnetic shims that can trim the E and B fields and eliminate the strong focussing characteristics of the device.[4] The

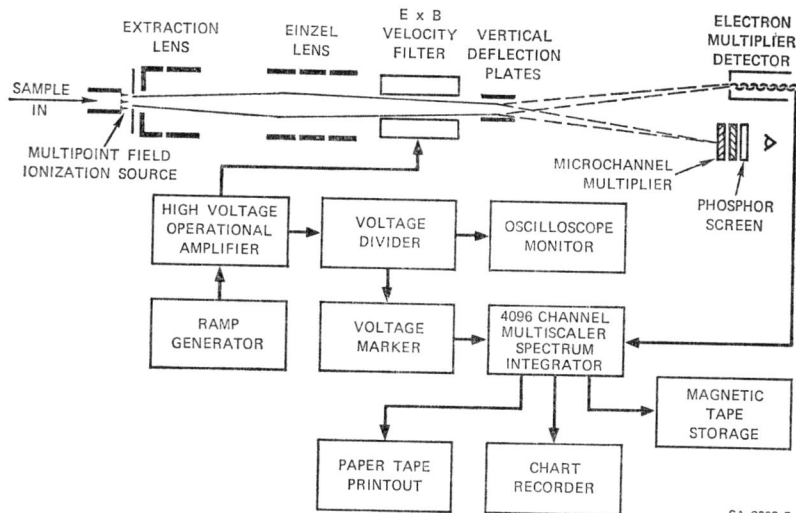

FIG. 3   Schematic of field ionization fingerprint apparatus.

velocity filter is operated with a constant magnetic field and mass scanning is accomplished by varying the electrostatic cross field. A pair of deflection plates situated immediately after the velocity selector provides for positioning the ion beam either on the entrance aperture of the multiplier detector or on the surface of an image intensifier. The image intensifier device consists of two 3·5-cm diameter series connected microchannel electron multiplier plates[5] backed by a fluorescent screen that is visible through a viewing port. A single ion impinging on the front surface of the image intensifier will produce about 10^8 electrons at the output of the second microchannel plate. This current is accelerated by about 1700 V to the fluorescent screen. Single ion impingements are thus made visible by this device which serves as an invaluable means for optimizing the ion beam optics and beam finding. The mass selected ion beam is detected through a 1 × 3 mm slit by a Bendix 4700 continuous dynode electron multiplier coupled with an SSR model 1120 amplifier/discriminator.

A mass scan typically covers a range of about 1000 mass units and is completed in less than 1 sec. Repetitive mass scans are performed by supplying

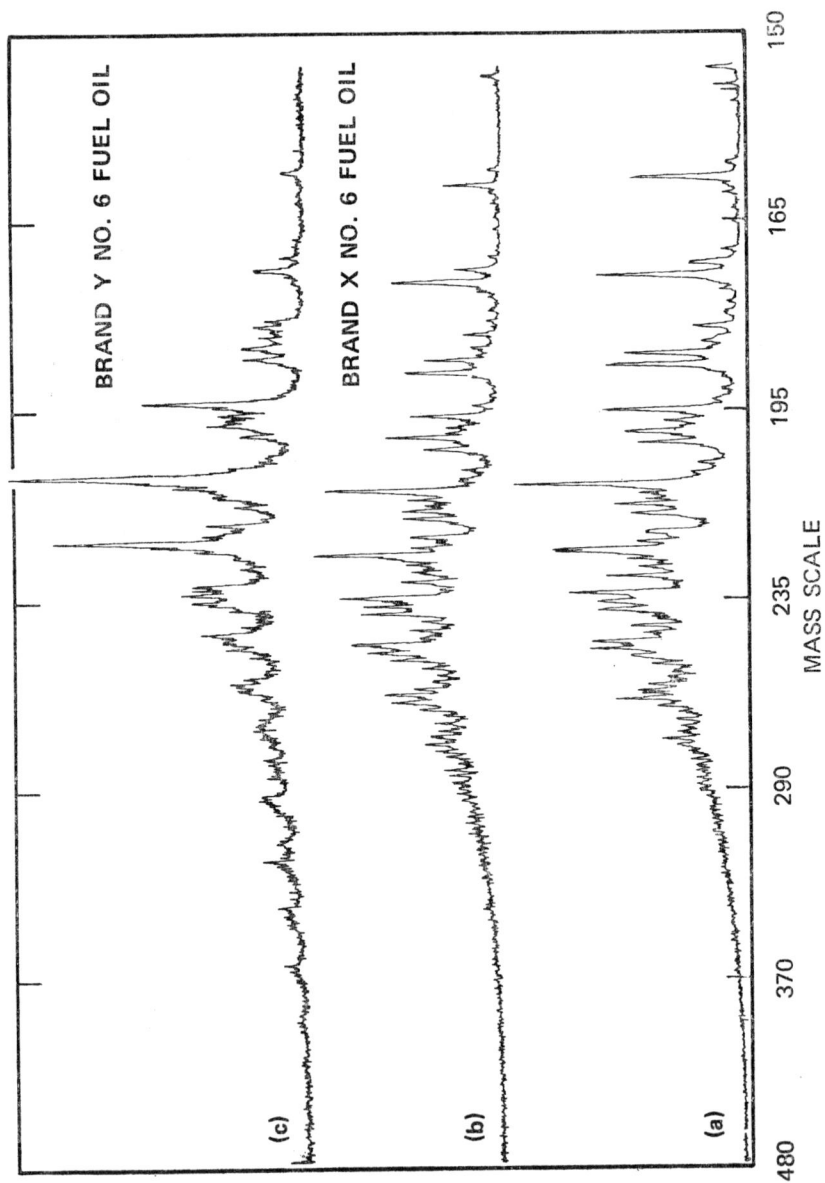

FIG. 4 Mass fingerprint comparisons of fuel oil samples.

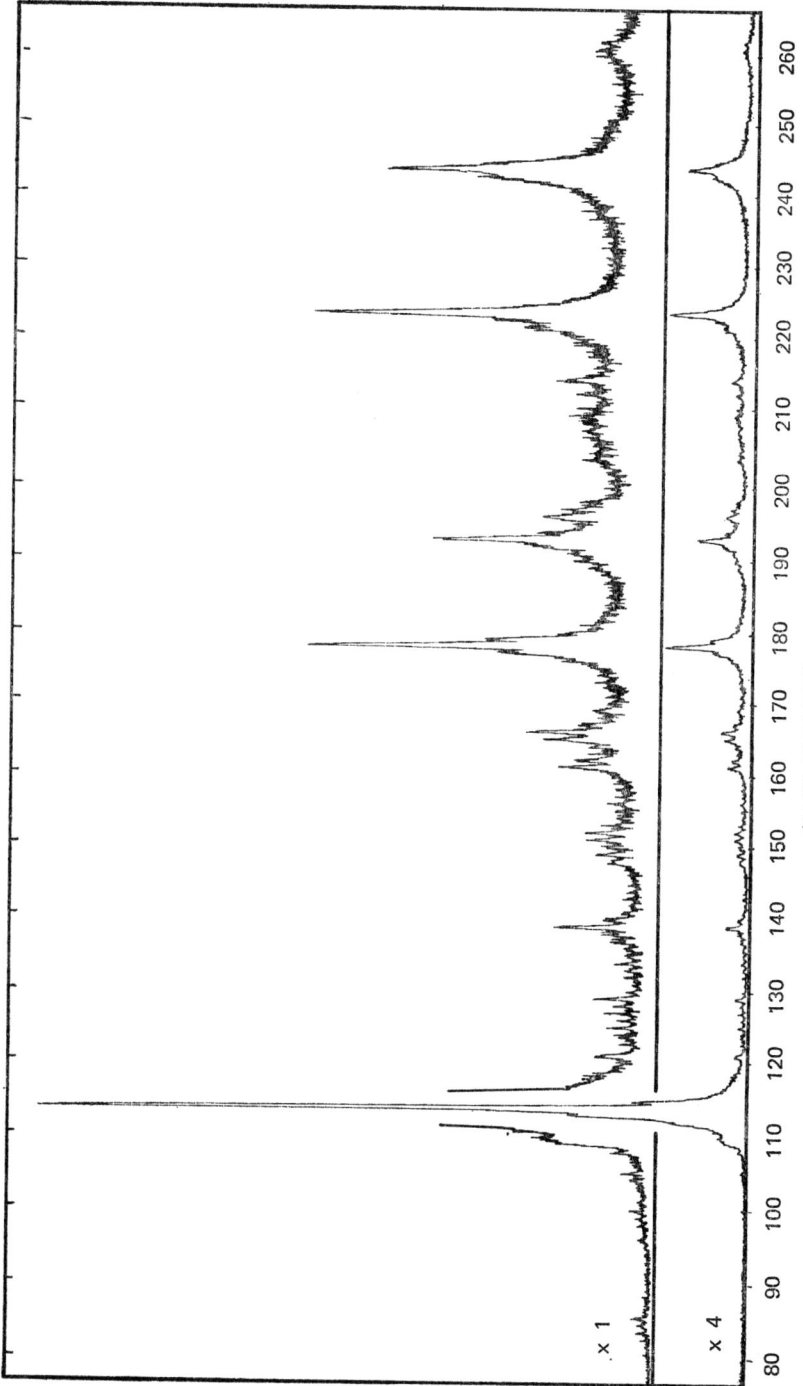

Fig. 5   Fingerprint of lyophilized urine.

the amplified ramp voltage from a sweep generator to the electrostatic deflecting plates of the Colutron filter. The mass range covered is obtained by adjusting the ramp voltage amplitude and d.c. offset. A pulse, synchronized to the ramp voltage, triggers a Model ND-2400 Nuclear Data 4096-channel multichannel analyser which operates in the multichannel scaling mode of data acquisition and which obtains its pulsed signal input from the amplifier/discriminator unit. This arrangement permits the storage of signals from a particular mass into a corresponding channel during repetitive mass scanning. The mass resolution of the system is determined by the ion beam optics, the stability of electronics, and the number of storage channels per mass unit in the multichannel analyser. Present resolution is limited by electronic stability and is in the range of 300 to 1000, depending on the mass number.

Figures 4 and 5 show fingerprint mass spectra which illustrate the performance of the instrument. Figure 4 is a group of three spectra, two of which (4(a) and 4(b)) were obtained from the same brand X No. 6 crude oil sample while the third spectrum (4(c)) was obtained from a brand Y No. 6 oil. The data for each of these spectra were accumulated over a period of about one-half hour during which time the sample temperature was gradually increased from room temperature to about 200°C. The degree of instrument reproducibility is indicated in spectra 4(a) and 4(b). The dissimilarity between these two spectra and that of 4(c) demonstrates the effectiveness of the fingerprinting technique. Although the envelope of mass peaks is similar for brand X and Y, the distribution of mass peak amplitudes within each homologous series is visibly different for these two brands.

The instrument is also being applied to urine analysis. Figure 5 shows a typical urine spectrum. Water was removed from a 10 ml urine sample by lyophilization prior to mass analysis. The sample was temperature programmed during the mass analysis from ambient to about 200°C.

Development effort in the immediate future will be directed towards improving both sensitivity and resolution of the instrument. This will involve further development of the multipoint source as well as increasing the stability of various crucial electronic components of the instrument.

REFERENCES

1. Aberth, W. H., Spindt, C. A. and Sperry, R. R., 21st Annual Conf. on Mass Spectrometry and Allied Topics, San Francisco, May 1973.
2. Colutron velocity filter, Model 300-6, Colutron Corp., Boulder, Colorado 80302.
3. Wahlin, L., *Nucl. Instr. Meth.*, 1964, 27, 55.
4. Seliger, R. L., 11th Symposium on Electron, Ion and Laser Beam Technology, Boulder, Colorado, San Francisco Press, Inc., 1971, p. 183.
5. Microchannel plates manufactured by Varian, Palo Alto Tube Division, Palo Alto, California 94303.

Discussion

Peter J. Derrick (University College, London, U.K.): I have doubts as to whether field ionization mass spectrometry can be useful for 'fingerprinting' complex mixtures. Field ionization can generally not be used for quantitative analysis, since the field ion current for any component in a mixture typically depends on other factors besides its partial pressure. Some of these factors are not understood, but polarization energies and ease of condensation are certainly important. The composition of a field ion mass spectrum of a mixture can vary drastically with temperature and with total pressure. It is possible for a component of a mixture to be suppressed completely merely by the presence of certain other components and so give no peaks at all in the mass spectrum. It is these objections which have prevented field ionization mass spectrometry from becoming the very powerful analytical tool it once promised to be.

W. Aberth: Fingerprint mass spectrometry requires that identical multicomponent samples will yield identical mass spectrum patterns. Quantitative evaluation of one sample constituent relative to another in the same sample is not important and consequently neither is the relative ionization efficiency of the different sample compounds important. However, quantitative comparison of a sample constituent relative to the same constituent in a different sample can still be made. The change in relative ionization efficiency with temperature and pressure experienced by some investigators may be due more to a change in surface interaction rates and sample composition than changing field ionization efficiency. It should also be pointed out that until now no apparatus was capable of producing true mass fingerprints and hence reproducible spectra of multicomponent samples were unobtainable.

49

An Interactive Display Oriented Data System for GC–MS

By VICTOR L. DaGRAGNANO and H. P. HOTZ

(Finnigan Corporation, Sunnyvale, California, U.S.A.)

THE use of small computers to control the data acquisition with gas chromatograph/mass spectrometers has become common. However, these systems have been rather inflexible in their handling of data. Data have had to be plotted or listed in order to be interpreted, a process that has required considerable time. We have designed a fully-integrated data system which provides live display of the spectra on an oscilloscope. Since the data displayed are resident in the computer's memory, it is straightforward to provide data manipulation directly. To avoid tedious dialogues via teletype, our computer functions are controlled principally by push buttons. The push buttons are read by the computer, which then executes the desired function. Outputs appear as alphanumerics on the oscilloscope or, optionally, typed on the teletype.

The organization of the data system and its relation to the gas chromatograph/mass spectrometer (GC–MS) is shown in Fig. 1. The data system controls a setting of the quadrupole mass spectrometer by means of mass-set voltage produced by a digital to analogue convertor (DAC). Data are acquired by taking the output of the electron multiplier through a preamp and amplifier/integrator to an analogue to digital convertor (ADC). These digitized amplitudes are stored in the computer's memory in the form of counts. Thus a mass spectrum consists of a series of counts for each channel, with each channel representing one atomic mass unit. The integration time is determined by the computer and is timed by the computer's real-time clock. Live display of spectra is maintained on the cathode ray tube (CRT) during acquisition. Past spectra and additional programs are stored in the disc memory. Hard copy data output may either be obtained on the teletype or on the digital plotter.

Core memory provides space for 2048 channels of data. During acquisition, this is divided into two halves, one for the gas chromatogram and one for the mass spectrum. Halves of these regions may be displayed at will and small segments may be chosen to be expanded across the entire CRT screen. Any two sections may be displayed simultaneously. After acquisition, one region of memory may be transferred to or stripped from any other region of memory. Thus, data can be manipulated quite freely and flexibly between the

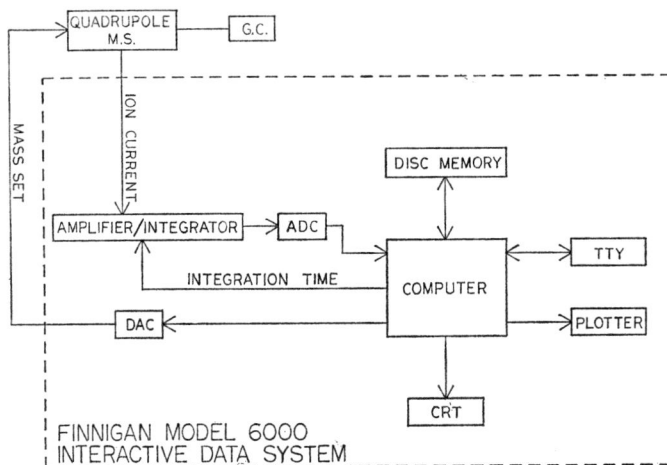

FIG. 1   Block diagram of the data system.

core memory divisions. This flexible handling of data in core memory provides the basis for all data handling in the system.

For data acquisition, one enters the parameters for the sample run into the system. Since a number of parameters must be entered, and alphanumeric information must be supplied, this is done by a means of a dialogue. Promptings are supplied by alphanumerics on the CRT and responses are provided by the operator on the teletype. An example of the information for a run is shown in Fig. 2. In general, the parts of the printing prior to the colon are the parts supplied by the computer and the parts following the colon have to be supplied by the operator. Once entered, the parameters remain in the machine and the operator only has to change the parameters he wishes. Figure 2 is an actual print-out by the computer of a set of current parameters for a run. Once the parameters have been entered, the computer will take control of the mass spectrometer and execute sequential mass scans recording the data as obtained. Upon the completion of each mass scan, the data are compacted and stored on the disc. The mass scan is integrated and one channel is added to the gas chromatogram in the appropriate region of core memory. One may select the live display region during acquisition just as at other times. Thus one can watch for a desired feature in a mass spectrum or the development of the gas chromatogram during acquisition. Two buttons are provided to control data acquisition. One is for acquire only and the data are not stored on the disc. One may then change to an acquire-and-save

```
FILE :UNC
TITLE:UNIV. OF NORTH CAROLINA SAMPLE BY GC-EI
MASS RANGE :30-199;200-400
INTEG. TIME:4,8
SECONDS PER SCAN :4
THRESHOLD:1
INST. RANGE SETTING:H
MAX. RUN TIME 66
```

FIG. 2   Example of run parameters listing.

mode by pushing a second button. Mass spectra will then be saved beginning with the one currently being acquired. The computer will not allow acquisition of data with a file name identical to that of data previously acquired. Thus there is no danger of confusing data in the disc files.

A second data acquisition mode is provided, which we call the mass fragmentography mode. In this mode, complete mass spectra are not acquired, but only mass fragmentograms are obtained. Up to four mass fragmentograms may be obtained simultaneously. Upon conclusion of data acquisition, these data are automatically written on disc for future reference. In this mode, the computer chooses the integration time so that maximum precision of the data is ensured.

After data have been acquired, data may be recalled from disc memory to core memory by entering the file name into the acquisition parameters. When data from a past file are recalled, all of the parameters of that run are restored to core so that they may be read or a similar run may be taken. Gas chromatograms, except mass fragmentograms, are not stored on disc. Recalling a gas chromatogram results in a reconstructed chromatogram. This may be either the full chromatogram as acquired or a limited mass gas chromatogram. The mass limitations are determined by intensifying a range as set by the lever wheel switches. Any mass spectrum of the run may be recalled by pushing the button with the number of that spectrum intensified. If more than one spectrum is intensified, the average of those spectra is obtained.

The ability to intensify is controlled by a push button and two lever wheel switches. One switch controls the start channel and the other the number of channels of span. The span may be in either direction from the start channel. The use of intensified channels provides a means of focussing the computer's attention to particular channels in the data, and this procedure is used very much as a light pen is used in larger interactive data systems. One simple use of these switches is to determine the size of a fragment lost in the mass spectrum. One might start with a certain peak, set the direction of intensification downward, and increment the span until it just reaches the next lower peak. The number of mass units lost from the first peak to get the second would then be one less than the reading on the span lever wheel switch.

We provide three push buttons allowing one to determine the amplitude of either mass spectra mass peaks or gas chromatograph peaks. The first of these is total intensity, which is particularly useful for mass spectra. When this is pushed, the total of the channels intensified is output. The output is either in the form of the absolute counts stored in the data system or it may be converted to read in volts output of the preamplifier, corrected for whatever integration time was used in its acquisition.

The area push button allows a background to be subtracted from the total. Three button pushings are required to obtain this output. On the first and second pushings, background is determined by fitting (least squares fit) to the channels intensified. The peak is then intensified and the button pushed for the third time. The result is then output and consists of the net total above the least squares straight line fit to the background, the calculated number in these peak channels below the least squares line fit line to the background, and a size. The size is the net area multiplied by a factor which is the ratio of a size chosen for a standard peak area to the total counts previously obtained in that standard peak. Thus a normalized, or calibrated answer may be obtained for quantification of the analysis.

The ratio of the totals of any two peaks may be obtained by two pushings of the ratio push button. The first peak is intensified and the button pushed. Then the second peak is intensified and the button pushed again. When the second pushing occurs, the counts in the intensified channels of the second peak are divided by the counts in the intensified channels of the first peak and the ratio output. These may of course be either single channels or multiple channels in either peak, and it may thus be used either for mass spectra or gas chromatograms.

Data output is provided either on the teletype or on the digital plotter. A labelling provision is provided so that one may input one line of text to be applied to the data output. This line of text, which will appear on the CRT, is put at the top of the listing or plot and is followed by a line of the title of the run. When one of the output buttons is pushed, the data output are those shown on the CRT display. Normalization is provided by normalizing either to the highest channel within the display region, or to the first channel intensified if one is intensified. The teletype output may also be in absolute counts. As an example of the flexibility of output one might go through the following sequence of steps: one might recall a mass spectrum from near the top of a gas chromatogram peak, and save it by transferring it to the GC memory region. One could then recall a mass spectrum from a background region adjacent to the gas chromatogram peak, and subtract this background spectrum from the peak spectrum to obtain a net spectrum. The interesting portion of this mass spectrum could be expanded to fill the full screen and the resulting spectrum plotted. Of course, if the plot does not show what one desired, one would not have to plot it, since one can preview it on CRT. Selected portions of the plot may be multiplied by any desired factor so that small features may be made more visible. The plot will automatically indicate these scale factors when this is done. If one does not wish to wait for the output at the time one sets it up, one may add it to a line-up for delayed output. This is done by pushing the delayed output push-button and then indicating the kind of output desired. Later, such as when one is ready to leave for the day, one may instruct the data system to perform all of the delayed outputs. Gas chromatograms may also be plotted. When mass fragmentography is used, the resulting plots are somewhat different. The four mass fragmentograms are plotted above each other.

Automatic comparison of spectra has come to be a useful function of computers. We provide a library search routine in our data system. The search routine provides for storing multiple libraries of spectrum codes on the disc. The operator may add to or delete from these libraries at will. He may also output or input a library from paper tape so that additional libraries may be exchanged with other users. The search scheme is one in which the mass of the most intense peak in each 14-atomic-mass unit range is listed. Fifty-two ranges, beginning with mass 34 are used. A spectrum obtained on the data system may be automatically coded, the code edited, and the code searched against the library codes. The ten best fits are then output, together with the number of mis-matches for each. Provision is made for altering parameters in the search so that missing peaks in the library are, or are not significant and unknown missing peaks in the spectrum are, or are not significant.

In conclusion, let us emphasize that the data system is display oriented. During acquisition, that which is displayed are the input data. On output, what you see is what will be reduced to hard copy. Data manipulation is of the displayed data and other data are called into display for use. The display orientation of the system thus allows significantly more rapid data manipulation and ensures that only significant data will be outputed.

50
Analysis of Ionization Efficiency Curves

By R. A. W. JOHNSTONE and B. N. McMASTER

(*The Robert Robinson Laboratories, The University, Liverpool, U.K.*)

AT the last International Mass Spectrometry Meeting in Brussels, we described[1] a computer-aided data acquisition system for obtaining electron-impact ionization efficiency curves. After mathematical smoothing, the data were processed by the EDD method[2] and we have termed the whole acquisition and interpretation of data as the IE/EDD method. The technique has been used to compare first ionization potentials with those obtained by photoelectron spectroscopy and generally good agreement was observed between the two methods.[3] Some appearance potentials were also measured[3] and the IE/EDD technique has been used to compare the appearance potentials of metastable and normal ions.[4] In practice, the IE/EDD method yields a well-defined appearance potential, but we have been conscious of the dependence of the measured appearance potential on the shape of the foot of the ionization efficiency curve. In this region, the curve-shape is dependent not only on the electron energy distribution but also on the ionization cross-section which can change rapidly at low energies. Accordingly, we examined the EDD method in more detail, and this led us to a new method for processing appearance potential data which removes almost all the subjectivity inherent in other methods.

CONSIDERATION OF THE EDD METHOD

This processing of ionization efficiency data is carried out[2] by defining a 'true' ionization current at an electron energy (E) by expression (1)

$$I(E) = i(V) - bi(V + \Delta V) \tag{1}$$

where i is the measured ion current at electron voltage (V) and b is a factor, generally about 0·9. We have found that the EDD method is really a critical slope approach. Rearranging eqn (1) yields (2) and in the limit $\Delta V \to 0$, eqn (3) is obtained. Then, if V_0 is the voltage at $I(E) = 0$,

$$I(E) = (1 - b)i(V) - b\,\Delta V\left(\frac{\Delta i(V)}{\Delta V}\right); \qquad \Delta i(V) = i(V + \Delta V) - i(V) \tag{2}$$

$$I(E) = (1 - b)i'(V)\left(\frac{i(V)}{i'(V)} - \frac{b\,dV}{(1-b)}\right) \qquad (3)$$

we have $i(V_0)/i'(V_0) = b\,dV/(1-b)$, assuming $(1-b)i'(V_0) \neq 0$.

Equation (3) shows that $I(E)$ depends critically on the behaviour of $i(V)/i'(V) = \{d[\ln i(V)]/dV\}^{-1}$ and may be considered a critical slope method.

The EDD method performs an exact deconvolution if the electron energy distribution is exponential $[m(U) = \exp(-U/kT)$, with $b = \exp(-\Delta V/kT)]$ but this is never the case in practice. Moreover, it is not easy to choose the best b-value for any given set of experimental data. For example, if the

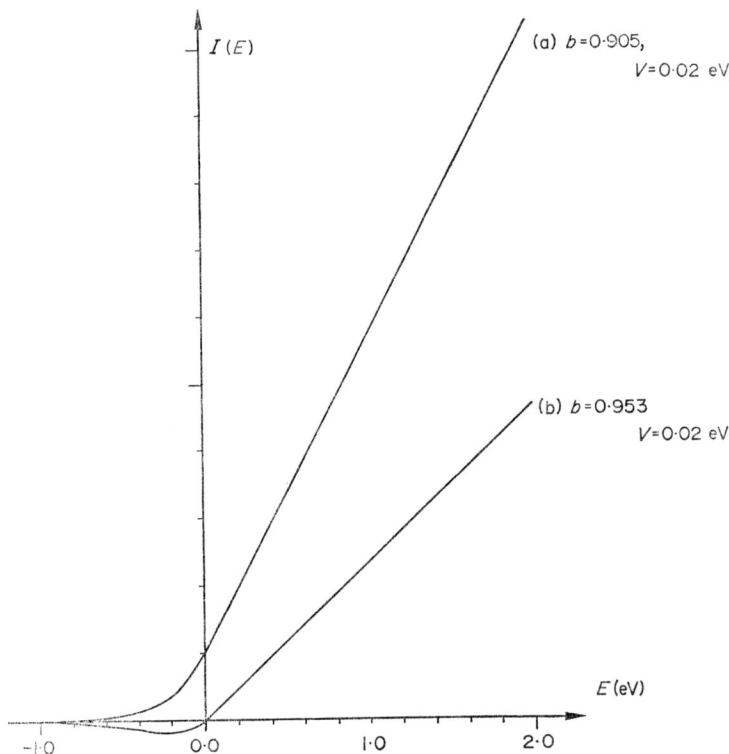

FIG. 1   EDD curves for Maxwellian electron energy distribution (with linear threshold law).

electron energy distribution is of a Maxwellian type $[m(U) = U\exp(-U/kT)]$ a b-value can be chosen so that $I(E)$ reaches zero asymptotically thereby removing the low energy tail but giving the wrong appearance potential (Fig. 1(a)); a suitably larger value of b will cause $I(E)$ to cut the abscissa at the true appearance potential before reaching a minimum and approaching zero asymptotically from a negative direction (Fig. 1(b)). Thus the appearance potential measurement is subjective and depends on a correct choice for b. Usually this is done by fitting b to the known appearance potential of a reference compound, but there is no reason to assume the threshold laws are the same for both the reference and the compound under investigation.

CONSIDERATION OF THE DOUBLE EDD METHOD

If the EDD method is applied twice, the expression (4) is obtained[5] and this is an exact inverse convolver for a Maxwellian type distribution, with $b = \exp(-\Delta V/kT)$.

$$I(E) = i(V) - 2bi(V + \Delta V) + b^2 i(V + 2\Delta V) \tag{4}$$

Although the electron energy distribution in the instrument is unlikely to be of the Maxwellian type, it may be close enough for this method to work satisfactorily. The double EDD method is certainly better than the single EDD process and has been used by us to compare with the critical slope curve matching (CSCM) developed below.

CONSIDERATION OF CRITICAL SLOPES

A function of the form, $y = x^n$, becomes $y/y' = (1/n)x$ after differentiation and gives the threshold law for any ionization efficiency curve except near threshold, where it also depends on the form of the electron energy distribution. As examples, we show in Fig. 2 the effect of convoluting a linear threshold law ($n = 1$) with three types of electron energy distributions and then

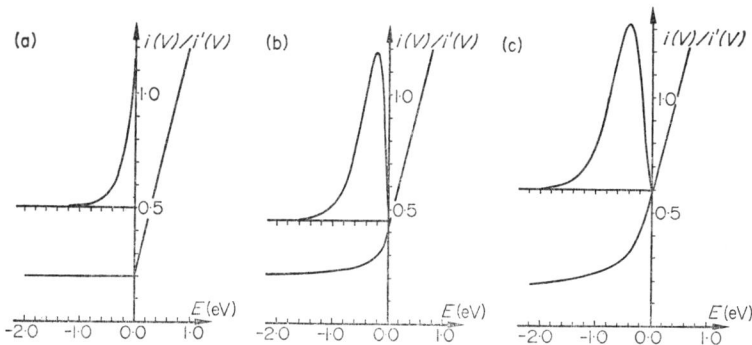

FIG. 2   Calculated $i(V)/i'(V)$ curves for different electron energy distributions (with linear threshold law). Insets show the reversed distributions, i.e. $i''(V)''$. (a) Exponential, $m(U) = \exp(-5U)$. (b) Maxwellian, $m(U) = U \exp(-5U)$. (c) Experimental, $m(U) = U^2 \exp(-5U^{1.1})$.

calculating the $i(V)/i'(V)$ curves. The first two distributions are standard, ideal ones, but the third is of a type $[m(U) = U^a \exp(-cU^b)]$ which can be fitted to the experimental data (as shown later) if sufficiently good data are available near threshold. It is apparent that this method can be used to distinguish both different threshold laws and different energy distributions.

CRITICAL SLOPE CURVE MATCHING (CSCM) METHOD

As shown in Fig. 2, for any particular electron energy distribution, the shape of the $i(V)/i'(V)$ curve can be predicted for various threshold laws. Accordingly, we have calculated $i(V)/i'(V)$ curves for various electron energy distributions, that is, the general formula given above has been used with different values of a, b, c. Generally, we have found that b is critical, since it alters the shapes of the curves markedly near threshold for small changes in its value. However, the exact values of a, c are less critical as these tend mostly to move the curves

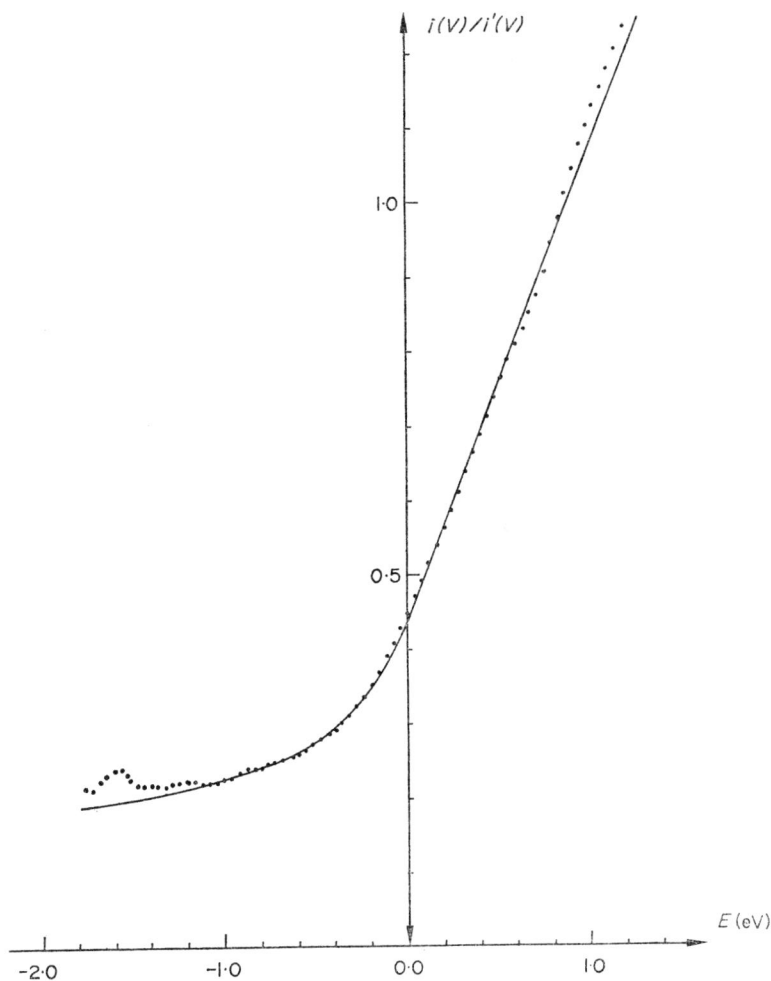

FIG. 3 Comparison of experimental and calculated $i(V)/i'(V)$ curves for argon. Dotted line = experimental points for Ar·+. Full line = calculated curve with $a = 2\cdot0$, $b = 1\cdot1$, $c = 5\cdot0$, $n = 1\cdot5$.

bodily along the ordinate. The shapes of $i(V)/i'(V)$ curves for different values of a, b, c, n were calculated and compared or 'matched' with the experimental $i(V)/i'(V)$ curves in the threshold region, where the shape depends primarily on the electron-energy distribution. The distribution giving the best fit to the data is then used to calculate (total) $i(V)/i'(V)$ curves for different threshold laws (n). These calculated $i(V)/i'(V)$ curves when matched with the experimental ones give the appearance potential (Fig. 3).

In use, the CSCM method is fairly rapid given a family of calculated $i(V)/i'(V)$ curves for matching with the experimentally determined curves. The method has a number of advantages (i) the experimental electron energy distribution for the ion source containing the compound to be investigated is determined each time, and does not rely on it being the same at some other time with other sources or compounds (ii) changes in the threshold laws for different substances can be compared (iii) the method is not subjective (except in getting the best fit of curves) and therefore does not rely on assumptions regarding the shapes of the ionization efficiency curves. The $i(V)/i'(V)$ curves require no arbitrary normalization like other 'absolute' techniques such as the critical slope, derivative, and deconvolution methods, but unlike other semi-empirical methods (e.g. semi-log, extrapolated voltage difference, vanishing current, etc.). The CSCM technique would appear to provide an excellent means of determining electron-impact appearance potentials and examples of its use are given below.

EXPERIMENTAL APPEARANCE POTENTIALS

Appearance potentials for the processes shown in Table I were determined and are compared with other previous data. For Argon and Xenon a value of $b = 0.905$ ($\equiv kT = 0.2$ eV) with the double EDD method gave curves which rose quite sharply from the abscissa and a difference in ionization potentials

TABLE I

Comparison of Appearance Potentials (eV)

Double-EDD	CSCM	Other[6]
1. I.P. (Ar)—I.P. (Xe)		
3·63 ± 0·05	3·56 ± 0·05	3·629
2. I.P. (C_2H_5OH)		
10·58 ± 0·05	10·45 ± 0·05	10·50 (PI), 10·48 (PI), 10·63 (PE)
3. $C_2H_5OH \rightarrow C_2H_5O^+ + H\cdot + e$		
10·76 ± 0·05	10·58 ± 0·05	11·12 (RPD), 10·9 (VC), 10·95 (LE)
4. $C_2H_5OH \rightarrow CH_3O^+ + CH_3\cdot + e$		
11·27 ± 0·1	11·07 ± 0·1	12·34 (RPD), 11·3 (VC), 11·6 (LE)

PI = photoionization,   PE = photoelectron spectroscopy,
RPD = retarding potential difference,   VC = vanishing current,
LE = linear extrapolation.

very close to that found by spectroscopic methods. The CSCM method gave ionization potentials at different positions on the $i(V)$ curves, but with a similar difference, and showed similar threshold laws for the two gases ($n = 1.5$). For ethanol, the double EDD and CSCM methods gave somewhat different appearance potentials. Other values will be reported.

REFERENCES

1. Johnstone, R. A. W., Mellon, F. A. and Ward, S. D., *Advances in Mass Spectrometry*, 1970, **5**, 334.
2. Winters, R. E. and Collins, J. H., *J. Chem. Phys.*, 1966, **45**, 1931.
3. Johnstone, R. A. W. and Mellon, F. A., *J. Chem. Soc., Faraday II*, 1972, **68**, 1209.
4. Bentley, T. W., Johnstone, R. A. W. and McMaster, B. N., *JCS Chem. Comm.*, in press.
5. Vogt, J. and Pascual, C., *Int. J. Mass Spectrom. Ion Phys.*, 1972, **9**, 441.
6. Franklin, J. L., Dillard, J. G., Rosenstock, H. M., Herron, J. T., Draxl, K. and Field, F. H., 'Ionisation Potentials, Appearance Potentials and Heats of Formation of Gaseous Positive Ions', NSRDS-NBS 26 (1969).

Discussion

A. J. C. Nicholson (C.S.I.R.O. Australia): How does your method work when the 'straight' portion of your curves extend across several vibrational levels of the ion or even across a region where there are poorly resolved autoionizing levels?

B. N. McMaster: The method is only expected to give a reliable appearance potential when the critical slope curve has a linear 'post-threshold' region extending over about 0·5 eV and gives a good fit to the calculated curve. We expect this to be the case for ionization efficiency curves of polyatomic fragment ions where structure is seldom observed, especially when using an unfiltered electron beam. However, if any structure is partially resolved, the critical slope curve will exhibit broad 'wobbles' in the post-threshold region rather than a straight line and a reliable appearance potential could not be derived. This situation is more likely to arise when measuring ionization potentials of small molecules and photo-electron spectroscopy would be much preferred in these cases. We note that this method thus provides its own check on the reliability of any measured appearance potential from the quality of fit of the calculated and experimental critical slope curves.

51

Coupling of a Liquid Chromatograph to a Mass Spectrometer

By R. E. LOVINS, S. R. ELLIS, G. D. TOLBERT
and C. R. McKINNEY

(*Department of Biochemistry, University of Georgia, Athens, Georgia, U.S.A.*)

INTRODUCTION

LIQUID chromatography has in recent years enjoyed a widespread resurgence in interest and analytical usage primarily as a result of an increasing number of chemicals which are either thermally labile and consequently unsuited for gas chromatographic analysis or which have other characteristics which to date have required alternative methods of analysis. The analysis of such compounds as nucleotides,[1] nucleic acid bases,[1] steroids,[2] analgesics,[3] vitamins and natural products,[4] as well as the identification of such environmental hazards as polychlorinated biphenyls and other pesticides,[5] has reportedly been accomplished using high speed liquid chromatography.

As a result of the increasingly widespread applications of liquid chromatography in analytical and bioanalytical procedures, the need for reliable methods of analysing and identifying the components generated by liquid chromatography has become apparent. This paper describes an interface which has been designed to couple a liquid chromatograph directly to a mass spectrometer thereby providing on-line mass spectrometric identification of liquid chromatographic effluent components.

EXPERIMENTAL

Apparatus

A duPont Model 830 Liquid Chromatograph equipped with a 254-nm wavelength UV detector was used for the liquid chromatography experiments. The liquid chromatographic equipment was the generous loan of the E. I. duPont de Nemours Instrument Products Division. A duPont Model 21–490 single focussing mass spectrometer equipped with linear scan, variable ionizing voltage, and a differential pumping system was used for the mass spectrometric analyses.

A schematic of the interface coupling the mass spectrometer to the liquid chromatograph is shown in Fig. 1. The interface was designed to accept a portion of the effluent from the liquid chromatograph corresponding to a component peak of the mixture being separated, remove the solute from the

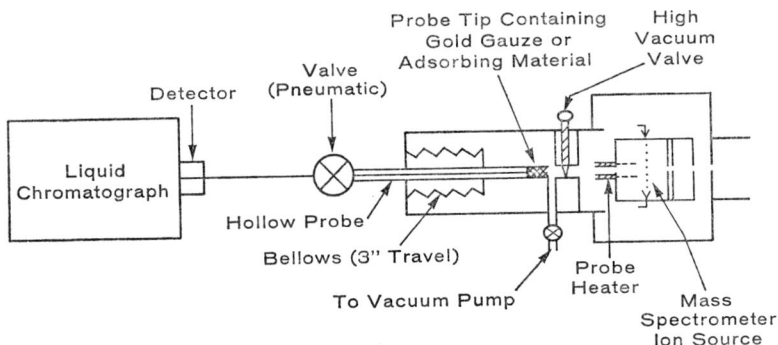

FIG. 1 Schematic of LC-MS interface showing the major components including the Capillary line from LC to pneumatic valve located on the hollow probe, the motor driven hollow probe on 3-in. bellows and the high vacuum inlet valve (motor driven).

mobile phase by either flash evaporation or selective adsorption and introduce the solute into the mass spectrometer ion source for analysis.

The interface is composed basically of a hollow probe which can be driven in and out of the mass spectrometer ion source and whose tip is packed with fine mesh gold gauze for flash evaporation or an active adsorbent such as charcoal. The liquid chromatograph is connected to the hollow probe by 3–4 ft of stainless steel capillary tubing (1 μlitre/in. capacity). The tubing is connected to a small Teflon reservoir (10 ml) which is part of a pneumatic valve assembly having zero dead volume attached to the end of the hollow probe. The reservoir/valve assembly was designed and fabricated in the University of Georgia Instrument Shop. The hollow probe ($12\frac{3}{8}$ in.) has an inside diameter of 0·032 in., with the centre tube running from the Teflon pneumatic valve to the removable probe tip. The removable probe tip has the dimensions 1·09 × 0·128 in. i.d. The probe is attached to the outer housing through a metal bellows (Metal Bellows Corp., Sharon, Mass.) having a 3 in. travel and is motor driven (Bodine type 729-WH K 1506 130 rev/min) in and out of the mass spectrometer. The high vacuum isolation valve separating the probe housing from the mass spectrometer ion source was designed by E. I. duPont Instrument Products Division and fabricated in the University of Georgia Instrument Shop. The isolation valve is also motor driven on a bellows having a 1 in. travel.

Procedure
The sample in the eluting solvent from the liquid chromatograph after detection is collected in the Teflon reservoir/valve assembly on the probe. After evacuation of the probe and probe housing (~ 10 sec), the pneumatic valve of the reservoir is opened and the sample is drawn into the probe tip. For flash evaporation of the solvent/solute the probe tip is maintained at a temperature of 80°–120°C [depending on the eluting solvent[6]]. After the solute material has been isolated on the gold gauze in the probe tip and the solvent has been removed, the probe and housing is evacuated ($\sim 10^{-6}$ torr), the isolation valve is opened and the probe is inserted into the ion source of the mass spectrometer for analysis. The complete cycle requires 5–7 min. In

order to collect all the peaks emerging from the liquid chromatograph for analysis, especially those which elute very close to each other, the liquid chromatograph must be operated in a stopped flow mode such that the solvent flow from the chromatograph is stopped while each emerging peak is being isolated and analysed through the interface as described.

RESULTS AND DISCUSSION

The manual isolation and mass spectral analysis of individual fractions from a multicomponent system in a liquid chromatograph, while not as involved as

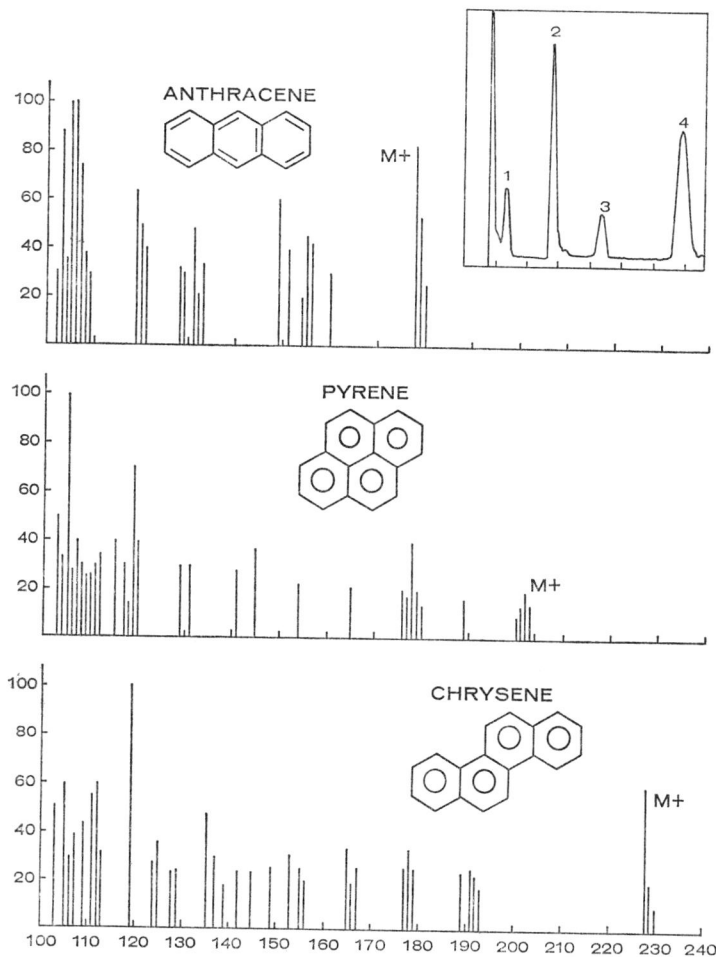

FIG. 2 Liquid Chromatogram of a mixture containing 1-naphthalene, 2-anthracene, 3-pyrene and 4-chrysene. Separated on a column having a stationary phase of octadecyl-silane and a mobile phase of 70% methanol/30% water at 1000 lb/in², ambient temperature (Inset) and the corresponding mass spectra of the components isolated on the LC-MS interface probe from peaks 2, 3 and 4 of the liquid chromatographic separation shown in the inset.

the trapping of components emerging from a gas chromatograph for mass spectral analysis, can nevertheless be time consuming. The interface coupling the mass spectrometer to a liquid chromatograph described above has been designed to eliminate much of the routine manual isolation procedures otherwise necessary in liquid chromatography.

The sample is transferred from the liquid chromatograph to the interface through capillary tubing (1 μlitre/in.) to minimize loss of material and cross-contamination. The samples presented in this paper have been isolated from the eluting solvent by flash evaporation. The probe temperature was maintained at 80–120°C at a pressure of $\sim 10^{-2}$ torr,[6] during sample introduction. The small orifice in the probe (0·32 in.) acts as a restriction thus maintaining a sufficient vacuum during sample transfer from the Teflon reservoir to the

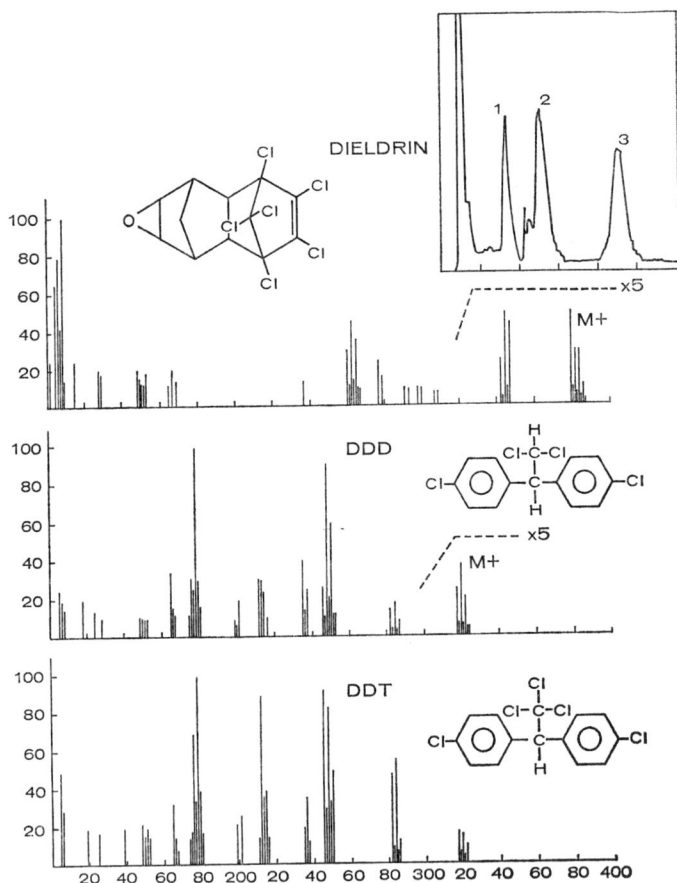

FIG. 3   The liquid chromatogram of a mixture of Dieldrin, DDD and DDT separated on a duPont ETH column using 70% heptane/30% isopropyl alcohol at 1000 lb/in.², ambient temperature (shown in inset) and the corresponding mass spectra of the components isolated using the LC-MS interface probe corresponding to the compounds in peaks 1, 2 and 3 in the liquid chromatograms shown in the inset.

probe tip to ensure efficient flashing of the solvent and consequent coating of solute on the gold gauze. Preliminary studies of percent retention of non-volatile solute during flash evaporation of solvent have shown the efficiency of the flashing evaporative separation technique to be 60–80 % depending on the ratio of solvent/solute vapour pressures. We have not undertaken, at present, an exhaustive study of the most favourable vapour pressure ratios for solvent/solute separation. We are however, investigating the use of high capacity adsorbent materials in the probe tip so that separation of solvent/solute can be made on the basis of selective adsorption rather than on vapour pressure differences thus increasing the number of volatile components which can be isolated through the interface. After isolation of the material on the probe tip, the probe is inserted into the ion source, the sample is volatilized and ionized. The separation of two multicomponent mixtures and the subsequent mass spectral analysis data obtained after isolating the components through the interface are presented in Figs. 2 and 3 and are included to demonstrate the utility and limitations of the apparatus. Figure 2 contains the liquid chromatogram (inset) and mass spectra of a mixture of polycyclic hydro-carbons (1-naphthalene, 2-anthracene, 3-pyrene, 4-chrysene).

Note that no mass spectral data is included for peak No. 1 (naphthalene). This was due to the fact that the vapour pressure of naphthalene is sufficient to cause the compound to be vaporized with the solvent under the conditions used rather than being retained on the gold gauze. Use of adsorbing materials for solute isolation should extend the use of the interface for isolation of such volatile components. Figure 3 contains the chromatogram (inset) of a mixture of chlorohydrocarbons (1-Dieldrin, 2-DDD, 3-DDT) and their corresponding individual mass spectra after isolation through the interface. Analyses were performed on 50–100 μg quantities of material.

The design of the LC–MS interface represents the first step toward the marriage of two analytical tools whose potential capabilities and utility as a combined analytical tool hopefully are as varied and broad as has been demonstrated by gas chromatography–mass spectrometry.

REFERENCES

1. Kirkland, J. J., *J. Chromatogr. Sci.*, 1970, **8**, 72.
2. Henry, R. A., Schmit, J. A. and Dieckman, J. F., *J. Chromatogr. Sci.*, 1971, **9**, 513.
3. Henry, R. A. and Schmit, J. A., *Chromatographia*, 1970, **3**, 116.
4. Schmit, J. A., Henry, R. A., Williams, R. C. and Dieckman, J. F., *J. Chromatogr. Sci.*, 1971, **9**, 645.
5. Byrne, S. H., Schmit, J. A. and Johnson, P. E., *J. Chromatogr. Sci.*, 1971, **9**, 592.
6. Lovins, R. E., Craig, J., Fairwell, T. and McKinney, C., *Anal. Biochem.*, 1972, **47**, 539.

Discussion

P. Johnson (Hoechst Pharmaceutical Research, Milton Keynes, U.K.): Presumably you pass into the mass spectrometer only those fractions which have shown detectable peaks on the LC detector. You are therefore presumably limited to compounds with a good UV response or to having sufficient material for a RI response?

R. E. Lovins: Yes, these are limitations at present.

R. Tanner (Allen & Hanbury's Research Ltd, Ware, U.K.): What volume of eluate do you pass through the probe tip during accumulation of the solute?

R. E. Lovins: We usually pass 0·5–1·0 ml of eluate through the probe tip during isolation of the solute.

R. Tanner: Have you found any evidence for a memory effect due to incomplete cleaning of the probe tip during automated analysis?

R. E. Lovins: The memory effect is minimized by baking the probe at 280°–300° for a short time after each analysis to remove residual solute. After prolonged use (*i.e.* 1 or 2 days) the probe is removed and cleaned by washing or baking at high temperature for longer periods of time.

R. Large (British Petroleum Co. Ltd, Great Burgh, U.K.): What is the effective response time of the LC/MS system and do you operate on a stop/flow basis?

R. E. Lovins: The effective response time or cycle time of the probe is 2–5 minutes. As a consequence of the long turn around time of the probe interface we do operate the LC on a stopped-flow basis which seems to work adequately.

52
Qualitative and Quantitative Analysis of Dissolved Gas by Gas Chromatography— Mass Spectrometry

By D. LEIGH and N. LYNAUGH

(*V.G. Micromass Ltd, Winsford, Cheshire, U.K.*)

INTRODUCTION

The analysis of dissolved gas both qualitatively and quantitatively is important in many industrial applications. Previous methods of hydrocarbon oil analysis have used gas chromatography and/or infra-red spectroscopy[1] but these methods have necessitated preliminary degassing of large volumes (up to 5 litres) of oil to reach the desired detection levels (1 ppm or less). This process can only be carried out in small quantities in reasonably sized degassing chambers and is somewhat cumbersome and time consuming. A method of analysis which can be performed by direct processing of small quantities of liquid sample is therefore to be preferred. The technique of gas chromatography–mass spectrometry is already well established and together with multipeak monitoring facilities presents a sensitive and rapid method of qualitative and quantitative analysis of dissolved gas.

INSTRUMENT DETAILS

The instrument consists of a Micromass 6 mass spectrometer, a single-focussing 6 cm radius 90° instrument with electron multiplier detection, electro-magnet and eight-channel programmed peak selector. The instrument is interfaced with a dual column Pye 104 chromatograph using a molecular separator of the Watson-Biemann type. Figure 1 shows a schematic layout of the complete system.

The mass spectrometer can be operated in the normal scanning mode or in the programmed peak selection mode. In the latter mode (Fig. 1) individual masses are selected at a constant magnetic field strength using the PP2(8) eight-channel programmed peak selector. Using precision ten-turn potentiometers a voltage is selected on the PP2(8) and fed to the programmed power supply where the accelerating voltage required to tune a particular mass is developed. Up to eight individual selections are then automatically cycled at a

463

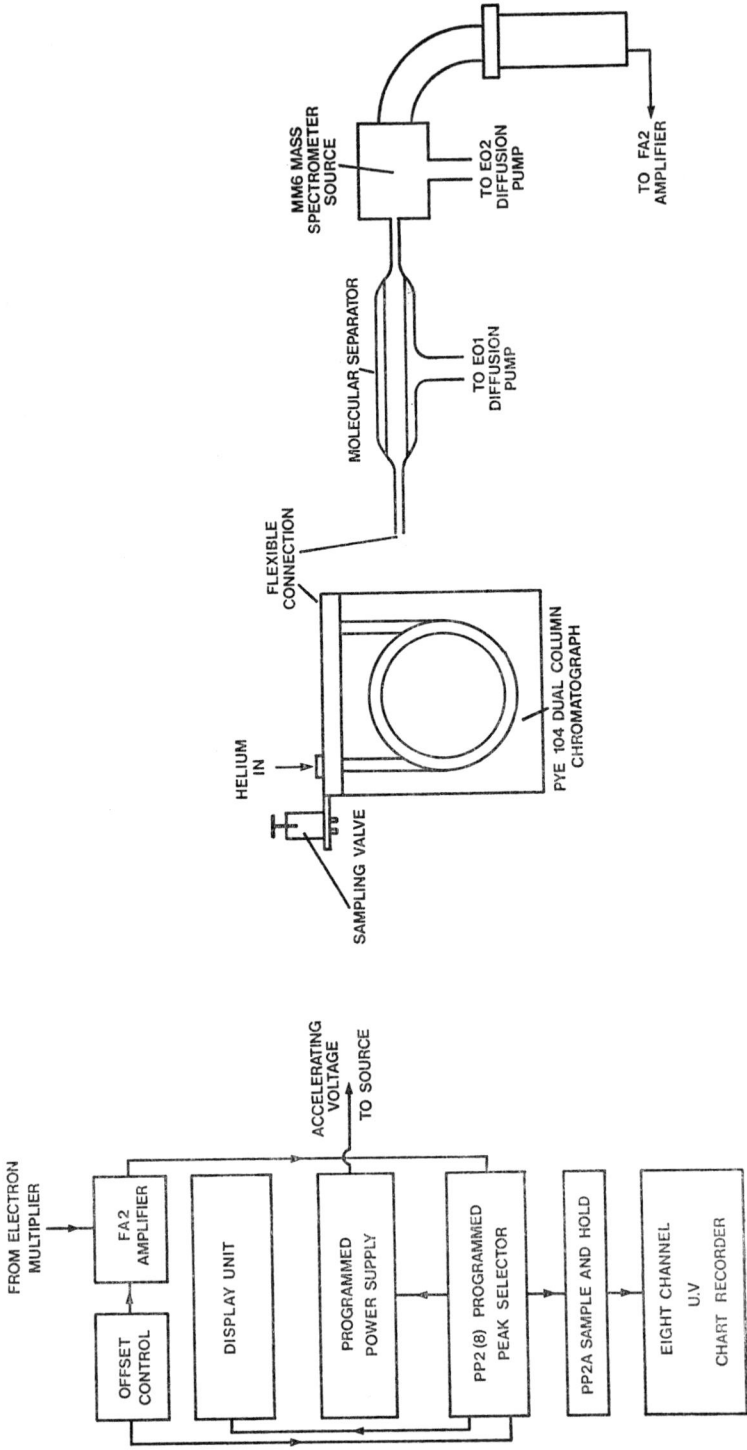

FIG. 1 Schematic diagram of the complete GC–MS system with multipeak monitoring.

rate of between 0·1 and 100 sec. The built-in amplifier in the PP2(8) incorporates an individual gain control for each channel allowing simultaneous display of masses with intensities in the ratio 100:1. In order to increase the overall dynamic range, the output from the electron multiplier is fed first to an FA2 fast amplifier which acts as a preamp for the PP2(8). An offset control on the input to the FA2 allows full background suppression on each channel individually. The maximum dynamic range thus achieved is 1 to 100 000, that is if 1 ppm is taken as the lower limit, analysis may be performed between 1 ppm and 10 % gas in liquid.

The output of the PP2(8) is fed to the PP2 A 'Sample-and-hold' unit which acts as an integrator with variable response between 1 and 3000 ms. The output from the unit is fed to an eight-channel UV chart recorder. A parallel output from the PP2(8) amplifier is also fed to an oscilloscope display unit.

The two columns used for the analysis are 5 ft. × ¼ in. o.d. glass columns packed with 85/100 mesh Activated alumina (Phase Separations Ltd) for analysis of C_2H_6, C_2H_4, C_3H_8, C_3H_6 and C_2H_2 and 85/100 mesh molecular sieve 5A (Phase Separations Ltd) for analysis of H_2, O_2, N_2, CH_4 and CO. The column exits are taken through a dummy Katharometer head unit to the rear of the chromatograph for connection to the mass spectrometer interface. The connection between column and interface is made with a short piece of 3 mm o.d., 1 mm i.d. silicone rubber which can easily be transferred from one column to the other.

By suitable choice of restriction between the column and interface the pressure at the column exit is arranged to be atmospheric at a volume flow rate of 20 ml/min helium.

The inlet system for the chromatograph consists of two, six-port gas

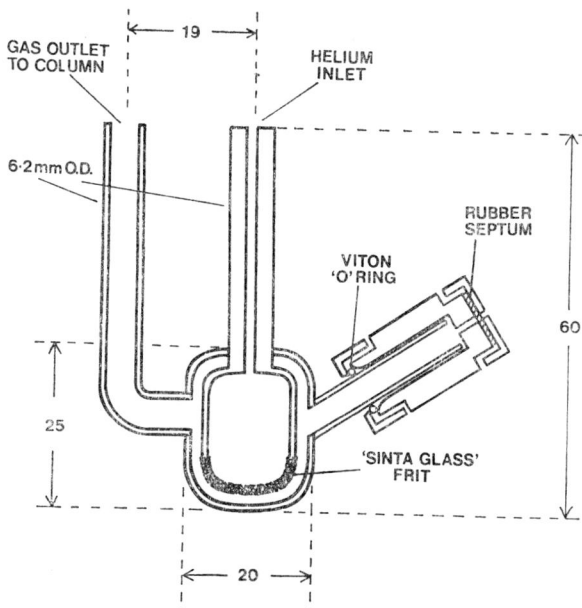

FIG. 2   Diagram of the oil degassing vessel (dimensions in millimetres).

sampling valves (Pye Unicam 12654), one for each column, which are used both for calibration and liquid sampling. In the calibration mode gas mixtures (prepared by Hilger and Watts, Westwood Industrial Estate, Margate, Kent, and guaranteed to 5%) containing 0·1% or 0·5% of each component in nitrogen are used together with several sample loops of known volume. The measured internal volume of the sampling valve is added to the loop volume and calibration based on peak area carried out before and after each liquid sample or batch of samples.

For the purpose of degassing the oil samples and transferring the gas to the chromatographic column, the apparatus of Fig. 2 has been constructed.[2] It consists of a 'Sinta glass' gas distribution tube (Gallenkamp GD802 porosity 2) enclosed in a Pyrex glass envelope giving an enclosed volume of approximately 3 ml. One side-arm carries a removable injection septum unit manufactured in stainless steel and held in position using a compression fitting with Viton 'O' ring. The other side-arm forms the gas outlet. The apparatus is connected between the normal sample loop ports of the gas sampling valve and is flushed out with helium. A sample of liquid is then injected via the rubber septum. The sample loop is then switched to the carrier stream. The carrier is made to bubble through the liquid to degas it and carries the free gas to the chromatographic column for analysis. The 'sinta glass' frit presents a negligible back pressure to the flow of carrier as judged by the pressure reading (Pirani gauge) in the interface pumping line.

THEORETICAL BACKGROUND

If 1 ml of liquid is chosen as the standard sample then for 1 ppm detection the problem is to detect and quantitatively estimate 10^{-6} ml of gas. This is equivalent to $1·9 \times 10^{-9}$ g of propane. If a quantity w(g) of sample gas produces a chromatographic peak which is t seconds wide at the base and gives N ions/second at the peak top and if the sensitivity of the instrument is Q C/μg then with the approximation that the peak is triangular

$$\tfrac{1}{2}Nt = \frac{Qw}{10^{-6}} \text{ coulombs}$$

$$= \frac{Qw}{1·6 \times 10^{-19} 10^{-6}} \quad \text{ions}$$

Therefore

$$N = \frac{1·25 \times 10^{25}}{t} Qw \quad \text{ions/sec}$$

If the peak selector has n channels and a cycle time T seconds for n channels then the number of ions/sampling (N_s) collected at mass m during sampling at the peak top is

$$N_s = \frac{NT}{n} = \frac{1·25 \times 10^{25}}{tn} QwT$$

If $N_s = 20$ ions/sampling for minimum detection then the minimum amount of gas detectable (W_{min}) is

$$W_{min} = \frac{20\,tn}{1\cdot25 \times 10^{25}\,QT}$$

If $Q = 10^{-11}$ C/μg (experimental value for propane mass 29) then with $n = 8$, $T = 2$ and $t = 20$ sec,

$$W_{min} = 2\cdot56 \times 10^{-11}\,g$$

This is equivalent to 0·014 ppm propane from a 1 ml sample. The value of Q will be different for each component due to variations in separator efficiency and ionisation efficiency. The range of values is approximately 10^{-11} C/μg propane to $5\cdot10^{-12}$ C/μg for hydrogen.

The above calculations assume that there is no background at mass m. In general, the minimum detectable mass of gas is increased by the presence of residuals. For example, if there is a residual ion current on mass 29 equivalent to say 10^6 ions/sec then the number of residual ions collected per sampling, cycling at 1 sec is $10^{-6}/8 = 125\,000$. The statistical noise (±3 standard deviations which gives a 99% confidence limit that a signal equal to the noise will be detected) on mass 29 would be $\pm3\,(125\,000)^{\frac{1}{2}} = \pm1060$. The detection level is thus raised by a factor $2120/20 = 100$. For propane therefore the level would be raised to $2\cdot56 \times 10^{-9}$ g, equivalent to 1·4 ppm.

As well as the detection limits imposed by the statistics of ion current detection there are practical limitations to residual ion current stability due to variations in pumping speeds and ion source stability. Variations in pumping speeds are particularly important when large amounts of helium are present in the system in GC–MS work. In general the practical limit of detection is of the order of 0·1% of the residual current. Thus, with residual current levels below approximately 10^6 ions/sec and at a cycling rate of 1 sec the statistical noise is the overriding factor. At residual levels greater than approximately 10^6 ions/sec the detection level is determined by the practical 0·1% factor.

EXPERIMENTAL

The mass spectrometer is operated with wide slits (0·5 mm source slit and 1·0 mm collector slit) giving flat-topped peaks with maximum sensitivity at a resolving power (10% valley definition) of 50. This is sufficient for the light hydrocarbon gases under study. In order to minimize the background ion current due to helium the source is operated with 30 eV electron energy. Other operating parameters were 100 μA trap current and +15 V ion repeller voltage.

Seven masses are selected as shown in Table I at a fixed magnetic field and are displayed continuously at a cycling rate of 1 sec. It is impractical to include Mass 2(hydrogen) together with masses up to 44 because of the loss of sensitivity over a wide range of accelerating voltage. Mass 2 is therefore tuned separately by changing the magnetic field strength and appears on channel 4.

It is reasonable to expect that the removal of gas from the oil using the apparatus of Fig. 2 will follow an exponential law giving rise to peak tailing

TABLE I

Selected Base Peaks For Analysis

Channel	Mass	Component
1	41	C_3H_6
2	32	O_2
3	28	N_2, CO, C_2H_6, C_2H_4
4	16 (2)	CH_4 (H_2 with field change)
5	26	C_2H_2, C_2H_6, C_2H_4
6	29	C_3H_8
7	44	CO_2

and consequent loss of chromatographic resolution. This is indeed the case when the chromatogram is obtained isothermally.

If, however, the analysis is performed with linear temperature programming of the column, the resulting chromatographic peaks are considerably sharpened. Figure 3 shows a multipeak chromatogram obtained from a sample of oil which had been exposed to a gas mixture containing 0·1% hydrocarbon gases in air. The program was started on injection and ran from 50°–150°C at 10°/min. Figure 3 demonstrates the advantages of multipeak programming in that ethane and ethylene can be determined without interference from the large amount of nitrogen present by tuning to mass 26 or 29. Also, nitrogen and oxygen can be determined by their response on masses 28 and 32 respectively, though they are chromatographically unresolved.

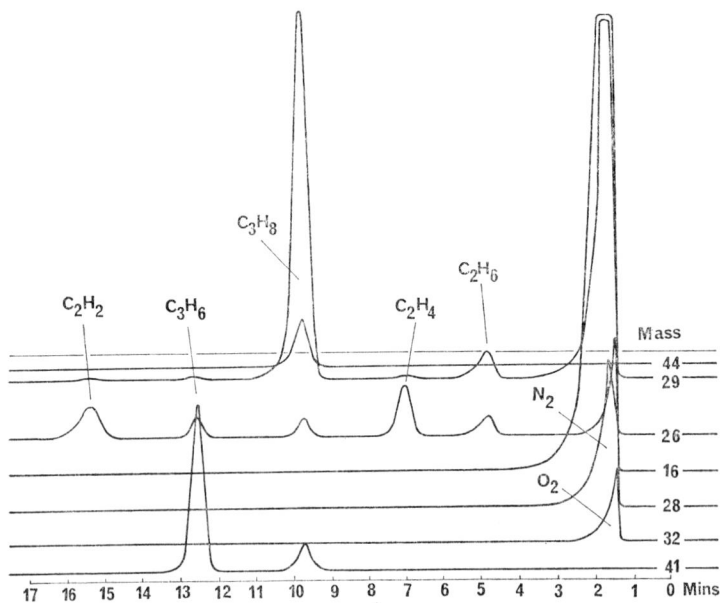

Fig. 3   Multipeak chromatogram obtained from a prepared oil sample.

RESULTS

Table II summarizes the detection levels obtained using a gas mixture with components present at a level equivalent to 100 ppm in a 1 ml oil sample.

Since the chromatographic peaks obtained from liquid samples exhibit some tailing the detection level for dissolved gas is raised by approximately 50% over the values shown in Table II.

TABLE II

Detection Levels Thus Far Established

Column	Mass	Component	Detection limit ppm
Activated Alumina	41	C_3H_6	0·1
	32	O_2	12
	28	N_2	48
	26	C_2H_2	1
	26	C_2H_6	0·8
	26	C_2H_4	0·5
	29	C_3H_8	0·4
	44	CO_2	1·6
Mol Sieve 5A	2	H_2	20
	32	O_2	16
	28	N_2	70
	28	CO	80
	15	CH_4	0·8

In order to test the efficiency of the oil degassing technique several oil samples containing quantities of propane were prepared. Propane was chosen as the test gas since it is the most soluble (1900% in oil) of the gases under study. If the process is efficient for propane therefore it may be judged to be so for the less soluble components. Table III summarizes the results obtained.

TABLE III

Analysis of Prepared Oil Samples Containing Propane

Nominal propane concentration (ppm)	Propane analysis (ppm)
2 500	2 566 (236)
1 000	830 (60)
500	570 (60)
250	245 (26)
125	124 (12)

Figures in parentheses are standard deviations.

DISCUSSION

The values shown in Table II for the detection level of hydrocarbon gases are approaching the theoretical limit of the present system. The values for atmospheric gases and CO are substantially greater than the theoretical limit due mainly to the high residual level of these gases in the system. The source of these residuals appears to be the chromatographic column and is independent of the type of column used. An attempt at purification of the helium stream using molecular sieve traps at liquid nitrogen temperature showed no improvement in the residual levels. The detection level of hydrogen appears rather high but is due mainly to the volume/volume method of measurement. The mass detection sensitivity is only a factor of 2 less than that for propane despite the expected reduction in the transfer efficiency of the interface for the lighter gas. The limit of detection for hydrogen can therefore only be improved by degassing larger volumes of liquid samples. The values in Table III indicate a standard deviation of 10% for propane in oil. This is probably the limit of injection reproducibility for oil with the type of syringe used. Further improvements in detection levels may be obtained with improvement in column performance and with reduction of residual levels at masses corresponding to atmospheric gases.

REFERENCES

1. Waddington, F. B. and Alan, D. J., *Electrical review*, May 1969.
2. Thibault, M. and Galand, J., Laboratoire Central Des Industries Electrique, March 1969, p. 885.

53
A New Chemical Ionization Mass Spectrometer

By E. M. CHAIT, CHARLES BLANCHARD and
VICTOR H. ADAMS

(*E. I. du Pont de Nemours & Co.* (*Inc.*), *Instrument Products Division,
Monrovia, California, U.S.A.*)

ONE of the problems with the conventional electron ionization source is that molecular ions are often produced which are so excited that no peak representing the molecular weight of the compound or the intact molecule is observed in the spectra. This lack of molecular ion creates a problem in sample identification because one must depend on deduction of structure from fragment ions alone. In addition, the electron ionization spectra tend to be complex and, therefore, difficult to interpret. In cases where only a rapid identification is desired, the complexity of the electron ionization spectra is unnecessary.

In chemical ionization, the reaction between the ions formed by ion molecular reactions of a high pressure reagent gas (1 torr) and sample molecules produce a simplified spectrum characterized by very little fragmentation and an abundant M + H ion. This gives an indication of molecular weight even for samples which do not give a molecular ion under electron ionization conditions.

Now that chemical ionization has become a well established technique in mass spectrometry,[1] it is necessary to design chemical ionization mass spectrometers that allow ease of operation while allowing a variety of sample inlets. This is especially true in the case of a chemical ionization mass spectrometer designed for GC-MS operation.

INSTRUMENTATION

A new chemical ionization mass spectrometer appropriate for GC-MS operation as well as probe and batch introduction has been created by the addition of a modified ion source to the Du Pont 21-490 series mass spectrometers. This ion source which is a dual chemical ionization/electron ionization source is shown schematically in Fig. 1. The conventional source has been modified by closing the electron aperture and exit slits of the source to allow the requisite high pressure for chemical ionization to be built up in the source

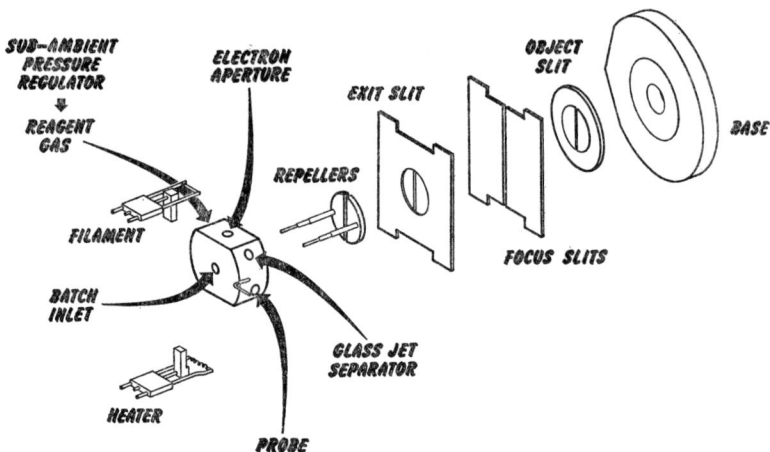

DUAL CI/EI SOURCE SCHEMATIC

FIG. 1　Schematic of Dual CI/EI ion source.

ionization chamber. The source is pressurized by the reagent gas through an opening in the ionization chamber with the reagent gas flow controlled by a subambient pressure regulator and needle valve. During chemical ionization operation, optimum sensitivities obtained at ionization chamber pressures of 0·4 to 0·8 torr. A high speed, high conductance pumping system has been incorporated as an integral part of the mass spectrometer. This pumping system accommodates the source gas load and maintains the analyser of the mass spectrometer at 3×10^{-7} torr while the ionization chamber is at 1 torr. GC effluents may be introduced into the source via a glass jet separator using helium as the carrier gas. This arrangement allows a unique approach to GC-MS chemical ionization in which full freedom of choice of reagent gases is allowed while the GC can be carried out under conditions of helium carrier gas which are familiar to chromatographers. In actual operating experience, such diverse reagent gases have been used as methane, isobutane, carbon monoxide, and water/argon mixtures.[2] Since the source is a dual chemical ionization/ electron ionization source, it may be converted back to electron ionization operation merely by cutting off the flow of reagent gas. Under these circumstances, it is possible to change modes of ionization within 15 sec, therefore allowing both types of spectra to be obtained during the same GC run.

CHEMICAL IONIZATION GC-MS

Since the source has been designed for optimum GC-MS performance, it is possible to obtain extremely high sensitivity for GC-MS in the nanogramme range. Figure 2 shows the sensitivity that can be expected from this source clearly indicating an abundant M + H peak for as little as 5 ng of methyl stearate injected on the column with isobutane used as the reagent gas. The

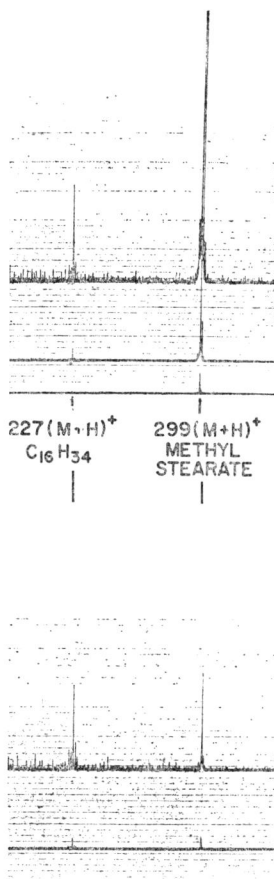

FIG. 2   200 ng and 5 ng of methyl stearate scanned at 4 sec/decade. Isobutane reagent at 0·5 torr pressure.

results on a total GC-MS separation are shown in Figs. 3 and 4. Figure 3 shows the separation of a mixture of barbiturate drugs; three of these drugs, pentobarbital, hexobarbital and secobarbital show no molecular ion under conventional electron ionization conditions. When run under chemical ionization conditions, however, using isobutane as a reagent gas, abundant M + H ions are observed for each one of these components. The application of the chemical ionization system for GC/MS is facilitated by the use of computerized mass spectrometer data system, such as the Du Pont 21-094. Under these conditions, because the chemical ionization source is also an electron ionization source, it is possible to calibrate the mass spectrometer with perfluorokerosene or other appropriate mass standards in the electron ionization mode and switch to the chemical ionization mode for analytical work.

Typical applications for chemical ionization, GC-MS include those compounds which do not give any molecular ions under electron ionization conditions making analysis very difficult, for example, in the case of the

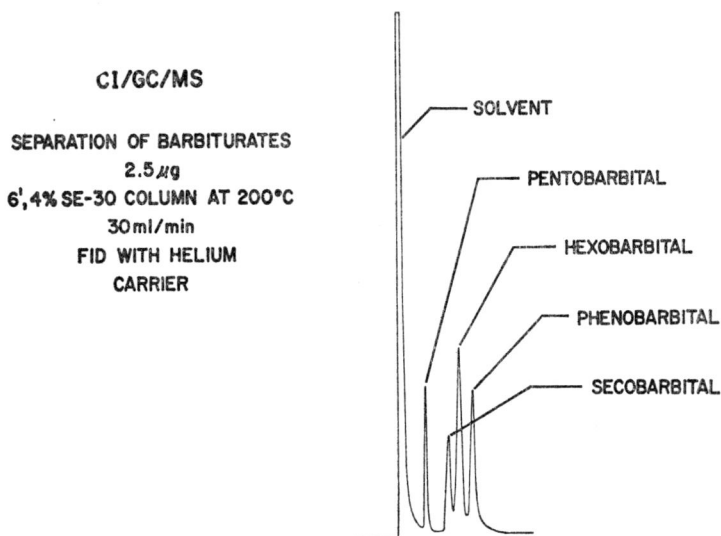

Fig. 3   GC separation of barbiturates.

trimethylsilyl ether derivatives of glycerols. Under these conditions, it is possible to obtain M + H ions for all components in a separation giving identification of the compounds.

DISTINCTION OF ISOMERS BY CHEMICAL IONIZATION

A unique application for the chemical ionization source is the distinction of isomers by the use of varying reagent gases.[3] Many organic compounds give very similar electron ionization spectra, particularly if they are isomers. Under these circumstances, it is difficult to distinguish the compounds; and methods which have been devised in order to distinguish isomers such as ion kinetic energy spectroscopy are ineffective in making this distinction on the high

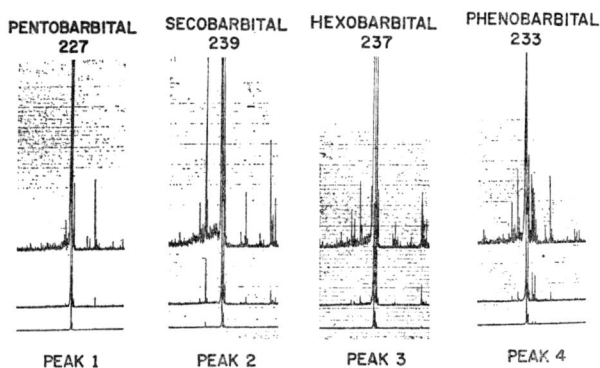

Fig. 4   M + H ions of separated barbiturates.

speed basis required by GC-MS analysis. Chemical ionization, however, offers this capability in conjunction with the GC-MS experiment. Because of the versatility of the chemical ionization source in using a variety of reagent gases, and maintaining GC conditions with helium as the carrier gas, it is possible to select the reagent gas that is appropriate to bring about the structure reactivity relationship necessary to distinguish isomers. Several cases in point represent the variety of capabilities of isomer distinction. Morphine and its isomer, morphone or Dilaudid give very similar electron ionization spectra. It is impossible to distinguish the compounds *a priori*. When these two compounds are run under chemical ionization conditions using methane as the reagent gas, two quite different spectra are obtained, however. Methane has an advantage over isobutane as the reagent gas since it produces a CH_5^+ which is a stronger acid and therefore promotes more fragmentation than the molecule while at the same time giving an abundant M + H ion. Under these conditions, the spectrum of morphine shows a characteristic loss of OH from the molecular ion as would be expected from the structure, whereas morphone shows a loss of 42 mass units or a loss of ketene ($CH_2C=O$) from the molecular ion, quite consistent with the carbonyl group on one of the rings replacing the hydroxyl. This ability to *a priori* distinguish keto enol isomerism seems to have far-reaching implications. In another case of isomerism, that of coumarin and isocoumarin, it is impossible to distinguish these compounds on the basis of their fragmentation in either electron ionization or chemical ionization with isobutane, methane or mixtures of argon and water. In the case of amobarbital and its isomer pentobarbital, however, it is possible through using an argon/water mixture as reagent gas to distinguish the two compounds, with the spectrum of pentobarbital favouring the fragmentation involving the hydrocarbon side chain.

These examples clearly indicate the value of doing chemical ionization in conjunction with electron ionization GC-MS experiments in order to obtain even more information from the sample, and at the same time to indicate the importance of being able to pressurize the ion source separately with reagent gas so that the conventional helium GC conditions can be used while maintaining complete flexibility in the selection of a reagent, as required by the experiment.

CHEMICAL IONIZATION MIXTURE ANALYSIS USING THE DIRECT INTRODUCTION PROBE

One of the characteristics of chemical ionization spectra is their simplicity, which makes analysis possible without prior separation.

It is possible, using the direct sample introduction probe, to analyse even the fairly complex mixture, using chemical ionization to produce essentially single peak spectra of each component, and then to interpret the composite spectrum of the mixture,[5] as shown in Fig. 5. This is a sample of an extract of the gastric contents of a drug overdose victim, and under these circumstances it was possible to obtain essentially single peak spectra for each one of the drugs ingested and to identify the drug. Note that in the case of acetylcarbromal, which contains one bromine, it was possible to observe the characteristic bromine isotope ratio for the M + H peak and the peak at two mass

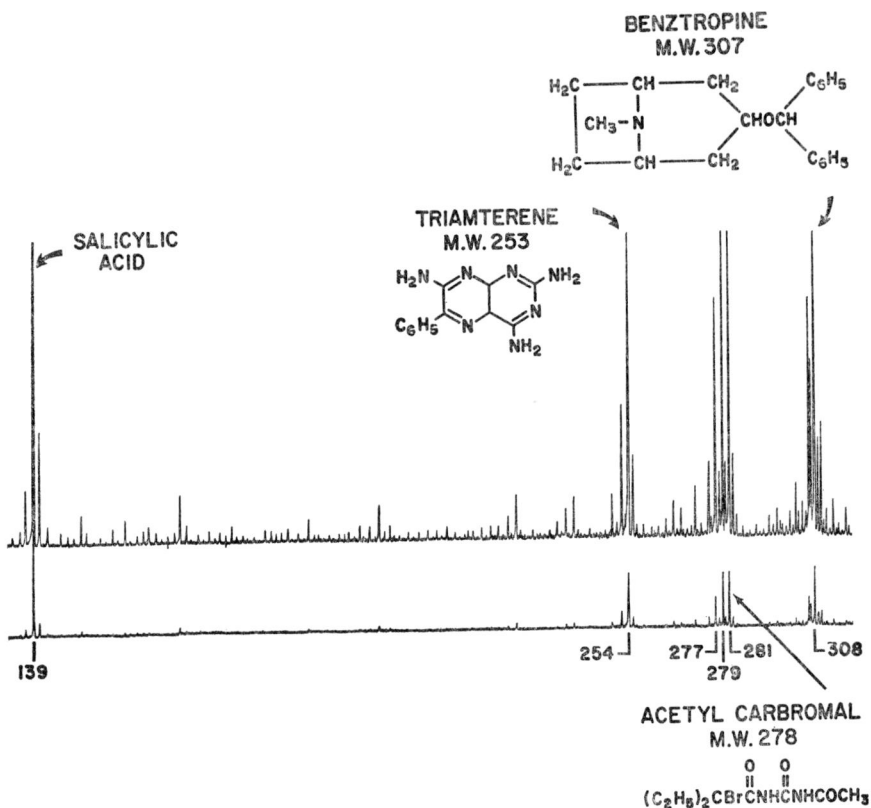

FIG. 5 Gastric contents extract analysed with direct sample introduction probe without prior separation.

units higher, thereby confirming the elemental composition as well as the molecular weight of the sample. Obviously, under these conditions, it makes little difference whether the compounds are capable of giving a molecular ion under electron ionization conditions. This type of analysis by chemical ionization without separation has proved to be extremely valuable in the analysis of illicit drugs and toxicological samples.

REFERENCES

1. Munson, M. S. B., *Anal. Chem.*, 1971, **43**, 28A.
2. Hunt, D. F. and Ryan, J. F., *Anal. Chem.*, 1972, **44**, 1306.
3. Chait, E. M. and Adams, V. H., ASMS Conference on Mass Spectrometry and Allied Topics, San Francisco, May 1973.
4. Chait, E. M. and Askew, W. B., ASMS Conference on Mass Spectrometry and Allied Topics, San Francisco, May 1970.
5. Milne, G. W. A., Fales, H. M. and Axenrod, T., *Anal. Chem.*, 1972, **44**, 1815.

54
A Computerized Mass Spectrometry Laboratory

By D. HENNEBERG, K. CASPER, B. WEIMANN
and E. ZIEGLER

(*Max-Planck-Institüt für Kohlenforschung, Mülheim a. d. Ruhr, Germany*)

COMPUTER SYSTEM

IN 1968 a DEC system-10 was installed in the MPI Mülheim and in the meantime has been enlarged to the configuration shown in Fig. 1. About a hundred scientists are working in the institute mainly in the field of organo-metallic chemistry, photo- and radiation chemistry and theoretical organic chemistry, supported by large analytical laboratories. The time-sharing computer system is a mixed system, in which real time acquisition and interpretation of data from chromatographs (27 lines for about 50 GC's, 5% CPU-time), conventional and Fourier-NMR instruments (10% CPU-time with spectra-simulations) and mass spectrometers (two fast and one slow scan, 5% CPU-time) is done simultaneously with off-line calculations such as X-ray structure analysis (35%) quantum chemistry and kinetic calculations (35%) and, among other applications, literature documentation.

In this system, which at the present time handles up to 32 jobs simultaneously within a multi-programming time-sharing environment, the MS-laboratory participates with about six jobs. This participation on a large, powerful system allows an almost complete computerization of nearly all tasks of the MS-laboratory.

In favour of a clearer survey of the present state of the MS part of this project details have been omitted. Some features and aspects have been published before[1,2] or will be published in future.

MASS SPECTROMETERS

For analytical purposes four (magnetic) mass spectrometers are used: one for GC-MS, two for qualitative analysis of liquids and solids, and one for quantitative analysis of gas mixtures.

The latter instrument is connected to the slow ADC (Fig. 1). It operates with slow scan. The whole process, beginning with the start of measurement through processing of the spectra and analysis by a regression method, up to the output of the final analysis report, is completely automated.

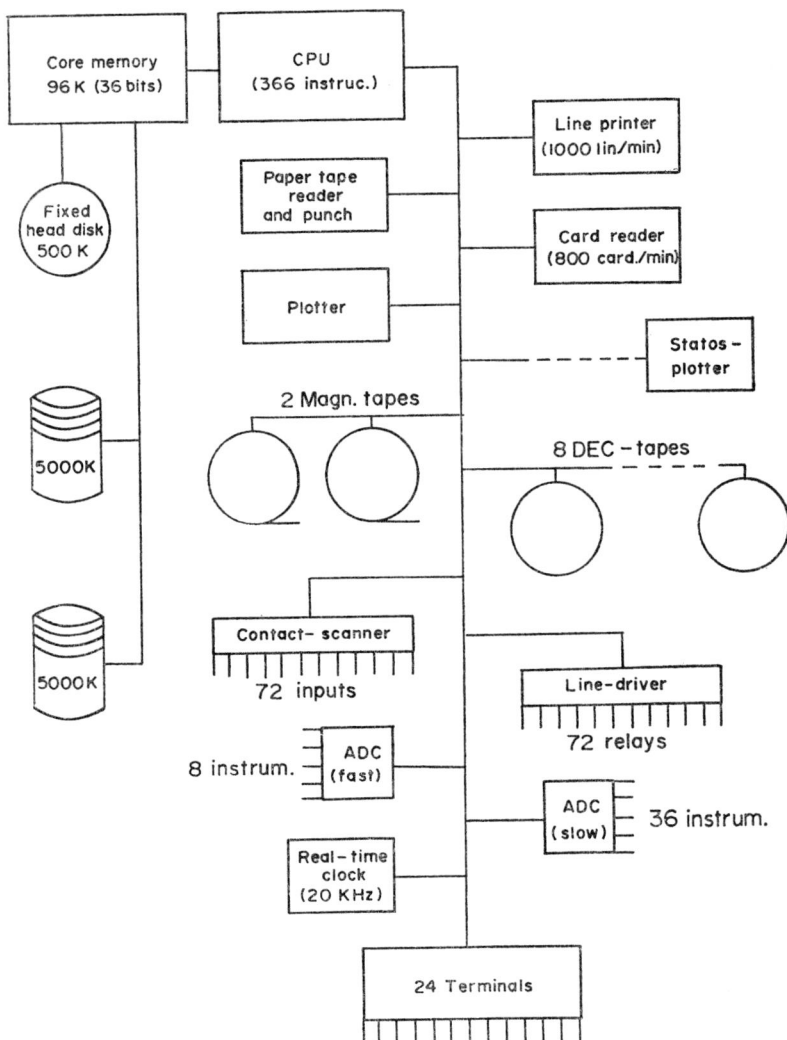

Fig. 1   Configuration of the computer system.

TECHNIQUE OF MEASUREMENT

For low resolution GC-MS analysis with the aid of computers the technique of taking a series of spectra in a certain time cycle was introduced in 1967 by Hites and Biemann[3] and is now usually applied for this purpose. The two main aspects of this technique are the inherent automation and the possibility of constructing mass chromatograms for any fragment, even if the interest in looking at a certain fragment arises after measurement and during interpretation of the analysis.

Principally this technique represents a quasi continuous method of measuring mass spectra as a function of time. The advantages of this method for GC-MS are obvious. We also apply the same technique of automatically taking series of spectra for most other applications of mass spectrometry. A great advantage is that one whole set of programs is applicable to a lot of quite different types of analytical problems. The second point which favours this method is the fact that the full potential of information to be gained out of a sample by mass spectral analysis normally cannot be included in one spectrum. A more or less great number of spectra is required mainly for direct inlet procedures to follow fractionated evaporation or decomposition of unstable compounds (*see* example in Figs. 6 and 7), but also for groups of spectra measured with a sample under varying conditions like energy (70 eV/low or complete breakdown-curves), sensitivity, pressure, background, reactant gas (D_2O), etc.

INTERFACING

At the moment two, fast-scan, low-resolution mass spectrometers are connected on line to the computer as shown in Fig. 2. Whereas the normal ion source output is fed to the fast ADC, another connection to the slow ADC allows one to sample slow signals, as for instance total ionization. Usually total ionization curves are reconstructed from total ionization values derived from the spectra, thus gaining one point of the curve each few seconds. But this is not sufficient, because in chromatograms with good separation efficiency

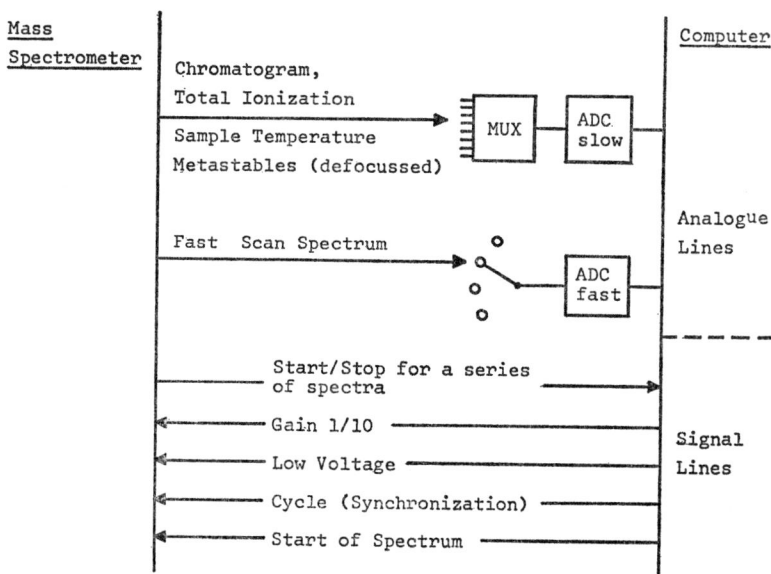

FIG. 2 Diagram of connections between mass spectrometer and computer.

(we almost exclusively use capillary columns) or in processes like irregular evaporation or decomposition of unstable compounds (in our case some transition metal complexes) time constants of a second and below occur, for which a sampling rate of at least two points per second is desirable.

Our fast ADC (Fig. 1) handles one channel at a time only. Therefore we established 30 sec periods subdivided into twelve time slots, each 2·5 sec long. Each time slot can be assigned by program to one of the fast ADC channels in order to acquire data. A typical slot assignment is given in Fig. 3: Channel Ø (GC-MS combination) acquires a spectrum all five seconds, channel 1 (for instance measurement of a solid via direct inlet) takes spectra all 10 seconds, each third one with the ion source switched to low voltage, channel 3 is given to a Fourier-NMR experiment or in future to a third mass spectrometer.

Channel	Slots												Experiment
Ø	N		N		N		N		N		N		GC/MS
1		N				N				L			MS
2			P				P					P	NMR or MS
	Ø			10			20			30			Seconds

FIG. 3   Typical slot assignment in a 30 sec period for three quasi-simultaneous experiments with high data rates. N: Normal 70 eV spectrum. L: Low voltage spectrum. P: Fourier-NMR or third MS.

The signal lines to the contact scanner or from the line driver (Fig. 1) start and stop automatic measurement of a series of spectra, start each single scan, switch the ionization voltage and switch the spectrometer sensitivity (electrometer resistor of the pre-amplifier) to one-tenth of its normal value. The latter happens, if the preceding spectrum exhibited peaks surpassing the voltage range of the ADC. This leads in the case of a longer period of very intense mass spectra to an alternation between normal- and reduced-intensity spectra. This feature provides not only a high dynamic range but it is also a requirement for automation, for instance if a GC-MS run contains peaks from main- and trace-components.

MEASUREMENT

A survey over the tasks of a set of programs, their sequence and co-operation during mass spectral analyses is given in Fig. 4. The upper left depicts the procedures for measurement. First the parameters are defined for a series of spectra: Channel number, cycle time, sequence of normal and low voltage spectra, automatic gain, a parameter to adjust for amplifier-noise, suppression of processing and storage of spectra during peak-free regions of chromatograms[4] and sampling rate of the ADC. Then, automatic measurement proceeds between 'start' and 'stop'.

Parallel to the acquisition- and prereduction-program another program processes the reduced data, outputs control values for each spectrum to allow for supervision of spectra and monitoring of the experiment, produces a

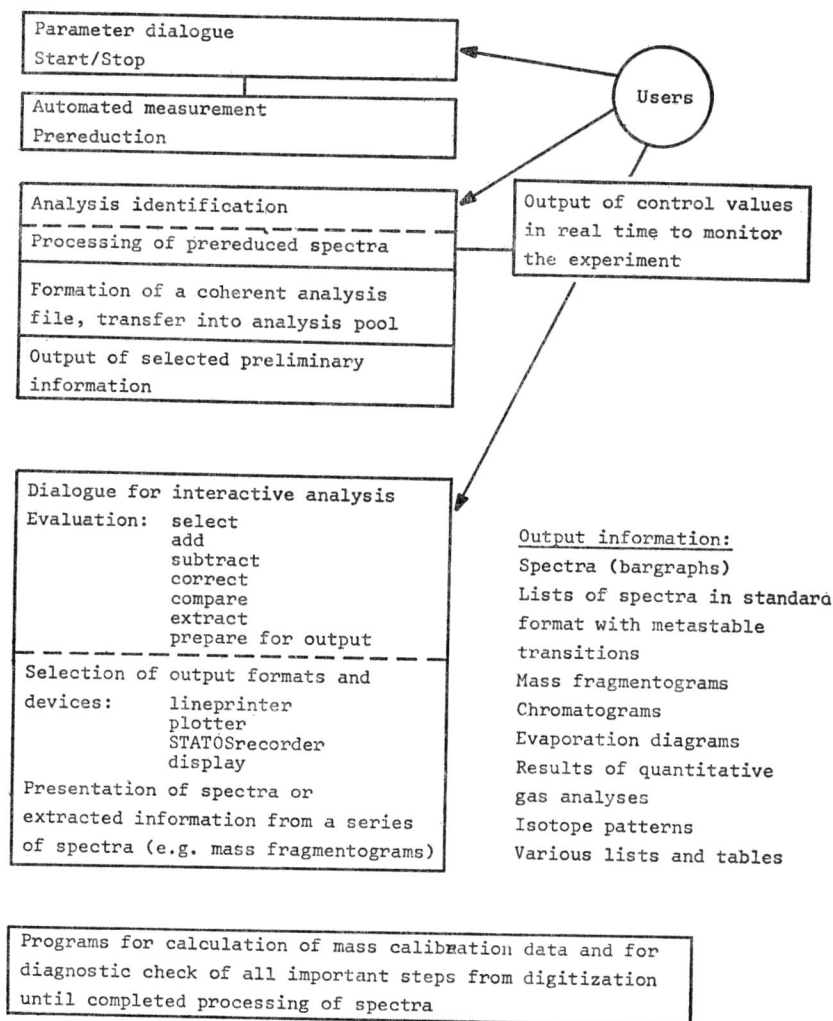

FIG. 4 Functions of existing MS-software.

complete file with all information belonging to the analysis and generates at the end automatically an output suitable for at least preliminary information for the analyst. In the case of a GC-MS run this information consists of (i) a plot of the chromatogram with positions, numbers and total ionization values of each spectrum and (ii) plots or tables of the mass spectra corresponding to peak maxima in the chromatogram.

GC-MS runs need no operator interaction as long as the control values for each spectrum indicate that the spectra are correct and have sufficient intensity. The same level of automation can also be achieved for measurements with the direct-introduction probe by adding an electronic control unit to the mass spectrometer, which adjusts the sample temperature corresponding to the sample pressure (total ion current).[5]

Main characteristics of the processing of the digitized spectra (*see also* Ref. 2) are:

   (i) The ratio between highest measurable and lowest detectable peak is 30 000 without and 300 000 with automatic gain switching.

   (ii) The mass scale is based on mass/time relationship and includes continuous correction dependent on integral mass values.

   (iii) Metastables are identified as far as they are recognizable in fast scan spectra.

In order to restrict the size of programs and the extent of disk-storage, two filtering algorithms are applied to each spectrum: (i) peaks smaller than 1/3000 of the base peak are suppressed and (ii) the number of peaks in a spectrum is limited at the moment to 500 by elimination of peaks which are chosen according to a low-mass low-intensity priority.

PRESENTATION AND MANIPULATION

Presentation of intermediate or final results in various forms on different output devices and manipulation (*i.e.* operations like averaging, correction, reduction, preparation of mass fragmentograms out of a series of spectra, etc.) are handled through a dialogue-program, indicated in the middle left of Fig. 4.

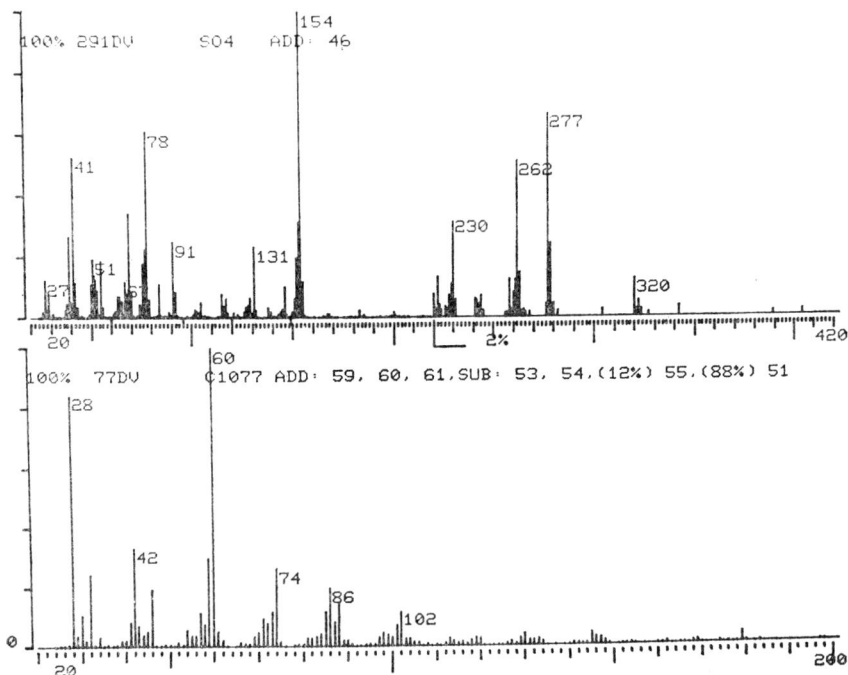

FIG. 5   Display output of two mass spectra. Note the headlines of the spectra with analysis number, size of the basepeak (1 DV equals 0·1 V) and notation of the manipulation of the displayed spectrum.

Figure 5 is a photo from the display. In the upper spectrum the scale has been changed at m/e 220, which is one out of numerous possibilities to arrange the output in desired form by means of a dialogue. The lower half shows the spectrum of a certain deuterated hydrocarbon species of formula $C_{12}D_{17}H$, which was eluted in GC-MS strongly overlapped with the same but perdeuterated species. Averaging and corrections for overlap and background have been applied to gain an optimal spectrum. The whole manipulation is ascertainable in the headline.

We offer this information in all our spectra-outputs (even more is included in the tabular form), because many of the spectra finally used for analytical purposes have undergone more or less complicated corrections. All corrections, which are not simply background-subtractions, are guided by thorough study of mass fragmentograms. A very useful special version of subtraction can be requested such as to make the intensity at a certain m/e zero. It then subtracts the right percentage of a denoted spectrum.

Figure 6 shows mass fragmentograms of the direct inlet of a thermally unstable water and air sensitive nickel-complex with triphenyl-phosphine ligand. Spectrum number 46 can be seen in the upper part of Fig. 5. M/e 320 is a primary fragment and follows a normal evaporation curve. M/e 262 (triphenylphosphine) has a double origin: As far as being synchronous to m/e

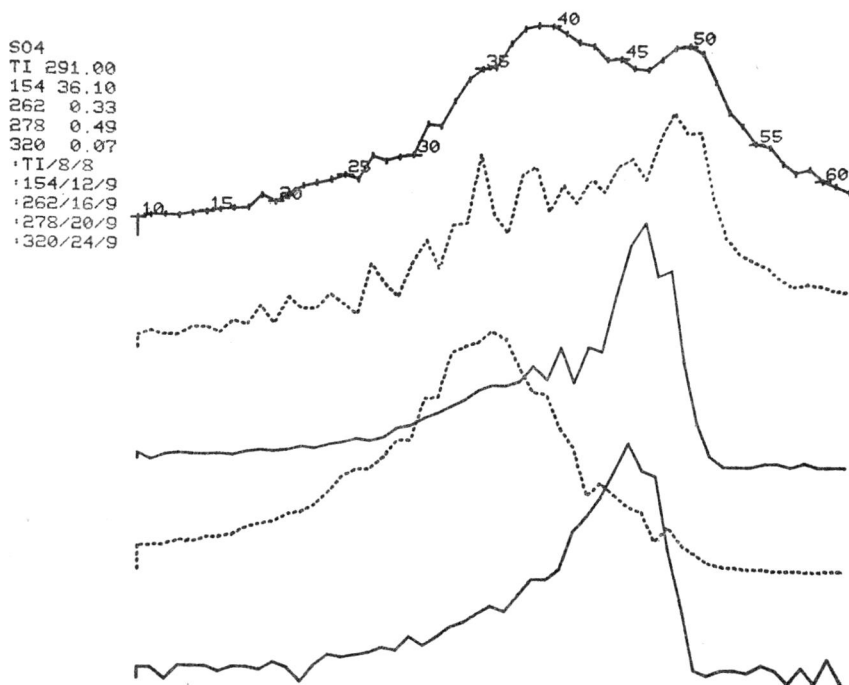

FIG. 6 Mass fragmentograms of the direct inlet of an unstable compound (*see* text). In the upper left are noted the absolute values of the maximal intensities of the corresponding fragments and below the dialogue to arrange baselines and amplitudes of the single fragmentograms.

320, it is a fragment of the complex, but a second maximum can be recognized just before that of m/e 320. M/e 154, the basepeak in the spectrum, exhibits an irregular shape, indicating decomposition of the sample. M/e 278 (triphenyl-phosphine oxide) appears before the complex, however, when added to the sample, it appears after the complex at higher evaporation temperatures. Our present explanation for this effect is, that the occurrence before the complex indicates evaporation of triphenylphosphine oxide as a reaction product, which covers the sample crystals. This is one of our most striking examples, which clearly demonstrates the need for taking series of spectra in order to get mass fragmentograms, which enable the correct interpretation of spectra and reactions which may take place in the mass spectrometer.

Other examples, as for instance plots of chromatograms or lists of spectra in standard format with metastable transitions have been published elsewhere.[2]

STORAGE REQUIREMENTS

Table I shows the five logically separated parts of the disk-storage for the MS laboratory. With the technique of spectra series a GC-MS run needs 10 to 50 k words, a qualitative analysis 2 to 10 k. With the storage capacity given in Table I the whole full-time production of one to two weeks of our laboratory

TABLE I

Structure of Data Storage on Directly Accessible Disk Packs

Purpose	Contents	Size (words)
Measurement	Programs Parameter sets Mass calibration data Prereduced spectra	300 k
Manipulation Presentation	Programs Output data	200 k
Collection of spectra (in preparation)	Spectra Signatures	2 000 k
Pool for analyses	Analyses to be evaluated	1 500 k
Storage for analyses	Recently evaluated analyses	1 000 k

can be held in the pool- and storage-parts for immediate access. Every day the operating system of the computer looks for the free space left in the storage-part and, if needed, initiates automatically the transfer of the whole storage-part to magnetic tape.

DOCUMENTATION

A reduction of the large amount of data in a series of spectra prior to final storage on tapes requires not only time but also a decision: which part of information will certainly be of no interest in the future. Therefore we store the complete analysis files. We prefer tapes instead of microfilm documentation because the information on tapes can be reprocessed.

DEVELOPMENT

At the moment a collection of mass spectra is being assembled together with appropriate search facilities.

The next developments are directed towards more assistance from the computer in daily interpretation tasks such as finding molecular weight, classification of compounds, library search and aiding the writing of analysis reports.

An interesting way of extracting peaks which are important for interpretation of a spectrum has been developed.[6] A program module looks at a 70 eV- and at a neighbouring low-voltage spectrum and extracts only peaks which in the low voltage spectrum have a higher intensity relative to total ionization than in the 70 eV spectrum. By this means an effective and reasonable reduction of information is achieved leaving all peaks of a spectrum, whose formation is energetically favoured.

CONCLUSIONS

It should be emphasized, that participation in a large time-sharing computer system with interactive capabilities allows one to handle all tasks in parallel, independently from each other, and it furthermore allows immediate access to all needed programs, to a large amount of measured data and to the spectra in a library. Fast interactive response is very important for computer-aided evaluation and interpretation. When sitting at a display terminal it is only a matter of seconds between the type-in of a request and the appearance of the answer on the screen.

These were the characteristic features of the system, which allowed a smooth and very flexible integration of the computer into the daily analytical work of our mass spectrometry laboratory.

REFERENCES

1. Ziegler, E., Henneberg, D. and Schomburg, G., *Anal. Chem.*, 1970, **42**, 51A.
2. Henneberg, D., Casper, K., Ziegler, E. and Weimann, B., *Angew. Chem. internat. Edit.*, 1972, **11**, 357.
3. Hites, R. A. and Biemann, K., 'Advances in Mass Spectrometry', 1968, Vol. 4, p. 37.
4. Henneberg, D., Casper, K. and Ziegler, E., *Chromatographia*, 1972, **213**, 209.
5. Franzen, J., Küper, H., Riepe, W. and Henneberg, D., *Int. J. Mass Spectrom. Ion Phys.*, 1972/3, **10**, 353.
6. Damen, H. and Henneberg, D., to be published.

55
Studies Upon Ion Beam Energies

By M. E. S. F. SILVA and R. I. REED

(*Chemistry Department, The University of Glasgow, Glasgow, U.K.*)

IONS with excess kinetic energy were first detected by Bleakney.[1] Methods used for their detection include:

(a) The study of the peak shape or peak intensity with the variation of some potential in the direction of propagation of the ions.[1,2]

(b) A study of the peak using deflection plates placed immediately after the source[3] or an electrostatic sector before[4] or after[5] the magnetic sector.

(c) The use of non-magnetic instruments such as the trochoidal path mass spectrometer,[6] the coincidence time of flight[7] and the time of flight.[8]

(d) The study of metastable peak widths and shapes.[9,10]

In the present study an instrument was built with the electrostatic sector following the magnetic sector. Similar instruments had previously been built to study ion-molecule reactions[5] and to improve the resolving power of a mass spectrometer.[11] While this work was proceeding three papers[12,13,14] were published reporting results with instruments having the electrostatic after the magnetic sector.

EXPERIMENTAL

The basic instrument to which the electrostatic sector was attached was an MS2 (Metropolitan Vickers, Series No. 11) single focussing model with a 90° of arc magnetic sector,[15] an electron impact Nier type ion source[16] and a Faraday cup as collector. In this type of instrument the ions were accelerated by a potential of 2 kV and focussed at the defining slit immediately in front of the Faraday cup. The position of this slit was taken as the source of the ions proceeding to the electrostatic sector.

The dimensions of the sector were chosen following the calculations of H. Ewald and H. Liebl[17] for a toroidal sector. The sector constructed was a spherical symmetric one of 152·4 mm radius and with an angle of deflection of 60° of arc. By choosing a spherical sector as opposed to a cylindrical one, focussing is obtained both in the radial and axial directions and a higher value is obtained for the velocity dispersion with the same radius.[18] The distance between the two plates was chosen to be 22·9 mm considering mainly the accuracy required in machining and positioning to obtain a precision of 1 in 1000. An ion with mass m_0, velocity v_0, entering the sector in the median plane will describe an orbit of radius $a_e = 152\cdot4$ mm and will be submitted to a centripetal force given by $m_0 v_0^2/a_e = eE$, where e is the electronic charge of the ion and E_0 the intensity of the radial electrostatic field between the plates and for a radius of a_e.

In the sector constructed, ions with 2000 V energy will be focussed by the electrostatic sector when it has $+300$ V on the concave and -300 V on the convex plate.

The two spherical plates were held in position and electrically insulated from the brass box in which they were contained by means of pyrophilite pieces which were machined to size and heat-treated to increase their mechanical strength. To correct for the fringing fields, grounded diaphragms were placed at the entrance and exit of the electrostatic sector at a distance calculated using the expression derived by Herzog.[19] Second-order effects were calculated, and at present the main limitation to the resolution is the spread in the ion beam energy obtained with the particular source used. The best focussing position was determined experimentally by moving the two bellows, one situated on either side of the electrostatic sector. The whole assembly is shown in Fig. 1. Between the magnetic and electrostatic sectors is

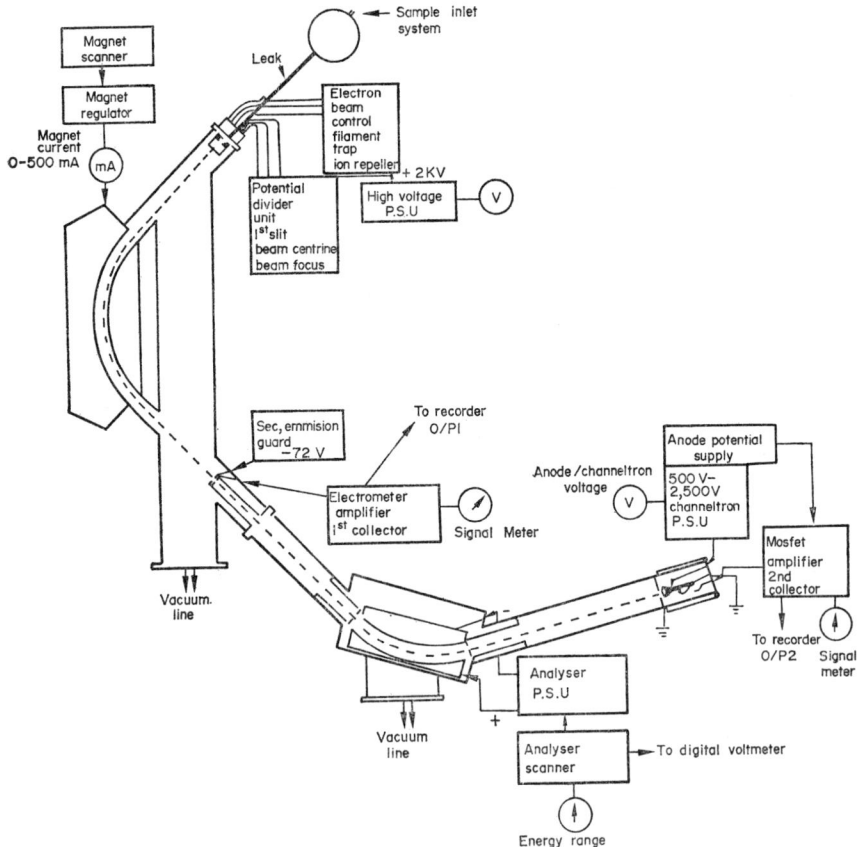

FIGURE 1

the first collector formed by two plates. The top one, which operates at -72 V defines the beam that reaches the second. This latter plate is connected externally to an electrometer which records the signal, while that part of the beam which passes through the slit reaches the electrostatic sector.

As a final collector, a Mullard channel electron multiplier was used. The final assembly is shown in Fig. 2. The 'channeltron' was used as a current amplifier and the electrons coming from the output end were attracted to a small positively biased trap.

"CHANNELTRON" ASSEMBLY

Scale 1:2.5

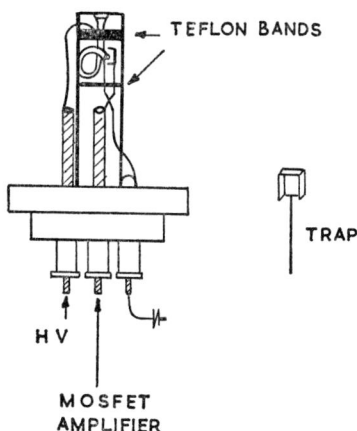

FIGURE 2

The behaviour of the 'channel electron multiplier' was a major source of concern. A very remarkable mass discrimination effect against lower masses was noticed similar to that observed in pulse counting.[20] The effect is particularly noticeable for masses under twenty-eight. A relative sensitivity to water differing markedly from that towards methane was also observed.

Part of the work was done towards the end of the life of the 'channeltron' indicated by a steady decay of its amplification. To increase this, higher applied voltages were used, but for voltages over 1·5 kV 'after pulses'[21] and field saturation effects[21,22] were observed.

Throughout these experiments the pressure in the apparatus, in the absence of a sample was about 10^{-6} torr.

RESULTS

The mass resolution normally obtained by the MS2 was doubled by the addition of the sector. The energy resolution obtained with it was of the order of 1 V in 1000 V.

Energetic peaks were detected by plotting the intensity obtained in the 'channeltron' against the potential on the plates of the electrostatic sector. Where necessary, corrections were introduced in the 'channeltron' reading to compensate for pressure changes by dividing its value by that of the same peak as measured at the first collector, which was assumed to be proportional to the pressure. While molecular ions gave rise to Gaussian curves characteristic of a Maxwell-Boltzmann distribution, for fragment ions some groups of higher kinetic energy appear superimposed on the Gaussian curve.

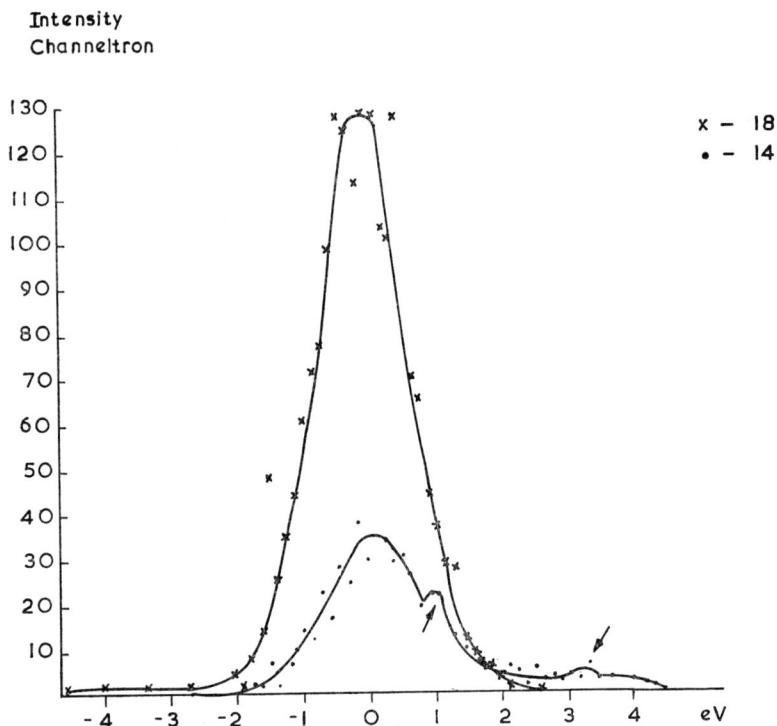

FIGURE 3

The following energetic peaks were detected:

(a) As shown in Fig. 3, peak m/e = 14 from nitrogen has two groups of energetic ions with approximately 1 eV and 3·3 eV of kinetic energy relative to the main thermal group. These results are in good agreement with the ones previously reported in the literature.[23]

(b) An asymmetric distribution was observed for m/e = 14,[24] in methane, but this could be due to the presence of some nitrogen.[25]

(c) A satellite with 1·9 eV kinetic energy was observed for peak 17 of methanol agreeing with the results in the literature.[26]

The method previously reported by Löhle and Öttinger[27] and called DADI[13] was used for the detection of metastables. The following transitions, previously reported[23] were observed.

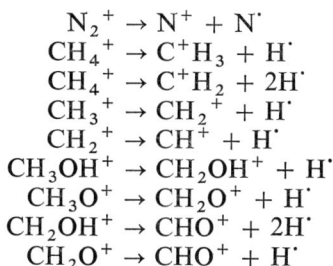

$$N_2^+ \to N^+ + N^{\cdot}$$
$$CH_4^+ \to C^+H_3 + H^{\cdot}$$
$$CH_4^+ \to C^+H_2 + 2H^{\cdot}$$
$$CH_3^+ \to CH_2^+ + H^{\cdot}$$
$$CH_2^+ \to CH^+ + H^{\cdot}$$
$$CH_3OH^+ \to CH_2OH^+ + H^{\cdot}$$
$$CH_3O^+ \to CH_2O^+ + H^{\cdot}$$
$$CH_2OH^+ \to CHO^+ + 2H^{\cdot}$$
$$CH_2O^+ \to CHO^+ + H^{\cdot}$$

All metastables were detected in saturation conditions, *i.e.* they were detected as pulses and as the intensity of the pulse has no direct connection with the abundance of the ion, no study of the variation of intensity with pressure could be attempted.

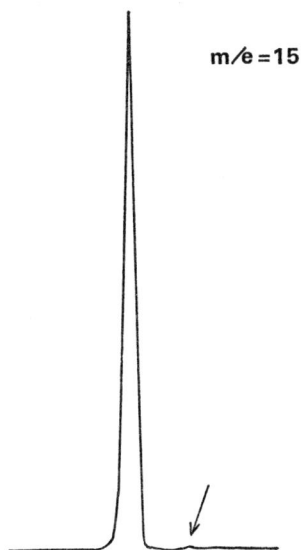

m/e = 15

FIGURE 4

An abnormal metastable was determined for peak fifteen from methane. It is shown in Fig. 4. It was attributed to a transition in the first field free region originating with an ion of nominal mass of 15·05,[23] which was not resolved from the main beam at m/e = 15 by the magnetic sector.

Using the expression

$$d \, (\text{Volts}) = 4V_1 \frac{m_2}{m_1} \left(\frac{\mu T}{\text{eV}} \right)^{\frac{1}{2}}$$

which for this instrument relates the metastable peak with the kinetic energy released in the process, where V_1 is the voltage of one plate; for the other letters, see Beynon.[9] An attempt was made to study the transition CH_2OH^+

$\rightarrow CHO^+ + 2H\cdot$. However, the peak profile was not available and the range over which pulses were counted was taken as the width: the results compare poorly with the ones previously obtained.[28,29]

ACKNOWLEDGMENT

One of us (M.E.S.F.S.) is indebted for the financial support of the Instituto de Alta Cultura, Lisbon, Portugal.

REFERENCES

1. Bleakney, W., *Phys. Rev.*, 1930, **35**, 1180.
2. Hagstrum, H. D. and Tate, J. T., *Phys. Rev.*, 1941, **59**, 354.
   Washburn, H. W. and Berry, C. E., *Phys. Rev.*, 1946, **70**, 559.
   Waldron, J. D., *Trans. Faraday Soc.*, 1954, **50**, 102.
   Taubert, R., *in* 'Advances in Mass Spectrometry', *ed.* J. D. Waldron, 1959, p. 483.
   Hipple, J. A., Fox, R. E. and Condon, E. U., *Phys. Rev.*, 1946, **69**, 347.
3. Berry, C. E., *Phys. Rev.*, 1950, **78**, 597.
   Durup, J. and Heitz, F., *J. Chimie Physique.*, 1964, **61**, 470.
4. Morrison, J. D. and Stanton, H. E., *J. Chem. Phys.*, 1958, **28**, 9.
5. White, F. A., Rourke, F. M. and Sheffield, J. C., *Appl. Spectr.*, 1958, **12**, 46.
6. Bleakney, W. and Hipple, J. A., *Phys. Rev.*, 1938, **53**, 521.
7. McCulloh, K. E., Sharp, T. E. and Rosenstock, H. M., *J. Chem. Phys.*, 1965, **42**, 3501.
8. Franklin, J. L., Hierl, P. M. and Whan, D. A., *J. Chem. Phys.*, 1967, **47**, 3148.
9. Beynon, J. H., Saunders, R. A. and Williams, A. E., *Z. Naturf.*, 1965, **20a**, 180.
10. Higgins, W. and Jennings, K. R., *Chem. Commun.*, 1965, p. 99.
    Flowers, M. C., *Chem. Commun.*, 1965, p. 235.
11. Gall', R. N., *Sov. Phys. Techn. Phys.*, 1969, **14**, 263. Translated from *Zhurnal Tekhnicheskoi Fiziki*, 1969, **39**, 360.
12. Löhle, U. and Öttinger, Ch., 17th. Ann. Conf. on Mass Spectrometry and Allied Topics, Dallas, 1969, Paper No. 29.
13. Maurer, K., Brunée, C., Kappus, G., Habfast, K., Schröder, U. and Schulze, P., 19th. Ann. Conference on Mass Spectrometry and Allied Topics, Atlanta, 1971, Paper K9.
14. Wachs, T., Bente III, P. F. and McLafferty, F. W., *Int. J. Mass Spec. Ion Phys.*, 1972, **9**, 33.
15. Blears, J. and Mettrick, A. K., *Proc. XIth. Intern. Congr. Pure and Appl. Chem.*, 1947, **1**, 333.
16. Nier, A. O., *Rev. Sci. Instr.*, 1940, **11**, 212.
17. Ewald, H. and Liebl, H., *Z. Naturforsch.*, 1957, **12a**, 28; 1959, **14a**, 129, also *in* 'Mass Spectrometry', *ed.* C. A. McDowell, p. 216, McGraw-Hill Book Co., 1963.
18. Silva, Miss M. E. S. F., Ph.D. Thesis.
19. Herzog, R., *Z. für Physik.*, 1935, **97**, 596.
20. Potter, W. E. and Mauersberger, K., *Rev. Sci. Instr.*, 1972, **43**, 1327.
21. Evans, D. S., *Rev. Sci. Instr.*, 1965, **36**, 375.
22. Andersen, R. D. and Page, D. E., *Nucl. Instr. Meth.*, 1970, **85**, 141.
23. Reed, R. I. and Silva, Miss M. E. S. F., 21st Ann. Conf. on Mass Spec. and Allied Topics, San Francisco 1973.
24. Fuchs, R. and Taubert, R., *Z. Naturforsch.*, 1964, **19a**, 494.
25. McDowell, C. A. and Warren, J. W., *Disc. Faraday Soc.*, 1951, **10**, 13.
26. Tsuchiya, T., *Bull. of Chem. Soc. Japan*, 1964, **37**, 1308.
27. Löhle, U. and Öttinger, Ch., *J. Chem. Phys.*, 1969, **51**, 3097.
28. Beynon, J. H., Fontaine, A. E. and Lester, G. R., *Int. J. Mass Spec. Ion Phys.*, 1968, **1**, 1.
29. Lifshitz, C., Shapiro, M. and Sternberg, R., *Israel J. of Chem.*, 1969, **7**, 391.

56
An UHV Secondary Ion Mass Spectrometer For Differential In-Depth Analysis

By J. MAUL, F. SCHULZ and K. WITTMAACK

(*Gesellschaft für Strahlen- und Umweltforschung mbH München,
Physikalisch-Technische Abteilung, Neuherberg, Germany*)

INTRODUCTION

Until recently, routine secondary ion studies were carried out by means of microprobe analysers with magnetic mass spectrometers (for a review see *e.g.* Refs. 1 and 2). The outstanding advantage of highly specialized instruments of this type is the availability of ion microscope images of the sputtered surface. The complexity in design and operation of these machines is a disadvantage in that non-imaging secondary ion mass spectrometry supplies sufficient information in specimen analysis. For this type of investigation quadrupole filters have been demonstrated to be the more appropriate mass separating tool.[3]

The only disadvantage of a quadrupole analyser in secondary ion studies results from the fact that it is not capable of preventing sputtered neutrals and high energy sputtered or backscattered ions from bombarding internal surfaces such as the quadrupole rods.[4] Ternary ions thereby produced cannot be filtered and may thus reach the detector even if the latter is positioned off-axis. The resulting background increases with increasing primary ion energy and allows trace analysis only at probe energies up to about 1 keV[3] whereas at 10 keV the background intensity may amount to more than 10% of the peak intensity.[4]

In a very recent study[4] we have investigated this effect and demonstrated that it may be completely eliminated by simply using an aperture bounded plate capacitor in front of the quadrupole analyser as an energy filter and a suppressor for sputtered neutrals. After optimization of the geometry of the plate capacitor an intensity ratio of peak to background of 2×10^8 could be obtained for an aluminium specimen.

The spectrometer has been equipped with a telefocus ion gun[5] which is able to produce trapezoidal beam profiles with a diameter of up to 5 mm at the target position. This allowed a determination of implantation profiles with high depth resolution (≈ 20 Å).[6-8] In the investigations the detection limit for certain impurities such as boron was limited by the presence of cracked products resulting from hydrocarbons in the residual gas. We therefore decided to rebuild our previous set-up[4] in an ultra high vacuum version. In this paper we will describe the construction and demonstrate its capabilities by some typical results.

APPARATUS

The construction of the UHV secondary ion mass spectrometer is shown in Fig. 1. The analyser chamber, the ion gun vessel and the components are all made of stainless steel and are bakeable. Copper sealing is applied throughout. A pressure step allows differential pumping of the system. A 70 litre/sec turbomolecular pump (Pfeiffer TVS 253) serves to evacuate the gun vessel and to pump the noble gas load (usually argon) introduced via the ion source outlet. Additionally it provides a good starting pressure for the 200 litre/sec sputter ion pump (Ultek D-1 pump) which is assisted by a titanium sublimation pump and a cryopanel. The ultimate residual gas pressure in the target chamber is in the 10^{-11} torr range, the argon partial pressure increases up to 10^{-8} torr during gun operation depending upon the ion beam current.

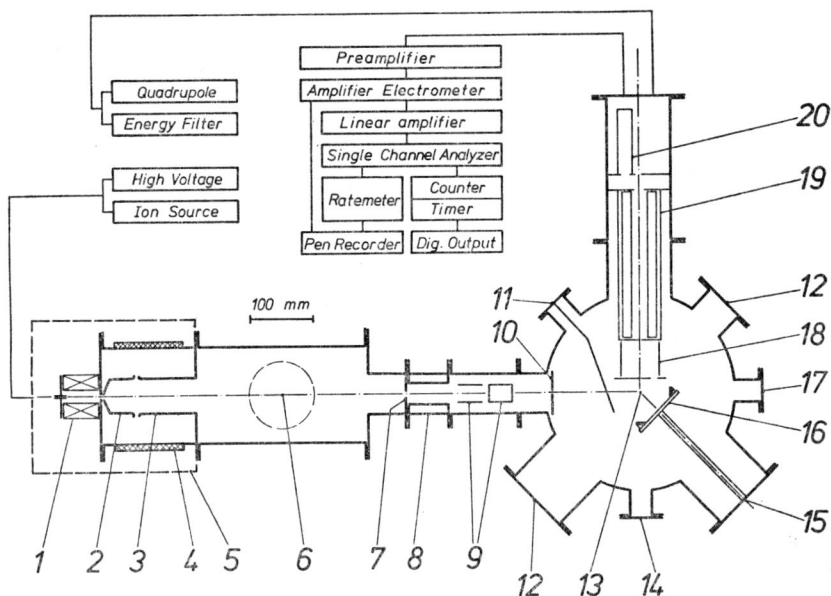

FIG. 1   Construction of the UHV secondary ion mass spectrometer.

1. Ion source
2. Extraction electrode
3. Acceleration electrode
4. Ceramic insulator
5. Protective cover
6. Turbomolecular pump
7. Pressure step
8. UHV valve
9. Deflection plates
10. Aperture
11. Beam monitor
12. Viewport
13. Ion pump
14. Variable leak
15. Rotary UHV feedthrough
16. Target holder
17. Current feedthrough
18. Energy filter
19. Quadrupole mass filter
20. Multiplier

Great care has been devoted to an optimum design[9,10] and operation[11] of the ion source with respect to high gas efficiency. The argon gas flow out of the source thus can be minimized. Reactant gases such as oxygen may be introduced into the target chamber through a variable leak, *e.g.* to increase the secondary ion yield.[12]

According to the implications in the use of secondary ion mass spectrometry to investigate shallow impurity concentration profiles in solids[6] the probe energy is limited to an upper value of 15 keV. For most targets this is

FIG. 2   Picture of the spectrometer (protective cover demounted).

sufficient to obtain a maximum sputtering yield.[13] The secondary ion mass spectrometer thus allows high speed differential in-depth analysis (DIDA) at high depth resolution and high sensitivity. The data acquisition system for signals from either single particle counting or direct current measurement is fully automated. The lay-out of the complete system (Fig. 2) meets commercial standards.

RESULTS

Figures 3 and 4 present results which demonstrate some specific properties of the arrangement. Figure 3 shows the spectrum of positive ions emitted from germanium upon bombardment with 15 keV argon ions. No special surface treatment was applied before taking the spectra. Therefore a variety of impurities is observed, the high intensity of Na^+ and K^+ being well-known.[3] The main result to be demonstrated is the high mass resolution, Δm (FWHM) = 0·3, which allows the germanium isotopes to be separated completely. This

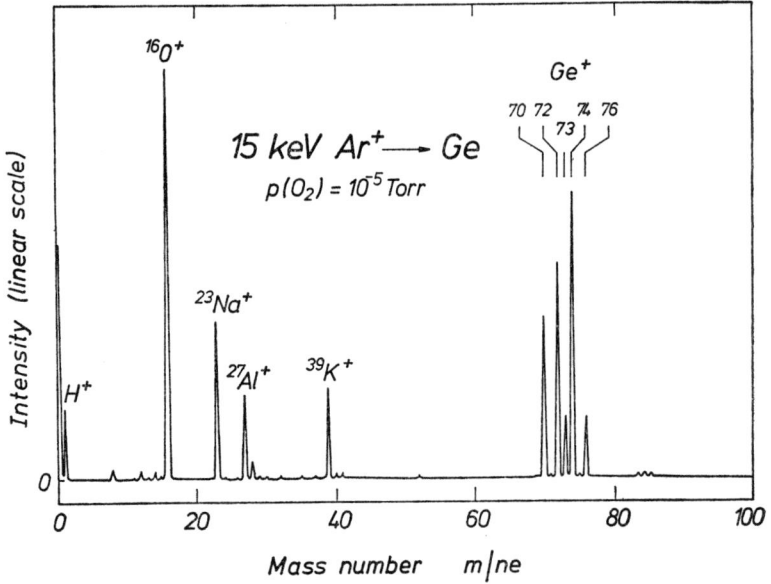

FIG. 3   Mass spectrum of germanium at enhanced oxygen partial pressure.

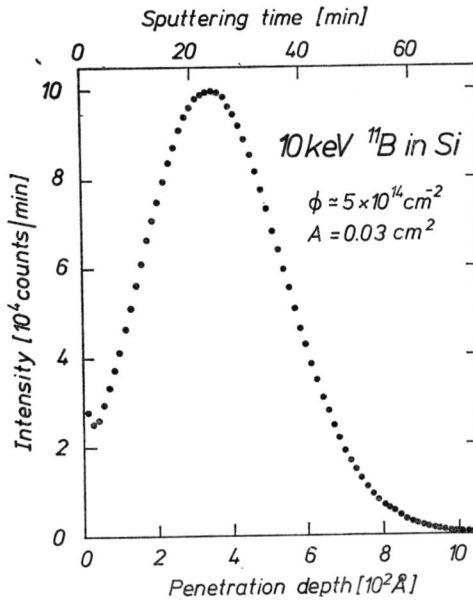

FIG. 4   Implantation profile of boron in amorphous (predamaged) silicon. Sputtering with
6 keV Ar$^+$ at 45°.

gain in resolution as compared to spectra obtained with non-energy selected secondary ions[3] is due to suppression of high energy ions.[4]

Figure 4 shows the range profile of 10 keV boron ions implanted in amorphous silicon. Each point corresponds to a sputtered layer of 15 Å. One can see that very shallow concentration profiles can be measured with high accuracy. More details concerning the energy dependence of boron range in silicon and the shape of the range profiles have been reported elsewhere.[6-8]

REMARKS

Shortly after our study concerning background reduction in the mass spectra[4] a more complicated ion optical approach to eliminate high energy secondary ions has been reported.[14] In that set-up a 180° segment of a cylindrical mirror analyser is used rather than a plate capacitor. An additional Einzel lens following the energy analyser serves to focus the secondary ion beam into the quadrupole filter. Signal-to-background ratios in excess of 10^6 have been achieved for a GaAs target bombarded with 2 keV argon ions.

Despite the more sophisticated construction of that set-up there does not seem to exist a pronounced advantage in comparison with a plate capacitor. The reason is that an optimum ion optical design would not require a focus at the entrance aperture of the quadrupole but matching of the emittance of the energy selected secondary ions and the acceptance of the mass filter.[15] In view of lacking knowledge with respect to the shape of the emittance and acceptance diagrams the simplest way to achieve high intensity in the mass spectra is ion deflection at low energy dispersion and the use of beam defining apertures as large as possible. An aperture bounded plate capacitor allows us to meet both demands very easily.

REFERENCES

1. Socha, A. J., *Surface Sci.*, 1971, **25**, 147.
2. Shinoda, G., Kohra, K. and Ichinokawa, T., *eds.*, 'Proc. VI. Int. Conf. X-ray Optics and Microanalysis', University of Tokyo Press, 1971, chapter V.
3. Benninghoven, A. and Loebach, E., *Rev. Sci. Instrum.*, 1971, **42**, 49.
4. Wittmaack, K., Maul, J. and Schulz, F., *Int. J. Mass Spectrom. Ion Phys.*, 1973, **11**, 23.
5. Wittmaack, K. and Schulz, F., *in* 'Proc. V. Int. Conf. Electron and Ion Beam Science and Technology', *ed.* R. Bakish. The Electrochemical Soc., Princeton, 1972, 181.
6. Schulz, F., Wittmaack, K. and Maul, J., 'Rad Effects', in press.
7. Wittmaack, K., Maul, J. and Schulz, F., 'Proc. Int. Conf. Ion Implantation in Semiconductors and Other Materials', New York, 1972, in press.
8. Wittmaack, K., Schulz, F. and Maul, J., *Phys. Letters*, 1973, **43A**, 477.
9. Wittmaack, K. and Schulz, F., *in* 'Proc. II. Int. Conf. Ion Sources', *ed.* F. Viehböck, H. Winter and M. Bruck, SGAE, Vienna, 1972, 863.
10. Wittmaack, K., to be published.
11. Wittmaack, K. and Schulz, F., 'XII Symp. Electron, Ion and Laser Beam Technology', Cambridge, Mass., 1973, to be published in *J. Vac. Sci. Techn.*
12. Benninghoven, A. and Mueller, A., *Phys. Letters*, **40A**, 1972, 169.
13. Sigmund, P., *Phys. Rev.*, 1969, **184**, 383.
14. Schubert, R. and Tracy, J. C., *Rev. Sci. Instr.*, 1973, **44**, 487.
15. Ruedenauer, F. G., *Int. J. Mass Spectrom. Ion Phys.*, 1971, **6**, 309.

Discussion

R. E. Honig (R.C.A., Princeton, U.S.A.): Could you explain in more detail how the flat-bottomed sputtered craters were produced?

J. Maul: They are due to a telefocus ion gun, developed especially for this purpose by Dr K. Wittmaack consisting of an extraction and an accelerating electrode. After optimization of the geometry and the operation parameters of the ion source and the ion gun—and indeed all of them had an influence on the beam profile—we learned what requirements have to be met, to obtain such trapezoidal beam profiles.

R. E. Konig: Were you able to obtain similar shapes at lower primary energies?

J. Maul: These crater shapes we get not only for primary ion energies of about 10 keV, but also for lower ones. For each primary ion energy they are achieved within a certain range of beam currents. According to the space charge law we get them at higher currents for higher primary ion energies.

H. Liebl (Max-Planck-Institut für Plasmaphysik, Germany): How did you confine the sampled area of the erosion crater?

J. Maul: In the case of the implantation profiles shown, the implanted region was sufficiently smaller in diameter than the sputtered area.

57
Field Desorption Mass Spectrometry Biochemical Applications

By H.-R. SCHULTEN and H. D. BECKEY

(*Institut für Physikalische Chemie der Universität Bonn, Bonn, Germany*)

INTRODUCTION

IT is no longer a 'work of art' to obtain Field Desorption (FD) mass spectra (MS) and the method, having now become a routine procedure, leads us to expect wide-spread fields of application. The two main advantages of the method, namely small sample consumption and high molecular ion intensities, displayed even by highly polar compounds, are the reason for a general trend towards analytical problems in biochemistry.[1] The direct and rapid determination of the molecular weight of chemical species in prepurified extracts is of extreme importance, especially in biomedicine and environmental research. With this purpose in mind, and in order to avoid thermal decomposition of the samples from biological sources, the sample molecules are normally desorbed at an optimal anode temperature.[2] This requires only the heat of ionic desorption, lowered by the extremely high electric field (~ 0.5 V/Å), which results in minimal thermal stress (no evaporation). These arguments hold for molecules with molecular ions in the range of $< m/e$ 1500, usually encountered in mass spectrometric investigations. Here FD-Mass Spectrometry (MS) has proved to be a useful tool for molecular weight determination of amino acids,[3] oligopeptides,[4] sugars,[5] oligosaccharides,[6] nucleosides and nucleotides[7] and other substances which are involatile and thermally unstable.

Considerably heavier molecules may be investigated by controlled thermal degradation, followed by ionization and identification of the smaller structural units in the mass spectrometer. Pyrolysis (Py)-MS of such material yields, of course, extremely complex mixtures of pyrolysis products. However, the qualitative determination of the components is greatly facilitated by the use of ionization methods displaying enhanced molecular ion intensities and minimal or no fragmentation.[8] A newly developed technique, Pyrolysis Field Desorption Mass Spectrometry (Py-FD-MS) widens the scope both of pyrolysis and mass spectrometry as demonstrated by promising applications to complex biological materials such as biopolymers,[9,10] bacteria, tissues and blood cells.

INVESTIGATION OF DRUGS AND DRUG METABOLITES

If the aim of the FD-MS study is the determination of the molecular weight (at the optimal anode temperature), this method offers some decisive advantages over others in the field of metabolic research:

(1) In contrast to all other methods of ion production for mass spectrometry the sample in FD is not evaporated from the introduction system, but is supplied to the emitter via the emitter dipping technique.[5] With FD-MS polar compounds of low volatility and/or thermal instability display high molecular ion intensities and little or no fragmentation.

(2) The rate of desorption of the sample is influenced by numerous parameters such as: class of compound, solvent,[11] sample concentration, length and distribution of the microneedles grown at high temperature on the surface of the emitter,[11,12,13] field strength and in particular by the emitter temperature. With the sensitivity of the presently available mass spectrometers, the upper and the lower limit of the sample consumption for one FD analysis lies in the range of 10^{-6} to 10^{-10} g adsorbed material on the emitter surface. This meets the requirements of submicro-analysis in life sciences very well.

(3) Since extracts of living cell material are difficult to purify completely when only small quantities of the substances are present, the high molecular ion intensities exhibited in FD-MS are of particular importance in metabolic studies. In the case of FD-MS the impurities also give only molecular ions. This results in a considerably simplified spectrum and therefore enables easier identification of the metabolites and/or non-metabolic decomposition products.

(4) The possibility of fast and direct detection of unprotected polar compounds obviates the necessity of derivatization, thus saving time and labour. Additionally the drawbacks of derivatization are avoided. These include the incompleteness of reaction leading either to a multiplicity of products or a less than quantitative yield, and the substantial increase in molecular weight which occurs upon derivatization of a polyfunctional compound. For example, with trimethylsilyl-(TMS) 72 mass units are added per functional group, leading to a molecular weight of m/e 707 for 5'-AMP-(TMS)$_5$,[14] while free 5'-AMP yields a signal at m/e 348 for the (M + 1) ion as the base peak of the FD spectrum.[7]

Most of the FD-MS work in drug metabolism has been done so far with anti-cancer drugs such as cyclophosphamide (Endoxan)[15] and sultam derivatives (mainly amino acid conjugates).[16] In the following we will demonstrate some of the significant differences of Electron Impact (EI) and FD-MS by means of a model compound and well-known drug, acetylsalicylic acid. For the determination of the molecular weight as mentioned above the sample is desorbed at the optimal anode temperature which is obtained by choosing the smallest emitter heating current necessary to produce detectable ion currents. The FD-MS of acetylsalicylic acid recorded in this manner is shown in Fig. 1(a). Only the molecular ions and the protonated molecules

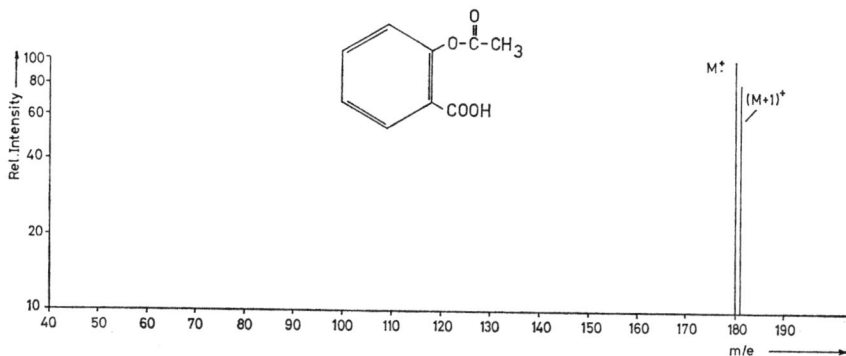

FIG. 1(a)   FD† mass spectrum of acetylsalicylic acid obtained at optimal anode temperature (∼8 mA emitter heating current for a 10-μm, high-temperature activated, tungsten wire emitter).

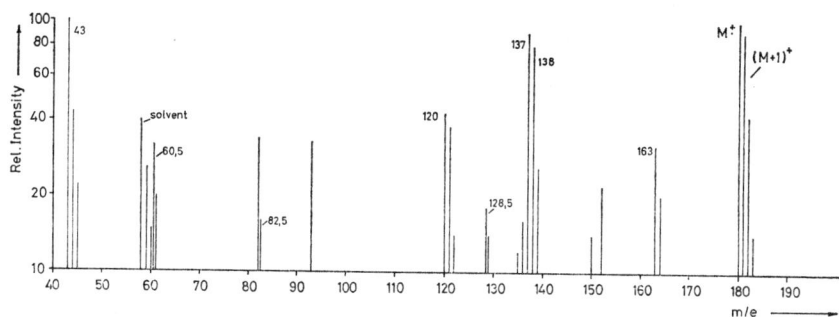

FIG. 1(b)   FD† mass spectrum of acetylsalicylic acid recorded within 6 min exposure time on one trace of a photoplate, raising the emitter heating current constantly from 0–40 mA during this time. The solvent is acetone.

appear with high intensities. The occurrence of the $(M + H)^+$ species in the FD-MS is enhanced by an increase of the acidity of the solvent[11] or of the acidic moiety of the sample molecule. Nevertheless FD-MS does not yield only 'one peak spectra', as may be seen in Fig. 1(b). For a thick layer of the adsorbed compound and a rapid rise in the emitter heating current more thermal energy is transferred. This may well result in a strong fragmentation giving spectra similar to some extent to typical EI spectra as demonstrated for substituted sultams.[16] The FD-mass spectrum of acetylsalicylic acid still displays the molecular ion as the base peak of the spectrum but is also marked by pronounced fragments. Again, some characteristic features of FD-MS such as fragmentation of molecular ions by simple bond rupture, elimination of neutral particles from protonated molecules, the formation of field- and surface-induced cluster ions and the occurrence of doubly charged ions appear as described previously.[16] The interpretation of such spectra is somewhat difficult, but yields valuable structural information. The EI-MS taken at

† All FD spectra were obtained with a CEC21-110B instrument, photographically recorded on vacuum evaporated AgBr-plates (IONOMET). Resolution >20 000 (10% valley definition). (This also refers to Figs. 1(b), 3(b) and 4.)

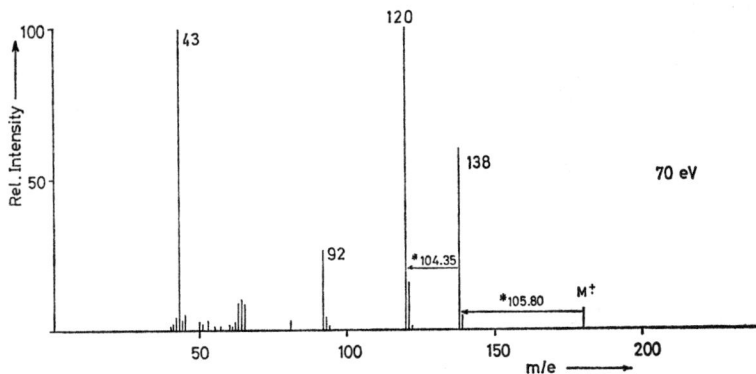

FIG. 2(a)   EI mass spectrum of acetylsalicylic acid. 70-eV electron energy, 60°C probe temperature.

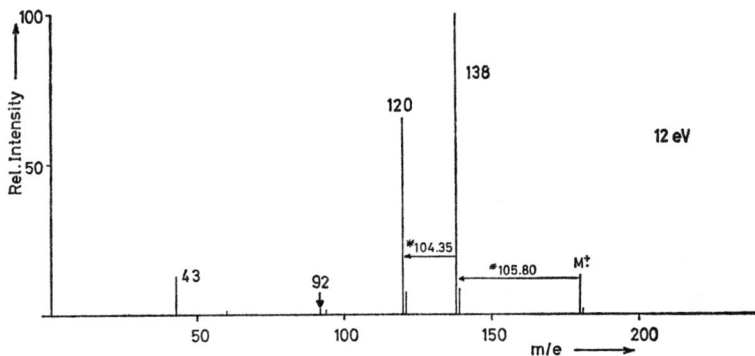

FIG. 2(b)   EI mass spectrum of acetylsalicylic acid. 12-eV electron energy, 60°C probe temperature.

60°C probe temperature show 6% rel. int. of the molecular ion for 70 eV electrons (Fig. 2(a)) and 12% rel. int. for 12 eV electrons (Fig. 2(b)). This reflects the phenomenon of enhanced relative molecular ion intensity at low electron energy despite a reduced absolute intensity. Although for many drugs Chemical Ionization (CI) spectra have $(M + 1)^+$ as base peak, for acetylsalicylic acid this ion has a relative intensity of only 17%, the base peak at m/e 121 being due to a loss of ketene following the water elimination from the $(M + 1)^+$ ion.[17]

APPLICATION TO ENVIRONMENTAL RESEARCH

The arguments used in the part dealing with drugs and drug metabolites are valid for pesticides and pesticide metabolites as well. Moreover, as it is a common observation that in a living organism a non-physiological compound is degraded to a metabolite of higher polarity and that this is the initial step for the more frequently occurring methods of detoxification, the advantages of

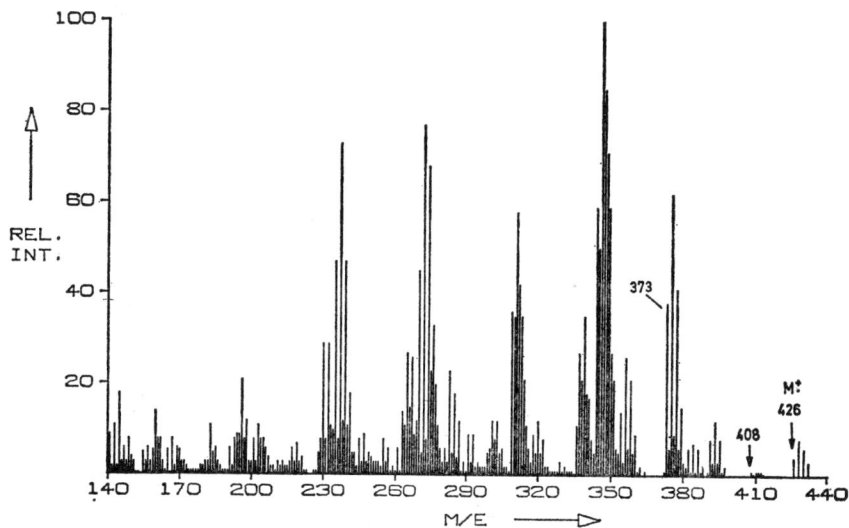

FIG. 3(a) EI mass spectrum of dihydrochlorden-1,3-dicarboxylic acid, 70-eV electron energy.

using FD-MS become even more pronounced as the polarity of the parent compound, pesticide or drug, increases. Further, since higher polarity corresponds to higher reactivity this may imply a higher hazard of the toxic environmental chemical to a living organism.

Most of the FD-MS work on pesticides so far has been done on bridged chlorinated compounds of the dimethano-naphthalene and methano-indene types.[18,19] We give here a typical example, dihydrochlorden-dicarboxylic acid: a main metabolite of aldrin in soil, plants and animals.[20] Figure 3(a)

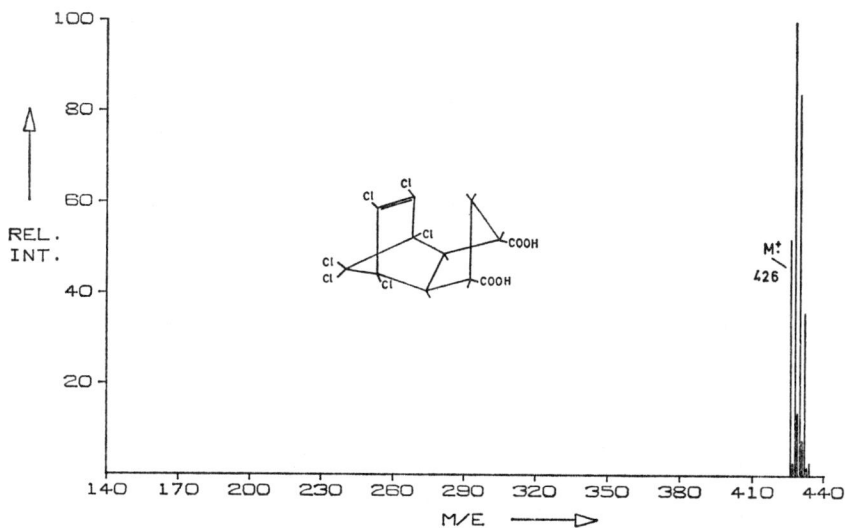

FIG. 3(b) FD† mass spectrum of dihydrochlorden-1,3-dicarboxylic acid, obtained at optimal anode temperature.

shows the EI-MS of this compound. The molecular ion appears with low intensity (3% rel. int.) and is difficult to detect in the case where the sample cannot be purified completely, for instance after an extraction process from biological material. In contrast the FD-MS (Fig. 3(b)) displays only the molecular ion plus the isotope peaks. In general it is found that where a small molecular peak is observed in the EI-MS, the FD-MS has this ion as the base signal, but FD-MS does not necessarily show the molecular ion as the base peak where it is not observable at all with EI-MS.

Extremely polar compounds such as salts of large organic acids are often of importance in environmental studies. These are inaccessible to the usual mass spectrometric investigation since they are virtually involatile. However, with FD-MS it has been possible to study the complete mass spectra of acetates[11] and phosphates.[21] Guaiacolsulphonic acid potassium salt for instance shows only two peaks at m/e 281, 283, due to cationization $(M + K)^{+}$.

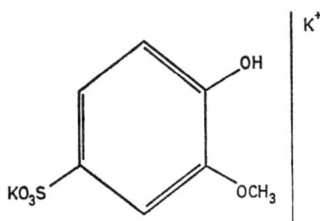

Investigations are presently being undertaken to determine parent pesticides and their metabolites in soil and marine samples. These will reveal the potential of FD-MS for the analysis of strongly contaminated extracts and for other trace analysis problems.

PYROLYSIS FIELD DESORPTION MASS SPECTROMETRY

Py-FD-MS is the direct submicro-pyrolysis on the surface of a high tempera-ture activated tungsten wire emitter. This method is necessary if one wants to obtain structural information about compounds with a molecular weight considerably exceeding m/e 1500. The larger the weight of the detected ions, the more valuable the information obtained. The observation of these large primary products is favoured by four important conditions:

First, minimal difference in time and place between pyrolysis and ionization; second, minimal sample quantities; third, minimal residence time of the free ions in the pyrolysis zone, and fourth, minimal excess energy transferred by ionization.

Py-FD-MS fulfils all these requirements to a high degree:

First, pyrolysis and ionization coincide in place and time; second, the sample quantity is 10^{-8} to 10^{-9} g; third, the residence time of the free ion in the pyrolysis zone is of the order of $<10^{-11}$ sec; fourth and finally, the average excess energy does not exceed 0·2 eV.

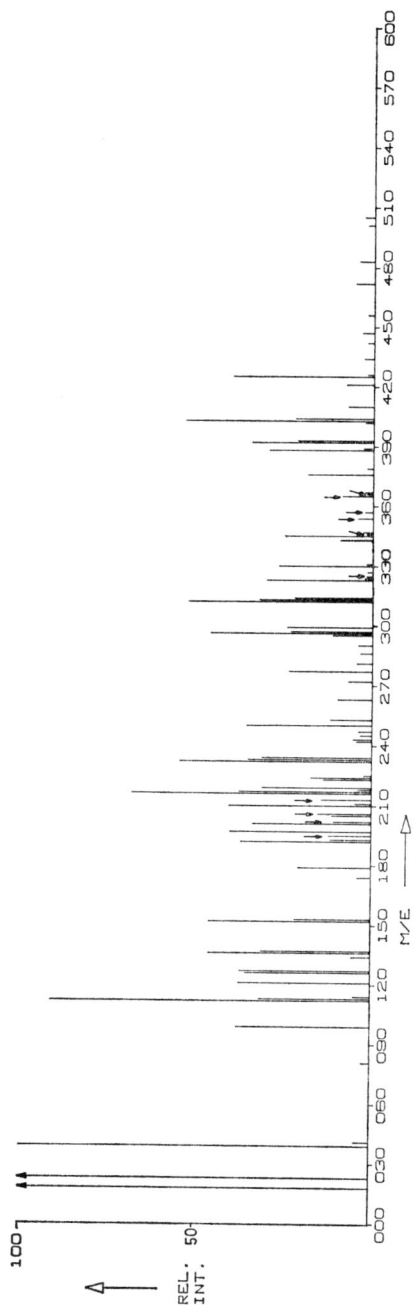

FIG. 4   Pyrolysis Field Desorption mass spectrum of herring DNA.

The first objects of our investigations were biopolymers, where high resolution ($>$20 000, 10% valley definition) enabled the determination of the elemental composition of large structure units.

Figure 4 shows the Py-FD mass spectrum of herring DNA. All five bases *i.e.* cytosine, methylcytosine, thymine, adenine and guanine as well as the corresponding nucleosides and nucleotides have been identified (the experimental conditions and results are discussed in detail in Ref. 9). As may be

TABLE I

Observed Doubly Charged Ions (indicated by $\downarrow$ in Fig. 5) Representing Structural Units of DNA

Ion structure	m/e
(deoxycytidine diphosphate + 2H)$^{++}$	195·019
(deoxythymidine diphosphate + 2H)$^{++}$	202·019
(Na-deoxycytidine diphosphate + 2H)$^{++}$	205·510
(Na-deoxythymidine diphospate + 2H)$^{++}$	213·010
(Na-dithymidine diphosphate + 2H)$^{++}$	325·050
(cytidine-thymidine triphosphate + 2H)$^{++}$	346·542
(dithymidine triphosphate + 2H)$^{++}$	354·042
(Na-cytidine-thymidine triphosphate + 2H)$^{++}$	357·533
(Na-dithymidine triphosphate + 2H)$^{++}$	365·033
(Na-^{13}C dithymidine triphosphate + 2H)$^{++}$	365·535

inferred from Table I some dinucleotides were detected which offer prospects for obtaining sequence information by Py-FD-MS.

To ascertain the proposed fragmentation pattern and ion structures investigations have been initiated on synthetic oligonucleotides of known structure.

CONCLUSION

The possibilities for the application of FD-MS in biochemistry, in particular to problems of drug and pesticide metabolism have been outlined and will certainly lead to further applications in pharmacology, toxicology and forensic sciences. Furthermore, the combination of column liquid chromatography and FD-MS[22] will probably point out new ways for dealing with problems in the above-mentioned area.

Py-FD-MS widens the scope of applicability in investigating bipolymers and complex biological material. The first steps into this field have been taken and gave valuable information about the building blocks of polynucleotides and polysaccharides.

Although FD-MS of organic substances is a relatively recent (1969) offspring of mass spectrometry it is well on its way to becoming an established and useful analytical tool for research in the delineated area. However, the authors are aware of the fact that until now the emphasis has been placed more on exploring and testing the potentialities of the method rather than on solving actual problems.

ACKNOWLEDGMENTS

The authors gratefully acknowledge the supply of the EI spectra by M. Jarman, Chester Beatty Research Institute, Institute of Cancer Research: Royal Cancer Hospital, London, Great Britain (Figs. 2(a), 2(b)) and W. Tomberg, Institut für Ökologische Chemie der Gesellschaft für Strahlen- und Umweltforschung, St Augustin, West Germany (Fig. 3(a)).

REFERENCES

1. Beckey, H. D., *in* 'Biochemical Applications of Mass Spectrometry', *ed.* George Waller, Wiley-Interscience, 1972, p. 810.
2. Beckey, H. D. and Schulten, H.-R., *Angew. Chemie* (Review), in preparation (1973).
3. Winkler, H. U. and Beckey, H. D., *Org. Mass Spectrom.*, 1972, **6**, 655.
4. Winkler, H. U. and Beckey, H. D., *Biochem. Biophys. Res. Commun.*, 1972, **46**, No. 2, 391.
5. Beckey, H. D., *Int. J. Mass Spectrom. Ion Phys.*, 1969, **2**, 500.
6. Krone, H. and Beckey, H. D., *Org. Mass Spectrom.*, 1971, **5**, 983.
7. Schulten, H.-R. and Beckey, H. D., *Org. Mass Spectrom* , 1973, **7**, 861
8. Schulten, H.-R., Beckey, H. D., Meuzelaar, H. L. C. and Boerboom, A. J. H., *Anal. Chem.*, 1973, **45**, 191.
9. Schulten, H.-R., Beckey, H. D., Boerboom, A. J. H. and Meuzelaar, H. L. C., *Anal. Chem.*, in press, 1973.
10. Schulten, H.-R., Symposium on Rapid Methods and Automation in Microbiology, Stockholm, Sweden, 4–8 June, 1973.
11. Schulten, H.-R. and Beckey, H. D., *Org. Mass Spectrom.*, 1972, **6**, 885.
12. Schulten, H.-R. and Beckey, H. D., *Messtechnik*, 1971, **78**, 196.
13. Beckey, H. D., Hilt, E. and Schulten, H.-R., *J. Phys. E: Sci. Instrum.*, 1973, **6**, 1043.
14. Lawson, A. M., Stillwell, R. N., Tacker, M. M., Tsuboyama, K. and McCloskey, J. A., *J. Amer. Chem. Soc.*, 1971, **93**, 1014.
15. Schulten, H.-R. *et al.*, *Anal. Biochem.*, in preparation, 1973.
16. Schulten, H.-R., Beckey, H. D., Eckhardt, G. and Doss, S. H., *Tetrahedron*, in press, 1973.
17. Milne, G. W. A., Fales, H. M. and Axenrod, T., *Anal. Chem.*, 1971, **43**, 1815.
18. Schulten, H.-R., Prinz, H., Beckey, H. D., Tomberg, W., Klein, W. and Korte, F., *Chemosphere*, 1973, **2**, 23.
19. Schulten, H.-R. and Beckey, H. D., *J. Agr. Food Chem.*, 1973, **21**, 372.
20. Gäb, S., Palar, H. and Korte, F., *Tetrahedron*, in press, 1973.
21. Schulten, H.-R., Beckey, H. D., Bessel, E. M., Forster, A. B., Jarman, M. and Westwood, J. H., J.C.S. *Chem. Comm.*, 1973, **13**, 416.
22. Schulten, H.-R. and Beckey, H. D., *J. of Chromatogr.*, 1973, **83**, 315.

58
A Quadrupole Mass Spectrometer for the Identification and Characterization of Neutral Fragments Formed by the Interaction of Electrons with Gaseous Molecules

By J. R. REEHER and H. J. SVEC

(*Ames Laboratory–USAEC and Department of Chemistry, Iowa State University, Ames, Iowa, U.S.A.*)

In a conventional mass spectrometer ion source, gaseous molecules interact with electrons in various ways to form many species, including positive ions, negative ions, and neutral radicals and molecules, all of which may be formed in ground or excited states. In order to completely characterize the behaviour of gaseous molecules under electron impact, all of these species must be identified and the energetics of their formation and ionization determined. The application of mass spectrometric methods to the detection of positive and negative ions and the determination of their energetics of formation are well documented. However, the neutral fragments formed during decomposition processes and the energetics of their formation and ionization are not so well known. Although indirect methods, such as metastable ion transitions, isotopic labelling, and deconvolution of ionization efficiency curves, are available for deducing the makeup of neutral fragments, methods for direct detection have had limited success. However, a quadrupole mass spectrometer with a modified ion source and ion detection system for characterizing the neutral fragments formed in a mass spectrometer ion source has been developed recently in our laboratory. This instrument has several advantages over developments described previously for this purpose, including those by Beck, Osberghaus and Niehaus,[1-5] Melton,[6-9] Saunders, Larkins and Saalfeld,[10] and Preston, Tsuchiya and Svec.[11] These advantages include:

(1) Greatly improved sensitivity for detecting neutral fragments.
(2) Convenient scanning of neutral fragment mass spectra.
(3) Convenient measurement of neutral fragment appearance potentials and vertical ionization potentials.

The ion source (Fig. 1) has a primary (1Y) reaction chamber with a pulsed electron beam for producing the neutral fragments and a secondary (2Y) ionization chamber with a d.c. electron beam for ionizing the neutral fragments formed in the first. The two chambers are separated by three tungsten mesh grids, each of which is 92% transparent and which preserve the integrity of the electric fields in the ion source. The distance between the centres of the two chambers is less than 8 mm.

Sample gases enter the primary reaction chamber through a nozzle perpendicular to the primary electron beam and are condensed on a liquid-nitrogen cooled surface immediately after traversing the primary ionization region.

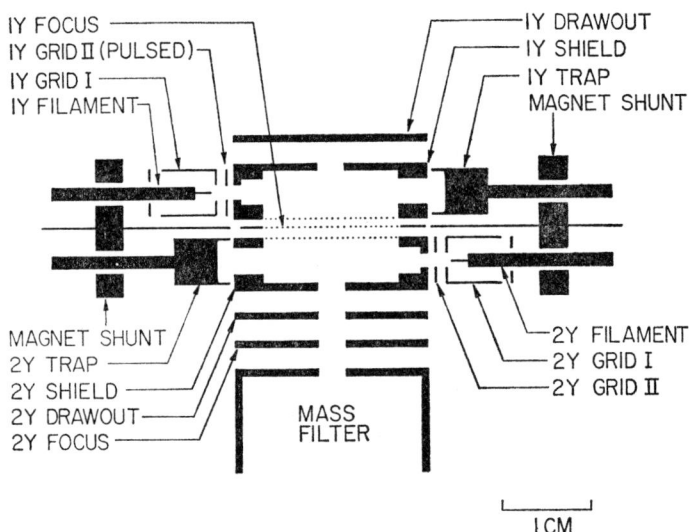

FIG. 1   Dual ionization chamber ion source.

The electron guns are pentodes with two grids between the filament and the shield or ionization chamber. The pulsed electron beam in the primary chamber is produced by pulsing the second grid of the primary electron gun. The grid closest to the filament collects the electrons emitted during the 'off' cycle of the pulsed electron beam, eliminating the problem of space charge building up in the vicinity of the filament. In the second chamber, the two grids act as a lens to focus the electron beam. Grounded plates between the components of the two electron guns reduce pickup of the a.c. pulse by the components of the secondary electron gun. Opposing the direction of current in the two electron guns also helps to eliminate this problem.

In general, maximum trap currents of 50 μA are utilized in the secondary electron beam to minimize ion trapping. However, in the primary reaction chamber, higher trap currents (up to 200 μA) may be utilized to produce neutral fragments, since they are not trapped in the field of the electron beam.

The processes which occur in the primary reaction chamber produce positive ions and neutral fragments at the frequency of the pulsed electron beam. The positive ions are attracted to the primary drawout plate, while the neutral fragments are allowed to diffuse through the tungsten mesh grids to the secondary ionization chamber.

In the secondary ionization chamber, two types of ion currents are produced. Those gas molecules which enter this chamber and interact with the secondary electron beam without having first interacted with the primary electron beam produce a d.c. ion current, which after mass analysis yields the normal mass spectrum of the sample gas. The particles which interact with the secondary electron beam after having interacted with the pulsed primary electron beam

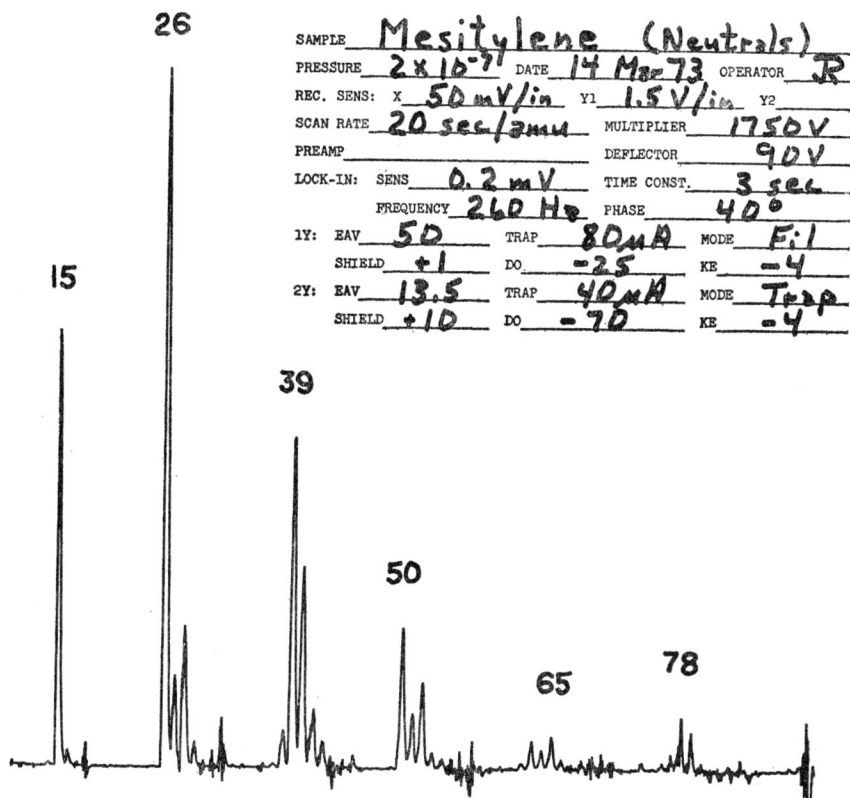

FIG. 2 Neutral mass spectrum of mesitylene.

produce an a.c. ion current, which after mass analysis yields the neutral mass spectrum of the sample gas.

The total ion current produced in the secondary ionization chamber is focussed into the quadrupole mass filter where it is mass analysed. Those ions which pass through the mass filter are deflected to the first dynode of a 17-stage Cu-Be electron multiplier. The function of the deflector is to eliminate any direct optical path between the ion source and the electron multiplier. A platinum black surface situated opposite the exit hole from the mass filter absorbs any photons which originate in the source region and traverse the

mass filter to the deflector region. After amplification by the electron multiplier, the output signal is applied to a pre-amplifier which separates the signal
into its a.c. and d.c. components and amplifies them. The d.c. component,
with a gain of 10^8 V/A, is recorded directly as the normal mass spectrum of the
sample. The a.c. component, with a gain of 10^9 V/A is processed by a
Princeton Applied Research lock-in amplifier whose output is recorded as the
neutral mass spectrum of the sample.

The neutral mass spectrum of mesitylene is shown as recorded from the
instrument in Fig. 2. In order to relate these raw data to the relative abundances
of the neutral fragments present, two major corrections must be made. First,
the spectrum must be corrected to compensate for the mass discriminations

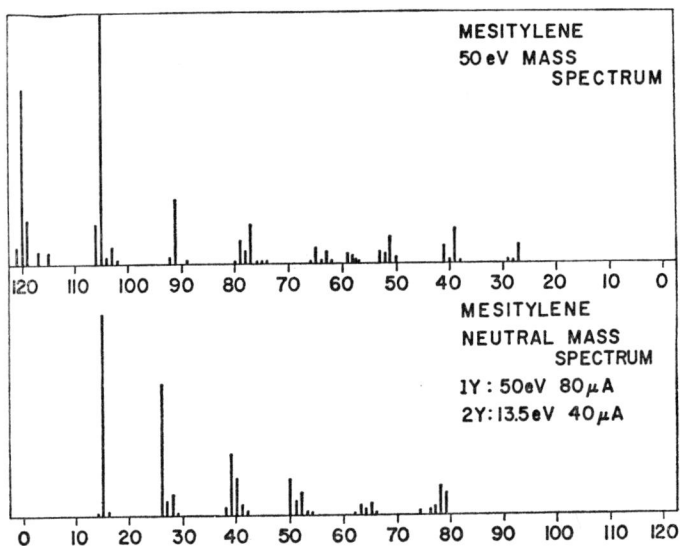

FIG. 3   Comparison of neutral mass spectrum of mesitylene with normal positive ion mass
spectrum.

inherent in the quadrupole mass filter and the electron multiplier. Second,
since each neutral fragment has a different cross-section for ionization, the
neutral mass spectrum must be corrected by dividing each peak intensity by
its corresponding cross-section. After making these two corrections, a
spectrum is obtained as shown in Fig. 3. Here, the neutral mass spectrum is
plotted with ascending mass while the normal positive ion mass spectrum is
plotted above it with descending mass, so that correlations between neutral
fragments and complementary positive ions can be observed more readily.
Although exact correlation is not evident, it should be noted that the shape of
the envelope of the two spectra plotted in this manner have the same general
appearance.

The neutral mass spectra of several methylsubstituted aromatic compounds
are compared in Fig. 4. Note the increase in methyl radical abundance as more

methyl groups are substituted on the ring, and the increase in abundance of higher mass neutrals from the heavier molecules. Note also that many of the same neutral fragments are observed from all of these compounds. This observation indicates that these compounds are randomizing to a great extent after interacting with an electron and before dissociating in the primary reaction chamber.

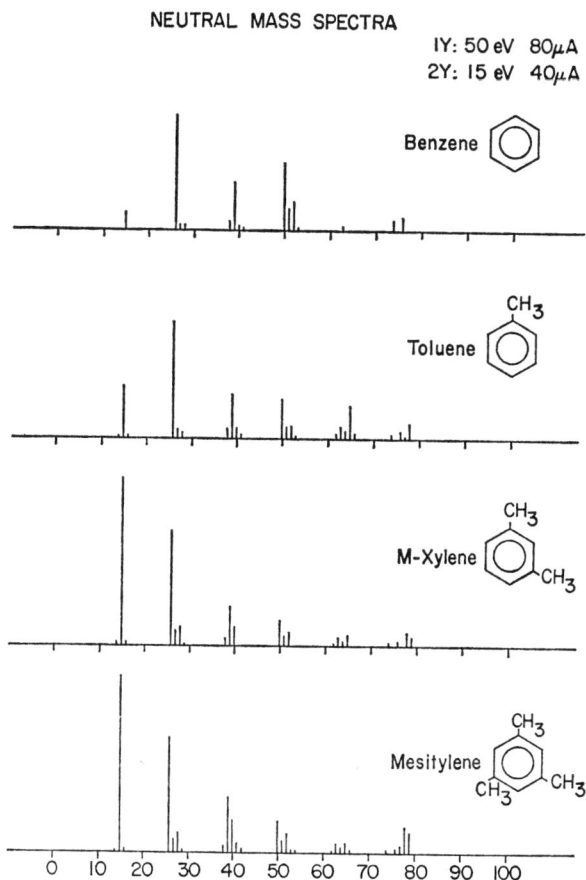

FIG. 4   Neutral mass spectra of benzene and some methyl-substituted benzenes.

Some uncorrected ionization efficiency curves of toluene neutral fragments being ionized in the secondary chamber are shown in Fig. 5. The energy scale was then calibrated with the parent ion from benzene. The data obtained from these curves are tabulated in numerical form in Fig. 6. The values determined are reproducible to about ± 0.06 V with good agreement between the linear extrapolation and extrapolated voltage difference methods of treating the data. Literature values for the vertical ionization potentials of C_3H_3 and C_4H_3 were not available. Measurement of neutral fragment ionization potentials such as these is essential to the determination of bond energies from

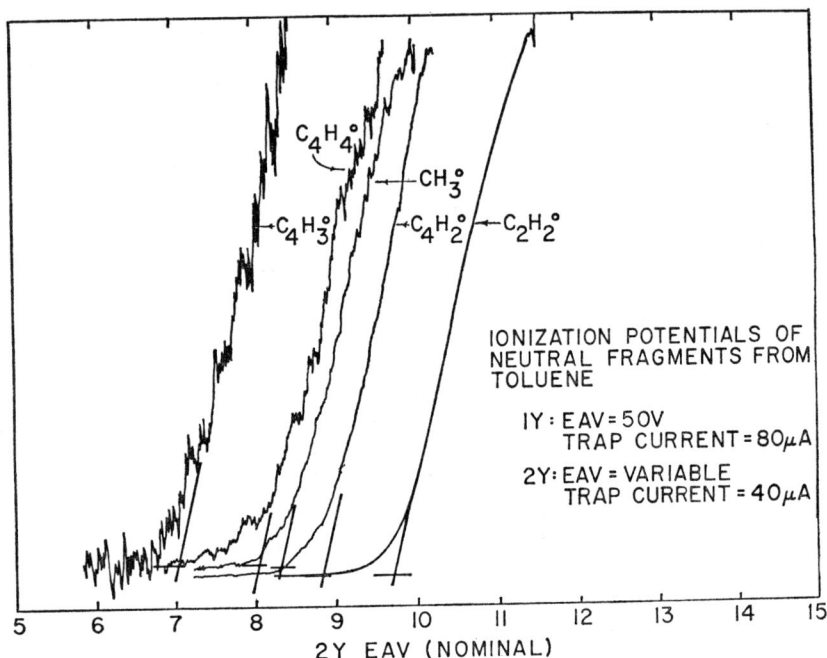

FIG. 5   Ionization efficiency curves of neutral fragments from toluene.

mass spectrometric data. Appearance potentials of neutral fragments can be determined similarly by scanning the energy of the primary electron beam.

In conclusion, this instrument development should open a new area of mass spectrometry which will lead to a better understanding of the processes which occur in a conventional mass spectrometer ion source.

IP'S OF NEUTRAL FRAGMENTS FROM TOLUENE

M/e	MOLECULAR FORMULA	LITERATURE	LE	EVD
15	CH_3	9.91	9.93	9.94
26	C_2H_2	11.40	11.38	11.37
39	C_3H_3	—	9.43	9.37
50	C_4H_2	10.20	10.47	10.37
51	C_4H_3	—	8.67	8.61
52	C_4H_4	9.87	9.68	9.70

FIG. 6   Ionization potentials of neutral fragments from toluene.

REFERENCES

1. Beck, D. and Osberghaus, O., *Z. für Physik*, 1960, **160**, 406.
2. Beck, D. and Niehaus, A., *J. Chem. Phys.*, 1962, **37**, 2705.
3. Beck, D., *Disc. Faraday Soc.*, 1964, **36**, 56.
4. Genzel, H. and Osberghaus, O., *Z. Naturf.*, 1967, **22a**, 331.

5. Niehaus, A., *Z. Naturf.*, 1967, **22a**, 690.
6. Melton, C. E., *J. Chem. Phys.*, 1966, **45**, 4414.
7. Melton, C. E., *J. Sci. Instr.*, 1966, **43**, 927.
8. Melton, C. E. and Rudolph, P. S., *J. Chem. Phys.*, 1967, **47**, 1771.
9. Melton, C. E., *Int. J. Mass Spec. Ion Phys.*, 1968, **1**, 353.
10. Saunders, R. A., Larkins, J. T. and Saalfeld, F. E., *Int. J. Mass Spec. Ion Phys.*, 1969, **3**, 203.
11. Preston, F. J., Tsuchiya, M. and Svec, H. J., *Int. J. Mass Spec. Ion Phys.*, 1969, **3**, 323.

Discussion

H. Hotop (University of Freiburg, Germany): Did you study the dependence of the neutral fragment mass spectra on the frequency of the pulsed primary electron beam? The reason for asking this question is that one may need higher frequencies in order to get rid of neutrals which have undergone wall collisions before ionization in the second chamber.

H. J. Svec: Yes, we studied frequencies over a range of about 1 kHz. The 260 Hz frequency we finally chose was a compromise which gave maximum neutral fragment signals for materials we knew were producing neutral fragments and at the same time had the appropriate phase dependence. Incidentally, excited rare gas atoms or C_2H_2 from benzene are very good neutral species to use in tuning the instrument.

H. Hotop: Why do you have a *great* loss in sensitivity—as you mentioned—when pulsing the secondary electron beam?

H. J. Svec: The loss was not *great*. We simply wanted every bit of sensitivity we could possibly get and even a factor of two was worthwhile eliminating. Pulsing only one electron beam also somewhat simplifies the overall instrumental problems. We've always maintained that one can solve problems by complicated means but our objective is however to always find the simplest possible way which produces a suitable result.

R. Botter (C.E.A., Saclay, France): Can you not have large discrimination effects in your apparatus for the neutral fragment measurements, if you have a beam of molecules directed perpendicular to the diffusion direction? In this case only those fragments which are formed with initial kinetic energy will diffuse out of the initial beam and therefore partly to the second ionization chamber. Under this condition they will not be condensed on the cold trap.

S. J. Svec: This may be true. I believe however that the apparent correspondence between the neutral fragment spectra we observe and that of the positive ions indicates that such discriminations are not serious. Perhaps they account for the imperfections that we observe in the correlations.

R. E. Honig (R.C.A., Princeton, U.S.A.): Where in the ion source were the liquid nitrogen surfaces located?

S. J. Svec: Directly opposite the gas inlet, immediately adjacent to the electron gun.

R. E. Honig: What is the explanation for the negative noise in your raw data?

S. J. Svec: Negative noise results when intense direct current peaks are processed in the lock-in-amplifier. They are electronic artifacts.

59

Reagent Gases for Chemical Ionization Mass Spectrometry

By DONALD F. HUNT

(*Department of Chemistry, University of Virginia,*
Charlottesville, Virginia, U.S.A.)

In chemical ionization mass spectrometry (CIMS) sample molecules are ionized by chemical reactions in the gas phase rather than by the conventional electron bombardment method.[1] Fundamental to the development of CIMS as a method for the identification and structure elucidation of organic compounds is the finding that the nature of the CI spectrum is dependent on the type of ion-molecule reaction employed to ionize the sample. Different structural information can be obtained with different reagent gases. In this paper we summarize the results of our research with deuterium oxide, argon–water, ammonia, and nitric oxide as reagents for CIMS.

Studies with deuterium oxide indicate that this reagent is ideally suited for the determination of active hydrogen in organic compounds on a nanogramme scale.[2] Electron bombardment of deuterium oxide at 1 torr affords ions at m/e 22 (D_3O^+), 42 ($[D_2O)_2D^+]$), 62 ($[D_2O]_3D^+$), 82 ($[D_2O]_4D^+$) and 102 ($[D_2O]_5D^+$). These ions, in turn, function as Brönsted acids in the gas phase and deuterate most organic compounds. In addition, as a consequence of the large number of collisions that occur between the sample and neutral deuterium oxide in the CI source, all of the acidic hydrogens on the organic sample are exchanged for deuterium. Determination of the number of active hydrogens is therefore easily accomplished by recording two CI spectra, one with either water, methane, or isobutane to establish the molecular weight of the sample and one with deuterium oxide to count the number of hydrogens that undergo exchange. Typical spectra are displayed in Table I.

In general, hydrogens attached to hetero-atoms in alcohols, phenols, carboxylic acids, amines, amides and mercaptans undergo essentially complete exchange in the ion source when deuterium oxide is employed as the reagent gas at 0·4 torr. Small amounts of deuterium incorporation ($<15–20\%$) occur with some ketones, aldehydes, esters and olefins, but this does not usually complicate the analysis. The above method is also applicable to the analysis of GC effluents since quantitative exchange is accomplished by using an inert gas as the GC carrier gas and adding sufficient deuterium oxide to the gaseous effluent to maintain its partial pressure in the CI source above 0·4 torr. As a consequence of its ability to participate in rapid and selective isotopic exchange

517

reactions, deuterium oxide is an excellent CI reagent for differentiating primary, secondary and tertiary amines, and other compounds having identical molecular weights but different functional groups.

Research with a mixture of argon and water as a CI reagent indicates that spectra obtained with this combination of gases exhibit all the features that are characteristic of both conventional electron impact (EI) and CI (CH_4) mass spectra.[3] Electron bombardment of Ar–H_2O at 1 torr affords H_3O^+, Ar^+, $Ar_2{}^+$, Ar^*, (metastable argon) and a large number of low energy electrons. As outlined above, H_3O^+ functions as a gaseous Brönsted acid and protonates

TABLE I
Deuterium Oxide Chemical Ionization Mass Spectra

Compound	Mol. wt.	$M^a + D^+$	$M^a + D^+ - D_2O$	$M^a + D_3O^+$	Other m/e	Hydrogens exchanged
		% Relative abundance				
Aniline	93	100	—	40	—	2
Di-n-pentylamine	157	100	—	—	—	1
Triethylamine	101	100	—	—	—	0
Benzamide	121	100	—	40	105 (5)	2
Fructose	180	5	100	8	147 (5)	5
Isoamyl ether	158	100	—	25	71 (20)	0
Acetophenone	120	100	—	10	105 (5)	0
Methyl methyacrylate	100	100	—	70	149 (10)	0
Digitoxigenin	374	100	20	—	338 (15)	2
Adenosine	267	100	—	—	273 (15),* 272 (2), 140 (25), 136 (15)	5
Triamterene	253	100	—	—	—	6

[a] M = Molecular weight of sample with all active hydrogens exchanged for deuterium.

most organics to give abundant M + 1 ions. Reaction of the sample with the remaining reagent species in the source by charge-exchange, Penning ionization, or low-energy electron impact produces a distribution of molecular and fragment ions which closely resembles that formed by the conventional electron bombardment technique. As shown in Table II, relative abundances of fragment ions produced by the EI and CI (Ar–H_2O) methods seldom differ by more than a factor of two or three.

On-line data processing of high resolution mass spectra can also be accomplished with the CI (Ar–H_2O) technique since the mass spectrum of the usual internal standard, perfluorokerosene, is identical to that obtained by electron impact. A comparison of the sample ion current generated using conventional EI and CI (Ar–H_2O, N_2–H_2O and He–H_2O) indicates that the sensitivity of CI (Ar–H_2O and N_2–H_2O) are almost identical and about 50 times greater than that obtained with either CI (He–H_2O)[4] or EI.

TABLE II
Electron Impact and Chemical Ionization (Argon–Water) Spectra

4-Decanone (mol. wt 156)			Di-n-pentylamine (mol. wt 157)			5α-Dihydrocorticosterone (mol. wt 348)		
m/e	$EI^{a,b}$	$CI(Ar-H_2O)^a$	m/e	$EI^{a,b}$	$CI(Ar-H_2O)^a$	m/e	$EI^{a,b}$	$CI(Ar-H_2O)^a$
157	—	84	158	—	93	349	—	50
156	7	4	157	8	5	348	0·5	1
113	35	20	156	2	20	331	—	50
86	30	7	142	0·8	4	330	7	2
85	20	6	128	2	1·5	317	100	15
71	45	46	114	2	3	313	—	25
58	39	17	100	100	100	299	5	10
57	25	10	71	6	6	295	—	15
43	100	100	70	8	2·5	271	56	60
			58	2	10	253	7	10
			57	4	6	161	75	100
			56	6	5			
			44	80	50			
			43	36	83			
			30	30	35			

[a] Relative abundance of sample ion peaks. Base peak equal 100.
[b] 70-eV electron impact spectrum.

Because the total ion current, that due to both sample and reagent, is frequently more than 100 times that produced by electron bombardment in our dual EI/CI source,[5] tuning our AEI MS–902 spectrometer for operation with narrow source and collector slits at resolutions between 10 000 and 20 000 is greatly simplified. As a consequence all high resolution service samples in our laboratory are now run by the CI technique using either Ar–H$_2$O or N$_2$–H$_2$O as the reagent gas. We find this method to be particularly valuable for the analysis of biological samples where the amount of material available for examination is frequently sufficient for only one experiment. Molecular weight and elemental composition data are obtained from the abundant M + 1 ion which is almost always present in the spectrum. In addition, valuable structural information is obtained from the electron impact type fragmentation that is also invariably observed.

In contrast to the situation with Ar–H$_2$O, almost no fragmentation occurs when ammonia is employed as a CI reagent gas. Electron bombardment of ammonia at 1 torr generates ions at m/e 18 (NH_4^+), 35 ($[NH_3]_2H^+$), and 52 ($[NH_3]_3H^+$), and these ions function as either electrophiles or weak Brönsted acids toward organic samples. Research to date indicates that ammonia is a useful CI reagent for identifying organic functional groups, for selectively

TABLE III

Ammonia Chemical Ionization Mass Spectra

Compound	Mol. wt	$M + H^+$	$M + NH_4^+$	$M + (NH_3)_2H^+$	Other
		% Relative abundance[a]			
Cyclohexylamine	99	60	100	—	—
N-Ethylaniline	121	100	95	—	—
N,N-Diethylaniline	149	100	—	—	—
Triethylamine	101	100	—	—	—
N-Ethylacetamide	87	30	100	5	—
1,5-Diphenyl-3-pentadienone	234	100	20	—	—
Lauric acid	200	—	100	5	—
Propyl propanoate	116	—	100	75	—
Cyclohexanone	98	—	100	25	—
Trans-Cyclohexane-1,2-diol	116	—	100	—	—
D(-) Ribose	150	—	100	12	—
2-Decanone	156	—	100[b]	23[b]	m/e 170 (83) $M + CH_3NH_3^+ - H_2O$
Benzamide	121	—	100[b]	45[b]	—

[a] Relative abundance of sample ions. Base peak equals 100.
[b] Recorded with CH_3NH_2 as the reagent gas. Ions correspond to $M + CH_3NH_3^+$ and $M + (CH_3NH_2)_2H^+$.

ionizing basic compounds in complex organic mixtures, and for determining the molecular weights of polyhydroxylic compounds.

As shown in Table III, amines, amides and α,β-unsaturated ketones are the only compounds examined that are sufficiently basic to accept a proton from the ammonium ion.[6] All of these compounds with the exception of tertiary amines also form M + 18 ions as a result of electrophilic attachment of the ammonium ion to the organic sample. Ketones, esters, ethers, and anhydrides suffer electrophilic attachment by NH_4^+ but are not protonated in the gas phase. Alkanes, aromatics, alkenes, alcohols, ethers, and nitro compounds fail to react with NH_4^+ when the ratio of reagent to sample is maintained at the usual 100/1 level.

Exceptions to this latter generalization occur when two or more oxygenated functional groups exist in close proximity in the same molecule. Abundant M + 18 ions are formed from diols such as trans-cyclohexane-1,2-diol and are presumably stabilized by hydrogen bonding between the two hydroxyl groups and the NH_4^+ ion. This result suggests that ammonia may find utility as a reagent for probing the geometrical relationship of functional groups in polyfunctional organic molecules. Electrophilic addition of NH_4^+ also occurs to underivatized sugar molecules.[7] Because the energy liberated in this reaction is relatively low, very few of the M + 18 ions suffer fragmentation. Consequently, ammonia is an excellent reagent gas for determining the molecular weight of compounds of this type.

An early report[8] that aldehydes lose water and form protonated Schiff bases under CI (NH$_3$) conditions has been confirmed in our laboratory. We find, however, that most, if not all, of the protonated Schiff base product is generated in a two-step process involving a neutral gas phase reaction between ammonia and the aldehyde and subsequent protonation of the Schiff base product by NH$_4^+$. Addition-elimination reactions involving ammonia are not observed with ketones, esters and amides. Not unexpectedly, ions due to protonated Schiff base products are also observed when methylamine is used as a CI reagent gas. Addition-elimination reactions with this reagent are considerably faster than those with ammonia and occur with both aldehydes and ketones. An analogous reaction, however, is not observed with esters and amides.

In addition to our research with deuterium oxide, argon–water, and ammonia, we have also recently examined the utility of nitric oxide as a CI reagent.[9] Electron bombardment of nitric oxide at 1 torr produces NO$^+$ in high abundance. This ion, in turn, functions as an oxidising agent, hydride

TABLE IV
Nitric Oxide Chemical Ionization Mass Spectra

Compound	Mol. wt	M-1	M-OH	M + NO$^+$	Other
		% Relative abundance[a]			
n-Pentyl acetate	130	—	—	100	—
Acetophenone	120	—	—	100	—
Heptanal	114	90	—	100	—
Di-n-amyl ether	156	100	—	—	—
Hexadecanoic acid	256	5	10	100	—
1-Pentanol	88	10	—	—	M-2 + 30 (100), M-3 (20)
2-Pentanol	88	25	10	—	M-2 + 30 (100)
2-Methyl-2-pentanol	88	—	100	—	—
n-Decane	142	100	—	—	85 (5), 99 (2)
p-Di-t-butylbenzene	190	—	—	100	190 (10)
Trans-2-decene	140	42	—	100	114 (9), 100 (6)
2,4,4-Trimethyl-1-pentene	112	10	—	100	—
2,4,4-Trimethyl-2-pentene	112	—	—	100	85 (10), 115 (35)
1-Hexene	84	5	—	100	86 (80)
1-Heptene	98	10	—	100	86 (80), 100 (40)
1-Octene	112	7	—	100	86 (70), 100 (60), 114 (30)
Cyclohexane	84	100	—	—	—
Anisole	108	—	—	10	108 (100)

[a] Relative abundance of sample ion peaks. Base peak equal 100.

and hydroxide abstractor, and electrophile toward organic molecules. Results of our investigation to date indicate that this reagent can be used to identify organic functional groups and to qualitatively assay components in complex hydrocarbon mixtures.

As shown in Table IV, CI (NO) spectra of esters and ketones exhibit a single ion which results from electrophilic addition of NO^+ to the carbonyl oxygen of these compounds. Aldehydes undergo the same reaction but also suffer hydride abstraction to give an abundant M-1 ion. CI (NO) spectra of ethers display an M-1 ion but fail to show an adduct ion. Acids undergo electrophilic addition as well as hydroxide abstraction in the presence of NO^+.

Of additional interest is the finding that CI with nitric oxide can be used to differentiate primary, secondary and tertiary alcohols. CI (NO) spectra of primary alcohols exhibit three ions corresponding to M-1, M-3 and M-2 + 30. These species are produced by hydride abstraction from the alcohol and oxidation of the alcohol followed by either hydride abstraction from, or electrophilic addition to the resulting aldehyde. Secondary alcohols are converted to M-1, M-2 + 30 and M-17 ions by NO^+. CI (NO) spectra of tertiary alcohols exhibit a single peak at M-17.

CI (NO) spectra of hydrocarbons are also of interest since little or no fragmentation accompanies the ionization step. As illustrated in Table IV, the dominant ion in the CI (NO) spectra of saturated hydrocarbons corresponds to M-1. Spectra of internal olefins generally show two ions corresponding to M-1 and M + 30 and are easily distinguished from cycloalkanes. Spectra of the latter type compounds contain an M-1 ion only. In contrast to the situation with disubstituted olefins, straight chain terminal alkenes exhibit M-1 and M + 30 ions as well as a series of abundant fragment ions at m/e 86 (C_4H_8NO), 100 ($C_5H_{10}NO$) and 114 ($C_6H_{12}NO$). Dienes undergo hydride abstraction, electrophilic addition and charge exchange with NO^+. Electron-rich aromatic compounds also suffer oxidation in the presence of NO^+. Electrophilic addition of NO^+ to the aromatic ring system is also usually observed.

REFERENCES

1. (a) Field, F. H., 'MTP International Review of Science', Physical Chemistry, Series I, Vol. 5, ed. A. Maccoll, Butterworth and Co. Ltd, Oxford and University Park Press, Baltimore, Md., 1972, chapter 5; (b) Field, F. H., 'Ion Molecule Reactions', Vol. I, ed., J. L. Franklin, Plenum Press, New York, N.Y., 1972, chapter 6; (c) Munson, M. S. B., Anal. Chem., 1971, **48**, 28A.
2. Hunt, D. F., McEwen, C. N. and Upham, R. A., Anal. Chem., 1972, **44**, 1292.
3. Hunt, D. F. and Ryan III, J. F., Anal. Chem., 1972, **44**, 1306.
4. Arsenault, G. P., J. Amer. Chem. Soc., 1972, **94**, 8241.
5. Manufactured by Scientific Research Instruments Corporation, Baltimore, Md.
6. (a) Dzidic, I. and McCloskey, J. A., Org. Mass Spectrom., 1972, **6**, 939; (b) Dzidic, I., J. Amer. Chem. Soc., 1972, **94**, 8333.
7. Hogg, A. M. and Nagabhushan, T. L., 'Twentieth Annual Conference on Mass Spectrometry and Allied Topics', Dallas, Texas, June 1972, Abstract No. P2.
8. Beggs, D. P., 'Twentieth Annual Conference on Mass Spectrometry and Allied Topics', Dallas, Texas, June 1972, Abstract No. N4.
9. Hunt, D. F. and Ryan III, J. F., J.C.S. Chem. Comm., 1972, p. 620.

60
Modification of an Ion Source for Improvement of the Electron Energy Resolution

By MIODRAG MILETIC,†
ALEKSANDAR STAMATOVIC,‡ RADOJKO MAKSIC†
and KIRO ZMBOV†

†('Boris Kidric' Institute, Vinca, Yugoslavia)
‡(Institute of Physics, Beograd, Yugoslavia)

INTRODUCTION

THE classical electron impact ion source for magnetic deflection mass spectrometry has an electron energy spread of the order of magnitude of 1 eV. Most of this energy spread originates from the electron source which is usually a hot filament made of high melting point metal. Additional electron energy spread, resulting from non-uniformity of electrostatic accelerating fields, could be eliminated, but even that does not give enough energy resolution to resolve most of the near threshold structure of the ionization efficiency curve and to determine accurately enough the critical potentials of the studied electron impact phenomena.[1] So, methods for reduction of the electron energy spread or its influence have to be applied in order to obtain more information from electron collision experiments.

Generally there are two ways of solving the problem, namely, using one of several types of electron monochromator in order to reduce the electron energy spread,[1] or by the application of different deconvolution techniques or methods[2] to eliminate the influence of the electron energy spread on the final result of measurements. In fact, it is possible to combine both the monochromatization of the electron beam and deconvolution of the data obtained by such a 'monochromatic' electron beam, but very little of that is done in practice. Deconvolution itself is applied more and more, owing to the availability of digital computer time, but it seems that the results are inferior to the ones obtained by the use of a 'monochromatic' electron beam.

EXPERIMENTAL

The mass spectrometer used was a 90° sector field magnetic analyser with a secondary electron multiplier as an ion current detector. The original ion source was a Nier-type electron bombardment ion source with the ions being extracted by a penetrating electrostatic field from outside the collision chamber. The gas inlet was so arranged to introduce gas directly into the collision region.

Before any modification of the ion source, different methods of electron beam monochromatization were considered and some advantages found for the trochoidal electron monochromator (TEM);[3] they were: very tolerable experimental conditions required for operation, simple construction and small size, insensitivity to small fringing fields of the main magnet of the mass spectrometer, and the presence of an electron beam confining magnetic field.

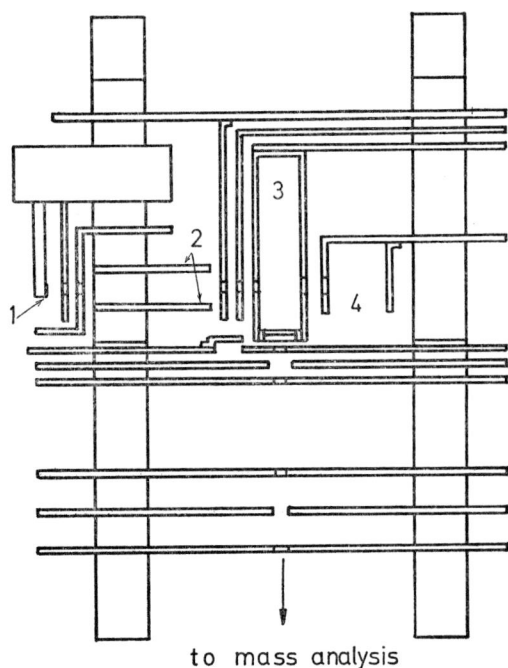

to mass analysis

FIGURE 1

In fact, under existing conditions RPD method[4] is the only electron mono-chromator applicable except for TEM, and TEM has some advantages such as better resolution and better signal-to-noise ratio. The choice of TEM has brought some disadvantages too, such as critical electrostatic shielding and rather low current in the electron beam. The latter disadvantage is especially inconvenient for the reason that the ion detection system sensitivity has to be increased by several orders of magnitude to compensate for the much smaller electron current, and that in many cases means that one has to go to an ion counting technique.

The adaptation of the ion source was performed only in the electron gun and collision chamber region, leaving the ion focus system unchanged. Figure 1 shows the modified ion source. The lack of space has limited the length of

TEM to approximately 10 mm. Electrons emitted from the filament (1) are accelerated and collimated by two electrodes with circular apertures 1 mm in diameter. The electron beam so formed is decelerated and dispersed in the analyser part of the TEM consisting of two parallel plates (2) at a distance of 3 mm. The following two electrodes have apertures of 1 mm diameter too, but with the centre axis at a distance of 4 mm from the axis of the incoming electron beam. The collision chamber (3) entrance aperture is also 1 mm in diameter while the exit aperture is 1·5 mm as well as the aperture at the front end of the electron collector system (4). All electrodes have been made of stainless steel (non-magnetic) and no plating of electrodes was tried (gold plating was excluded because of the mercury diffusion pumps).

The electrode structure in the monochromator and collision chamber part of the ion source was shielded, and the shield held at the potential of the filament. This was necessary because the ion source structure is relatively close to the walls of the vacuum chamber and at the same time is held at about 2–3 kV above ground potential. This caused, in the first experiments, severe field penetration in the region of the TEM, and operation under these conditions was quite difficult. It appeared to be hard to make the shield so effective as to avoid any penetration at all, so that some differences in TEM potentials settings with the high voltage on and off exist even when the source shield is applied. This is easy to understand because the energy of the electrons in the deflection region of the TEM is very low (lower than 0·01 eV).

RESULTS

The TEM characteristics have been checked by retarding potential analysis of the electron beam coming from the electron monochromator. In Fig. 2 a

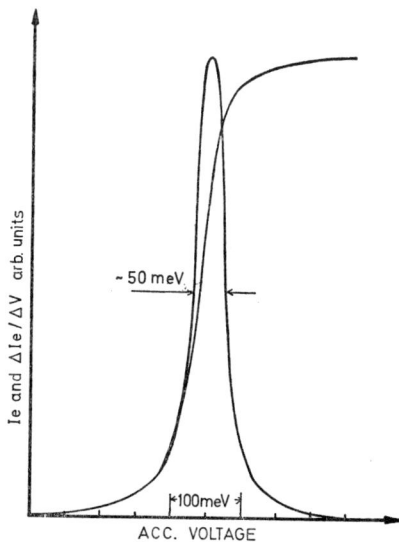

FIGURE 2

typical electron beam retarding curve and its derivative are shown. It has been found that, in spite of the unbaked system and the plain stainless steel electrodes, the results are fairly good. Namely, the half-widths of about 50 meV are easily obtainable, but with smaller currents than expected (about 10^{-9} A). With the ion source at the high voltage needed for the operation of the mass spectrometer, the retuning of the TEM was necessary in order to get the same current and resolution as with the high voltage off. As with the earlier use of the TEM in connection with quadrupole mass spectrometers,[5] where the shielding from the r.f. field was critical, this is the only disadvantage of using the TEM instead of the RPD technique.

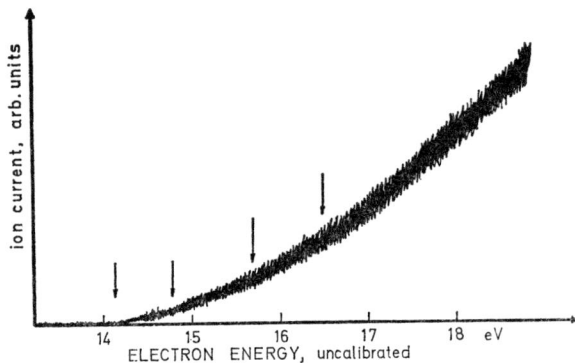

FIGURE 3

The preliminary results of the modified ion source application for study of ionization process in Kr are shown in Fig. 3. The structure seen in the figure is in rather good agreement with the results already published.[6]

The study of negative ion formation is impossible at the moment, owing to necessary modification of the mass spectrometer, but we do intend to study the structure of the ionization efficiency curve for negative ions too.

CONCLUSIONS

From the preliminary results obtained with the modified ion source it has been concluded that there are no principal difficulties for the use of the TEM in the ion source of classical sector magnetic field mass spectrometers. As it was anticipated the stray magnetic field of the mass spectrometer has no effect on the operation of the TEM. Good electrostatic shielding has been shown to be essential, and it seems to be the only critical requirement for proper operation of the TEM. Simplicity of construction and operation as well as better resolution (compared to RPD) showed that the choice of TEM for the electron monochromator is correct. The lack of space in the source region prevented the use of some other TEM geometries suggested in an earlier paper[3] that would probably have given better electron current to resolution ratio. Gold plating or platinum black plating of electrodes would give better results too, but in this case it was not applicable.

REFERENCES

1. Stamatovic, A., 'On the significance of the energy resolution in the experimental study of the electron-atom and electron-molecule collisions', V Yugoslav Symposium on Physics of Ionized Gases, Herceg Novi, 1970.
2. Giessner, B. G. and Meisels, G. G., *J. Chem. Phys.*, 1971, **55,** 2269.
3. Stamatovic, A. and Schulz, G. J., *Rev. Sci. Instrum.*, 1970, **41,** 423.
4. Fox, R. E., Hickam, W. M., Grove, D. J. and Kjeldaas, Jr, T., *Rev. Sci. Instrum.*, 1955, **26,** 1101.
5. Stamatovic, A. and Schulz, G. J., *J. Chem. Phys.*, 1970, **53,** 2663.
6. Brion, C. E., Frost, D. C. and McDowell, C. A., *J. Chem. Phys.*, 1966, **44,** 1034.

61

An Improvement of the Pulse Counting Method for Mass Spectroscopic Trace Analysis

By HIROSHI MIYAKE and MASAMI MICHIJIMA

(*Department of Nuclear Engineering, Kobe University of Mercantile Marine, Kobe, Japan*)

INTRODUCTION

THE detection technique for very small ion currents is one of the most important factors deciding the accuracy of the isotopic analysis of very small samples. In recent years, the development of ion detectors has made it possible to count single slow ions. Nevertheless, it is difficult to measure ion currents below 10^{-20} A. Measurements for long intervals may be disturbed by fluctuations of the ion counting efficiency which may be caused by the gain drift of the secondary electron multiplier and by the shift of the pulse height discrimination level. It is also required that the bursts of noise pulses induced from external sources are discriminated.

The multi-channel scaling of the ion current is an effective method to eliminate these difficulties.[1,2] The recurrent multi-channel scaling synchronized with the scanning of the ion beam gives refined mass spectra, for which the effect of the fluctuation of the ion counting efficiency is minimized.[3,4] False peaks produced by accidental bursts of noise pulses can easily be discriminated and rejected from the mass spectra in the calculation of peak heights.

In this paper, measurements of partial pressures below 10^{-15} torr for isotopic analyses of very small rare gas samples are described.

EXPERIMENTAL

The mass spectrometer used in this study was of the 90° magnetic deflection type with 100 mm radius of curvature. Its main parts were made of stainless steel. The analyser tube, bakeable valves and an oilless UHV pumping system which consisted of a sputter-ion pump (16 litre/sec) and a titanium sublimation pump were outgassed by baking them at 300°C in an oven, using an oil diffusion pump system. Total pressures were measured with a modulated Bayard–Alpert gauge mounted in the metal envelope. The operating temperature of the gauge was reduced by use of a thoria-coated tungsten filament. Pressures in the range $3–6 \times 10^{-11}$ torr were achieved after a 10-hour bake.

The ion source of the analyser was of the Nier-type[5] which was modified in such a way that the electrode surfaces could be heated by electron bombardment. The use of a (BaSrCa)O-coated cathode reduced the operating temperature of the ion source and reduced the background hydrocarbons markedly. Although the outgassing of carbon dioxide from the cathode increased the operating pressures slightly, the cracking pattern of carbon dioxide did not disturb the isotopic analyses of noble gases except for neon.

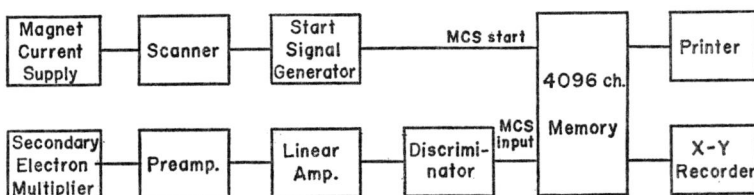

FIG. 1   Block diagram of the electronics for the recurrent multi-channel scaling.

The mass resolution of about 140 was obtained with a 0·1 mm source slit in conjunction with a 0·5 mm collector slit. The background ions scattered from the analyser tube wall were reduced by a factor of more than 10 by use of a baffle slit of 2 mm width placed at 5 cm apart from the collector slit.

The block diagram of the electronics for the measurement in the recurrent multi-channel scaling mode is shown in Fig. 1. The ion detector was an Allen-type 12 stage secondary electron multiplier with Cu–Be dynodes which had a gain of about 10^5 with a supplied voltage of 2 kV. It was connected to a low-noise charge-sensitive pre-amplifier of which the equivalent noise charge was 5×10^2 electrons.

The memory system has a size of 4096 channels and a count capacity of 10^6. The maximum input rate in the multi-channel scaling mode is 10 MHz. The built-in timer is switch-adjustable to set the address dwell time from 10 μsec to 0·9 sec. The stored data are read out with a digital printer and an X–Y recorder.

The analyser magnet current was scanned with a motor-driven potentiometer. The interval of one way scanning was switch-selectable from 5 to 120 min. The starting of multi-channel scaling was triggered by the automatic returning signal of the scanner. The memory size and the dwell time were chosen as the channel advance stopped before the next turning of the scanning. The stored memory was partly printed out during the waiting interval in order to check the time variation of the accumulated counts.

RESULTS AND DISCUSSION

It is necessary for the exact measurement of very low partial pressures that the noise counting rate is very low and that the fluctuation of the peak positions in the mass spectrum is very small for the recurrence of the multi-channel scaling.

The pulse heights for single ions increased with the voltage applied to the secondary electron multiplier. On the other hand the noise counting rate was markedly reduced by lowering the applied voltage to the multiplier. When the discriminating level was set at the half of the average pulse height for Ar^+ ions, the noise counting rate was 0·005 counts/sec at an applied voltage of 2·4 kV and 0·02 counts/sec at 2·8 kV. For measuring an ion current below 10^{-21} A it was necessary to set the discriminating level high although the counting loss increased with increasing level.

The recurrence of the multi-channel scaling synchronized with the scanning of the analyser magnet current gave a stable superposition of mass spectra. A 16-hours test operation in which the peak positions were recorded in every run verified that the fluctuation of the peak positions corresponded to less than 0·02 amu in the mass range from 28 to 44. The increase of the peak width measured at one hundredth of the peak height was only 0·03 amu at m/e = 28 for a 12 days test run. The analysing magnetic field was changing along a hysteresis loop, the magnet current following a triangular time

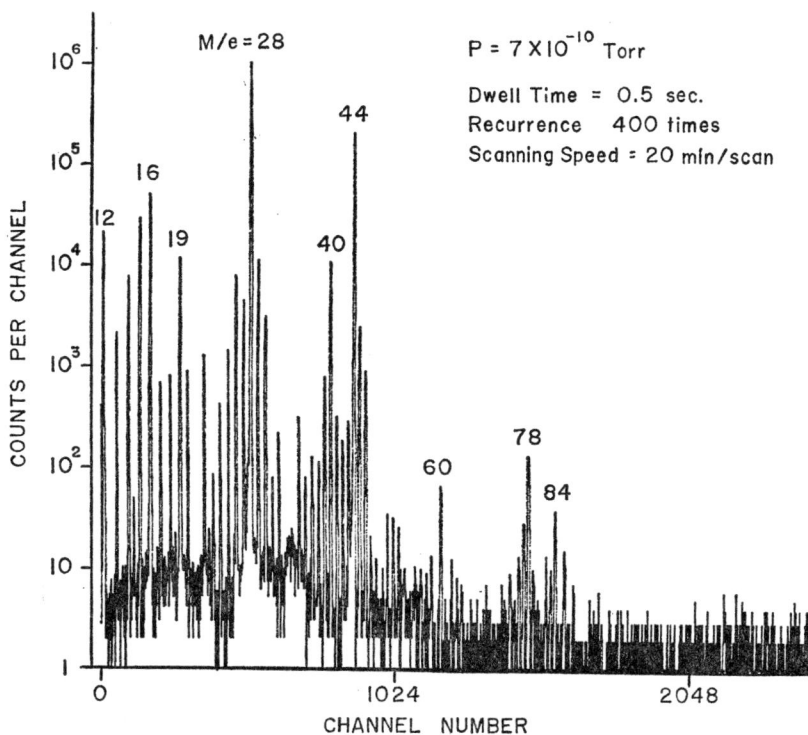

FIG. 2   Mass spectrum of residual gases for mass range 12–138. Total counting interval per channel is 200 sec. Electron emission current is 100 μA. Ion accelerating voltage 1·0 kV. Applied voltage to the secondary electron multiplier 2·4 kV. Pulse height discriminating level corresponds to the multiplier output of 8×10^4 electrons. One count per channel corresponds to 1×10^{-16} torr. The baffle slit was not used in this run.

function. The spectral shift due to the hysteresis effect was eliminated by scanning several times before the recurrent multi-channel scaling was started. An example of the mass spectrum of residual gases is shown in Fig. 2.

A computer programmed analysis of the mass spectra with rejection of noise channels was carried out. Since the peaks produced by the bursts of noise pulses were mostly formed in a short interval, they persisted in only one or two channels if the dwell time was longer than 0·1 sec. These noise peaks or noise channels could easily be discriminated by comparing the second order differential coefficient with the expected statistical error. The fitting of the Gaussian distribution for each peak was suitable for the calculation of peak heights and peak positions.

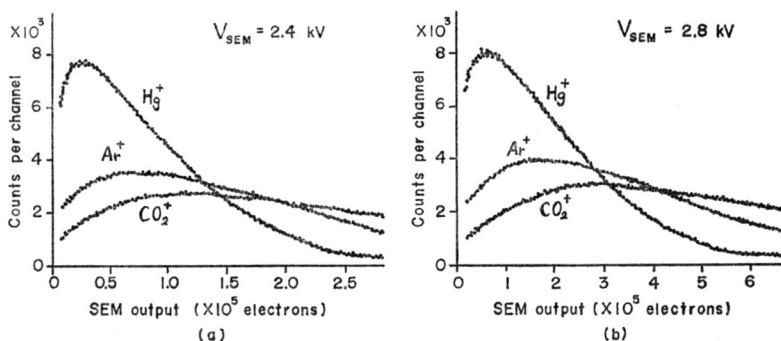

FIG. 3 Pulse height distributions of the output of the secondary electron multiplier (SEM) for Hg^+, Ar^+ and CO_2^+. (Ion accelerating voltage × mass number = 40 keV amu.)

The accuracy of the measurement depended strongly on the pulse height distribution of the secondary electron multiplier output. The counting efficiency for each ion had to be determined for each operation, because the pulse height distribution depended markedly on the operating condition of the multiplier and on the kind of ions. This is shown in Fig. 3. It is noticeable that the shape of the pulse height distribution changes only slightly with the voltage applied to the multiplier although the average pulse height changes by a large factor. This means that the pulse height distribution depends mostly on the distribution of the number of secondary electrons ejected by the impinging ions. The counting loss rates for Hg^+, Ar^+ and CO_2^+ ions estimated from the analysis of the pulse height distributions were about 50%, 8% and 3%, respectively, in the case shown in Fig. 3.

REFERENCES

1. Blauth, E. W. et al., J. Vac. Sci. Technol., 1971, **8**, 384.
2. Ihle, H. R. and Neubert, A., J. Mass Spectrom. Ion Phys., 1971, **7**, 189.
3. Miyake, H. and Michijima, M., Shitsuryo Bunseki (Mass Spectroscopy), 1971, **19**, 110 (in Japanese).
4. Miyake, H. and Michijima, M., Shinku, 1972, **15**, 283 (in Japanese).
5. Miyake, H., Otsuji, T. and Michijima, M., Shitsuryo Bunseki (Mass Spectroscopy), 1968, **16**, 339.

Discussion

W. Aberth (Stanford Research Institute, California, U.S.A.): I would like to comment that in our laboratory we are using two mass spectrometers with multi-channel analysers for data acquisition. One of these machines uses a Wien type velocity filter for mass analysis and the other uses a quadrupole. Since they both mass scan electrostatically, the scanning rate can be very fast—less than one second per scan as contrasted to 10 minutes per scan for the magnetic deflection type instrument.

M. Michijima: The present object of our study is to apply the recurrent multi-channel scaling method to improve ion detection in an ordinary magnetic deflection type mass spectrometer for trace analyses. Accordingly the most suitable speed of magnetic scanning, *e.g.* 10 or 20 minutes per scan was selected.

L. D. Nguyen (Institute de Physique Nucléaire, Orsay, France): I have some comments concerning the speed of scanning and the sensitivity. For the speed of scanning, I think you can obtain a higher speed by high voltage scanning. In our case (Int. Journal of Mass Spec. and Ion Physics 11, 1973), we obtain a speed of 20 μs per channel.

For the sensitivity we obtain a noise level less than one ion per minute *i.e.* a current of about 10^{-21} A. I think you can improve your sensitivity by decreasing the noise of your system particularly by decreasing the high voltage applied on the dynodes of your multiplier within certain limits because the noise produced by electrical discharges between different dynodes is an exponential function of the applied voltage.

M. Michijima: The reduction of the noise counting rate is necessary to get high sensitivity. As you comment, the noise could be markedly reduced by decreasing the applied voltage to the multiplier. We used an applied voltage of 2·4 kV taking into account the relation of the noise and the gain of the multiplier and obtained 0·005 counts/sec of the noise counting rate.

62
Static Mass Spectrometers with Axial Symmetry

By H. LIEBL

(*Max-Planck-Institüt für Plasmaphysik, Euratom-Association,*
Garching, Germany)

INTRODUCTION

Conventional static mass spectrometers accept ions which travel near the entrance axis. This is the most appropriate way of designing a mass spectrometer as long as the ions to be analysed can be accelerated to an energy which is high compared to their initial energy, so that they fill a small solid angle after acceleration. However, if the ions are to be analysed at their initial energy without post-acceleration, collection efficiency becomes very poor. If a small sample area emitting n_0 charged particles with a cosine distribution (Fig. 1), is assumed, the fraction of particles emitted within the

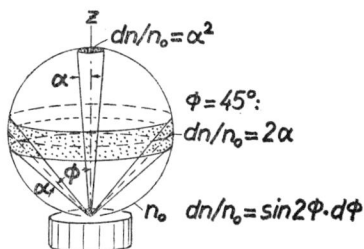

FIG. 1   Sample emitting n_0 particles with a cosine distribution.

small half-angle α around the normal to the surface is $dn/n_0 = \alpha^2$. On the other hand, if the same half-angle is opened on both sides of a conical surface with the colatitude φ, the fraction becomes $dn/n_0 = 2 \sin 2\varphi \,.\, \alpha$, and for $\varphi = 45°$ we get $dn/n_0 = 2\alpha$. This is a factor $2/\alpha$ more than in the former case (*e.g.* a factor 50 for $\alpha = 2\cdot3°$).

Particle spectrometers capable of accepting this latter solid angle thus have considerably higher transmission compared with the former case, whereby similar imaging properties are to be expected, α being the same.

This fact has long been utilized in beta-ray spectroscopy and in recent years also in electron spectroscopy for designing deflection-type analysers with axial symmetry which accept particles within a certain solid angle range confined by two conical surfaces around the axis, at the apex of which the source is located.

In this paper examples are given of how this principle can be applied in the design of mass spectrometers with angular and energy focussing. Furthermore, useful applications for this type of mass spectrometer are discussed.

EXAMPLES

Figure 2 shows an example, where of the particles emitted by sample O a conical hollow beam, confined by an annular aperture A (α-stop), is deflected by a spherical condenser as energy analyser E, so that ions having a certain energy eU_0 leave it parallel to the z-axis. An electrostatic annular lens L_1 forms an annular focus at the focal distance f_1. Another annular aperture B (β-stop) confines the energy bandwidth. A second annular lens L_2, located at

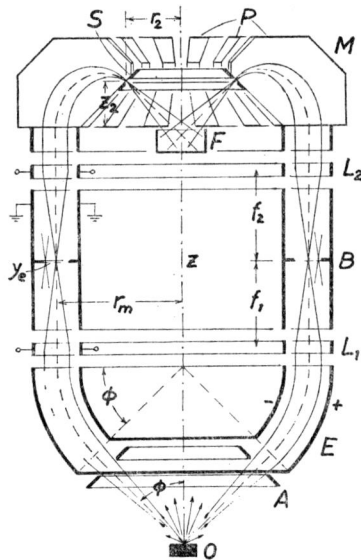

FIG. 2 First example of double-focussing mass spectrometer with axial symmetry. O, sample; A, angular stop; E, energy analyser; $L_{1,2}$, annular lenses; B, energy stop; M, magnet; P, pole pieces; S, exit slit; F, Faraday cup.

its focal distance f_2 from B, makes the trajectories of the ions with energy eU_0 parallel to the z-axis again. The magnet M performs the mass separation. The magnetic field is formed between wedge-shaped pole pieces P. The field-forming planes of the pole pieces go through the z-axis (see Fig. 2(a)). The field lines are circular arcs around the z-axis. The field strength is inversely proportional to the distance from the z-axis. Such fields have been used in beta spectrometry[1,2] and in mass spectrometry.[3-6] (A field of the same kind can be generated inside a toroidal iron-free coil with windings contoured like the pole pieces and with free space in between to let the ions pass through. However, the field strengths attainable this way are much lower.) The entrance boundaries are perpendicular to the z-axis. The ions follow trochoidal trajectories in planes through the z-axis. In the calculated case sketched in Fig. 2 there is angular focussing of the beam entering parallel, on a circle about the z-axis having the radius $r_2 = 0.43r_m$ at a distance $z_2 = 0.38r_m$ from the entrance plane, the angle of deflection being 135°. There an annular exit slit S is located, which is interrupted where the windings W (see Fig. 2(a)) require it. The mass separated ions are collected in a Faraday cup F.

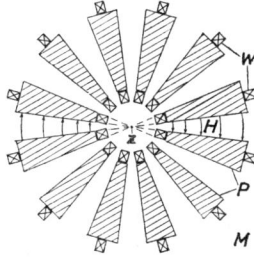

FIG. 2(a)   Cross section through magnet M. P, pole pieces; W, windings; H, field lines.

In order to achieve energy focussing, the energy dispersion brought about by the spherical condenser at B has to be compensated by the energy dispersion of the magnetic field. Let ions of energy eU_0 leave the energy analyser E parallel to the z-axis, then the trajectories of the ions of energy $e(U_0 + \Delta U)$ are inclined against the z-axis by the small angle $\gamma_e = L_e \Delta U/U_0$, L_e being the energy dispersion coefficient of the energy analyser E.[7] The lens L_1 focusses these ions at a distance y_e from the central trajectory (distance r_m from z-axis) given by

$$y_e = \gamma_e f_1 = L_e f_1 \frac{\Delta U}{U_0} \tag{1}$$

Similarly, the energy dispersion of the magnetic field, calculated back from the exit slit S, is

$$y_m = N_m f_2 \frac{\Delta U}{2U_0} \tag{2}$$

where N_m denotes the momentum dispersion coefficient of the magnetic field. The condition for energy focussing is $y_e = y_m$, and hence

$$L_e f_1 = \frac{N_m f_2}{2} \tag{3}$$

For the spherical condenser we have $L_e = \sin \varphi$, and thus with $\varphi = 45°$ we get $L_e = 0.707$. The coefficient N_m is found to be $N_m = 1.325$ for the magnetic field of Fig. 2. With these values we get from eqn (3) $f_2 = 1.07f_1$.

In the second example (Fig. 3), instead of a spherical condenser a half of a 'cylindrical mirror analyser'[8,9] serves as energy analyser. The electrostatic field is generated between concentric metal cylinders. Such an analyser is easy to build. Its ion optical quality, however, can be superior to that of a

spherical condenser if the geometry is appropriately chosen. The example in
Fig. 3 shows $\varphi = 45°$, the inner cylinder radius has the value $r_1 = 0.516r_m$,
and the distance between beam entrance and exit is $z_1 = 1.08r_m$. At the exit
a third thin-walled metal cylinder is placed and connected to the potential
prevailing at the same radius in the remainder of the field. This keeps the
fringe field small. The energy dispersion coefficient of such a field is found to
be $L_e = 0.855$ if the ions travel after leaving the field on the same potential
as before entering it. Using the same magnetic field as in the first example,
energy focussing is achieved, from eqn. (3), for $f_2 = 1.29f_1$. Figure 3 shows a

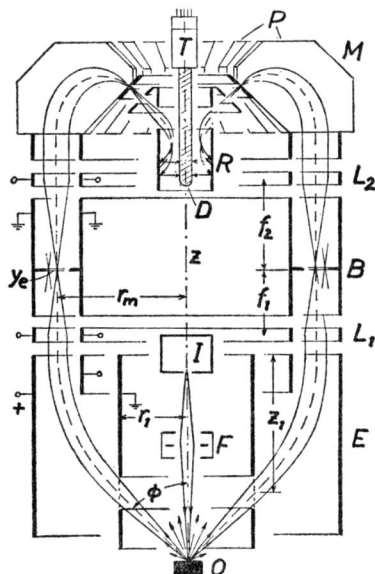

FIG. 3 Second example. I, primary ion source; F, focussing optics; R, ion-electron-
converter tube; D, scintillator rod; T, photomultiplier tube. Other notations as in Fig. 2.

detector for positive ions, replacing the collector of Fig. 2, that has a geometry
especially well adapted to this coaxial arrangement. It works according to the
ion-electron-converter principle,[10,11] but has axial symmetry. The ions are
post-accelerated, after passing the exit slit S, to 10 to 15 keV and enter the
electrostatic field between an aluminium tube R and an aluminized scintillator rod D. The ions hit the inside of the tube R and release electrons which
in turn are accelerated towards the scintillator D. The light generated there
is guided to a photomultiplier tube T.

Figure 4 shows a third example. Prior to entering the energy analyser, the
ions here are accelerated between two conical grids N_1 and N_2 up to an
energy eU_1 which is high compared to their initial energy eU_0. Thereby the
side-rays are refracted so that they enter the energy analyser nearly parallel
to the central rays. The energy analyser has the same geometry as in the
previous example (Fig. 3). Here, however, because of the parallel beam entry,
a directional focus occurs at the exit boundary. The calculation yields an

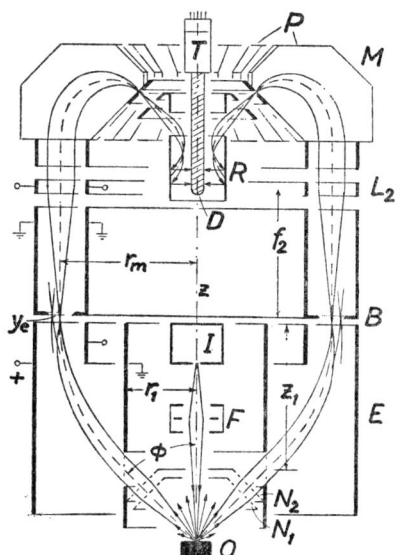

FIG. 4   Third example. $N_{1,2}$, conical grids. Other notations as in Figs. 2 and 3.

energy dispersion of $y_e = 0.6615 r_m \, \Delta U / U_1$. The condition for energy focussing $y_e = y_m = N_m f_2/2$, using the same magnet as in the previous examples ($N_m = 1.325$), is now met by the relation $f_2 = r_m$.

The magnet geometry used in these examples is not the only solution to achieve directional focussing. There are other solutions with different deflection angles, coupled with different values of r_2 and z_2. The values of N_m, that have to be used in order to meet the condition for energy focussing are thereby also different. Furthermore, instead of the symmetric lenses L_1 and L_2 electrostatic immersion lenses can be used, whereby the beam refraction has to be taken into account to satisfy the condition of energy focussing.

POSSIBLE APPLICATIONS

Owing to the multi-gap magnet design, only relatively low magnetic fields can be achieved with conventional technology (excluding superconducting coils). Therefore, ion energies must be low compared with what one is used to in static mass spectrometry.

One useful application would be the analysis of secondary ions released from a solid sample by ion bombardment. Figures 3 and 4 show an ion source I with a focussing system F, arranged coaxially inside the mass analyser. The mean energy of secondary ions ranges from a few electron volts to some tens of electron volts, depending on whether they come from surface layers or from the bulk. This energy range could be handled by such a mass spectrometer having handy dimensions.

Another application could be the analysis of sputtered neutrals, which up to now has been hampered by lack of efficient ionization and interference with the residual gas.[12] The sputtered neutrals could be ionized by electron impact anywhere along their way within the field-free volume between the sample and the energy analyser (or grid N_1). Thus, a large ionization volume would afford increased ionization probability. Ionized residual gas would be discriminated against, since no drawing field exists within the ionization volume.

The great advantage of SAMS (secondary atom mass spectrometry) over SIMS (secondary ion mass spectrometry) would be the elimination of the matrix effects, which make quantitative SIMS so difficult.

REFERENCES

1. See 'Alpha-, Beta- and Gamma-Ray Spectroscopy', *ed.* K. Siegbahn, North-Holland Publ. Comp., Amsterdam, 1965, Vol. I, p. 126–35.
2. O'Connell, J. S., *Rev. Sci. Instr.*, 1961, **32**, 1314–16.
3. Poschenrieder, W. and Warneck, P., *J. Appl. Phys.*, 1966, **37**, 2812–20.
4. Liebl, H., *J. Appl. Phys.*, 1967, **38**, 5277–83.
5. Ruedenauer, F. G., *Int. J. Mass Spectrom. Ion Phys.*, 1970, **4**, 181–201.
6. Liebl, H., *Int. J. Mass Spectrom. Ion Phys.*, 1971, **6**, 401–12.
7. Comp. Liebl, H., *in* 'Recent Developments in Mass Spectroscopy', Univ. of Tokyo Press, 1970, p. 188–93.
8. Zashkvara, V. V. *et al.*, *Sov. Phys.-Tech. Phys.*, 1966, **11**, 96–9.
9. Sar-el, H. Z., *Rev. Sci. Instr.*, 1967, **38**, 1210–16.
10. Schuetze, W. and Bernhard, F., *Z. Phys.*, 1956, **145**, 44–7.
11. Daly, N. R., *Rev. Sci. Instr.*, 1960, **31**, 264–7.
12. Smith, A. J., Cambey, L. A. and Marshall, D. J., *J. Appl. Phys.*, 1963, **34**, 2489–90.

63
Mass Spectrometric Studies of Low Temperature Flames

By J. R. WYATT,† J. J. DeCORPO, M. V. McDOWELL
and F. E. SAALFELD

(*U.S. Naval Research Laboratory, Washington, D.C., U.S.A.*)

INTRODUCTION

THERE is a great and continuing interest in many reactions occurring in gas phase combustion. Knowledge of these reactions is important for the more efficient utilization of fuels for heating and propulsion, their relationship to air pollution and environmental quality, and, when uncontrolled, their potential danger to men and materials. To obtain a better understanding of these reactions, analyses must be made of the actual species existing during the reactions. Although mass spectrometry has been employed for gas analysis for more than thirty years, the problem of sampling reactive species in flow systems at atmospheric pressure is still largely unsolved.

This paper describes a simple method of sampling material from a flow system at atmospheric pressure which is applicable to many areas of study. It is currently being used in the study of low temperature combustion processes, particularly those producing cool flames[1] in a dynamic system. The advantages of this method of sampling are: the gases are sampled adiabatically directly into the ionizing electron beam of a mass spectrometer; the time of extraction and analysis is less than 10 μsec; no elaborate, costly vacuum system is required for the pressure reduction; only a minute amount of material is required and beam modulation is unnecessary.

EXPERIMENTAL

The sampling system depicted in Fig. 1, uses a 30 mm o.d., 120 cm length, Vycor reactor tube. In order to generate and stabilize hydrocarbon cool and blue flames, in the reactor tube, it is necessary to have a temperature gradient along the tube. This is accomplished by having an electrically conducting tin oxide coating[2] on the exterior of the tube. The system used required 2 A at

† National Research Council Postdoctoral fellow, 1972–73.

REACTOR--ION SOURCE ASSEMBLY

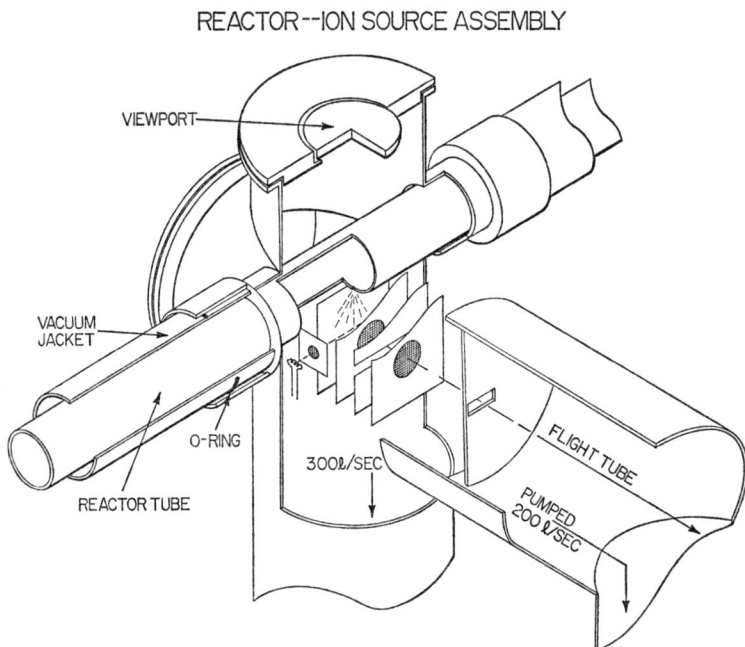

FIG. 1 Cutaway drawing of the mass spectrometer and reactor tube combination. The ionization region has been simplified for clarity.

300 V to heat a 75-cm section of the tube to 500°C, while a typical hydrocarbon mixture (n-C_4H_{10}, 125 ml/min; O_2 100 ml/min; He 500 ml/min) was passed through it.

A sampling orifice was drilled in the wall of the reactor tube using a 250 W, cw CO_2 laser. The reactor tube was evacuated with a mechanical pump and the laser beam was focussed on the wall of the reactor tube with a water-cooled germanium lens (focal length 25 cm). The laser was pulsed for a period (20 msec) sufficient to drill about 80% through the tube wall. The remaining drilling was accomplished using 6 to 10 short (3 msec) laser pulses. Penetration of the reactor tube wall was determined by observing a rapid pressure increase in the tube.

A cross-section of a laser-drilled hole is shown in Fig. 2. The reactor tube was carefully cut and ground in order to examine the orifice. Since it was not possible to do this without shattering the glass at the smallest portion of the orifice, the exact dimensions and shape of the orifice (10^{-3} cm diameter) are somewhat uncertain. However, the listed dimension is consistent with the orifice size calculated from effusion data. In addition, photographic evidence supports the 10^{-3} cm diameter orifice and further suggests that the orifice is circular in cross section.

As shown in Fig. 1, the reactor tube is inserted through the ion source housing of a Bendix Model 12-107 Time-of-Flight mass spectrometer such that the sampling orifice is over the centre of the ion source less than one cm from the electron beam. This mounting position required several minor ion

CROSS SECTION OF A
LASER DRILLED HOLE IN 30MM
VYCOR TUBING

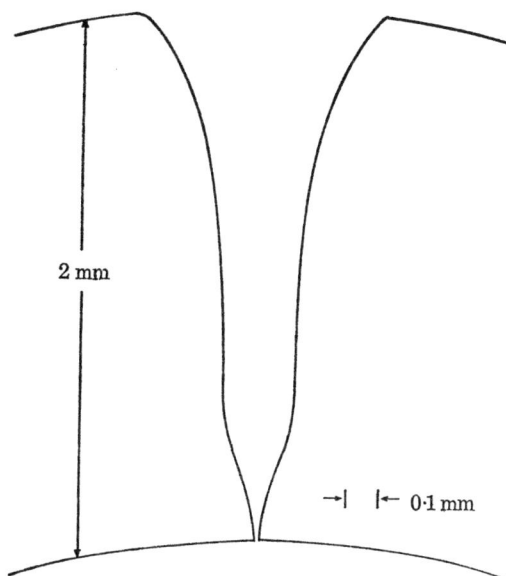

FIG. 2   Cross-section of a laser-drilled hole in 30 mm Vycor tubing.

source modifications. The open source of the Time-of-Flight mass spectrometer is particularly suitable for this application because, with a differentially pumped ion source, less than 10% of the effusing molecules pass through the ionizing region more than once. Thus, modulation of the effusing beams is not required for most applications.[3]

RESULTS AND DISCUSSION

Plots of the ion current intensities for the $C_4H_{10}^+$ and $C_4H_8^+$ as a function of the position of the cool flame front relative to the sampling orifice (point 0 on the abscissa) are shown in Fig. 3. A similar plot for the ion currents of CH_3O^+, CH_2O^+ and H_2O^+ is shown in Fig. 4. From these data, it is apparent that in the preignition zone (before the cool flame front reaches the orifice) little C_4H_{10} has reacted and only small amounts of other hydrocarbon species have been formed; however, when the cool flame front is moved directly over the sampling orifice, there is a precipitous decrease in $C_4H_{10}^+$ current accompanied by a sharp increase in the $C_4H_8^+$, CH_3O^+, CH_2O^+, and H_2O^+ ion currents, indicating that the butane has undergone oxidation producing C_4H_8, CH_3OH, CH_2O and H_2O among other products. When the post cool flame front is sampled, the various ion currents remain relatively constant. These observations are consistent with the hypothesis

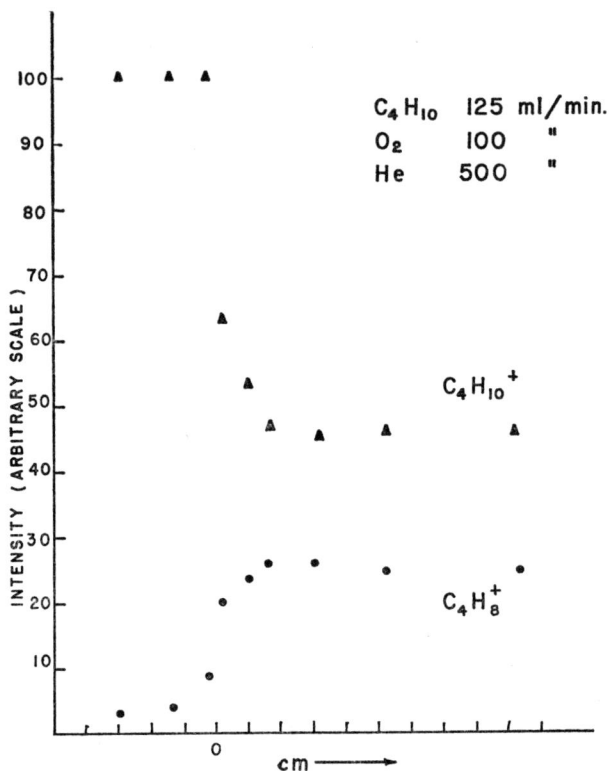

FIG. 3 Ion intensities of $C_4H_{10}^+$ and $C_4H_8^+$ versus position of the cool flame front of *n*-butane relative to the orifice position, (0). The flame front was moved to the indicated positions by varying the temperature of the tube.

that equilibrium between the gases adjacent to the reactor tube walls and the interior of the tube has been established prior to the sampling process.

The identities of the species obtained from a mass spectrometric analysis of the material effusing from the cool and blue flames are cited in Table I, and agree with gas chromatographic analyses of material sampled by conventional microprobe techniques.[4,5] In addition, several species which are difficult to

TABLE I

Species Identified in the *n*-Butane Cool and Blue Flames

Cool flame		Blue flame	
n-C_4H_{10}	H_2CO^a	*n*-C_4H_{10}	$C_2H_2{}^a$
O_2	CH_3OH^a	O_2	CH_2CO^a
He	$C_4H_8{}^a$	He	$C_2H_4{}^a$
H_2O^a	$C_4H_9{}^a$	$CH_4{}^a$	

^{a} Indicates that the species is a product of the cool or blue flames.

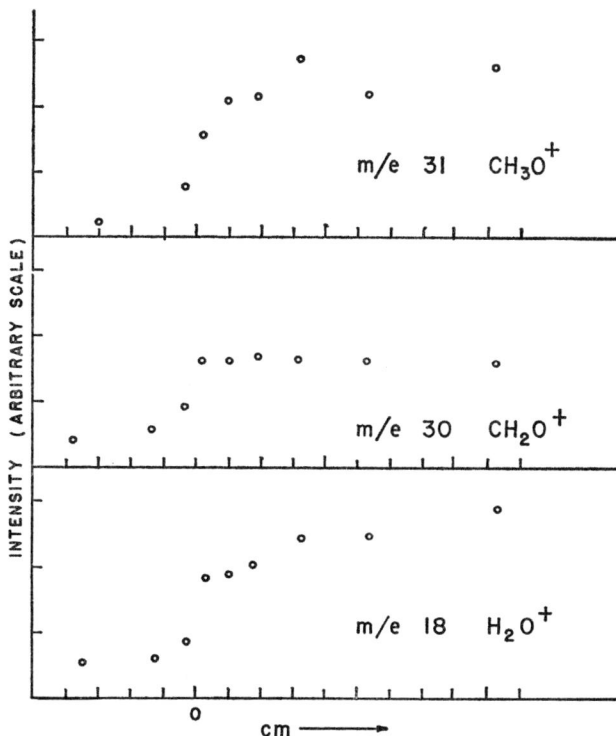

Fig. 4   Ion intensities of CH_3O^+, CH_2O^+, H_2O^+ versus position of the cool flame front of n-butane relative to the orifice position, (0). The flame front was moved to the indicated positions by varying the temperature of the tube.

detect using conventional sampling methods (*e.g.* H_2O, CH_2O), also have been readily observed.

CONCLUSION

The use of a reactor tube with a laser-drilled orifice in combination with a mass spectrometer has been shown to be a powerful, yet simple, technique for the sampling and analysis of flow systems at atmospheric pressure. For flow system at lower pressures, the orifice diameter can be increased by increasing the laser drilling time. Since these leaks are as strong as the tube wall, they can be employed also for the analysis of a system in which the pressure is much greater than one atmosphere. This method should prove useful for the analysis of many systems such as lasers, plasmas, flames, and atmospheric contaminants.

ACKNOWLEDGMENTS

We would like to thank Mr Carmen Carosella of the Solid State Division of this Laboratory for his assistance with the 250 W, cw laser. We appreciate the collaboration of Drs R. Sheinson and F. Williams of the Chemistry Division on design of the flow reactor system and interpretation of the flame data.

REFERENCES

1. (a) Bradley, J. N., 'Flame and Combustion Phenomena', Methuen & Co., London, 1969, chapter 6.
   (b) Fish, A., *Ang. Chem.*, 1968, 7, 45.
2. Manufactured by Corning Glass Work, Corning. N.Y.
3. Kistiakowsky, G. B. and Kidd, P. H., *J. Amer. Chem. Soc.*, 1957, **79**, 4825.
4. Williams, F. W. and Sheinson, R. S., *Comb. Sci. and Tech.*, in press.
5. Williams, F. W. and Stumpf, W. L. Jr, *Anal. Chem.*, 1972, **44**, 1829.

Discussion

H. J. Svec (Iowa State University, U.S.A.): How do you initiate the flame?

F. E. Saalfeld: The flames are initiated by heating the reactor tube. By selecting the proper reactor tube temperature and gas flow conditions, the flame zones can be produced at any position within the reactor tube.

H. J. Svec: What other gases have you experimented with?

F. E. Saalfeld: We have looked at a series of n-Hydrocarbons, butane, pentane and hexane. One of the most interesting flames is acetaldehyde which my co-worker Dr Wyatt will discuss at the forthcoming Combustion Society Meeting.

A. J. C. Nicholson (C.S.I.R.O., Australia): Does the orifice ever get blocked up with soot or dust?

F. E. Saalfeld: No, we have not had any problems of this type. I believe the reason our orifice has never blocked is that our flames are not sufficiently hot, therefore no C_2 radicals are produced in these flame stages.

D. Price (University of Salford, U.K.): Have you considered the effect of boundary layer on the pinhole sampling technique you have described?

F. E. Saalfeld: Yes, we have been very concerned with the boundary layer with this approach. However, with the 'cool' and 'blue' flames, the wall of the reactor tube is hotter than the flame. This causes turbulence in the flame stage and prevents the formation of laminar flow in the reactor tube. With 'hot' flames the boundary layer problem would be very serious and to sample these flames we would have to alter the reactor tube in a manner to prevent laminar flow at the boundary.

PLENARY LECTURE

Mass Spectrometric Studies in High Temperature Chemistry

By J. DROWART and S. SMOES

(*Laboratorium voor Fysische Scheikunde, Vrije Universiteit, Brussels, Belgium*)

It is regretted that the manuscript for this paper was not to hand at the time of printing. The full text will now be printed elsewhere.

INORGANIC APPLICATIONS

Chairmen

G. DIJKSTRA
(*State University, Utrecht, The Netherlands*)

J. FRANZEN
(*Institüt für Spectrochemie, Dortmund, Germany*)

D. PRICE
(*University of Salford, U.K.*)

64
Application of Mass Spectrometry to Inorganic Flames

By D. PRICE, J. H. LIPPIATT and R. W. BROWN

(*Department of Chemistry and Applied Chemistry,*
University of Salford, Salford, U.K.)

and D. C. A. IZOD

(*Ministry of Defence, Royal Armament Research and Development Establishment,*
Sevenoaks, Kent, U.K.)

MANY investigations have been carried out on flames in which the chemical reactants, fuel and oxidizer, are present in the gaseous state but comparatively little when they are present as solids. A pyrotechnic flame[1] is an example of a self-sustaining exothermic inorganic flame in which the reactants are present as solids.

Pyrotechnics find extensive use in both military and civil applications because of their simplicity, reliability, compactness and low cost. The military use is much more extensive than the civil use, which is mainly confined to the signal application for maritime, railway and motorway purposes.

Perhaps the single most important military application is that of illumination on the night battlefield. The compositions used for this purpose consist in the main of a mixture of magnesium and sodium nitrate powders which are consolidated into a cylindrical form at high pressing loads. During combustion, the flame reaches a temperature of about 2500°K and the combustion products are mainly solid magnesium oxide, in a fine particulate form, and sodium vapour. Due to the high density of the sodium vapour and its high oscillator strengths, radiation from the $^3P_{1/2}$ and $^3P_{3/2}$ states to the ground state ('D' lines) is imprisoned in the flame. As a result, the spectral power distribution shows extensive broadening similar to that found in the high pressure sodium vapour lamp. This process makes the magnesium/ sodium nitrate system the most efficient light emitter of all pyrotechnic compositions.

Signal compositions which consist of magnesium/barium nitrate and magnesium/strontium nitrate mixtures together with a chlorine donor such as potassium perchlorate produce the green and red colours respectively. The green colour is produced by the BaCl band between 506 nm and 532 nm. The red colour is produced by a mixture of the SrCl and SrOH bands between 600 nm and 700 nm.

A further class of pyrotechnics which are of considerable interest are the infra-red emitting flares. These fall into two categories. Combustion of a consolidated mixture of magnesium and PTFE powders results in a flame which is rich in incandescent carbon particles. The spectral power distribution is grey body in shape having a correlated colour temperature of about 1800°K. An alternative source of infra-red is based on a thermite type of reaction. By using a titanium/manganese dioxide mixture a solid white-red hot slag is produced. The emission is again grey body in shape but the correlated colour temperature is around 1200°K.

Despite the great antiquity of the art of pyrotechny little is known about the detailed combustion processes occurring in the flame. Also, there is little information available on the thermo-chemical data of common pyrotechnic ingredients. In this situation the development of new compositions has proceeded by 'trial and error' methods which are expensive and time-consuming. In recent years, there has been a determined effort to put development onto a more scientific footing. Spectroscopic techniques, particularly emission spectroscopy, have been used to gain information as to the species present in pyrotechnic flames. These techniques are limited due to difficulty of identification and also because they are unable to give concentration gradients through the flame due to its heterogeneous nature. Both these difficulties can be overcome by successful mass spectrometric monitoring.

The work described in this paper arises from an attempt to study pyrotechnic systems using a Bendix MA-2-time-of-flight mass spectrometer. Experience gained during development of a molecular beam system to sample directly from the pyrotechnic flame will be described. In addition, application of a Knudsen cell-mass spectrometer combination to investigate the high temperature behaviour of species relevant to pyrotechnics is discussed.

MOLECULAR BEAM–MASS SPECTROMETER SYSTEM

The prerequisite of this experiment is the withdrawal and quenching of representative samples from the pyrotechnic flame for analysis by the time-of-flight mass spectrometer. Since the flare is in the region of atmospheric pressure, a suitable interface is required to transfer the sample into the ion source which is at 10^{-9} atmospheres pressure. At such high sample inlet pressures it is not possible to use a simple pinhole leak into the mass spectrometer.[2,3] Sampling must be via a molecular beam system which decreases the pressure through a series of orifices and pumping chambers. The system, used to gain preliminary experience in sampling from pyrotechnic flames, is shown in Fig. 1.

The sampling probe must withstand the effects of a highly oxidising flame at temperatures in excess of 2000°K. Difficulties were encountered in fabricating, from commercially available ceramic tubing, a suitable probe which would allow rapid expansion of flare material into the first chamber. For this reason quartz probes, orifice diameter 0·5 mm, have been used. Although quartz is inferior to ceramics in withstanding the pyrotechnic flame conditions, the probes used have been found able to withstand several experiments, after which they are easily replaced.

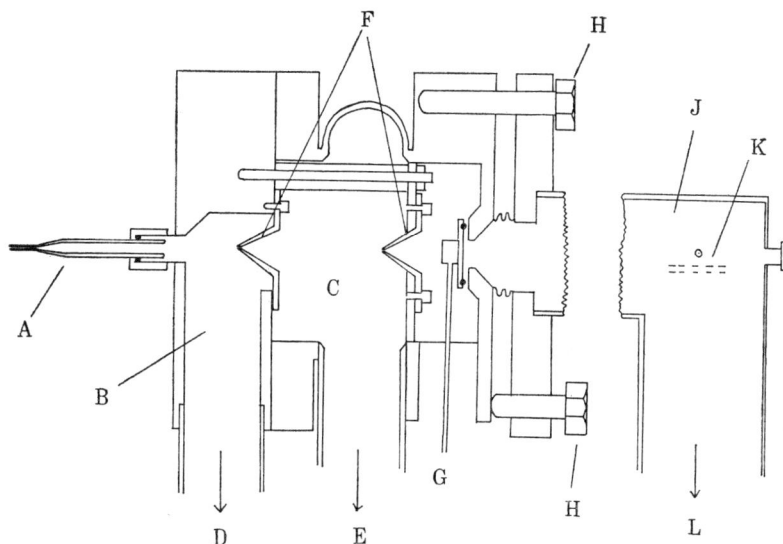

FIG. 1 (A) Quartz probe. (B) First chamber. (C) Second chamber. (D, E, L) Pumps. (F) Conical skimmers. (G) Gate valve in closed position. (H) Adjusting bolts. (J) Mass spectrometer. (K) Ionization region. Optical window is at extreme right.

The first chamber is evacuated to a pressure of 100–200 N m^{-2} by means of a Geryk model 4SIA rotary pump. Sample passes into the second chamber via a 0·4-mm diameter orifice in the conical skimmer. This chamber is pumped to a pressure of 0·5 N m^{-2} by a 2-in. diffusion pump which uses Santovac 5 oil in order to minimize the hydrocarbon background contribution to the spectrum. The sample finally passes into the ion source (0·5 × 10^{-3} N m^{-2}) via a second skimmer. The probe and skimmer orifices are aligned with the ionization region by means of the adjustable bolts while sighting through the optical window. This alignment has been found to be critical because of the small cross-sections of the intersecting molecular and electron beams. The specially designed gate valve enables the molecular beam system to be removed for any reason without the mass spectrometer being let up to atmospheric pressure.

Initial experiments were conducted with a conventional gas flame. Changes were observed in the peak heights and their ratios for the species carbon dioxide (m/e 44), water (18) and methane (16) as different regions of the flame were sampled. Thus the molecular beam–mass spectrometer system is capable of detecting changes in major species' concentration in a gas flame.

When the system was applied to pyrotechnic flames it was found that the probe became blocked by solid deposit at times between 3 sec and 11 sec after ignition. These times were obtained by observing the pressure in the first chamber; a sudden drop in this pressure indicated probe blockage. These times should be sufficient for the rapid scanning time-of-flight mass spectrometer to provide an analysis of samples taken from the flame. Experiments have shown that the larger particles can be removed from the flame region by spinning the pyrotechnic at *ca.* 10^3 rev sec^{-1}. This technique can be used to extend the sampling time.

The spectra observed, when sampling from the pyrotechnic flame, were dominated by the nitrogen (m/e 28, 14) and oxygen (m/e 32, 16) peaks due to air sampled with the flame. The only observed changes to the pre-burn spectra were increases at m/e 44 (CO_2) and 18 (H_2O); there were no peaks characteristic of the pyrotechnic. The carbon dioxide and water observed were presumably the result of combustion of the container cartridge and the linseed oil or lithographic varnish binders of the flare.

On dismantling the beam system after the firing of a flare, it was found that the first cone had become heavily coated with a grey deposit. Thus a large quantity of flare material had entered the first chamber. Subsequent experiments, in which deposit was collected on a polished aluminium disc placed at various positions through the sampling system, established that flare material was reaching the electron beam in the ion source. The intensity of material deposited on the disc decreased rapidly as the distance from the inlet probe increased.

These experiments have shown that a beam of particles can be abstracted from a pyrotechnic flame and fed into the ionization region of the mass spectrometer before the inlet probe is blocked by solid deposit. The question which remains is why a mass spectrum of the sample was not observed. It appears probable that the amount of sample reaching the ionization region was only a small fraction of the air, carbon dioxide and water whose signals dominate the observed spectra. It should be noted that the flare sample only passes through the electron beam once since it will condense on collision with a surface. The gaseous species can make several passes through the ionization region before being removed.

The beam system is being modified, by enlarging the various orifices and increasing pumping speeds, in order to enhance the relative amount of flare material reaching the ionization region. Experiments are also being conducted in which the pyrotechnic will be ignited at reduced pressure in order to reduce the background sampled with the flare material.

KNUDSEN CELL–MASS SPECTROMETER SYSTEM

A Bendix Knudsen cell system[4] coupled to the MA-2 time-of-flight is being used to study the high temperature behaviour of pyrotechnic materials and to obtain thermodynamic data for relevant species. Attempts will also be made to produce and monitor pyrotechnic reactions within the Knudsen cell.

Magnesium (m.p. 923°K) has been found to sublime, under reduced pressures, at *ca.* 640°K by observing the isotopes at 24, 25 and 26 amu. No polymers of magnesium were observed at any temperature up to 1200°K.

The mode of decomposition of sodium nitrate has been investigated in the Knudsen cell. Decomposition commenced at *ca.* 570°K, peaks being observed at m/e 30 and 44 indicating the production of nitric and nitrous oxides respectively. As the temperature was raised, the signal due to nitric oxide increased and above 670°K peaks due to nitrogen and oxygen were observed. Similar results were obtained by Charsley *et al.*[5] in a DTA–mass spectrometry study of the thermal decomposition of magnesium nitrate hexahydrate. They found nitric oxide as the initial decomposition product at 400°K and that nitrous oxide and oxygen are formed in addition above 600°K.

REFERENCES

1. Shidlovsky, A. A., 'Fundamentals of Pyrotechnics', Rpt. AD-462 474, Picatinny Arsenal, Dover, N.J.; Ellern, H., 'Military and Civilian Pyrotechnics', Chemical Publishing Co. Inc, N.J., 1968.
2. Greene, F. T., Brewer, J. and Milne, T. A., *J. Chem. Phys.*, 1964, **40**, 1488.
3. Homann, K. H., *in* 'New Experimental Techniques in Propulsion and Energetics Research', *ed.* D. Andrews and J. Surugue, AGARD Conf. Proc. No. 38, Technivision Services, Slough, England, 1970, p. 371.
4. Bowles, R., *in* 'Time-of-Flight Mass Spectrometry', *ed.* D. Price and J. E. Williams, Pergamon Press, Oxford, 1969, p. 211.
5. Aspinal, M. L., Madoc-Jones, H. J., Charsley E. L. and Redfern, J. P., *in* 'Proc. 3rd Int. Conf. on Thermal Analysis' *ed.* H. G. Wiedemann, Birkhäuser Verlag, Basel, 1972, Vol. 1, p. 303.

65

Application of a Molecular Beam Source Mass Spectrometer to the Study of Reactive Fluorides

By M. J. VASILE, G. R. JONES† and W. E. FALCONER

(Bell Laboratories, Murray Hill, New Jersey, U.S.A.)

INTRODUCTION

MASS spectroscopic study of high oxidation state fluorides and oxyfluorides is complicated by the high reactivity and thermal instability of many of these molecules. Conventional mass spectrometric inlets and ion sources are in general not well suited for the study of these reactive species, since they are frequently thermostatted at $\sim 250°C$ and often permit contact of the samples with glass, stainless steel, traces of water or organic materials, or surface oxide layers with which many fluorides react rapidly. Mass spectra of reactive fluorides have been obtained in systems where collisions with the walls of the ion source take place after extensive prefluorination and seasoning with the compound under study.[1,2] Also, cooled inlets[3,4] for the introduction of highly reactive fluorides have been described which are satisfactory for molecules sufficiently volatile to effuse into the ion source at reduced temperatures.

We required an apparatus to study, mass spectrometrically, the phenomenon of association of inorganic pentafluorides in their vapours. Spectroscopic and electron diffraction studies of several pentafluoride systems gave conflicting results concerning the presence of polymeric species in the gas phase.[5-9] Mass spectrometric[10,11] studies using conventional inlet systems have shown dimeric ions in SbF_5 and BiF_5 in small amounts relative to the monomeric ions. The determination of the extent of association in the vapour phase by mass spectrometry requires inlet systems that operate over a wide range of temperature and a method of sample introduction to the ion source which minimizes collisions of the molecules with the source walls. The species subject to ionization should be those that most accurately represent the composition of the vapour phase which is in equilibrium with either the solid or liquid phase. A molecular beam source with interchangeable inlets was designed and constructed to satisfy these requirements.

† Present address: Royal Radar Establishment, Malvern, England.

APPARATUS

A diagram of the molecular beam–mass spectrometer is shown in Fig. 1. Effusion takes place either from a Monel oven heated by resistive windings, which is used for involatile solid samples, or from Monel gas sources over a range of temperatures. End plates with orifice diameters of 0·012, 0·025, 0·051 and 0·10 cm and wall thicknesses of 0·025 cm were fabricated for the Monel ovens.

A 2-inch diameter liquid nitrogen cooled surface protrudes into the source chamber to increase the pumping speed for condensable gases. A Monel collimating plate with a 0·051-cm diameter knife-edge aperture separates the source chamber from the analyser chamber. The source chamber can be isolated from the rest of the apparatus by gate valves so that samples and inlets may be changed conveniently.

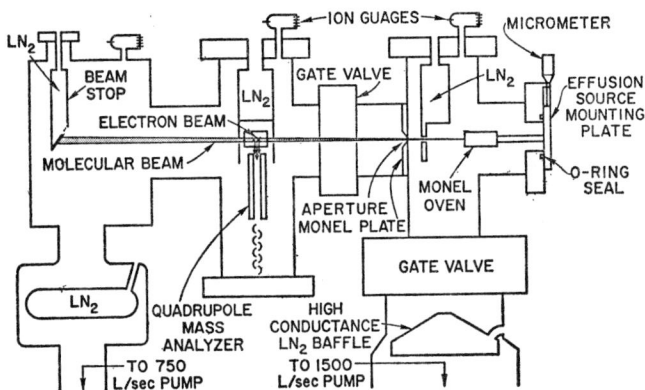

FIGURE 1

The effusion source to collimating plate distance is 5·7 cm, and the distance from the collimating plate to the centre of the ion source is 33·1 cm. The ion source of the mass analyser is surrounded by a liquid-nitrogen cooled OFHC copper surface. A 0·051-cm diameter collimating aperture and a 0·025-cm diameter source (which was most commonly employed) should produce a molecular beam with a penumbra 0·46 cm in diameter and an umbra 0·19 cm in diameter[12] at the outer surface of this cold shield. This shield contains a square opening 0·3 cm on edge through which the beam passes and which blocks the ion source filaments and walls from the beam. The electron beam, the molecular beam, and the resulting ion beam are on mutually perpendicular axes. After emerging from the ion source the molecular beam impinges on a liquid nitrogen cooled OFHC copper beam stop.

The mass analyser is a modified E.A.I. Quad 250B with 0·635-cm diameter × 12·7-cm long poles. Modifications to the ion source include using a true Pierce-type electron gun,[13] blocking the axial orifice in the Faraday cage, and improving the mounting on the quadrupole housing. The mass filter is powered by an Extranuclear Laboratories quadrupole power supply with a type E high Q-head. With this particular mass filter, the power supply produces a maximum r.f. voltage of 4 kV at 1·76 MHz, corresponding to a maximum mass of 2300 amu. The entire mass scale is scanned on one range.

EXPERIMENTAL

Solid samples are loaded into prefluorinated Monel ovens in a dry, oxygen-free helium atmosphere (O_2 + H_2O concentration less than 1 part in 10^6). A tight-fitting Teflon end-cap is used to cover the effusion source during transfer to the apparatus. The source chamber is back-filled with dry nitrogen, and a vigorous flow of dry nitrogen is maintained while inserting the effusion source. Upon evacuation of the source chamber to a pressure less than 10^{-6} torr, the gate valve to the analyser chamber is opened. The Monel oven is then gradually heated until a molecular beam is observed. The position of the source is adjusted for maximum signal, and the source temperature is adjusted to give the necessary beam intensity.

Mass spectra are usually taken using an electron emission of 450 μA with an energy of 70 V. Ions are injected into the quadrupole with a nominal energy

TABLE I
Associated Species in the Vapour Phase of Pentafluorides

Compound	Source temp. (°K)	Vapour[a] pressure (torr)	Orifice diameter (cm)	Oligomer found		
				Dimer	Trimer	Tetramer
VF_5	263	1	0·025	—	—	—
NbF_5	339	1	0·025	+	+	+
			0·050			
TaF_5	332	1	0·025	+	+	—
			0·050			
CrF_5	298	n.a.	0·025	—	—	—
MoF_5	323	1	0·10	+	+	+
WF_5[b]	348	n.a.	0·050	—	—	—
ReF_5	329	0·5	0·025	+	+	+
OsF_5	348	1	0·050	+	+	—
IrF_5[c]	338	n.a.	0·012	+	+	+
RuF_5	383	6	0·10	+	+	+
RhF_5	363	n.a.	0·025	+	+	+
			0·050			
PtF_5	368	n.a.	0·025	+	+	—
PF_5	161	120	0·025	—	—	—
AsF_5	191	115	0·025	—	—	—
SbF_5[c]	298	2	0·025	+	+	+
BiF_5	360	2	0·025	+	+	—

[a] Vapour pressures are calculated from equations given in Refs. 14, 15.
[b] Extensive disproportionation occurred.
[c] Trace amounts (1 part in 10^6) of pentameric ions detected.
+ Species detected.
− Not detected.
n.a. Not available.

of 15 V, with the poles biased 6 V above ground. The mass filter power supply was set up to scan in the constant ΔM mode. The linearity of the mass scale and the mass scale calibration were checked periodically with perfluoro-kerosene-H which was admitted through a heated gas inlet. The gas inlets simply replaced the entire solid source on the apparatus.

RESULTS

Mass spectrometric data on association for the pentafluorides of sixteen elements are collected together in Table I. Vapour pressures have been esti-mated from vapour pressure–temperature relationships,[14,15] where known. Conditions were such as to ensure vapour saturation in equilibrium with either a liquid or solid phase. Ions characteristic of dimers and trimers of the pentafluorides were found for eleven of the elements listed, and six of these showed tetrameric ions. It is probable that the pentafluorides of V, Cr, P and As are not associated in gas phase, at least to the detection limit of dimer ions, about 1 part in 10^5. The pentafluoride of tungsten disproportionates to the hexafluoride and tetrafluoride to such an extent that nothing definitive can be said about the vapour phase composition.

Table II contains mass spectra of two representative pentafluorides, NbF_5 and RhF_5. The spectrum of NbF_5 is typical of the pentafluorides in which the metal atom is in its highest formal oxidation state. Metals which are capable of forming stable hexafluorides gave spectra similar to that of RhF_5. Nominal 16-eV electron energy spectra for RhF_5 are also given in Table II. At 70-eV electron energy, oligomers constitute 31% of the total ions in the NbF_5 spectrum, and 35% of the total ions in the RhF_5 spectrum. In the case of RhF_5 the oligomers account for 87% of the total ions when the electron energy is reduced to a nominal value of 16 eV. The dominance of dimer and trimer ions at low electron energies is consistent either with the vapour being principally composed of associated species which crack to give monomeric fragments at 70 eV, or with a significant difference in ionization cross section between monomeric and associated species. A clear distinction between these two possibilities cannot be made with the apparatus in its present form.

DISCUSSION

To draw meaningful conclusions from the mass spectrometric observations of this study, it is necessary to consider the importance of three possible perturbations which can obscure the interpretation of the data: (1) the possibility that the higher molecular weight ions observed arise from ion-molecule reactions in the ion source, (2) the possibility that neutral agglomer-ates are formed by condensation at the molecular beam source due to free jet expansion, and (3) the correction for the known decrease in transmission of the quadrupole mass filter with increasing mass when the filter is operated in the constant ΔM Mode.

TABLE II

Source temperature (°C) electron energy	NbF_5 66° 70 eV	RhF_5 90° 70 eV	RhF_5 90° 16 eV
M^+	1	39	—
MF^+	1	29	—
MF_2^+	5	80	—
MF_3^+	16	62	—
MF_4^+	100	100	—
MF_5^+	—	21	22
M_2^+	—	0·7	—
M_2F	—	2·0	—
M_2F_2	—	0·7	—
M_2F_3	—	2·0	—
M_2F_4	—	3·0	—
M_2F_5	—	5·0	—
M_2F_6	—	2·0	—
M_2F_7	0·4	10	—
M_2F_8	1·4	3·6	3·2
M_2F_9	52	100	100
M_2F_{10}	—	10	18
M_3F_{11}	—	0·4	—
M_3F_{12}	—	2·0	—
M_3F_{13}	0·09	11	2·5
M_3F_{14}	1·5	21	19
M_3F_{15}	—	6	13
M_4F_{18}	—	0·2	—
M_4F_{19}	—	0·2	—
M_4F_{20}	—	0·2	—

Mass spectra are reported relative to the most intense peak = 100, and corrected for quadrupole discrimination effects.
Peaks not detected above noise level —.

Ion-Molecule Reactions

An upper limit estimate of the yield of secondary ions was made assuming an entirely monomeric beam of NbF_5 and a reaction cross-section of 10^{-14} cm^2 for eqn (1).

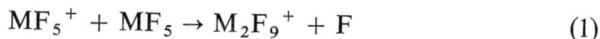

$$MF_5^+ + MF_5 \rightarrow M_2F_9^+ + F \qquad (1)$$

The ratio of secondary ions to primary ions was computed to be 2×10^{-5}. Consistent with this upper limit estimate, no CD_5^+ could be detected from eqn (2), which has a cross-section[16] of $\sim 10^{-14}$ cm^2.

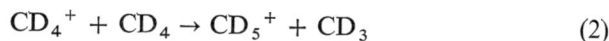

$$CD_4^+ + CD_4 \rightarrow CD_5^+ + CD_3 \qquad (2)$$

It is highly improbable that ion-molecule reactions are contributing significantly to the observation of oligomer ions.

Condensation in Free-Jet Expansion

If free-jet expansion were responsible for the association observed, the degree of association should increase with increasing pressure or orifice size.[17]

Measurements on IrF_5 showed the degree of association to be independent of orifice size over the range 0·012 cm to 0·10 cm dia. for a source temperature of 65°C. The results for NbF_5, TaF_5, RhF_5, and BiF_5 always showed a reduction in the extent of association when the source temperature/pressure was increased. Deliberate attempts to obtain AsF_5 oligomers by isentropic expansion from the cold source (Temp. = 195°K) with driving pressures as high as 300 torr did not produce oligomers. On the contrary, HF at 60 torr from the same source at room temperature yielded mass spectra of $(HF)_n$, $n = 1$–4, in accord with published results.[18]

Quadrupole Mass Discrimination

Decreasing ion transmission with increasing mass or at the same mass with increasing resolution is a well known characteristic of quadrupole mass filters.[19-21] Since an order of magnitude decrease in ion transmission relative to that of magnetic mass spectrometers might be expected for a scan from 60 to 600 amu,[22] the intensities observed for the associated ions relative to the monomer ions would be considerably less than their true abundances. Accordingly, the relative transmission of the quadrupole mass filter was measured using several techniques. Least squares semi-log fits of the transmission data yielded eqn (3) which was used to correct the intensities of oligomer ions relative to the monomer ions.

$$\ln\left(\frac{T(R_1)}{T(R_2)}\right) = -0.018\,(R_1 - R_2) \tag{3}$$

R_1 and R_2 are the 10% maximum resolutions measured at mass M_1 and M_2, and $T(R_1)/T(R_2)$ is the relative transmission. For this particular instrument the transmission correction must be applied to the raw data to obtain a realistic estimate of the abundances of associated ions.

CONCLUSIONS

The mass spectra of a large number of inorganic pentafluoride systems show that associated species exist in the gas phase. The experimental evidence indicates that the origin of the association is not due to ion-molecule reactions or clustering during isentropic expansion. The associated ions observed may therefore be associated either with their neutral counterparts, or with fragmentations from electron impact on higher agglomerates. This fragmentation precludes the possibility of quantitatively determining the relative abundances of neutral agglomerates from the mass spectra.

Despite the absence of quantitative measurements of the abundances of the neutral agglomerates, it is apparent that association in the vapour phase is the rule rather than the exception for pentafluorides beyond the first transition series. Those pentafluorides found to be associated in the vapour phase in this study are known to be associated in the solid phase,[14] either as tetrameric rings or in infinite chains. By implication, association of the liquids is very likely. The presence or absence of associated species in the vapour and liquid phases of NbF_5, TaF_5, and MoF_5 has been the subject of some controversy.[5-9] The present results clearly support extensive association for the saturated vapours of these three species. A considerably more detailed account of this study has recently appeared in the literature.[23]

ACKNOWLEDGMENT

The authors are indebted to Mr W. A. Sunder for synthesizing the fluorides required for this study.

REFERENCES

1. Begun, G. M. and Compton, R. N., *J. Chem. Phys.*, 1969, **51**, 2367.
2. Svec, H., *in* 'NATO Advanced Study Institute on Mass Spectrometry, Glasgow, 1964', *ed*. R. I. Reed, Academic Press, London, 1965, p. 233.
3. Cristy, S. S. and Mamantov, G., *Int. J. of Mass Spec. and Ion Phys.*, 1970, **5**, 309.
4. Holzhaur, J. K. and McGee, H. A. Jr, *Anal. Chem.*, 1969, **41**, 24A.
5. Romanov, G. V. and Spriridonov, V. P., *Siberian Chemical Journal*, 1968, p. 126.
6. Selig, H., Reis, A. and Gasner, E. L., *J. Inorg. Nucl. Chem.*, 1968, p. 2087.
7. Alexander, L. E., *Inorg. Nucl. Chem. Letters*, 1970, p. 1053.
8. Ouellette, T. J., Ratcliffe, C. T. and Sharp, D. W. A., *J. Chem. Soc. (A)*, 1969, p. 351.
9. Beattie, I. R., Livingston, K. M. S., Ozin, G. A. and Reynolds, D. J., *J. Chem. Soc. (A)*, 1969, p. 958.
10. Müller, A., Roesky, H. W. and Böhler, D., *Z. Chem.*, 1967, **7**, 469.
11. Lawless, E. W., *Inorg. Chem.*, 1971, **10**, 2084.
12. Ramsey, N. F., 'Molecular Beams', Oxford University Press, 1956.
13. Pierce, J. R., 'Theory and Design of Electron Beams', Van Nostrand, 1954.
14. Canterford, J. H. and Colton, R., 'Halides of the Second and Third Row Transition Metals', John Wiley and Sons, 1968.
15. Colton, R. and Canterford, J. H., 'Halides of the First Row Transition Metals', John Wiley and Sons, 1968.
16. McDaniel, E. W., Cermak, V., Dalgarno, A., Ferguson, E. E. and Friedman, L., 'Ion Molecule Reactions', John Wiley and Sons, 1970.
17. Gordon, R. J., Lee, Y. T. and Herschbach, D. R., *J. Chem. Phys.*, 1971, **54**, 2393.
18. Dyke, T. R., Howard, B. J. and Klemperer, W., *J. Chem. Phys.*, 1972, **56**, 2442.
19. Karasek, F. W., *Research and Development*, Nov. 1970, p. 55.
20. Brubaker, W. M., *Adv. Mass Spectrometry*, 1968, **4**, 293.
21. Ehlert, T. C., *J. Physics*, 1970, **E3**, 237.
22. Finnigan, J. L., private communication.
23. Vasile, M. J., Jones, G. R. and Falconer, W. E., *Int. J. Mass Spectrom. and Ion Phys.*, 1972/3, **10**, 457.

66
Electrohydrodynamic Ionization Mass Spectrometry†

By B. N. COLBY and C. A. EVANS, Jr

(Materials Research Laboratory, University of Illinois, Urbana, Illinois, U.S.A.)

ELECTROHYDRODYNAMIC (EH) ionization is based on the interaction of a conducting liquid meniscus with an electric field. As the field strength is increased, the liquid meniscus forms a sharply pointed cone. When a sufficiently high electric field is reached, ions are produced from the liquid tip by field evaporation. Unlike solid field ion emitters, the liquid tip is self-forming, self-replenishing, and is readily operated at pressures up to 5×10^{-5} torr with ion current in excess of 200 μA.

A cross-sectional representation of the ion source is shown in Fig. 1. The liquid metal sample is contained in the $2\frac{1}{2}$ ml syringe. The liquid meniscus is formed at the tip of the modified number 27 hypodermic needle by operating the plunger with a mechanical drive (not shown) exterior to the vacuum system. The electric field is applied between the needle and the extractor electrode. The position of the needle in the extractor electrode aperture is critical to stable source operation. The needle tip must protrude into the plane of the extractor electrode but must not extend through it. If this criteria is not met, charged droplet formation rather than ionization will result. The collector cup serves a dual purpose; it reduces contamination of the source housing and, when connected to an oscilloscope, provides an indication of proper source operation.

The electronics used to operate the EH ion source are shown in Fig. 2. The 50 MΩ resistor in series with the needle and its power supply (Spellman High Voltage Power Supply) provides the necessary current limitation for ion source operation. With a significantly lower resistance ionization is unsteady at best and much larger resistance greatly reduces the magnitude of ion production. The rest of the circuit provides the extractor electrode potential. Since the ion's kinetic energy is determined by the potential of the needle with respect to ground, instrument transmission will be lost if IR drop across the 50 MΩ resistor is not held constant. This is achieved by automatically increasing and decreasing the extractor electrode potential to maintain a constant ion current. The needle potential is set to $+8$ kV, the maximum

† This research was supported in part by the National Science Foundation under Grant GH-33634.

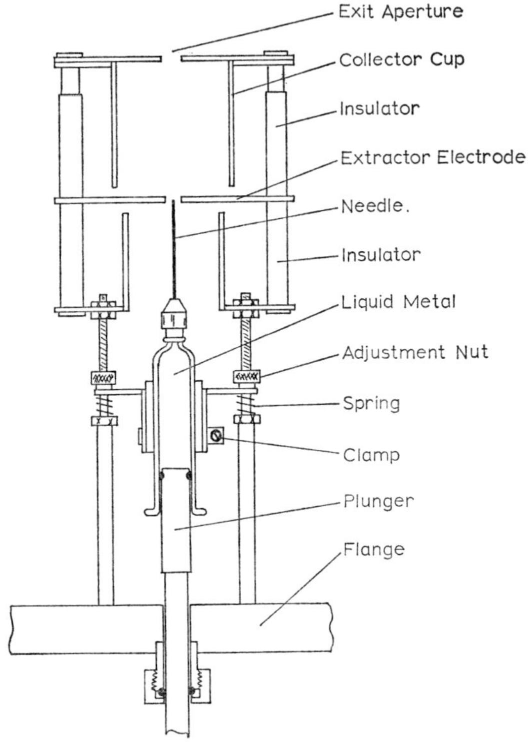

FIG. 1   Cross-sectional representation of electrohydrodynamic ion source.

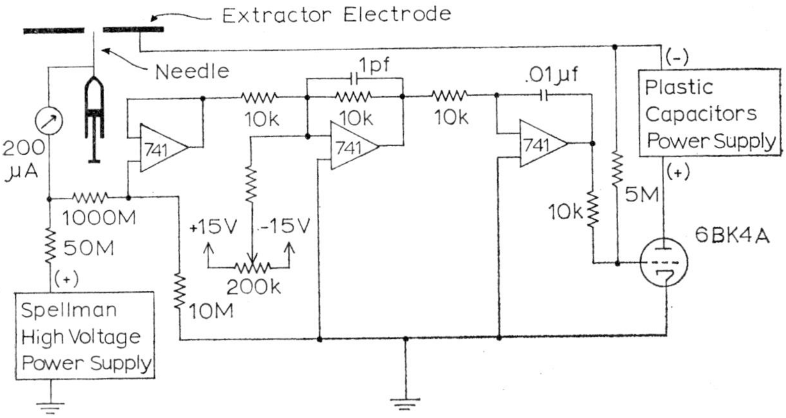

FIG. 2   Ion source electronics.

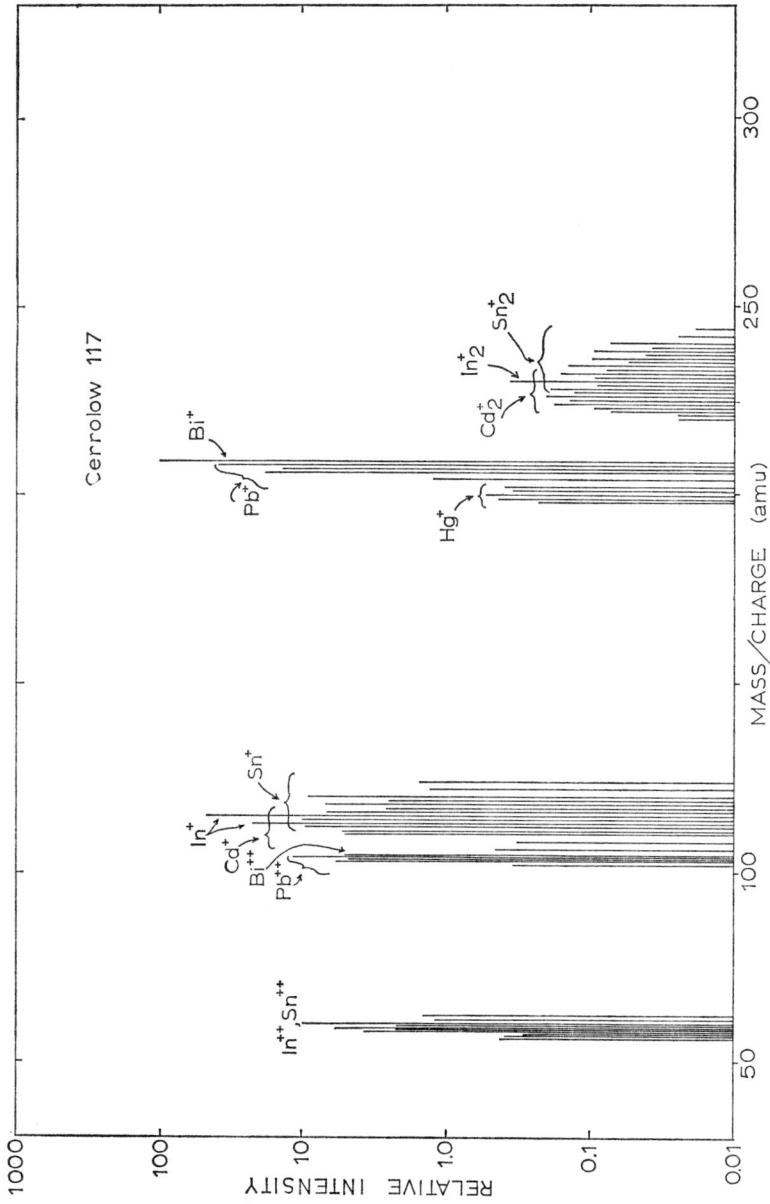

FIG. 3 Electrohydrodynamic mass spectrum of Cerrolow 117.

accelerating voltage of the A.E.I. MS-902 double focussing mass spectro-meter employed in this study, and the extractor electrode is allowed to vary between −500 and −5000 V.

Figure 3 represents an EH mass spectrum of Cerrolow 117, a quinternary eutectic alloy of Bi, Pb, Sn, Cd and In which also contained a 6000 ppm impurity of Hg. Note that except for Sn the predominant species in the

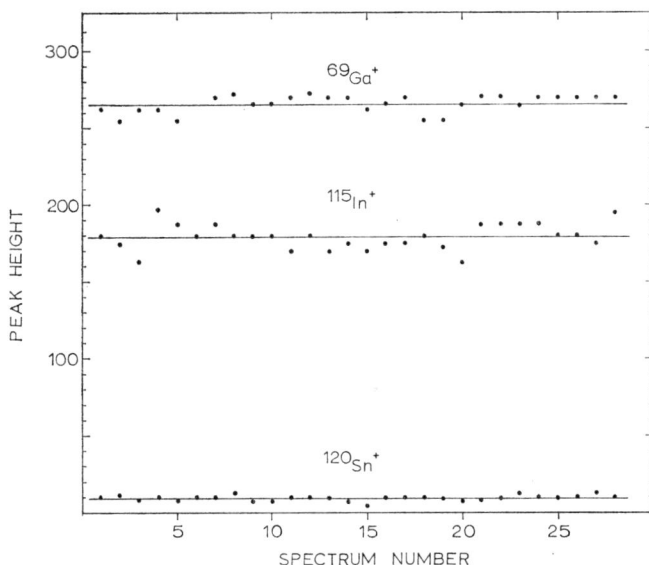

FIG. 4   Peak height reproducibility with GaInSn.

spectrum are the singly-charged monoatomic ions. Also the polyatomic peaks are at least an order of magnitude less intense. This particular spectrum was obtained at an elevated temperature since the sample melting point was 47°C. The reproducibility of peak heights was determined to be the order of 5% relative standard deviation. This is shown in Fig. 4 for 28 successive spectra of a GaInSn eutectic alloy taken over a 30-min period.

While the peak heights or ion yields are reasonably constant for a given set of operating conditions, sample temperature did affect the relative ion yields of several elements. This is shown in Table I for the six elements of Cerrolow 117. The Bi^+ increases with temperature, In^+ and Pb^+ remain constant, Sn^+ and Cd^+ decrease, and Hg^+ is observed only at the lower temperature.

TABLE I

Variation of Ion Yield with Temperature for Six Elements in Cerrolow 117

Relative ion yield	Bi^+	In^+	Pb^+	Sn^+	Cd^+	Hg^+
54°C	0·550	2·07	0·672	0·449	0·798	2·17
70°C	0·722	2·04	0·622	0·271	0·541	—

The elements in three samples were determined quantitatively using peak heights corrected only for isotopic abundance. Figure 5 shows the apparent concentrations plotted against known atomic concentrations for the elements in Cerrolow 117, GaInSn eutectic and GaInSn containing 1% by weight of Cerrolow 117. The elemental concentrations range from several hundred ppm atomic for Cd to about 70% atomic for Ga. Since In^+ was used as an internal standard for the doped GaInSn and is known to have a high relative ion yield by this technique, the values for Cd, Pb, and Bi appear low for this sample.

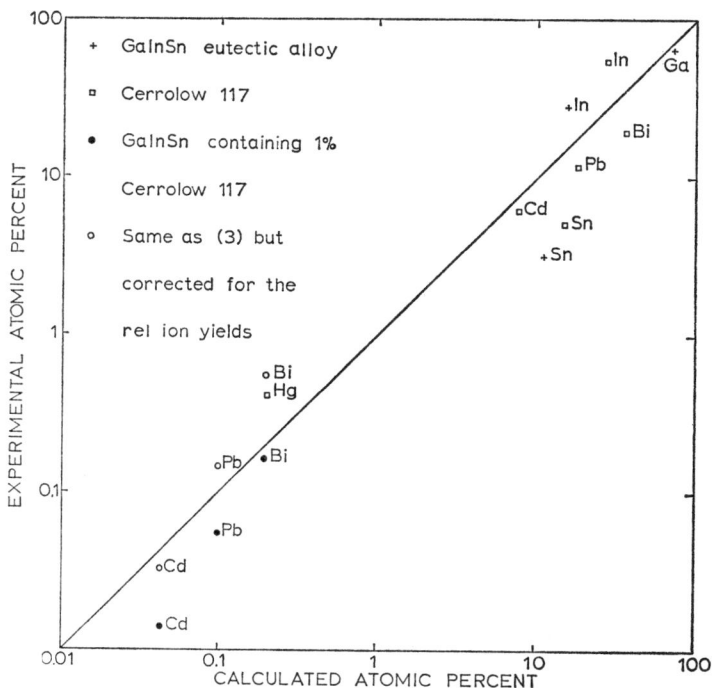

FIG. 5   Quantitative aspect of electrohydrodynamic mass spectrometry.

When these values are corrected for the relative ion yield of In^+ in GaInSn and each element's relative ion yield in Cerrolow 117, the points fall much closer to the ideal 45° line. In this instance the Bi appears high due to a Ga_3^+, interference at mass 209. This figure shows that all these elements are determined to within a factor of two of the known concentrations. The use of relative sensitivity factors for a specific element in a specific matrix would be expected to greatly improve overall accuracy.

Discussion

H. J. Svec (Iowa State University, Ames, U.S.A.): What is the size of the needle?

C. A. Evans: The work described in this paper was performed using needles from 200 to 250 micrometres internal diameter.

H. J. Svec: How much excess kinetic energy do the ions have?

C. A. Evans: No experiments have been done to date on the initial kinetic energy spread of the ions produced by the EH ion source.

H. J. Svec: Have you tried the technique for water solutions? If so, what have been the results?

C. A. Evans: Water solutions have not been run in the ion source due to the vapour pressure/ vacuum system problems.

67
Condensation and Vaporization Coefficients of Graphite[†]

By M. HOCH, D. RAMAKRISHNAN
and T. VERNARDAKIS

(*Department of Materials Science and Metallurgical Engineering,
University of Cincinnati, Cincinnati, Ohio, U.S.A.*)

SYMBOLS

I	ion intensity; *e.g.* I_{25} is the ion intensity of the $C^{12}C^{13}$ vapour species.
M	graphite prepared by Monsanto.
p_x	equilibrium pressure of species x; *e.g.* p_2 equilibrium pressure for C_2.
R_2	concentration of $C^{12}C^{13}$ mole fraction, $$R_2 = \frac{C^{12}C^{13}}{C_2{}^{12} + C^{12}C^{13} + C_2{}^{13}}$$
R_3	concentration of $C_2{}^{12}C^{13} + C^{12}C_2{}^{13}$ mole fraction, $$R_3 = \frac{C_2{}^{12}C^{13} + C^{12}C_2{}^{13}}{C_3{}^{12} + C_2{}^{12}C^{13} + C^{12}C_2{}^{13} + C_3{}^{13}}$$
S	geometric surface area 'exposed to gas' in square centimetres.
T	Temperature in °K.
U	UF4F graphite.
$\alpha_1, \alpha_2, \alpha_3$	vaporization coefficient of C_1, C_2 and C_3, respectively.
α_M, α_U	vaporization coefficient from graphite$_M$ and graphite$_U$, respectively.
$\beta_1, \beta_2, \beta_3$	condensation coefficients of C_1, C_2 and C_3.

† This research has been supported in part by the Air Force Materials Laboratory, Air Force Systems Command, United States Air Force, WPAFB, Ohio 45433, in contract No. F 33615-69-C-1531.

INTRODUCTION

THE vaporization and condensation of graphite and its vapour has received a great deal of attention in recent years because graphite shows promise as a structural material for aerospace applications. Studies by a number of investigators[1-8] on the vaporization of graphite in the temperature range 2200–2700°K have shown this to be quite a complicated process. Studies, on the other hand, concerned with the composition of the equilibrium carbon vapour over graphite generally agree that this vapour is composed of atomic C species as well as polyatomic or molecular species such as C_2, C_3, C_4, and C_5,[9-12] with the C_3 molecule being the most abundant species.

The vaporization coefficient of graphite has been determined by comparing the rate of vaporization under non-equilibrium conditions (free vaporization) to that under equilibrium conditions (Knudsen vaporization). The reported experimental values for various grades of graphite varied between 0·001 and 0·3. Thorn and Winslow[5] reported values for vaporization coefficients of C_1, C_2, and C_3 for non-pyrolytic graphite at 2450°K as $\alpha_1 = 0.37$, $\alpha_2 = 0.34$, and $\alpha_3 = 0.08$. Burns and co-workers[9] studied the vaporization of pyrolytic graphite originating either from basal planes or from edges. Vaporization coefficients were determined for each face with the following results at 2500°K: for the basal planes (c-face) $\alpha_1^c = 0.14$, $\alpha_2^c = 0.26$, and $\alpha_3^c = 0.03$; for the edges (a-face) $\alpha_1^a = 0.23$, $\alpha_2^a = 0.38$, and $\alpha_3^a = 0.04$. Zavitsanos's[10] results using pyrolytic graphite for vaporization from the basal plane (c-face) under equilibrium and non-equilibrium conditions at about 2700°K gave values for α^c as follows: $\alpha_1^c = 0.24$, $\alpha_2^c = 0.50$, and $\alpha_3^c = 0.023$.

The vaporization coefficients obtained by Zavitsanos for the c-face are in better agreement with those determined by Burns et al. for the a-face rather than the c-face. This problem was partially resolved by Wachi and Gilmartin[13] who suggest that the vaporization process for carbon species vaporizing from the a- and c-face is identical for the two surfaces since identical activation energies of vaporization were determined for these species. The predominant process for c-face vaporization is considered to be the formation of the C_2, C_3, and C_5 molecules which appear to evaporate mainly from the edges of the basal plane (a, b-plane) of the crystallites at the grain boundaries from other discontinuities and from surface defects. The process for a-face vaporization is thought to be due to carbon evaporation from a new 'white' allotropic form of carbon.

THEORY AND EXPERIMENTAL SETUP

The isotopic exchange technique in combination with the Time-of-Flight mass spectrometer has been used to determine the condensation coefficients of AlN,[14] BN,[14] and NiO.[15] The theoretical model used for studying the exchange reactions is based on the following assumptions:

(1) The exchange reaction between the gas molecules and the solid takes place only on the surface of the solid.
(2) The vaporization of solid takes place only from a monolayer of the surface.

The theoretical treatment derived for AlN[14] applies directly to the exchange reactions between C_2 gas and C(s), whereas a different equation[16] applies to those between C_3 gas and C(s). The two equations used to determine the condensation coefficients are:

$$\beta_2 = \frac{h}{S} \frac{R_2}{(\frac{1}{2} - R_2)} \tag{1}$$

$$\beta_3 = \frac{h}{S} \frac{R_3}{(\frac{1}{2} - \frac{2}{3}R_3)} \tag{2}$$

The isotopic exchange technique employs a Knudsen cell which contains two samples of the solid which are to be vaporized, each containing a high concentration of a particular isotope. In this work the solids are C^{12} and C^{13}. A schematic sketch of the tungsten Knudsen cell and the tantalum sample holder is shown in Fig. 1.

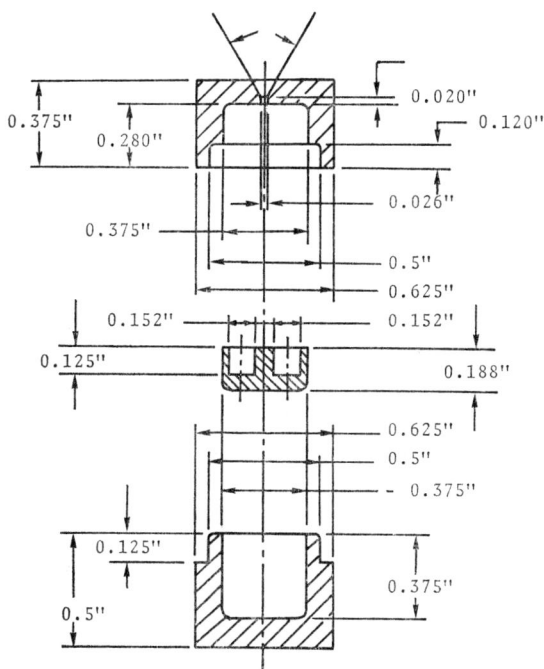

FIG. 1 Tungsten Knudsen cell and tantalum insert for condensation coefficient measurement.

In the second type of experiment the study of difference in vaporization coefficients between various types of graphite was undertaken. The first experiment was to study the difference between the vaporization of the spectrographic grade UF4F graphite and the fine crystalline C^{13} powder. For this purpose the tungsten Knudsen cell equipped with a graphite liner (made of UF4F) was used into which pellets, P_1, P_2, and P_3, could be introduced (Fig. 2).

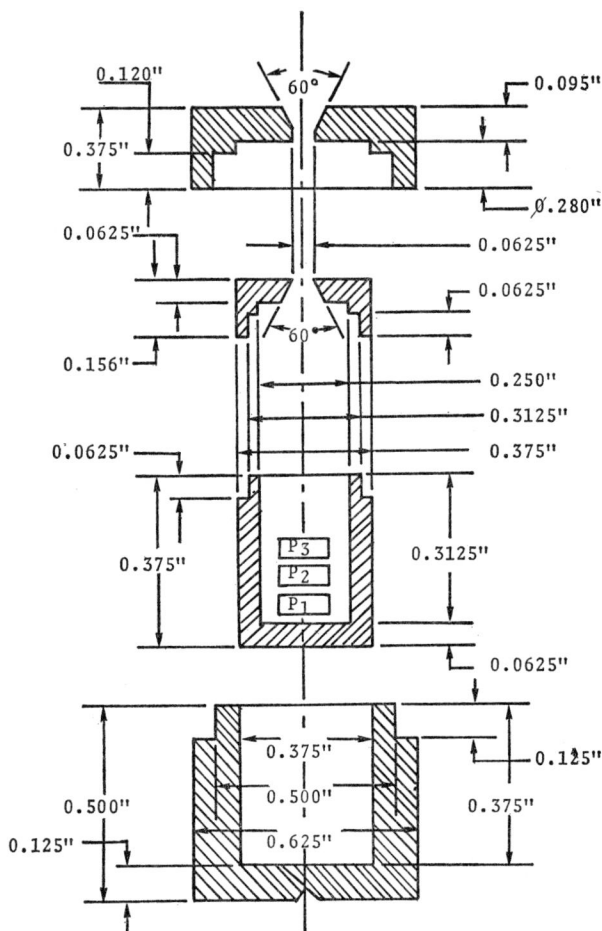

FIG. 2 Tungsten Knudsen cell and graphite liner for comparison of vaporization coefficients.

MATERIALS AND EXPERIMENTAL PROCEDURE

The isotopic exchange technique requires carbon samples containing very high concentration of C^{12} and C^{13} isotopes. 99·93% pure C^{13} and C^{12} powders were obtained from Monsanto Research Corporation, Mound Laboratory, Miamisburg, Ohio. The C^{13} powder contained 94·3% C^{13} and 5·7% C^{12} isotopes. Table I lists the spectrochemical analysis data of the powder samples used in this investigation.

The powders were pressed into small pellets using a 0·152-inch diameter steel die and a load of 500 pounds. The compact thus obtained had a density of 1·03 g/ml. This material is identified with the letter M. The other graphite material used was spectrographic grade, UF4F, rod. Its properties are also given in Table I. This material is identified with the letter U.

After degassing of the components separately the experimental runs were

TABLE I

Properties of Graphite

(a) Carbon 13 (Monsanto) Analysis: Spectrochemical Data

Iron	0·05 weight %
Silicon	0·01 weight %
Magnesium	0·000 5 weight %
Aluminium	<0·000 5 weight %
Nickel	<0·005 weight %
Molybdenum	0·001 weight %
Hydrogen	<0·1 weight %

(b) Ultra Purity Spectrographic Electrodes, Grade UF4F

Ash Content	10 ppm
Grain size	max. 0·008 inch
Bulk density	1·76 g/ml
Porosity	22·7%
Electrical resistivity	0·000 45 ohm in

carried out at 2700°K and the ion intensities of the various species recorded. For each species four to six readings each were taken with shutter closed, shutter open, and each experiment was completed in about $1\frac{1}{2}$ hours. Measurements were taken at 25, 30, 35 and 40 eV ionization energies.

EXPERIMENTAL RESULTS AND DISCUSSION

The concentration of a species will be proportional to the peak intensity, as obtained by the mass spectrometer, *e.g.*

$$(C_2{}^{12}C^{13}) \simeq \frac{Z \cdot I_{37}}{Z \cdot I_{36} + Z \cdot I_{37} + Z \cdot I_{38} + Z \cdot I_{39}}$$

Z contains the various mass spectrometer constants, and is the same for one species, irrespective of the isotopic distribution. Thus ion intensities can be used instead of concentrations.

Condensation Coefficients

Table II contains the data of the condensation coefficient measurements.

Run 2 was made at a higher temperature and thus the data obtained in this run are more reliable. Based on these results we get $\beta_2 = 4 \times 10^{-2}$ and

TABLE II

Condensation Coefficients of C_2 and C_3 on Graphite$_M$

Exchange Run	Configuration	Temp. (°K)	h (cm^2)	S_M (cm^2)	R_2	R_3	β_2	β_3
1	No insert	2 613	$8·1 \times 10^{-3}$	0·72	0·18	0·23	$6·4 \times 10^{-3}$	$7·5 \times 10^{-3}$
2	Ta insert	2 663	$3·4 \times 10^{-3}$	0·234	0·37	0·38	$4·1 \times 10^{-2}$	$2·3 \times 10^{-2}$
3	Ta insert	2 623	$3·4 \times 10^{-3}$	0·234	0·2	0·42	$9·8 \times 10^{-3}$	$2·8 \times 10^{-2}$

TABLE III

Vaporization Coefficient of Graphite$_M$ and Graphite$_U$

Run	Configuration P_1 P_2 P_3	S_U (cm^2)	S_M (cm^2)	Temp. $(°K)$	I_{12}	I_{13}	I_{24}	I_{25}	I_{26}	I_{36}	I_{37}	I_{38}	I_{39}	D_1	D_2	D_3	$\left(\dfrac{\alpha_U}{\alpha_M}\right)_1$	$\left(\dfrac{\alpha_U}{\alpha_M}\right)_2$	$\left(\dfrac{\alpha_U}{\alpha_M}\right)_3$
1	O O O	2·43	0	2 668	31·6		30·5			311									
2	M M M	2·25	0·89	2 663	35·0	77·7	14·2	49·0	81·2	41·6	223	544	585	0·450	0·366	0·364	0·178	0·145	0·144
3	M M M	2·25	0·89	2 660	42·0	95·8	16·7	74·8	94·7	58·3	273	634	579	0·439	0·410	0·413	0·173	0·162	0·163
4	M M U	2·84	0·475	2 668	10·9	13·4	4·7	11·7	9·1	17·7	66·4	88·8	42·8	0·813	0·704	0·735	0·136	0·118	0·123
5	M M O	2·25	0·653	2 670	45·2	62·4	23·0	55·3	61·8	59·4	285	471	291	0·725	0·568	0·581	0·210	0·165	0·169
																	(Avg.: 0·157 ± 0·020)		

$\beta_3 = 2\cdot 3 \times 10^{-2}$ for the condensation coefficients of C_2 and C_3 respectively. The condensation coefficient of C_2 obtained in this investigation is lower than the vaporization coefficient of C_2 reported by Thorn and Winslow,[5] Zavitsanos,[10,12] and Burns et al.[9] But the condensation coefficient of C_3 is in good agreement with the vaporization coefficient of C_3 reported by the same authors.

Comparison of Vaporization Coefficients

Table III contains the experimental results on the comparison of the vaporization coefficients of $graphite_U$ and $graphite_M$.

Let us define D_x as the ratio of the carbon isotopes in each gaseous species, i.e.

$$D_1 = \frac{I_{12}}{I_{13}} \tag{3}$$

$$D_2 = \frac{I_{24} + \frac{1}{2}I_{25}}{I_{26} + \frac{1}{2}I_{25}} \tag{4}$$

and

$$D_3 = \frac{I_{36} + \frac{2}{3}I_{37} + \frac{1}{3}I_{38}}{I_{39} + \frac{2}{3}I_{38} + \frac{1}{3}I_{37}} \tag{5}$$

If the vaporization coefficients for $C_1(g)$, $C_2(g)$, and $C_3(g)$ are equal from the same surface but different from one surface and the other (e.g. different vaporization coefficients from $graphite_U$ than from $graphite_M$) then we would have

$$D_1 = D_2 = D_3 \tag{6}$$

Inspecting Table III we see that $D_1 \leq D_2 = D_3$, showing that within experimental error the vaporization of C_1, C_2 and C_3 is similar.

The data in Table III indicate that the intensities of the species coming from the surface of $graphite_M$ are greater than those coming from the surface of $graphite_U$ even though $S_U > S_M$. Also the increase in intensity of the species coming from S_M is not due to a 'line of sight' effect, i.e. the fact that the $graphite_M$ pellet is opposite the orifice, because in run 4 where a $graphite_U$ pellet was opposite the orifice the intensities containing the C^{13} isotope only are larger than those containing C^{12} only. Thus we conclude that the vaporization coefficient of $graphite_M$ is larger than that of $graphite_U$.

Using the geometrical surface areas one can relate the intensity I_x to the surface area S_x and vaporization coefficients α_x

$$I_x \simeq p_x S_x \alpha_x \tag{7}$$

by expressing the ratios of two intensities and rearranging

$$\left(\frac{\alpha_U}{\alpha_M}\right)_1 = \frac{I_{12}}{I_{13}} \cdot \frac{S_M}{S_U} = D_1 \cdot \frac{S_{13}}{S_{12}} \tag{8}$$

$$\left(\frac{\alpha_U}{\alpha_M}\right)_2 = D_2 \cdot \frac{S_M}{S_U} \tag{9}$$

and

$$\left(\frac{\alpha_U}{\alpha_M}\right)_3 = D_3 \cdot \frac{S_M}{S_U} \tag{10}$$

The last three columns of Table III show that the vaporization coefficient for C_1, C_2 and C_3 of graphite$_U$ is six times less than for graphite$_M$.

Further experiments are being carried out where the graphite liner in the tungsten cell is being eliminated; various pellets P_1, P_2, P_3, are placed in a tungsten Knudsen cell. In P_3 graphite$_M$ is introduced and in P_1 and P_2 various types of graphite, such as glassy-carbon, Pocographite, etc. In this way the latter's vaporization coefficients are compared with that of graphite$_M$ and thus their relative vaporization rates are obtained.

REFERENCES

1. Brewer, L., Gilles, P. W. and Jenkins, F. A., *J. Chem. Phys.*, 1948, **16**, 797.
2. Doehaerd, T., Goldfinger, P. and Waelbroeck, F., *J. Chem. Phys.*, 1952, **20**, 757.
3. Brewer, L. and Kane, J. S., *J. Phys. Chem.*, 1955, **59**, 105.
4. Hoch, M., Blackburn, P. E., Dingledy, D. P. and Johnston, H. L., *J. Phys. Chem.*, 1955, **59**, 97.
5. Thorn, R. J. and Winslow, G. H., *J. Chem. Phys.*, 1955, **23**, 1369; 1957, **26**, 186.
6. Honig, R. E., *J. Chem. Phys.*, 1954, **22**, 126.
7. Chupka, W. A. and Inghram, M. G., *J. Phys. Chem.*, 1953, **21**, 371; 1953, **21**, 1313.
8. Drowart, J., Burns, R. P., DeMaria, G. and Inghram, M. G., *J. Chem. Phys.*, 1959, **31**, 1131.
9. Burns, R. P., Jason, A. J. and Inghram, M. G., *J. Chem. Phys.*, 1964, **40**, 1161.
10. Zavitsanos, P. D., 'The Vaporization of Pyrolytic Graphite', G. E. Report R66SD31, May 1966.
11. Palmer, H. B. and Shelef, M., *in* 'Chemistry and Physics of Carbon', *ed.* P. L. Walker Jr, Marcel-Dekker, Inc., New York, 1968, pp. 85–135.
12. Zavitsanos, P. D., 'Experimental Study of the Sublimation of Graphite at High Temperatures, Final Report', G. E. Report, December 1969, Revised June 1970.
13. Wachi, F. M. and Gilmartin, D. E., *Carbon*, 1970, **8**, 141.
14. Hoch, M. and Ramakrishnan, D., *J. Electrochem. Soc.*, 1971, **118**, 1204–11.
15. Ramakrishnan, D. and Hoch, M., *J. Electrochem. Soc.*, 1971, **118**, 1212–16.
16. Ramakrishnan, D. and Hoch, M., 'The Condensation Coefficients of C_2 and C_3', Technical Report AFML-TR-70-66, Air Force Materials Laboratory, Wright-Patterson Air Force Base, Ohio, February 1971.

68
The Behaviour of Evaporation of the Compounds Belonging to the Binary System BaO–Al₂O₃

By K. HILPERT, H. BESKE, H. W. NURNBERG

(*Zentralinstitüt für Analytische Chemie, Jülich, Germany*)

A. NAOUMIDIS

(*Institüt für Reaktorwerkstoffe der Kernforschungsanlage, Jülich, Germany*)

INTRODUCTION

THE fuel elements of high temperature gas-cooled reactors[1] contain coated nuclear fuel particles.[2] As in the future the temperature is to be raised further with respect to applications in the process heat field safety requires to improve the retention ability of the coated particles for solid fission products (*e.g.* ^{137}Cs, ^{89}Sr, ^{90}Sr and ^{140}Ba). One possible new way to do this is by the addition to the nuclear fuel of certain refractory oxides such as Al_2O_3, which can react with the solid fission products. If the compounds being formed are comparatively stable it becomes possible in this way to diminish the fission product release. In connection with tests and studies[3-5] related to this approach of fission product retention the hitherto unknown fundamental data on the behaviour of evaporation of the compounds belonging to the binary system BaO–Al₂O₃; the clarification of the involved chemical reactions and solid phases; and the determination of the corresponding thermodynamic parameters were required.[6]

According to Purt[7] as well as to Toropow and Galachow[8] the system BaO–Al₂O₃ consists of the compounds $BaAl_{12}O_{19}$, $BaAl_2O_4$ and $Ba_3Al_2O_6$. These authors determined the liquidus curve of the system. Additionally, the dimensions of the unit-cell and the Debye–Scherrer patterns of the barium aluminate compounds have been determined. However, no data about the evaporation of these compounds were available.

EXPERIMENTAL

Instrument

The measurements were carried out with the commercially available Knudsen cell attachment to the VARIAN MAT CH-5 mass spectrometer. The vapour beam effusing from the cell is directed towards an electron impact ion source. The ions produced are magnetically deflected over an angle of 90° and are detected either by a multiplier or a Faraday cage.

The cylindrical molybdenum Knudsen cell was heated by heat radiation and electron bombardment. Its inner diameter and height were 8 mm respectively. The cell could be adjusted from outside. The cutter-shaped effusion gap had a 0·4 mm radius. The sample was put into a molybdenum crucible of 4 mm height and of 0·5 mm wall thickness. The temperature was measured with an optical pyrometer on the bottom of a black body cavity laterally placed close to the bottom of the cell. Its accuracy, checked on the basis of the melting points of silver and platinum, was better than $\pm 1\%$. Temperature measurements in additionally arranged lateral cavities and investigations according to a method reported by Storms[9] have proved a homogeneous temperature distribution within measuring accuracy.

Samples

The samples were prepared by heating of $BaCO_3$ and Al_2O_3. Despite constant temperature, measurements of exactly stoichiometric samples showed fluctuations in vapour pressure, even though the samples consisted of pure compounds according to the Debye–Scherrer patterns. For this reason, under- and over-stoichiometric samples of the compositions 0·95 BaO, 6·05 Al_2O_3; 0·9 BaO, 1·1 Al_2O_3 and 1·1BaO, 0·9Al_2O_3 were investigated.

To prepare these samples, mixtures of $BaCO_3$ and Al_2O_3, supplied by E. Merck, Darmstadt, Germany, were pressed into pellets which were heated in a tantalum crucible for 16 hours at 1823°K under flowing argon. Subsequently, parts of the pellets were melted by an oxy-hydrogen torch to ascertain that the samples had completely reacted. The heated samples as well as those that were additionally melted have been checked by X-ray diffraction according to the Debye–Scherrer method.

RESULTS

Gaseous Phase

Investigating the vapour pressure of the molten and the merely heated samples no other Ba-containing ions could be observed than Ba^+ and BaO^+. With respect to the different composited samples, the intensity of Ba^+ exceeds the intensity of BaO^+ by a factor 1·7 to 3·3. It can be derived from ionization efficiency measurements and from the resulting appearance potentials that Ba and BaO exist in the gaseous phase and that Ba^+ and BaO^+ result from simple ionization.

While the BaO pressures are only temperature-dependent and have practically the same values for samples having the same composition, according to the Debye–Scherrer patterns, the vapour pressures of Ba in some cases are clearly different. This observation and a Ba^+ ion current varying with time, which could be observed sometimes at the beginning of a measurement after an especially fast heating-up process, indicate that Ba is formed by reduction of BaO. This conclusion is confirmed by mass-spectrometric Knudsen cell measurements carried out with BaO by Hilpert and Gerads[10] applying graphite and Mo cells as well as Mo cells lined with Pt or Al$_2$O$_3$. Although, according to the results obtained with Al$_2$O$_3$ lined cells by us and according to the studies of Inghram, Chupka and Porter,[11] BaO sublimes congruently, the results obtained with Mo cells showed almost the same Ba/BaO ratios as those being observed in this work.

Vapour Pressure

The vapour pressure was evaluated with the following equation

$$p(X) = I(^iX^+) . T(^iX^+) . E . \frac{\sigma(Ag)}{\sigma(X)} . \frac{\gamma(^{107}Ag^+)}{\gamma(^iX^+)} . \frac{h(^{107}Ag)}{h(^iX)} \qquad (1)$$

where the sensitivity E represents a calibration value defined by

$$E = \frac{p(^{107}Ag)}{I(^{107}Ag^+)T(^{107}Ag^+)} \qquad (2)$$

The meaning of the symbols used in both equations is as follows:

$p(X)$, $p(^{107}Ag)$	vapour pressure of the molecule or atom X or of ^{107}Ag respectively.
$I(^iX^+)$, $I(^{107}Ag^+)$	intensity for the isotope of the mass i in the same unit which was used for ^{107}Ag.
$T(^{107}Ag^+)$, $T(^iX^+)$	cell temperature belonging to the intensity of $^{107}Ag^+$ or $^iX^+$ respectively.
$\dfrac{\sigma(Ag)}{\sigma(X)}$	ratio of the ionization cross-sections of Ag and X at 22 eV.
$\dfrac{\gamma(^{107}Ag^+)}{\gamma(^iX^+)}$	correction for the different amplification factors of the multiplier for $^{107}Ag^+$ or $^iX^+$ respectively.
$\dfrac{h(^{107}Ag)}{h(^iX)}$	ratio of the isotopic abundancies of ^{107}Ag and ^iX.

For an ionization energy of 22 eV the following σ-ratios were calculated by linear interpolation using the maximum ionization cross-sections published by Mann:[12]

$$\frac{\sigma(Ag)}{\sigma(Ba)} = 0.39 \qquad \frac{\sigma(Ag)}{\sigma(BaO)} = 0.42$$

The sensitivity E was determined according to the following methods:

(1) Quantitative vaporization of a known amount of silver.
(2) Use of known vapour pressures of silver as given by Honig.[13]

At first, in the course of one measurement the sensitivity was determined according to methods 1 and 2. During the first 20 min of such a measurement the intensity of $^{107}Ag^+$ was measured at seven different temperatures and the sensitivity could be calculated according to method 2. Then the silver was completely evaporated, and by means of method 1 the sensitivity E could be determined using the Hertz–Knudsen equation. In this way, in the course of five different measurements in each case the sensitivity could be evaluated according to the two methods. The values of E calculated by means of method 2 differed by a mean factor $f = 0.65$ from those determined by quantitative evaporation.

The calibration for the measurement of the BaO vapour pressure was carried out according to method 2. The sensitivity being determined in this way was subsequently multiplied by the previously calculated factor f. Thus, the computation of the vapour pressure was performed on the basis of the sensitivity obtained by method 1, which represents an absolute calibration. The advantage of the outlined procedure was that a calibration according to method 2 takes only some minutes, and it is not necessary to weigh out the silver before putting it into the cell.

In this way the BaO vapour pressure curves of all samples have been determined by a least-squares fit. The curves of the samples 0.95 BaO, 6.05 Al_2O_3 (molten) and 0.95 BaO, 6.05 Al_2O_3 (heated) were in good agreement. According to the Debye–Scherrer patterns the two samples consist of $BaAl_{12}O_{19}$. Thus, the measured BaO vapour pressures are the dissociation pressures of this compound.

According to the interpretation of the Debye–Scherrer patterns, the samples 0.9 BaO, 1.1 Al_2O_3 (molten) and 0.9 BaO, 1.1 Al_2O_3 (heated) consist of $BaAl_2O_4$ and $BaAl_{12}O_{19}$ whereas the sample 1.1 BaO, 0.9 Al_2O_3 (molten) is composed of $BaAl_2O_4$ only. The vapour pressures of these samples coincide within measurement accuracy and are higher than the BaO pressure of $BaAl_{12}O_{19}$. They represent the dissociation pressures of $BaAl_2O_4$.[14]

The BaO pressure of the sample 1.1 BaO, 0.9 Al_2O_3 (heated) exceeds the dissociation pressure of $BaAl_2O_4$ by a factor 6.8×10^3. Beside $BaAl_2O_4$ this sample probably consists of $Ba_3Al_2O_6$ as could be proved by means of the Debye–Scherrer patterns. Hence, the corresponding vapour pressure curve represents the dissociation pressure of $Ba_3Al_2O_6$.

According to the Debye–Scherrer patterns and the vapour pressure curves

TABLE I

BaO Vapour Pressures of the Barium Aluminate Compounds and BaO

Compound	BaO vapour pressure (p in torr, T in °K)	Temperature range (°K)	p_{BaO} at 1 818 °K (torr)
$Ba_3Al_2O_6$	$\log p = -2.459 \times \dfrac{10^4}{T} + 11.25$	1 448–1 711	5.3×10^{-3}
$BaAl_2O_4$	$\log p = -2.814 \times \dfrac{10^4}{T} + 9.38$	1 835–2 094	7.9×10^{-7}
$BaAl_{12}O_{19}$	$\log p = -2.830 \times \dfrac{10^4}{T} + 9.14$	1 897–2 170	3.8×10^{-7}
BaO	$\log p = -2.173 \times \dfrac{10^4}{T} + 10.07$	1 681–1 332	1.31×10^{-2}

the sample $1 \cdot 1$ BaO, $0 \cdot 9$ Al$_2$O$_3$ (heated) in contrast to the sample $1 \cdot 1$ BaO, $0 \cdot 9$ Al$_2$O$_3$ (molten) contains Ba$_3$Al$_2$O$_6$. This discrepancy might be explained by fractionation, which happens during the melting process of a part of a pellet, due to the different melting points of Ba$_3$Al$_2$O$_6$ and BaAl$_2$O$_4$.

The equations for the dissociation pressures are given in Table I. If for one compound more than one vapour pressure curve has been determined average values were calculated. Additionally, in this table the vapour pressure curve of BaO is given from the measurements by Hilpert and Gerads.[10] From the corresponding relations, the vapour pressures at 1818°K were calculated. They are given also in Table I and enable a better comparison of the results

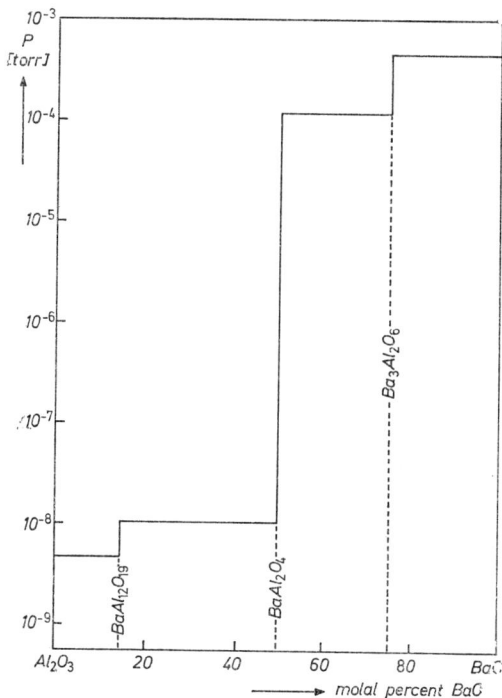

FIG. 1   BaO vapour pressure versus composition at 1620°K.

for the various compounds. After the determination of the vapour pressures for the complete representation of the behaviour of the BaO–Al$_2$O$_3$ system it is possible to employ a right-angle space model with three axes to represent pressure p, temperature and composition c. A section of this model parallel to the cp-plane yields the vapour pressure as a function of the composition. This diagram is shown in Fig. 1 for a constant temperature of 1620°K, which represents the expected temperature of the reactor fuel. Figure 1 indicates that by formation of a barium aluminate compound the BaO vapour pressure can be decreased to an unusually great extent if BaAl$_2$O$_4$ or BaAl$_{12}$O$_{19}$ is being formed. This means, with respect to fission product retention, that the quantity of Al$_2$O$_3$ added to the nuclear fuel at least should be such that all fission products eligible for reaction can react forming BaAl$_2$O$_4$.

Reactions of Evaporation

The compounds $Ba_3Al_2O_6$, $BaAl_2O_4$ and $BaAl_{12}O_{19}$ evaporate according to the following relations:

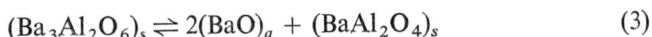

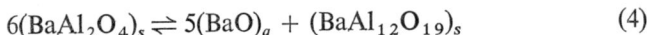

$$(Ba_3Al_2O_6)_s \rightleftharpoons 2(BaO)_g + (BaAl_2O_4)_s \tag{3}$$

$$6(BaAl_2O_4)_s \rightleftharpoons 5(BaO)_g + (BaAl_{12}O_{19})_s \tag{4}$$

and

$$(BaAl_{12}O_{19})_s \rightleftharpoons (BaO)_g + 6(Al_2O_3)_s \tag{5}$$

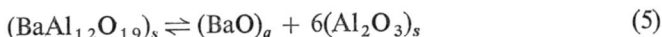

To prove these equations, the phase law will be applied to the systems given by the equilibria (eqns (3), (4), (5)). The systems consist of two components (BaO and Al_2O_3) and three phases (($BaO)_g$ and two solids). Substituting these quantities into the phase law yields one degree of freedom. Consequently at constant temperature the BaO vapour pressure is fixed. This is in agreement with the experimental results. If there were one or three solid phases in equilibrium with the gas phase the number of degrees of freedom would become 2 or 0 respectively. This is inconsistent with the results of the measurements. Thus, on evaporating a barium aluminate compound a second solid phase must be formed additionally. The composition of this solid phase follows from the two facts of (I) the incongruent evaporation of the barium aluminate compounds in which gaseous BaO is formed and (II) from the prerequisite that only adjoining phases of a phase system can exist side by side in an equilibrium system. Hence, the solid phases, which are allowed to exist in equilibrium with gaseous BaO, are $Ba_3Al_2O_6$ and $BaAl_2O_4$, $BaAl_2O_4$ and $BaAl_{12}O_{19}$ or $BaAl_{12}O_{19}$ and Al_2O_3.

Thermodynamic Parameters

The equilibrium constants $K_{(Ba_3Al_2O_6)}$, $K_{(BaAl_2O_4)}$ and $K_{(BaAl_{12}O_{19})}$ for the reactions of evaporation given by eqns (3), (4) and (5) respectively are defined by the following expressions:

$$K_{(Ba_3Al_2O_6)} = \frac{a^2_{(BaO)_g} a_{(BaAl_2O_4)_s}}{a_{(Ba_3Al_2O_6)_s}} \tag{6}$$

$$K_{(BaAl_2O_4)} = \frac{a^5_{(BaO)_g} a_{(BaAl_{12}O_{19})_s}}{a^6_{(BaAl_2O_4)_s}} \tag{7}$$

and

$$K_{(BaAl_{12}O_{19})} = \frac{a_{(BaO)_g} a^6_{(Al_2O_3)_s}}{a_{(BaAl_{12}O_{19})_s}} \tag{8}$$

where a is the activity.

For Knudsen cell measurements one generally assumes that the vapour pressure is equal to the fugacity because only low vapour pressures can be measured. On the other hand the activity of compounds existing as a pure phase equals 1. Substituting this into eqns (6), (7) and (8) yields the relations

$$K_{(Ba_3Al_2O_6)} = p^2_{(BaO)} \tag{6a}$$

$$K_{(BaAl_2O_4)} = p^5_{(BaO)} \tag{7a}$$

$$K_{(BaAl_{12}O_{19})} = p_{(BaO)} \tag{8a}$$

One obtains then with the van't Hoff equation for the reaction enthalpies corresponding to eqns (3), (4) and (5) 225, 645 and 130 kcal mole^{-1}. These values refer to 1580°K, 1963°K and 2036°K respectively.

ACKNOWLEDGMENT

The authors are grateful to Dr G. Wolff for helpful discussions.

REFERENCES

1. Schulten, R., *Atomwirtschaft*, 1966, **11**, 218; Büker, H., *Naturwissenschaften*, 1969, **56**, 112.
2. Nickel, H., *KFA-Report*, Jül-687-RW (1970).
3. Nickel, H., Förthmann, R. and Hamesch, M., *KFA-Report*, Jül-846-RW (1972).
4. Hamesch, M., *KFA-Report*, Jül-730-RW (1971).
5. Förthmann, R., Gyarmati, E., Hilpert, K. and Nickel, H., *Atomwirtschaft*, 1972, **17**, 265.
6. Hilpert, K., Thesis, University Bonn (1973).
7. Purt, G., *Radex-Rundschau*, 1960, **4**, 198.
8. Toropow, N. A. and Galachow, F. J., *Bulletin of the Academie of Sciences S.S.S.R.*, 1952, **82**, No. 1.
9. Storms, E., *High Temp. Sci.*, 1969, **1**, 456.
10. Hilpert, K. and Gerads, H., 'Mass Spectrometric Investigation of the Vaporization of BaO Using Knudsen Cells Made of Mo, Pt, Graphite and Al$_2$O$_3$', to be published.
11. Inghram, M. G., Chupka, W. A. and Porter, R. F., *J. Chem. Phys.*, 1955, **24**, 2159.
12. Mann, J. B., *J. Chem. Phys.*, 1967, **46**, 1646.
13. Honig, R. E., *in* 'The Characterization of High-Temperature Vapors', *ed.* J. L. Margrave, J. Wiley, New York, 1967, p. 478.
14. Hilpert, K., Naoumidis, A. and Wolff, G., 'Mass Spectrometric Study of the Behaviour of Evaporation of BaAl$_2$O$_4$', submitted to *High Temp. Sci.*

69
A Mass Spectrometric Study of the Vaporization of CdTe and $Hg_{1-x}Cd_xTe$ Compounds

By EDMUND J. ROLINSKI

(*Air Force Materials Laboratory, Wright-Patterson Air Force Base, Ohio, U.S.A.*)

DUANE E. EARLEY

(*University of Dayton Research Institute, Dayton, Ohio, U.S.A.*)

and THOMAS E. JOYCE

(*Texas Instruments, Inc., Dallas, Texas, U.S.A.*)

INTRODUCTION

ELECTRONICALLY useful group IIB chalcogenides have been studied to provide thermodynamic properties. However, in the case of CdTe, thermal analysis studies of deNobel[1] and Lorenz[2] were primarily aimed to determine equilibrium compositions for crystal growth studies. The development of the phase diagram for CdTe was further studied by Brebrick and Strauss[3] by optical absorption methods. Knudsen cell experiments were performed by Korneeva *et al.*[4] assuming a composition of the vapour species. Mass spectrometric measurements were first obtained by Drowart and Goldfinger[5] and Goldfinger and Jeunehomme.[6] Their measurements indicated that the vapour above the subliming solid was primarily Cd and Te_2. Further mass spectrometric studies were made by DeMaria *et al.*[7] and Ivanov and Vanyukov[8] to determine relative values for electron impact ionization cross-sections. The purpose of this study was to systematically analyse the vaporization of CdTe with the intent to extend this work to the $Hg_{1-x}Cd_xTe$ system.

EXPERIMENTAL

This investigation of the vaporization of CdTe and $Hg_{1-x}Cd_xTe$ was conducted using the mass spectrometric technique. The details of such techniques as applied to the area of high temperature phenomena have been previously discussed.[9] The instrument used in this study is a Nuclide Analysis Associates

587

12-90-HT mass spectrometer. This instrument is a 90°-sector, 30·5 cm radius of curvature, first-order direction focussing mass spectrometer equipped with a high temperature Knudsen cell sample system.

The sample system consists of a cylindrical Knudsen cell centered on the axis of a bifilarly wound tungsten helix which serves as a heating element. Both the cell and heater are surrounded by radiation shielding, and cell heating is accomplished by radiative heat transfer. The entire furnace complex is mounted on an assembly which in turn is driven by a bellows drive arrangement, allowing the investigator to adjust the position of the furnace assembly when under vacuum.

The vacuum housing around the furnace assembly is water jacketed, and water cooling is also provided to some of the support members of the furnace. The housing contains viewports enabling the experimenter to sight the top, side and bottom of the Knudsen cell in order to make optical temperature measurements, and it has provision for thermocouple feedthroughs. A mechanically operated 'shutter' is located above the cell and is driven by a micrometer-bellows assembly. This movable collimator between the sample system and the ion source serves to differentiate between species in the molecular beam from the cell and those present as background gases in the vacuum system. The 'shutter' assembly also provides information concerning the intensity profile of the molecular beam.

The Knudsen cell used in this investigation consists of a tantalum jacket surrounding a POCO graphite cell. The graphite cell has a 0·1 cm diameter orifice. The channel length of the orifice in this design is 0·15 cm and the ratio of internal cell area to orifice area is 810 to 1. The tantalum jacket encapsulating the graphite cell serves as a holder, provides a convenient place for thermocouple attachment and helps to ensure isothermal cell conditions.

The sample temperatures in this investigation were determined using Chromel–Alumel thermocouples which were peened into blind holes in the bottoms of the cells or cell holders. The legs of the thermocouples were led out of the vacuum envelope through Conax fittings. The thermocouples were fabricated from calibrated 24 gauge Chromel and Alumel wire. The hot junctions were formed by arc-welding in an inert atmosphere. A typical hot junction had a 1·25 mm bead when made in this way. Prior to installation, the calibration of selected thermocouples was checked against a standard thermocouple. After installation, the calibration was again checked against the fusion point of zinc. A separate calibration run was made to check the sample temperature and the bottom-holder temperature. Where appropriate, corrections to actual sample temperatures were made. We take the error in our temperature measurements to be $\pm 1·0°C$ based on our observations. The thermal emf's generated by these thermocouples were referenced to the ice point and were measured using a Leeds and Northrup Type K-5 potentiometer in conjunction with a L & N d.c. null detector, Model 9834-1.

The ion source on the Nuclide instrument is a Nier-type electron-impact source. During the course of these studies, data were taken using 18 eV electrons except where noted. The emission current was always held at 0·5 mA. The ions resulting from electron impact ionization processes occurring in the source were accelerated through 5000 volts and focussed on the entrance slit to the magnetic analyser which was set at 0·153 mm width. The exit slit from the analyser field was set at the same width. With such a geometry, a resolution of 1/1500 was obtained.

The detector in use for this study was a 16-stage electron multiplier. This multiplier has Cu–Be dynodes. A series of experiments using ionic species of a range of atomic weights was carried out using an insertable Faraday cup collector to measure the multiplier gain. The dependence of multiplier gain on the inverse square root of the ionic mass, which has previously been reported,[10,11,12] was not observed. Within the scope of our experiments, the gain appears to be quite insensitive to the mass of the impinging ion. Such behaviour is consistent with the findings of Gingerich.[13] In view of this observation, we took the gain of the multiplier to be 10^6 for all species under these conditions.

The CdTe samples used in this investigation were polycrystalline materials made by Gould Laboratories, Cleveland, Ohio, starting from Ventron 'ultrapure' CdTe, Lot 4F-248 which had a listed purity of 99·9999 %. Mass spectrographic analysis was performed on this material and listed element impurities were 65 ppm C, and 7 ppm Oxygen. The CdTe sample material was used as received following several hours of degassing at about 150°C. The $Hg_{1-x}Cd_xTe$ samples were obtained from Honeywell[14] as type N, 90 V, Nos. 121–131 with $x = 0·27$. The $Hg_{0·73}Cd_{0·27}Te$ was degassed at about 60°C prior to elevated heating runs.

RESULTS

Initial experiments were carried out to identify the ionic species resulting from the vaporization of the CdTe system. Mass spectra were recorded under isothermal conditions of 550°C and mass numbers assigned to the peaks by peak counting and by magnetic field measurements. The peaks of the ionic species originating from the Knudsen cell were differentiated from background peaks by their shutterability. The ionic species were identified by comparing the experimentally determined relative intensities within each peak group with calculated isotopic abundances for each ion and by appearance potential determinations for each species. The appearance potentials were determined by applying the linear extrapolation method to plots of the ionization efficiency curves for each species. The energy of the ionizing electrons was fixed by measurement of the appearance potential of Hg, which is present as a background gas. The ionic species identified and their appearance potentials were in good agreement with those of Franklin et al.[15] for the first appearance potentials of Cd^+, Te^+, Te_2^+. The appearance potential for Te_3^+ was determined to be 8·5 ± 0·5 eV.

The ionization efficiency curve for Te^+ displayed a sharp break. The first AP was taken to represent the simple ionization of the neutral Te. The second AP of 12·0 ± 0·5 eV marked by the discontinuity of the slopes of the ionization efficiency curve, was taken to represent the dissociative ionization of the Te_2 molecule substantiating the results of Drowart and Goldfinger.[5]

Calculations of the degree of dissociation at different pressures and temperatures indicate that at the temperatures of this investigation, dissociation is negligible. Measurements obtained indicate actual concentrations of

Te^+ and Te_3^+ in the gas phase to be less than $2\cdot0\%$ and $0\cdot2\%$, respectively, of that of Te_2^+. These findings justify the assumption, used in all of the thermodynamic calculations in this investigation, that the vaporization of CdTe is congruent and gives mainly the species $Cd_{(g)}$ and $Te_{2(g)}$. The presence of $Te_{(g)}$ and $Te_{3(g)}$ are minor and have no significant effect on the thermodynamic properties.

The ion current intensity of each of the ions identified in the mass spectrum was recorded as a function of Knudsen cell temperature. It has been shown[16] that the partial pressure, P_i, of a neutral species is directly proportional to the product of the cell temperature and the intensity measured for an ion resulting from ionization of the neutral in question, thus

$$P_i = kI_i^+ T$$

In order to relate the intensity of a given species as measured by the mass spectrometer to the pressure of the neutral precursor of that species inside the

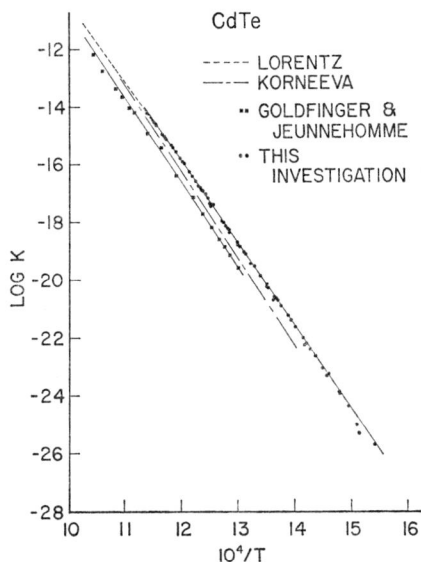

FIG. 1   Vaporization of CdTe.

equilibrium enclosure, it is necessary to carry out some type of calibration procedure. In the case of the present work such calibration was carried out by a quantitative vaporization of a quantity of CdTe. The rationale involved in the treatment of the data resulting from these quantitative vaporizations has been discussed in part previously.[9,17]

Having established the partial pressures for P_{Cd} and P_{Te_2} (Table I), the calculations for the equilibrium constant can be carried out using the expression:

$$K = P^2_{Cd(g)} \cdot P_{Te_2(g)}$$

Heats of reaction were calculated using the third law method, since:

$$-RT \ln K = \Delta H°_{298} + T\Delta\left(\frac{(G° - H°_{298})}{T}\right)$$

The free energy functions were calculated using C_p information for $Cd_{(g)}$ and $Te_{2(g)}$ from Stull and Sinke[18] and $CdTe_{(s)}$ expression from Goldfinger and Jeunehomme.[6] A comparison was made with the second law value

$$\Delta H°_T = -R\frac{d \ln K}{d\,(1/T)}$$

which is obtained from a plot of log K versus $1/T$ (see Fig. 1). The second law value of $\Delta H°_{763} = 133$ kcal/mole is in good agreement with the value of $\Delta H°_{874} = 134$ of Goldfinger and Jeunehomme.[6]

Third law calculations of $\Delta H°_{298}$ of the congruent vaporization gave the values listed in Table I whose average is $139 \cdot 2 \pm 0 \cdot 5$ kcal and is in excellent agreement with the value of $140 \cdot 7$ kcal reported by Goldfinger and Jeunehomme.[6] Values were calculated for relative electron impact ionization cross-sections and were in agreement with previous investigations,[7,8] and those of Otvos and Stevenson[19] assuming additivity.

Several experimental runs were made on $Hg_{0.73}Cd_{0.27}Te$. Based on the limited runs made it was determined that this system vaporizes incongruently and seems to be diffusion limited above 300°C. The $Hg_{0.73}Cd_{0.27}Te$ samples developed an outside scale of significant thickness during vaporization. This

TABLE I

Vaporization of CdTe $2CdTe_{(s)} = 2Cd_{(g)} + Te_{2(g)}$

$T(°K)$	P_{Cd} (atm)	P_{Te_2} (atm)	$-log\,K$	$\Delta H°_{298}$ (kcal)
649	$3 \cdot 16 \times 10^{-9}$	$1 \cdot 79 \times 10^{-9}$	$25 \cdot 748$	$139 \cdot 2$
649	$3 \cdot 13 \times 10^{-9}$	$1 \cdot 71 \times 10^{-9}$	$25 \cdot 776$	$139 \cdot 4$
661	$3 \cdot 84 \times 10^{-9}$	$3 \cdot 29 \times 10^{-9}$	$25 \cdot 314$	$140 \cdot 5$
663	$4 \cdot 87 \times 10^{-9}$	$3 \cdot 43 \times 10^{-9}$	$25 \cdot 090$	$140 \cdot 1$
669	$9 \cdot 52 \times 10^{-9}$	$4 \cdot 66 \times 10^{-9}$	$24 \cdot 374$	$139 \cdot 2$
676	$1 \cdot 15 \times 10^{-8}$	$6 \cdot 36 \times 10^{-9}$	$24 \cdot 075$	$139 \cdot 6$
677	$1 \cdot 32 \times 10^{-8}$	$7 \cdot 10 \times 10^{-8}$	$23 \cdot 908$	$139 \cdot 2$
677	$1 \cdot 09 \times 10^{-8}$	$6 \cdot 63 \times 10^{-9}$	$24 \cdot 104$	$139 \cdot 9$
686	$2 \cdot 24 \times 10^{-8}$	$1 \cdot 15 \times 10^{-8}$	$23 \cdot 239$	$138 \cdot 9$
687	$2 \cdot 12 \times 10^{-8}$	$1 \cdot 13 \times 10^{-8}$	$23 \cdot 294$	$139 \cdot 2$
690	$2 \cdot 51 \times 10^{-8}$	$1 \cdot 32 \times 10^{-8}$	$23 \cdot 080$	$139 \cdot 3$
696	$3 \cdot 55 \times 10^{-8}$	$1 \cdot 81 \times 10^{-8}$	$22 \cdot 642$	$139 \cdot 0$
701	$4 \cdot 28 \times 10^{-8}$	$2 \cdot 25 \times 10^{-8}$	$22 \cdot 385$	$139 \cdot 0$
704	$5 \cdot 12 \times 10^{-8}$	$2 \cdot 60 \times 10^{-8}$	$22 \cdot 166$	$139 \cdot 0$
706	$5 \cdot 00 \times 10^{-8}$	$2 \cdot 69 \times 10^{-8}$	$22 \cdot 172$	$139 \cdot 4$
706	$4 \cdot 77 \times 10^{-8}$	$5 \cdot 51 \times 10^{-8}$	$22 \cdot 243$	$139 \cdot 6$
708	$5 \cdot 57 \times 10^{-8}$	$3 \cdot 00 \times 10^{-8}$	$22 \cdot 031$	$139 \cdot 2$
714	$7 \cdot 55 \times 10^{-8}$	$4 \cdot 33 \times 10^{-8}$	$21 \cdot 608$	$139 \cdot 0$
719	$9 \cdot 16 \times 10^{-8}$	$4 \cdot 72 \times 10^{-8}$	$21 \cdot 402$	$139 \cdot 2$
721	$1 \cdot 02 \times 10^{-7}$	$5 \cdot 42 \times 10^{-8}$	$21 \cdot 249$	$139 \cdot 1$

TABLE I—*continued*

$T(°K)$	P_{Cd} (atm)	P_{Te_2} (atm)	$-\log K$	$\Delta H°_{298}$ (kcal)
727	$1·38 \times 10^{-7}$	$7·06 \times 10^{-8}$	20·871	139·0
730	$1·47 \times 10^{-7}$	$7·97 \times 10^{-8}$	20·764	139·2
730	$1·45 \times 10^{-7}$	$7·85 \times 10^{-8}$	20·782	139·3
732	$1·57 \times 10^{-7}$	$8·22 \times 10^{-8}$	20·693	139·2
732	$1·65 \times 10^{-7}$	$8·80 \times 10^{-8}$	20·621	139·0
734	$1·54 \times 10^{-7}$	$8·21 \times 10^{-8}$	20·711	139·6
740	$2·24 \times 10^{-7}$	$1·17 \times 10^{-7}$	20·231	139·2
741	$2·21 \times 10^{-7}$	$1·17 \times 10^{-7}$	20·243	139·4
742	$2·28 \times 10^{-7}$	$1·23 \times 10^{-7}$	20·194	139·3
748	$3·00 \times 10^{-7}$	$1·60 \times 10^{-7}$	19·842	139·3
753	$3·73 \times 10^{-7}$	$1·95 \times 10^{-7}$	19·567	139·2
757	$3·90 \times 10^{-7}$	$2·12 \times 10^{-7}$	19·492	139·7
762	$5·31 \times 10^{-7}$	$2·77 \times 10^{-7}$	19·107	139·1
764	$5·61 \times 10^{-7}$	$2·79 \times 10^{-7}$	19·056	139·2
768	$6·59 \times 10^{-7}$	$3·48 \times 10^{-7}$	18·821	139·2
770	$7·16 \times 10^{-7}$	$3·70 \times 10^{-7}$	18·722	139·2
779	$9·79 \times 10^{-7}$	$4·95 \times 10^{-7}$	18·324	139·3
781	$1·08 \times 10^{-6}$	$5·49 \times 10^{-7}$	18·194	139·1
781	$1·02 \times 10^{-6}$	$5·36 \times 10^{-7}$	18·254	139·3
783	$1·12 \times 10^{-6}$	$5·88 \times 10^{-7}$	18·132	139·2
785	$1·24 \times 10^{-6}$	$6·49 \times 10^{-7}$	18·001	139·2
787	$1·27 \times 10^{-6}$	$6·80 \times 10^{-7}$	17·960	139·2
797	$1·99 \times 10^{-6}$	$1·04 \times 10^{-6}$	17·385	138·8
797	$1·95 \times 10^{-6}$	$9·61 \times 10^{-7}$	17·437	139·1
798	$1·97 \times 10^{-6}$	$9·74 \times 10^{-7}$	17·423	139·1
800	$2·05 \times 10^{-6}$	$1·04 \times 10^{-6}$	17·359	139·3
800	$1·99 \times 10^{-6}$	$1·03 \times 10^{-6}$	17·389	139·4
803	$2·34 \times 10^{-6}$	$1·19 \times 10^{-6}$	17·186	139·1
806	$2·57 \times 10^{-6}$	$1·32 \times 10^{-6}$	17·060	139·1
809	$2·88 \times 10^{-6}$	$1·43 \times 10^{-6}$	16·926	139·1
810	$3·02 \times 10^{-6}$	$1·55 \times 10^{-6}$	16·850	139·1
814	$3·25 \times 10^{-6}$	$1·69 \times 10^{-6}$	16·748	139·3
818	$3·78 \times 10^{-6}$	$1·94 \times 10^{-6}$	16·557	139·1
821	$4·29 \times 10^{-6}$	$2·16 \times 10^{-6}$	16·401	139·2
825	$4·80 \times 10^{-6}$	$2·48 \times 10^{-6}$	16·243	139·1
825	$4·88 \times 10^{-6}$	$2·48 \times 10^{-6}$	16·229	139·2
831	$6·02 \times 10^{-6}$	$3·12 \times 10^{-6}$	15·947	138·9
831	$6·04 \times 10^{-6}$	$3·13 \times 10^{-6}$	15·942	138·9
833	$6·27 \times 10^{-6}$	$3·19 \times 10^{-6}$	15·902	139·1
837	$7·06 \times 10^{-6}$	$3·60 \times 10^{-6}$	15·746	139·1
841	$8·34 \times 10^{-6}$	$4·10 \times 10^{-6}$	15·545	138·9
842	$8·12 \times 10^{-6}$	$4·28 \times 10^{-6}$	15·549	139·1
845	$9·06 \times 10^{-6}$	$4·61 \times 10^{-6}$	15·422	139·2
847	$9·62 \times 10^{-6}$	$4·83 \times 10^{-6}$	15·350	139·2
853	$1·14 \times 10^{-5}$	$5·64 \times 10^{-6}$	15·135	139·2
855	$1·25 \times 10^{-5}$	$6·20 \times 10^{-6}$	15·014	139·1
867	$1·73 \times 10^{-5}$	$8·63 \times 10^{-6}$	14·588	139·1
879	$2·33 \times 10^{-5}$	$1·16 \times 10^{-5}$	14·201	139·4

scale was easily flaked off leaving a clean surface of the base sample material. During the vaporization only Hg and Te_2 were detected to any significant amounts. Microprobe analysis of the scale indicated that it was primarily CdTe, confirming the depletion of Hg and Te from the compound. Further experiments below 300°C could not be made since Hg is a background gas in the mass spectrometer.

REFERENCES

1. deNobel, D., *Philips Res. Repts.*, 1959, **14**, 361–99.
2. Lorenz, M. R., *J. Phys. Chem. Solids*, 1962, **23**, 939.
3. Brebrick, R. F. and Strauss, A. J., *J. Phys. Chem. Solids*, 1964, **25**, 1441–5.
4. Korneeva, I. V., Belyaev, A. V. and Novoselova, A. V., *Russ. Jour. Inorg. Chem.*, 1960, **5**, 1–3.
5. Drowart, J. and Goldfinger, P., *J. Chim. Phys.*, 1958, **55**, 721–32.
6. Goldfinger, P. and Jeunehomme, M., *Trans. Farad. Soc.*, 1963, **59**, 2851–67.
7. DeMaria, G., Goldfinger, P., Malaspina, L. and Piacente, V., AFML TR-64-331, WPAFB, Ohio (1965).
8. Ivanov, Yu. M. and Vanyukov, A. V., *Pribory i Tekhnika Eksperimenta*, 1968, No. 3, 145–7.
9. Grimley, R. T., 'Mass Spectrometry', *in* 'The Characterization of High Temperature Vapors', John Wiley and Sons, Inc., New York, 1967.
10. Inghram, M. G., Hayden, R. and Hess, D., *Nat'l Bur. Std. (U.S.), Circ.*, 1951, **522**.
11. Inghram, M. G. and Hayden, R., Nuclear Science Series Report 14, 1954.
12. Akishin, P. A., *Usp. Fiz. Nauk.*, 1958, **66**, 331.
13. Gingerich, K. A., 'Mass Spectrometry in Inorganic Chemistry', Advances in Chemistry Series 72, ACS, Washington, D.C., 1968, pp. 291–9.
14. Lin, J., Honeywell Research Center, Minneapolis, Minnesota, private communication, Contract F33615-72-C-1612.
15. Franklin, J. L., Dillard, J. G., Rosenstock, H. M., Herron, J. T., Draxl, K. and Field, F. H., 'Ionization Potentials, Appearance Potentials, and Heats of Formation of Gaseous Positive Ions', NSRDS-NBS 26, U.S. G.P.O., Washington, D.C., June 1969, pp. 223–4.
16. Honig, R. E., *J. Chem. Phys.*, 1954, **22**, 126.
17. Grimley, R. T. and Joyce, T. E., *J. Phys. Chem.*, 1969, **73**, 3047.
18. Stull, D. R. and Sinke, G. C., 'Thermodynamic Properties of the Elements', Advances in Chemistry Series, No. 18, Amer. Chem. Soc., 1956.
19. Otvos, J. W. and Stevenson, D. P., *J. Amer. Chem. Soc.*, 1956, **78**, 546.

70

Mass Spectrometric Study of the Vaporization of Perrhenates

By K. SKUDLARSKI and W. LUKAS

(*Institute of Inorganic Chemistry and Metallurgy of Rare Elements,
Technical University of Wroclaw, Poland*)

INTRODUCTION

VAPORIZATION processes and the mass spectrometric investigation of oxy-acids salts are described in some reviews.[1-3] Recently papers have been published on the vaporization of thallium sulphate,[4] sodium sulphate,[5] rubidium and caesium sulphates,[6] copper (II), mercury (II), thallium (I), rubidium and caesium nitrates,[7] and perrhenates of magnesium and barium.[8]

The present paper presents a mass spectrometric study on the vaporization or sublimation of perrhenates of group IA, IB elements and thallium (I). Preliminary results on the vaporization of zinc and mercury (I) perrhenates are also given.

The presented investigations were undertaken under J. Drowart's direction with G. Exsteen and A. Vander Auwera Mahieu collaborating (Laboratoire de Chimie Physique Moléculaire, Université Libre de Bruxelles, Belgique). The vapour pressures of sodium and potassium perrhenates and the enthalpies of vaporization of sodium perrhenate and the sublimation of potassium perrhenate have been determined.[9] The investigations concerned with lithium, rubidium, caesium, copper (I), silver (I), thallium (I), zinc (II) and mercury (I) perrhenates have been continued at the Technical University of Wroclaw, Poland. The mass spectra of sodium and potassium perrhenates were also measured.

EXPERIMENTAL

Materials

$LiReO_4$ and $NaReO_4$ were obtained from ammonium perrhenate by ion exchange. $KReO_4$ was a commercial product: Purissimus, of Fulka AG, Buchs SG, Switzerland. $RbReO_4$ and $CsReO_4$ were obtained by the neutralization of hydroxides with perrhenic acid. $AgReO_4$ and $TlReO_4$ were obtained by precipitation from silver and thallium nitrates solutions with ammonium perrhenate. $CuReO_4$ was obtained by the reduction of $Cu(ReO_4)_2$ with metallic copper in a vacuum. The intermediate product $Cu(ReO_4)_2$ was obtained from NH_4ReO_4 by ion exchange. $Zn(ReO_4)_2$ was obtained by dissolution of metallic zinc in $HReO_4$, H_2O_2 solution. $HgReO_4$ was obtained by thermal decomposition of $Hg_2O . Hg_2(ReO_4)_2$, and $Hg_2O . Hg_2(ReO_4)_2$ from $HgNO_3$ water solution by precipitation with $HReO_4$. All these substrates were dried under vacuum and stored in a water-free atmosphere.

Apparatus

The experiments were carried out in a 200 mm radius curvature, 60° sector, single focussing mass spectrometer type MI-1305 specially adapted for thermodynamic investigations at high temperatures.[10] It was equipped with an oven assembly with single or double[11] effusion cells, and with an electron multiplier SI-2. The Knudsen effusion cells made of silica glass were inserted in a tantalum container and heated in a pyrophylite oven with a tungsten spiral. The temperatures were measured with three or four thermocouples clamped to tantalum Pt/PtRh.

The accelerating voltage was 2000, 1000 or 600 V depending on the measured mass spectra range.

RESULTS AND DISCUSSION

The mass spectra of the examined perrhenates are presented in Table I. The highest are the intensities of M^+ ions, the intensities of all other ions being considerably lower.

The mass spectra of alkali metal perrhenates show that the electron impact decomposition increases according to the atomic weight of the alkali element. The same observations were made for alkali metal halogenides,[12-14] borates[15] and hydroxides.[16] In order to determine the origin of all observed ions, the ionization efficiency curve and the approximate appearance potentials of all the ions have been estimated. The potentials of the ion formation increased according to the decrease in the oxygen atom number in the $MReO_x^+$ ions ($x = 0$ to $x = 4$). The intensities of different ions were highly influenced by the deflecting potential of electrodes situated close behind the ion source exit slit. The most deflected were the MO^+ and MRe^+ ions, the less deflected $MReO_4^+$ ions. The above-mentioned examination and the comparison with the ions formed from alkali metal halides,[14,17,18] hydroxides[22-24] and borates[15,20,21] entitle us to conclude that the M^+, $MReO_x^+$, MO^+ are formed from monomer and dimer molecules, and the $M_2ReO_4^+$ ions from dimer molecules.

As can be seen from Table I, the changes of ion intensities with increasing atomic weight of alkali elements are different for the particular ion form. The intensities of $MReO_3^+$, $MReO_2^+$ ions decrease just as the $MReO_4^+$ ion intensities decrease. The diminishment of $MReO^+$ ion intensity is less pronounced and the MRe^+ intensity increases considerably as compared with $MReO_4^+$ intensity. The $Re_2O_7^+$ ions observed in copper (I), silver (I), thallium (I), mercury (I) and zinc (II) perrhenate vapour mass spectra are parent ions. They originate from the Re_2O_7 which is the product of the thermal decomposition of perrhenate. The fragment ions $Re_2O_x^+$ ($x = 4$ to $x = 6$), ReO_y^+ ($y = 0$ to $y = 3$), also originating from Re_2O_7, can be

TABLE I

The Mass Spectra of Perrhenates. Ionizing Electron Energy 4·8 MJ/mol (50 eV)

(1)	(2) $LiReO_4$ 920°K	(3) $NaReO_4$ 833°K	(4) $KReO_4$ 833°K	(5) $RbReO_4$ 820°K	(6) $CsReO_4$ 840°K	(7) $CuReO_4$ 800°K	(8) $AgReO_4$ 900°K	(9) $TlReO_4$ 680°K	(10) $HgReO_4$ 640°K	(11) $Zn(ReO_4)_2$ 770°K
M^+	100	100	100	100	100	100	100	100	100	100
MO^+	<1	0·4	0·6	0·2	0·5	1·1	—	0·45	—	—
MRe^+	0·04	0·09	0·06	0·12	0·12	a	0·5	0·43	—	—
$MReO^+$	0·1	0·05	0·02	0·02	0·018	0·2	0·7	0·11	—	—
$MReO_2^+$	0·1	0·026	0·013	0·01	0·006	0·5	0·4	0·05	—	—
$MReO_3^+$	0·6	0·33	0·07	0·05	0·03	0·7	2	0·10	—	2
$MReO_4^+$	3·0	1·4	0·27	0·24	0·10	0·5	17	0·8	—	—
M_2^+	—	—	—	—	—	3·3	—	—	—	—
M_2O^+	—	—	—	—	—	7·5	—	—	—	—
$M_2ReO_2^+$	—	—	—	—	—	0·14	—	—	—	—
$M_2ReO_3^+$	—	—	—	—	—	0·4	—	—	—	—
$M_2ReO_4^+$	10	1·90	0·47	0·31	0·25	45	18	0·02	20	20
$M_2(ReO_4)_2^+$	—	—	—	—	—	12	—	—	—	25
$M(ReO_4)_2^+$	—	—	—	—	—	0·04	—	—	—	5
$M_3(ReO_4)_2^+$	—	—	—	—	—	0·02	—	—	—	—
Re^+	0·1	—	—	—	a	b	b	0·5	b	b
ReO^+	0·3	0·07	0·06	0·04	a	b	b	a	b	b
ReO_2^+	0·7	a	0·13	0·06	0·09	b	b	0·5	b	b
ReO_4^+	0·4	0·48	0·29	0·13	a	b	b	a	b	b
ReO_3^+	0·5	0·19	0·15	0·05	0·06	b	b	0·1	b	b
$Re_2O_7^+$	~0·1	1·0	0·24	0·08	—	380	500	—	100	1 300
O_2^+	—	—	—	—	—	—	200	—	—	—

a Not measured.

b The intensities of ions ReO_x^+ ($x = 0$–4) and $Re_2O_y^+$ ($y = 4$–6) coming from ionization of Re_2O_7 are not given.

observed in the previously mentioned perrhenate mass spectra as well. Their intensities were not given in Table I since they are proportional to the $Re_2O_7^+$ ion intensities and characteristic of the Re_2O_7 spectrum.

The mass spectrum of copper (I) perrhenate differs considerably from the alkali metal perrhenates mass spectra. It shows a high intensity of parent dimer $(CuReO_4)_2^+$ ions, which are not found in other perrhenate spectra. The examination of the preheated copper perrhenate vapour mass spectra proves that the vapour phase is mainly dimeric. The existence of parent dimer ions suggests that the binding of Cu atoms in a copper perrhenate dimer molecule is covalent, contrary to alkali metal perrhenate molecules showing the mainly ionic character of M–O bond. There is a considerable resemblance between copper perrhenate bond fragmentation and that of copper salt molecules, and between alkali metal perrhenate fragmentation and that of alkali metal salts.

The mass spectrum of silver perrhenate molecules differs considerably from the copper perrhenate spectrum. The dimer perrhenate parent ions were not observed. The spectrum is comparable to the lithium perrhenate spectrum, but the number of fragment ions is considerably smaller.

The mass spectrum of thallium (I) perrhenate is very similar to the mass spectra of alkali metal perrhenates. It suggests that the degree of ionicity of the bond M–O is much the same.

The preliminary examination of zinc perrhenate indicates a high degree of thermal decomposition of $Zn(ReO_4)_2$. Its mass spectrum resembles the mass spectrum of copper (I) perrhenate. The ions of the masses higher than $(ZnReO_4)_2$ were not observed, but their presence cannot be excluded. The vapour composition of zinc perrhenate has not been determined yet, but the possible presence of the gaseous zinc (I) perrhenate must be taken into consideration.

The preliminary examinations of mercury (I) perrhenate also show a considerable thermal decomposition but it is lower than that of zinc perrhenate. Only the ions coming from dimer or larger molecules appear in the mass spectra, as is the case with copper (I) perrhenate, but the existence of the parent $(HgReO_4)_2^+$ ions could not be verified because their mass is beyond the mass spectrometer measuring range.

The fragmentation pattern of dimer and monomer molecules was calculated using the method described by Gorokhov.[25] It consists of the measurement of the mass spectra of vapours preheated in a double effusion cell. Great care was taken to ensure a constant temperature in the effusing cell during the heating of the lower part of the cell. A special calibration procedure was undertaken. Due to the very high intensity of the M^+ ions in the mass spectra it was necessary to calculate which part of these ions originated from the dimer, and which from monomer molecules. It was determined that at temperatures 920°K 30% of Li^+ ions, at 850°K nearly 10% of Rb^+ and Cs^+ ions and at 762°K 15% of Tl^+ originate from the dimer, and the rest come from monomeric molecules. The creeping of silver perrhenate on the Knudsen cell walls made this type of measurement impossible.

The relation of the effective ionization cross-sections of dimer to monomer molecules was determined for $CsReO_4$ and $KReO_4$, using the double Knudsen cell method described by Gorokhov,[26] and Berkowitz.[13]

The measurements of the partial vapour pressures of perrhenates were performed by the Knudsen integral method. The results are given in the form of equations:

$$\log \frac{p}{\text{Pa}} = -\frac{A}{T} + B$$

The constants A and B are given in Table II. In the temperature range of our measurements, the partial vapour pressures of the dimeric forms are near the monomer vapour pressure for lithium, sodium and potassium perrhenates. The pressures of the monomeric forms of rubidium, caesium and thallium perrhenates are 1–2 orders higher than those pressures for the dimeric form.

TABLE II

Vapour Pressure of Perrhenates. Coefficients A and B in Equation:

$$\log \frac{p}{\text{Pa}} = -\frac{A}{T} + B$$

Compound	Temperature range $(T/^\circ K)$	Monomer		Dimer	
		A	B	A	B
$LiReO_4$	750–950	10 350	11·42	10 520	11·61
$NaReO_4$	660–780	9 930	11·60	12 400	15·84
$KReO_4$	685–815	10 030	11·39	14 350	16·99
$RbReO_4$	710–863	9 750	11·33	13 450	14·79
$CsReO_4$	750–850	9 453	11·67	13 440	14·79
$CuReO_4$	650–780	—	—	9 613	11·29
$TlReO_4$	550–700	8 934	12·78	11 910	15·07

Worth noting is the fact that perrhenates of lighter alkali metals, sodium and lithium, which have lower melting points show vapour pressures of the same order, or even higher than the heavier alkali metal perrhenates melting at considerably higher temperatures.

The concentration of the dimeric forms increase from caesium to lithium. The direction of this change is the same as for alkali metal borates,[15] and hydroxides.[27]

The thermal decomposition of the alkali metal perrhenates could be observed to a small extent only for $LiReO_4$. This is in accordance with the decreasing stability of alkali metal carbonates[28] and sulphates[5,29] according to the atomic weight of alkali elements.

It should be emphasized that the thermal stability of gaseous molecules increases according to the increase in the atomic mass of alkali elements, and contrary to the decrease in their stability for electron impact decomposition. This suggests that the ionic bonds, being more susceptible to electron bombardment decomposition are responsible for the stability of gaseous alkali metal salt molecules.

The enthalpies of vaporization or sublimation were calculated for dimers from the temperature dependence of the $M_2ReO_4{}^+$ ions, which originate

TABLE III

Enthalpy of Vaporization or Sublimation of Perrhenates Calculated from II Law and Enthalpy of Dissociation of Dimers
(Uncertainty: ± 25 kJ/mol)

	$LiReO_4$	$NaReO_4$	$KReO_4$	$RbReO_4$	$CsReO_4$	$CuReO_4$	$AgReO_4$	$TlReO_4$
	ΔH° (850°K, vap.)	ΔH° (750°K, vap.)	ΔH° (750°K, sub.)	ΔH° (780°K, sub.)	ΔH° (775°K, sub.)	ΔH° (700°K, vap.)	ΔH° (810°K, vap.)	ΔH° (630°K, sub.)
	kJ/mol	kJ/mol	kJ/mol	kJ/mol	kJ/mol	kJ/mol	kJ/mol	kJ/mol
$MReO_4(S,L) = MReO_4(g)$	198	190	192	187	183	—	220	171
$2MReO_4(S,L) = M_2(ReO_4)_2(g)$	201	237	274	257	257	184	225	228
	ΔH° (850°K, dis.)	ΔH° (750°K, dis.)	ΔH° (750°K, dis.)	ΔH° (780°K, dis.)	ΔH° (775°K, dis.)	ΔH° (700°K, dis.)	ΔH° (810°K, dis.)	ΔH° (630°K, dis.)
	kJ/mol	kJ/mol	kJ/mol	kJ/mol	kJ/mol	kJ/mol	kJ/mol	kJ/mol
$M_2(ReO_4)_2(g) = 2MReO_4(g)$	195	142	109	116	109	—	215	114

from dimeric molecules only. The enthalpies of sublimation or vaporization for the monomeric form were also calculated and are presented in Table III. In order to enable the mutual comparison of sublimation enthalpies for alkali metal perrhenates, the sodium and lithium perrhenates enthalpies of melting were estimated, based on the melting points for other salts of these metals. For $LiReO_4$ 25 kJ/mol and for $NaReO_4$ 13 kJ/mol have been accepted.[30]

It can be seen that the sublimation enthalpy of the monomeric form of perrhenates decreases gradually according to the decrease of the atomic weight from Cs to Na and for Li the sublimation enthalpy is nearly the same as for Na. The sublimation enthalpy of the dimeric form has a maximum value for sodium perrhenate. The lower value of sublimation enthalpy, as in the case of lithium perrhenate dimer was observed for lithium borate[21] and halogenides[30] as well.

The dimer-to-monomer decomposition enthalpies increase from caesium to lithium perrhenates just like those for borates[21] and halogenides.[31-34]

The highest dimer dissociation enthalpy was observed for silver perrhenate. The decomposition enthalpy of the thallium perrhenate dimer is nearly equal to that of rubidium perrhenate.

REFERENCES

1. Büchler, A. and Berkowitz-Mattuck, J. B., 'Gaseous Ternary Compounds of the Alkali Metals' *in* 'Advances in High Temperature Chemistry', Academic Press, New York, 1967, p. 95.
2. Semenov, G. A., 'Probl. sovrem. khim. koordinac. soedin.', Leningrad Univ., 1970, 3, 16.
3. Gorokhov, L. N. and Semenov, G. A., *Adv. Mass Spectr.*, 1971, 5, 349.
4. Cubicciotti, D., *High Temp. Sci.*, 1970, 2, 389.
5. Cubicciotti, D. and Kaneshea, F. J., *High. Temp. Sci.*, 1971, 4, 32.
6. Cubicciotti, D., *High Temp., Sci.*, 1971, 3, 349.
7. Lander, T., Towler, C. and Wuth, T., *Adv. Mass Spectr.*, 1971, 5, 379.
8. Semenov, G. A., Nikolaev, E. N. and Opendak, I. G., *Zh. Neorg. Khim.*, 1972, 17, 1819.
9. Skudlarski, K., Drowart, J., Exteen, G. and Van der Auvera-Mahieu, A., *Trans. Faraday Soc.*, 1967, 63, 1146.
10. Skudlarski, K., *Pribory i Tekhn. Eksperim.*, 1971, No. 2, 268.
11. Lukas, W., to be published.
12. Berkowitz, J. and Chupka, W. A., *J. Chem. Phys.*, 1958, 29, 653.
13. Berkowitz, J., Tasman, H. A. and Chupka, W. A., *J. Chem. Phys.*, 1962, 36, 2170.
14. Schoonmaker, R. C. and Porter, R. F., *J. Chem. Phys.*, 1959, 30, 283.
15. Makarov, A. V. and Nikitin, O. T., *Teplofiz. Vysokikh Temperatur Akad. Nauk S.S.S.R.*, 1971, 9, 1073.
16. Schoonmaker, R. C. and Porter, R. F., *J. Chem. Phys.*, 1959, 30, 830.
17. Friedman, L., *J. Chem. Phys.*, 1955, 23, 477.
18. Milne, T. A. and Klein, H. M., *J. Chem. Phys.*, 1960, 33, 1628.
19. Porter, R. F. and Schoonmaker, R. C., *J. Chem. Phys.*, 1958, 29, 1970.
20. Büchler, A. and Berkowitz-Mattuck, J. B., *J. Chem. Phys.*, 1963, 39, 286.
21. Gorokhov, L. N., Gusarov, A. V. Makarov, A. V., and Nikitin, O. T., *Teplofiz. Vysokikh Temperatur Akad. Nauk S.S.S.R.*, 1971, 9, 1173.
22. Gorokhov, L. N. and Gusarov, A. V., *Zh. Fiz., Khim.*, 1970, 44, 269.
23. Gusorov, A. V. and Gorokhov, L. N., *Zh. Fiz. Khim.*, 1968, 42, 860.
24. Schoonmaker, R. C. and Porter, R. F., *J. Chem. Phys.*, 1958, 28, 454.
25. Gorokhov, L. N., *Vestn. Mosk. Univ. Ser. Mat. Fiz. Khim.*, 1958, No. 6, 231.

26. Akishin, P. A., Gorokhov, L. N. and Sidorov, L. N., *Vestn. Mosk. Univ.*, *Ser. Mat. Fiz. Khim.*, 1959, No. 6, 194.
27. Glemser, O. and Wendlandt, H. G., *Adv. in Inorg. Chem. and Radiochem.*, Vol. 5, p. 215–58.
28. Gusarov, A. V., Gorokhov, L. N. and Efimova, A. G., *Teplofiz. Vysokikh Temperatur Akad. Nauk S.S.S.R.*, 1967, **5**, 783.
29. Ficalora, P. J., Uy, O. M., Muenow, D. W. and Morgrawe, J. L., *J. Amer. Ceram. Soc.*, 1968, **51**, 574.
30. Bauer, S. H. and Porter, R. F., 'Metal Halide Vapors: Structures and Thermochemistry', *in* 'Molten Salt Chemistry', Interscience Pub., J. Wiley, New York, 1964.
31. Eisenstadt, M., Rothberg, C. M. and Kush, P., *J. Chem. Phys.*, 1958, **29**, 797.
32. Margulescu, I. G. and Topor, L., *Rev. Roum. Chem.*, 1970, **15**, 997.
33. Rudakoff, G., *Z. Chem.*, 1965, **5**, 292.
34. Polyachenok, O., *Zh. Fiz. Khim.*, 1970, **44**, 2414.

71
The Direct Determination of Thermodynamic Activity of Ni in Iron-Rich Fe–Ni and Fe–Cr–Ni Alloys

By G. G. CAMERESI and G. MAZZEI

(*Centro Sperimentale Metallurgico, Rome, Italy*)

INTRODUCTION

SEVERAL investigations have been published to date concerning the thermodynamic activity of the metal components in alloys. Determinations for Fe–Ni melts have been made by several different methods: transpiration;[1] Knudsen cell effusion with analysis of the condensate,[2] and Knudsen cell effusion with measurements of relative component intensities by mass spectrometric methods.[3-6]

In this work the multiple Knudsen cell technique[7,8,9] in which the effusing vapours from each cell are alternately analysed in the mass spectrometer has been used. In this technique one cell contains the alloy sample and the others pure metals as standard state. The direct determination of the activity by the ratio of the ion currents permitted an investigation of the iron-rich Fe–Ni liquid alloys.

Measurements made indicate that the solute activities are slightly greater than ideal below 2·5% by weight and the activity coefficient of Ni in dilute solution in liquid iron at 1570°C, $\gamma°_{Ni}$, is 1·82. Experimental activity data in the higher concentrations for Ni are in good agreement with published values.

EXPERIMENTAL

This work was undertaken to determine the activity of Ni at low concentration in binary Fe–Ni and in ternary Fe–Cr–Ni melts by a direct method previously described.[8-9] The Knudsen cell assembly is shown in Fig. 1 and was developed in the laboratory where its performance was severely tested. It consists of a movable plate supporting a tantalum shield case thermally and electrically insulated by vitreous silica spacers.

Inside the case is placed a refractory metal block surrounded by two rounded tungsten ribbons. These are spot welded onto two tungsten holders fitted on the plate and insulated by ceramic spacers.

The metal block inside the case is supported by a tungsten tripod; it was

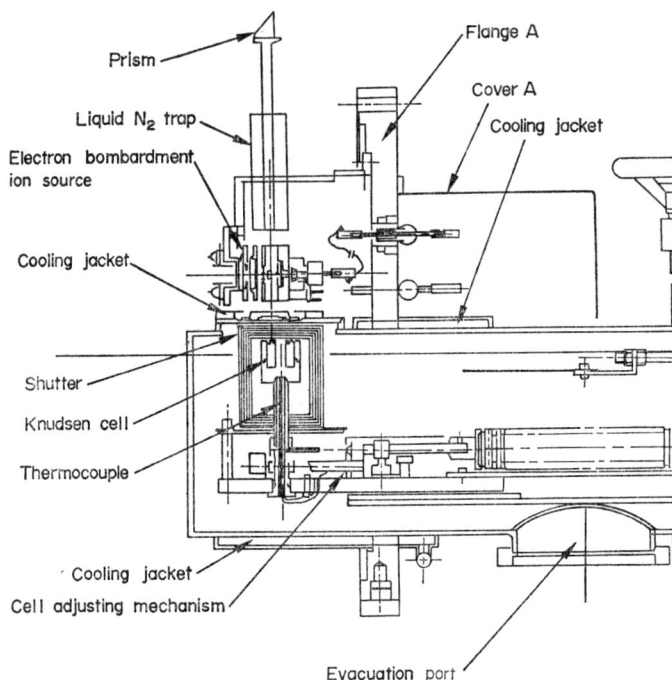

FIG. 1   Side view of Knudsen cell assembly.

designed as a container for two or three Knudsen cells so as to allow the best isothermal conditions.

The material used for the Knudsen cells (15 mm height and 6 mm o.d.) was pure Al_2O_3 at zero porosity and pure ThO_2; the sample diffusion for both materials was negligible up to 1570°C. The lids were machined with an ultrasonic drill to obtain knife-edged holes and the best Knudsen conditions. A further examination using optical systems permitted measurement of the hole diameters at ± 0.01 mm. The effective areas of effusion were calibrated by comparing the ion current of effusing silver vapours at constant temperature. Translational and rotational movements of the isothermal block were performed by means of two external knobs. The rotation angle was more than 270°, so it was possible to align three cells at 120° to each other with the shutter slit and ionizing chamber. A fine alignment of the effusion holes with the ionization region could be obtained by combining translational and rotational movements. Rough adjustments were carried out looking into a prism and an optical pyrometer through a window placed at the top of the ionization region.

The radiation heating up to 1300°C was sufficient; higher temperature, over 2000°C, could be reached by electron bombardment of the refractory metal block connected to ground. Bombarding ribbons and shield case were maintained at negative potential up to 1500 V.

A liquid-nitrogen trap immediately above the ion source lowered the pressure in this region to less than 1×10^{-7} torr (1.33×10^{-5} N/m^2) and stopped effusing vapours.

The temperature was measured by a W–Re 5%/W–Re 25% thermocouple fitted at the bottom of the isothermal block. Every thermocouple used in the runs was calibrated at the melting points of pure Au, Ni and Fe: the resulting accuracy was ±2°C. The response time of the assembly was determined by comparing the observed temperature with the recorded signals for such elements continuously changing by increasing and decreasing the temperature. The greatest difference in the observed temperature for the same ion current intensity was less than 3°C at a temperature variation of 5°C per minute.

Automatic variation of the temperature was achieved by a pilot device. A PID system provided a fine control of the temperature stability better than ±2°C up to 1700°C. The mass spectrometer used was a Mattauch–Herzog type double focussing instrument by Jeol (Japan Electronic Optics Laboratories) mod.01UB equipped with Knudsen accessory units or with spark source for analytical purposes.[10] The electron bombardment ion source was a classical type open structure. The minimum resolution fixed in the experiments was 3000. The lowest partial pressure observed was less than 1×10^{-8} torr ($1 \cdot 33 \times 10^{-6}$ N/m^2).

Electrical and photoplate detection systems were available. Alignment of the cells was previously performed by electrical detection when photoplates were used in the experiments.

The alloy samples, average weight approximately 0·400 g, were prepared in two ways: by the direct melting in Knudsen cells of weighed pure components or by preliminary melting of larger quantities of the same components in pure Al$_2$O$_3$ crucibles in a furnace heated by a tungsten wire. In the

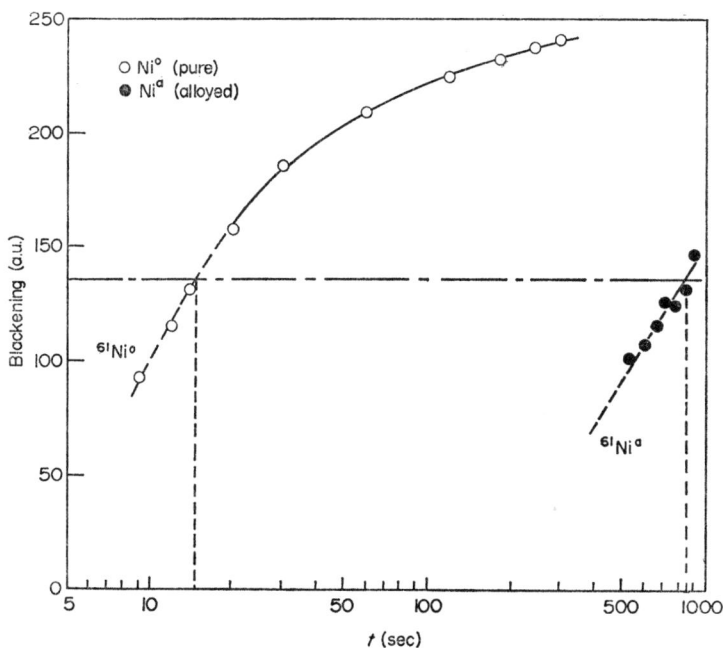

FIG. 2   Variation of blackening with the exposure times for ^{61}Ni related to the alloy sample and pure Ni.

G. G. CAMERESI AND G. MAZZEI

latter case the atmosphere was pure argon at first, then the samples were remelted under vacuum. All these operations were standardized and the alloy ingots were cut in two or more pieces and examined to control homogeneity and possible contamination.

Two ways were followed to determine thermodynamic activity of Ni: (a) by the ratio of the ion currents electrically detected for alloyed and pure Ni respectively; (b) by the ratio of the exposure times related to the same isotope for alloyed and pure Ni at the same blackening value of the photoplate line. This last way is based on the reciprocity law which holds for Ilford Q_2 plates used in the runs. So, the same blackening at constant temperature corresponds to the same charge for alloyed and pure Ni in the linear range of the sensitivity curves. These were obtained by plotting blackening versus the exposure times as shown in Fig. 2: therefore the ratio of the exposure times related to the same isotope of pure Ni on alloyed Ni is the thermodynamic activity:

$$Q_{Ni^\circ} = Q_{Ni}a = I^+_{Ni}a \cdot t_{Ni}a = I^+_{Ni^\circ} \cdot t_{Ni^\circ} \qquad \frac{I^+_{Ni}a}{I^+_{Ni^\circ}} = \frac{t_{Ni^\circ}}{t_{Ni}a} = a_{Ni}$$

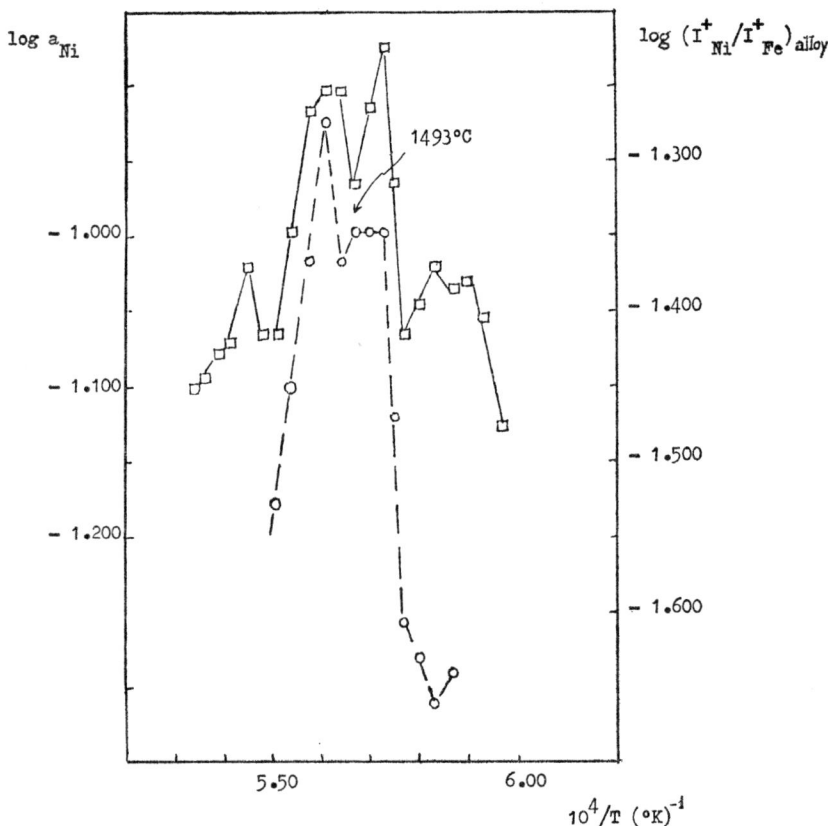

FIG. 3   Variation of $\log a_{Ni}$ (□) and $\log (I^+_{Ni}/I^+_{Fe})$ (○) with $1/T$ in °K^{-1} at the composition $N_{Ni} = 0{\cdot}115$ for Fe–Ni alloy. The plots were obtained by continuously decreasing the temperature. The temperature of 1493°C yields the liquidus point at this composition.

Photoplate detection presents the following advantages: evaluation of an averaged signal derived from a charge integration system; opportunity to detect and identify all atomic or molecular species in equilibrium with the condensed phase; high sensitivity, not influenced by noise or eventual trouble of SEM or any electrical device; the possibility of controlling the eventual change of the alloy composition in comparing the spectra related to the alloy sample, at the same exposure time, before and after the run. Accuracy can be increased by placing the spectra as close as possible. In such a way every typical influence of the various parameters involved in photoplate detection may be considerably reduced.

Some experiments were carried out with Fe–Ni alloy samples by slowly decreasing the temperature and continuously recording ion currents related to the effusing vapours. In this case the Ni activity was calculated by the ratio of the ion currents of alloyed and pure Ni measured by a successive alignment of the two Knudsen cells.

The plots of log a_{Ni} and log $(I^+{}_{Ni}/I^+{}_{Fe})$ versus I/T shown in Figs. 3 and 4 appear strictly correlated. Generally speaking, this way of treating the

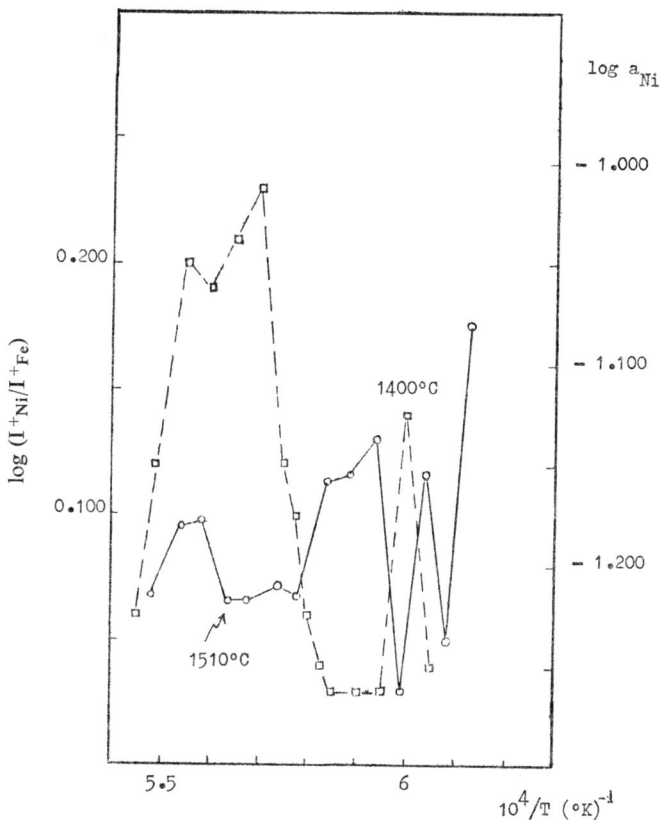

FIG. 4   Variation of log a_{Ni} (□) and log $(I^+{}_{Ni}/I^+{}_{Fe})$ (○) with $1/T$ in °K^{-1} at the composition $N_{Ni} = 0.041$ for Fe–Ni alloy. The temperature of 1510°C yields the liquidus point at this composition. The points at 1400°C may represent the transition point $(T_{\delta-\gamma})$ for iron.

experimental data seems to offer unsuspected chances of getting sophisticated information on the equilibrium diagrams of the system to be investigated. Solidification points determined in correspondence with the slopes change[11] agree very well with published values. Of course, the performance of the whole system and the control of temperature must be guaranteed. Moreover, the experimental data have to be as great as possible to ensure a statistical validity to the observed slopes.

RESULTS

Thermodynamic activity data of Ni for binary Fe–Ni and ternary Fe–Cr–Ni alloys are presented in Table I. Within the experimental error they are self-consistent and in good agreement with reported values in published works at Ni concentrations greater than 10% by weight. The evaluable accuracy is better than 5%.

TABLE I
Experimental Activities and Activity Coefficients of Ni in Fe–Ni and in Fe–Cr–Ni Alloys at 1 560°C

Mole fraction of Ni	Mole fraction of Cr	Activity of Ni	Activity coefficient of Ni
0·012	—	0·017	1·430
0·022	—	0·026	1·162
0·041	—	0·033	0·797
0·057	—	0·034	0·593
0·115	—	0·050	0·440
0·163	—	0·062	0·380
0·326	—	0·172	0·520
0·502	—	0·330	0·658
0·093	0·050	0·050	0·537
0·025	0·050	0·030	1·200

Activities were calculated by elaborating the experimental data obtained by the use of the electrical and the photoplate detection at a constant temperature of 1560°C. In the experiments carried out the maximum working temperature was limited to 1570°C to reduce diffusion phenomena and contamination by decomposition of the crucible material.

Each activity value represents the average of at least two independent determinations. Figure 5 shows the experimental data related to the Fe–Ni alloy as two straight line plots of $\log \gamma_{Ni}$ versus N_{Fe}^2. The lines have the following equations:

$$\log \gamma_{Ni} = 4\cdot391 N_{Fe}^2 - 4\cdot131 \tag{1}$$
$$\log \gamma_{Ni} = 0\cdot529 N_{Fe}^2 - 0\cdot047 \tag{2}$$

By using eqn (1), valid for the iron-rich compositions, the activity coefficient of Ni at infinite dilution, γ°_{Ni}, gives a result of 1·82. Thus, at $N_{Ni} < 3\%$ by weight the solute activity is slightly greater than ideal. In the range of

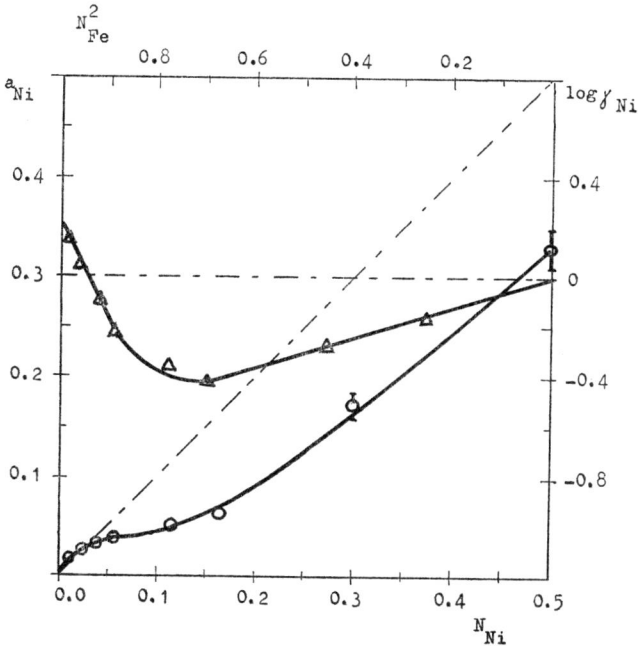

FIG. 5   Variation of a_{Ni} (O) with N_{Ni} and of log γ_{Ni} (Δ) with N_{Fe}^2 in binary Fe–Ni alloy.

higher concentrations of nickel a negative deviation from ideality is observed. It would be interesting to confirm by some other technique the behaviour shown with the iron-rich composition where there is a lack of previous experimental results. The activity of Ni up to $N_{Ni} = 0\cdot10$ in the ternary system Fe–Cr–Ni at $N_{Cr} = 0\cdot05$ appears not to be influenced by the presence of Cr as shown from the experimental data of Table I. Possible small differences for the Ni activity in Fe–Ni and in Fe–Cr–Ni may be included into the experimental error.

REFERENCES

1. Zellars, G. R., Payne, S. L., Morris, J. P. and Kipp, R. L., *Trans A.I.M.E.*, 1959, **215**, 181–5.
2. Speiser, R., Jacobs, A. and Spretnak, J. W., *Trans. A.I.M.E.*, 1959, **215**, 185–92.
3. Belton, G. R. and Fruehan, R. J., *J. Phys. Chem.*, 1967, **71**, 5, 1403–9.
4. Belton, G. R. and Fruehan, R. J., *Trans. A.I.M.E.*, 1969, **245**, 113–17.
5. Belton, G. R. and Fruehan, R. J., *Met. Trans.*, 1970, **1**, 781–7.
6. Neckel, A. and Wagner, S., *Ber. Bunsenges. Physik Chem.*, 1969, **73**, 210–17.
7. Büchler, A. and Stauffer, J. L., 'Symposium on Thermodynamics—Proceedings Series', Vol. I IAEA, Vienna 1966.
8. Cameresi, G. G., De Maria, G., Gigli, R. and Piacente, V., *Ric. Sci.*, 1967, **37**, 1092–7.
9. De Maria, G., Piacente, V., Cameresi, G. G. and Gigli, R., 'Proc. of X Congresso Nazionale della Soc. Chimica Italiana', 1968.
10. Watanabe, E. and Naito, M., 'Proc. of the Int. Conference on Mass Spectr.', Sept. 1969, p. 249.
11. Gokcen, N. A., Chang, E. T. and Marx, P. C., 'Proc. of Third Int. Symp. on High Temperature Technology', U.S.A. 1967.

72

Mass Spectrometric Investigations for Geological Age Determinations by the Potassium/Calcium Method

By K. G. HEUMANN and E. KUBASSEK

(Fachbereich Anorganische Chemie und Kernchemie,
Technische Hochschule, Darmstadt, Germany)

INTRODUCTION

GEOLOGICAL age determinations can be realized by the analytical determination of radiogenic produced isotopes. The best known methods in this field are the uranium/lead, rubidium/strontium and potassium/argon methods. In all these cases isotope ratio measurements with a high precision are indispensable. Therefore mass spectrometry has an important analytical application in this field.

Potassium/argon dating is a well known method, but in many cases this method is useless because of the loss of argon from the investigated material. Therefore, sometimes the potassium/calcium method is to be preferred.

Common potassium contains about 0.012% of the radioactive isotope ^{40}K. ^{40}K undergoes a dual decay, by electron capture (ε) to ^{40}Ar and by β^- emission to ^{40}Ca:

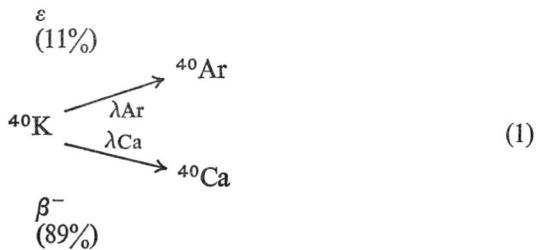

$$
\begin{array}{c}
\varepsilon \\
(11\%) \\
\nearrow \quad ^{40}Ar \\
^{40}K \quad {\scriptstyle \lambda Ar} \\
\quad {\scriptstyle \lambda Ca} \searrow \quad ^{40}Ca \\
\beta^- \\
(89\%)
\end{array}
\qquad (1)
$$

If one can assume that there has been no calcium loss from the very beginning of the formation of the investigated geological material, a quantitative analysis of the radiogenic produced isotope ^{40}Ca can give the age of the sample.

The following equations (2)–(5) show how one can determine the age of

611

geological materials by the potassium/calcium method:

$$N_{40_{Ca,rad}} = \frac{\lambda_{Ca}}{\lambda_{Ar} + \lambda_{Ca}} \cdot N_{40_K} \cdot \{\exp[(\lambda_{Ar} + \lambda_{Ca}) \cdot t] - 1\} \tag{2}$$

$$N_{40_{Ca,t}} = N_{40_{Ca,acc}} + N_{40_{Ca,rad}} \tag{3}$$

$$N_{44_{Ca,t}} = N_{44_{Ca,acc}} \tag{4}$$

$$\left(\frac{N_{40_{Ca}}}{N_{44_{Ca}}}\right)_t = \left(\frac{N_{40_{Ca}}}{N_{44_{Ca}}}\right)_{acc} + \frac{\lambda_{Ca}}{\lambda_{Ar} + \lambda_{Ca}} \cdot \frac{N_{40_{K,t}}}{N_{44_{Ca,t}}} \cdot \{\exp[(\lambda_{Ar} + \lambda_{Ca}) \cdot t] - 1\} \tag{5}$$

λ constant of radioactive decay
N Number of atoms
t age of the material
(index t concentration today; acc accessory non-radiogenic calcium; rad radiogenic calcium)

For the use of the potassium/calcium method it is necessary that the potassium content of the geological material should be high and the non-radiogenic calcium part should be low ($N_K : N_{Ca} \geq 100$). This condition is given for a number of sylvites and lepidolites and in some cases also for feldspar.

To obtain a precise age determination by the potassium/calcium dating method the following analytical problems must be solved:

(1) Precise isotope ratio measurements of calcium.
(2) Separation of small calcium amounts.
(3) Quantitative analysis of small calcium amounts, for example using isotope dilution analysis.

ISOTOPE RATIO MEASUREMENTS OF CALCIUM

Precise isotope ratio measurements of calcium are one of the most important presuppositions for the application of the described dating method. Table I gives a summary of calcium isotope measurements in the literature and the obtained errors (standard deviation). In all cases the isotope ratio $^{44}Ca/^{40}Ca$ was determined with more precision than the $^{48}Ca/^{40}Ca$ ratio. That is the reason why eqn (5) is related to ^{44}Ca and not to ^{48}Ca.

The precision of earlier calcium isotope ratio measurements was only high enough to use the potassium/calcium method for a few geological samples. Therefore, Backus[9] determined only minerals with an age of more than 1000 million years and the error he obtained was rather high.

In the case of our isotope ratio measurements with calcium carbonate and calcium chloride samples we observed an error of 0·4 and 0·2% for the $^{44}Ca/^{40}Ca$ ratio and of 0·7 and 0·6% for the $^{48}Ca/^{40}Ca$ ratio (Table I). These values are more precise than the other results described in the literature.

To improve the isotope ratio measurements of calcium the parameters must be known which influence the isotope ratio determination in the ion source

TABLE I

Isotope Ratio Measurements of Calcium by Thermal Ionization

Sample	Filament	Isotope ratio	Error (%)	References
CaO, Ca(NO$_3$)$_2$	Single and	^{44}Ca/^{40}Ca	1	Turnbull, 1963[1]
CaCO$_3$, CaSO$_4$	Triple	^{48}Ca/^{40}Ca	1·5–2	
CaC$_2$O$_4$,	Single	^{44}Ca/^{40}Ca	1·8	Backus et al., 1964[2]
Ca(NO$_3$)$_2$		^{48}Ca/^{40}Ca	2·9	
		^{48}Ca/^{40}Ca	0·8	Hirt and Epstein, 1964[3]
CaI$_2$	Triple	^{44}Ca/^{40}Ca	1	Artemov et al., 1966[4]
CaCl$_2$/ZnCl$_2$	Triple	^{44}Ca/^{40}Ca	0·5	Miller et al., 1966[5]
Ca(NO$_3$)$_2$	Triple	^{44}Ca/^{40}Ca	1	Letolle, 1968[6]
CaCl$_2$	Double	^{44}Ca/^{40}Ca	0·6	Stahl, 1968[7]
		^{48}Ca/^{40}Ca	1·2	
CaCO$_3$	Double	^{44}Ca/^{40}Ca	0·4	Heumann et al., 1970[8]
		^{48}Ca/^{40}Ca	0·7	
CaCl$_2$	Double	^{44}Ca/^{40}Ca	0·2	Heumann et al., 1973[8]
		^{48}Ca/^{40}Ca	0·6	

of a mass spectrometer. Following dependences of the measured calcium isotope ratios in a double filament thermal ion source have been found:

(1) The dependence on the temperature of the ionization filament.
(2) The mass discrimination effect during the evaporation of the sample.
(3) The dependence on the anion of the salt used as sample.
(4) The dependence on the size of the crystals using a calcium carbonate sample.

About the points (1)–(3) we have reported in detail.[8] What follows are some investigations concerning point (4).

In the case of large carbonate crystals (10–50 μ) we measured a higher isotope ratio in relation to small carbonate crystals ($<5\ \mu$): 2·6% higher for ^{44}Ca/^{40}Ca and 5·4% higher for ^{48}Ca/^{40}Ca.

Two explanations for the difference in the isotope ratios between large and small calcium carbonate crystals are possible:

(1) During the formation of the large carbonate crystals there could have been taking place an enrichment of the heavier calcium isotopes at the surface of the crystals. If only the surface of the crystals is analysed the measured isotope ratios ^{44}Ca/^{40}Ca and ^{48}Ca/^{40}Ca increase in relation to the standard value.

(2) In the double filament ion source there can take place different evaporation and dissociation processes for small and large carbonate crystals.

To prove an enrichment of the heavier calcium isotopes at the crystal surface, layer by layer of the large crystals was taken off by solving the calcium carbonate with diluted acetic acid. This process was controlled by a microscope. Table II gives the results.

After solving by steps 75% of the whole calcium carbonate no variation of the isotope ratio was observed within the limits of error. Therefore one can

TABLE II

Calcium Isotope Analysis of Different Layers in CaCO$_3$ Crystals with a Size of 30–50 μ

Dissolved CaCO$_3$ (%)	Isotope ratio		Size of the CaCO$_3$ crystals (μ)
	$(^{44}Ca/^{40}Ca) \times 10^5$	$(^{48}Ca/^{40}Ca) \times 10^6$	
0	2 168 $\pm$ 4	1 955 $\pm$ 2	30–50
0·1	2 160 $\pm$ 13	1 952 $\pm$ 8	30–50
1	2 170 $\pm$ 7	1 964 $\pm$ 10	30–50
10	2 165 $\pm$ 20	1 973 $\pm$ 24	10–30
20	2 169 $\pm$ 6	1 967 $\pm$ 26	10–30
50	2 155 $\pm$ 9	1 952 $\pm$ 6	10–30
75	2 162 $\pm$ 12	1 941 $\pm$ 8	10–20
90	2 147 $\pm$ 19	1 912 $\pm$ 18	5–10
95	2 115 $\pm$ 5	1 875 $\pm$ 8	<5

exclude an enrichment of the heavier calcium isotopes at the crystal surface. When the crystals become smaller than 5 μ by the solving process the observed isotope ratio is identical with that obtained from small crystals.

With increasing temperature CaCO$_3$ dissociates:

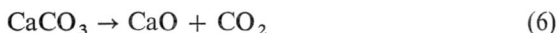

$$CaCO_3 \rightarrow CaO + CO_2 \qquad (6)$$

It is known that the dissociation pressure of CO$_2$ depends on the size of the CaCO$_3$ crystals.[10] To control whether different evaporation and dissociation processes in the ion source are responsible for the various isotope ratios in the case of small and large crystals an ion source was built where the evaporation filament could be used in combination with an electron bombardment system. With this system both kinds of crystals were heated step by step at the evaporation filament up to 970°C. This temperature was adequate to that used for double filament thermal ionization. The CO$_2$ peak was measured; the obtained results show that the small CaCO$_3$ crystals dissociate rather quickly before the temperature of 970°C is reached. If the temperature rises beyond 400°C only CaO is available at the evaporation filament. Contrary to the small crystals there is still a CO$_2$ emission in the case of the large crystals at a temperature of 970°C. Therefore it can be assumed that the large carbonate crystals are still not all converted into CaO. The different isotope ratio can be explained by the assumption that in the case of the large crystals not only CaO molecules are evaporated but for example also CaCO$_3$ molecules. In the case of an evaporation of CaCO$_3$ molecules on the one hand and of CaO on the other hand the dissociation mechanism at the ionization filament must be different for large and small calcium carbonate crystals, and this explains the variation in the isotope ratio.

ISOTOPE DILUTION ANALYSIS WITH A ^{44}Ca SPIKE SOLUTION

For the isotope ratio measurements of calcium in a thermal ion source one must be sure that the calcium sample is free of potassium. Because of the high ionization sensitivity of potassium in a thermal ion source the isotope ^{40}K can adulterate the intensity of ^{40}Ca. For the separation of calcium from potassium and other elements, in principle, two methods can be used:

(1) Precipitation of calcium as $CaCO_3$.
(2) Separation of calcium by cation ion exchangers.

By the first method one needs at least some milligrammes of calcium for the precipitation of $CaCO_3$; in addition to calcium some other elements can be precipitated as carbonate. But this method has the advantage that the separation procedure is very simple and that after recrystallizing the sample will be nearly free of potassium. The second method takes more time and is more difficult as to the experimental procedure; but several elements can be separated from calcium at the same time.

To test the precision of the calcium isotope dilution analysis in connection with a calcium separation a ^{44}Ca spike solution was taken to determine the calcium content of some pure chemicals. In all cases calcium was isolated as $CaCO_3$ and recrystallized for purification.

By eqn (7) one can calculate the calcium content:

$$N_2 = \frac{N_1 \cdot (h_{44}{}^{spike} - R_3 \cdot h_{40}{}^{spike})}{R_3 \cdot h_{40}{}^{sample} - h_{44}{}^{sample}} \qquad (7)$$

$N_{1,2}$ calcium content in the spike solution and sample
$h^{spike,\ sample}$ abundance of the isotope in the spike solution and sample
R isotope ratio $^{44}Ca/^{40}Ca$ after the isotope dilution

The calcium isotope abundances of the samples are identical with 'common' calcium, because there are no or only very small isotope effects during chemical procedures.[11]

Table III gives the results of the isotope dilution analysis:

(1) Calcium amounts of ≥ 1 ppm can be analysed with a precision better than 5%.
(2) In the case of calcium amounts <1 ppm the separation of calcium as $CaCO_3$ is not significant because the value of R_{spike} becomes similar to that of R_3.

TABLE III

Isotope Dilution Analysis of Pure Chemicals

Run No.	Sample	Isotope ratio R_3 $^{44}Ca/^{40}Ca$	Calcium content in the sample (ppm)
1	KCl	0·073 35	3·18
2	KCl	0·073 58	2·85
3	KCl	0·073 34	3·13
1	KNO$_3$	0·074 88	1·02
2	KNO$_3$	0·074 87	1·04
1	NaNO$_3$	0·075 94	$\leq$0·2
2	NaNO$_3$	0·075 46	
	Spike	0·075 70 $\pm$ 0·000 30	

To determine amounts in the ppb-range the calcium content of the spike solution must be about 1–100 μg. Therefore, after the isotope dilution only a separation of calcium by ion exchangers is possible. Figure 1 shows the separation curve of a solution containing Fe, Al, K, Na and Ca by a column filled with Dowex 50-X12. The element mixture in the solution can be compared with the composition of a feldspar. The calcium curve in Fig. 1 is drawn on a fifty times scale.

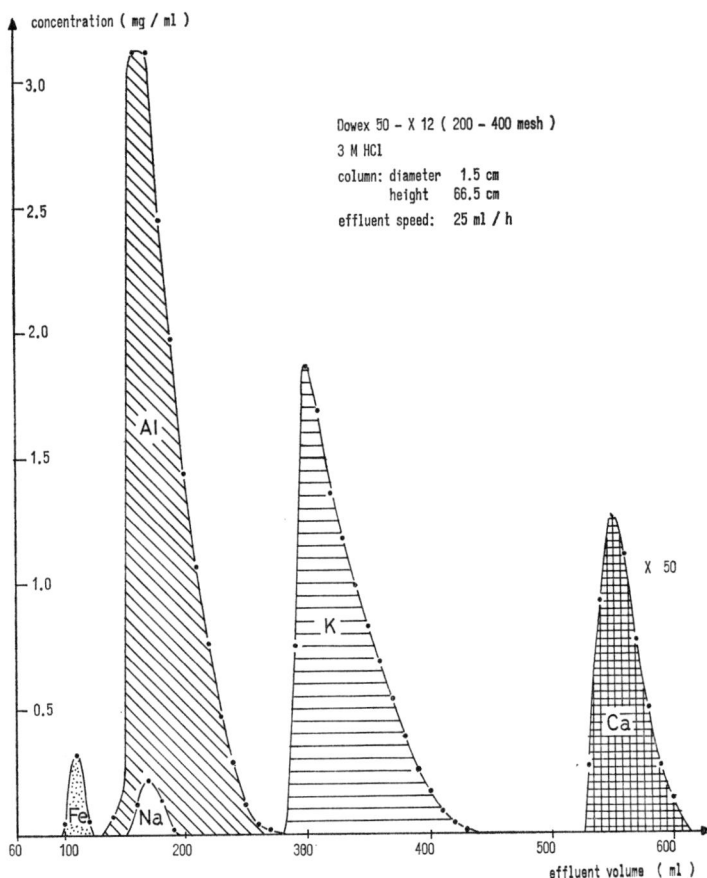

FIG. 1 Separation of calcium in a Dowex 50-X12 filled column.

To realize calcium analysis in the ppb-range it is essential to work in plastics apparatus and with suprapure chemicals. For a precise calcium isotope ratio measurement 1 μg CaCl$_2$ is enough. So it should be possible by using a small quantity of calcium spike to determine some ppb calcium with the method of isotope dilution. The error of this method is mainly given by the error of the isotope ratio measurement as can be seen from eqn (7).

POTASSIUM/CALCIUM DATING OF GEOLOGICAL MATERIAL

Figure 2 gives the difference between the isotope ratio $^{40}Ca/^{44}Ca$ of potassium containing material today and of accessory calcium in dependence on the potassium/calcium ratio. Age determinations out of the limits of error (in

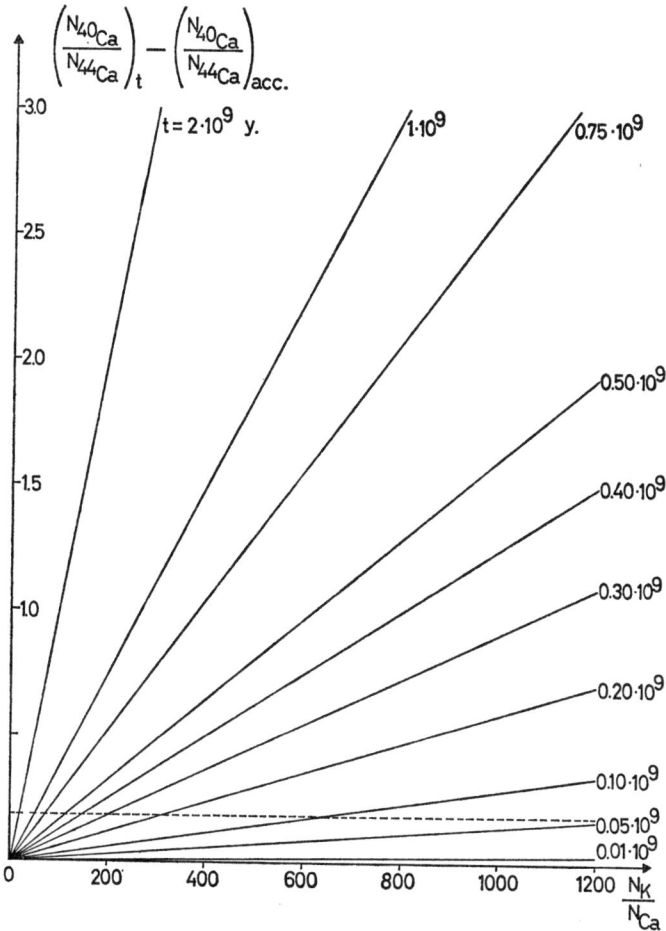

FIG. 2   The variation of the $^{40}Ca/^{44}Ca$ ratio in relation to accessory calcium in dependence on the potassium/calcium ratio and on the age of a geological material.

Fig. 2 is marked an error of 0·4% for the ratio $^{44}Ca/^{40}Ca$) are only possible with geological materials where the conditions given above the marked line are realized. Recently, Heumann et al.[8] have been able to improve the precision of the calcium isotope ratio measurement $^{44}Ca/^{40}Ca$ to 0·2% (Table I). Therefore the application of the potassium/calcium dating method should increase. For example, it should be possible to determine many feldspars by the potassium/calcium method.

REFERENCES

1. Turnbull, A. H., 'Surface Ionisation Techniques in Mass Spectrometry', Report AERE-R 4295, Harwell, 1963, p. 9.
2. Backus, M. M., Pinson, W. H., Herzog, L. F. and Hurley, P. M., *Geochim. Cosmochim. Acta*, 1964, **28**, 735.
3. Hirt, B. and Epstein, S., *Helv. Phys. Acta.*, 1964, **37**, 179.
4. Artemov, Y. M., Knorre, K. G., Strizkov, V. P. and Ustinov, V. I., *Geochemistry (U.S.S.R.)* (English Transl.), 1966, **3**, 1082.
5. Miller, Y. M., Ustinov, V. I. and Artemov, Y. M., *Geochemistry (U.S.S.R.)* (English Transl.), 1966, **3**, 929.
6. Letolle, R., *Earth Planet. Sci. Lett.*, 1968, **5**, 207.
7. Stahl, W., *Earth Planet. Sci. Lett.*, 1968, **5**, 184.
8. Heumann, K. G., Lieser, K. H. and Elias, H., 'Recent Developments in Mass Spectroscopy', University of Tokyo Press, Tokyo, 1970, p. 457.
9. Backus, M. M., Thesis, Massachusetts Institute of Technology, 1955.
10. 'Gmelins Handbuch der Anorganischen Chemie', 8th edn, 'Calcium', Verlag Chemie, Weinheim, 1961, Teil B, p. 854.
11. Heumann, K. G. and Lieser, K. H., *Z. Naturforsch.*, 1972, **27b**, 126.

73

Investigation by Mass Spectrometry of Thermal Phases During the Formation of the Moon and the Earth

By L. D. NGUYEN, M. DE SAINT SIMON

(Institut de Physique Nucléaire, Orsay, France)

G. PUIL, Y. YOKOYAMA

(Centre des Faibles Radioactivités, C.N.R.S., Gif-sur-Yvette, France)

and F. ARBEY

(Laboratoire de Sédimentologie et de Géologie Dynamique, Orsay, France)

INTRODUCTION

THE aim of our work is to develop a method which allows us to determine with accuracy the rare earth elements (REE) contained in very small quantities of lunar material (<1 mg), particularly in mineral phases in order to investigate their cosmochemical and geochemical behaviour. In this work, using the isotopic dilution method we determine ten REE: La, Ce, Nd, Sm, Eu, Gd, Dy, Er, Yb and Lu in Luna 20 samples. A thermodynamical interpretation of the results is discussed.

ANALYTICAL PROCEDURES

Identification and Separation of the Mineral Phases of Luna 20 Samples
In a first step, optical microscopic examination is performed and allows us to distinguish four main mineral phases: basic plagioclases, olivine, pyroxene and regolites (fragments of the first minerals and glass). We then use the X-ray analysis (Guinier device) to improve the determination of the mineralogic composition.

The size of the grains of Luna 20 soils are a few hundred micrometres and their weight is near 100 μg. To determine with sufficient accuracy the REE contained in such small quantities of lunar samples we have improved the isotopic dilution method with surface ionization mass spectrometry by increasing its sensitivity and accuracy.

Mass Spectrometric Analyses

Sensitivity—We increase the sensitivity of the mass spectrometer and decrease the chemical contamination.

(*a*) Sensitivity of the mass spectrometer. We use an electronic system developed in our laboratory[1] for detecting ions arriving at the collector. By improving electronic circuitry and carefully choosing individual components we obtain a final noise level of less than one ion per minute, *i.e.* a

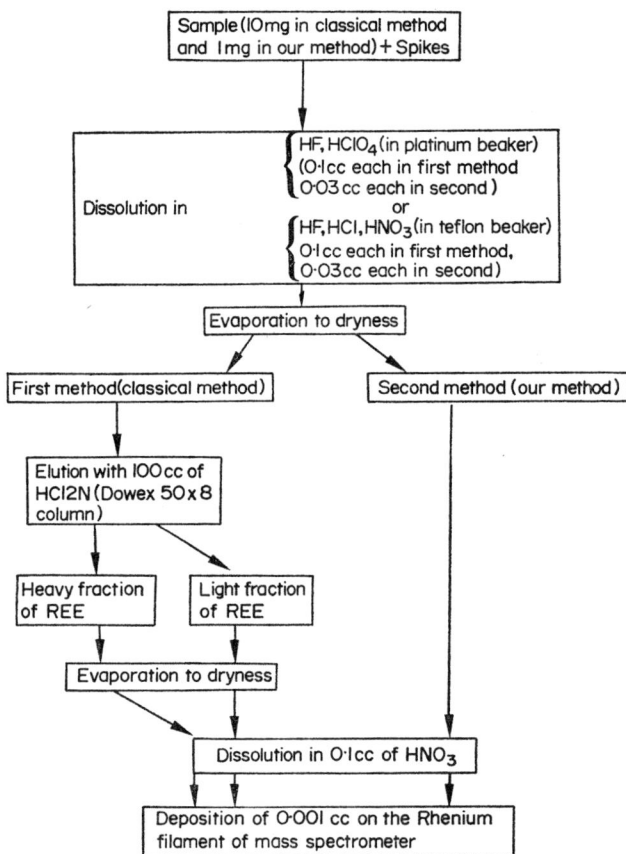

FIG. 1 Chemical procedure of REE (comparison of classical method with our method).

current of 10^{-21} A. In addition the mass scanning is carried out by a staircase-shaped variation of accelerating voltage of the ion source synchronized with the gating of the multiscaler.

(*b*) Chemical contamination. A chemical procedure is used for a 1 mg sample, as shown in Fig. 1 (comparison of classical method with our method). The acids are commercial Merck Suprapure grade. Some of them are further purified in our laboratory. Only platinum, quartz, teflon and polyethylene vessels are used.

We notice that in our method the number of chemical operations and quantities of acids used are greatly reduced, and only about 1/100 of the initial REE contained in the sample is deposited on the rhenium filament of the mass spectrometer. Using either the classical method or our method we obtain the same purity of mass spectra with a minimum interference between REE and by barium compound, provided that we use the procedure described in the following paragraph.

TABLE I

Blank Determination of REE

Elements	Philpotts et al.[2] $10^{-9} g$	Our results	
		Dissolution in HF, $HClO_4$ $10^{-9} g$	Dissolution in HF, HCl, HNO_3 $10^{-9} g$
La		0·90	
Ce	8·0	1·10	
Nd	4·67	0·20	0·38
Sm	0·81	0·050	0·07
Eu	1·01	0·028	0·02
Gd	1·01	0·20	0·099
Dy	0·49	0·61	0·033
Er	0·32	0·03	0·02
Yb	0·26	0·08	
Lu		0·13	
Quantities of samples to be analysed in the same conditions (g)		10^{-3}	10^{-3}

We show in Table I the comparison of our results with reported blank determination.[2] For a smaller sample we think that it is possible to improve our method by using a smaller quantity of purer acids.

Accuracy—We try to obtain a high accuracy in our measurements by increasing the purity of the mass spectrum and the accuracy of the spike calibration.

(*a*) Purity of mass spectrum. For each REE we record the mass spectrum of M^+ (metallic ions) or MO^+ (monoxide ions) at a well defined temperature of the filament according to our previous study (Fig. 2).[3] With this method we obtain the same purity of mass spectrum of the final products in the two chemical procedures indicated in Fig. 1. But in the second method, our method, the quantities of acids used were three orders of magnitude less than in the first method.

(*b*) Accuracy in spike calibration. Spikes are calibrated against standard solutions of normal isotopic composition which are prepared from 'specpure' oxides purchased from Johnson Matthey and Co. Ltd. Before using, these

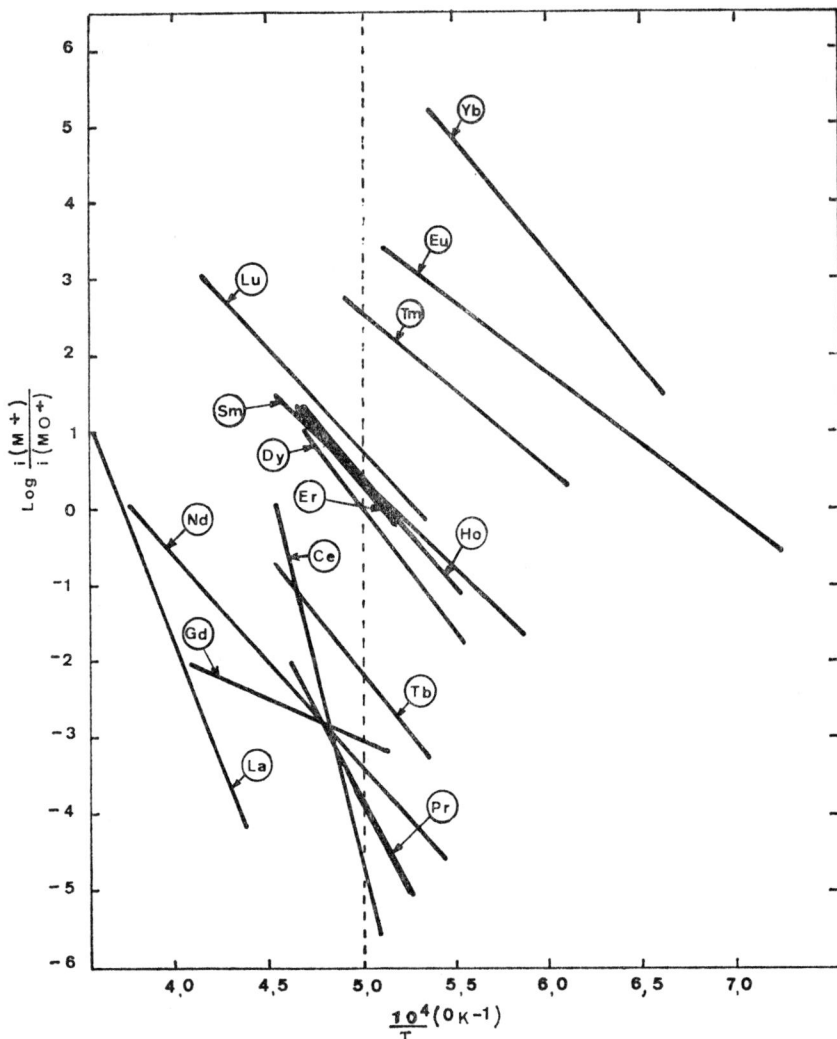

FIG. 2   Log $i(M^+)/i(MO^+)$ versus $10^4/T$ for lanthanides deposited on rhenium filament of the mass spectrometer (i represents the intensity of ion beam and T the absolute temperature).

products are submitted to the following analyses:[4]

(i) thermal balance under vacuum and dried air, oxygen, helium and argon for thermogravimetric study;

(ii) thermal balance, coupled with mass spectrometric analysis, for determining the nature of released gas;

(iii) X-ray diffraction for determining the true nature of these products.

We note that the anionic impurities (CO_2 and/or H_2O) can reach 15% of the weight of these products (Fig. 3).

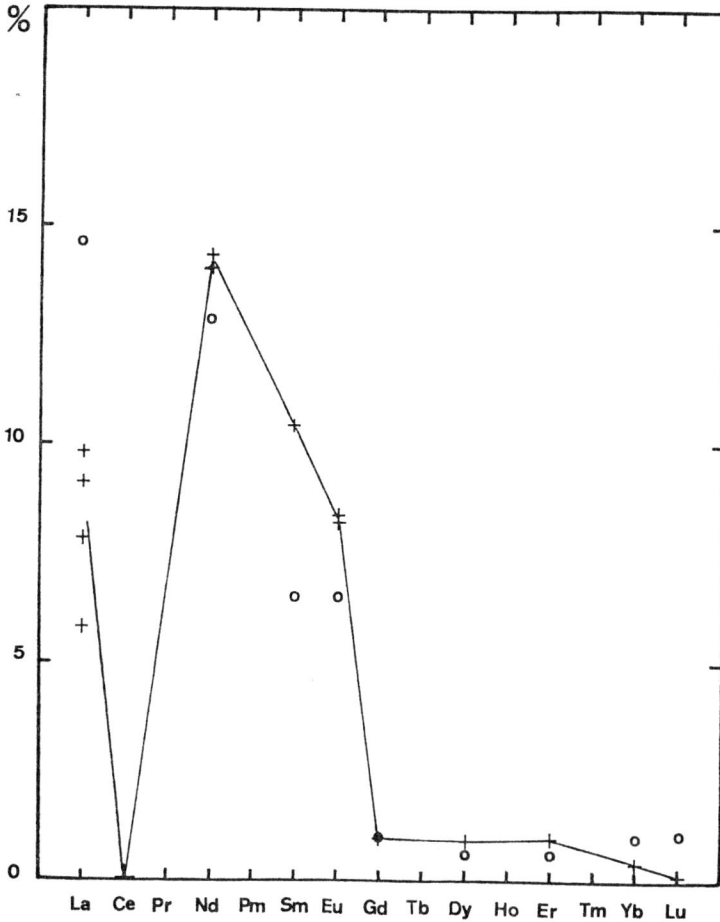

FIG. 3   Weight loss of our standard REE oxide powder after thermal cycles (cross) and the difference between the gravimetric calibration and the titrimetric one (circle) of the standard solutions, for Gast *et al.* (*Science*, 1970, **167**, 485).

RESULTS

We have tested our method by determining REE contained in 1 mg of BCR1 and our results (the average of six analyses) are in good agreement with those reported.[5] Our results of Luna 20 samples[6,7,8,9] (total rock) are given in Fig. 4. They correspond to the average of analyses of samples with different weights as following: 0·703, 0·800, 1·267, 1·644, 1·657 and 3·489 mg. For each sample we performed three mass spectrometric analyses. We observed a good agreement between our different measurements. In the same figure the results obtained by 14 MeV neutron activation of Surkov *et al.*[10] and those of Loubet *et al.*[11] using classical mass spectrometry with samples in the same slice of the carrot of Luna 20 are compared with ours. Our results are in good agreement with those reported in spite of the much smaller quantities of sample used in our determination.

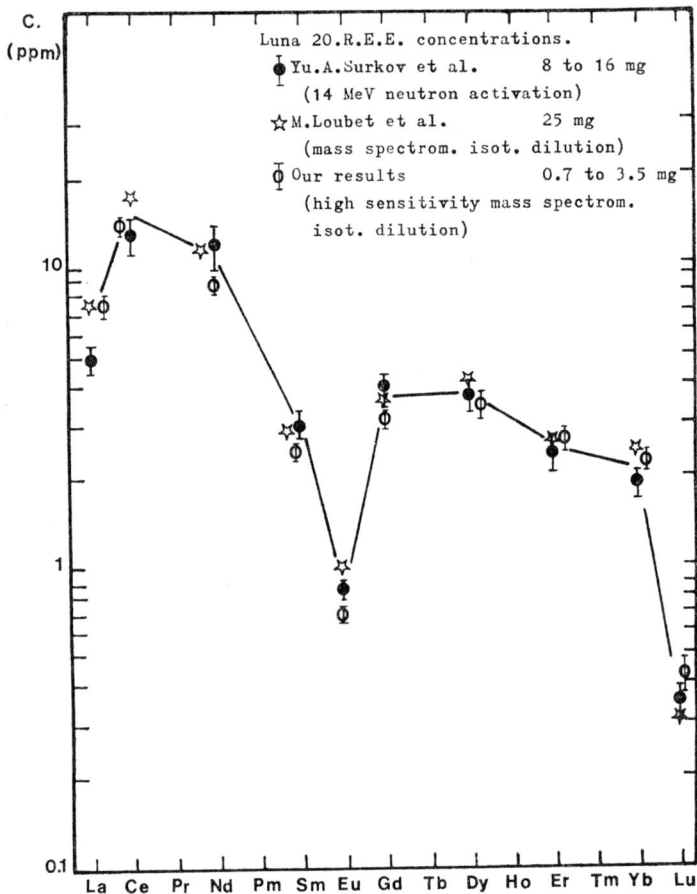

FIG. 4   REE concentrations in Luna 20.

The results of our investigation on REE concentrations in Luna 20 samples are compared with those reported on Apollo 11 to 15 and Luna 16[12] in Fig. 5.

We notice that there is some similarity between Luna 20 and Apollo 12 mare basalt in REE pattern but with smaller depletion of europium in Luna 20 than Apollo 12. The smaller Eu depletion of Luna 20 can be explained by the fact that highland material contains much plagioclase. The results of the determination of REE concentrations in different mineral phases will be exposed at the conference.

DISCUSSION

One of the striking features of different samples returned by the Apollo and Luna 20 missions is the depletion of europium in total rock in contrast with earth samples. The Eu depletion found in Luna 20 soils, a highland sample, seems to indicate that the moon as a whole is depleted in Eu and that it is

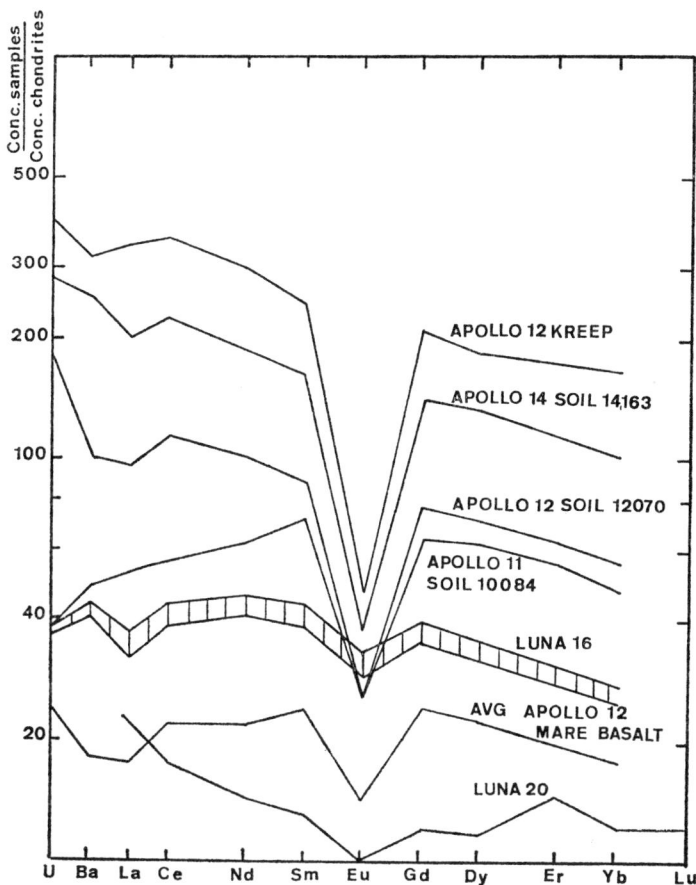

FIG. 5 Comparison of our results on REE concentrations normalized to chondrites in Luna 20 sample with those reported on Apollo 11 to 15 and Luna 16.

necessary to suppose another mechanism, in addition to partial melting or fractional crystallization,[13] to explain this phenomenon. In 1970, L. D. Nguyen and Y. Yokoyama proposed a mechanism in which the depletion in Eu may be explained by the difference of volatilization of REE from their oxides at high temperature and in vacuum, presumably at an early high temperature stage of the sun, Hayashi phase.[14]

Previous papers[3,15] have shown that the thermal decomposition of REE oxides heated under vacuum forms generally monoxide and metallic vapours, particularly Eu which forms metal vapours with a great vapour pressure. It is interesting to note that the presence of air drastically decreases the decomposition volatility of Eu which tends to be no longer different from adjacent elements Nd and Sm.[16] Depletion of Eu by the mechanism of volatilization is therefore expected only in a reducing system, that is the case of the moon formation, and it is a less pronounced event in terrestrial rocks.

In the following discussion we assume that the moon was formed by an accretion of small grains of solid which have a chondritic abundance pattern of REE. We calculate the evaporation rate of REE oxides according to a theory due to A. N. Nesmeyanov[17] and the vapour pressure data of Shchukarev et al.[15] for a heating period of 100 years and a temperature of 1400°K. This temperature corresponds to that expected at the orbit of the earth for the first 100 years of the Hayashi phase.[18] An example of this calculation of relative depletion of Eu is shown in Fig. 6. The calculated

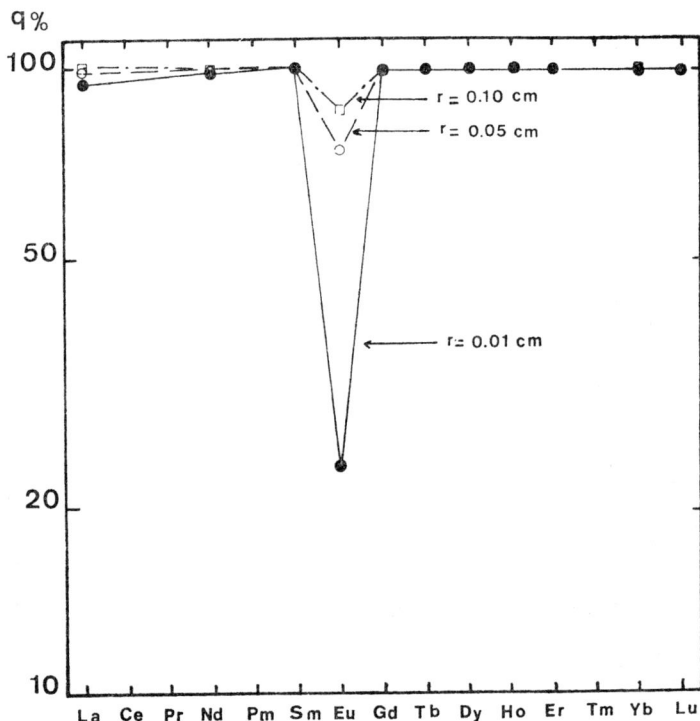

FIG. 6   Relative depletions of REE calculated by using the vapour pressure data of oxides under vacuum[15] with different radius r(cm) of grains.

curves are for a radius of the grains r of 0·01, 0·05 and 0·10 cm with a density of 1 g/cm^3. We note that for $r = 0·01$ cm the calculation gives the same Eu depletion as that found in Apollo 11 rocks and for $r = 0·05$ cm the same Eu depletion as that of rock 12 035 which is supposed[19] to have a closer REE composition to the original lunar material.

CONCLUSION

With high sensitivity mass spectrometry the abundance of REE can be determined in a sample of 1 mg or much less.

The study of the Luna 20 sample by this method shows that europium is also depleted in such highland samples, although, before the mission, it was expected, in view of liquid–solid partition theory, that the lunar highlands would be rich in europium. The most simplified interpretation of this discovery may be that the moon as a whole is depleted in europium to some extent. If it is so, the Eu depletion of lunar samples can be, in part, attributed to its evaporation loss at the moment of the formation of the moon. The determination of the relative importance of the evaporation process to that of the liquid–solid partition process will be an interesting subject of further studies to understand the early history of the moon and that of the solar system.

REFERENCES

1. Nguyen, L. D., Goby, G. and Rosenbaum, B., *Int. J. Mass Spectrom. Ion phys.*, 1973, **11**, 205.
2. Philpotts, J. A., Schnetzler, C. C., Bottino, M. L., Schuhmann, S. and Thomas, H. H., *Earth Planet. Sci. Letters*, 1972, **13**, 429.
3. Nguyen, L. D. and de Saint Simon, M., *Int. J. Mass Spectrom. Ion Phys.*, 1972, **9**, 299.
4. Nguyen, L. D., de Saint Simon, M., Foex, M., Coutures, J. P. and Gerdanian, P., Tenth Rare Earth Research Conference, Carefree, Arizona (U.S.A.) 30 April–3 May 1973, 1056.
5. Philpotts, J. A. and Schnetzler, C. C., *Proceedings of the Apollo* 11 *Lunar Science Conference*, 1970, **2**, 1471.
6. Nguyen, L. D., Puil, G., de Saint Simon, M. and Yokoyama, Y., Symposium on Cosmochemistry, Cambridge, Mass. (U.S.A.) 14–16 August 1972, T5.
7. Nguyen, L. D., de Saint Simon, M., Puil, G. and Yokoyama, Y., Fourth Lunar Science Conference, Houston, Texas (U.S.A.) 5–8 March 1973, 563.
8. Nguyen, L. D., Puil, G., de Saint Simon, M. and Yokoyama, Y., Réunion annuelle des Sciences de la Terre, Paris (France) 19–22 March 1973, 317.
9. Nguyen, L. D., de Saint Simon, M., Puil, G. and Yokoyama, Y., Tenth Rare Earth Research Conference, Carefree, Arizona (U.S.A.) 30 April–3 May 1973, 1068.
10. Surkov, Yu. A., Kirnozov, F. F., Kolezov, G. M., Ivanov, I. N., Rivkin, R. N. and Chpanov, A. P., C.N.R.S., International Colloquium on the Activation Analysis, C.E.N. Saclay (France) 2–6 October 1972.
11. Loubet, M., Birck, J. L., Allegre, C. J., *C.R. Ac. Sci.*, 1972, **275D**, 1075.
12. Hubbard, N. J., Nyquist, L. E., Rhodes, J. M., Bansal, B. M., Wiesmann, H. and Church, S. E., *Earth Planet. Sci. Letters*, 1972, **13**, 423.
13. Haskin, L. A. and Allen, R. O., *Proceedings of the Apollo* 11 *Lunar Science Conference*, 1971, **2**, 1213.
14. Nguyen, L. D. and Yokoyama, Y., 33rd Annual Meeting of the Meteoritical Society, Skyland, Virginia (U.S.A.) 27–30 October 1970; *Meteoritics*, 1970, **5**, 214.
15. Shchukarev, S. D. and Semenov, S. A., *Dokl. Akad. Nauk*, 1961, **141**, 652.
16. Benezech, G. and Foex, M., *C.R. Ac. Sci.*, 1969, **268**, 2315.
17. Nesmeyanov, A. N., 'Vapor pressure of the chemical elements', Elsevier Publishing Company, Amsterdam/London/New York, 1963.
18. Larimer, J. W. and Anders, E., *Geochim. Cosmochim. Acta*, 1967, **31**, 1239.
19. Masuda, A. and Tanaka, T., *Contr. Mineral. and Petrol.*, 1972, **34**, 336.

Discussion

K. G. Heumann (Technische Hochschule, Darmstadt, Germany): Can you explain the analytical reason why you have used the isotope dilution analysis for the determination of the rare earths and not for example such a sensitive method as activation analysis with thermal neutrons?

L. D. Nguyen: It is well known that for absolute determination of rare earth elements (REE) the classical method of isotopic dilution analysis is the more sensitive method (*see*

Philpotts *et al.* paper in Analytical Chemistry). Besides by increasing the sensitivity of the mass spectrometer and by decreasing the contamination we improve this classical method by many orders of magnitude. In the last 'Fourth Lunar Science Conference held at Houston, March 1973', we compared our method with the others such as slow neutron, 14 Mev neutron activation etc. . . . (performed by different authors) with the analysis of REE contained in the same lunar sample and it is shown that our method is the most sensitive.

K. G. Heumann: You spoke about the accuracy of your results. Please explain what accuracy means in this case.

L. D. Nguyen: By the term 'to improve the accuracy' I mean 'to minimize the absolute experimental errors'. For this purpose it is necessary to eliminate or to decrease different contributions of these errors: mass discrimination, interference between different REE and Barium compounds etc. and particularly the biggest, produced by the spike calibration (error up to 15%). The aim of our work is to eliminate or to minimize these errors by different described improvements.

74

Ion Microanalyser Observation of Samples from the Natural Reactor of Oklo: Preliminary Results

By G. SLODZIAN

(*Laboratoire de Physique des Solides associé au C.N.R.S.,*
Université Paris-Sud, Orsay, France)

and A. HAVETTE

(*Laboratoire de Pétrographie Volcanologie,*
Université Paris-Sud, Orsay, France)

INTRODUCTION

THE attention of the scientific community was focussed on the uranium bed of Oklo when it became known that the isotope ratio of $^{235}U/^{238}U$ exhibited unusual values.[1] The impoverishment in ^{235}U and the presence of rare earth elements with abnormal isotopic ratios founded the idea that, in the remote past, fission reactions occurred in given areas of the ore so that presently, we can observe a true fossil atomic reactor.

In order to study the different stages followed by the reactor, several projects are in progress. Among other methods, localized analysis may bring a significant contribution. From such a method, one can expect not only distribution maps of the different elements but also *in situ* measurements of isotopic ratios.

The microanalyser[2-4] seems to be well adapted for that double goal. Let us recall the features of the instrument which are necessary for a good understanding of what follows. A primary ion beam (A^+ or O_2^+ or O^-) strikes the polished surface of the sample. The target is sputtered and among the emitted particles are found ions characteristic of the elements present in the sample. An optical system (including among other things the equivalent of a mass spectrograph) allows the direct imaging of the surface: images made of ions of only one kind can be observed on the fluorescent screen or recorded on a photographic film. The spacial resolution limit is about 1 μm, the mass-resolving power is 300 and the imaged field is 250 μm. It should be noted that an aluminium grid is deposited on insulating samples in order to eliminate charge effects. The contrast obtained on the distribution images is essentially due to local variations of concentration. However, it should be remembered

629

that geometrical defects on the surface can bring shadow effects and that the sputtering rates can vary from one mineral to another. When this occurs, one area may look enriched in a given element: strictly speaking, the relative intensities measured at the same point are the only things to be considered.

DISTRIBUTION MAPS

Lead and Uranium Distribution

Uranium is detected by means of U^+ or UO^+ ions; the ratio of UO^+ to U^+ is about 20. On the micrographs (Figs. 1 to 4), it is seen that uranium is distributed in grains which are irregular in shape and sometimes crossed by cracks.

Lead is detected by means of Pb^+ ions; PbO^+ and $PbOH^+$ ions are also found, but their ratio to Pb^+ ions is only 0·04 and 0·20. Since it is very likely that the ionization probabilities of uranium and lead are different, the ratio $^{238}U^+/^{206}Pb^+$ is not equal to the ratio of the respective atomic concentrations. However, knowing the disintegration constant and the approximate decay time, it is possible to estimate that the ratio of the ionization probabilities is about 1.

From observation of the micrographs, one can infer that, roughly speaking, there are two types of areas:

(1) Grains containing uranium and a small concentration of lead (the ratio $^{238}U^+/^{206}Pb^+$ is about 7·7). Uranium is combined with oxygen but we did not try to determine which type of oxide was formed; we just know from other studies[6] that it could be uraninite.

(2) Embedding the grains, a region which looks like containing more lead. In these areas, elsewhere[5] identified as clay, along with lead, we found other elements: Al, Si, Na, K, Mg, Fe, Ba. The chemical state of lead is far from understood. In some cases, electron probe measurements and optical observations allowed us to identify a sulphide.

The cracks observed in the grains very often seem to contain more lead. Surprisingly, this observation is also true for normal types of minerals such as pitchblend from Canada or uraninite from Katanga.

On the micrographs (Fig. 4) we can see distribution maps which are somewhat different from the previous arrangement. Some areas have similar concentrations of uranium and lead, whereas others exhibit a depletion of lead. Moreover, the distribution map of potassium shows that most of the lead is no longer located in the clay. Besides that, it should be noted that on a fine scale, the lead distribution is heterogeneous and correlates to the calcium distribution.

Calcium, Strontium and Barium Distribution

Ca, Sr and Ba have a remarkable distribution (Figs. 1 and 3). Calcium is highly concentrated in 'the uranium oxide' grains. The measured intensities suggest that the atomic concentration of calcium is at least as high as the uranium one! Further, from the different micrographs, and specially from Fig. 4, it appears that calcium and lead tend to replace each other. It is

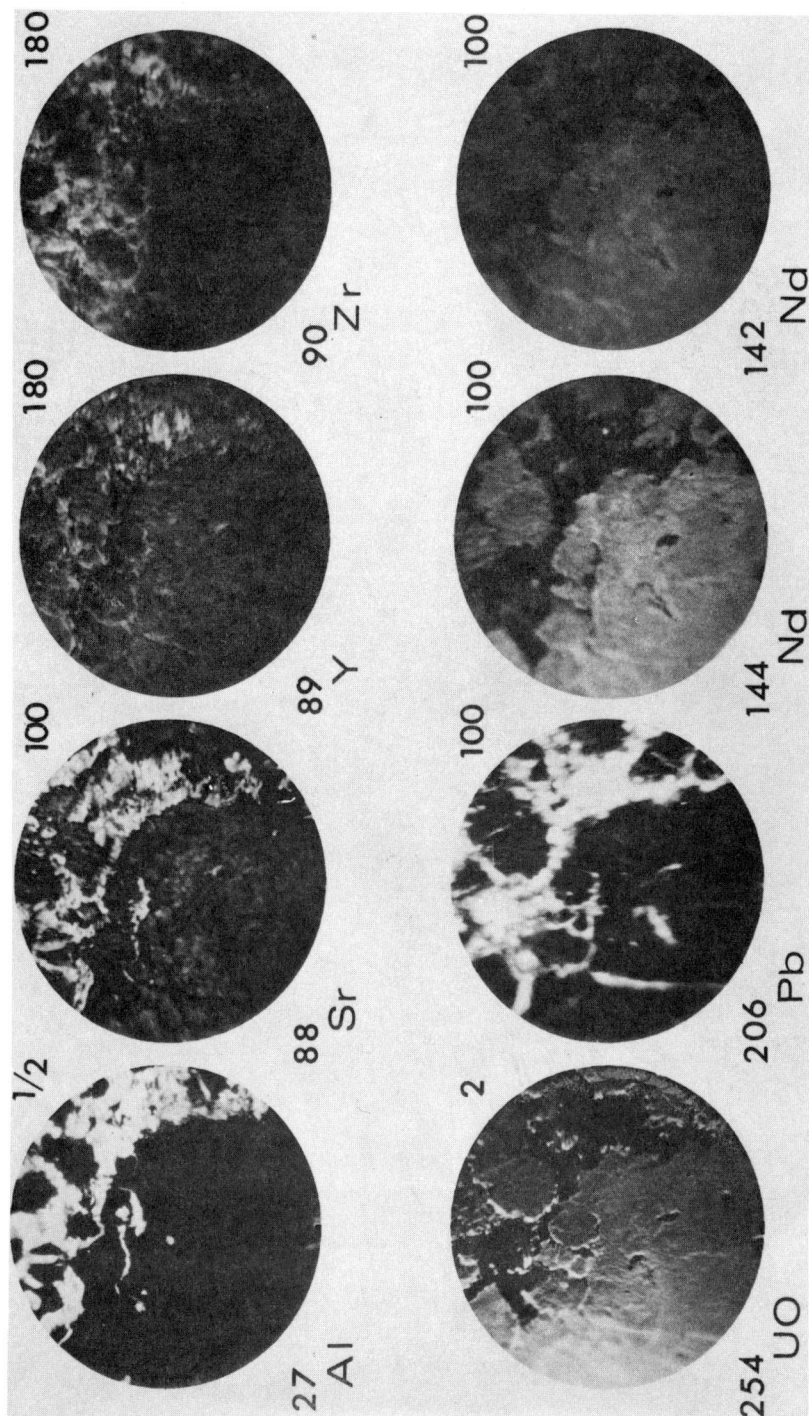

FIGS 1, 2, 3 and 4   The viewing field is 250 μ. The number on the upper, right-hand side of each micrograph gives the exposure time in seconds.

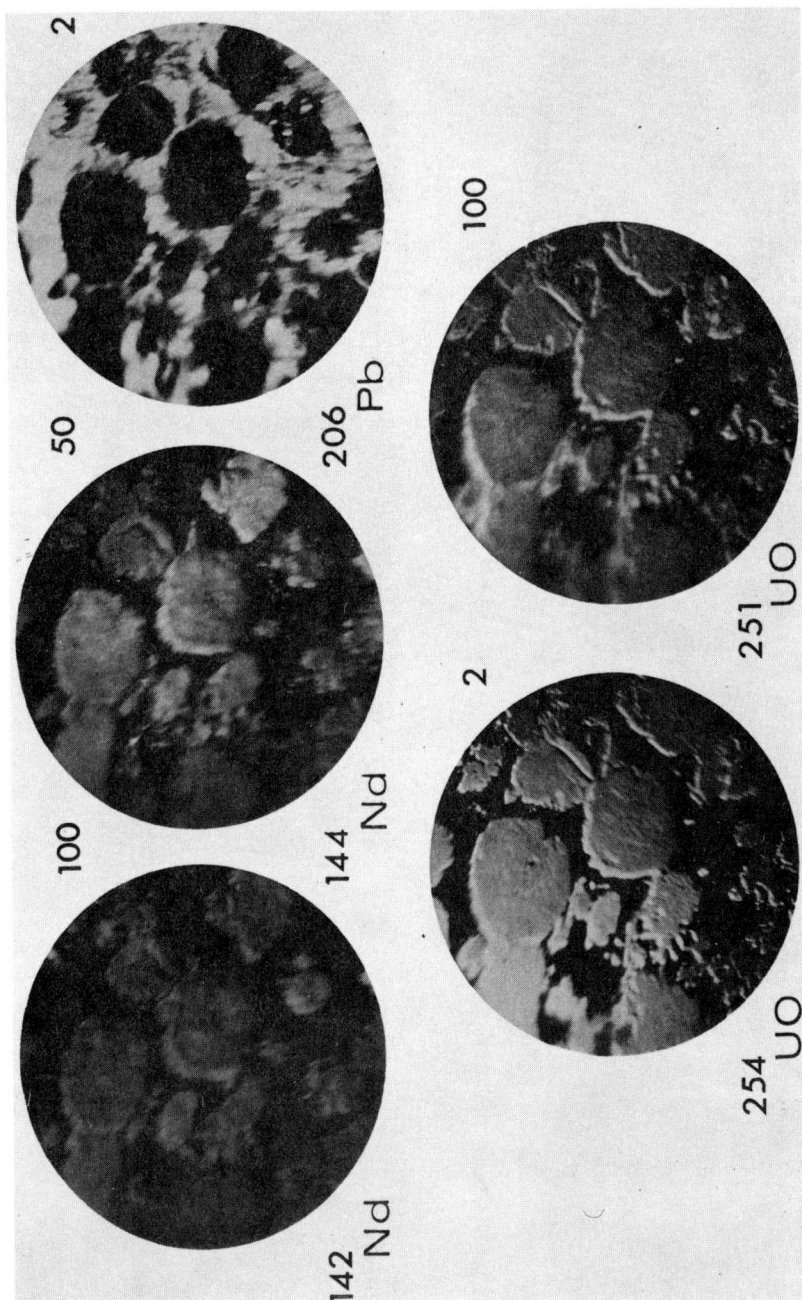

FIGURE 2

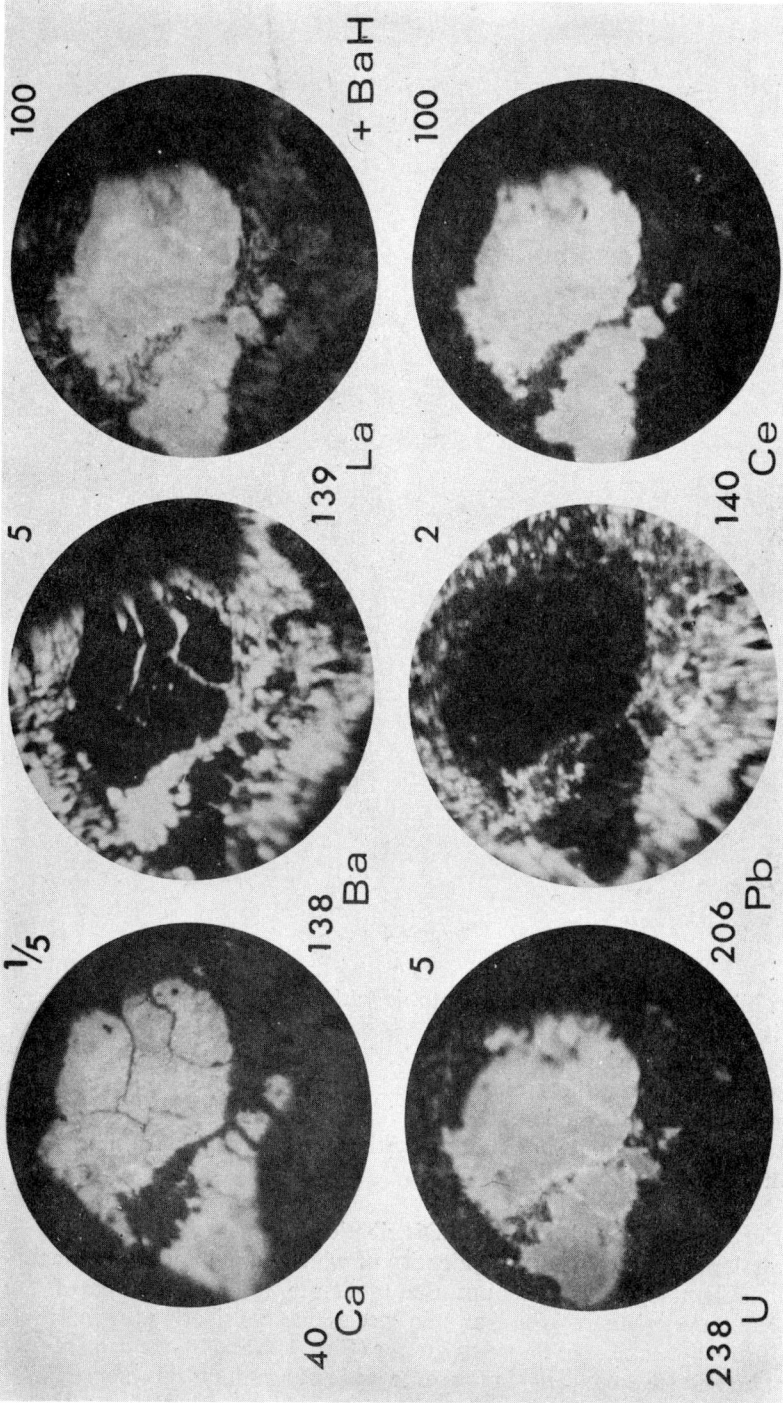

FIGURE 3

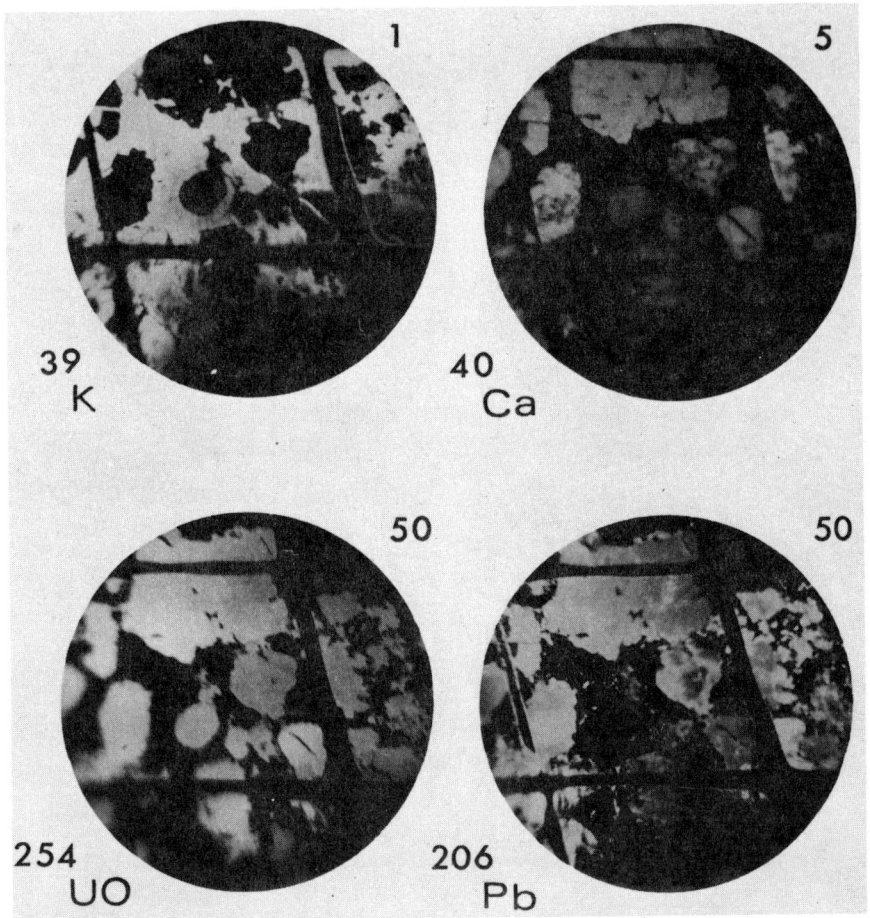

FIGURE 4

worthwhile to mention that the isotopic ratio $^{44}Ca/^{40}Ca$ measured on a grain is normal.

Strontium and barium have similar distributions characterized by a higher concentration outside the grains. We did not thoroughly examine the fine structure shown on the micrographs inside the grains where barium might be of radiogenic origin. It should be noted that there is roughly ten times more barium than strontium.

Rare Earth Elements and Zirconium Distribution

We only looked for elements the identity of which would not be reasonably questionable taking into account the moderate resolving power of the instrument, the existence of polyatomic species and considering the information already available. On the micrographs (Figs. 1, 2 and 3) one can see that Y, Nd, La and Ce are located in the same grain as uranium; this fact suggests that they have no tendency to migrate. The distributions of ^{144}Nd and ^{142}Nd

show that their isotopic ratio is unusual. In fact most of the so-called ^{142}Nd is probably ^{142}Ce which is a fission product just like ^{144}Nd. Further, the distribution of ^{139}La suggests that lanthanum migrated outside the grain; it is much more likely to find a small amount of ^{138}Ba^{1}H$^-$ ions corresponding to the region rich in barium, outside the grain.

IN SITU MEASUREMENTS OF ISOTOPIC RATIOS

From densitometric measurements on ^{235}U^{16}O, ^{238}U^{16}O micrographs, one could in principle deduce isotopic ratios. Such measurements are difficult to realize on samples exhibiting a rough topography and besides, they often are less accurate.

However, if isotopic heterogeneities are strong enough, distribution images may offer a quick method for showing them and give their location in the sample.

Another operating procedure is to replace the photographic film by a particle detector and to limit the viewing field by means of a diaphragm suitably inserted; it is then possible to pick out the ions emitted by a small area of the sample, the location of which can be exactly chosen according to the distribution image. At first, our measurements were limited to a 20 μm diameter area well suited to the kind of sample being analysed.

We investigated only the isotopic ratios of uranium and lead in a small number of spots. We observed that: ^{235}U/^{238}U was varying from place to place, let us say between 4 and 5·10^{-3}; and that ^{207}Pb/^{206}Pb was much lower inside the grains than outside (6·10^{-2} and 9·10^{-2}).

As an example, let us give the above mentioned two ratios as measured at the same spot: we found 4·4 10^{-3} for uranium and 5·8 10^{-2} for lead.

From these numbers, one can compute the datation of the grain: 1·6 10^9 years. Of course, it would be of great interest to extend these measurements to a number of grains. Such an extension requires some kind of automation and a careful sampling through the uranium ore. It is worthwhile mentioning that in the region recorded in Fig. 4 where uranium and lead can be found together, we measured 7·10^{-2} for the ratio ^{207}Pb/^{206}Pb and 0·38 10^{-2} for the ratio ^{235}U/^{238}U.

DISCUSSION AND CONCLUSION

Distribution images should be of great help to determine the various sources of the elements present in the sample and infer what kind of migration they suffered. For instance, from the micrographs one can make sure that the rare earth elements do not appreciably migrate. It is clear that such information is useful for the interpretation of more precise analytical data obtained with non-local methods. Moreover, 'unexpected' distributions may be discovered, providing new information and contributing to a better understanding of the formation and working conditions of the reactor.

The isotopic ratios given in this paper should be accepted with great care because it should not be forgotten that the results given here are preliminary and from only a small number of measurements. Also it should be recalled that in some cases polyatomic ions can make the measurement meaningless. That would be the situation if the intensity of $^{206}Pb^1H^+$ ion happened to be strong enough to perturb the measurement of $^{207}Pb^+$. A general study of the emission of polyatomic ions by mineral samples suggests that MH^+ are emitted in appreciable amounts when MOH^+ is also strong; in our samples $PbOH^+$ has a low intensity. Besides, the datation of the sample gives values which are a little lower than those usually accepted[5] suggesting that there is no significant perturbation at mass number 207 (at the precision of our measurements). However, these arguments still leave place for undesirable 'surprises' so that some facetious investigator might ask: 'are you sure it's lead?' The answer is that we checked for the presence of lead with the electron probe and that the isotopic ratios are on the average the same as those measured with non-local mass-spectrometry.

In conclusion, the potentialities of localized analysis by secondary ion emission seem to be promising enough on Oklo samples to justify more systematic investigations.

ACKNOWLEDGMENTS

The authors are indebted to Dr Long-Den N'Guyen for suggesting this work, to Professor R. Brousse for encouraging it, to R. Naudet for providing the samples and stimulating discussions, to R. Dennebouy whose skill in operating the instrument was of an irreplaceable help and to H. Bizouard for measurements with the electron probe.

This work was sponsored by the C.N.R.S. (N. A 6580061).

REFERENCES

1. Neuilly, M. and Bussac, J. *et al.*, *C.R. Acad. Sc., Paris*, 1972, p. 275, 23 Oct.
2. Castaing, R. and Slodzian, G., 'Proc. Em. Reg. Conf. on electron microscopy', Delft, 1960, p. 169.
3. Jerome, D. and Slodzian, G., *Bul. Soc. Fr. Min. Cris.*, 1971, p. 94.
4. Havette, A., *Proc. Sc. de la Terre*, 1973.
5. Naudet, R., C.E.A. Saclay, Internal Report, June 1973.
6. Naudet, R., private communication.

75
Isotope Ratio Analysis of Neodymium and Gadolinium Fractions Obtained from Ion Exchange Runs

By JORMA AALTONEN

(*Department of Radiochemistry, University of Helsinki, Finland*)

INTRODUCTION

In the studies of isotopic effects taking place in chemical exchange reactions a mass spectrometer is almost always needed for the isotope ratio analysis. In this work the mass spectrometer MS 7 from A.E.I. (United Kingdom) was used for the analysis of some neodymium and gadolinium fractions obtained from two different ion exchange runs. From the results of the isotope ratio analysis the isotope separation factor for neodymium and gadolinium in the ion exchanger-complexing agent system can be calculated.

Neodymium and gadolinium both have seven stable isotopes. The mass numbers of the neodymium isotopes are 142, 143, 144, 145, 146, 148 and 150. The mass numbers of the gadolinium isotopes are 152, 154, 155, 156, 157, 158 and 160. The useful radionuclides as tracers for neodymium and gadolinium are [147]Nd (half-life 11·1 days) and [159]Gd (half-life 18·6 hr).

ION EXCHANGE RUNS

In the isotope separation study of neodymium and gadolinium a recycle ion exchange technique was used. The starting material was neodymium oxide and gadolinium oxide from Lindsay, U.S.A. In order to make it easier to follow the movement of the neodymium and gadolinium band in the column, the starting material was irradiated for 25 hours in TRIGA reactor in Otaniemi, Finland. The migration of a band through the column was detected with a NaI scintillation counter.

The size of the ion exchange column was 0·7 × 100 cm. The column was packed with Dowex 50 X8 (400 mesh) cation exchange resin. The eluent was ammonium α-hydroxyisobutyrate, pH 4·45. The concentration of the eluent was 0·32 M for neodymium and 0·20 M for gadolinium. The flow of the eluent was controlled by a peristaltic pump connected to the column. The number of cycles in the neodymium and gadolinium ion exchange runs was 7 and 6 respectively. Elution curves are given in Figs. 1 and 2.

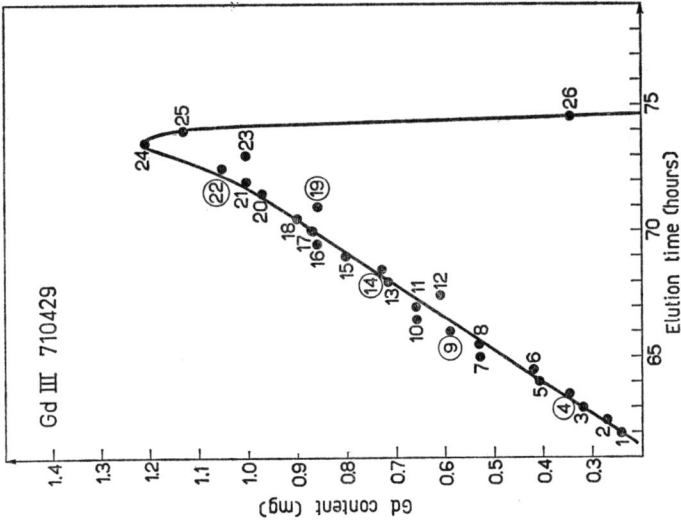

Fig. 2 Elution curve of the gadolinium ion exchange run after 6 cycles.

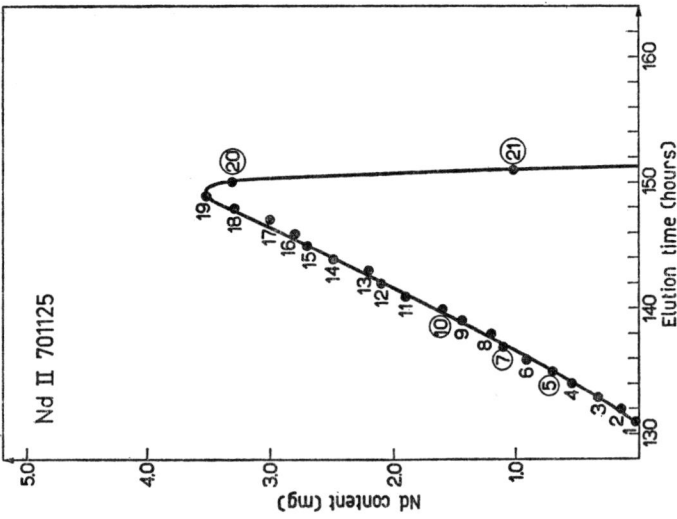

Fig. 1 Elution curve of the neodymium ion exchange run after 7 cycles.

At the beginning of an ion exchange run the resin in the column was in equilibrium with ammonium α-hydroxyisobutyrate of the same concentration and pH as the eluent used in a particular run.

The activated neodymium oxide and gadolinium oxide were first dissolved in 0·1 M nitric acid, then mixed with the eluent solution. Two 5-ml solutions,

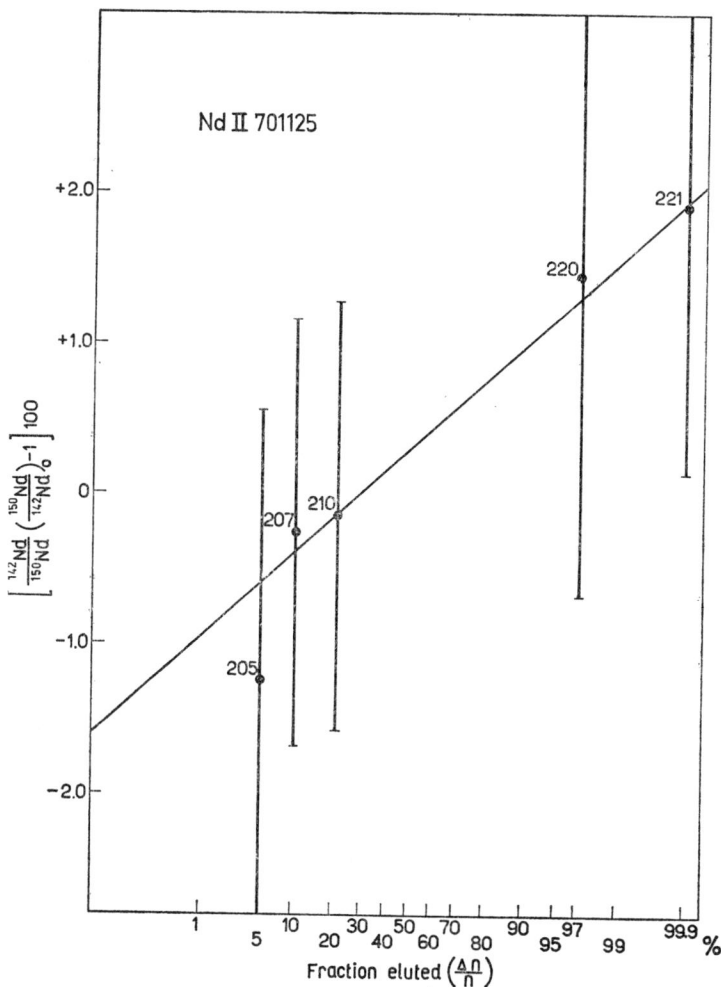

FIG. 3   Local enrichment percentages (R-1) 100 of the samples obtained in the neodymium ion exchange run. The proportion of the amount of neodymium $\Delta n/n$ is plotted as a function of (R-1) 100.

one containing 30 mg neodymium and the other 20 mg gadolinium, were used for two ion exchange runs. The fractions from the runs were collected at 30- to 60-minute intervals with a fraction collector.

The activity corresponds to the amount of the material in the fraction. The proportion ($\Delta n/n$) of the activity of an individual fraction Δn to the total

JORMA AALTONEN

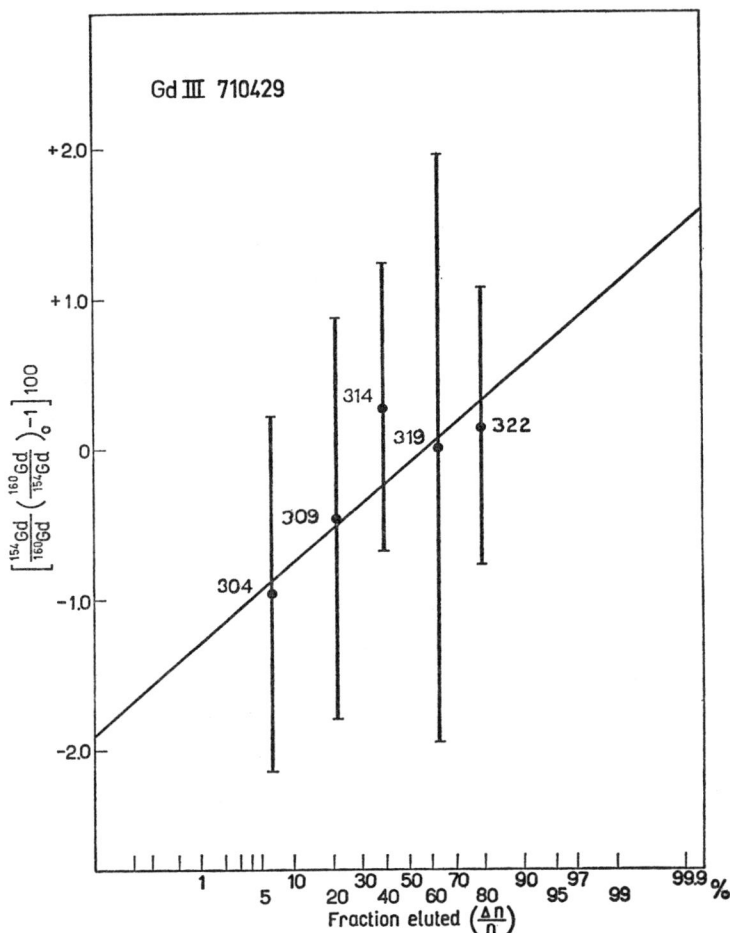

FIG. 4 Local enrichment percentages (R-1) 100 of the samples obtained in the gadolinium ion exchange run. The proportion of the amount of gadolinium $\Delta n/n$ is plotted as a function of (R-1) 100 on the probability graph paper.

activity n, was calculated. Local enrichment percentages for neodymium and gadolinium are given in Figs. 3 and 4 respectively.

Table I shows the values of parameters (elution time, migration rate, fractionation interval, number of cycles and number of theoretical plates) in the neodymium and gadolinium ion exchange runs.

TABLE I

Parameters of Neodymium and Gadolinium Separation Experiments

Experiment	Elution time (hr)	Migration rate cm/hr	Fractionation interval (min)	Number of cycles	Number of theoretical plates
Nd II 701125	148	4·7	60	7	1 000
Gd III 710429	74	8·1	30	6	350

TABLE II

Isotopic Ratios $^{144}Nd/^{142}Nd$, $^{146}Nd/^{142}Nd$, $^{148}Nd/^{142}Nd$ and $^{150}Nd/^{142}Nd$ in Samples 205, 207, 210, 220 and 221 of the Separation Experiment Nd II 701125

Sample No.	$\Delta n/n$ (%)	Isotopic ratio $\dfrac{^{144}Nd}{^{142}Nd} \times 10^4$	Isotopic ratio $\dfrac{^{146}Nd}{^{142}Nd} \times 10^4$	Isotopic ratio $\dfrac{^{148}Nd}{^{142}Nd} \times 10^5$	Isotopic ratio $\dfrac{^{150}Nd}{^{142}Nd} \times 10^5$	Local enrichment factor $\dfrac{^{142}Nd}{^{150}Nd}\left(\dfrac{^{150}Nd}{^{142}Nd}\right)_0$
205	5·0	8 764 ± 27	6 301 ± 32	2 109 ± 12	2 048 ± 21	0·987 8
207	10·4	8 721 ± 38	6 269 ± 35	2 084 ± 12	2 028 ± 12	0·997 5
210	21·8	8 728 ± 28	6 243 ± 37	2 086 ± 16	2 026 ± 12	0·998 5
220	97·2	8 658 ± 49	6 187 ± 55	2 062 ± 17	1 994 ± 25	1·014 5
221	99·9	8 683 ± 35	6 209 ± 29	2 046 ± 14	1 985 ± 18	1·019 2
Standard = 0		8 712 ± 37	6 246 ± 40	2 096 ± 14	2 023 ± 17	

TABLE III

The Isotopic Ratios $^{154}Gd/^{160}Gd$, $^{155}Gd/^{160}Gd$, $^{156}Gd/^{160}Gd$, $^{157}Gd/^{160}Gd$ and $^{158}Gd/^{160}Gd$ in Samples 304, 309, 314, 319 and 322 of the Separation Experiment Gd III 710429

Sample No.	$\Delta n/n$ (%)	Isotopic ratio $\dfrac{^{154}Gd}{^{160}Gd} \times 10^5$	Isotopic ratio $\dfrac{^{155}Gd}{^{160}Gd} \times 10^4$	Isotopic ratio $\dfrac{^{156}Gd}{^{160}Gd} \times 10^4$	Isotopic ratio $\dfrac{^{157}Gd}{^{160}Gd} \times 10^4$	Isotopic ratio $\dfrac{^{158}Gd}{^{160}Gd} \times 10^4$	Local enrichment factor $\dfrac{^{154}Gd}{^{160}Gd}\left(\dfrac{^{160}Gd}{^{154}Gd}\right)_0$
304	6·6	10 133 ± 80	6 821 ± 50	9 443 ± 71	7 189 ± 51	11 441 ± 55	0·990 4
309	20·3	10 184 ± 89	6 800 ± 43	9 493 ± 69	7 220 ± 61	11 475 ± 73	0·995 4
314	38·9	10 259 ± 60	6 885 ± 50	9 526 ± 42	7 254 ± 43	11 475 ± 33	1·002 7
319	62·5	10 232 ± 153	6 839 ± 85	9 471 ± 274	7 195 ± 79	11 501 ± 98	1·000 1
322	79·5	10 246 ± 52	6 886 ± 37	9 538 ± 42	7 227 ± 22	11 456 ± 30	1·001 5
Standard = 0		10 231 ± 46	6 828 ± 29	9 500 ± 46	7 194 ± 34	11 495 ± 34	

ISOTOPE RATIO ANALYSIS

Five fractions from the runs Nd II 701125 and Gd III 710429 were selected for mass spectrometric analysis. Neodymium and gadolinium were precipitated with oxalic acid. The oxalate precipitates were heated in small beakers in an oven at 800°C. The created oxide samples were dissolved in 0·1 M nitric acid and about 10 μg of neodymium and gadolinium were transferred onto the side filaments of triple filament beads.

Before the start of a mass spectrometric run, each filament was preheated in the ion source with a current of 3 A for at least three hours. The cyclic scanning of the magnetic field was started after a stable ion current was reached. The mass spectra of neodymium and gadolinium samples were recorded with a Hitachi–Perkin Elmer 159 strip chart recorder. From the neodymium mass spectra the peak heights corresponding to the mass numbers 142, 144, 146, 148 and 150 were measured. The isotopic ratios of these masses were calculated with regard to mass 142. In the case of gadolinium the mass numbers measured were 154, 155, 156, 157, 158 and 160. The isotopic ratios of gadolinium isotopes were calculated with regard to mass 160. At least twenty sets of mass spectra were measured from every sample. The mean values and the standard deviations of the mean were also calculated for the isotope ratios (Tables II and III).

The isotope separation experiments of neodymium and gadolinium by ion exchange resulted in the separation factors $S_{150}^{142}(Nd) = 1\cdot000\ 27 \pm 0\cdot000\ 05$ and $S_{160}^{154}(Gd) = 1\cdot000\ 42 \pm 0\cdot000\ 10$.

DISCUSSION

In the isotope ratio analysis of the neodymium samples the standard deviation of the mean lies between 0·3% to 1%. In the case of gadolinium samples it is somewhat higher, about 0·5% to 1%. In order to achieve an accuracy of at least 1% for the analysis of a sample, a set of twenty mass spectra have to be made.

The isotope separation factor of neodymium is within the limits of error relatively the same as that of samarium $S_{154}^{144}(Sm) = 1\cdot000\ 19 \pm 0\cdot000\ 06$ (Aaltonen, J., *Ann. Acad. Sci. Fenn.*, 1967, **AII**, 137.) The isotopic effect found in the ion exchange runs of neodymium and gadolinium is 1% to 2% as compared to the starting material.

In the case of gadolinium, the separation factor seems to be too high. This may be caused by the uncertainty in the determination of the number of theoretical plates in the ion exchange run Gd III 710429.

Discussion

H. J. Svec (Iowa State University, U.S.A.): How did you eliminate mixing in the tubing of the recirculation system?

J. Aaltonen: After every completed cycle the selector valve of the recycle system was opened so that fresh eluent could flow through the peristaltic pump and the possible tailing of the band was cut off. The tubing of the pump was of Tygon with no detectable adsorption of the rare earths.

H. J. Svec: What resin did you use and in what form was it prior to loading with the rare earths? What was the eluting agent?

J. Aaltonen: The resin in the column was Dowex 50 X8 400 mesh in ammonium form. The eluent agent was ammonium α-hydroxyisobutyrate which is commonly used in the separation of the rare earths.

L. D. Nguyen (Institut de Physique, Nucleaire, Orsay, France): I think in the case of the gadolinium the purpose of your experiments was to determine in the future the dose of the irradiation of meteoritic or lunar samples by slow neutrons by determining the abundance ratios of gadolinium. In this case have you compared your results with those obtained by Wasserburg *et al.*?

J. Aaltonen: The primary idea of this study was to find out the magnitude of the isotopic effect taking place in the ion exchange runs of neodymium and gadolinium.

L. D. Nguyen: Have you observed in the mass spectrum interference with Barium compounds?

J. Aaltonen: Before the sample loading, the beads were heated in the ion source for about three hours to be sure there were no impurities on the filaments. During the measurements no interference was found to come from the sample in the mass ranges of neodymium and gadolinium.

76
Mass Spectrometric System for Locating Fuel Cladding Failures in Fast Neutron Reactors

By R. HAGEMANN, S. TISTCHENKO, P. LOHEZ

(*Département de Recherche et Analyse,
Service d'Analyse et d'Etudes en Chimie Nucléaire et Isotopique,
Gif-sur-Yvette, France*)

and G. NIEF

(*Département de Recherche et Analyse, Gif-sur-Yvette, France*)

FUEL cladding failures in a fast neutrons reactor lead to the presence in the reactor cover gas of fission gases coming from the fuel. Mass spectrometric analysis of the krypton and xenon isotopes which are present in the cover gas allows detection of the cladding failures. In addition, the measurement of isotopic ratios permits the determination of the age of the fuel which leads in principle to a partial localization of the cladding failures.

One possible method for obtaining a more definite localization is to tag the fuel elements with xenon and krypton having different isotopic compositions for each fuel assembly and enriched with light-stable isotopes not produced by fission.

When a cladding failure occurs, the spike is transferred into the sodium and then into the cover gas, and the failure is located from the isotopic analysis of the spike. This method has been studied for the 'Phenix' reactor.

The spikes are a series of mixtures of xenon isotopes, each spike being defined by its ratio $^{124}Xe/^{129}Xe$. The first spike is natural xenon, and the variation of the above ratio between two successive spikes is made to be 1·25. To double the number of available spikes, krypton enriched with ^{78}Kr is added to the xenon spikes in one half of the rods.

The concentration of the spike in the cover gas after a cladding failure will be very low. As a matter of fact, the amount of spike which can be introduced in each fuel rod is limited to a volume of about 1 cm^3 and after a cladding failure this amount is diluted by a factor larger than 10^7 by the cover gas. Moreover, because the spike may not be quantitatively transferred from the rod through the sodium to the cover gas, the concentration of the spike in the gas to be analysed may be as low as 10^{-9}.

To make possible the isotopic analysis of the spike, it is therefore necessary, as a first step, to concentrate the xenon and krypton present in several litres of the cover gas. The accuracy of the measurements of the ratios ^{124}Xe/^{129}Xe has to be better than 10% to be able to discriminate between two adjacent spikes in the series.

In this paper the analytical system which could be set up in the 'Phenix' reactor is described. The set up includes a sampling line in which xenon and krypton are trapped and separated from argon which is the main constituent of the cover gas. The sampling line is connected to the inlet line of a mass spectrometer.

THE SAMPLING LINE

The diagram of the sampling line is shown in Fig. 1.

The separation of xenon and krypton from argon is performed by chromatography on activated charcoal at $-80°$C, using three columns in series. Up to 20 litres of the cover gas are passed through a filter to retain radioactive aerosols, then sent into the first column which contains about 20 g of activated-charcoal at $-80°$C; xenon and krypton are preferentially trapped. Then argon is eluted using a flow of helium purified on activated-charcoal at liquid nitrogen temperature. To achieve a higher separation of argon from the krypton and xenon without losing the latter gases, the argon elution is carried out with two more columns containing respectively about 8 g and 4 g of activated-charcoal.

The gases are transferred from one column to the next by heating at a temperature of 250°C. After the argon elution is finished in the third column at $-80°$C, helium is pumped off, then the column is heated to 250°C and xenon, krypton and residual argon are transferred into the reservoir of the inlet line of the mass spectrometer. Impurities which would be present in the cover gas are recovered at the end of the above cycle with the spike. Any impurities of CS_2, C_6H_6 and C_2HCl_3 would give mass spectra which could interfere with those of krypton and xenon. Because of this possibility, it is necessary to purify the gas coming from the third column before its introduction into the mass spectrometer, otherwise it would be necessary to perform the isotopic analysis using a mass spectrometer able to discriminate between rare gases and hydrocarbons of the same nominal masses.

The purification is done by cracking any hydrocarbon impurities on a hot wire in the reservoir of the mass spectrometer. The pyrolysis products which are mainly H_2, CH_4, CO are eliminated by a getter pump connected to the reservoir. It has been checked by mass spectrometric analysis, using a tungsten wire heated to 1800°C, that CS_2, C_6H_6 and C_2HCl_3 are completely cracked into lighter compounds which are eliminated by the getter pump.

The sampling line is designed to operate completely automatically, using a small computer.

The measured performance using 20 litres of argon containing 10^{-3} vpm of xenon and krypton is as follows.

The yields of xenon and krypton which are recovered are respectively 60% and 80%. Argon is almost completely eliminated as the ratio xenon/argon is increased by a factor of about 10^8. The recovered gases are free of hydrocarbons and compounds which could interfere with the mass spectra

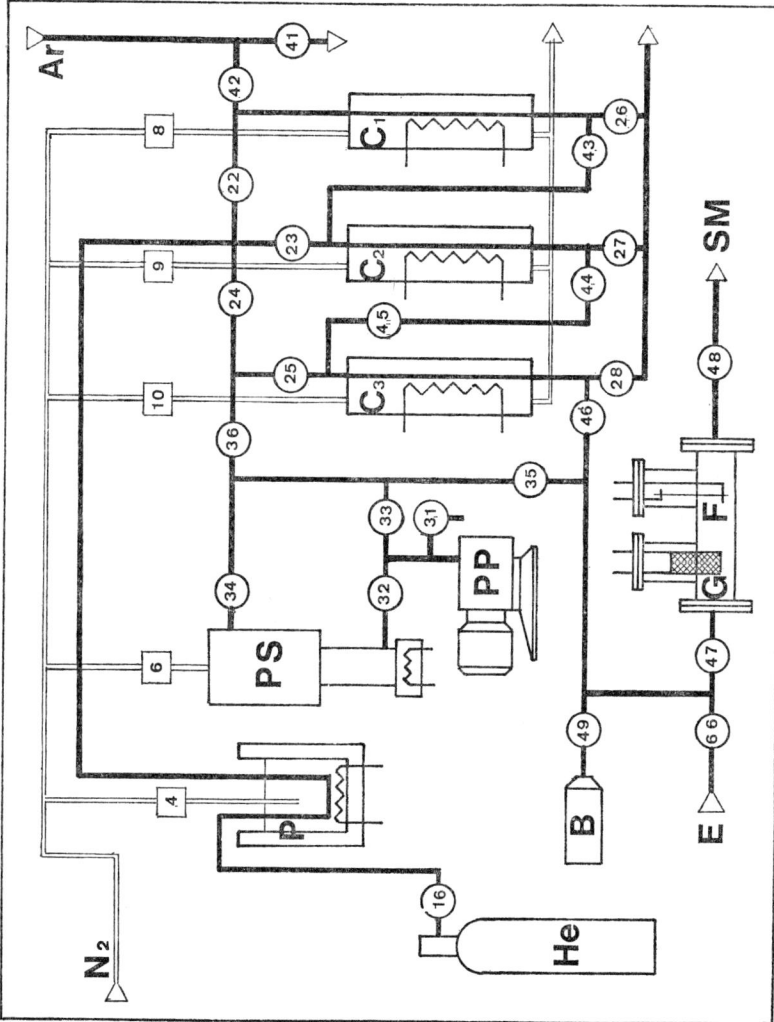

FIG. 1 Schematic diagram of the sampling line. C_1, C_2, C_3: chromatographic columns; P: cold trap for helium purification; PS: mercury diffusion pump; PP: mechanical pump; G: getter pump; F: hot filament; E: inlet for reference and standard samples; B: cylinder used to stock the processed cover gas sample if the mass spectrometric analysis is delayed; SM: mass spectrometer. (The numbers correspond to automatic valves).

of xenon and krypton and therefore a high resolution mass spectrometer is not needed.

When the sampling line is running automatically, it takes one hour for a complete analysis including the processing.

THE MASS SPECTROMETER

The sampling line is connected to the inlet line of a quadrupole mass spectrometer. The resolution is 150 at mass 130. The pumping system has been specially designed to eliminate residual gases which could interfere with the krypton and xenon spectra.

The pumping system includes two mercury diffusion pumps in series (150 litres and 10 litres/s), a vacuum reservoir of 5 litres, a zeolite filter and a mechanical pump. The pressure in the 5-litre reservoir is regulated, so that the fore pump is only needed a few minutes a day. The mass spectrometer is pumped by a 200-litre/s ion pump when the sampling line is not used.

The same computer controls the processing line procedure and the mass spectrometer peak switching and in addition is used for the acquisition of the data and for the calculations of isotopic ratios. The different masses are scanned by a peak switching technique, using a 0 to 10 V d.c. voltage supply programmed within 1 mV to the correct masses by the computer.

The ionic currents are measured by the counting technique, using a 17-stage electron multiplier and a 100 MHz scaler.

The measuring sequences are the following:

(1) Introduction of 0.1 cm^3 of helium containing 500 vpm of natural xenon, argon, krypton, and determination of the exact positions (in volts) of the maximum of each peak of interest.
(2) Measurement of the isotopic ratios on the same gases to check the sensitivity and calibrate the mass discrimination factors.
(3) Analysis of the actual processed cover gas sample (if the measured intensities indicate the presence of krypton and xenon coming from a cladding failure the isotopic ratios are calculated and the results printed).
(4) Confirmation of any spike detection is carried out by analysis of a reference sample which has a $^{124}Xe/^{129}Xe$ ratio nearest that of the detected spike-gas.

The results obtained in the laboratory show that with 10^{-5} cm^3 STP of natural xenon in front of the leak, the quadrupole mass spectrometer can measure the $^{124}Xe/^{129}Xe$ and $^{126}Xe/^{129}Xe$ ratios with an accuracy better than 10% (2σ). For all the others isotopic ratios measured relative to ^{129}Xe, the accuracy is 3%.

CONCLUSIONS

The analytical system which has been described will operate fully automatically. It will be possible to obtain every hour an isotopic analysis of any spike eventually present in the cover gas of the 'Phenix' Reactor. A sample of 20 litres of the cover gas having a concentration of xenon and krypton as low as 10^{-9} is sufficient to make measurements of the isotopic ratios $^{78}Kr/^{82}Kr$ and $^{124}Xe/^{129}Xe$ with an accuracy (2σ) better than 10%.

77
Influence of the Compatibility Between Impurities and Matrices on the Accuracy of Results in Spark Source Mass Spectrometry: The Case of the Matrices of Uranium and Plutonium

By J. P. BILLON

(*Commissariat à l'Energie Atomique, Centre d'Etudes de Bruyères-le-Chatel, Montrouge, France*)

IN most meetings about spark source mass spectrometry two kinds of people can be found. On one side the analysts, who, with the best working conditions, try to get the best from their spectrometer in the same way as with any other apparatus. On the other, the theoreticians whose aim seems only to improve their 'toy'. Working on radioactive samples, mainly plutonium, forbids any modification, and our laboratory then belongs to the first category.

Up to now, for convenience in sample preparation, most of our work has been done on properly prepared and cleaned solid samples. Quantitative analysis requires the use of relative sensitivity factors (RSF) but the apparent lack of reproducibility of the method increases the difficulties both to determine and use RSF's. Even when RSF's are computed from results on known samples and analysed in given working conditions, they may not be used without care. Generally their values are between 0·3 and 3.

In day-to-day analysis, standard deviation, which is roughly 20%, may reach 100% and even more. In such cases the use of RSF seems of little interest. It is necessary therefore to make replicate analysis and a statistical computation in order to obtain results in agreement with other methods. This is costly and time-consuming.

Here we intend to show some of our results to illustrate points with:

(1) A statistical computation which has been made on results obtained from known uranium samples. This computation allowed us some approach to RSF.[1]
(2) A comparison between three laboratories which analysed the same sample using the same method.[2,3]

NICKEL

A	B	C	D	E	F	G	H	I	J	K	L	Sample
2·1	1·5	1·6	1·3	1·15	1·8	1·15	1·25			1·5	1·7	R.S.F.
1·5→2·7	1·0→20	1·1→2·1	0·9→1·7	0·8→1·5	1·3→22	0·8→1·5	0·9→1·6			1·0→2·0	1·2→2·2	Range

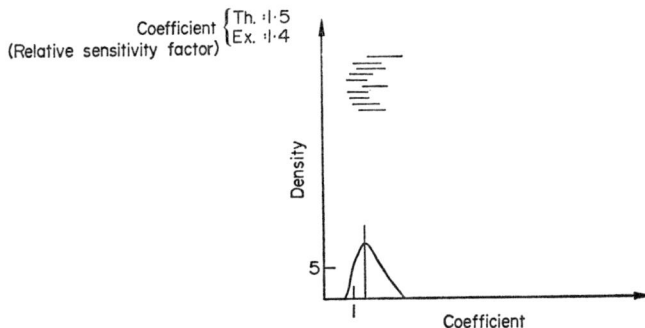

Coefficient {Th. :1·5 / Ex. :1·4}
(Relative sensitivity factor)

FIGURE 1

AN APPROACH TO RSF

We have made a statistical computation on a large number of results from twelve uranium samples. Concentrations covered a rather large range. A RSF with a given standard deviation of 30 % was applied to each element in each sample. As can be seen in Figs. 1–4, the ranges obtained were super-imposed to determine a kind of density. This method allowed the assimilation of a large number of results. Details are given in Reference 4. Experimental values obtained here are in good agreement with those computed previously by Goshgarian.[5]

IRON

A	B	C	D	E	F	G	H	I	J	K	L	Sample
2·7→5	1·85	1·1→27	1·5	0·8→14	1·2→24	1·3	0·9→20	1·25	1·0→3·5	1·2	1·7	R.S.F.
1·9→6·5	1·3→2·4	0·7→3·5	1·0→20	0·6→1·8	0·8→3·0	0·9→1·7	0·6→26	0·9→1·6	0·7→4·5	0·9→1·5		Range

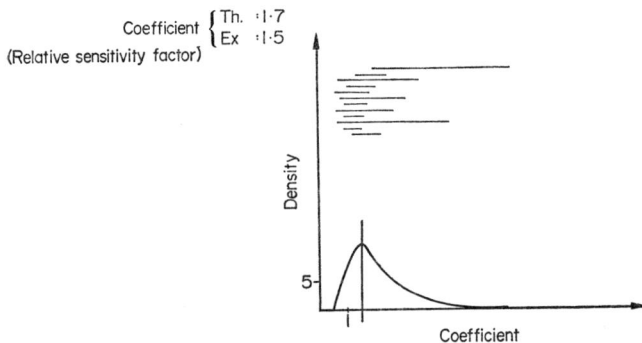

Coefficient {Th. :1·7 / Ex :1·5}
(Relative sensitivity factor)

FIGURE 2

COPPER

A	B	C	D	E	F	G	H	I	J	K	L	Sample
2·7	1·6	2·1→3·7	1·8→2·4	3	2·9	3·3	3	1·1	3·9→7·7	2·5→3·1	2·3	R.S.F.
1·9→3·5	1·1→2·1	1·5→4·8	1·3→3·1	2·1→3·9	2→3·8	2·3→4·3	2·1→3·9	0·8→1·4		1·8→4·0	1·6→3·0	Range

Coefficient { Th. :2 Ex. :2·4

(Relative sensitivity factor)

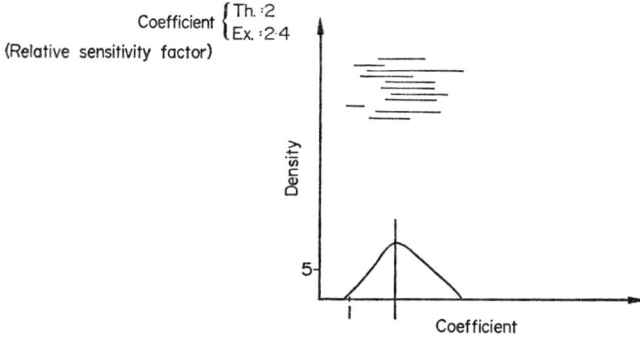

FIGURE 3   The RSF value under sample J refers to a non-homogeneous material.

Because of the lack of standard samples for Li, Na, K and Mg, Ca, Sr we were not able to determine RSF and the problem remains for such elements.

No problem arises when the matrix is plutonium. Uranium being a common impurity in plutonium, we made a statistical study of observed RSF for U in Pu and, afterwards, verified that

$$\frac{Ke}{U} = \frac{Ke/Pu}{Ke_U/Pu}$$

The experimental values for RSF are between 0·3 and 3. They are highly

MANGANESE

A	B	C	D	E	F	G	H	I	J	K	L	Sample
obs 1	2·4	3·2	2·7	1·3	2·1	2·2	3·5	1·6	4·5	2·0	2·1	R.S.F.
	0·7→3·1	2·2→4·2	1·9→3·5	0·9→1·7	1·5→2·7	1·5→2·9	2·4→4·6	1·1→2·1	3·2→5·8	1·4→2·6	1·5→2·7	Range

Coefficient { Th. :2·7 Ex. : 2·3

(Relative sensitivity factor)

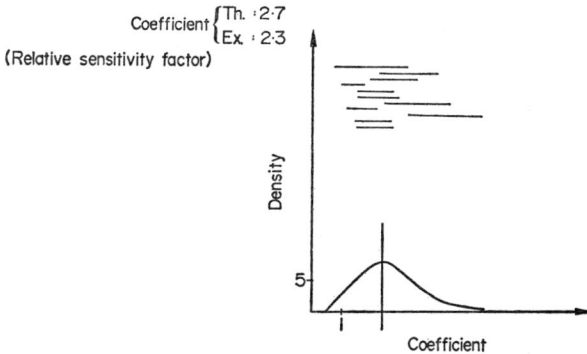

FIGURE 4

dependent on working conditions, mainly related to how optical density profiles on photoplates are measured. Great care is necessary when comparing RSF from various laboratories.

COMPARISON BETWEEN LABORATORIES

The same uranium sample has been analysed by three laboratories using the *same* method.

Even for a specially chosen sample, observed standard deviations are often large. However, it is important to notice that zirconium, which is an element known to be 'soluble' in uranium—in a metallurgical meaning—gives the smallest SD (15%). A statistical computation on means and variances (Table I) shows that results from the three laboratories are in good agreement.[6] Then, at least in this case, although the three spectrometers are not identical instrumental parameters may be neglected and a common RSF may be used.

DISCUSSION

All these results show, when the considered element is 'soluble' in the matrix, that a 15% standard deviation can be obtained. We think this is a limit which includes all the experimental and instrumental parameters and that the apparent lack of accuracy and reproducibility comes from the sample itself which is not homogeneous at the level of the small portions taken by the spark. This lack of homogeneity appears with every microanalysis method and causes difficulty when one wishes to compare spark source mass spectrometry results obtained on a single photoplate with those of spectrophotometry for example. It has been seen that determination of RSF requires a large number of results.

In the case of solid samples—metals—when concentrations are higher to a 'limit of solubility' in the matrix, inclusions appear and erratic results are obtained.[7] One can find impurities which behave in the same way. It seems that these elements precipitate together.[6] With the kind of metallic matrix we analyse, we think we cannot have high accuracy since with replicate analysis and computation from individual results, observed standard deviation is not only dependent on the method but also on the composition of the sample. Thus each determination is only representative of the small portion of sample used.

Microanalysis can be done but the spark being so erratic cannot be compared with ionic secondary emission.

When one needs an 'average analysis', samples must be homogenized. Two methods may be used. (*a*) The metal can be converted into the oxide although particles are generally still too large and there is no important improvement in reproducibility and (*b*) the metal may be taken into solution which is dried on graphite from which electrodes are made. First results using this latter method are promising although some elements are lost.

TABLE I

Comparison Between Laboratories

Distributions study internal standard: Uranium 235

Laboratories	Distribution parameter	Al	Si	Cr	Mn	Fe	Ni	Cu	Zr
1	Max. of distribution	290	155	20	29	180	53	51	27
	Computed mean	265	158	20	24	157·5	55	53·5	26
	Standard deviation $\frac{s}{M} \times 100$	34%	38%	17·5%	15%	34%	20%	20%	15%
2	Max. of distribution	300	187	21	24	147	66	49	30
	Computed mean	194	166	19	26	122·5	74·5	61	28·5
	Standard deviation $\frac{s}{M} \times 100$	29%	31%	26%	29%	29%	32%	29·5%	14%
3	Max. of distribution	150	201	18·5	19	200	66	31	35
	Computed mean	142	207	21·5	20	205	77	39	34
	Standard deviation $\frac{s}{M} \times 100$	45%	36%	23%	30%	32%	26%	24%	18%

654 J. P. BILLON

REFERENCES

1. Neuilly, M. and Leclerc, J. C., CEA-R-3911.
2. Stefani, R., Theorie et technique de la spectrométrie de masse à étincelles', Rapport CEA-R-3445.
3. Desjardins, M., 'Essai d'analyse absolue par SME. Thèse de docteur ingenieur de la Faculté des Sciences de l'Université de Grenoble', No. 459, 1964.
4. Billon, J. P., CEA-R-3883.
5. Goshgarian and Jensen, 'Interatomic properties of solids, demonstrated by spark con', Source. léè Ann. Conf.
6. Cetama, Groupe 12, CEA-R-4348.
7. Bonversox, D. F. and Leary, J. A., *Journal of Nuclear Materials*, 1968, **27**, 181–6.

78

An Analytical Method for the Investigation of Thin Films by Spark Source Mass Spectrography

By V. LIEBICH and H. MAI

(*Zentralinstitut für Festkörperphysik und Werkstofforschung, Dresden, Germany*)

INTRODUCTION

DURING the last decade very thin films became more and more important in research and technical application. The preparation of films having identical quality depends on the reproducibility of their physical and chemical composition. The most important factors regarding the chemical composition are a guarantee of correct matrix element concentrations and suppression of impurity elements. Therefore the analytical characterization includes the precise single element determination as well as the survey analysis for impurity elements in order to check the effectivity of the preparation process.

At present no universal highly sensitive method is available for the examination of thin films (thickness below 1 μm) simultaneously for all elements of the periodic table. Spark source mass spectrographs with low voltage arc or r.f. spark were mainly used for survey analyses of films having a thickness ≤ 1 μm.[1-10] In Table I critical experimental parameters and dimensions of obtained spark craters are surveyed for a number of publications which have appeared during the last few years. In most cases the penetration depth of the spark discharge has exceeded 1 μm. Applying those techniques to the analysis of films < 1 μm the resulting contribution of the substrate would cause large blanks in the mass spectra.

The object of the experiments described below was to work out a method for the examination of a special group of samples, *i.e.* conducting film on insulating substrate, having a thickness range between 0·01 and 1 μm. The spark penetration depth should be limited almost to the thin film to avoid serious blanks due to target excitation. For that reason a good in-depth resolution and optimum limits of detection had to be obtained.

EXPERIMENTAL

A spark source mass spectrograph MS 7 (A.E.I., Manchester) has been used to carry out the necessary experiments. The spark discharge has been initiated

TABLE I

Comparison of Methodical Details of Some Papers Concerning Thin Film Analysis by SSMS

Reference	Electrode configuration	Sample type	Film thickness (μm)	Sparked volume Depth (μm)	Sparked volume Width (μm)	Probe electrode	Remarks
(1)	▲● ☐● ⊕●	c	0, 1, 3–5	0·2–10	0·4–100	Fe, Au, Si	
(2)	☐● ▲●	c, c/n	0·4–0·5 ≦5	3–200	600	Au	
(3)	▲●	n	—	0·04	300	Au	Grain boundary
(4)	▲●	n, c	—	0·3–5	—	—	Grain boundary
(5)	☐●	c	3–300	0·2–10	25–300	Re, Ta, Rh, Al	Al surface coating
(6)	▲●	c	—	10	500	Pt, W	
(7, 8)	▲●	c	—	≦1–15	30–300	Pt, W	d.c. arc
(9, 10)	⊕●	c, c/n	0·1–50	0·2–20	50–150	Rh, Ta, Graphite	

● — fixed probe electrode   ☐ — sample electrode, motor xy-scan   n — non-conducting
▲ — sample electrode, local analysis   ⊕ — rotating sample electrode   c/n — conducting film on non-conducting substrate
☐ — sample electrode, manual xy-scan   c — conducting

between the 'large area' sample electrode and a thin probe electrode. The tip of that probe electrode was positioned in the ion-optical axis of the ion source. The size of the resulting spark craters normally depends on the following main factors:

(1) Energy consumption of the discharge.
(2) Material and structure of sample electrode.
(3) Material and shape of probe electrode.

Rough experiments showed that for the interesting type of sample the second point is the most important factor. Hence the assumption was made that due to the action of the insulating substrate the penetration depths could be limited almost to the film if a sufficiently low energy consumption in the discharge could be realized. The following precautions were met to control the spark energy.

(a) Generation of single discharges.
(b) Reduction of spark voltage.
(c) Prevention of double excitation of the same area element of the sample.

Some changes of the r.f. generator of the MS 7 as external triggering and an additional capacitive adjustment of the oscillator gave in contrast to the original version real single r.f. pulses.

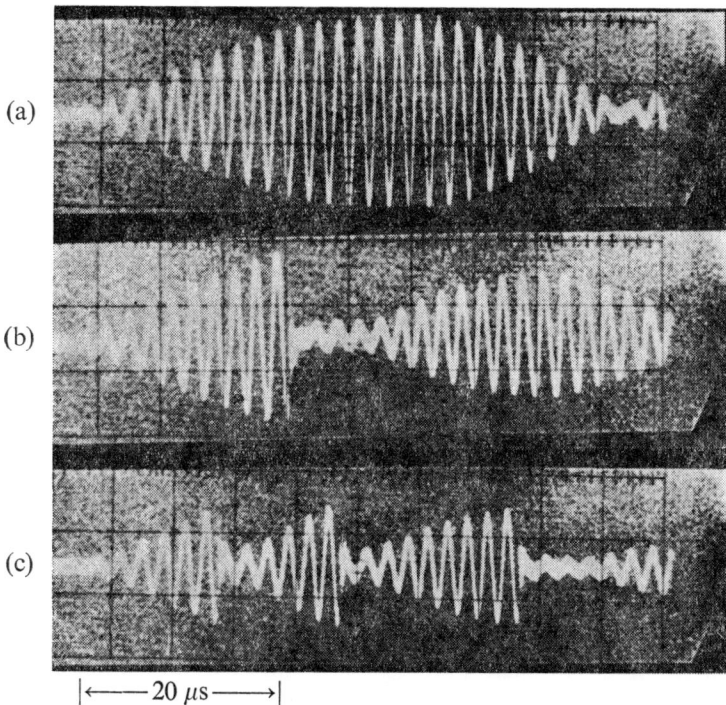

FIG. 1   Oscilloscopic representation of spark voltage. (a) no spark, (b) single spark, (c) multiple spark.

The gap between probe and sample electrode was chosen so that the break-down voltage was roughly equal to the maximum amplitude of the r.f. voltage. Under these conditions it was possible to obtain single sparks because in one r.f. pulse the breakdown voltage could not be reached a second time. A certain decrease of the electrode distance due to electrode adjustment or change in microstructure of electrode surface can result in further breakdowns within the same r.f. pulse.

Figure 1 demonstrates the behaviour of the r.f. voltage for (a) no spark, (b) single spark, and (c) multiple spark.

The reduction of spark energy has been realized by the choice of low r.f. amplitudes. This results in very small electrode gaps. The lower limit of the gap and in this way also of the spark voltage is determined by the repro-ducibility of the micro-manipulator adjustment.

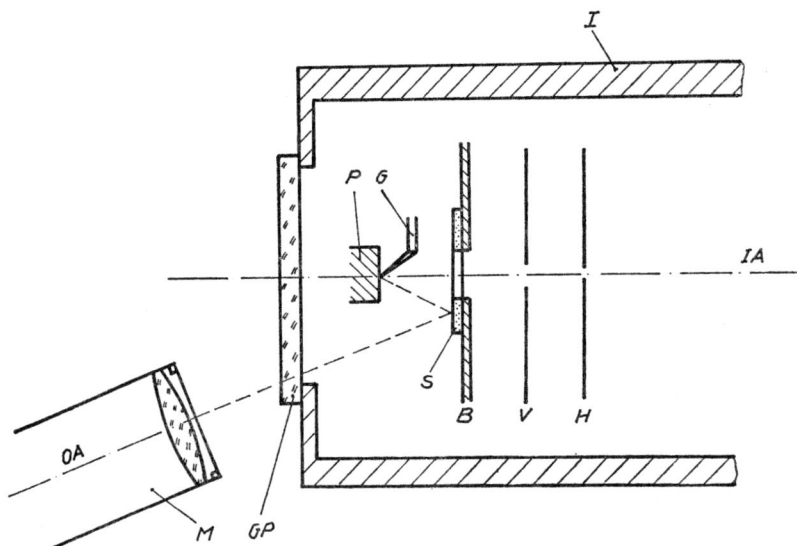

FIG. 2 Sample arrangement in the ion source. M–microscope, P–sample; G–probe; S–mirror; GP–front plate; B–accelerating electrode; V–slit No. 1; H–main slit, I–source housing; IA–ion optical axis, OA–optical axis.

The third precaution has been fulfilled by the installation of a stereo-microscope (SM XX, VEB C. Zeiss, Jena). Then a controlled electrode adjustment was possible with the aid of a mirror mounted inside the ion source (Fig. 2).

LIMITS OF DETECTION

Provided all other parameters are kept constant the limits of detection[11] of spark source mass spectrography are determined by the exposure of the mass spectrum with respect to the amount of sample analysed. The demands for improvement of the spatial resolution are accompanied by the reduction of the crater volume. Thus a compromise between both factors has to be found for every analytical problem.

A way to avoid a serious deterioration of the limits of detection without loss of spatial resolution could be to integrate over a larger number of sparks put on identical parts of the sample. In thin film analysis where a homogeneous area distribution could be presumed, normally the area resolution is not of interest. Therefore an optimum in-depth resolution can be achieved without loss in detection power provided a sufficient sample area is available. Limits of detection near 1 ppm/atm have been estimated for a film of 0·01 μm thickness and a sample area of 5 × 5 mm^2.

RESULTS AND DISCUSSION

The experimental parameters are shown in Table II. Under these conditions the examination of spark penetration has been carried out on synthetic samples—vacuum coated copper films (0·005–2 μm thick)—on glass substrates of optical quality.

TABLE II

Experimental Parameters

Sample electrode	Manual xy-scan
	Scanned area $f_s = 4 \times 4$ mm^2
Probe electrode	High purity niobium, tip diameter $d_t \approx 0·07$ mm, fixed position
Electrode distance	$d_e \leqq 0·08$ mm
R.f. oscillator	R.f. voltage according to 25 per cent mains input voltage, trigger pulse length $t_p = 20$ μs, external triggering
Accelerating voltage	$U_b = 20$ kV
Magnet current	$I_m = 250$ mA
Main slit width	$d_m = 0·04$ mm
Photoplate	Ilford Q2
Development	ID 13, 3 min at 20°C

Figure 3 shows spark craters obtained by one r.f. pulse each in copper films of different thickness. It can be seen that in the area of the crater bottom the copper film is almost completely evaporated whereas the substrate remains almost unaffected. In Fig. 3(a) for a film thickness of 0·09 μm the spark creates, in addition to the crater, a lifting of the surrounding film from the surface. Well-defined crater walls are shown in Fig. 3(b) (film thickness 0·03 μm) whereas for 0·008 μm film thickness the crater shape could not be reproduced so clearly. It shows one of the smallest penetrations obtained so far.

The measurements of crater depth and film thickness have been carried out on an interference microscope (Epival interphako, VEB C. Zeiss, Jena). The samples were vacuum coated with a 0·03 μm copper film before the measurements for the compensation of phase shifts due to reflection of light on metal and insulator surface respectively. It is hoped that the crater shape will not be seriously affected by the coating.

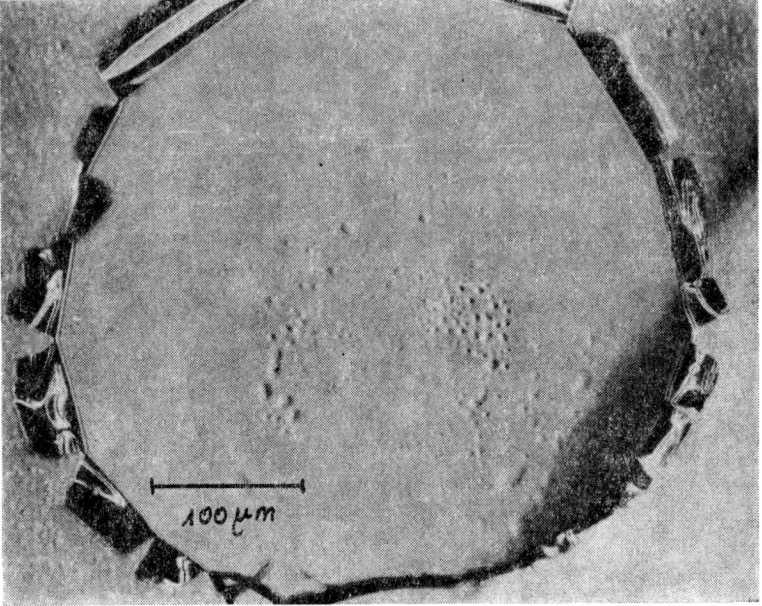

FIG. 3(a)

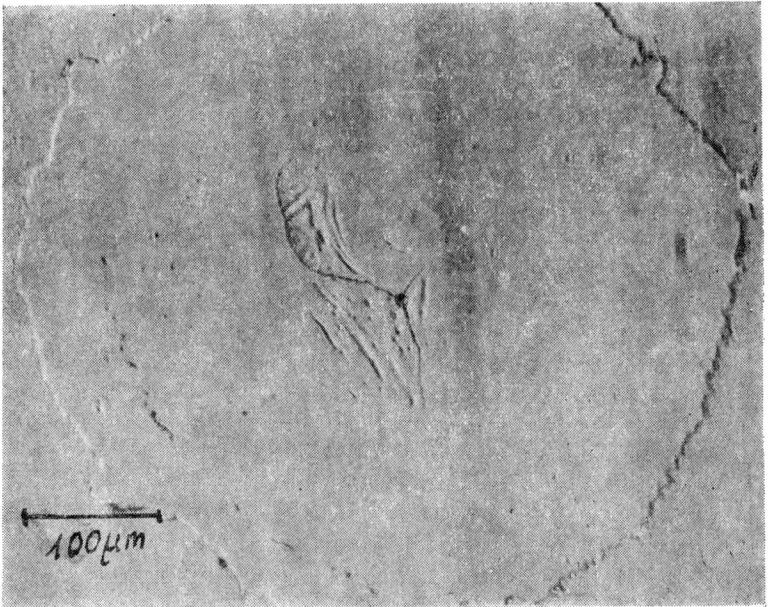

FIG. 3(b)

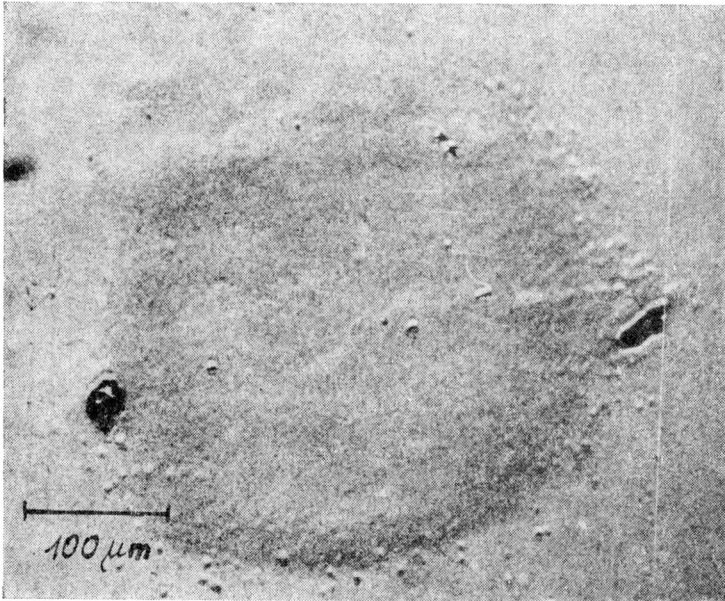

FIG. 3(c)

FIG. 3   Spark craters produced by one r.f. pulse (a) 0·09 μm-film, (b) 0·03 μm-film, (c) 0·008 μm-film.

The results of crater measurements are shown in Fig. 4. Curve No. 1 represents the relation between film thickness and spark penetration depth. In the range below 1 μm the crater depth is nearly equal to the film thickness whereas near 1 μm and above the single spark can no longer reach the substrate.

The errors of the measurements of crater depth could be estimated to a mean error factor of 1·3. In critical cases, *i.e.* when craters having mis-shaped walls had to be measured, it could reach a value of 2.

Curve No. 2 demonstrates the crater volume as a function of film thickness. The volume has been calculated from measured crater depths and diameters— assuming a cylindrical crater shape. The maxima of both curves are not, at present, fully understood; an understanding is hoped for after completion of the present experiments. Representative mean values of crater dimensions and film thickness are collated in Table III.

Mass spectra obtained from these copper films show in every case copper as the matrix element. The contributions of the matrix elements of the

TABLE III

Representative Results of Optical Measurements

Film thickness [μm]	1·8	0·65	0·16	0·090	0·044	0·032	0·007 7	0·004 8
Crater depth [μm]	0·11	0·63	0·18	0·086	0·045	0·054	0·007 3	0·005 0
Crater diameter [μm]	110	90	250	400	630	570	340	570

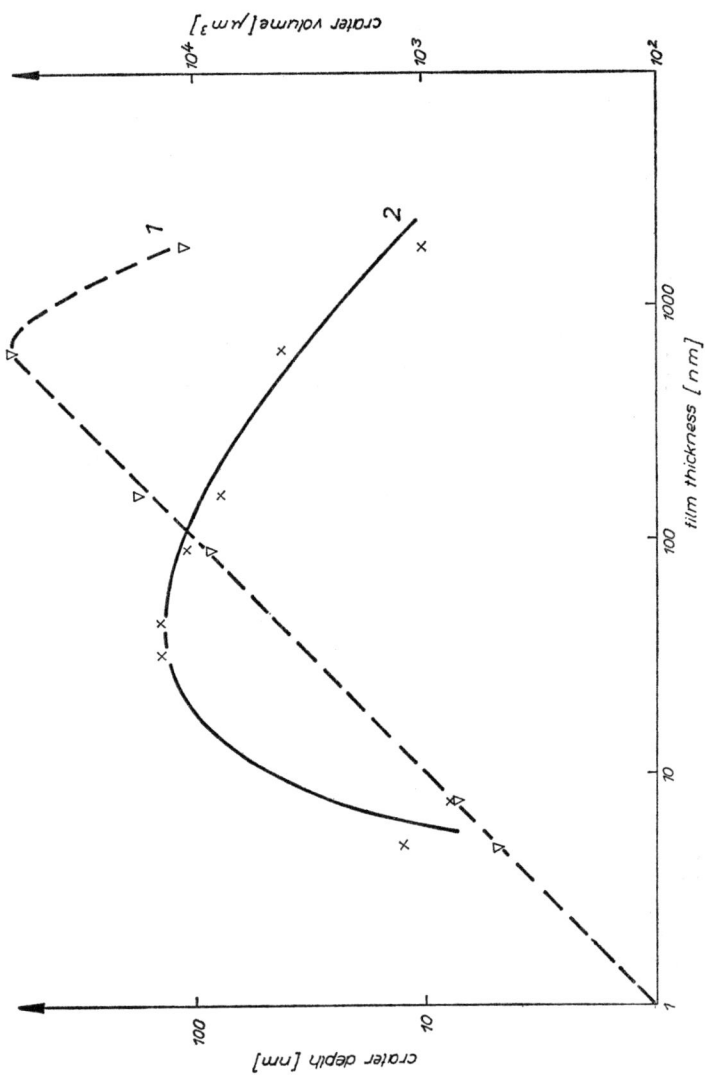

FIG. 4   Depths and volumes of spark craters as a function of film thickness.

substrate remain below 30%/atm in nearly all experiments of the copper signal. Thus in addition to the optical measurements from the mass spectra the conclusion can be also drawn that the penetration of the r.f. spark was restricted nearly almost to the metallic film.

The values of relative crater depths—*i.e.* the relation of crater depth to film thickness—have been calculated from copper and substrate element concentrations and compared with the values from interferometric measurements (Table IV, columns 3 and 4). The very good agreement of the results

TABLE IV

**Relative Crater Depths Obtained by Optical Measurements,
Calculated from the Mass Spectra and for an Assumed Penetration of 1 μm**

Film thickness (μm)	Relative crater depth		
	Optical	Mass spectra	Assumed penetration of 1 μm
0·005	1·1	—	200
0·032	1·7	1·3	31
0·090	1·0	1·1	11
0·16	1·2	1·2	6·3
0·65	1·0	1·4	1·5
1·8	0·06	—	0·6

of both methods seems to be a chance hit. A much larger deviation could be explained by taking into account the error factor of optical measurements and relative sensitivity factors defined to 1. Column 4 finally shows calculated values of the relative crater depth assuming that the penetration in the different films examined would amount to 1 μm: a value hitherto published in a large number of papers for conducting solids. For films <1 μm it is a measure for the reduction of blanks due to the substrate obtained by single spark in comparison to multiple spark excitation.

The method described here has been successfully applied to the identification and analysis of various metal and semiconductor thin films (0·01–0·5 μm) on glass and other insulators.

The application of the method is not restricted to this sample type. It is useful also for the checking of coating processes for metallic substrates, etc. In that case the substrate should be replaced by an insulator for the preparation of model samples.

CONCLUSIONS

Thin conducting films on insulating substrates have been analysed by the single spark method. In the film thickness range between 0·005 and 1 μm the spark penetration could be restricted nearly almost to the film. Due to the combined effects of single discharges and the insulating substrate the smallest crater depths achieved are in the range between 50 and 100 Å whereas the crater diameter amounts to values between 100 and 600 μm. The limits of detection obtained for a scanned area of about 25 mm^2 are near 1 ppm/atm.

REFERENCES

1. Hickam, W. M. and Sweeney, G. G., *in* 'Mass Spectrometric Analysis of Solids', *ed*. A. J. Ahearn, Elsevier Publ. Co., New York, 1966, chapter V.
2. Malm, D. L., *in* 'Progress in Analytical Chemistry, Physical Measurements and Analysis of Thin Films', Plenum Press, New York, 1969, Vol. 2, p. 148.
3. Desjardins, M. and Williams, J. P., *J. Amer. Ceram. Soc.*, 1968, **51**, 296.
4. Tong, S. S. C. and Williams, J. P., ibid, 1970, **53**, 58.
5. Cupachin, M. S., Krjuckova, O. I. and Ramendik, G. I., 'Analytical Methods of Spark Source Mass Spectrometry', Atomizdat, Moscow, 1972.
6. Yamaguchi, N., Suzuki, P. and Kammori, O., *Bunseki Kagaku*, 1969, **18**, 3; ibid., 1969, **18**, 370.
7. Sato, K. and Yamaguchi, N. *et al.*, *Nippon Kinzoku Gakkaishi*, 1970, **34**, 610.
8. Sato, K. and Yamaguchi, N. *et al.*, *J. Iron Steel Inst. of Japan*, 1972, **58**, 1495.
9. Roberts, J. A. and Millett, E. J., *Proc. 6th Ann. MS 7 Users Conf.* 1966, p. 39.
10. Clegg, J. B., Millett, E. J. and Roberts, J. A., *Anal. Chem.*, 1970, **42**, 713.
11. Mai, H., *in* 'Spurenanalyse in hochschmelzenden Metallen VEB Deutscher Verlag für Grundstoffindustrie', Leipzig, 1970, p. 70.

79

Mass Spectrometric Study of the Vacuum Arc

By D. STÜWER

(*Institut für Spektrochemie, Dortmund, Germany*)

FOR a long period, the r.f.-spark was the only ion source in solid state mass spectrometry; nowadays its domain is shared by the vacuum arc more and more. The vacuum arc discharge is unipolar and stationary, thus leading to better reproducibility and higher precision in analytical results. At a first glance, the most striking feature of vacuum arc spectra is that they show many more lines than r.f.-spectra. A more detailed look reveals the reason: higher ionization states are much more populated in the vacuum arc. This phenomenon was first discussed by Franzen and Schuy[1] on the basic assumption of local thermodynamic equilibrium. Though only being a postulate not yet proved, vacuum arc phenomena later on were often discussed on this basis—as for example by Venkatasubramanian and Rajagopalan.[2]

As the populations of ionization states represent the phenomenon under investigation, sufficient information on the processes involved must be encoded in these distributions. The problem is an adequate strategy for decoding. Thus we investigated, by thoroughly measuring the population distributions of the various ionization states, a systematic set of samples:

(1) pure metals as general reference,
(2) alloys of several components for studying the influence of a main constituent on alloying constituents, and
(3) binary alloys for investigating the influence of an alloying constituent on the ionization distribution of the main constituent.

The vacuum arc as an ion source was attached to an A.E.I.-MS 7 spectrometer. The evaluation of the photoplates was performed by an automatic evaluation system which has been described before.[3]

Typical distributions for the population frequencies of the various ionization states are shown in Fig. 1 containing some results for pure metals. The diagrams show the decadic logarithm of the concentration ratios c_i—with the sum normalized to unity—plotted as a function of the number i of ionic charges. For reason of comparison, it is commendable to characterize such distributions quantitatively by one numerical value. For this we have chosen

$$Q = \sum_i i^2 c_i$$

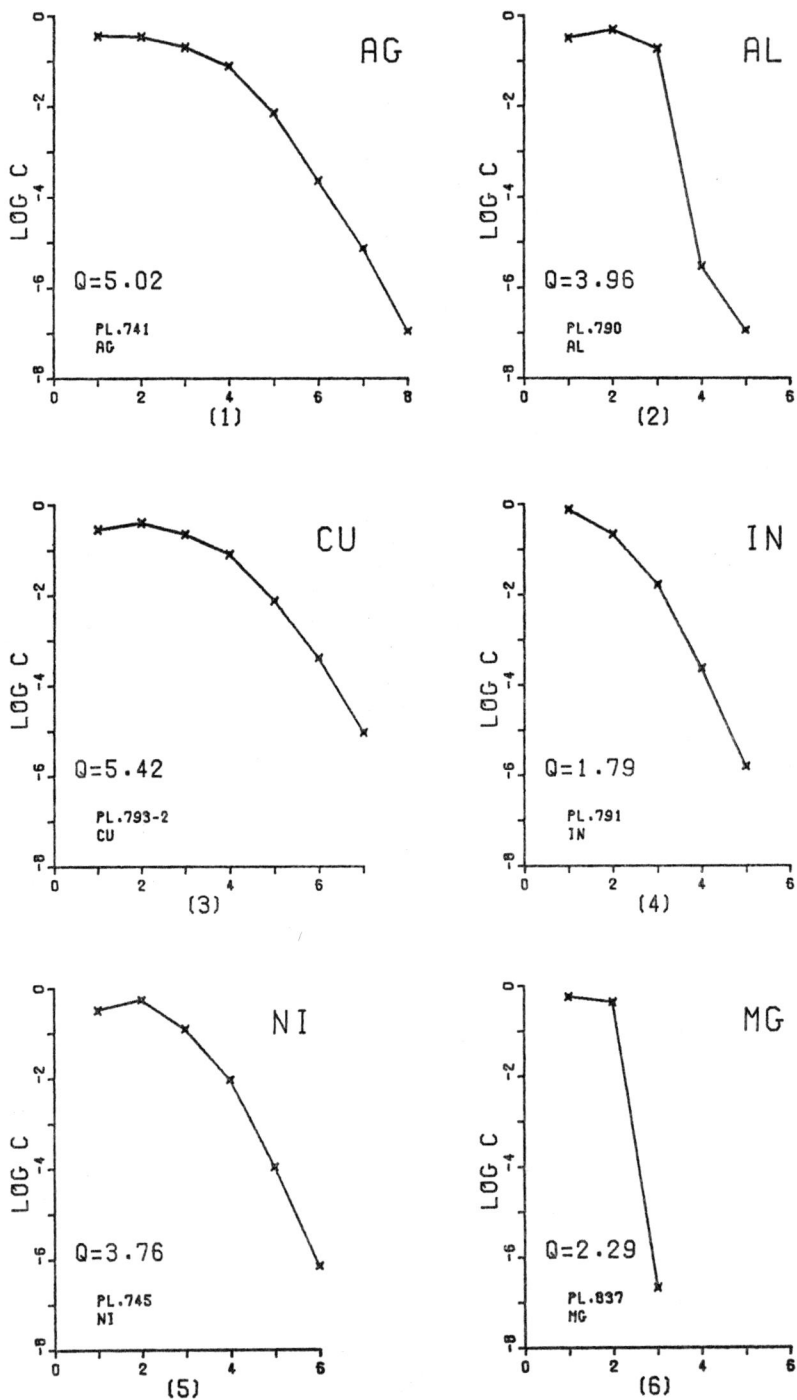

FIG. 1 Population of ionization states—pure metals.

This value Q is called 'centre of charge' because it reflects—together with temperature and density of the electrons—the Debye-radius as the similarity parameter of a plasma.

The most conspicuous feature is the sudden drop in the distributions of Mg and Al. In both cases there is a corresponding step in the sequence for the ionization energies because the next electron is to be taken from a closed shell. Another remarkable fact is the wide spread of Q-values although the ionization energies of the elements are comparable. This must be caused by different electron temperatures in the ionization region of the plasma which, for its part, depends on the properties of the cathode material. This dependence is especially demonstrated by the fact that pressed Ag powder leads to a Q-value of 4·44 whereas for vacuum molten Ag a Q of 5·02 is found in Fig. 1. In addition to the elements of Fig. 1, the Q-values were measured for pure In, Sn, Sb, Pb and Bi. They appear well correlated to the conductivity of the cathode material, the electrical as well as the thermal. For electrically and thermally good conducting materials we get high electron temperatures in the ionization region of the vacuum arc plasma.

This statement can be checked by investigating high-alloy samples. When alloying elements yield a rather 'cold' plasma into a matrix with a rather 'hot' plasma, the centres of charge of the alloying constituents should be shifted towards higher values. The examples in Fig. 2 demonstrate this effect. The alloying constituents—from left to right—are Sn and Sb, the main constituents—from top to bottom—are Cu, Ag and In. The ionization distributions of pure Sn and Sb are not shown, because here the triply charged state is already the highest detectable one. According to expectation the centres of charge for both elements—Sn and Sb—are shifted towards higher values when embedded into matrices of higher conductance.

This dependence just demonstrated must be a mutual one: when the centres of charge for the alloying constituents are shifted towards higher values, that of the main constituent must be shifted towards a lower value. Investigating this conjecture, Fig. 3 shows the distributions for the components of some binary alloys: Ni in Cu, Al in Ag and Mg in Ag. The Q-values for the pure components are known from Fig. 1. Summarizing the results there is a somewhat surprising feature: whereas the centre of charge of Cu is lowered by Ni and that of Ni is—vice versa—raised by Cu as expected, in the case of Al in Ag both centres of charge are going down. Nearly the same is the case for Mg in Ag. This fact must now be discussed.

In both samples the alloying constituents are of much lower atomic weight than the main constituent, in contrast to the previously discussed samples. The vacuum arc represents a plasma effluent out of a reaction region of high particle density into an ambient vacuum. The expression for the velocity of this particle propagation contains a diffusion component and this is mass-dependent. The lighter an element the faster it escapes from the high-density reaction region of the plasma. Thus the mass spectrum does not only demonstrate the lowering of the centre of charge for the main constituent, it also shows a lowering or at least no rising of the centre of charge for the alloying constituent. All particles take part in the energy balancing processes with the electron gas, e.g. ionization and recombination, for only a short time and thus particles escaping faster take less part in the heating-up process. The

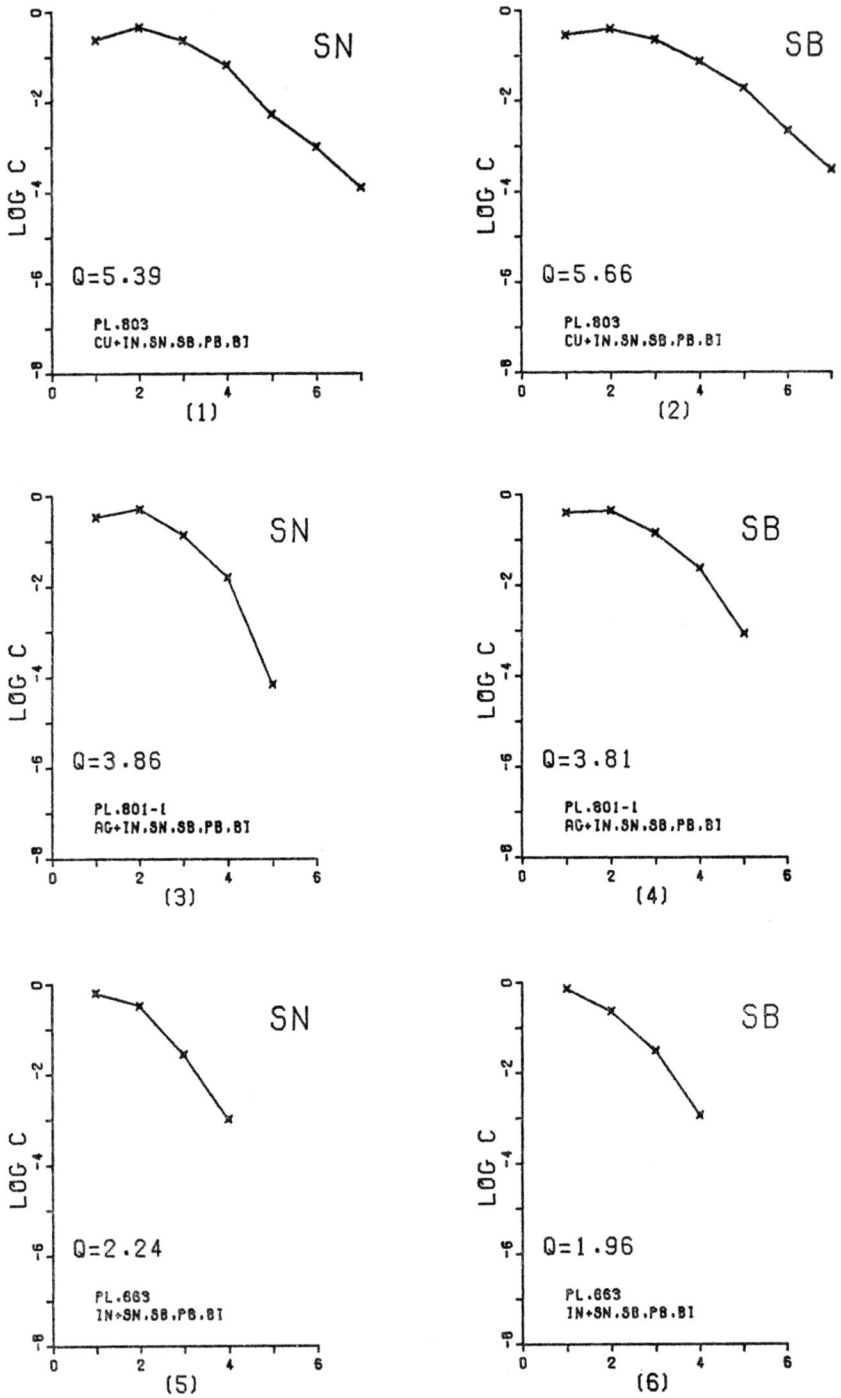

FIG. 2   Population of ionization states—alloys.

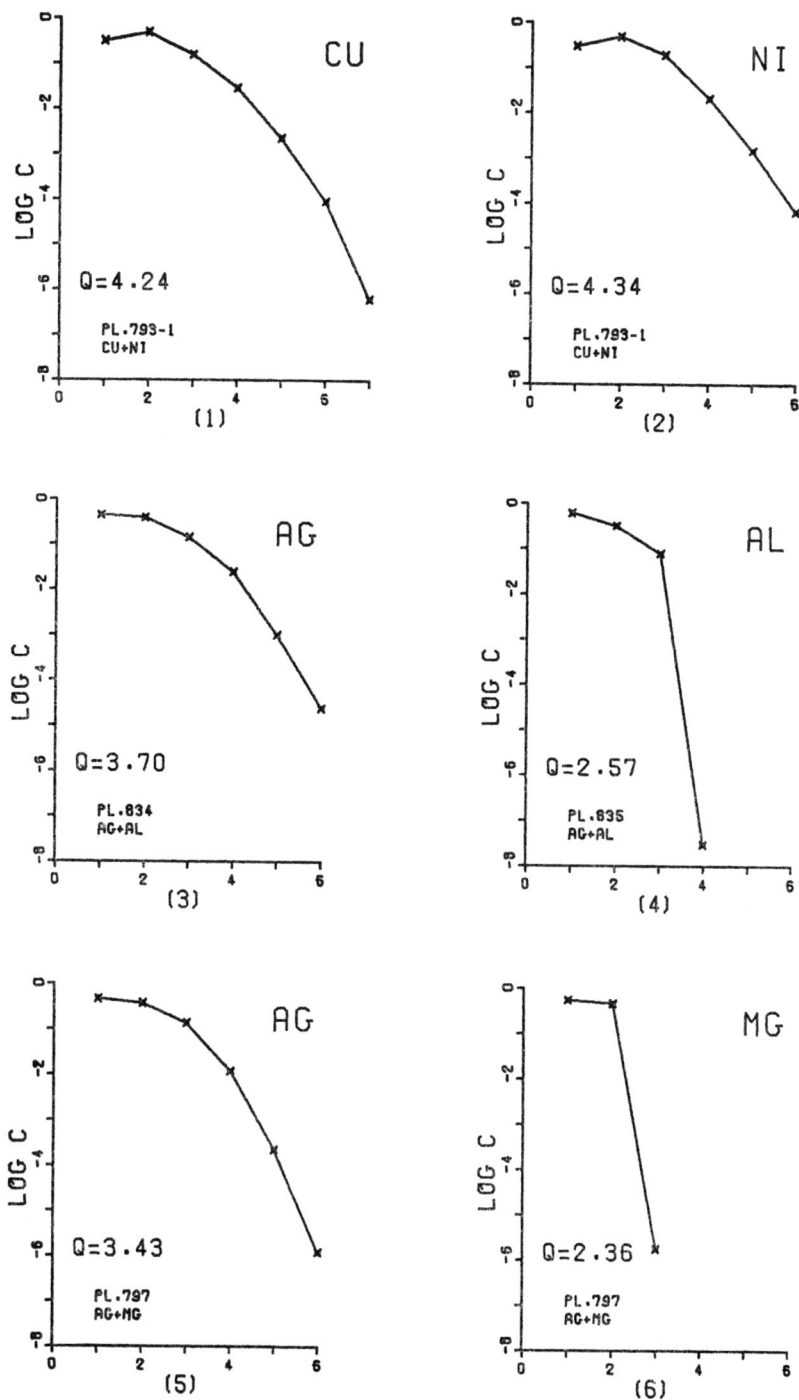

FIG. 3 Population of ionization states—binary alloys.

temperature of the electron gas is left unbalanced with that of the heavier species: there is no thermal equilibrium in the vacuum arc.

Let us now consider this statement from a more theoretical point of view. In the vacuum arc we have a potential hump of about 70 eV in the vicinity of the cathode, as pointed out by Davis and Miller.[4] The relaxation time for thermalization of electrons with an initial kinetic energy of 70 eV can be calculated as in the order of 10^{-12} sec, the electron density assumed to be about 10^{18} cm^{-3}. Thus the electrons can be treated as thermalized and we may speak of an 'electron temperature'. For getting an estimate of this electron temperature we remember a relation often used in emission spectroscopy:

$$10 \, kT \lesssim E_j$$

where E_j represents the ionization energy of the highest detectable ionization state. Supposing the detection power of mass spectrometry two orders of magnitude higher, we have—with respect to the Maxwellian distribution— to use

$$14 \, kT \lesssim E_j$$

Thus we find the electron temperature somewhat above 100 000°K. This means that the mean kinetic energy of the thermalized electrons is about one-seventh of the initial kinetic energy corresponding to the potential hump.

The electron temperature having been estimated, we can now calculate the relaxation times for ionization and recombination by their reaction cross-sections and their frequencies respectively. We get a relaxation time in the order of 10^{-8} sec for ionization, and in the order of 10^{-6} sec for recombination. The relaxation time for recombination is markedly higher, because the low-energy electrons are predominantly contributing to recombination, and their relative density is becoming higher with the decrease in electron temperature during the heating-up of the ionic and neutral gas. This increase of recombination with decreasing electron temperature has still another consequence. In the effluent plasma of the vacuum arc, with its decrease of electron density and temperature, the population of ionization states can only be shifted towards lower ionization because of the increasing recombination. Compared with plasmas in thermal equilibrium, the vacuum arc shows a surplus of highly ionized states. This surplus cannot be a consequence of the propagation into the ambient vacuum, it represents the non-equilibrium nature of the vacuum arc favouring ionization against recombination.

In the effluent discharge plasma of the vacuum arc there are two concurrent processes: the energy-exchange of the hot electron gas with the heavier components; and the flux into the ambient vacuum, the latter rapidly lowering the density of particles. Thus the rate of energy-exchange processes is vanishing, some state of unbalanced energy-exchange reactions is frozen in. Ionization is still much favoured against recombination because of the higher relaxation time for recombination. A marked surplus of ionization processes is yielding population distributions of the ionization states differing markedly from those of equilibrium plasmas.

REFERENCES

1. Franzen, J. and Schuy, K. D., 'Phenomena in Ionized Gases', Int. Conf. Belgrade, 1965, Vol. 3, p. 242.
2. Venkatasubramanian, V. S. and Rajagopalan, P. T., *Proc. of the Nucl. Phys. and Sol. State Phys. Symp.*, Bombay 1970, Vol. II, p. 547.
3. Franzen, J., Schönfeld, W. and Stüwer, D., 'Advances in Mass Spectrometry', The Institute of Petroleum, 1970, Vol. 5, p. 322.
4. Davis, W. D. and Miller, H. C., *J. Appl. Phys.*, 1969, **40,** 2212.

80
Fingerprint Spectra in Secondary Ion Mass Spectrometry

By H. W. WERNER, H. A. M. DE GREFTE
and J. VAN DEN BERG

(Philips Research Laboratories, Eindhoven, The Netherlands)

INTRODUCTION

MASS spectrometry can be used to analyse gases, liquids and solids. The specimens may be in the form of elements or compounds.

The electron-bombardment ion source has long been used in the analysis of gases.[1] The spectra of gases in the atomic state generally consist of singly and multiply charged ions of the gas being analysed. In the case of gases in the molecular state or of chemical compounds, the ions consist not only of the types just mentioned but also of ions produced by decomposition of the compounds before or after ionisation. The frequency with which the individual fragmentary ions occur depends on the binding energy of the original molecule and is therefore characteristic of the molecule concerned (fingerprint spectrum). Such fingerprint spectra have long been used for the qualitative and quantitative analysis of gases and mixtures of gases. The mass spectrum obtained from a gas mixture is a linear superposition of the contributions of the individual components. Knowing the fingerprint spectra of the individual components, their concentration can therefore be calculated from the measured mass spectrum of the mixture.[2,3]

Secondary-ion mass-spectrometry (SIMS) is becoming increasingly important for the analysis of thin layers.[4–30] In SIMS the specimen to be analysed is placed in the mass spectrometer and bombarded with a beam of primary ions. As bombardment is continued, layer after layer of the sample is peeled off. The sputtered neutral particles[9] or the positive or negative[10–12] ions characterize the original specimen and can be used for qualitative analysis and, after calibration of the abundance of particular types of ions, for quantitative analysis.[12,13,28] In analogy to fingerprint spectra in gas mass-spectrometry the concept of fingerprint spectra in secondary-ion mass-spectrometry was previously introduced.[30,37] In this latter case the relation between the fingerprint spectrum and the 'structure' of the sample under investigation is more complex than for gas mass-spectrometry as there are much more essential parameters involved in the formation of secondary ions from a solid than in the generation of ions from a gas molecule. These essential parameters, typically of a solid, such as the band structure and in particular work function, sublimation energy, crystal orientation etc., are somehow reflected in the secondary-ion fingerprint-spectrum. Therefore the secondary-ion fingerprint-spectrum should be typical for a given chemical compound. To prove this, several chromium oxides have been investigated in the present work.

Regarding the analysis of solids by means of secondary-ion mass-spectrometry, the situation is similar to the above mentioned mode of gas analysis when the sample consists of a mixture of different compounds of two elements. An analysis of the concentration as a function of depth can be carried out in that case, by means of fingerprint spectra, if the measured mass spectrum is produced by a linear superposition of the different fingerprint spectra. In the present paper the validity of this superposition theorem will be proven for the case of thin chromium oxide films.

SIMS-FINGERPRINT SPECTRA OF CHROMIUM OXIDES

Experimental
Samples of different chromium oxides have been prepared by pressing chemically prepared powders of CrO_2, Cr_2O_3 and CrO into pellets. These samples, as well as discs of CrO_3 and vacuum-melted chromium were analysed in a secondary-ion mass-spectrometer (Cameca S.A., Type IMS 300) by means of Ar^+ ion bombardment (angle of incidence 45°, energy of the argon ions: 5·5 keV when measuring positive secondary-ions, 14·5 keV when measuring negative secondary-ions). The residual gas pressure in front of the target during bombardment amounted to 10^{-7} torr (Nitrogen equivalent) $\approx 10^{-5}$ Pa (1 Pa = 1 N/m^2).

Results and Discussion
Positive and negative secondary-ion currents obtained from these samples are given in Table I. From these spectra one can make the following observations: The spectra obtained from the chromium oxide samples, with a ratio r of the oxygen/chromium content $r = n_0/n_{Cr}$ in the sample varying between 3 and 0, are different.† This is illustrated in greater detail in Figs. 1a, 1b and 1c.

From these results one may conclude that a decrease in the oxygen content from the CrO_3 sample towards the Cr sample is reflected in the characteristic mass-spectra as a decrease in the intensity of ions with high oxygen content ($r^* \geq 1$) and an increase in the intensity of ions with a low oxygen content ($r^* \leq \frac{2}{3}$). This conclusion is similar to the one obtained for thin films of Cu oxides[30] and other oxides.[28] Regarding our CrO sample one must bear in mind that the existence of CrO is not generally accepted.[31-34] Some peaks in our mass spectra from the Cr_2O_3 and CrO samples, however, show a slight, yet systematic difference, indicating the existence of two different

† The chromium sample, supposed to contain no oxygen, is in fact slightly oxidized, as can be seen from the chromium oxygen clusters.

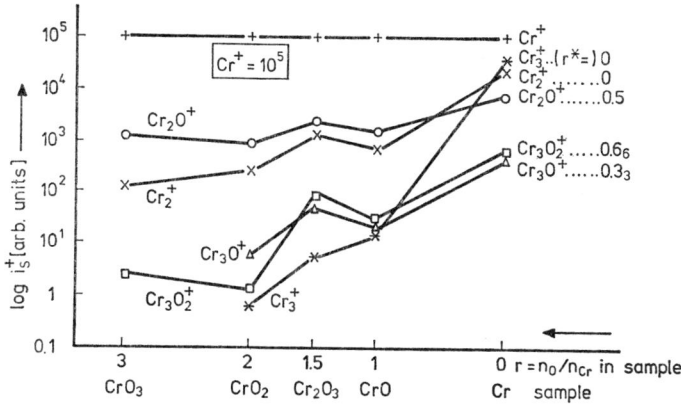

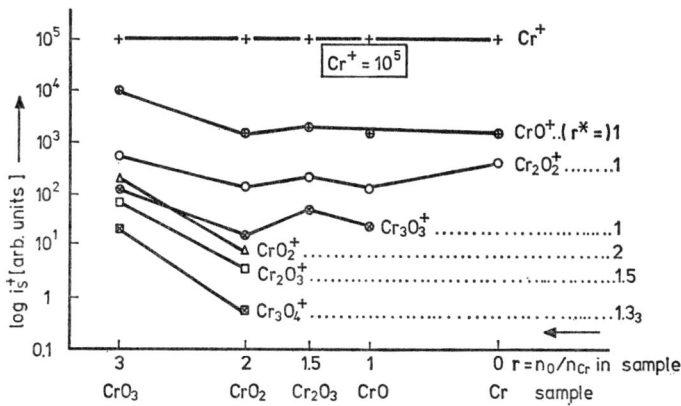

FIG. 1(a) and (b)   Logarithm of the positive secondary ion currents $\log i_s^+$ from various chromium oxide samples plotted against the ratio r = (number of oxygen atoms/number of chromium atoms) = n_O/n_{Cr} in the samples. The values of the (O/Cr)-ratios r^* in the various ion species are also given.

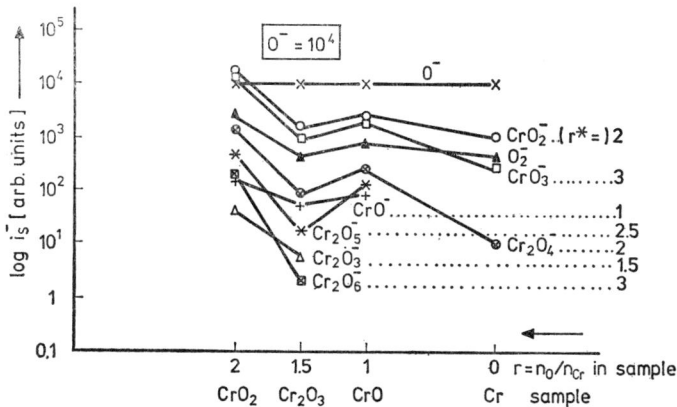

FIG. 1(c)   Logarithm of the negative ions.

chromium oxides. This result is confirmed by recent thermodynamic considerations of Venema[35] showing the possibility of CrO formation under suitable conditions.

THE APPLICATION OF FINGERPRINT SPECTRA FOR A DEPTH ANALYSIS OF THIN CHROMIUM OXIDE LAYERS

Experimental

Technically pure, vacuum-melted chromium was heated in an oxygen atmosphere so as to produce a chromium oxide layer whose oxygen content decreases with depth z. (We wish to thank Dr Kwestroo and Mr Hall for the preparation of this and the other samples.) This layer was analysed with the same instrument as described above.

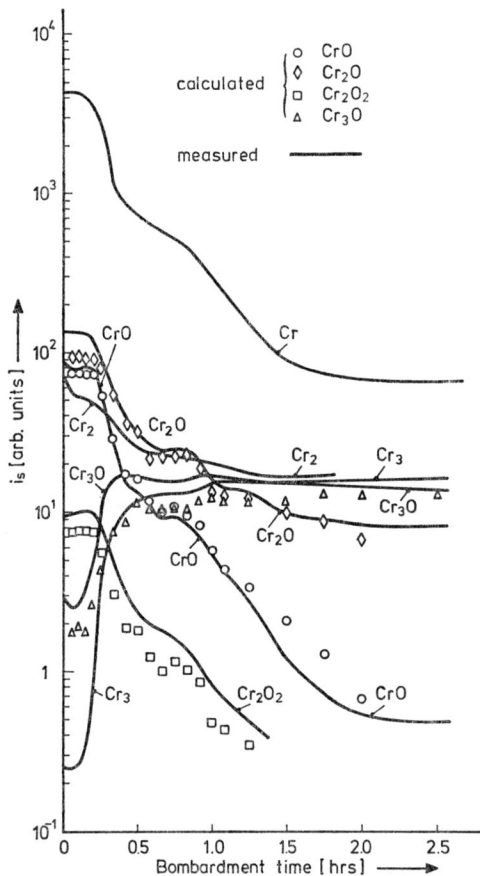

FIG. 2 Secondary ion currents i_s from a thin chromium oxide layer corrected for charging of the surface as a function of bombardment time t or depth z. (Sputtering speed was approximately 6 μm/h.) Best fits through these corrected experimental points are given by solid lines. The corresponding calculated points, assuming three phases are given by symbols.

Results and Discussions

The positive secondary-ion currents $i^+{}_s$, obtained as a function of bombardment time t ($\sim$ depth z) and corrected for charging of the surface are given in Fig. 2. The abundance of the different ion species at $t = 0$ is found to agree within the experimental error with the fingerprint spectrum obtained from pure Cr_2O_3 (Table I). From this agreement one may conclude that the topmost layer consists mainly of Cr_2O_3.

With increasing bombardment time t one penetrates into zones with decreasing degree of oxidation. This can be shown as follows: from Fig. 2 one can derive as a function of bombardment time t the ratios of the different secondary ion currents i_s to the chromium ion current i_{Cr} (normalized fingerprint spectra Fig. 3). From a comparison of these values with the ones given in Fig. 1a or Table I one can conclude that as a function of bombardment time t the measured spectra match best first with Cr_2O_3, than with the less oxygen bearing CrO and finally with Cr.

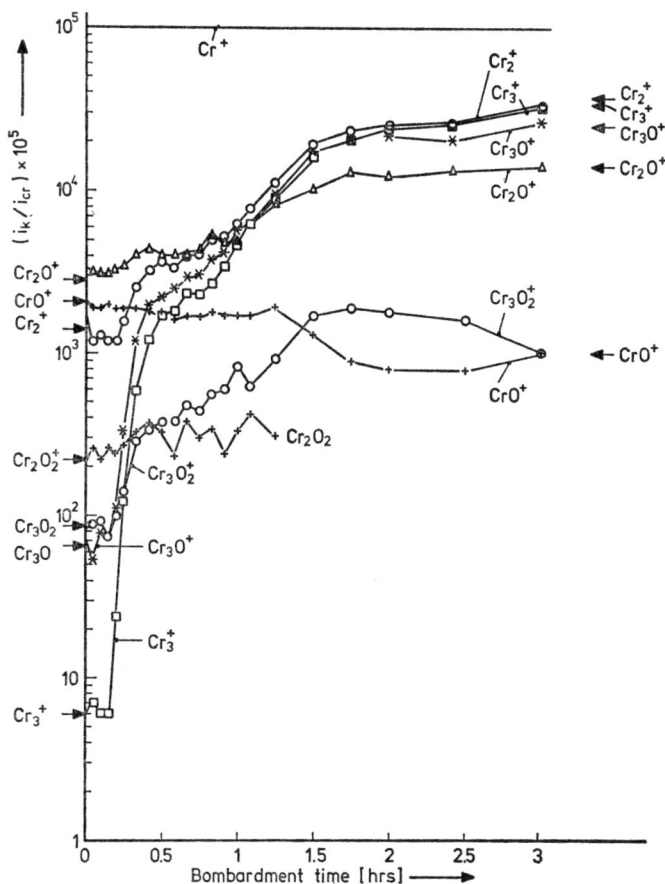

FIG. 3 Ratio of different secondary ion currents i_s to the chromium ion current i_{Cr}, as a function of time (normalized fingerprint spectra). The fingerprint spectra of pure Cr_2O_3 and of Cr are indicated by arrows on the left hand and right hand ordinates.

TABLE I

Abundances of Different Ion Species (Fingerprint or Characteristic Spectra) from Different Chromium Oxide Samples Obtained Under Slightly Different Conditions, of Primary Ion Current Density and Residual Gas Pressure, as for Measurements Given in Fig. 2

Sample	CrO_3		CrO_2		Cr_2O_3		CrO		Cr	
Ion	i_s^+	i_s^-	i_s^+	i_s^-	i_s^+	i_s^-	i_s^+	i_s^-	i_s^+	i_s^-
O	$<1 \times 10^1$		$4 \cdot 0 \times 10^1$	$1 \cdot 0 \times 10^4$	$1 \cdot 6 \times 10^2$	$1 \cdot 0 \times 10^4$	$1 \cdot 0 \times 10^2$	$1 \cdot 0 \times 10^4$	$1 \cdot 1 \times 10^2$	$1 \cdot 0 \times 10^4$
O_2	$<1 \times 10^1$		$9 \cdot 0$	$2 \cdot 4 \times 10^3$	$6 \cdot 0 \times 10^7$	$5 \cdot 6 \times 10^2$	$4 \cdot 0 \times 10^1$	$5 \cdot 5 \times 10^2$	$<6 \times 10^1$	$6 \cdot 0 \times 10^2$
Cr	$1 \cdot 0 \times 10^5$		$1 \cdot 0 \times 10^5$	<5	$1 \cdot 0 \times 10^5$	<1	$1 \cdot 0 \times 10^5$	$<6 \times 10^1$	$1 \cdot 0 \times 10^5$	$1 \cdot 8 \times 10^1$
CrO	$9 \cdot 9 \times 10^3$		$1 \cdot 3 \times 10^3$	$1 \cdot 9 \times 10^2$	$2 \cdot 1 \times 10^3$	$5 \cdot 8 \times 10^1$	$1 \cdot 6 \times 10^3$	$1 \cdot 1 \times 10^2$	$1 \cdot 0 \times 10^3$	$1 \cdot 0 \times 10^2$
CrO_2	$1 \cdot 4 \times 10^2$		7	$1 \cdot 5 \times 10^4$	≤ 8	$1 \cdot 7 \times 10^3$	<3	$1 \cdot 8 \times 10^3$	$<8 \times 10^1$	$1 \cdot 3 \times 10^3$
CrO_3	$<1 \times 10^1$		<1	$1 \cdot 8 \times 10^4$	<4	$9 \cdot 2 \times 10^2$	<3	$1 \cdot 6 \times 10^3$	$<8 \times 10^1$	$2 \cdot 7 \times 10^2$
CrO_4	$<1 \times 10^1$		<1	$5 \cdot 0 \times 10^4$	<4	<1	<3	$3 \cdot 6 \times 10^1$	$<8 \times 10^1$	$<2 \times 10^1$
Cr_2	$1 \cdot 1 \times 10^2$		$3 \cdot 0 \times 10^2$	<5	$1 \cdot 4 \times 10^3$	<1	$9 \cdot 7 \times 10^2$	$\leq 6 \times 10^1$	$3 \cdot 4 \times 10^4$	$<2 \times 10^1$
Cr_2O	$1 \cdot 0 \times 10^3$		$9 \cdot 7 \times 10^2$	<5	$2 \cdot 9 \times 10^3$	<1	$1 \cdot 6 \times 10^3$	$\leq 2 \times 10^1$	$1 \cdot 4 \times 10^4$	$<2 \times 10^1$
Cr_2O_2	$7 \cdot 3 \times 10^2$		$1 \cdot 5 \times 10^2$	<5	$2 \cdot 2 \times 10^2$	<1	$1 \cdot 6 \times 10^2$	$\leq 4 \times 10^1$	$3 \cdot 2 \times 10^2$	$<2 \times 10^1$
Cr_2O_3	$1 \cdot 0 \times 10^2$		$5 \cdot 0$	$4 \cdot 6 \times 10^1$	<4	$4 \cdot 0$	≤ 5	$\leq 2 \times 10^1$	$<8 \times 10^1$	$<2 \times 10^1$
Cr_2O_4	$\leq 1 \cdot 4 \times 10^1$		<1	$1 \cdot 2 \times 10^3$	<4	$7 \cdot 7 \times 10^1$	<3	$3 \cdot 5 \times 10^2$	$<8 \times 10^1$	$<2 \times 10^1$
Cr_2O_5	$<1 \times 10^1$		<1	$6 \cdot 4 \times 10^2$	<4	$2 \cdot 3$	<3	$1 \cdot 6 \times 10^2$	$<8 \times 10^1$	$<2 \times 10^1$
Cr_2O_6	$<1 \times 10^1$		<1	$1 \cdot 3 \times 10^2$	<4	<1	<3	$<1 \times 10^1$	$<8 \times 10^1$	$<2 \times 10^1$
Cr_3	$<1 \times 10^1$		≤ 1	<5	$6 \cdot 0$	<1	$5 \cdot 0$	2×10^1	$3 \cdot 2 \times 10^4$	$<2 \times 10^1$
Cr_3O	$\leq 1 \cdot 4 \times 10^1$		$7 \cdot 0$	<5	$6 \cdot 8 \times 10^1$	<1	$3 \cdot 7 \times 10^1$	6×10^1	$2 \cdot 6 \times 10^4$	$<2 \times 10^1$
Cr_3O_2	$2 \cdot 8 \times 10^1$		$1 \cdot 6 \times 10^1$	<5	$8 \cdot 4 \times 10^1$	<1	$3 \cdot 7 \times 10^1$	7×10^1	$1 \cdot 0 \times 10^3$	$<2 \times 10^1$
Cr_3O_3	$1 \cdot 1 \times 10^2$		$1 \cdot 6 \times 10^1$	<5	$5 \cdot 2 \times 10^1$	<1	$3 \cdot 2 \times 10^1$	3×10^1	$<8 \times 10^1$	$<2 \times 10^1$
Cr_3O_4	$2 \cdot 8 \times 10^1$		<1	<5	<4	<1	<3	1×10^1	$<8 \times 10^1$	$<2 \times 10^1$
Cr_3O_5	$<1 \times 10^1$		<1	$3 \cdot 1 \times 10^2$	<4	<1	<3	$3 \cdot 7 \times 10^1$	$<8 \times 10^1$	$<2 \times 10^1$

The measured secondary ion currents i_s can thus be considered to originate from a superposition of ion currents from three chromium containing phases viz.: phase I (Cr_2O_3) with concentration $c_I(t)$, phase II (CrO) with concentration $c_{II}(t)$ and phase III (relatively pure chromium) with concentration $c_{III}(t)$.

For any mass line i_k ($k = 1, 2, \ldots n$), one can therefore write, assuming that these 3 phases are independent of each other:

$$i_1(t) = c_I(t)i_{1I} + c_{II}(t)i_{1II} + c_{III}(t)i_{1III}$$
$$i_2(t) = c_I(t)i_{2I} + c_{II}(t)i_{2II} + c_{III}(t)i_{2III}$$

$$\text{etc.}$$

$$i_n(t) = c_I(t)i_{nI} + c_{II}(t)i_{nII} + c_{III}(t)i_{nIII}$$

As c_I, c_{II} etc. are equal to the surface coverage[28] by phases I, II and III we have the additional relation: $c_I + c_{II} + c_{III} = 1$. The values i_{1I}, $i_{2I} \ldots$ $i_{kI} \ldots i_{nI}$ are the fingerprint spectra of phase I. They are obtained by measuring a sample of pure Cr_2O_3 (see Table I) for which by definition $c_I = 1$ and $c_{II} = c_{III} = 0$. The fingerprint spectra of phases II and III, i_{kII} and i_{kIII} ($k = 1, 2 \ldots n$) can be found in an analogous way from pure samples (see Table I). The values of i_{kI} and i_{kIII} can also be obtained from $i_k(0)$ and $i_k(3 \text{ hrs})$ respectively, assuming $c_{II}(0)$ and $c_{III}(3 \text{ hrs})$ to be zero.

From three lines $i_1(t)$, $i_2(t)$ and $i_3(t)$ for which we have chosen: Cr^+, $Cr_2{}^+$ and $Cr_3{}^+$, the concentrations of the three phases as a function of time were calculated (Fig. 4) using $i_k(0)$, $i_k(3 \text{ hrs})$, as fingerprint spectra i_{kI} and i_{kIII} and the CrO-values from Table 1 as i_{kII}. One can see that in this case a third phase—CrO—is found.

From these calculated values of c_I, c_{II} and c_{III} and the fingerprint spectra the time dependence of some other mass lines was calculated. These calculated values are indicated by symbols in Fig. 2.

As the existence of the CrO phase is not generally accepted the same calculations were carried out as before but now for two phases only viz. Cr_2O_3 and Cr with concentrations c_I and c_{III} respectively. In this case,

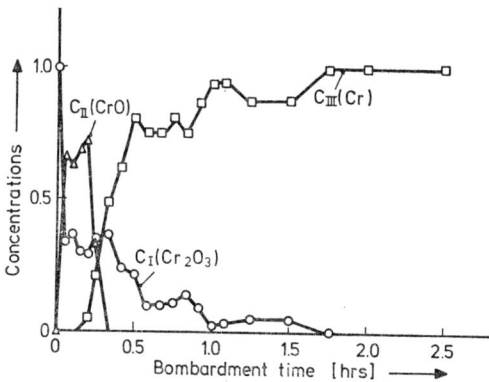

FIG. 4   Calculated concentrations $c_I(t)$, $c_{II}(t)$ and $c_{III}(t)$ of Cr_2O_3, CrO and Cr respectively as a function of bombardment time.

however, negative values for the concentration of the second phase ($c_{III} = -1\cdot5$) were found.

The occurrence of various oxidation states of chromium was also found by Benninghoven[36] with the aid of static-SIMS when analysing mono-molecular chromium oxide layers. Bearing in mind that the transmission of the two instruments may be different, our fingerprint spectra are in good agreement with Benninghoven's. Since, however, Benninghoven, in each case measured only the outermost layer, it follows from the agreement between the spectra that the cluster ions we measured originated also from the topmost layers.

CONCLUSIONS

(1) It was shown that the different chromium oxides also give different fingerprint spectra. In particular a correlation between the intensity of the oxygen bearing cluster ions and the oxygen content in the sample was found.

(2) An oxidized chromium layer was shown to consist of three chromium oxide/chromium phases, whose concentrations as a function of depth were calculated by means of fingerprint spectra.

(3) From a comparison of our measurements with static SIMS it was concluded that the cluster ions originate from the topmost layers.

REFERENCES

1. Ewald, H. and Hintenberger, H., 'Methoden und Anwendungen der Massenspektroskopie', Verlag Chemie, Weinheim/Bergstrasse, 1953, pp. 35–42.
2. Barnard, G. P., 'Modern Mass Spectrometry', The Institute of Physics, London, 1953, pp. 192–230.
3. Kienitz, H., Hrsg, 'Massenspektrometrie', Verlag Chemie, Weinheim/Bergstrasse, 1968, pp. 233–86.
4. Herzog, R. F. K. and Viehböck, F., *Phys. Rev.*, 1949, **76**, 855.
5. Herzog, R. F. K., Poschenrieder, W. P., Rüdenauer, F. G. and Satkiewicz, F. G., 15th Annual Conf. Mass Spectr. and Allied Topics (ASTM E-14), Denver, Col., 1967, p. 301.
6. Benninghoven, A., *Ann. Physik*, 1965, **15**, 113.
7. Beske, H., Dissertation, Univ., Mainz, 1966.
8. Werner, H. W., *Philips Tech. Rdsch.*, 1966, **27**, 346.
9. Smith, A. J., Marshall, D. J., Cambey, L. A. and Michael, J., *Vacuum*, 1964, **14**, 263.
10. Benninghoven, A., *Zs. f. Phys.*, 1967, **199**, 141.
11. Evans, Ch. A. Jr, 'Advances in Mass Spectrometry', Vol. 5, *ed.* A. Quayle, The Inst. of Petroleum, London, 1971, p. 436.
12. Werner, H. W., *in* 'Dev. Appl. Spectrosc.', Vol. 7A, E. L. Grove and A. J. Perkins, *eds.* Plenum Press, New York, 1969, pp. 243, 247.
13. Beske, H. E., *Zs. f. Natf.*, 1967, **22a**, 459.
14. Benninghoven, A., *Zs. f. Natf.*, 1967, **22a**, 841.
15. Werner, H. W. and de Grefte, H. A. M., *Vakuumtechnik*, 1967, **17**, 37.
16. Fogel, Ya. M., *Soviet Phys. Usp.*, 1967, **10**, 17.
17. Andersen, C. A., *Int. J. Mass Spectrom. Ion Phys.*, 1969, **2**, 61.
18. Andersen, C. A., *Int. J. Mass Spectrom. Ion Phys.*, 1970, **3**, 493.
19. Castaing, R. and Hennequin, J. F., 'Advances in Mass Spectrometry', vol. 5, *ed.* A. Quayle, The Institute of Petroleum, London, 1971, p. 419.
20. Castaing, R. and Slodzian, G., *J. de Microscopie*, 1962, **1**, 395.

21. Liebl, H. J. and Herzog, R. K. F., *J. Appl. Phys.*, 1963, **34**, 2893.
22. Liebl, H. J., 'Advances in Mass Spectrometry', Vol. 5, *ed*. A. Quayle, The Institute of Petroleum, London, 1971, 433.
23. Gaukler, K. H. Paper given at Conference on Electronmicroscopy and SIMS, held at KFA-Jülich, October 1972.
24. Benninghoven, A., Paper given at Conference on Electronmicroscopy and SIMS, held at KFA-Jülich, October 1972.
25. Beske, H., Paper given at Conference on Electronmicroscopy and SIMS, held at KFA-Jülich, October 1972.
26. Werner, H. W., *Vacuum*, 1972, **22**, 613.
27. Evans, C. A. Jr, *Analytical Chem.*, 1972, **44**, 67A.
28. Werner, H. W. and de Grefte, H. A. M., *Surface Sci.*, 1973, **35**, 458.
29. Rüdenauer, F. G., *Int. J. Mass Spectrom. Ion Phys.*, 1971, **6**, 309.
30. Werner, H. W., de Grefte, H. A. M. and van den Berg, J., Proceedings of the 1972 Garching-Conference on Ion-Surface Interaction-Sputtering, published in Rad. Effects, 1973, **18**, 269.
31. Gulbranssen, E. A. and Andrew, K. F., *J. El. Chem. Soc.*, 1952, **99**, No. 10, p. 402.
32. Hansen, M. and Anderko, K., 'Constitution of binary alloys', McGraw-Hill, New York, 1958, p. 546.
33. Shunk, F. A., 'Constitution of binary alloys', second supplement, McGraw-Hill, New York, 1969, p. 275.
34. Maier, C. G., U.S. Bur. Mines, Bull, Nr. 436, 1972.
35. Venema, A., Proc. 5th Czechoslovak Conf. on Electronics and Vacuum Physics, October 1972.
36. Benninghoven, A., Frühjahrstagung Regensburg 1972 der Deutschen Physikalischen Gesellschaft.
37. Werner, H. W., Frühjahrstagung Regensburg 1972 der Deutschen Physikalischen Gesellschaft.

Discussion

K. G. Heumann (Technische Hochschule Darmstadt, Germany): I have a chemical question: CrO_3 and Cr_2O_3 samples are easy to prepare, but how have you prepared pure CrO_2 and CrO samples?

H. W. Werner: The CrO_2 sample we obtained from our magnetic department. The CrO was especially prepared for us by Dr Kwestroo of our preparative chemical department according to a special method which includes different heating cycles of critical length.

Moreover, in the paper submitted to the editor for printing we have discussed the fact that the existence of CrO is not generally accepted. (Due to lack of time I have omitted this discussion in today's presentation.) Notwithstanding different opinions on whether CrO does exist or not, we have the fact that our SIMS fingerprint spectra of our 'CrO' sample are slightly different from the other oxides. Keeping in mind however the doubts on its existence we intend to repeat these measurements and hope to be able to prove (a) the long term reproducibility of the CrO-fingerprint spectra and (b) to show that the technique of fingerprint-spectra is a relatively simple method for the identification of a specific oxide and (extrapolating therefrom) in general for chemical compounds.

G. Blaise (Universite d'Orsay, France): Comment about the 'Superposition theorem'. In a complex sample like a metal containing inclusions of an oxide, the 'fingerprint' is composed of ions coming from the metal, from the bulk of the inclusions and from the boundaries between the metal and the inclusions.

If the size of the inclusions is large enough so that ion emission coming from the boundaries can be neglected, the superposition theorem evidently applies.

Now, if the size of the inclusion decreases the ion emission coming from the boundaries prevails and it becomes very difficult to apply the superposition theorem.

At least in perfectly disordered solid solutions the ionization efficiency of sputtered atoms depends on the concentration through the electronic structure of the solutions. So, the superposition theorem cannot be applied in that case.

H. W. Werner: Yes, I quite agree. Fortunately our sample (as well as the copper oxides in ref. 30) seems to prove the necessary assumption of 3 independent phases, as could moreover be verified by microscopic studies of the sample after various times of bombardment. We have seen relatively large crystals of different colours (green Cr_2O_3 in the beginning and brown (CrO?) at intermediate stages) so that influence from interface-structures may be neglected. Formula (1) fully written should have been:

$$i_1(t) = c_I i_{II} + c_{II} i_{III} + c_{III} i_{IIII} + c_{I,II} i_{II,II} + c_{I,III} i_{1,I,III} + c_{II,III} i_{III,III}.$$

In our case however the mixed terms, due to the contribution of the interface regions can be neglected as the areas ($\sim c_{I,II}$, $c_{I,III}$, $c_{II,III}$) covered by the interfaces are small compared to the areas ($\sim c_I$, c_{II}, c_{III}) covered by the pure phases due to the relatively large crystal size.

In the extreme case of a solid solution c_I, c_{II} and c_{III} are zero and only $c_{I,II}$, $c_{I,III}$, etc. are non-vanishing. If it were possible to find the (concentration dependent) 'fingerprint spectra' also in this case one could proceed as in the case of independent phases.

The effect of surface roughness (large crystals) as described above may also be supported by an incompletely flat bottom to the crater.

Another point to mention is the existence of so called Wadsley defects (Ref: see articles of J. S. Anderson in 'Chemistry of Extended Defects in Non-Metallic Solids', *eds*. L. Eyring and M. O'Keeffe, North Holland Publishing Company, 1970 and in 'Problems of Non-stoichiometry', *ed*. A Rabenau, North Holland Publishing Company, 1970) which explains the existence of non-stoichiometric compounds by a structure consisting of several layers each 10–20 Å thick of stoichiometric compounds say Cr_2O_3 and CrO whose frequency of occurrence may vary with depth so as to give varying m and n values in the formula found for the average composition Cr_mO_n as a function of depth.

81
Secondary Ion Mass Spectrometric Investigations of Monomolecular Layers and Thin Films

By W. K. HUBER, E. LÖBACH
and G. RETTINGHAUS

(Balzers Aktiengesellschaft für Hochvakuumtechnik und Dünne Schichten, Balzers, Liechtenstein)

INTRODUCTION

SECONDARY ion mass spectroscopy (SIMS) is principally a method for analysing the very first atomic monolayer of a solid and therefore is frequently associated with sorption and related phenomena only. However, a number of bulk properties also are reflected with high sensitivity by the surface state. It is the virtually absolute depth resolution which makes SIMS particularly selective for such phenomena. Thanks to the high sensitivity of SIMS, very low primary ion currents can be applied. Thus undesired alterations of the surface to be investigated can be prevented almost completely.

Besides that, SIMS is suited to revealing structural properties of a solid which are connected with internal surfaces; so-called interfaces. This requires that the layers of the material are removed successively by adequate means (sputtering by increasing the primary ion intensity, controlled evaporation or sublimation, etc.).

The principles of SIMS and the apparatus used here have been exhaustively described elsewhere.[1,2,3] In what follows typical examples will be discussed which demonstrate the range covered by SIMS.

SURFACE BEHAVIOUR OF STAINLESS STEEL DURING HEATING IN UHV

A stainless steel surface was analysed during thermal treatment. The experiment was part of a programme to optimize the degassing technique to be used in ultra high vacuum. The specimen, a sheet of X5 Cr–Ni 18-9 stainless steel 1 mm thick, was ground and washed in Freon. No further pretreatment was applied. It was heated directly by passing an electric current through it, in steps up to 900°C during SIMS analysis under UHV conditions. Typically,

683

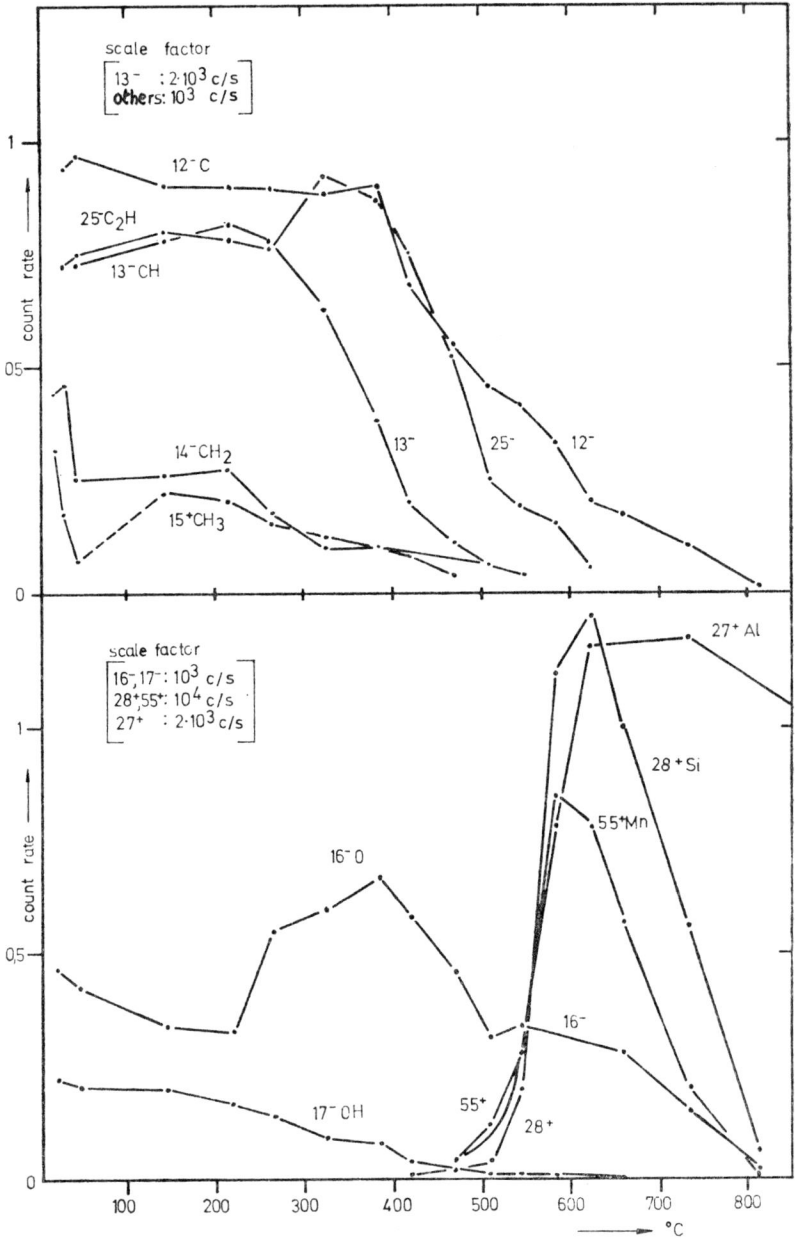

FIG. 1 Behaviour of characteristic SIMS peaks from a stainless steel sample during heating. Primary argon ion current 2×10^{-10} A. c/s (counts per second) means pulse rate counted at the output of the multiplier, corresponding to the rate of incoming ions.

the temperature was kept constant for more than 10 min for each step. After reaching the new temperature equilibrium, the surface conditions remained rather stable during the measuring time. Figure 1 shows the response of some characteristic peaks during the first temperature cycle applied to the fresh specimen.

Primary interest was devoted to the behaviour of the hydrocarbon contamination generally present at such surfaces before bakeout. Two general types of hydrocarbon contamination could be distinguished, characterized apparently by their binding energy:

(1) The loosely bound one delivers 'cracking patterns' of positive secondary ions which appear similar to those found with hydrocarbons in the gas phase, whereas the contribution to negative ions is less characteristic. Thermal desorption takes place rapidly at slightly increased temperatures (not shown in Fig. 1). Very high depletion cross-sections under ion bombardment have been observed.

(2) The strongly bound type delivers more negative than positive ions, both at lower masses. Typical ions, among others, are 13^-, 14^-, 25^-, 15^+, and with reservations because of the carbon content of the material, 12^+, 12^- and 24^-. The 'cracking pattern' seems not to be characteristic of the individual hydrocarbon species adsorbed.

The depletion of the different peaks to be traced without doubt to hydrocarbons does not proceed in parallel, thus reflecting a rather complicated behaviour in adsorption, fractionation and cracking which will not be discussed further here. Depletion of the strongly bound hydrocarbons begins at temperatures higher than 350°C.

Other types of surface impurities have been observed which also may severely influence the UHV behaviour. At moderate temperatures, the most predominant ones are the ubiquitous alkali metals Na and K. Because of their notoriously high ionization cross-sections, Na and K are highly over-represented by their peak heights in the SIMS spectra. Also it is highly significant that after prolonged treating at 900°C these ions virtually disappear and do not reappear either as secondary or thermic ions during subsequent treatment by sputtering or moderate heating. This means that diffusion of these species from the bulk to the surface does not proceed any longer.

During the first cycle of stepwise temperature increase, between 500 and 600°C, a sudden increase of Al^+, Si^+ and Mn^+ occurred. The virtually simultaneous occurrence indicates that one common cause is responsible for these processes, the more as the metallurgical data of austenitic steels undergo deep changes in this temperature range.

At higher temperatures Al^+, Si^+ and Mn^+ behaved in quite different ways: Si^+ and Mn^+ disappeared fast at 800°C. Al^+ decreased only very slowly but steadily at 900°C (not shown in Fig. 1). Al^+ reappeared again during subsequent heating between 500° and 600°C, whereas Si^+ did so only after sputtering away about 10 monolayers. Mn^+, having once disappeared, failed to reappear again. It is hard to believe that the supply of Mn has been exhausted completely by the first process, including heating at 900°C for about 2 hours. Alternatively, it must be considered that segregation in the monolayer range can be influenced by the state of the surface, for example by the presence of certain adsorption layers such as oxygen.

From the metallurgical viewpoint it may be interesting to study such segregation phenomena in more detail under conditions better defined with respect to the special problem to be studied.

After prolonged heating at 900°C mainly Cr^+ and Fe^+ ions, further Cr_2^+, $CrFe^+$, Fe_2^+, and minor constituents of the steel were detected, with the residual gas background kept low enough to avoid rapid contamination

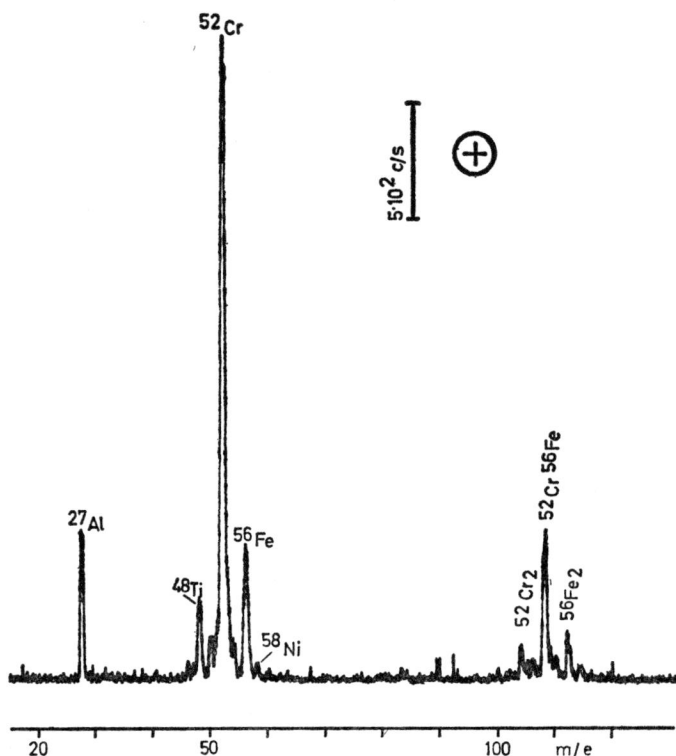

FIG. 2 Positive spectrum of a nearly clean stainless steel surface. In the negative spectrum only a small sulphur peak (32) and a not clearly identified peak at mass 26 could be detected, both as low as 200 counts per second.

from the gas phase. Cr^+ remained enriched in relation to Fe^+, when compared to a freshly sputtered surface. Figure 2 shows the spectrum of such a nearly clean surface. The conditions which govern the ratio Cr^+/Fe^+ are not revealed as yet.

The surface behaviour becomes complicated by oxidation phenomena when, for example, oxygen is admitted. Figure 3 shows the spectra of a surface cleaned by heating at 900°C and by sputtering, then oxidized and heated again to 530°C, leading to Al and Si segregation once more.

As is well known, in the positive spectrum the first steps of oxidation are generally characterized by an increased yield of metal ions rather than by

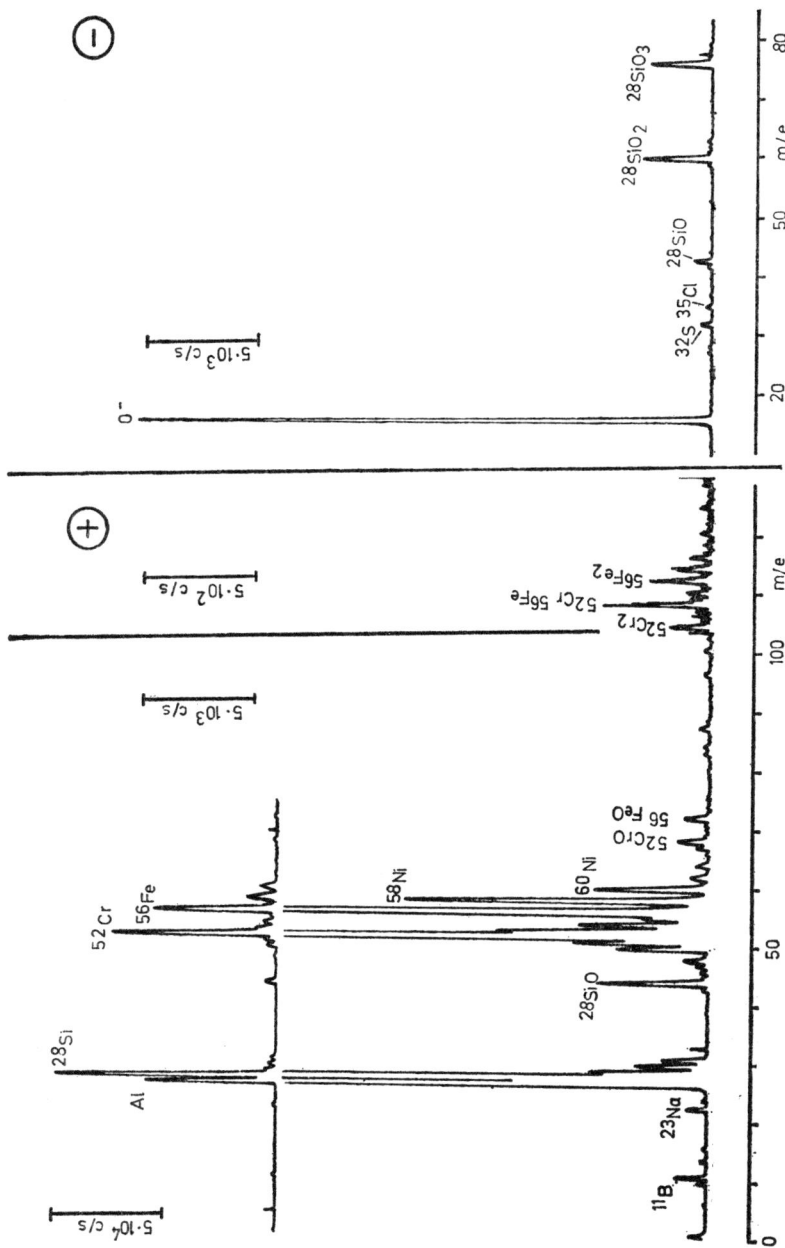

FIG. 3 Positive and negative spectra of a partly clean stainless steel surface after subsequent oxidation. For details see text.

appearance of oxide ions (which, however, are present), whereas O^- (and OH^-, if water vapour is involved) readily appears in the negative spectrum. Figure 3 shows in addition SiO_2^- and SiO_3^-.

During the first heating process the original oxides and hydroxides which are due to previous exposure to atmosphere were removed at 800° and 500–600°C respectively, as can be seen from the representative ions O^- and OH^- (Fig. 1).

The experiment discussed here gives an impression of the variety of problems involved simultaneously. Final answers cannot be expected from such a survey but only from a deeper investigation. Nevertheless, the conclusions to be drawn for UHV-technology are obvious already.

DEPTH PROFILES

The measurement of depth profiles is demonstrated by Figs. 4 and 5. Figure 4 is taken from the oxide layer of an anodized tantalum sample. The sample was prepared by subsequent anodization first in concentrated H_3PO_4 and then in dilute $Na_2SO_4 + KF$ solution. During the measurement the oxide film was successively sputtered away by increasing the primary argon ion current to 10^{-6} A. It is established that the phosphorus containing layer is in the centre of the oxide layer, thus demonstrating that tantalum and oxygen must have migrated through the phosphorus-rich zone in which the phosphorus atoms are immobile. Because the thickness of the oxide film was

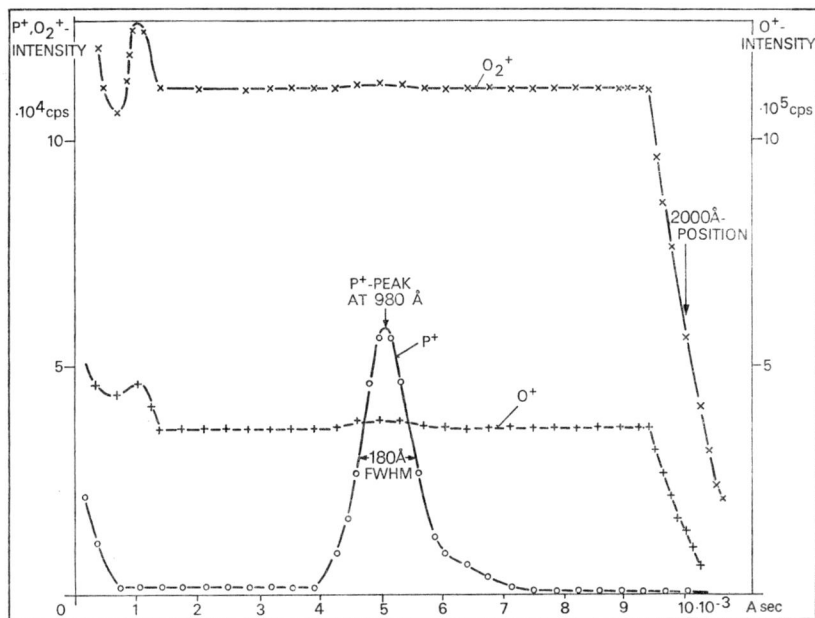

FIG. 4 Depth profile of a tantalum oxide film with a narrow phosphorus layer in its centre.

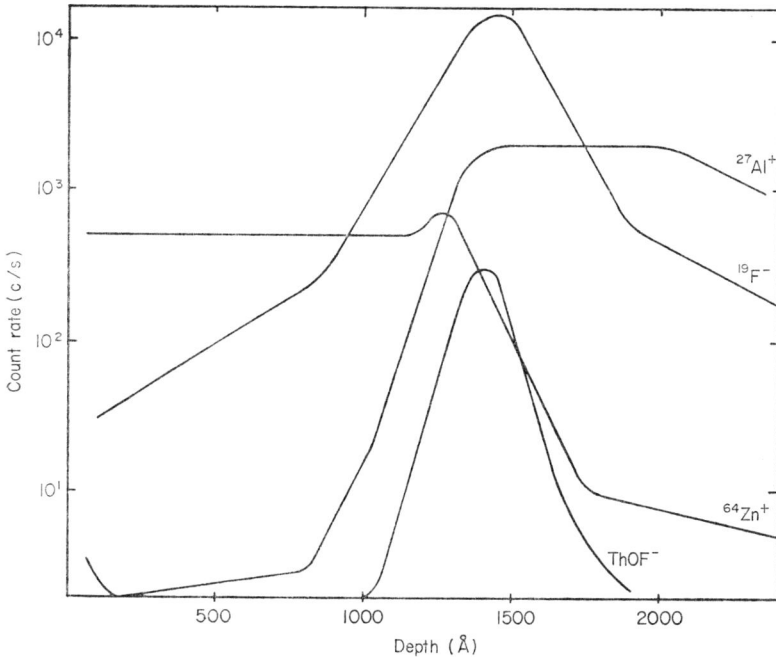

FIG. 5   Interface between a film of ZnS (1400 Å) and the substrate (cleaned by glow discharge with oxygen in the presence of Al and Thoriumfluoride).

known to be 2000 Å, the measured thickness of the phosphorus layer is 180 Å.

The depth resolution achieved in such measurements depends on the material to be removed and is typically limited to about 10% of the thickness of the removed layer. This is not only due to lateral non-uniformities of the ion beam but is given mainly by the peculiarities of the sputtering process itself. In Fig. 5 the problem of the interface between the substrate and the evaporated film is illustrated. The substrate has been 'cleaned' in a glow discharge prior to evaporation of the ZnS. The vacuum system was used before for the evaporation of Thoriumfluoride. The inner walls of the system were protected by aluminium foil and the sputter gas was oxygen.

Only a few of the many species present are shown in Fig. 5. The Zn ion is representative of the film. The other three types of ions (out of numerous others) clearly show the presence of contamination at the interface. The amount and kind of contamination strongly depends on the working conditions.

REFERENCES

1. Benninghoven, A., *Surface Sci.*, 1971, **28**, 541.
2. Huber, W. K., Selhofer, H. and Benninghoven, A., *J. Vac. Sci. Technol.*, 1972, **9**, 482.
3. Huber, W. K. and Löbach, E., *Vacuum*, 1972, **22**, 605.

Recent Progress in the Understanding of the Mechanisms of Ion–Molecule Reactions

By J. DURUP

(*Laboratoire des Collisions Ioniques,*† *Université de Paris-Sud, Orsay, France*)

INTRODUCTION

A continuously increasing amount of work has been devoted in the past twenty years to the study of ion–molecule reactions. The purpose of the present paper is, within the limits of the subfield of the mechanisms of the simplest ion–molecule reactions, to review still extant problems rather than well-established results.

To begin with, a few basic considerations pertaining to the language of chemical dynamics have to be given. A simple abstraction reaction such as $A + BC \rightarrow AB + C$ generally takes place on a potential hypersurface, which is usually represented by contour curves which are lines of constant potential energy. Typical curves are shown on Fig. 1, where the co-ordinates are the internuclear distances R_{AB} and R_{BC}. There is still one co-ordinate missing, viz. the angle between AB and BC, so that the curves of Fig. 1 may be considered as pertaining, *e.g.* to linear ABC configurations only.

The right-hand figures represent sections of the potential hypersurface along the reaction co-ordinate which is the co-ordinate along the easiest reaction path.

The upper figure is an example of a potential surface with a potential barrier. This always gives rise to so-called direct trajectories for the reactive system as represented by the dashed line.

The lower figure is an example for a potential surface with a potential well. This may in some cases lead to so-called complex trajectories, where the system needs some time before decomposing into products.

The type of trajectory is best evidenced by the measurement of the distribution of the end-point of the velocity vector of the product ion, compared to the velocity of the centre-of-mass of the whole reacting system. Direct trajectories give rise either to forward peaking or to backward peaking of this

† Part of the Laboratoire de Physico-chimie des Rayonnements, associated to the C.N.R.S.

$$A + BC \longrightarrow AB + C$$
$$(\text{linear } ABC)$$

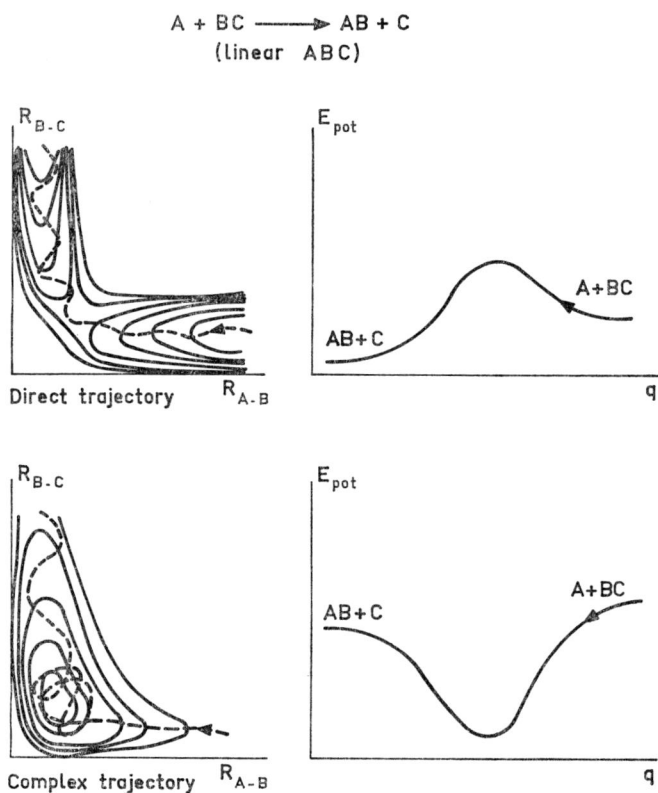

Direct trajectory

Complex trajectory

FIG. 1   Schematic representations of potential surfaces and reaction trajectories (see text).

(a) $CO^+ + D_2 \rightarrow COD^+ + D$
           $T_1 = 1.02$ eV

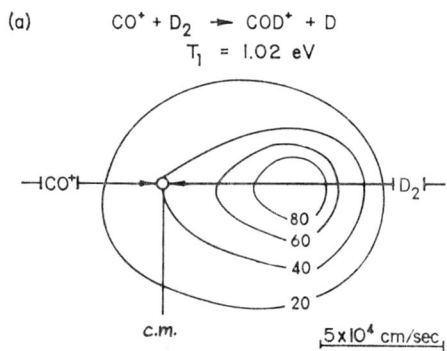

FIG. 2   Product intensity contour map for $CO^+ + D_2 \rightarrow COD^+ + D$ (reproduced from Ref. 28, Fig. 1a).

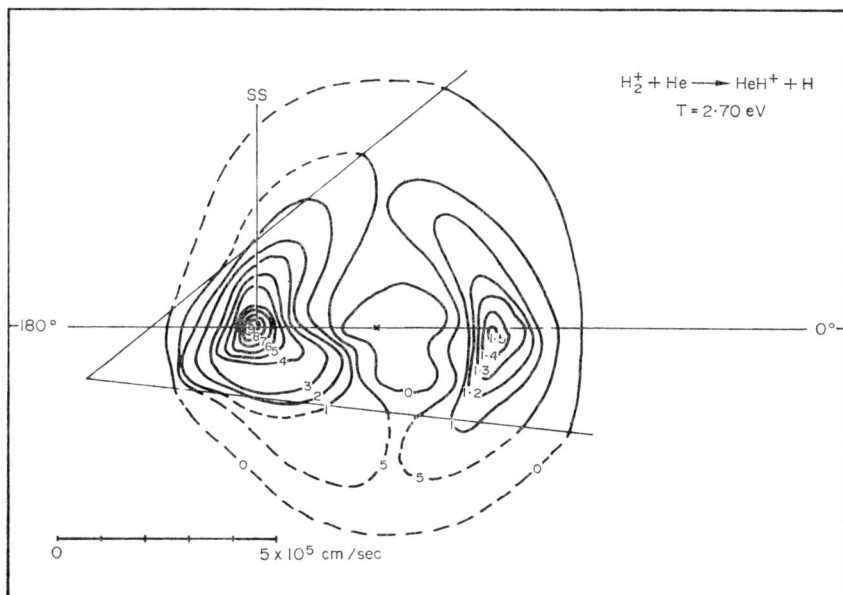

FIG. 3 A contour map in a case of preferential backward scattering (reproduced from Ref. 57). The velocity of centre-of-mass is marked by the cross in the centre of the pattern. The point marked SS corresponds to the spectator stripping model, *i.e.* to no velocity change of the product H atom.

distribution, as shown, *e.g.* on the contour maps represented on Figs. 2 and 3, respectively. Complex trajectories may be defined phenomenologically as those where the lifetime of the collision complex is large enough with respect to its rotation period for the contour map to be symmetrical with respect to the plane perpendicular to the relative velocity of the reactants and passing by their centre-of-mass, as exemplified on Fig. 4.

Such contour maps are drawn from experimental angular and velocity distributions. Types of apparatus for such measurements will not be described here and may be found in the recent literature.[1]

COMPARISON BETWEEN NEUTRAL–NEUTRAL AND ION–NEUTRAL REACTIONS

Figure 5 shows potential curves (along the reaction co-ordinate) for the simplest neutral–neutral and ion–neutral abstraction reactions:

$$H + H_2 \longrightarrow H_2 + H \tag{1}$$

$$H^+ + H_2 \longrightarrow H_2 + H^+ \tag{2}$$

Three very general features have to be noted.

(i) All neutral–neutral abstraction reactions go over a potential barrier, whereas most ion–neutral reactions go through a potential well.

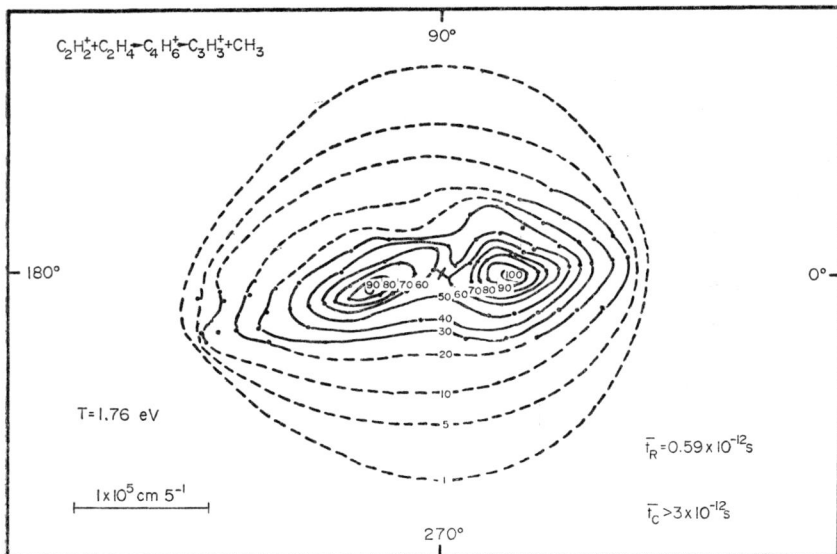

FIG. 4   A contour map in a case of complex trajectory (from Ref. 44, Fig. 16a).

(ii) In neutral–neutral reactions the long-range forces are weak (usually the potential in $-R^{-6}$ due to Van der Waals forces), whereas in ion–neutral reactions there is always the long-range ion-induced dipole potential $-\alpha e^2/2R^4$. The latter is responsible for the very large cross sections of many ion–molecule reactions. It also leads to reactive collisions where the total angular momentum may be very large; whether in that case the products retain this angular momentum mainly as rotational angular momentum of either product or of angular momentum of relative motion is not clear at the present time either from theory or from experiment.

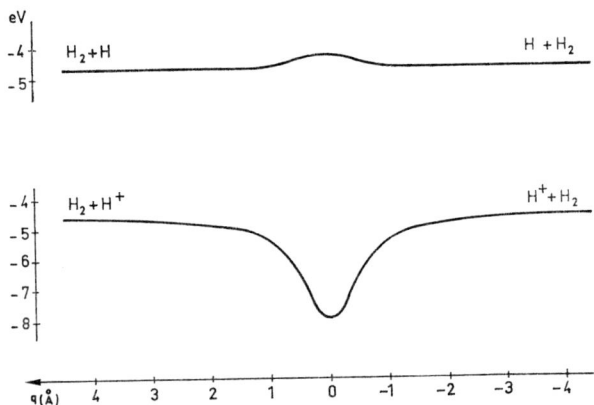

FIG. 5   Potential curves for the simplest neutral–neutral and ion–neutral reactions, approximately replotted from calculations of Refs. 2 and 3 for the linear configuration of the nuclei.

(iii) (Not apparent on Fig. 5.) Most usual neutral–neutral reactions take place in the ground electronic state of the whole system, which is often separated by a large energy gap from the first excited state which then need not be considered for the understanding of the reaction dynamics at low energies. In contrast, for all ion–molecule reactions there are at least two electronic states close to each other, which are for example:

$$A^+ + BC \text{ correlated with } AB^+ + C$$

and

$$A + BC^+ \text{ correlated with } AB + C^+$$

In any ion–molecule reaction the second state (and possibly others) may be reached from the reactants either when they have a small excess energy, or even without any excess energy if the potential curve of the excited state is partly inside the potential well of the ground state. Therefore most ion–molecule reactions may be fully understood only by taking into account more than one electronic state of the whole system. Examples will be given in the following three sections.

THE STUDIES OF $H^+ + H_2$ AND $Ar^+ + H_2$ BY THE TRAJECTORY SURFACE HOPPING METHOD

As a first example the most salient results of the study of the $H^+ + H_2$ system by Tully, Preston, Krenos and Wolfgang[3–6] and of the $Ar^+ + H_2$ system by Chapman and Preston[7] are summarized.

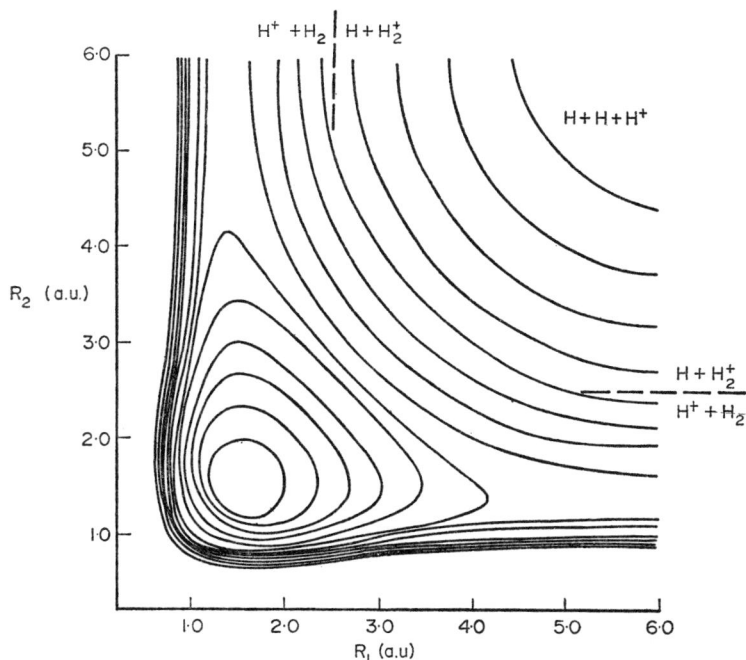

FIG. 6   The lowest potential surface for collinear $H_3{}^+$ (reproduced from Ref. 4, Fig. 1a).

A schematic correlation diagram for the H_3^+ system is the following, where isotopic species are introduced for the sake of clarity, and where the reactants are supposed to be $H^+ + D_2$:

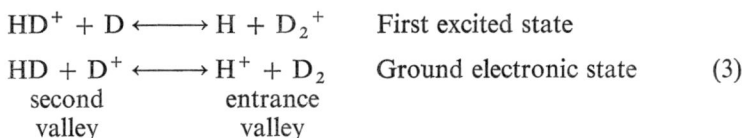

$$HD^+ + D \longleftrightarrow H + D_2^+ \qquad \text{First excited state}$$

$$HD + D^+ \longleftrightarrow H^+ + D_2 \qquad \text{Ground electronic state} \qquad (3)$$

$$\underset{\text{valley}}{\text{second}} \qquad \underset{\text{valley}}{\text{entrance}}$$

The potential surfaces of these two states were calculated by Preston and Tully[3,4] for the linear configuration of the nuclei. Figure 6 reproduces Fig. 1a of Ref. 4. A section of the potential surfaces by a plane of constant, large R_1 (or R_2) gives the curves of Fig. 7 (Fig. 1 of Ref. 3), which are the well-known potential curves of the diatomics H_2 and H_2^+. There is a crossing which occurs at an internuclear distance (in the molecule) larger than either the equilibrium distance of H_2 or of H_2^+; this crossing will move with varying $H-H_2$ distance to give a line, but at finite $H-H_2$ distances the crossing is avoided because the two H_3^+ states belong to the same $^1\Sigma^+$ symmetry species (see later): therefore there will be an avoided crossing 'seam', the location of which appears as dashed lines on Fig. 6.

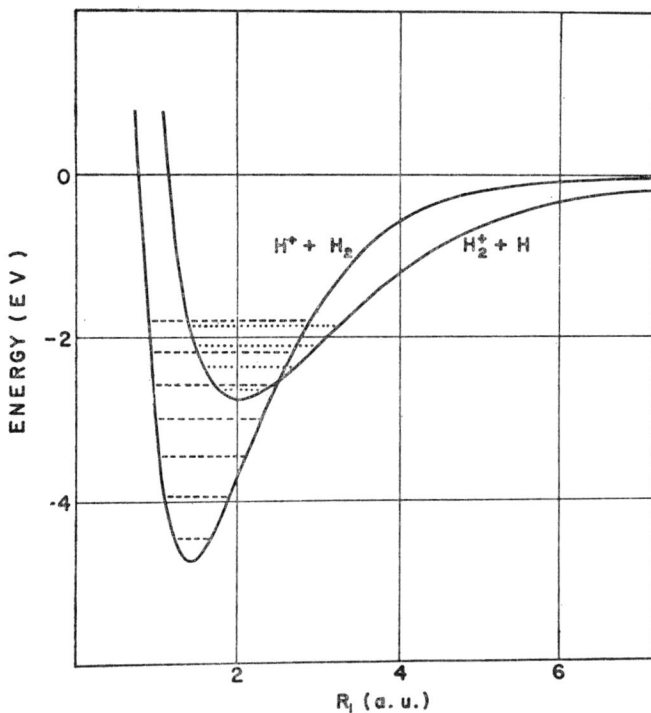

FIG. 7   Potential curves of the ground states of H_2 and H_2^+ drawn with the same asymptote corresponding to the energy of $H^+ + H + H$ at infinity (reproduced from Ref. 3, Fig. 1).

Now the method developed by Tully and Preston[5] consists of calculating classical trajectories on the potential surfaces, with some probability of jumping from one surface to the other when passing by the avoided crossing seam. A similar method was developed later by Wolf and Birkinshaw to generate reaction trajectories shown as computer-animated films.[8]

The results of Tully and Preston's *ab initio* calculations showed an excellent agreement with experiment, regarding partial cross sections as well as angular and velocity distributions of products.[4,5] Similar agreement was achieved by Chapman and Preston[7] for the ArH_2^+ system.

Two important conclusions of these studies, which are probably of general significance, are the following:

(i) The avoided crossing region is reached only by a motion perpendicular to the reaction co-ordinate; thus if the $D^+ + H_2$ reaction takes place with vibrationally unexcited H_2 molecules there will be no hopping in the entrance valley; only when the products are leaving through the exit valley, if the reactants had enough incident translational energy (2 eV in this case) this energy may be transformed into vibration along the co-ordinate perpendicular to the reaction co-ordinate, allowing for the avoided crossing seam to be reached and surface-hopping to occur;

(ii) the final state of the products depends on the features of each trajectory in a complicated way (odd or even number of surface hoppings, and choice of the exit valley); therefore it makes no sense to refer, *e.g.* to the 4 possible exit channels of reaction $D^+ + H_2$ (see earlier, reactions (3)), as due to 4 different mechanisms (D^- – transfer, charge transfer, D – transfer, elastic scattering).

THE N_2O^+ SYSTEM

This will be taken as an example of a more complicated system. The reactions which take (or not) place through an N_2O^+ intermediate state are listed below along with their main features as known from experiment.

$$O^+(^4S_u) + N_2(\tilde{X}\ ^1\Sigma_g^+) \longrightarrow NO^+(\tilde{X}\ ^1\Sigma^+) + N(^4S_u) \qquad (4)$$

This reaction has under thermal conditions a very low rate constant ($1 \cdot 2 \times 10^{-12}$ cm^3/sec) with a negative temperature coefficient.[9] This rate constant increases strongly with vibrational excitation of N_2.[10] At translational energies higher than 1 eV the cross section also rises rapidly to a maximum at about 10 eV, without much dependence on vibrational energy.[11–13]

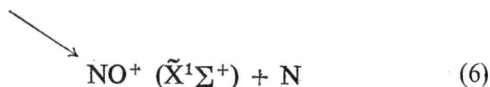

$$O^+(^2D_u) + N_2(\tilde{X}^1\Sigma_g^+) \longrightarrow N_2^+(\tilde{A}^2\Pi_u) + O\ (^3P_g)^{14} \qquad (5)$$

$$\searrow$$

$$NO^+\ (\tilde{X}^1\Sigma^+) + N \qquad (6)$$

Reaction 6 takes place only with vibrationally excited N_2.[15]

J. DURUP

$$N^+ (^3P_g) + NO(\tilde{X}^2\Pi) \longrightarrow NO^+(\tilde{X}^1\Sigma^+) + N (^4S_u) \qquad (7)$$

$$\searrow$$

$$N_2^+ (\tilde{B}^2\Sigma_u^+) + O(^3P_g) \qquad (8)$$

Reaction 7 has a rate constant of 8×10^{-10} cm³/sec.[16] Reaction 8 is the first ion–molecule reaction in the proper sense which was shown to give rise to excited products.[17] It was observed at a few eV incident energy and its cross section decreases with decreasing energy, which led us to assume that the resultants were in their ground state, the reaction then being endothermal. However other reasons were recently developed by the authors of the work which would favour a reaction with $N^+(^1D_g$ or $^1S_g)$.[18]

$$N_2^+ + O(^3P_g) \longrightarrow NO^+ + N \ (1.4 \times 10^{-10} \text{ cm}^3/\text{sec}) \ [19] \qquad (9)$$

$$\searrow$$

$$O^+ + N_2 \ (<10^{-11} \text{ cm}^3/\text{sec}) \ [20] \qquad (10)$$

These observations may be understood from the knowledge of the potential hypersurfaces of the various electronic states of N_2O^+. *Ab initio* calculations of potential curves of N_2O^+ were performed by Lorquet and Cadet[21] and by Pipano and Kaufman.[22] In Fig. 8 we have approximately replotted some of Pipano and Kaufman's data in such a way as to show both dissociation valleys on the same graph. Their calculations were performed for the linear configuration only, to which the term species indicated in the middle of the figure are pertinent. However since in most cases the intermediate N_2O^+ will not be linear we also indicated the term species in the planar symmetry

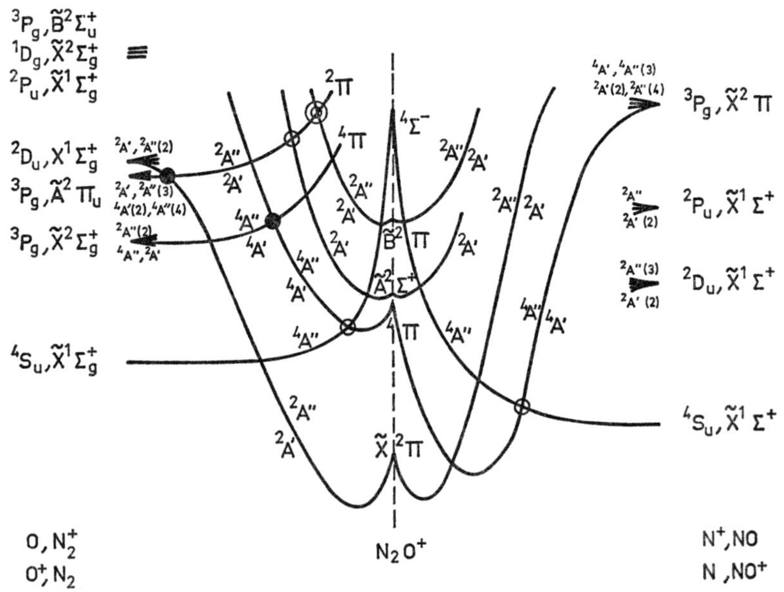

FIG. 8 Schematic potential diagram for N_2O^+ replotted in an approximate way from Ref. 22.

(C_s group). We also listed the other states correlated with each state of the reactants. The crossings marked with dots are actually avoided crossings in all configurations. The circled ones are avoided only in non-linear configurations. Both may be considered as points where transitions are likely to occur.

From her data Kaufman explained all the features of reactions 4–6[22] and the other reactions may be regarded in a similar way.

Reaction 4 is very slow under thermal conditions because it requires a spin-changing transition from a $^4\Sigma^-$ state to the X $^2\Pi$ state of N_2O^+. The negative temperature coefficient may be explained by an increase in the probability of this transition when the collision duration increases. The strong increase of the cross section with N_2 vibrational energy or at translational energy higher than 1 eV arises from the possibility to reach the first $^4\Pi$ state, to which the transition takes place out of the linear configuration. The mechanism of this transition was successfully treated in the case of N_2O^+ by O'Malley.[23]

Reaction 5 clearly takes place at the avoided crossing at a distance well into the entrance valley; reaction 6 needs some vibrational energy to reach a crossing with a state leading to $N(^2D_u)$ or $N(^2P_u)$.

Reaction 7 is similar to 5. The occurrence of reaction 8 probably indicates that the $^4\Pi$ first excited state of N_2O^+ is connected with $O(^3P_g) + N_2^+(\tilde{B}\ ^2\Sigma_u^+)$ on the left of Fig. 8, unless the reactant is an excited N^+. A more detailed knowledge of the potential surfaces would be necessary to understand why reaction 9 and not 10 takes place.

CORRELATION DIAGRAMS IN THE GENERAL CASE

A system such as N_2O^+ where extensive *ab initio* calculations have been performed is obviously an exception. Most other systems may be understood only from qualitative correlation diagrams. The latter may be either adiabatic or diabatic.

Adiabatic correlation diagrams are easily drawn from the knowledge of the spectroscopic states of reactants and products, and from the application of the non-crossing rule for states of the same species, which seems to be valid as well for polyatomic as for diatomic states.[24]

Unfortunately qualitative adiabatic correlations say nothing about avoided crossings, and it is impossible, *e.g.* to decide *a priori* whether one is in situation (a) or (b) of Fig. 9. Here (b) represents adiabatic potential curves pertaining to the same physical situation as the diabatic curves (c). Adiabatic states are exact eigenstates of the electronic Hamiltonian of the whole system at fixed configurations of the nuclei. Diabatic states have no precise definition and are essentially states obtained from a wavefunction varying smoothly with internuclear distances, thus leading also to smoothly varying energies.

In case (b) or (c), if one has good calculations it is relatively unimportant whether they are adiabatic or diabatic. In the former case one knows that transitions between adiabatic states I and II are able to be induced by the relative motion of the nuclei; in the latter case transitions between diabatic states 1 and 2 are able to be induced by some neglected part of the electronic Hamiltonian. Transition probabilities may be estimated from Landau–Zener

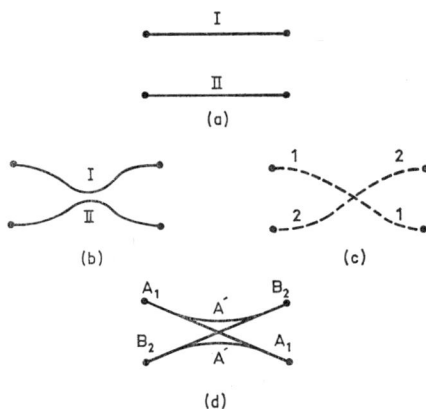

FIG. 9　Schematic correlation diagrams (see text).

formula. One may also in a simpler way consider that from the fourth uncertainty relation a transition between adiabatic states such as I and II of Fig. 9(b) will be likely to occur as soon as the energy gap ΔE is no more than $\hbar/\Delta t$ where Δt is the time needed by the system for travelling through the region where I and II are near each other (*e.g.* with O^+ ions accelerated to 1 eV by the ion-induced dipole attraction, ΔE will be 0·3 eV if the transition may take place over 0·2 Å). If ΔE is much smaller than $\hbar/\Delta t$ the transition will always take place.

In this context two results of Chapman and Preston's calculations[7] on $Ar^+ + H_2$ may be of general application:

    (i) transitions under the conditions of ion–molecule reactions actually take place only at avoided crossings, within about 0·2 Å;

    (ii) in cases as shown on Fig. 9(d), where under nuclear configurations of relatively high symmetry (here C_{2v}) two states are allowed to cross (here A_1 and B_2) whereas for lower symmetry (here C_s) the corresponding states (here both A') undergo an avoided crossing, the most probable event even in the configuration of lower symmetry turns out to be the *diabatic* way, that is to say the one following a dashed line in Fig. 9(c) and therefore representing a jump from one adiabatic state to the other.

From the above discussion it is clear that it is extremely useful to be able to derive diabatic correlations. The simplest way, in principle, to get diabatic states is to build them from single configurations of the valence electrons on the molecular orbitals, if these orbitals themselves are smooth functions of the internuclear distances. But the latter condition usually is not met, since the orbitals also may undergo avoided crossings (see, *e.g.* Ref. 25). Therefore one should know also how to get diabatic correlations of orbitals, *i.e.* correlations allowing for a smooth variation of the orbitals with varying internuclear distances.

This may be done easily for diatomic systems by using correlations derived from those of one-electron molecules (such as H_2^+) which have three good quantum numbers; such correlations have been extremely successful for the understanding of many phenomena in atomic collisions at high or moderate energies.[26] However there exists at the present time no such prescriptions for polyatomic systems. Therefore the best way to derive diabatic orbital correlations for ion–molecule reactive systems has been to use chemical intuition, as exemplified by the work of Mahan.[27]

We shall illustrate his approach with the example of the reaction

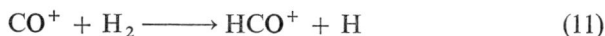

$$CO^+ + H_2 \longrightarrow HCO^+ + H \qquad (11)$$

which occurs through a direct trajectory, as shown in Fig. 2 reproduced from Kerstetter and Wolfgang,[28] in spite of the expectation that the stable intermediate H_2CO^+ would have a deep potential well with several states in it allowing for vibronic relaxation leading to a complex trajectory.

Now the orbital correlation diagram proposed by Mahan[27] for the geometry corresponding to that stable H_2CO^+ (i.e. the C_{2v} symmetry group) is shown in Fig. 10. The important point is that the $\sigma_g(H_2)$ orbital on the left which will have the a_1 symmetry in C_{2v} cannot be correlated in a 'natural'

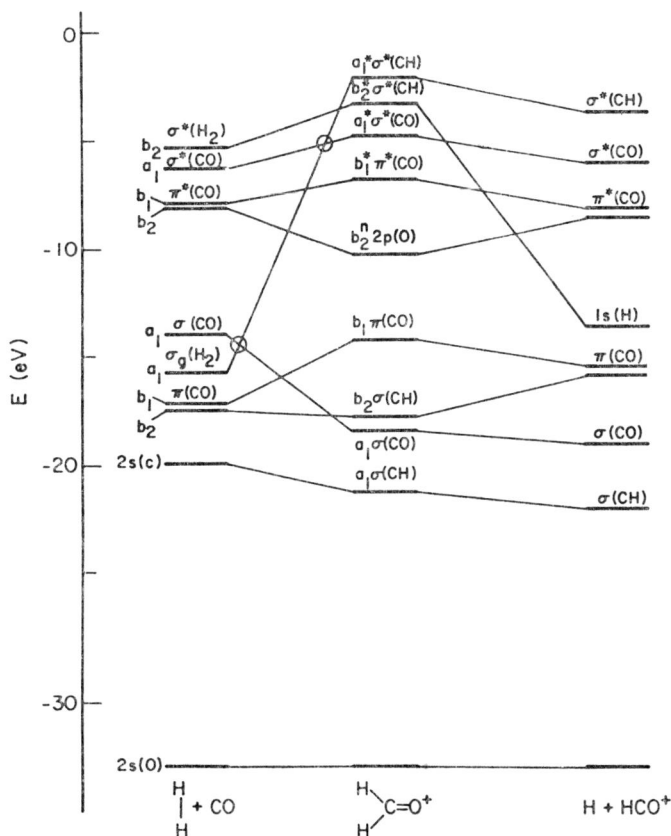

FIG. 10 Orbital correlation diagram for $CO^+ - H_2$ collisions in the C_{2v} symmetry group (reproduced from Ref. 27, Fig. 4).

way to the $a_1\sigma(CO)$ orbital (to which it would be adiabatically correlated) nor to the $a_1*\sigma*(CO)$ orbital, since these orbitals are localized on the CO bond. It has to be correlated with the very high $a_1*\sigma*(CH)$ orbital. Now when we populate the orbitals by the 11 valence electrons of the system we see that at least one electron (out of the two electrons of the $\sigma_g(H_2)$ orbital and the one of the $\sigma(CO)$ orbital) will have to go to this high-energy $a_1*\sigma*(CH)$ orbital and therefore the whole system will go into an excited state instead of going into the potential well of the stable H_2CO^+ ion. In other words the reaction will not take place at thermal energy, except if the nuclear configuration is different from that of the stable H_2CO^+, *e.g.* if it is linear. Then the correlation diagrams show that the reaction may take place, but in a direct way without any potential well.[27]

FORBIDDEN TRANSITIONS

Ion–molecule reactions may be forbidden in two ways: spin-forbidden or symmetry-forbidden (we do not consider here the energy requirements, which may always be met if the reactants or the products have translational energy in the convenient range).

Spin-forbidden Reactions

In most cases it has been shown that spin-forbidden transitions, in ion–molecule reactions as well as in other types of collision phenomena, are for light atoms about 10^2 times less probable than similar but spin-allowed transitions.

The following examples are found in the field of ion–molecule reactions.

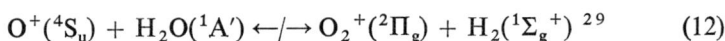

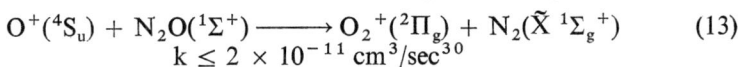

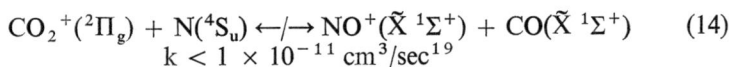

$$O^+(^4S_u) + H_2O(^1A') \longleftarrow/\longrightarrow O_2^+(^2\Pi_g) + H_2(^1\Sigma_g^+) \quad [29] \tag{12}$$

$$O^+(^4S_u) + N_2O(^1\Sigma^+) \longrightarrow O_2^+(^2\Pi_g) + N_2(\tilde{X}\ ^1\Sigma_g^+) \tag{13}$$
$$k \leq 2 \times 10^{-11}\ cm^3/sec\,[30]$$

$$CO_2^+(^2\Pi_g) + N(^4S_u) \longleftarrow/\longrightarrow NO^+(\tilde{X}\ ^1\Sigma^+) + CO(\tilde{X}\ ^1\Sigma^+) \tag{14}$$
$$k < 1 \times 10^{-11}\ cm^3/sec\,[19]$$

$$O^+(^4S_u) + N_2(\tilde{X}\ ^1\Sigma_g^+) \longrightarrow NO^+(\tilde{X}\ ^1\Sigma^+) + N(^4S_u) \tag{4}$$

$k = 1\cdot2 \times 10^{-12}\ cm^3/sec$ (when occurring through the $^2\Pi$ state) see above section on N_2O^+ system

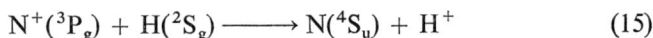

$$N^+(^3P_g) + H(^2S_g) \longrightarrow N(^4S_u) + H^+ \tag{15}$$

$k = 8 \times 10^{-13}\ cm^3/sec$ (when occurring through the $\tilde{X}\ ^2\Pi$ state), independently of incident translational energy in the range of $0\cdot0025$ to $0\cdot25$ eV.[31]

There are however two exceptions: the ion–molecule reaction

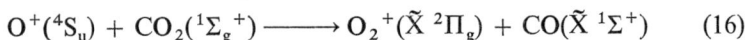

$$O^+(^4S_u) + CO_2(^1\Sigma_g^+) \longrightarrow O_2^+(\tilde{X}\ ^2\Pi_g) + CO(\tilde{X}\ ^1\Sigma^+) \tag{16}$$

$k = 1\cdot1 \times 10^{-9}\ cm^3/sec$,[32-35] and the collision-induced excitation

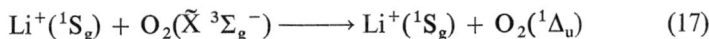

$$Li^+(^1S_g) + O_2(\tilde{X}\ ^3\Sigma_g^-) \longrightarrow Li^+(^1S_g) + O_2(^1\Delta_u) \tag{17}$$

at 400 eV and scattering angle $>3°$.[36]

The high rate constant of reaction 16 is very strange especially when compared with that of the isoelectronic reaction 13. It was ascertained by Fehsenfeld et al.,[33,37] that the reacting species in reaction 16 is effectively ground-state $O^+(^4S_u)$.

Symmetry-forbidden Reactions

An ion–molecule reaction in the proper sense involves at least a triatomic system, where the only symmetry element which exists in all configurations is the plane of the three atoms.

No ion–molecule reaction which would require a change of the symmetry of wave function with respect to that plane has ever been tested. Such a reaction would require very particular excited states, e.g.:

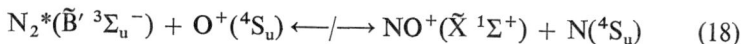

$$N_2^*(\tilde{B}'\,^3\Sigma_u^-) + O^+(^4S_u) \longleftarrow/\longrightarrow NO^+(\tilde{X}\,^1\Sigma^+) + N(^4S_u) \qquad (18)$$

However, we observed the non-occurrence of such symmetry-forbidden processes in double charge transfer experiments at high energy and zero scattering angle:[38,39]

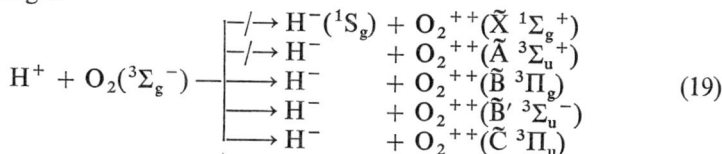

$$H^+ + O_2(^3\Sigma_g^-) \begin{array}{l} \rightarrow/\rightarrow H^-(^1S_g) + O_2^{++}(\tilde{X}\,^1\Sigma_g^+) \\ \rightarrow/\rightarrow H^- \quad + O_2^{++}(\tilde{A}\,^3\Sigma_u^+) \\ \longrightarrow H^- \quad + O_2^{++}(\tilde{B}\,^3\Pi_g) \\ \longrightarrow H^- \quad + O_2^{++}(\tilde{B}'\,^3\Sigma_u^-) \\ \longrightarrow H^- \quad + O_2^{++}(\tilde{C}\,^3\Pi_u) \end{array} \qquad (19)$$

The first two processes (19) are respectively spin-forbidden and symmetry-forbidden, and do not appear, whereas the other ones, which are spin and symmetry-allowed, do appear. A more detailed discussion of such symmetry-forbidden processes is given elsewhere.[40]

COMPLEX TRAJECTORIES AND THE LIFETIME OF THE INTERMEDIATE COMPLEX

Complex trajectories mean more or less long-lived intermediate complexes. These are usually associated with some chemical stability of the intermediate species, e.g. $C_4H_8^+$ in the reaction

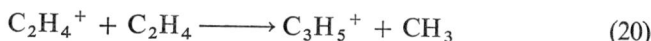

$$C_2H_4^+ + C_2H_4 \longrightarrow C_3H_5^+ + CH_3 \qquad (20)$$

However in bimolecular reactions the intermediate complex always has enough internal energy to dissociate and in particular when an exothermal reaction occurs a long lifetime of the intermediate complex requires extensive vibrational relaxation as observed in reaction 20.[41,42]

Experiments on intermediate complex lifetimes are of different kinds and give only estimates.

(i) Various authors by studying an ion–molecule reaction as a function of reactant translational energy observed a continuous transition between direct trajectories (at relatively large energy) and complex trajectories (at relatively small energy). Such evidence was reviewed by Henglein.[43] A recent example is shown on Figs. 4 and 11 from Herman and Birkinshaw's work[44] on the reaction

$$C_2H_2{}^+ + C_2H_4 \longrightarrow C_3H_3{}^+ + CH_3 \qquad (21)$$

Figure 4 shows an example of complex trajectory whereas in Fig. 11 there is a dominant forward peaking with already some backward peaking, indicating a lifetime $\bar{t}_C$ of the intermediate complex of the order of the rotation period $\bar{t}_R$.

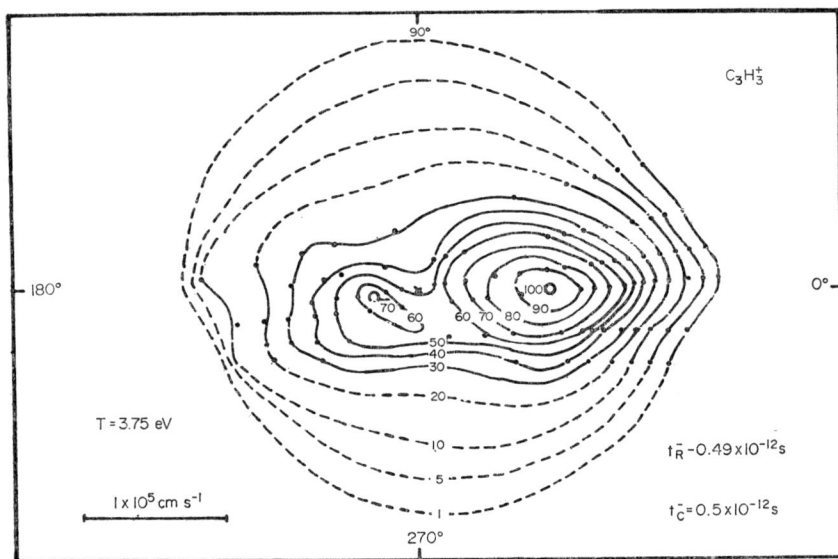

FIG. 11   A contour map in the transition region between direct and complex trajectories (from Ref. 44, Fig. 16b).

(ii) An apparatus (Fig. 12) for the direct measurement of intermediate complex lifetimes in the nanosecond to microsecond range was recently built by Marie Durup and colleagues.[45] The primary ions are slowed to a few eV in a very narrow collision region (typically 30 μm) defined by two grids, where they meet the target gas. Complexes moving out of this region are accelerated by a strong electric field (1 V/μm) and the moment they dissociate is indicated by the energy of the product ions which are mass and energy-analysed by a magnetic sector. This technique is similar to that used by Ottinger[46] for the determination of metastable ions lifetimes. Measurements are in progress.

(iii) Lifetimes in the 10^{-8} to 10^{-4} sec range for intermediate complexes formed under thermal conditions have been deduced by various authors from the study of the competition between dissociation and collision-induced deactivation.[47] Thus various estimates of the

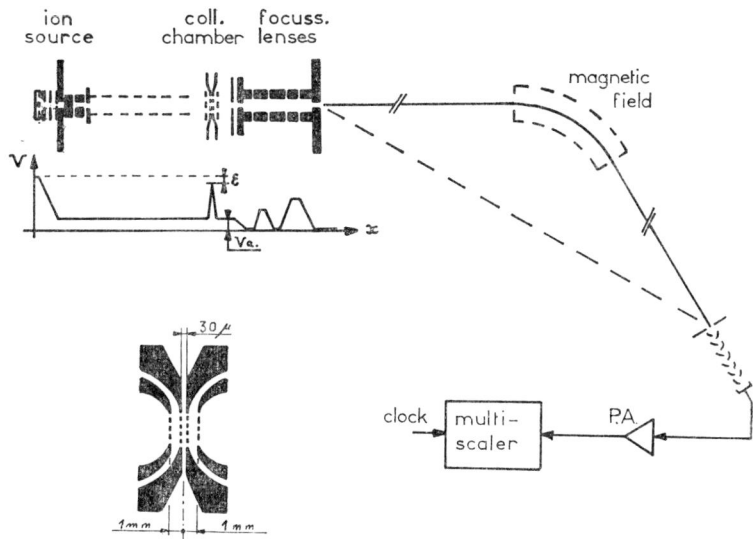

Fig. 12 An apparatus for the direct measurement of intermediate complex lifetimes.[45]

lifetime of the intermediate complex in reaction 20 could be given (see *e.g.* Meisels and Tibbals[47]). This of course rests upon assumptions about the collisional deactivation cross-sections. A better knowledge of the latter can now be obtained from the study of the variation of the reaction cross-section of excited ions during their collisional cooling.[48] It should be kept in mind however that collisions under thermal conditions may lead not only to stabilization but also to dissociation of metastable states.[49]

ISOTOPIC SCRAMBLING

Scrambling phenomena, *i.e.* a more or less statistical distribution of isotopic product ions from reactants with isotopic atoms in non-equivalent positions, have often been considered as an indication of a long-life complex. Recent work contradicts such a conclusion, and we shall briefly discuss the kind of evidence which scrambling phenomena may bring.

The occurrence or non-occurrence of simple scrambling reactions, such as

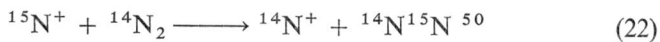

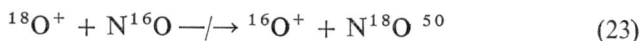

$$^{15}N^+ + {}^{14}N_2 \longrightarrow {}^{14}N^+ + {}^{14}N^{15}N \quad {}^{50} \tag{22}$$

$$^{18}O^+ + N^{16}O \longrightarrow\!\!\!/\!\!\longrightarrow {}^{16}O^+ + N^{18}O \quad {}^{50} \tag{23}$$

simply indicates the absence or the presence of an energy barrier between both equivalent valleys. In a reaction such as (23), where the occurrence of a barrier is also apparent from the correlation diagram,[51] the height of that barrier would be deduced from the observation of an energy threshold for the scrambling reaction.

Scrambling effects in less simple reactions, such as

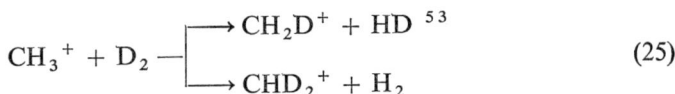

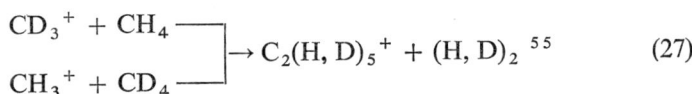

$$^{16}O^- + N_2{}^{18}O \longrightarrow NO^- + NO \quad {}^{52} \tag{24}$$

$$CH_3{}^+ + D_2 \begin{cases} \longrightarrow CH_2D^+ + HD \quad {}^{53} \\ \longrightarrow CHD_2{}^+ + H_2 \end{cases} \tag{25}$$

$$C_2H_4{}^+ + C_2D_4 \longrightarrow C_3(H, D)_5{}^+ + C(H, D)_3 \quad {}^{54} \tag{26}$$

$$\left. \begin{array}{l} CD_3{}^+ + CH_4 \\ CH_3{}^+ + CD_4 \end{array} \right\} \rightarrow C_2(H, D)_5{}^+ + (H, D)_2 \quad {}^{55} \tag{27}$$

may be fully understood only by a knowledge of the velocity distribution of products. Thus it was shown in particular by Huntress[54] that reaction 26 occurs with extensive isotope scrambling in a large range of incident energies including the range where, from the velocity distributions measured by Herman et al.,[42] the trajectories are direct, as well as the range where they are complex. Similarly, the contour maps obtained by Weiner et al.[55] for the various isotopic products of reaction 27 show that there is a kinetic competition between scrambling and dissociation but that both are more rapid than one rotation of the complex.

In cases of reactions such as 24, 26 and 27 it would also be useful to know the height of the energy barrier (if any) between the various equivalent exit valleys. This would require appearance potential measurements, best obtained by photo-ionization of specifically labelled compounds identical with the intermediate complex, e.g. 1,1,2,2-tetradeuteromethylcyclopropane for the understanding of reaction 26, which was shown to proceed probably through a methylcyclopropane intermediate.[56]

OTHER PROBLEMS

Other interesting problems, not considered here, include the following:

(i) effects of fine structure on ion–molecule reaction cross section (e.g. $Ar^+ + H_2$ with Ar^+ in the $^2P_{3/2}$ or $^2P_{1/2}$ state);

(ii) possibility of observing quantum mechanical interference effects on the angular and energy distribution of ion–molecule reaction products;

(iii) discussion of the conditions for the occurrence of complex trajectories.

ACKNOWLEDGMENTS

The author is grateful to Drs Richard K. Preston, Joyce J. Kaufman, Zdenek Herman, Fred C. Fehsenfeld, David L. Smith, Victor Sidis, Michel Barat and Professor Bruce H. Mahan for sending him preprints of their not yet published work.

REFERENCES

1. See *e.g.* McDaniel, E. W., Čermák, V., Dalgarno, A., Ferguson, E. E. and Friedman, L., 'Ion-Molecule Reactions', Wiley Interscience, New York, 1970, pp. 107–58.
2. Shavitt, I., Stevens, R. M., Minn, F. L. and Karplus, M., *J. Chem. Phys.*, 1968, **58**, 1925.
3. Preston, R. K. and Tully, J. C., *J. Chem. Phys.*, 1971, **54**, 4297.
4. Krenos, J., Preston, R. K., Wolfgang, R. and Tully, J. C., *Chem. Phys. Lett.*, 1971, **10**, 17.
5. Tully, J. C. and Preston, R. K., *J. Chem. Phys.*, 1971, **55**, 562.
6. Preston, R. K. and Cross, R. J., in press.
7. Chapman, S. and Preston, R. K., in press.
8. Wolf, F. A. and Birkinshaw, K., VIII ICPEAC, Belgrade 1973, 154.
9. Ferguson, E. E., Bohme, D. K., Fehsenfeld, F. C. and Dunkin, D. B., *J. Chem. Phys.*, 1969, **50**, 5039.
10. Schmeltekopf, A. L., Ferguson, E. E. and Fehsenfeld, F. C., *J. Chem. Phys.*, 1968, **48**, 2966.
11. Giese, C. F., *Adv. Chem. Ser.*, 1966, **58**, 20.
12. Stebbings, R. F., Turner, R. B. and Rutherford, J. A., *J. Geoph. Res.*, 1966, **71**, 771.
13. Cohen, R. B., *J. Chem. Phys.*, 1972, **57**, 676.
14. Rutherford, J. A. and Vroom, D. A., *J. Chem. Phys.*, 1971, **55**, 5622.
15. Neynaber, R. H. and Magnuson, G. D., *J. Chem. Phys.*, 1973, **58**, 4586.
16. Goldan, P. D., Schmeltekopf, A. L., Fehsenfeld, F. C., Schiff, H. I. and Ferguson, E. E., *J. Chem. Phys.*, 1966, **44**, 4095.
17. Brandt, D., Ottinger, Ch. and Simonis, J., Max Planck Institut für Stömungsforschung Bericht 119, 1973.
18. Brandt, D. and Ottinger, Ch., ibid. Bericht 131, 1973.
19. Fehsenfeld, F. C., Dunkin, D. B. and Ferguson, E. E., *Planet. Space Sci.*, 1970, **18**, 1267.
20. Ferguson, E. E., Fehsenfeld, F. C., Goldan, P. D., Schmeltekopf, A. L. and Schiff, H. I., *Planet. Space Sci.*, 1965, **13**, 823.
21. Lorquet, J. C. and Cadet, C., *Int. J. Mass Spectrom. Ion Phys.*, 1971, **7**, 245.
22. Pipano, A. and Kaufman, J. J., *J. Chem. Phys.*, 1972, **56**, 5258; Kaufman, J. J., to be published.
23. O'Malley, T. F., *J. Chem. Phys.*, 1970, **52**, 3269.
24. Naqvi, K. R., *Chem. Phys. Lett.*, 1972, **15**, 634.
25. Barat, M., Dhuicq, D., François, R. and Sidis, V., in press.
26. Barat, M. and Lichten, W., *Phys. Rev. A.*, 1972, **6**, 211.
27. Mahan, B. H., *J. Chem. Phys.*, 1971, **55**, 1436.
28. Kerstetter, J. and Wolfgang, R., *J. Chem. Phys.*, 1970, **53**, 3765.
29. Chiang, M. H., Gislason, E. A., Mahan, B. H., Tsao, C. W. and Werner, A. S., *J. Phys. Chem.*, 1971, **75**, 1426; reverse reaction also never observed to our knowledge.
30. Derwish, G. A. W., Galli, A., Giardini-Guidoni, A. and Volpi, G. G., *J. Chem. Phys.*, 1964, **40**, 3450; the quoted value is a maximum since the reactant O⁺ may be in a metastable state.
31. Melius, C. F., VIII. ICPEAC, Belgrade 1973, 809 (theoretical calculation).
32. Paulson, J. F., Mosher, R. L. and Dale, F. L., *J. Chem. Phys.*, 1966, **44**, 3025.
33. Fehsenfeld, F. C., Ferguson, E. E. and Schmeltekopf, A. L., *J. Chem. Phys.*, 1966, **44**, 3022.
34. Dunkin, D. B., Fehsenfeld, F. C., Schmeltekopf, A. L. and Ferguson, E. E., *J. Chem. Phys.*, 1968, **49**, 1365.
35. Mosesman, M. and Huntress, W. T., *J. Chem. Phys.*, 1970, **53**, 462.
36. Aberth, W. and Lorents, D. C., *Phys. Rev.*, 1969, **182**, 162.
37. Fehsenfeld, F. C., Schmeltekopf, A. L. and Ferguson, E. E., *Planet. Space Sci.*, 1965, **13**, 219.
38. Appell, J., Durup, J., Fehsenfeld, F. C. and Fournier, P., *J. Phys. B.*, 1973, **6**, 197.
39. Fournier, P., unpublished results.
40. Durup, J., *Chem. Phys.*, in press.
41. Durup, M. and Durup, J., *Adv. Mass Spectrom.*, 1968, **4**, 677.
42. Herman, Z., Lee, A. and Wolfgang, R., *J. Chem. Phys.*, 1969, **51**, 452.
43. Henglein, A., *J. Chem. Phys.*, 1970, **53**, 458.

44. Herman, Z. and Birkinshaw, K., *Ber. Bunsenges. Phys. Chem.*, 1973, **77**, 566.
45. Durup, M., Durup, J., Gitton, B. and Ozenne, J. B., to be published.
46. Ottinger, Ch., *Zts f. Naturf.*, 1967, **22a**, 19.
47. Meisels, G. G. and Tibbals, H. F., 15 *Annual Conf. Mass. Spectrom.*, Denver, Colorado, 1967, 11.
48. Smith, D. L. and Futrell, J. H., in press.
49. Bowers, M. T., paper 96 presented at the present Conference.
50. Fehsenfeld, F. C., Albritton, D. L., Busch, Y. A., Fournier, A., Govers, T. R. and Fournier, J., to be published.
51. Tully, J. C., Herman, Z. and Wolfgang, R., *J. Chem. Phys.*, 1971, **54**, 1730.
52. Tiernan, T. O. and Clow, R. P., paper 33 presented at the present Conference.
53. Harrison, A. G. and Keyes, B. G., *Can. J. Chem.*, 1973, **51**, 1265.
54. Huntress, W. T., *J. Chem. Phys.*, 1972, **56**, 5111.
55. Weiner, J., Smith, G. P. K., Saunders, M. and Cross, R. J., *J. Amer. Chem. Soc.*, 1973, **95**, 4115; Constantin, E., private communication.
56. Weiner, J., Lee, A. and Wolfgang, R., *Chem. Phys. Lett.*, 1972, **13**, 613.
57. Pacák, V., Birkinshaw, K. and Herman, Z., VIII ICPEAC, Belgrade 1973, 106.

ION-MOLECULE REACTIONS

Chairmen

T. AST
(University of Belgrade, Yugoslavia)

P. KNEWSTUBB
(University of Cambridge, U.K.)

D. H. WILLIAMS
(University of Cambridge, U.K.)

82

Reactions of $C_4H_9^+$ Ions Formed by Electron Impact Ionization of Isomeric Butyl Halides

By J. R. HASS and K. R. JENNINGS

(*Department of Molecular Sciences, University of Warwick, Coventry, U.K.*)

INTRODUCTION

THERE has recently been considerable interest in the structure of relatively simple gaseous hydrocarbon ions.[1,2] Mass spectrometric methods of determining ion structures have been based on fragmentation reactions which occur in the ion source or in a field free region of a magnetic deflexion instrument. These methods suffer from the possible complication that the internal energy necessary to cause fragmentation may be more than enough to cause isomerization so that the two processes may compete, thereby reducing the usefulness of the method. The continued development of gas phase ion molecule chemistry has now made it possible to investigate the reactions of ions of much lower internal energies and much longer lifetimes. If, at the moment of formation, the ions contain insufficient energy to isomerize to a common structure, it is to be expected that ions generated from isomeric neutral species will undergo different bimolecular reactions with added neutral species and so allow one to distinguish the structure of the ions. This technique has been successfully used to distinguish structures of $C_2H_5O^+$, $C_3H_6O^{\cdot+}$ and $C_3H_7^+$ ions by observing their reactions in an ion cyclotron resonance (ICR) mass spectrometer.[1,3,4]

The present work was designed to try to obtain structural information about $C_4H_9^+$ ions generated by electron impact ionization from isomeric butyl halides. Williams and co-workers[5] have studied the fragmentation of specifically deuterium labelled butyl ions in the first field free region of a double focussing mass spectrometer. They conclude that in all cases isomerization to a common structure or mixture of structures precedes decomposition so that no structural information can be inferred. Ausloos and co-workers[2] have used pulse radiolysis followed by end product analysis to study the reactions of butyl ions produced from various hydrocarbons. For example, end product analysis of products produced from the radiolysis of *n*-hexane indicated the presence of *n*-butyl, *sec*-butyl and *tert*-butyl ions, but no evidence was found to suggest the presence of *iso*-butyl ions. Similarly, although 2-methylbutane might be expected to yield some *iso*-butyl ions, evidence was obtained only for the presence of *sec*-butyl and *tert*-butyl ions, and it was concluded that under their experimental conditions, the *iso*-butyl ion must rearrange in $\sim 10^{-10}$ s to the more stable *tert*-butyl ion structure.

Isomeric butyl chlorides, bromides and iodides were chosen as sources of $C_4H_9^+$ ions in the present study since their appearance potential is little above the ionization potential of the halides so that complications arising from reactions of other ions were minimized. In addition, reasonable yields of $C_4H_9^+$ ions could be obtained by working at electron energies of less than 1 eV above the appearance potential for these ions, thereby minimizing effects arising from the internal excitation of the ions. Under these conditions further fragmentation of the $C_4H_9^+$ ion was negligible. A search has been made for systematic differences in the ion chemistry of the $C_4H_9^+$ ions produced from isomeric species so that they could be correlated with differences in either structure or internal energy. The relative importance of these two factors was investigated by varying the total pressure and the composition of the reaction mixture.

EXPERIMENTAL

All ICR spectra were recorded on a Varian ICR V-5900 instrument modified as previously described.[6] Reaction mixtures were made up by measurement of partial pressures of the gases on a fixed volume and were estimated to be accurate to $\pm 3\%$. Butyl halides were obtained from B.D.H. Laboratories Ltd and purified by spinning band column distillation and preparative GLC. GLC-MS was used to show that there was only one source of $C_4H_9^+$ ions in each sample. Other reagents were of analytical reagent purity and were used without further purification.

RESULTS AND DISCUSSION

Gross[1] has reported that whereas the $C_3H_7^+$ ion generated from *sec*-C_3H_7Cl undergoes a condensation reaction with furan to give the $C_7H_{11}O^+$ ion, no corresponding reaction is observed with $C_3H_7^+$ ions generated from *n*-C_3H_7Cl. Reactions of $C_4H_9^+$ ions generated from various precursors with furan were therefore investigated.

In equimolar mixtures of C_4H_9X and furan, it was found that all butyl ions, generated from any butyl halide, underwent reaction 1 to form the expected condensation product:

$$C_4H_9^+ + C_4H_4O \rightarrow C_8H_{13}O^+ \quad (m/e = 125) \quad (1)$$

However, in the case of $C_4H_9^+$ ions formed from *n*- and *sec*-butyl halides, an additional product ion was observed at m/e = 107, and the $C_4H_9^+$ ion was shown to be the sole precursor by means of the ion cyclotron double resonance (ICDR) technique. This is interpreted in terms of the following reaction sequence:

$$C_4H_9^+ + C_4H_4O \rightarrow C_8H_{13}O^{+*} \rightarrow C_8H_{11}^+ + H_2O \qquad (2)$$

No qualitative difference was observed in the reactivity of butyl ions generated from different halides of a given isomeric butyl group. Reaction 2 therefore enables one to distinguish $C_4H_9^+$ ions generated from all n- and sec-butyl halides from those generated from iso- and $tert$-butyl halides.

Since acetyl compounds have been shown to be reactive neutral species in certain systems,[7] reactions of the type shown below were studied as a possible means of distinguishing the $C_4H_9^+$ ions generated from different isomeric species:

$$C_4H_9^+ + CH_3CO-X \rightarrow C_6H_{12}O^+ + X \cdot \qquad (3)$$

$$\rightarrow C_6H_{11}O^+ + HX \qquad (4)$$

When X = Cl, butyl ions from all the halides undergo the reaction

$$C_4H_9^+ + CH_3COCl \rightarrow C_6H_{11}O^+ + HCl \qquad (5)$$

and the hydrogen atom lost comes exclusively from the $C_4H_9^+$ ion since no DCl is lost when $C_4H_9^+$ reacts with CD_3COCl. Since these reactions did not distinguish between the butyl ions formed from different precursors, biacetyl was investigated as a less reactive neutral species. In fact, this proved to be too unreactive in that no $C_4H_9^+$ ion would undergo reactions analogous to 3 and 4 with biacetyl. The case in which X = OH, $i.e.$ acetic acid, was therefore investigated.

In the absence of any butyl halide, a number of ion–molecule reactions occur in the acetic acid system, and these may be summarized as follows:

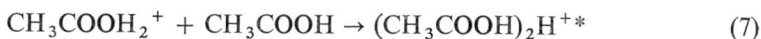

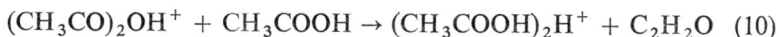

$$CH_3COOH^{\cdot+} + CH_3COOH \rightarrow CH_3COOH_2^+ + CH_3COO \cdot \qquad (6)$$

$$CH_3COOH_2^+ + CH_3COOH \rightarrow (CH_3COOH)_2H^{+*} \qquad (7)$$

$$(CH_3COOH)_2H^{+*} \rightarrow (CH_3CO)_2OH^+ + H_2O \qquad (8)$$

$$(CH_3COOH)_2H^{+*} \xrightarrow{+M} (CH_3COOH)_2H \qquad (9)$$

$$(CH_3CO)_2OH^+ + CH_3COOH \rightarrow (CH_3COOH)_2H^+ + C_2H_2O \qquad (10)$$

These reactions give rise to product ions at m/e = 61, 103, and 121, and their occurrence was confirmed both by ICDR negative signals and by investigating the relative ion abundance as a function of pressure (Fig. 1). The relative importance of reactions 9 and 10 in forming $(CH_3COOH)_2H^+$ ions could not be deduced from the data obtained.

In C_4H_9X/CH_3COOH mixtures, in no case did a butyl ion undergo the expected condensation reaction with the elimination of water and no new ions appeared to be formed. However, the ICDR technique showed that the $C_4H_9^+$ ions generated from all n- and iso-butyl halides contributed to the m/e = 61 and m/e = 103 ions whereas no $C_4H_9^+$ ion derived from a sec- or $tert$-butyl halide did so. The ICDR results indicate the occurrence of the

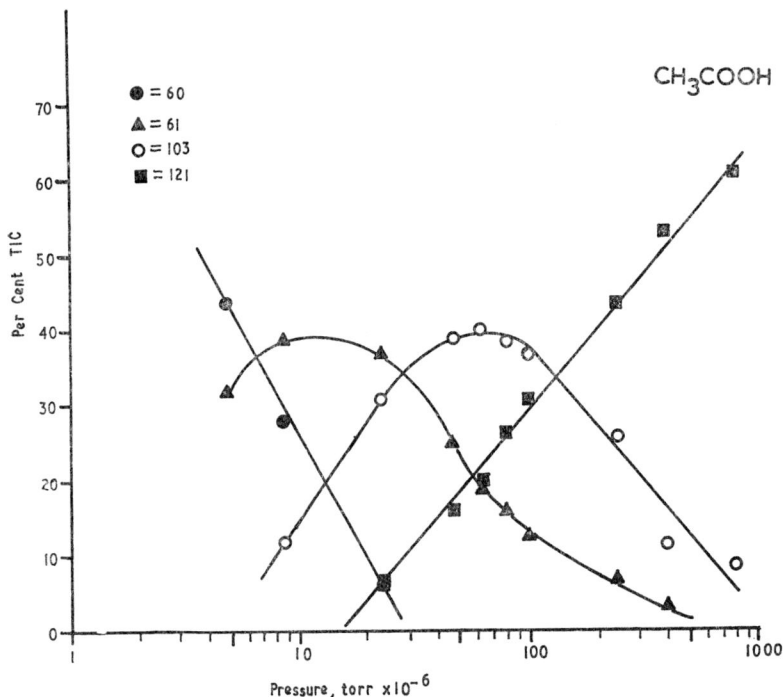

FIG. 1   Relative abundance of ions as a function of pressure in the acetic acid system at ∼15 eV.

following reactions:

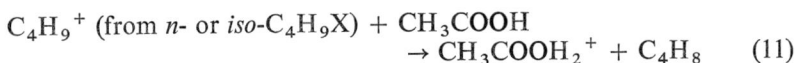

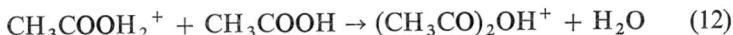

$$C_4H_9^+ \text{ (from } n\text{- or } iso\text{-}C_4H_9X) + CH_3COOH$$
$$\rightarrow CH_3COOH_2^+ + C_4H_8 \quad (11)$$

$$CH_3COOH_2^+ + CH_3COOH \rightarrow (CH_3CO)_2OH^+ + H_2O \quad (12)$$

Experiments using C_4H_9X/CD_3COOD mixtures showed that the $C_4H_9^+$ ion contributed only a proton to the $m/e = 61$ and 103 ions.

No double resonance signal could be found connecting the $C_4H_9^+$ ion with the $(CH_3COOH)_2H^+$ ion, $m/e = 121$, at pressures up to $\sim 1 \times 10^{-4}$ torr, and above this pressure, space charge effects made the technique unreliable.[8] The lack of a signal might suggest that an excited species is involved in the reaction sequence, causing the collision complex between $CH_3COOH_2^+$ and CH_3COOH to be unstable with respect to the elimination of a molecule of water. In this case, the intensity ratio $I(121^+)/I(103^+)$ should increase as the pressure of an inert collision gas is increased, and this was tested using the iso-butyl bromide/acetic acid system. Since $C_4H_9^+$ ions do not react further with iso-butyl bromide, it forms a suitable collision gas, and the ratio $I(121^+)/I(103^+)$ was investigated as a function of total pressure for different ratios of neutral reactants, with the results shown in Fig. 2. For comparison, the same ratio is shown for the acetic acid system in the absence of a butyl halide. It can be seen that at pressures below about 2×10^{-4} torr, the ratio

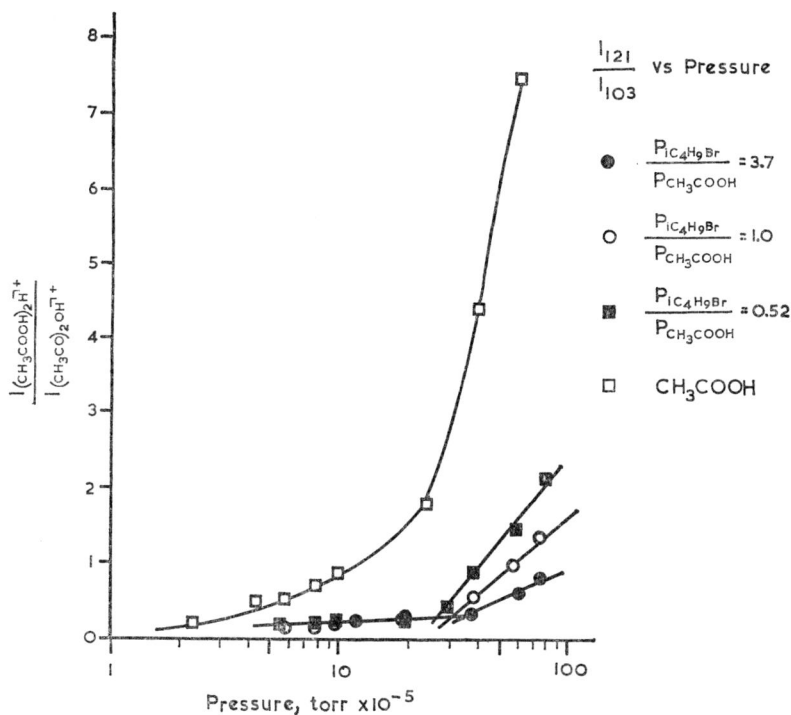

FIG. 2   Plot of $I(121^+)/I(103^+)$ as a function of pressure.

$I(121^+)/I(103^+)$ shows very little pressure dependence and the results do not support the above hypothesis. The results at higher pressures are unreliable because of the possible effects of space charge and further reactions.

The above reactions provide a method of distinguishing $C_4H_9^+$ ions generated from different isomeric butyl halides as the following table shows:

Source of $C_4H_9^+$	$+ C_4H_4O \rightarrow 107^+$	$+ CH_3COOH \rightarrow 61^+, 103^+$
n-C_4H_9X	Yes	Yes
sec-C_4H_9X	Yes	No
iso-C_4H_9X	No	Yes
$tert$-C_4H_9X	No	No

The difference in reactivity may be due to the presence of structurally different ions or to the presence of structurally similar ions containing different amounts of internal energy. The heats of formation of n-, sec-, iso-, and $tert$-$C_4H_9^+$ ions are respectively 912, 803, 858 and 736 kJ mol^{-1} and at the very low pressures used in ICR experiments, it is unlikely that the newly formed $C_4H_9^+$ ions will lose energy before reacting. Consequently if n-$C_4H_9^+$ ions are formed initially from n-C_4H_9X, any isomerization will be reversible because of the lack of stabilizing collisions and the experiments described above would not distinguish between n-$C_4H_9^+$ ions in their ground vibrational state and s-$C_4H_9^+$ ions containing 109 kJ mol^{-1} of vibrational energy, for example. On the other hand, the lack of stabilising collisions does allow one to look for reactivity differences between ions which are structurally different at the instant of formation, even if they rearrange to vibrationally hot isomeric ions before reacting.

REFERENCES

1. Gross, M. L., *J. Amer. Chem. Soc.*, 1971, **93**, 253.
2. Lias, S. G., Rebbert, R. E. and Ausloos, P., *J. Amer. Chem. Soc.*, 1970, **92**, 6430 and Ausloos, P., Chapter 3, *in* 'Progress in Reaction Kinetics' Vol. V, *ed.* G. Porter, Pergamon Press, 1970.
3. Beauchamp, J. L. and Dunbar, R. C., *J. Amer. Chem. Soc.*, 1970, **92**, 1477.
4. Eadon, G., Diekman, J. and Djerassi, C. J., *J. Amer. Chem. Soc.*, 1970, **92**, 6205.
5. Davis, B., Williams, D. H. and Yeo, A. N. H., *J. Chem. Soc. (B)*, 1970, 81.
6. O'Malley, R. M., Jennings, K. R., Bowers, M. T. and Anicich, V. G., *Int. J. Mass Spec. & Ion Phys.*, 1973, **11**, 89.
7. Bursey, M. M., Elwood, T. A., Hoffman, M. K., Lehman, T. A. and Tesarek, J. M., *Anal. Chem.*, 1970, **42**, 1370.
8. Goode, G. C., Ferrer-Correia, A. J. and Jennings, K. R., *Int. J. Mass Spec. & Ion Phys.*, 1970, **5**, 229.

83
Chemical Applications of New Developments in Ion Cyclotron Resonance Spectroscopy

By J. L. BEAUCHAMP

(*Department of Chemistry, Noyes Laboratory,
California Institute of Technology, Pasadena, California, U.S.A.*)

INTRODUCTION

ION trapping techniques, which permit ions to be confined to a well-defined spatial region for a variable and controlled period of time, have application to the study of a wide range of phenomena involving the production and study of ions in the gas phase. Trapped ion cyclotron resonance techniques, originally introduced by McIver,[1] have been developed in our laboratory for use with the conventional ICR cell,[2] making a single instrument versatile for a wide range of experiments. The basis of the trapped ion method is shown in Fig. 1, which represents a side view of the ICR cell. The electron beam is pulsed to form ions in the source region of the cell. During the trapping period, the source drift voltages are left off and a negative bias (for positive ions) is applied to the resonance region drift plates. The centres of the ion

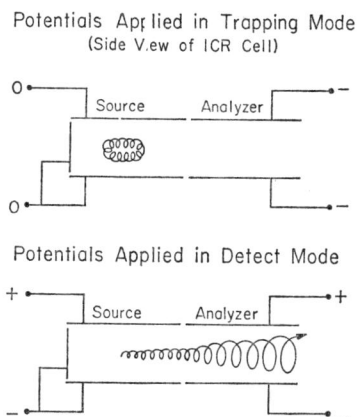

FIG. 1   Potentials applied to the ICR trapped ion cell in trapping and detect modes. Charged particle motion during the trapping and detect modes is illustrated.

717

orbits remain on an equipotential of the electrostatic field which close on themselves with the field configuration illustrated in Fig. 1. After a suitable delay, ions are sampled by switching the electrode potentials for the normal drift-mode operation. Ion trapping without appreciable loss of up to 10 sec at 10^{-6} torr can be easily achieved. This permits $\sim 10^3$ collisions of ions with neutrals before sampling occurs. The long ion residence times in the source region also facilitate the study of processes associated with photo-excitation and photodecomposition of ions.[3]

The major application of ICR trapped ion techniques has been to the study of processes involving fast electron and atomic ion transfer reactions. Equilibria observed in atomic ion transfer reactions permit an accurate assessment of relative neutral and ion thermochemical properties such as proton affinities[4,5] and carbonium ion stabilities.[6] To illustrate the power of the technique for probing reaction kinetics the reactions occurring in electron impact ionized HD serve as an example.[7]

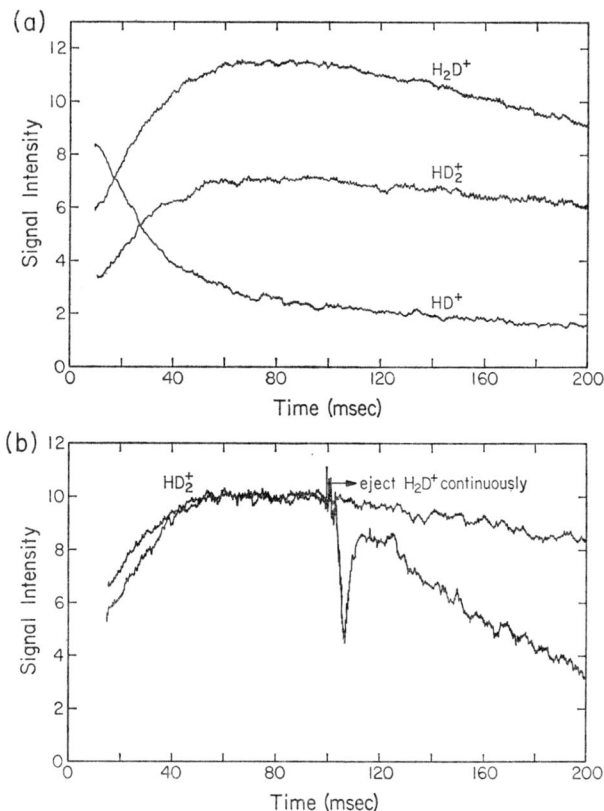

FIG. 2 (a) Variation of ion abundance with time in HD. Ions were formed by a 10 msec pulse of the electron beam at 70 eV electron energy. The curve labelled HD^+ also represents a small H_3^+ signal due in part to a H_2 impurity in the sample. (b) Repeat of HD_2^+ scan with and without continuous ejection of H_2D^+ after 100 msec. The anomalous feature in the HD_2^+ decay curve results from transients generated by turning on the irradiating oscillator in the timing sequence.

FAST PROTON AND DEUTERON TRANSFER
PROCESSES IN HD

Figure 2(a) shows the variation of ion abundance with time in HD. The decay of HD^+ is accompanied by the formation of H_2D^+ and HD_2^+ in accordance with reactions 1 and 2. Even though at long times the ratio of H_2D^+ and HD_2^+

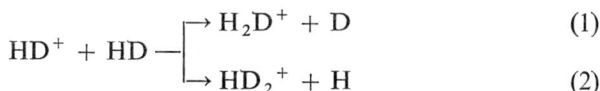

$$HD^+ + HD \longrightarrow \begin{array}{l} \rightarrow H_2D^+ + D \qquad (1) \\ \rightarrow HD_2^+ + H \qquad (2) \end{array}$$

becomes constant, the deuteron and proton transfer reactions (3) and (4) are occurring, being the chain propagation steps in the conversion of HD to a

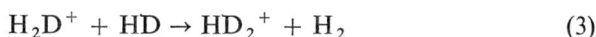

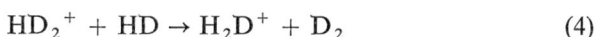

$$H_2D^+ + HD \rightarrow HD_2^+ + H_2 \qquad (3)$$

$$HD_2^+ + HD \rightarrow H_2D^+ + D_2 \qquad (4)$$

mixture containing H_2 and D_2. The kinetics of reactions (3) and (4) can be explored using a technique unique to ICR trapped ion methods. In Fig. 2 (b), the time dependence of the HD_2^+ signal from Fig. 2(a) is twice reproduced. For one of the curves, the species H_2D^+ is continuously ejected after 100 msec by an r.f. field applied at its cyclotron frequency to the source region of the ICR cell. This causes reaction (4) to proceed entirely to the right, leading to the decay of HD_2^+ and permitting the extraction of the rate constant for process 4. The rate constant for reaction 3 can similarly be determined, and the results are summarized in Table I along with the rate constants for a variety of other fast electron and proton transfer reactions.[7] Also included

TABLE I

Kinetics of Fast Electron and Proton Transfer Reactions

Reaction	Rate constant (10^{-10} cm^3 molecule^{-1} sec^{-1})		
	Experimental	Statistically corrected total rate	Langevin
$H_2D^+ + HD \rightarrow HD_2^+ + H_2$	$2\cdot4 \pm 0\cdot3$	$14\cdot4$	$15\cdot9$
$HD_2^+ + HD \rightarrow H_2D^+ + D_2$	$3\cdot4 \pm 0\cdot3$	$20\cdot4$	$15\cdot2$
$CH_3D_2^+ + CH_2D_2 \rightarrow CH_2D_3^+ + CH_3D$	$0\cdot3 \pm 0\cdot1$	$1\cdot5$	$13\cdot1$
$CH_2D_3^+ + CH_2D_2 \rightarrow CH_3D_2^+ + CHD_3$	$0\cdot3 \pm 0\cdot1$	$1\cdot5$	$13\cdot1$
$^{14}NH_4^+ + {}^{15}NH_3 \leftrightarrows {}^{15}NH_4^+ + {}^{14}NH_3$	$6\cdot0 \pm 0\cdot6$	$12\cdot0$	$11\cdot6^a$
$^{12}CO^+ + {}^{13}CO \leftrightarrows {}^{13}CO^+ + {}^{12}CO$	$4\cdot1 \pm 0\cdot3$	$8\cdot2$	$8\cdot6^a$
$^{12}CO_2^+ + {}^{13}CO_2 \leftrightarrows {}^{13}CO_2^+ + {}^{12}CO_2$	$3\cdot7 \pm 0\cdot3$	$7\cdot4$	$8\cdot0$
$^{14}N_2^+ + {}^{15}N_2 \leftrightarrows {}^{15}N_2^+ + {}^{14}N_2$	$6\cdot1 \pm 0\cdot5$	$12\cdot2$	$8\cdot2$

[a] Ion-dipole interactions may be important, thus invalidating the Langevin model as an estimate of the encounter rate.

in Table I are the statistically corrected total encounter rates and the rates calculated using the Langevin model.[8] The statistically corrected rates are generated from the experimental rates with an assumed model for the reaction. For example, in the case of reaction (4), the two reaction intermediates I and II are considered possible. Ignoring isotope effects, complex II,

$$
\begin{array}{cc}
\underset{\displaystyle \underset{D}{|} \overset{\displaystyle H}{|}}{\overset{\displaystyle H}{\underset{D}{}} {\,}^{+}\!\!{\cdots}} & \qquad \qquad \underset{\displaystyle \underset{D}{|} \overset{\displaystyle H}{|}}{\overset{\displaystyle D}{\underset{D}{}} {\,}^{+}\!\!{\cdots}} \\
\text{I} & \qquad \qquad \text{II}
\end{array}
$$

which is formed in 33 % of the encounters, will dissociate to give H_2D^+ 50 % of the time. Hence, the rate for reaction 4 represents one-sixth of the total encounter rate which is predicted by the Langevin model. In the case of HD the agreement between the Langevin rate and the statistically corrected rate is quite good. In the case of methane, however, the measured rate is much slower, indicating that proton transfer occurs with a very low probability. For the electron transfer reactions cited in Table I it is apparent that the Langevin model satisfactorily explains the charge transfer rate in CO and CO_2, but that in nitrogen electron transfer occurs outside the orbiting impact parameter.

DETERMINATION OF ION THERMOCHEMICAL PROPERTIES

Studies of ion thermochemical properties using ICR techniques have been concerned primarily with an examination of the generalized reaction 5 in which the relative binding energies of a reference acid A to two bases B_1 and B_2 are determined.[4, 5] This includes studies of base strengths and acidities of neutrals. Of more recent interest is an examination of the thermochemical properties of ions and neutrals which can be determined from an examination of reaction (6) in which the relative binding energies of a reference base B to two acids A_1 and A_2 are determined. An example is the fluoride transfer reaction (7) in which F^- is the reference base and the difluoromethyl and

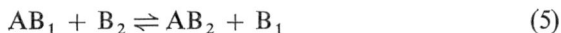

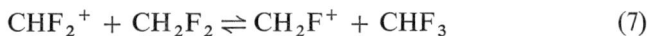

$$AB_1 + B_2 \rightleftharpoons AB_2 + B_1 \tag{5}$$

$$A_1B + A_2 \rightleftharpoons A_2B + A_1 \tag{6}$$

$$CHF_2{}^+ + CH_2F_2 \rightleftharpoons CH_2F^+ + CHF_3 \tag{7}$$

fluoromethyl carbonium ions are the gas phase Lewis acids. Studies of the preferred direction and equilibrium in reaction (7) can yield accurate carbonium ion stabilities.[6] Summarized in Fig. 3 are recently determined fluoride affinities (RF heterolytic bond dissociation energies) for a series of fluorinated hydrocarbons.[9] A low value of $D(R^+ - F^-)$ indicates a more stable carbonium ion. As can be seen from Fig. 3 fluorine substituents have unusual effects on carbonium ion stabilities which can be understood in terms of dative π-bonding and σ-withdrawal effects when fluorine is bound to the carbonium ion centre and ion-dipole interactions when fluorine is

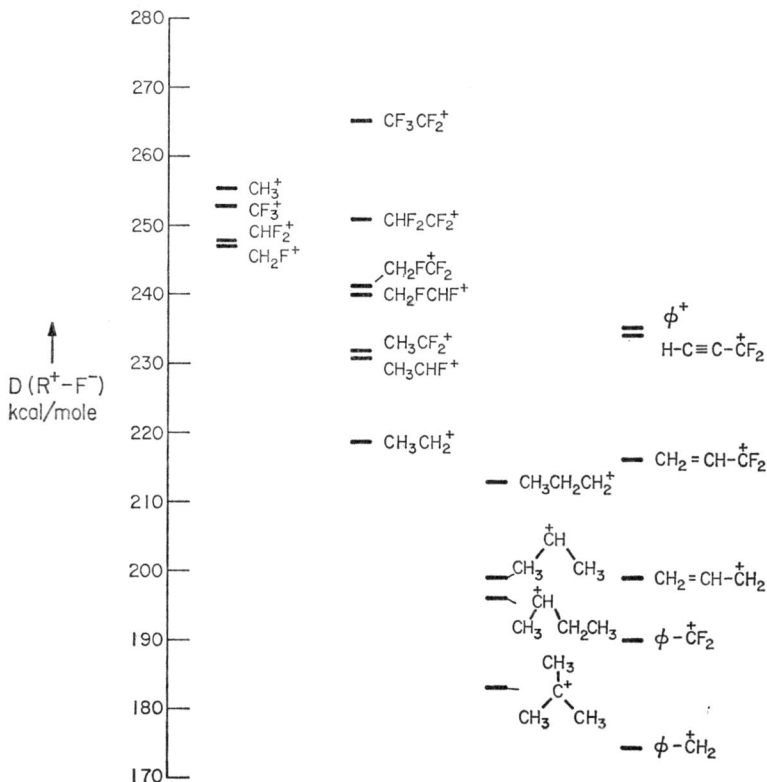

FIG. 3   Fluoride affinities (R$^+$–F$^-$ heterolytic bond dissociation energies) of carbonium ions.

located apart from the carbonium ion centre. Interestingly, different reference bases give different orderings for carbonium ion stabilities. In addition, data such as that illustrated in Fig. 3 can be used to determine ionic heats of formation. From the measured fluoride affinity, the heat of formation of the phenyl cation $(C_6H_5{}^+)$ is calculated to be 270 ± 4 kcal/mole, which is 15 kcal/mole less than the currently accepted value.[10]

DIRECT OBSERVATION OF ELECTRONS IN ICR EXPERIMENTS

In addition to and in conjunction with trapped ion techniques, a second new development will permit application of ICR to the study of processes involving production and reaction of free electrons in gases. In the ICR cell, charged species oscillate in the trapping well at a frequency proportional to the square root of the ratio of the trapping voltage to the particle mass. By utilizing a radiofrequency detector with an electric field component parallel rather than perpendicular to the primary field, the oscillation of particle motion in the trapping field can be excited and the particles detected. Using the 'Q-meter'

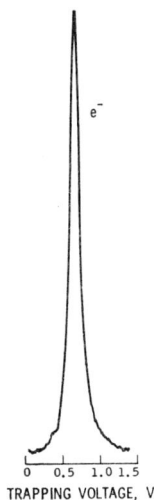

FIG. 4 Resonant power absorption by electrons due to excitation of their oscillatory motion in the trapping field. The resonance is observed by sweeping the trapping voltage. Other conditions were: detector frequency, 4·6 MHz; r.f. amplitude 5 mV; electron current 5×10^{-11} A from scattering of the electron beam off CO_2 at an impact energy of 4·0 eV.

circuit developed by Huntress and Simms,[11] the power absorption of electrons at a frequency of 4·6 MHz is shown in Fig. 4, with the trapping voltage being scanned to observe the resonance. The direct observation of electrons can be utilized to study the kinetics and energetics of a variety of processes, including electron attachment reactions, associative detachment reactions, inelastic excitation by electron impact, and electron-neutral collision frequencies through a study of pressure broadening of the resonance absorption curve.

ACKNOWLEDGMENT

This research was supported by the United States Atomic Energy Commission under Grant No. AT(04-3)767-3.

REFERENCES

1. McIver, R. T. Jr, *Rev. Sci. Instr.*, 1970, **41**, 1953.
2. McMahon, T. B. and Beauchamp, J. L., *Rev. Sci. Instr.*, 1972, **43**, 509.
3. Dunbar, R. C., *J. Amer. Chem. Soc.*, 1973, **95**, 472. Recent experiments in our laboratory (B. S. Freiser and J. L. Beauchamp, unpublished results) have utilized the ICR trapped ion analyser described in Reference 2 to advantage in studying the photodissociation of ions.
4. Beauchamp, J. L., *Ann. Rev. Phys. Chem.*, 1971, **22**, 527.
5. Henderson, W. G., Taagepera, M., Holtz, D., McIver, R. T. Jr, Beauchamp, J. L. and Taft, R. W., *J. Amer. Chem. Soc.*, 1972, **94**, 4728.
6. McMahon, T. B., Blint, R. J., Ridge, D. P. and Beauchamp, J. L., *J. Amer. Chem. Soc.*, 1972, **94**, 8934.
7. McMahon, T. B., Miasek, P. G. and Beauchamp, J. L., unpublished results.

8. Henchman, M., 'Rate Constants and Cross Sections', *in* 'Ion-Molecule Reactions', *ed*. J. L. Franklin, Plenum Press, New York, 1972, chapter 5.
9. McMahon, T. B., Ridge, D. P., Park, J. Y. and Beauchamp, J. L., unpublished results.
10. Franklin, J. L. *et al.*, NSRDS-NBS 26, 'Ionization Potentials, Appearance Potentials, and Heats of Formation of Gaseous Positive Ions', U.S. Government Printing Office, Washington, D.C., 1969.
11. Huntress, W. T. Jr and Simms, W. T., *Rev. Sci. Instr.*, submitted for publication.

Discussion

J. B. Hasted (Birkbeck College, London, U.K.): Are the symmetrical resonance charge transfer data consistent with ionic mobility data?

J. L. Beauchamp: Where meaningful comparisons can be made, the charge transfer rate constants are consistent with mobility data for molecular ions moving in their parent gases. It is essential that the comparison be made at E/P values appropriate for thermal energy ions. The agreement in the case of N_2, where charge transfer occurs outside the orbiting impact parameter, is particularly good.

84

On the Relative Gas Phase Basicities of some (Un)substituted Piperidines and 1-Aza-adamantanes, Determined by Ion Cyclotron Resonance

By C. B. THEISSLING and N. M. M. NIBBERING

(*Laboratory for Organic Chemistry, University of Amsterdam, Amsterdam, The Netherlands*)

INTRODUCTION

IN various recent publications it has been shown that ion cyclotron resonance spectroscopy (ICR)[1a-g] and chemical ionization mass spectrometry[2a-b] have been successfully applied to ascertaining gas phase properties of organic molecules.

In many cases these intrinsic properties, *i.e.* free of any interfering solvation effects, appeared to be considerably different from the properties determined in solution. For example, the acidity order of simple alcohols in the gas phase is the reverse of the order in solution.[3] The basicity order of simple aliphatic amines in the gas phase has been studied extensively both in a qualitative[4,5] and in a quantitative way.[6-8]

These investigations[4-8] have demonstrated that the basicity increases upon an increasing length of the alkyl chain and upon a higher degree of alkyl substitution at the nitrogen atom. Using these results a complete thermodynamic analysis of the 'anomalous order' of amine basicities in solution (NH_3 < primary < secondary > tertiary) has been presented.[9]

The basicity order of some *N*-nitroso-amines in the gas phase however parallels the order in solution.[10]

Recently, the relative solution basicities, ultraviolet, infra-red and ^{13}C NMR spectra of some (un)substituted 1-aza-adamantanes have been measured in our laboratory.[11] This study has been undertaken to look for 'through bond' and 'through space' interactions in some 1-aza-adamantane derivatives, compounds very well suited for this purpose (*vide infra*). As solvation effects might obscure the intrinsic properties of compounds, as mentioned above, it seemed to us desirable to establish the relative gas phase basicities of the title compounds using the ICR technique.

725

RESULTS

The basicities of some (un)substituted piperidines and 1-aza-adamantanes, determined in aqueous solution,[11] have been compiled in Table I.

In measuring the relative order of gas phase basicities of these compounds by ICR, it must be recognized that some of their molecular ions generated in the ICR cell may suffer ring opening by α-cleavage, a well-known reaction in conventional mass spectrometry.[12a,b] This would be a serious problem, because the ring-opened molecular ions, being odd-electron species, may abstract a hydrogen atom from a neutral molecule in the collision process

TABLE I

Base Strengths Measured in Aqueous Solution at 22°C

Compound	pK_b	Compound	pK_b
piperidine	2·88[a]	1-aza-adamantane	2·96 ± 0·1
1-methylpiperidine	3·92[a]	1-aza-4-methylene-adamantane	4·16 ± 0·1
1-methyl-4-piperidone	6·01 ± 0·1	1-aza-adamant-4-one	5·43 ± 0·1

[a] 'Handbook of Chemistry and Physics', 48th ed. (values at 25°C).

to give protonated molecules with a ring-opened structure. The latter structures could lead to a wrong order of relative basicities. The ring-opened ions, however, may also give rise to protonated molecules with a ring-closed structure in an additional collision process with a neutral molecule, when the possible proton transfer reaction is exothermic or thermoneutral.

The following experimental conditions have been chosen in order to suppress the potentially possible ring opening of the molecular ions of the present compounds as much as possible:

(1) The electron energy has been adjusted at such a value (10–10·5 eV) that in the ICR spectra exclusively $M^{+\cdot}$- and $(M + H)^+$-peaks are present (cf. Table II).

(2) The total pressure has been held at $7-9 \times 10^{-5}$ torr in order to guarantee sufficient collisions between ions and neutral molecules in the ICR cell and to deactivate those ions possessing excess of internal energy (cf. Table II).

(3) The total ion current has been maintained at the 10^{-11} A range, as recommended by K. R. Jennings et al.[13] in particular for double resonance experiments, but also to have as much as possible neutral molecules in excess of ions (see also point (2)).

Using these conditions, the relative gas phase basicities of the title compounds have been studied by observing the preferred direction of proton transfer in reaction (1),[4] applying the powerful double resonance technique:

$$B_1H^+ + B_2 \rightleftharpoons B_1 + B_2H^+ \tag{1}$$

In this way each of the compounds has been tested against most of the other members of the series; this procedure provides for a check on the consistency of the relative order of the gas phase basicities.

TABLE II

Relative Intensities of the Primary and Secondary Ions in the ICR Spectra of Some of the Mixtures Studied

Reaction[a]	Compound I		$M^{+\cdot}$ m/e (RI)[b]	MH^+ m/e (RI)[b]	Compound II		$M^{+\cdot}$ m/e (RI)[b]	MH^+ m/e (RI)[b]
1	1-aza-adamantane	(I)	137 (100)	138 (14)	quinuclidine	(IV)	111 (62)	112 (20)
2	quinuclidine	(IV)	111 (97)	112 (29)	1-aza-4-methylene-adamantane	(II)	149 (100)	150 (13)
3	1-aza-4-methylene-adamantane	(II)	149 (48)	150 (100)	1-methylpiperidine	(VI)	99 (61)	100 (71)
4	1-methylpiperidine	(VI)	99 (67)	100 (20)	1-aza-adamant-4-one	(III)	151 (100)	152 (15)
5	1-aza-adamant-4-one	(III)	151 (80)	152 (100)	piperidine	(V)	85 (31)	86 (76)
6	piperidine	(V)	85 (60)	86 (98)	1-methyl-4-piperidone	(VII)	113 (100)	114 (45)

[a] See Table III.
[b] RI = mass uncorrected relative intensities at a total pressure of 7–9 × 10⁻⁵ torr and at an electron energy of 10–10·5 eV.

TABLE III

Double Resonance Results for Forward (F) and Reverse (R) Proton Transfer Reactions at 10–10·5 eV[a,b]

I X = H,H
II X = CH$_2$
III X = O

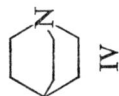

IV

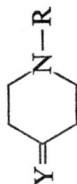

V R = H; Y = H, H
VI R = CH$_3$; Y = H, H
VII R = CH$_3$; Y = O

No.	Reaction	F[c]	R[c]
1	$C_9H_{15}N$ (I) $+ C_7H_{13}NH^+ = C_9H_{15}NH^+ + C_7H_{13}N$ (IV)	—	0
2	$C_7H_{13}N$ (IV) $+ C_{10}H_{15}NH^+ = C_7H_{13}NH^+ + C_{10}H_{15}N$ (II)	—	0
3	$C_{10}H_{15}N$ (II) $+ C_6H_{13}NH^+ = C_{10}H_{15}NH^+ + C_6H_{13}N$ (VI)	—	0
4	$C_6H_{13}N$ (VI) $+ C_9H_{13}NOH^+ = C_6H_{13}NH^+ + C_9H_{13}NO$ (III)	—	+
5	$C_9H_{13}NO$ (III) $+ C_5H_{11}NH^+ = C_9H_{13}NOH^+ + C_5H_{11}N$ (V)	—	−
6	$C_5H_{11}N$ (V) $+ C_6H_{11}NOH^+ = C_5H_{11}NH^+ + C_6H_{11}NO$ (VII)	—	+

Double Resonance Results for Forward (F) and Reverse (R) Charge Transfer Reactions at 10·5, 15 eV[a]

No.	Reaction	10·5 eV		15 eV	
		F[d]	R[d]	F	R
1[e]	$C_9H_{15}N$ (I) $+ C_7H_{13}N^{+\cdot} = C_9H_{15}N^{+\cdot} + C_7H_{13}N$ (IV)	0	0	+	+
2[e]	$C_7H_{13}N$ (IV) $+ C_{10}H_{15}N^{+\cdot} = C_7H_{13}N^{+\cdot} + C_{10}H_{15}N$ (II)	0	0	+	+

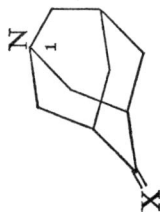

		F	R	F	R
3	$C_{10}H_{15}N$ (II) $+ C_6H_{13}N^{+\cdot} = C_{10}H_{15}N^{+\cdot} + C_6H_{13}N$ (VI)	+	+	0	+
4[f]	$C_6H_{13}N$ (VI) $+ C_9H_{13}NO^{+\cdot} = C_6H_{13}N^{+\cdot} + C_9H_{13}NO$ (III)	+	-	+	-
5[g]	$C_9H_{13}NO$ (III) $+ C_5H_{11}N^{+\cdot} = C_9H_{13}NO^{+\cdot} + C_5H_{11}N$ (V)	0	0	0	0
6[g]	$C_5H_{11}N$ (V) $+ C_6H_{11}NO^{+\cdot} = C_5H_{11}N^{+\cdot} + C_6H_{11}NO$ (VII)	0	+	0	+

[a] Each of the compounds has been tested against most of the other members of the series. Only the reactions reflecting the relative order of gas phase basicities are given.

[b] In most of the cases, the molecular ion also appears to transfer a proton to the neutral molecule of the stronger base. The intensity ratio of the double resonance signals for proton transfer by the molecular ion and the protonated molecule decreases when the irradiation is applied to the analyser region instead of the source region. Thus, some α-cleavage of the molecular ions seems to occur, even at the applied electron energies where fragment ions could not be detected anymore; another indication for ring-opened ions follows from the observation that in some reactions, part of the protonated molecules participates in charge transfer reactions (see further text).

[c] A negative sign (−) is generally associated with an exothermic or thermoneutral reaction, a positive sign (+) with endothermic reactions. A zero (0) indicates that the reaction was investigated but no signal change was observed. In conjunction with a forward (−) this suggests that the reverse reaction was not proceeding measurably.[4]

[d] In charge-transfer reactions a positive sign (+) is generally associated with an exothermic, thermoneutral or endothermic reaction. For the significance of a zero (0), see note c.

[e] The (+) signals, observed both for the F- and R-charge transfer reactions prevent deduction of the relative order of IP's. In the F-reactions however, sweep-out of $M^{+\cdot}_{IV}$ c.q. $M^{+\cdot}_{II}$ in the ion source region by use of the irradiating oscillator reduces the abundances of $M^{+\cdot}_{I}$ c.q. $M^{+\cdot}_{IV}$; sweep-out of $M^{+\cdot}_{I}$ c.q. $M^{+\cdot}_{IV}$ in the R-reaction has no influence upon the abundances of $M^{+\cdot}_{IV}$ c.q. $M^{+\cdot}_{II}$. These observations point to a relative order of IP's I < IV < II.

[f] The negative double resonance signal in the forward reaction cannot be explained by us.

[g] The observed double resonance signals imply that IP(piperidine) > IP(1-methyl-4-piperidone), in agreement with the stabilizing effect of alkyl groups.[4] Yet, the gas phase basicity of 1-methyl-4-piperidone appears to be lower than that of piperidine (see text).

In addition to proton transfer reactions charge transfer reactions (2) have been studied in this manner, because some papers have shown a correlation between gas phase basicities and ionization potentials, *i.e.* the basicity will increase with decrease of the ionization potential:[7,8]

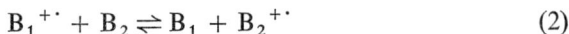

$$B_1^{+\cdot} + B_2 \rightleftharpoons B_1 + B_2^{+\cdot} \tag{2}$$

Thus, the latter reaction can provide an additional check on the relative order of the gas phase basicities of the present compounds.

The results of the double resonance experiments have been compiled in Table III.

From the data in Table III the following relative order of gas phase basicities of the title compounds can be deduced: 1-methyl-4-piperidone < piperidine ≈ 1-aza-adamant-4-one < 1-methylpiperidine < 1-aza-4-methylene-adamantane < quinuclidine < 1-aza-adamantane.

It is interesting to note that the same order is found at 15 eV and at the same pressure range (7–9 × 10^{-5} torr). This would further support that the order actually refers to intrinsic basicities of the compounds studied.

DISCUSSION

The 1-aza-adamantane derivatives are ideal systems to study 'through bond' and 'through space' interactions.

Optimal through bond interaction can be expected when a substituent such as carbonyl or methylene is present at position 4; in that case the lone pair orbital on nitrogen is almost parallel to the π-orbital of the substituent and to the C^2–C^3 sigma bond (cf. Fig. 1).

Through bond and through space interactions have been the subject of extensive research during the last years.[14a,b,c] These interactions are however, difficult to separate. Nevertheless, some years ago Heilbronner *et al.* were able using photo-electron spectroscopy[14a] to show in 1,4-diazabicyclo[2.2.2]-octane, that the through bond interaction dominates the through space interaction.

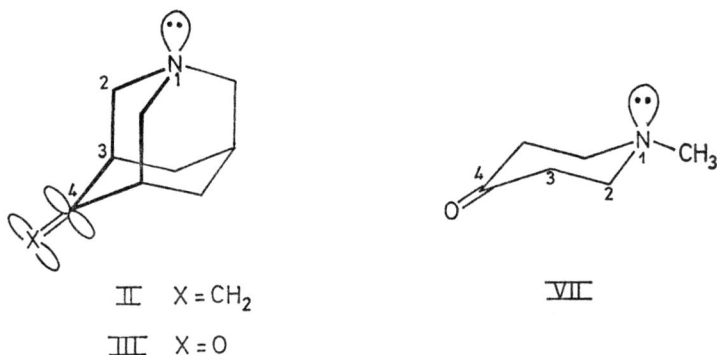

II   X = CH₂

III   X = O

VII

FIG. 1   1-aza-4-methylene-adamantane (II), 1-aza-adamant-4-one (III) and 1-methyl-4-piperidone (VII). In the left-hand figure the path of sigma-coupling has been indicated schematically.

The same author has recently demonstrated that through space and through bond interactions in bicyclo[3.2.2] nona-6,8-diene almost cancel each other.[14b]

Through bond interaction in the electronic ground-state of the present 1-aza-adamantane derivates could not be detected with certainty in our laboratory,[11] by application of absorption, emission, infrared and ^{13}C NMR spectroscopy and by the study of the relative basicities in solution (cf. Table I). The gas phase basicities of the present 1-aza-adamantane derivatives do in fact parallel those in solution (cf. Table I); hence, through bond interaction seems not to be the most likely explanation for the observed order of gas phase basicities. Any analogy from solution phase studies however, must be drawn with great caution. For example, a threefold to fourfold attenuation of substituent effects in pyridines in the gas phase has been observed as compared with those in solution.[15] Similarly, the stabilizing effect of alkyl groups on the ammonium ions in the gas phase has been found to be six times stronger than in solution.[7]

Yet, we prefer to suggest a through space interaction in the 1-aza-adamantane compounds as the dominating effect, in particular because of the ^{13}C NMR studies in our laboratory; ^{13}C NMR shifts are known to be very sensitive to changes in charge density, but very distinct shifts, to be expected in the case of through bond interaction, have not been observed.[11]

The through space interaction can account for the basicity order, found for the 1-aza-adamantane derivatives, when it is realized that the substituents at position 4 represent dipoles. The dipole moment of the methylene group and the larger one of the carbonyl group at position 4 both point to such a direction, that the positive charge will be more concentrated at the carbon atom 4, nearest to the nitrogen atom.† The lone-pair electrons at the nitrogen will therefore experience a net positive charge by an interaction through space, effecting a lower basicity in the observed order. Such a through space interaction is the only reasonable explanation for the lower basicity of 1-methyl-4-piperidone with respect to that of 1-methylpiperidine, an order also found in solution (cf. Table I); a through bond interaction is impossible because of the axial position of the nitrogen lone pair which is perpendicular to the $C^{2(5)}$–$C^{3(6)}$ bonds[16] (cf. Fig. 1). The lower basicity of piperidine versus 1-methylpiperidine and of quinuclidine versus 1-aza-adamantane again demonstrates the stabilizing effects of alkyl groups.[4,7,10]

Unfortunately, the discussed effects cannot be interpreted in a quantitative way as long as accurate quantitative relative gas phase basicities of the title compounds are unknown.

EXPERIMENTAL

A Varian-V5903 ICR spectrometer (serial number 126) equipped with a standard flat three-section cell and a dual inlet system was used in this study.

---

† The ^{13}C NMR spectrum of 1-aza-4-methylene-adamantane shows a downfield shift of the carbon atom 4 ($\delta = 155.1$ ppm rel. to TMS) with respect to the methylene carbon atom ($\delta = 100.8$ ppm rel. to TMS). The chemical shift of carbon atom 4 in 1-aza-adamant-4-one is 214.3 ppm and in 1-aza-adamantane 36.6 ppm rel. to TMS. For further details see Reference 11.

The electron energy was set using a digital voltmeter and pressures were determined from the ion pump current meter.

Single resonance spectra were recorded by field modulation (amplitude 20G) with a sweep rate of 10 min, response 0·3 sec. In the ion cyclotron double resonance experiments square wave modulation was employed and the spectra were recorded with a sweep rate of 10 min, response 3 sec, sweep width 50 kHz. The irradiating oscillator (the amplitude used 0·5 × 0·1) was applied to the source and analyser region, the latter giving the best results.

During the scans, the total ion current was also monitored[13] by connecting the output of the electrometer, which measured the total ion current, to a digital voltmeter. It appeared that the total ion current dropped less than 1 % at the peaks. The static drift potentials applied to source and analyser were usually less than 0·5 V and the electron collector voltage was zero.

The present compounds (solids and liquids) were introduced into the reaction cell through the standard inlet system except for 1-aza-adamant-4-one. For this compound a direct insertion probe manufactured by Varian was used. This probe allows the introduction of the sample via a vacuum lock into the ionization chamber close to the filament. In this position the probe can be heated to volatilize the sample. Very careful control of temperature was necessary to obtain a reasonable working pressure and to hold the vapour pressure rather constant, while at the same time preventing too fast evaporation and loss of sample. The probe temperature was ∼50°C.

The chemicals used in this study were obtained from commercial sources except for the 1-aza-adamantane derivatives which were available in our laboratory. The syntheses of these compounds have been reported elsewhere.[11]

The ^{13}C NMR spectra were measured on a Varian XL-100 instrument.

ACKNOWLEDGMENTS

The authors wish to thank Dr J. W. Verhoeven for helpful discussions, Mr C. Kruk who measured the ^{13}C NMR spectra and Dr A. W. J. D. Dekkers for the samples of the 1-aza-adamantane derivatives. We are also grateful to Dr W. N. Speckamp and Professor Dr Th. J. de Boer for their interest in this work and to the Netherlands Organization for Pure Research (SON/ZWO) who provided the grant for the purchase of the ICR spectrometer.

REFERENCES

1. (a) Baldeschwieler, J. D., *Science*, 1968, **159**, 263.
   (b) Goode, G. C., O'Malley, R. M., Ferrer-Correia, A. J. and Jennings, K. R., *Nature*, 1970, **227**, 1093.
   (c) Goode, G. C., O'Malley, R. M., Ferrer-Correia, A. J. and Jennings, K. R., *Chemistry in Britain*, 1971, **7**, 12.
   (d) Baldeschwieler, J. D. and Sample-Woodgate, S., *Accounts Chem. Res.*, 1971, **4**, 114.
   (e) Futrell, J. H., *in* 'Dynamic Mass Spectrometry', Vol. 2, *ed.* D. Price, Heyden and Son Ltd, London, 1971, p. 97.
   (f) Gray, G., *Advan. Chem. Phys.*, 1971, **19**, 141.

(g) Drewery, C. J., Goode, G. C. and Jennings, K. R., *in* 'M.T.P. International Review of Science (Physical Chemistry)', Vol. 5, *ed.* A. Maccoll, Butterworth, London, 1972, p. 183.
2. (a) Field, F. H., *Accounts Chem. Res.*, 1968, **1**, 42.
   (b) Munson, B., *Anal. Chem. Rev.*, 1971, **43**, 28.
3. Brauman, J. I. and Blair, L. K., *J. Amer. Chem. Soc.*, 1970, **92**, 5986.
4. Brauman, J. I., Riveros, J. M. and Blair, L. K., *J. Amer. Chem. Soc.*, 1971, **93**, 3914.
5. Dzidic, I., *J. Amer. Chem. Soc.*, 1972, **94**, 8333.
6. Bowers, M. T., Aue, D. H., Webb, H. M. and McIver, R. T., *J. Amer. Chem. Soc.*, 1971, **93**, 4314.
7. Aue, D. H., Webb, H. M. and Bowers, M. T., *J. Amer. Soc.*, 1972, **94**, 4726.
8. Henderson, W. G., Taagepera, M., Holtz, D., McIver, R. T., Beauchamp, J. L. and Taft, R. W., *J. Amer. Chem. Soc.*, 1972, **94**, 4728.
9. Arnett, E. M., Jones, F. M. III, Taagepera, M., Henderson, W. G., Beauchamp, J. L., Holtz, D. and Taft, R. W., *J. Amer. Chem. Soc.*, 1972, **94**, 4724.
10. Billets, S., Jaffé, H. H. and Kaplan, F., *Org. Mass Spectrom.*, 1973, **7**, 431.
11. Dekkers, A. W. J. D., Verhoeven, J. W. and Speckamp, W. N., *Tetrahedron* in press.
12. (a) Budzikiewicz, H., Djerassi, C. and Williams, D. H., 'Mass Spectrometry of Organic Compounds', Holden-Day, Inc., San Francisco, 1967.
    (b) Palecek, J. and Mitera, J., *Org. Mass Spectrom.*, 1972, **6**, 1353.
13. Goode, G. C., Ferrer-Correia, A. J. and Jennings, K. R., *Int. J. Mass Spec. Ion Phys.*, 1970, **5**, 229.
14. (a) Heilbronner, E. and Muszkat, K. A., *J. Amer. Chem. Soc.*, 1970, **92**, 3818.
    (b) Goldstein, M. J., Natowsky, S., Heilbronner, E. and Hornung, V., *Helv. Chim. Acta*, 1973, **56**, 294. (See also references cited therein.)
    (c) Sasaki, T., Eguchi, S., Kiriyama, T. and Sakito, Y., *J. Org. Chem.*, 1973, **38**, 1648.
15. Taagepera, M., Henderson, W. G., Brownlee, R. T. C., Beauchamp, J. L., Holtz, D. and Taft, R. W., *J. Amer. Chem. Soc.*, 1972, **94**, 1369.
16. Cookson, R. C., Henstock, J. and Hudec, J., *J. Amer. Chem. Soc.*, 1966, **88**, 1061.

85
ICR-ESR and Optical Spectroscopy Study of Electron Impact Excitation and Ionization in Ammonia

By R. MARX, G. MAUCLAIRE, M. WALLART and A. DEROULEDE

(*Laboratoire de Physico-Chimie des Rayonnements (associé au C.N.R.S.), Université de Paris-Sud, Orsay, France*)

EXCITATION ionization and ion-molecule reactions in molecular gases irradiated with low energy electrons (0 to 100 eV) have been studied in a specially designed ICR cell.

This cell may be used either with an ESR spectrometer to study the free radicals resulting from excitation and ion-molecule reactions, or with optical detection of the light emitted by electronically excited molecules or fragments.

This experimental set-up has been used to study NH_3. It has been shown that ground state $\dot{N}H_2$ are produced by electron impact above 6·2 eV and excited $\dot{N}H_2{}^*$ above 8·5 eV. At 12 eV the $\dot{N}H_2$ radicals observed by ESR are produced half by ion-molecule reaction and half by excited or superexcited molecule decomposition.

EXPERIMENTAL METHODS

The Apparatus

Figure 1 shows a schematic drawing of the ICR cell used with an ESR spectrometer or an optical detection device. A detailed description of the ICR spectrometer has been given elsewhere[1,2] but a special cell has been designed for these experiments.

The filament producing the electron beam is differentially pumped in order to prevent the free radicals produced by pyrolysis being trapped in the ESR cavity.

The cell is shorter than a conventional one because both the ESR cavity and the ICR cell must be located in a homogeneous magnetic field between the pole pieces of the electromagnet. It has only three short sections and therefore it cannot be used to measure ion-molecule reaction rate constants.

Depending on the energy of the electron beam, one or both of the following

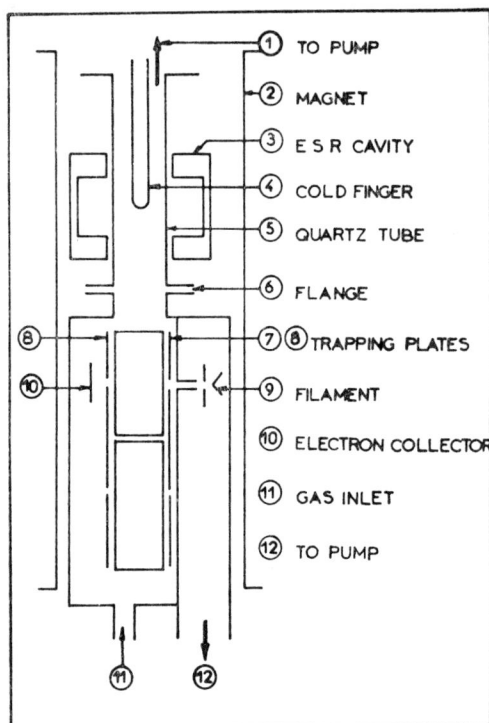

FIGURE 1

processes occur in the ion source

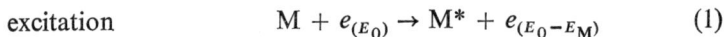

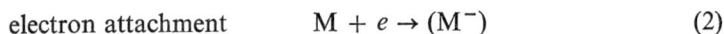

excitation $$M + e_{(E_0)} \rightarrow M^* + e_{(E_0 - E_M)} \qquad (1)$$

electron attachment $$M + e \rightarrow (M^-) \qquad (2)$$

ionization $$M + e \rightarrow M^+ + 2e \qquad (3)$$

The primary species produced in the ion source may then undergo further reactions.

So, this ICR cell has been used: to determine positive and negative ion abundance curves and threshold excitation spectra; to produce free radicals and electronically excited species which may be analysed by ESR or optical spectroscopy. Finally an attempt was made to determine the contribution of ion-molecule reactions to the production of free radicals.

Positive and Negative Ions Abundance Curves

To observe positive ions the trapping voltage is positive (~ 0.5 V), so that negative charges are ejected by the trapping plates and only positive ions are trapped and drifted along the axis of the cell to be collected on the end plates. The total ion current (TIC) is measured with a picoammeter and recorded versus the energy of the electron beam.

The ion current may be mass analysed using a frequency swept r.f. generator connected to the drift plates of the analyser.

To collect negative ions the trapping voltage is reversed ($V_T \sim -0.5$ V). Moreover, the low energy electrons which would otherwise be trapped and drift along the cell with the negative ions have to be ejected using a 3 MHz generator connected to the source trapping plates.

Negative ions may be mass analysed in the same way as positive ions.

Threshold Excitation Spectra

To obtain threshold excitation spectra below the ionization potential one would have to measure the 'zero-energy' electrons produced in process 1. However, in an ICR cell one traps and measures all the negative species of energy up to the trapping potential which is in our experiments at least 0.1 eV. Of course, to measure only electrons the negative ions have to be ejected in the ion source.

Free Radicals Abundance Curves

Free radicals are produced in the ICR cells by various processes following excitation and ionization. They are pumped through the ICR cell into the cavity of an ESR spectrometer where they are trapped on a cold finger in a matrix of unreacted molecules. Because ESR is not a very sensitive method, the radicals have to be accumulated for a rather long time. The time needed to obtain measurable ESR spectra depends on pressure, electron beam intensity and, of course, on the efficiency of the free radical producing process.

Electron beam intensities and pressure are usually higher than in an ICR experiment $I_e = 1$ to $100~\mu$A; $P \geq 5 \cdot 10^{-4}$ torr.

Moreover, the delay between production and trapping of the free radicals being about 10^{-3} sec, only long-lived radicals may be observed.

Finally, because of the high magnetic field no charged species can reach the ESR cavity.

Optical Emission Spectra of Electronically Excited Molecules or Fragments

To observe the emission spectrum of electronically excited species produced in the ion source of the ICR cell, the ESR cavity is replaced by a quartz window (Fig. 2). The light coming out of the cell is focussed on the entrance slit of a monochromator (or an interference filter), analysed and measured with a photomultiplier.

The electron beam is pulsed to allow phase sensitive detection which improves the signal-to-noise ratio and prevents the light coming from the heated filament to perturb the measurements.

With this experimental set-up one may record fluorescence spectra of the emitting species at constant energy of the exciting electron beam. Recording the intensity of an emission line as a function of the energy of the electron beam gives the fluorescence excitation curve for the species emitting the line.

Contribution of Ion-Molecule Reactions to Free Radical Formation Above the Ionization Potential

The ICR cell may be used in two different ways to produce free radicals.

(a) If the ions are trapped in the cell, the contribution of the ion-molecule

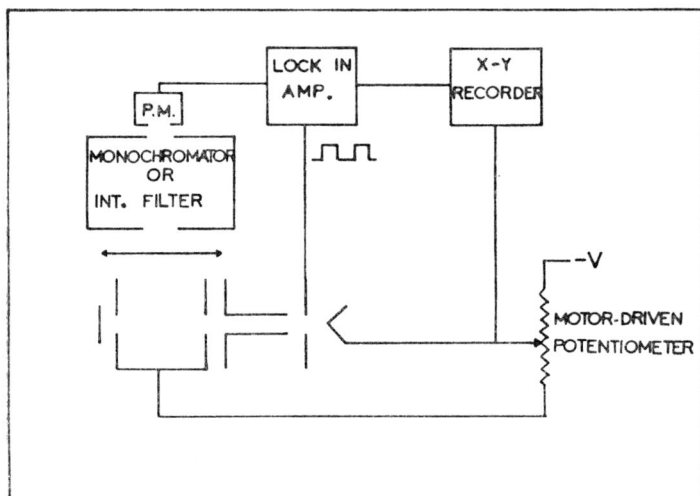

FIGURE 2

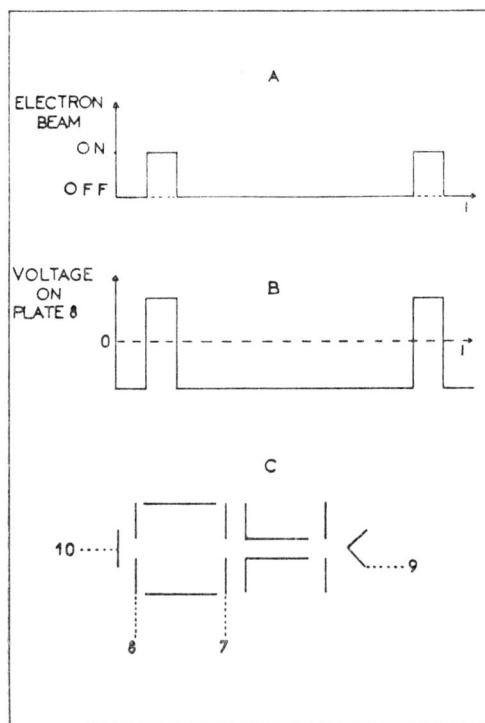

FIGURE 3

reaction is enhanced. In fact, at the relatively high pressure we have to use $(P \geq 5 \cdot 10^{-4}$ torr), the usual trapping time $(\sim 10^{-3}$ sec) is sufficient to allow a reactive collision for all the primary ions.

(b) If the ions are ejected in a time short enough to prevent any ion-molecule reaction before neutralization, one would observe only free radicals coming from excited molecule decomposition.

The time allowed for ejection, to have a negligible contribution of ion-molecule reactions, depends on the rate constant of this reaction and on the pressure.

Since cyclotron ejection, with a r.f. field between the drift plates, takes too much time, the ions are ejected with a 'detrapping' negative potential on one of the trapping plates of the ion source.

In order to prevent the ejection voltage perturbing the energy of the ionizing electron beam, we use the pulsing sequences shown in Fig. 3.

The ions are produced during the electron beam pulse $(10^{-6}$ sec) while the trapping potential is $+V_T$, the same as in the experiment without ejection. Then the ions are ejected by applying the detrapping pulse as shown in Fig. 3. The energy of the ionizing electrons is then the same with or without ejection of the ions and it is possible to compare the ESR spectra obtained with or without ejection of the ions, all other experimental conditions being the same.

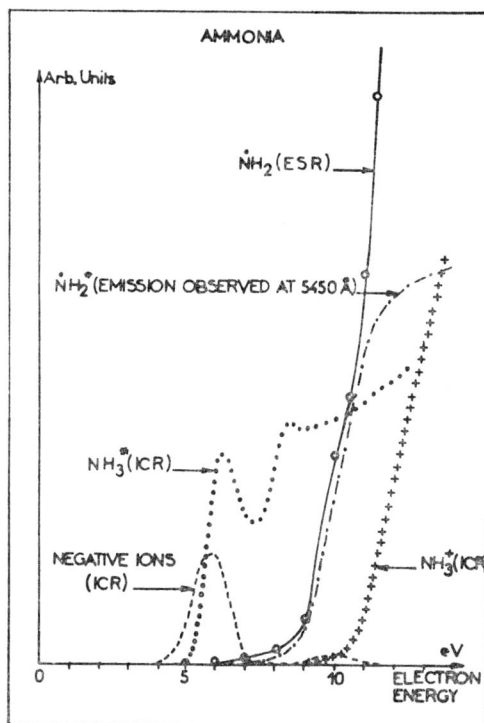

FIGURE 4

EXPERIMENTAL RESULTS AND DISCUSSION
FOR AMMONIA NH_3

Figure 4 shows the different curves obtained for ammonia with electrons of energy up to 14 eV.

For each experiment the energy scale has been calibrated using the appearance potential of $NH_3^{\cdot+}$ (10·4 eV) and the first negative ions peak (5·65 eV).

The intensity scales are given in arbitrary units and there is no normalization factor between the different experiments.

Processes Occuring Below the Ionization Potential of NH_3

Negative ion curve and threshold excitation spectrum—Our results are in good agreement with the curves published in the literature and obtained by different methods.[3,4,5]

The first peak of the negative ion curve contains NH_2^- and H^- coming from

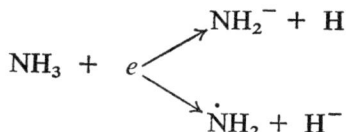

$$NH_3 + e \underset{\searrow}{\overset{\nearrow}{}} \begin{array}{l} NH_2^- + H \\[1em] \dot{N}H_2 + H^- \end{array}$$

In the second one there is also NH^- but the total intensity is 20 times lower. The first peak of the excitation spectrum corresponds to the transition $(\tilde{X} \to \tilde{A}\ {}^1A''_2 v = 6)^5$ and the second one to $(\tilde{X} \to \tilde{B}\ {}^1E'' v = 5)$. The other peaks are very poorly resolved.

Fluorescence excitation curve of NH_2^*—The emission spectrum of $\dot{N}H_2$ consists of a set of poorly resolved bands starting at about 5000 Å. The excitation curve shown in Fig. 4 was obtained at $\lambda = 5450$ Å.

The O–O transition for $\dot{N}H_2\tilde{A}\ {}^2A_2 \to \dot{N}H_2\tilde{X}\ {}^2B_1$ would give a fluorescence at 9762 Å. So $\lambda = 5450$ Å corresponds to a rather high vibrational level of $NH_2\tilde{A}\ {}^2A_2$.

$\dot{N}H_2$ abundance curve—As shown previously $\dot{N}H_2$ is the only free radical observed by ESR in NH_3. The appearance potential of $\dot{N}H_2$ is around 6 eV.

Near threshold the curve increases very slowly, a rather sharp increase starts at 8–9 eV and a second one at 10–11 eV near the appearance potential of NH_3^+.

Discussion—These results lead to the following conclusions:

Dissociative electron attachment has a very low contribution to $NH_2^{\cdot}$ production: no radical is observed at the maximum of the negative ion abundance curve (5·65 eV).

The first excited state of NH_3 (max. at 6·2 eV) gives only ground state $NH_2^{\cdot}$ observed by ESR and no excited $\dot{N}H_2^*$.

The second excited state of NH_3 (max. at 8·5 eV) produces excited NH_2^* observed by fluorescence and also probably ground state $\dot{N}H_2$. The trapped $\dot{N}H_2$ observed by ESR in this energy range may be produced directly as ground state $\dot{N}H_2$ and (or) as excited $\dot{N}H_2^*$ which are de-excited before being trapped.

This is in good agreement with the observations of Okabe and Lenzi[6] on photo-excited ammonia: in spite of the fact that there is enough energy to produce excited $\dot{N}H_2{}^*$ with 6·5 eV photons or electrons, $\dot{N}H_2{}^*$ are produced only above 8·2 eV.

Processes Occurring Above the Ionization Potential

It has been shown that above the ionization potential two ion-molecule reactions produce $\dot{N}H_2$ radicals:

between 10·5 and 15·7 eV only:

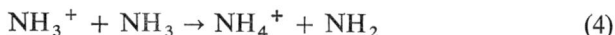

$$NH_3{}^+ + NH_3 \rightarrow NH_4{}^+ + NH_2 \qquad (4)$$

above 15·7 eV there is also:

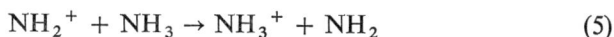

$$NH_2{}^+ + NH_3 \rightarrow NH_3{}^+ + NH_2 \qquad (5)$$

We have tried to determine the contribution of reaction (4) to the production of $\dot{N}H_2$ at 12 eV by ejecting $NH_3{}^+$ ions. The result of this experiment is not very accurate for the following reasons:

Because we have to pulse the electron beam the mean intensity is only a few μA. Therefore to have a measurable ESR spectrum we need a rather high pressure of at least 10^{-3} torr. Moreover, the mean time required in our cell to eject the $NH_3{}^+$ ions is 10^{-5} sec. The rate constant of reaction (4) being $k = 1·4 \times 10^{-9} cm^3 mol^{-1} s^{-1}$ (Ref. 2) 40% of the $NH_3{}^+$ ions undergo a reactive collision before being ejected. So the decrease in the ESR spectrum observed when the ions are ejected was only 15% to 20% corresponding to a contribution of ion-molecule reactions of 50% $\pm$ 10%.

REFERENCES

1. Mauclaire, G., Thesis, Orsay, 1973.
2. Marx, R. and Mauclaire, G., *Int. Mass Spectrom. Ion Phys.*, 1973, **10**, 213.
3. Compton, R. N., Stockdale, J. A. and Reinhardt, P. W., *Phys. Rev.*, 1969, **180**, 111.
4. Sharp, T. E. and Dowell, J. T., *J. Chem. Phys.*, 1969, **50**, 3024.
5. Skerbele, A. and Lassettre, E., *J. Chem. Phys.*, 1965, **42**, 395.
6. Okabe, H. and Lenzi, M., *J. Chem. Phys.*, 1967, **47**, 5241.

Discussion

J. Durup (Université de Paris-Sud, Orsay, France): Isn't it possible, once a qualitative understanding of the observed phenomena is achieved, to use processes where products are monitored in two different ways (*e.g.* emission of $NH_2{}^*$, and ESR spectrum of $NH_2{}^{\cdot}$, etc.) for a relative calibration of your detecting devices with respect to one another?

R. Marx: In principle yes. But it would be very difficult to get any absolute value from the ESR spectra, first because the collection efficiency of $NH_2{}^{\cdot}$ on the cold trap would be very difficult to evaluate and also because the sensitivity calibration of the spectrometer for this very peculiar sample is not obvious.

The calibration of the optical detection would also introduce a considerable error.

J. B. Hasted (Birkbeck College, London, U.K.): If sufficient ESR spectra of molecular ions are known or can be collected, then this technique might contribute to the study of the difficult problem of electron-ion recombination at surfaces.

R. Marx: A few ESR spectra of molecular ions are known: NH_4^+ in ionic solid matrices for example.

In our apparatus the ions cannot be collected in the ESR cavity, they cannot reach the cold finger because of the high transverse magnetic field. However one can imagine some modification which would allow this kind of experiment.

86
Mass Spectrometric Ion Sampling from Reactive Plasmas

By M. J. VASILE and G. SMOLINSKY

(*Bell Laboratories, Murray Hill, New Jersey, U.S.A.*)

DURING the past ten years, several reports of mass spectrometric sampling of r.f. or d.c. discharges have appeared in the literature.[1-9] The objectives of most of these studies were to better define the ion sampling process or to elucidate ion-molecule reactions and energy transfer mechanisms in discharges or afterglows involving atomic or simple molecular systems. Our requirements for a diagnostic procedure for reactive r.f. plasmas evolved from our work with mixtures of argon and volatile organosilicon compounds,[10] which, when subjected to a glow discharge, form polymeric films possessing unique optical properties.[11]

From an examination of the literature[10,12-14] it is apparent that there has been a considerable amount of speculation about the mechanisms by which polymers are produced in a gas discharge. Very few attempts have been made to obtain direct experimental data regarding the chemistry of such reactive discharges. Therefore we have constructed an apparatus which samples either the ionic or neutral species arriving at the walls of the discharge vessel. We report our findings on discharges of mixtures of argon and vinyltrimethylsilane (VTMS, structure 1).

$$CH_2=CHSi(CH_3)_3$$

$$(1)$$

APPARATUS

The apparatus was designed to allow only a minimal exposure of the mass filter to the flux of reactive molecules, since the polymer films produced from organic compounds are insulating. Figure 1 is a diagram of the discharge tube, cylinder lenses, and mass filter and shows the differential pumping arrangement. The sampling orifice is a 12-μ diameter hole which was laser machined in a 5·1-cm diameter aluminium end-plate. The dimensions of the orifice were chosen so as to avoid clustering effects due to free jet expansion and to minimize disturbances of the equipotentials in the sheath.

The mass filter is a commercially available E.A.I. Quad 300 equipped with

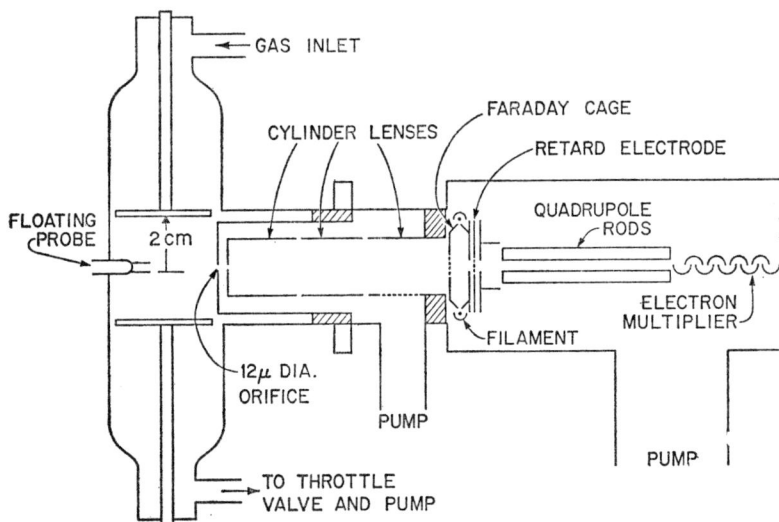

FIGURE 1

a dual filament cross-beam ionizer. The discharge tube (Fig. 1) was constructed from nominal 58 mm i.d. Pyrex tubing, with appropriate connections to accommodate electrode supports, gas inlet and flow control, evacuation, and a floating double probe. Pressure in the discharge tube was measured with a calibrated capacitance manometer. R.f. power was supplied to one of the electrodes from a commercially available 13·56 MHz generator.

EXPERIMENTAL

Gases and gas mixtures were stored in a 3-litre stainless steel tank and admitted to the discharge tube through a variable leak valve. The discharge tube was continuously pumped through a throttle valve by a liquid nitrogen trapped pump, which was independent of either of the pumping systems previously described. Pressure in the reactor tube was varied between 0·1 and 1·0 torr. The r.f. potential applied to one of the electrodes was monitored by a high-impedance 10-MHz oscilloscope.

The potentials of the outer support cylinder and sampling orifice were allowed to float in order that they might attain the wall potential. Potentials on the cylinder lenses were adjusted to obtain maximum signal intensity, viz., approximately -390 V on the first lens, -80 V on the second, and -100 V on the third. The potential on the Faraday cage was also adjusted between 0 and $+5$ V for maximum ion intensity. The retarding grid following the Faraday cage in the mass spectrometer was maintained at ground potential unless it was in use, when a ramp from 0 to $+150$ V was applied to it.

For the studies on pure argon, the floating double probe was used to measure the electron temperature and estimate the number density of charge carriers and the plasma-to-wall potential difference. The floating double probe could not be used in a reactive plasma because of the almost instantaneous deposition of an electrically insulating film.

RESULTS AND DISCUSSION

Mass spectra of pure argon discharges at 150 V applied r.f. potential showed the low abundance isotopes of argon at m/e 36 and 38, as well as a trace amount of Ar^{+2} at m/e 20. The balance of the spectrum consisted of ions that result from contaminants, the most prominent of which occur at m/e 12 (C^{+}), m/e 28 (N_2^{+}, CO^{+} or $C_2H_4^{+}$), and m/e 41 (ArH^{+}). The total impurity ions account for 2·6% of the argon ion intensity, while ArH^{+} accounts for 1·1%. The Ar_2^{+} ion was also observed, at approximately the same intensity level as ArH^{+}. The argon ion signal exhibited a very broad maximum over the pressure range studied. Floating probe measurements showed that the electron temperature varied between 2·3 and 1·9 eV over the pressure range 0·1 to 0·9 torr.

Spectra of 1% VTMS–argon discharges were characteristic of an ion-molecule reaction dominated regime. The dependence of the total ion signal with pressure was very close to that of pure argon, while the sum of the ion currents of each species in the 1% mixture was a factor of two lower than that obtained in pure argon. There were only two ionic species which had significant intensities, Ar^{+} and ArH^{+}. These ions accounted for 96% of the total at 0·12 torr, and decreased somewhat with pressure to 78% of the total at 0·93 torr.

Mass spectra of the ions originating from the discharge of the 13% mixture for two pressures, 0·1 and 0·54 torr, are shown in Fig. 2. At the lower pressure, the principal ions are H_3^{+}, CH_3^{+}, $C_2H_2^{+}$, Ar^{+} and ArH^{+}, none of which contain silicon, while at the higher pressure the principal ions are H_2SiMe^{+}, $HSiMe_2^{+}$ and $SiMe_3^{+}$, all of which contain silicon. The transition from a situation where the dominant ions contain no silicon to one where they do is accompanied by a sharp decrease in total ionization for the 5, 8 and 13% mixtures.

A conspicuous feature observed in discharges of VTMS at all concentrations and pressures studied is the lack of correlation between the relative intensities of the ions formed in the discharge and those obtained from low energy electron impact.[15] VTMS has an ionization potential of 9·8 eV,[16] which is about 6 eV below that of argon, and a cross-section four times that of argon[15] between 30 and 70 eV. Since the electron temperature of the discharge is at most a few electronvolts, it was expected that ionization of VTMS by electron impact would be far more probable than ionization of argon. In the electrode-to-plasma sheath, where the electrons are more energetic, VTMS ions should be produced at a rate approximately four times that of argon ions.

Electron impact induced fragmentation produces $C_2H_3SiMe_2^{+}$ (m/e 85) as the dominant ion for electron energies between 9·8 eV and 23 eV. It was reported[15] that the mass 72 fragment ion has about the same appearance potential as the mass 85 fragment. However, at electron energies of 15 to

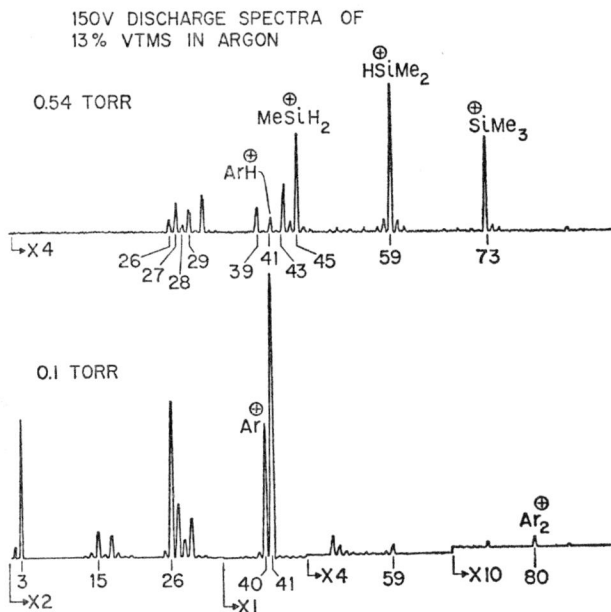

150 V DISCHARGE SPECTRA OF
13% VTMS IN ARGON

FIGURE 2

16 eV, the intensities of the mass 72 as well as those of the mass 59 fragment and the parent ion are about an order of magnitude less than that of the mass 85 fragment. The mass 26, 27, 43 and 45 fragments do not even appear from electron impact until energies of about 14 eV are reached. In the discharge, the mass 26 and 27 ions are evident at low pressures (0·1 to 0·3 torr), the mass 43 and 45 fragments reach a maximum at 0·3 torr, and the mass 59 and 73 ions dominate the spectrum above about 0·6 torr. In contrast, the mass 85 fragment was never seen in the discharge at any significant level for any of the mixtures over the pressure range studied. The change in distribution of ionic products that occurs with increasing pressure can also be obtained by simply increasing the power input to the discharge. Figure 3 illustrates this effect with an 8% VTMS–argon mixture at 0·6 torr.

The neutral species originating from discharges in the 13% mixture were sampled as a function of pressure, using 25 eV electrons in the ion source with the lens potentials arranged to block ions from the discharge. Neutral fragments could not be observed for the 1 and 5% mixtures with the discharge on, although the sensitivity of the spectrometer was more than adequate to sample neutral molecules with the discharge off. It is surprising to find that the discharge is so efficient at affecting the monomer. At low pressures, the dominant neutral fragments observed were H_2, C_2H_2 and C_2H_4. These species were also produced at high pressures, along with $HSiMe_2$ (m/e = 59) and $SiMe_3$ (m/e = 73).

Considering the results obtained from the pressure and power variations of discharges in pure argon and the VTMS–argon mixtures, it is tempting to formulate an hypothesis by which some of these results can be accounted for. At low number densities of VTMS, or with sufficiently high power, electrons in the discharge can attain energies necessary to ionize Ar and H_2, which can then react to give the ion-molecule reaction products observed, or enter into dissociative charge transfer with VTMS. Dissociative charge transfer on VTMS by argon ions should yield high appearance potential fragment ions, one of which is abundantly observed ($C_2H_2^+$, AP = 14·1 eV). As the number density of VTMS increases, inelastic processes that occur below the ionization potential of VTMS become dominant, and reduce the electron energy to a point where ionization of Ar or H_2 is insignificant. If the electrons

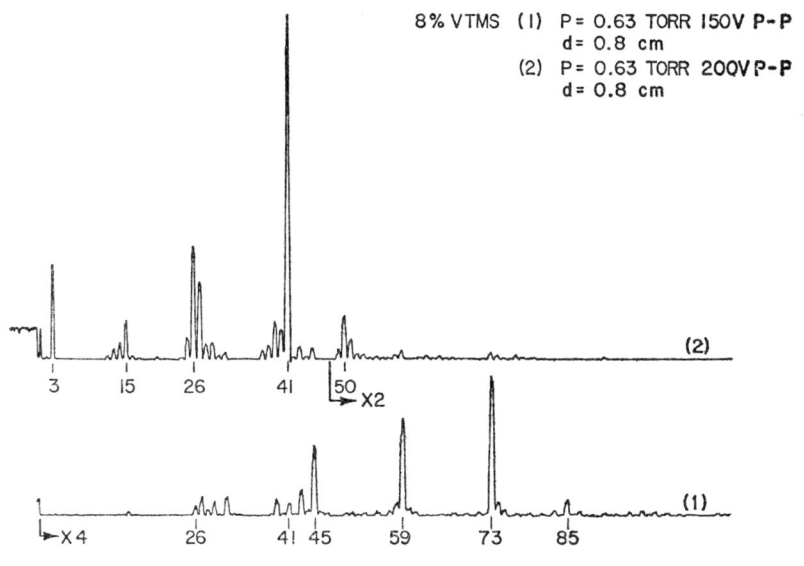

8% VTMS (1) P = 0.63 TORR 150V P-P
d = 0.8 cm
(2) P = 0.63 TORR 200V P-P
d = 0.8 cm

FIGURE 3

have sufficient energy to excite argon to the ^3P metastable levels, then energy transfer from the metastables to VTMS should yield fragment ions with appearance potentials of 11·6–11·7 eV. Such fragment ions are observed at masses 59 and 73.

The lack of primary ionization fragments of VTMS is at variance with the above scheme. For example, electrons that are sufficiently energetic to excite argon to its ^3P metastable levels or to its ion also have more than enough energy to cause primary ionization of VTMS. The key to this problem may be the unspecified inelastic processes that cause the electron energy to drop with increasing number density of VTMS.

A more detailed account of these studies will appear as a two part paper in the literature (*Intl. J. Mass. Spectrom. and Ion Phys.*).

REFERENCES

1. Knewstubb, P. F. and Tickner, A. W., *J. Chem. Phys.*, 1962, **36**, 674.
2. Franklin, J. L., Studniarz, S. A. and Ghosh, P. K., *J. Appl. Phys.*, 1968, **39**, 2052.
3. Studniarz, S. A. and Franklin, J. L., *J. Chem. Phys.*, 1968, **49**, 2652.

4. Kohout, F. C. and Neiswender, D. D., *Intl. J. Mass Spec. and Ion Phys.*, 1970, **4**, 21.
5. Whiting, H. L., *J. Appl. Phys.*, 1969, **40**, 236.
6. Sullivan, J. J. and Buser, R. G., *J. Vac. Sci. and Tech.*, 1969, **6**, 103.
7. Seguin, J. G., Dugan, C. H. and Goodings, J. M., *Intl. J. Mass Spec. and Ion Phys.*, 1972, **9**, 203.
8. Fehsenfeld, F. C., Schmeltekopf, A. L., Goldan, P. D., Schiff, H. I. and Ferguson, E. E., *J. Chem. Phys.*, 1966, **44**, 4087.
9. Gilkinson, J. L., Held, H. and Chanin, L. M., *J. Appl. Phys.*, 1969, **40**, 2350.
10. Vasile, M. J. and Smolinsky, G., *J. Electrochem. Soc.*, 1972, **119**, 451.
11. (a) Tien, P. K., Smolinsky, G. and Martin, R., *Appl. Optics*, 1972, **11**, 637.
    (b) Thompson, L. F. and Smolinsky, G., *J. Appl. Poly. Sci.*, 1972, **16**, 1179.
12. Yasuda, H. and Lamaze, C. E., *J. Appl. Polymer Sci.*, 1971, **15**, 2277.
13. Westwood, A. R., *European Polymer J.*, 1971, **7**, 363.
14. Denaro, A. R., Owens, P. A. and Crawshaw, A., *European Polymer J.*, 1968, **4**, 93; 1969, **5**, 471; 1970, **6**, 487.
15. Smolinsky, G. and Vasile, M. J., *Org. Mass Spectrom*, 1973, **6**.
16. Bock, H. and Seidl, H., *J. Organometal. Chem.*, 1968, **13**, 87.

87
Product Distribution in Ion–Neutral Collisions

By F. S. KLEIN, G. D. LEMPERT, E. MURAD†
and A. PERSKY

(*Isotope Department, Weizmann Institute of Science, Rehovot, Israel*)

REACTIVE scattering measurements of ion–neutral collisions have been made in an instrument which is shown schematically in Fig. 1. The principal parts of the apparatus are:

(1) a monoplasmatron ion source,[1] which can be run also in an electron–impact mode. The emerging ions are accelerated into

(2) a two-dimensionally focussing 90° magnetic sector mass separator. The mass-separated ion beam, of circular cross-section, is decelerated and focussed by a

(3) three cylinder lens system. The ion beam was operated at energies between 5 and 50 eV. The total primary beam ion current is monitored by

(4) a grid-amplifier system and scattered by

(5) a crossed neutral beam, emerging from a fused glass multichannel array source at 90° to the ion beam.

Reactant and product ion energy distributions are observed by

(6) the retarding grid method, mass analysed by

(7) a quadrupole mass filter and detected by

(8) an electrostatic electron multiplier (channeltron).

The observed signal-to-noise ratio is greatly enhanced by passing the signal plus background through a narrow band amplifier and phase sensitive detector circuitry (P.A.R. 127 lock-in-amplifier).

The signal is tuned with the aid of

(a) a strip chart recorder, which displays the lock-in-amplifier output directly and

(b) an on-line Varian 620 mini-computer, programmed to print out digital information as well as plot ion intensity distributions as a function of the programmed parameter scattering angle or energy of the mass-separated ions. The computer accumulates measured signals to the required accuracy before proceeding to the next parameter setting.

† Senior Guest Scientist 1972/73. Permanent address: Air Force Cambridge Research Labs., Hanscom Field, Bedford, Massachusetts 01730, U.S.A.

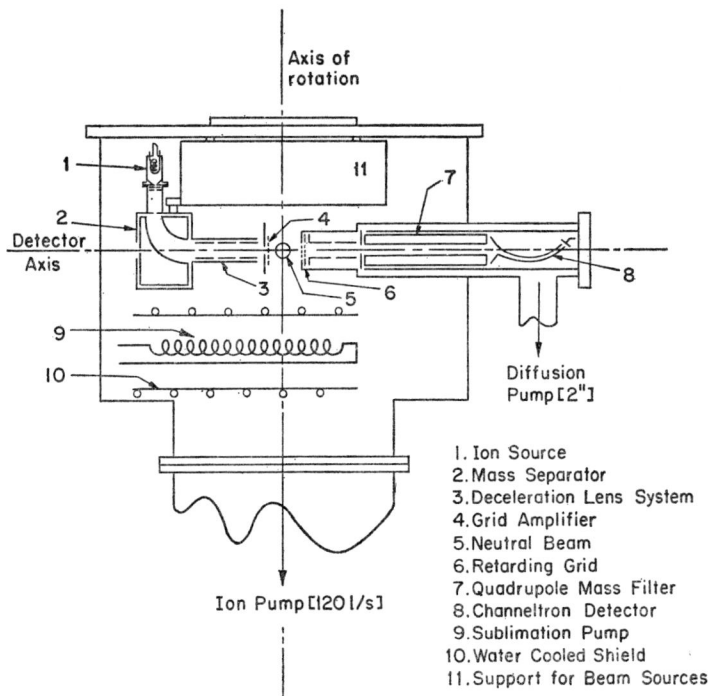

FIG. 1   Schematic diagram of the apparatus.

Items 1 to 5 are fixed with respect to each other and rotatable with respect to the detection system, items 6 to 8, within an angular range of $-20°$ to $+110°$ with respect to the primary ion beam. The whole system is continuously pumped by a 120 litre/sec ion pump and during an experiment by a water-cooled-baffled-titanium sublimation pump. The background pressure is approximately 10^{-7} torr before experiment and approximately 10^{-6} torr when both beams are flowing. Items 6 to 8 are differentially pumped by a liquid nitrogen trapped oil diffusion pump (2") system.

This apparatus was tested by the reaction

$$CO^+ + D_2 \rightarrow COD^+ + D \qquad (1)$$

which has been carefully studied by Kerstetter and Wolfgang[2] and by Ding et al.[3] We found in agreement with these studies a narrow forward-directed angular product distribution and relatively high product energies as summarized below:

$E(CO^+)$	$E(COD^+)$ eV, lab energy $\pm0\cdot1$ V
14·5	13·5
7·5	7·3
5·3	5·0

These observations are in agreement with expectations for an essentially stripping mechanism. Specimen results of this reaction are shown in Fig. 2.

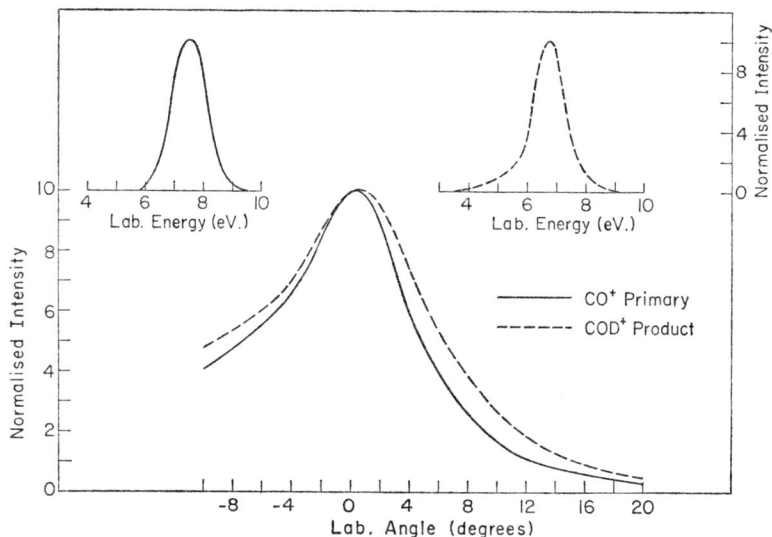

FIG. 2 Angular and energy distributions of primary and product ions of the reaction $CO^+ + D_2 \rightarrow COD^+ + D$.

The systems we are studying at present are the reactions:

$$CO^+ + CO_2 \rightarrow CO_2^+ + CO \qquad \text{charge exchange} \qquad (2)$$

and

$$C^+ + CO_2 \rightarrow CO^+ + CO \qquad \text{atom transfer} \qquad (3)$$

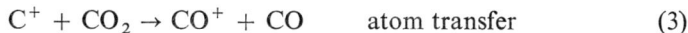

Schildcrout and Franklin[4] reported rate constants of 2·0 and 1·4 × 10^{-9} cm^3 molecule^{-1} sec^{-1} respectively for these reactions. ICR experiments in our laboratory found[5] 2·8 and 2·0 × 10^{-9} cm^3 molecule^{-1} sec^{-1} respectively, measured in a slightly different energy region. At higher centre of mass energies of the reactants (3 eV and higher) the rate of the second reaction is much slower than that of the first, an observation indicated by ICDR results from our laboratory.[5] We have observed the atom transfer reaction (3) as a function of scattering angle with C^+ ion energies of 5 to 10 eV, but as yet only results of poor signal-to-noise ratios have been recorded, due to small product signals and also instabilities. We shall therefore concentrate on the first reaction.

The apparatus was recently modified in the region of the scattering centre to reduce stray electric fields. A narrower and more symmetric primary beam angular distribution was then observed (Fig. 3). The corresponding angular and energy distributions of the product ion of the reaction (2) are also shown in Fig. 3. This reaction was repeated using an isotopically labelled primary beam, $C^{18}O^+$. All the product signal, within the limits of experimental error, was observed at mass 44, and is thus due to a charge transfer reaction.

In order to operate at lower primary beam energies, down to approximately 0·3 eV, with narrow angular and energy beam distributions, a Herman modification of the Lindholm deceleration lens[6] has been built and is presently being installed in the beam apparatus.

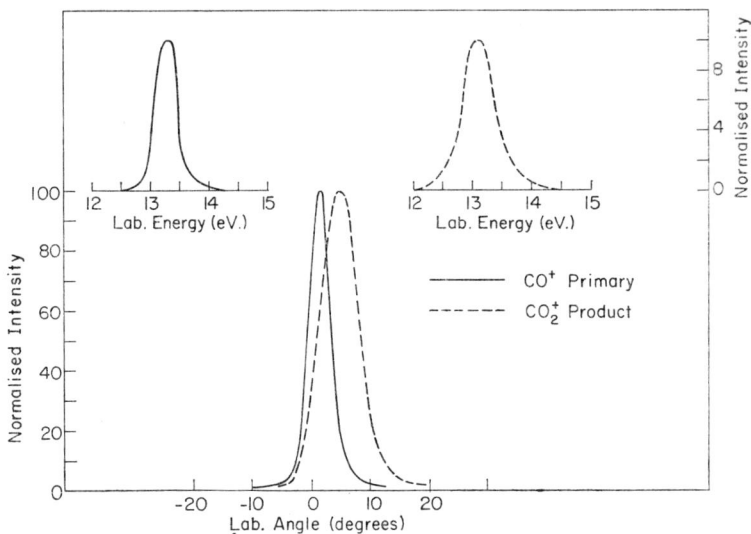

FIG. 3   Angular and energy distributions of primary and product ions in the reaction $CO^+ + CO_2 \rightarrow CO_2^+ + CO$.

REFERENCES

1. Menzinger, M. and Wahlin, L., *Rev. Sci. Instr.*, 1969, **40**, 102.
2. Kerstetter, J. and Wolfgang, R., *J. Chem. Phys.*, 1970, **53**, 3765.
3. Ding, A., Henglein, A., Hyatt, D. and Lacmann, K., *Z. Naturforschung*, 1968, **23a**, 2084.
4. Schildcrout, S. M. and Franklin, J. L., *J. Chem. Phys.*, 1969, **51**, 4055.
5. Jaffe, S. and Klein, F. S., *J. Chem. Phys.*, in press.
6. Herman, Z., Kerstetter, J., Rose, T. and Wolfgang, R., *Rev. Sci. Instr.*, 1969, **40**, 538.

Discussion

J. Durup (Université de Paris-Sud, Orsay, France): Did you also perform experiments with the last set-up using isotopically labelled species?

F. S. Klein: Yes. We obtained the same results as with the first set-up.

J. Durup: It, therefore, appears that the charge-transfer produced CO_2^+ would be your 'high-energy' component.

F. S. Klein: We did not draw this conclusion but are looking for a better answer.

88

Positive Ion Clustering Reactions in a Photoionization Mass Spectrometer

By LARRY I. BONE

(*Department of Chemistry, East Texas State University, Commerce, Texas, U.S.A.*)

FOR quite some time we have been conducting ion clustering experiments in hopes of determining the structures and forces involved in ion cluster formation. Our first study in this series[1] involved the hydration of NO^+ and suggested to us that hydrogen bonding was important in those clusters. Further studies reported herein involve the clustering of NO^+ with CH_3OH and H_2S.

It is quite clear that hydrogen bonding is not a prerequisite for ion cluster formation since many types of clusters have been observed which involve molecules not capable of hydrogen bonding. Neutrals such as SO_2, CO_2, N_2, O_2 and even the rare gases have been observed to cluster with various positive and negative ions. In fact it could probably be safely said that ion clusters evolving about any neutral molecule can be observed if your instrumentation is sensitive enough.

It is interesting to note that no study to date has reported ion clusters containing more neutral molecules than would normally be expected in the first solvation sphere, except for clusters involving water and ammonia.[2] Clampitt and Jefferies[3] have observed $M(H)_6^+$ and $M(Ne)_6^+$ where M is Na, K and Rb but their data do not show any larger clusters even though $M(X_6)^+$ is quite abundant. In contrast, Kebarle[4] has reported $H(H_2O)_8^+$ and Castleman and Tang[5] have reported clusters as large as $Pb(H_2O)_{10}^+$. In these examples, as well as numerous others, not only can more water be added than would normally be expected in a first hydration sphere, but the values of the rate constants or the equilibrium constants for the successive addition of water do not show a discontinuity at any hydration number. Although other explanations can be offered,[4] this suggests that clusters involving molecules which can hydrogen bond may have a different structure or there may be different forces involved in the cluster formation when hydrogen bonding is possible.

EXPERIMENTAL

The work described here was carried out in a photoionization quadrupole mass spectrometer which has been described in detail elsewhere.[1,6] A diagram

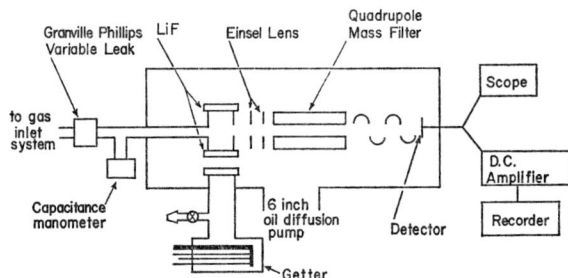

FIG. 1   Photoionization mass spectrometer.

of the instrument is shown in Fig. 1. Nitric oxide and the polar reagent are premixed in the gas inlet and then bled into the reaction chamber through a variable leak. The reaction chamber, which serves as the source of the mass spectrometer, is gas tight except for a 1/25-in. ion exit hole. The exit hole is covered with a fine mesh screen and the entire source is at the same potential allowing the ions to react in a field free region.

NO^+ is produced by photoionization from a Krypton resonance lamp with a CaF_2 window. This lamp passes only one line, 10·03 eV, and thus nitric oxide is the only ion produced from any of the mixtures. Ions diffuse from the source and the ion distribution is determined as a function of the pressure in the source.

RESULTS AND DISCUSSION

By studying the abundance of the various ions as a function of mole fraction and pressure in the source we[1] were able to deduce the following mechanism for the hydration of NO^+ which is consistent with that observed by other authors using different techniques.[7-9]

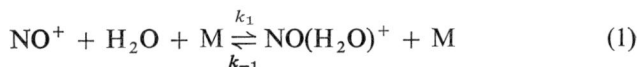

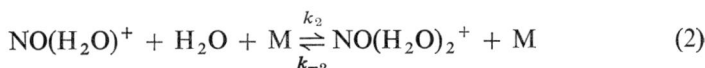

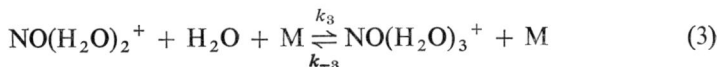

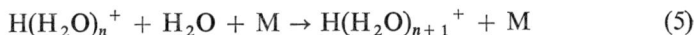

$$NO^+ + H_2O + M \underset{k_{-1}}{\overset{k_1}{\rightleftharpoons}} NO(H_2O)^+ + M \qquad (1)$$

$$NO(H_2O)^+ + H_2O + M \underset{k_{-2}}{\overset{k_2}{\rightleftharpoons}} NO(H_2O)_2^+ + M \qquad (2)$$

$$NO(H_2O)_2^+ + H_2O + M \underset{k_{-3}}{\overset{k_3}{\rightleftharpoons}} NO(H_2O)_3^+ + M \qquad (3)$$

$$NO(H_2O)_3^+ + H_2O \overset{k_4}{\rightarrow} HONO + H(H_2O)_3^+ \qquad (4)$$

$$H(H_2O)_n^+ + H_2O + M \rightarrow H(H_2O)_{n+1}^+ + M \qquad (5)$$

The values of k_1, k_2 and k_3 are, depending on the third body, near $1·3 \times 10^{-28}$, $1·2 \times 10^{-27}$ and 2×10^{-27} cm^6 sec^{-1} respectively. The values of k_{-1}, k_{-2} and k_{-3} are very small, varying from 10^{-12} to 10^{-14} cm^3 sec^{-1}.

Our studies involving the clustering of NO^+ with CH_3OH, while quite similar to those with water, show some differences. Figure 2 shows the abundance of various product ions as a function of pressure in the source for a 0·5 mole fraction mixture of CH_3OH and NO.

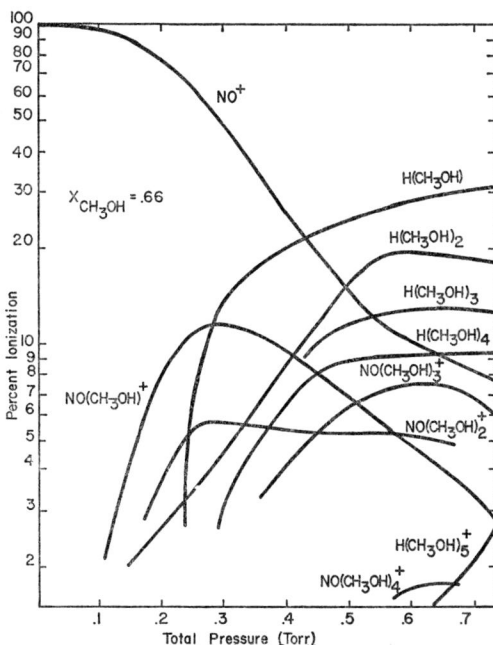

FIG. 2   Reaction of NO^+ with CH_3OH.

The following mechanism has been deduced from such data taken at a series of different mole fractions.

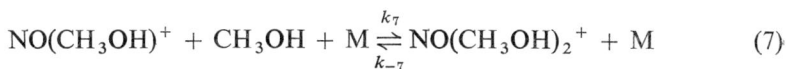

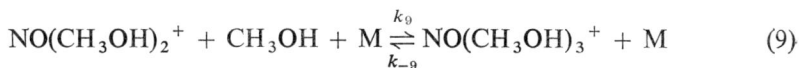

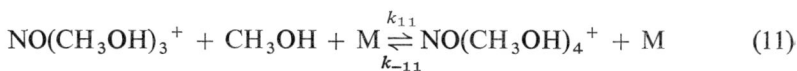

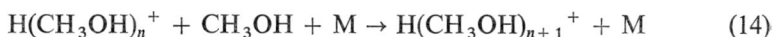

$$NO^+ + CH_3OH + M \overset{k_6}{\underset{k_{-6}}{\rightleftharpoons}} NO(CH_3OH)^+ + M \qquad (6)$$

$$NO(CH_3OH)^+ + CH_3OH + M \overset{k_7}{\underset{k_{-7}}{\rightleftharpoons}} NO(CH_3OH)_2{}^+ + M \qquad (7)$$

$$NO(CH_3OH)^+ + CH_3OH \overset{k_8}{\rightarrow} CH_3ONO + H(CH_3OH)^+ \qquad (8)$$

$$NO(CH_3OH)_2{}^+ + CH_3OH + M \overset{k_9}{\underset{k_{-9}}{\rightleftharpoons}} NO(CH_3OH)_3{}^+ + M \qquad (9)$$

$$NO(CH_3OH)_2{}^+ + CH_3OH \overset{k_{10}}{\rightarrow} CH_3ONO + H(CH_3OH)_2{}^+ \qquad (10)$$

$$NO(CH_3OH)_3{}^+ + CH_3OH + M \overset{k_{11}}{\underset{k_{-11}}{\rightleftharpoons}} NO(CH_3OH)_4{}^+ + M \qquad (11)$$

$$NO(CH_3OH)_3{}^+ + CH_3OH \overset{k_{12}}{\rightarrow} CH_3ONO + H(CH_3OH)_3{}^+ \qquad (12)$$

$$NO(CH_3OH)_4{}^+ + CH_3OH \overset{k_{13}}{\rightarrow} CH_3ONO + H(CH_3OH)_4{}^+ \qquad (13)$$

$$H(CH_3OH)_n{}^+ + CH_3OH + M \rightarrow H(CH_3OH)_{n+1}{}^+ + M \qquad (14)$$

Mathematical analysis of the pressure dependence of the different NO^+ cluster ions for various mole fractions shows that they are produced by a

third-order mechanism and react by competitive second and third-order
processes. If k_{-6}, k_{-7}, k_{-9} and k_{-11} are very small as they are in the water
system, the appearance of proton clusters, $H(CH_3OH)_n{}^+$, at relatively low
pressures suggests the mechanism as proposed. The following rate constants
have been determined:

$$k_6 = 8 \cdot 0 \times 10^{-28} \text{ cm}^6 \text{ sec}^{-1} \qquad k_9 = 5 \times 10^{-27} \text{ cm}^6 \text{ sec}^{-1}$$
$$k_7 = 1 \cdot 9 \times 10^{-27} \text{ cm}^6 \text{ sec}^{-1} \qquad k_{10} = 10^{-11} \text{ cm}^3 \text{ sec}^{-1}$$
$$k_8 = 3 \cdot 6 \times 10^{-12} \text{ cm}^3 \text{ sec}^{-1}$$

Comparison of the clustering rate constants for water and methanol shows
that methanol clusters some six times faster for the first step but that the rates
approach each other as the cluster grows. This is consistent with the fact that
methanol is more polarizable. This enhances the attraction to small clusters
where the field strength is large but falls off as the cluster grows. Water, on

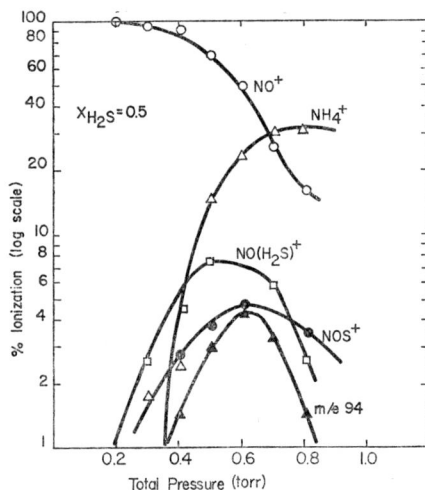

FIG. 3   Reaction of NO^+ with H_2S.

the other hand, has a larger permanent dipole and a lower polarizability,
causing the reaction rates to be less sensitive to cluster size. Because of the
smaller size of the proton, reactions (8), (10), (12) and (13) in the methanol
systems compete better with the clustering reactions than does reaction (4)
in the water system.

In addition to the fact that the clustering reactions are faster in methanol
than water, we were also able to observe NO^+ clusters with up to four
methanols while only three waters have been observed. By analogy with
water and methanol clustering data, similar experiments involving NO^+ and
H_2S should show multiple clustering with comparable rates.

We have conducted a series of experiments involving mixtures of NO and
H_2S. Sample ion abundance data are shown as a function of pressure in
Fig. 3.

A possible mechanism for this system is:

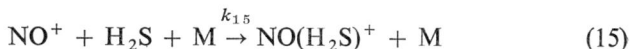

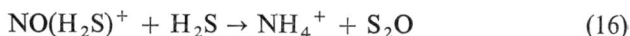

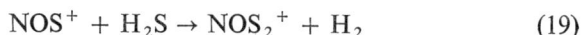

$$NO^+ + H_2S + M \xrightarrow{k_{15}} NO(H_2S)^+ + M \tag{15}$$

$$NO(H_2S)^+ + H_2S \rightarrow NH_4^+ + S_2O \tag{16}$$

$$NO(H_2S)^+ + NO + M \rightleftharpoons (NO)_2(H_2S)^+ + M \tag{17}$$

$$NO^+ + H_2S \rightarrow NOS^+ + H_2 \tag{18}$$

$$NOS^+ + H_2S \rightarrow NOS_2^+ + H_2 \tag{19}$$

k_{15} has been measured and is found to be near 1.0×10^{-28} cm^6 sec^{-1} or about 0·6 times as fast as the comparable reaction in water.

The most interesting observation from these experiments is that we do not observe $NO(H_2S)_2^+$ or any larger clusters. We also do not observe $H(H_2S)_n^+$ for any n. This observation at first suggests that NO^+ will not cluster twice with H_2S and the reason is not the familiar reaction producing the solvated proton (reactions 4, 8, 10, 12 and 13).

The non-observation of $H(H_2S)_2^+$ is consistent with our suggestion in Ref. 1 that NO^+ clusters in water and methanol in hydrogen bonded chain-like structures

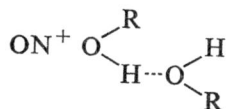

and that H_2S can not cluster in this way because of sulphur's near inability to hydrogen bond. This argument is also consistent with a recent observation by Kebarle[10] that the equilibrium constant for the addition of one dimethyl ether to a proton is comparable to the addition of one water, but addition of two dimethyl ethers requires a pressure comparable to that required for the addition of eight waters.

The proof of our hydrogen bonded structure by the non-observation of $NO(H_2S)_2^+$ is far from conclusive because of the observation of NH_4^+ in our NO–H_2S experiments. The entire reason why $NO(H_2S)_2^+$ is not observed could be explained by reaction (16) or possibly even the reactions which produce either NOS^+ or NOS_2^+, which we do not yet understand.

ACKNOWLEDGMENTS

I would like to acknowledge Mary Jane McAdams, David Lyle Turner and Donald K. Riddle who carried out the majority of this work in conjunction with East Texas State University's Undergraduate Honours Programme. I would like to acknowledge the financial support of the Robert A. Welch Foundation and the Faculty Research Committee at East Texas State University.

REFERENCES

1. McAdams, M. J. and Bone, L. I., *J. Chem. Phys.*, 1972, **57**, 2173.
2. Kebarle, P., *Adv. in Chem. Series*, 1968, **72**, 24.

3. Clampitt, R. and Jefferies, D. K., *Nature*, 1970, **226**, 141.
4. Kebarle, P., Searles, S. K., Zolla, A., Scarborough, J. and Arshad, M., *J. Am. Chem. Soc.*, 1967, **89**, 6393.
5. Castleman, A. W. and Tang, I. N., discussion at The International Nucleation Theory Workshop, Clark College, Atlanta, Georgia, 10–12 April 1972.
6. Hopkins, J. M. and Bone, L. I., *J. Chem. Phys.*, 1973, **58**, 1473.
7. Fehsenfeld, F. C., Mosesman, M. and Ferguson, E. E., *J. Chem. Phys.*, 1971, **55**, 2120.
8. Howard, C. J., Rundle, H. W. and Kaufman, F., *J. Chem. Phys.*, 1971, **55**, 4772.
9. Puckett, L. J. and Teague, M. W., *J. Chem. Phys.*, 1971, **54**, 2564.
10. Yamdaani, R., Davidson, B. and Kebarle, P., discussion at the 165th National Meeting of The American Chemical Society, Dallas, Texas, 8–13 April, 1973.

Discussion

J. H. Futrell (University of Utah, U.S.A.): The switching reactions you have described provide a way of generating solvated protons which bypass the energy relaxation problems associated with equilibrium studies which start with the monosolvated proton. Can you deduce equilibrium constants from your experiments? Do you have any k's for the water and methanol systems?

L. I. Bone: We have been able to deduce equilibrium constants for other systems where the constants are not as large as they are for the solvated protons. At room temperature we do not see any reverse reactions. We plan to raise the source temperature which will reduce the equilibrium constants such that the reverse reactions will occur. We should then be able to measure very good equilibrium constants for both the water and methanol systems.

J. Durup (Université de Paris-Sud, Orsay, France): How is it possible to state that a 'switching' reaction such as reaction 8 occurs in one step? It seems that it is under these relatively high pressures indistinguishable (from the stoichiometric and from the kinetic viewpoints) from a sequence of reactions composed of the reverse of the clustering reactions (reactions 6, 7, 9 and 11) and the reaction of the unclustered ion (reactions 8, 10 and 12).

L. I. Bone: Proton clusters $H(ROH)_n^+$ must be produced from a reaction which initially involves NO^+ since this is the only primary ion produced under 10·03 eV photoionization. At pressures of 0·3 to 0·4 torr we are able to show by a steady state treatment that $NO(ROH)^+$ disappears by combination of a second order and a third order process. At these pressures $H(ROH)^+$ is growing-in rapidly and therefore must be a major portion of the second order rate constant we measure. The reverse of reactions 7, 9 and 11 could also contribute to this rate but their rate constants must be very low at room temperature because the equilibrium for $NO(ROH)_n^+ + ROH + M \leftrightharpoons NO(ROH)_{n+1}^+ + M$ must certainly be large. Such a reaction has also been observed in the NO^+-water system by a number of authors and was found to be second order.

89

Long-Range Interactions and Momentum Transfer Collisions in the Dissociative Charge Transfer Reactions between Acetylene and Ar$^+$

By R. S. LEHRLE and R. S. MASON†

(*Chemistry Department, University of Birmingham, Birmingham, U.K.*)

INTRODUCTION

LONG-RANGE interactions in which there is little or no transfer of kinetic energy are a significant feature of many near-resonant simple charge transfer reactions.[1] This is especially the case for near-resonant reactions involving atoms and atomic ions in which the modulus of the energy defect ($|\Delta E|$) is small; ΔE is the difference between the Recombination Energy (RE) of the reactant ion and the Appearance Potential (AP) of the product ion. These reactions occur at relative translational energies up to many kiloelectronvolts. Since there are vibrational and rotational as well as many more electronic energy levels available in polyatomic molecules it is reasonable to expect that a reactant ion with a sufficiently high Recombination Energy is in near resonance with at least one of these energy levels. One might therefore also expect such reactions to proceed via a long distance interaction, although presumably considerations such as the Franck–Condon principle should be taken into account. If the polyatomic ion produced is sufficiently excited it may undergo unimolecular decomposition. The effect of both the RE of the reactant ion and its translational energy on unimolecular decomposition have been the subject of previous studies in this laboratory, particularly involving propane.[2] In these studies long-distance interactions have been assumed to be predominant. Unfortunately, although apparent evidence of charge transfer into isolated electronic states of the parent product ion was obtained, so far there has been insufficient knowledge of the internal energy states of the polyatomic ion for those states involved in the reaction to be identified. Acetylene is a simpler molecule than propane for which more information is available concerning the internal energy states of its positive ion $C_2H_2{}^+$. By studying acetylene it was hoped to obtain more specific details of the energy transfer process involved in dissociative charge transfer (DCT) reactions.

† Present address: School of Chemistry, University of Leeds, Leeds, U.K.

DCT reactions of acetylene with the reactant ions Xe^+, Kr^+, Ar^+, and Ne^+ have been studied.[3] In this paper the results mainly of the Ar^+/C_2H_2 reaction are discussed.

The investigation utilized Total Charge Collection (TCC) combined with mass analysis of the product ions, and Time Resolved Mass Spectrometry (TRMS) as before,[2] although a different approach to the interpretation of the TRMS results has been used. Briefly the experimental set-up comprised a continuous beam of Ar^+ ions (containing both the $^2P_{3/2}$ and the $^2P_{1/2}$ species) which was accelerated through the ionization region, called 'reaction zone', of the Bendix Time of Flight Mass Spectrometer at a translational energy selected between 100 and 2000 eV. Acetylene gas was introduced into the reaction zone where Dissociative Charge Transfer with the Ar^+ ions occurred. The Total Charge Transfer Reaction Cross-Section Q_T was measured by TCC as a function of translational energy. In a separate experiment the product ions were mass analysed by the time of flight technique. In a third experiment the Ar^+ beam was pulsed and the product ions were allowed to reside in the reaction zone for controlled times before applying the pulse which withdraws them for mass analysis. The abundance of each product ion could thereby be measured as a function of this residence time T_r. The Ar^+ ion pulse width was *ca.* 0·2 μs.

A pressure of acetylene below *ca.* 2×10^{-5} torr ($2·66 \times 10^{-3}$ Nm^2) was maintained so that single collision conditions existed in the reaction zone.

RESULTS AND DISCUSSION

The Collection Curves (Fig. 1) measured during TCC experiments were obtained by applying a small symmetrical collection potential across the two parallel plates which serve as a boundary to the reaction zone. 'Slow' positive product ions are collected on the negative 'collector' plate. Ions with a maximum kinetic energy of up to 3 eV would be expected to be completely collected with a collection potential above 3 V and the current I_s measured on the collector should reach a plateau if all ions are being collected. However, the present curves display quite a steep limiting gradient, especially at the lower reactant ion translational energies. This indicates that product ions with quite a high kinetic energy may be being formed in momentum transfer collisions, and that possibly a higher degree of momentum transfer occurs at the lower end of the translational energy range.

The product ion residence time results are presented (Fig. 2) in terms of 'Delay Curves', in which product ion intensity is plotted as a function of T_r. The scatter in the results is due primarily to the comparatively low signal to noise ratio which occurs when pulsed ion beams are used. Within the scatter the $C_2H_2^+$ curve is horizontal for the first few microseconds, whereas all the other product ions show decay. Decay of a particular ion intensity must be due either to unimolecular decomposition into daughter fragments, or disappearance of the ion from the reaction zone due to excess kinetic energy (no significant decrease in the intensity of thermally energetic ions is expected for about five or six microseconds), or both. An ion intensity decaying due to unimolecular decomposition would be accompanied by the

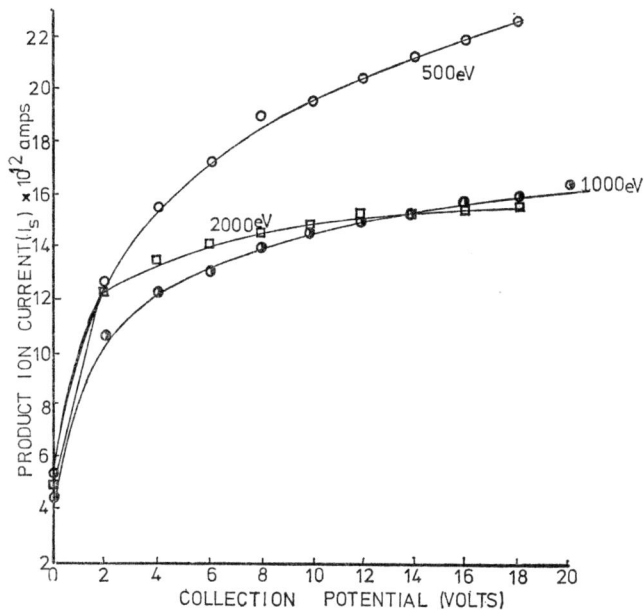

FIG. 1   Collection curves obtained during TCC experiments for Ar^+/C_2H_2 DCT.

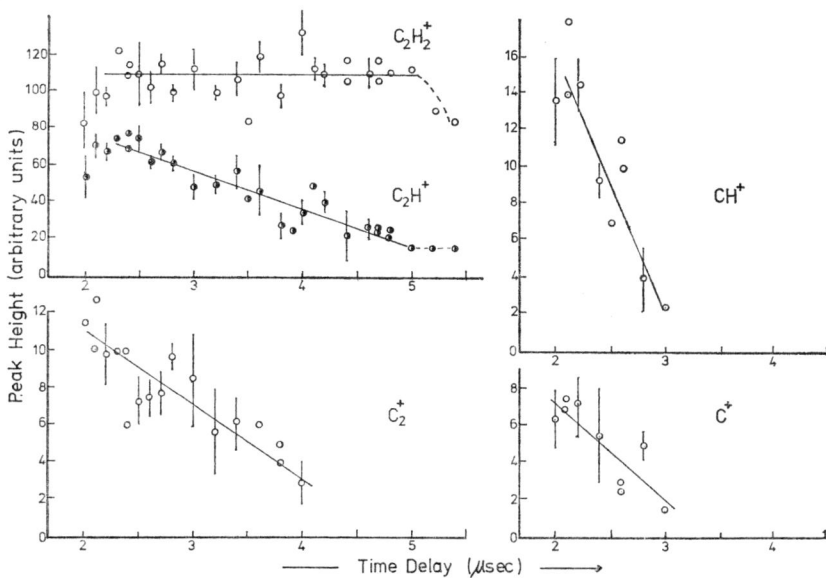

FIG. 2   Delay curves Ar^+/C_2H_2.

corresponding appearance of its daughter ion. Since no such phenomena are observed in this instance it is concluded that whereas the observed $C_2H_2^+$ ions are only thermally energetic, the fragment product ions have excess kinetic energy which is most probably due to their being formed in momentum transfer collisions. It is also possible that fragment ions may have excess energy due to dissociation; however it is unlikely that such ions would have more than 2 or 3 eV energy of dissociation.[3] Evidence from the collection curves in Fig. 1 suggests that ions of considerably higher energy are being formed.

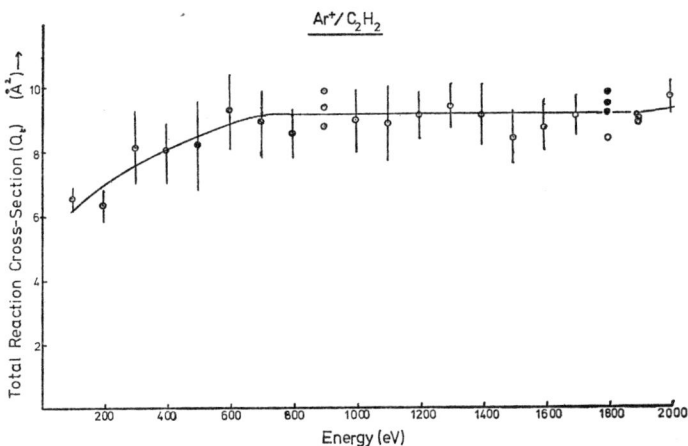

FIGURE 3

The curve of Q_T against reactant ion translational energy (TE) (shown in Fig. 3) although it includes large error bars shows an increase in Q_T to a maximum around 700 eV. However, any other curve structure, if it exists, is masked by the uncertainty in the measured points. Since the values of Q_T here correspond to zero momentum transfer processes, then those values are a measure of the reaction cross-section for the formation of the observed $C_2H_2^+$ ions.

The mass analysis results are represented as plots of Relative Abundance (RA) versus TE to form a Translational Energy Curve for each product ion (Fig. 4). The $C_2H_2^+$ TE curve is distinctly different from those of its fragment ions. It shows two distinct maxima at 700 and ca. 1650 eV respectively. The maximum at 700 eV is in agreement with the apparent rise of Q_T to a maximum in Fig. 3, therefore the second maximum at 1650 eV in Fig. 4 could also correspond to a second maximum in the reaction cross-section for the formation of the observed $C_2H_2^+$ ions, which is masked in Fig. 3 by the high degree of uncertainty. The 'Near Adiabatic Theory' of Massey[4] can be applied to these maxima in a manner similar to its application in simple charge transfer studies.[1] Thus from the position of the maxima, values of $|\Delta E|$ can be calculated using the formula: $|\Delta E| = 0.57 E_T^{\frac{1}{2}}/m^{\frac{1}{2}}a$ (eV). ΔE is the energy defect of the reaction giving rise to the observed maximum at a translational energy E_T (measured in eV); a is called the near adiabatic parameter and is assumed here to have a value of 7 Å, which was the average

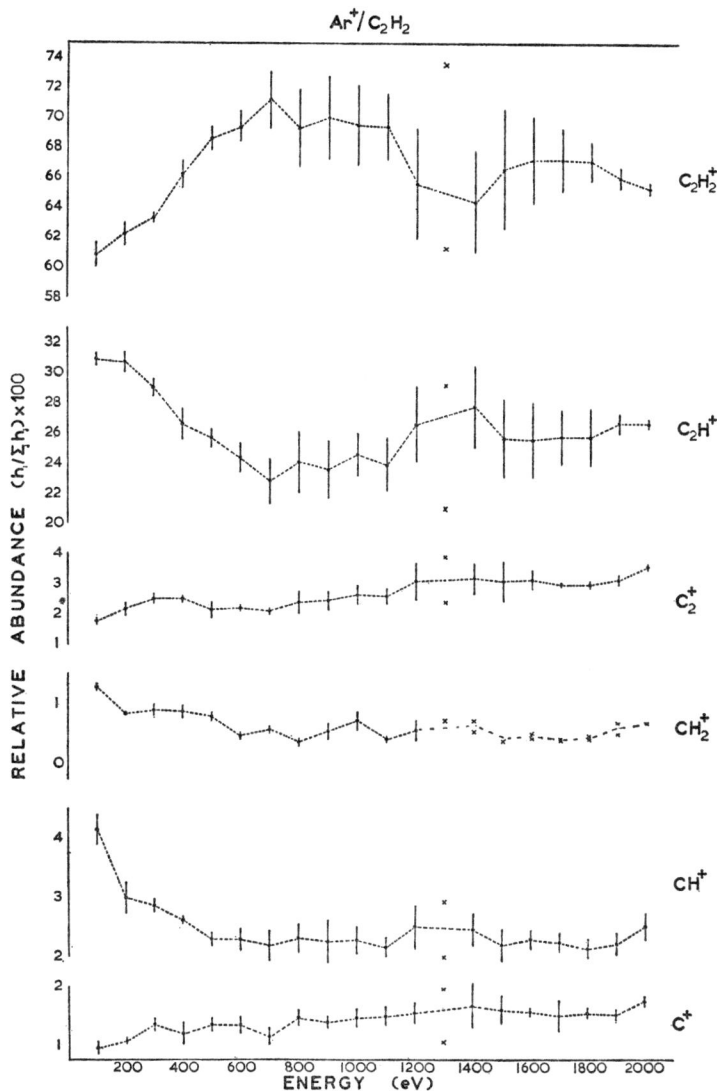

FIG. 4 Mass analysis: translational energy curves.

of many values found by Hasted[1] to be applicable to simple charge transfer reactions. The maxima at 700 and 1650 eV give rise to $|\Delta E|$ values of 0·34 eV and 0·52 eV respectively.

The second IP for C_2H_2 has been reported as 16·27 eV and corresponds to the formation of $C_2H_2^+$ in the $^2\Sigma_g^+$ state (presumably the $v = 0$ level) which is localized mainly along the HC–CH co-ordinate.[5,6] The Ar^+ RE for the $^2P_{3/2}$ and $^2P_{1/2}$ states is 15·76 and 15·94 eV respectively.[7] Thus for the reaction

$$Ar^+(^2P_{3/2} \text{ and } ^2P_{1/2}) + C_2H_2 \rightarrow Ar + C_2H_2^+(^2\Sigma_g^+)$$

ΔE would be -0.32 and -0.51 eV respectively for the two states of Ar^+ involved.

The surprisingly good agreement of the latter predicted energy defects with those 'found' by experiment leads us to believe that the observed $C_2H_2^+$ ions are formed in the $^2\Sigma_g^+$ ($v = 0$) level.

By comparing the TE curves in Fig. 4 and separating them into groups of similar behaviour, three groups emerge:

(a) $C_2H_2^+$
(b) C_2H^+, (CH^+), (CH_2^+?)
(c) C_2^+, C^+

In accordance with the arguments put forward previously,[2] ions of the same group might have the same precursor ion which is different to the precursor of ions in a different group. This precursor could be a particular excited state of the parent ion $C_2H_2^+$ for instance.

Comparison of the delay curves in Fig. 2 shows that the rate of loss from the reaction zone of CH^+ is much higher than that for C_2H^+, C_2^+ and C^+ all of which seem to have a similar rate of loss. This suggests that CH^+ is formed in a collision involving much more momentum transfer than the collisions leading to the formation of C_2H^+, C_2^+ and C^+ and therefore CH^+ cannot belong to group (b), and therefore in this instance does not have the same precursor as C_2H^+.

In conclusion, we have provided direct evidence that long-range interactions in charge transfer reactions involving polyatomic molecules can occur in a similar manner to simple charge transfer. In addition we are able to discern between product ions emerging from momentum transfer and non-momentum transfer collisions. In the Ar^+/C_2H_2 reaction the AP of all possible fragment ions is above the RE of Ar^+, and it is not surprising that, in order to provide the necessary transfer of energy for the formation of these ions, the ion and molecule must be involved in a 'close' interaction. However, in the Ne^+/C_2H_2 reaction,[3] in which the formation of C_2H^+, $C_2^+(+ H_2)$, and CH_2^+ are exothermic with respect to the RE of Ne^+ and CH^+ formation has a ΔE of only -0.1 eV, all these ions were found to have been formed in momentum transfer collisions except possibly for CH_2^+. The implication is that, although long-distance interactions do occur, momentum transfer collisions may be more important in Dissociative Charge Reactions than has previously been thought to be the case.

REFERENCES

1. Hasted, J. B., 'Physics of Atomic Collisions', Butterworths, 2nd edn, 1971.
2. Lehrle, R. S., Parker, J. E. and Robb, J. C., 'Advances in Mass Spectrometry', Vol. 5, 1970.
3. The results of the Xe^+/C_2H_2, Kr^+/C_2H_2, and Ne^+/C_2H_2 reactions are to be published.
4. Massey, H. S. W. and Burhop, E. H. S., 'Electronic and Ionic Impact Phenomena', Oxford University Press, Oxford, 1952.
5. Al-Joboury, M. I., May, D. P. and Turner, D. N., *J. Chem. Soc.*, 1965, p. 616; Baker, C. and Turner, D. N., *Chem. Comm.*, 1967, p. 797.

6. Fiquet-Fayard, F., *J. Chim. Phys.*, 1967, **64**, 320.
7. Ionization Potentials, Appearance Potentials, and Heats of Formation of Gaseous Positive Ions, N.S.R.D.S., N.B.S., 1968, 26.

Discussion

J. B. Hasted (Birkbeck College, London, U.K.): I believe it would be preferable to analyse these data in terms of avoided crossings rather than in terms of the adiabatic maximum rule.

I would also suggest that it is possible to derive quantitative momentum distribution functions from your condenser experiment $I^+(E)$ functions.

R. S. Mason: The adiabatic maximum rule provides a direct confirmation of our conclusion that the observed $C_2H_2^+$ ions are formed by long-range interaction, without momentum transfer. It also confirms that this process can be interpreted in terms of known energy states of the ion.

Your second suggestion is an interesting one, and could lead to an approximate indication of the magnitude of the kinetic energy of the various momentum-transfer products.

90
Fast Ion and Atom Impact Mass Spectrometry of Simple Molecules†

By W. G. GRAHAM, C. J. LATIMER, R. BROWNING and
H. B. GILBODY

(*Department of Pure and Applied Physics,
The Queen's University of Belfast, Belfast, U.K.*)

INTRODUCTION

In recent experiments in this laboratory[1,2,3] we have studied the ionization and dissociation of simple molecular gases by 5–45 keV hydrogen and helium beams using a mass spectrometric technique which ensured a high and uniform collection efficiency for all the secondary product ions despite the wide spread in their energies of formation and initial directions. These results shed new light on the relative importance of the various processes leading to ionization and dissociation of the target molecules. Dissociative processes proceeding by charge transfer and ionization make a large and often dominant contribution to the molecular ionization, but it is not yet clear to what extent such processes may be expected to hold for other projectiles.

The present paper describes studies of the ionization and dissociation of CO, N_2, O_2 and H_2 by 5–45 keV Ne^+, Ne and Na^+ projectiles. Cross-sections for the formation of particular secondary ions have been determined and general mechanisms for the dissociative processes are presented. The results are compared with those previously obtained using beams of hydrogen and helium over the same energy range.

EXPERIMENTAL APPROACH

The specially designed mass spectrometer and measuring procedure has previously been described in detail.[1,2,3] Cross-sections σ_f corresponding to the formation of a particular fragment ion f are related to the mass spectrometer output signal S_f by

$$S_f = k\sigma_f I_0 p$$

where I_0 is the primary beam flux, p is the target gas pressure and k is a constant which we have shown is substantially independent of the species of

† This research was supported by the Science Research Council.

secondary ion and projectile. Absolute cross-sections σ_f were obtained by comparing the mass spectrometer output signals S_f with those observed using a H^+ beam, the cross-sections for which are known from our previous work.[1] The total cross-section σ_+ for the production of secondary ions may be deduced from the values of σ_f using the relation

$$\sigma_+ = \Sigma_f n_f \sigma_f$$

Projectile charge changing cross-sections σ_{10} (charge transfer) in the case of ion impact or σ_{01} (stripping) in the case of atom impact were measured using the growth curve and equilibrium fraction methods.[4]

RESULTS AND DISCUSSION

General Features
Cross-sections σ_f for the formation of the various secondary ions produced by 5–45 keV Ne^+, Ne and Na^+ projectiles in CO, N_2, O_2 and H_2 have been measured. Although random errors associated with the relative values of σ_f are within $\pm 7\%$, our normalization procedure may introduce uncertainties in absolute magnitudes of up to 15%.

The results of several other measurements[5,6,7,8] are available for comparison with the present data and in general show good agreement in the case of σ_+. However the low efficiency of secondary ion collection obtained by other workers precludes a quantitative comparison of the dissociative processes with the present data.

The data obtained with CO, N_2 and O_2 reveal a number of similar features which are not strongly dependent on the projectile. For ion impact in these cases the secondary ions are formed mainly through dissociative processes and comparison of σ_+ with σ_{10} and the total ionization cross-section $\sigma_e = \sigma_+ - \sigma_{10}$ shows that, despite having a larger energy defect, dissociative ionization is at least as probable as dissociative charge transfer. In the case of Ne atom impact $\sigma_+ \simeq \sigma_{01}$ showing that dissociative ionization processes are dominant.

Doubly Charged Products
Figure 1 shows the results for a CO target. Cross-sections $\sigma(CO^{++})$ for ion impact are considerably larger than those for Ne atom impact and the curves show evidence of structure similar to that observed in our earlier work[1,2] with H^+ and He^+ projectiles (Fig. 2). This type of energy variation suggests that at the lower impact energies the transfer ionization process

$$X^+ + CO \rightarrow X + CO^{++} + e$$

is dominant, while the pure double ionization process

$$X^+ + CO \rightarrow X^+ + CO^{++} + 2e$$

makes an increasing contribution at higher impact energies. Indeed this explanation has already been confirmed[9] in the case of N_2^{++} formation by 5–50 keV protons in N_2.

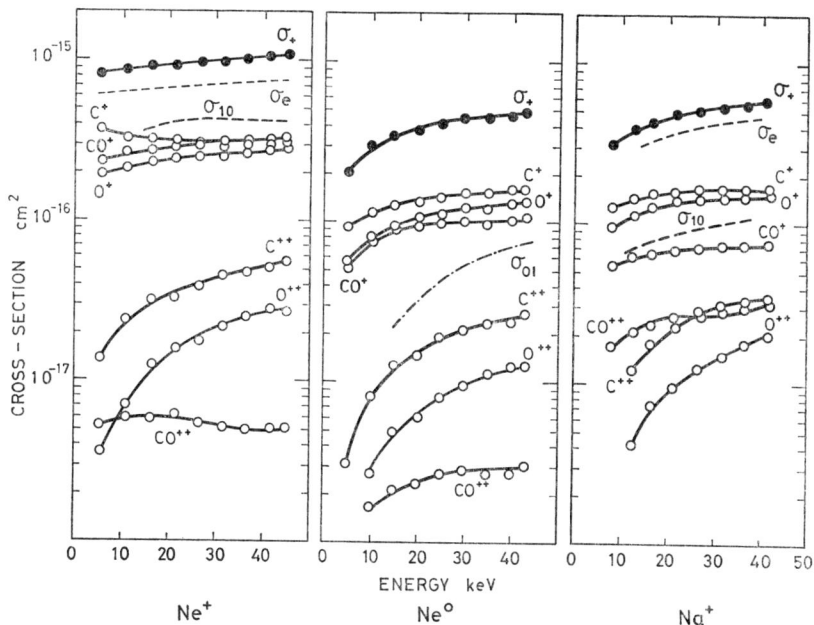

FIG. 1   Cross-sections for secondary ion formation in collisions of Ne $^+$, Ne and Na $^+$ with CO.

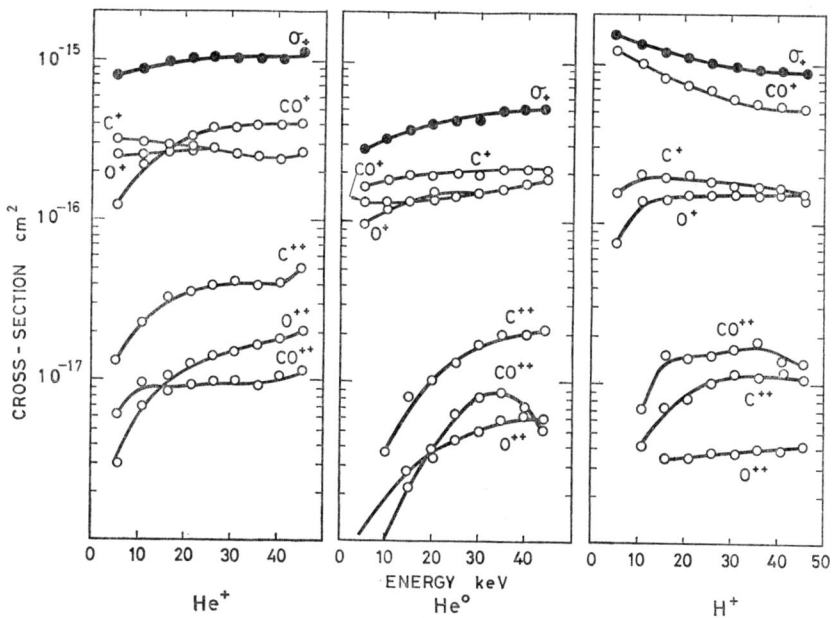

FIG. 2   Cross-sections for secondary ion formation in collisions of He $^+$, He and H $^+$ with CO.

Cross-sections for the formation of C^{++} and O^{++} from CO attain values which in our energy range are of the same order or greater than those for CO^{++}. Cross-sections for the formation of N^{++} from N_2 and O^{++} from O_2 are similar in magnitude to $\sigma(C^{++}) + \sigma(O^{++})$ from CO. The mechanism we propose for the formation of these doubly charged fragment ions is that they arise mainly from a two-step process in which firstly a vacancy in the lowest molecular orbital of the valance shell of the doubly charged molecular ion is created, followed by predissociation of this precursor ion. The binding energy of this bonding electron may then be made available for the production of C^{++} and O^{++} ions. Thus for pure ionization the first step is of the type

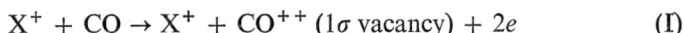

$$X^+ + CO \rightarrow X^+ + CO^{++} \text{ (1σ vacancy)} + 2e \qquad \text{(I)}$$

processes which have energy defects estimated from the data of Seigbahn et al.[10] and El-Sherbini and van der Wiel[11] to be about 70 eV. The CO^{++} ion so formed may then undergo predissociation according to

$$CO^{++} \rightarrow C^{++} + O \ (+\ 23 \text{ eV approx.}) \qquad \text{(IIa)}$$

or

$$CO^{++} \rightarrow C + O^{++} \ (+\ 10 \text{ eV approx.}) \qquad \text{(IIb)}$$

The energy defects associated with the first step would of course be reduced if they proceeded by transfer ionization and cannot be precisely defined because electronic and vibrational states may be excited simultaneously. If the initial ionization process created a vacancy in a less tightly bound orbital (e.g. 2σ) some of the predissociative channels would be energetically inaccessible. However, some C^{++} ions could be formed in this way and the molecule could fragment to form C^+ and O^+ ions simultaneously.

In support of this mechanism we observe that the ratio $\sigma(C^{++})/\sigma(O^{++})$ from CO remains constant virtually independent of projectiles (see Figs. 1 and 2). This would be expected if this ratio reflects the branching ratios for predissociative processes (IIa) and (IIb). Further, a comparison of (IIa) and (IIb) shows that extra energy must always be available in C^{++} production and so a much wider variety of final states is accessible in this case, leading to more C^{++} than O^{++} production in accordance with our observations.

Since the electronic structure and energies of the 1σ and 2σ orbitals in N_2 and O_2 correspond[10] to those in CO, it might be expected that the ions N^{++} from N_2 and O^{++} from O_2 would arise from the analogous processes leading to

$$\sigma(N^{++}) \simeq \sigma(O^{++}) \simeq \sigma(C^{++}) + \sigma(O^{++})$$

as we have already noted.

The proposed mechanism would suggest that C^+ and O^+ ions should also be formed in the predissociative process

$$CO^{++} \rightarrow C^+ + O^+ \ (+\ 31 \text{ eV approx.}) \qquad \text{(IIc)}$$

Photoionization data[12,13] indicate that this may be the dominant decay channel for CO^{++}. However, C^+ and O^+ ions may be formed simultaneously following a single primary ionization of CO (see (IVa) below) and without more detailed analysis the two decay schemes would be indistinguishable.

Singly Charged Products

The cross-sections for the production of singly charged molecular ions exhibit some features which are not readily explained. These ions may arise from ionizing collisions in addition to endothermic charge transfer in the case of Ne^+ impact. In the case of targets of CO, N_2 and O_2 the molecular ion formation cross-sections do not exhibit well-defined peaks but pass through broad maxima. These formation cross-sections therefore do not appear to be greatly influenced by the nature of the projectiles or the energy defects of the processes involved. This behaviour suggests that the formation of singly charged molecular ions may be influenced by competing inelastic channels some of which, we shall show, may also lead to fragment ion formation.

The secondary ions produced most abundantly in CO, N_2 and O_2 are the singly charged fragment ions. The ionization cross-sections σ_e are all of similar magnitude and large values of σ_e at the higher energies generally accompany proportionately large yields of fragment ions. We suggest that a similar mechanism to that for the production of the doubly charged fragments is responsible for most of the fragmentation into singly charged ions. If, for example, C^+ and O^+ ions are formed firstly by the creation of a 1σ vacancy in CO

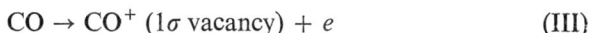

$$CO \rightarrow CO^+ \ (1\sigma \text{ vacancy}) + e \qquad \text{(III)}$$

involving an energy defect[10] of about 38 eV, the CO^+ ion so formed may predissociate

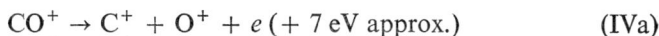

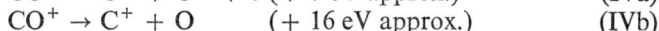

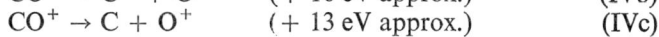

$$
\begin{array}{lll}
CO^+ \rightarrow C^+ + O^+ + e & (+ \ 7 \text{ eV approx.}) & \text{(IVa)} \\
CO^+ \rightarrow C^+ + O & (+ \ 16 \text{ eV approx.}) & \text{(IVb)} \\
CO^+ \rightarrow C + O^+ & (+ \ 13 \text{ eV approx.}) & \text{(IVc)}
\end{array}
$$

The creation of a 2σ or 1π vacancy in CO ($\Delta E \simeq -20$ eV) would apparently preclude dissociation and the molecular ion would result. Nevertheless, a few eV of extra energy transferred during the collision into electronic or vibrational energy would allow predissociation to proceed. Hence there are a large number of processes which may contribute to the overall fragme ztation. A number of arguments support the proposed process.

(*a*) The fragmentation of H_2 is generally less pronounced than in CO, N_2 and O_2 or in any other case we have studied. The mechanism given above is of course precluded in H_2 since there are no subshell vacancies to be created. Further, in the proposed mechanism fragmentation is accompanied by electron production cross-section hence, as is observed, the electron production cross-section is markedly smaller in collisions with H_2.

(*b*) The cross-sections $\sigma(O^+)$ from O_2 are larger than $\sigma(N^+)$ from N_2. This is to be expected since a vacancy created in the 2σ orbitals in O_2 can cause predissociation directly[10] to $O^+ + O$ but a further excitation of the molecule must accompany the creation of the corresponding vacancies in N_2 for the molecular ion to predissociate.

(*c*) Similar arguments to those given for the doubly charged ions from CO apply to the relative numbers of C^+ and O^+ ions observed. The ratio $\sigma(C^+)/\sigma(O^+)$ would not be expected to be strongly influenced by the projectile species. One might also expect the ratio to be nearer unity because of the influence of channels (IVa) and (IIc). This is observed.

CONCLUSION

The ionization and dissociation processes involved in collisions of the type described are clearly most complex. Nevertheless, our results appear to be consistent with a simple model which might be applicable to fragmentation in other collision combinations. However, a more complete understanding of the cross-sections for a particular target/projectile combination clearly requires a more detailed knowledge of the interaction, particularly at the lower impact energies where, as might be expected, our model tends to be less satisfactory. A better test of the model would involve an accurate measurement of the energy loss spectrum obtained with the primary and analysed secondary ion measured in coincidence.

REFERENCES

1. Browning, R. and Gilbody, H. B., *J. Phys. B. (Proc. Phys. Soc.)*, (2), 1968, **1**, 1149–56.
2. Browning, R., Latimer, C. J. and Gilbody, H. B., *J. Phys. B: Atom. Molec. Phys.*, 1969, **2**, 534–40.
3. Browning, R., Latimer, C. J. and Gilbody, H. B., *J. Phys. B: Atom. Molec. Phys.* 1970, **3**, 667–75.
4. Graham, W. G., Latimer, C. J., Browning, R. and Gilbody, H. B., *J. Phys. B: Atom. Molec. Phys.*, to be published.
5. Gusev, V. A., Polyakova, G. N., Erko, V. F., Zats, A. V., Oksyuk, A. A. and Fogel, Ya. M., *Sov. Phys. JETP*, 1971, **33**, 863–6.
6. Gusev, V. A., Polyakova, G. N., Fogel, Ya. M., *Sov. Phys. JETP*, 1969, **28**, 1126–30.
7. Ogurtsov, G. N. *et al.*, *Sov. Phys.-Tech. Phys.*, 1966, **11**, 84–8 and 362–6.
8. Polyakova, G. N., Gusev, V. A., Erko, V. F., Fogel, Ya. M. and Zats, A. V., *Sov. Phys. JETP*, 1970, **31**, 637–42.
9. Afrosimov, V. V., Leiko, G. A., Mamaev, Yu. A., Panov, M. N. and Fedorenko, N. V., 'Proc. Intern. Conf. Phys. Elec. and At. Collisions', Massachusetts, U.S.A., 1969, pp. 114–17.
10. Seigbahn, K., Nordling, C., Johannson, G., Hedman, J., Heden, P. F., Hamrin, K., Gelius, U., Bergmark, T., Werme, L. O., Manne, R. and Baer, Y., 'ESCA Applied to Free Molecules', 1969, North Holland, Amsterdam.
11. El-Sherbini, T. M. and van der Wiel, M. J., *Physica*, 1972, **59**, 433–52.
12. Carlson, T. A. and Krause, M. O., *J. Chem. Phys.*, 1972, **56**, 3206–9.
13. Van Brunt, R. H., Powell, F. W., Hirsch, R. G. and Whitehead, W. D., *J. Chem. Phys.*, 1972, **57**, 3120–9.

Discussion

W. Aberth (Stanford Research Institute, California, U.S.A.): Does the cross-section represented by the curve of Ne$^+$ + CO producing C$^+$, which is increasing with decreasing collision energy, represent a chance energy resonance?

R. Browning: Our lowest energy point is at 5 keV and a peak in the cross-section may occur at a lower energy—there is no exact resonance in this case.

91
Mass Spectrometer Studies of Low Temperature Gas Films

By R. H. PRINCE and G. R. FLOYD

(Department of Physics, York University, Toronto, Canada)

1. INTRODUCTION

THE condensation of gaseous species clearly provides a method of concentrating molecules to densities not attainable in the gas phase. Ionic species produced by subsequent electron bombardment can yield information concerning ion-molecular binding energies and reactive surface collisions having applications in atmospheric studies and astrophysics. Some recent results obtained by this technique are reviewed and implications for mass spectroscopic analysis discussed.

2. APPARATUS

The apparatus used in these studies is shown in Fig. 1, and is based on a Vacuum Generators ultra-high vacuum system equipped with an 80 litre/sec ion pump and a titanium sublimation unit. Sorption pumps are used for primary evacuation to avoid hydrocarbon contamination, a point which has great importance in the reactive studies using condensed acetylene reported in Section 3(c). A copper target, cryogenically cooled, provides a substrate for gaseous deposition from a reduced pressure storage reservoir, equipped with a high vacuum bypass valve for rapid evacuation. In addition, a provision for target heating is provided. The electron source is a Pierce type electron gun designed for parallel flow at the 2-mm anode orifice, so that no attempt is made to focus the beam at the target, where the beam diameter is typically 5 mm. This is desirable from the point of view of power dissipation in the condensed film, and also to maximize the beam current. The latter has values up to several hundred microamperes at 100 eV, using a 4-watt oxide cathode.

Ions created by the beam-target interaction are focussed into a quadrupole mass filter by means of an 'Einzel lens' immediately before it. Ion detection is by means of a channel electron multiplier and single channel analyser.

A second electron-impact ionizer is mounted into the lens plates so that routine partial pressure analysis may be performed.

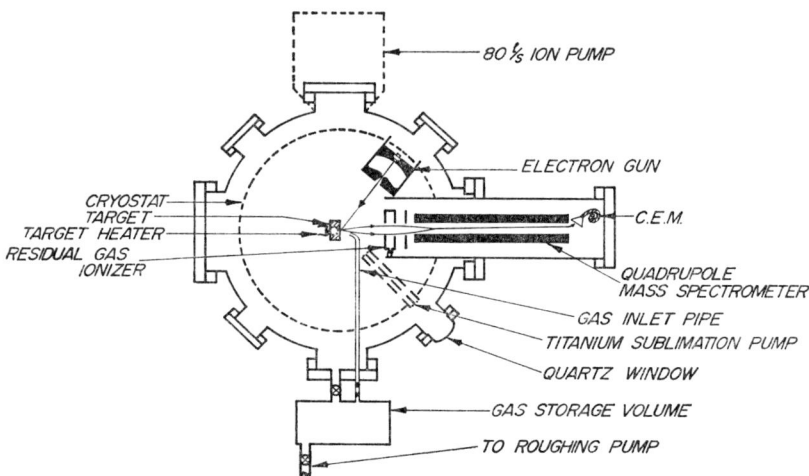

FIG. 1   Schematic diagram of apparatus.

3(a)   Production of Ionized Clusters

As an example of this process, we refer to recent work[1] on the system $H^+(H_2O)_n$, and analysis[2] of the system $H^+(H_2)_n$ reported earlier by Clampitt and Gowland.[3]

Electrons at 90 eV are directed at the target on which water is being continuously deposited at a low rate after establishing a film of adequate thickness (100–1000 nm) to eliminate substrate interactions. Untreated results for experiments performed at 193°K are shown in Fig. 2, wherein mass peaks are observed at m/e values of 55, 73, 91, 109, 127, 145 corresponding to $3 \leq n \leq 8$. Experiments conducted at a lower temperature of 153°K yielded similar results although the intensities were reduced, an effect attributed to surface charging. A model may be constructed whereby the kinetic energy distribution of the proton fragment[4] from the reaction

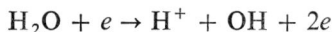

$$H_2O + e \rightarrow H^+ + OH + 2e$$

is utilized to overcome the attractive surface potential. The distribution in observed mass peaks may then be used to determine the cluster–crystal interaction potential. As anticipated, a minimum occurs, since two types of binding exist. The first is an ion-induced dipole attraction which decreases with increasing cluster size and the second is a neutral–neutral binding, assigned to hydrogen bonding for water ice, and a much weaker Lennard–Jones binding for solid hydrogen, which increases *linearly* with increasing n. The minimum occurs at a value of n very close to the number of nearest neighbours for the lattice point, so that the cluster size distribution is determined essentially by the nature of the lattice and not by gas kinetic considerations. This procedure accounts for the most probable species being $H^+(H_2O)_5$ for the water lattice, and $H^+(H_2)_8$ for experiments with solid hydrogen.[3]

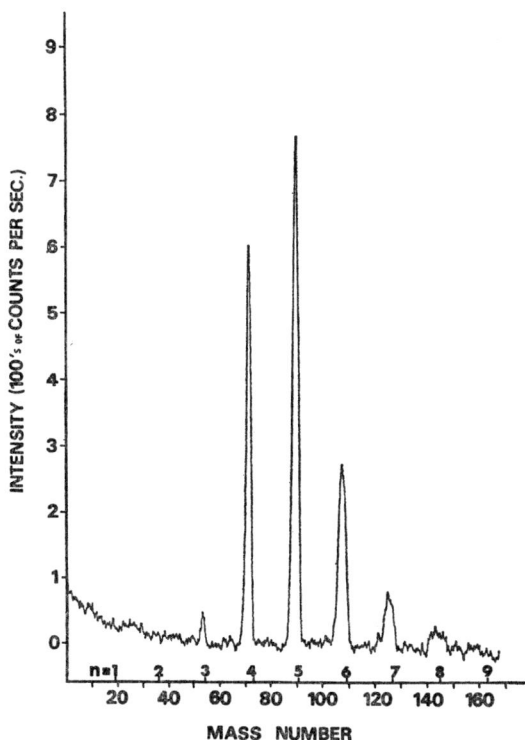

FIG. 2   Observed mass distribution of species $H^+(H_2O)_n$ produced by bombardment of water ice at 193°K by 80 eV electrons.

3(b)   Interaction with Adsorbed Atoms

The ejection of ions by electron impact of physisorbed and chemisorbed species is well known in ion source operation at low background pressure, whereby species liberated from source electrodes contribute significantly to the ion spectrum. In magnetic instruments this leads, of course, to 'ghost' peaks since such ions are frequently quite energetic. It appears likely that bound species are excited to an antibonding excited state and gain kinetic energies of the order of 1 to 20 eV, sufficient to escape. Redhead[5] has suggested that Auger neutralization is an effective process such that perhaps less than 0·01% of ions escape without undergoing neutralization. His studies estimated ion kinetic energies without mass analysis for the oxygen–tungsten system. We have on numerous occasions observed this phenomenon for nitrogen. In this case, experiments are performed with electrons of about 100 eV and a target at room temperature or above. The electron excitation presumably leads to the ejection of N_2^+ and N^+ by dissociative ionization, and in the presence of multilayers a family of ions of the type $N_2^+(N)_n$ is observed (Fig. 3). The production of this sequence is in some respects analogous to a sequence $H_3^+(H_2)_n$ reported by Clampitt for H_2 layers,[6] with the exception that nitrogen dissociates on chemisorption.

The distribution of cluster size is seen to decrease in an exponential

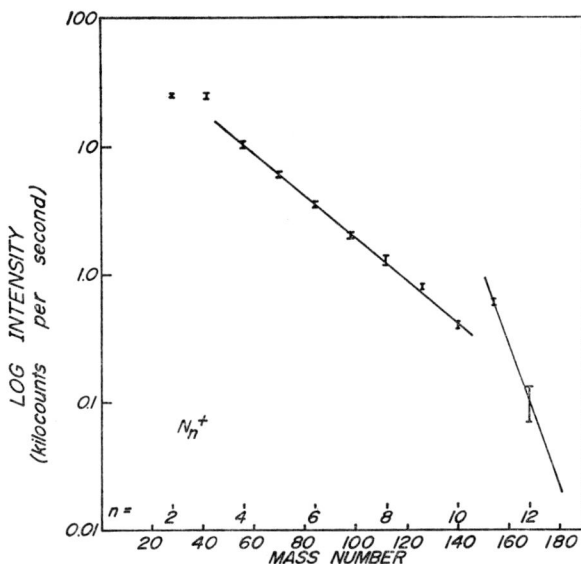

FIG. 3   Nitrogen ion sequence produced by low energy electron bombardment of gold
substrate.

fashion, with a discontinuity occurring beyond m/e = 140 beyond which the
spectrum suggests a second sequence. If we assume N_2^+ as nucleation centre,
then the first sequence terminates at $N_2^+(N)_8$. This is again in accordance
with observations of Clampitt[7] on the system $Li^+(Ne)_n$ and with the theory
of Magee and Funabashi[8] on the clustering of non-polar molecules on ions,
based on 'shell' structures.

3(c)   Reactive Processes

In the last few years, a large number of complex organic molecules have been
observed in the radio spectrum of the interstellar medium.[9] Of the molecules
reported, several would appear to be related to acetylene-cyanoacetylene,
methylacetylene, CH^+ and possibly C_2H as 'xogen'. Acetylene, itself however,
has no permanent dipole moment and is thus not detectable in the microwave
region. The presence of such acetylene derivatives strongly supports arguments for the occurrence of the parent molecule in at least some regions of
the interstellar medium; furthermore, thermodynamic calculations show that
acetylene is a likely component of some (carbon-rich) stellar atmospheres.

We now consider molecules formed by radiation-induced reactions in solid
acetylene as a simulation of an interstellar grain process in which reactions
occur in a condensed gas mantle.

The apparatus is identical with that used in previous experiments reported
above. A preliminary mass analysis of species on the target at 55°K indicated
no detectable contamination other than species commonly found in a residual
gas analysis. To discriminate against contamination, deuterated acetylene
was prepared in our laboratory by the reaction of heavy water with calcium
carbide. Multiple distillation was used in purification and analysis by an

Hitachi Perkin-Elmer RMU-6 mass spectrometer indicated no detectable impurities.

The acetylene was then deposited to produce a film of several thousand ångströms thickness, and bombarded with electrons of 150 eV energy, and product ions analysed. This spectrum is highly complex since it contains radicals stabilized by the low temperature used. A second spectrum of reaction products was also taken by isolating the system and warming the

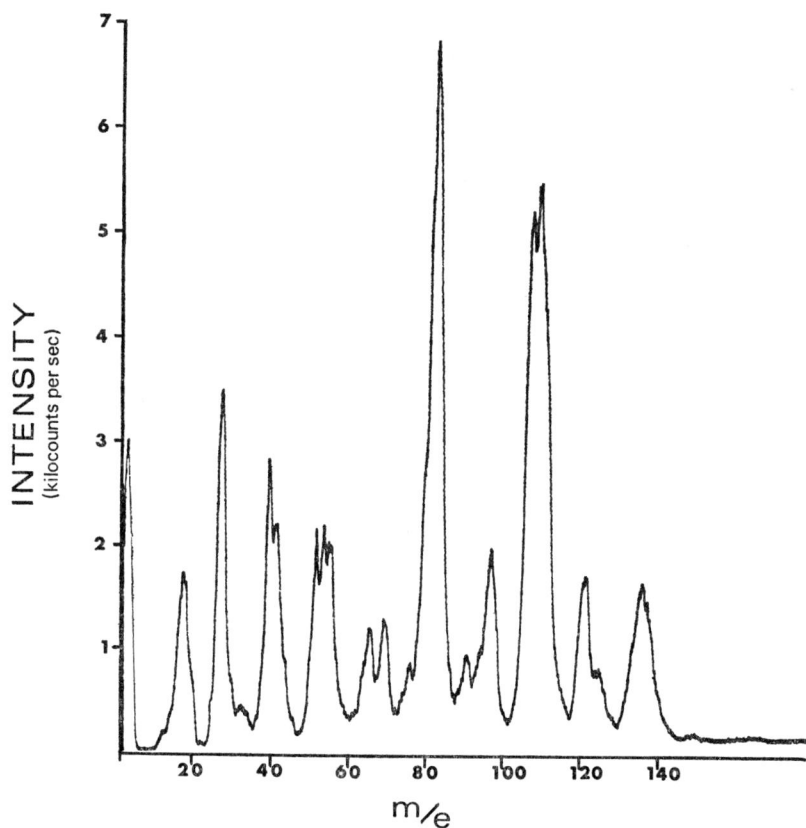

FIG. 4   Spectrum of reaction products from condensed C_2D_2 at 55°K.

target to room temperature. These (most stable) products in a typical mass scan are shown in Fig. 4. Radiation induced polymerization of gaseous acetylene has been studied by numerous workers.[10,11] Diacetylene, benzene and polymers of benzene have been observed. Aromatic or ring-type compounds are thus expected in this experiment, and in fact three major categories of ions are observed, as shown in Table I, and each will be discussed in turn. Our identification of mass peaks is not unique, but it is felt that these assignments are justified on three grounds. Firstly, acetylene is known to polymerize to benzene, and the presence of an aromatic family is plausible. Secondly, we feel that the molecules we have chosen are the most stable compounds with the required masses. Thirdly, using such assignments, all of the positive aromatic ions and fragments are present up to and including those with ten π electrons.

(i) *Aromatic Ions*—Using the Huckel rule, the first aromatic should have 2π electrons. The cyclopropenium cation is a three-carbon ring with 2π electrons and a mass of 42 amu as observed in Fig. 4.

The second aromatic should have six π electrons, such as benzene with a

TABLE I

IONS PRODUCED BY LOW ENERGY ELECTRON BOMBARDMENT OF ACETYLENE

AT 55°K

M/E	MASS NUMBER	RELATIVE INTENSITY	IDENTIFICATION	
AROMATIC SPECIES				
42	42	M	CYCLOPROPENIUM CATION (2π ELECTRONS)	$C_3D_3^+$
84	84	L	BENZENE CATION {6π ELECTRONS FOR PARENT}	$C_6D_6^+$
98	98	M	TROPYLIUM CATION (6π ELECTRONS)	$C_7D_7^+$
136	136	M	NAPTHALENE CATION {10π ELECTRONS FOR PARENT}	$C_{10}D_8^+$
RESONANCE STABILIZED IONS				
52	52	M	DIACETYLENE CATION D-C=C-C≡C-D	$C_4D_2^+$
54	54	M	D-C=C-C≡C-D / D	$C_4D_3^+$
56	56	M	D-C=C-C=C-D / D D	$C_4D_4^+$
108	108	L	PHENYLACETYLENE C≡C-D	$C_8D_6^+$
122	122	M	INDENIUM CATION	$C_9D_7^+$
OTHER IONS				
4	4	L	DEUTERIUM CATION D_2^+	
18	18	M	METHYL CATION CD_3^+	
28	28	L	ACETYLENE CATION $C_2D_2^+$	
40	40	M	ARGON CATION AR^+	

six-carbon ring at mass 84. This ion is dominant in the spectrum and shifts appropriately when using C_2H_2. A five-carbon ring, negatively charged, would also satisfy the 6π electron requirement, but we are concerned here with positive ions only. The tropylium ion at mass 98 also has six π electrons and is thought to be a rearrangement ion for substituted benzene compounds. The largest aromatic molecule present has ten π electrons, identified as napthalene at mass 136.

(ii) *Resonance Stabilized Ions*—A number of the peaks in the spectrum have been identified as resonance stabilized ions, particularly a family of three peaks at masses 52, 54, 56 shown as $C_4D_2^+$, $C_4D_3^+$, $C_4D_4^+$ in Table I. All of these species have negative enthalpies of formation,[11] and the reactions involved are

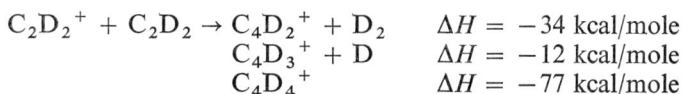

$$C_2D_2^+ + C_2D_2 \rightarrow C_4D_2^+ + D_2 \quad \Delta H = -34 \text{ kcal/mole}$$
$$C_4D_3^+ + D \quad \Delta H = -12 \text{ kcal/mole}$$
$$C_4D_4^+ \quad \Delta H = -77 \text{ kcal/mole}$$

The first two reactions have been well-documented in gas phase studies,[11] whereas the third is less extensively reported.

The remaining resonance stabilized ions occur at masses 108 and 122, identified as phenylacetylene and indene, respectively. The former is a likely reaction in the present experiment, since benzene is formed in a large excess of acetylene. Indene is possibly a product of a reaction between benzene and the propenium cation.

(iii) *Other Ions*—The peaks at masses 4, 12 and 18 are D_2^+, C^+ and CD_3^+ respectively, fragments of the parent ion $C_2D_2^+$ at mass 28. Argon is used as a calibration peak.

The ramifications of such grain processes in liberating complex hydrocarbons into the interstellar medium are considerable, and under consideration. Recent upward revisions in the D/H ratio would now appear to make feasible radio searches for C_2HD and C_6H_5D.

ACKNOWLEDGMENTS

It is a pleasure to acknowledge the helpful discussions with W. W. Duley and C. Leznoff, and correspondence with R. Clampitt of the Culham Laboratory. We further acknowledge the financial support of the National Research Council of Canada and one of us (G.R.F.) the award of a postgraduate scholarship.

REFERENCES

1. Floyd, G. R. and Prince, R. H., *Nature Physical Science*, 1972, **240**, 11.
2. Prince, R. H., *Nature Physical Science*, 1973, **242**, 127.
3. Clampitt, R. and Gowland, L., *Nature*, 1969, **223**, 815.
4. Appell, J. and Durup, J., *Intern. J. Mass Spectr. Ion Phys.*, 1973, **10**, 247.
5. Redhead, P. A., *Can. J. Phys.*, 1964, **42**, 886.
6. Clampitt, R., 'Proc. 2nd Symp. Adsorption–Desorption Phenom.', Florence 1971, Academic Press.

7. Clampitt, R., *Nature*, 1970, **226**, 141.
8. Magee, J. L. and Funabashi, K., *Rad. Res.*, 1959, **10**, 622.
9. Solomon, P. M., *Physics Today*, 1973, **26**, 32.
10. Lind, S. C., 'Radiation Chemistry of Gases', Reinhold, New York, 1961.
11. Szabo, I. and Derrick, P. J., *Intern. J. Mass Spectr. Ion Physics*, 1971, **7**, 55.

92
A Crossed-Beam Apparatus for Investigation of Ion Molecule Reactions

By M. VESTAL, C. BLAKLEY, P. RYAN and
J. H. FUTRELL

(*Department of Chemistry, University of Utah, Salt Lake City, Utah, U.S.A.*)

INTRODUCTION

INTERSECTING ion and neutral beams have been employed by several investigators interested in the chemical dynamics of ion-molecule reactions.[1] In the ideal beam experiment the quantum state of both reactants and products should be determined.[2] The instrument described in this report is an attempt to approach more closely to this ideal than has been done in earlier work.

The layout of the instrument is shown in Fig. 1. While superficially not unlike other crossed-beam instruments, this particular machine incorporates several unique design features. These include a high pressure chemical ionization source, an apertureless exponential retarding lens, and a secondary ion mass/energy analyser that can be scanned through nearly a full octant of a sphere.

ION SOURCE

The ion source is mounted in a 20-cm diameter chamber which is pumped by a 2400 litres/sec diffusion pump.[3] High energy electrons (1–3 keV) enter the source through a 0·3-mm diameter by 1-mm capillary and ions exit through a 1-mm diameter aperture. The ion exit aperture is covered with 35% transparent 60 line/mm nickel mesh.[4] This is done in an attempt to maintain a sharp transition from viscous to molecular flow at the aperture even at relatively high source pressures (up to about 5 torr). The electron beam enters the source transverse to the ion beam at a distance of 6 mm from the ion exit. At a source pressure of 1 torr the average number of collisions that an ion formed in the electron beam undergoes in diffusing to the ion exit is about 10^6.[5]

Ions exiting the source are accelerated at *ca.* 750 eV in a cathode lens and focussed onto the object aperture of the magnetic mass spectrometer by an einzel lens. Cylindrically symmetric electrostatic ion optics are used throughout the lens system. A 1·5-mm diameter aperture in the last element of the lens system serves both as the differential pumping aperture and the object for the mass analyser.

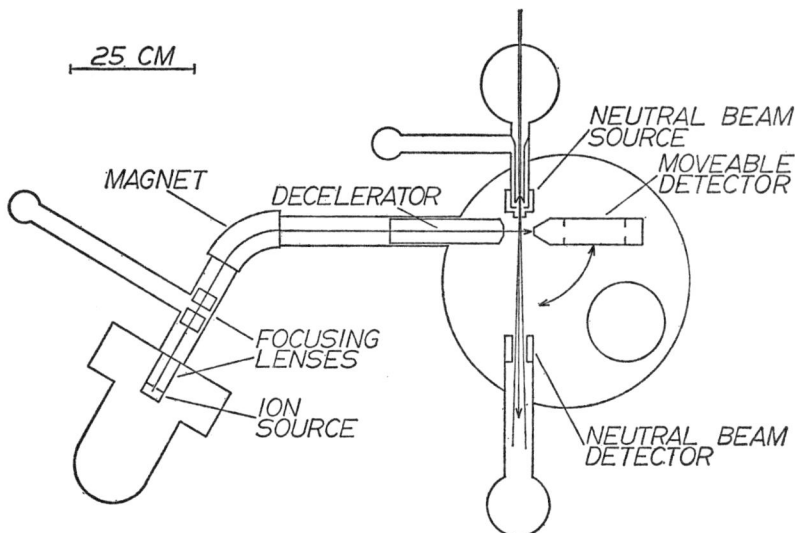

FIG. 1   Layout of the ion-neutral crossed-beam apparatus.

Located in the analyser tube, 3 cm from the object aperture is a two-element electrostatic quadruple lens. This lens was designed according to the graphical procedure given by Enge for astigmatic operation.[6] The diverging plane of the lens is the plane perpendicular to the magnetic field. In this plane the lens parameters are calculated to give a virtual image at the object slit of the mass analyser. Thus in this plane the ion beam is essentially unchanged except for a small decrease in the angular divergence. In the converging plane, parallel to the magnetic field, the parameters are adjusted to produce an image at the analysing aperture (0·5 mm diameter). The lens voltages found empirically to give the most intense ion beam are in good agreement with the values calculated according to the method of Enge.

PRIMARY MASS ANALYSER

The primary mass analyser is a 12·5-cm radius, 60-degree magnetic sector salvaged from an Hitachi RMS-4.[7] The flight tube was modified to electrically isolate it from ground by the installation of Kovar-Pyrex-Kovar seals at both ends of the tube and the insertion of teflon sheets between the magnet poles and the tube. This allows the collision region to be at ground potential. With this arrangement the final nominal ion energy is the same as the ion source voltage relative to ground. The analyser is pumped by a 100 litre/sec diffusion pump[3] with liquid nitrogen baffle.

DECELERATING LENS

After exiting from the analyser tube through the 0·5-mm diameter aperture, the ion beam enters the decelerating lens. Horizontal and vertical deflectors and an einzel lens are provided at the entrance to the deceleration lens to align the beam with the lens. The decelerating lens consists of a stack of 42 identical stainless steel plates, each with a 1·25-cm aperture, spaced 0·5 cm apart by ceramic spacers and connected to an exponentially decreasing resistive voltage divider. The first two and last three elements are connected to variable resistors which are empirically adjusted to correct for end effects.

With this arrangement the retarding potential is given by

$$V = -V_0 \exp(-\alpha x) \tag{1}$$

where V_0 is the energy of the ions at $x = 0$ and x is the distance along the axis of the field. In the paraxial approximation[8] the trajectories of ions in such a field are given by

$$r(x) = C \exp\left(\frac{\alpha x}{4}\right) \sin\left[(3)^{\frac{1}{2}}\left(\frac{\alpha x}{4} + \varphi\right)\right] \tag{2}$$

where r is the distance from the axis and

$$C = \frac{2}{\alpha}\left(\frac{40_0{}^2 - 2\theta_0 r_0\alpha + r_0{}^2\alpha^2}{3}\right)^{\frac{1}{2}} \tag{3}$$

$$\varphi = \sin^{-1}\left(\frac{r}{C}\right)$$

r_0 is the distance from the axis at the entrance and θ_0 is the entrance angle. The first focus of the lens occurs at

$$X = \frac{4\pi}{(3)^{\frac{1}{2}}\alpha} \tag{4}$$

The magnifications are given by

$$M_\theta = \left(\frac{\theta_f}{\theta_0}\right)_{r_0 - 0} = -\exp\left(\frac{\pi}{(3)^{\frac{1}{2}}}\right) \approx -6\cdot1 \tag{5}$$

$$M_r = \left(\frac{r_f}{r_0}\right)_{\theta_0 - 0} = -\exp\left(\frac{\pi}{(3)^{\frac{1}{2}}}\right) \approx -6\cdot1 \tag{6}$$

At the focus the energy of the ions is

$$V = V_0 \exp\left(\frac{-4\pi}{(3)^{\frac{1}{2}}}\right) \approx 7 \times 10^{-4} \tag{7}$$

The Abbe sine condition[9] requires that

$$M_\theta{}^2 M_r{}^2 = \frac{V_0}{V} \tag{8}$$

which restricts the minimum magnifications that can be attained by any retarding lens. The exponential lens has the desirable property of making the magnifications equal. It is not possible to maintain purely exponential fields at the ends, but by empirically trimming the potential on elements near the ends, results in good agreement with the theory have been obtained.

NEUTRAL BEAM SOURCE

The neutral beam source uses a differentially pumped free-jet similar to one described in the literature.[10] The nozzle is a 0·1 mm capillary 3 cm long. The skimmer is a 60-degree cone with a 0·5-mm aperture. The nozzle to skimmer distance may be adjusted from 0·1 to 6 mm by a micrometer screw outside the vacuum housing. The beam is modulated by a rotating disc chopper at frequencies from 15 to 150 Hz by selecting various gear ratios connecting the chopper shaft to a 30 Hz synchronous motor. The beam is further defined by collimating apertures of 1-mm diameter at 1 cm downstream from the skimmer and 3-mm diameter, 3 cm downstream. The nozzle chamber is pumped by a 2400 litres/sec diffusion pump[3] and the chopper and collimator chambers are differentially pumped by 150 litres/sec diffusion pumps.[5]

The neutral beam intersects the ion beam at a distance of 1 cm from the final collimating aperture. After crossing the reaction region the beam is detected by a flow-through ionization gauge and trapped in a separate chamber pumped by a 1200 litres/sec diffusion pump.[3] The beam source gives total intensities of approximately 10^{16} molecules/sec with a beam diameter of 4 mm at the collision centre and an angular divergence of 6 degrees. Molecular velocity distributions are obtained by using the highest chopper speed and measuring the time-of-flight between the chopper and the detector.

SECONDARY ION ANALYSER

The secondary ion analyser and beam intersection is housed in a 45-cm diameter glass bell jar. The chamber is pumped by a 2400 litres/sec pump[3] equipped with a high conductance liquid nitrogen baffle. The secondary ion analyser consists of a series of collimating apertures, a filter lens energy analyser, a quadrupole mass filter[11] and an electron multiplier.[12] With the 3-mm collimating apertures presently used the analyser acceptance is 4 degrees. The detector is mounted so that it may be rotated about an axis through the collision centre by 110 degrees. This axis itself may be rotated independently through an angle of 90 degrees relative to one of the primary beams while remaining perpendicular to the other primary beam. Thus the detector may be scanned through somewhat more than a full octant of a sphere. This motion is effected by gear trains inside the vacuum housing connected to shafts passing through rotating seals to the outside. The angles are read directly on vernier dials mounted directly on the apparatus inside the bell jar. The out of plane motion allows the primary beam divergences out of the nominal collision plane to be determined experimentally and allows the detector to reach some centre of mass angles which would otherwise entail scattering the neutral beam from the detector.

DATA HANDLING

The data handling system is shown schematically in Fig. 2. The pulses from the electron multiplier are amplified and counted using the multi-scaler mode of a 256 channel memory-core analyser.[13] A digital ramp generator, designed and built in our laboratories, provides the retarding potential to the secondary ion energy analyser. Each half cycle of the beam chopper a 256-step digital ramp is produced which both advances the address of the multi-channel analyser as well as increasing (or decreasing) the retarding potential. In this way a complete integral energy spectrum (over a selected energy range) is recorded each half cycle of the beam chopper. During the time that the molecular beam is on, the signals are added into the memory and during the

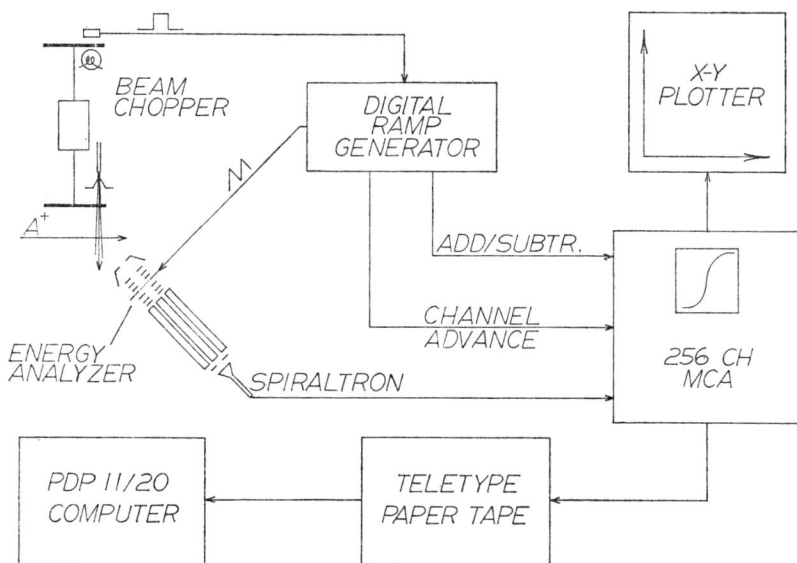

FIG. 2   Block diagram of the data handling system.

time the beam is off the signals are subtracted. Thus both signal averaging and automatic background subtraction are accomplished. After a statistically significant number of counts have been accumulated the integral energy spectrum may be plotted on an X–Y recorder, printed by a teletype, or punched onto paper tape for processing off-line on the PDP 11/20 computer.

PERFORMANCE DATA

The performance of the ion optical system is illustrated in Figs. 3 and 4. Figure 3 shows a typical result for ion current as a function of final ion energy. These results were obtained on the $C_4H_9^+$ beam with 1 torr of isobutane in the ion source and used a Faraday cup with a 3-mm diameter aperture located at the collision centre. As can be seen from the figure the transmission of the retarding lens is essentially flat down to about 1 eV nominal energy and falls off only about a factor of 2 down to thermal energies.

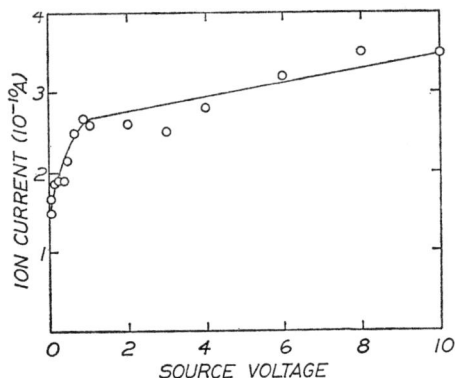

FIG. 3   Ion current as a function of ion source potential, or nominal ion energy at the collision centre. $C_4H_9^+$ from isobutane at 1 torr, 150°C, measured with 3-mm aperture Faraday cup at the collision centre.

The surprisingly high intensity at zero source voltage may be understood by reference to Fig. 4, where differential energy spectra obtained at a source temperature of 150°C and a source pressure of 1 torr are shown. The ion source was at 1·0 V. These spectra clearly are not the thermal distributions anticipated but rather show significant excess kinetic energy. This excess

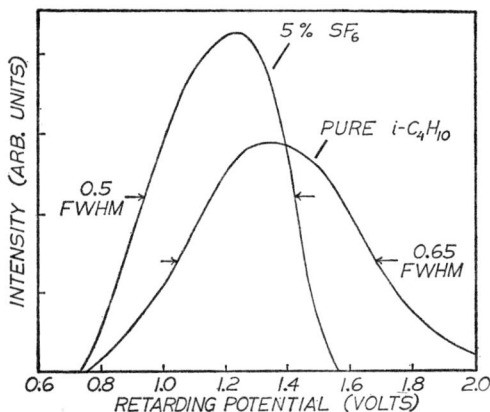

FIG. 4   Differential energy distributions for $C_4H_9^+$ from isobutane (1 torr, 150°C) at 1·0 eV nominal energy.

energy is believed to be due to excess positive ion space charge. The distributions are significantly narrowed by adding a small amount of SF_6 to the isobutane. A detailed discussion of this effect is beyond the scope of this report.

The angular distributions for the primary ion beam vary from 4° FWHM at 1 eV. The distributions are in good agreement with the theoretical expectations for the exponential retarding lens.

APPLICATIONS

This apparatus is currently being used for studies on reactive scattering. Preliminary results have been obtained for the reactions

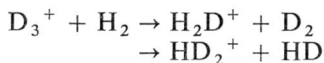

$$D_3{}^+ + H_2 \rightarrow H_2D^+ + D_2$$
$$\rightarrow HD_2{}^+ + HD$$

and on the reaction

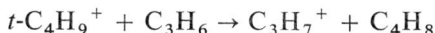

$$t\text{-}C_4H_9{}^+ + C_3H_6 \rightarrow C_3H_7{}^+ + C_4H_8$$

These reactions all appear to occur by a long-lived complex at low initial kinetic energies.

REFERENCES

1. (a) Henglein, A., Lacmann, K. and Jacobs, G., *Ber. Bunsenges. Phys. Chem.*, 1965, **69**, 279.
   (b) Fink, R. D. and King, J. S., *J. Chem. Phys.*, 1967, **47**, 1857.
   (c) Gentry, W. R., Gislason, E. A., Mahan, B. H. and Tsao, Chi-Wing., *J. Chem. Phys.*, 1968, **49**, 3058.
   (d) Champion, R. L., Doverspike, L. D. and Bailey, T. L., *J. Chem. Phys.*, 1966, **45**, 4377.
   (e) Turner, B. R., Fineman, M. A. and Stebbings, R. F., *J. Chem. Phys.*, 1965, **42**, 4088.
   (f) Herman, Z., Kerstetter, J. D., Rose, T. L. and Wolfgang, R., *Rev. Sci. Instr.*, 1969, **40**, 538.
2. Herman, Z. and Wolfgang, R., *in* 'Ion-Molecule Reactions', *ed.* J. L. Franklin, Plenum Press, New York, 1972, Vol. 2, chapter 12.
3. The 2400 litre/sec pumps are NRC Model VHS-6, the 1200 litre-sec pump is NRC Model VHS-4, and the 150 litre/sec pumps are NRC Model SHS-2, from NRC/Varian, Newton, Mass. 02161. The 100 litre/sec pump is a 2 in. Hitachi supplied with the RMS-4.
4. Electroformed nickel mesh from Buckbee-Mears Corp., St Paul, Minn.
5. Calculated from equations given by S. Dushman, 'Scientific Foundations of Vacuum Technique', John Wiley and Sons, New York, 1949, p. 81ff.
6. Enge, H. A., *Rev. Sci. Instr.*, 1959, **30**, 248.
7. Hitachi Ltd, Tokyo, Japan.
8. Pazkowski, B., 'Electron Optics', Elsevier, New York, 1968.
9. Ibid, p. 45.
10. Parson, J. M. and Lee, Y. T., *J. Chem. Phys.*, 1972, **56**, 4558.
11. Finnigan Instruments Corp., Sunnyvale, California.
12. Bendix Electr-Optics Div., Gallileo Park, Sturbridge, Mass., Model 4219.
13. Northern Scientific, Inc., Middleton, Vis., Model NS-600.

93
Arrival Time Distribution Measurements in Ion-Molecule Reaction, Chemical Ionization, and Ion Equilibrium Studies

By G. G. MEISELS, G. J. SROKA and R. K. MITCHUM

(*Department of Chemistry, University of Houston, Houston, Texas, U.S.A.*)

INTRODUCTION

THE study of ion behaviour in ion sources operating at pressures ranging from a few torr to an atmosphere[1] has become important to analytical applications of chemical ionization mass spectrometry,[2] to measurements of ion-molecule reaction rates,[3,4,5] and to the evaluation of ionic equilibria, including the associated determination of heats of reaction and of intrinsic acidities and basicities in simple systems.[1,6-10] Particularly in the last two areas, some knowledge of ion residence times is required either to assure constancy of residence times for kinetic measurements, or to establish that ionic equilibrium is indeed reached under conditions accessible by experiment. The field has received a particularly strong stimulus as a result of the development of chemical ionization sources for analytical applications, and because of the convenience of their existence these have been employed extensively for reaction rate and equilibrium studies, with residence times only inferred indirectly. We have addressed ourselves to the development of an understanding of ion drift properties, energies, and residence time distributions in such sources using as the principal approach the direct measurement of arrival time distributions.[5,11-13] While we address ourselves primarily to measurements of physical and kinetic parameters, it is of considerable interest to seek a reconciliation of one of the most perplexing discrepancies currently in the literature: the discordant values of heats and free energies for the sequence of steps which constitute the hydration of the proton.

Initial experimental measurements concerning the solvation of the proton in the gas phase employed equilibrium studies in a field-free source at near-atmospheric pressure[6] and collisional detachment.[14] Results were in reasonable agreement with each other and with a set of bond energies calculated by *ab initio* methods.[15] However, in an extensive investigation using chemical ionization sources where pressure is of the order of a few torr and a small extraction field is maintained to improve sensitivity, equilibrium constants were found to differ by orders of magnitude from those determined earlier, and free energies and enthalpies of reaction were a factor of two smaller than previously suggested.[8,9]

In thermodynamic studies, it is essential that ion sampling reflects equilibrium rather than a steady state. Cunningham et al.[10] recently demonstrated that equilibrium could be achieved only after about 0·3 msec. This is nearly an order of magnitude longer than average residence times in chemical ionization sources as measured earlier by an arrival time distribution method.[5] Young and Falconer[16] determined rate coefficients; they noted their dependence on E/P and the absence of equilibrium. While there was thus clear evidence for lack of complete equilibrium Cunningham et al.[10] were unable to account for the pressure independence of data obtained in chemical ionization sources, and for the linearity of Van't Hoff Plots in such systems, which were initially taken as evidence for the attainment of equilibrium. They could also not rationalize the discrepancies in the free energy and heat of hydration of H_3O^+.

We present here a partial resolution of the reported discrepancies.

EXPERIMENTAL

Arrival time distributions were determined using a modified Atlas CH-4 mass spectrometer as described previously.[5,11-13] Briefly, water vapour was introduced into the ion source at a pressure of 1 torr and held constant by a servovalve controlled by a Baratron Capacitance manometer whose sensing arm was attached directly to the source. A 0·1 μsec pulse of electrons having an energy of 220 eV and entering the source in a plane 0·7 cm from the single ion exit and extraction slit ionized the vapour. Since the position of the equilibrium and the rate of at least the forward reaction are strongly dependent on E/P, field strength is maintained at 1 V/cm. The electron pulse is used to gate a time-to-pulse height converter whose ramp is terminated by the ion arrival pulse. The corresponding output pulses are sorted on a 400-channel pulse height analyser, each channel corresponding to an interval of 1·25 μsec. Arrival time distributions of each mass are measured by accumulating pulses for a few minutes. The memory of the analyser is then transferred onto tape. After the collection of the required number of arrival time distributions the tapes are processed on a Univac 1108 computer using the time-sharing mode. Numerical evaluations of model systems are performed in the same manner.

RESULTS AND DISCUSSION

Figure 1 shows illustrative arrival time distributions of H_3O^+, $H_5O_2^+$ and $H_7O_3^+$, obtained over various periods taken directly from the output of the multichannel analyser. They are shifted slightly to take into account the variation in transit time in the analyser as a result of the difference in mass, using t (transit) $= 1·5 (m/e)^{\frac{1}{2}} \mu$sec.[11] It is apparent that equilibrium is not established over most of the range of times accessible in a typical chemical ionization source. If equilibrium were established, the normalized arrival

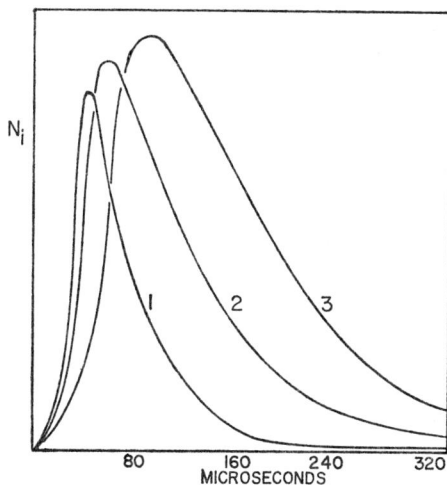

FIG. 1   Arrival time distributions of H_3O^+, $H_5O_2^+$ and $H_7O_3^+$ (curves 1, 2 and 3) at a field strength of 1 V/cm, 440°K, and 1 torr water.

time distributions would be identical for all species involved in the equilibrium.[11] This is not the case; this is recognized clearly by examining the ion current ratio obtained by dividing the distribution for H_3O^+ by that for $H_5O_2^+$ (Fig. 2). Only at times in excess of 0·2 msec does the ratio approach constancy.

Arrival time distributions of H_3O^+ are dominated by times of less than approximately 150 μsec, and those of $H_5O_2^+$ by times less than 200 μsec; these periods are much shorter than those required for the establishment of equilibrium. In continuous experiments these portions of the distribution dominate the measured ion current; the observed ion current ratio is a time average

$$\frac{[H_5O_2^+]}{[H_3O^+]} = \int_{t=0}^{t=\infty} \left(\frac{i_{H_5O_2}(t)}{i_{H_3O}(t)} \right) dt \tag{I}$$

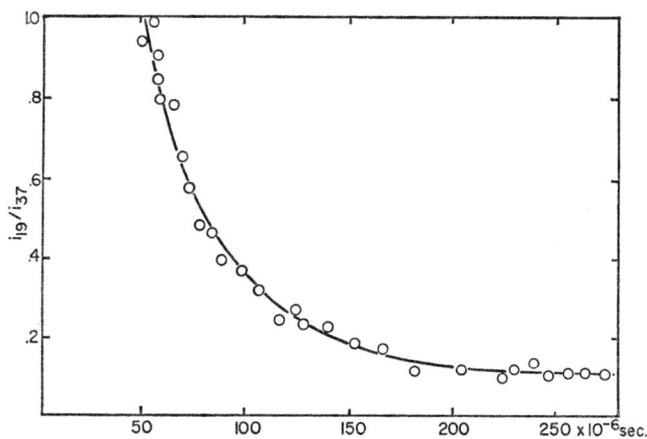

FIG. 2   Dependence of ratio of ion currents of H_3O^+ to $H_5O_2^+$ on time.

which can approach the equilibrium value only if the areas under the arrival time distribution curves at times less than those required for the attainment of equilibrium are negligible. Even when the average residence time is in excess of that required for the attainment of equilibrium, the ion current ratio may include significant time regions where equilibrium is not established.

QUASITHERMODYNAMIC PARAMETERS DERIVED UNDER CONDITIONS OF KINETIC CONTROL

The earlier work of Cunningham et al.[10] as well as Fig. 2 clearly demonstrate that at relatively low temperatures and at small water concentrations the reaction is far from equilibrium but that relative ion intensities are determined instead by the kinetics of the approach to equilibrium. The reaction sequence representing the first few hydration steps is

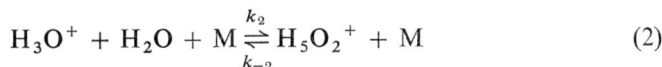

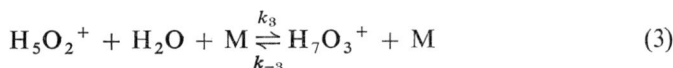

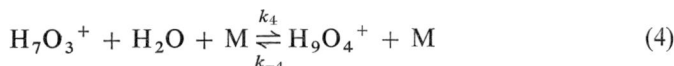

$$MH^+ + H_2O + M \underset{k_{-1}}{\overset{k_1}{\rightleftharpoons}} M + M + H_3O^+ \tag{1}$$

$$H_3O^+ + H_2O + M \underset{k_{-2}}{\overset{k_2}{\rightleftharpoons}} H_5O_2^+ + M \tag{2}$$

$$H_5O_2^+ + H_2O + M \underset{k_{-3}}{\overset{k_3}{\rightleftharpoons}} H_7O_3^+ + M \tag{3}$$

$$H_7O_3^+ + H_2O + M \underset{k_{-4}}{\overset{k_4}{\rightleftharpoons}} H_9O_4^+ + M \tag{4}$$

In pure water, MH^+ is merely H_2O^+; with methane as the major drift gas, MH^+ is CH_5^+ and yet another reaction step must be added.

$$CH_4^+ + CH_4 \overset{k_0}{\rightarrow} CH_5^+ + CH_3 \tag{5}$$

In addition to reactions (1)–(4), each ion is also lost by diffusion to the walls, by extraction from the source, and by neutralization. As long as the diffusion and recombination coefficients are of similar magnitude for all ions involved, these ion loss processes should affect all ions nearly equally. Differences arise from the kinetic behaviour which converts initial ions earlier in time so that the mean time during which each ion exists in the source increases as the number of steps required to produce it increases. For simplicity, we neglect this complication here.

The greatest disparity in reported heats, free energies, and entropies of reaction exists for the second equilibrium, reaction (2). We confine our further considerations to this step.

The operational definition of the equilibrium constant is

$$K_a = \frac{[H_5O_2^+]}{[H_3O^+] \cdot [H_2O]} \tag{II}$$

The subscript a is used to indicate that the quantity calculated via equation II is not necessarily the equilibrium constant K but an experimental observable whose relationship to the equilibrium constant is to be established.

The first approach is based on a model in which reactions (1), (2) and (3) proceed only in the forward direction; consequently, the reverse reactions do not significantly contribute to specific ion concentrations. At the same time, it must be assumed that reaction (O), the formation of reagent MH^+, occurs instantaneously. Because of the large concentration of M this is a reasonable assumption, and M^+ is typically almost completely converted to MH^+ within 10^{-7} sec, short in comparison to mean residence times of 10^{-5} sec.[11,12] The solution to the linear first-order differential equations describing the time dependent behaviour of ion concentrations under these assumptions is readily available[7,17] and leads to

$$K_a = \frac{k_2[M]}{\exp\left(-k_1[M][H_2O]\right) - \exp\left(-k_2[M][H_2O]\right)}$$
$$\times \left(\frac{\exp\left(-k_1[M][H_2O]t\right) - \exp\left(-k_3[M][H_2O]t\right)}{(k_3 - k_1)[M][H_2O]} \right.$$
$$\left. - \frac{\exp\left(-k_2[M][H_2O]t\right) - \exp\left(-k_3[M][H_2O]t\right)}{(k_3 - k_2)[M][H_2O]} \right) \qquad \text{(III)}$$

In the limit where all exponents are near zero, for convenience denoted as time approaching zero, one can write the relationship

$$\lim_{t \to 0} K_a = \left(\frac{k_2}{2}\right)[M]t \qquad \text{(IV)}$$

This quantity is clearly independent of the concentration of water in the system.

The second explicit solution in closed form is based on the assumption that all steps other than reaction (2) are at equilibrium. The approach to equilibrium of two opposing quasi-first order reactions is given by

$$K_a = K \frac{1 - \exp -\{(k_2[M][H_2O] + k_{-2}[M])t\}}{1 + K[H_2O]\exp\{-(k_2[M][H_2O] + k_{-2}[M])t\}} \qquad \text{(V)}$$

This equation clearly shows that K_a is always less than K unless time approaches infinity. In the limit where the sum of the product of time and the forward and reverse reaction frequencies is very small (*i.e.* under kinetic control)

$$\lim_{t \to 0} K_a = k_2[M]t \qquad \text{(VI)}$$

This relationship is again independent of water concentration, and differs from (IV) by a factor of 2 as a result of the assumption that step 1 is at equilibrium.

Comparison of the preceding equations with experiments can be made by inserting rate constants determined under ideal conditions where all ions are indeed at their equilibrium concentrations, using experimental arrival times. In the absence of arrival time distribution measurements, mean drift times given by drift theory may be employed, since these have been shown to lead to calculated residence times within a factor of two of those experimentally

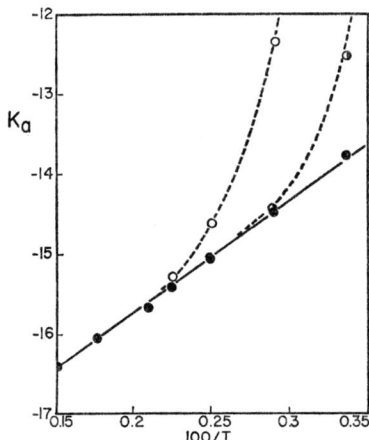

FIG. 3   Van't Hoff Plot of quasi-equilibrium constant K_a calculated from eqn (V). The slope of the line is that defined by eqn (VI) for complete kinetic limitation and no reverse reaction. Solid circles: water pressure 0·001 torr; semi-filled circles: 0·01 torr, open circles: 0·1 torr.

observed.[5,11,12] The mean residence time $\bar{t}$ is conveniently given by[12]

$$\bar{t} = \left(\frac{z}{T}\right)\left(\frac{P}{E}\right)(\alpha\mu)^{\frac{1}{2}} \times 10^{-2} \text{ sec} \qquad \text{(VII)}$$

where z is the drift length in centimetres, T is in °K, the pressure P of [M] in torr and the repeller or extraction field E in V/cm; α and μ are the polarizability and reduced mass in atomic units.

Unfortunately, the rate constants k_1, k_{-1}, k_3 and k_{-3} are not known sufficiently well to evaluate the temperature dependence of K_a through eqn (III). From Kebarle's recent investigations it is possible to extract $k_2 = 3·16 \times 10^{-30} \exp(4900/RT) \text{ cm}^6 \text{ sec}^{-1}$ and $k_{-2} = 5·19 \times 10^{-6} \times \exp(-26\,700/RT) \text{ cm}^3 \text{ sec}^{-1}$. Figure 3 summarizes the calculated quasi-Van't Hoff plot based on eqns (V) and (VII), these rate constants, a drift length z of 1 cm and a field strength of 1 V/cm at 1 torr methane as the drift gas. It is apparent that results are independent of water concentration and reasonably linear at higher temperatures.

The linearity of Van't Hoff Plots under non-equilibrium conditions can be understood easily by consideration of eqns (IV) or (VI) and (VII), and by the variation of log T with $1/T$. The latter is very nearly a linear function of temperature between 300° to 700°K and, over this range, can be represented by $\log T = 3·147 - 2·22 \times 10^2/T$ with a deviation of less than 0·5%. In the limit of complete control by kinetic parameters, we may therefore write, beginning with (VI) and (VII) and the ideal gas law

$$\lim_{t \to 0} \log K_a = -7·881 + \log\left(\frac{A_2(\alpha\mu)^{\frac{1}{2}}}{k}\right) + \log\left(\frac{P^2 z}{E}\right) - \frac{E^2}{(2·303\,RT)} - 2\log T \qquad \text{(VIIIa)}$$

$$= -14·175 + \log\left(\frac{A_2(\alpha\mu)^{\frac{1}{2}}P^2 z}{kE}\right) - \frac{(E_2 - 2025)}{(2·303\,RT)} \qquad \text{(VIIIb)}$$

where A_2 and E_2 are the pre-exponential factor and activation energy of reaction (2), respectively. The apparent heat of reaction ΔH_a obtained from a Van't Hoff plot is $E_2 - 2025 = -6{\cdot}9$ kcal/mole if Kebarle's value[10] for E_2 is employed. A line corresponding to this equation is drawn in Fig. 3 and represents the asymptote to the full equation at higher temperatures.

Our calculated value of ΔH_a is in almost exact agreement with $\Delta H_a = 7$ kcal/mole reported by Beggs and Field[8] when they employed methane as the major gas. This agreement is probably fortuitous because we have assumed that drift is entirely dictated by gross drift considerations. Since it is customary to adjust the repeller potential at will, it is well possible that in other investigations a lesser potential was employed where mean residence times are more nearly described by diffusion considerations, thereby changing the functional dependence of mean residence times on temperature and with it the limiting slope of Van't Hoff plots.

CONCLUSIONS

The analysis based on kinetic limitation of reactions presented here demonstrates that independence of pressure and linearity of Van't Hoff plots does not assure that equilibrium has been achieved. It leads to a calculated value of the apparent heat of reaction which depends only on the activation energy of the rate controlling step and the functional form of the temperature dependence of mean residence times. As long as applied fields are sufficient to assure the applicability of (VII) the resultant ΔH_a will be independent of instrumental parameters, drift lengths, field strengths, etc. and will be approximately 2 kcal/mole less than the activation energy of the limiting step. These factors will, however, affect the intercept of Van't Hoff plots and thereby exercise considerable influence on the apparent values of free energies and entropies of reaction. This is consistent with earlier observations.[8,9]

ACKNOWLEDGMENTS

This investigation was supported in part by the Robert A. Welch Foundation, grant number E-210, and in part by the United States Atomic Energy Commission under contract AT-(40-1)-3606. We are sincerely grateful for this assistance.

REFERENCES

1. Kebarle, P. and Godbole, E. W., *J. Chem. Phys.*, 1963, **39**, 1131; also see Kebarle, P., in 'Ion-Molecule Reactions', *ed.* J. L. Franklin, Plenum Press, New York, 1972, chapter 7.
2. Fales, H., in 'Mass Spectrometry: Techniques and Applications', *ed.* G. A. W. Milne, Wiley-Interscience, New York, 1971; and Field, F., in 'Mass Spectrometry', *ed.* A. MacColl, Butterworth, London, 1972.
3. Field, F. H., *J. Amer. Chem. Soc.*, 1969, **91**, 2827.
4. Vredenberg, S., Wojcik, L. and Futrell, J. H., *J. Phys. Chem.*, 1971, **75**, 590.
5. Sroka, G., Chang, C. and Meisels, G. G., *J. Amer. Chem. Soc.*, 1972, **95**, 1052.

6. Kebarle, P., Searles, S. K., Zolla, A., Scarborough, J. and Arshadi, M., *J. Amer. Chem. Soc.*, 1967, **89**, 6393.
7. Good, A., Durden, D. A., Kebarle, P., *J. Chem. Phys.*, 1970, **52**, 212.
8. Beggs, D. P. and Field, F. H., *J. Amer. Chem. Soc.*, 1971, **93**, 1567 and 1576.
9. Bennett, S. L. and Field, F. H., *J. Amer. Chem. Soc.*, 1972, **94**, 5186.
10. Cunningham, A. J., Payzant, J. D. and Kebarle, P., *J. Amer. Chem. Soc.*, 1972, **94**, 7627.
11. Chang, C., Sroka, G. J. and Meisels, G. G., *J. Chem. Phys.*, 1971, **55**, 5154.
12. Chang, C., Sroka, G. J. and Meisels, G. G., *Int. J. Mass Spechrom. Ion Phys.*, 1973, **11**, 367.
13. Chang, C., Meisels, G. G. and Taylor, J. A., *Int. J. Mass Spectrom. Ion Phys.*, accepted for publication.
14. DePaz, M., Leventhal, J. J. and Friedman, L., *J. Chem. Phys.*, 1969, **51**, 3748.
15. Newton, M. D. and Ehrenson, S., *J. Amer. Chem. Soc.*, 1971, **93**, 4971.
16. Young, C. E. and Falconer, W. E., *J. Chem. Phys.*, 1972, **57**, 918.
17. For example, see Hildbrand, F. B., 'Advanced Calculus for Applications', Prentice-Hall, Englewood Cliffs, N.J., 1962, p. 7; or Margenau, H. S. and Murphy, G. M., 'The Mathematics of Physics and Chemistry', Van Nostrand, New York, 1943, p. 41.

Discussion

J. B. Hasted (Birkbeck College, London, U.K.): Is it possible to extract velocity distributions from these data? We have found by retardation analysis that such distributions have Maxwell–Boltzmann tails, whilst the mean velocities are represented by the Wannier equation.

G. G. Meisels: These data do not permit an unfolding of the distributions, and a simple modification of our procedure is planned which will permit us to measure at least the tail of the velocity distribution on exit by a time-of-flight method. However, such measurements yield information only on the longitudinal or one-dimensional distribution, and a separate measurement such as by a deflection technique is required for the evaluation of the transverse velocity distribution. Recent independent but essentially identical theoretical analyses by ourselves (Int. J. Mass Spectrom. Ion Phys. **12**, (1973)) and by Woo *et al.* suggest elliptical or eggshaped velocity distributions characterized by one-dimensional Maxwell–Boltzman distributions with different mean energies in the transverse and longitudinal directions.

94
Reactions of O⁻ Ions with Some Unsaturated Hydrocarbons

By GILL C. GOODE and K. R. JENNINGS†

(*Department of Chemistry, The University, Sheffield, U.K.*)

INTRODUCTION

THE formation and reactions of negative ions are of importance in understanding the ion chemistry of the ionosphere and other planetary atmospheres and of combustion processes and reactions of interest in atmospheric pollution studies. Attempts have been made to measure electron affinities and rate constants both for the formation of negative ions and for their reactions with atoms and molecules.[1−7] The flowing afterglow technique has been used extensively to study negative ion-molecule reactions at thermal energies of ions found in flames and the ionosphere, such as O^-, O_2^- and O_3^-, and rate constants for reactions of these ions with a variety of molecules have been determined.[2,3,8]

The main objective of the present work was the obtaining of both qualitative and quantitative information on possible reactions of O^- ions in hydrocarbon flames, and the results of a study of their reactions with acetylene, ethylene and propylene are presented.

EXPERIMENTAL

The work was carried out using a Varian V-5900 ion cyclotron resonance mass spectrometer as described previously,[9] but with the modifications mentioned below. The high sensitivity of the technique makes it particularly suitable for studies of negative ions. For most experiments, standard magnetic field modulation was used. Since the trap current regulation was unsatisfactory at electron energies below 4 eV, filament current regulation was employed. A five-section, 'flat' cell was constructed, having source, ejection, reaction, analyser and collector regions. When operated in the negative ion mode, 98% of the recorded 'total ion current' (TIC) was shown to be due to the collection of scattered electrons and only 2% was due to negative ions. Since it is necessary to know the total negative *ion* current for quantitative

† Present address: Department of Molecular Sciences, University of Warwick, Coventry CV4 7AL, Warwickshire.

work,[10] the electrons were ejected from the cell, immediately after they had left the source region, by applying a r.f. field to the trapping plates.[11] The frequency varied with operating conditions but was within the range 6–11 MHz. The TIC was monitored using a Keithley Model 602 electrometer.

Reactants were standard reagent grade gases (B.O.C. Ltd and Cambrian Chemicals Ltd) and were used without further purification. Deuterated compounds were obtained from Merck, Sharpe and Dohme Company, Canada.

RESULTS

The primary ions produced from acetylene, ethylene and propylene and their deuterated analogues by dissociative attachment were of a much lower abundance than the secondary hydrocarbon ions formed by reactions of O^- ions with hydrocarbon molecules and their maximum yield occurred at electron energies different from those used to produce O^- ions from various molecules. Reliable quantitative measurements of secondary ion currents could therefore be made without interference.

Reaction with Acetylene

Nitrous oxide was used to produce O^- ions by dissociative electron attachment at an electron energy of 1·5 eV (uncorrected). In the presence of acetylene, the major reactions were:

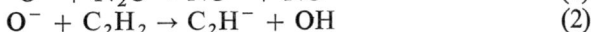

$$O^- + N_2O \rightarrow NO^- + NO \qquad (1)$$
$$O^- + C_2H_2 \rightarrow C_2H^- + OH \qquad (2)$$
$$O^- + C_2H_2 \rightarrow C_2OH^- + H \qquad (3)$$

Comparison of the ionization efficiency curves of NO^-, C_2H^- and C_2OH^- with that of O^- shows them to have similar shapes and to have maxima at the same electron energy. The occurrence of the three reactions was also confirmed using the ion cyclotron double resonance (ICDR) technique.[12]

At a pressure of about 5×10^{-6} torr, the percentage conversion was sufficiently low for relative rate constants of these reactions to be obtained, using both ion cyclotron resonance (ICR) and total ion current (TIC) methods.[10] Good agreement was obtained between the two methods and led to values of $k_2/k_1 \simeq 14$ and $k_2/k_3 \simeq 17$.

A qualitative investigation of the dependence of k_2/k_1 on the translational energy of the O^- ion was made using the ICR technique by irradiating the O^- ions with a second r.f. field. The ratio k_2/k_1 was observed to increase as the translational energy of O^- increased; since, in the absence of acetylene, k_1 was found to fall only slightly as the translational energy of O^- increased, this suggests that k_2 rises with increasing O^- translational energy. This was confirmed by observing k_2/k_1 for reactions of O^- ions produced from both N_2O and CO in a mixture of $N_2O/CO/C_2H_2$. The O^- ions produced from N_2O are thought to have a translational energy of 0·38 eV, whereas those produced from CO have near thermal translational energies.[13,14] By working with electron energies of $\sim 1·5$ eV and 9·5 eV (uncorrected) and by careful control of the N_2O/CO ratio in the mixture, it was possible to produce approximately equal yields of O^- ions from N_2O and CO respectively, and the ratios of k_2/k_1 were approximately 14 and 9 respectively. Under these conditions and at low conversions, the ratio of the ion currents O^-/NO^- was approximately the same for each source of O^- ions so that these observations again suggest that k_2 increases with increasing translational energy of the O^- ion. This is also consistent with the results obtained from a mixture of $O_2/N_2O/C_2H_2$. At an electron energy of ~ 7.0 eV (uncorrected), O^- produced from O_2 with a translational energy of 1.64 eV[15] gave $k_2/k_1 \simeq 28$.

Owing to the sensitivity problems associated with the use of CO as a source of thermal O^- ions, the use of NO_2 as an alternative source was investigated. This has a much larger cross-section for the production of O^- ions, and their translational energy is believed to be close to thermal.[1] In a $CO/NO_2/C_2H_2$ mixture, O^- ions were produced from NO_2 (1.2 eV) and CO (9.5 eV) and the major reactions observed were (2) and (3) together with (4)

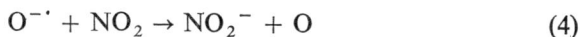

$$O^{-\cdot} + NO_2 \rightarrow NO_2^- + O \tag{4}$$

Using both ICR and TIC techniques, $k_4/k_2 \simeq 1.4$ for both sources of O^- ions, strongly suggesting that NO_2 does indeed give thermal O^- ions. This was therefore used as a source of O^- ions in subsequent experiments.

Reaction with Ethylene

In a mixture of NO_2 and C_2H_4 the major reactions observed were (4) and (5), with (6) also occurring to a minor extent

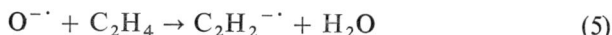

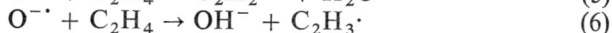

$$O^{-\cdot} + C_2H_4 \rightarrow C_2H_2^{-\cdot} + H_2O \tag{5}$$
$$O^{-\cdot} + C_2H_4 \rightarrow OH^- + C_2H_3\cdot \tag{6}$$

Comparison of ionization efficiency curves and the use of ICDR confirmed that O^- ions were the precursors of NO_2^-, $C_2H_2^-$ and OH^-. At a pressure of $\sim 5 \times 10^{-5}$ torr, the pressure was sufficiently low to obtain relative rate constants, and k_4/k_5 was found to be $\simeq 4$. The identity of the ion at $m/e = 26^-$ was shown to be $C_2H_2^-$ rather than CN^- by working with the mixture NO_2/C_2D_4, when the ion moved to $m/e = 28^-$, $C_2D_2^-$. In this mixture, $k_4/k_5 \simeq 7$, indicating that O^- reacts more slowly with C_2D_4 than with C_2H_4.

In order to establish which hydrogen atoms are removed in reaction (5), the mixtures NO_2/CH_2CD_2 and NO_2/CH_2CF_2 were investigated. In the former mixture, secondary ions were observed at $m/e = 26^-$, 28^- and 46^-, but not at 27^-. The peaks at $m/e = 26^-$ and 28^- were of approximately equal intensity and are identified with $C_2H_2^-$ and $C_2D_2^-$ respectively, suggesting that the two hydrogen atoms are removed from the same carbon atom of the ethylene molecule, e.g.

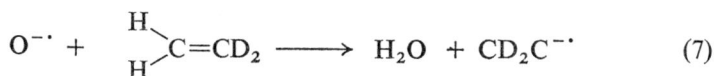

$$O^{-\cdot} + \begin{matrix} H \\ H \end{matrix}\!\!\diagdown\!\!C\!=\!CD_2 \longrightarrow H_2O + CD_2C^{-\cdot} \tag{7}$$

Support for this mechanism is given by the observation on a peak at $m/e = 62^-$ in the NO_2/CH_2CF_2 system, which is believed to be formed in an analogous reaction

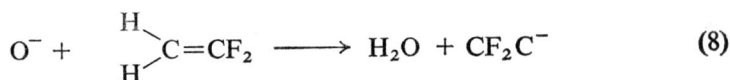

$$O^- + \quad \begin{matrix} H \\ \diagdown \\ \diagup \\ H \end{matrix} C{=}CF_2 \quad \longrightarrow \quad H_2O + CF_2C^- \qquad (8)$$

If the $C_2H_2^-$ ion has the structure CH_2C^-, this would explain the fact that it cannot be formed directly from acetylene.

Reaction with Propylene

In order to study the effect of substitution on reaction (5), the reactions of O^- with propylene were studied in a mixture of NO_2/CH_3CHCH_2 at a nominal electron energy of 1·2 eV. In addition to reaction (4), the two major reactions were

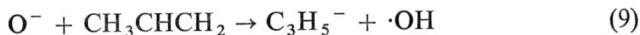
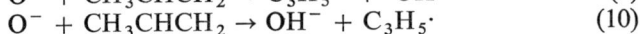
$$O^- + CH_3CHCH_2 \rightarrow C_3H_5^- + \cdot OH \qquad (9)$$
$$O^- + CH_3CHCH_2 \rightarrow OH^- + C_3H_5\cdot \qquad (10)$$

where $k_4/k_9 \simeq 6\cdot5$. In order to determine which hydrogen atoms are involved in these reactions, the NO_2/CD_3CHCH_2 and $NO_2/CH_3CH{=}CD_2$ systems were investigated. In the former system, the major secondary ions formed were at $m/e = 17^-(OH^-)$, $18^-(OD^-)$, $46^-(NO_2^-)$ and 43^-. The observation of both OH^- and OD^- ions formed in a reaction analogous to reaction (10) indicates that in this reaction, the O^- ion is unselective in removing a hydrogen or deuterium atom. The ion at $m/e = 43^-$ could be due either to $C_3H_3D_2^-$ (OD lost) or $C_3HD_3^-$ (H_2O lost), although the former is to be expected by analogy with reaction (9). This would indicate that the hydrogen atom of the OH radical comes entirely from the methyl group of the propylene molecule. This was confirmed by observing the $m/e = 43^-$ ion in the $NO_2/$ $CH_3CH{=}CD_2$ system. In this case, the peak can only be ascribed to the $C_3H_3D_2^-$ ion, arising from the exclusive loss of a hydrogen atom from the methyl group. As in the reaction of O^- ions with ethylene, there is no evidence of any H,D randomization in the reaction complex.

Absolute Rate Constants for Reactions of Thermal O^- Ions

The relative rate constants obtained in this work can be put on an absolute basis by referring them to an absolute value for k_4 from the literature.[5] Taking $k_4 = 1\cdot2 \times 10^{-9} \; cm^3 \; molecule^{-1} \; sec^{-1}$, the following absolute rate constants were calculated for reactions of thermal O^- ions (all in units of $cm^3 \; molecule^{-1} \; sec^{-1}$)

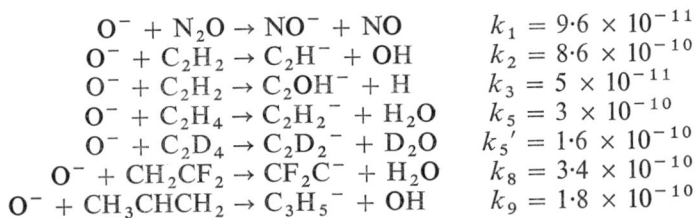

$$O^- + N_2O \rightarrow NO^- + NO \qquad k_1 = 9\cdot6 \times 10^{-11}$$
$$O^- + C_2H_2 \rightarrow C_2H^- + OH \qquad k_2 = 8\cdot6 \times 10^{-10}$$
$$O^- + C_2H_2 \rightarrow C_2OH^- + H \qquad k_3 = 5 \times 10^{-11}$$
$$O^- + C_2H_4 \rightarrow C_2H_2^- + H_2O \qquad k_5 = 3 \times 10^{-10}$$
$$O^- + C_2D_4 \rightarrow C_2D_2^- + D_2O \qquad k_5' = 1\cdot6 \times 10^{-10}$$
$$O^- + CH_2CF_2 \rightarrow CF_2C^- + H_2O \qquad k_8 = 3\cdot4 \times 10^{-10}$$
$$O^- + CH_3CHCH_2 \rightarrow C_3H_5^- + OH \qquad k_9 = 1\cdot8 \times 10^{-10}$$

DISCUSSION

The results obtained for the reaction of O^- ions with acetylene are in good agreement with those obtained previously using widely differing techniques. In studies using a tandem mass spectrometer,[7] in which the O^- ions have an average translational energy of 0·3 eV, reactions (2) and (3) were found to be the two major reactions with $k_2/k_3 \simeq 10$. In a time-of-flight instrument[1] in which the O^- ions were believed to be thermal, $k_2/k_3 \simeq 33$ and other reactions were again of minor importance. In the present work, the O^- ions produced from N_2O probably had an average translational energy of about 0·38 eV,[13] since the pressure was too low for any thermalizing collisions to occur, and $k_2/k_3 \simeq 17$. Owing to the low yield of C_2OH^- ions, this ratio was not studied using other sources of O^- ions.

The variation of k_2/k_1 with translational energy of the O^- ion is in qualitative agreement with results in the literature. There is now a body of evidence to suggest that k_1 falls with increasing O^- translational energy;[14] for thermal O^- ions, $k_1 = 2·5 \times 10^{-10}$ cm³ molecule⁻¹ sec⁻¹, falling to $\sim 7 \times 10^{-11}$ cm³ molecule⁻¹ sec⁻¹ for a translational energy of 0·38 eV.[14] On the other hand, k_2 has been shown to increase with increase in O^- translational energy,[1] so k_2/k_1, should certainly rise as the translational energy rises. However, if one combines the data from References 1 and 14, k_2/k_1 (thermal) $\simeq 5·6$ and k_2/k_1 $(E_{tr}(O^-) = 0·38$ eV$) = 37$, so that the ratio increases by a factor of $\sim 6·6$ over this energy range. The present results give a factor of only 1·6, arising from the much slower decrease in k_1 observed in this study compared with that found in Reference 14.

The absolute value of k_2 obtained in this work, $8·6 \times 10^{-10}$ cm³ molecule⁻¹ sec⁻¹, is in good agreement with the values of $1·07 \times 10^{-9}$ and $1·4 \pm 0·3 \times 10^{-9}$ cm³ molecule⁻¹ sec⁻¹ obtained by other techniques.[1,7] The value found for k_3 is in excellent agreement with the result obtained using at ime-of-flight instrument,[1] but is considerably lower than that obtained in a tandem mass spectrometer.[7]

Although an early study of the O^-/C_2H_4 system using the flowing afterglow technique appeared to suggest that no secondary ions are formed,[3] the results obtained in the present work are in agreement with more recent results obtained in time-of-flight[1] and tandem mass spectrometers.[7] The results of the deuterium labelling experiments indicate clearly that $C_2H_2^-$ is formed and that its structure at the moment of formation is probably CH_2C^-. The value of $k_5/k_6 = 7$ found in the present work may be compared with the value of 4 obtained in a tandem instrument, using ions of 0·3 eV translational energy. The absolute value of k_5 is again in excellent agreement with values previously obtained of $2·4 \times 10^{-10}$ and $\leq 4·2 \times 10^{-10}$ cm³ molecule⁻¹ sec⁻¹. The substantially lower value found for k_5' in the O^-/C_2D_4 system cannot be compared with a literature value, but the lower value found on deuteration is consistent with the trend observed in the O^-/C_2H_2 and O^-/C_2D_2 systems for reaction (2).[1] Similarly, there are no other values of k_8 in the literature, and although one might have expected k_8 to be less than k_5 on grounds of symmetry, this appears to be more than compensated for by the polar nature of the molecule.

In the O^-/propylene system, both reactions (9) and (10) were observed, but the obtaining of relative rate constants for these two reactions was not attempted owing to the large difference in masses of the product ions and the uncertainty of the corrections involved. An estimate of k_9 was obtained, however, by comparing the peaks at m/e = 41^- and 46^- in the NO_2/ CH_3CHCH_2 system, yielding $k_9 = 1{\cdot}8 \times 10^{-10}\,cm^3$ molecule^{-1} sec^{-1}. This is substantially higher than the value of $\sim 5 \times 10^{-11}\,cm^3$ molecule^{-1} sec^{-1} obtained using the flowing afterglow technique,[3] and considerably higher than a value of $\sim 1 \times 10^{-11}\,cm^3$ molecule^{-1} sec^{-1} obtained in a tandem instrument.[7] It is possible that part of the differences arises from variations in the translational energies of the O^- ions, but it is unlikely that this effect alone can explain the differences.

We thank the Science Research Council for financial support during the course of this work.

REFERENCES

1. Stockdale, J. A. D., Compton, R. N. and Reinhardt, P. W., *Int. J. Mass Spectrom. Ion Phys.*, 1970, **4**, 401.
2. Bohme, D. K. and Fehsenfeld, F. C., *Can. J. Chem.*, 1969, **47**, 2717.
3. Bohme, D. K. and Young, L. B., *J. Amer. Chem. Soc.*, 1970, **92**, 3301.
4. Neuert, H., Rackwitz, R. and Vogt, D., *Advances in Mass Spectrometry*, 1968, **5**.
5. Ferguson, E. E., *Can. J. Chem.*, 1969, **47**, 1815.
6. Paulson, J. F., 'Ion Molecule Reactions', *ed.* J. L. Franklin, Butterworths, London, 1972, chapter 4.
7. Futrell, J. H. and Tiernan, T. O., Ibid, chapter 7.
8. Ferguson, E. E., *Advances Electron. Electron Phys.*, 1968, **24**, 1.
9. Goode, G. C., O'Malley, R. M., Ferrer-Correia, A. J. and Jennings, K. R., *Chem. Brit.*, 1971, **7**, 12.
10. Goode, G. C., O'Malley, R. M., Ferrer-Correia, A. J., Massey, R. I., Jennings, K. R., Futrell, J. H. and Llewellyn, P. M., *Int. J. Mass Spectrom. Ion Phys.*, 1970, **5**, 393.
11. Beauchamp, J. L. and Armstrong, J. T., *Rev. Sci. Inst.*, 1968, **39**, 1772.
12. Goode, G. C., Ferrer-Correia, A. J. and Jennings, K. R., *Int. J. Mass Spectrom. Ion Phys.*, 1970, **5**, 229.
13. Chantrey, P. J., *J. Chem. Phys.*, 1969, **51**, 3369, 3380.
14. Marx, R., Mauclaire, G., Fehsenfeld, F. C., Dunkin, D. B. and Ferguson, E. E., *J. Chem. Phys.*, 1973, **58**, 3267.
15. Chantrey, P. J. and Schulz, G. J., *Phys. Rev.*, 1967, **156**, 134.

Discussion

J. Durup (Université de Paris-Sud, Orsay, France): If actually the $C_2H_2^-$ ion is not stable in the CH–CH geometry, one has a unique method for determining unambiguously the structure of positive ions such as $C_2H_2^+$: we have shown that double charge transfer of positive molecular ions leading to negative molecular ions by high-energy collisions at zero-scattering angles takes place without any change in geometry (Franck-Condon transition). Therefore in such double-charge-transfer experiments a CH_2C^+ ion would lead to stable CH_2C^- whereas a $CHCH^+$ ion would lead to no $C_2H_2^-$ ion. This technique might be of very wide use.

K. R. Jennings: We interpret our results as suggesting that, at the moment of formation, the $C_2H_2^-$ ion has the structure CH_2C^-, and we know of no observation where the symmetrical structure, $CHCH^-$, has been supported. One cannot rule out the possibility that the symmetrical ion may exist under other experimental conditions but on present experimental evidence, the experiment which you suggest should be valid.

C. Lifshitz (Hebrew University, Jerusalem, Israel): It has been assumed by Stockdale *et al.* that O^-/NO_2 is formed with a considerable amount of translational energy. This is obtained if the excess energy, in the dissociative electron capture, is distributed between the fragments as in a quasi-diatomic molecule. Would your results suggest, therefore, that on the contrary, most of the energy ends up as internal energy of the NO?

K. R. Jennings: Stockdale *et al.* suggest that if one uses the first peak (1·9 eV) for O^- production from NO_2, the O^- translational energy is $\leq 0·3$ eV, and they discuss their results in terms of thermal O^- ions. (Int. J. Mass Spectrom. Ion Phys (1970) **4**, 401.) Our results obtained with the $CO/NO_2/C_2H_2$ system appear to support this and any excess energy must therefore, as you suggest, end up primarily as internal energy of the NO molecule.

95
Gas-Phase Ion-Molecule Interactions: Energy Transfer and Primary Product Ion Excitation in Dissociative Charge-Transfer Reactions

By ROY S. LEHRLE

(*Chemistry Department, University of Birmingham, Birmingham, U.K.*)

and JOHN E. PARKER

(*Chemistry Department, Heriot-Watt University, Edinburgh, U.K.*)

INTRODUCTION

DISSOCIATIVE charge-transfer reactions considered here involve the interaction of a translationally energetic ion with a polyatomic molecule. An electron exchange process converts the polyatomic molecule into a polyatomic ion, which may be internally excited, and which may dissociate in various ways to produce fragment ions. The objective in the present work is to obtain information about the energy transfer and fragmentation processes in such dissociative charge-transfer reactions.

Previous work on methane[1] indicated that for a specified reactant ion the most probable fragmentation of the methane is that for which the appearance potential of the fragment is closest to the recombination energy of the reactant ion. This type of behaviour may be described as 'internal energy resonant'. For the methane systems the translational energy of the reactant ion has little influence upon the fragmentation channels, except for endothermic channels which exhibit a translational energy threshold.[1] However, the cross-sections for charge-transfer reactions are often strongly dependent on the relative translational energy of the reactants. This dependence has been interpreted in terms of the 'Near-adiabatic theory',[2] which has been extensively applied to simple charge-transfer systems.[3] A model developed by Rosenstock *et al.*[4] may apply when interpreting the fragmentation of larger target molecules. This 'Quasi-Equilibrium' theory assumes that the initial energy transfer process (*e.g.* charge transfer) produces a polyatomic ion of a particular internal energy for which many excited states of the ion can exist, each with the same total energy but with the energy distributed in different ways. The assumption is made that the primary product ions are

formed with a random distribution amongst these states, and that the population of ions in each state is unaffected when those states corresponding to
decomposition channels disappear to give product ions. Thus there is a
quasi-equilibrium between the states in each ion.

The present work sets out to (a) assess any contribution of internal energy
resonance, (b) to examine the energetics of the observed fragmentations
with a view to obtaining information about the internal energy distribution,
and (c) to investigate the influence of translational energy in an attempt to
understand the energy transfer processes.

EXPERIMENTAL

A reactant ion beam of specified translational energy (100–2000 eV) interacts
with reactant gas in the source region of a modified Bendix time-of-flight
mass spectrometer. The following reactant ions were used: $Xe^+(^2P_{3/2}$,
12·13 eV; $^2P_{1/2}$, 13·44 eV), Kr^+ ($^2P_{3/2}$, 14·0 eV; $^2P_{1/2}$, 14·67 eV) and
CO^+ ($^2\Sigma^+$, 14·0 eV). Similar experiments were performed using a translationally energetic beam of He atoms (600–2000 eV). The helium neutral
beam was formed by charge transfer of a He^+ ion beam with thermal energy
helium gas, any ions remaining in the beam being then removed by a transverse electric field. The helium beam contained not only ground state atoms
($1^1 S_0$), but also metastable excited atoms He (2^3S_1, 19·82 eV) and He
(2^1S_0, 20·61 eV). (The figures in the brackets refer to the recombination energy
of the ions and the excitation energy of the neutral atom.)

The source region of the mass spectrometer contained propane gas at
thermal energies, and the length of the reaction zone was defined by the
geometry of the total charge collection system.[5] The reactant ion currents
were of the order of 10^{-8} amp and the propane pressures were ca. 10^{-5} torr;
under these conditions the product ion currents were first order in reactant
gas pressure and reactant ion current. At each specified reactant ion translational energy the total reaction cross-section (Q) for the formation of all
product ions was measured by the total charge collection technique.[3] The
apparatus was designed so that the slow product ions could alternatively be
swept out of the reaction zone and mass analysed by the time-of-flight
method.

The electron impact energy used for the production of all reactant ions
was 20 eV. The reproducibility of the total reaction cross-section was ca.
$\pm4\%$ for the Kr^+/C_3H_8 and CO^+/C_3H_8 systems and ca. $\pm7\%$ for the $Xe^+/$
C_3H_8 system. The mass analysed relative abundances were reproducible to
ca. $\pm3\%$ for CO^+/C_3H_8 and Xe^+/C_3H_8 and ca. $\pm6\%$ for the Kr^+/C_3H_8
system.

RESULTS AND DISCUSSION

The appearance potentials (eV) for propane and its fragments are as follows:
$C_3H_8^+$, 11·2; $C_3H_7^+$, 12·1; $C_3H_6^+$, 12·2; $C_3H_5^+$, 13·7; $C_3H_4^+$, 14·8;
$C_3H_3^+$, 17·0; $C_2H_5^+$, 12·0; $C_2H_4^+$, 11·7; $C_2H_3^+$, 14·0; $C_2H_2^+$, 14·5;

CH_3^+, 15·2. (All values are taken from Reference 6 except those for $C_3H_4^+$ and CH_3^+, which were taken from References 7 and 8 respectively.)

Using these values, and the recombination energy values listed in the previous section, graphs of relative abundance against $\Delta\varepsilon(=$RE–AP$)$ were plotted for all fragments from all the reaction systems. A peak at $\Delta\varepsilon = 0$ in such a plot would indicate that internal energy resonance is the dominant energy transfer and fragmentation mechanism.[1] No indication of such a peak was obtained; the graph showed a great scatter in the $\Delta\varepsilon =$ positive region, and low values in the $\Delta\varepsilon =$ negative region. On this basis it can be concluded that simple internal energy resonance is not a useful model in interpreting these propane dissociative charge transfer systems.

Graphs of fragment relative abundance against corresponding fragment appearance potential were plotted for all fragments for each of the reaction systems. (Relative translational energy 2000 eV in each case.) Each of these curves (including that for the helium neutral system) displayed maxima at energies 0·8, 2·8 and $\geq$5·8 eV above the ionization potential of C_3H_8. The same maxima were evident in corresponding plots of electron-impact results obtained with both 70 eV and 20 eV electrons. The relative heights of the three peaks depended on the RE of the ion, the excitation energy of the neutral, or the kinetic energy of the electron, but the locations of the peaks were always identical on the AP axes. The decomposition channels clearly lie in three discrete energy regions and the results show that the predominant fragmentation process in each region is the same for each of the reaction systems. Although these findings could be discussed in terms of a simple quasi-equilibrium model, it may alternatively be proposed that the primary product ion $C_3H_8^+$ is being formed in three predominant electronic states which may be populated to varying degrees by different energy transfer species. Variation of the relative translational energy appeared to have little influence on this distribution pattern for internal energy. This was demonstrated by plotting similar graphs for all the systems at various relative translational energies in the range 100–2000 eV; the observed changes in population of the levels were very small (ca. $\pm 2\%$) and thus the general picture was unchanged.

On the other hand the total reaction cross-section (Q) displayed significant variation with translational energy; indeed for each of the systems plots of Q against translational energy showed several maxima. The translational energies corresponding to these maxima are listed below:

Kr^+/C_3H_8: 300, 500, 850, 1600, $\geq$2000 eV
CO^+/C_3H_8: $\leq$100, 300, 800, 1200, 1500, 1800 eV
Xe^+/C_3H_8: 300, 950, 1350, 1900 eV

Thus the total probability of dissociative charge transfer is enhanced at certain relative translational energies, although (as concluded in the previous paragraph) the internal energy distribution pattern changes insignificantly with translational energy. This indicates that the energy *transfer* process is influenced by translational energy even though the predominant *distribution* of energy is not. In an attempt to understand this situation it has been assumed that C_3H_8 can accept energy at any level corresponding to its ionization potential or the appearance potentials of its fragments. The near-adiabatic theory[2,3] has been applied to each of the above maxima, and in every case it has been possible to define one of the C_3H_8 energy acceptance levels which corresponds to a value in the range 6–8 Å for the near-adiabatic parameter a. Thus all maxima can be interpreted in terms of energy transfer processes leading to the $C_3H_8{}^+$ ion in known excited (decomposition) states. However, since the present results provide no evidence that the relative decomposition modes are strongly influenced by translational energy, it must be concluded that considerable randomization of energy occurs subsequent to the transfer process.

CONCLUSIONS

This work indicates that simple internal energy resonance is not important in determining the fragmentation in dissociative charge transfer reactions of propane. The observations indicate that three types of fragmentation reaction, with discrete internal energy requirements, are predominant whichever reactant ion (or bombarding particle) is used. It is suggested that this could arise if the ion-molecule interaction processes tended to form $C_3H_8{}^+$ in three predominant electronic states which are populated to varying degrees by the different energy transfer species. The influence of translational energy on the total reaction cross-sections has been interpreted in terms of the near-adiabatic theory, and it is concluded that translational energy thus converted must be randomized since it does not strongly favour particular fragmentation modes.

REFERENCES

1. Homer, J. B., Lehrle, R. S., Robb, J. C. and Thomas, D. W., *Trans. Farad. Soc.*, 1966, **62**, 619.
2. Massey, H. S. W. and Burhop, E. H. S., 'Electronic and Ionic Impact Phenomena', Clarendon, Oxford, 1952, p. 441.
3. Hasted, J. B., *Adv. Electron. Electron Phys.*, 1960, **13**, 1.
4. Rosenstock, H. M., Wallenstein, M. B., Wahrhaftig, A. L. and Eyring, H., *Proc. Natl. Acadm. Sci.*, *U.S.*, 1952, **38**, 667.
5. Lehrle, R. S., Parker, J. E., Robb, J. C. and Scarborough, J., *J. Mass Spectrom. Ion Phys.*, 1968, **1**, 455.
6. Pettersson, E. and Lindholm, E., *Arkiv Fysik*, 1963, **24**, 49.
7. Field, F. H. and Franklin, J. L., 'Electron Impact Phenomena', Academic Press, New York, 1957.
8. Franklin, J. L., Dillard, J. G., Rosenstock, H. M., Herron, J. T., Draxl, K. and Field, F. H., National Bureau of Standards, NSRDS-NBS 26, Washington, 1969.

96

The Formation and Reaction of Electronically Excited $(N_2^{+\cdot})^*$ Ions

By P. R. KEMPER and M. T. BOWERS

(*Department of Chemistry, University of California,*
Santa Barbara, California, U.S.A.)

INTRODUCTION

THE charge transfer reaction between $He^{+\cdot}$ and N_2, reaction (1),

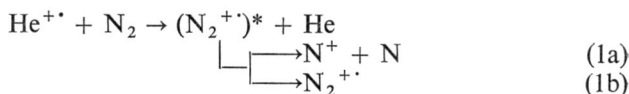

$$He^{+\cdot} + N_2 \rightarrow (N_2^{+\cdot})^* + He$$
$$\longrightarrow N^+ + N \tag{1a}$$
$$\longrightarrow N_2^{+\cdot} \tag{1b}$$

must be one of the most studied ion molecule reactions.[1-12] Interest has centred on the competition between pathways (a) and (b) in (1). There has been considerable speculation with regard to the details of the reaction mechanism, in particular the nature of the excited state(s) of $(N_2^{+\cdot})^*$ initially formed in the reaction as well as the dissociative state(s) of $N_2^{+\cdot}$ that eventually lead to the N^+ product. Most early experimental data had been consistent with the notion that the $C^2\Sigma_u^+$ state of $N_2^{+\cdot}$ is formed by the charge transfer and competition exists between emission to the $X^2\Sigma_g^+$ ground state of $N_2^{+\cdot}$ and dissociation to the $N^+(^3P)$ and $N(^4S)$ products via the $^4\Pi_u$ state of $N_2^{+\cdot}$.[13] Emission from the $C^2\Sigma_u^+$ state has been observed by a number of workers.[11,14,15] The $^4\Pi_u$ state was chosen as the most likely candidate leading to dissociation because early approximate potential curves by Gilmore[16] indicated it intersected the $C^2\Sigma_u^+$ state at approximately the recombination energy of $He^{+\cdot}$. Both processes are expected to occur within 10^{-8} sec after formation of $(N_2^{+\cdot})^*$. This mechanism suggests the k_{1a}/k_{1b} ratio reflects only the intramolecular properties of $(N_2^{+\cdot})^*$ and subsequent collisions of $(N_2^{+\cdot})^*$ with both gases should play little or no role in the measured ratio. Measured ratios of k_{1a}/k_{1b} have varied from *ca.* $1 \cdot 0^4$ to $2 \cdot 2$.[8] The total rate constant, $k_1 = k_{1a} + k_{1b}$, has been measured numerous times[1,4,12,17,18] with values between $1 \cdot 2$ and $1 \cdot 7 \times 10^{-9}$ cm³/s indicating charge transfer is facile and occurs on nearly every collision.

In this paper, a report on the dependence of k_{1a}/k_{1b} on pressure is given. These data suggest a mechanism entirely different from the previously accepted mechanism must dominate reaction (1). A much more thorough report including the dependence of k_{1a}/k_{1b} on kinetic energy and on reaction time is given elsewhere.[19]

EXPERIMENTAL

The experiments were performed on a laboratory built drift cell ICR spectro-meter previously described.[20,21] Pressures were measured on a MKS Baratron Capacitance Manometer or on an Ion Gauge calibrated against the Baratron. The details of the experimental technique used are given elsewhere.[19]

RESULTS

The total thermal rate constant for reaction (1) has been measured a large number of times in our laboratories yielding $k_1 = 1.5 \pm 0.3 \times 10^{-9}$ cm^3/s. This value is in good agreement with other workers who report values from 1.2–1.7×10^{-9} cm^3/s.[1,4,12,17,18] The theoretical capture rate constant is 1.65×10^{-9} cm^3/s indicating reaction occurs on nearly every collision.

The $N^+/N_2^{+\cdot}$ ratio resulting from reaction (1) for $^{14}N^{14}N$ has been

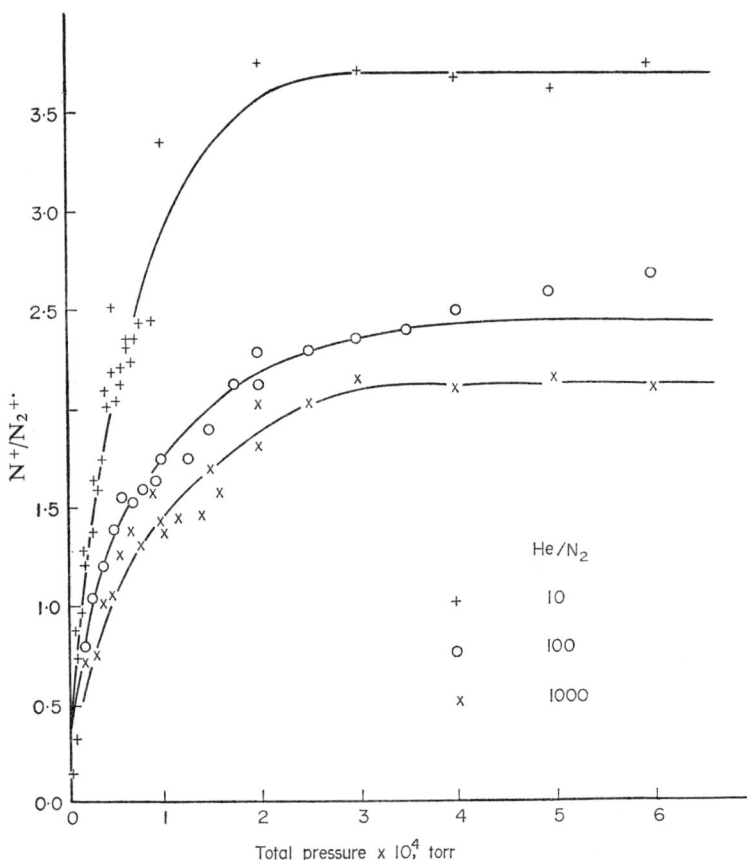

FIG. 1   Plot of the experimental $N^+/N_2^{+\cdot}$ ratio for $^{14}N^{14}N$ as a function of total pressure for three separate values of the He/N$_2$ mixing ratio.

measured as a function of total pressure and He/N$_2$ mixing ratio. The results are presented in Fig. 1 (see also Table I). Complementary data have been reported for $N^+/N_2^{+\cdot}$ *versus* reaction time.[19]

We have also observed similar pressure dependent results for He$^{+\cdot}$

TABLE I

Rate Parameters Used to Fit the Curves in Fig. 1

k_q	$(1\cdot2-1\cdot5) \times 10^{-9}$ cm^3/s
k_d	$(5-10) \times 10^{-11}$ cm^3/s
k_1	$(0\cdot5-1\cdot5) \times 10^{-9}$ cm^3/s
k_2	$(3-6) \times 10^{-10}$ cm^3/s
$k_{2'}$	$(1-2) \times 10^{-10}$ cm^3/s
k_3	$(1-10) \times 10^8$ sec^{-1}
k_t	$(1-3) \times 10^2$ sec^{-1}
k_c	0

reacting with $^{15}N^{15}N$. The absolute value of the N^+/N_2^+ ratio is significantly different, however, as the data in Fig. 2 indicate. Over the range of He/N$_2$ ratios of 10 to 1000 the ratio of the N^+/N_2^+ signals for $^{14}N^{14}N$ is consistently larger than the $^{15}N^{15}N$ data by a factor of $1\cdot55 \pm 0\cdot05$.

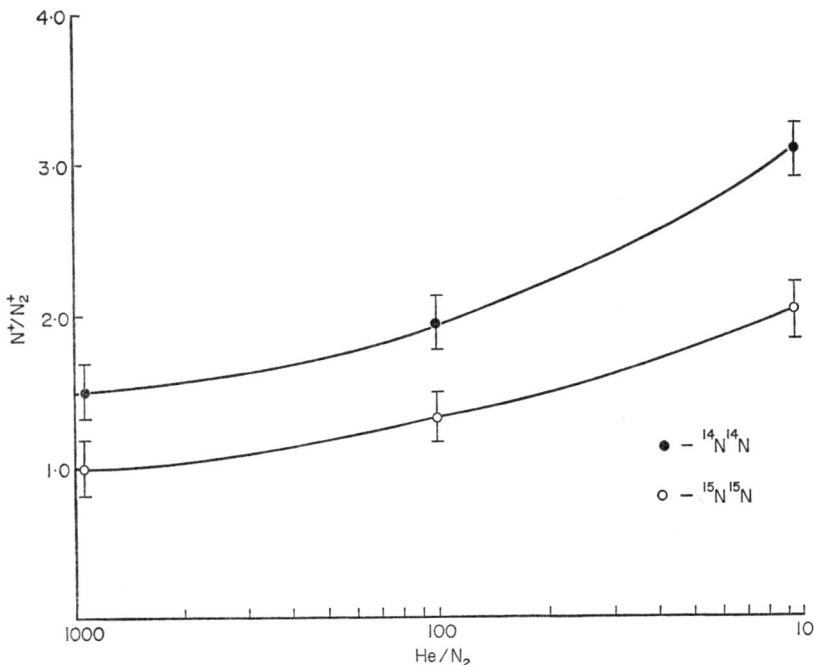

FIG. 2   A plot of the ratio of $N^+/N_2^{+\cdot}$ as a function of He/N$_2$ mixture ratio for both $^{14}N^{14}N$ and $^{15}N^{15}N$. The total pressure was *ca.* 2×10^{-4} torr for all measurements.

DISCUSSION

The experimental data suggest

(1) the distribution of $(N_2^+)^*$ ions initially formed from $He^{+ \cdot}$. Charge transfer is dominated by electronic states with lifetimes against either emission or dissociation of at least 10^{-4}–10^{-3} sec;

(2) the substitution of $^{15}N^{15}N$ for $^{14}N^{14}N$ has a considerable quantitative effect on the N^+/N_2^+ ratio but the qualitative behaviour is similar.

Our interpretation of these phenomena is that the quartet manifold of states, shown in Fig. 3b, receives the majority of the $N_2^{+ \cdot}$ ions from the charge transfer reaction while the $C^2\Sigma_u^+$ state is formed in, at most, 10% of the collisions. The following mechanism is suggested and the appropriate differential equations have been solved.

This mechanism very nicely fits the experimental data for the range of parameters given in Table I. These parameters in all cases are in agreement with those calculated from simple collision theory[20] or from those measured by experiment.[23-26]

An interesting consequence of these results is that spin does not seem to be conserved. Reaction (2) seemingly dominates over reaction (3),

$$He^{+ \cdot}(^2S) + N_2(^1\Sigma_g^+) \rightarrow N_2^{+ \cdot}(^4\Pi_u) + He(^1S) \tag{2}$$

$$He^{+ \cdot}(^2S) + N_2(^1\Sigma_g^+) \rightarrow N_2^{+ \cdot}(C^2\Sigma_u^+) + He(^1S) \tag{3}$$

a phenomenon not usually observed in ion molecule collisions.[27] Little is known about spin conservation in thermal energy charge transfer reactions, however, and this interesting point deserves further study. There does appear to be at least one other case in which spin does not appear to be conserved, the similar system of $Ne^{+ \cdot}$ undergoing charge transfer with N_2 where the product $N_2^{+ \cdot}$ state appears to be in the quartet manifold.[28]

The existence of long lived $(N_2^+)^*$ states with lifetimes greater than 1×10^{-3} sec has been clearly demonstrated by the observation of reaction (4) in the ICR spectrometer[26,29]

$$(N_2^+)^* + N_2 \rightarrow N_3^+ + N \tag{4}$$

where the reactant $(N_2^+)^*$ ions have 5–6 eV internal energy. These results are discussed in detail elsewhere[26] but serve to support the interpretation of the charge transfer mechanism suggested here.

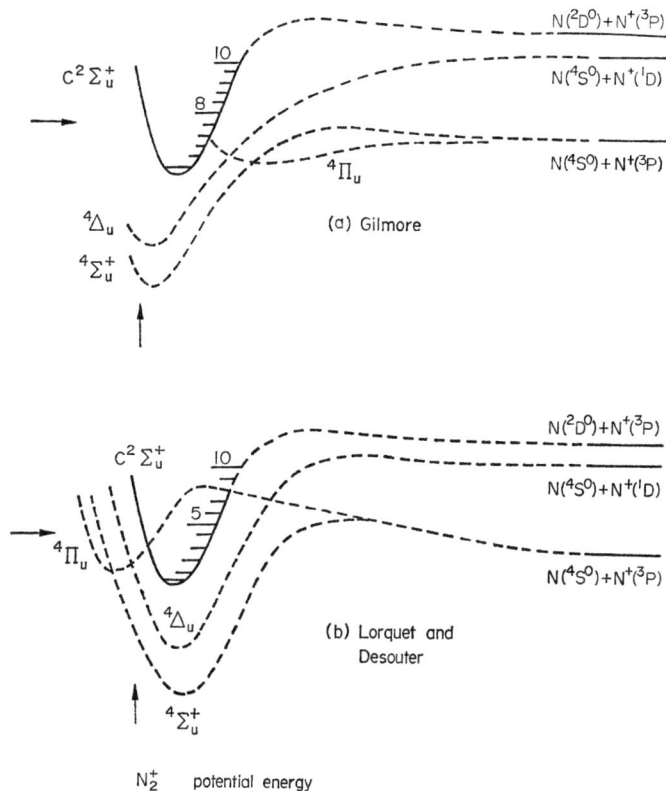

FIG. 3   Potential energy curves for selected excited states of $N_2^{+\cdot}$. The horizontal arrows correspond to the recombination energy of He$^{+\cdot}$. The vertical arrows correspond to the equilibrium ground state N_2 separation. The curves in (a) are taken from Gilmore, Ref. 16—the dashed lines indicate uncertainty. Those in (b) are taken from Lorquet and Desouter, *Chem. Phys. Letters*, 1972, **16**, 136.

ACKNOWLEDGMENT

The support of the National Science Foundation under Grant GP-15628 is gratefully acknowledged.

REFERENCES

1. Sayers, J. and Smith, D., *Disc. Farad. Soc.*, 1964, **37**, 167.
2. Ferguson, E. E., Fehsenfeld, F. C., Schmeltekopf, A. L. and Schiff, H. I., *Planet Space Sci.*, 1964, **12**, 1169.
3. Fehsenfeld, F. C., Schmeltekopf, A. L., Goldan, P. D., Schiff, H. I. and Ferguson, E. E., *J. Chem. Phys.*, 1966, **44**, 4087.
4. Warnek, P., *J. Chem. Phys.*, 1967, **47**, 4279.
5. Moran, T. F. and Friedman, L., *J. Chem. Phys.*, 1966, **45**, 3837.
6. Heimerl, J., Johnson, R. and Biondi, M. A., *J. Chem. Phys.*, 1969, **51**, 5041.
7. Stebbings, R. F., Smith, A. C. H. and Ehrhardt, H., *J. Chem. Phys.*, 1963, **39**, 968.
8. Ferguson, E. E., *Adv. Electronics and Elect. Phys.*, 1968, **24**, 1.

9. Schmeltekopf, A. L., Ferguson, E. E. and Fehsenfeld, F. C., *J. Chem. Phys.*, 1968, **48**, 2966.
10. Stebbings, R. F., Rutherford, J. A. and Turner, B. R., *Planet. Space Sci.*, 1965, **13**, 1125.
11. Inn, E. C. Y., *Planet. Space Sci.*, 1967, **15**, 19.
12. Dunkin, D. B., Fehsenfeld, F. C., Schmeltekopf, A. L. and Ferguson, E. E., *J. Chem. Phys.*, 1968, **49**, 1365.
13. For a discussion of the development of this mechanism, see Ref. 11.
14. Carroll, P. K., *Can. J. Phys.*, 1959, **37**, 880, and references therein.
15. Ferguson, E. E. and Fehsenfeld, F. C., private communication.
16. Gilmore, F. R., *J. Quant. Spectr. Radiative Transfer*, 1965, **5**, 369.
17. Farragher, A. L., *Trans. Faraday Soc.*, 1970, **66**, 1411.
18. Laudenslager, J. B. and Bowers, M. T. (to be published).
19. Kemper, P. R. and Bowers, M. T., *J. Chem. Phys* (in press).
20. Anicich, V. G. and Bowers, M. T., *Int. J. Mass Spectrom. Ion Phys.*, 1973, **11**, 329.
21. Anicich, V. G., Ph.D Thesis, University of California at Santa Barbara, Santa Barbara, California.
22. Gioumousis, G. and Stevenson, D. P., *J. Chem. Phys.*, 1958, **29**, 294.
23. Cress, M. C., Becker, P. M. and Lampe, F. W., *J. Chem. Phys.*, 1966, **44**, 2212.
24. Asundi, R. K., Schultz, G. J. and Chantry, P. J., *J. Chem. Phys.*, 1967, **47**, 1584.
25. Ryan, K. R., *J. Chem. Phys.*, 1969, **51**, 570.
26. Bowers, M. T. and Kemper, P. R., to be published.
27. Durup, J., Sixth International Mass Spectrometry Conference, Edinburgh, Scotland (1973), p. 691.
28. Ferguson, E. E., Paper presented at the Twenty-fifth Gaseous Electronics Conference, London, Ontario, Canada, October 1972.
29. Jaffe, S., Karpas, Z. and Klein, F., *J. Chem. Phys.*, 1973, **58**, 2190.

Discussion

J. Durup (Université de Paris-Sud, Orsay, France): It would be rather unexpected that the spin-conservation rule would be violated in He^+—or—N_2 charge transfer. This cross section would decrease if one slightly heated the He^+ ions, since the lifetime of any collision complex would become very short and a spin change very unlikely. Can this test be done? In addition the same phenomenon (decrease of N^+/N_2^+ ratio with decreasing pressure) was reported many years ago by Comes and Lessmann using a photo-ionization source, where no quartet N_2^+ would be produced.

M. T. Bowers: It is our opinion that the factors that govern thermal energy charge transfer reactions are not well understood. It is, to us, conceivable that spin is not conserved in these reactions. A second reaction, that of $Ne^{+\cdot}$ with vibrationally excited N_2 molecules, has been reported that also appears to violate the spin-conservation rule (Ferguson and co-workers, 25th Gaseous Electronics Conference, London, Ontario, Canada, Oct. 1972; *J. Chem. Phys.*, Spring, 1973). We have observed little charge in the total $He^{+\cdot} + N_2$ charge transfer rate constant at energies to 1 eV c.m. There is a small but measureable increase in the $N^+/N_2^{+\cdot}$ ratio over the energy range 0–1 eV. Thus, no dramatic effects are observed at low translational energies.

The photoionization experiments of Comes and Lessmann (*Z. Naturforsch*, 1962, 1964) bear little relation to the charge transfer studies presented here. They observed a linear increase of the $N^+/N_2^{+\cdot}$ ratio from zero to *ca.* 4% as the pressure in their same increased. This increase was ascribed to a collisional dissociation of the sort

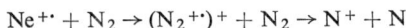

$$Ne^{+\cdot} + N_2 \rightarrow (N_2^{+\cdot})^+ + N_2 \rightarrow N^+ + N$$

where the excess energy is derived from the relative translational energy of the reactants. In our experiments, $N^+/N_2^{+\cdot}$ varies from *ca.* 0·1 to *ca.* 3·0 as a function of pressure under conditions where the ionic species have nearly thermal translational energies.

97

High Energy Ion Molecule Reactions: Charge Exchange 20/11

By T. AST, R. G. COOKS and J. H. BEYNON

(*Department of Chemistry, Purdue University, W. Lafayette, Indiana, U.S.A.*)

Ion kinetic energy spectrometry (IKES)[1] may be applied to the study of both unimolecular and bimolecular ionic reactions. The latter include collision-induced dissociation reactions as well as various charge transfer reactions occurring at high (500 eV–10 000 eV) translational energies. This paper is concerned with charge exchange reactions of doubly charged ions. That is 20/11 reactions[2] of the type

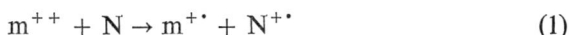

$$m^{++} + N \rightarrow m^{+\cdot} + N^{+\cdot} \qquad (1)$$

where m^{++} is a rare gas ion and the collision partner is a monatomic or diatomic neutral species. Our interest in this reaction stems in part from the fact that it forms the basis for determining the peaks due only to doubly charged ions ($2E$ spectra) in the mass spectra of complex molecules.[3]

Experiments were done using the Hitachi/Perkin–Elmer RMH-2 mass spectrometer modified for ion kinetic energy work.[4] The collision gas was introduced into the first field-free region of the instrument at a pressure of approximately 4×10^{-5} torr. The electric sector voltage was set to twice the value, E, normally used for plotting mass spectra. At the electric sector setting $2E$, only singly charged ions produced from doubly charged ions in the charge-exchange reaction shown in (1) above are recorded. All such reactions do involve a small change in the translational energy of the product ion relative to that of the reactant ion so that if a narrow energy-resolving (β) slit is used, ions are not transmitted *exactly* at the value $2E$. An alternative method that enables these small energy differences to be observed involves scanning the accelerating voltage over a narrow range in order to provide the product ions with just the energy required for them to be transmitted by the electrostatic sector when set at the value $2E$. The plots of ion abundance versus ion accelerating voltage give, to a good approximation, the kinetic energy loss (or gain) during reaction.[1] This presupposes (i) that the translational energy of the reactant ion is high compared to the energy differences being observed and (ii) that the angle of scattering in the reaction is small.

The experimental techniques used in the present study as well as the assumptions used in interpreting the kinetic energy distributions owe much to a related study on the charge stripping reaction (10/20) of rare gas ions.[5]

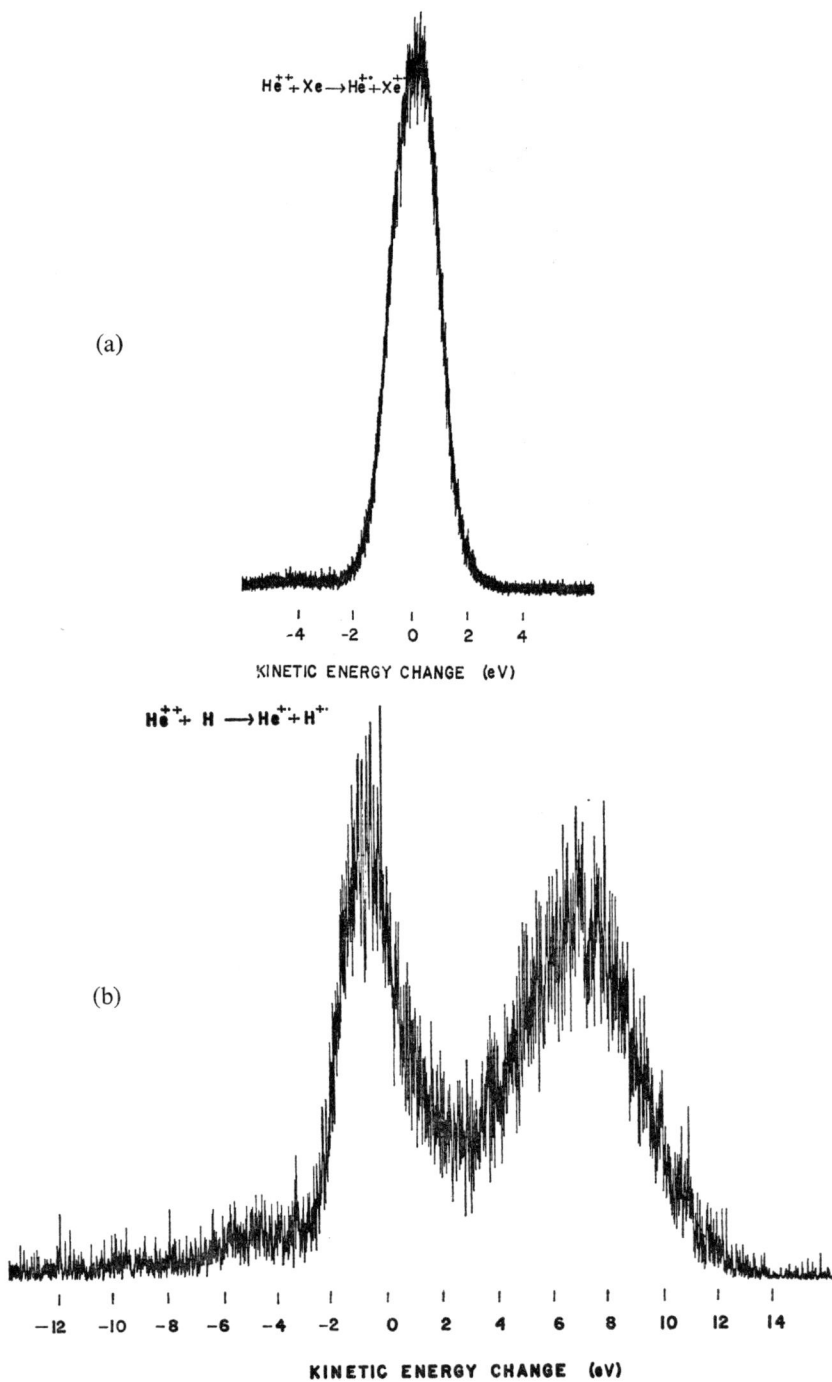

FIG. 1　Kinetic energy spectrum of He$^{+\cdot}$ ions produced by charge exchange of He^{++} ions with (a) Xe and (b) H$_2$.

The charge exchange reactions studied showed complex kinetic energy loss (gain) curves, often with several maxima. Here we present the relevant data and discuss in detail the origin of typical peaks. Figure 1 illustrates the sensitivity obtained on a single scan and the peak shapes observed for charge exchange of He^{++} with two different target species.

The measured voltage corresponding to zero change in translational energy can be affected by any small contact potential variations on the electric sector walls. During the course of an experiment, the measured electric sector voltage necessary to transmit the main ion beam was plotted as a function of accelerating voltage. Any small departure from a linear relationship could be used to calculate the true value of electric sector voltage. The observed variation in the zero position was ± 0.2 eV and this is the major source of uncertainty in the kinetic energy measurements.

TABLE I[a]

Kinetic Energy Loss Peaks $He^{++} \to He^{+\cdot}$ ($^2S_{1/2}$)

Collision gas	Position[b]	Energy available[c]	IP[d]
He	-10.3^e	24.0	24.6
Ar	-1.7	15.4	15.5, 15.7
Kr	-0.8	14.5	14.0, 14.7
Xe	0.0	13.7	12.1, 13.4
H_2	-1.5	15.2	15.4

[a] All data in electron volts.
[b] Peak centre.
[c] Minus second column + 13.7 eV.
[d] Two values given when ground state is a doublet.
[e] A very weak, broad peak.

Results for helium are given in Table I. This case is typical and shows many of the features seen in experiments using the other rare gases. He^{++}, having no orbiting electrons, has no excited states and, in general, the processes observed in all the rare gases can be explained as occurring from the ground states of the reactant doubly charged ions. This is not surprising in view of the method of ionization which uses electrons of 120 eV energy, but it contrasts with the results obtained on charge stripping[5] where long-lived excited states of singly charged ions made a substantial contribution to the reactions observed after ~ 10 μsec. The energy liberated by the helium ion on charge exchange is either 54.3 eV or 13.7 eV[6] depending on whether the product $He^{+\cdot}$ is formed in the ground or first excited state. All higher states of $He^{+\cdot}$ make too little energy available to ionize the collision gas.

An extra source of energy is the translational energy of the reactant ion. This can supplement the electronic energy available from the He^{++} in such a way as to provide an exact match with the energy required to drive the charge transfer process. It is found that the heights of the peaks due to charge transfer for various collision gases increase as the amount of supplementary energy necessary decreases. Many of the energy loss peaks observed for He^{++}

are due to the reaction

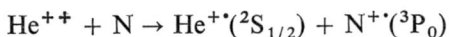

$$He^{++} + N \rightarrow He^{+\cdot}(^2S_{1/2}) + N^{+\cdot}(^3P_0)$$

and in the case where Xe is the collision gas, the energy loss for this process is near zero, so that the observed transition is comparatively intense. This is shown in Fig. 1(a). It is also noteworthy that the widths of these energy loss peaks vary inversely with the magnitude of the energy loss. In the case of Xe referred to above, the width of the peak at half height is only 2·0 eV compared with the corresponding width of the main beam of 1·0 eV.

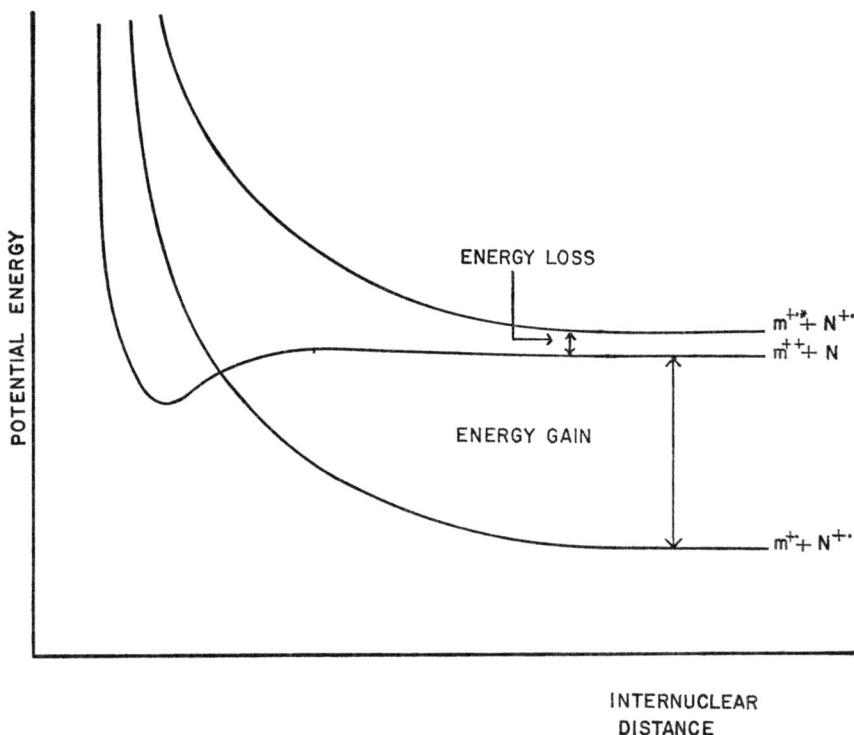

FIG. 2   Mechanisms of charge exchange leading to energy loss and to energy gain peaks.

When a diatomic or larger molecule is used as collision gas a much more intense peak is generally seen located exactly at zero energy loss, within experimental error. The much closer lying energy levels in such molecules are apparently accessible to act as a sink for any excess energy available and to provide, in most cases, a very good energy match without the necessity of converting any translational energy of the reactant ion into internal energy of the products.

The mechanism of a charge exchange reaction in which kinetic energy is lost is illustrated in Fig. 2. Curve crossing from the reactant state ($m^{++} + N$) to the product state ($m^{+\cdot*} + N^{+\cdot}$) requires that an amount of kinetic energy equal to the differences between the enthalpies of these states be provided by the reactant ion to drive the reaction. This energy loss can be measured directly from the energy of the high velocity product, $m^{+\cdot *}$. Figure 2 also illustrates the fact that charge exchange can occur when the product state $(m^{+\cdot} + N^{+\cdot})$ is lower in energy than the reactant state $(m^{++} + N)$. In this situation the excess internal energy must be released as kinetic energy. For reasons analogous to those presented elsewhere for energy loss reactions,[1] the high energy product $(m^{+\cdot})$ will acquire virtually all of this kinetic energy. It will therefore be recorded as an energy gain peak, the position of which measures the exothermicity of the reaction directly. It must be emphasized that there will generally be numerous product states $(m^{+\cdot} + N^{+\cdot})$ of energy both greater and smaller than the energy of the reactant and that Fig. 2 only shows two such states. Clearly the assignment of energy gain peaks becomes difficult when the ground state of the products is much more stable than that of the reactants. This is the case for charge exchange reactions of helium, including the reaction illustrated in Fig. 1(b).

Energy gain peaks are often more intense than energy loss peaks. Their shapes are not understood in detail and further work is in hand to obtain more knowledge in this area. However, as an illustration of the appropriateness of the mechanism just outlined, consider the case of charge exchange of Kr^{++} using neon as collision gas. This reaction shows no energy loss peak and a single intense energy gain peak centered at $+1\cdot1$ eV with a width at half height of $1\cdot6$ eV. The reaction leading to ground state $Kr^{+\cdot}$ makes available $23\cdot8/24\cdot5$ eV and of this some $21\cdot6$ eV is required to ionize neon, leaving a small (~ 2 eV) energy excess which gives rise to the observed energy gain peak.

With few exceptions, the doubly charged rare gas ions undergo transitions only to the ground state or the first excited state of the corresponding singly charged ion in processes giving rise to energy loss peaks as typified by the reactions of helium already discussed. Some Ar^{++} reactions are, however, exceptional. The $Ar^{++} + Kr$ reaction illustrates this and it also illustrates another feature, the possibility that the collision gas can become doubly ionized, viz. that the reaction be of the 20/12 rather than the 20/11 type. The kinetic energy distribution (Fig. 3) for the bombardment of krypton with doubly charged argon ions shows three peaks centred at $-12\cdot2$, $-3\cdot4$ and $-0\cdot4$ eV respectively. The peak corresponding to the largest energy loss can be assigned to the 20/12 reaction, involving the ground states of all species. The $Ar^{++}/Ar^{+\cdot}$ energy difference is $27\cdot5$ eV. Therefore, the total energy transferred to the collision gas is $27\cdot5 + 12\cdot2 = 39\cdot7$ eV. This agrees satisfactorily with the double ionization potential of krypton ($38\cdot5$ eV) considering the width and the low abundance of the peak being measured. The peak at $-3\cdot4$ eV corresponds to the reaction in which the ground state Ar^{++} ions undergo transitions to the second excited state of $Ar^{+\cdot}$. This process contributes $11\cdot2$ eV to the energy required to ionize neutral krypton. The translational energy of $-3\cdot4$ eV means that a total of $14\cdot6$ eV has been transferred. The ionization potential of Kr is $14\cdot0/14\cdot6$ eV (the ground state is a doublet). The third peak at $-0\cdot3$ eV corresponds to a transition of Ar^{++} to the first excited state of $Ar^{+\cdot}$ that lies $14\cdot1$ eV below the energy of the ground state of Ar^{++}. The total energy transferred ($14\cdot5$ eV) again agrees closely with that required to ionize krypton.

$$Ar^{++} + Kr \longrightarrow Ar^{+\cdot} + Kr^{+\cdot}$$

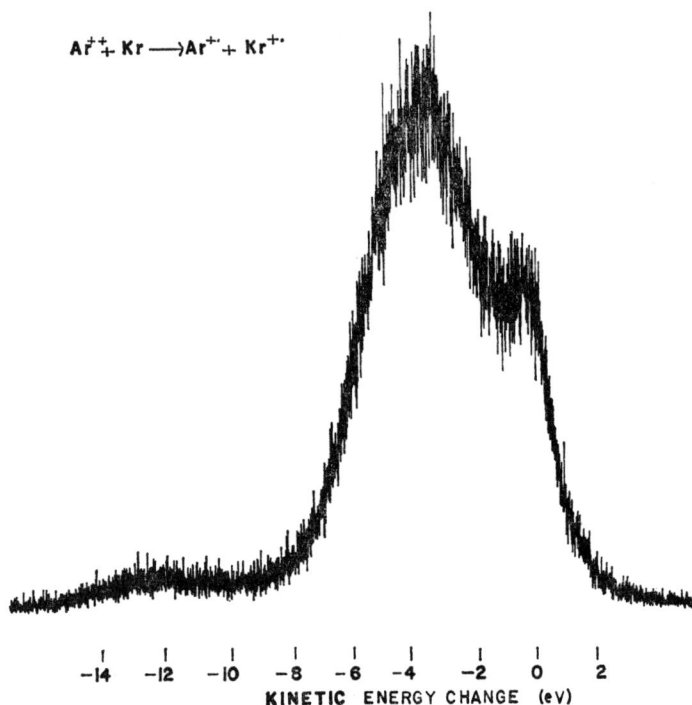

FIG. 3 Kinetic energy spectrum of $Ar^{+\cdot}$ ions formed by charge exchange of Ar^{++} with krypton.

The type of analysis made above has been applied to many other charge exchange reactions of rare gas doubly charged ions. Examples of some of the assignments made are given in Table II. Comparisons of the values in the last two columns provide an indication of the agreement achieved.

TABLE II[a]

**Assignments of Energy Loss Peaks for
Charge Exchange Reactions of Kr^{++}, Xe^{++}, Ar^{++} and Ne^{++}**

Reactant ion	Target	Product ions	Energy loss[b]	Energy available	IP[c]
Kr^{++}	Kr	$Kr^{+\cdot}(^2S) + Kr^{+\cdot}(^2P)$	3·2	14·2	14·0/14·7
Xe^{++}	H_2	$Xe^{+\cdot}(^2P) + H_2^{+\cdot d}$	0·0	19·9/21·2	15·4[d]
Xe^{++}	Kr	$Xe^{+\cdot}(^2S) + Kr^{+\cdot}(^2P)$	4·3	14·2	14·0/14·7
Ne^{++}	Xe	$Ne^{+}(^2S) + Xe^{+\cdot}(^2P)$	0·5	13·6–14·6	12·1/13·4
Ar^{++}	Ne	$Ar^{+\cdot}(^2S) + Ne^{+}(^2P)$	8·2	22·3	21·6
Ar^{++}	Xe	$Ar^{+\cdot}(^2S) + Xe^{+\cdot}(^2P)$	0·0	14·1	12·1/13·4
Kr^{++}	Ar	$Kr^{+\cdot}(^2S) + Ar^{+\cdot}(^2P)$	4·6	15·6	15·5/15·7

[a] All data in electron volts.
[b] Peak centre.
[c] Two values given for $^2P_{3/2}$ and $^2P_{1/2}$ doublets.
[d] $H_2^{+\cdot}$ is presumably formed in a vibrationally excited state.

ACKNOWLEDGMENT

We thank the National Science Foundation for financial support.

REFERENCES

1. Cooks, R. G., Beynon, J. H., Caprioli, R. M. and Lester, G. R., 'Metastable Ions' Elsevier Publishing Company, Amsterdam, 1973.
2. Hasted, J. B., 'Physics of Atomic Collisions', Butterworths, London, 1964, p. 3.
3. Ast, T., Beynon, J. H. and Cooks, R. G., *Org. Mass Spectrom.*, 1972, **6**, 749.
4. Beynon, J. H., Baitinger, W. E., Amy, J. W. and Komatsu, T., *Int. J. Mass Spectrom. Ion Phys.*, 1969, **3**, 47.
5. Ast, T., Beynon, J. H. and Cooks, R. G., *J. Amer. Chem. Soc.*, 1972, **94**, 6611.
6. This value and all other thermochemical data used are taken from
   (a) Moore, C. E., 'Atomic Energy Levels', NBS Circ. 467, 1949.
   (b) Franklin, J. L., Dillard, J. G., Rosenstock, H. M., Herron, J. T., Draxl, K. and Field, F. H., 'Ionization Potentials, Appearance Potentials and Heats of Formation of Gaseous Positive Ions', Nat. Stand. Ref. Data Series, National Bureau of Standards, U.S.A., No. 26, 1969.

98
Ion–Molecule Reactions of Organometallic Complexes

By J. MÜLLER

(*Anorganisch-Chemisches Laboratorium der Technischen Universität, München, Germany*)

SCHUMACHER and Taubenest were the first to observe bimetallic ions in the mass spectra of mononuclear organometallic complexes and to identify the species as products of ion–molecule reactions (IMR).[1,2] From these as well as from other investigations on this subject[3–9] it turned out that IMR of organometallic complexes are characterized by rather high cross-sections and involve a much greater variety of reaction pathways than do IMR of organic molecules. Therefore they can be easily studied by use of conventional mass spectrometers if only the sample pressures are slightly increased and the repeller or the drawing-out-plate potentials are adjusted to be equal to the potential of the collision chamber in order to increase the residence time of the ions in the ion source.

In our work we have investigated two principal types of IMR which will be discussed using selected examples. All secondary ions mentioned in this paper have been identified as products of IMR by studying the dependence of their intensities on pressure and on drawing-out-plate potentials.

FORMATION OF BIMETALLIC IONS FROM MONONUCLEAR COMPLEXES

The general equation for such processes is as follows:

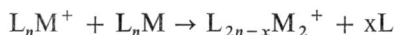

$$L_n M^+ + L_n M \rightarrow L_{2n-x} M_2^+ + xL$$

The intensities of the secondary ions $L_{2n-x}M_2^+$ depend on the square of the pressure of L_nM. The primary collision complex $(L_nM)_2^+$ has not been observed in any case; it is stabilized by loss of one or more ligand molecules L which can take away excess energy from the complex as kinetic energy. The secondary ion $L_{2n-x}M_2^+$ may undergo further fragmentation processes which are often accompanied by the corresponding metastable peaks. Usually the primary stabilizing loss of ligands as well as the succeeding fragmentations of secondary ions obey the same rules as the fragmentation of 'normal'

Scheme 1

Formation of Secondary Ions in the Mass Spectrum of CpV(CO)$_4$

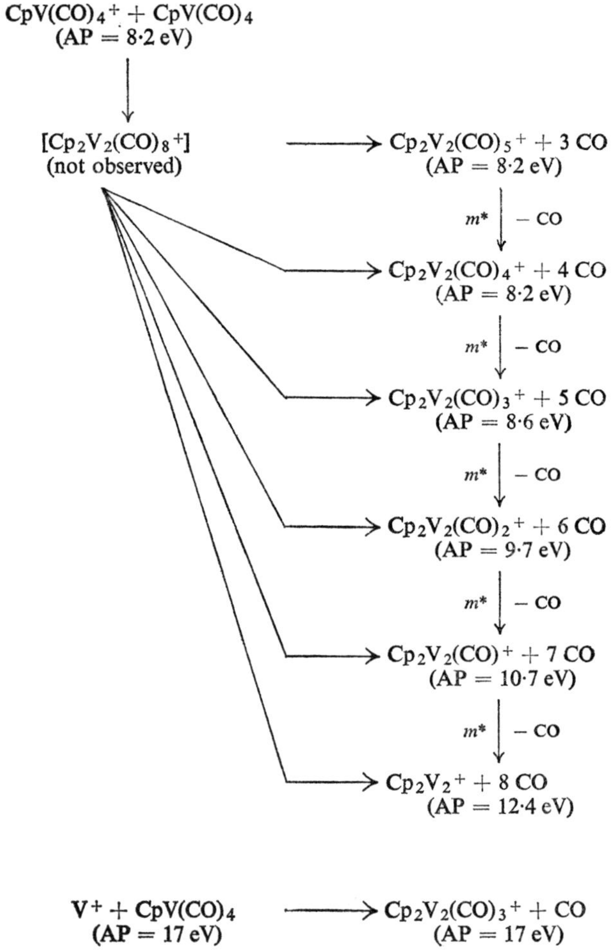

CpV(CO)$_4^+$ + CpV(CO)$_4$
   (AP = 8·2 eV)

$\downarrow$

[Cp$_2$V$_2$(CO)$_8^+$] $\longrightarrow$ Cp$_2$V$_2$(CO)$_5^+$ + 3 CO
(not observed)   (AP = 8·2 eV)

                          m^* $\Big|$ − CO

                          $\longrightarrow$ Cp$_2$V$_2$(CO)$_4^+$ + 4 CO
                                (AP = 8·2 eV)

                          m^* $\Big|$ − CO

                          $\longrightarrow$ Cp$_2$V$_2$(CO)$_3^+$ + 5 CO
                                (AP = 8·6 eV)

                          m^* $\Big|$ − CO

                          $\longrightarrow$ Cp$_2$V$_2$(CO)$_2^+$ + 6 CO
                                (AP = 9·7 eV)

                          m^* $\Big|$ − CO

                          $\longrightarrow$ Cp$_2$V$_2$(CO)$^+$ + 7 CO
                                (AP = 10·7 eV)

                          m^* $\Big|$ − CO

                          $\longrightarrow$ Cp$_2$V$_2^+$ + 8 CO
                                (AP = 12·4 eV)

V$^+$ + CpV(CO)$_4$ $\longrightarrow$ Cp$_2$V$_2$(CO)$_3^+$ + CO
   (AP = 17 eV)   (AP = 17 eV)

ions.[10] Thus, in the IMR of cyclopentadienyl metal carbonyls the CO ligands are most easily lost. As an illustration Scheme 1 shows several (not all!) reactions leading to some of the secondary ions which can be observed in the mass spectrum of cyclopentadienyl vanadium tetracarbonyl [CpV(CO)$_4$; Cp = C$_5$H$_5$] under suitable conditions.

The influence of the excitation energy of the primary ion CpV(CO)$_4^+$ on the cross-section of the formation of the secondary ion Cp$_2$V$_2$(CO)$_4^+$ is shown by Fig. 1. This diagram exhibits a maximum of the ratio I[Cp$_2$V$_2$(CO)$_4^+$]/ I[CpV(CO)$_4^+$] at an electron energy which is some electron volts higher than the AP of the secondary ion. Such resonance phenomena have been observed for organometallic IMR in several cases.[3,4,7,8]

Other examples for organometallic IMR involving formation of bimetallic

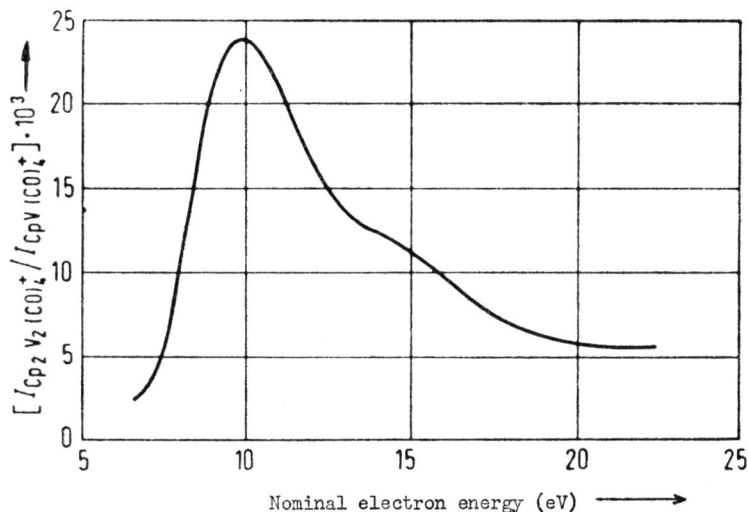

FIG. 1   The ratio of the intensities of $Cp_2V_2(CO)_4^+$ and $CpV(CO)_4^+$ as a function of electron energy for the reaction $CpV(CO)_4^+ + CpV(CO)_4 \rightarrow Cp_2V_2(CO)_4^+ + 4\,CO$.

ions are presented by the following equations:

Chromium hexacarbonyl:

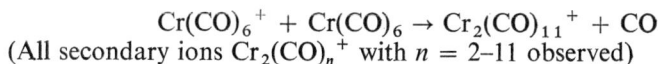

$$Cr(CO)_6^+ + Cr(CO)_6 \rightarrow Cr_2(CO)_{11}^+ + CO$$
(All secondary ions $Cr_2(CO)_n^+$ with $n = 2\text{--}11$ observed)

Tetrakis(trifluorophosphine)nickel:

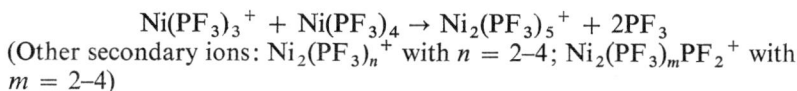

$$Ni(PF_3)_3^+ + Ni(PF_3)_4 \rightarrow Ni_2(PF_3)_5^+ + 2PF_3$$
(Other secondary ions: $Ni_2(PF_3)_n^+$ with $n = 2\text{--}4$; $Ni_2(PF_3)_m PF_2^+$ with $m = 2\text{--}4$)

Dibenzene chromium:

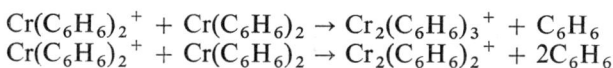

$$Cr(C_6H_6)_2^+ + Cr(C_6H_6)_2 \rightarrow Cr_2(C_6H_6)_3^+ + C_6H_6$$
$$Cr(C_6H_6)_2^+ + Cr(C_6H_6)_2 \rightarrow Cr_2(C_6H_6)_2^+ + 2C_6H_6$$

Cyclopentadienyl chromium dicarbonyl nitrosyl:

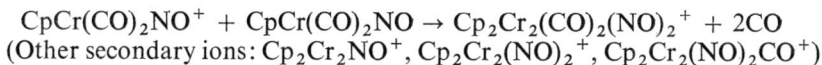

$$CpCr(CO)_2NO^+ + CpCr(CO)_2NO \rightarrow Cp_2Cr_2(CO)_2(NO)_2^+ + 2CO$$
(Other secondary ions: $Cp_2Cr_2NO^+$, $Cp_2Cr_2(NO)_2^+$, $Cp_2Cr_2(NO)_2CO^+$)

In some cases and under special conditions (higher pressures and low electron energies) trimetallic ions have also been observed, for example in the mass spectrum of benzene chromium tricarbonyl. Obviously such ions arise from collisions between bimetallic secondary ions and neutral molecules, e.g.

$$(C_6H_6)_2Cr_2(CO)_3^+ + C_6H_6Cr(CO)_3 \rightarrow (C_6H_6)_2Cr_3(CO)_6^+ + C_6H_6$$

FORMATION OF NEW METAL–LIGAND BONDS
BETWEEN A COMPLEX ION AND A LIGAND MOLECULE

Such processes may occur according to the following general equation:

$$L_nM^+ + L' \rightarrow L_{n-x}L'M^+ + xL$$

IMR of this type have been studied for $L_nM^+ = CpNiNO^+$, $CpCo(CO)_2{}^+$, $CpMn(CO)_3{}^+$, and $CpV(CO)_4{}^+$ with a large number of σ and π donor molecules L', such as H_2O, R_2O, NH_3, NHR_2, H_2S, HCl, CH_3Cl, NF_3, PF_3, AsF_3, SbF_3, ethylene, acetylene, butadiene, benzene, cyclic mono- and oligo-olefins.

The experiments have been carried out by use of a heated double gas inlet system, one reservoir containing the complex, the other one containing the ligand sample. Usually the ionization energies of organometallic complexes are lower than those of the ligands L' studied; therefore it is possible to adjust the electron energy of the ion source to values which allow ionization of only the complex molecules. The secondary ion intensities have been found to be linearly dependent on the pressures of both the complex and the ligand sample.

The following IMR have been observed:

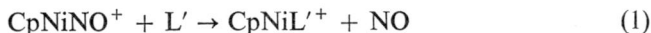

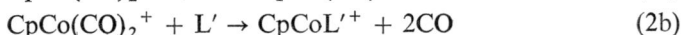

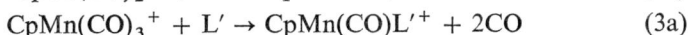

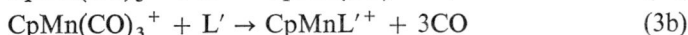

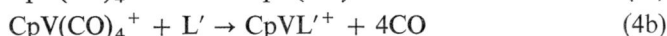

$$CpNiNO^+ + L' \rightarrow CpNiL'^+ + NO \tag{1}$$
$$CpCo(CO)_2{}^+ + L' \rightarrow CpCo(CO)L'^+ + CO \tag{2a}$$
$$CpCo(CO)_2{}^+ + L' \rightarrow CpCoL'^+ + 2CO \tag{2b}$$
$$CpMn(CO)_3{}^+ + L' \rightarrow CpMn(CO)L'^+ + 2CO \tag{3a}$$
$$CpMn(CO)_3{}^+ + L' \rightarrow CpMnL'^+ + 3CO \tag{3b}$$
$$CpV(CO)_4{}^+ + L' \rightarrow CpV(CO)L'^+ + 3CO \tag{4a}$$
$$CpV(CO)_4{}^+ + L' \rightarrow CpVL'^+ + 4CO \tag{4b}$$

The secondary ions arising from eqns (2a), (3a) and (4a) have much lower intensities than those formed by processes (2b), (3b) and (4b). With only few exceptions the secondary ions $CpML'^+$ have the same appearance potentials as the ionic collision partners $CpML_n{}^+$, and it must be concluded that the reactions require no or almost no activation energies.

In the cases of (1) and (2b) even IMR with L' = saturated hydrocarbons (for example cyclohexane) occur; in these reactions L' loses an H_2 molecule and the resulting olefine molecule is co-ordinated to the metal atom, e.g.

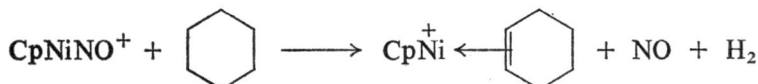

It follows from appearance potential measurements that for such reactions activation energies of about 1·5 eV are necessary.

Most of the secondary ions $CpML'^+$ show a series of fragmentation reactions of the complexed L' ligand. In many of these cases new double bonds which can be co-ordinated to the central atom are formed either by loss of H_2 molecules or by ring cleavage processes:

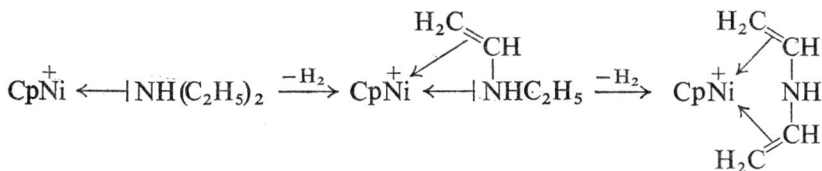

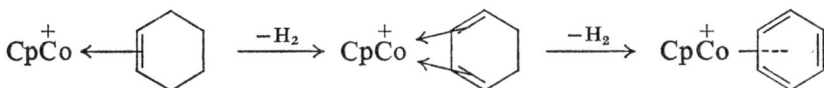

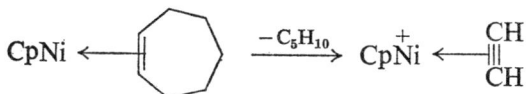

$$\overset{+}{CpNi} \longleftarrow NH(C_2H_5)_2 \xrightarrow{-H_2} \overset{+}{CpNi} \longleftarrow NHC_2H_5 \xrightarrow{-H_2} \overset{+}{CpNi}\ NH$$

$$\overset{+}{CpCo} \longleftarrow \bigcirc \xrightarrow{-H_2} \overset{+}{CpCo} \longleftarrow \bigcirc \xrightarrow{-H_2} \overset{+}{CpCo} \longleftarrow \bigcirc$$

$$CpNi \longleftarrow \bigcirc \xrightarrow{-C_5H_{10}} \overset{+}{CpNi} \longleftarrow \overset{CH}{\underset{CH}{|||}}$$

In order to compare the results of the IMR of this type relative cross-sections Q_r have been determined in the following way. For each $CpML_n^+/L'$ system a value $Q_{L'}$ was calculated from the sum of the intensities of all secondary ions ($CpML'^+$ as well as fragments arising from this ion) of the system at 20 eV and normalized with respect to the intensity I of the primary ion $CpML_n^+$ and the reservoir pressure p of the ligand L':

$$Q_{L'} = \frac{I(\text{secondary ions})}{I(CpML_n^+) \cdot p(L')}$$

The value Q_r for $L' = O(C_2H_5)_2$ is set to unity, and Q_r is defined as

$$Q_r = \frac{Q_{L'}}{Q_{Et_2O}}$$

From the Q_r values listed in Table I various qualitative influences can be recognized which affect the cross-sections of the IMR, although in some cases a detailed interpretation is rather complicated. The most important parameter seems to be the relative donor character of the ligand L'. Thus, the very weak donor molecules NF_3, HCl, and CH_3Cl do not react, and the strong acceptor ligand PF_3 yields secondary ions of very low intensities. The increase of the Q_r values with increasing donor properties of L' is reflected in the series $H_2O < C_2H_5OH < (C_2H_5)_2O$ as well as $NH_3 < NH(C_2H_5)_2$.

A second influence affecting Q_r is the number of co-ordination centres of L', in terms of which the increase of Q_r in the series

$$(C_2H_5)_2O < H_3COC_2H_4OCH_3$$

or cyclohexene < 1,3-cyclohexadiene can be explained.

The third parameter seems to be the molecular size of L' which should affect the collision cross-section of the IMR. This influence can be recognized for example in the series butadiene < 1,3-cyclohexadiene < 1,5-cyclo-octadiene.

TABLE I

**Relative Cross-Sections Q_r for Ion–Molecule Reactions
in the Systems $CpML_n^+/L'$**

L'	$CpNiNO$	$CpCo(CO)_2$	$CpMn(CO)_3$	$CpV(CO)_4$
PF_3	0·00	0·05	0·01	0·01
H_2S	—	0·25	0·26	0·48
H_2O	0·01	0·45	0·95	0·91
C_2H_5OH	0·59	—	—	—
$(C_2H_5)_2O$	1·00	1·00	1·00	1·00
$H_3COC_2H_4OCH_3$	1·32	1·08	1·24	1·71
NH_3	0·63	0·78	0·66	0·81
$NH(C_2H_5)_2$	0·99	0·90	0·63	1·57
butadiene	0·55	0·78	0·55	0·76
cyclohexane	0·77	0·73	0·00	0·00
cyclohexene	0·96	1·05	0·40	1·33
1,3-cyclohexadiene	1·14	1·18	0·42	0·86
benzene	0·87	0·88	0·53	1·10
cyclo-octene	1·79	0·88	0·55	1·38
1,5-cyclo-octadiene	2·13	1·73	0·68	1·48
1,3,5-cyclo-octatriene	1·81	1·15	0·58	1·57
cyclo-octatetraene	1·49	1·48	0·92	1·81

It follows from these investigations that IMR of organometallic complexes provides valuable information on the formation and stability of metal–metal and metal–ligand bonds.

REFERENCES

1. Schumacher, E. and Taubenest, R., *Helv. Chim. Acta*, 1964, **47**, 1525.
2. Schumacher, E. and Taubenest, R., *Helv. Chim. Acta*, 1966, **49**, 1447.
3. Müller, J. and Fenderl, K., *Chem. Ber.*, 1970, **103**, 3141.
4. Gilbert, J. R., Leach, W. P. and Miller, J. R., *J. Organometal. Chem.*, 1971, **30**, C41.
5. Kraihanzel, C. S., Conville, J. J. and Sturm, J. E., *Chem. Commun.*, 1971, 159.
6. Foster, M. S. and Beauchamp, J. L., *J. Amer. Chem. Soc.*, 1971, **93**, 4924.
7. Müller, J. and Fenderl, K., *Chem. Ber.*, 1971, **104**, 2199.
8. Müller, J. and Fenderl, K., *Chem. Ber.*, 1971, **104**, 2207.
9. Müller, J. and Goll, W., *Chem. Ber.*, 1973, **106**, 1129.
10. Müller, J., *Angew. Chem. Internat. Edit.*, 1972, **11**, 653.

99

Ion–Molecule Reactions in CS₂ and COS Studied by 'Time of Flight' Mass Spectrometry

By M-TH. PRAET and J. P. DELWICHE†

(*Institut de Chimie, Université de Liège, Liège, Belgium*)

THE data available on the ion–molecule reactions in CS_2 are now rather old. Herman and Čermák[1] and Henglein[2] showed the existence of ions with a m/e ratio higher than the m/e ratio of the parent ion. These ions are $C_2S_2^+$ (m/e = 88), CS_3^+ (m/e = 108), and $C_2S_3^+$ (m/e = 120). As the pressure range used by these authors was rather narrow, we decided to reinvestigate the ion–molecule reactions occurring in CS_2 with an experimental set up allowing the use of high pressures (up to 1 torr).

The mass spectrometer used was a Bendix model 14 TOF fitted with a modified ion source.[3] The highest ionization chamber pressure reached in these experiments was 1 torr.

The mass spectra of CS_2 were recorded with 50 eV electrons and with an extracting field of 10 V/cm. The electron beam intensity was kept in the 5 to 7×10^{-8} A range to reduce the space charge effects while the whole ion source was heated at 440 °K in order to avoid spurious effects due to absorption of gases on the ionization chamber walls.

In addition to the usual ions of the mass spectra of CS_2, we found by increasing the pressure in the ionization chamber the ions $C_2S_2^+$, CS_3^+, and $C_2S_3^+$ already observed and also, at high pressure, two new ions: CS_4^+ (m/e = 140) and $C_2S_4^+$ (m/e = 152).

The relative abundance of the C^+, S^+, CS^+ ions, if the pressure is increased, tends to decrease exponentially as is usual for reactive primary ions. (Figures 1(c), (b).)

Figure 1(a) shows that the relative abundance of the CS_2^+ ions goes through a maximum indicating that the parent ions can also be created by ion–molecule reactions, most probably by charge transfer between the C^+, S^+ and/or CS^+ primary ions and the neutral molecules. Another mechanism of formation for the CS_2^+ ions must also be proposed because we observe that, with electrons having an energy lower than the appearance energies of the fragment ions, the relative abundance of CS_2^+ ions does not show the

† 'Chercheur qualifié' of the Belgian 'Fonds National de la Recherche Scientifique'.

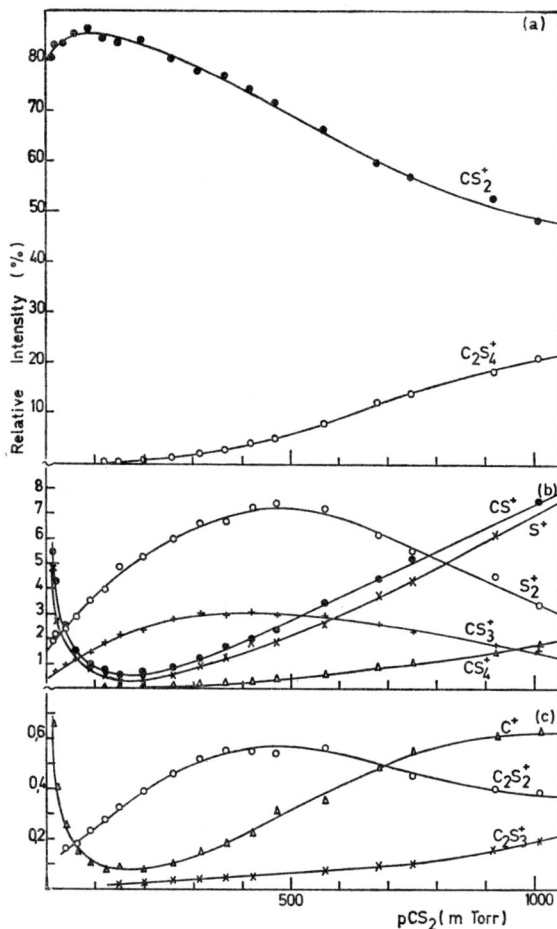

FIG. 1 Pressure dependence of ion intensities in the mass spectrum of CS_2 at electron energy of 50 eV.

expected exponential decrease but decreases very slowly. The formation of CS_2^+ ions in these conditions results from charge transfer reactions between slightly vibrationally excited CS_2^+ ions and neutral molecules.

The S_2^+ ion relative abundance shows an initial increase followed by a decrease with increasing pressure as shown in Fig. 1(b), therefore this ion might be a product as well as a reactive ion in the ion–molecule reactions.

In order to find the precursors of the secondary ions, we measured the appearance energies of all the ions; for the primary ions C^+, CS^+, and S^+ we found 20.0 ± 0.1 eV, 16.2 ± 0.1 eV and 14.8 ± 0.1 eV respectively; the appearance energies of $C_2S_2^+$, CS_3^+ and $C_2S_3^+$ have been measured to be between 13.6 and 13.9 eV, in good agreement with the Henglein's value of 13.3 eV.[2] These energies are lower than the appearance energies of the primary ions.

We can conclude that the primary ions C$^+$, CS$^+$ and S$^+$ react only by charge transfer to produce CS$_2{}^+$ secondary ions.

The disappearance rate constants for the primary ions have been measured and we found:

$$k_d C^+ = 13.5 \pm 2.0\ 10^{-10}\ \text{cm}^3\ \text{molecule}^{-1}\ \text{s}^{-1}$$
$$k_d S^+ = 7.5 \pm 1.5\ 10^{-10}\ \text{cm}^3\ \text{molecule}^{-1}\ \text{s}^{-1}$$
$$k_d CS^+ = 8.7 \pm 1.7\ 10^{-10}\ \text{cm}^3\ \text{molecule}^{-1}\ \text{s}^{-1}$$

Henglein[2] attributed the formation of C$_2$S$_2{}^+$, CS$_3{}^+$ and C$_2$S$_3{}^+$ ions to reactions between CS$_2{}^+$ ions in their first excited electronic state $^2\Pi_u$ and neutral molecules. This state was supposed to be at 13.6 eV according to RPD measurements of Collin.[4] However, recent measurements in photoelectron spectrometry[5] showed that the $^2\Pi_u$ state of CS$_2{}^+$ ion is located at 12.69 eV (adiabatic transition) and that the vibrational structure of the electronic band extends up to 13.2 eV. This excludes the possibility of reaction between CS$_2{}^+$ ions formed by direct ionization in the $^2\Pi_u$ state and neutral molecules to give heavy secondary ions. It has been shown, however, by photoelectron spectrometry that high vibrational levels can be populated through autoionization mechanisms. In addition, photoionization and RPD measurements of Momigny and Delwiche[6] showed the existence of strong autoionization structures in the ionization efficiency curves of CS$_2{}^+$ at 13.6 eV. This led us to conclude that C$_2$S$_2{}^+$, CS$_3{}^+$ and C$_2$S$_3{}^+$ result from collisions between vibrationally excited CS$_2{}^+$ ions in their $^2\Pi_u$ state formed by autoionization and neutral molecules.

The CS$_4{}^+$ and S$_2{}^+$ ions that have appearance energies in the same energy range (13.6 to 13.9 eV) are also formed in the same type of reactions.

In contrast, the C$_2$S$_4{}^+$ ions appear at 10 eV and therefore can be formed by parent ions created by a direct ionization process. As seen in Fig. 1(a), the curve of the C$_2$S$_4{}^+$ intensity as a function of pressure shows an upward curvature indicating a high kinetic order for the reaction producing this ion; we have found a value of 4 and we suggest a complex mechanism similar to that observed in C$_2$H$_3$Cl[7] and C$_2$H$_3$F[8] involving a collision stabilization process between CS$_2{}^+$ ions and neutral molecules.

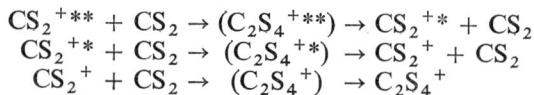

$$CS_2{}^{+**} + CS_2 \rightarrow (C_2S_4{}^{+**}) \rightarrow CS_2{}^{+*} + CS_2$$
$$CS_2{}^{+*} + CS_2 \rightarrow (C_2S_4{}^{+*}) \rightarrow CS_2{}^{+} + CS_2$$
$$CS_2{}^{+} + CS_2 \rightarrow (C_2S_4{}^{+}) \rightarrow C_2S_4{}^{+}$$

This is sustained by the fact that the C$_2$S$_4{}^+$ ions are only present in the spectrum for pressures higher than 100 mtorr and that, at the same pressures, the relative abundance of the CS$_2{}^+$ ions starts to decrease.

The disappearance rate constant of the CS$_2{}^+$ ion above 100 mtorr is 1.76 10^{-11} cm^3 molecule^{-1} s^{-1}.

When the pressure is increased above 100 mtorr in the ionization chamber the relative abundance of C$^+$, S$^+$, and CS$^+$ ions increases rapidly. The explanation for this phenomenon is still uncertain but it seems likely that it can be produced by dissociative collisions of CS$_2{}^+$ ions or ions having a higher mass; the number of such collisions per unit of time increasing if the pressure in the ionization chamber increases.

In a study of the ion–molecule reactions occurring in COS, Džidić *et al.*[9] using a Nier type mass spectrometer observed the following ions: S_2^+ (m/e = 64), CS_2^+ (m/e = 76), and S_3^+ (m/e = 96). Our study of the same subject in the pressure range 0–1 torr with the same experimental conditions as for CS_2 led us to observe $C_2O_2S_2^+$ (m/e = 120) ions and ions with m/e = 112 for which we cannot attribute a definite formula, their very low relative intensity precludes their identification by measurement of isotopic ratios. We have also observed that the SO^+ ions can be formed by ion–molecule reactions and can themselves react at high pressure with neutral molecules (Fig. 2(b)).

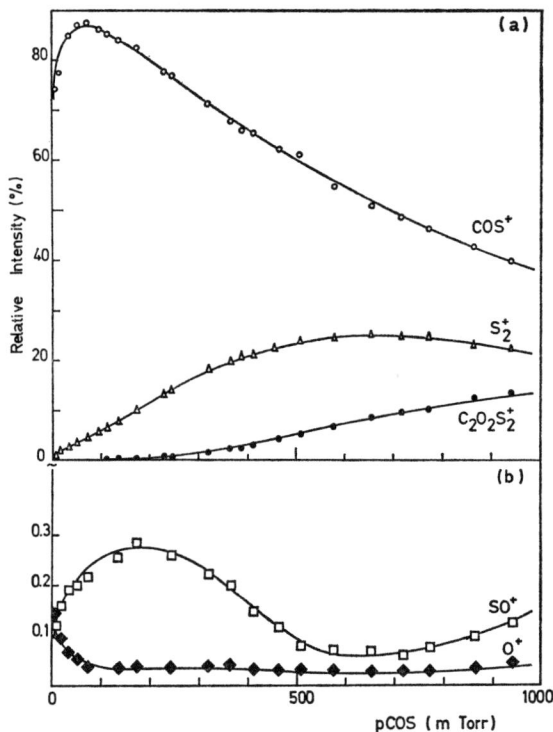

FIG. 2   Pressure dependence of ion intensities in the mass spectrum of COS at electron energy of 50 eV.

Figures 2(b) and 3(a), (b) show that the fragment ions disappear in an exponential manner with increasing pressure while the COS^+ ions are also formed by charge transfer especially in the low pressure range (Fig. 2(a)).

The appearance energies of $C_2O_2S_2^+$, S_2^+, and CS_2^+ coincide with the ionization energy of COS. As shown in Figs. 2(a), 3(a) the S_2^+ and CS_2^+ ions are present at low pressures while the $C_2O_2S_2^+$ ions appear only for pressures higher than 100 mtorr (Fig. 2(a)).

The formation order of $C_2O_2S_2^+$ is 4 indicating a complex series of reactions involving stabilizing collisions similar to those in CS_2 for the formation of $C_2S_4^+$.

The formation orders of S_2^+ and CS_2^+ ions are not integrals and vary with the pressure. At the present time we are not able to explain this behaviour.

Džidić *et al.*[9] suggested that the S_2^+ ions could be formed by COS^+ and S^+. Our measurements however do not show any correlation between S_2^+ and S^+. In addition these authors proposed that the CS^+ ion could be a precursor of CS_2^+ but this was ruled out by our experiments because we observed CS_2^+ ions at the appearance threshold of COS^+.

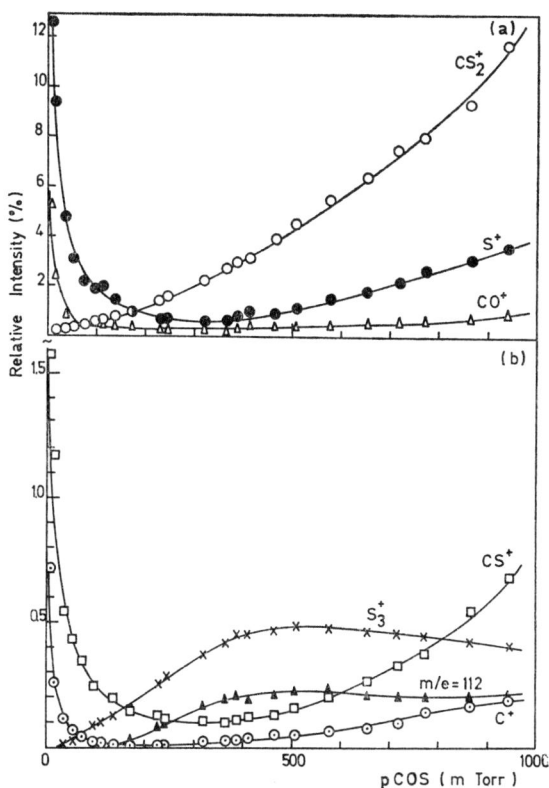

FIG. 3   Pressure dependence of ion intensities in the mass spectrum of COS at electron energy of 50 eV.

Some COS^+ ions are also formed in charge transfer processes between neutral molecules and primary fragment ions C^+, O^+, CO^+, and CS^+ for electron energies higher than the appearance energies of these ions, and for lower electron energies the charge transfer reactions must be of the type

$$COS^{+*} + COS \rightarrow (COS)_2^{+*} \rightarrow COS^+ + COS$$

A similar type of reaction was encountered in CS_2.

The total rate constant for the disappearance of COS^+ is $2 \cdot 63 \ 10^{-11} \ cm^3$ molecule^{-1} s^{-1} while the disappearance rate constants for the primary ions are

$$k_d C^+ = 24{\cdot}0 \pm 4{\cdot}0 \ 10^{-10} \ cm^3 \ molecule^{-1} \ s^{-1}$$
$$k_d O^+ = 14{\cdot}0 \pm 3{\cdot}0 \ 10^{-10} \ cm^3 \ molecule^{-1} \ s^{-1}$$
$$k_d S^+ = 10{\cdot}2 \pm 3{\cdot}0 \ 10^{-10} \ cm^3 \ molecule^{-1} \ s^{-1}$$
$$k_d CO^+ = 15{\cdot}7 \pm 4{\cdot}0 \ 10^{-10} \ cm^3 \ molecule^{-1} \ s^{-1}$$
$$k_d CS^+ = 10{\cdot}3 \pm 4{\cdot}0 \ 10^{-10} \ cm^3 \ molecule^{-1} \ s^{-1}$$

This last value agrees very well with the Džidić's value[9] of $7{\cdot}7 \ 10^{-10} \ cm^3$ molecule^{-1} s^{-1} obtained by a pulsed method.

Interesting is the case of the SO^+ ion because its appearance energy is about 15·5 eV. No primary fragment ions have an appearance energy near this value, but in this energy range two excited electronic states of COS^+ are known by photoelectron spectrometry:[5] the A $^2\Pi$ state at 15·08 eV with vibrational structure extending up to 16 eV and the B $^2\Sigma^+$ state at 16·04 eV. It is therefore highly probable that the secondary SO^+ ions are formed by collision of excited COS^+ ions with neutral molecules. Above 200 mtorr the relative intensity of SO^+ tends to decrease and we note the appearance of an ion with a m/e = 112 (Fig. 3(b)). The sum of the relative intensities of these two ions remains constant over a wide pressure range indicating that SO^+ is a possible precursor for the m/e = 112 ion; the relative intensities of these two ions are however very low (Figs. 2(b), 3(b)).

Above 100 mtorr the relative abundance of C^+, O^+, CO^+, S^+, CS^+ tends to increase. The reason for this is probably the occurrence of dissociative collisions as it was for CS_2.

REFERENCES

1. Čermák, V. and Herman, Z., *J. Chim. Phys.*, 1960, **57**, 717.
2. Henglein, A., *Z. Naturforsch.*, 1962, **17a**, 37.
3. Futrell, J. H., Tiernan, T. O., Abramson, F. P. and Miller, C. D., *Rev. Sci. Instrum.*, 1968, **39**, 340.
   Miller, C. D., Tiernan, T. O. and Futrell, J. H., *Rev. Sci. Instrum.*, 1969, **40**, 957.
4. Collin, J., *J. Chim. Phys.*, 1960, **57**, 717.
5. Turner, D. W., Baker, C., Baker, A. D. and Brundle, C. R., *in* 'Molecular Photoelectron Spectroscopy', Wiley Interscience, London, 1970.
6. Momigny, J. and Delwiche, J., *J. Chim. Phys.*, 1968, **65**, 1213.
7. Herman, J. A., Myher, J. J. and Harrison, A. G., *Can. J. Chem.*, 1969, **47**, 647.
   Hughes, B. M., Tiernan, T. O. and Futrell, J. H., *J. Phys. Chem.*, 1969, **73**, 829.
8. Herman, J. A. and Harrison, A. G., *Can. J. Chem.*, 1969, **47**, 957.
   O'Malley, R. M. and Jennings, K. R., *Int. J. Mass Spectrom. Ion Phys.*, 1969, **2**, 441.
9. Džidić, I., Good, A. and Kebarle, P., *Can. J. Chem.*, 1970, **48**, 664.

The Determination of Ion Structure

By J. H. BEYNON and R. G. COOKS

(*Department of Chemistry, Purdue University, W. Lafayette, Indiana, U.S.A.*)

INTRODUCTION

IT may be argued that the development of any material science begins with an understanding of structure. Nowhere has this been more true than in chemistry, where questions of structure (atomic, molecular and crystal) have spawned so many achievements in synthesis and kinetics. To appreciate the comparable effect in physics, one needs only to consider the tremendous advances that followed in the wake of Rutherford's experiments on atomic structure. These included the development of all branches of spectroscopy. So mass spectrometry is not unique in that a proper understanding of the structures of the species with which it deals is vital to its full development.

Organic chemists have now been using mass spectrometry for about thirty years to determine the molecular structures of unknown organic compounds. In some of the earliest work[1,2] generalizations were made connecting fragmentation patterns and molecular structure and as more and more compounds were examined, many such empirical correlations were noted. The success of the mass spectrometer in qualitative analysis has rested largely upon these correlations and also upon the ability to determine the formulae of the molecular and fragment ions.[3] In the case of the molecular ion this leads to direct information concerning molecular structure. The number of rings and double bonds can be deduced and the relative abundance of this ion also gives crude information concerning its stability and hence, indirectly, its structure.

This review is concerned entirely with methods of determining the structures of charged species and is restricted to work on organic ions. Much of the understanding of the empirical correlations referred to above has been based upon postulates of ion structure. These structures have been assumed to be similar to the well-characterized structures that occur in the ground state chemistry of neutral species. The activated states of the ions from which fragmentation occurs have often been identified with structures in which the charge is localized on a particular atom. A prominent pioneer of this work was McLafferty[4] and later a vast amount of information was collated by Djerassi[5] who, by careful 'electron book-keeping' and the use of isotopically labelled compounds, was able to prove which atoms took part in particular fragmentation reactions and to suggest plausible structures for the reactant and product species.

It has sometimes been argued that none of the structures postulated have any real justification. True, there is an indirect aspect to each of the methods used to probe gaseous ion structures, but this is no different in kind to those associated with the usual methods of determining molecular structures of organic compounds. There is now a very large amount of evidence that in the majority of cases, the few simple assumptions made concerning analogies between the structures and fragmentation reactions of organic ions and the behaviour of ground state organic molecules are justified. These, coupled with the concept of charge localization in the activated complex, have led to a broad understanding of the origins of the main peaks in mass spectra. It is indeed fortunate that the basis of this understanding was painstakingly laid, often in the face of fierce criticism that saw the individual weaknesses without recognizing the overall strength. For it is surely as absurd to dismiss the general approach that has been used so successfully to deduce unknown molecular structures as it is to claim in individual cases that a postulated structure for an ion is necessarily correct because it seems to be the only 'reasonable' structure.

This review will deal only briefly with the early work; not because it is less important in the light of later research, but because it has been described in detail in many previous reviews.[6] The classical work of Meyerson[7] especially on the structure of $C_7H_7^+$ ions is also so well known that it will only be treated briefly. Other early work will be mentioned only where it seems to have made a similar lasting contribution to our ideas on ion structure. The main body of this paper will concern techniques which are currently being applied, including ion–molecule and ion–photon interactions, metastable peak shapes, kinetic methods, thermochemical and isotopic labelling studies.

It is worth emphasizing at this stage that the structures of ions encountered in mass spectrometry may vary depending upon their internal energy. Also, in the case of molecular ions, the structure will not necessarily coincide with that of the neutral molecules from which they were derived. For similar reasons, it cannot be assumed that the structures of the stable ions that traverse the flight path in the mass spectrometer and that give rise to the normal mass peaks are necessarily the same as the structures of the reactive ions (containing more internal energy) that decompose either in the ionization chamber or during their journey towards the detector. It is of paramount importance in considering the various methods of gleaning information on ion structures to distinguish between those that lead to information concerning the mechanism and energetics of the *formation* of stable molecular or fragment ions and those that lead to structural information based on the rates, energetics or mechanisms of *fragmentation* reactions. These latter will, of course, refer only to reactive ions. Ion structure is a function of both internal energy and of time so that a complex series of molecular ion isomerizations might occur even in ions with insufficient internal energy to undergo any spontaneous fragmentation. Ion enthalpy measurements refer to stable ions having the minimum internal energy possible subject to experimental limitations. Other methods of studying stable ions deal with more poorly defined ions and investigations of the timing and energy requirements for intramolecular rearrangements of stable ions still lie in the future. These complexities, so much more acute than those met in studying analogous species in solution, arise because a gaseous ion forms an isolated system. They are magnified by the broad distribution of internal energies represented in virtually all experiments on ion structure.

It is also worth remembering that mass spectrometry differs from most of the other techniques used for determining molecular structures in that the sample molecules are destroyed during their examination. The instrument is, in fact, a reactor that is used to study ionic reactions in the gas phase. So it is vitally necessary to ensure that the sample molecules introduced into this reactor do not undergo any change before ionization. Typically, a molecule will undergo some fifty or so collisions with the walls of the ionization chamber before ionization occurs and this can produce gross changes in structure when the ionization chamber walls are at their usual temperature of the order of 200°C. In this respect, it is only necessary to mention the work of Spiteller[8] who showed that the mass spectra of most of the paraffinic hydrocarbons that had been published at that time were, in fact, spectra of thermally modified species.

ION PROPERTIES

Although there are many cases where bond strengths in the neutral molecule seem to bear little relationship to the bonds actually broken during fragmentation, there are also many cases where bond strengths do seem to play a part in determining the fragmentation pathway, especially in hydrocarbons. Lester[9] used a method based on the hypothesis that the frequency of fracture of a particular bond was related to its bond strength to predict the abundances of the $(M-15)^+$ peaks in the mass spectra of octane isomers with good success. In other cases, too, the structure of the decomposing ion seems to mirror the structure of the neutral molecule from which it was formed.

An early rule concerned with fragmentation behaviour was due to Stevenson[10] and states that when an ion $R_1R_2{}^+$ fragments, the charge will be retained on the fragment (R_1, say), having the lower ionization potential. This rule is a general one and means (in the absence of any reverse activation energy for the reaction) that the total enthalpy of the products will be as low as possible. The rate of any fragmentation reaction increases as the excitation energy available in the reactant ion increases above the minimum for detectable reaction products to be formed. Since product ions $R_1{}^+$ can be formed at lower energy, at energies just sufficient to form $R_2{}^+$, the reactant $R_1R_2{}^+$ ions will be decomposing rapidly to give $R_1{}^+$ and so this ion will predominate. Another way of stating this same rule would be to say that simple bond cleavage tends to give stable products. Both statements of the rule imply, of course, that the structure of the ion $R_1{}^+$ is the same as that of the radical $R_1{}^{\cdot}$.

These concepts gradually went further than Stevenson's rule and implied that stability of products was a necessary and often a sufficient condition that prominent peaks due to this reaction would be seen in the mass spectrum. The observed abundance of any ion will depend not only upon its rate of formation, but also upon its rate of subsequent fragmentation and this will be affected by its stability. But the observation of an extremely abundant ion must always mean a high rate of formation whatever the subsequent fragmentation rate; and in any case there are many striking examples where abundant (and not particularly stable) ions are accompanied by neutral species that can exist in a stable form. The fact that a reaction is accompanied by a neutral moiety having the elements of H_2O, HCN, CO_2, CH_4, etc. is not proof that the neutral is eliminated as a molecule of water, hydrocyanic (or isocyanic) acid, carbon dioxide, methane, etc. But on the other hand, the very large number of cases that have been noted in which the elements of a stable compound are eliminated when an abundant ion is formed or where a particularly stable ionic structure can be written for the daughter ion must be due to more than just coincidence. Indeed, there is now a method by which the structure of the neutral can be inferred so that confirmation that a stable structure has been formed can be obtained. The neutral fragment can be extracted from the ion chamber and characterized by its ionization efficiency curve.[11]

In the absence of any collision process, the energy transmitted to the molecular ion at the moment of fragmentation or the energy contained in a fragment ion at the moment of its formation represents the maximum available for subsequent reactions. The ion may lose energy by radiation but there is no source from which extra energy can be acquired. If the various possible ion structures are enumerated then that structure in which the most easily removed electron has been lost from the molecule to form the ion will be the structure in which the greatest amount of energy is available as vibrational and other forms. This is, therefore, the structure from which fragmentation is most likely. This concept was introduced by Djerassi[5] who has used it widely to rationalize the fragmentation patterns of many classes of organic compounds. The concept is of a charge localized, for example, on a heteroatom such as O, N or S from each of which reactions can occur. The classic example is the McLafferty rearrangement in ketones.

Localization of charge in the product ion can also be considered to promote reaction. Such localization is equivalent to forming the products in their most stable states. In a sense then, charge is localized throughout the sequence: reactant → activated complex → products, and so the potential energy of the system is continuously minimized. Stability of the products provides the 'drive' or selectivity for the reaction to proceed. In moving towards the activated complex, many possible conformations will be obtained on a statistical basis. If, in one of these, atoms are brought into relative positions closer to those which they would occupy in the products the 'localized stability' so produced means that more energy is available to drive the reaction. There is thus a cascading effect, increasing the reaction rate in the *appropriate* direction. Again, it must be emphasized that these structures with localized charges are not necessarily the only, or indeed the most numerous, structures present. They merely represent the structures of particularly reactive species. Generally speaking, any structural feature that aids in delocalizing the positive charge over the whole ion structure will increase the stability and reduce the amount of fragmentation that occurs. Thus in many aromatic systems the ion of lowest enthalpy may correspond to one in which one of the π-electrons has been removed. Such a structure may lead to concentration of the excess energy then available for fragmentation in a side chain and there may be a high probability of fragmentation in this fashion. The ring itself can, however, behave as an efficient energy sink and fragmentation in the ring is much less likely.

A question arises as to whether the atoms, as they are located in the product ion, have rearranged from their positions in the reactant ion. A useful guide to the detection of such rearrangements is the comparison of spectra taken at high and low bombarding electron energies. Rearrangement ions are relatively more prominent at the lower electron energies, the reason being that their lower frequency factors are compensated for by their lower activation energies.[12]

GROUNDWORK STUDIES

Spectral Comparison

An early example of a fragmentation in which suggestions were made for the structure of the activated complex was the loss of neutral CO from metastable molecular ions of anthraquinone.[13] In this case, fragmentation was assumed to take place from the low enthalpy molecular ion in which the charge is localized on an oxygen atom. The atoms of carbon and oxygen that ultimately form the neutral molecule of carbon monoxide are already close to their optimum relative positions for forming carbon monoxide in the anthraquinone molecule and it only requires a change in the hybridization of the carbon from the trigonal form towards a digonal arrangement to allow a more stable form due to increased strengths in the σ and π bonds of the carbonyl group. An intermediate state is visualized in which neutral carbon monoxide has almost separated from the ion. In this form the carbon

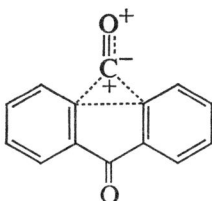

hybridization has changed, the polarization induced in bonds adjacent to the CO group has drawn negative charge towards the carbon and any movement of the two rings towards one another will lead to increased stability and also help to 'drive' the reaction.

In the elimination reaction discussed above, the fragment ion formed is postulated to assume the structure of the molecular ion of fluorenone. If this is so, and in addition the loss of CO from the molecular ion of anthraquinone is the dominant mode of fragmentation of this ion, one would expect the mass spectrum of fluorenone to be similar to that of anthraquinone. Indeed from mass 180 (the molecular weight of fluorenone) downwards, the spectra are almost identical. In this case, one assumes the structure of the fluorenone molecular ion to be essentially the same as that of the neutral molecule, but this assumption cannot always be made and indeed, comparisons of spectra have been used to deduce that some ion structures markedly different from the corresponding neutral species can be formed. The method of spectral comparisons was extended to include consideration of metastable peaks by Brown and Djerassi[14] who applied it in showing that the ion formed by loss of CO_2 from methyl phenyl carbonate was apparently identical to the molecular ion of anisole.

Use of Labelled Isotopes in Structure Determination

Some of the earliest studies in this general area were those of Meyerson and Rylander[15] and were concerned with the structure of the $C_7H_7^+$ ions that form the base peak in the mass spectrum of toluene. Calculations had shown that although the benzyl radical is more stable than the tropylium radical, the situation is reversed when the even-electron positive ions are compared, the tropylium ion having the lower enthalpy. Meyerson and Rylander labelled toluene with an atom of ^{13}C in the α-position and looked at the fragmentation reactions of the $(M-1)^+$ ions. They found that when the elements of acetylene were eliminated, the chance of losing the ^{13}C atom in the neutral moiety was precisely 2/7 showing that all seven carbon atoms had become equivalent, and suggesting that a seven-membered ring had formed to give the tropylium ion structure. This was the first example of the method of characterizing the structure of an ion by its fragmentation reactions. This method is now widely used and further examples of it are given below.

Another early example of work using labelled materials that added considerably to the ideas concerning ion structure used *tert*-butylbenzene, the labelled material being 1-^{13}C-(1,1-dimethyl)ethylbenzene.[16] The molecular ion of this compound fragments with loss of one of the methyl groups and interest centred upon the structure of the fragment ion so formed. This dissociates with loss of two carbon atoms and there is a metastable peak corresponding to the loss of the elements of ethylene to give a $C_7H_7^+$ ion. There are no C_8 peaks as might be expected if single carbons could be lost in succession. The isotopic distribution found in the C_7 ions was such as would be expected if the three side chain carbons of the first-formed fragment ion had become indistinguishable. It was proposed that this was a consequence of the formation of a substituent symmetric with respect to the ring, namely a cyclopropane co-ordinated with the phenyl cation.

Much work has been carried out using isotopically labelled compounds in order to determine which particular atoms take part in a fragmentation, and from this work it has sometimes proved possible to infer the structure of the fragmenting ion. A pioneer of this approach has been Djerassi. For example, 1-keto steroids can show a prominent ion at m/e 124 which can be visualized as arising by simple ring cleavage:

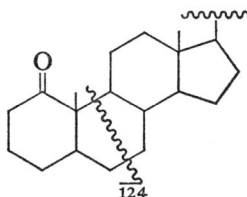

However, deuterium labelling has shown, among other results, that the C-5 hydrogen is lost. Hence this reaction must involve hydrogen transfer in two directions (reciprocal hydrogen transfer). The following mechanism was therefore suggested:[17]

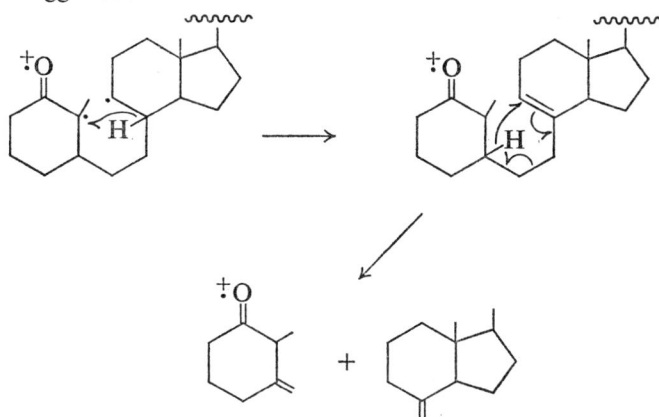

Thermochemical Measurements

It has been appreciated since the earliest days of mass spectrometry that appearance potential measurements of fragment ions can be used as a basis for the determination of the enthalpies of these ions and hence to infer their structure. Much of the early work was concerned with the enthalpies of hydrocarbon ions, for example the $C_4H_9^+$ and $C_3H_7^+$ ions that are formed in very high abundance in the mass spectra of most alkanes. It was assumed that the very large numbers of these ions formed was an indication of their stability. The basis of the method is indicated in Fig. 1.

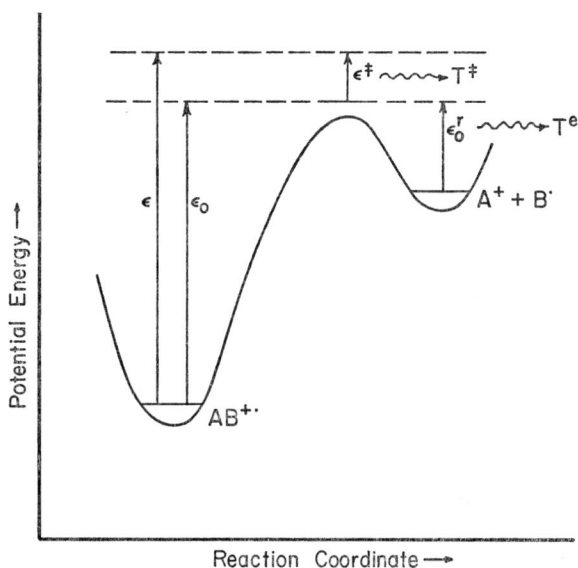

Reaction Coordinate →

FIGURE 1

The enthalpy of the ion $AB^{+\cdot}$ in the transition state is equal to the sum of the enthalpy of this ion in its ground state and the appearance potential. This was equated to the sum of the enthalpy of the neutral fragment $B^{\cdot}$ and that of the ion A^+ to obtain a value for the enthalpy of this product ion. The method was of very limited value because of the failure to realize that a kinetic shift operates in such cases[18] leading to an excess energy $\varepsilon^{\ddagger}$ in the activated state and also to an inability to estimate the value of the reverse activation energy of the process which must be corrected for. Discussion later in this paper will show how these difficulties are currently beginning to be overcome.

MODERN METHODS

Approaches Based upon Ion-Molecule Reactions

Reactions at low kinetic energy—Much is currently being learned about ion structures by several methods which depend upon the study of ion–molecule reactions. These reactions can be assumed to take place from ions which are stable to unimolecular fragmentation, thus stable ion structures are characterized. It does not necessarily follow that the structures found are those with which the ions are formed at threshold since ion–molecule reactions have cross-sections which may show a pronounced dependence upon the internal energy of the ion. This factor is particularly important if the relative kinetic energy of the reactants is low. This is so in studies by ion cyclotron resonance (ICR), as well as in many tandem mass spectrometer and ion source experiments.

Typical of the ICR studies is the demonstration that the product ions $C_3H_6O^{+\cdot}$ formed by McLafferty's rearrangement of the 2-hexanone molecular ion or by double rearrangement in 5-nonanone have the same structure as the enolic ions generated from 1-methylcyclobutanol.[19] Three reactions were found which were undergone by each of the above ions; in addition three other reactions, undergone by the keto ion generated by ionization of acetone, were not observed for any of the enolic ions. The great advantage of ICR in studies such as these is the fact that, using double resonance techniques, the masses of the species in the reaction can be completely specified. In these experiments a particular daughter ion is chosen for observation and the effect upon its abundance of irradiating at the frequencies of possible parent ions is tested.

The importance of keto-enol tautomerism in the chemistry of neutral molecules has led to much interest in the possibility that keto and enol ions may interconvert. The ICR experiments by Dickman *et al.* described above indicated that the $C_3H_6O^{+\cdot}$ species do not so interconvert on the usual ICR time scale. Haas *et al.*[20] have shown that the enolic $C_3H_6O^{+\cdot}$ ion formed from 2-hexanone does not ketonize even when the time scale is extended to ~ 0.1 sec. However, on this time scale, the ion formed from 2-propylcyclopentanone gives reactions typical of the keto ion in contrast to its behaviour after 10^{-3}–10^{-4} sec. Apparently this cyclic product ion can, given sufficient time, ketonize. Moreover, ketonization can be prevented by increasing the number of collisions that the ion undergoes in the ICR cell, perhaps because such collisions reduce its internal energy.

Although the masses of the reactant and product ions are determined by these ICR experiments, only a certain feature of the ion structure is probed. If, for example, in the 2-propylcyclopentanone case, allylic cleavage were to occur it could lead, by the path shown below, to a keto ion other than cyclopentanone molecular ion.

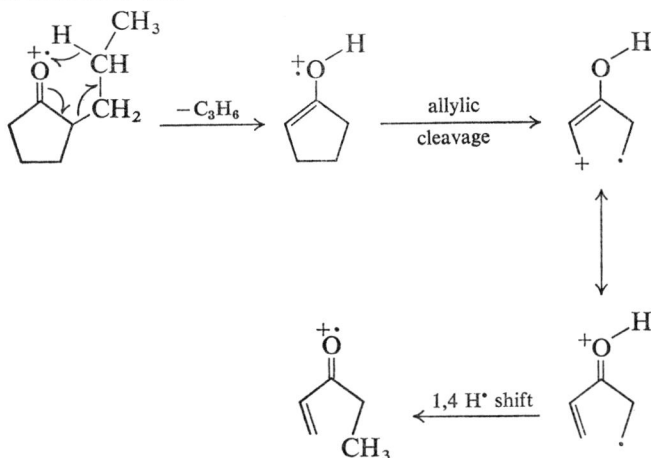

$$ \text{(reaction scheme: } 2\text{-propylcyclopentanone molecular ion} \xrightarrow{-C_3H_6} \text{cyclopentanone ion} \xrightarrow{\text{allylic cleavage}} \text{ring-opened ion} \downarrow \xrightarrow{1,4\ H^\bullet\ \text{shift}} \text{keto ion with } CH_3 \text{)} $$

This reaction does not correspond to keto-enol tautomerism and hence no general deduction concerning the tautomerism is possible. Because the ion molecule reactions used to determine ion structures are relatively unspecific, they cannot distinguish between the alternative keto ion structures. New reactions which might do so could be sought.

Differences in reactivity of ions of identical structure but different internal energies have been met in several recent ICR studies and have caused some difficulty. This has been particularly true of studies on the $C_3H_7^+$ ion.[21,22] In the light of these difficulties, it seems that use of multiple sources of the ion in question and as many probe ion–molecule reactions as possible must be practised.

The tandem mass spectrometer is frequently used for the study of ion–molecule reactions at low kinetic energy. It can be applied to problems of ion structure using conventional product ion abundance data, analogous to that obtained in ion source experiments but, in addition, accurate measurements on the relative rates of charge or proton transfers can be made and thermochemical thresholds can be measured and used in structural characterization. In spite of this experimental flexibility, problems are met in attempting to characterize ions by these techniques because of the effects of internal energy on reactivity. This may preclude definite conclusions about structure, even in careful work of the type done on the $C_4H_8^{+\bullet}$ ion.[23]

While studies of ion–molecule reactions within an ion source have generally concentrated on small ions and problems other than ion structure, the revival of this field in the guise of chemical ionization, is expected to produce much data bearing upon our topic. As an illustration of the type of information obtained from ion source reactions consider the results on $C_4H_9^+$ generated from n-butane and iso-butane.[24] When allowed to interact with various alkenes the former is much more reactive and it is concluded that it yields the sec-butyl cation which does not rearrange on the time scale of the experiment to the more stable tertiary isomer generated from iso-butane. The observed reactivity differences are in this instance unlikely to be due to differences in internal energy although this possibility has not been completely eliminated.

Somewhat related to ion source studies is a recently developed method[25] in which ion structures are inferred from consideration of the *neutral products* of ion–molecule reactions. The method has obvious limitations in terms of sensitivity, inability to distinguish individual steps in the sequence leading to the end product and the lack of control over ion lifetime. Isomerization in the ion prior to reaction is not readily distinguished from isomerization in an intermediate excited neutral species. In spite of these difficulties, results on a number of hydrocarbon ions, not always in agreement with mass spectrometric data, have been obtained.

Reactions at high kinetic energy—High energy ion–molecule reactions are considered to be those in the energy range from several hundred to more than 10 000 eV. The high energy reactions which give structural information are collisional excitation, followed by fragmentation, and also various charge transfer processes including charge stripping.

Ion–molecule reactions which lead to dissociation of the products can also occur at low relative translational energies if the ion is internally excited. Such processes, which apparently proceed via long-lived complexes, are mechanistically quite distinct from the high energy collisional processes considered here. In these latter reactions there is little momentum transfer, the reaction cross-section is large and the ion is excited by interaction with the electrons of the target species the nature of which has only secondary effects on the reaction. For scattering angles close to zero and at energies of several thousand electron volts almost all the energy lost by the fast moving ion serves to increase its internal energy; very little, in any form, is transferred to the neutral target.[26]

After collision, the excited ions may fragment in the field-free region in which collision occurred, giving products which can be analysed by kinetic energy and mass to characterize the reaction completely. This characteristic, as well as the experimental methods used, is reminiscent of metastable ions. Collision-induced decompositions, however, are sharply distinguished from metastable ion reactions in many ways, not the least important being the fact that the energized ions may have a large range of internal energies and may fragment almost immediately upon impact or as much as several microseconds later. The method shares one most significant advantage with metastable ion studies—because of the high velocity of the fragmenting ion the kinetic energy release accompanying the reaction is amplified in the laboratory co-ordinate system. Using an instrument of good energy resolution it is thus possible to measure this energy release, no matter how small it is. In characterizing ions by their collision-induced dissociations it is possible to measure several distinct properties. The most readily measured is the relative abundance of the various products, which can be determined even on low energy resolution instruments. Such measurements have been made for complex ions by Jennings[27] and later by McLafferty and co-workers.[28] As one example of the conclusions possible $C_7H_8^{+\cdot}$ ions have been studied by this technique and it is found that the ions generated from n-butyl benzene have the same structure as the cycloheptatriene molecular ion which appears to be different from that of the toluene molecular ion.[28] In addition to the abundance measurement, the kinetic energy release is derivable from the collision-induced dissociation peak shapes. Moreover, an accurate measure of peak position gives the kinetic energy lost in each process.[29] Since these energy losses may be 10 eV or more, they provide yet another distinguishing feature of the reactant ion. There are then, three measurements which can be used in studying ion structure by collision-induced dissociation. Of these, relative abundance measurements are now being made routinely but kinetic energy releases and kinetic energy losses have only been determined in a few cases.[26]

The most notable feature of collision-induced decompositions is that they are expected to be largely independent of the internal energy of the reactant ion (but not of the internal energy acquired after excitation). It is possible that cases may exist in which there is more than one structure represented in the stable ion beam entering the collision chamber (usually some microseconds after ion formation). Such a mixture of structures is expected to be a rare complication in what is clearly one of the most promising techniques of ion structure determination. Certainly the full range of information available on peak positions and shapes is not yet being utilized.

An experimental point of some note concerns the advantages of using a reversed sector instrument, that is a mass-analysed ion kinetic energy (MIKE) spectrometer, for studies of collision-induced dissociation and other high-energy ion–molecule reactions. In this technique all reactions proceeding from a given ion can be studied under the same ion source conditions by mass selecting the reactant ion and then scanning a kinetic energy analysing device such as an electrostatic sector. No high-voltage scanning is necessary and a simple computer-compatible system can be built. It is also possible, using exactly the same equipment, to study other high-energy ion–molecule reactions such as charge exchange of doubly-charged ions, charge stripping and charge inversion. ((1)–(3), below, where N is the target species.)

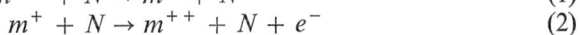

$$m^{++} + N \rightarrow m^+ + N^+ \tag{1}$$
$$m^+ + N \rightarrow m^{++} + N + e^- \tag{2}$$
$$m^+ + N \rightarrow m^- + N^{++} \tag{3}$$

The kinetic energy loss (and in some cases gain) accompanying these reactions provides a new source of structure-specific thermochemical data. The methods allow the determination of double ionization potentials and electron affinities, as well as characterizing states of different internal energy represented in an ion beam of given kinetic energy and a single mass to charge ratio.[26] Although a start has been made in kinetic energy measurements on the products of these reactions their cross-sections are almost entirely unexplored.

Ion–Photon Interactions

There has, because of experimental difficulties, been far less work on ion–photon than on ion–molecule interactions. The detailed information potentially available through ion–photon reactions, however, is now the object of a growing amount of research. Of the work which has been done, little is yet structural in the sense employed here. For example, the photodissociation studies of Durup and co-workers[30] have taken $H_2^{+\cdot}$ as their subject. The methods being developed in this work however, should be applicable to more complex ions and already Dunbar and Fu[31] have shown that when $C_7H_8^{+\cdot}$ ions are photodissociated in an ICR cell and the residual $C_7H_8^{+\cdot}$ ion current is plotted as a function of wave-length, three entirely distinct curves are obtained depending upon whether toluene, cycloheptatriene or norbornadiene is used as the source of $C_7H_8^{+\cdot}$. Clearly three structurally distinct and stable types of $C_7H_8^{+\cdot}$ ions are formed in the three experiments. Gross differences in ion internal energy would of course affect the shapes of these curves, but the method has the advantage, common to all experiments done some time after ion formation, that some internal energy selection is achieved.

Using conventional mass spectrometric instrumentation, photo-excitation has been employed in the study of such organic ions as $C_5H_{12}^{+\cdot}$.[32] This approach is inherently flexible in that irradiation can be carried out either in a field-free region or within the ion source and the effect on the mass spectrum recorded. The photon flux from tunable lasers is sufficiently high to excite efficiently ions already produced by electron bombardment, although the quantum energy may be too small to permit direct photo-ionization.

Absorption and emission spectra of some simple ions have been recorded by Herzberg.[33] Such data give highly specific information on molecular structure but application to complex ions under mass spectrometric conditions seems likely to be difficult.

The last of the photometric methods, which like all the others is still in an early stage of development, is photodetachment. This experiment involves measurement of the photon energy required to convert a negative ion to the neutral molecule so that the electron affinity is obtained directly, without complications due to ionic solvation.[34]

A great deal of structural information including electronic and vibrational levels in small ions is now being obtained by photo-electron spectroscopy. We shall not attempt to cover this work which has already been the subject of several reviews[35] and which only borders on the field of mass spectrometry as does the complementary subject of Penning ionization electron spectroscopy.[36] We do, however, note that photo-electron spectroscopy has been used to provide information concerning the geometry of triatomic ions.[37] In addition, applications to more complex molecules, made in conjunction with photoionization and theoretical calculations, have led to detailed interpretations of electronic structure in such molecules as pyridine.[38]

Thermochemical Methods

As already stated, a sufficiently accurate measurement of the standard heat of formation of an ion should be one of the best methods of distinguishing between isomeric structures. The problem of course lies with the determination of these quantities which has frequently been made by inaccurate methods and more importantly, with insufficient consideration for the subtleties of the measurement including the excess energy terms involved. Recent reviews[39,40] have emphasized the presence of these terms, the kinetic shift and the reverse activation energy, as well as the difficulties in estimating them. It should be noted that the same terms appear in photoionization as in electron impact measurements. Indeed, with the development of the electron energy selection technique,[41] some of the most important disadvantages of electron impact ionization have been removed and values of appearance potentials in good agreement with photoionization results are routinely obtained by this method. The next step in any ion enthalpy determination should be the evaluation of these excess energy terms. A fraction of the kinetic shift can be determined from the difference in appearance potentials between metastable and normal daughter ions.[42] This method has not been widely applied although it would seem to have much to offer. Another important step in correcting for the excess energy terms comes from measurements of the translational energy distribution of the product ions. In principle, such measurements, taken over a range of ion life times, could be used to determine both the reverse activation energy and the excess energy of the activated complex. The latter term controls the rate of fragmentation and is, from all the available evidence, statistically partitioned between translation energy of separation and internal excitation of the products. Hence, the experimentally determined kinetic energy release is composed of two terms, only one of which is time dependent. Measurements over a range of life times allow the separation of the two terms. Both excess energy terms can be determined from the separate components of the kinetic energy release if the mode of energy partitioning is known in each case. Statistical partitioning of the excess energy of the activated complex may be assumed but the manner in which the reverse activation energy is distributed between translational and internal energy of the products is much more complex and data on ions are only just beginning to be acquired.[43] Measurements on metastable ions can provide the necessary kinetic energy values and even at this early stage in the development of these techniques, considerable improvement in enthalpy values is possible merely by correcting for the entire measured kinetic energy release (the actual correction must be at least this large). For example, the heats of formation[44] of $C_5H_7^+$ ions formed from a series of C_5H_8 molecules generally lie approximately 8 kcal mole^{-1} above those measured for $C_5H_7^+$ ions formed from C_6H_{10} molecules. Although this is *prima facie* evidence for two different structures this conclusion is called in question by the kinetic energy release data (referring to metastable ions, not necessarily exactly the same energy ions as referred to in the appearance potential experiments). Cyclopentene loses H⋅ to give $C_5H_7^+$ with an average kinetic energy release of 3 kcal mole^{-1} while cyclohexene releases 0·3 kcal mole^{-1} in forming the same ion by methyl loss. Since the kinetic energy release is unlikely to represent the entire excess energy (it is for example about 60% for H⋅ loss from the $C_7H_8^{+⋅}$ molecular ion)[45] these results suggest that the apparent enthalpy difference between the two $C_5H_7^+$ ions may lie only in the excess energy term and that their structures may be identical.

Before leaving this subject, it is worth emphasizing that the magnitude of the excess energy correction may be very large. The loss of formaldehyde from the anisole molecular ion to give ionized benzene is accompanied by an average kinetic energy release of 0·32 eV and this represents only about 15% of the excess energy.[46] Clearly differences in ion structure may be completely masked in the presence of terms of this size, unless corrections are made for these terms.

Isotopic Labelling Studies

Isotope effects—This section will deal both with reactive ions and with the kinds of stable ions already discussed above. Isotope effects upon ion abundance as well as upon kinetic energy release contain implicit information on ion structure and can be accurately measured. We shall, however, only consider kinetic isotope effects.

Although measurements of these effects can readily be carried out for reactions occurring in the ion source (giving rise to integral mass peaks in the mass spectrum) there are considerable advantages in studying reactions of metastable ions. These include (i) the specificity with which the process is defined so there is no possibility of contributions from other reactions to the measured ion abundance, (ii) the lower internal energies of metastable ions which make the isotope effects much larger than for the high energy ion source reactions.[47] For simple cleavage reactions it is possible to find very large isotope effects in metastable ions; hydrogen atom loss from iso-butane has a value for H'/D' of over 1000. It has been proposed that a valuable indication of whether a hydrogen atom elimination occurs by simple bond cleavage or by rearrangement (hence giving an indication of ion structure) is provided by the magnitude of the kinetic isotope effect upon H'(D') loss.[48]

An example of the application of kinetic isotope effects to problems of ion structure comes from examination of the $(M-C_2H_4)^{+\cdot}$ ion formed from substituted acetanilides.[49] This ion may have either the anilinic structure or the isomeric imino structure.

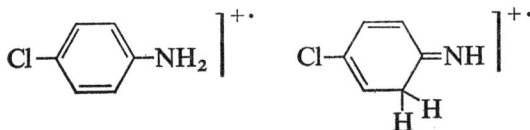

p-Chloroaniline shows metastable peaks for both Cl' loss and HCN loss from the molecular ion. *N*-deuteration decreases the metastable peak for HCN loss, relative to the reference reaction, by a factor of about two. If *p*-chloro-acetanilide yields the imino ion directly it is proposed that HCN loss should follow without conversion to the anilinic form and there should be no primary isotope effect on the metastable ion abundance. In fact an isotope effect of about 2 was again observed, a result that indicates the intermediacy of the anilinic ion in the reaction sequence rather than direct formation of the imino ion.

Isotopic distributions—Consideration of the route of labelled atoms through a fragmentation sequence is a well-established procedure for determining ionic fragmentation mechanisms and ionic structures. The possible complications arising from the exchange of groups between equivalent positions must be carefully considered. Complete randomization of the hydrogen atoms in numerous aromatic ions has been found in metastable ions.[50] Since there is only one experimental set of data—the label distribution in the product ion—it may be difficult to distinguish between molecular ion isomerization *before* fragmentation (for each process there will be some characteristic activated complex which may make certain atoms or groups equivalent) and any isomerization which is implicit in attaining the activated complex for fragmentation.

Carbon labelling of benzene establishes that the ring atoms can interchange positions and so, independently, can the hydrogen atoms which are bonded to these carbons.[51,52] This experimental result can only be achieved by the examination of fragment ions and yet it gives no information about the structure of the activated complex for the fragmentation reaction itself. What it does indicate is that highly symmetrical isomeric ions such as prismane are present at some stage of the exchange/fragmentation sequence.

Recent studies have provided direct evidence for the exchange of substituents around an aromatic nucleus. For example in 2-phenylthiophene[53] the phenyl group can migrate independently of the carbon atom to which it is attached. It seems probable that such substituent exchange occurs in an acyclic form of the molecular ion in contrast to ring atom rearrangement which may well involve a cyclic ion structure.

Kinetic Methods

Metastable ion abundances—An ion may be characterized by the set of all the reactions that it undergoes. Thus the fragmentation pattern arising from a given ion serves to fingerprint the structure of that ion. Mass spectra do not provide data of this type since the reactions leading to a particular ion are not completely defined. However, if reactions of metastable ions are studied both parents and daughters of all transitions are determined. It is most convenient to list all the reactions that any particular ion will undergo, and experimentally this is most easily done by using a reversed sector mass spectrometer. Even the simple listing of all the reactions that a particular ion can undergo serves, in almost every case, to differentiate between all possible isomers. It is, however, so simple a matter to measure the abundances of the products of fragmentation of metastable ions that most of the early work employed such measurements.[54] A further extension to the techniques of identification, concerned with measuring the shapes of metastable peaks, is treated in the next section.

The interpretation of measurements of relative abundances initially met with some difficulties associated with the effects of internal energy. It is now established that the internal energy indeed affects abundance ratios, but the magnitudes of these effects and the factors controlling them are established.[55,56] It is thus possible to differentiate between different structures even in the face of the differences in internal energy which arise when ions are prepared by different methods. To illustrate the magnitude of the internal energy effects and the ease with which isomeric ions may be distinguished, consider the data on $C_3H_7O^+$ given by Tsang and Harrison.[56] The four isomeric metastable ions $C_2H_5O^+=CH_2$, $(CH_3)_2C=O^+H$, $C_2H_5CH=O^+H$ and $CH_3O^+=CHCH_3$ each generated from several sources lose H_2O, C_2H_4 and CH_2O in distinct ratios that are very different for the four species. For example the ratios for loss of H_2O and C_2H_4 give average values of 3·0, 0·5, 1·7 and 0·01 respectively.

It should be noted that special precautions must be taken when there is a large disparity in intensity between the two metastable peaks being considered.

Small errors translate into large ratio differences and there is also the ever-important question of whether the weaker peak is actually due to a metastable ion decomposition rather than a collision-induced dissociation. The method may be misleading if an ion isomerizes to a second structure before fragmentation occurs and the isomerization rather than the fragmentation is the rate limiting step. Uccella and Williams[57] have apparently uncovered a case of such behaviour in the $C_9H_{11}{}^+$ ion. A large number of compounds were examined and although they fell into two classes in terms of metastable ion abundance ratios a single reacting ion structure is thought to be responsible. It seems to us that this situation represents a real difficulty in applying this technique. Appreciable differences in abundance ratios arise because the isomerization step may introduce appreciable internal energy differences that are not filtered out as they normally are for reactions of metastable ions.

Another technique that has advantages in studying fragmentation mechanisms and ion structure is field ionization. Since ions move rapidly through a high field, the position, and hence time, of fragmentation can be determined by measurements of the kinetic energy of product ions. Unimolecular reactions occurring in times of the order of 10^{-11} sec can be followed. This enables the structures of fragmenting ions to be followed during the course of internal rearrangements. A recent example is concerned with the loss of various deuterated methyl radicals from labelled cyclohexene.[58] These results show that the mechanism involves vinylic cleavage and that the time for randomization of the deuterium atoms in the ionic structure is of the order 10^{-9} sec.

Another application of field ionization is to determine whether or not a fragment ion is formed via a rearrangement reaction. This is done by comparison of electron impact and field ionization spectra, rearrangement ions tending to give stronger peaks following field ionization.[59]

Kinetic Energy Release

The last parameter with which we shall deal that gives information about ion structure is the translational energy released when an ion undergoes fragmentation. We shall only discuss such measurements on metastable ions since these have the following advantages: (i) When metastable ions are sampled there is energy 'filtering' resulting in a low and controlled internal energy, (ii) there is no question as to the identity of the reactant ion when a particular product ion is examined and (iii) the kinetic energy release in the centre-of-mass system is amplified in the laboratory system by the high velocity of the ion. This means that very small kinetic energy releases could be detected (estimated to be as small as 10^{-6} eV) if such processes existed, while the entire range observed in practice extending from 10^{-4} eV to 10 eV is covered, high reproducibility being achievable.[26]

An advantage of using kinetic energy release to characterize a reaction, rather than to use the relative abundances of the ions formed, is due to the fact that the kinetic energy contains considerably more detailed information concerning the reaction. As well as giving an energy release measured from the width of the peak observed, the relative probabilities of releasing various amounts of energy is contained in the *shape*. Fine structure in the shapes of metastable peaks adds further information and is discussed later.

A simple example of the application of the above principles to a problem of ion structure concerns the $C_8H_8O^{+\cdot}$ ion formed from butyrophenone. The question to be answered is whether this fragment ion has the same structure as the molecular ion of acetophenone. Both ions fragment further to give $C_7H_5O^+$, the reaction being accompanied by release of 7 meV in the case of acetophenone and 46 meV in the other case.[60] It is therefore concluded that the $C_8H_8O^{+\cdot}$ ions have different structures, as shown below.

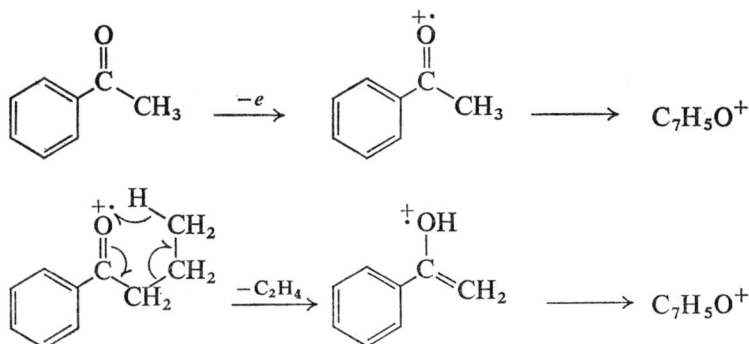

A recent isotope labelling study[61] has demonstrated that methyl loss from the McLafferty rearrangement product ion involves one ring hydrogen atom, in agreement with the above conclusion.

The sources of the translational energy measured above are the reverse activation energy and the non-fixed energy of the activated complex. Provided that ions having a broad distribution of internal energies are formed in the ion source, the instrument will 'select' for examination only ions having relatively small internal energies covering a rather narrow range. Hence changes in the internal energy distribution of ions formed in the source will have only a small effect upon the kinetic energy release, even when the contribution to this energy from the reverse activation energy is negligible. Of course, when the reverse activation energy makes the major contribution to the energy released, internal energy effects become even less important.[62,63] As already discussed in connection with ion abundance measurements, there is one set of circumstances that can affect the conclusions drawn from any data on metastable ions, and this is concerned with a high energy isomerization occurring prior to fragmentation.

Unique structural information can be obtained by study of charge separation reactions of doubly-charged ions. Many of these ions are metastable and the process is necessarily accompanied by the release of considerable energy due to coulombic repulsion between the charges. The energy release translates directly into an approximate intercharge distance in the decomposing ion.[64] In the case of small ions, the precise energy release can be used to infer energy levels[65] of the products and reactants while in more complex ions it can be used to follow the rearrangement of substituents on aromatic rings, the localization of charge on particular atoms and reactions accompanied by gross changes in conformation, such as ring opening.[66]

Consider, for example, the doubly charged molecular ion of aniline which fragments as shown:

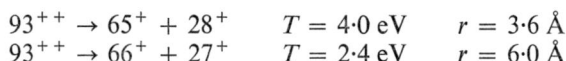

$$93^{++} \rightarrow 65^+ + 28^+ \qquad T = 4 \cdot 0 \text{ eV} \qquad r = 3 \cdot 6 \text{ Å}$$
$$93^{++} \rightarrow 66^+ + 27^+ \qquad T = 2 \cdot 4 \text{ eV} \qquad r = 6 \cdot 0 \text{ Å}$$

Formation of 27^+ ($C_2H_3^+$) clearly occurs from a different structure than formation of 28^+ (H_2CN^+) and the intercharge distances suggest linear and cyclic reacting structures respectively. By way of contrast, entirely distinct molecules may generate a single reacting ion as shown by kinetic energy release data. Thus 1,4-diaminobenzene and 1,4-diamino-2,5-dimethylbenzene both yield the ion 108^{++}, which fragments to give $80^+ + 28^+$ with an energy release of $2 \cdot 60$ eV ($r = 5 \cdot 51$ Å) and $2 \cdot 55$ eV ($r = 5 \cdot 6$ Å) respectively.[66]

Using these measurements on doubly-charged ions it is sometimes possible to 'watch' a cyclic ion ring-opening as hydrogens are removed from it. The three isomeric xylenes yield doubly-charged molecular ions which behave identically suggesting that they have isomerized to a common structure. One transition undergone by these ions is loss of CH_3^+ for which the intercharge distance in the reacting ion, calculated from the kinetic energy release, is $5 \cdot 5$ Å. The $(M-H_2)^{++}$ ions undergo the analogous reaction and the inter-charge distance is measured to be the same, indicating that loss of H_2 has not been accompanied by skeletal rearrangement. However, for the $(M-H_2-H_2)^{++}$ ions CH_3^+ elimination is associated with an intercharge distance of $7 \cdot 7$ Å, suggesting that ring opening accompanies removal of the second hydrogen molecule.

The best method for studying abundances of doubly-charged ions is to plot 2E mass spectra, that is, spectra in which doubly-charged ions formed in the ion-source are charge exchanged in the field-free region using a collision gas and in which only the products of these reactions are transmitted by the electric sector and detected. By combining data on the fragmentation pathways of doubly-charged ions with ion abundance and kinetic energy release data it is possible to infer ion structures. Of particular interest is the fact that certain ions apparently have exceptional stability and occur quite generally, for example, in the mass spectra of hydrocarbons.[67] The ions $C_mH_2^{++}$ ($m = 2$–5) and $C_nH_6^{++}$ ($n = 4$–10) apparently have particular stability and the types of reaction they undergo and the kinetic energy release accompanying these reactions, suggest linear structures bearing only terminal hydrogen atoms. Such structures include $H-C^+=C=C=C^+-H$ and $H_3C-C^+=C=C=C=C=C^+-CH_3$.

It has recently been found that some metastable peaks possess composite structures[68] and this observation has proved to be a further powerful probe into ion structures. Even when such a feature is not completely resolved, new ion structural information may be obtainable. For example, metastable $C_7H_7^+$ ions formed from a variety of sources, undergo C_2H_2 loss with release of between 29 meV and 45 meV of kinetic energy.[69] There is a correlation between the presence of a benzylic group in the molecular ion and the larger energy releases. These results are most simply interpreted in terms of fragmentation from two distinct structures, benzylic and tropylium. An independent study[70] of the same reactions has shown that the peak shapes can be synthesized by mixtures of functions representing the two extreme types associated with benzylic and tropylium groups in the fragmenting ions.

In cases in which composite metastable peaks are readily resolved, as for example in NO˙ loss from nitrobenzenes, it has always been possible to suggest alternative structures for the daughter ions. It is possible to study substituent effects upon the relative abundance of each component of the metastable peak and upon the energy release in each reaction. Such a study[71] showed clearly that the accepted three-membered cyclic transition state for nitrobenzene fragmentation was indeed involved, the magnitude of the energy release, as well as the peak abundance, increasing with the electron donating power of the substituent. In addition, this type of data indicated that the competitive process probably occurs by oxygen migration to the ortho position as shown below.

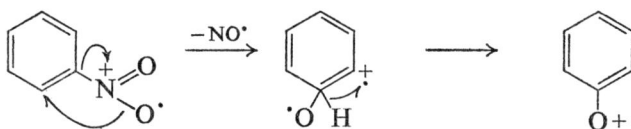

$$\text{(benzene ring with } N^+O_2 \text{ group)} \xrightarrow{-NO^{\cdot}} \text{(benzene ring with } ^{\cdot}O\,H) \longrightarrow \text{(benzene ring with } O+)$$

As a final illustration of the use of metastable peak shapes, let us consider its application for determining ion structures and fragmentation mechanisms, not in the field free region, but in the ion source.[72] It is known that anisoles give composite metastable peaks for formaldehyde elimination and that the peak shapes for p- and m-chloroanisole are distinctive. If a reaction occurring in the ion source generates the m- or p-chloroanisole molecular ion then the subsequent fragmentation of this metastable ion can be used to determine its structure. Application of this principle (which should be quite general) to the

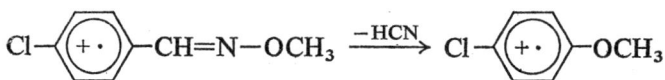

$$\text{Cl}\!-\!\!\langle +\cdot \rangle\!-\!\text{CH}=\!\text{N}\!-\!\text{OCH}_3 \xrightarrow{-\text{HCN}} \text{Cl}\!-\!\!\langle +\cdot \rangle\!-\!\text{OCH}_3$$

reaction showed that p-chloro oxime ether fragmented to give an ion having the m-chloroanisole molecular ion structure and vice versa. The oxime ether reaction in the ion source therefore occurs via a five- rather than a four-centered activated complex.

CONCLUSION

The subject of ion structure has developed slowly and the current ideas contain contributions from a variety of fields of knowledge. A distillation of these incoming methods and ideas has been difficult to achieve and the subject is still advancing as our appreciation of the basis for each of the individual methods is clarified.

A continuing problem, encountered to a greater or smaller extent in each method of determining ion structures, is the separation of the effects of internal energy from effects due to ion structural differences. It is hoped that charge exchange will come to be more commonly used in problems of ion structure.

In summary, this article has sought to provide some appreciation for the way in which our present ideas on ion structure have developed and to present, within a logical framework, a description of the current status of the subject. The problem is sufficiently large that it will continue to reward those who develop fresh approaches to it. There remains much to be done in extending and testing the methods discussed here.

ACKNOWLEDGMENT

We thank the National Science Foundation for its support of our work described in this article.

REFERENCES

1. Washburn, H. W., Wiley, H. F., Rock, S. M. and Berry, C. E., *Ind. Eng. Chem.*, Anal. Ed., 1945, **17**, 74.
2. Mohler, F. L., *J. Wash. Acad. Sci.*, 1948, **38**, 193.
3. Beynon, J. H., *Nature*, 1954, **174**, 735.
4. McLafferty, F. W., *in* 'Mass Spectrometry of Organic Ions' *ed.* F. W. McLafferty, Academic Press, 1963.
5. Budzikiewicz, H., Djerassi, C. and Williams, D. H., 'Interpretation of Mass Spectra Compounds', Holden-Day, Inc., 1964.
6. See, for example, Meyerson, S. and McCollum, J. D., *Advan. Anal. Chem. Instrum.*, 1963, **2**, 179.
7. Grubb, S. and Meyerson, S., *in* 'Mass Spectrometry of Organic Ions', *ed.* F. W. McLafferty, Academic Press, 1963.
8. Spiteller, G., *in* 'Some Newer Physical Methods in Structural Chemistry', *ed.* R. Bonnett and J. G. Davis, United Trade Press, London, 1967.
9. Lester, G. R., *in* 'Advances in Mass Spectrometry', Pergamon Press, London, 1959.
10. Stevenson, D. P., *Disc. Faraday Soc.*, 1951, **10**, 35.
11. Svec, H. J., personal communication.
12. Williams, D. H. and Cooks, R. G., *Chem. Commun.*, 1968, p. 663.
13. Beynon, J. H., 'Mass Spectrometry and its Applications to Organic Chemistry', Elsevier Publishing Company, Amsterdam, 1960.
14. Brown, P. and Djerassi, C., *J. Amer. Chem. Soc.*, 1967, **89**, 2711.
15. Meyerson, S. and Rylander, P. N., *J. Chem. Phys.*, 1957, **27**, 901.
16. Meyerson, S. and Rylander, P. N., *J. Amer. Chem. Soc.*, 1956, **78**, 5799.
17. Powell, H., Williams, D. H., Budzikiewicz, H. and Djerassi, C., *J. Amer. Chem. Soc.*, 1964, **86**, 2623.
18. Chupka, W. A., *J. Chem. Phys.*, 1959, **30**, 191.
19. Dickman, J., MacLeod, J. K., Djerassi, C. and Baldeschwieler, J., *J. Amer. Chem. Soc.*, 1969, **91**, 2069.
20. Haas, J. R., Bursey, M. M., Kingston, D. G. I. and Tannebaum, H. P., *J. Amer. Chem. Soc.*, 1972, **94**, 5095.
21. McAdoo, D. J., McLafferty, F. W. and Bente, III P. F., *J. Amer. Chem. Soc.*, 1972, **94**, 2027.
22. Gross, M. L., *J. Amer. Chem. Soc.*, 1971, **93**, 253.
23. Futrell, J. H. and Tiernan, T. O., *in* 'Ion-Molecule Reactions', *ed.* J. L. Franklin, Plenum Press, New York, 1972, pp. 515–19.
24. Munson, M. S. B., *J. Amer. Chem. Soc.*, 1967, **89**, 1772.
25. Ausloos, P. and Lias, S. G., *in* 'Carbonium Ions', *ed.* J. L. Franklin, Plenum Press, New York, 1972, Vol. 2, chapter 16.
26. Cooks, R. G., Beynon, J. H., Caprioli, R. M. and Lester, G. R., 'Metastable Ions', Elsevier, Amsterdam, 1973.
27. Jennings, K. R., *Int. J. Mass Spectrom. Ion Phys.*, 1968, **1**, 227.
28. McLafferty, F. W., Kornfeld, R., Haddon, W. F., Levsen, K., Sakai, I., Bente, III P. F., Tsai, S.-C. and Schudemage, H. D. R., *J. Amer. Chem. Soc.*, 1973, **95**, 3886.

29. Beynon, J. H., Bertrand, M., Jones, E. G. and Cooks, R. G., *Chem. Commun.*, 1972, 341.
30. Durup, J., presented at the 21st Annual Conference on Mass Spectrometry and Allied Topics, San Francisco, 1973.
31. Dunbar, R. C. and Fu, E. W., *J. Amer. Chem. Soc.*, 1973, **95**, 2716.
32. Ellefson, R. E., Denison, A. B. and Weber, J. H., presented at the 20th Annual Conference on Mass Spectrometry and Allied Topics, Dallas, Texas, 1972.
33. Herzberg, G., 'The Spectra and Structures of Simple Free Radicals', Cornell University Press, Ithaca, 1971.
34. Smyth, K. C. and Brauman, J. I., *J. Chem. Phys.*, 1972, **56**, 5993.
35. See, for example, Frost, D. C , *in* 'Mass Spectrometry', *ed.* A. Maccoll, MTP International Review of Science, University Park Press, Baltimore, 1972.
36. Hotop, H. and Niehaus, A., *Int. J. Mass Spectrom. Ion Phys.*, 1970, **5**, 415.
37. Potts, A. W. and Price, W. C., *Proc. Roy Soc. Lond. A*, 1972, **326**, 181.
38. Jonsson, B. O. and Lindholm, E., *Int. J. Mass Spectrom. Ion Phys.*, 1969, **3**, 385.
39. Franklin, J. L., *in* 'Carbonium Ions', *ed.* G. A. Olah and P. von R. Schleyer, Interscience, New York, 1968, Vol. 1.
40. Harrison, R. G., *in* 'Topics in Organic Mass Spectrometry', *ed.* A. L. Burlingame, Wiley Interscience, New York, 1970.
41. Maeda, K., Semeleuk, G. P. and Lossing, F. P., *Int. J. Mass Spectrom. Ion Phys.*, 1968, **1**, 395.
42. Hertel, I. and Ottinger, Ch., *Z. Naturforsch.*, 1967, **22a**, 40.
43. Cooks, R. G., Setser, D. W., Jennings, K. and Jones, S., *Int. J. Mass Spectrom. Ion Phys.*, 1971, **7**, 493.
44. Lossing, F. P., personal communication, 1972.
45. Bertrand, M., Beynon, J. H. and Cooks, R. G., *Int. J. Mass Spectrom. Ion Phys.*, 1972, **9**, 346.
46. Cooks, R. G., Bertrand, M., Beynon, J. H., Rennekamp, M. E. and Setser, D. W., *J. Amer. Chem. Soc.*, 1973, **95**, 1732.
47. See, for example, Ottinger, Ch., *J. Chem. Phys.*, 1967, **47**, 1452.
48. Bertrand, M., Beynon, J. H. and Cooks, R. G., *Org. Mass Spectrom.,* 1973, **7**, 193.
49. Uccella, N., Howe, I. and Williams, D. H., *Org. Mass Spectrom.*, 1972, **6**, 229.
50. Howe, I., *in* 'Mass Spectrometry', *ed.* D. H. Williams, A Specialist Periodical Report, The Chemical Society, London, 1971, Vol. I.
51. Beynon, J. H., Caprioli, R. M., Perry, W. O. and Baitinger, W. E., *J. Amer. Chem. Soc.*, 1972, **94**, 6828.
52. Dickinson, R. J. and Williams, D. H., *J. Chem. Soc. B*, 1971, 249.
53. Rennekamp, M. E., Perry, W. O. and Cooks, R. G., *J. Amer. Chem. Soc.*, 1972, **94**, 4985.
54. Shannon, T. W. and McLafferty, F. W., *J. Amer. Chem. Soc.*, 1966, **88**, 5021.
55. Occolowitz, J. L., *J. Amer. Chem. Soc.*, 1969, **91**, 5202.
56. Tsang, C. W. and Harrison, A. G., *Org. Mass Spectrom.*, in press.
57. Uccella, N. and Williams, D. H., *J. Amer. Chem. Soc.*, 1972, **94**, 8778.
58. Derrick, P. J., Falick, A. N. and Burlingame, A. L., *J. Amer. Chem. Soc.*, 1972, **94**, 6794.
59. Beckey, H. D., *in* 'Recent Topics in Mass Spectrometry', *ed.* R. I. Reed, Gordon and Breach Science Publishers, New York, 1971.
60. Beynon, J. H., Caprioli, R. M. and Shannon, T. W., *Org. Mass Spectrom.*, 1971, **5**, 967.
61. Tomer, K. B. and Djerassi, C., *Org. Mass Spectrom.*, 1972, **6**, 1285.
62. Jones, E. G., Bauman, L. E., Cooks, R. G. and Beynon, J. H., *Org. Mass Spectrom.*, 1973, **7**, 185.
63. Jones, E. G., Beynon, J. H. and Cooks, R. G., *J. Chem. Phys.*, 1972, **57**, 2652.
64. Beynon, J. H., Caprioli, R. M., Baitinger, W. E. and Amy, J. W., *Org. Mass Spectrom.*, 1970, **3**, 455.
65. Beynon, J. H., Caprioli, R. M. and Richardson, J. W., *J. Amer. Chem. Soc.*, 1971, **93**, 1852.
66. Ast, T., Ph.D. Thesis, Purdue University, 1972.
67. Ast, T., Beynon, J. H. and Cooks, R. G., *Org. Mass Spectrom.*, 1972, **6**, 749.
68. Beynon, J. H., *Advan. Mass Spectrom.*, 1968, **4**, 123.

69. Cooks, R. G., Beynon, J. H., Bertrand, M. and Hoffman, M. K., *Org. Mass Spectrom.*, in press.
70. Holmes, J. L. and Weese, G. M., *Canad. J. Chem.*, in press.
71. Beynon, J. H., Bertrand, M. and Cooks, R. G., *J. Amer. Chem. Soc.*, 1973, **95**, 1739.
72. Beynon, J. H., Bertrand, M. and Cooks, R. G., *Org. Mass Spectrom.*, in press.

METASTABLES

Chairman

K. R. JENNINGS
(University of Warwick, Coventry, U.K.)

100
A Novel Method of Ion Structure Identification

By P. BRUCK, J. TAMAS, G. CZIRA

(*Central Research Institute for Chemistry of the Hungarian Academy of Sciences, Budapest, Hungary*)

and K. KORMENDY

(*Research Group for Peptide Chemistry of the Hungarian Academy of Sciences, Budapest, Hungary*)

IDENTIFICATION of ion structures is one of the cardinal problems of organic mass spectrometry.[1] The most important advancement in this field was the introduction of the method of metastable characteristics, suggested first by Shannon and McLafferty.[2]

Competing metastable transitions are only observed if two competitive decomposition reactions of similar activation energies occur. This limits the applicability of the method.

We were faced with this limitation in the course of the structure identification of a rearranged fragment ion, resulting from 2,3-dihydroimidazo[2,1-a]-phtalazin-6(5H)-one (Compound I) and its 5N-methyl derivative (Compound II) upon electron impact (Fig. 1).[3]

In the mass spectrum of I—besides M and M-1—the only abundant ion appears at m/e 130 (Fig. 2, upper part). This fragment occurs via two competitive fragmentation routes:

The M-1 ion can lose $C_2H_3^{15}N$, *i.e.* the 'lower part' of the imidazole ring, leading to the m/e 145 ion, which decomposes to m/e 130 by NH elimination.

In a competitive process, the M-1 ion loses $HC^{15}N$, resulting in an ion at m/e 159. This fragment eliminates CH_3N, and again the m/e 130 ion is formed. Elimination of CH_3N from m/e 159 is not possible without the previous rearrangement of its structure. It is reasonable to assume that in the first step a new bond is formed between the methylene group and the N atom at position 5, which is followed by the cleavage of the CO–N bond.

The fragmentation of Compound II—the 5N-methyl derivative of I—is completely analogous to the decomposition of the base compound (Fig. 2, lower part).

HCN loss of the M-1 ion leads to a fragment at m/e 173, which is the homologue of the m/e 159 ion of Compound I. The m/e 173 ion eliminates C_2H_5N, and the m/e 130 is formed. This process is analogous to the

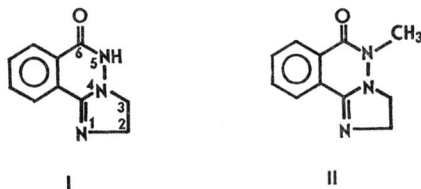

FIGURE 1

$159 \rightarrow 130$ route of Compound I, and requires a rearrangement, similar to the case of the m/e 159 ion in I.

On the other hand, the M-1 ion of II also eliminates C_2H_3N, resulting in the 5N-methyl homologue of the m/e 145 ion (Compound I) at mass 159. This fragment loses CH_3N, giving rise to the m/e 130 ion, exactly as in the case of I.

It is supposed that both m/e 159 ions have the same structure. However, the 5N–C bond is already present in the intact molecule of II, while in the case of I its formation requires a rearrangement.

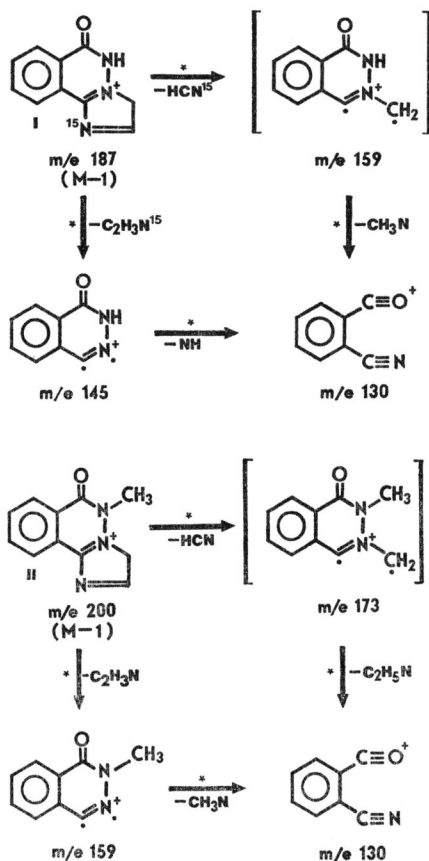

FIGURE 2

In order to prove the structural identity of the two m/e 159 ions, it is necessary to apply some more exact and more quantitative method, in addition to the evidence provided by their similar decomposition routes.

The method of metastable characteristics is not applicable to this case, since no other reaction competes with the 159 → 130 process. On the other hand, it is possible to characterize the structure of the m/e 159 ion by the activation energy of its further decomposition reaction.

Let us consider two molecules, M_1 and M_2, decomposing upon electron impact to X^+. If the ions X^+ have identical structure and state in both cases, the activation energy ($\Delta E^\ddagger$) of their further decomposition ($X^+ \rightarrow Y^+$) must be equal. The activation energy of the $X^+ \rightarrow Y^+$ process can be approximated by the difference of the mass spectrometric appearance potentials of the appropriate ions:

$$\Delta E^\ddagger \approx AP[Y^+] - AP[X^+] \tag{1}$$

However, the application of this procedure is restricted by certain concurrent processes, leading to the same ion, Y^+ (*e.g.* 173 → 130 in the case of Compound II).

To eliminate this effect, we suggest calculating the activation energies of metastable $X^+ \rightarrow Y^+$ processes instead of normal ones, using the equation

$$\Delta E^\ddagger \approx AP[m^*(X^+ \rightarrow Y^+)] - AP[X^+] \equiv \Delta AP^* \tag{2}$$

where $AP[m^*(X^+ \rightarrow Y^+)]$ is the appearance potential of the metastable ions, corresponding to the $X^+ \rightarrow Y^+$ process.

Let us suppose, that for ions X^+, having identical structure and state, the deviation of ΔAP^* from the thermodynamic value of activation energy ($\Delta E^\ddagger$) is equal, independently of the origin of X^+ (*vide infra*).

Then, if the ΔAP^* values for X^+, originated from M_1 and from M_2, respectively, are equal, one may conclude, that the structure and the state of ions X^+ of both origins are identical.

The applicability of the suggested method was tested on some model compounds (III–VIII).

(1) The molecular ion of aniline (III) and the M-42 fragment ion of acetanilide (IV) are known to have the same structure (Fig. 3).[4] The ΔAP^* value for the HCN expulsion from aniline is 4·02 eV, while it is 4·06 eV for the same reaction of the M-42 ion in the case of IV.

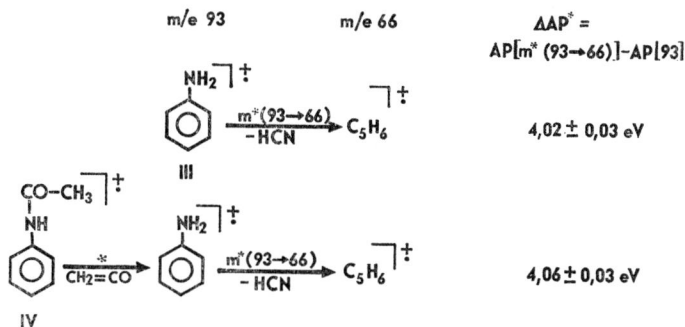

FIGURE 3

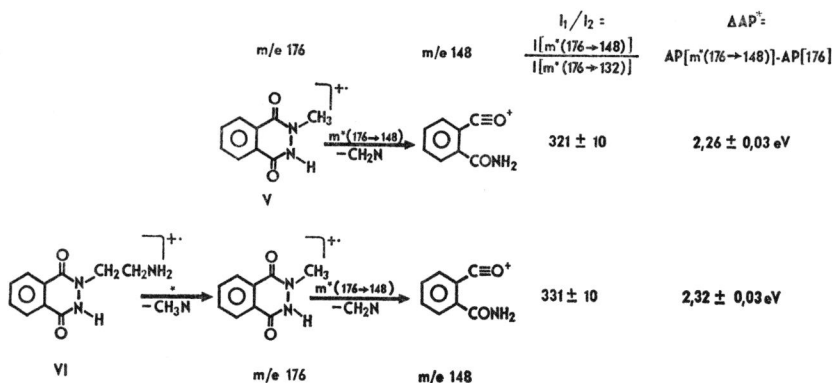

FIGURE 4

(2) The molecular ion of 3-(2-aminoethyl)-1,4(2H)-phthalazindione (Compound VI, Fig. 4) loses CH_3N, resulting in an ion at m/e 176. This fragment has the same elemental composition as the molecular ion of 3-methyl-1,4(2H)-phthalazindione (Compound V). Both m/e 176 ions eliminate H_2CN, giving rise to m/e 148 ions.

The ΔAP^* values of the 176 → 148 reactions are 2·26 eV for Compound V, and 2·32 eV for Compound VI, respectively, supporting the suggestion that the structures of these m/e 176 ions are identical. The method of metastable characteristics led to the same conclusions (Fig. 4).

(3) We studied the decomposition of two, homologous ω-hydroxyalkyl-amino-phthalazinone-derivatives (VII and VIII).[5] In a primary fragmentation process, the molecular ions lose the hydroxyalkyl side-chain via McLafferty rearrangement, giving rise to m/e 161 ions in both cases. The only decomposition reaction of the latter ions results in the m/e 130 fragment by N_2H_3 elimination. The ΔAP^* values for these reactions are equal within the experimental error (Fig. 5), demonstrating the structural identity of the m/e 161 ions, independently of their origin.

Now, let us return to our original problem, i.e. to the structure identification of the m/e 159 ions. In Fig. 6, we summarized the fragmentation of four

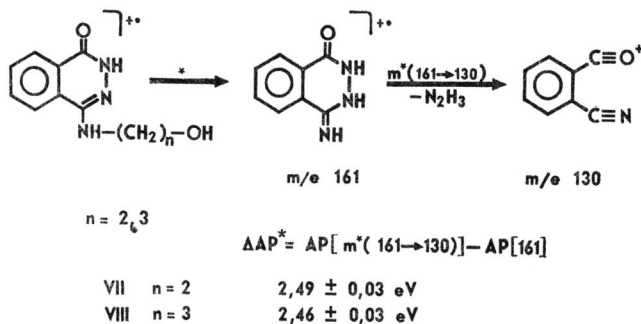

FIGURE 5

compounds (I, II, IX and X), which results in m/e 159 ions of the same elemental composition. All of these m/e 159 ions possess only a single decomposition route, which leads to the m/e 130 ion. However, the ΔAP^* values of the $159 \rightarrow 130$ reactions are practically equal for Compounds I and II, while they are very significantly different either for IX, or for X (Fig. 6).

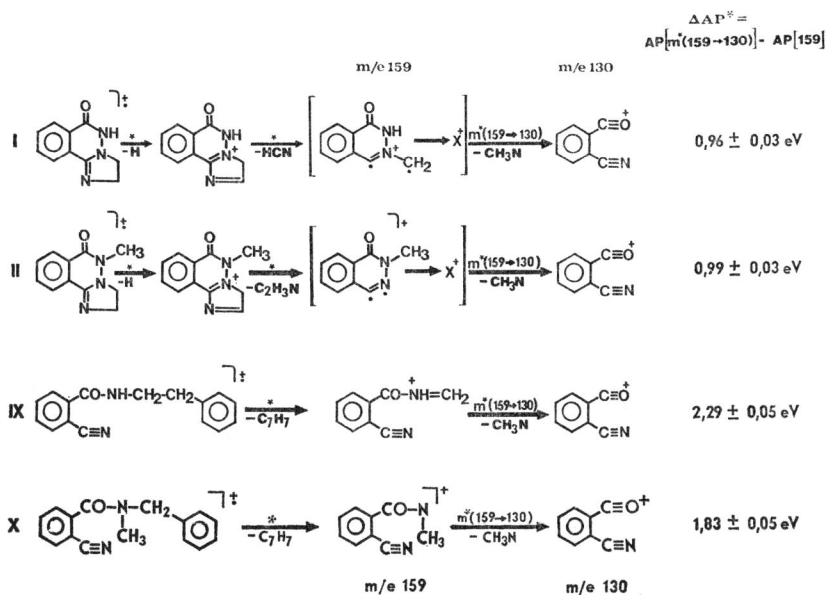

FIGURE 6

These data definitely support the proposal that the decomposition of the m/e 159 ions of Compounds I and II proceeds from a common structure, X^+.

Although the method, applied above, does not give an exact answer to either the structure, or to the mechanism of formation of the m/e 159 ion, it seems to be very likely that the rearrangement of m/e 159 in Compound I involves a methylene insertion reaction,[6] leading to the formation of an N–CH$_3$ group.

Finally, it is necessary to discuss some effects influencing the applicability of the method. At least two sources of excess internal energy must be considered:†

(a) The value of the kinetic shift may be significant,[8] and in such cases the application of AP data often leads to poor approximation of the thermodynamic data. However, when the ΔAP^* values of two decomposition processes are related, the effect of the kinetic shift decreases since a significant part of the deviation is cancelled out.

---

† We neglect the effect of internal thermal energy.[7]

(b) Let us now consider the case, when the reverse reaction

$$X^+ + N \rightarrow M_1{}^+$$

requires some activation energy ($\Delta E_r{}^\ddagger$), which contributes to the excess energy of the dissociation products, X^+ and N.[9]

When the reverse reactions of the formation of X^+ from M_1 and from M_2 require different activation energies, X^+ will contain different amounts of excess energy, even if its structure is the same.

If a part of this excess energy appears, as internal energy, then the higher is the difference in the activation energy of the reverse reactions, the higher is the deviation of the ΔAP^* values. The most significant differences in $\Delta E_r{}^\ddagger$ are expected, when X^+ is a molecular ion on the one hand ($M_1 = X$), while X^+ is formed from $M_2{}^+$ by rearrangement, on the other hand. Even in such cases (see Figs. 3 and 4), the deviation of the ΔAP^* values does not exceed the experimental error.

A possible explanation of this result is that in the case of the above rearrangement reactions, $\Delta E_r{}^\ddagger$ is not significantly higher than the reproducibility of the AP measurements.

Recently, a similar assumption was made to explain the lack of flat-topped metastable peaks in the cases of some rearrangement reactions.[9] The slight activation energy of ion–molecule reactions is in accordance with the above hypothesis.

The experiments were carried out applying an MS 902 mass spectrometer. For the AP determinations, the metastable ions were refocussed by decreasing the voltage of the ESA.[10] ΔAP^* values were obtained by the application of the semilog–plot technique, without the use of a reference sample. The data were reproducible several weeks later to $\pm 0\cdot03$–$0\cdot05$ eV. The temperatures of the source and the inlet systems were 100–150°C.

REFERENCES

1. Williams, D. H., in 'Determination of Organic Structures by Physical Methods', ed. F. V. Nachod and J. J. Zuckerman, Academic Press, 1971, Vol. 3.
2. Shannon, T. W. and McLafferty, F. W., J. Am. Chem. Soc., 1966, **88**, 5021.
3. Bruck, P., Tamás, J. and Körmendy, K., to be published.
4. Hammerum, S. and Tomer, K. B., Org. Mass Spectrom., 1972, **6**, 1369.
5. Bruck, P., Tamás, J. and Körmendy, K., in preparation.
6. Kirmse, W., 'Carbene Chemistry', Academic Press, 1964.
7. Chupka, W. A., J. Chem. Phys., 1959, **30**, 191.
8. Vestal, M. L., in 'Fundamental Processes in Radiation Chemistry', ed. P. Ausloos, Wiley-Interscience, 1968.
9. Williams, D. H. and Howe, I., 'Principles of Organic Mass Spectrometry', McGraw-Hill, 1972, p. 85.
10. Hickling, R. D. and Jennings, K. R., Org. Mass Spectrom., 1970, **3**, 1499.

101
Shapes and Relative Abundances of Metastable Ion Peaks as an Aid to Ion Structure Determination

By J. L. HOLMES

(*Chemistry Department, University of Ottawa, Ottawa, Ontario, Canada*)

It has become commonplace to use the relative abundance of separated first field-free region metastable ion peaks, generated from a given dissociation of a labelled precursor molecule, as a tool to elucidate ion fragmentation mechanisms. If the process involves stereospecific selection of label atoms, then a structure for the transition state of the dissociating ion can be accurately inferred. Quite often the atoms are found to have apparently lost their positional identity and are randomly distributed among the products of the fragmentation; in such cases structural assignments generally are not possible. It should be stressed that it is transition state structure that is inferred from these experiments and not the ground state structure of the precursor ion.

We report here some studies of the shapes of the so-called Gaussian metastable ion peaks, which are those most commonly observed. In particular two dissociations will be discussed in detail, namely

$$C_7H_7{}^+ \rightarrow C_5H_5{}^+ + C_2H_2 \qquad (1)$$

and

$$C_6H_{10}{}^{+\cdot} \rightarrow C_5H_7{}^+ + CH\cdot{}_3 \qquad (2)$$

In the first detailed discussion of metastable ion peak shapes[1] it was concluded that the commonly observed Gaussian-type peaks could not be associated with single-valued energy release, and thus a kinetic energy distribution function of some kind was needed to describe their shape. One reason for slow progress in this field has been the observation of metastable ion peaks under a wide variety of poorly defined experimental conditions. Indeed, without a β-slit, many instruments only produce broad, diffuse metastable ion peaks with relatively little distinction possible between Gaussian and flat-top peaks. In this study an A.E.I. G.E.C. MS902S mass spectrometer, equipped with an energy resolving β-slit and an ion detector based on a scintillator photo-multiplier system[2] was used. Good energy resolution was achieved by decreasing the source slit to 0·001 in., the β-slit to 0·004 in. and the collector slit to 10% transmission. It was found that further reduction of these parameters did not change the metastable ion peak profile, as has also been reported by Jones *et al.*[3] Under these conditions the energy spread of the main ion beam was minimal, and so corrections for main beam width were very small. The acceleration potential for transmission of the metastable ion was 7500 V and the electron energy was 70 V.

In the present study, peak profiles have been calculated assuming a simple exponential distribution of energies (ε^*) among those ions whose unimolecular dissociation rate constants (k_1) allow them to be observed in the metastable lifetime region. It is also assumed that this internal energy distribution will be reflected in that which is released as translational kinetic energy. The unimolecular dissociation rate constants will themselves be some function of the energy content (ε^*), which at the lower limit may be close to threshold for some ions, and the number of degrees of freedom participating in the activated state. In general the greater the number of degrees of freedom involved, the wider the range of internal energies which will be observed through the metastable 'window'.

Thus the shape of a metastable ion peak for which a continuous energy distribution function applies can be defined by an equation of the form

$$h = \exp\left(-\beta w^n\right) \qquad n = 1\cdot6 \to 2\cdot0$$

where h is the peak height, w its corresponding width; the value of the exponent n depends upon the number of participating degrees of freedom in the transition state, *i.e.* the curvature of the k_1 versus ε^* plot for that ion. The energy release calculated from the half height width ($T_{0.5}$) of these peaks represents the average energy release.[4]

It should be possible therefore, by comparing observed metastable peak shapes with calculated profiles, to obtain information as to the shape of the k_1 versus ε^* curves. Furthermore, by adjusting instrument parameters it should be possible to displace the metastable 'window' to transmit ions of longer lifetimes. For dissociations in which a distribution of energies are released, a displacement of the window to longer lifetimes should result in

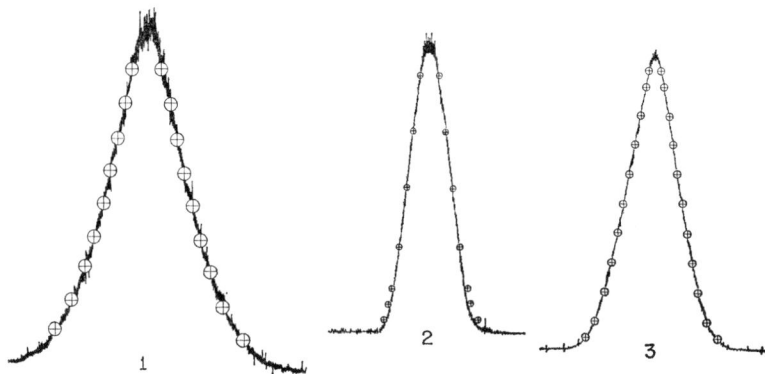

FIG. 1   Class I Gaussian Metastable Ion Peaks.

1. $C_6H_5COOH^{+\cdot} \to C_6H_5CO^+ + OH^\cdot$ $n = 1\cdot65$
2. $C_5H_6^+ \to C_5H_5^+ + H^\cdot$ $n = 2\cdot0$
3. $C_6H_6^{+\cdot} \to C_6H_5^+ + H^\cdot$ $n = 1\cdot80$

smaller $T_{0.5}$ values. For the benzene molecular ion dissociation by H atom loss we have observed a small but significant reduction in $T_{0.5}$ from 0·077 to 0·061 eV on decreasing the acceleration potential from 7000 V to 1700 V.

We propose that Gaussian type metastable peaks can be divided into two classes. The first, Class I, are peaks whose profiles can be reproduced by the mathematical expression shown above. These peaks then, arise from the dissociation of a single structure or electronic state of the precursor ion.

Three Class I peaks are shown in Fig. 1, the points calculated for the indicated value of n agree very well with the experimental curves.

The second group, Class II, are those peaks having relatively broad skirts, which cannot be wholly fitted to the simple mathematical expression. We find that these peaks can be easily resolved into two Gaussian components and we suggest that they arise from the concurrent dissociation of two or more ions having different structures and/or occupying different electronic states. $T_{0.5}$ in this case is not simply related to an average energy release, but depends on the proportions of each dissociating species present.

The best example of this second class of metastable ion peaks and a fragmentation of considerable interest, is the $C_7H_7^+$ dissociation (1). It has been widely accepted[5,6] that the $C_7H_7^+$ ion produced from toluene and ethyl benzene has the symmetrical ring expanded tropylium structure. Recently, the hypothesis that $C_7H_7^+$ ions have exclusively a tropylium structure has been questioned; gas phase radiolysis studies of toluene[7] have indicated that only about one third of the $C_7H_7^+$ ions are formed through a symmetrical intermediate and that both tropyl and benzyl cations were reacting with the substrate.

We have examined the metastable ion peak for reaction (1), generating $C_7H_7^+$ ions from ten precursor molecules. The peak profiles depended on the

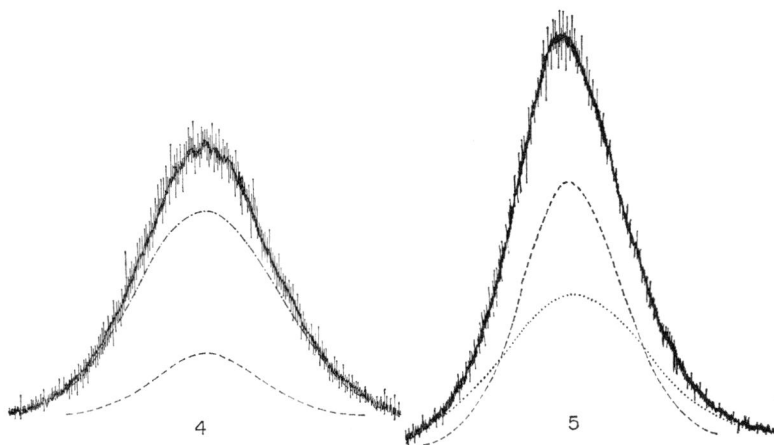

FIG. 2   Class II Gaussian Metastable Ion Peaks.

4. $[C_6H_5CH_2Cl^{+\cdot}] \rightarrow C_7H_7^+ \rightarrow C_5H_5^+ + C_2H_2$
5. $[C_6H_5CH_3^{+\cdot}] \rightarrow C_7H_7^+ \rightarrow C_5H_5^+ + C_2H_2$

Dotted lines are the common Gaussian components which when added, produce the experimental curves.

precursor molecule; two of them are shown in Fig. 2. All the peaks however, could be generated by summation of the *same two Gaussian peaks* in appropriate proportions. For simplicity, pure Gaussian ($n = 2 \cdot 0$) profiles were chosen, and the toluene peak was first constructed by trial and error procedure. It should be stressed that having subtracted one pure Gaussian from the observed peak, the remainder was also of Gaussian form. The appropriate $T_{0.5}$ values for the two basic Gaussians are given in Table I. Each basic

TABLE I

The Relative Contributions from Two Ionic Species in the Dissociation $C_7H_7^+ \rightarrow C_5H_5^+ + C_2H_2$ and their Dependence on Precursor Molecule

Precursor molecule	Narrow Gaussian[a]	Wide Gaussian[a]
Ortho-iodo toluene	64	36
Toluene	56	44
Cycloheptatriene	56	44
Bitropyl	56	44
Norbornadiene	56	44
Benzyl benzoate	50	50
Ethyl benzene	41	59
Bibenzyl	22	78
Benzyl chloride	18	82
Diphenyl methane	18	82

[a] $T_{0.5}$ narrow Gaussian $= 0 \cdot 029$ eV, $T_{0.5}$ wide Gaussian $= 0 \cdot 059$ eV. Contributions represent relative peak areas.

Gaussian represents a $C_7H_7^+$ species and it is concluded that the same two $C_7H_7^+$ species are produced by all these precursors, but in varying amounts. Benzyl chloride for example produces largely the wide Gaussian generating species. The energy release difference is small but at present we have no way of relating $T_{0.5}$ to ε^*. Tentatively we would describe the dissociating species as being different structures rather than electronic states of $C_7H_7^+$. As a test for structural differences we examined process (1) using $C_6H_5CD_2CD_2C_6H_5$ as the precursor molecule. There was no evidence for any positional scrambling of hydrogen and deuterium atoms in the molecular ion prior to its central cleavage to produce a $C_7H_5D_2^+$ ion, either from examination of the daughter ion or metastable ion peak abundances. The metastable ion peaks for the dissociations

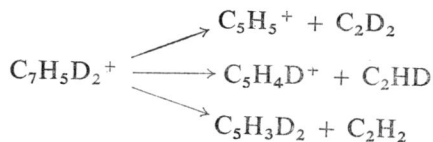

$$C_7H_5D_2^+ \begin{cases} \longrightarrow C_5H_5^+ + C_2D_2 \\ \longrightarrow C_5H_4D^+ + C_2HD \\ \longrightarrow C_5H_3D_2 + C_2H_2 \end{cases}$$

were of identical shape (the same as that observed for unlabelled bibenzyl) and had abundances (peak areas) in the ratio 1:10:10 respectively, *i.e.* that required for complete randomization of hydrogen and deuterium atoms prior to dissociation. Thus *if two structural* forms of $C_7H_7^+$ are present they cannot be distinguished by labelling experiments.

A second example of variable Class II Gaussian peaks are the metastable ion peak profiles for the process

$$C_6H_{10}^{+\cdot} \rightarrow C_5H_7^+ + CH_3^{\cdot} \qquad (2)$$

as a function of the precursor ion's structure and origin. For cyclohexene, the metastable ion peak for the above dissociation was resolved into two components A and B; $T^A_{0.5} = 0.012$ eV and $T^B_{0.5} = 0.046$ eV. In the case of cyclohexene-1d_1 both components in the M–CH_3 and M–CH_2D peaks had abundance ratios consistent with complete hydrogen atom scrambling in the molecular ions prior to dissociation, namely $2.33:1$ respectively. The behaviour of 3-methylcyclopentene was similar insofar as two components were needed to reproduce the observed peak profile. One was component A, the second, C, was appreciably broader than B ($T^C_{0.5} = 0.062$ eV). In 3-methylcyclopentene 1-d_1 the relative abundances of C in the CH_3 and CH_2D loss metastable ion peaks were in the random statistical ratio but nearly all of component A appeared in the CH_3 loss peak. There are thus three $C_6H_{10}^{+\cdot}$ species which competitively dissociate in the first field free region.

Had only the gross peak areas been considered (as is current practice) it would have been concluded that cyclohexene molecular ions rearrange to an intermediate with complete hydrogen scrambling prior to dissociation by methyl loss and that 3-methylcyclopentene molecular ions do not. It is possible that instances of such behaviour will be common and that many labelling experiments will need reinterpretation following a study of peak shapes. Other isomers are being studied and further results will be reported shortly.

REFERENCES

1. Beynon, J. H. and Fontaine, A. E., *Z. Naturforsch*, 1967, **22a**, 334.
2. Daly, N. R., McCormick, A. and Powell, R. E., *Rev. Sci. Inst.*, 1968, **39**, 1163.
3. Jones, E. G., Bauman, L. E., Beynon, J. H. and Cooks, R. G., *Org. Mass. Spectrom.*, 1973, **7**, 185.
4. Beynon, J. H., Caprioli, R. M., Baitinger, W. E. and Amy, J. W., *Org. Mass Spectrom.*, 1970, **3**, 661.
5. Grubb, H. M. and Meyerson, S., *in* 'Mass Spectrometry of Organic Ions', *ed.* F. W. McLafferty, Academic Press, Inc., New York, 1963.
6. Rinehart, K. L., Buchholz, A. C., Van Lear, G. E. and Cantrill, H. L., *J. Amer. Chem. Soc.*, 1968, **90**, 2983.
7. Yukio Yamamoto, Setsuo Takamuku and Hiroshi Sakurai, *J. Amer. Chem. Soc.*, 1972, **94**, 661.

102
A Direct Measurement of the 'Kinetic Shift' in Benzene

By CHAVA LIFSHITZ, ARTHUR MacKENZIE PEERS,†
MATANYA WEISS and MORRIS J. WEISS

*(Department of Physical Chemistry, The Hebrew University
of Jerusalem, Israel)*

INTRODUCTION

THE Quasi Equilibrium Theory (QET)[1-3] of mass spectra assumes that the rate coefficient $k(E)$, for the unimolecular dissociation of a polyatomic ion, is a monotonically increasing function of its internal energy, E above the minimum necessary for dissociation. The larger the number of degrees of freedom of the ion and/or the greater the activation energy, E_0 for the process, the lower is the rate at threshold and the slower the rise of $k(E)$ above threshold.

The residence time of an ion in an ordinary magnetic sector mass spectrometer is of the order of microseconds. The 'kinetic shift' is that excess energy above threshold, necessary according to the theory,[4] for the reaction to be observed in the mass spectrometer. Vestal,[2] has calculated for example, for the reaction:

$$C_6H_6{}^+ \rightarrow C_6H_5{}^+ + H \qquad (1)$$

in benzene, assuming an activation energy $E_0 = 3.9$ eV, that an excess energy $(E-E_0)$ of 1·50 eV is needed for a reaction rate constant $k_1(E) = 10^5 \text{ sec}^{-1}$.

The benzene system is of particular interest, since an entirely different approach assumes that dissociations of the benzene ion take place from non-interacting separate electronic states and there is recent experimental evidence[5] corroborating this view. If a non-interacting excited electronic state of the benzene ion has to be reached, for reaction (1) to set in, then the excess energy necessary for observing reaction (1), should be independent, or only slightly dependent, on the time available for decomposition.

There have been various attempts in the past to measure 'kinetic shifts'. The width of the 'metastable' breakdown curve in the breakdown graph of the molecule, as obtained by photoionization–mass spectrometry, is a measure

† Visiting Research Associate from Laboratoire Curie, Institut de Physique Nucléaire, 11, Rue Pierre et Marie Curie, Paris 5.

of the 'kinetic shift'.[6] Electron impact methods have relied, in one way or another, on measuring—by conventional methods—the appearance potential of the fragment at various times following the formation of the parent ion. Appearance potentials of 'metastable ions' were thus compared with those of the normal fragment ions.[7,8] Appearance potentials obtained, following long residence times in an ICR instrument were compared with those obtained, for the same fragment ions, in an ordinary mass spectrometer.[9] The approach taken here was to trap the ions in an electron space charge. Ions may thus be trapped for periods up to milliseconds.[10,11,12] This method has not been employed, until now, for the purpose of measuring 'kinetic shifts'. The possible advantage that we foresaw, in using the trapped-ion method for this purpose, was that the residence times of the parent ions could be varied continuously over several orders of magnitude. Furthermore, the appearance potentials of the fragment ions could be measured in the same instrument, by the same technique and under comparable conditions.

EXPERIMENTAL

The Varian MAT CH4, single-focussing 60° magnetic sector mass-spectrometer was employed. The AN4 ion-source was modified and a repeller electrode was introduced. The trapping technique used was similar to that used by Bourne and Danby.[13] A continuous electron beam of ~ 5 eV energy and 8–10 μA trap current, provided by thermionic emission from a directly heated rhenium filament, was used to trap the ions produced when a voltage pulse, variable in height, negative with respect to the ionization chamber and of ~ 1 μsec duration, was applied to the filament. The nominal electron energy scale was set by the height of the voltage pulse on the filament. Our system, which is very similar to the one employed by Herod and Harrison,[12] in terms of the pulsing scheme, trapping conditions, pressures employed, etc., was checked by obtaining very similar results to theirs for the trapping of Ar^+ ions and for ion–molecule reactions in the methane system.

Appearance potentials were obtained by measuring ionization efficiency curves at various delay times following the ionizing pulse. The logarithm of the ion current was plotted as a function of the nominal electron energy.[7,14] The ionization efficiency curve of the benzene parent ion was employed as a reference standard (I.P.(C_6H_6) = 9·246 eV),[15] to calibrate the electron energy scale and to take into account any possible variations with delay time. $C_6H_6^+$ ions do not undergo any ion–molecule reactions with benzene molecules under our conditions.[16] The contribution of ion–molecule reactions of $C_6H_5^+$ ions is also negligibly small and any possible effect these might have on the measured ionization efficiency curves, was neglected in the present study.

RESULTS AND DISCUSSION

Figure 1 represents the results of one experiment at short and long delay times. These are represented in the form of semi-logarithmic plots of the ion currents of $C_6H_6^+$ and $C_6H_5^+$, as a function of the nominal electron

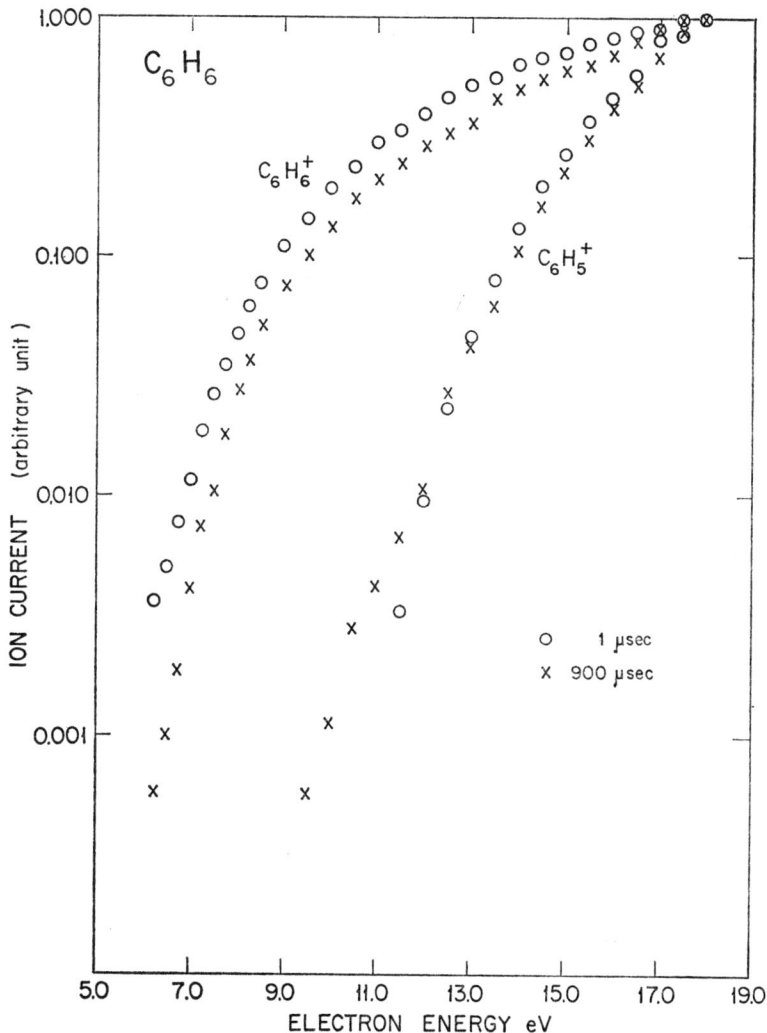

FIG. 1   Semilogarithmic plots of $C_6H_6^+$ and $C_6H_5^+$ ionization efficiency curves from benzene at short (1 μsec) and long (900 μsec) delay times following the ionizing pulse.

energy, at 1 μsec and 900 μsec following the ionizing electron pulse. The ion currents were normalized at 18 V (nominal). These curves yield

$$\text{A.P.}(C_6H_5^+)_{1\mu sec} = 14 \cdot 3 \pm 0 \cdot 1 \text{ eV}$$

and

$$\text{A.P.}(C_6H_5^+)_{900\mu sec} = 12 \cdot 8 \pm 0 \cdot 2 \text{ eV}$$

The RPD (Retarding Potential Difference) method gave, using the same instrument[16] a value: A.P.$(C_6H_5^+) = 14 \cdot 1$ eV to be compared with the photoionization value[15] of A.P.$(C_6H_5^+) = 13 \cdot 8$ eV. These last values of the appearance potential coincide with a known excited electronic state of the benzene ion,[17] on the one hand and with the onset of $C_6H_5^+$ formation by dissociative charge exchange,[5] with $k \simeq 3 \times 10^5$ sec^{-1}, on the other.

The endothermicity of reaction (1) is still not very well known. While Vestal[2] adopted a value of $E_0 = 3.9$ eV, Klots[18] assumed $E_0 = 4.55$ eV, where E_0 is the activation energy for reaction (1). Since reaction (1) involves a simple bond cleavage, E_0 may be equated with its endothermicity. Let us further assume that the ground electronic state of the benzene ion is involved.

$$E_0 = \Delta H_f(C_6H_5) + \text{I.P.}(C_6H_5) + \Delta H_f(H) - \Delta H_f(C_6H_6^+) \quad (2)$$

All of the thermochemical values necessary for calculating E_0 according to eqn (2) are known.[19] The main uncertainty lies in the ionization potential of the phenyl radical. The accepted value,[19] $\text{I.P.}(C_6H_5) = 9.20$ eV, corresponds to the vertical ionization potential of the radical, as obtained by the semi-logarithmic method.[20] It leads to $E_0 = 4.5$ eV. A more recent value,[21] obtained by photoionization, $\text{I.P.}(C_6H_5) = 8.1 \pm 0.1$ eV leads to $E_0 = 3.4$ eV.

If the reaction proceeds through the ground electronic state of the benzene ion, then our preliminary results would indicate that $E_0 = (\text{A.P.}(C_6H_5^+) - \text{I.P.}(C_6H_6)) \leq 3.6 \pm 0.2$ eV. Furthermore, our results show a 'kinetic shift' of ≥ 1.5 eV for $k(E) \simeq 10^5$ sec^{-1}.

Clearly, more experiments are needed along these lines. We intend to do those in the near future and to include some other systems as well, e.g. C_6H_5CN, which has been the subject of several previous studies by other methods.[5,7]

REFERENCES

1. Rosenstock, H. M., Wallenstein, M. B., Wahrhaftig, A. L. and Eyring, H., *Proc. Natl. Acad. Sci. (U.S.)*, 1952, **38**, 667; Rosenstock, H. M., and Krauss, M., *in* 'Mass Spectrometry of Organic Ions', *ed.* F. W. McLafferty, Academic Press, New York, 1963, pp. 2–64; Rosenstock, H. M., *Advan. Mass Spectry.*, 1968, **4**, 523.
2. Vestal, M. L., *in* 'Fundamental Processes in Radiation Chemistry', *ed.* P. Ausloos, Wiley, New York, 1968, pp. 59–118.
3. Wahrhaftig, A. L., *in* 'Mass Spectrometry, MTP International Review of Science, Physical Chemistry Series One, Volume 5', *ed.* A. Maccoll, Butterworths, London, 1972, pp. 1–24.
4. Chupka, W. A., *J. Chem. Phys.*, 1959, **30**, 191.
5. Andlauer, B. and Ottinger, C., *J. Chem. Phys.*, 1971, **55**, 1471; *Z. Naturforsch*, 1972, **27a**, 293.
6. Chupka, W. A., Berkowitz, J. and Miller, S. I., 20th American Society for Mass Spectrometry Meeting, 4–9 June, 1972, Dallas, Texas.
7. Hertel, I. and Ottinger, Ch., *Z. Naturforsch.*, 1967, **22a**, 40.
8. Hickling, R. D. and Jennings, K. R., *Org. Mass Spectrom.*, 1970, **3**, 1499.
9. Gross, M. L., *Org. Mass Spectrom.*, 1972, **6**, 827.
10. Baker, F. A. and Hasted, J. B., *Phil. Trans. Roy. Soc.*, 1966, **261**, 33; Hasted, J. B., *in* 'Some Newer Physical Methods in Structural Chemistry', *ed.* R. Bonnett and J. G. Davis, United Trade Press, London, 1967.
11. Redhead, P. A., *Can. J. Phys.*, 1967, **45**, 1791.
12. Herod, A. A. and Harrison, A. G., *Int. J. Mass Spectrom. Ion Phys.*, 1970, **4**, 415.
13. Bourne, A. J. and Danby, C. J., *J. Sci. Instrum.*, Ser. 2, 1968, **1**, 155.
14. Lossing, F. P., Tickner, A. W. and Bryce, W. A., *J. Chem. Phys.*, 1951, **19**, 1524.
15. Brehm, B., *Z. Naturforsch.*, 1966, **21a**, 196.
16. Lifshitz, C. and Reuben, B. G., *J. Chem. Phys.*, 1969, **50**, 951.

17. Samson, J. A., *Chem. Phys. Lett.*, 1969, **4**, 257.
18. Klots, C. E., *Z. Naturforsch.*, 1972, **27a**, 553.
19. Franklin, J. L., Dillard, J. G., Rosenstock, H. M., Herron, J. T., Draxl, K. and Field, F. H., 'Ionization Potentials, Appearance Potentials, and Heats of Formation of Gaseous Positive Ions', NSRDS-NBS 26, U.S. Government Printing Office, 1969.
20. Fisher, I. P., Palmer, T. F. and Lossing, F. P., *J. Am. Chem. Soc.*, 1964, **86**, 2741.
21. Sergeev, Yu. L., Akopyan, M. E. and Vilesov, F. I., *Opt. Spectr.*, 1972, **32**, 121.

Discussion

G. G. Meisels (University of Houston, U.S.A.): Harrison at Toronto has reported that the average translational energy of the trapped ions is about 0·3 to 0·4 eV; with such long times, do you find evidence for collision processes (ion–molecule reactions)?

C. Lifshitz: The parent ($C_6H_6^+$) ions do not react with benzene molecules to form dimers under our (fairly high) temperatures. The rate coefficient for the formation of the adduct $C_6H_5^+ \cdot C_6H_6$, by the phenyl ion is also low ($\sim 10^{-12}$–10^{-11} cc/molec sec). Charge exchange is unlikely since I.P. (C_6H_5) < I.P. (C_6H_6). We have so far neglected to take into account contributions of collision processes or ion–molecule reactions to the ionization efficiency curves of $C_6H_6^+$ and $C_6H_5^+$ at long times. We plan to measure these at several different pressures and to extrapolate to zero pressure.

R. Botter (C.E.A.—Saclay, France): Will the trapped parent ion not be excited by the low energy electron during the trapping time? This is especially critical if they would be vibrationally excited, since it would lower the appearance potential of the fragment ion.

C. Lifshitz: Vibrational excitation of neutral molecules by low energy electron impact is a resonance process going via a temporary negative ion state. Vibrational excitation by a non-resonant process is believed to have low cross-sections for neutrals, as well as for ions. We hope to test the importance of such processes by measuring the time dependence of appearance potentials of fragments for which no kinetic shift is to be expected (*e.g.* CH_3^+ from CH_4) and by varying the energy of the trapping electron beam.

103
Field Ionization Kinetic (FIK) Studies of the Unimolecular Gas-Phase Reactions of Organic Radical-Cations at 10^{-11} to 10^{-5} Sec

By P. J. DERRICK,† A. M. FALICK
and A. L. BURLINGAME

(*Space Sciences Laboratory, University of California, Berkeley, California, U.S.A.*)

THE field ionization kinetics (FIK) technique allows the rates of unimolecular gas-phase ionic reactions to be measured over a time range extending from 10^{-11} (or in some cases 10^{-12}) to 10^{-5} sec (Refs. 1–3). Thus FIK provides a *time-resolved* view of the reactions induced by ionization, whereas other mass spectrometric techniques (*e.g.* electron impact (EI), photoionization (PI), chemical ionization) typically afford an *integrated* view of all events occurring within some lengthy and largely arbitrary time interval of the order of microseconds. The FIK technique can be regarded as adding a new dimension, namely time, to mass spectrometry. The one-dimensional peak in a conventional mass spectrum is replaced in FIK by a two-dimensional curve of rate of reaction against time. Considering a hypothetical situation in which the internal energy distribution within a molecular ion is the same for FI as for EI or PI, the area under the FIK curve of rate against time for a particular fragment corresponds to the peak intensity of that fragment in the EI or PI mass spectrum (considering a primary fragment which does not itself fragment). We have sought to apply the FIK technique to the general problem of reaction mechanisms in gas-phase ion chemistry. The present mechanistic understanding of unimolecular gas-phase ionic reactions must be adjudged as limited when compared to the level of mechanistic understanding achieved in some other areas of chemistry. Many of the reaction mechanisms encountered in organic mass spectrometry rest on the most tenuous evidence. Often the molecular weights of the reactant and the charged product are the sole bases. Yet the unimolecular reaction systems set up by gas-phase ionization have a perhaps unparalleled potential for elucidating principles of chemical reactivity. A variety of reaction types is coupled with an essential

† Present address: Department of Chemistry, University College London, 20 Gordon Street, London WC1H 0AJ.

simplicity resulting from the lack of solvent and collision effects. We feel that an important cause of the comparatively limited understanding of these reaction systems has been a lack of genuine kinetic data. The FIK technique goes a long way toward fulfilling this need for kinetic data.

One of our major interests continues to be the complex rearrangement and fragmentation induced by ionization of alkenes. In earlier papers,[4,5] we have described the study of specifically deuterated cyclohexene and 2-methyl-propene using the FIK technique. It would appear that the cyclohexene carbon skeleton can remain intact while the hydrogens undergo rapid shifts (within 10^{-12} to 10^{-11} sec) around the ring. The hydrogen rearrangement can be completely rationalized by 1,3 hydrogen transfers. 2-Methylpropene also undergoes rapid rearrangement following ionization, which again can be completely explained in terms of 1,3 hydrogen transfers. There is inde-pendent evidence to suggest that 2-methylpropene does not undergo skeletal rearrangement following low energy ionization.[6] The possibility of 1,2 hydrogen transfers within the 2-methylpropene and cyclohexene ions can certainly not be completely dismissed, but there is no need to invoke such transfers to explain the FIK results. In the present paper, we describe the extension of our FIK studies of alkenes to 1-butene. The 1-butene system is considerably more complicated than the 2-methylpropene. The 1-butene ion rearranges to 2-butene and 2-methylpropene structures even at very low ionizing energies.[6b,7] In addition, there is considerable hydrogen rearrange-ment.[8] However, it should be possible to gain a good understanding of the behaviour of the 1-butene ion by making sufficient FIK measurements on carefully chosen specifically isotopically labelled species. As a first step we have studied 1-butene-1,1-d_2 (1) and 1-butene-4,4,4-d_3 (2). FIK results on these species will define the complexity of the problem more precisely, while hopefully allowing certain of the hydrogen transfers to be identified.

Figures 1 and 2 show the ion currents of fragments m/e 41 (M–CD$_2$H),

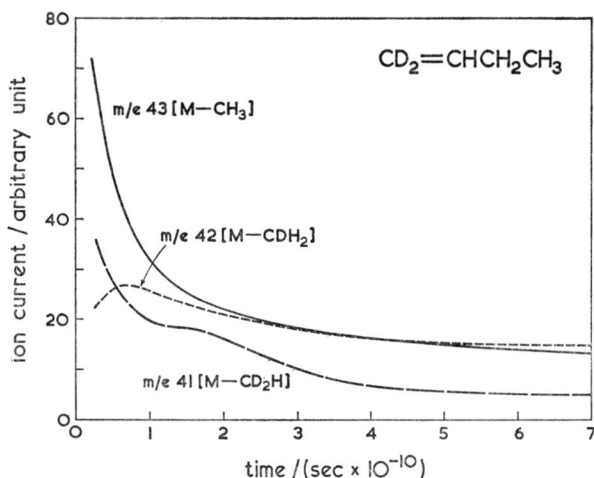

FIG. 1   Ion currents due to m/e 41 (M–CD$_2$H), m/e 42 (M–CDH$_2$), and m/e 43 (M–CH$_3$) as a function of reaction time following FI of 1-butene-1,1-d$_2$.

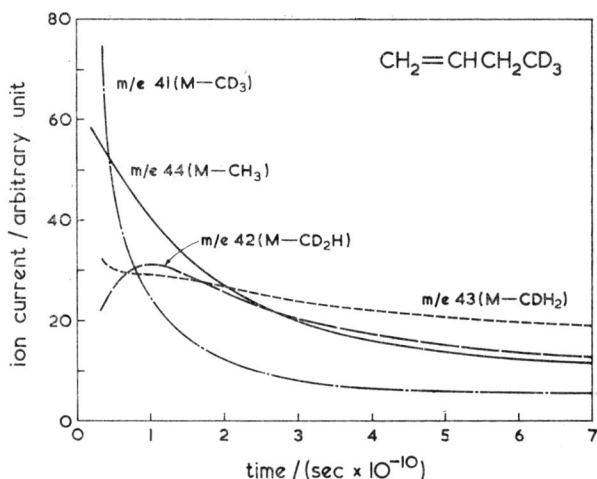

FIG. 2   Ion currents due to m/e 41 (M–CD$_3$), m/e 42 (M–CD$_2$H), m/e 43 (M–CDH$_2$) and m/e 44 (M–CH$_3$) as a function of reaction time following FI of 1-butene-4,4,4-d$_3$.

m/e 42 (M–CDH$_2$), and m/e 43 (M–CH$_3$) from 1-butene-1,1-d$_2$ (**1**) and m/e 41 (M–CD$_3$), m/e 42 (M–CD$_2$H), m/e 43 (M–CDH$_2$), and m/e 44 (M–CH$_3$) from 1-butene-4,4,4-d$_3$ (**2**) formed at times between 3×10^{-11} sec and 7×10^{-10} sec following FI. All these fragments correspond to a single fragment (M–CH$_3$)$^+$ in the undeuterated species. The ion currents in Figs. 1 and 2 can be readily converted to rates of reaction or rate constants.[3,4] For the purposes of our discussion, however, this is unnecessary, since for fragment ions of similar masses (as in Figs. 1 and 2) the ratios of their ion currents at any time can be considered to be equal to the ratios of their rates of formation at that time.[3] The particulars of our experimental technique,[3,4] which utilizes a double focussing FI mass spectrometer, and our method[3,9] of transforming the measurements into kinetic data (Figs. 1 and 2) have been described. Table I shows the ratios of the metastable intensities due to loss of methyl groups from C$_4$D$_2$H$_6$$^{\cdot+}$ and C$_4$D$_3$H$_5$$^{\cdot+}$ species in the first field-free region between the ion source and the electrostatic analyser (ESA). These metastables, representing reaction occurring at about 10^{-6} sec following ionization were measured by a standard technique[10] involving varying the ESA voltage. Calculated ion currents assuming complete H/D randomization and zero isotope effect are shown for comparison.

The formation of three fragments with 1-butene-1,1-d$_2$ and four fragments with 1-butene-4,4,4-d$_3$ corresponding to a single fragment with the un-deuterated species is convincing proof that extensive rearrangement occurs either prior to or during fragmentation. It is evident that the relative ion currents of the fragments from the deuterated species are sensitive functions of time from 3×10^{-11} to 7×10^{-10} sec (Figs. 1 and 2). We see this dependence on time as an accurate indication of the progress of the various rearrangements. The most interesting portions of the curves in both Figs. 1 and 2 are those containing structure between 3×10^{-11} and 3×10^{-10} sec. After 3×10^{-10} sec the ratios of the ion currents slowly approach the

TABLE I

Relative Ion Currents for Loss of Methyl Groups from the
$C_4D_2H_6^{\cdot+}$ and $C_4D_3H_5^{\cdot+}$ Species at Various Times Following FI

	1-*Butene*-1,1-d_2		
	m/e 41	m/e 42	m/e 43
	($M–CD_2H$)	($M–CDH_2$)	($M–CH_3$)
Measured (time in sec)			
3×10^{-11}	29	19	52
3×10^{-10}	15	42	43
5×10^{-10}	15	43	42
7×10^{-10}	14	45	41
$\sim 10^{-6}$ (metastables)	10 ± 3	52 ± 3	38 ± 3
Calculated			
complete H/D randomization	11	54	36

	1-*Butene*-4,4,4-d_3			
	m/e 41	m/e 42	m/e 43	m/e 44
	($M–CD_3$)	($M–CD_2H$)	($M–CDH_2$)	($M–CH_3$)
Measured (time in sec)				
3×10^{-11}	45	11	16	28
3×10^{-10}	11	29	33	27
5×10^{-10}	11	27	37	25
7×10^{-10}	10	27	40	23
$\sim 10^{-6}$ (metastables)	3 ± 3	25 ± 3	52 ± 3	19 ± 3
Calculated				
complete H/D randomization	2	27	54	18

values predicted on the basis of completely random selection of the three H and/or D atoms to be lost as methyl (Table I). At $\sim 10^{-6}$ sec the ratios of the fragment ion currents are equal to the predicted values for completely random loss. There may be several structural isomers (*e.g.* 1-butene$^{\cdot+}$, 2-butene$^{\cdot+}$, 2-methylpropene$^{\cdot+}$) among the ions decomposing at $\sim 10^{-6}$ sec. We suggest that within each of these species the H and D are completely randomized. This would certainly be expected to be the case for 2-methylpropene ions.[5]

Consider now the ion currents at 3×10^{-11} to 3×10^{-10} sec (Figs. 1 and 2). We suggest that the high ion currents of m/e 43 ($M–CH_3$) with 1 and m/e 41 ($M–CD_3$) with 2 at the shortest times represent fragmentation of the un-rearranged 1-butene-1,1-$d_2^{\cdot+}$ and 1-butene-4,4,4-$d_3^{\cdot+}$. The ion currents of both these fragments decline steadily with time reflecting the conversion of the 1-butene-1,1-$d_2^{\cdot+}$ and 1-butene-4,4,4-$d_3^{\cdot+}$ ions to other species. With both **1** and **2** there are two other ions, m/e 41 ($M–CD_2H$) and m/e 44 ($M–CH_3$), whose ion currents decline with increasing time (Table I). We suggest that some of the original 1-butene ions isomerize to 2-butene and that the 2-butene ions may fragment to form m/e 41 ($M–CD_2H$) and m/e 44 ($M–CH_3$) (*see* eqns (1) and (2)). The 1-butene ions must be able to rearrange to 2-butene ions (as in eqns (1) and (2)) in less than 3×10^{-11} sec. If our time-resolution were better we would expect to observe a maximum in the m/e 41 ($M–CH_3$) ion

$$CD_2{=}CHCH_2CH_3^{\overset{+}{\cdot}} \longrightarrow CH_3^{\overset{\cdot}{+}} + C_3D_2H_3^{\overset{+}{\cdot}}$$
$$m/e\ 43$$

$$CD_2HCH{=}CHCH_3^{\overset{+}{\cdot}}$$

$$\longrightarrow CH_3^{\cdot} + C_3D_2H_3^{+}$$
$$m/e\ 43$$

$$\longrightarrow CD_2H^{\cdot} + C_3H_5^{+}$$
$$m/e\ 41$$

$$(1)$$

$$CH_2{=}CHCH_2CD_3^{\overset{+}{\cdot}} \longrightarrow CD_3^{\overset{\cdot}{+}} + C_3H_5^{+}$$
$$m/e\ 41$$

$$CH_3CH{=}CHCD_3^{\overset{+}{\cdot}}$$

$$\longrightarrow CD_3^{\cdot} + C_3H_5^{+}$$
$$m/e\ 41$$

$$\longrightarrow CH_3^{\cdot} + C_3D_3H_2^{+}$$
$$m/e\ 44$$

$$(2)$$

currents at some time $< 3 \times 10^{-11}$ sec. At times of 3×10^{-11} sec and longer, the 2-butene ions $CD_2HCH{=}CHCH_3^{\cdot+}$ and $CH_3CH{=}CHCD_3^{\cdot+}$ are rearranging further to other species, so that the observed proportions of $m/e\ 41$ (M–CD_2H) and $m/e\ 44$ (M–CH_3) fall with time, reflecting the progress of these further rearrangements (Table I). Another possible explanation for the formation of $m/e\ 41$ (M–CD_2H) and $m/e\ 44$ (M–CH_3) would involve skeletal rearrangement. Consider 1-butene-1,1-$d_2^{\cdot+}$. A shift of the C-4 methyl to C-2 followed by a 1,2 hydrogen shift from C-2 to C-1 would yield a 2-methylpropene ion, which could lose CD_2H as required. We consider that such a relatively complicated rearrangement involving a methyl shift is unlikely to be important in the time scale under consideration (3×10^{-11} sec and less). Direct 1,2 and 1,4 hydrogen transfers in the original 1-butene ions can be dismissed as far as the formation of $m/e\ 41$ (M–CD_2H) and $m/e\ 44$ (M–CH_3) at these short times is concerned; 1,2 hydrogen shift from C-2 to C-1 would produce an energetically unfeasible structure and 1,4 transfer in $CH_2{=}CHCHCD_3^{+}$ cannot lead to the loss of CH_3 as required.

The formation of the fragments $m/e\ 42$ (M–CDH_2) from 1 and $m/e\ 42$ (M–CD_2H) from 2 can be rationalized in terms of 1,3 hydrogen shifts. The formation of $m/e\ 43$ (M–CDH_2) with 2 cannot. So hydrogen shifts other than 1,3 or skeletal rearrangements are certainly operative. These cannot, however, be securely identified from the present results with 1 and 2 (Figs. 1 and 2).

Further interpretation of these results should be possible when the kinetics of the skeletal rearrangements of the 1-butene ion are better understood.

We wish in the remainder of this paper to focus our attention on the proposed isomerization of the 1-butene ion to 2-butene. We envisage the isomerization as proceeding via a direct 1,3 hydrogen shift. Two consecutive 1,2 hydrogen shifts is another possibility. The attractive feature of the 1,3 mechanism is that it can be considered electrocyclic (*see* eqn 3). The open shell can retain its delocalized π characteristic throughout the rearrangement. Comparison is unavoidably provoked with the 1,3 hydrogen shift within the stable carbonium ion 2,4-dimethyl-2-pentyl[11] (*see* eqn (4)). The carbonium

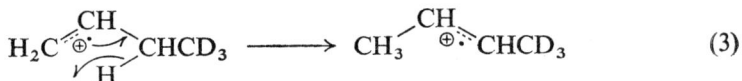

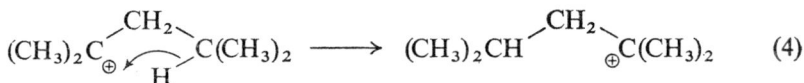

$$H_2C \overset{CH}{\underset{H}{\oplus}} CHCD_3 \longrightarrow CH_3 \overset{CH}{\oplus} CHCD_3 \tag{3}$$

$$(CH_3)_2C \overset{CH_2}{\underset{H}{\oplus}} C(CH_3)_2 \longrightarrow (CH_3)_2CH \overset{CH_2}{\underset{\oplus}{}} C(CH_3)_2 \tag{4}$$

ion does not undergo 1,2 hydrogen shifts and rearrangement does not proceed via protonated cyclopropane.[11] It is also worth noting that no proposed 1,2 hydrogen shift in a free radical has been substantiated.[12] We feel that the present FIK study provides good support for our earlier suggestion[5] that 1,3 shift of allylic hydrogen across an ionized olefinic bond is a facile and ubiquitous unimolecular gas-phase reaction. The results suggest that the frequency factor for this reaction via a four-membered cyclic transition state is close to 10^{-11} sec^{-1}. To *securely* establish any unimolecular gas-phase rearrangement of a radical-cation would be valuable. We feel that with the FIK results on cyclohexene,[4] 2-methylpropene,[5] and 1-butene we are close to substantiating 1,3 allylic hydrogen shifts.

ACKNOWLEDGMENTS

The financial support of the National Aeronautics and Space Administration (Grant NGL 05-003-003) and the National Science Foundation (Grant NSF GP-38389X) for this work is gratefully acknowledged.

REFERENCES

1. (a) Beckey, H. D., *Z. Naturforsch. A*, 1961, **16**, 505; (b) Beckey, H. D., *Z. Naturforsch. A*, 1966, **21**, 1920; (c) Beckey, H. D., Hey, H., Levsen, K. and Tenschert, G., *Int. J. Mass Spectrom. Ion Phys.*, 1969, **2**, 101.
2. Derrick, P. J. and Robertson, A. J. B., *Proc. Roy. Soc. Ser. A*, 1971, **324**, 491.
3. Falick, A. M., Derrick, P. J. and Burlingame, A. L., *Int. J. Mass Spectrom. Ion Phys.*, in press.
4. Derrick, P. J., Falick, A. M. and Burlingame, A. L., *J. Amer. Chem. Soc.*, 1972, **94**, 6794.
5. Derrick, P. J. and Burlingame, A. L., submitted to *J. Amer. Chem. Soc.*

6. (a) Lias, S. G. and Ausloos, P., *J. Res. Nat. Bur. Stds.*, 1971, **75A**, 591; (b) Sieck, L. W., Lias, S. G., Hellner, L. and Ausloos, P., *J. Res. Nat. Bur. Stds.*, 1972, **76A**, 115.
7. Meisels, G. G., Park, J. Y. and Giessner, B. G., *J. Amer. Chem. Soc.*, 1969, **91**, 1555.
8. (a) Bryce, W. A. and Kebarle, P., *Can. J. Chem.*, 1956, **34**, 1249; (b) Millard, B. J. and Shaw, D. F., *J. Chem. Soc. B*, 1966, 664.
9. Pfeifer, J.-P., Falick, A. M. and Burlingame, A. L., *Int. J. Mass Spectrom. Ion Phys.*, in press.
10. See for a description of the technique, Hills, L. P. and Futrell, J. H., *Org. Mass Spectrom.*, 1971, **5**, 1010.
11. Saunders, M. and Stofko, J. J., Jr, *J. Amer. Chem. Soc.*, 1973, **95**, 252.
12. Wilt, J. W., *in* 'Free Radicals', *ed.* J. K. Kochi, John Wiley & Sons, Inc., 1973, **1**, 378.

Discussion

F. W. McLafferty (Cornell University, U.S.A.): You postulate a 1,3-H shift mechanism which for the $CH_2=CHCH_2CD_3$ should give $CH_3CHCHCD_3$ as the intermediate. Should not the decomposition of this intermediate by methyl loss give nearly equal amounts of $C_3H_7^+$ and $C_3H_4D_3^+$? In contrast, your data show that the $C_3H_4D_3^+$ production increases with time to become substantially *greater* than that for $C_3H_7^+$.

Also, your abstract states that this method provides information on the entropy of activation for the rearrangement reaction $C_5H_{11}CHO^{+\cdot} \rightarrow C_4H_8 + CH_2CH-OH^{+\cdot}$ which indicates that this is a *concerted*, not a stepwise process. However, the overall entropy for a 2 step process, in which one step is a tight complex rearrangement reaction and the other is a loose complex reaction, should be little different from a concerted process which is simply the simultaneous occurrence of these same two processes. Can you comment on this?

P. J. Derrick: We suggest that $CH_2CHCH_2CD_3^{+\cdot}$ can rearrange very rapidly (10^{-12}– 10^{-11} sec) to $CH_3CHCHCD_3^{+\cdot}$ and that m/e 44 (M–CH_3) formed at a few $\times 10^{-11}$ sec originate in large part from $CH_3CHCHCD_3^{+\cdot}$. Other rearrangements of $CH_2CHCH_2CD_3^{+\cdot}$ do, however, certainly occur in times of the order of 10^{-11} sec and $CH_3CHCHCD_3^{+\cdot}$ will rearrange further in the same time frame. Formation of m/e 42 (M–CD_2H) and m/e 43 (M–CDH_2) is evidence of this. It is the fragmentation of isotopic and structural isomers other than $CH_2CHCH_2CD_3^{+\cdot}$ and $CH_3CHCHCD_3^{+\cdot}$ which is responsible for the high intensity of m/e 44 (M–CH_3) relative to that of m/e 41 (M–CD_3) at times longer than a few $\times 10^{-11}$ sec. We do see the reaction system set up ionization of 1-butene as being extremely complex, but we do believe the complexity can be unravelled by making further FIK measurements on ^{13}C- and D-labelled compounds.

To move on to your second question, we use the terms 'tight' and 'loose' to refer to extreme types of activated complex (see for example G. M. Wieder and R. A. Marcus, J. Chem. Phys., **37**, 1835 (1962)). We envisage an activated complex in the proposed concerted mechanism which is closer to the 'loose' than the 'tight' extreme. The important feature of our concerted mechanism is that the reaction is considered to be a dissociation (although a hydrogen shift occurs simultaneously). The two fragments have begun to move apart at the activated complex. The activation energy for the reverse reaction is probably low. The proposed activated complex therefore conforms to the usual meaning of 'loose'. One vibration will have been converted to a translation along the reaction co-ordinate. Perhaps other vibrations have been converted (to some degree) to the free rotations of the incipient fragments. By contrast we envisage the activated complex in the rearrangement step of the two-step mechanism as conforming rather closely to the definition of a 'tight' activated complex.

104
Metastable Transitions in Isotopically Substituted Ethylenes

By I. NENNER, H. NGUYEN NGHI
and R. BOTTER

(*Département de Recherche et Analyse, CEN Saclay, Gif sur Yvette, France*)

INTRODUCTION

METASTABLE ion dissociations observed in mass spectrometry are of invaluable importance in all the studies of monomolecular dissociation mechanisms, and are often a direct test of the fundamental hypothesis of the Quasi Equilibrium Theory of mass spectra (QET).[1]

The subject has been reviewed recently by Holmes and Benoit[2] and by Wahrhaftig[3] in relation with theory. These authors have shown the importance of the study of isotopically substituted molecules.

In the course of the study of metastable transitions in deuterium substituted acetylene and ethylene[4] a very large isotope effect has been found for the loss of HD and D_2 from C_2HD_3 (loss of HD/loss of $D_2 \approx 25$). Such large isotope effects have also been observed in other partially substituted molecules such as the partially deuterated methanes[5] and $(CD_3CHDCH_3)^{+}$[6] in which the observed loss of H is 300 times larger than the loss of D. Ottinger,[6] and Vestal also observed an isotopic effect favouring the loss of H_2 rather than HD from $C_3H_5D_2{}^{+}$.[6,7]

These effects are all far from statistical. For CH_3–CD_3 Löhle and Ottinger[8] and Lifshitz and Sternberg[9] observe mainly the loss of HD and almost no H_2 or D_2 from the metastable parent ion. These results show that the H_2 from the C_2H_6 ion is due to the loss of one hydrogen atom from each carbon[8] and that no or very little scrambling of the hydrogen takes place before dissociation at low internal energy. However these authors and D'or et al.[10] concluded from the mass spectra that at higher energy reshuffling occurs before dissociation. From the isotopic ratio in the metastable dissociation of the $C_2H_2D_2{}^{+}$ fragment ion, Löhle and Ottinger state that in this latter only partial rearrangement has occurred. However, their results on the relative intensities of the metastable peaks corresponding to the loss of H_2, HD and D_2 from $C_2H_2D_2{}^{+}$ seem to disagree with those of Lifshitz and Sternberg.

These observations, the results on the isotopic effect in C_2D_3H, and the relation of metastable transitions with the QET led us to undertake a study of the metastable ion dissociation in the deuterated ethylenes. All the substituted ethylenes, except the *cis* $C_2H_2D_2$ have been analysed. The isotope effect in the metastable reaction $C_2X_4^+ \rightarrow C_2X_2^+ + X_2$ (where X = H or D) were carefully measured and the experimental results compared with calculated values in the frame of the QET.

Preliminary results were communicated at the Atlanta ASTM meeting.[11]

EXPERIMENTAL SECTION AND RESULTS

The experiments were performed on a single focussing 20-cm radius, 60° deflection mass spectrometer. The ionizing electron energy was about 50 eV and the ion accelerating voltage 5 kV. The ion source and the analyser were pumped separately. The tube pressure was always less than 10^{-7} torr.

The deuterium labelled compounds were prepared in the laboratory. The chemical purity was always better than 99·9% for all the compounds. The isotopic purities were as follows: C_2H_4 better than 99·9%, C_2D_4: 98% *trans* and geminal $C_2H_2D_2$: 92%. Both CH_3D and CD_3H were obtained as a mixture of about 10% in C_2H_4 and C_2D_4 respectively. Although the presence of C_2H_4 and C_2D_4 did not allow the precise measurement of the mass spectra of CH_3D and CD_3H, they did not disturb the determination of the metastable transitions in these compounds.

Mass Spectra

The mass spectra of C_2H_4, *trans* CHD–CHD, CH_2–CD_2 and CD_4 are given in Table I. Fragment ions below mass 24 have been neglected; they represent only a few percent of the total ionization. The isotopic contributions (especially ^{13}C) have been subtracted. The mass spectra are in good agreement with those of Delfosse and Hipple[12] and of Dibeler *et al.*[13]

A slight isotope effect can be observed in the mass spectra of C_2H_4 and C_2D_4. The spectra of *trans* and geminal $C_2H_2D_2$ are quite similar. This

TABLE I

Mass Spectra of C_2H_4 1,2 *trans* and 1,1 D_2 Ethylene and C_2D_4

m/q	C_2H_4		$C_2H_2D_2$			C_2D_4	
	Fragments	RA	Fragments	trans	Geminal	Fragments	RA
32						$C_2D_4^+$	100
31							
30			$C_2H_2D_2^+$	100	100	$C_2D_3^+$	52·1
29			$C_2HD_2^+$	31·0	36·2		
28	$C_2H_4^+$	100	$C_2D_2^+$, $C_2DH_2^+$	36·1	36·1	$C_2D_2^+$	55·1
27	$C_2H_3^+$	54·5	C_2DH^+	33·8	40·0		
26	$C_2H_2^+$	52·8	C_2D^+, $C_2H_2^+$	10·6	11·0	C_2D^+	7·0
25	C_2H^+	8·6	C_2H^+	2·5	2·7		
24	C_2^+	2·1	C_2^+	1·7	−1·8	C_2^+	1·5

proves that H and D undergo scrambling before dissociation. However, differences in the relative intensities of the C_2HD^+ fragment show that the HD loss is larger in the 1,1 D_2 compound and this would indicate that randomization of H and D is not complete before dissociation.

Metastable Peaks

The metastable transitions observed in C_2H_4 for the loss of H or H_2 from the C_2 ions are summarized in Table II together with the kinetic energy of the metastable fragment ions. The first reaction, loss of H_2 from the parent ion,

TABLE II
Metastable Reaction and Kinetic Energy Release in $C_2H_4^+$†

Reaction		Apparent mass	Intensity‡	Kinetic energy (eV)
$C_2H_4^+$	$C_2H_2^+ + H_2$	24·14	0·031	0·010 ± 0·005
$C_2H_3^+$	$C_2H_2^+ + H$	25·04	0·001	
C_2H^+	$C_2^+ + H$	23·04	0·001 5	0·20 ± 0·05

† Values taken from Reference 4.
‡ Percent of the parent $C_2H_4^+$ peak.

is the most important metastable transition. The kinetic energy release during decomposition is very low and the metastable peak shape is nearly the same as that of a normal peak.

The metastable reaction $C_2X_4^+ \rightarrow C_2X_2^+ + X_2$ has also been studied in all the substituted ethylenes (except *cis* dideutero). The results are given in Table III. Since the kinetic energy release remains very low in all these cases, the observed isotope effects are certainly not of instrumental origin.

Two kinds of isotope effect may be seen. First, the loss of D_2 from $C_2D_4^+$ is larger than the loss of H_2 from $C_2H_4^+$. Secondly, in each partly substituted ethylene, where two or three competitive reactions can take place; the loss of the lighter neutral fragment is always favoured over the heavier one. For the dideutero compounds, both the *trans* and géminate isomers give the same results for the metastable intensities indicating that scrambling takes place before dissociation of the metastable ion. Such an effect is completely opposite to that observed in CH_3–CD_3 where little scrambling has been observed at low internal energy.[8,9]

The observed relative intensities corresponding to the loss of H_2, HD and D_2 in the partially substituted ethylenes are in all cases far from the statistical values. We have tried to verify if such large isotopic effects may be explained by the Quasi Equilibrium Theory.

CALCULATION OF THE MASS SPECTRA

The general expression of the rate constant for the monomolecular dissociation of ions as given in the QET is

$$k_{(E)} = \frac{\alpha}{h} \frac{W^\ddagger(E - E_a)}{\rho(E)}$$

TABLE III

Isotopic Effect in the Metastable Transition $C_2X_4^+ \rightarrow C_2X_2^+ + X_2$

Molecule		Reactions		m^*	Relative abundance		
					Loss of H_2	Loss of HD	Loss of D_2
C_2H_4	(28)	$C_2H_4^+ \rightarrow C_2H_2^+ + H_2$		24·14	$3·0 \times 10^{-2}$		
C_2H_3D	(29)	$C_2H_3D^+ \rightarrow C_2H_2^+ + HD$		23·31		$<0·06 \times 10^{-2}$	
		$\rightarrow C_2HD^+ + H_2$		25·14	$4·5 \times 10^{-2}$		
$C_2D_2H_2$	(30)	$C_2D_2H_2^+ \rightarrow C_2H_2^+ + D_2$	gém.	22·53			$<0·01 \times 10^{-2}$
		$C_2D_2H_2^+ \rightarrow C_2H_2^+ + D_2$	trans	22·53			$<0·01 \times 10^{-2}$
		$\rightarrow C_2HD^+ + HD$	gém.	24·30		$0·85 \times 10^{-2}$*	
		$\rightarrow C_2HD^+ + HD$	trans	24·30		$0·9 \times 10^{-2}$	
		$\rightarrow C_2D_2^+ + H_2$	gém.	26·13	$7·0 \times 10^{-2}$		
		$\rightarrow C_2D_2^+ + H_2$	trans	26·13	$7·4 \times 10^{-2}$		
C_2D_3H	(31)	$C_2D_3H^+ \rightarrow C_2HD^+ + D_2$		22·52			$<0·1 \times 10^{-2}$
		$\rightarrow C_2D_2^+ + HD$		25·29		$3·0 \times 10^{-2}$	
C_2D_4	(32)	$C_2D_4^+ \rightarrow C_2D_2^+ + D_2$		24·50			$3·5 \times 10^{-2}$

where $W^{\ddagger}(E - E_a)$ is the number of states of the activated complex with energy less than or equal to $E - E_a$ and E_a the activation energy. $\rho(E)$ is the density of states of the ion with internal energy E; α is the number of equivalent paths for the reaction and h Planck's constant.

One of the greatest difficulties encountered in using this formula is being able to accurately enumerate the density of states. This problem has been reviewed in detail by Rosenstock[14] and more recently, by Wahrhaftig.[3]

We have used Haarhoff's method[15] for the enumeration of states in the present calculations. Rate constants can be readily evaluated with a computer from Haarhoff's final expression. These rate constants have been calculated for all the primary reactions. For the second generation of fragment ions we assume that all the first generation ions having internal energies larger than the activation energy for the given reaction will dissociate.

The relative abundance of the ions can then be calculated from the $k(E)$.

Set of Reactions

The set of reactions assumed for the dissociation of $C_2H_4^+$ is shown in Fig. 1. Reaction 5 has been reduced from Szabo's[16] charge exchange experiments on C_2H_4. The set of reactions for the deuterated compounds has been deduced by analogy with the scheme of Fig. 1. For $C_2H_2D_2$ we have assumed complete randomization. The same activation energies have been used in the case of partially and totally deuterated compounds for all reactions except reaction 3.

Rate Constant

Frequencies were estimated for the ion and the intermediate complex (values are given in Ref. 17). For the ground state ion we have taken the frequencies of the molecule for the C–H stretching and for one bending vibration. The C–C stretching, symmetric bending and twisting vibrations have been obtained from photoelectron spectroscopy measurements.[18,19] Values of other frequencies have been taken lower in the ion than in the neutral molecule. For the activated complex frequencies having about the same values have been chosen. The reaction co-ordinate for the loss of H is evident.

For H_2 loss from $C_2H_4^+$ two intermediate complexes have been considered.

$$
\begin{array}{c}
\text{H} \,\text{-}\,\text{-}\, \text{H} \\
\text{\footnotesize |} \;\; (+) \;\; \text{\footnotesize |} \\
\end{array}
$$

First a symmetric one of the form $\text{H—C} \stackrel{=}{=} \text{C—H}$ where the two hydrogens are lost from the two carbon atoms. This form has been used for the calculation of mass spectra.

Secondly an asymmetric structure in equilibrium with the ground state.

$$CX_2 = CX_2^+ \rightleftharpoons CX - X_3^+ \rightleftharpoons (CX - CX_3^+)^{\ddagger} \rightarrow CX \equiv CX^+ + X_2$$

This type of dissociation has been suggested by Prasil and Frost[20] and by Lorquet and Lorquet.[21] These last authors have calculated that the asymmetric form is the first excited state of the ion and is only 0·8 eV above the ground state. Moreover, the rotational barrier should be only 0·3 eV (1·08 eV

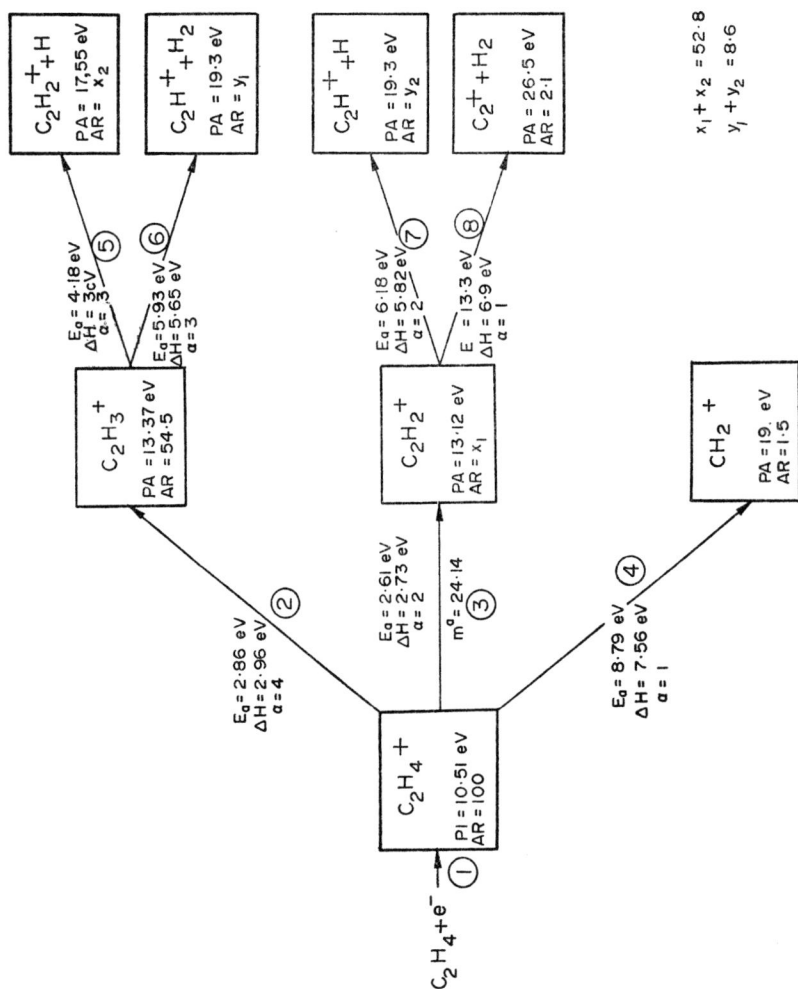

FIG. 1   Reaction scheme for the dissociation of the $C_2H_4^+$ ion.

in the molecule) according to these authors. Free rotation has been considered in our calculations. An activation energy of 0·1 eV for the reverse reaction $(CX - CX_3^+ \rightarrow CX_2 = CX_2^+)$ has been chosen.

Differences in activation energy for the loss of H_2, HD and D_2 from $C_2H_2D_2^+$ of 0·05 eV, of the order of the differences in zero point energy of the fragments, have been taken.

RESULTS AND DISCUSSION

The calculated breakdown curves are shown in Fig. 2 for $C_2H_4^+$, $C_2H_2D_2^+$ and $C_2D_2^+$ together with the experimental results of Szabo for the first two ions. The agreement between the experimental points and the calculated curves is good although some small discrepancies may be observed. The mass spectra for C_2H_4 and $C_2H_2D_2$ have been calculated from these breakdown curves and by using the photoelectron spectrum[19] as an internal energy distribution function. The results are shown in Table IV for both the experimental and calculated breakdown curves together with the mass spectra obtained by electron impact (50 eV energy) and by photoionization (He 21·21 eV resonance line).[22] The two experimental spectra are different. This

TABLE IV

Comparison of the Experimental and Calculated Mass Spectra of C_2H_4 and $C_2H_2D_2$

M/q	Ions	Photoionization 21·21 eV 22	Electron impact (50 eV)	Convolution of the breakdown curves and the distribution function obtained from photoelectron spectroscopy	
				Exper. breakdown curves	Calcul. breakdown curves
		C_2H_4			
28	$C_2H_4^+$	100	100	100	100
27	$C_2H_3^+$	137	54·5	147	160
26	$C_2H_2^+$	76	52·8	82	78
25	C_2H^+		8·6		4
24	C_2^+		2·1		1
		$C_2H_2D_2$			
30	$C_2H_2D_2^+$	100	100	100	100
29	$C_2HD_2^+$	93	31·0	83	99
28	$C_2H_2D^+$, $C_2D_2^+$	85	36·1	65	73
27	C_2HD^+	59	33·8	43	42
26	$C_2H_2^+$, C_2D^+	15	10·6	9	26
25	C_2H^+		2·5		4
24	C_2^+		1·7		1

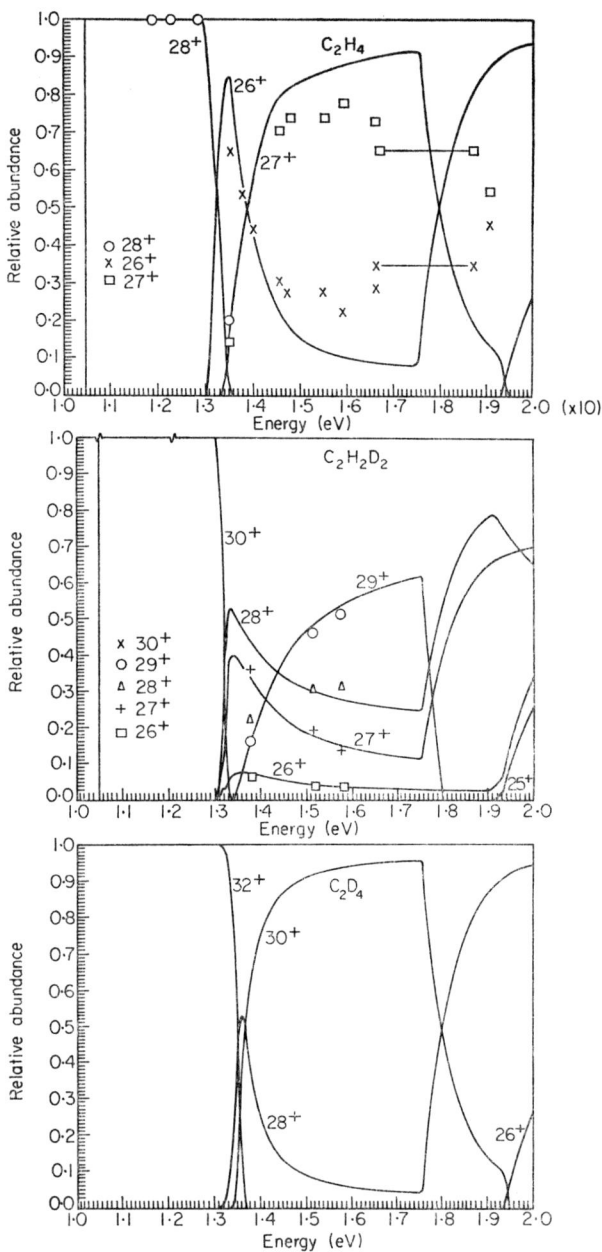

FIG. 2   Breakdown curves for $C_2H_4^+$, $C_2H_2D_2^+$ and $C_2D_4^+$.

difference is certainly due to autoionization and to forbidden transitions in the case of electron impact. The agreement between the experimental photo-ionization mass spectra and the two calculated spectra is good. In this case the internal energy distribution functions are exactly the same. This shows that the fragmentation of $C_2H_4^+$ for the most important reactions, is well described by the QET.

Metastable Transitions

The rate constants for the various reactions $C_2X_4^+ \rightarrow C_2X_2^+ + X_2$ have been calculated more accurately near the minimum in order to compare the calculated isotope effect with the relative intensities of the isotopic metastable peaks. The rate constants found for C_2H_4 are larger than those for C_2D_4 and are in agreement with the observed effect (metastable intensity larger

TABLE V

Rate Constants for the Reaction

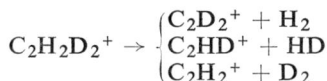

$$C_2H_2D_2^+ \rightarrow \begin{cases} C_2D_2^+ + H_2 \\ C_2HD^+ + HD \\ C_2H_2^+ + D_2 \end{cases}$$

Energy	Symmetric complex[a] $k \times 10^{-5}$			Asymmetric complex[b] $k \times 10^{-5}$		
	k_{H_2}	k_{HD}	k_{D_2}	k_{H_2}	k_{HD}	k_{D_2}
13·06	5·61			2·2		
13·11	4·99	9·56		1·8	3·1	
13·16	12·8	7·92	4·18	5·0	2·9	1·5
13·21	26·7	20·2	5·37	9·2	6·5	2·6
13·26	49·5	41·8	16·9	21	9·8	3·5
13·31	85·1	77·2	40·5	38	15·5	8·1

[a] Reaction path $H_2C = CH_2^+ \leftrightarrows (H-\overset{\overset{H--H}{|+|}}{C}-C-H)^{\ddagger} \rightarrow C_2H_2^+ + H_2$

[a] Reaction path $H_2C = CH_2^+ \leftrightarrows HC-CH_3^+ \leftrightarrows (HC-CH_3^+)^{\ddagger} \rightarrow C_2H_2^+ + H_2$.

for C_2D_4 than for C_2H_4, *see* Table III). The results obtained for $C_2H_2D_2$, with the two configurations for the intermediate complexes, are summarized in Table V. The intensities calculated with these rate constants and for a time range of 10^{-6} sec to $3 \cdot 10^{-6}$ sec are, for the symmetric complex, $H_2/HD \approx 5$ and $D_2/H_2 \approx 160$, for the asymmetric complex $H_2/HD \approx 2 \cdot 3$ and $D_2/H_2 \approx 20$. If we take a time range from 3×10^{-6} sec to 5×10^{-6} sec these ratios become respectively 50, 10^5, 3 and 120.

The calculated rate constants can very well account for the observed isotope effect. Moreover, an increase in the time limit increases the isotope effect by a very large factor. The time range in which the metastable reaction may occur can vary from one mass spectrometer to another; this can easily explain the differences in the isotope effects observed by different authors for this type of transition. Decreasing the observation time will reduce the isotope effect but also increase the relative intensities of the metastable peaks. This is in agreement with the results of Löhle, Lifshitz and the present authors.

The differences in the activation energies for the loss H_2, HD and D_2 used in the calculation may be larger than the differences in the zero point energies by a factor of 2, but this does not change the present conclusions. Moreover, the relative rate constants k_{H_2}, k_{HD} and k_{D_2} depend on the frequencies chosen and the relative values of k_{H_2} and k_{HD} at a given energy may be reversed.

The internal energy distribution may also have an effect on the relative intensities of the isotopic metastable peaks. If this distribution is not quasi continuous it may reduce or enhance one of the isotopic fragments. Tunnelling effects should equally be taken into consideration.

New experiments have been undertaken for the measurements of the metastable ion intensities produced by photoionization with the He 21·21 eV resonance line. The metastable transition $C_2H_4^+ \rightarrow C_2H_2^+ + H_2$ has been observed, but the intensity of the metastable peak is at present too low to make any accurate measurements on the substituted molecules.

REFERENCES

1. Rosenstock, H. M., Ph.D. Thesis, University of Utah, 1952.
2. Holmes, J. L. and Benoit, M. B., *in* 'Mass Spectrometry', Phys. Chemistry Series One, *ed.* A. Maccoll, Butterworth, London, 1972, Vol. 5.
3. Wahrhaftig, A. L., same refer.
4. Botter, R., Hagemann, R., Khodadadi, G. and Rosenstock, H. M., *in* 'Recent Developments in Mass Spectrometry', *ed.* K. Oyata and T. Hayahawa, Univ. of Tokyo Press, 1970.
5. Ottinger, Ch., *Z. Naturforsh*, 1965, **20a**, 1232.
6. Ottinger, Ch., *J. Chem. Phys.*, 1967, **47**, 1452.
7. Vestal, M. and Futrell, J. H., *J. Chem. Phys.*, 1970, **52**, 978.
8. Löhle, U. and Ottinger, Ch., *J. Chem. Phys.*, 1969, **51**, 3097.
9. Lifshitz, C. and Sternberg, R., *Int. J. of Mass Spectrom. and Ion Phys.*, 1969, **2**, 303.
10. D'Or, L., Collin, J. E. and Longree, J., *Bull. Classe Sci. Acad. Roy. Belg.*, 1966, **52**, 518.
11. Baumel, I., Hagemann, R. and Botter, R., XIXth An. Conf. on Mass Spect. and Allied Topics, Atlanta, 1971.
12. Delfosse and Hipple, *Phys. Rev.*, 1937, **52**, 843.
13. Dibeler, V. H., Mohler, F. L. and Hemptine, M. de, *J. of Res. of the N.B.S.*, 1954, **53**, 107.
14. Rosenstock, H. M., 'Advances in Mass Spectrometry', The Institute of Petroleum, London, Vol. IV, p. 523.
15. Haarhoff, P. C., *Mol. Phys.*, 1963, **6**, 337 and 1963–4, **7**, 101.
16. Szabo, I., *Arkiv. für Fysik.*, 1966, **31**, 287.
17. Nenner, I., *These de 3è cycle*, 1972.
18. Baker, A. D., Baker, C., Brundle, Cr. and Turner, D. W., *Int. J. of Mass Spectrom. and Ion Phys.*, 1968, **1**, 285.
19. Branton, C. R., Frost, D. C., Marika, T., McDowell, C. A. and Stenhouse, I. A., *J. Chem. Phys.*, 1970, **52**, 802.
20. Prasil, Z. and Frost, W., *J. Chem. Phys.*, 1967, **71**, 3166.
21. Lorquet, A. J. and Lorquet, J. C., *J. Chem. Phys.*, 1968, **49**, 4955.
22. Botter, R., Dibeler, V. H., Walker, J. A. and Rosenstock, H. M., *J. Chem. Phys.*, 1966, **45**, 1298.

105

Charge-Exchange and Metastable Transition Continua in the 180° Magnetic Mass Spectrometer

By J. M. McCREA

(*Monroeville, Pennsylvania, U.S.A.*)

HIPPLE,[1] Coggeshall[2] and Viney[3] have all given extensive treatments of metastable peak contours in the 180° mass spectrometer, and each has assumed that the primary ion beam is a line beam. The author's evaluations of charge-exchange continua in sector and Mattauch–Herzog instruments also utilize a line-beam assumption with the additional restriction that beam intensity does not decrease with distance travelled.[4,5,6] Although the point has not been stressed, the calculated contours have shown an infinite intensity for every case in which the secondary ions approach the detector in a direction parallel to that of the primary beam. Thus there are infinity anomalies at both ends of the contours for the 180° instrument. Coggeshall has shown that incremental velocity effects can explain a broadened contour,[2] and it seems reasonable to expect that the assumption of a primary beam of finite width would predict a broadened contour without any point of infinite intensity. This report, based on a primary beam of rectangular shape and specified width, confirms this expectation. Because the line-beam case is basic to the treatment of a beam of finite width, it is also possible to report previously unpublished expressions for the complete contour applicable to either metastable-transition or charge-exchange processes.

In a constant magnetic field free of electric field, the ion speed remains constant and the path is an arc of a circle with its radius proportional to the mass and inversely proportional to charge and field strength once the speed has been fixed by external acceleration. Let r, m and p be radius, ion mass and ion charge respectively before transition, and let $'$ superscripts denote the subsequent values. Also let m^* be the mass of a singly charged ion that travels a full semicircle of radius r^* without change and let $F = (m'/p')/(m/p)$. Then $r' = Fr$, and from the usual mass spectrometer formula it can be shown that $r^* = fr$ where $m^* = f^2 m/p$ or $f^2 = (m^*p/m)^{\frac{1}{2}}$. Charge exchange is described by F over unity with no mass change. If a metastable ion decomposes into charged and uncharged fragments, F is below one and the geometry is as illustrated in Fig. 1. The cosine and sine laws for triangle SCC_1 give

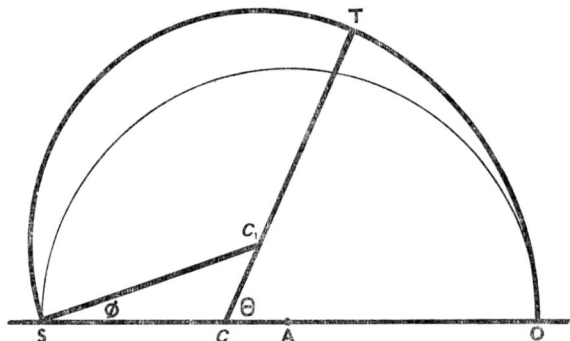

FIG. 1   Geometrical Relationships for a Metastable Transition. The metastable primary ion enters the magnetic field normally at O and sweeps out a circular arc subtending angle θ at centre C until the transition occurs at T. The charged fragment then sweeps out an arc of shorter radius about centre C_1 until it hits the collector S on the line SCO from a direction inclined at an angle φ to the normal.

relations F, f, θ and φ as follows:

$$\cos \theta = \frac{2f(1-f)-(1-F)}{(2f-1)(1-F)} \tag{1}$$

$$F \sin \varphi = (1-F) \sin \theta \tag{2}$$

Equation 1 is strictly equivalent to Hipple's equation (14)[1] and Coggeshall's equation (15),[2] and, with allowances for typographical omissions in a defining relation and the equation itself, to Viney's equation (44)[3] also.

By the author's technique of resolving displacements parallel and perpendicular to SAO,[6]

$$I = \frac{[-\mathrm{d}i(\theta)/\mathrm{d}t]}{(\cos \theta \tan \varphi - \sin \theta)(|1-F|)} \tag{3}$$

where I is the intensity of secondary ions at S, $-\mathrm{d}i(\theta)/\mathrm{d}t$ is the production rate of secondary ions from primary ions arriving at T with a rate $i(\theta)$ and t is transit time. If instrument residual pressure is constant, $i(\theta) = i(0) \exp(-kt)$ applies to both metastable and charge-exchange transitions; in the first case k equals ln 2 divided by the half life, and in the second the product of exchange cross-section and pressure. Time t is obtained by dividing $r\theta$, the distance from O to T by the ion speed v given by $\frac{1}{2}mv^2 = pV$, where V is the external ion accelerating potential. The result can be simplified by the mass spectrometer equation involving magnetic field H. Two cases arise from normal instrument operation, V constant in magnetic scanning and H constant for either electrical scanning or ion-sensitive plate detection. In the first, $t = (m/2Vp)^{\frac{1}{2}}\theta l/f$ where the design radius l is inserted for r^*, and in the second, $t = \theta m/Hp$. Evaluation of $-\mathrm{d}i(\theta)/\mathrm{d}t$ for the second case and substitution in eqn (3) gives

$$I = \frac{k \exp(-km\theta/Hp)i(0)}{(\cos \theta \tan \varphi - \sin \theta)(|1-F|)} \tag{4}$$

Equations 1, 2 and 4 and the defining relation $m^* = f^2 m/p$ permit an exact calculation of I as a function of m^* for the allowed values of f between F

and one. A similar result follows for magnetic scanning. Infinity anomalies are implicit in these line-beam treatments, because I becomes infinite when either $\theta = 0°$ or $\theta = 180°$. Similar anomalies are implicit in all earlier investigations,[1-6] although the present compact formulation of the result was not achieved in them.

In principle, average intensity $\bar{I}$ at a point on the collector for a uniform primary beam of width w is given by $\int_0^w (I/w)\,dx$ and a second integration over collector width w' would give the average $\bar{\bar{I}}$. Even if these integrations were easy, the infinity anomalies would complicate matters at and near the peaks where experimental interest is greatest. As a realistic alternative, the range of θ that allows collection of secondary ions corresponding to a specific part of the source slit was determined, and the collector ion yield was then evaluated by integration over all source positions. Equation 1 can be transformed in terms of displacements from the position of a mean beam with one side as a power series in θ and the other as a power series in displacements. An accurate series for θ itself results from neglecting θ^6 and higher terms, solving the remaining quadratic in θ^2 and using the binomial expansion for the square root twice. Particularly for small θ the range of θ admitted by the collector slit restriction varies over different portions in the source slit and the relative times of transit from that slit vary. The decrease in intensity of the segment of primary beam can be evaluated in terms of θ and then in terms of displacement by expanding the exponential, and the actual secondary ion collection for a segment of the source determined by integrating over the accepted range. The total collection follows by integrating over source position, and is expressed in terms of a power series in displacements.

A more instructive form that involves the approximation that the average of the exponentials is the exponential of the average θ is obtained by treating the primary intensity as constant over the small allowed ranges of θ and evaluating the average θ separately. The expression for collector yield then becomes a product of constant factors, a geometrical term and an exponential term related to half time or cross-section. The geometrical term, which converges well for usual slit geometries, is given numerically by substituting small $z = $ displacement$/l$ in

$$(1 + \tfrac{1}{2}z)^{\frac{1}{4}} \left[\begin{array}{l} (z + s + \alpha s)^{3/2} - (z - s + \alpha s)^{3/2} \\ \qquad - (z + s - \alpha s)^{3/2} + (z - s - \alpha s)^{3/2} \\[6pt] +A\{(z + s + \alpha s)^{5/2} - (z - s + \alpha s)^{5/2} \\ \qquad - (z + s - \alpha s)^{5/2} + (z - s - \alpha s)^{5/2}\} \\[6pt] +B\{(z + s + \alpha s)^{7/2} - (z - s + \alpha s)^{7/2} \\ \qquad - (z + s - \alpha s)^{7/2} + (z - s - \alpha s)^{7/2}\} \\[6pt] + \text{terms of order 9/2, 11/2 and higher} \end{array} \right]$$

where $s = w/2l$ and
$\alpha = w'/w$ are two simplifying ratios
$A = (F^2 + 3F - 3)/20(1 - F)(1 - 2F)(1 + \tfrac{1}{2}z)$
$B = (7F^4 + 90F^3 - 171F^2 + 90F - 9)/672(1 - F)^2(1 - 2F)^2$
$\qquad \times (1 + \tfrac{1}{2}z)^2$

The geometrical term is independent of F to the order of terms in the square of slit width ratios, and within this approximation differentiating and equating the result to zero gives the condition for the maximum of the term as a z given by

$$z = \frac{s[4(1 - \alpha + \alpha^2)^{\frac{1}{2}} - 1 - \alpha]}{3}$$

Since $z = 0$ corresponds to the usual formula $m^* = m'^2/m$ for the metastable peak, the geometric term tends to shift the peak by a fraction of the average slit width. The effect of the exponential is to offset some of the geometric shift.

A similar procedure to that employed for evaluation of the geometric term from eqn (1) and auxiliary relations is used to obtain $\bar{\theta}$, the average θ for use in the exponential term. The result, expressed correctly to terms in z^2, is

$$\frac{3F}{4[2(2F - 1)(1 - F)]^{\frac{1}{2}}} \times$$

$$\left[\frac{(z + s + \alpha s)^2 - (z - s + \alpha s)^2 - (z + s - \alpha s)^2 + (z - s - \alpha s)^2}{(z + s + \alpha s)^{3/2} - (z - s + \alpha s)^{3/2} - (z + s - \alpha s)^{3/2} + (z - s - \alpha s)^{3/2}} \right]$$

In both the expression for the geometric term and the expression for $\bar{\theta}$, zero is to be inserted for any expression of z, s and αs that becomes negative as numerical substitution proceeds. This rule for inserting zero correctly accounts for changes in boundary conditions as z changes between $-(s + \alpha s)$ and $+(s + \alpha s)$. As a result, $\bar{\theta}$ is proportional to $(z + s + \alpha s)^{\frac{1}{2}}$ for z between $-(s + \alpha s)$ and $-|s - \alpha s|$ but eventually becomes proportional to $z^{\frac{1}{2}}$ with a different proportionality constant as z becomes appreciably greater than $s + \alpha s$, as might be expected for an approach to the line-beam approximation.

A number of calculations have been made for the charge exchange $2+$ to $1+$ with experimentally practical values of s (0·000 05, 0·0005, 0·005) and α (0·4, 0·6, 0·8, 1·0, 1·4, 2·0, 3·0 and 5·0). The geometrical term displays a striking variation in its contribution to peak intensity and peak shape as s is changed, along with a significant shift in the location of the position of maximum intensity. This is illustrated in Fig. 2. The variation of s alters the relative widths of both source and collector slits simultaneously and produces an effect on intensity that is greater than the square of the relative change in widths. Increase of α at constant s increases both peak width and height slowly, but there is relatively little shift in the position of the maximum until α exceeds 1·4. At $s = 0·0005$, the increase of peak height with α is slower than a linear rise because the values of the maximum divided by α are 1·16, 1·11, 1·06, 1·00, 0·91, 0·81, 0·69 and 0·55 on a relative basis for the sequence of α values given at the start of the paragraph. These calculations clearly demonstrate that taking finite beam width into account eliminates the infinity anomalies implicit in eqn (4) for the line-beam case. Alternative modifications can achieve the same result; it may be inferred from the author's work for sectors with different angles[6] that a departure from a perfect 180° will eliminate the infinities, and Coggeshall has shown that changes in kinetic energy during transition will do likewise.[2]

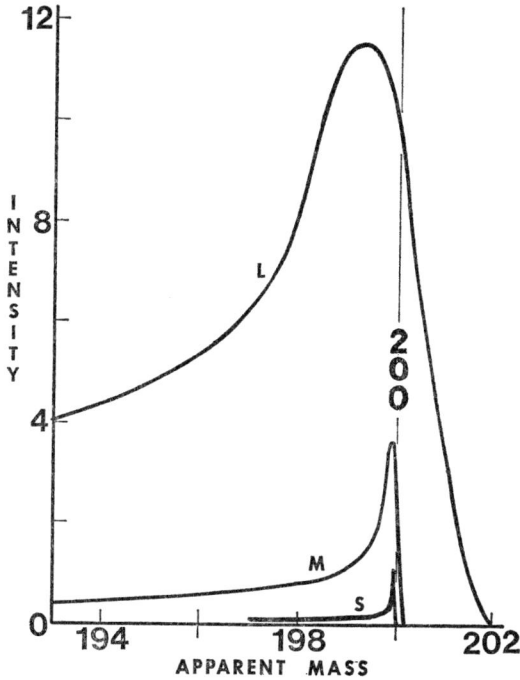

FIG. 2   Peak contours for the 2+ to 1+ charge exchange of an ion of mass 100. (L, large slits with $s = 0.005$; M, medium slits with $s = 0.0005$; S, small slits with $s = 0.000\,05$; source and collector slits of equal width. The ordinates for the contour S have been increased a hundred times and those for the contour M tenfold to obtain a visually meaningful comparison.)

From the present work, it does seem that slit widths should be definitely considered in experiments relating to ion lifetimes and kinetic energy straggling. The author hopes subsequently to compare results for the rectangular beam with those for either a triangular or a Gaussian beam as a prelude to determining lifetimes for some metastable ion species.

REFERENCES

1. Hipple, J. A., *Phys. Rev.*, 1947, **71**, 594.
2. Coggeshall, N. D., *J. Chem. Phys.*, 1962, **37**, 2167.
3. Viney, B. W., *Int. J. Mass Spectrom. Ion Phys.*, 1972, **8**, 417.
4. McCrea, J. M., ibid., 1970, **5**, 381.
5. Ibid., 1972, **9**, 167.
6. Ibid., 1973, **11**.

The Mass-Spectrometric Monitoring of Ions in Plasma

By J. B. HASTED

(*Birkbeck College, University of London, London, U.K.*)

To an increasing extent mass-spectrometers are being used to monitor the densities of different ionic species in plasma; in electrical discharges, afterglows and flames, in ionized regions of the upper atmosphere, and in plasma produced by nuclear radiation, by photons and by chemi-ionization. In this article we consider the confidence with which such monitoring experiments can be regarded; this must depend on the theory of the ion effusion or extraction process and on the efficiency of the mass-spectrometer.

The ion extraction problem is also fundamental to the operation of ion sources in which plasma is generated.

The mass-spectrometer sees the plasma region through an orifice in a metal plate, across which a considerable gas pressure gradient develops. Since the ion population density varies spatially through plasma, it is sometimes necessary that the entire monitoring system be movable, for example, radially or axially within a cylindrical plasma chamber. In Fig. 1 is shown a diagram of a monitor, consisting of an orifice through which ions pass via a lens system into a mass-analysis region and thence onto a detector. The mass-spectrometer chamber is differentially pumped to a pressure sufficiently low for the ion mean free path to be much longer than the distance between orifice and detector. When a flame plasma at atmospheric pressure is monitored, an intermediate pumping chamber is necessary, but for plasma at pressures in the range 0·1–10 torr, no such precaution is required. Orifice diameters are normally of the order of a few microns. The orifice is pierced in a flat metal plate, but for monitoring flowing plasma a nose-cone is sometimes used.

THEORY OF ION EXTRACTION THROUGH AN ORIFICE

The theory of ion extraction through an orifice follows the theory of the collection of charged particles at a plane metal probe which was originally worked out by Langmuir.[1] Indeed the first post-war mass-spectrometric monitors, built by Boyd,[2] were themselves miniaturized like probes. Subsequently Pahl[3] and also Knewstubb and Tickner[4] extended this type of

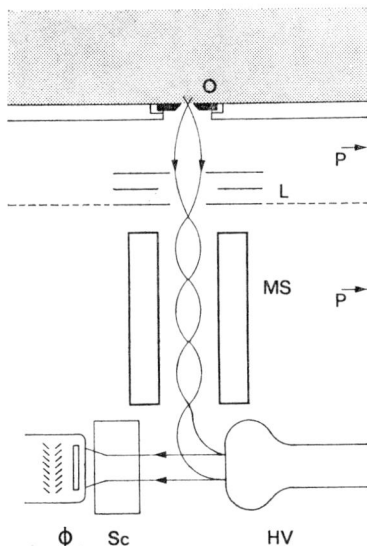

FIG. 1   Schematic diagram of mass-spectrometric monitoring of ions in plasma. O, orifice; L, ion optical lens; MS, quadrupole mass-spectrometer; HV, high voltage electrode biassed 25 kV negative; Sc, scintillator; ϕ, photomultiplier, P, diffusion pumps.

research, and fast pumping allowed much longer path lengths to be used in the mass-spectrometer region.

In the plane-probe treatment it is assumed that the orifice collects the ion current which would be collected by a metal probe in the plane of its front surface. We thus assume that the potential of the plane of the plasma side of the orifice plate is uniform across the hole as is illustrated in Fig. 2(c). We shall see that ion optical effects challenge the validity of this assumption.

Normally the orifice is maintained at the wall potential of the discharge, that is, at the equilibrium potential which a floating wall† takes up with respect to the plasma, so as to allow equal fluxes of positive and negative particles to arrive at it. When the positive ions in the plasma are at temperature T_+ and the electrons at temperature T_e, their mean velocities $\bar{v}_+$ and $\bar{v}_e$ are necessarily different. Except in electronegative gases, the macroscopic number densities of electrons and positive ions in any region of the plasma are equal, $n_+ = n_e$; the faster species (nearly always the electrons) would arrive at the wall at a greater rate were they not prevented from doing so by the attraction of the space charge of the species remaining in the plasma. Thus a potential gradient is built up between the wall (potential V_w) and the bulk of the plasma (potential V_{pl}); it is called the sheath, and the potential across it will be denoted by $V_s = V_w - V_{pl}$. From kinetic theory the current i_+ of positive ions flowing to an area A of wall is

$$i_+ = \frac{An_+\bar{v}_+}{4} \tag{1}$$

† The German literature, perhaps more appropriately, describes the floating wall potential as 'schwimmende'.

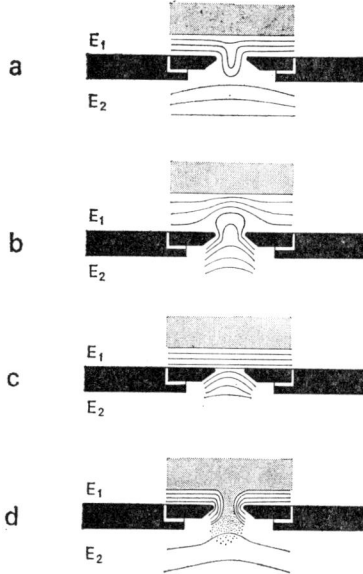

FIG. 2 Equipotential planes in the region of the orifice under four different conditions: (a) $E_2 < E_1$; (b) $E_2 > E_1$; (c) $E_2 = E_1$; (d) collapse of the sheath due to orifice diameter exceeding d.

where n_+ is the ion density at the plasma side of the sheath. It is assumed that this is the ion current which would flow through an orifice of area A. When the electrons possess a Maxwellian velocity distribution of mean energy $\bar{E}_e$, the electron and ion flow rates are related as follows:

$$\frac{n_+\bar{v}_+}{4} = \frac{n_e\bar{v}_e}{4} \exp\left(\frac{-eV_s}{\bar{E}_e}\right) \tag{2}$$

so that

$$V_s = \bar{E}_e(2)^{\frac{1}{2}} \ln\left(\frac{m_+\bar{E}_e}{m_e\bar{E}_+}\right) \tag{3}$$

since $\bar{E} = \frac{1}{2}m\bar{v}^2$. Effects due to the presence of negative ions have been neglected.

A negatively biassed probe or orifice, at potential V with respect to the plasma potential ($V_{pl} = 0$), will collect the ion current which is appropriate to the behaviour of the sheath as a space-charge-limited planar diode. The thickness d of the sheath is consistent with the equation

$$\frac{en_+\bar{v}_+}{4} = \frac{1}{9\pi}\left(\frac{2e}{m_+}\right)\frac{V^{3/2}}{d^2} \tag{4}$$

provided that collisions with gas molecules in the sheath are neglected. The calculation of sheath thickness is necessary because the extraction process depends upon d lying within certain limits. It must be smaller than the ionic mean free path

$$1_{f+} = \frac{1}{\langle n_0\sigma\rangle} \tag{5}$$

within a gas consisting of molecules of number density n_0; the total collision cross-section is σ and an average must be made over a range of impact velocity. But d must not be much smaller than the orifice diameter, since otherwise the sheath would fail to extend across the orifice plane, as is shown in Fig. 2(d).

Since in normal operation V is made as small as possible with $V \simeq V_s$, d will also be small, and will be difficult to estimate. Therefore the orifice diameter must be made as small as possible, usually a few microns or even less. Multiple perforations are sometimes used, and the orifice must be knife-edged.

For large negative bias the ion current is to some extent voltage-independent, but because of the operation of eqn (4), the sheath thickness increases; when it becomes greater than the ionic mean free path, inelastic ionic collisions take place within the sheath, as was shown in the experiments of Böhme and Goodings.[5] Therefore the temptation to apply such bias must be resisted, and the orifice must be allowed to take up approximately the true wall potential.

A further condition that must be satisfied is that the surrounding wall of the plasma chamber also takes up a true wall potential. This can only be so if the charged particle diffusion is ambipolar, i.e. with the motion of the faster group of particles hindered by the Coulomb attraction of the slower group. Under these conditions a single ambipolar diffusion coefficient

$$D_a = \frac{D_+ K_- + D_- K_+}{K_+ + K_-} \tag{6}$$

is appropriate; here D represents diffusion coefficient and K the mobility (drift velocity in unit electric field). Diffusion will be not ambipolar, but free, unless the Debye lengths for both species are much smaller than the dimensions of the plasma chamber. The Debye length is given by:

$$\lambda_D = \left(\frac{kT}{8\pi ne^2} \right)^{\frac{1}{2}} \tag{7}†$$

Normally this condition is not difficult to satisfy, but in the very late afterglow the small values of n raise the Debye length to a value sufficiently large to cause anomalous monitoring efficiency. Replenishment of the electrons in stationary afterglows has sometimes been considered necessary.[6]

CHARACTERISTICS OF BIASSED MONITORING ORIFICES

When the monitoring orifice is biassed to a potential V with respect to plasma potential $V_{pl} = 0$, the transmitted positive ion current shows a variation which is known as the characteristic. Its form depends upon the ion collection efficiency, but when precautions are taken to make this as large as possible and independent of V, then the characteristics will resemble those in Fig. 3. These data are taken from the work of Lindinger,[7] and the dimensions and details of the experiment are shown in the caption to the figure.

† This equation is written in Gaussian units, as are all others in this article.

It will be noticed that the characteristic is drawn with the negative bias to the right so that it resembles an electron probe characteristic in form.

But in the characteristics of Fig. 3 the ion current also increases with increasing *positive* bias (to the left). There should of course be no positive ions able to escape through the orifice in this region; the observed positive ions arise from a secondary effect, being formed by impact ionization of effusing gas molecules by the accelerated electrons. Observations in this region have been used by Howorka[8] to monitor the density of neutral gas in the plasma chamber.

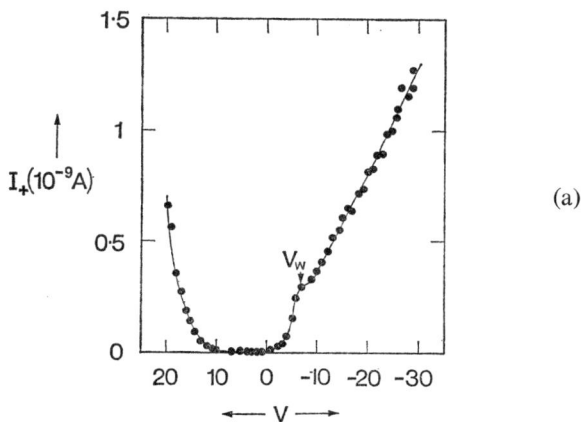

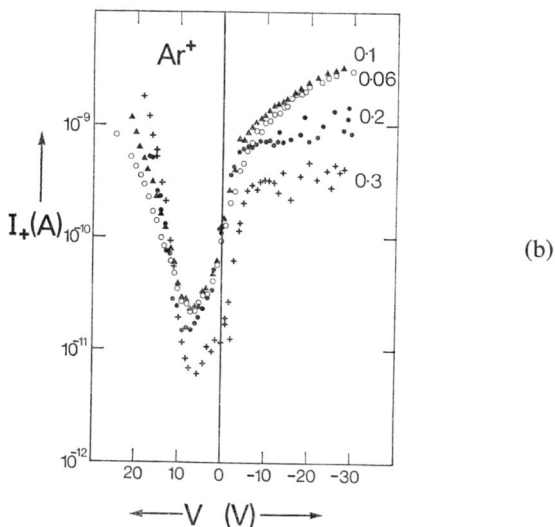

FIG. 3   Characteristics of mass-spectrometric monitoring system due to Lindinger. Orifice diameter, 10 μm; thickness 10–20 μm. Directly behind the orifice is an accelerating orifice lens at -900 to -1200 V potential, followed by further lenses. (a) Total extracted positive ion current from hollow cathode discharge of 2 mA current in argon at pressure 0·2 torr. (b) Ar$^+$ current, with discharge current 5 mA in argon at pressures 0·06 torr (open circles), 0·1 torr (triangles), 0·2 torr (closed circles) and 0·3 torr (crosses).

When the orifice is close to plasma potential it is only the most energetic of the ions in the distribution of energies which can surmount the space-charge barrier and penetrate the orifice. As the negative orifice bias $-V$ is increased a greater proportion are able to penetrate, and the orifice current density j_+ rises sharply according to the equation

$$j_+ = \frac{n_+ e \bar{v}_+}{4} \exp\left(\frac{-eV}{kT_+}\right) \tag{8}$$

The equivalent exponential rise in probe electron current can sometimes be used to provide a measure of the electron temperature, and it is not impossible that the same technique be applied here. This sharp rise appears clearly in the region 0 to -7 V on Fig. 3(a) and $+5$ V to -4 V on Fig. 3(b). But for stronger negative bias the entire energy distribution of positive ions can penetrate the sheath, and the ion current is space-charge-limited, being given by eqn (4). The characteristic rises relatively slowly, and not proportionately to $V^{3/2}$, since there is variation in sheath thickness d. Ion optical and possibly other effects contribute to this rise, which is not understood quantitatively. In this idealized model, ion gas molecule collisions, ion-optical effects, and negative ions have been neglected.

ION-OPTICAL EFFECTS

Behind the orifice are situated an ion-optical lens system, a mass-spectrometer and a detector. The ions pass from the electric field $E_1 = V/d$ of the sheath into the accelerating field E_2, and if these fields are both uniform except in the orifice region, then the orifice is itself a single aperture or Calbick lens,[9] whose focal length f is given approximately by

$$f = \frac{-(E_2 - E_1)}{4V} \tag{9}$$

where V is the orifice potential with respect to plasma potential $V_{pl} = 0$. This focal length would be inconveniently long if the accelerating field were of the strength necessary to prepare the ions for a sector mass-spectrometer. Although sector instruments are often used for plasma monitoring,[4,7] the strong acceleration is rendered unnecessary by the use of a quadrupole mass-spectrometer. A weak acceleration of less than 50 V is all that is required for this instrument, so that focal lengths of a few centimetres can be achieved. The accelerating field can be applied by means of a grid or by a cylinder or aperture lens. Under the normal operating conditions, that is with the orifice at wall potential, the action of the orifice lens should be such that the characteristic has a reasonably flat maximum at a low accelerating potential. We shall see that because of sheath collisions and other effects the characteristic is rather different for a negatively biassed orifice. An example of an unbiassed orifice characteristic is shown in Fig. 4; it is taken from the orifice sampling of an ion drift tube at different ratios of field strength to pressure.[10] The energy distributions of the effusing ions are approximately known from

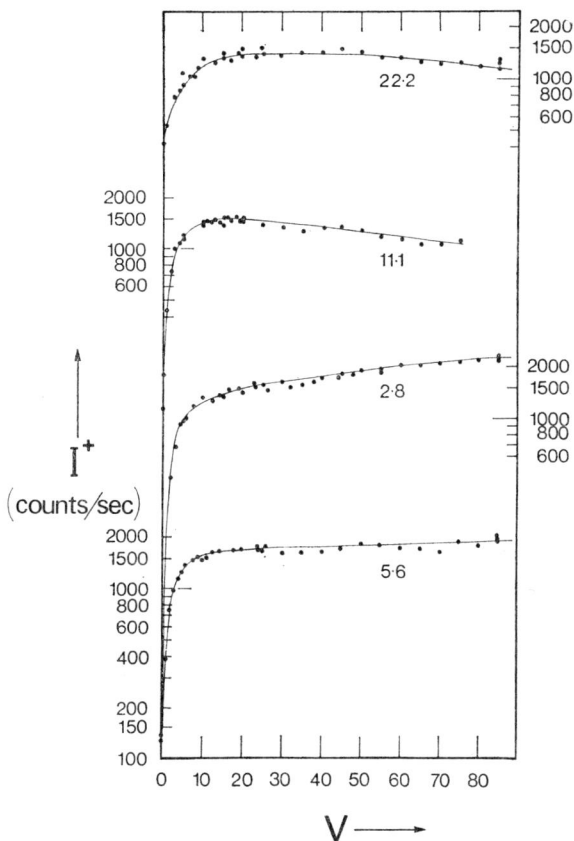

FIG. 4   Characteristics of a drift tube ion sampling system for O^+ in helium gas. Orifice diameter 0·05 mm, ratios of field strength to pressure E/p, (V cm^{-1} torr^{-1}) denoted for each characteristic.

both theory and experiment, and the relatively flat characteristics are a test of the absence of ion-gas-molecule collisions in the accelerating region.

Detailed tests of the characteristics of a plasma monitoring system can be made in terms of two potentials, that of the orifice and that of the accelerating lens. An example of biassed orifice characteristics, due to Zwirner,[11] is shown in Fig. 5. It is seen that there is an optimum accelerating potential for each value of orifice potential, and their mutual dependence can be compared with orifice extraction theory.

When a potential V is applied to the orifice, thus forming a sheath with $V = V_s$, then for a given plasma with $n_+ \bar{v}_+$ constant, eqn (4) leads to a proportionality $V^{3/2} \propto d^2$. Therefore $d \propto V^{3/4}$, and the extraction field $E_1 \propto V^{1/4}$. But when this proportionality is used in conjunction with eqn (9) to analyse the data of Fig. 4 the test is inadequate because of the failure of probe theory at high gas pressures. Fette and Hesse[12] applied the high pressure probe theory of Schulz[13,25] to ion extraction, obtaining expressions for

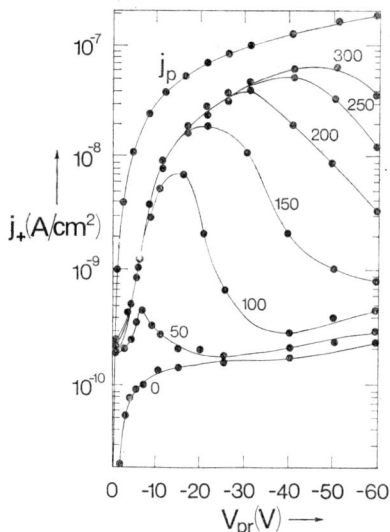

FIG. 5 Characteristics of biassed orifices for various accelerating lens potentials (V) marked against each curve. j_p represents the total orifice current density, and j_+ represent total positive ion collector current densities, in the nitrogen pink afterglow at pressure 4 torr. Orifice system dimensions in text.

the sheath field E_1

$$E_1 \simeq E_0 \left(\frac{5}{2} c^2 \frac{V}{\bar{E}_e}\right)^{2/5} \tag{10}$$

and the sheath thickness

$$d \simeq \frac{5}{2} \left(\frac{V}{\bar{E}_e}\right)^{3/5} \frac{2}{3} c^{1/5} \lambda_D \left(\frac{n_+}{n_{+0}}\right)^{\frac{1}{2}} \tag{11}$$

where c is a constant such that $(2)^{\frac{1}{2}} \le c \le 2\cdot 32$, $\bar{E}_e$ is the mean electron energy in electron volts, λ_D the Debye length for electrons, and

$$E_0 = \frac{\bar{E}_e}{c\lambda_D} \tag{12}$$

$$n_{+0}{}^2 E_0 = n_+{}^2 E \tag{13}$$

Zwirner applied the theory to his data without being able to obtain agreement; we propose instead the following: when the orifice is taken to be as a Calbick lens, eqn (10) can be combined with eqn (9) to give

$$\frac{V_a - V(I_{max})}{4d_a V(I_{max})} - \frac{\bar{E}_e}{4c\lambda_D V(I_{max})} \left(\frac{5}{2} c^2 \frac{V(I_{max})}{\bar{E}_e}\right)^{2/5} = \text{constant} \tag{14}$$

where V_a is the potential of the accelerating lens, which is situated at a distance d_a behind the orifice.

When the observed current is a maximum and reaches its maximum value I_{max}, the focal length will have been adjusted to its optimum value. All the parameters in eqn (14) are known or can be estimated approximately. The theory is tested by calculating the constancy of the expression from the observed data. When estimates are made of Zwirner's parameters as follows: $d_a = 5$ cm, $V_e = 1$ eV, $n_e = 10^{10}$ cm^{-3}, $c = 2$, $\lambda_D = 700$, $(V_e/n_e)^{\frac{1}{2}} = 0.007$ cm, the constancy for the data of Fig. 5 is within 15%, there being no very systematic variation. It should be possible with well-understood plasmas to obtain passable agreement with the high pressure orifice theory outlined above, but, of course, collisional effects interfere with experiments with biassed orifices.

It is possible, in these experiments as in others, that the finite thickness of the orifice plate enhances collisional effects. The orifice thickness in Zwirner's experiments was 5 μm, which is smaller than the orifice diameter 10 μm. The gas kinetic mean free path was of order 2·5 μm—an example of an experiment well designed by present-day standards.

Because of collisional effects, biassed orifices should be avoided, but even with unbiassed orifices, the ion accelerating system should always be tuned for maximum signal. For afterglows and plasma with electron and ion temperatures close to 300°K, only minimal acceleration will be applied; but for hotter plasma, the electron temperature will usually be in excess of the ion temperature; $V_w - V_{pl}$ will be sizeable and rather greater acceleration will be required, since the orifice operates as a Calbick lens in the manner described above.

FURTHER REQUIREMENTS OF MONITORING SYSTEMS

Some articles relevant to mass-spectrometric monitoring[13,14] and to probe theory[15] may be mentioned at this point. It is noted by Märk[14] that a mass-spectrometer suitable for plasma monitoring must possess:

(1) minimal ion energy discrimination,
(2) high source pressure capability,
(3) large dynamic range,
(4) absence of critical injection geometry.

Although the quadrupole instrument might seem to be weak on (4), it is nevertheless currently in the highest favour for plasma monitoring.

The ion detector for a monitoring system must be as sensitive as possible, so that the minimum orifice diameter can be used. In single particle counting the signal–noise ratio depends upon the background count, and it is worth mentioning that the lowest backgrounds (0·02 counts/sec) have been obtained[16] by an adaptation of the Daly secondary emission scintillator detector.[17]

A monitoring system will often require calibration, both absolute and mass-discriminatory. The orifice and accelerating lens should not show mass-discrimination, but a quadrupole mass-spectrometer may do so if the operating conditions are not correct.

A mass-discriminatory test of a system can be made in a flowing afterglow, using interconversion of ions in the plasma by addition of a small flow rate of reactive gas. As the flow rate is increased the parent ion density decreases exponentially, and the product ion density rises asymptotically to a steady value[18] as shown in Fig. 6. Although calculation of the reaction rates from

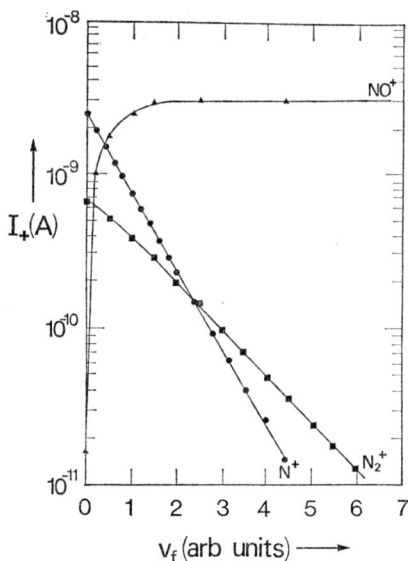

FIG. 6   Variation with nitric oxide flow rate v_f of ion currents sampled from flowing after-glow system. The dominant ion NO^+ is formed by the processes

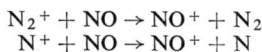

$$N_2{}^+ + NO \rightarrow NO^+ + N_2$$
$$N^+ + NO \rightarrow NO^+ + N$$

these changes depends only on measurement of reactive gas partial pressure and not on relative proportions of ions, it is also possible to test for mass-discrimination by comparison of the ion currents. It will be seen from Fig. 6 that these particular data do not add up (*i.e.* the total ion current is pressure-dependent); this is because no correction has been applied for diffusion losses. The axial ion density depends upon the ratio of ambipolar diffusion coefficient to square of diffusion length, and the ambipolar diffusion coefficients for different ions vary with the mobility, hence with the mass; the apparent ambipolar diffusion coefficient also depends upon the parabolic velocity distribution for the laminar carrier gas flow.[19] It would be possible to apply these corrections, but a simple alternative would be to work in a flow tube of sufficient diameter (say 20 cm) for the radial diffusion correction to be negligible. The flow velocity should still be maintained sufficiently high for axial diffusion to be negligible. If suitable conditions cannot be maintained, then there is no alternative to an independent check on mass-discrimination using a separately mounted electron impact source, containing gases whose ionization cross-sections are accurately known.

The absolute calibration of a mass-spectrometric monitor can only be made against a measurement of electron or ion density in the plasma. Rather than use ion cyclotron resonance, one may diagnose the electron density by means of microwave permittivity[20] or by Langmuir or similar probe. Although good absolute calibrations have been made in this way, and the results used in electron–ion recombination rate measurement,[21] it is significant that no quantitative test of orifice extraction theory has yet been made; it is important that this omission be repaired. Obviously the use of electronegative gases introduces the disadvantage of having to take negative ion densities and measurements into account.

In probe theory in an electronegative gas, as developed by Boyd and Thompson,[22] the dominating factor is the ratio of negative ion to electron density, $\beta = n_-/n_e$. For $\beta \leq 2$ no very serious errors are involved in using conventional probe theory, but for $\beta > 2$, a treatment of the spherical probe leads to

$$i_+ = 4\pi r_s^2 \frac{n_+ e}{4} \left(\frac{8kT_+}{\pi m_+} \right)^{\frac{1}{2}} \tag{14}$$

where the sheath radius

$$r_s^2 = \frac{16(\pi)^{\frac{1}{2}} \lambda_+^2 \eta^{3/2}}{9A} \tag{15}$$

$$\lambda_+ = \left(\frac{kT}{8\pi n_+ e^2} \right)^{\frac{1}{2}} \tag{16}$$

and

$$A \simeq \left(\frac{r_s - r_p}{r_p} \right)^{3/2} \tag{17}$$

r_p is the probe radius and $\eta = V/\bar{E}_e$. Provided the plasma is sufficiently dense for a sheath to be maintained, the positive ion density n_+ is approximately proportional to the square of the positive ion current i_+. Langstroth[23] reported that these conditions were achieved in the very late afterglow of electronegative gases, when it could be expected that electron attachment processes would ensure $\beta > 2$. He found that the slope of the logarithm of the time-dependence of i_+ changed by a factor of 2 as the negative ion density built up at the expense of electron density. But removal of the negative bias from the orifice caused the effect to disappear, by reducing the sheath thickness. This constitutes another argument against using biassed orifices. Treatment of orifice extraction in spherical geometry is only valid when the accelerating field penetrates deeply into the sheath (Fig. 1(d)).

In some early experiments[26] the monitoring of negative ions from an afterglow was found to be unsuccessful. The failure to observe negative ions was due to their failure to surmount the negatively charged barrier presented by the sheath of faster-diffusing electrons. However, in the late afterglow of an electronegative gas, the electrons have disappeared by attachment to form negative ions, whose diffusion coefficients are of the same order of magnitude as those of positive ions. The disappearance of the barrier causes a dramatic increase in the observed negative ion current, which shows clearly in Fig. 7, due to Puckett and Lineberger.[27]

Another difficulty that arises in the sampling of negative ions is their removal by associative detachment collisions with radicals, such as

$$Cl^- + H \rightarrow HCl + e$$

It is believed by Parkes that Cl^- cannot be sampled in fuel-rich flames because this reaction equilibrium moves to the right near the orifice where the electrons are lost, although it is balanced in the bulk of the plasma.

In addition to the space charge barrier, to which we have devoted most of

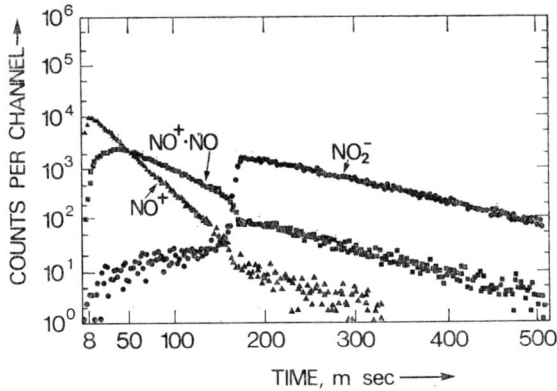

Fig. 7   Positive and negative ion count rates in the NO_2 afterglow.[27]

this discussion, there can arise three other barriers which affect mass-spectrometric monitoring: (i) that due to macroscopic electric fields existing in the
plasma, (ii) the cool boundary layer at the edge of a flame, and (iii) the
convective barrier existing at high pressures.

(i) Parkes[24] has discussed the situation arising from the drift of ions in
the direction of the orifice, due to the application of electric field. There is
competition between convective flow and the drift, which favours the
extraction of heavy ions. It has been confirmed[28] that the discrimination is,
as predicted, reduced by minimization of the orifice size.

(ii) In a high gas temperature system the cooler boundary layer near the
vessel wall can change ion populations, by clustering and by shifting equilibria. Knewstubb and Tickner's[4] hydrates may have been formed in this
layer.

(iii) In the convective regime in which the orifice diameter exceeds the mean
free path, cooling occurs on expansion of the gas flowing through the orifice.
Ion populations freeze at a temperature much lower than those characteristic
of the plasma, and clustering reactions can occur. Hayhurst and Telford[29]
have given an excellent theoretical treatment of the problem.

The absolute calibration of a monitoring system would in all probability
demonstrate the importance of convective flow. The simplified treatment due
to Kebarle[30] is as follows.

The convective flow F of gas (including ions) through a thin-walled orifice
of area A mm^2 is approximately

$$F \simeq 0 \cdot 2 \, A \text{ litre sec}^{-1} \qquad (19)$$

Consider a co-ordinate system with origin at the orifice centre and radius
variable r, the distance perpendicular to the orifice plane being denoted by z.
The convective flow velocity

$$v = \frac{F}{2\pi r^2} \qquad (20)$$

The ion will only be sampled if the convective flow time τ, through orifice of radius r_o,

$$\tau = \int_{r_o}^{r} 2\pi r^2 \frac{dr}{F} \tag{21}$$

is smaller than the drift or free fall time

$$\tau = \frac{z}{v_d} \tag{22}$$

The effective radius r_m of the sampling orifice is therefore different from the actual radius r_o, being given by

$$r_m{}^2 = r_o{}^2 \left[R^2 - \left(\frac{F}{2\pi R v_d r_o} \right)^2 \right] \tag{23}$$

where R is the real positive root of

$$R^4 - R - 3 \left(\frac{F}{2\pi v_d r_o{}^2} \right)^2 = 0 \tag{24}$$

Convective effects are of course minimized by the use of large draw-out potentials, but the disadvantages of these have been discussed earlier; the reduction of orifice radius of course extends the range of pressure over which monitoring is feasible.

REFERENCES

1. Langmuir, I. and Mott-Smith, H. M., *Phys. Rev.*, 1926, **28**, 727.
2. Boyd, R. L. F., *Nature*, 1950, **165**, 142; Tüxen, O., *Z. Physik*, 1936, **103**, 463.
3. Pahl, M. and Weimer, U., *Z. Naturforsch.* 1958, **13a**, 50, 745; 1957, **12a**, 926; *Naturwiss.*, 1957, **44**, 487.
4. Knewstubb, P. F. and Tickner, A. W., *J. Chem. Phys.*, 1962, **37**, 2941; 1963, **38**, 1031.
5. Bohme, D. K. and Goodings, J. M., *Rev. Sci. Inst.*, 1966, **37**, 362–6; *J. App. Phys.*, 1966, **37**, 4261.
6. Court, G. R., Thesis, University of Birmingham 1953.
7. Lindinger, W., Dissertation, University of Innsbruck, 1971 Proc. II Int. Conf. on Ion Sources, SGAE, Vienna 1972, p. 85.
8. Howorka, F., Proc. II Int. Conf. on Ion Sources, SGAE, Vienna 1972, p. 92.
9. Calbick, C. J., *Phys. Rev.*, 1931, **38**, 585; 1932, **42**, 580.
10. Kosmider, R. and Hasted, J. B., Unpublished Data 1973.
11. Zwirner, W., Proc. II Int. Conf. on Ion Sources, SGAE, Vienna 1972, p. 160.
12. Fette, K. and Hesse, J., *Z. Naturforsch.*, 1970, **25a**, 518.
13. Studniarz, S. A., 'Electrical Discharges', *in* 'Ion Molecule Reactions', *ed.* J. L. Franklin, Butterworths, London, 1973, chapter 14.
14. Märk, T. D., Europhysics Study Conference on atomic and molecular physics of ionized gases, C.E.N. Saclay, 1973, p. 41.
15. Schott, L., 'Electrical Probes', *in* 'Plasma Diagnostics', *ed.* W. Lochte-Holtgreven, North Holland, Amsterdam, 1968.
16. Werner *et al.*, quoted in Reference 14.
17. Daly, N. R., *Rev. Sci. Inst.*, 1960, **31**, 264.
18. Goldan, P. D., Schmeltekopf, A. L., Fehsenfeld, F. C., Schiff, H. I. and Ferguson, E. E., *J. Chem. Phys.*, 1966, **44**, 4095.
19. Bolden, R. C., Hemsworth, R. S., Shaw, M. J. and Twiddy, M. D., *J. Phys. B*, 1970, **3**, 45, 61; Huggins, R. J. and Cahn, J. H., *J. App. Phys.*, 1967, **38**, 180.
20. Weller, C. S. and Biondi, M. A., *Phys. Rev.*, 1968, **172**, 198.
21. Mahdavi, M. R., Hasted, J. B. and Nakshbandi, M. M., *J. Phys. B*, 1971, **4**, 1726–37.
22. Boyd, R. L. F. and Thompson, J. B., *Proc. Roy. Soc. A*, 1959, **252**, 102.

914   J. B. HASTED

23. Langstroth, G. F. O., Thesis, University of London.
24. Parkes, D., *Trans. Far. Soc.*, 1971, **64,** 711.
25. Schulz, G., *Z. Phys.*, 1965, **183,** 51.
26. Fite, W. L. and Rutherford, J. A., *Dis. Faraday Soc.*, 1964, **37,** 192.
27. Puckett, L. J. and Lineberger, W. C., *Phys. Rev. A*, 1970, **1,** 1635–41.
28. Bouby, L. (Orsay), Private communication to D. Parkes.
29. Hayhurst, A. N. and Telford, N. R., *Proc. Roy. Soc. Lond.*, 1971, **A322,** 483–507.
30. Kebarle, P., *Adv. Chem.*, 1966, **58,** 210.

THEORY AND FUNDAMENTALS

Chairmen
A. J. H. BOERBOOM
(FOM Laboratory, Amsterdam, The Netherlands)

S. FACCHETTI
(CEE Euratom, Italy)

106

Predissociation of Triatomic Ions Studied by Photoelectron–Photoion Coincidence Spectroscopy

By J. H. D. ELAND

(*Physical Chemistry Laboratory, Oxford, U.K.*)

INTRODUCTION

THE study of unimolecular ionic decompositions is complicated by the fact that both electron and photon impact produce molecular ions in a range of energy states, and the energy deposition function is not usually known. Since all the characteristics of such reactions are functions of the internal energy of the molecular ions, any mass spectrometric measurement of a rate constant, branching ratio or energy release is a complicated average, whose meaning in terms of molecular reaction dynamics is hard to determine. This difficulty is overcome in the photoelectron–photoion coincidence technique, where the reactions of ions formed with known initial energies are examined.

Ionization by photons of a single fixed energy produces photoelectrons whose kinetic energy is equal to the photon energy less the energy needed to form ions in some particular electronic, vibrational and rotational state:

$$KE = h\nu - E^*$$

The photoelectron spectrum of a molecule is thus a spectrum of the energy states of the molecular ions accessible by photon impact. When electrons of kinetic energy KE are detected molecular ions with energy E^* have been produced simultaneously. A knowledge of the electron and ion flight times then allows the pairs of particles formed in individual ionization events to be identified by the coincidence technique. The fates of molecular ions in known initial states are studied by this means; the two principal results are the breakdown diagram, obtained from the mass spectrum at each initial energy, and the kinetic energies released in fragmentation, obtained from the time-of-flight (TOF) coincidence spectra.

The earliest applications of the coincidence method[1] were determinations of the breakdown diagrams for some small molecules, and they led to the conclusion that certain molecular ions, particularly CH_4^+ and CD_4^+, are stable within a range of energies above the threshold for formation of fragments. It has since been discovered that this conclusion was in error,[2] as has also been shown by Stockbauer's independent work using a related coincidence technique.[3] More attention has recently been paid to the kinetic energies released in fragmentation, which can be measured simultaneously with the breakdown diagram provided the electron energy analyser is of the differential type. This paper summarizes findings made by this method on the decays of COS^+, N_2O^+, CO_2^+, CS_2^+ and SO_2^+, and presents new results on the decomposition of D_2O^+. These triatomic species have been chosen for study because their simplicity offers some hope that the molecular dynamics involved in their decay may eventually be understood in detail.

GENERAL OBSERVATIONS

With the exceptions of the $\tilde{A}$ states of N_2O^+ and COS^+, which are discussed below, all ionic states which lie above dissociation limits are found to be fully dissociated. This comment is valid for all the molecular ions so far studied by the coincidence method and not just triatomic ions; it indicates that ionic fluorescence processes leading to stable states are rare. The only other common characteristic shared by all the triatomic ions studied is that their fragmentation patterns (or branching ratios) and kinetic energy release distributions are in no case adequately explained either by the QET model or by models of direct dissociation. This is not at all surprising, but it underlines the need for a more satisfactory theory of unimolecular decay.

RESULTS

Dissociations of COS^+ and N_2O^+

In their $\tilde{A}(^2\Pi)$ and $\tilde{B}(^2\Sigma^+)$ states COS^+ ions can dissociate by a spin-forbidden pathway producing $S^+(^4S_u)$ ions, while in $\tilde{B}$ and vibrationally excited levels of $\tilde{A}$ above (3, 0, 0) they can also decompose to $S^+(^2D_u)$ by a spin-allowed process. The coincidence measurements[4] show that the spin-allowed process progressively takes over from the spin-forbidden one as the excitation energy available exceeds the threshold for $S^+(^2D)$. In the (0, 0, 0) vibrational level of $\tilde{A}$ the COS^+ ions fluoresce;[5] the coincidence measurements show that only about 6% of them do so, while the remainder dissociate to $S^+(^4S_u)$. The rate of the predissociation increases with increasing vibrational energy in the $COS^+(\tilde{A})$ ions, so that fluorescence no longer competes at higher energies. The kinetic energy released in forming $S^+(^4S_u)$ from $COS^+(\tilde{A})$ has a sharp distribution[6] about the large CM energy of 0·7 eV.

The decomposition of N_2O^+ ions from the $\tilde{A}^2\Sigma^+$ states resembles that of $COS^+(\tilde{A})$ closely. This decomposition was long believed not to occur at all because of the existence of the well known $N_2O^+(\tilde{X} \leftarrow \tilde{A})$ emission band. It has now been shown[4] that $N_2O(\tilde{A}\ ^2\Sigma^+)$ ions do decompose in competition with the emission of fluorescence, but the dissociation is detectable only when the N_2O^+ ions also possess vibrational excitation energy. The presence of vibrational energy seems to accelerate the predissociation, exactly as in $COS^+(\tilde{A})$, even though both decays are spin-forbidden and cannot be vibrational predissociations in any usual sense. The kinetic energy released in forming NO^+ from $N_2O^+(\tilde{A})$ is again sharply distributed[6] about a large value (0·8 eV in CM), which suggests that the NO^+ product is formed predominantly in the $v' = 5$ or 6 vibrational level. It is interesting to note that the appearance potential of NO^+ from N_2O found by Dibeler *et al.*[14] corresponds exactly to a vibrationally excited Rydberg level of N_2O with $N_2O^+(\tilde{A}\,^2\Sigma^+$ as 1, 0, 0) as its core. Dissociation might conceivably precede autoionization in this case.

Dissociations of CO_2^+ and CS_2^+

Of the four states of CO_2^+ formed in photoionization of CO_2 at 584 Å only $\tilde{C}\,^2\Sigma_g^+$ represents ions with sufficient energy to dissociate. The predissociation of $CO_2^+(\tilde{C})$ to $O^+(^4S_u) + CO(^1\Sigma_g^+)$ is found to be complete,[7] even though this process is spin-forbidden in exactly the same way as the decays of $COS^+(\tilde{A})$ and $N_2O(\tilde{A})$. This seems a surprising observation, since spin-orbit coupling should be less efficient in CO_2^+ than in COS^+.

It is less surprising that $CS_2^+(\tilde{C}\,^2\Sigma_g^+)$ and $CS_2^+(\tilde{D}\,?)$ are fully predissociated;[8] ions in the $\tilde{C}$ state can yield both CS^+ and S^+ in their ground state, while from the $\tilde{D}$ state S_2^+ ions and electronically excited S^+ and CS^+ are also accessible. The observation[8] that $CS_2^+(\tilde{D})$ decays lead to all three energetically possible products indicates that the parent ions survive for some time before dissociating, despite the continuous contour of the $\tilde{D}$ band in the photoelectron spectrum. The kinetic energies released in the dissociations of $CS_2^+(\tilde{D})$ are not, however, continuously distributed as dissociation following a statistical model would require, but have sharply peaked structures, which may match the different electronic states in which the products can be formed. The kinetic energy release distributions in formation of S^+ and CS^+ from $CS_2^+(\tilde{C})$ are similar,[8] and again the spin-forbidden S^+ formation is accompanied by a large, sharp energy release.

Dissociation of SO_2^+

The third band in the 584 Å photoelectron spectrum of SO_2 includes two, or possibly three, near-degenerate electronic states of SO_2^+, all of which are energetically capable of decomposing to either S^+ or SO^+. Only SO^+ is actually formed, and this lack of an expected competition suggests that the mechanism might be more direct than statistical. The kinetic energies released in formation of SO^+, on the other hand, are unique among the energies yet found in triatomic decays in that they have broad continuous distributions.[9] The average energies are also surprisingly close to the predictions of QET. The SO^+ product ions must be formed in a range of vibrational states, even when the SO_2^+ is initially formed in a single vibrational level.

Dissociation of D_2O^+

An examination of the ionic decomposition of water by the coincidence method was suggested to the author by A. J. and J. C. Lorquet, who have made a full theoretical study of the system following the original work of Fiquet-Fayard and Guyon.[10] The heavy isotope had to be used rather than H_2O because of the low resolution of the TOF mass spectrometer at Oxford.[11] Decomposition takes place from the $\tilde{B}\,^2B_2$ state of D_2O^+ and as this state covers a range of

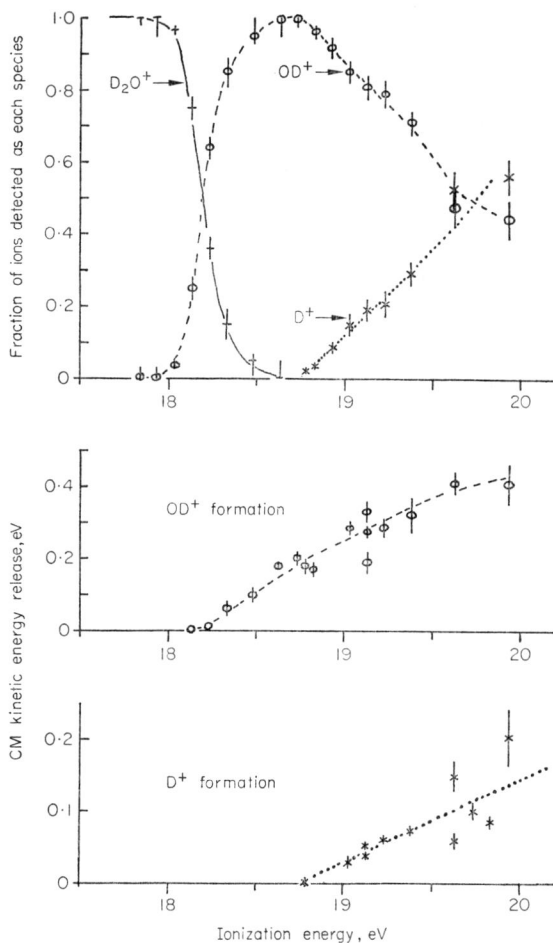

FIG. 1　Breakdown diagram of D_2O^+ in the region of the $\tilde{B}\,{}^2B_2$ state, with the CM kinetic energies released in formation of the OD^+ and D^+ fragments.

energies and decomposes competitively into OD^+ and D^+ the results are best presented in the form of a breakdown diagram (Fig. 1). The kinetic energies released in the fragmentations are also plotted in the figure; in formation of OD^+ the energy distributions are sharp, approximating the release of single energies, whereas in formation of D^+ they are broad, so the energies quoted are averages. The kinetic energy releases occasion some systematic uncertainty in the construction of the breakdown diagram since corrections must be made to allow for their effect on the ion collection efficiencies. The corrections are made using numerical ion trajectory calculations which fit the experimental TOF peak shapes, and it seems unlikely that the errors could be more than 20% in the most unfavourable OD^+/D^+ ratio. Apart from the effect of kinetic energies the apparatus has no detectable mass discrimination between m/e 2 and m/e 20.

In cases such as this where fragment ion thresholds fall within a photo-electron band the photoelectron–photoion coincidence method can be used to determine appearance potentials. The crossover point between D_2O^+ and OD^+ occurs at $18\cdot19 \pm 0\cdot03$ eV, and this can be taken as the appearance potential of OD^+. The cross-over region is no wider than expected from the instrumental resolution of $0\cdot13$ eV. Extrapolation of the D^+ curve to zero yields an appearance potential of $18\cdot75 \pm 0\cdot05$ eV for D^+. The uncertainty in both these figures is due in part to uncertainty in the calibration of the energy scale against the ionization potential of N_2 to $N_2^+(\tilde{B}\ ^2\Sigma_u^+)$ in D_2O/N_2 mixtures.

The mass spectrum of D_2O at 584 Å can be predicted by integrating the product of the breakdown diagram with the photoelectron spectrum of D_2O. The predicted intensities are D_2O^+ $1\cdot00$, OD^+ $0\cdot36$, D^+ $0\cdot07$; an assumption has been made here that the partial cross-sections for production of D_2O^+ $\tilde{X}$, $\tilde{A}$ and $\tilde{B}$ are equal. The D^+ ion from D_2O is predicted to be much more abundant[12,13] than H^+ from H_2O at 584 Å, and it will be interesting to see whether this is confirmed by direct measurement.

According to Fig. 1 the kinetic energy releases are smooth functions, within experimental error, of the excess energies above threshold. Such a smooth variation is very unlikely to be real since the excess energy must go mainly into vibration of the OD^+ or OD product, where the quanta are large. Fine structure on these curves is currently being sought in the breakdown of H_2O^+ where the larger quanta should be easier to see. The decays of H_2O^+ should in fact offer a case where decomposition from individual vibration levels of the molecular ion can be studied rather easily.

REFERENCES

1. Brehm, B. and v. Puttkamer, E., 'Advances in Mass Spectrometry', Vol. IV, *ed.* E. Kendrick, Inst. Petroleum 1968, p. 591.
2. Brehm, B. and Fuchs, V., private communication.
3. Stockbauer, R., *J. Chem. Phys.*, 1973, **58**, 3800.
4. Eland, J. H. D., *Int. J. Mass Spectrom. Ion Phys.*, in press.
5. Horani, M., Leach, S., Rostas, J. and Berthier, G., *J. Chim. Phys.*, 1966, **63**, 1015.
6. Brehm, B., Eland, J. H. D., Frey, R. and Küstler, A., to be published.
7. Eland, J. H. D., *Int. J. Mass Spectrom. Ion Phys.*, 1972, **9**, 397.
8. Brehm, B., Eland, J. H. D., Frey, R. and Küstler, A., *Int. J. Mass Spectrom. Ion Phys.*, in press.
9. Brehm, B., Eland, J. H. D., Frey, R. and Küstler, A., *Int. J. Mass Spectrom. Ion Phys.*, in press.
10. Fiquet-Fayard, F. and Guyon, P. M., *Mol. Phys*, 1966, **11**, 17.
11. Eland, J. H. D. and Danby, C. J., *Int. J. Mass Spectrom. Ion Phys*, 1972, **8**, 153.
12. Dibeler, V. H., Walker, J. A. and Rosenstock, H. M., *J. Res. Nat. Bur. Standards*, 1966, **70A**, 459.
13. Cairns, R. B., Harrison, H. and Schoen, R. I., *J. Chem. Phys.*, 1971, **55**, 4886.
14. Dibeler, V. H., Walker, J. A. and Liston, S. K., *J. Res. Nat. Bur. Standards*, 1967 **A71**, 371.

Discussion

C. E. Klots (Oak Ridge National Laboratory, Tennessee, U.S.A.): Do you correct your coincidence rates for the variation of collection efficiency with ion kinetic energy? If so, how?

J. H. D. Eland: Yes. The TOF peak shapes are first interpreted to yield kinetic energy release distributions, then once the kinetic energy distributions are known the collection efficiencies can be calculated numerically. The calculation examines every ion trajectory in terms of the apparatus geometry and fields, to determine which of all the possible initial velocity vectors lead to ion detection.

P. M. Guyon (C.N.R.S., Lure, Orsay, France): Have you observed any metastable ions in your technique? It will be very interesting indeed to see if one observes a difference using a photoionization mass spectrometry technique and a coincidence technique with the 584 Å He line. Since in the first case one expects to reach states of the excited ions via indirect processes such as autoionization, it might lead to metastable decomposition that might not be obtained via direct ionization.

J. H. D. Eland: I have not yet observed any metastable ions by the coincidence technique, though this is mainly a question of sensitivity. V. Fuchs has found metastables in coincidence experiments at Freiburg using a retarding field electron energy analyser, which makes the sensitivity greater. I agree that in the case of small molecules like N_2O or COS the metastables familiar from electron impact mass spectra arise mainly from ions in states populated by autoionization.

107
Predissociation Processes in Triatomic Ions

By J. MOMIGNY, H. WANKENNE, G. MATHIEU,
J. P. FLAMME

(*Laboratoire d'Etude des Etats Ionisés, Université de Liège, Liège, Belgium*)

and M. A. ALMOSTER-FERREIRA

(*Faculdade de Ciências, Lisbon, Portugal* and *Laboratório Calouste Gulbenkian de Espectrometria de Massa, Lisbon, Portugal*)

INTRODUCTION

THE experimental study and theoretical interpretation of metastable transitions arising from small molecular ions is of leading importance for the understanding of dissociative ionization mechanisms.

Many non-collision and collision-induced metastable transitions have been studied since 1969 in diatomic and some triatomic molecular ions: $N_2^+,$[1-8] $NO,$[1,3,9] $CO^+,$[1] $CH^+,$[10,11] NO_2^+ and $N_2O^+,$[12] H_2S^+ and $H_2O^+.$[13] We have also studied the DBr^+ metastable ion whose results will not be discussed in this paper owing to their recent publication.[14]

We will therefore describe with some details the occurrence of the metastable transitions:

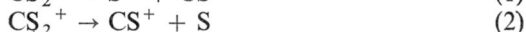

$$CS_2^+ \rightarrow S^+ + CS \qquad (1)$$
$$CS_2^+ \rightarrow CS^+ + S \qquad (2)$$

and also

$$CH_2^+ \rightarrow C^+ + H_2 \qquad (3)$$

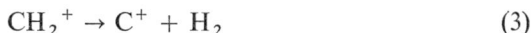

the latter observed in the mass spectra of CH_4, C_2H_4 and CH_2CO. Simultaneously the appearance of the fast decaying CS_2^+ and CH_2^+ ions will be considered for the three mechanisms.

EXPERIMENTAL

The experimental method used in this paper has been previously described.[1,2,7] It involves a double focussing Nier-Johnson mass spectrometer used in the focussing mode with appropriate counting techniques and with automatic recording of both the metastable peak shape and the ionization efficiency curves. Occasionally resonant charge exchange experiments with noble gas

TABLE I

Characteristics of the Slow (Metastable) Dissociation Processes of the CS_2^+ Ion

Dissociation process	Appearance potential (eV)	Mode of fragmentation	Kinetic energy released (eV)	Charge exchange reactions	
				Primary ion	Relative intensity
(1) $CS_2^+ \rightarrow S^+ + CS$	13–14	Collision-induced only	?	—	—
	(15 ± 1)	Non-collision-induced + possibly collision-induced	<0·25	Kr^+	Strong[a]
	(17 ± 1)	Non-collision-induced + possibly collision-induced	<0·25	—	—
(2) $CS_2^+ \rightarrow CS^+ + S$	13–14	Collision-induced only	?	—	—
	$(16·3 \pm 1)$	Non-collision-induced + possibly collision-induced	<0·4	Ar^+	Weak

[a] A similar effect is observed for the fast rate appearing S^+ ions.

ions have been used, in order to state more accurately the thresholds of the
processes studied.

Metastable CS_2^+ Ions

The experimental results collected about processes (1) and (2) are summarized
in Table I.

From the best values for $\Delta U_f(CS_2) = 1\cdot21$ eV; $\Delta U_f(S^+) = 13\cdot25$ eV;
$\Delta U_f(CS) = 3\cdot04$ eV[15] and for (IP)$_1$ and (IP)$_2$ of CS respectively equal to
$11\cdot33$ eV[16,17] and to $11\cdot39$ eV,[15] as for the position in energy of the ground
and first excited states of CS,[18] we have calculated the successive dissociation
limits for processes (1) and (2); a tentative correlation of the dissociation
products with the known electronic states of CS_2^+ as deduced from PES[19–21]
is given in Fig. 1 using the $C_{\infty v}$ symmetry.

In order to take into account a continuous band observed by PES[21]
between $16\cdot5$ and $17\cdot5$ eV, an additional potential energy curve has been
drawn (D state). In this Fig. 1, the energy position of breaks in the ionization
efficiency curves of CS^+ and S^+ given by RPD method on a t.o.f. instrument[22] as well as the photoionization threshold values[23] have been drawn.
From Fig. 1 it is possible to give the following picture of the dissociative
phenomena suffered by CS_2^+.

Appearance Processes of S^+—(a) The previous EI[22] and PI[23] observations
pointed out the appearance of S^+ ions resulting from fast rate decaying
CS_2^+ around the first dissociation limit at $15\cdot08$ eV.

In the same energy range (15 ± 1) eV-S^+ ions produced by slow rate
decaying metastable CS_2^+ ions are observed and the KE release during this
process is lower than $0\cdot25$ eV (*see* Table I). The first dissociation limit
$S^+(^4S) + CS$ $(X^1\Sigma^+)$ only correlates with a $^4\Sigma^-$ repulsive state which

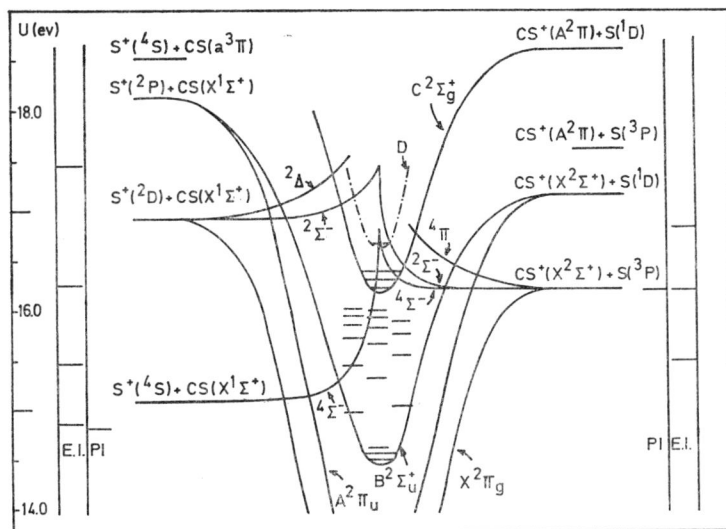

FIG. 1 Suggested set of potential energy curves in the $C_{\infty v}$ symmetry for CS_2^+.

necessarily crosses the $X^2\Pi_g$, $A^2\Pi_u$ and $B^2\Sigma_u^+$ and $C^2\Sigma_g^+$ states. Therefore the following interpretation of the described situation may be proposed:

(i) A predissociation mechanism of the $X^2\Pi_g$ or $A^2\Pi_u$ states by the $^4\Sigma^-$ state (*see* Fig. 1) can explain the fast rate appearing S^+ ions at 15·08 eV. This predissociation can take place in quite a short time owing to the important spin-orbit interaction due to the occurrence of S atoms in the molecule.

(ii) The slow rate appearance of S^+ ions around 15 eV can be interpreted as a predissociation of the $B^2\Sigma_u^+$ state by the $^4\Sigma^-$ state. The selection rule $\Sigma^+ - \Sigma^-$ forbidding this predissociation may be slightly violated when $\Delta S \neq 0$ and for a large spin-orbit coupling, as well as via vibronic coupling.[13] This process can therefore take place within a time comparable with the metastable ions lifetime.

It is to be pointed out that such a slow predissociation can favourably compete with the radiative lifetime of the $B^2\Sigma_u^+$ state of CS_2^+ which is ranging about $3 \cdot 10^{-7}$ sec.[24]

Some of the data reported here seem to be in contradiction with the fact that by direct Franck–Condon transitions the upper levels populated by the $X^2\Pi_g$, $A^2\Pi_u$ and $B^2\Sigma_u^+$ states lie below the energy threshold of the slow or fast rate appearing S^+ ions.

Therefore we must admit that under EI the predissociation levels of the $X^2\Pi_g$, $A^2\Pi_u$ or $B^2\Sigma_u^+$ states are populated from some auto-ionized Rydberg states lying around 15 eV (*see* Fig. 1) through non-radiative transitions to these states.

Moreover the slow or fast appearance of S^+ ions originating from the slightly endothermic dissociative charge exchange with Kr^+ ions (*see* Table I) shows that the Franck–Condon principle does not hold any more in this case: this has already been suggested by Hasted[25] and Champion.[26]

(b) The observation at (17 ± 1) eV of a second threshold for the slow rate appearing S^+ ions is probably to be correlated with the observation between 16·5 and 17·5 eV of a low continuum in the PES of CS_2. This can correspond to a forbidden predissociation which is able to explain the occurrence of the metastable transition. Therefore we suggest that in this energy range the $C^2\Sigma_g^+$ or another stable electronic state of CS_2^+ (drawn tentatively in Fig. 1) is populated through autoionization processes[27] and is predissociated by the $^2\Sigma^-$ state correlated with the dissociation limit $S^+(^2D) + CS(X^1\Sigma^+)$ (*see* Fig. 1).

(c) It is to be pointed out that EI appearance efficiency curves of S^+ ions obtained by the RPD method reveal two additional breaks between the first two dissociation limits leading to S^+ ions at 15·08 and 16·92 eV (*see* Table I and Fig. 1).

These breaks situated at 15·5 eV and 16·25 eV are not observed in our experiments on metastable ions; we propose to explain this as follows:

(i) The break at 15·5 eV is more than probably due to the population in this energy range of the repulsive $^4\Sigma^-$ state through autoionization processes (*see* Fig. 1).

(ii) A predissociation mechanism of the $C^2\Sigma_g^+$ by the $^4\Sigma^-$ (see Fig. 1) may explain the fast or slow rate appearance of the S^+ ions at 16·25 eV but the amount of KE released during the process makes it difficult to observe the eventual slow process in the experimental conditions used.

Appearance Processes of CS^+—(a) The lowest value detected by the RPD method has been shown to be due to an ion-pair process such as $CS^+ + S^-$. We shall not discuss this further.[22,28]

(b) At around 16·16 eV, fast rate appearing CS^+ ions have been detected both by EI[22] and PI;[23] we think that these ions appear through the predissociation of the $C^2\Sigma_g^+$ by the $^4\Sigma^-$ state correlated with the first dissociation limit $CS^+(X^2\Sigma^+) + S(^3P)$. The large spin-orbit coupling in this molecule favours this process.

The appearance potential of the slow rate decaying CS_2^+ ions lies a little higher in energy and may be partly due to the same process and to the probably slower predissociation process of the $C^2\Sigma_g^+$ by the $^2\Sigma^-$ state, in agreement with the maximum KE release reported in Table I.

Collision-induced Processes—The appearance potentials of the collision-induced metastable transitions (1) and (2) at about 13 to 14 eV leads us to conclude that the predissociation states implied in these processes can be populated by collision-induced electronic transitions from the $A^2\Pi_u$ or $B^2\Sigma_u^+$ states of the CS_2^+ ions, in agreement with previous works.[7,19−21]

Metastable CH_2^+ Ion

Electron impact on CH_4, C_2H_4 and CH_2CO shows the occurrence of the metastable transition $CH_2^+ \rightarrow C^+ + H_2$. The total KE released in the non-collision and collision-induced process, estimated from the metastable peak shape, is not higher than 0·8 eV.

The appearance efficiency curves of both the metastable ion and the normal C^+ ion have been recorded for the three substances. The results are presented in Fig. 2 where the appearance energies of collision or non-collision-induced processes are indicated for the metastable transition and for the production of normal C^+ ions.

From Fig. 2 we easily conclude that purely collision-induced components of the metastable transition $CH_2^+ \rightarrow C^+ + H_2$ occur in each substance: the first one appears within a 1 eV interval above the first dissociation limit $-C^+(^2P_u) + H_2(^1\Sigma_g^+)-$; the second one appears near the dissociation limit $-C^+(^4P_g) + H_2(^1\Sigma_g^+)-$. A purely non-collision-induced metastable transition appears for the three substances in a 1 eV interval energy around the dissociation limit $-C^+(^2D_g) + H_2(^1\Sigma_g^+)$. In addition, it appears in Fig. 2 that in each substance, the normal C^+ ions detected are produced first by collision-induced and at higher energies by non-collision-induced processes.

Although the bimolecular component of these normal C^+ ions could possibly be due to faster decaying collision-induced metastable CH_2^+ ions, in the case of non-collision-induced components, the higher energy threshold of the metastable CH_2^+ ions with respect to the normal C^+ threshold prevents us from identifying the appearance of these ions to the same slow or fast process.

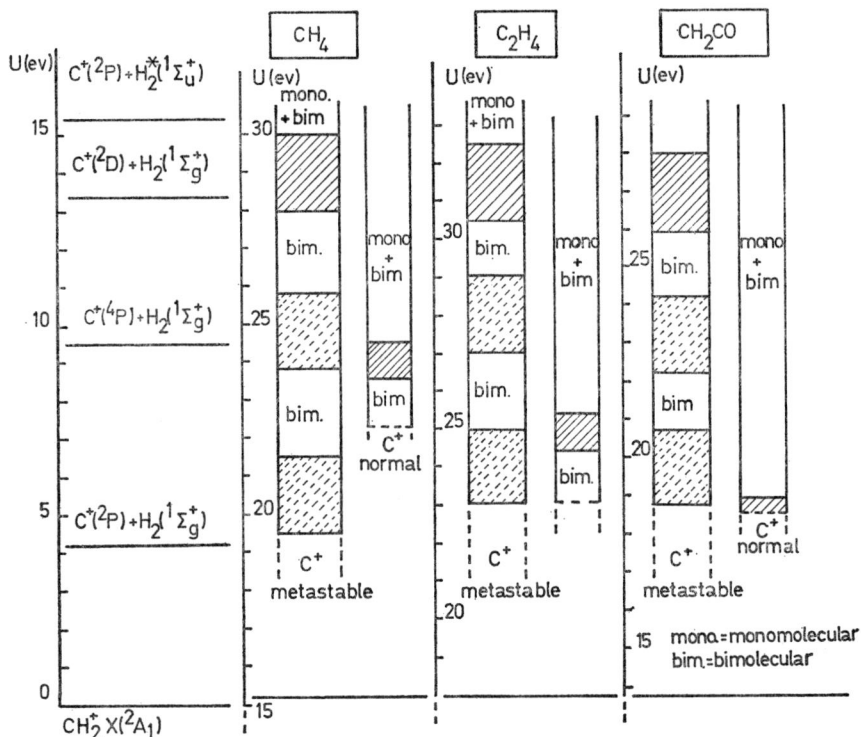

FIG. 2   Appearance potentials of the $CH_2^+ \rightarrow C^+ + H_2$ metastable transition and of the C^+ fast decaying ions from CH_4, C_2H_4 and CH_2CO.

The parallelism of the experimental situation in the three substances induces us to describe in some detail the case of CH_4.

CH_2^+ Metastable Ion from CH_4

We show in Fig. 3 the energy position of the successive dissociation limits of $CH_2^+ \rightarrow C + H_2^+$ on the left side and of $CH_2^+ \rightarrow C^+ + H_2$ on the right side; in each case the electronic states correlated with these limits are given in the C_{2v} symmetry. The limits are drawn with respect to the ground state $X(^2A_1)$ of CH_2^+. As it was pointed out by Lindholm,[29] the CH_2^+ ion at its threshold in CH_4 would be in the $\tilde{A}(^2B_1)$ state; it is known, however, that the energy interval between the $\tilde{A}$ and $\tilde{X}$ states does not exceed ~ 0.2 eV.[30]

The energy scale on the left side is given with respect to the photoionization AP of CH_2^+ from CH_4.[31]

The situation described in Fig. 2 allows us to propose tentative interpretations for the first collision-induced mechanism appearing around 20 eV (a) and for the non-collision-induced mechanism appearing around 29 eV (b).

(a) *Ionization phenomena in CH_4 around* 20 eV *and the appearance of the collision-induced process* $CH_2^+ \rightarrow C^+ + H_2$—In the T_d symmetry the electronic structure of CH_4 is $(2a_1)^2(2f_2)^6 (3a_1)$; at 12·6 eV the ground state CH_4^+ ion is produced in the $^2F_2$ state; around 22·4 eV and $^2A_1$ state

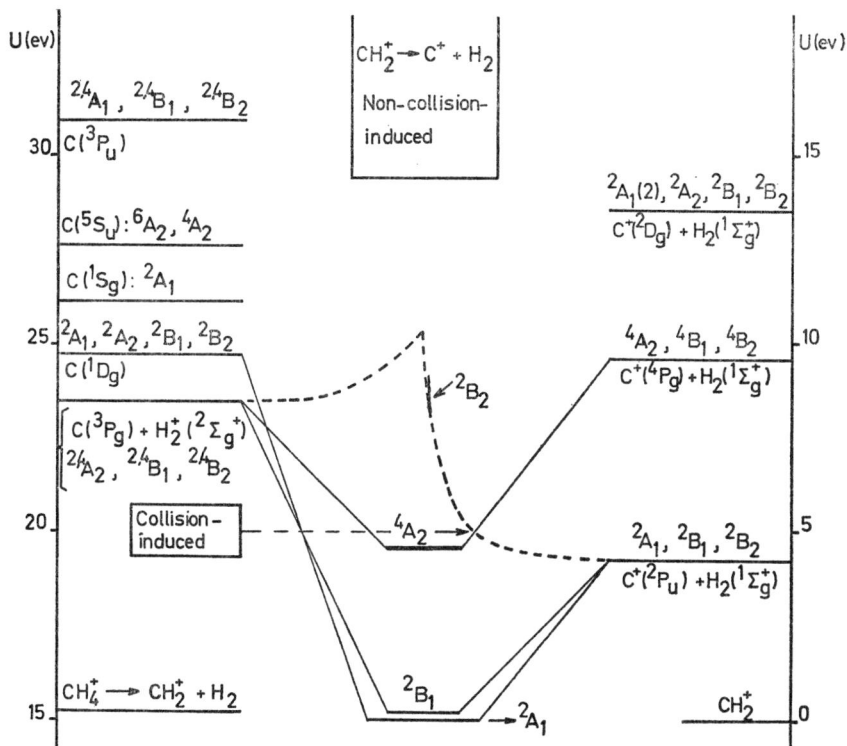

FIG. 3   Suggested set of potential energy curves in the C_{2v} symmetry for CH_2^+.

appears. In the ionization efficiency curves obtained by EI a break appears around 19·3 eV;[32] a similar structure is also present in the appearance efficiency curves of CH_2^+ from CH_4.[33] None of these structures are detectable with photoionization[31] or photoelectron spectroscopic methods.[34]

By charge exchange with Ne^+ ions however, an important increase in the cross-section for CH_2^+ ions from CH_4 is observed.[35] This seems to us to support the possibility that an electronic excited configuration of CH_4^+ is formed (by EI or CE) around 20 eV and is also able to decay in some excited electronic configuration of $CH_2^+ + H_2$.

This electronically excited configuration of CH_4^+ could be induced in CH_4 simultaneously with the $^{1,3}F_2$ and $^{1,3}A_1$ superexcited states of CH_4 discovered by Ehrhardt around 20 eV.[32] We suggest that this doubly excited electronic configuration of CH_4^+ could be: $(1a_1)^2(2f_2)^4(3a_1)^1(^4F_1)$. If we turn our attention to the electronic configuration of CH_2 which is in C_{2v} symmetry: $(2a_1)^2(1b_2)^2(3a_1)^1(1b_1)^1(4\overline{a_1})$, the lowest electronic configurations of CH_2^+ will possibly be such as given below:

$(2a_1)$	$(1b_2)$	$(3a_1)$	$(1b_1)$	$(4a_1)$	
2	2	1	0	0	$\tilde{X}(CH_2^+)^2A_1$
2	0	1	0	0	$\tilde{A}(CH_2^+)^2B_1$
2	1	1	1	0	$\tilde{B}(CH_2^+)^2A_2 - {}^4A_2$
2	1	2	0	0	$^2B_2$

It is naturally impossible to decide without elaborate *a priori* quantum calculations whether the $^2B_2$ or the $^4A_2$ states could be the lowest in energy. Owing to our hypothesis about the energy position of the $^4F_1$ state of CH_4^+ we will propose that the $^4A_2$ state of CH_2^+ lies also around 20 eV.

The $^4F_1$ state reduced into the C_{2v} symmetry gives the $^4A_2 + {}^4B_1 + {}^4B_2$ states of CH_4^+. Under these states the $^4A_2$ is able to be correlated with $CH_2^+(^4A_2) + H_2(^1A_1)$. On this basis we propose to present a tentative explanation of our observation of the collision-induced metastable $CH_2^+ \rightarrow C^+ + H_2$ around 20 eV; this is given in Fig. 3 where the successive dissociation asymptote of $CH_2^+ \rightarrow C^+ + H_2$ and $CH_2^+ \rightarrow C + H_2^+$ is given.

In this interpretation we propose that the collision-induced metastable transition around 20 eV is due to the perturbation by collision of the strongly forbidden predissociation of the $^4A_2$ state of CH_2^+ by the $^2B_2$ state.

(b) *The non-collision-induced metastable transition* $CH_2^+ \rightarrow C^+ + H_2$ *around 29 eV*—If we remember that the maximum kinetic energy released in this metastable transition is around 0·8 eV, the predissociation mechanism giving rise to the non-collision-induced metastable transition can only take place at the third asymptote $C^+(^2D_g) + H_2 ({}^1\Sigma_g^+)$.

It is premature to decide whether the predissociation is a spin-orbit forbidden one or a tunnelling process through a potential energy hill.

In our opinion, however, the spin-orbit forbidden mechanism is more than probable as suggested by the strong perturbation upon the non-collision-induced process when using high atomic number target gases like Xe, in agreement with our observations in the N_2^+ metastable transition.[2,7]

ACKNOWLEDGMENTS

We thank the Fonds de la Recherche Fondamentale Collective and the Patrimoine de l'Université de Liège for their financial support to two of us, as also for research funds.

REFERENCES

1. Wankenne, H. and Momigny, J., *Chem. Phys. Letters*, 1969, **4**, 132.
2. Momigny, J. and Wankenne, H., 'Advances in Mass Spectrometry', Vol. 5, 1970, p. 70.
3. Newton, A. S. and Sciamanna, A. F., *J. Chem. Phys.*, 1969, **50**, 4868; 1970, **52**, 327.
4. Albritton, D. L., Schmeltekopf, A. L. and Ferguson, E. E., 6th Int. Conf. on the Phys. of Electronic and Atomic Collisions (Cambridge, Mass, 1969), p. 331.
5. Fournier, P., van de Runstraat, C. A., Govers, T. R., Schopman, J., de Heer, F. J. and Los, J., *Chem. Phys. Letters*, 1971, **9**, 426.
6. Fournier, P., Ozenne, J. B. and Durup, J., *J. Chem. Phys.*, 1970, **53**, 4095.
7. Wankenne, H and Momigny, J., *Int. J. Mass Spectrom. Ion Phys.*, 1971, **7**, 227.
8. Lorquet, J. C. and Desouter, M., *Chem. Phys. Letters*, 1972, **16**, 136.
9. Pham Dong. and Bizot, M., *Int. J. Mass Spectrom. Ion Phys.*, 1972/3, **10**, 227.
10. Lorquet-Julien, A., Lorquet, J. C., Momigny, J. and Wankenne, H., *J. Chim. Phys.*, 1970, **67**, 64.
11. Lorquet, A. J., Lorquet, J. C., Wankenne, H., Momigny, J. and Lefebvre-Brion, H., *J. Chem. Phys.*, 1971, **55**, 4053.
12. Newton, A. S. and Sciamanna, A. F., *J. Chem. Phys.*, 1970, **52**, 327.

13. Jones, E. G., Beynon, J. H. and Cooks, R. G., *J. Chem. Phys.*, 1972, **57**, 3207.
14. Mathieu, G., Wankenne, H. and Momigny, J., *Chem. Phys. Lett.*, 1972, **17**, 260.
15. Hildenbrand, D. L., *Chem. Phys. Lett.*, 1972, **15**, 379.
16. Jonathan, N., Morris, A., Okuda, M., Smith, D. J. and Ross, K. J., *Chem. Phys. Lett.*, 1972, **13**, 334.
17. King, G. H., Kroto, H. W. and Suffolk, R. J., *Chem. Phys. Lett.*, 1972, **13**, 457.
18. Tewarson, A. and Palmer, H. B., *J. Mol. Spectroscopy*, 1968, **27**, 246.
19. Turner, D. W. and Al Joboury, M. I., *J. Chem. Phys.*, 1962, **37**, 3007.
20. Eland, J. H. D. and Danby, C. J., *Int. J. Mass Spectrom. Ion Phys.*, 1968, **1**, 111.
21. Brundle, C. R. and Turner, D. W., *Int. J. Mass Spectrom. Ion Phys.*, 1969, **2**, 195.
22. Momigny, J. and Delwiche, J., *J. Chim. Phys.*, 1968, **65**, 1213.
23. Dibeler, V. H. and Walker, J. A., 'Adv. in Mass Spectrometry', Vol. 4, 1968, p. 767.
24. Hayden Smith, W., *J. Chem. Phys.*, 1961, **51**, 3431.
25. Hasted, J. B., 'Physics of Atomic Collisions', Butterworths, London, 1964.
26. Champion, R. L. and Doverspike, L. D., *J. Chem. Phys.*, 1968, **49**, 4321.
27. Tanaka, Y., Jursa, A. S. and Le Blanc, F. J., *J. Chem. Phys.*, 1960, **32**, 1205.
28. Locht, P., Ph.D. Thesis, Université of Liège, 1971.
29. Lindholm, E., *Arkiv för Fysik*, 1968, **37**, 37.
30. Bender, C. F. and Schaeffer, III H. F., *J. Mol. Sp.*, 1971, **37**, 423.
31. Chupka, W. A., *J. Chem. Phys.*, 1968, **48**, 2337.
32. (a) Frost, D. C. and McDowell, C. A., *Proc. Roy. Soc. London*, 1957, **A249**, 194.
    (b) Collin, J. E., *Mem. Soc. Roy. Sc. Liège*, 1966, **XIV**, No. 1.
    (c) Ehrhardt, H., Linder, F. and Meister, G., *Z. Naturf.*, 1965, **20a**, 989.
33. Al Joboury, M. I. and Turner, D. W., *J. Chem. Soc.*, 1964, 4434; *J. Chem. Soc.*, 1967, 373.
34. Sjögren, H., *Arkiv för Fysik*, 1966, **31**, 159.

Discussion

R. J. Donovan (University of Edinburgh, U.K.): Concerning the 'spin forbidden' predissociations in $CS_2{}^+$, I would like to point out that the spin–orbit coupling in CS_2 (neutral) is sufficiently strong that the spin quantum number is no longer strictly a good one. There are several pieces of evidence which show this, and two which immediately spring to mind are:

(i) Photo-excitation of the $\tilde{A}$ state results in the formation of $S(3\,^3P_J)$ ('spin forbidden') and not $S(3\,^1D_2)$ ('spin allowed').
(ii) Photo-fragmentation of CS_2 following absorption in several Rydberg bands again leads to $S(3\,^3P_J)$ and *not* $S(3\,^1D_2)$ (H. Okabe, J. Chem. Phys., **56** (1972), 4381).

In the face of this evidence would it not be better to reconstruct your correlation diagram using double groups, thus taking account of spin-orbit coupling?

J. Momigny: I would entirely agree with Dr Donovan's remark if we had considered only spin-forbidden predissociations in order to explain the occurrence of metastable $CS_2{}^+$ ions. However, it has to be pointed out that our explanation was based on the degree to which Σ^+, Σ^- predissociations are forbidden.

This suggestion nevertheless is a useful one and I would like to consider its impact on the subject, as soon as possible.

May I add that the occurrence of spin-forbidden processes in the photochemistry of CS_2, the rate at which the processes are taking place is not known; or this rate would be the most useful piece of information in order to make a comparison with the occurrence of the $CS_2{}^+$ metastable ions.

108
Mass Spectrometric Studies of the Photoionization of Simple Gases in the Extreme Ultraviolet

By R. BROWNING, J. FRYAR† and R. CUNNINGHAM

(*Department of Pure and Applied Physics, The Queen's University of Belfast, Belfast, U.K.*)

INTRODUCTION

ALTHOUGH dissociative photoionization processes may produce significant numbers of certain ionic species within planetary ionospheres, and despite the fundamental nature of the processes involved, little progress has yet been made in their elucidation. Theoretically, few calculations can yet be made because of the complexity of the processes for even the simplest molecules. Experimentally such processes may be investigated using a mass spectrometer, but low signal strengths and kinetic energy discrimination effects arising from molecular dissociation, when taken in conjunction with the short wavelengths at which such processes become important, make accurate investigations particularly difficult. We describe here measurements in which we have met these problems and applied a specially designed mass spectrometric system to the study of the photoionization of some simple molecules.

METHOD

Basically the apparatus consisted of a monochromator coupled to a mass spectrometer. A $\frac{1}{2}$m Seya-Namioka monochromator dispersed the line radiation produced by a specially constructed duoplasmatron light source, and the light intensity was monitored using a photomultiplier viewing a glass plate coated with sodium salicylate. The light passed through a stainless steel chamber biased 11 kV above ground containing the gas under investigation at a pressure $<10^{-4}$ torr. A representative proportion of the photoions formed along the path of the beam was swept by a field in excess of 1 kV cm^{-1} into a wide-aperture, 60° magnetic mass spectrometer and detected as

---

† Now at Department of Physics, University of Windsor, Windsor, Ontario, Canada.

single particles. The complete system will be described in a future publication,[1] but the uniform transmission characteristics of the system may be judged from the fact that the heights of the flat-topped peaks in the mass spectra (a) fully saturated with the field sweeping the photoions into the mass spectrometer (b) became independent of the width of the photoion detector at sufficiently large apertures and (c) could be summed as partial cross-sections to reproduce independently measured total photoionization cross-sections.[1,2]

RESULTS AND DISCUSSION

General

The ratio of the peak heights recorded mass spectrometrically was determined from the integrated count at each peak, normalizing to a fixed photon flux. Small corrections[1] were applied to correct the measured ratios to eliminate the effect of spurious signals due to collisions occurring during the extraction of the photoions and to take account of the non-uniform response of the detector to ions of different mass. From the measured ratios it was possible to deduce partial cross-sections (dissociative and non-dissociative) with a knowledge of the total photoionization cross-sections. However, the results described below have not been so reduced, as most of the significant features can be appreciated simply by observing the variation in the ratio of the peak heights of the mass spectra with photon energy. The ratios reported here are believed to be accurate to within 8%.[1]

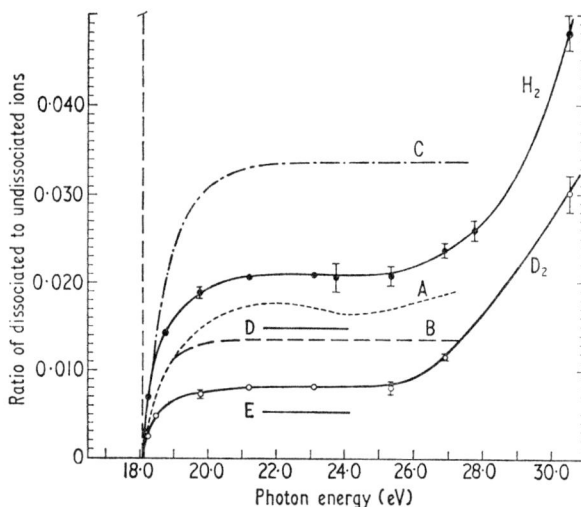

FIG. 1   The ratio of the cross-sections for dissociative to non-dissociative photoionization of H_2 and D_2 as a function of photon energy: —●—, —○— Present results for H_2 and D_2; ────── A: curve calculated using the H_2 ground vibrational wave function and electronic matrix element variation given by Flannery and Opik;[5] ───── B: as curve A but without the electronic matrix element variation; —·—·—· C: simple harmonic oscillator approximation (for comparison); ──────── D, E: Plateau values for H_2, D_2 as given by Villarejo.[6]

Hydrogen

The results for H_2 are shown in Fig. 1, with the exception of a result at 304 Å (40·8 eV).[3] Over much of the energy range shown (<26 eV) the photoions are formed solely through the $1s\sigma_g$ state of H_2^+, and this permits approximate calculations to be made as described fully by Browning and Fryar.[4] As the photon energy increases beyond the threshold for dissociative photoionization at 18·08 eV, the H_2 molecule, which is in its ground vibrational state, may be photoionized to the repulsive part of the $H_2^+(1s\sigma_g)$ state which subsequently dissociates. In this region the dissociative photoionization fraction H^+/H_2^+ may be described and calculated using the reflection approximation. At sufficiently large photon energies (above ~21 eV) a short plateau in the H^+/H_2^+ ratio is observed: here the ratio may also be calculated using a sum rule extending over the associated Franck–Condon factors for H_2^+. Both types of calculation may be corrected using the variation of the electronic matrix elements as given by Flannery and Opik,[5] and curve A is the most complete calculation of this type yet available. The plateau value (D) of Villarejo,[6] who has made a most complete and exact calculation of the Franck–Condon factors for H_2 and D_2, agrees well for H_2 with the plateau shown on curve B, to which it corresponds. However, it cannot be said that the agreement between experiment and theory is yet sufficiently satisfactory, and improved calculations would seem to be worthwhile.

The rise in H^+/H_2^+ from the plateau value ($0·0208 \pm 0·0017$) may be readily interpreted as arising from transitions, initially from large internuclear separations in the vibration of H_2, to the repulsive $2p\sigma_u$ state of H_2^+. The ratio H^+/H_2^+ reaches $0·108 \pm 0·024$ at 304 Å, a value which indicates that dissociative photoionization may be expected to be an important source of protons in the ionosphere of the outer planets.[9]

Oxygen

The corresponding results for oxygen are shown in Fig. 2. Since oxygen is too complex a molecule for calculation of the type described above to be expected to be worthwhile, an interpretation of the results most usefully proceeds[1] by comparing the results with the potential curves of Gilmore[10] and the results of photoelectron spectroscopy.[7,8]

The initial rise to a short plateau in O^+/O_2^+ at 19–20 eV may be likened to that described above in H_2; in this case O^+ ions would appear to arise from transition to repulsive part of the $A^2\Pi_u$ state of O_2^+ above the dissociative threshold to $O^+(^4S) + O(^3P)$ which occurs at 18·73 eV. The next rise between 20–21 eV would seem to arise as a result of predissociation of the $^2\Sigma_g$ state of O_2^+.[1,11,12]

Between 24 and 28 eV photoionization of a $\sigma_u 2s$ electron, which has anti-bonding character, may occur (see Fig. 2). However the few results obtained show no striking fall in the O^+/O_2^+ ratio as might be expected if stable O_2^+ ions resulted. Some predissociation in this region may therefore be expected. The one high energy point at 40·8 eV (304 Å) indicates that the release of a strongly bonding $\sigma_g 2s$ electron leads to dissociation, and it would appear that the removal of an electron from this orbital is involved in

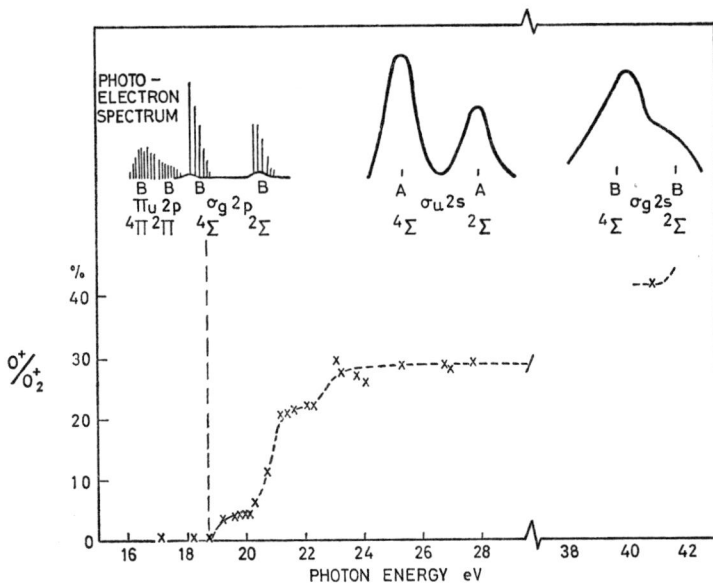

FIG. 2   The ratio O^+/O_2^+ of the cross-sections for photoionization of oxygen. (The dotted line through the experimental points serves to guide the eye: rapid variations may arise through predissociation. The photoelectron spectrum from the $\pi_u 2p$ and $\sigma_g 2p$ molecular orbitals is taken from Turner et al.[7] using 584 Å radiation. The $\sigma_u 2s$ and $\sigma_g 2s$ peaks at much lower resolution are taken from Siegbahn et al.[8] The letters A and B indicate the antibonding or bonding character of the electron in the orbital designated.)

the explanation of the large degree of fragmentation which results from heavy particle impact.[13]

Other Simple Molecules

Nitrogen, carbon monoxide and nitric oxide have been examined in less detail, but it is already apparent that in N_2 and CO, the dissociative ionization fractions are generally smaller than in O_2. For example, in N_2 the ratio N^+/N_2^+ remains at about 4% for the first few volts from threshold, rising to about 25% at 40·8 eV. This can be understood by comparing the photoelectron spectra with the thresholds for dissociative ionization, bearing in mind the results reported above. The peaks in the photoelectron spectra[8] lie largely below the threshold for dissociative ionization in these cases, except for the lowest lying molecular orbitals which occur just below 40 eV. In NO, however, the release of an electron from higher orbitals (2σ) just above threshold can also lead to dissociation, in much the same way as in O_2; it is therefore not surprising that in this case the dissociative ionization fractions are larger, and similar to those reported above for O_2.

ACKNOWLEDGMENTS

We wish to thank the Science Research Council for the provision of a research grant. One of us (R.C.) is also indebted to the Ministry of Education, Northern Ireland, for a Research Studentship.

REFERENCES

1. Fryar, J. and Browning, R., *J. Phys. B: Atom. Molec. Phys.*, to be published.
2. Samson, J. A. R. and Cairns, R. B., *J. Geophys. Res.*, 1964, **69**, 4583.
3. Fryar, J. and Browning, R., *Planet. Space. Sci.*, 1973, **21**, 709, 1080.
4. Browning, R. and Fryar, J., *J. Phys. B: Atom. Molec. Phys.*, 1973, **6**, 364.
5. Flannery, M. R. and Opik, U., *Proc. Phys. Soc.*, 1965, **86**, 491.
6. Villarejo, D., *J. Chem. Phys.*, 1968, **49**, 2523.
7. Turner, D. W. *et al.*, 'Molecular Photoelectron Spectroscopy', Wiley–Interscience, 1970.
8. Siegbahn, K. *et al.*, 'ESCA Applied to Free Molecules', North Holland Publishing Company, 1969.
9. McConnell, J. C., Private communication.
10. Gilmore, F. R., *J. Quant. Spectry. Radiative Transfer*, 1965, **5**, 369.
11. Doolittle, P. H., Schoen, R. I. and Schubert, K. E., *J. Chem. Phys.*, 1968, **49**, 5108.
12. Danby, C. J. and Eland, J. H. D., *Int. J. Mass Spectrom. Ion Phys.*, 1972, **8**, 153.
13. Graham, W. G., Latimer, C. J., Browning, R. and Gilbody, H. B., *J. Phys. B: Atom. Molec. Phys.*, to be published.

Discussion

A. J. C. Nicholson (C.S.I.R.O. Australia): Many years ago Dr Morrison and I measured O^+ from O_2 by electron impact and obtained ions from the ion pair dissociation. Should not you have a small peak in your graph at about 17·5 eV?

R. Browning: Simultaneous ion pair production has indeed been observed by others in the photoionization of O_2 over a rather narrow spectral region. Evidence for this process (just below the first main rise in O^+/O_2^+ at 18·7 eV) is present in our results too, which suggest a value of $O^+/O_2^+ \approx 0.2\%$ in this region.

J. Eland (Oxford University, U.K.): It certainly seems surprising that the O^+/O_2 curve remains flat in the 25 eV region, particularly in view of Edquist and Lindholm's 304 Å photoelectron spectrum of O_2, which clearly shows the $\tilde{C}$ state as predissociating. Could you comment further on this point?

R. Browning: There are rather few data points in the 25 eV region and a much more complete understanding would be available if a continuous spectrum (from a synchrotron for example) could be used instead of the line radiation used here. Certainly our results suggest a rather constant branching ratio in this region, but contributions from several states are included.

109
Metastable Transitions in Small Molecular Ions: Predissociation Processes in CO^+, CO_2^+ and COS^+

By M. A. ALMOSTER-FERREIRA and M. M. PIRES

(*Faculdade de Ciências, Lisbon, Portugal*
and
Laboratório Calouste Gulbenkian de Espectrometria de Massa, Lisbon, Portugal)

1. INTRODUCTION

METASTABLE transitions have proved to be of great importance in the understanding of the fundamental processes of dissociation of gaseous ionized matter. The occurrence of such phenomena on triatomic molecular ions such as CS_2^+, COS^+ and CO_2^+ obtained by electron impact are being thoroughly investigated, as well as the possibility that bimolecular positive fragment ions resulting from the decomposition of the molecular ions undergo the same process of dissociation. Concerning this aspect, fast and slow rate decaying processes of diatomic molecular ions are also being investigated.

The metastable transitions are explained by predissociation processes, and it has been possible to distinguish between pure and collision-induced predissociation phenomena.

Some of the results already obtained are reported.

2. EXPERIMENTAL TECHNIQUES

All the experimental studies have been performed on a single-focussing, 90°, Mass Spectrometer (A.E.I.-MS2 SG) which has been modified in several ways.

To study the decompositions of metastable ions occurring in the field-free region between the ion source exit slit and the entrance slit of the magnetic field, the defocussing technique[1,2] has been used. The adaptation of a counting system to the mass spectrometer has largely improved the results.

The appearance potential measurements have been performed either by the semilogarithmic plot method or the vanishing current method.

939

3. EXPERIMENTAL RESULTS AND THEIR INTERPRETATION

The following metastable transitions have been studied, with more or less detail, according to the experimental possibilities.

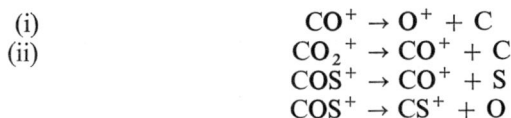

(i) $CO^+ \rightarrow O^+ + C$
(ii) $CO_2{}^+ \rightarrow CO^+ + C$
                          $COS^+ \rightarrow CO^+ + S$
                          $COS^+ \rightarrow CS^+ + O$

Results for $CS_2{}^+$ have been already presented.[3,4]

3.1. The occurrence of the metastable transition $CO^+ \rightarrow O^+ + C$ in the CO molecule when submitted to electron impact, has been reported in an earlier paper[5] as unimolecular and with an appearance potential of about 26 eV, without any further interpretation.

Considering the fact that this same transition is of great interest for the study of the metastable decomposition of the CO^+ ion coming from both $CO_2{}^+$ and COS^+, its study in the CO molecule was undertaken first. The defocussing technique has been used in all the experiments for this transition.

Determinations of the appearance potential of the slow rate appearing O^+ ions formed from CO under electron impact have shown that the efficiency curve presents changes of slope which, calculated from several runs, correspond to energies at $26\cdot0 \pm 0\cdot5$ eV (small cross-section), $28\cdot0 \pm 0\cdot5$ eV (higher cross-section) and $30\cdot0 \pm 0\cdot5$ eV (very high cross-section), and which were assigned to three possible different processes.

The influence of increasing pressure of CO upon the transition indicates that below 30 eV a collision induced transition occurs superimposing itself on some monomolecular process that may simultaneously occur, and that above 30 eV a unimolecular process together with a collision-induced one is present, the non-collision-induced one being more intense.

The rare gas target effects upon the non-collision induced metastable transition intensity have been investigated with the following results: no effect was obtained when using Ne^+ and Ar^+; an increase of 68 % with He^+ and a decrease of 75 % with Kr^+ with respect to the null effect with Ne^+ and Ar^+.

The enhancement of the metastable ions by He^+ can be explained as a consequence of a resonance between a recombination energy of He^+ and the threshold energy of the unimolecular metastable transition.[6] This assumption seems to be supported by the fact that recombination energies for He^+ are reported at energies around 26 eV and 32 eV.[7] The negative effect of Kr^+ can be understood providing that the atomic weight of krypton is high enough to perturb a possibly not too strong spin-orbit coupling, that is, sufficiently high to cause the decay rate of the metastable transition to increase enough to allow the decrease of the metastable intensity to be observable.

Under our working conditions it was not possible to measure the energy released as kinetic energy, because the metastable peak associated with the transition falls in a nominal mass range where it was difficult to experiment.

3.1.1. From spectroscopic observations,[8] three states are well known for the CO^+ ion, the $\tilde{X}^2\Sigma^+$, the $\tilde{A}^2\Pi$ and the $\tilde{B}^2\Sigma^+$ states, the corresponding potential curves having been calculated by Krupenie and Weissman.[9] A $^2\Delta$

state has also been observed spectroscopically by J. Marchard *et al.*[10] These states, the ground state excepted, are not of interest in interpreting our experimental results which involve a higher range of energies. A tentative study has been made to have an insight into the potential energy curves for possible higher states of the CO$^+$ ion.

The existence of another state (not known spectroscopically) can be predicted from the carbon monoxide electron configuration by Mulliken[11] at about $25 \cdot 5 \pm 0 \cdot 5$ eV and was observed by photoionization by Weissler *et al.*[12] with an onset at about 25·5 eV. More recently its existence has been suggested by El-Sherbini[13] and Siegbahn *et al.*[14,16]

On the other hand, extensive quantum mechanical calculations have been performed by Schaefer[15] on the radical CN (which is isoelectronic with CO$^+$ and N$_2^+$), for 59 molecular states.

Considering that the position of the known states of CO$^+$ in relation to the

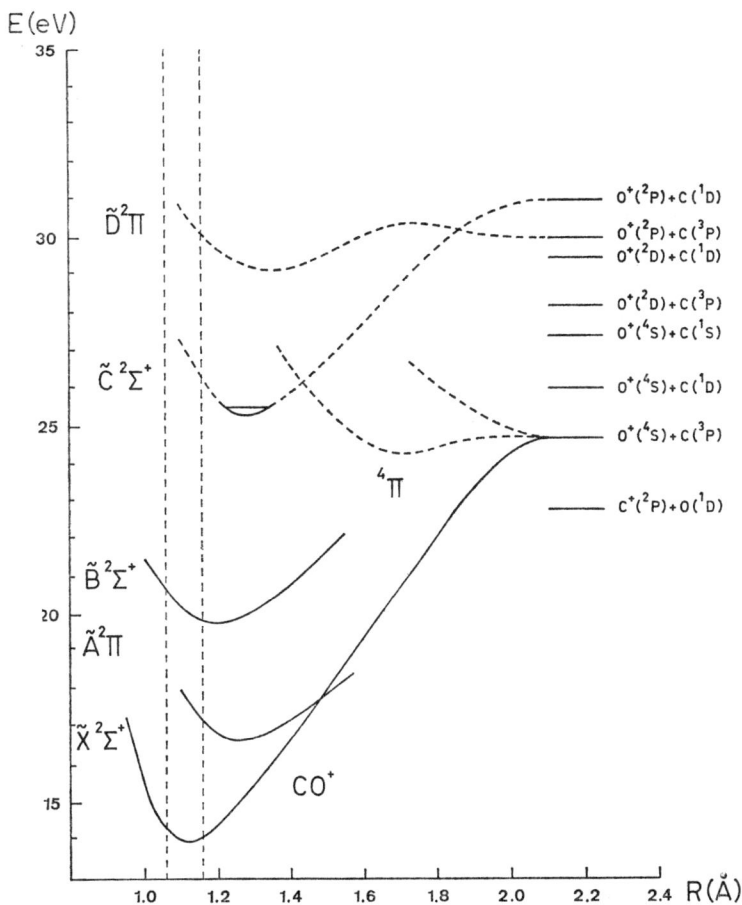

FIG. 1 Experimental (solid lines) and predicted (dashed lines) potential energy curves for the CO$^+$ ion.

corresponding calculated ones for CN are comparable and in fair agreement, an attempt was made to obtain a more complete diagram of the potential energy curves of the CO^+ ion, including the already known curves and some others, as shown in Fig. 1. We do not pretend it to be accurate, but we consider it to be correct in its main features.

The solid lines represent the known states and the dashed lines those states the position and form of which are presumed from the calculated potential energy curves for radical CN, as well as what is considered as the $\tilde{C}^2\Sigma^+$ state of CO^+, by comparison with the same state in CN and N_2^+.

The following interpretations of experimental results may be considered.

The slow appearance rate of O^+ ions at 26 eV may be explained by a predissociation mechanism of the $\tilde{C}^2\Sigma^+$ state of CO^+ (which may have been populated by collision-induced electronic transitions from lower states), by the $^4\Pi$ state which correlates with the first dissociation limit $O^+(^4S) + C(^3P)$.

The same kind of mechanism could explain the process observed at 28 eV accepting the possibility that some other stable state of the CO^+ ion exists in the Franck–Condon region at higher energy, which even if not observed spectroscopically may be induced by electron impact. On the other hand, the fact that a higher cross-section is observed for this process than for the previous one may suggest that higher vibronic levels of the $\tilde{C}$ state can be involved in a predissociation process by some state correlated to one of the dissociation asymptotes above the first one, if a situation is to be found for the $\tilde{C}^2\Sigma^+$ state of CO^+ similar to that observed by De Heer[17] for the $\tilde{C}^2\Sigma^+$ state of N_2^+ in which ratios of probability for predissociation to that for radiative decay increase for higher vibronic levels.

For the unimolecular process observed around 30 eV the following interpretation can be proposed. One of the potential curves calculated by Krupenie for the $^2\Pi$ states for the radical CN which correlates with a dissociation limit about 3 to 4 eV above the first dissociation limit, presents a potential maximum (probably due to an avoided crossing) and a shallow well. Since a $^2\Pi$ state can correlate with both the dissociation limits $O^+(^2P) + C(^3P)$ and $O^+(^2P) + C(^1D)$ for the CO^+ ion at energies about 30 eV and 31 eV, this strongly suggests the possibility of avoided curve crossing around these energies. If it is accepted that a $^2\Pi$ state for the CO^+ ion exists in the Franck–Condon region, as has been tentatively drawn (see Fig. 1), the unimolecular metastable transition observed at 30 eV could be explained by a predissociation taking place by a tunnelling process through the energy barrier around 30 eV.

The lack of knowledge of accurate potential energy curves for higher excited states of the CO^+ ion prevents one presenting more than a hypothesis to explain the experimental observations, which nevertheless is theoretically possible.

3.2. The occurrence of metastable transitions for the triatomic molecular ions CO_2^+ and COS^+ formed by electron impact has not been mentioned before.

The predissociation of CO_2^+ has been studied by photoelectron–photoion coincidence spectroscopy by Eland[18] with incidence of the study of the process leading to the formation of O^+. Some results about the metastable transition $CO_2^+ \rightarrow CO^+ + O$ are being reported and they refer to the study of this fragmentation process in the region between the exit slit of the ion source and the entrance slit of the magnetic field.

The intensity of the resulting CO$^+$ ions is very small. On the efficiency curve for the transition CO$_2$$^+$ $\rightarrow$ CO$^+$ + O two thresholds could be detected, one at 20 $\pm$ 1 eV and another at 30 $\pm$ 1 eV.

From measurements of the effect of increasing pressure of CO$_2$ in the ion source upon the metastable ions it was concluded that at both thresholds a monomolecular decay process was occurring, but below 30 eV the increasing pressure was favouring at the same time the occurrence of a collision-induced process.

From the shape of the metastable peak associated with the metastable transition, the kinetic energy release was measured using methods known from the literature.[19,20] The influence of rare gas target effects was also investigated and the results obtained show a negative effect for Kr$^+$ explained in the same way as for CO$^+$, and a favourable influence of Ne$^+$ in the enhancement of the metastable transition which can also be explained by a resonance process as mentioned before. The fact that recombination energies of 21·5 and 21·6 eV are known for Ne$^+$ seems to corroborate the value of 20 $\pm$ 1 eV for the appearance of a low rate decaying process of the CO$_2$$^+$ ion at this energy.

The study of slow rate decaying processes for the fragmentation of COS$^+$ ion, at low pressure, using the defocussing technique, has shown that in the efficiency curves for the decomposition COS$^+$ $\rightarrow$ CS$^+$ + O after the acceleration region, a threshold at 14 $\pm$ 1 eV and another at 18 $\pm$ 1 eV could be detected.

The process occurring at 14 $\pm$ 1 eV is left for further study since it is believed to be due to some more complex process than just the electron impact effect upon the molecule.[21,22] The existence of a unimolecular decay process for the COS$^+$ leading to CS$^+$ ions at energies above 18 eV was verified by means of the defocussing technique. For the metastable peak (nominal mass 32·06) the kinetic energy release was measured and found to be practically negligible. The results for this transition have similar characteristics to those observed for CS$_2$$^+$ $\rightarrow$ CS$^+$ + S.[4,5]

Only a few results could be obtained for the metastable transition COS$^+$ $\rightarrow$ CO$^+$ + S and they must be looked at as preliminary.

A collision-induced fragmentation process was observed at energies considerably above 20 eV for which a kinetic energy release of about 0·25 eV could be calculated. The defocussing technique was not performed upon this transition.

3.2.1. To interpret the experimental results obtained on the metastable decomposition of the triatomic molecular ions, the CO$_2$$^+$ ion will be considered first.

Figure 2 shows a diagram of potential energy curves for the CO$_2$$^+$ ion where onsets of ionization were taken from the values given by Eland,[18] as well as the energies of the two considered dissociation limits and the vibronic levels of the $\tilde{C}^2\Sigma_g^+$ state. Only the first of these levels is highly populated by photoionization, the other two having a low probability of population.

Considering that the first dissociation asymptote related to the ground state of the ion is CO$^+$($^2\Sigma_g^+$) + O(^3P_g) at an energy of 19·45 eV as calculated by use of the Dibeler's[23] photoionization dissociation energy D(OC–O), and that

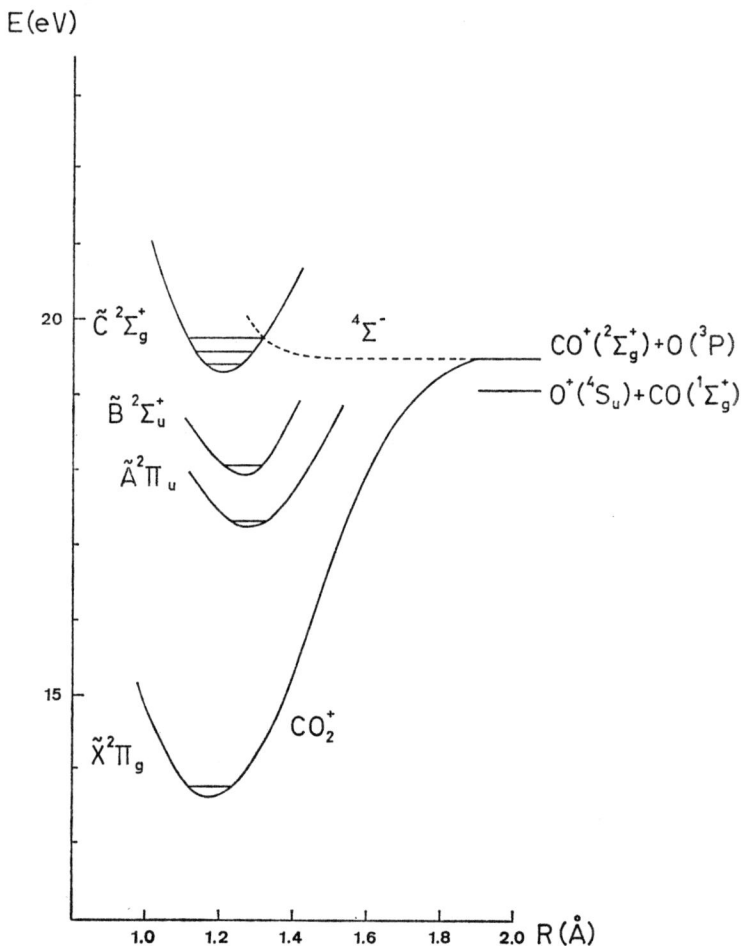

FIG. 2 Potential energy curves for the CO_2^+ ion.

the kinetic energy released in the transition is about 0·4 eV, a probable mechanism for the formation of CO^+ ions from a low rate decaying process seems to be excitation to the dissociative $\tilde{C}^2\Sigma_g^+$ state followed by predissociation to the $CO^+(^2\Sigma_g^+) + O(^3P)$ dissociation limit probably by a repulsive state $^4\Sigma^-$ which correlates with this dissociation limit. The symmetries of the two states involved fulfil the conditions when, according to Herzberg[8b] the selection rule $\Sigma^+ \leftarrow\!\!\mid\!\!\rightarrow \Sigma^-$ can be violated if spin-orbit coupling is large. Very probably the low vibronic levels of the $\tilde{C}^2\Sigma_g^+$ state are not involved in this transition, a low intensity of CO^+ ions being expected, in agreement with experimental observations.

The same predissociation mechanism can explain the appearance of CO^+ ions formed by a fast rate process, a lower vibronic level of the $\tilde{C}^2\Sigma_g^+$ state of the CO_2^+ ion being probably involved.

The slow fragmentation of the COS^+ ion in $CS^+ + O$ and $CO^+ + S$ will now be considered. Taking into account Turner's[24] values for onsets of excited states for the COS^+ ion, the metastable transition $COS^+ \rightarrow CS^+ + O$ taking place in a monomolecular way and with practically no kinetic energy release can be explained considering that after excitation to the $COS^+ \tilde{C}^2\Sigma_g^+$ state it predissociates by some state which could correlate with the first dissociation limit leading to the formation of CS^+ ions. If the $\tilde{C}^2\Sigma_g^+$ state is the one responsible for the predissociation—and it seems to be—to be consistent with the fact that the appearance of the transition is observed around 18 eV with practically no kinetic energy released, the vibrational level of the $\tilde{C}$ state involved in the process should be the first one.

For the slow fragmentation $COS^+ \rightarrow CO^+ + S$ the appearance potential for the metastable transition is not known, but since it should not be less than 15·6 eV (appearance potential of the CO^+ ion obtained by fast dissociation of CO_2 by electron impact), and considering that about 0·25 eV of kinetic energy are released, the same $\tilde{C}^2\Sigma_g^+$ state of the molecular ion should be able to explain the appearance of CO^+ ions found by slow fragmentation of the CO_2^+ ion, since it is also above the dissociation level at which CO^+ ions can be formed in the ground state. Some other possibilities are nevertheless to be considered for both metastable transitions involving the COS^+ ion.

REFERENCES

1. Barber, M., Jennings, K. R. and Rhodes, R., *Z. f. Naturf.*, 1967, **22a**, 15.
2. Barber, M. and Elliot, R. M., ASTM Committee E-14, 12th Annual Conference on Mass Spectrometry, Montreal, June 1969.
3. Momigny, J., Mathieu, G., Wankenne, H. and Almoster-Ferreira, M. A., *Chem. Phys. Letters* (in publication).
4. Momigny, J., Mathieu, G., Wankenne, H., Flamme, J. P. and Almoster-Ferreira, M. A., this Congress, Predissociation Processes in Triatomic Ions, Paper 107.
5. Wankenne, H. and Momigny, J., *Chem. Phys. Letters*, 1969, **4**, 132.
6. Albritton, D. C., Schmeltekopf, A. L. and Ferguson, E. E., VIth International Conference on Physics of Electronic and Atomic Collisions, 1969.
7. Gustafsson, E. and Lindholm, E., *Ark. f. Fysik*, 1960, **18**, 219.
8. Herzberg, G., (a) 'Molecular Spectra and Molecular Structure', Vol. I, 'Spectra of Diatomic Molecules', Van Nostrand Co., London, 1950; (b) 'Molecular Spectra and Molecular Structure', Vol. III, 'Electronic Spectra and Electronic Structure of Polyatomic Molecules', Van Nostrand Co., London, 1969.
9. Krupenie, P. H. and Weissman, S., *J. Chem. Phys.*, 1965, **43**, 1529.
10. Marchand, J., D'Incan, J. and Janin, J., *Spectrochimica Acta*, 1969, **25A**, 605.
11. Mulliken, R. S., (a) *Rev. Mod. Phys.*, 1932, **4**, 1; (b) *Phys. Rev.*, 1934, **46**, 549.
12. Weissler, G. L., Samson, J. A. R., Ogawa, M. and Cook, G. R., Research Report on Photoionization Analysis by Mass Spectroscopy, 1958, University of Southern California, Los Angeles.
13. El-Sherbini, Th. M. and Van Der Wiel, M. J., *Physica*, 1972, p. 433.
14. Siegbhan, K. *et al.*, 'ESCA Applied to Free Molecules', North Holland Publ. Co., Amsterdam, 1969.
15. Schaefer III, H. F. and Heil, T. G., *J. Chem. Phys.*, 1971, **54**, 2573.
16. Gaydon, A. G., 'Dissociation Energies and Spectra of Diatomic Molecules', 3rd ed., Chapman and Hall Ltd., London, 1968.
17. Govers, T. R., Van Runstraat, C. A. and De Heer, F. J., *J. Phys. B: Atom. Molec. Phys.*, 1973, **6**, 173.
18. Eland, J. H. D., *Int. J. Mass Spectrom. Ion Phys.*, 1972, **9**, 397.
19. Beynon, J. H., Saunders, R. A. and Williams, A. E., *Z. f. Naturf.*, 1965, **20a**, 180.

20. Newton, A. S. and Sciamanna, A. F., *J. Chem. Phys.*, 1966, **44**, 4327.
21. Almoster-Ferreira, M. A. and Costa, M. L., *Rev. Port. Quim.*, 1972, **14**, 21.
22. Smith, W. H., *J. Chem. Phys.*, 1969, **51**, 3410.
23. Dibeler, V. H. and Walker, J. A., 'Advances in Mass Spectrometry', Vol. 4, *ed.* E. Kendrick, Elsevier Publ. Co., London, 1968.
24. Turner, D. W., Baker, C., Baker, A. D. and Brundle, C. R., 'Molecular Photoelectron Spectroscopy', Wiley–Interscience, London, 1970.

Discussion

J. Durup (Universite de Paris-Sud, Orsay, France): I think it was suggested by S. Leach that the $\tilde{A}$ and/or $\tilde{B}$ states of CO_2^+ in their high vibrational levels might also be pre-dissociated. It seems that you cannot from your data exclude this possibility.

M. A. Almoster-Ferreira: Our results are very much in favour of the proposed interpretation, but, nevertheless, do not exclude the possibility of the $\tilde{A}$ or $\tilde{B}$ states being also involved. At the moment we are trying to determine whether autoionization phenomena are simultaneously contributing to the experimental observations.

110
Translational Energy of Ions by the Deflection Method: Heats of Formation of Several Free Radicals

By J. L. FRANKLIN and D. K. SEN SHARMA†

(*Department of Chemistry, Rice University, Houston, Texas, U.S.A.*)

INTRODUCTION

It has been shown by Haney and Franklin[1] that appearance potentials measured by electron impact often include excess energy which must be corrected for if valid thermochemical results are to be obtained from the measurement. Further, Haney and Franklin showed that the excess energy at onset, E^*, can be computed from the translational energy in the centre of mass, $\bar{\varepsilon}_t$, by means of the equation

$$E^* = \alpha N \bar{\varepsilon}_t \tag{1}$$

where N is the number of vibrational degrees of freedom and α is an empirically determined constant having a value of 0·44. Using this technique Haney and Franklin[2] obtained very satisfactory values for the heats of formation of several ions and free radicals.

The present study was undertaken to apply this technique to the measurement of the heats of formation of several hydrocarbon and amine free radicals. However, whereas Haney and Franklin[1,2] employed a time-of-flight mass spectrometer and determined the translational energy of the ions from peak shape analysis, it was necessary for us to employ a sector-field mass spectrometer which we modified to permit measurement of translational energy by the deflection method of Taubert.[3]

EXPERIMENTAL

The mass spectrometer employed in these studies is a 12-inch radius, 60° sector-field, electron impact instrument equipped with a conventional source and ion gun when measuring appearance potentials. When translational

† Present address: Department of Chemistry, Michigan State University, East Lansing, Michigan 48823.

energies were measured a shortened ion gun was employed and the ion beam was swept across the collector slit by a variable electric field applied between two deflection plates mounted just inside the analyser. ε_i was obtained from the deflection data by means of the equation

$$\varepsilon_i = \frac{U_d{}^2}{K^2} \qquad (2)$$

where ε_i is the translational energy of the ion, U_d is the potential across the deflection plates and K is a constant obtained by a standard calibration procedure. The average translational energy, $\bar{\varepsilon}_i$, was obtained by graphical integration of the curve of ε_i against relative ion intensity. In all cases the translational energy was measured as close as possible to the appearance potential and in no case more than one volt above it. $\bar{\varepsilon}_t$ was obtained from $\bar{\varepsilon}_i$ by use of the equation

$$\bar{\varepsilon}_t = \frac{m_i + m_n}{m_n} \bar{\varepsilon}_i - \frac{m_i}{m_n} (\tfrac{3}{2}kT) \qquad (3)$$

where m_i and m_n are the masses of the ion and neutral fragments, respectively. The source temperature, T, was measured by means of a chromel-alumel thermocouple attached to the outer wall of the ionization chamber. As a means of evaluating our method we determined the translational energies of several ions studied by Haney and Franklin[1] with generally satisfactory results, as shown in Table I.

Appearance potentials were evaluated by the extrapolated voltage difference method of Warren.[4] Ionization efficiency curves of $N_2{}^+$ and Ar^+ were employed to calibrate the voltage scale.

The chemicals used in this study were obtained from Matheson Co., Inc., and from Matheson, Coleman and Bell and were used without further purification.

TABLE I

Comparison of Translational Energies by the Deflection Method with those of Haney and Franklin[1] by Peak Shape Analysis

	$\bar{\varepsilon}_t$, kcal/mole		α	
	Haney and Franklin	This study	Haney and Franklin	This study
$CH_3NO_2{}^+ \rightarrow CH_3{}^+ + NO_2$	4·2	5·0	0·45	0·39
cyclo-$C_3H_6{}^+ \rightarrow CH_2{}^+ + C_2H_4$	12·1	10·2	0·45	0·53
cyclo-$C_3H_6{}^+ \rightarrow C_2H_2{}^+ + CH_4$	1·6	2·8	0·45	0·23
cyclo-$C_3H_6{}^+ \rightarrow C_2H_3{}^+ + CH_3$	2·8	2·9	0·36	0·35
$C_2N_2{}^+ \rightarrow CN^+ + CN$	4·8	4·1	0·57	0·66
$C_2H_2{}^+ \rightarrow CH^+ + CH$	6·4	5·9	0·43	0·46
$CH_3CN^+ \rightarrow CH_2{}^+ + HCN$	2·8	2·4	0·50	0·59
$CH_2Cl_2{}^+ \rightarrow CH_2{}^+ + Cl_2$	8·3	8·5	0·48	0·47
$CH_3NH_2{}^+ \rightarrow CH_3{}^+ + NH_2$	5·7	4·1	0·39	0·54
$CH_3NH_2{}^+ \rightarrow NH_2{}^+ + CH_3$	2·8	3·7	0·45	0·34

TABLE II

Appearance Potentials of Ions and ΔH_f of Hydrocarbon Free Radicals and Ions (kcal/mole)

Process	$\bar{\varepsilon}_t$	AP		E^*	ΔH_0	$\Delta H_f(R)$	
		This work	Lit.			This work	Lit.
$CH_3-C_2H \rightarrow CH_3^+ + C_2H$	3·6	369	355[6]	23	346	130	112,[6] 130,[5] 114,[7] 116[8]
$1\text{-}C_4H_8 \rightarrow CH_3^+ + C_3H_5$	2·0	326	—	26	304	41	40,[10] 38,[11] 39·6 ± 1·5[12]
$(CH_3)_3C-CH=CH_2 \rightarrow CH_3^+ + C_5H_9$	2·9	354	—	61	293	19	—
$C_2H_5C\equiv CH \rightarrow CH_3^+ + C_3H_3$	4·4	348		46	302	82	75,[13] 80·7,[14] 86[15]
$(CH_3)_3C-C\equiv CH \rightarrow CH_3^+ + C_5H_7$	2·5	340		46	294	59	—
$(CH_3)_2C=CH_2 \rightarrow CH_3^+ + C_3H_5$	4·6	378		61	317	53	58[16]

RESULTS AND DISCUSSION

The purpose of this study was to determine the heats of formation of the following free radicals: C_2H, allyl, propargyl, 1,1-dimethyl allyl, 1,1-dimethyl propargyl, $CH_3C\!=\!CH_2$, CH_3NH, $(CH_3)_2N$, CH_2NH_2, $(C_2H_5)NHCH_2$ and $(C_2H_5)_2NCH_2$. All were determined by measuring the appearance potential and the translational energy of methyl ions in the reaction

$$e + CH_3R \rightarrow CH_3{}^+ + R + 2e \qquad (4)$$

The results obtained for the hydrocarbon radicals are given in Table II and those for the various nitrogen-containing radicals in Table IV.

TABLE III
Bond Strengths and Stabilization Energies (kcal/mole)

Bond	Bond strength	Stabilization energy
$n\text{-}C_3H_7\!-\!CH_3$	84	—
$CH_2\!=\!CH\!-\!CH_2\!-\!CH_3$	74	10
$C_2H\!-\!CH_2\!-\!CH_3$	75	9
$(CH_3)_3C\!-\!CH_3$	79	—
$C_2H_3\!-\!\underset{\underset{CH_3}{\vert}}{\overset{\overset{CH_3}{\vert}}{C}}\!-\!CH_3$	66	13
$C_2HC\!-\!\underset{\underset{CH_3}{\vert}}{\overset{\overset{CH_3}{\vert}}{}}CH_3$	67	12

Our value (130 kcal/mole) for $\Delta H_f(C_2H)$ agrees exactly with that of Wyatt and Stafford[5] which is the most recent and, in our opinion, the most accurate of the literature values. The previous values[6-8] are all considerably smaller and are probably less reliable. Wyatt and Stafford's measurement by a high-temperature thermodynamic method can hardly be in error by more than one or two kcal/mole and is thus much more accurate than previous methods.

Our values for the heats of formation of allyl and propargyl are in good agreement with recent high values by others. The previously accepted heat of formation of allyl of 32 kcal/mole[9] seems now to be definitely superseded by a value around 40 kcal/mole. Similarly, the more recent determinations of ΔH_f (propargyl) by Tsang[14] and by Walsh[15] are somewhat greater than the older value of Collin and Lossing.[13] Our determination is in close agreement with that of Tsang[14] using a shock tube method and is only 4 kcal/mole less than that which Walsh[15] determined from studies of the iodine catalysed

TABLE IV

Appearance Potentials of Ions and Heats of Formation of Free Radicals from Amines (kcal/mole)

Process	$\bar{\varepsilon}_t$	AP		E^*	ΔH_0	$\Delta H_f(R)$	
		This work	Lit.			This work	Lit.
$CH_3NH_2 \rightarrow CH_3^+ + NH_2$	4·1	334	339[1]	27	307	41	41,[16] 43·3 $\pm$ 3,[12] 40,[11] 47·2[18]
$(CH_3)_2NH \rightarrow CH_3^+ + CH_3NH$	3·1	341	—	33	308	43·6	37,[17] 34·3 $\pm$ 2,[11] 41·7,[12] 45·2[18]
$(CH_3)_3N \rightarrow CH_3^+ + (CH_3)_2N$	2·6	343	—	38	305	39	34,[17] 29·3 $\pm$ 2,[11] 37·4,[12] 38·2[18]
$C_2H_5NH_2 \rightarrow CH_3^+ + CH_2NH_2$	4·4	360		46	314	43	
$(C_2H_5)_2NH \rightarrow CH_3^+ + CH_2N(C_2H_5)H$	2·0	355		37	318	37	
$(C_2H_5)_3N \rightarrow CH_3^+ + CH_2N(C_2H_5)_2$	3·0	386		79	307	23	

isomerization of propyne. In view of the widely different methods employed, the agreement of Tsang's, Walsh's and our values is surprisingly good.

We also determined $\Delta H_f(CH_3—C\equiv CH_2)$ by measuring the appearance potential and translational energy of CH_3^+ from isobutene. Our value of 53 kcal/mole leads to D(RH) in propene of 100 kcal/mole. This is about 5 kcal/mole less than $D(C_2H_3—H)$ as might be expected.

Heats of formation of the dimethyl allyl and dimethyl propargyl radicals have not previously been determined. The values appear reasonable, however. Values of the $CH_3—R$ bond strengths computed from our radical heats of formation are given in Table III and show stabilization energies relative to that of the similar bond in a paraffin hydrocarbon that are in quite good agreement. This would be expected and thus lends support to our values for the dimethyl allyl and dimethyl propargyl radicals.

Our determinations of the heats of formation of the various radicals containing nitrogen are given in Table IV. Our heats of formation of CH_3NH and $(CH_3)_2N$ agree within about 2 kcal/mole with those recommended by Benson and O'Neal[12] and to the somewhat higher values of Golden et al.[18] and thus appear to be satisfactory. Further, the bond dissociation energies obtained by subtracting the ionization potential of methyl (227 kcal/mole) from the ΔH_0 in Table IV are 81 and 78, respectively, for CH_3NH and $(CH_3)_2N$ and are in the range expected.

Our value for $\Delta H_f(NH_2)$ is in fair agreement with the previously accepted value but disagrees seriously with the new higher value of Golden et al.[18] The latter determination appears to yield the most reliable value and thus our measurement seems to be erroneously low. We do not understand why our value is too low but we suspect that our appearance potential may be at fault. If our value for E^* is combined with Haney and Franklin's[1] appearance potential, $\Delta H_f(NH_2)$ would be computed as 46 kcal/mole, in good agreement with Golden et al.[18]

The heats of formation of the three radicals formed by breaking a C–C bond in each of the three ethyl amines are not very satisfactory, although there are no measurements with which they may be compared. However, if we again subtract the ionization potential of methyl from our ΔH_0, we find the strengths of the C–C bonds to be 87, 91 and 80 kcal/mole, respectively, for the mono-, di- and tri-ethyl amines. The progression is certainly not the expected one. Although we cannot be certain which values are the most nearly correct ones, it seems likely that for the diethyl compound is somewhat too high and that for the triethyl compound is somewhat too low. The large number of vibrational modes in these compounds magnify any error in the translational energy excessively and thus result in either over- or under-correction of the appearance potential.

ACKNOWLEDGMENT

The authors wish to express their appreciation to the Robert A. Welch Foundation for the financial support of this work.

REFERENCES

1. Haney, M. A. and Franklin, J. L., *J. Chem. Phys.*, 1968, **48**, 4093.
2. Haney, M. A. and Franklin, J. L., *Trans. Faraday Soc.*, 1969, **65**, 1794.
3. Taubert, R., *Z. Naturforsch.*, 1964, **199**, 484; Taubert, R., *in* 'Advances in Mass Spectrometry', Vol. 1, *ed.* J. D. Waldron, Institute of Petroleum, London, 1959, p. 489; Bracher, J., Ehrhardt, H., Fuchs, R., Osberghaus, O. and Taubert, R., *in* 'Advances in Mass Spectrometry', Vol. 2, *ed.* R. M. Elliot, Institute of Petroleum, London, 1963. p. 285.
4. Warren, J. W., *Nature*, 1950, **165**, 810.
5. Wyatt, J. R. and Stafford, F. E., *J. Phys. Chem.*, 1972, **76**, 1913.
6. Coats, F. H. and Anderson, R. C., *J. Amer. Chem. Soc.*, 1957, **79**, 1340.
7. JANAF Thermodynamic Tables, *ed.* D. R. Stull (Clearing House for Federal Scientific and Technical Information; Springfield, Va.), 1968, Document No. PB-168, 370. Tables issued to Jan., 1971.
8. Tsang, W., Bauer, S. H. and Cowperthwaite, M., *J. Chem. Phys.*, 1962, **36**, 1768.
9. McDowell, C. A., Lossing, F. P., Henderson, I. H. S. and Farmer, J. B., *Can. J. Chem.*, 1956, **34**, 345.
10. Lossing, F. P., *Can. J. Chem.*, 1971, **49**, 357.
11. Kerr, J. A., *Chem. Rev.*, 1966, **66**, 465.
12. Benson, S. W. and O'Neal, H. E., 'Kinetic Data on Gas Phase Unimolecular Reactions', Vol. 21, Nat. Standards Ref. Data Service, Nat. Bureau Standards (U.S.), U.S. Dept. of Commerce, 1970.
13. Collin, J. and Lossing, F. P., *J. Amer. Chem. Soc.*, 1957, **79**, 5848.
14. Tsang, W., *Int. J. Chem. Kinetics*, 1970, **2**, 23.
15. Walsh, R., *Trans. Faraday Soc.*, 1971, **62**, 2085.
16. Franklin, J. L., Dillard, J. G., Rosenstock, H. M., Herron, J. T., Draxl, K. and Field, F. H., 'Ionization Potentials, Appearance Potentials, and Heats of Formation of Gaseous Positive Ions', Nat. Stand. Ref. Data Ser., Nat. Bur. Stand. (U.S.), Vol. 26, U.S. Dept. of Commerce, 1969.
17. Gowenlock, B. G., Jones, P. P. and Majer, J. R., *Trans. Faraday Soc.*, 1961, **57**, 23.
18. Golden, D. M., Solly, R. K., Gac, N. A. and Benson, S. W., *J. Amer. Chem. Soc.*, 1972, **94**, 363.

Discussion

J. Eland (Oxford University, U.K.): Have you developed a method of transforming from the deflection curves to the kinetic energy release distribution? Such a method would also be useful in interpreting TOF peak shapes, which is a closely related problem.

If not, does your derivation of average translational energy releases involve the assumption of a particular form of energy release distribution?

J. L. Franklin: Most of our data gave distributions that are nearly Gaussian. However, some do not and, as a consequence, we plot our deflection curves in terms of energy and use graphical integration to obtain the average energies. In some instances, this results in deviations from the Gaussian results of as much as 20–30 per cent.

With the time-of-flight instrument, when employing polyatomic molecules, we have normally found our distributions to be quite close to Gaussian. In a few instances, we have performed graphical integrations of the time-of-flight data and compared it to the values obtained from the peak width at half height and obtained very close agreement. There are exceptions to this. With certain diatomics, we have found some ions to be formed with very large translational energies. For example, the N^{++} ion from N_2 has a translational energy of 5 volts. These curves are in no sense Gaussian and must be treated in a quite different fashion.

111
Evaluation and Application of Energy Deposition Functions in Electron-Impact Mass Spectrometry

By G. G. MEISELS, R. H. EMMEL

(*Department of Chemistry, University of Houston, Houston, Texas, U.S.A.*)

S. E. SCHEPPELE, R. K. MITCHUM,†
K. F. KINNEBERG and J. H. DRAEGER

(*Departments of Chemistry and Biochemistry, Oklahoma State University, Stillwater, Oklahoma, U.S.A.*)

INTRODUCTION

APPLICATION of quasi-equilibrium theory to the interpretation and prediction of mass spectra produced by 70-eV electrons requires knowledge of both unimolecular dissociation rate constants and their dependence on internal energies and structure, and of the internal energy distribution function present after electron impact.[1] The nature of the latter has not been extensively evaluated, nor has its dependence on molecular structure and temperature. Internal energy distribution functions are not readily measureable directly. It is commonly assumed that they can be estimated by some combination of the energy deposition function, the energy in excess of the ionization potential which is transferred to the molecular ion by electron impact, and the internal (thermal) energy distribution of the precursor neutral molecule. This approach assumes that the *measured* energy deposition function is not itself dependent on gas temperature. In this communication we address ourselves primarily to the veracity of this statement.

Among the possible methods of estimating energy deposition functions other than the direct measurement by two-electron coincidence spectrometry,[2] those based on the optical approximation appear superior.[1,3] Measurements required for their application based on threshold photoelectron–photoion coincidence mass spectrometry[4] are underway in our laboratory, but they do not readily lend themselves to the study of complex molecules which are of interest for other reasons,[5] nor to an investigation of temperature effects. Consequently, we have chosen to employ an older and less attractive albeit

† Presently at the University of Houston.

better developed method, the second derivative technique,[6] in an attempt to determine the energy deposition function, $P(ED)$, applicable to the 70-V mass spectrum of 1,2-diphenylethane-1,2-dione (I) and to evaluate the effect of ring substituents on $P(ED)$ for (I). The molecules studied to date are (I), the 4-methyl (II), the 4-methoxy (III) and the 3-trifluoromethyl (IV) homologues:

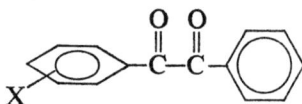

$$X = H(I),\ 4\text{-}CH_3(II),\ 4\text{-}CH_3O(III),\ 3\text{-}CF_3(IV)$$

The assumption that near threshold the cross-section for single ionization to a given state by electron impact increases linearly with excess energy and is proportional to the relative transition probabilities to that state leads to the result that the second derivative of the total ionization efficiency (SDIE) curve corresponds to the band envelope of these relative transition probabilities.[6a] Although a linear threshold law may not attain in all instances—especially for energies considerably in excess of the threshold value—the technique should as a first approximation yield reasonable estimates of $P(ED)$ as long as the transition probability curves for individual states have similar shapes.[1,6b] Autoionization which could introduce some error is expected as a first approximation to affect somewhat the shape of the total SDIE curve rather than its total area.[6b] The requirement that the SDIE curves be obtained for all ions represents a pragmatic problem. To circumvent this difficulty the 1,2-diphenylethane-1,2-dione system was chosen for study since the 70-eV mass spectra of (I) and (II) are dominated by and below 20 eV consist essentially only of ions formed by competitive rupture of the C-1–C-2 bond of their respective molecular ions followed by loss of CO from these primary fragment ions. Subject to the following modifications this kinetic scheme is applicable to the mass spectra of (III) and (IV): (a) the 4-methoxybenzoyl ion formed from the molecular ion of (III) fragments with loss of CO to form m/e 107 which in turn loses H_2CO to form m/e 77 and (b) cleavage of a C–F bond is competitive with rupture of the C-1–C-2 bond in the molecular ion of (IV).

EXPERIMENTAL

The experimental method[5] is essentially that described by Meisels and co-workers.[7] Measurements were made on an LKB-9000 mass spectrometer. To qualitatively evaluate the effect of thermal energy of the neutral on $P(ED)$ data were taken at two ion source temperatures.

Samples were introduced into the ion source from the gas chromatograph oven via the molecular separators. These two temperatures were respectively 123° and 155° for (I), 135° and 150° for (II), 152° and 155° for (III), and 113° and 150° for (IV). Considering the dimensions of the inlet system the mean molecular temperature at the ion source entrance should be similar at a given ion source temperature for these compounds. The high ion source temperature was 310°. The low temperature was 230° for (II) and 250° for the other compounds. Studies on (I) at 310° are currently in progress.

RESULTS

The $P(\text{ED})$ functions applicable to the 70-eV mass spectra of (II) at ion source temperatures of 230° and 310° are reproduced in Figs. 1 and 2 respectively as representative examples. These functions were obtained by summing the SDIE curves for the molecular ion of (II), m/e 224, the 4-methylbenzoyl ion, m/e 119, the C_7H_7 ion, m/e 91, and the C_6H_5 ion, m/e 77.

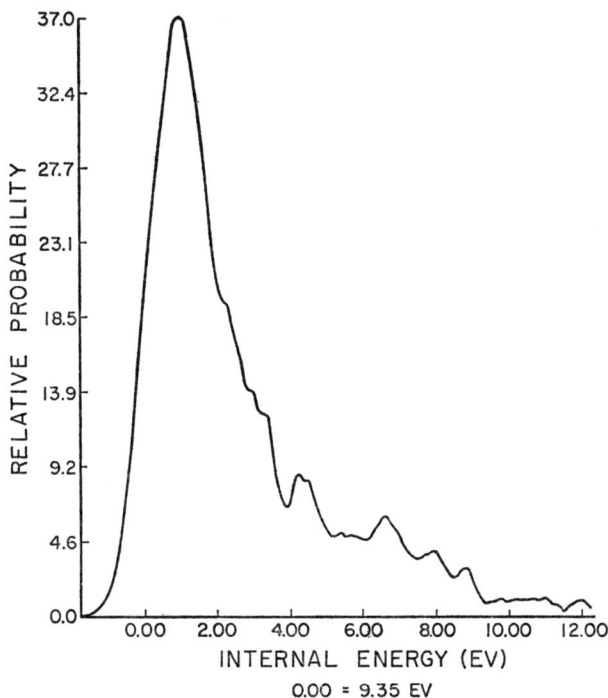

FIG. 1   Energy deposition function applicable to the 70 eV mass spectrum of (II) at an ion source temperature of 230°.

These ions account for 79·9 and 75·6% of the total ion abundance in the 70-eV mass spectra at ion source temperatures of 230° and 310° respectively. The ions investigated in the mass spectra of (III) and (IV) at 310° and of (I) at 250° account for 82·3, 88·2 and 85·8% of the total ion abundance at 70 eV respectively. Thus, even if the second derivative technique yielded the exact energy deposition functions, these $P(\text{ED})$ functions are incomplete for 70-V electrons owing to the neglect of processes occurring at ionizing energies of about 19–20 eV or more.

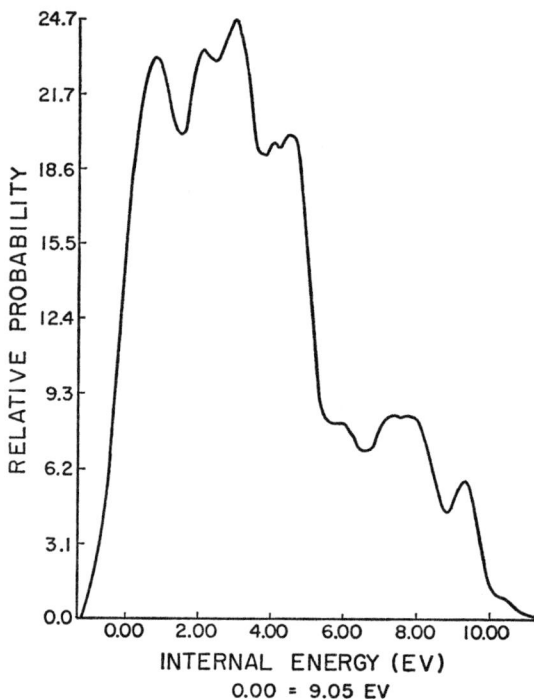

FIG. 2 Energy deposition function applicable to the 70 eV mass spectrum of (II) at an ion source temperature of 310°.

DISCUSSION

The $P(ED)$ functions for (I) through (IV) demonstrate that molecular ions are formed with considerable excess energy, e.g. at an ion source temperature of 250° (230°) they are produced with 2 to 3 eV of internal energy. A similar result was deduced for a series of ring-substituted 1,2-diphenylethanes based on photoelectron spectra.[8] The molecular ions from these compounds are formed with a wider range of internal energies than are those from ethane,[9] propane and butane.[6b,9] Although data are available for only a limited number of compounds, it appears that the excitation probability is generally greater for aromatic compounds than for aliphatic hydrocarbons.

As seen in Figs. 1 and 2 the thermal energy of (II) is reflected in the distribution of internal energies of the molecular ions; as the average temperature of (II) is increased the probability of forming higher energy molecular ions increases relative to the lower energy ones. A similar result attains for (III). However, for the trifluoromethyl homologue the $P(ED)$ function is altered negligibly upon increasing the ion source temperature. These results are in agreement with earlier evidence[5,10] that the thermal energy of the neutral must be considered in a detailed analysis (prediction) of mass spectra.

The ratio of the area under the $P(ED)$ function at the lower ion source temperature to the area under the one at the higher ion source temperature is 0·93 for (II), 0·94 for (III), and 1·08 for (IV). These deviations from unity may not be outside experimental uncertainty. If they are real, they may be the result of changes in the weighting factors of the band envelopes for the transition probabilities from vibronic states of the neutral to states in the molecular ion with changes in the thermal energy of the neutral. However, changes in conformational populations with change in temperature for these compounds (*vide infra*) also have to be considered.

TABLE I

Contribution of Individual Ions to the Area under the Energy Deposition Functions

Compound	m/e	Ion type	Percent contribution to P(ED)	
			250°	310°
I	210	SM[a]	2·9	
	105	PF	78·6	
	77	SF	18·5	
II	224	SM	2·6[b]	0·8
	119	PF	72·2[b]	66·8
	91	SF	11·0[b]	17·2
	105	PF	8·8[b]	11·2
	77	SF	5·4[b]	4·0
III	240	SM	1·0	0·3
	135	PF	83·4	78·5
	107	SF	4·9	6·4
	105	PF	2·9	4·3
	77	SF	7·8[c]	10·5
IV	278	SM	0·4	0·2
	259	PF	0·4	0·6
	173	PF	0·4	0·6
	145	SF	6·1	4·7
	105	PF	67·0	72·3
	77	SF	21·2	17·7

[a] SM, PF and SF represent stable molecular, primary fragment, and secondary fragment ions respectively.
[b] Ion-source temperature 230°.
[c] Preliminary data for (III)-d$_5$ indicate that CO loss from m/e 107 and from m/e 110 contributes 7·8 and 4·2% to P(ED). The value of their sum (12·0) is in fair agreement with the value 7·8 for (III).

The effect of thermal energy of the neutral can also be considered in terms of the percentage contribution to the area under $P(ED)$ from molecular ions which (*a*) are stable, (*b*) fragment to produce stable primary ions, and (*c*) react to produce secondary fragment ions. As seen in Table I the contribution of stable molecular ions, as expected, decreases with an increase in the thermal energy of the neutrals.

The effect of thermal energy of the neutral on the fragmenting molecular ions is seen to be complex. Increasing the molecule's thermal energy decreases the percentage of molecular ions which fragment to produce stable sub-stituted benzoyl ions, *i.e.* m/e 119, 135 and 173. Alternatively, the percentage of molecular ions which fragment to produce stable benzoyl ions, m/e 105, increases with increasing molecular temperature. However, increasing the molecular temperature increases the fraction of molecular ions which frag-ment by competitive rupture of the C-1–C-2 bond. The corresponding variation is also observed for those molecular ions from (IV) which fragment via loss of fluorine.

Except for (IV) the percentage of molecular ions accounted for by fragmen-tation of the substituted benzoyl ions increases with increasing ion source temperature. For (II) and (IV) but apparently not for (III), the fraction of molecular ions with sufficient excess energy to fragment to produce stable ions of mass 77 is seen to decrease with increasing source temperature.

However, it may well be possible[5,11] that the threshold behaviour for the 77 ion is non-linear and that for (II) and (IV) the deviation from linearity increases with increasing temperature. A complicating factor for (III) may be that m/e 77 is formed from both m/e 107 via m/e 135 and from m/e 105. This point is under investigation.

Finally, the SDIE curves for the fragment ions broaden with an increase in temperature. This result is at least qualitatively consistent with the expected existence of a broader range of internal energies at each nominal deposition energy.

We believe that these observations are caused in part by the temperature dependence of the rotamer populations on the hypersurface of these mole-cules.[12] This phenomenon would clearly complicate interpretation of the observed thermal effects and its importance is to some extent corroborated by an analysis of the temperature dependence of the UV and NMR spectra.

ACKNOWLEDGMENTS

This investigation was supported in part by the Robert A. Welch Foundation and in part by the United States Atomic Energy Commission at the University of Houston, and by the Oklahoma State University Research Foundation. We are sincerely grateful for this assistance.

REFERENCES

1. Meisels, G. G., Chen, C. T., Giessner, B. G. and Emmel, R. H., *J. Chem. Phys.*, 1972, **56**, 793.
2. Ehrhardt, H., *Adv. in Mass Spectrometry*, 1971, **5**, 81.
3. Meisels, G. G. and Emmel, R. H., *Int. J. Mass Spectrom. Ion Phys.*, 1973, **11**, 455.
4. Stockbauer, R., *J. Chem. Phys.*, 1973, **58**, 3800.
5. Scheppele, S. E., Mitchum, R. K., Kinneberg, K. F., Meisels, G. G. and Emmel, R. H., *J. Amer. Chem. Soc.*, in press.
6. (a) Morrison, J. D., *Revs. Pure Appl. Chem.*, 1955, **5**, 22; (b) Chupka, W. A. and Kaminsky, M., *J. Chem. Phys.*, 1961, **35**, 1991.
7. Meisels, G. G., Park, J. Y. and Giessner, B. G., *J. Amer. Chem. Soc.*, 1970, **92**, 254.
8. McLafferty, F. W., Wacks, T., Lifshitz, C., Innorta, O. and Irving, P., *J. Amer. Chem. Soc.*, 1970, **92**, 6867.

9. Chupka, W. A. and Berkowitz, J., *J. Chem. Phys.*, 1967, **47**, 2921.
10. Chupka, W. A., *J. Chem. Phys.*, 1959, **30**, 191.
11. Mitchum, R. K., Ph.D. thesis, Oklahoma State University, 1973.
12. For leading references pertaining to and a discussion of conformation effects in 1,2-dicarbonyl compounds, see Reference 5.

Discussion

M. T. Bowers (University of California, Santa Barbara, California, U.S.A.): Have you looked at total ion current second derivatives and compared with the sum of the second derivatives of the individual fragment ions? It seems to me such measurements are very useful for molecules with a large number of fragment ions.

G. G. Meisels: We have done so for only one other compound because the time constant of the total ion current monitor is too long to permit convenient measurements. In that instance there was good agreement. However the relative contributions of the different ions to the energy deposition functions do not change greatly (Table I) and discrimination such as differences in collection efficiencies cannot be large enough to account for shifts in $P(ED)$ as large as that shown in Figs. 1 and 2.

112
Vibrational Anharmonicity and the Fragmentation of Ions

By Z. PRÁŠIL

(*Institute of Research, Production and Application of Radioisotopes, Prague, Czechoslovakia*)

LE-KHAC HUY and W. FORST

(*Department of Chemistry, Université Laval, Québec, Canada*)

IONS prepared by charge exchange, and by electron or photon impact, are often very highly excited as evidenced by their decay which proceeds by several parallel high-energy channels. As long as the decay of the excited species proceeds from some well-defined bound state and not an electronically repulsive state, the decay should be calculable by the statistical (or quasi-equilibrium) theory of mass spectra. The theory asserts that, at a given energy, any internal state of the fragmenting system is just as likely as another (subject to certain restrictions), and this assumption (characteristic of a microcanonical ensemble) then leads to the appearance of a density of states term in the microcanonical rate constant.[1a]

Vibrational anharmonicity of the fragmenting system enters into the calculated rate constant in two ways: one implicit and the other explicit. Implicitly, it is the basis of the otherwise unspecified mechanism responsible for making 'one state as likely as another', *i.e.* for the statistical redistribution of energy among the internal degrees of freedom of the system, while the explicit involvement of vibrational anharmonicity concerns the way it affects the number or density of states.[1b]

This communication addresses itself to the explicit effect; explicit because an appropriate parameter (the anharmonicity constant) must be specified for each degree of freedom. Relative to the usual independent harmonic oscillator representation, the anharmonic case presents two complications:[2,3] (i) the spacing of vibrational levels decreases with increasing energy, causing the anharmonic integrated density of states (see below for definition) to be higher than its harmonic counterpart when total energy E is not too high, but (ii) when E becomes high enough, the anharmonic integrated density tends to a constant value for there are no more bound states to accommodate the rising energy. The reason is that the number of bound energy levels of an anharmonic oscillator is finite, as it should be for a real molecular vibration, whereas the harmonic oscillator 'never dissociates', *i.e.* contains an infinite number of energy levels, and therefore the harmonic integrated density of states is a monotonically increasing function of energy.

The most general method available for the calculation of densities of states is one based on the inversion of the partition function. We have[1c]

$$I_k[N(E)] = \mathscr{L}^{-1}\left\{\frac{Q(s)}{s^k}\right\} = \frac{1}{2\pi i}\int_{c-i\infty}^{c+i\infty}\frac{Q(s)\exp{(sE)}\,ds}{s^k} \tag{1}$$

where $I_k[N(E)]$ is the kth integral of $N(E)$, $Q(s)$ is the partition function for the species in question, written as a function of the transform parameter $s = 1/kT$ (k = Boltzmann's constant, T = temperature), and $\mathscr{L}^{-1}\{\ \}$ represents the inverse Laplace transform, given explicitly in terms of the inversion integral on the right of eqn (1). For $k = 0$ the inversion procedure leads to the density of states $N(E)$; for $k = 1$, the inversion procedure leads to $\int_0^E N(E)\,dE = G(E)$, the integrated density, and for $k = 2$ the inversion procedures leads to $\int_0^E G(E)\,dE$, the second integral of the density that appears in the calculations of the average translational energy of fragment ions, for instance.[1d,4]

Evaluation of the inversion integral in eqn (2) is best done by the method of steepest descents. We have described the method a number of times[1c,5,6,7] and therefore we give only the result:

$$I_k[N(E)] = \frac{Q(\theta)}{(\ln\theta^{-1})^k\theta^E[2\pi\theta^2\varphi''(\theta)]^{\frac{1}{2}}} \tag{2}$$

Here, $\varphi(z) = \ln{[Q(z)]} - k\ln{(\ln z^{-1})} - E\ln z$, $z = \exp{(-s)}$, θ is the value of z which is the solution of $-z\varphi'(z) = 0$, and prime and double prime signify first and second derivatives, respectively. The calculation of $I_k[N(E)]$ therefore involves only the partition function, its logarithm and logarithmic derivatives, and solution of the equation for θ.

We shall represent the reactant species by a collection of n independent Morse oscillators. The energy levels of the ith Morse oscillator, in excess of ground state, are given by

$$\varepsilon_i = \alpha_i[v_i(1 - x_i) - v_i^2 x_i] \qquad v_i = 0, 1, \ldots, m_i$$

where $\alpha_i = hc\omega_i$, ω_i being the normal frequency of the ith oscillator in cm^{-1}, x_i is its anharmonicity coefficient, and v_i its quantum number. The maximum value of v_i is $m_i = (x_i^{-1} - 1)/2$, so that there are $m_i + 1$ levels in all. The one-oscillator partition function is therefore

$$Q_i(z) = \sum_{v_i=0}^{m_i}\exp\left(\frac{-\varepsilon_i}{kT}\right) = \sum_{v_i=0}^{m_i}z^{\varepsilon_i} \tag{3}$$

and for n such oscillators

$$Q_n(z) = \prod_{i=1}^n Q_i(z) = \prod_{i=1}^n\left(\sum_{v_i=0}^{m_i}z^{\varepsilon_i}\right)$$

The summations over v_i cannot be obtained in closed form for arbitrary α_i and x_i, and therefore must be done term by term, which requires some machine time. $I_k[N(E)]$ of eqn (2) calculated in this fashion shall be referred to as 'Method A'.

An alternative treatment of the Morse oscillator partition function is possible,[8] originally due to Kubo.[9] If we let $u = \alpha/kT$, then (dropping the subscript i for the moment) the one-oscillator energy levels are

$$\frac{\varepsilon}{kT} = u[v(1 - x) - v^2x] = uv - uvx - uv^2x$$

Hence

$$Q(u) = \sum_{v=0}^{m} [\exp(-uv) \exp(uvx + uv^2x)] \qquad (4)$$

If we expand the second exponential to first order, $\exp(uvx + uv^2x) \approx 1 + uvx + uv^2x$, this has roughly the same effect as a cut-off on v; therefore the summation in eqn (4) can be extended to infinity without too much error:

$$Q(u) \approx \sum_{v=0}^{\infty} [\exp(-uv) \times (1 + uvx + uv^2x)] \qquad (5)$$

The advantage of (5) over (3) is that the summation over v can now be obtained in closed form, thus saving machine time compared with (3). After some manipulation, eqn (5) summed over all v and transcribed into the z-formulation becomes

$$Q(z) \approx \frac{1}{1 - z^\alpha}\left(1 - \frac{2xz^\alpha \ln z^\alpha}{(1 - z^\alpha)^2}\right)$$

which is, in essence, the harmonic partition function multiplied by a correction factor. We can further simplify the logarithm of $Q(z)$ by taking $\ln(1 - y) \approx -y$. For the ith oscillator, Q, α and x will each have subscript i, and the total partition function (or its logarithm) is then obtained in the usual way by taking a product of the $Q_i(z)$'s (or a sum of the $\ln Q_i(z)$'s). $I_k[N(E)]$ of eqn (2) calculated in this fashion shall be referred to as 'Method B'.

The two versions of the steepest-descent method for independent Morse oscillators are compared in Table I which gives the integrated density $G(E)$ of cyclopropane, for which exact results have become recently available.[10] Method B comes off rather poorly in the comparison, due in part to an additional (and unnecessary) approximation made by the authors of Reference 8: they take $k = 0$ in eqn (2), i.e. calculate $N(E)$ first, and then proceed to approximate $G(E) \approx N(E)/\ln \theta^{-1}$. This approximation can be easily avoided by the simple expedient of taking $k = 1$ in eqn (2) at the start of the calculations, as we have done. Method A (with $k = 1$) is seen to give essentially the exact count (within about 2%), a result we have obtained before on another system.[3] This establishes Method A as a standard against which the other methods may be compared.

Cyclopropane is a rather large molecule which is not suitable for investigating the high-energy behaviour of the density of states, i.e. at energies high enough to dissociate some or all of the oscillators in the molecule. We have used previously[3] a 'small molecule' model where the energy required to

TABLE I

Comparison of Anharmonic $G(E)$ for Cyclopropane[a]

Energy (kcal/mole)	Exact[b]	Method A[c]	Method B[d]
10	8·17 (2)	8·29 (2)	9·25 (2)
20	9·29 (4)	9·33 (4)	1·13 (5)
30	3·45 (6)	3·50 (6)	4·53 (6)
50	9·20 (8)	9·30 (8)	1·35 (9)
100	1·21 (13)	1·22 (13)	2·19 (13)
150	8·66 (15)	8·72 (15)	1·18 (16)
200	1·45 (18)	1·47 (18)	3·35 (18)
	(Multiplier in powers of ten in parentheses)		

[a] Vibrational frequencies and anharmonicity constants as in Reference 2.
[b] Reference 9.
[c] This work.
[d] Reference 8, Table III.

dissociate all oscillators is a relatively modest $123 \times 10^3 \, cm^{-1}$ (352 kcal/mole). Figure 1 shows the energy dependence of $G(E)(Morse)/G(E)(harmonic)$, *i.e.* the anharmonicity correction factor, for this 'small molecule'. Curve (a) represents Method A and shows the expected behaviour: it starts at unity at $E = 0$, goes through a maximum, and then declines asymptotically to zero, since $G(E)(harmonic)$ increases monotonically with energy, while $G(E)(Morse)$ reaches a constant value. Curve (c) represents Method B; it behaves correctly up to the maximum, but then starts to diverge, which shows that extending the summation in eqn (5) to infinity begins to introduce an appreciable error at high energies. Nevertheless Method B is acceptable at lower energies if

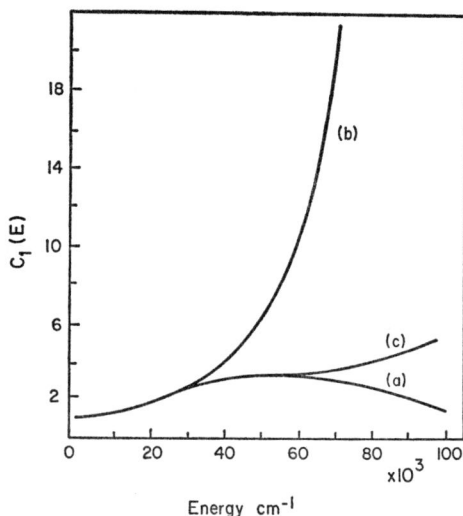

FIG. 1 Anharmonicity correction factor $C_1(E) = G(E)(Morse)/G(E)(harmonic)$ as a function of energy for a 'small molecule' model. (a) Method A; (b) Method due to Haarhoff; (c) Method B. Energy necessary to dissociate all oscillators is $125 \times 10^3 \, cm^{-1}$ in this case.

economy of machine time is the prime consideration and high accuracy is unimportant. Curve (b) represents an earlier formula due to Haarhoff;[11] it is seen to give a gross overestimate of anharmonicity at essentially all energies.

To investigate the effect of the various density of states approximations on the energy dependence of the calculated microcanonical rate constant, we use as model reactions two decomposition channels of the ethane ion, $C_2H_6^+ \rightarrow C_2H_5^+ + H$ (channel 1) and $C_2H_6^+ \rightarrow CH_3^+ + CH_3$ (channel 2). The microcanonical rate constant for decomposition into channel i at total parent energy E is given by[1b]

$$k_i(E) = \frac{\alpha_i G_i^*(E - E_{0i})}{hN(E)}$$

$N(E)$ is the density of states of $C_2H_6^+$; we have assumed that the vibrational frequencies of the ion were the same as those of the neutral ground state molecule. These frequencies, as well as all other parameters were given previously.[12] $G_i^*(E - E_{0i})$ is the integrated density of the transition state for the appropriate reaction channel. For simplicity, and to avoid ambiguities in the assignment of transition state parameters, we have assumed that the transition state for every channel is a species in every respect similar to the reactant $C_2H_6^+$ except that one molecular vibration is 'missing', i.e. has

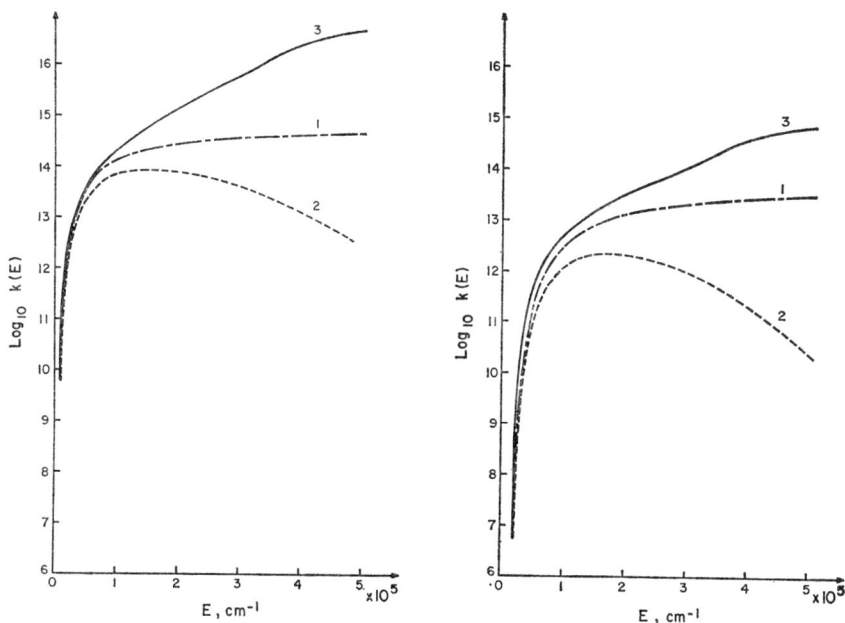

FIG. 2 Microcanonical rate constant $k(E)$ as a function of energy for two decomposition channels of the ethane ion.

Left: $C_2H_6^+ \rightarrow C_2H_5^+ + H$ (channel 1)
Right: $C_2H_6^+ \rightarrow CH_3^+ + CH_3$ (channel 2)

1: harmonic, 2: anharmonic, Haarhoff's method; 3: anharmonic, Method A. Energy necessary to dissociate all oscillators in $C_2H_6^+$ is $1\cdot5 \times 10^5$ cm^{-1}.

become the reaction co-ordinate. The reaction co-ordinate is one C–H stretch at 3175 cm^{-1} for channel 1, and a C–C stretch at 1016 cm^{-1} for channel 2.

The results are shown in Fig. 2. Curve 1 is the rate constant calculated on the harmonic model; it reaches a constant value at high energy, the expected classical behaviour. Curve 2 represents the harmonic $k(E)$ corrected for anharmonicity using the formula due to Haarhoff. It shows a high-energy *decline* in the rate constant, which we have noticed before,[12] but which, in view of the present results of Fig. 1, is entirely spurious. This is one case where a correction makes things worse. Curve 3 is calculated by Method A and can be considered as essentially exact for the assumed models. Method B would give results practically indistinguishable from Curve 3 up to about $1·5 \times 10^5$ cm^{-1} (on account of the logarithmic $k(E)$ scale), which makes the method quite adequate for the calculation of breakdown graphs, for instance, where only relative rate constants are involved. At higher energies, however, the expense of the summations in eqn (3) of Method A cannot be avoided if reasonably accurate rate constants are desirable.

REFERENCES

1. Forst, W., 'Theory of Unimolecular Reactions', Academic Press, New York, 1973. (a) p. 258ff. (b) p. 46ff. (c) p. 106ff. (d) p. 327ff.
2. Forst, W. and Prasil, Z., *J. Chem. Phys.*, 1970, **53**, 3065.
3. Huy, Le-Khac, Forst, W. and Prasil, Z., *Chem. Phys. Lett.*, 1971, **9**, 476.
4. Lau, K. H. and Lin, S. H., *J. Phys. Chem.*, 1971, **75**, 2458.
5. Forst, W. and Prasil, Z., *J. Chem. Phys.*, 1969, **51**, 3006.
6. Prasil, Z., *Advan. Mass Spectrom.*, 1971, **5**, 53.
7. Forst, W., *Chem. Rev.*, 1971, **71**, 339.
8. Hoare, M. R. and Ruijgrok, T. W., *J. Chem. Phys.*, 1970, **52**, 113.
9. Kubo, R., 'Statistical Mechanics', North Holland Publishing, Amsterdam, 1965, p. 218.
10. Stein, S. E. and Rabinovitch, B. S., *J. Chem. Phys.*, 1973, **38**, 2438.
11. Haarhoff, P. C., *Mol. Phys.*, 1963, **7**, 101.
12. Prasil, Z. and Forst, W., *J. Phys. Chem.*, 1967, **71**, 3166. (To reduce machine time in the present work, the anharmonicity coefficient of the ethane vibration at 822 cm^{-1} has been increased by a factor of 10.)

Discussion

C. E. Klots (Oak Ridge National Laboratory, Tennessee, U.S.A.): Anharmonicities are clearly important but I think the model of independent anharmonic oscillators must be poor. When the reaction coordinate becomes anharmonic, the bending vibrations start turning into rotations—*i.e.* the anharmonicities are coupled.

W. Forst: We obtain the density of states by inverting the partition function. The independent oscillator model is used here merely for convenience, since the partition function for the whole molecule is then simply a product of one-oscillator partition functions. If we knew how to write the molecular partition function with all the internal couplings taken into account, we could, in principle, obtain the corresponding density of states by the same inversion technique. However if, in addition, we wished to consider also the coupling of the *reaction coordinate* to other degrees of freedom, this would require a modification of transition state theory (see, for example, M. Jungen and J. Troe, *Ber. Bunsenges.* (1970) **74**, 276).

113
Theory of Ionic Fragmentations: Recent Developments†

By C. E. KLOTS

*(Health Physics Division, Oak Ridge National Laboratory,
Oak Ridge, Tennessee, U.S.A.)*

INTRODUCTION

SINCE its inception, the quasi-equilibrium theory (QET) of ionic fragmentations has suffered from a credibility gap. This is scarcely unique these days. What is remarkable is that this gap should be so persistent when, at last, a number of definitive experiments are suggesting strongly that the theory is often valid. Similarly a number of recent conceptual developments are serving to eliminate from the theory much of its remoteness.

The QET has always suffered from an aura of remoteness. In 1952, when the theory was first enunciated,[1] mass spectrometry was an exotic art, seemingly unrelated to the usual province of the chemist. For example, that same year a formulation of the kinetics of thermal decompositions of molecules was published,[2] yet it seems to have been several years before the two treatments were generally recognized as equivalent. Evidently, even in 1952, chemists were too specialized. Today the mass spectrometer is an accepted research tool and the grounds for this remoteness have largely disappeared.

Nevertheless a number of obstacles, often manufactured by QET practitioners themselves, have detracted from the theory's effectiveness, and served to cast doubt on its usefulness. We propose to discuss several of these obstacles and indicate the progress being made in effecting their removal.

ENERGY-CONTENT DISTRIBUTION

The earliest applications of the QET were in calculations of mass spectrometer cracking patterns. In order to do the calculations it was necessary to guess at the internal energy distributions of the molecular ions formed by electron impact. The distributions required in order to fit experiment, even approximately, were very broad—often implying tens of electron volts of energy

† Research sponsored by the U.S. Atomic Energy Commission under contract with Union Carbide Corporation.

residing in the parent ion. These distributions seemed unreasonable, and indeed were demonstrated to be so.[3]

A reconciliation was soon effected. The traditional QET expression for the first-order dissociation rate constant of a species with total energy $(E + E_0)$ is

$$k(E + E_0) = \frac{1}{h} \frac{\displaystyle\sum_{x=0}^{x=E} g_i^{\ddagger}(x)}{\rho(E + E_0)} \tag{1}$$

where $\rho(E + E_0)$ is the density of states of the parent species; the numerator is a sum over the degeneracy factors $g_i^{\ddagger}$ of the accessible levels of the transition states of energy x. In evaluating these expressions, some very poor arithmetical approximations were being used. After this had been recognized,[4] and better arithmetic employed, the inordinately broad internal energy distributions were no longer needed to account for the observed cracking patterns.

This experience, nevertheless, left something of a bad taste. It re-emphasized the fact that internal energy distributions had to be guessed at, with little available in the way of guidelines. Secondly, the new methods for evaluating densities of states and sums of degeneracies led to expressions which were imposingly complex and uneasily assimilated. The frequent use of computers in their evaluation did not ameliorate this remoteness. In any case, the calculation of a mass spectral cracking pattern does not pose a stringent test of the quasi-equilibrium theory. The necessary averaging over an assumed, or even well-known, internal energy distribution renders less convincing whatever apparent success might be achieved.

This source of ambiguity no longer exists. Three distinct techniques now exist for preparing ionic species with a reasonably well-defined internal energy. These techniques of (1) charge exchange, (2) monoenergetic electron attachment, and (3) photo-electron coincidence measurements have already provided a considerable bulk of largely still-unassimilated data. Likewise, the appearance[5] of a simple, transparent, and accurate method for obtaining densities of states dispels any remoteness associated with the calculations themselves.

PROPERTIES OF THE TRANSITION STATE

The transition state is a surface in phase space, passage through which is viewed as constituting a chemical reaction. It is, accordingly, quite unsusceptible to experimental study outside of the context of the chemical reaction itself. Spectroscopic examination, for example, is quite unavailable. Yet it is precisely such spectroscopically-defined parameters as the normal co-ordinate frequencies which are necessary for an evaluation of eqn (1).

There is yet another difficulty inherent in the transition-state formulation of chemical kinetics. The formalism may very well describe the equilibrium flux through the transition state. It has nevertheless long been recognized that to go from this to the *net* flux requires the inclusion of additional transmission factors. In the present context, for example, eqn (1) should be written rather as

$$k = \frac{1}{h} \frac{\sum\limits_{x=0}^{x=E} g_i^{\ddagger}(x)}{\rho(E + E_0)} \cdot \left(\frac{k_1 \cdot k_{12}}{N_1}\right) \tag{2}$$

where the additional factors are defined as follows:

$k_1 =$ the fraction of the instantaneous flux genuinely originating from the reactant(s).

$k_{12} =$ the fraction of these forward crossings which ultimately proceed to products.

$N_1 =$ the average number of forward crossings per genuine trajectory.

The necessity for inclusion of the third of these components seems not to have been hitherto recognized. A role for all three components may nevertheless be discerned in some recent model-trajectory computations.[6]

From considerations of microscopic reversibility one can derive the relations

$$\frac{k_1 k_{12}}{N_1} = (k_1 + k_2 - 1) = \frac{k_2 k_{21}}{N_2} \tag{3}$$

where k_2, k_{21} and N_2 are the parameters, analogous to those defined above, but pertaining to the reverse of the reaction under consideration. These two relations do not suffice to determine the six parameters involved. The *a priori* evaluation of transmission coefficients of a chemical reaction remains then an unsolved problem.

Quite apart from these several difficulties one encounters in applying the transition-state formalism to unimolecular decompositions, there is an important defect, intrinsic to the formalism, which is fatal for many purposes. Although the theory might serve to predict a rate constant, it is mute on the ultimate distribution of excess energy among the several degrees of freedom of the products. This is a topic of great current interest, and, indeed, is of great importance in mass spectrometry itself where the extent of secondary decompositions will depend critically on the disposition of excess energy from the primary fragmentation. The difficulty was recognized several years ago, in an early attempt to calculate the average kinetic energies of the separating fragments.[7] It was recognized then that an auxiliary assumption, arising from outside the theory itself, was needed for this purpose.

REFORMULATION OF THE THEORY

It has been recognized for a long time that the rate constant for a chemical reaction could be obtained directly from that of the reverse reaction together with the equilibrium constant. The only stipulation is that the reactants be, in each case, in thermal equilibrium. The quasi-equilibrium hypothesis constitutes the analogous stipulation with respect to unimolecular decompositions. Within the framework of this hypothesis it should then be possible to derive fragmentation rate constants from the details of the bimolecular association reaction. This is just what Rice[8] recognized some years ago, when he noted that fast bimolecular association reactions imply large frequency factors for the reverse dissociation.

The microscopic formulation of this approach to a unimolecular rate constant is just

$$k(E + E_0) = \frac{1}{h} \frac{\sum_{x=0}^{x=E} g_i(\sigma/\pi \dot{\chi}^2)_i}{\rho(E + E_0)} \tag{4}$$

where the notation is as in eqn (1), but with g_i the degeneracy factor of the separated states of internal energy x, and σ_i is the cross-section for their association to form the activated species. This formulation, and its relation to that of the transition state, has been discussed elsewhere.[9] It is clear, for example, that transmission coefficients and their attendant uncertainties do not intrude in this formulation.

In order to make use of this description one must nevertheless supply cross-sections for the bimolecular association. The simplest model for this process which one might entertain is the Langevin picture of spiralling collisions dominated by the long-range potential. This model has been investigated rather thoroughly, and very simple expressions for fragmentation rate constants obtained.[10] Although we shall have occasion, below, to observe some instances where the Langevin model is clearly inadequate, these will only serve to emphasize the central role of this model as a starting point for further developments. Thus, it has already served nicely to identify and illuminate several aspects of unimolecular decompositions. Rotational energy, for example, has always played a sort of orphan's role in discussions of chemical reactions. In 1968 Chupka[11] nevertheless demonstrated the efficiency of rotational energy in effecting the decomposition of the methane ion. The Langevin model shows that this should be quite generally true, with one minor proviso.

This single proviso is in itself of interest. In the neighbourhood of the threshold for fragmentation, molecular rotation will give rise to a centrifugal barrier which can impede decomposition. Tunnelling through this barrier can occur and, if it does, will exhibit an anomalous mass dependence.[12] This mechanism seems to account for the observed[13] intensities of metastable decompositions of methane and its isotopic variants.

The Langevin model predicts, finally, that rate constants should go through a maximum as a function of energy. Since the model has been shown to provide an upper limit to quasi-equilibrium rate constants, it is probable that this result will be retained in its subsequent refinements. While of dubious practical content, this startling prediction poses a challenge to the conventional wisdom of chemical kinetics.

DISTRIBUTION OF EXCESS ENERGY AMONG REACTION PRODUCTS

The formulation of quasi-equilibrium theory, contained in eqn (2), is cast directly in terms of the final states of the separated fragments. It follows, then, that when supplemented with some model for the association cross-sections, the distribution of the reaction products among the several vibrational and rotational states is directly obtainable. This has been investigated, so far, only within the context of the Langevin collision model. Again the results are very simple.[10]

The energy E, in excess of that required for the decomposition, can be used to define a temperature, via

$$E = \left(\frac{R+1}{2}\right) kT + \sum_v \varepsilon_v \left[\exp\left(\frac{\varepsilon_v}{kT}\right) - 1\right]^{-1} \tag{5}$$

where R is the number of rotational degrees of the products and ε_v/h the frequency of one of the final vibrational degrees of freedom. This temperature governs the dispersion of excess energy among reaction products. It makes, for example, the calculation of secondary fragmentation probabilities an especially simple matter.

The kinetic energy release in a decomposition is of much interest, being susceptible to measurement with a high degree of precision. The Langevin model predicts that the decomposition fragments will separate with a kinetic energy distribution very nearly that of a two-dimensional Boltzmann gas. The average energy is predicted to be just the kT of eqn (5).

Examples are known where the observed distributions are not of this form and where an activation energy for the association reaction is indicated.[14] This is not unexpected in complex four-centre fragmentations. Recently, however, a more remarkable type of anomaly has been noted. The fragmentation, for example, of benzonitrile

$$C_6H_5CN^+ \rightarrow C_6H_4^+ + HCN$$

is accompanied by an extremely small kinetic energy release. This, in itself, is noteworthy since the reaction, as a concerted process, would violate the Woodward–Hoffman orbital symmetry selection rules. What is especially noteworthy is that, for this reaction and several others,[15] the final kinetic temperature is clearly *less* than predicted by eqn (5). Accordingly, the temperature of the internal degrees of freedom must then be correspondingly higher.

This result is not understood, except in a very general way. It must be a reflection, via eqn (4), of the cross-sections of the bimolecular association reaction. It is thus consonant with the by-now familiar propensity of vibrational activation to expedite chemical reactions. Hence, if we are not yet able to calculate fragmentation rate constants in an *a priori* way, contact has at least been achieved with the main-stream of chemical kinetics. We may anticipate, then, a healthy symbiotic relationship between the chemistry of principle concern to this conference and that which, in 1952, seemed to comprise the historic domain of the chemist.

REFERENCES

1. Rosenstock, H. M., Wallenstein, M. B., Wahrhaltig, A. L. and Eyring, H., *Proc. Natl. Acad. Sci. U.S.*, 1952, **38**, 667.

2. Marcus, R. A. and Rice, O. K., *J. Phys. and Colloid. Chem.*, 1951, **55**, 894; Marcus, R. A., *J. Chem. Phys.*, 1952, **20**, 359.
3. Chupka, W. A. and Kaminsky, M., *J. Chem. Phys.*, 1961, **35**, 1991.
4. Rosenstock, H. M., *J. Chem. Phys.*, 1961, **34**, 2182.
5. Hoare, M. R. and Ruijgrok, Th. W., *J. Chem. Phys.*, 1970, **52**, 113; Hoare, M. R. ibid., 1970, **52**, 5695; ibid., 1971, **54**, 3058.
6. Morokuma, K. and Karplus, M., *J. Chem. Phys.*, 1971, **55**, 63.
7. Klots, C. E., *J. Chem. Phys.*, 1964, **41**, 117.
8. Rice, O. K., *J. Phys. Chem.*, 1961, **65**, 1588.
9. Klots, C. E., *J. Phys. Chem.*, 1971, **75**, 1526.
10. Klots, C. E., *Z. Naturforsch.*, 1972, **27a**, 553.
11. Chupka, W. A., *J. Chem. Phys.*, 1968, **48**, 2337.
12. Klots, C. E., *Chem. Phys. Letters*, 1971, **10**, 422; Abstracts, 19th Annual Conference on Mass Spectrometry and Allied Topics, Atlanta, 1971.
13. Ottinger, Ch., *Z. Naturforsch.*, 1965, **20a**, 1232; Hills, L. P., Vestal, M. L. and Futrell, J. H., *J. Chem. Phys.*, 1971, **54**, 3834.
14. Jones, E. G., Beynon, J. H. and Cooks, R. G., *J. Chem. Phys.*, 1972, **57**, 2652.
15. Klots, C. E., *J. Chem. Phys.*, 1973, **58**, 5364.

Discussion

J. Eland (Oxford University, U.K.): Are the distributions of kinetic energy released in fragmentation as important as we would like to think as probes of the mechanism of dissociation? How do these distributions come out in the Langevin theory?

C. E. Klots: Such measurements are very important. The Langevin model predicts a two-dimensional Boltzman distribution and deviations from this then tell a lot about the reverse reaction.

C. Lifshitz (Hebrew University, Jerusalem, Israel): In the usual formulation of the Quasi Equilibrium Theory, simple bond cleavages and rearrangement reactions are treated by using 'loose' and 'rigid' transition states, respectively. While this approach is rather arbitrary, it is appealing intuitively in that it can say something about the relative statistical weights of two such reactions. How is this problem taken into account in your development of the theory, where the statistical weights are those of the final product states?

C. E. Klots: The Langevin model is equivalent to the loose (the Eyring–Hirshfelder–Taylor) transition state picture. Deviations then imply that the transition state is tighter, or that transmission coefficients are non-unity, for example. A precise equivalence cannot be traced. Note, however, that the failure of the Langevin picture for the benzene decomposition shows that the 'transition state' for this 'simple bond cleavage' is not loose.

114

The Sociological Interpretation of Scientific Progress: The Case of Mass Spectrometry

By D. R. PANTON and B. G. REUBEN

(*Departments of Sociology and Chemistry,
University of Surrey, Guildford, U.K.*)

In the sociology of science we are faced with the problem of adequately accounting and providing an explanation for the processes of scientific advance and innovation.

Major contributions in this field have been carried out by Thomas Kuhn[1] and Robert Merton.[2]

As in many other disciplines, widely differing hypotheses are put forward to account for the various phenomena under investigation, in this case the accumulation of contributions to science which constitute scientific progress.

However, many sociologists of science have been unable to obtain useful results because they were not sufficiently conversant with scientific practice to understand and appreciate the subject of their investigations.

In other words, they have been faced with a 'Black-Box' problem, a situation where it is possible to study and measure the inputs and outputs of a system without coming to terms with the causes.

Consequently, one could argue that theories of scientific innovation are often constructed on the basis of the work of a few scientists concerning problems which in many cases are somewhat peripheral to scientific advance.

If one is to attempt to provide a more realistic explanation of scientific advance, one must try to understand the process of intellectual innovation in the scientific community which eventually appears as a formal paper in a scientific journal, and this requires not only sociological analysis but scientific background.

This work therefore is concerned with assembling and interpreting the history of Mass-Spectrometry, from its earliest days under Sir J. J. Thomson to its present-day activities, via study of the literature, interviews with scientists and questionnaires.

There are currently two or three differing theories in this area of sociology of science.

The functionalist approach[2] to the study of the social organization of research, as propounded by Merton, maintains that certified knowledge will accumulate automatically as a consequence of reasonable conformity to a set of institutional norms of behaviour which are held to be binding on the scientist and are legitimized in terms of institutional values.

The assumption here is that growth can only take place within 'open' communities, where these values are upheld—values such as 'open-mindedness' and the 'quest for truth'—and as science has undoubtedly developed more rapidly than intellectual movements, so the scientific community must be more open than other social groups.

However, in Kuhn's view,[1] it is quite possible to see scientific growth as a product of intellectual and social closure—the basis of this assertion being the view that for most of the time scientific activity consists in the 'attempt to force Nature into conceptual boxes supplied by the professional education'.

This view holds that scientists develop a strong commitment to a particular theoretical–methodological tradition and there are consequently powerful conservative forces within science working to limit the possibility and acceptance of innovation.

For Kuhn, Mertonian-type explanation adequately accounts for the activities of 'normal' science, where the existing accepted paradigms of scientific method are sufficient guides for research, and there is a common acceptance of a framework of ideas.

During these periods, there is a hostility to alternative rival theories, and scientists are not usually willing to give a non-biased evaluation of them, and Kuhn suggests that this is in the interests of efficiency, since science progresses rapidly when no-one is fretting over fundamental ideas.

A third approach to the explanation of the process of scientific discovery has used the concept of 'serendipity—the art of making happy and unexpected discoveries.

Many so-called 'accidental' discoveries such as penicillin or X-rays are extremely common in science, and sociologists have studied various situations where the following-up of a chance observation made in an experiment directed towards a different end has led to an important discovery.[17]

By reconstruction of the elucidation of a central problem in science, we hope it will be possible to see to what extent the detailed historical picture fits these rather quasi-philosophical theories of scientific development.

To take a specific case, if we look at the early history of the Mass-Spectrograph and its application to the discovery of isotopes we can see quickly that this was an area of science where a large number of incorrect results found their way into the literature.

At the same time it was a period of flux in science. The achievements of nineteenth-century science had been completed—classical thermodynamics, the 'billiard-ball' atomic theory, electromagnetic theory, etc., and as late as the eighteen-nineties many scientists considered that all the great discoveries had been made. Atomic physics, with the structure of the atom and the quantization of energy was just beginning.

One of the earliest experimental breakthroughs was Thomson's discovery of the electron, for which he was awarded one of the first Nobel prizes in 1906. By 1912 he was at the height of his fame and was established in the Cavendish chair of Physics at Cambridge. In that year too, he published the main observations made on his new 'parabola' mass analyser which he had invented.[3] Positive rays from a discharge tube were allowed to pass through a very narrow channel and were then analysed by electric and magnetic fields such that positive rays containing the ions of similar mass-to-charge ratio formed a parabola on a fluorescent screen placed in their path. This was a direct and obvious extension of his work on the electron. Ions of different m/e values fell on different parabolas. Thomson examined a wide variety of materials in this instrument, and it is important to note that the sharpness of the parabolas obtained was the first experimental proof that atoms of the same material were of even approximately similar masses.

Thomson obtained three parabolas from a sample of hydrogen, corresponding to m/e values of 1, 2 and 3. He attributed them respectively to hydrogen atoms, hydrogen molecules, and a new polymeric form of hydrogen H_3. This last he originally considered to occur 'only under certain conditions of pressure and current'[4] but later he changed his mind and suggested it was 'more stable than ozone' and could combine with both oxygen and mercury under the influence of an electric discharge.[3] He pointed out that it was not possible to reconcile the existence of this substance with the ordinary conception of valency if hydrogen is always regarded as monovalent.

Shortly afterwards, he obtained results for neon and recorded not only the expected parabola at m/e = 20, but another, fainter one at m/e = 22. This was shown not to be doubly charged CO_2 and, on the basis of Mendeleef's Periodic Law, there was no room for a new element with this atomic weight. Soddy had already proposed the idea of isotopes among the radio-elements[5] but there was a general feeling that isotopes were necessarily linked with radio-activity. Thomson originally thought the line was due to a new element[3] but later, possibly under the influence of Lindeman,[6] he preferred to consider that it was due to the compound $NeH_2{}^+$, by analogy with hydrogen.

In 1919, Dempster showed the hydrogen parabola at m/e = 3 was indeed $H_3{}^+$, that it was not a stable species, and that it was formed only under conditions in which hydrogen was dissociated.[7] Dempster noted especially the variation with pressure in the proportion of $H_3{}^+$ formed.

This result was confirmed by Aston in 1920 by accurate atomic weight measurements,[8,9] and the work of Smyth,[10] and Hogness and Lunn[11] correctly suggested that the mode of formation was via the ion-molecule reaction $H_2 + H_2{}^+ \rightarrow H_3{}^+ + H$. It was the first ion-molecule reaction of this kind to be postulated.

Thomson's views on hydrogen were thus more or less confirmed. The same could not be said for neon. In 1920 Aston reported accurate atomic weight measurements of his newly-built mass spectrograph at the Cavendish Laboratory which showed that the chemical atomic weight of neon lay between the real atomic weight of the two constituents detected on the mass spectrograph but corresponded to neither of them.[8,9] He also observed a faint line at m/e = 21 but had difficulty in reproducing it, and though he allowed of the possibility that it was due to a third isotope, he considered on balance that it was likely to be $^{20}NeH^+$. It was not until 1928 that Hogness and Kvalnes[12] repeated this experiment under conditions of greater sensitivity and pointed out that if the line at m/e = 21 were due to $^{20}NeH^+$, then there should be a corresponding line at m/e = 23 due to $^{22}NeH^+$. It being absent, they concluded that they had identified a third isotope of neon.

In spite of Aston's work, J. J. Thomson was reluctant to believe in isotopes. In 1916, he himself only obtained a single parabola from chlorine (*sic.*)[13] and in 1921 he opened a Royal Society discussion on isotopes with a detailed attack on Aston's data.[14] He suggested that the line at mass 37 in Aston's chlorine mass spectrum was due to H_2Cl^+ and that the neon hydrides were responsible for the additional lines in the neon spectrum. He pointed out quite reasonably that water gave a line at $m/e = 19$ which was generally agreed to be due to H_3O^+, and suggested, with considerable insight, that neon ions, having only seven outer electrons might well be a reactive species.

The evidence for the existence of isotopes gathered weight so that even Thomson's reputation was not sufficient to stem it, and Aston duly received a Nobel Prize in 1922. It was another 36 years before the neon hydride ion NeH^+ was identified,[15] and 41 years before Moran and Friedman[16] showed that it was formed in the reaction:

$$Ne + H_2^+ \rightarrow NeH^+ + H$$

From a sociological standpoint, this chapter in the chequered history of Mass-Spectrometry sheds a good deal of light on the theories already mentioned.

In terms of the normative behaviour of scientists—the steady-state theory of scientific advance—we can see in the attitude of Thomson to Aston something very much removed from the idea of open-mindedness.

Thomson, with all the undoubted attributes of a great man of science, strongly resisted the attempts to alter his own particular scientific theories in this respect and remained unconvinced about isotopes for some time after Aston had provided what others regarded as conclusive proof.

Yet this establishment-type attitude to young scientists and their novel theories, borne out by many people's experience would seem almost to be the rule rather than the exception, and it would therefore appear that Merton's conceptualization of the behaviour of scientists is an ideal towards which scientific practitioners only pay lip-service. In other words, what Merton would regard as deviate is in fact a perfectly acceptable form of behaviour.

Kuhn's theory of periods of normal and revolutionary science far more closely follows the actual pattern of scientific advance, and provides a fairly satisfactory account of the processes whereby new ideas are accepted into science, but he fails to adequately analyse the structural sources of innovation with respect to the generation of ideas.

Sociology of science necessarily implies a view of what scientists do and create, and therefore must examine why a scientist decides in what area to work next, and *ceteris paribus*, tackles a problem on the basis of rational criteria.

As far as theories of serendipity are concerned, the history of the discovery of isotopes was much confused by the occurrence of reactions between gaseous ions and neutral molecules. Thomson was correct in thinking that the peak at $m/e = 3$ in hydrogen was due to H_3 but wrong in attributing the $m/e = 22$ in neon to NeH_2, and so through his genius the discovery of isotopes was delayed for five years.

The majority of isotopes had been identified by the end of the 1930s—serendipity was absent to such an extent that no-one at all bothered to study ion-molecule reactions for their own inherent interest until the middle of the 1950s.

Serendipity leads as often as not into blind alleys, and many eminent and distinguished scientists, in all honour and integrity, have followed the wrong paths. The advance of science much more resembles the 'drunkards walk' than the ordered progress in which the lay-man and the University undergraduate are taught to believe.

REFERENCES

1. Kuhn, T. S., 'The Structure of Scientific Revolutions', Univ. Chicago Press, 1962.
2. Merton, R. K., 'Social Theory and Social Structure', Free Press, 1967, pp. 550–61.
3. Thomson, J. J., 'Rays of Positive Electricity', Longmans, London, 1913.
4. Thomson, J. J., *Phil Mag.*, 1912, **24**, 234.
5. Soddy, F., *Ann. Rep. Chem. Soc.*, 1910, p. 285.
6. Thomson, G. P., 'J. J. Thomson and the Cavendish Laboratory', Nelson, London, 1964.
7. Dempster, A. J., *Phil Mag.*, 1916, **31**, 438.
8. Aston, F. W., *Phil Mag.*, 1920, **39**, 449.
9. Aston, F. W., *Proc. Camb. Phil. Soc.*, 1920, **19**, 317.
10. Smyth, H. D., *Phys. Rev.*, 1925, **25**, 452.
11. Hogness, T. R. and Lunn, E. G., *Phys. Rev.*, 1925, **26**, 44.
12. Hogness, T. R. and Kvalnes, H. W., *Phys. Rev.*, 1928, **32**, 942.
13. Thomson, J. J., *Engineering*, 29 March 1918, p. 345.
14. Thomson, J. J., *Proc. Roy. Soc.*, 1921, A99, 87.
15. Stevenson, D. P. and Schissler, D. O., *J. Chem. Phys.*, 1958, **29**, 282.
16. Moran, T. and Friedman, L., *J. Chem. Phys*, 1963, **39**, 2491.
17. Barber, B. and Fox, R. C., *American J. Sociol.*, 1958, **64**, 128.

Information Theory and Mass Spectrometry

By T. L. ISENHOUR and J. B. JUSTICE

(*University of North Carolina, North Carolina, U.S.A.*)

IN this paper, we shall attempt to review some of the ways in which information theory has been applied to mass spectrometry. However, in fairness, we must note that most mass spectrometry interpretation, while often resorting to sophisticated techniques, has made very little use of information theory. This is actually quite reasonable in that information theory itself is still in a very early stage of development. Furthermore, information theory has found, thus far, only a limited range of applications; although, most who have encountered it at all, feel information theory should be a very fundamental approach to a variety of problems. The exception to this statement is in the area of communications where information theory has already made major contributions.

In this presentation we shall discuss some of the various methods of interpretation which have been applied to mass spectrometry, give a brief introduction to information theory, show a few information theory calculations on mass spectrometry, and conclude with some speculations as to possible applications of information theory to mass spectrometry.

METHODS OF INTERPRETATION

For our discussion we will describe three general methods which have been used to interpret mass spectrometry. These are not necessarily exclusive, nor parallel classifications, but serve conveniently for the purposes of this presentation.

Library Comparisons

The comparison of spectra of known compounds to unknown compounds is a general method of interpretation which is widely applied. While the comparison method is a purely empirical approach, it is perhaps the most widely used and successful method of interpretation of complex data collections in all fields. Surely the telephone directory is a better method of storing and retrieving telephone numbers (by comparison of names) than any general algorithm based on names, location, etc.

In order to make a comparison of an unknown spectrum with the known spectra in a library, some form of search must be done. Considerable effort has gone into devising storage schemes and search strategies to improve the efficiency of spectral comparison.

A great number of mass spectrometry search systems have been developed—particularly since the digital computer became widely available in the 1960s. We will not attempt to review this subject, but rather present some particular ideas which move in the direction of our interest in the application of information theory. In 1964, Tal'rose, Raznikov and Tantsyrev considered the problem of determining the minimum data sufficient to identify individual organic substances by coincidence of their mass spectral lines.[1] It was shown for 900 spectra, intensity ratios of only two mass values (39 and 41) were sufficient to establish that a given compound was one of twelve possible compounds within the library. Three ratios were shown sufficient for unequivocal identification assuming a 5% error in the measurement. The interest here is that information theory may be frequently applied to show the minimum information necessary to answer such a question. Later on we shall discuss the point of relation between the theoretical minimum information and that which is often available.

Other workers developed various search methods in which often only a few intense peaks were used, spectra were divided into 14 mass unit positions keeping one or two peaks, etc.[2-4] These references serve to show that often only a small amount of the available information is necessary for an unequivocal identification.

Grotch, doing correlation studies using only the presence or absence of peaks, showed that from a library of 3200 spectra, only 18 perfect matches existed among non-identical compounds.[5] Therefore, Grotch concluded that compressing the spectra by removing all intensity information still caused little loss in information necessary for identification from a library. Further work of this sort[6] showed that 6700 mass spectra could be further reduced by the combining of peaks for the purpose of maximizing useful information per channel. (This point will be discussed later.) Other references[7-10] cover various sophisticated techniques for identification of mass spectra by comparison methods. Finally, reference (11) offers an excellent review of compound identification methods by computer matching of mass spectra.

Quasi Equilibrium Theory

The second method of mass spectral interpretation is that of Quasi Equilibrium Theory (QET). This approach is in complete contrast to the previous discussion in that theoretical methods are used to interpret and identify the mass spectrum. QET was proposed in 1952 by Rosenstock, Wallenstein, Wahrhaftig and Eyring.[12] Examples of applications are given in references (13–15). While we shall not discuss Quasi Equilibrium Theory, it being considerably off the subject of information theory applications to mass spectrometry, it must be noted that this approach does contrast with direct comparison methods by applying known or suspected chemical and physical relations of molecules in an attempt to elucidate the actual mechanism involved in the production of the spectrum.

Computer Aided Interpretation

The third general heading in our discussion is Computer Aided Interpretation. The first section on search systems depends, of course, very heavily on computer operations. However, these are the rather standard operations of direct comparison. The methods we wish to discuss under this heading are those that call on more unusual algorithms or mathematical approaches, which must be computer implemented, rather than those in typical search systems.

One of the early such attempts was the development of Dendral.[16] Dendral incorporates rules by which all conceivable structures for a mass spectral fragment can be generated. This approach includes heuristic procedures, some of which are based on chemical knowledge, which limit the possible list size; otherwise, it is shown that lists quickly become too large to be manageable or even computed. Another heuristic approach is that of pattern recognition, which was discussed in some detail at a recent meeting also sponsored by the Institute of Petroleum.[17] Crawford and Morrison have developed *ab initio* computer programs for the interpretation of mass spectra below molecular weight 200.[18] The unknown spectrum is integrated to yield molecular mass presence of functional groups and finally a molecular skeleton. Details are stored in structure matrices.

Chapman offers a good review of computer aided interpretation of mass spectrometry through 1969.[19] Applications such as references (20–22) give examples of computer interpretation of high resolution mass spectra of peptides.

For detailed reviews of mass spectral interpretation references (23–25) should be consulted.

INFORMATION THEORY

In information theory the basic unit of information is the bit—usually represented as an 'on-off' situation or 1,0 value mathematically. A string of bits can be thought of as a message from a source to a receiver. The information transferred from sender to receiver is referred to as the bit capacity of a channel and is a direct measure of the message's descriptive ability. For example, n bits have 2^n possible combinations and, therefore, may represent 2^n possible distinguishable states or messages. Hence it is logical that information is related logarithmically to the number of possible arrangements since

$$\text{information} = \log_2 2^n = n \text{ bits} \qquad (1)$$

This relationship holds even though the number of levels may be more than 2. However, computations are simplest when the number of levels for each channel of the message is some integral power of 2. For example, in an 8 level system, given m measurements

$$\text{information} = \log_2 8^m = m \log_2 8 = 3m \text{ bits} \qquad (2)$$

and since it requires 3 bits to describe each 8 level message it is consistent to require 3m bits. (Good introductions to information theory are references (26–29).)

The unfortunate problem of information theory is that while it is frequently easy to compute the *information* of a message, it is often very difficult to determine the *meaning* of the message—*meaning* being determined by the 'usefulness' of the information in the message. Meaning, or semantic information, unfortunately, is culturally or contextually dependent even to the extent that the same message has different interpretations to different persons. Hence, information may be a directly measured property of the transmitting channel but semantic information is a function of both the transmitter and receiver, and their past experiences, interactions, etc.

For example, the telegraph company charges for messages based on their information; that is, the number of characters transmitted. Consider, however, the following pair of messages.

Message 1: ARRIVE LISBON MONDAY NOON STOP WILL YOU
JOIN ME FOR LUNCH STOP R. BURTON
Message 2: NO STOP E. TAYLOR

Now it is clear that message 1 takes more bits to transmit than message 2. However, the amount of semantic information is about the same, *i.e.*, Richard wants to have lunch with Elizabeth—Elizabeth doesn't want to have lunch with Richard. However, notice that message 2 is only significant in the context of message 1. Without some preceding message, message 2, saying only 'NO' would probably have zero semantic information.

It was, of course, necessary to the development of the theory that the measure of information used be independent of context, so that the method would be widely applicable. Hence, information theory can tell you how many bits are necessary to ensure unambiguous interpretation of a message, but not how to interpret it.

In mass spectrometry we can do some rather interesting information calculations to show what the theoretical limits of the technique are. However, we have some problem in showing the practical limitations of these calculations.

For example—how many compounds could be represented by a mass spectrum? Let's take a rather limited case of a low resolution mass spectrometer with a mass range of 200, unit mass resolution, and intensity resolution of 1 %. Since 2^7 is about 100 (actually 128) we can say each mass position has 7 bits of information. So we have about 1400 bits in a mass spectrum. That's a lot of information!

Liz can refuse lunch one day at a time for the next 3·84 years with that many bits. Furthermore $2^{1400} \approx 10^{420}$ compounds can potentially be represented by our simple low resolution mass spectrum. That's a lot of compounds!

The present number of compounds known is somewhere between 2 and 5×10^6. Furthermore, if everyone in the world (about 4×10^9 people) isolated a new molecule every month for 100 years it would only amount to 5×10^{12} compounds. (We don't recommend this, by the way. There are already enough synthetic chemists about.) But let's consider the result of reversing the calculations. If there were 5×10^{12} compounds ($5 \times 10^{12} \approx 2^{42}$) we would only need about 42 bits of information, or a mass spectrometer with unit intensity resolution (*i.e.* peak/no peak) and only 42 resolvable mass positions; that is, if every possible combination of mass positions occurred among our spectra. And there's the rub. Even with the almost daily invention of new types of mass spectrometer sources, most of us are willing to agree that most theoretically possible mass spectra will never represent a real compound. Hence the problem is to find the semantic information of the spectrum.

An estimation of the information content of a channel can be made by measuring the so called 'entropy' of the channel. The relation used (which will be derived in another context later) has the logical basis that the closer a one-bit channel is to being 'on' exactly half the time, the more things it can represent. Obviously if a given channel is always 'on' or always 'off' it carries no information. The relation of information to probable results, where H is information, and p_i the probability of a given result in a given channel (for n possible results) is

$$H = -\sum_n p_i \log_2 p_i \qquad (3)$$

Notice that since $0 < p_i < 1$ the value of H is always positive. (Values of $p_i = 0$ or $= 1$ result, of course, in zero information.)

In the binary case the relation becomes

$$H = -p \log_2 p - (1 - p) \log_2 (1 - p)$$
(where p is the probability of a peak) (4)

and is symmetric about the maximum of $p = 0.5$. In general, the relation is a maximum where all values of $p_i = (1/n)$.

In the previously mentioned search system, Wangen et al.[6] used the general relation (eqn 3) to maximize entropy in each channel, thereby minimizing the necessary number of channels for identification. From 6700 mass spectra he found that combining peaks which were highly correlated allowed a decrease from 352 to 80 mass positions while increasing the number of spectra which were identical from 451 to 556. (It should be noted that most matches involved identical compounds thereby having the result that most compression removed noise rather than useful information.)

Another interesting application of information theory to mass spectrometry is an attempt to determine how much information is available for distinguishing among compounds. This approach, using a derivation for determining the information available to distinguish between two messages, leads to the relation stated in eqns 3 and 4.

Given n equally likely outcomes of a measurement, let n_1 be from compound 1 and n_2 from compound 2. ($n = n_1 + n_2$, of course.) Now the total information from the mass spectral measurement is $\log_2 n$. The information about each compound is, however, $\log_2 n_1$ and $\log_2 n_2$ for compounds 1 and 2, respectively. Message 1 occurs at a frequency (or with a probability, p) of n_1/n, and, $p_2 = n_2/n$. Thus the information available to tell them apart is:

$$H = \log_2 n - \frac{n_1}{n} \log_2 n_1 - \frac{n_2}{n} \log_2 n_2 \qquad (5)$$

$$H = \log_2 n - p_1 \log_2 n_1 - p_2 \log_2 n_2 \qquad (6)$$

since $n_1 = p_1 n$
and $n_2 = p_2 n$

$$H = \log_2 n - p_1 \log_2 p_1 n - p_2 \log_2 p_2 n \quad (7)$$

rearranging

$$H = \log_2 n - (p_1 + p_2) \log_2 n - p_1 \log_2 p_1 - p_2 \log_2 p_2 \quad (8)$$

since $p_1 + p_2 = 1$

$$H = -p_1 \log_2 p_1 - p_2 \log_2 p_2 \quad (9)$$

This may then be generalized to give our earlier expression (eqn 3) for the entropy of a measurement.

To determine the information in a small set of mass spectra, we examined a data set of 630 mass spectra consisting of 119 mass positions with 2% intensity resolution. The maximum information possible would occur if all intensities were equally probable. This would yield an H of 640 bits. Using the actual distribution of intensities resulted in H equal to 139·8 bits. Neglecting intensity entirely (*i.e.* peak/no peak data) yields $H = 58·7$ bits. Either result is still far more than enough to represent all our compounds.

However, such calculations assume independence of the mass positions which is certainly not the case in mass spectrometry. One application, in which we are very interested, is the possibility of determining, and hence removing, redundancy from spectral representations. A realistic limit will be reached if this can be accomplished.

In conclusion one must say that to date applications of information theory to mass spectrometry and to most methods of chemical analysis have been few—one notable exception being Hadamard spectroscopy. However, perhaps this is not to remain the case. One possible application, which we have tentatively discussed, but have too little accomplished to present results at this point, is the resolution of mixtures in mass spectrometry. In spite of the incredible amount of computer treatment of mass spectra, with only a few exceptions, and usually those of well behaved cases, unknown mixtures have been avoided.

Is this an impossible problem? Perhaps information theory can tell us if that's the case. If not, perhaps information theory can lay down some limits on what may be accomplished. We hope to find out.

The support of the National Science Foundation is gratefully acknowledged.

REFERENCES

1. Tal'rose, V. L., Raznikov, V. V. and Tantsyrev, G. D., *Dokl. Akad. Nauk. S.S.S.R.*, 1964, **159**, 182.
2. Knock, B. S., Smith, I. C., Wright, D. E., Ridley, R. G. and Kelly, W., *Anal. Chem.*, 1970, **42**, 1516.
3. Hites, R. A. and Biemann, K., *Adv. Mass Spectrom.*, 1968, **4**, 37.
4. Pettersson and Ryhage, *Arkiv. Kemi.*, 1967, **26**, 293.
5. Grotch, S. L., *Anal. Chem.*, 1970, **42**, 1214.
6. Wangen, L. E., Woodward, W. S. and Isenhour, T. L., *Anal. Chem.*, 1971, **43**, 1605.
7. Unruh, G. V., Spiteller-Friedmann, M. and Spiteller, G., *Tetrahedron*, 1970, **26**, 3039.
8. Reimendal, R. and Sjöval, J., *Anal. Chem.*, 1971, **43**, 2057.
9. Smith, D. H., *Anal. Chem.*, 1972, **44**, 536.
10. Gillis, R. G., *Org. Mass Spectrom.*, 1971, **5**, 79.

11. Ridley, R. B., *in* 'Biochemical Applications of Mass Spectrometry', *ed*. G. R. Waller, 1972, chapter 6.
12. Rosenstock, H. M., Wallenstein, M. B., Wahrhaftig, A. L. and Eyring, H., *Proc. Nat. Acad. Sci. U.S.*, 1952, **38**, 667.
13. Vestal, M., *J. Chem. Phys.*, 1965, **43**, 1356.
14. Friedman, L., *J. Chem. Phys.*, 1957, **27**, 613.
15. Kropf, A., *J. Chem. Phys.*, 1960, **32**, 149.
16. Bucks, A., Delfino, A. B., Djerassi, C., Duffield, A. M., Buchanan, B. G., Feigenbaum, E. A., Lederberg, J., Schroll, G. and Sutherland, G. L., *in* 'Advances in Mass Spectrometry', *ed*. A. Quayle, Vol. V, Institute of Petroleum, London, 1971, p. 314.
17. Isenhour, T. L. and Jurs, P. C., *in* 'The Applications of Computer Techniques in Chemical Research', sponsored by the Institute of Petroleum, Manchester, England, 1971.
18. Crawford, L. R. and Morrison, J. D., *Anal. Chem.*, 1971, **43**, 1790.
19. Chapman, J. R., *Chem. Brit.*, 1969, **5**, 563.
20. Senn, M. and McLafferty, F. W., *Biochem. Biophys. Res. Commun.*, 1966, **23**, 4.
21. Biemann, K., Cone, C., Webster, B. R. and Arsenault, C. P., *J. Am. Chem. Soc.*, 1966, **88**, 5598.
22. Barber, M., Powers, P., Wallington, M. J. and Wolstenholme, W. A., *Nature*, 1966, **212**, 5064.
23. Bursey, J. T., Bursey, M. M. and Kingston, D. G. I., *Chem. Rev.*, 1973, **73**, 191.
24. 'Mass Spectrometry: A Specialist Periodical Report', Vol. 1, *ed*. D. H. Williams, The Chemical Society, London, 1971.
25. 'Biochemical Applications of Mass Spectrometry', *ed*. Waller, G. R., Wiley, New York, 1972.
26. Bell, D. A., 'Information Theory and Its Engineering Applications', Pitman Publishing Corp., New York, 1962.
27. Raisbeck, G., 'Information Theory', M.I.T. Press, 1963.
28. Brillouin, L., 'Science and Information Theory', Academic Press, New York, 1962.
29. Shannon, C. E. and Weaver, W., 'The Mathematical Theory of Communication', University of Illinois Press, 1964.

COMPUTER SOFTWARE

Chairmen

A. L. BURLINGAME
(University of California, U.S.A.)

C. MERRITT
(U.S. Army Natick Laboratories, U.S.A.)

115

Structure Determination of Organic Molecules by Statistical Learning Methods

By J. FRANZEN and H. HILLIG

(*Institut für Spektrochemie, Dortmund, W. Germany*)

LEARNING Machines, *i.e.* computers programmed to be capable of 'learning', were first applied to the determination of structures of organic molecules from their mass spectra by Jurs and co-workers.[1] The special iterative learning method chosen by this group as the learning process, however, shows some disadvantages. Firstly, the need for convergence restricts the number of spectra in the training set, *i.e.* the computer cannot learn with an arbitrarily large number of spectra. This is a principal foible of the method chosen. Secondly, the learning procedure tends to need excessive computer time and storage, and elaborate methods have to be invented to cut down time and costs. Thirdly, the whole process of learning is not very clear, not very understandable for the user, and the geometrical interpretation of dichotomizing a collection of points in a ($n + 1$)-dimensional space by a n-dimensional hyperplane—although correct—does not help to reduce the distrust of the chemist confronted with this method.

As, however, we think the original idea to be extremely promising, we developed from decision theory another type of learning method which does not show the drawbacks outlined above.

Learning machines are based, in general, on binary decisions.[2] The whole problem of structure determination, therefore, has to be split up into a set of questions which can be answered by binary decisions. In general, these questions each will ask for the presence (or absence) of distinct structural details, such as 'Does the compound contain oxygen?', or 'Is it aromatic?' For complete structure determination, a series of different binary decisions has to be performed, arranged to form a strategic 'decision tree' or 'decision network'.

The work presented here is restricted to the investigation of such binary decisions. These are performed by mathematical decision functions applied to the mass spectrum of an unknown compound which is considered as a vector V. The general form of a decision function is

$$S = f(\mathbf{V}, \kappa_i) \tag{1}$$

where the scalar S is the 'result' of the decision. The decision function must be constructed so that $S > 0$ if the structural detail is present, and $S < 0$ if the structural detail is absent. κ_i is a set of function parameters which, for each type of decision, possesses characteristic values determined during the learning process.

For the derivation of a special decision function, let us consider, in concretum, a distinct binary decision, say the presence of at least one aromatic ring within the structure. Given is a large set of known spectra which can be divided into two classes, one containing all aromatic, and the other all non-aromatic compounds. We now draw our attention to the intensities of a

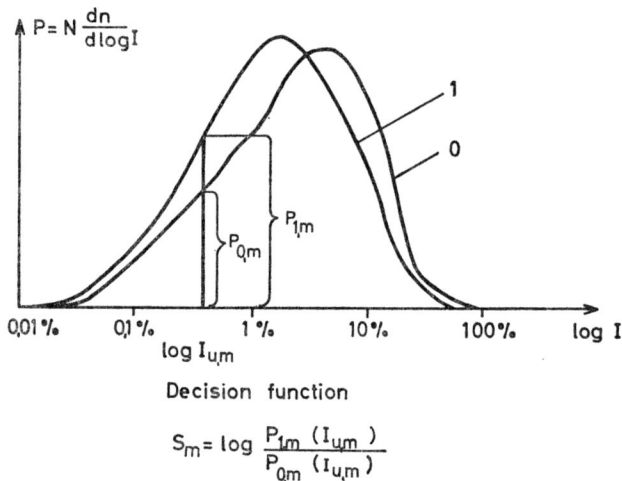

FIG. 1 Example of normalized frequency distributions of the intensities of all learning set spectra at mass m. The two curves, designated with 1 and 0, represent the spectra with presence or absence of the structural detail considered. The method is based upon the fact that the two curves are not congruent, giving, for most intensities, noticeable probabilities for or against the presence of the structural detail.

single mass only, say m, which forms the mth component of the vector $\mathbf{V}$. The statistical investigation of the intensities of this mass m throughout all spectra, separately for both structural classes, reveals, in general, two different frequency distributions, as presented in Fig. 1. The difference in these frequency distributions now may help to classify an unknown mass spectrum. Let the intensity of mass m of this unknown spectrum be $I_{u,m}$. From our frequency distributions for this mass m, we can read for the intensity $I_{u,m}$ a certain ratio $r = p_1/p_0$ of probability densities which give a statistical indication of whether our unknown spectrum stems from an aromatic compound or not. If r is larger than 1, our compound is, to some probability, aromatic, and vice versa. A simple form of a decision function thus reads

$$S_m = \log r_m = \log \frac{p_{1,m}(I_{u,m})}{p_{0,m}(I_{u,m})} \tag{2}$$

where the logarithm is introduced to fulfil the requirements assigned to eqn (1).

To make our decision surer, we have to consider other masses, too, giving other probability ratios r. If we assume, in a first approximation, that the intensities at different masses are mutually statistically independent, the overall probability for all masses is the product of the individual probabilities, which leads to the complete decision function

$$S = \sum_m \log r_m \tag{3}$$

considering now all masses.

To perform such decisions, we have to derive, from spectral data, two complete frequency distributions for each mass which then have to be stored. This can be done in two different ways: the distributions may be approximated by analytical functions, the parameters of which are stored solely, or the frequency distributions (better: the ratios $r_m(I_{u,m})$) are stored point by point at suitable intervals. We have investigated both methods with respect to their success in making correct decisions.

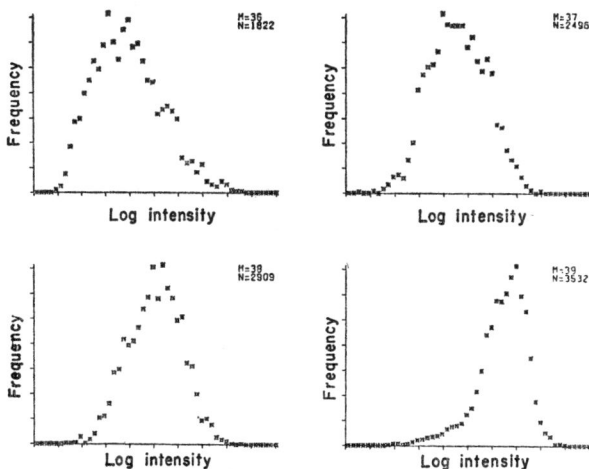

FIG. 2 Shape of experimental frequency distributions. The spectra are not split up into two structural classes as in Fig. 1 (masses 36 to 39).

Figure 2 represents computer-plotted distribution curves on a logarithmic intensity scale, as derived from masses 36 to 39. These curves are not separated into structural classes; they are shown here only to give a general impression of their shape. From these distributions we learn that they may roughly be described by Gaussian distribution functions, which are fully described by two parameters only, the mean μ and the variance σ^2. If we introduce the Gaussian distribution into eqn (2), the decision function S_m simplifies to a quadratic function of the form

$$S_m = \log r_m = a_m \cdot (\log I_{u,m})^2 + b_m \cdot \log I_{u,m} + c_m \tag{4}$$

where a_m, b_m and c_m can be easily related to $\mu_{1,m}$, $\mu_{0,m}$, $\sigma_{1,m}^2$, and $\sigma_{0,m}^2$. The storage requirements are thus greatly reduced since only the a_m, b_m and $\sum c_m$ are to be stored. Special provisions, however, have to be made for zero

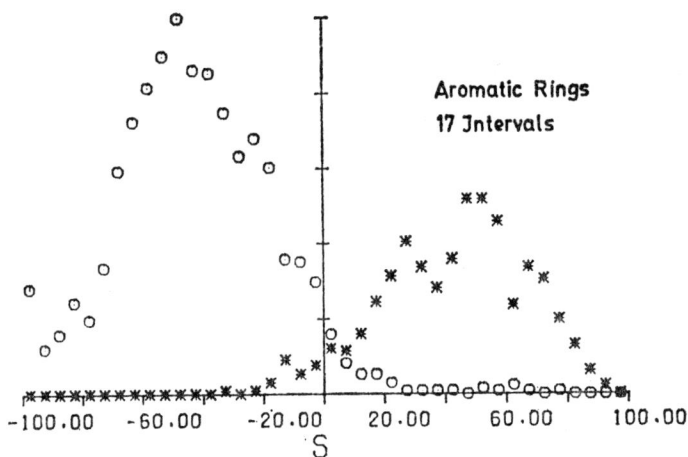

Aromatic Rings
17 Jntervals

-100.00   -60.00   -20.00   20.00   60.00   100.00

S

(a)

intensities which prevailingly appear in the mass spectra. By this, the number of coefficients to be stored is enlarged again.

If the Gaussian distributions for both structural classes both have equal width the number of parameters to be stored can be further reduced, because the decision functions S_m become linear

$$S_m = \log r_m = b_m \cdot \log I_{u,m} + c_m \qquad (5)$$

This type of function is the basis of the work of Jurs and co-workers, though their method of determining the coefficients b_m and c_m is completely different.

For the point-by-point storage of distribution functions, the intensity scale is, for each mass separately, divided into n intervals containing an equal number of intensity values each, except for the first interval which contains

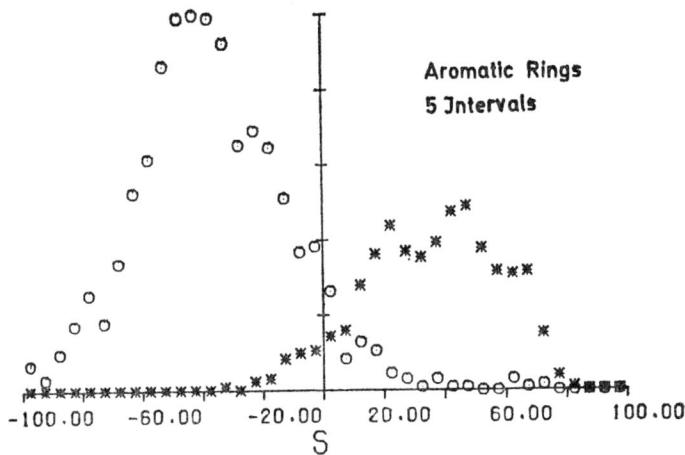

Aromatic Rings
5 Jntervals

-100.00   -60.00   -20.00   20.00   60.00   100.00

S

(b)

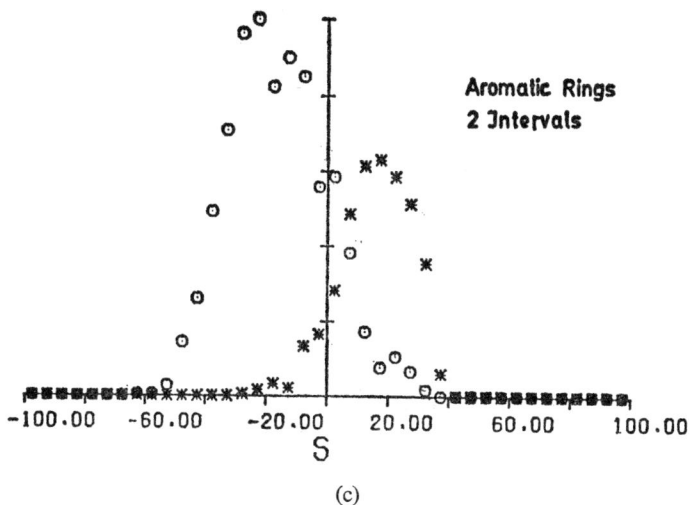

(c)

FIG. 3 Statistical investigations of the success rate of decision. The structural detail considered is the presence of aromatic rings. Frequency distributions of the resulting scalar S of the decision function are plotted against the magnitude of S, for both structural classes respectively. Asterisks mark aromatic compounds, circles describe non-aromatic samples.

(a) The total intensity range was divided, for each mass separately, in seventeen intervals of equal population. The overall success is 95·4%.

(b) Success rate of decision on aromatic rings with five intervals only. The overall success is 93·3%.

(c) Success rate 'aromatic rings' with only two intervals. Overall success is 86·6%.

all zero intensities. Different numbers n of intervals are investigated for their yield in correct decisions.

Experiments were performed with a subset of the MSDC collection of mass spectra.[3] Compounds have been restricted to molecular weights below $m = 500$ amu, and to the elements C, H, O, N, S, F, Cl, Br and I, only. To homogenize the data material, only spectra are admitted which start below mass 30 and show some intensities below 0·3%. All spectra are cut off below mass $m = 30$ and intensity $I = 0·1\%$. The remaining 4400 spectra are divided into a learning set (2200 spectra) and a testing set for the investigation of success rates.

The results exhibited the surprising fact that in all cases investigated so far, the point-by-point storage of distribution curves revealed higher success rates for decisions than the storage of the parameters of the Gaussian distributions, for equal storage place for both methods. In the following, therefore, we limit the presentation to the description of the distributions by storing the probability ratios r of suitably chosen intervals.

Figure 3 presents a statistical investigation of the success of the decision on the presence of aromatic rings, for 17, 5 and 2 intensity intervals each. The parameters of the decision function at first are 'learned' by measuring (counting) the probability ratios r within the intervals. Then the decision function is applied to 2200 spectra, and the frequencies of the resulting scalars

J. FRANZEN AND H. HILLIG

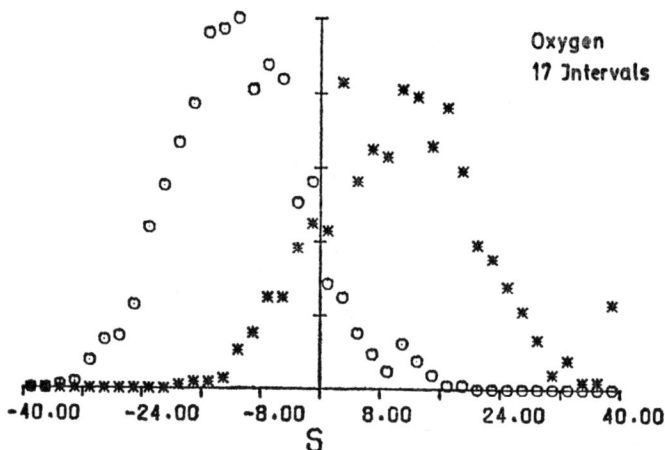

FIG. 4 Decision on the presence of oxygen. Seventeen intervals. Success 86·7%.

S are plotted against the magnitude of S. These frequency distributions must not be confounded with those of Fig. 1.

From Fig. 3 we see that the success rate drops only slightly if we confine the number of intensity intervals from 17 to 5, thus drastically reducing the storage space required. The Figs. 4, 5 and 6 show success rates for three other arbitrarily chosen structural characteristics.

From Figs. 3 to 6 we can easily learn that it is possible to assign a probability value to each individual decision for being correct. If a large absolute value of S is found, the probability of the decision to be correct is high. This probability value for the accuracy of individual decisions may be used later on within a network of decisions.

Decision functions have been applied, too, to modulo-14-spectra and 80

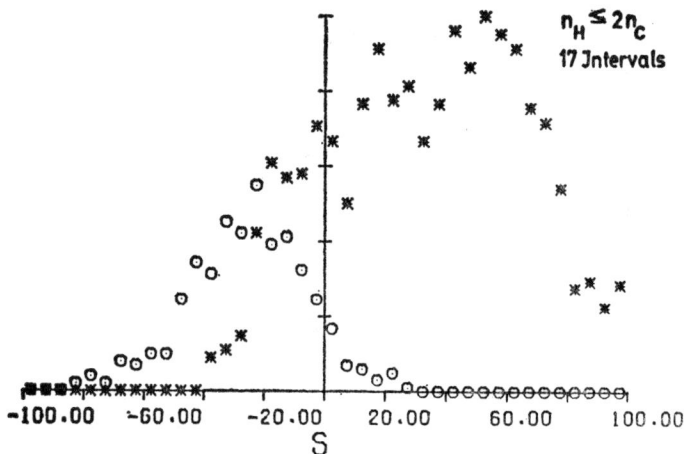

FIG. 5 Decision on the number of hydrogen atoms to be equal or less two times the number of carbon atoms. Seventeen intervals. Success 86·2%.

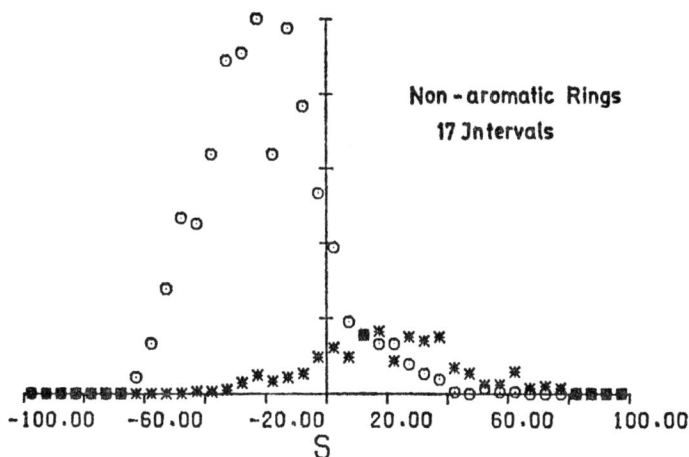

FIG. 6 Decision on the presence of non-aromatic rings. Seventeen intervals. Success 83·6%.

component periodicity-spectra.[4] The success rates turned out to be quite similar to those found with untransformed mass spectra. However, considerable computer time and storage is saved, because these spectra are much shorter.

Furthermore, a reduction of the number of masses which are considered is possible, too. If the 25 to 30 most significant masses are selected (which can be done automatically by the program), the drop in success rate is almost negligible. The automatic component selection calculates weighted average decision values $\bar{S}_m$ in both structural classes and sorts the masses with respect to the magnitude of $\bar{S}_{m,1} - \bar{S}_{m,0}$. The resulting statistically most significant masses agree well with our expectations from present knowledge in mass spectrometry. Therefore, this method seems a promising one with which to further enlarge this knowledge.

REFERENCES

1. Jurs, P. C., Kowalski, B. R. and Isenhour, T. L., *Anal. Chem.*, 1969, **41**, 21.
   Jurs, P. C., Kowalski, B. R., Isenhour, T. L. and Reilley, C. N., *Anal. Chem.*, 1969, **41**, 690.
   Kowalski, B. R., Jurs, P. C., Isenhour, T. L. and Reilley, C. N., *Anal. Chem.*, 1969, **41**, 695.
   Kowalski, B. R., Jurs, P. C., Isenhour, T. L. and Reilley, C. N., *Anal. Chem.*, 1969, **41**, 1945.
   Jurs, P. C., Kowalski, B. R., Isenhour, T. L. and Reilley, C. N., *Anal. Chem.*, 1969, **41**, 1949.
   Jurs, P. C., Kowalski, B. R., Isenhour, T. L. and Reilley, C. N., *Anal. Chem.*, 1970, **42**, 1387.
   Jurs, P. C., *Anal. Chem.*, 1971, **43**, 22.
   Isenhour, T. L. and Jurs, P. C., *Anal. Chem.*, 1971, **43**(9), 20A.
   Wangen, L. E., Frew, N. M., Isenhour, T. L. and Jurs, P. C., *Appl. Spectrosc.*, 1971, **25**, 203.
2. Nilsson, N. J., 'Learning Machines', McGraw-Hill, New York, 1965.
3. United Kingdom Atomic Energy Authority, Atomic Weapons Research Establishment, Mass Spectrometry Data Centre, Aldermaston, Reading, England.
4. Crawford, L. R. and Morrison, J. D., *Anal. Chem.*, 1968, **40**, 1469.

116
A Self-Training Interpretive and Retrieval System for Mass Spectra[1]

By F. W. McLAFFERTY, R. VENKATARAGHAVAN,
K.-S. KWOK and G. PESYNA

(*Department of Chemistry, Cornell University, Ithaca, New York, U.S.A.*)

A UNIQUE advantage of mass spectrometry for the determination of molecular structure is the unusual quantity and specificity of the data found in the mass spectrum of a complex molecule. Unfortunately, the skill and time required to use these data apparently constitute such a substantial obstacle that the chemist usually makes use of only a small percentage of these data for structure elucidation of an average unknown. The use of the modern digital computer is an obvious potential answer to this problem and, to this end, a variety of systems have been proposed.[1-4]

Three general types of systems have been developed. The first[2] matches the unknown mass spectrum against those in a reference file, and is thus mainly effective for those compounds for which reference spectra are available. The second class of systems is exemplified by the 'learning machine' approach of Isenhour.[3-5] For this the computer first correlates the spectra in the reference file separately for each specific structural feature sought; these 'pattern classifiers' are then used by the computer to determine whether it is probable that each of these structural features is present in the unknown compound. Bender, Shepherd and Kowalski[6] point out that the development of reliable linear pattern classifiers using all peaks up to m/e 300 requires the use of a minimum of 900 compounds; 'even with the most sophisticated computational methods and facilities, this poses a formidable data processing problem'. These authors propose that this effort can be reduced substantially by specific contractions of the data base.[6] Practical applications of such systems have been limited to date.

A third general class of systems attempts to utilize the present large body of knowledge concerning the mass spectral behaviour of organic molecules by programming the computer to follow more or less the same interpretive reasoning employed by the chemist utilizing known mass spectral correlations.[2-4] Probably the more extensive effort in this regard is the 'Artificial Intelligence' method developed by the Stanford group. An excellent example of the potential of this method is its recent application to the mass spectra of oestrogens.[7] A large amount of time is necessary to develop these programs, as human effort is required to correlate the mass spectra as well as to reduce this knowledge to a computer program. Further, as new reference spectra or other new mass spectral information become available, additional effort of a similar sort is necessary to incorporate this new information into the program.

We have recently described a 'Self-Training Interpretive and Retrieval System' (STIRS) which seeks to combine the advantageous features of each of these general approaches; a more complete description of the system can be found in the original literature.[1,4,8] To interpret the mass spectrum of an

TABLE I

Spectral Data Classes and Match Factors Utilized by STIRS

Class of spectral data	Match factor
1. Ion series	MF1
2. Low mass characteristic ions ($<$m/e 90)	MF2
2B. Overlapping characteristic ions[a] (m/e 67–118)	MF12
3. Medium mass characteristic ions (m/e 90–149)	MF3
3B. Overlapping characteristic ions[a] (m/e 119–175)	MF13
4. High mass characteristic ions ($>$m/e 150)	MF4
5. Neutral loss series[a]	MF14
6. Small primary neutral losses	MF5
6B. Overlapping primary neutral losses[a] (-33 to -90)	MF15
7. Large primary neutral losses	MF6
8. Secondary neutral losses from the most abundant odd mass loss	MF7
9. Secondary neutral losses from the most abundant even mass loss	MF8
Class 9 data of the unknown spectrum matched against Class 5 data of the reference spectrum	MF9
10. Fingerprint ions	MF10
Overall match factor	MF11

[a] Proposed new match factor currently under experimental evaluation.

unknown compound with STIRS it is not necessary for the chemist to correlate the spectral behaviour of related compounds; this training is done by the computer for each unknown. However, previous knowledge of mass spectral correlations is utilized to identify a variety of spectral data classes which are most indicative of particular types of molecular structure features (Table I). STIRS then searches the file of reference compounds to find those which match most closely for each of the selected data categories; if a particular structural feature is found to be common to a number of the selected compounds this is indicative of the presence of this feature in the unknown. Complete identification of the unknown is then attempted by combining the structural features indicated by each of the data classes with molecular weight, elemental composition, and any other available information.

A critical factor in the effectiveness of such a system is obviously the extensiveness and accuracy of the data file. A large effort over the last few years in this regard has resulted[1] in our present file of over 25 000 reference spectra. All of these spectra have been checked for errors by a variety of human and computer techniques. In the case of compounds for which more than one reference spectrum is available, these have been compared and the best spectrum selected. The 19 000 'best' spectra have been ordered by molecular weight, elemental composition, and chemical compound type.[1] The latter sort is made possible because the compounds for each reference spectrum have been coded in Wiswesser Line Notation (WLN). In the examination of an unknown mass spectrum by STIRS, a 'match factor' (MF) is calculated for each reference mass spectrum in each data class. For each class of spectral data the ten compounds in the best spectrum file of highest match factor are printed out, followed by any other spectra in the remaining duplicate file which have match factors above the match factor found for the tenth compound selected.

THE STIRS CONCEPT

The basic conceptual uniqueness of STIRS can be illustrated by the use of 'ion series' data (MF1, Table I). The usefulness of such sums of the intensities of particular ions to indicate particular structural classes has been employed in 'type analysis'[9,10] as well as 'ion series' of homologous ions[11-15] (also called 'reduced mass spectra').[13] For example, in an extensive study, Smith[15] has determined the average abundances of the 14 homologous ion series for 50 classes of compounds and has shown that these data can be used for an unknown mass spectrum by the computer to determine the most probable class of the unknown compound. However, he points out the difficulty in defining rigorous compound classifications; further subclassifications to solve this problem are often not possible due to lack of representative reference spectra. 'It may be necessary and/or useful to relax this definition of class depending on available reference spectra and the particular chemical problem at hand.'[15]

In contrast, for STIRS no previous arbitrary classifications of structural types need be set up in order to utilize the ion series data. In the calculation of MF1, STIRS determines the resemblance of the low mass ion series of the unknown and the particular reference compound. Computer selection of the reference spectra with the highest MF1 values automatically identifies those compounds whose 'reduced mass spectra'[13] most closely resemble that of the unknown. Thus, Smith defined separately classes of methyl n-alkanoates and dimethyl n-alkandioates;[15] although the data are similar, as expected, he finds useful differences in the reduced mass spectra of these compounds. In a test of STIRS using the mass spectrum of methyl nonane-1,9-dioate, the reference spectra found with highest MF1 and centroid match factors were methyl heptane-1,7-dioate and methyl octane-1,8-dioate; the first nine compounds selected were all methyl esters of aliphatic acids.[4] Further, the fourth most abundant ion series was the so-called 'low aromatic' series (m/e 38, 39, 50, 51, 63, 64, 75, 76); the computer ignored the human prejudice of this nomenclature in correlating the ion series data. Thus identifications by STIRS using ion series data are not limited to a predetermined set of structural classes for which the average reduced mass spectra have already been determined. STIRS should be capable of identifying any class of compounds represented in the reference file by a relatively few mass spectra with sufficiently similar and distinctive ion series data.

STIRS IDENTIFICATION OF ION SERIES

The ion series used for MF1 already include groups of peaks in addition to the 14 possible homologous ion series.[4] However, in many cases MF2 identifies structural features in addition to those found by MF1, despite the fact that MF2 actually uses a more limited set of peaks ($<m/e\ 90$). For example, for the mass spectral data of pregnenolone acetate, three of the ten reference compounds of highest MF1 values contained the steroidal skeleton, while this was true for 7 of the 10 compounds of highest MF2 values. Thus in using the 'characteristic ions' (MF2, MF3, MF4) it is not necessary to define either the classes of structural types or the most suitable ion series. However, by designating the ion series in advance (MF1), it is sometimes possible to identify a structural feature giving an ion series of relatively low abundance in the presence of a feature giving more abundant ions which do not fall in one or two of the designated ion series.

REFERENCE FILE

Progress has been made in a number of areas of development and application of STIRS since the original description of this system.[1,4,8] Doubling the size of the data base has greatly increased the power of the system, as expected. In a test of ten spectra whose identities were completely unknown to us,[16] STIRS identified the compound correctly in seven cases and predicted closely related isomers in two others (Table II). (MF values are given for cases in

TABLE II

Results of Unknowns Supplied by Dr T. L. Pugh

True structure	STIRS identification	MF10	MF11
		—	—
3,6-dimethyluracil	3,6-dimethyluracil	982	998
$CHF_2CF_2(CH_2)_2CF_2CHF_2$	$CHF_2(CHF)_4CHF_2$	—	—
n-tetradecane	n-tetradecane	737	889
benzyl formate	benzyl formate	906	809
		—	—
pregnenolone acetate	pregnenolone acetate	855	833
cis-decalin	cis-decalin	917	977
1,3-indandione	1,3-indandione	989	1 000
p-nitrodiphenylamine	p-nitrodiphenylamine	1 000	1 000

which a reference spectrum of the compound was present in the data base.) In the case of the imine, STIRS incorrectly indicated the presence of a saturated ring containing nitrogen (MF1), oxygen (MF1 and MF2) and –CO–N–CO– (MF5); we are refining our criteria in hopes of improving this. However, the criteria are based on the *probability* that a particular functionality will be present.

INSENSITIVITY OF STIRS TO QUANTITATIVE VARIATIONS IN SPECTRAL DATA

The capability of both the 'fingerprint' match factor (MF10) and the overall match factor (MF11) for retrieval of a reference mass spectrum from the data base is illustrated by the results of two mass spectra of cholesterol taken under very different instrumental conditions.[17] In the first spectrum the molecular ion was the base peak in the spectrum; the second had more than twenty peaks in the low mass region whose abundances were greater than that of the molecular ion. For both spectra the first six compounds selected by STIRS as having the highest MF11 values were all spectra of cholesterol. These were not the same spectra in the same order; the reference file contains eight spectra of cholesterol. The results using the 'fingerprint' match factor (MF10) for these two spectra showed a similar tolerance for their wide abundance differences.

INFORMATION FROM MORE COMPLEX MOLECULES

Probably the most elegant application of the 'Artificial Intelligence' program has been to the interpretation of the high-resolution mass spectra of oestrogens.[7] The performance of STIRS with such structures was tested utilizing the spectra of a number of these compounds as unknowns. The reference data used in this study were kindly supplied to us by Professor Djerassi, and the corresponding low resolution spectra incorporated into our data base. In general, the basic oestrogenic skeleton was identified by STIRS with almost no ambiguity, with the characteristic ion match factors (MF2, MF3, MF4, MF10) and the overall match factor (MF11) being the most indicative. The results utilizing the mass spectral data of equilenin acetate are typical of those observed. MF2, MF3 and MF10 indicated equilenins; the selections of MF11 are shown in Table III. Such information is not sought in the Artificial Intelligence program, which is applied only to those spectra which are known to be of oestrogenic compounds.

Identification by STIRS of the functional groups of the oestrogens was not completely satisfactory; carbonyl and unsaturated moieties were erroneously indicated in particular cases. This is probably due to the overlapping data from the large number of functionalities present. To improve these identifications we have designed specific series of neutral losses, analogous to the low mass ion series, to provide much more specific identification of the common small functional groups (Table I). However, in many cases STIRS provided reliable characterization of such groups. For example, the

TABLE III

Overall Match Factor Selections for Equilenin Acetate

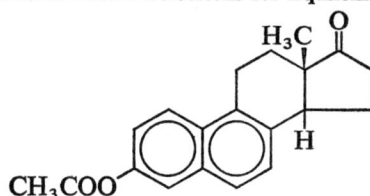

Compound	MF11
equilenin	536
13β-ethyl-18-nor-equilenin-3-methyl ether	496
1,2-dimethyl-D6-estrone-3-acetate	453
N-acetyl-O-methylcrotonosine	440
apovincamine	435
D6-estrone-3-acetate	432
D6-1-methylestradiol-3,17-diacetate	411
hinokione acetate	411
1-methyl-D6-estrone	400

MF5 selections for equilenin acetate, Table IV, fulfil the criteria for the presence of the acetate functionality. In addition, the mass (or elemental composition) of the molecular ion can be used with knowledge of the structural skeleton to confirm the postulated functional group identities.

In these trials STIRS did not do as well at locating the positions of the functional groups on the skeleton as did the Artificial Intelligence program. The performance in this should be improved by using the 'shift technique' program of STIRS[4] which is basically similar to the approach employed by the Artificial Intelligence technique. We plan to investigate the feasibility of characteristic ion comparisons in which the peak masses for both the reference and unknown mass spectra are shifted by the masses of the functional groups

TABLE IV

Neutral Loss Match Factor Selections for Equilenin Acetate

Compound	MF5
2-acetyl-2-azaquinuclidine-3-one	947
phenyl n-propyl ether	900
D6-estrone-3-acetate	900
D6-1-methylestradiol-3,17-diacetate	900
1,2-dimethyl-D6-estrone-3-acetate	900
para-acetophenetidide	900
tri-isopropylphosphine	900
N-methyl-4-acetoxy-6-methyl-2-pyridone	888
N-(3-methoxy-4-acetoxybenzyl)acetamide	888
p-chlorophenyl acetate	842

present in each. Even more comparable results should be possible through modification of STIRS to utilize the high resolution data.

The capabilities of STIRS for structure elucidation were also investigated for terpene hydrocarbons and their derivatives because the effect of structural details on the spectra of these compounds is particularly obscure.[18] The spectra of monoterpene hydrocarbons consist mainly of peaks of the same set of mass values which are often even of similar intensities.[19] This was

TABLE V

Overall Match Factor Selections for Fenchyl Alcohol

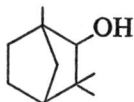

Name	Structure	MF11
elemol		607
(—)-*cis*-caran-*trans*-2-ol		602
iso-pulegol		589
(+)-*trans*-caran-*trans*-2-ol		562
1-terpineol		560
3,7-dimethyl-2,6-octadiene-1-ol		545
α-terpineol		536
thujyl alcohol isomer		531

reflected in the STIRS results for allo-ocimene which gave little specific structural information beyond the distinctive classification as a monoterpene hydrocarbon. A similarly disappointing lack of specificity was found in the STIRS results for the mass spectrum of 2-bornanone, although the reference spectrum giving the highest value for the overall match factor was that of 3-bornanone. Monoterpene alcohols show much more substantial variations in peak abundances;[20] however, identification of the specific arrangement of the hydrocarbon skeleton is still ambiguous, as shown by the overall match factor results for fenchyl alcohol, Table V. Note, however, that one of the selected compounds is a sesqueterpene alcohol and that all the rest are monoterpene alcohols. Further, since it is known that the data base contains a large variety of such compounds, these results indicate that correlation of the exact skeletal arrangement with the mass spectrum is difficult. It may be possible to find characteristic ion or neutral-loss series for such compounds using learning machine techniques.

MOLECULAR WEIGHT DETERMINATION

It has been pointed out that STIRS can often aid in a problem which plagues mass spectrometrists: establishing the molecular weight of compounds whose mass spectra do not show a molecular ion.[4] For example, for the mass spectrum of pregnenolone acetate, run as a complete unknown (Table II),[16] the peaks of highest masses were at m/e 315 (0·3%), 316 (0·1%), and 317 (0·05%). The choice of molecular weight for the unknown only affects the calculation of the neutral loss match factors, MF5 and MF6, as the neutral losses are determined as the difference between the mass of the particular ion and the postulated molecular weight. Using trial molecular weights of 315, 316 and 317 gave strong indications of a compound similar to pregnenolone acetate. For example, for mol. wt = 317, the highest MF11 values were given by pregnenolone acetate, pregnenolone acetate-20-ethylene ketal, 5α-pregnane-3β, 17α-diol-20-one-3-acetate, 6α-epoxyallopregnane-3-β-ol-20-one-3-acetate, 5β,6β-epoxypregnane-3β-ol-20-one-3-acetate, 16α-methylpregnenolone acetate, and pregnenolone. (Because this was a true unknown, reference spectra of the same compound were not eliminated from the data base.) However, for these three trial molecular weights the neutral loss match factors (MF5 and MF6) gave no structural features fulfilling the predetermined criteria, and the selected compounds bore little or no relationship to the types of compounds indicated so strongly by the overall match factor. This is compelling evidence that the peaks at masses 315, 316 and 317 are actually due to fragment ions, and that the molecular ion is of negligible abundance. The value of 358 was next tried for the molecular ion mass as suggested by the compound showing the highest MF11 value; the MF5 results are shown in Table VI. This overwhelming indication of the presence of acetate is also given by the compounds found to have the highest MF6 values (large neutral losses); again the first eight compounds selected are acetates. Nine of the first ten compounds selected by MF9 (secondary neutral loss after even mass loss) are also acetates. Thus these data give definitive

TABLE VI

**Low Mass Neutral Loss Match Factor Selections for
Pregnenolone Acetate Using Mol Wt = 358**

Compound	MF5
carvomenthyl acetate	909
2-(2H)-carvomenthyl acetate	888
1-oleo-2,3-diacetin	888
desmosterol acetate	888
cholesteryl acetate	888
3-β-acetoxy-pregn-δ(5,16)-dien-20-one	888
3β,21-acetoxy-pregn-δ(5,16)-dien-20-one	888
17α-methyl-17β-carbomethoxyandrost-5-en-3β-ol acetate	888
pseudoivalin acetate	888
3β-acetoxy-24-ethyl-5,22-cholestadiene	888
C19-formyl cholesterol acetate	833
ethylene diacetate	833
2-octyl acetate	833
pregnenolone acetate	833
menthyl acetate	833

evidence that the molecular weight of the unknown is 358, as well as establishing the presence of the acetate functionality. The new MF11 values are even more strongly indicative of the true structure; eleven of the first fifteen compounds selected are pregnenolone acetate derivatives.

FURTHER DEVELOPMENTS

Currently we are developing programs in an attempt to have the computer recognize the structural features common to the reference compounds of highest values of particular match factors utilizing the Wiswesser Line Notation (WLN) which is included for each compound in the data base. Computer interpretation of the WLN allows the corresponding structures to be displayed on the cathode ray tube (CRT) by the computer. We also hope to develop CRT-lightpen techniques for assembling trial structures using the postulated structural features, molecular weight, and other structural information.

MAN–MACHINE INTERACTION

Although increasingly the computer can take over many of the tasks of the human interpreter, and even excel his performance in some of these tasks, we feel that optimum result for complex molecules can only be obtained by using STIRS as an aid to the chemist. As Nilsson has said on the use of computer learning machines,[21]

Unfortunately, there is very little theory to guide our selection of measurements. At worst this selection process is guided solely by the designer's intuitive ideas about which measurements play an important role in the classification to be performed. At best the process can make use of known information about some measurements that are certain to be important. . . . We shall henceforth assume that the d measurements yielding the pattern to be classified have been selected as wisely as possible while remembering that the pattern classifier cannot itself compensate for a careless selection of measurements.

Bender and Kowalski[22] have stated

It would be erroneous to infer that pattern recognition removes the scientist from data analysis. Man–machine interaction is currently in vogue in chemistry and other fields for a good reason: man is the best pattern recognizer known today. . . . Computer techniques should be used but carefully supervised by the scientist.

ACKNOWLEDGMENTS

The authors are grateful to Drs Marianna Busch and J. W. Serum in connection with the recent expansion of the data base, and to the National Institutes of Health (GM 16609) and to the Environmental Protection Agency (R801106) for generous support of this work.

REFERENCES

1. 'Computer-Aided Interpretation of Mass Spectra', V, For Part IV, see McLafferty, F. W., Busch, M. A., Kwok, K.-S., Meyer, B. A., Pesyna, G., Platt, R. C., Sakai, I., Serum, J. W., Tatematsu, A., Venkataraghavan, R. and Werth, R. G., *in* 'Mass Spectrometry and NMR Spectroscopy in Pesticide Chemistry', *ed*. F. J. Biros and R. Haque, Plenum Press, New York, 1973.
2. McLafferty, F. W. and Gohlke, R. S., *Anal. Chem.*, 1959, **31**, 1160; Ridley, R. G., *in* 'Biochemical Applications of Mass Spectrometry', *ed*. G. R. Waller, Wiley–Interscience, New York, 1972, p. 177.
3. Ward, S. D., *in* 'Mass Spectrometry', Vol. 2, *ed*. D. H. Williams, Specialist Periodical Reports, the Chemical Society, London, 1973.
4. Kwok, K.-S., Venkataraghavan, R. and McLafferty, F. W., *J. Amer. Chem. Soc.*, 1973, **95**, 4185.
5. Jurs, P. C., Kowalski, B. R. and Isenhour, T. L., *Anal. Chem.*, 1969, **41**, 21.
6. Bender, C. F., Shepherd, H. D. and Kowalski, B. R., *Anal. Chem.*, 1973, **45**, 617.
7. Smith, D. H., Buchanan, B. G., Engelmore, R. S., Duffield, A. M., Yeo, A., Feigenbaum, E. A., Lederberg, J. and Djerassi, C., *J. Amer. Chem. Soc.*, 1972, **94**, 5962.
8. Kwok, K.-S., Ph.D. Thesis, Cornell University, 1973.
9. Hood, A., *in* 'Mass Spectrometry of Organic Ions', *ed*. F. W. McLafferty, Academic Press, New York, 1963, p. 597.
10. Seifert, W. G. and Teeter, R. M., *Anal. Chem.*, 1970, **42**, 750.
11. McLafferty, F. W., *in* 'Determination of Organic Structures by Physical Methods', *ed*. F. C. Nachod and W. D. Phillips, Academic Press, 1962, p. 93.
12. McLafferty, F. W., 'Interpretation of Mass Spectra', 1st *ed*., W. A. Benjamin, Inc., New York, 1966.
13. Crawford, L. R. and Morrison, J. D., *Anal. Chem.*, 1968, **40**, 1469.
14. Venkataraghavan, R., McLafferty, F. W. and Van Lear, G. E., *Org. Mass Spectr.*, 1969, **2**, 1.
15. Smith, D. H., *Anal. Chem.*, 1972, **44**, 536.

16. Supplied by Dr T. L. Pugh for the Workshop on Computer Matching of Unknown Mass Spectra, American Society for Mass Spectrometry Meeting, San Francisco, May 1973.
17. Bonelli, E. J., Knight, J. B., Skinner, R. F. and Taylor, D. M., *Finnigan Spectra.*, 1973, **3**, No. 2.
18. Weinberg, D. S. and Djerassi, C., *J. Org. Chem.*, 1966, **31**, 115.
19. Ryhage, R. and von Sydow, E., *Acta Chim. Scand.*, 1963, **17**, 2025.
20. Ibid., 1963, **17**, 2504.
21. Nilsson, N. J., 'Learning Machines', McGraw-Hill, New York, 1965.
22. Bender, C. F. and Kowalski, B. R., *J. Amer. Chem. Soc.*, 1972, **94**, 5635.

117

Diagnostic Functions in the Codification of Mass Spectral Data

By D. H. ROBERTSON and C. MERRITT, Jr

(Pioneering Research Laboratory, U.S. Army Natick Laboratories, Natick, Massachusetts, U.S.A.)

A particular characteristic of mass spectrometry which leads to difficulty for investigators in the field is that of overgenerous amounts of data; it has therefore been said often that mass spectra are over-determined; that there is more data produced than is required for unique identification of a particular compound. The advent of small laboratory-sized computers has established the potential for processing the data from a mass spectrometer or from the very common combination of mass spectrometer and gas chromatograph and it is now quite within the technical and financial grasp of most organizations to purchase and maintain a mini-computer system for use in an on-line configuration for the processing of mass spectral data. However, it is necessary to consider new ways of handling these data in order to make possible the best utilization of a small computer system. Traditionally all mass and intensity data have been employed in the creation of library files, and search algorithms have been employed which compare the 'unknown' with the library spectra on a peak by peak basis, using both mass and intensity values to define a match. In order to utilize effectively the characteristics of a mini-computer system, it is necessary to conceive ways of condensing or abbreviating the data which results from a mass spectrometer so that it will be diagnostic in nature as well as capable of being handled by a small system with limited core memory and limited mass storage facilities.

Natick Laboratories has been concerned with approaches to the handling of mass spectral data which would in the ideal case, involve the use of an irreducible minimum of information from the spectrum, capable of uniquely describing the compounds in the data library. It should be emphasized that the needs of each individual laboratory must be taken into consideration when one is in the process of designing a codification scheme and of developing search algorithms for comparing the data of an unknown with the data of a library of known mass spectra. There are often special analysis requirements for a limited collection of compounds which are found to be present in connection with a particular investigation. One may, for example, be concerned with natural product analysis wherein the volatile components of a particular sample type must be determined. The components of the sample are separated by a gas chromatograph into pure components, for each of which a mass spectrum is obtained. After all the components have been identified in a given sample type it is possible to create a library of reference compounds which may be used in all future investigations of samples of this type. In a typical case a library will contain 200–300 compounds, a size which is very amenable to the testing of codification schemes and search algorithms in their preliminary stages.

In-house requirements at Natick Laboratories involve frequent assignments of this nature from which a standard library of compounds results in each case and therefore there has been a good basis for testing these factors. It was recognized that some techniques might be effective for limited libraries while in the more general case they might fail *i.e.* they might not maintain sufficient uniqueness of character when applied to a large library such as the Wiley Data File or the mass spectra file provided on magnetic tape by the Mass Spectrometry Data Centre in Aldermaston, England.

A major difficulty in coming to grips with an effective solution to data handling problems in the field of mass spectrometry is that of determining what significance any particular approach may have when applied to the general case of a large library of mass spectral data. Of course the notorious heterogeneity of the existing data libraries exerts the restriction that any data handling routine must be independent of the inconsistencies in data which may be caused by such factors as different operating (instrumental) parameters, presence of trace impurities in the sample and inaccurate recording of the analogue signal which represents mass and intensity information.

Considering the mass range of the average mass spectrometer, there is no great difficulty in recording these values in an accuracy of unit resolution over a range which is satisfactory for most applications, especially in tandem gas chromatography–mass spectrometry configurations. In addition, this application produces an alarmingly large amount of data.

The first of the two general approaches which have been utilized for purposes of simplifying mass spectral data, involves the calculation of diagnostic functions utilizing mass and intensity data as they are obtained from the mass spectrometer but resulting, from appropriate manipulation of the data, in a single valued diagnostic function which is used in construction of a data file library; such a library provides for very rapid search and identification of unknown compounds once their individual functions have been calculated. The first of these calculations to be used was based on the principles of information theory. The desired goal is the application of a basic numerical definition which represents the amount of information that it conveys about the event of interest. A simple relationship is employed often known as the Khinchine entropy function.[1]

$$-\eta = \sum_{i=1}^{n} p_i \log p_i \tag{1}$$

In eqn (1), i is the index of units corresponding to the individual ions and p is the probability of each ion in the spectrum. If only one event has a probability of unity and all others are zero, the value of the entire function is zero; if all probabilities are equal, the function K will have a maximum value.

There lies in the continuum between zero and a maximum value, a collection of values which has been shown to be diagnostic for a wide range of functional group types, when applied to a library of organic compounds.

Table I lists a sample grouping of organic compounds and their corresponding Khinchine entropy function. In this particular case, it is easily possible to distinguish among the members of the group by using the Khinchine value. In actual practice it has been possible to utilize this function for libraries up to approximately 200 compounds and still retain unique diagnostic characteristics. When expansion is effected to extensive data libraries (*e.g.* the Wiley File) the non-uniformity of recorded spectral data in these libraries

TABLE I

Examples of Khinchine Function Values

Compound	Khinchine function
n-butane	0·927
2-methylpropene	0·812
t-2-butene	1·042
3-methyl-1,2-butadiene	1·216
1,3,5-hexatriene	1·340
1,5-hexadiyne	1·604
3-heptyne	1·380

results in inconsistencies among the Khinchine values which render them less than ideally useful for the unique identification of every compound in the library. Although not universally applicable, this approach is eminently practical with limited libraries, especially those which have been created from data which have been generated under the same instrumental conditions as data for the unknown compounds. A file of single-valued functions, one for each compound in the library, can be stored in the minimum configuration of core storage for most currently-produced mini-computer systems. Search algorithms are extremely simple for the comparison of an unknown with the members of the library; in total, the codification and search process is essentially trivial in nature and can be easily executed on an inexpensive mini-system.

The initial concept of this work was based on mathematical manipulation of mass spectral data with little specific consideration of the nature of the compounds whose spectra were being studied. Mathematical manipulation is a more purely academic exercise; consideration of the compound type being studied provides a more immediately useful approach, especially in applications which utilize on-line real-time data acquisition techniques. It was with this more practical approach in mind that other special codification schemes[2-5] have been investigated; one of the more successful of these has been named octal coding.[2] As such it represents a type of codification which falls into the general category of data compression. Such an operation is essential to the success of a mass spectral processing system from (a) the standpoint of coding individual compounds in a unique manner as well as in (b) the case of search algorithms which determine how the code for an unknown compound is compared with the members of a library of mass spectra similarly encoded. Table II shows how selective binary coding of characteristic peaks is accomplished by arbitrarily dividing the mass range of interest into multiple groups of seven. The number corresponding to the spectrum peak having the highest intensity is then encoded as a three bit binary number. Thus, the peak of the sixth position is encoded in the first grouping, the 2nd peak in the second and so on; zero is used to denote the absence of a peak within the grouping, thereby giving a total of 8 possible values; hence the term octal coding.

TABLE II

Example of Octal Coding

Mass ranges	23–29	30–36	37–43	44–50
M/e to be encoded	28	31	43	0
Position of m/e in octet	6	2	7	0
Binary code	110	010	111	000
Octal code	6	2	7	0

Representation of an octal number within the computer requires three bits; thus, in a 16 bit machine such as the Hewlett–Packard 2116B used first in setting up this system, five octal characters can be stored in each computer word with one bit left over. Thereby a single computer word can represent 35 amu, using 15 bits. The bit number 16 is used as a pointer to signal if the computer word in question is the last word being used for codification.

The searching of mass spectral libraries, when these libraries consist of a full tabulation of mass and intensity pairs for every ion produced by each compound, is a time and core storage consuming operation. In the case of octal coding the values required may be extracted readily during on-line data acquisition, thereby leading directly to on-line creation of a library without a requirement for storage of all mass and intensity data. In actual practice a library would be created by coding the mass spectra for a group of compounds expected in a certain analysis. Subsequently, during mass spectrometric analysis of unknown compounds, codification of the spectrum would occur under conditions much more likely to reproduce those utilized in creating the original library of known spectra.

The variability in the values of relative ion abundances with variation in mass spectrometer design and parameters of operation produces considerable uncertainty in the reliability of a diagnostic based on measurement of spectral intensity factors, and strongly points out the desirability of using a coding technique which does not depend upon the reproducibility of an intensity factor. Several variations on this basic code whereby a mass spectrum is represented in binary coded octal have been presented. They have in common the fact that sufficient combinations and permutations exist to code extensive data libraries (*e.g.* 2^{15} is equivalent to 32 768 possible combinations) in a unique manner.

It would seem axiomatic that the application of specific mass spectral information to the codification process would result in more highly diagnostic binary patterns which may be 'recognized' by the computer and correlated with like patterns in the similarly coded data library. For example, a practising mass spectroscopist would feel that starting mass for the groups or windows would have an effect upon the success of the search algorithms in differentiating among spectra, insomuch as there exists a series of ions which are routinely used as diagnostics for specific functional groups. Thus, a detailed understanding of mass spectral data is necessary before statistics can be successfully applied to them (*see* Table III).

TABLE III

Number of Ones Coded

Start	Transition			
	0·01	1·0	5·0	10·0
20	17·5	13·0	5·9	8·1
21	17·5	13·3	6·3	8·6
22	17·5	13·3	6·4	8·7
23	17·4	12·6	5·8	7·9
24	17·2	12·2	5·7	7·7
25	17·0	12·0	5·5	7·4
26	16·9	12·5	5·9	8·0

Since the number of binary bits or 'ones' coded in the library represents its information content, such representation is frequently employed. Considering a range of starting masses for the first window from 20–26, it can be seen for the intensity found at each transition (*i.e.* threshold) level the number of 'ones' coded remains essentially constant. The information content of the spectrum as an entity remains constant. Such a conclusion is obvious; but also misleading because it imputes to the code the ability to select the most diagnostic mass in each group of seven when it is actually selecting the most intense mass in each group of seven amus. A look at some statistical values based on the total 6880 compound library illustrates the point as suggested in Table IV.

Each pair of masses in this Table (selected from the statistical distribution of 'ones' when the starting mass is 23) represents respectively the last mass or position 7 in one window and the first or position 1 mass in the next higher adjacent window or group. It can be readily appreciated that if the scale is moved left or right to accommodate any starting mass that any one of these mass values would be moved into the adjacent window where it might or might *not* be coded as the most intense peak, depending of course upon the intensity of the other peaks in the window under consideration. In other words, for this group of adjacent mass pairs or indeed for any adjacent mass pair, a shift of starting mass used in the code may eliminate the boundaries which allow each mass to be coded in a separate window. Appearance of adjacent masses in a spectrum is in many instances a highly diagnostic feature. Commonly known examples would be the P and P-1 peaks and the isotope distribution peaks. A specific example has occurred with our library search routine aimed at the identification of alkyl benzenes on the basis of

TABLE IV

M/e Values at Window Boundary, Octal Code, 1 in 7

50–51	211–212	225–226
253–254	281–282	309–310

their m/e values at 91 and 92. Unless the starting mass is adjusted to allow each of the masses to fall in a separate group or window, the search routine is not successful.

In addition to the question of what starting masses to use in filling the groups or windows, one must address himself to the problem of selecting the codification scheme whereby these selected masses are represented by a binary code within the computer.

In Table V we see a comparison of coding schemes, using, from left to right, one peak in 14, one peak in 7 and one peak in 14 plus 2 bits for coding intensity. The first scheme requires 64 bits per spectrum, while the remaining two schemes each require 96 bits per spectrum. The term confusion refers to the number of compounds which agree as well or better than the right answer.

TABLE V

Comparison of Codes

Confusion	14 amu	7 amu	14 amu + 2
0	64·8	57·6	79·2
1	72·8	74·4	89·6
2	76·8	79·2	94·4
3	82·4	80·8	95·2
4	87·2	83·2	95·2
5	88·0	84·8	95·2

The values listed under each code are percentages of unique matches for the confusion factors with which they correspond. When one considers the absolute percentages for a single unique identification, *i.e.* for a confusion factor of zero, it is obvious that use of an intensity factor substantially increases the percentage of matches at all confusion levels.

However, a comparison of the values for one in 14 amu versus 1 in 7 amu differ much less and in practical work prove to be less important than the absolute difference in these values might indicate. Since most mini-systems for laboratory applications use a 16 bit word, a group of seven, each requiring three bits can be more easily coded, *i.e.* 5 groups of 3 bits with one left over as a pointer or indicator. Moreover, working with groups of 7 amu provides twice the opportunity for coding a mass which is diagnostic for a specific functional group. No intensity factor is coded, making possible more rapid on-line acquisition of gas chromatography–mass spectrometry data.

In most practical cases the investigator is in possession of some ancillary information about the compound or compounds he expects to identify; so the fact that more than one answer is given by a search algorithm does not mitigate against its utility.

In terms of practicality as regards computer power needed to do the job and ease with which the chosen technique provides on-line real-time processing of mass spectral data, it certainly need not be a unique compound. The conditions of mass spectral data libraries is such that it would seem unrealistic to hope for a coding and search technique that is always able to abstract the single right answer. There has been, for example, in these laboratories, a higher degree of success when working with special restricted libraries of compounds expected to appear in particular sample types; these libraries contain spectra coded from the mass spectrometer on which the 'unknown' is being run. Therefore greater internal self-consistency of data is achieved. When one goes to larger libraries, the spectra in which come from numerous sources, the internal self-consistency factor is reduced, often dramatically. If progress in purification of data file libraries can be encouraged at a rate near that at which codification and search techniques are proliferating at the present time, there is indeed promise for major advance in the general area of mass spectral data processing.

REFERENCES

1. Robertson, D. H. and Reed, R. I., Proc. of 19th Annual Conference on Mass Spectrometry and Allied Topics, Atlanta, Georgia, 1971, p. 68.
2. Robertson, D. H., Cavagnaro, J., Holz, J. and Merritt, C. Jr, Proc. of the 20th Annual Conference on Mass Spectrometry and Allied Topics, Dallas, Texas, 1972, p. 359.
3. Grotch, S. L., *Anal. Chem.*, 1970, **42**, 1214.
4. Grotch, S. L., ibid, 1971, **43**, 1362.
5. Grotch, S. L., ibid, 1973, **45**, 2.

118
Computer Interpretation of Mass Spectra

By D. H. K. KOO and R. D. SEDGWICK

(*Chemistry Department, University of Manchester Institute of Science and Technology, Manchester, U.K.*)

AN ever-increasing number of mass spectrometers are now linked directly to computerized data systems. These can produce high or low resolution mass spectra at speeds which are compatible with working on-line to gas chromatographs. In consequence an abundance of high quality spectra are produced at a rate which can be an embarrassment to the laboratory responsible for their interpretation. The spectra are, fortunately, in a digital format, which is an ideal starting point for beginning the automatic assignment of structure.

Considerable progress has been made already in library search routines[1-4] to identify compounds whose spectra have already been tabulated. However, these systems are at a disadvantage when presented with new compounds of quite simple structure. This point can be underlined when we consider that a large library may contain 10 000 spectra of all kinds whereas $C_{14}H_{31}N$ has 48 865 isomers. The alternative is to use the computer to interpret the mass spectra in much the same way as a human spectroscopist. This involves formalizing the basic rules governing mass spectral fragmentation mechanisms and using them to deduce structural features from a mass spectrum.

Other workers[5-10] have shown that this approach can be successful when applied to acyclic compounds, and this work represents a simplified approach to this problem. We noted that in the Dow index of mass spectra 54% of entries are acyclic compounds and of this majority less than 1% contain more than two secondary or tertiary atoms in their carbon skeleton and we have accepted this as an arbitrary limit in our initial work.

To represent molecular formulae in the computer requires a linear or matrix notation. We have chosen to use the Wiswesser Line Notation,[11] (WLN) which has found wide acceptance by chemists and is now used to encode formulae in A.P.I. mass spectra.

Briefly this notation uses letters to represent functional groups; Q represents –OH while V represents –CO–, and numbers to represent alkyl chains, thus $2Q$ is ethyl alcohol. Branches in a carbon skeleton are symbolized by Y for a secondary carbon

(–CH–)
|

while X represents a tertiary branch point

$$(-\overset{\textstyle |}{\underset{\textstyle |}{C}}-)$$

The two or three chains following these branch points are separated by ampersands, &, thus $2VX1\&1\&1$ is t butyl ethyl ketone.

The program is written in Fortran and has been run on the U.M.R.C.C. computer system.†

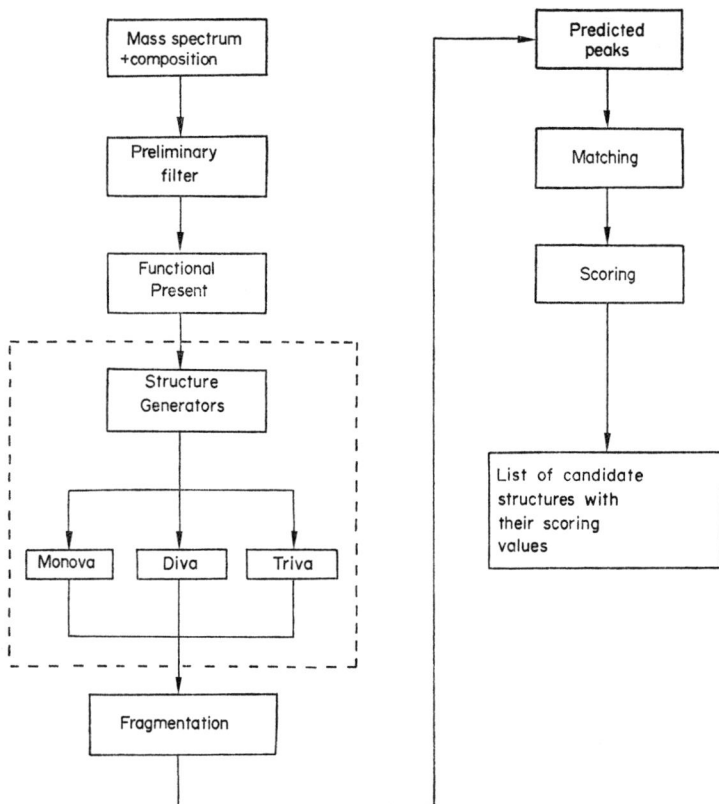

FIG. 1   General design of the program.

The general layout of the program is shown in Fig. 1. The input is the mass spectrum of an unknown compound obtained at either high or low resolving power. If low resolution spectra are used then we must also input the atomic composition of the molecular ion.

† The University of Manchester Regional Computer Centre provides access to their CDC 7600 computer through an ICL 1906A front end. Within this system the program requires 36 000 words for storage and execution, with about 4 seconds being sufficient time to process each spectrum of the complexity reported here.

THE PRELIMINARY FILTER

This first segment is used to determine the probable limits to the structural possibilities of the compound. The efficiency of this stage is a vital factor in keeping the overall computation and output to manageable proportions. The object is to determine the functional groups present and to place upper limits on carbon chain lengths. Consideration of the molecular atomic composition together with the rings plus double bonds, 'R value', limits the functional group types. These broad possibilities are retained or rejected by searching the spectrum for ions resulting from fragmentation involving loss of the proposed group. Further tests are then applied to position the functional group in the molecule.

Thus a compound $C_nH_{2n}O$ could be a ketone, aldehyde, unsaturated alcohol or ether. The compounds with a monovalent functional group, (–OH and –CHO) can be distinguished by the presence or loss of this group (or H_2O in the former case). The ketone and unsaturated ether can be distinguished by considering the partition of the molecule thus:

For a ketone R_1COR_2 we should find two ions R_1CO^+ and R_2CO^+, where R_1 and R_2 are alkyl radicals, such that

$$R_1CO + R_2CO = M_{ion} + CO$$

where M_{ion} = mass of the molecular ion. Application of this technique to

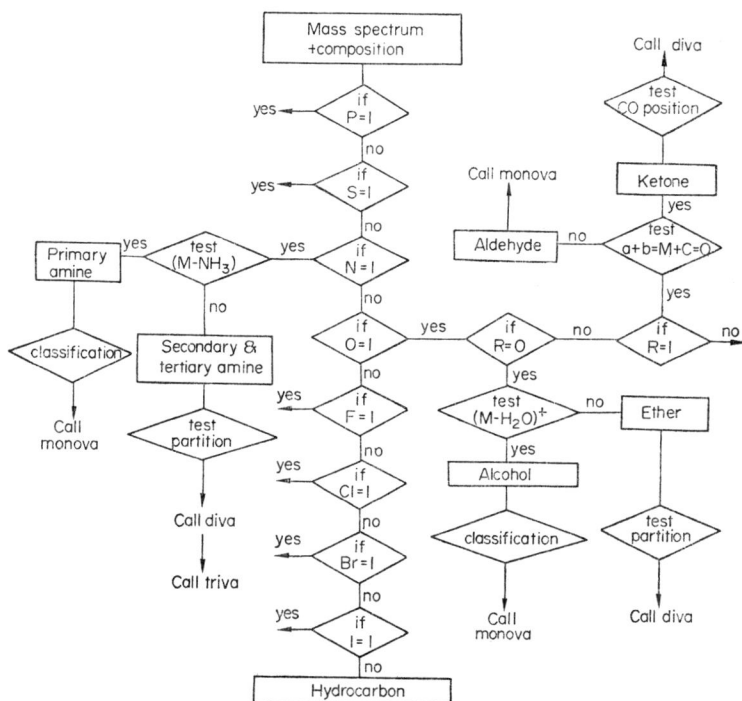

FIG. 2   Flow chart design of preliminary filter.

find a best fit not only confirms the functional group but also sets limits to the size of the two alkyl groups also present.

Similar arguments can be used to distinguish between primary, secondary and tertiary alcohols and amines.

A simplified outline of this segment is shown in Fig. 2. The limits set up at this stage are passed to the next segment to limit the extent of structures to be generated as possible candidates.

THE STRUCTURE GENERATOR

The functional groups present are classified as mono, di- or trivalent and are substituted by $-CH_3$, $-CH_2-$, or

$$-\overset{|}{C}H-$$

respectively in separate routines to give equivalent hydrocarbon structures. The isomeric possibilities of the equivalent hydrocarbons are then generated within any constraints previously determined. The functional groups are substituted back in the hydrocarbon isomers to give a set of possible structures. Some redundancy results at this stage and is eliminated by a number of symmetry tests. Thus a list of candidate structures is produced. This list is further edited using information deduced in the first segment to eliminate those generated isomers which are deemed improbable. Thus the list passed to the next stage is often kept quite short.

COMPUTER FRAGMENTATION

Each candidate structure is subjected in this segment to a number of fragmentation processes to produce a number of probable fragment ions. Odd fragments are generated by breaking the structure at all branch points. This is particularly easy using the WLN which accentuates these features. We also cause fragmentation of any linear carbon chains and add ions resulting from rearrangement processes specific to the functional group. Each candidate structure thus leads to a set of fragment ions of known mass (M). Since the fragmentation relies heavily on the presence of branch points in the structure, isomers of similar structure often lead to a different number (P) of predicted fragments.

MATCHING AND SCORING

The final segment involves comparing the masses in the experimental spectrum with those sets of masses produced as fragments from the candidate structures. Any coincidence of masses in the two lists is noted and is used to build a score for each candidate. Each structure will give N coincident masses ($N \leq P$) which will be the peaks used to score for this structure. In determining the score we use the coincident mass (M) and the corresponding experimental

STRUCTURE IS ALCOHOL

SECONDARY ALCOHOL

- -

ISOMERS	A	B	P	M
• Q Y 1 & 4	0.159263E 05	0.129401E 05	16	13
3 Y Q 2	0.736220E 04	0.414124E 04	16	9
Q Y 1 Y 1 & 2	0.174959E 05	0.803308E 04	14	9
Q Y 1 & 1 Y 1 & 1	0.140971E 05	0.969176E 04	16	11
2 Y Q Y 1 & 1	0.429380E 04	0.178908E 04	12	5
Q Y 1 X 1 & 1 & 1	0.126111E 05	0.810714E 04	14	9

2-HEXANOL
Q Y1 &4

STRUCTURE IS ALCOHOL

SECONDARY ALCOHOL

- -

ISOMERS	A	B	P	M
Q Y 1 & 4	0.229663E 05	0.186601E 05	16	13
3 Y Q 2	0.322146E 05	0.241610E 05	16	12
Q Y 1 Y 1 & 2	0.156320E 05	0.111657E 05	14	10
Q Y 1 & 1 Y 1 & 1	0.109964E 05	0.149973E 05	16	12
2 Y Q Y 1 & 1	0.226066E 05	0.150711E 05	12	8
Q Y 1 X 1 & 1 & 1	0.112950E 05	0.726107E 04	14	9

3-HEXANOL
2YQ3

FIG. 3   (a) 2-Hexanol QY1 & 4. (b) 3-Hexanol 2YQ3.

relative intensity (I). A number of scoring methods have been used but to date the most successful has been

$$\text{Score} = \frac{N}{P} \sum (M \times I) = B$$

The use of the products $M \times I$ is used quite arbitrarily in the belief that the higher the mass M of the matched peaks the more significant is the ion in determining the molecular structure. The weighting factor N/P is used to prevent the degree of branching influencing the score value. At this stage the use of accurate masses from high resolution spectra are a distinct advantage in determining the correct matching and lead to more satisfactory scores than nominal mass values from low resolution spectra. Finally, the candidate structures and their scores are printed and hopefully the structure with the highest score is that of the unknown.

OUTPUT

A typical printout is shown in Fig. 3(a). Here the input was 2-Hexanol, $C_6H_{14}O$.

The first line printed shows that the preliminary filter has detected the hydroxyl group and has classified the compound as an alcohol.

The next line indicates that as the result of the partition tests in the first segment the conclusion is that we have a secondary alcohol.

The possible secondary alcohol isomers are then generated and each structure is fragmented and scored by the method B outlined above. The resulting structures and their scores are then printed. The highest score is shown by the first structure, which is the correct inference.

Figure 3(b) shows a similar correct conclusion for 3-Hexanol.

RESULTS

We report here a test of the program using 140 mono-functional acyclic compounds. These included alcohols, ethers, ketones and amines. The correct structure was found in first place in more than 50% of all cases considered, in second place in 19% of cases and in third place in 11% of cases. Thus in 80% of cases the correct structure was found in the first three highest scoring places. This compares favourably with other published success rates for this type of program.[6]

Table I shows the score ranking for a number of alcohols.

It is interesting to note that when the correct structure is not found in the first place, the highest scoring structure has many structural features in common with the actual structure. Often the only difference is in the position of a methyl group, e.g. the mass spectrum of 3,4-dimethyl-3-hexanol is inferred to be 3-methyl-3-heptanol. Both are tertiary alcohols containing methyl, ethyl and butyl groups but in the latter structure we have n-butyl instead of 1-methylpropyl. The correct structure is found in second place.

TABLE I

Results for Alcohol: Mass Spectra

HRP spectra ionizing energy 70 eV
source temperature 200°C

Compound	Number of $C_nH_{2n+2}O$ isomers	Number of inferred isomers	Score ranking
2,2-dimethyl-3-butanol	32	6	4
3-hexanol	32	6	1
2-hexanol	32	6	1
2,3-dimethyl-2-butanol	32	3	2
2,3-dimethyl-3-pentanol	72	6	1
2-methyl-2-hexanol	72	6	1
2-heptanol	72	13	1
4-heptanol	72	13	1
2,4-dimethyl-2-pentanol	72	6	2
4-methyl-3-hexanol	72	13	2
1-heptanol	72	16	2
3,4-dimethyl-3-hexanol	171	13	2
2-ethyl-1-hexanol	171	34	1
2,2,4-trimethyl-1-pentanol	171	34	19
5-methyl-3-heptanol	171	22	3
2-methyl-2-heptanol	171	13	1
4-methyl-3-heptanol	171	22	3
4-methyl-4-heptanol	171	13	1
2-octanol	171	22	1
4-octanol	171	22	1

The operation of the preliminary filter which ensures that the output is kept to manageable proportions is accurate in 97% of cases. This means that in all these cases the inferred structure is at least an isomer of the actual structure, the success rate is comparable with that of most spectrum matching programs using library search routines.[2]

REFERENCES

1. Crawford, L. R. and Morrison, J. D., *Anal. Chem.*, 1968, **40**, 1464.
2. Knock, B. A., Smith, I. C., Wright, D. E. and Ridley, R. G., *Anal. Chem.*, 1970, **42**, 1516.
3. Grotch, S. L., *Anal. Chem.*, 1971, **43**, 1362.
4. Hertz, H. S., Hites, R. A. and Biemann, K., *Anal. Chem.*, 1970, **43**, 681.
5. Pettersson, B. and Ryhage, R., *Anal. Chem.*, 1967, **39**, 790.
6. Crawford, L. R. and Morrison, J. D., *Anal. Chem.*, 1971, **43**, 1790.
7. Venkataraghavan, R., McLafferty, F. W. and Van Lear, G. E., *Org. Mass Spectrom.*, 1969, **2**, 1.
8. Biemann, K., McMurray, W. and Fennessey, P. V., *Tetrahedron Letters*, 1966, 3997.
9. Lederberg, J., Sutherland, G. L., Buchanan, B. G., Feigenbaum, E. A., Robertson, A. V., Duffield, A. M. and Djerassi, C., *J. Amer. Chem. Soc.*, 1969, **91**, 1973.
10. Delfino, A. B. and Buchs, A., *Helv. Chim. Acta*, 1972, **55**, 2017.
11. Smith, E. G., 'The Wiswesser Line Formula Notation', McGraw-Hill, New York, 1968.

Discussion

F. W. McLafferty (Cornell University, U.S.A.): Could you compare the basic steps of your approach to that of the 'Artificial Intelligence' method developed by Djerassi, Lederberg, *et al.*?

R. D. Sedgwick: The basic philosophy is the same but we are in a much earlier stage of development. I find it difficult to comment on the details of their work since in common with all publications in this field, they are necessarily brief.

119

Dynamic Man–Computer Interactive Data Processing for High and Low Resolution Mass Spectrometry Applied to Sequence Analysis of Peptide Mixtures

By H. A. VAN'T KLOOSTER,
J. S. VAARKAMP-LIJNSE and G. DIJKSTRA

(*Laboratory for Analytical Chemistry, University of Utrecht,
Utrecht, The Netherlands*)

INTRODUCTION

THE critical phase in mass spectrometry data processing procedures is the acquisition of the primary signal. Particularly with high resolution techniques even slight fluctuations in the conditions of the instruments may ruin the analysis, unless fast corrections can be carried out.

On-line data acquisition is therefore most effective if a fast output presents results in such a form that the operator may evaluate this output within a few seconds, with instantaneous response of the system. These requirements are met by a CRT visual display-with-keyboard unit. In such a terminal the high-speed output of the CRT is combined with an equally fast feed-back of information to the computer. Computer control of the critical functions of the mass spectrometer is an alternative method to deal with instrumental fluctuations, as recent developments indicate.[1,2]

There are, however, other factors affecting the quality of the data obtained which can be dealt with effectively only by human intervention, such as the behaviour of samples being offered for MS-analysis as 'slightly contaminated' which eventually turn out to be plain mixtures. Further, a sample is rarely presented for MS-analysis without any preliminary information, *e.g.* concerning the origin, techniques of synthesis or isolation and other spectral data. This information is not only used for initial instrumental settings, but may also provide a check on data acquisition performance.

The use of a CRT display-with-keyboard terminal allows a quick first evaluation of several scans from one sample, *e.g.* as it evaporates during the temperature rise of a direct insertion probe.

A study of the ergonomics of spectrometric analysis and subsequent experience have shown that early presentation of well-edited data followed by human intervention greatly reduces the amount of data handled and presented for scrutiny by the system, thus reducing the total effort.

In building up the integrated method of analysis we have developed programs for both high and low resolution on-line procedures, with partial use of commercially available software.[3]

We wish to report here, as an example, the application of a high resolution procedure to sequence analysis of peptides. We aim at an automated method for the sequence analysis of small peptides in simple mixtures, based on the high resolution EI mass spectra of appropriate derivatives.

Although some kind of separation is in most cases feasible, *e.g.* by fractional vaporization, we reckon with the superposition of spectra of two or more peptides. We demonstrate that even in such a case the method provides unequivocal identification of the peptides present. A brief comparison of high and low resolution results is given.

INSTRUMENTATION

The main components of our system are an A.E.I. MS 902 mass spectrometer and a Ferranti Argus 500 computer. The basic Argus 500 has 8 K of core store, 24 bit words and a cycle time of 1 μsec. The main component of the interface is an Adage Model VT 13-AB analogue–digital converter with a resolution of 13 bits + sign in 4 μsec, which enables a maximum sampling rate of 125 kHz for the overall system. The noise signal is rejected by a hardware threshold. A direct store access system enables processing of peak information as the scan proceeds. The total ion current is monitored at regular intervals during the scan period, allowing normalization of peak-intensity values on completion of the scan.[4]

Peripherals in the basic configuration include an ICL TR 6 paper-tape reader (300 characters per second), a Facit 1500 paper-tape punch (150 char./sec) and a Teletype ASR33 teleprinter (output 10 char./sec). We have extended the system with a Ferranti WD103 terminal display, in order to implement the philosophy set out above, with a Data Products 2310 line printer (360 char./sec) to increase analytical capacity and with another 8 K of core store for integration of various data processing procedures and storage of GC–MS results, anticipating the use of a disc as a backing store. The WD103 is a keyboard and CRT arrangement which provides high-speed communication in alphanumeric characters. The display accepts data at a rate of 480 char./sec. For normal input–output subroutines no more core store is required than using a teletype.

The entire system allows on-line high resolution mass measurement with a scan speed of 16 sec per mass decade, giving an average accuracy of 5 ppm in the mass range 70–700. With low resolution conditions a scan speed of 8 sec per decade is normally used.[3]

ON-LINE HIGH RESOLUTION MASS MEASUREMENT

The on-line high resolution procedures integrate the automated calibration of the mass/time function, using perfluorokerosene as a reference, and the accurate measurement of masses.

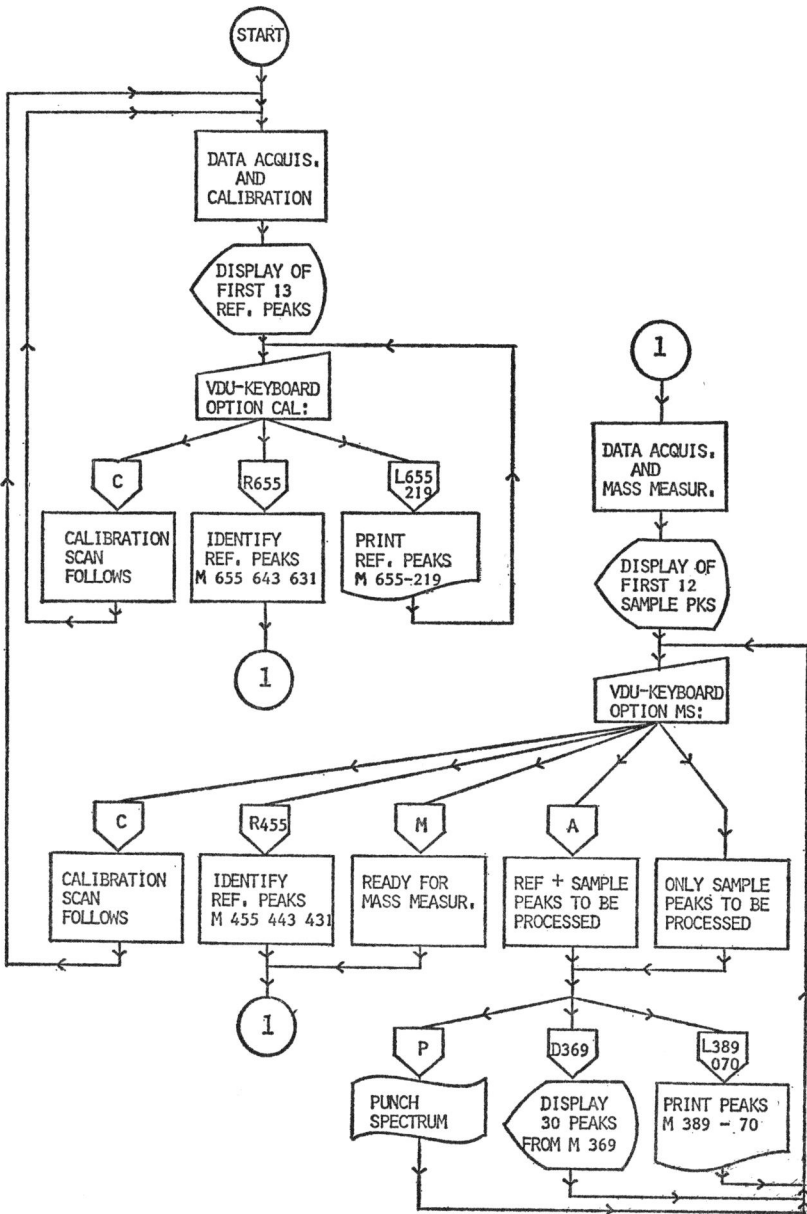

FIG. 1   On-line high resolution calibration and mass measurement.

TOTAL NO. OF PEAKS	580	MASS	INT	SAMPLESΔ
AV. SAMPLES PER PEAK	13.7Δ			
SCAN TIME	26.0	420.26761	0.54	9Δ
TOTAL ION CURRENT	4579.0	403.26298	0.18	5Δ
		387.22915	0.36	7Δ
		386.22967	1.61	13Δ
		375.23142	0.36	7Δ
		374.22323	1.43	13Δ
		358.23479	1.79	13Δ
		324.27412	1.25	12Δ
		323.27224	7.89	21Δ
EXIT CODE 0		322.26499	4.48	18Δ
		321.25948	0.18	5Δ
NO. OF SAMPLE PEAKS	403	321.22814	0.54	10Δ
MASS BASE PEAK	142.08757Δ			
INTENSITY BASE PEAK	558	KEYBOARD OPTION MS: P̲		

FIG. 2 On-line results of high resolution mass measurement. A few seconds after the scan has been completed (scan speed: 16 sec per mass decade; total scan time: 26 sec) the first 12 peaks in the high mass region of the spectrum are being displayed, preceded by a summary of scan data. The average peak width (av. samples per peak) gives an indication of the dynamic resolving power. The total number of peaks includes 403 peaks due to the peptide mixture, the rest being PFK-peaks. Peak intensities are calculated relative to the intensity of the base peak, the displayed value of which (558) is indicative of the quality of the spectrum.

A block diagram is given in Fig. 1, showing the subroutines and their activation by means of the visual display keyboard, using various option codes. These options include fast recalibration, the processing of mass deficient samples (in which case reference and sample peaks are not separated) and hard copy production by fast punch or line printer. Figure 2 shows a typical result as a first output, the analysed sample being a mixture of small peptides.

SEQUENCE ANALYSIS OF SMALL PEPTIDES IN MIXTURES

A key-factor in peptide sequencing by mass spectrometry is the chemical method employed to protect the polar functions of the peptide: the peptide derivative should be sufficiently volatile and the mass spectrum should contain sufficiently high 'sequence-peaks'. Acetylation and permethylation are usually applied. Kamerling et al. have demonstrated the utility of ethoxycarbonylpeptidemethylesters.[5] The electron impact mass spectra of these derivatives show series of both N-terminal and C-terminal sequence peaks. The same applies to permethylated ethoxycarbonylpeptidemethylesters.[6] In Fig. 3 the N- and C-terminal sequence ions of a schematic tripeptide are shown. This experience has formed a starting point in the development of our peptide sequencing method described below.

$$\overleftarrow{A_1}\quad \overleftarrow{B_1}\qquad \overleftarrow{A_2}\quad \overleftarrow{B_2}\qquad \overleftarrow{A_3}\quad \overleftarrow{B_3}\qquad \overleftarrow{M}$$

$$C_2H_5O \!-\! \underset{\underset{}{\overset{O}{\|}}}{C} \!-\! \underset{H}{N} \!-\! \underset{\underset{R_1}{\overset{CD_3}{|}}}{C} \!-\! \underset{\overset{O}{\|}}{C} \!-\! \underset{H}{N} \!-\! \underset{\underset{R_2}{\overset{CD_3}{|}}}{C} \!-\! \underset{\overset{O}{\|}}{C} \!-\! \underset{H}{N} \!-\! \underset{\underset{R_3}{\overset{CD_3}{|}}}{C} \!-\! \underset{\overset{O}{\|}}{C} \!-\! OCD_3$$

$$\overrightarrow{M}\quad \overrightarrow{Z_3}\ \overrightarrow{Y_3}\qquad \overrightarrow{Z_2}\ \overrightarrow{Y_2}\qquad \overrightarrow{Z_1}\ \overrightarrow{Y_1}$$

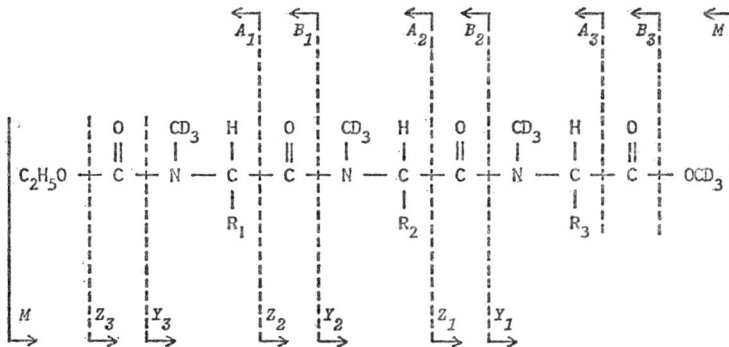

FIG. 3   N- and C-terminal sequence ions of a schematic perdeutero-permethylated ethoxycarbonyltripeptidemethylester.

A program along the lines of Biemann's peptide sequencing algorithm[7] using high resolution spectra, has been developed at the University of Liverpool, for use with an MS902–Argus 500 system.[8] This program generates possible sequences using the accurate masses of N-terminal amine- and aminoacyl-ions (*see* Fig. 3, ions A and B) as a primary criterion. In the case of alternative sequences the relative abundance of the corresponding sequence ions is used as a secondary criterion. Low-intensity peaks are ignored by programmed intensity thresholds.

Starting from this program we have developed a new algorithm: sequences are being reconstructed from N- towards C-terminus and vice versa. Consistency between an N → C and a C → N sequence forms the decisive criterion for a correct sequence.

The less reliable abundance of N-terminal sequence peaks as a criterion breaks down entirely when in peptide mixtures (*e.g.* from hydrolysates) some components are present in low concentration.

We take into account all peaks down to a threshold as low as the noise level. The criteria applied in the sequencing program were derived from the high resolution mass spectra of some 30 ethoxycarbonylpeptidemethylesters (di-, tri- and tetrapeptides), most of them also having been perdeuteroper-methylated.[9] For the identification of an N-terminal amino acid both the A_1- and B_1-peaks (*see* Fig. 3) must be present, whereas subsequent amino acid residues are identified by the presence of subsequent B-peaks. Identification of a C-terminal amino acid requires the presence of either a Y_1- or a Z_1-peak and the further reconstruction of a C → N sequence requires subsequent Z-peaks of higher order. A minimum overlap of two amino acids is required for consistency between an N → C and a C → N sequence being found initially. Consequently the presence of a molecular peak is not essential and unambiguous results were obtained without it.[9] A flow diagram of the program is given in Fig. 4.

The decision not to search for an overlap is made by the operator when the sample is suspected to contain minor concentrations of some peptides, for which at least partial sequences may then be deduced by manual interpretation of the initial results.

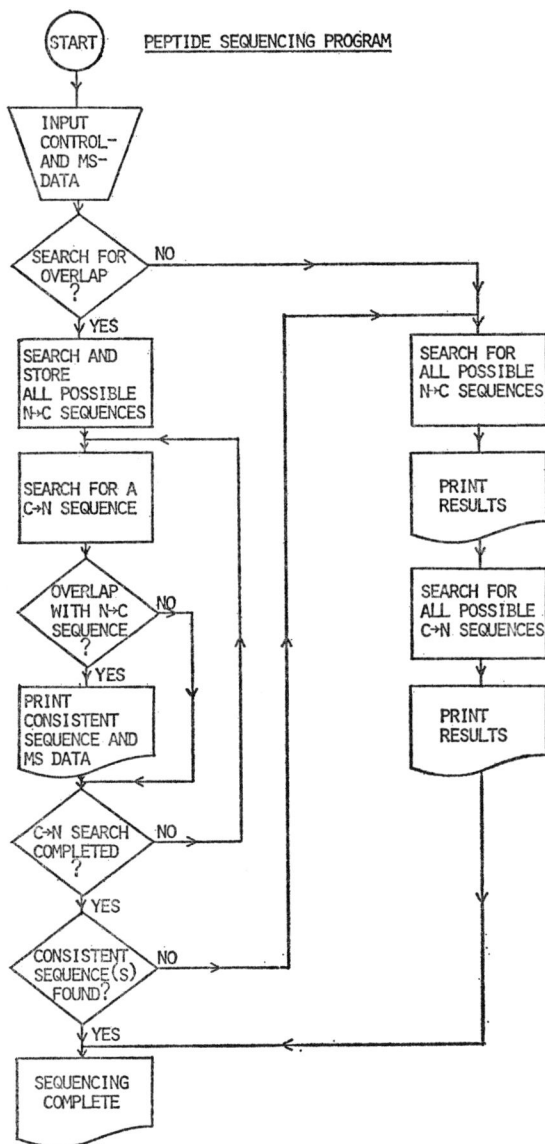

FIG. 4   Flow diagram of the peptide sequencing program.

As an example we give the analysis of a synthetic mixture of *ca.* 0·1 μmole Pro-Val, 0·05 μmole Asp-Phe and 0·05 μmole Gly-Leu-Tyr, being ethoxy-carbonylated and perdeuteropermethylated. Using a common direct insertion probe fractional vaporization has been achieved by introducing the probe-tip into the ion source at regular time intervals while increasing the source temperature. (An independently heated probe was not operational at the time.) During the analysis some 20 spectra were scanned, the results being

PEPTIDE SEQUENCE PROGRAM

DATE:	14-05-73	DERIVATIVE:	
SAMPLE CODE:	337PD-R2	N-TERMIN. :	C2H5OCO
TOTAL NO. OF PEAKS:	403	C-TERMIN. :	OCD3
MASS FIRST PEAK:	420.2676	AMIDE-N :	CD3
MASS BASE PEAK:	142.0876		
INTENSITY BASE PEAK:	558	AMINOACID ANALYSIS:	
		NOT AVAILABLE	
REL. INT. THRESHOLD:	0.01 %		
MASS TOLERANCE:	4.0 MMU		
SEARCH FOR OVERLAP:	YES		

- - - - - -

POSSIBLE SEQUENCE 1: - PRO - VAL

SEQUENCE PEAKS:

MASS MEASURED	ERROR MMU	INTENSITY REL./B.P.	ION-TYPE N-TERMIN.	ION-TYPE C-TERMIN.
320.2226	0.8	3.76	M	
286.1838	-0.8	0.36	B2	
275.1874	-0.4	2.51		Z2
258.1907	1.0	0.90	A2	
247.1926	-0.3	3.41		Y2
178.1348	-0.2	4.12		Z1
170.0815	-0.2	1.08	B1	
150.1402	0.1	17.74		Y1
142.0876	0.8	100.00	A1	

- - - - - -

POSSIBLE SEQUENCE 2: - ASP - PHE

SEQUENCE PEAKS:

MASS MEASURED	ERROR MMU	INTENSITY REL./B.P.	ION-TYPE N-TERMIN.	ION-TYPE C-TERMIN.
420.2676	2.7	0.54	M	
386.2297	1.9	1.61	B2	
375.2314	0.5	0.36		Z2
358.2348	2.0	1.79	A2	
226.1348	-0.2	0.90		Z1
222.1246	-0.2	15.05	B1	
194.1293	-0.7	46.42	A1	

- - - - - -

SEQUENCING COMPLETE

FIG. 5 Output of the peptide sequencing program. A summary of scan data and specified limits is first printed. The program allows processing of various peptide derivatives and the input of an amino acid analysis. As results in this case two peptides have been identified, based on the MS data printed underneath. The codes for the ion types correspond to those in Fig. 3.

displayed as described above. At a source temperature of 90°C a first spectrum was acquired, the displayed part showing a highest mass of 321·227 76. In the following displays no significant change in the high mass region was observed up to 110°C. At this temperature the results shown in Fig. 2 were obtained. A different spectrum again was acquired at 130°C, with a highest mass of 508·375 00.

The complete spectra of these three runs were stored on paper tape by simply pressing a 'P' on the keyboard of the visual display terminal. Processing by the peptide sequencing program resulted in:

> run 1: Pro-Val
> run 2: Pro-Val and Asp-Phe
> run 3: Gly-Leu-Tyr

The output for run 2 is shown in Fig. 5. The processing of these three spectra took less than 5 minutes altogether. It should be pointed out that no amino acid analysis was specified, so that seventeen amino acids were taken into account.

In order to evaluate explicitly the requirement of high resolution mass measurement we have converted the masses of the three HR spectra to unit resolution. These 'low resolution' spectra have been run with the same sequencing program with the same control data, except of course the mass tolerance. The results, summarized for both the HRP- and the 'LRP'-data are given in Table I. It should be emphasized that these 'LRP'-spectra contain a reduced number of peaks compared to real low resolution spectra, which consequently give rise to more artefacts.

In a detailed report a number of factors will be discussed concerning data reduction and evaluation both for identification and confirmation purposes, including the role of amino acid analysis and the use of high versus low resolution data.[9] Also a variety of other MS data processing procedures is

TABLE I

Number of Sequences Found

	Amino acid analysis specified				No amino acid analysis specified			
	$N \rightarrow C$	$C \rightarrow N$	Consistent	Artefacts	$N \rightarrow C$	$C \rightarrow N$	Consistent	Artefacts
HRP run no.								
1	1	5	1	0	2	5	1	0
2	4	6	2	0	5	8	2	0
3	2	5	1	0	2	6	1	0
'LRP' run no.								
1	1	11	1	0	4	26	4	3
2	13	17	5	3	31	55	6	4
3	5	10	2	1	10	33	5	5

described in a forthcoming paper[3] and work is in progress to prepare on-line identification procedures, using time sharing software and a disc as a secondary memory.

ACKNOWLEDGMENTS

This work was supported in part by the Netherlands Foundation for Chemical Research (SON) with financial aid from the Netherlands Organization for the Advancement of Pure Research (ZWO). We are indebted to Dr J. F. G. Vliegenthart for helpful discussions, to Mrs L. R. Smaling for the chemical derivatization of the peptide mixture and to Mr C. Versluis for technical assistance.

REFERENCES

1. 'Biochemical Applications of Mass Spectrometry', *ed.* G. Waller, John Wiley & Sons, New York, 1972.
2. McLafferty, F. W., Michnowicz, J. A., Venkataraghavan, R., Rogerson, P. and Giessner, B. G., *Anal. Chem.*, 1972, **44**, 2282.
3. van't Klooster, H. A., Vaarkamp-Lijnse, J. S. and Dijkstra, G., *Org. Mass Spectrom.*, in press.
4. Bowen, H. C., Chenevix-Trench, T., Drackley, S. D., Faust, R. C. and Saunders, R. A., *J. Sci. Instrum.*, 1967, **44**, 343.
5. Kamerling, J. P., Heerma, W. and Vliegenthart, J. F. G., *Org. Mass Spectrom.*, 1968, **1**, 351.
6. Vliegenthart, J. F. G. and Dorland, L., *Biochem. J.*, 1970, **117**, 31P.
7. Biemann, K., Cone, C., Webster, B. R. and Arsenault, G. P., *J. Amer. Chem. Soc.*, 1966, **88**, 5598.
8. Upham, R. A., Ward, S. D., Johnstone, R. A. W. and Kenner, G. W., unpublished results.
9. van't Klooster, H. A., Vaarkamp-Lijnse, J. S., Heerma, W., Vliegenthart, J. F. G., Upham, R. A., Ward, S. D., Johnstone, R. A. W. and Kenner, G. W., manuscript in preparation.

Discussion

P. Roepstorff (Danish Institute of Protein Chemistry, Copenhagen, Denmark): Why does your program by reduction of the spectrum from high to low resolution omit the correct sequences? A number of artefacts should be expected to show up but the sequence peaks of the correct sequences should still be present and thus the correct sequences should be deduced.

H. A. van't Klooster: Our present program considers deduced series of N → C and C → N sequence-peaks with maximum length only, *e.g.* if A, A–B and A–B–C are found respectively, only A–B–C is stored, unless for say A–B also a molecular ion has been identified, in which case A–B is stored too. If all possible partial sequences are taken into account the correct sequences are deduced indeed, but also a number of artefacts is thus being generated, even with high resolution data. On the other hand, incomplete series of either N → C or C → N sequence-peaks may still lead to a correct sequence, provided that no interference occurs by spurious sequence-peaks. These are to be eliminated from high rather than from low resolution data. For example, from the third high resolution spectrum of the discussed test-mixture the program has deduced Gly-Leu as one of the possible N → C sequences, and Tyr-Leu-Gly as a possible C → N sequence, thus concluding to Gly-Leu-Tyr as a correct sequence, although no B$_3$-peak was present. Converted to unit resolution, however, a peak at m/e 398 was identified as the B$_3$-peak of Gly-Leu-Ser, and the program accepts this artefact as a possible N → C sequence, rejecting Gly-Leu. Since no overlap occurs in this case Gly-Leu-Tyr is thus omitted, being a non-consistent sequence. The accurate mass of peak 398, contained in the high resolution data however, was 398·262 5, whereas the calculated exact mass of the B_3-ion of Gly-Leu-Ser is 398·304 9. The difference of 42 mmu, being far beyond the mass tolerance of 4 mmu, has prevented the generation of this artefact, in the case of processing high resolution data.

120

An Experimental International Conversational Mass Spectral Search System

By S. R. HELLER, H. M. FALES, G. W. A. MILNE
R. J. FELDMANN

(*National Institutes of Health, Bethesda, Maryland, U.S.A.*)

N. R. DALY, D. C. MAXWELL and A. McCORMICK

(*The Mass Spectrometry Data Centre, A.W.R.E., Aldermaston, Reading, U.K.*)

An interactive, conversational mass spectral search (MSS) system, available over ordinary telephone lines using teletypewriter terminals has been used by over 200 scientists in the U.S.A. and Canada since 1971. The system is used an average of 25 times per day and was originally located on a PDP-10 computer in the Division of Computer Research and Technology (D.C.R.T.) at the National Institutes of Health (N.I.H.) in Bethesda, Maryland, and supported by the National Heart and Lung Institute (N.H.L.I.).

The MSS is one component of a Chemical Information System being developed at D.C.R.T. The system has recently been transferred to the worldwide General Electric (G.E.) timesharing network which allows the system to be used for the cost of a subscription fee, the computer charge, and a local telephone call to any of over 300 cities in Japan, United States, Canada, Great Britain and nine other countries on the European continent. Further local service telephone facilities are expected. Most cities have both low (10 Hz) and high (30 Hz) speed telephone service available. At this meeting the system will be demonstrated by calling a local Edinburgh telephone number.

Details of the system have been presented elsewhere[1-6] and the use, value, and future of the system will be highlighted here. The present program options on the system operating on the G.E. network include:

1. Peak and intensity search
2. Molecular weight search
3. Complete and partial molecular formula search
4. Peak and molecular weight search
5. Peak and molecular formula search
6. Molecular weight and formula search

7. Dissimilarity index comparison
8. Spectrum printout
9. Automatic and manual microfiche retrieval
10. CRAB–comments and complaints
11. HARVEST–entering of new data
12. NEWS–news of the system
13. MSDC code list

At present, consideration is being given to implementing the display of spectra on graphics terminals. This option, available at N.I.H., was not transferred to the G.E. system due to software and system incompatibilities. In addition, a reverse spectrum search, that is, search for losses from the parent ion from 0 (Parent Ion) to −100, is under development. M.S.D.C. codes and Chemical Abstracts service (CAS) Registry Numbers (REGN) are to be added to the file for future searching by structural and functional groups.

PROGRAM: MASS SPEC PEAK AND INTENSITY SEARCH

USER: INTENSITY RANGE FACTOR FOR THIS SEARCH IS: 2

TYPE PEAK, INT
CR TO EXIT, 1 FOR ID #/NAMES

USER: 85,100

PROGRAM: FOUND 1365 SPECTRA WITH M/E PEAK: 85

```
# REFS     M/E PEAKS

  187        85
```

TYPE PEAK, INT
CR TO EXIT, 1 FOR ID #/NAMES

USER: 128,20

PROGRAM: FOUND 857 SPECTRA WITH M/E PEAK: 128

```
# REFS     M/E PEAKS

   8        85  128
```

TYPE PEAK, INT
CR TO EXIT, 1 FDR ID #/NAMES

USER: 3

ID#	MW	MF	NAME
1722	314	C19.H38.O3	METHYL 4-HYDROXYOCTADECANDATE
2085	198	C13.H26.O	DIHEXYL KETONE
2543	170	C10.H18.O2	GAMMA-DECALACTONE
2560	254	C16.H30.O2	GAMMA-PALMITOLACTONE
2561	282	C18.H34.O2	GAMMA-STEAROLACTONE
6505	128	C7.H12.S	6-TRIABICYCLO (3.2.1)OCTANE
8060	246	C15.H18.O3	6-EPI-ALPHA-SANTONIN
8470	309	C21.H27.N.O	6-DIMETHYLAMINO-4,4-DIPHENYL-3-HEPTANONE (METHADONE)

FIGURE 1

In general, the response to the system has been favourable and this positive reaction is the prime reason which has encouraged us to make the system available on such a scale to the scientific community. The main comments which users have made about the system during its trial period on the N.I.H. computer have been concerned with the size and nature of the spectral file. There are a large number of exact replicate spectra, and in addition there are in some cases several similar spectra of the same substance obtained from different sources. For example, there are 7 benzene spectra, 3 hydrogen spectra, 3 thiophene spectra and 7 acetone spectra. Obviously it is desirable to delete exact replicates. However, the consensus of opinion at a workshop session on matching systems held at the A.S.M.S. meeting in San Francisco was that holding several spectra of the same compound obtained under different conditions could be useful, particularly for more complex molecules, *e.g.* cholesterol, whose spectra may be quite sensitive to instrumental parameters. In addition to replication of spectra there are in some cases errors in peak location and intensity. Many spectra were recorded starting at m/e 40 or even as high as m/e 60; these spectra are often lost as possible answers because of this fact. The file is admittedly inadequate in such areas as drugs, steroids, pesticides, organometallics and other biochemical materials in general (amino acids, lipids, sugars, etc.). It is hoped that many of the deficiencies of the data file will be put right in the very near future and that subscribers to the search system will assist in still further improving it by submitting new spectra either through the on-line HARVEST option or by sending spectra to the Mass Spectrometry Data Centre at Aldermaston. It should be pointed out that maintaining and improving a data base of this size and complexity is an expensive procedure. In addition to producing the data base there are costs involved in loading and storing it on the G.E. computer system. It is for these reasons that it is necessary to charge a fairly high subscription rate for use of the system.

The remainder of this paper will be directed towards examples of some of the search options. Figures 1 and 2 are examples of the peak and intensity search option and are designed to show that only a few peaks are usually needed to narrow the number of possible answers down to just a few. Indeed in Fig. 2, three peaks lead to only one answer.

Figure 3 shows the molecular weight search for a molecular weight of 151, and indicates the presence of possible duplicate spectra.

An example of one of the combination search options, the molecular weight and peak search is shown in Fig. 4. In this search only one peak along with the molecular weight was needed to narrow the possible answers to three.

The complete molecular formula search option is shown in Fig. 5. Again the presence of more than one spectrum for the same compound is evident throughout the list.

After obtaining answers from searches such as those illustrated, most users wish to see the spectrum from the file for visual comparison. Since the spectra are all stored on-line on direct access discs, this is a simple matter, and Fig. 6 shows a sample spectrum printout for the simple spectrum of HBr. In addition to this on-line spectrum printout, microfiches of the file have been computer generated for viewing and are expected to be made available at cost to registered users of the system.

PROGRAM: MASS SPEC PEAK AND INTENSITY SEARCH

USER: INTENSITY RANGE FACTOR FOR THIS SEARCH IS: 4

TYPE PEAK, INT
CR TO EXIT, 1 FOR ID #/NAMES, 3 FOR ID, MW, MF, NAMES

USER: 86,100

PROGRAM: FOUND 507 SPECTRA WITH M/E PEAK: 86

REFS   M/E PEAKS

80   86

TYPE PEAK, INT
CR TO EXIT, 1 FOR ID #/NAMES, 3 FOR ID, MW, MF, NAMES

USER: 44,50

PROGRAM: FOUND 713 SPECTRA WITH M/E PEAK: 44

REFS   M/E PEAKS

11   86   44

TYPE PEAK, INT
CR TO EXIT, 1 FOR ID #/NAMES, 3 FOR ID, MW, MF, NAMES

USER: 74,10

PROGRAM: FOUND 804 SPECTRA WITH M/E PEAK: 74

REFS   M/E PEAKS

1   86   44   74

TYPE PEAK, INT
CR TO EXIT, 1 FOR ID #/NAMES, 3 FOR ID, MW, MF, NAMES

USER: 3

ID#	MW	MF	NAME
207	159	C8.H17.N.O2	LEUCINE ETHYL ESTER

FIGURE 2

TYPE MOLECULAR WEIGHT
CR TO EXIT

USER: 151

PROGRAM: FOUND 12 SPECTRA WITH MW: 151

PROGRAM: PRODUCE REFERENCES YES OR NO?

USER: YES

ID#	MW	MF	NAME
541	151	C8.H9.N.O2	ETHYL NICOTINATE
1894	151	C8.H9.N.O2	ETHYL NICOTINATE
3640	151	C7.H5.N.O3	P-NITRO-BENZALDEHYDE
3641	151	C8.H9.N.O2	METHYL P-AMINOBENZOATE
3642	151	C7.H5.N.O3	M-NITRO-BENZALDEHYDE
7283	151	C8.H9.N.O2	P-ACETAMIDOPHENOL
7284	151	C8.H9.N.O2	O-ACETAMIDOPHENOL
7638	151	C10.H17.N	1-(1-PYRROLIDINYL)-1-CYCLOHEXENE
8240	151	C8.H9.N.O2	ACETOAMINOPHENE
8316	151	C9.H11.O2	ACETAMINOPHEN METHYL ESTER
8439	151	C8.H9.N.O2	4'-HYDROXY-ACETANILIDE (ACETAMINOPHEN)
8532	151	C8.H9.N.O2	PHENYL N-METHYLCARBAMATE

FIGURE 3

PROGRAM: MOLECULAR WEIGHT AND PEAK SEARCH
CR TO EXIT

PLEASE GIVE MOLECULAR WEIGHT FIRST, THEN PEAKS

USER: THE MW IS: 156

PROGRAM: PEAK SEARCH

USER: INTENSITY RANGE FACTOR FOR THIS SEARCH IS: 5

TYPE PEAK, INT
CR TO EXIT
1 FOR ID/NAMES

USER: 100,50

PROGRAM: FOUND 3 REFERENCES TO THAT COMBINATION OF MW AND PEAKS

TYPE PEAK, INT
CR TO EXIT
1 FOR ID/NAMES

USER: 1

ID#	NAME
216	2,5-DIMETHYL-3-NORM-PROPYL-PIPERAZINE
1275	CAPROATE, ALLYL
6630	TRANS-2-THIABICYCLO(4.4.0)DECANE (TRANS-1-THIADECAHYDRONAP)

FIGURE 4

PROGRAM: MOLECULAR FORMULA SEARCH

TO SEARCH FOR A COMPLETE MF, TYPE ALL
TO SEARCH FOR A PARTIAL MF, TYPE IMBED
CR TO EXIT

USER: ALL

PROGRAM: TYPE IN MOLECULAR FORMULA
(ORDER OF ATOMS: C,H,N,O,S)
USER: C6.H6.O2

PROGRAM: FOUND 11 REFERENCES TO MOLECULAR FORMULA C6.H6.O2

PROGRAM: PRODUCE REFERENCES YES OR NO?

USER: YES

ID#	MW	MF	NAME
625	110	C6.H6.O2	METHYLFURYL KETONE
759	110	C6.H6.O2	5-METHYL-2-FURFURAL
1025	110	C6.H6.O2	CATECHOL
1026	110	C6.H6.O2	RESORCINOL
1027	110	C6.H6.O2	HYDROQUINONE
1539	110	C6.H6.O2	PYROCATECHOL
3084	110	C6.H6.O2	PYROQUINONE
3086	110	C6.H6.O2	PYROCATECHOL
3088	110	C6.H6.O2	RESORCINOL
3089	110	C6.H6.O2	FURYL METHYL KETONE
3093	110	C6.H6.O2	2-FURYL METHYL KETONE

FIGURE 5

PROGRAM: TYPE ID# FOR PEAKS/INTENSITIES
CR TO EXIT

USER: 100

ID#	MW	MF	NAME
100	80	H.BR	HYDROGEN BROMIDE

PROGRAM: THERE ARE 7 PEAKS IN SPECTRUM #100

PROGRAM: PRODUCE PEAKS/INTENSITIES YES OR NO?

USER: YES

PROGRAM: PLEASE GIVE RANGE FOR PEAKS
AND INTENSITY LEVEL

USER: THE MIN PEAK IS 0

USER: THE MAX PEAK IS: 100

USER: TYPE ALL PEAKS WITH INTENSITY > 0

M/E	INT
2	1
40	3
41	3
79	45
80	100
81	44
82	98

FIGURE 6

In summary, the MSS offers a number of options for searching a large on-line data base, now available 24 hours a day, seven days a week on the G.E. computer network. With the support of mass spectrometrists the system should improve in depth, providing a valuable tool for the mass spectral information needs of the worldwide scientific community.

REFERENCES

1. Heller, S. R., 'Conversational Mass Spectral Retrieval System and Its Use as an Aid in Structure Determination', *Anal. Chem.*, 1972, **44**, 1951.
2. Heller, S. R., Fales, H. M. and Milne, G. W. A., 'An Interactive Mass Spectral Search System', *J. Chem. Ed.*, 1972, **49**, 725.
3. Heller, S. R., Fales, H. M. and Milne, G. W. A., 'A Conversational Mass Spectral Search and Retrieval System. II. Combined Search Options', *Org. Mass Spectrom.*, 1972, **7**, 107.
4. Heller, S. R., Koniver, D. A. and Milne, G. W. A., 'A Conversational Mass Spectral Search System. III. Display and Plotting of Spectra and Dissimilarity Comparisons', *Anal. Chem.*, submitted.
5. Heller, S. R., Feldmann, R. J., Fales, H. M. and Milne, G. W. A., 'A Conversational Mass Spectral Search System. IV. The Evolution of a System for the Retrieval of Mass Spectral Information', *J. Chem. Soc.*, submitted.
6. Heller, S. R., 'DCRT/CIS Mass Spectral Search System User's Manual', November 1972, D.C.R.T., N.I.H., Bethesda, Maryland.

121
A User-Designed GC-MS Computer System

By B. HEDFJÄLL, R. RYHAGE and Å. ÅKERLIND

(Karolinska Institutet, Stockholm, Sweden)

It has now been ten years since a demonstration of a gas chromatograph connected to a mass spectrometer via a molecular separator was made. At that time a complex mixture of methylesters of fatty acids of butterfat was run (Fig. 1). It was possible to identify all the components of the mixture and by using rapid scan (3 seconds for mass range 10–500) several mass spectra from unresolved GC peaks could be identified. This is illustrated in Fig. 2 where scans I, J and K represent three C_{14} isomers with fragmentations much in agreement with their reference compounds.

The GC-MS method has been frequently used in accelerating voltage alternator (AVA) applications. In the beginning only two masses were studied (1965). Since then the number of masses studied during a GC run has increased to three or four. The sensitivity of the instrument has also been increased and samples can be studied in the picogramme range. This method is called mass fragmentography or mass chromatography and has been successful in the identification and qualitative determination of drug metabolites and for similar applications. The improvements in ion registration from the mass spectrometers output have played an important role in the analysis of substances.

The connection of a computer to the combined GC-MS instrument has been a powerful tool for reducing the manual work on mass spectrometric data. Figure 3 shows how one of the GC-MS instruments in our laboratory has been connected to a small computer. It is a block diagram of the GC-MS computer system. The GC is connected to the MS via a jet separator. In addition to supplying the signal for UV recording the output of the MS is fed through a special interface to a PDP-11/20 processor with an 8 K core memory. Other hardware units are DEC tape, display terminal and an electrostatic printer/plotter manufactured by Varian. When used as a printer an A4 page can be printed in one second and as a plotter the speed is about 5 seconds per mass spectrum.

Figure 4 shows a schematic diagram of the mass marker. The magnetic field is measured by a transducer which directly reads the nominal mass numbers. It is undoubtedly an advantage to use a mass marker both for increment and continuous scan as compared to the use of a time-dependent

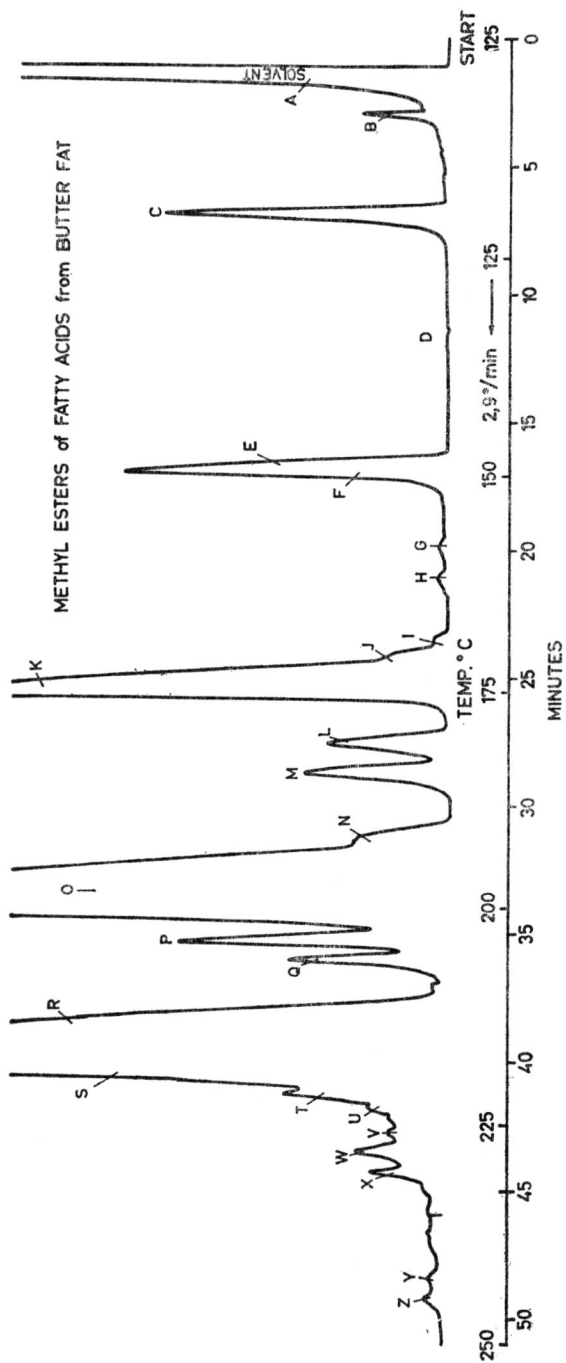

Fig. 1 Chromatographic separation of methyl esters of fatty acids from butterfat.

scan. We have found that a rapid and accurate mass marker increases the performance of the GC-MS computer combination. The mass marker used for this application is manufactured by LKB-Produkter.

The transducer consists of two conductors mechanically coupled together but electronically insulated. An oscillator is producing the necessary current through the primary circuit of the transducer. The frequency chosen is 320 Hz. The reference voltage over the condenser is 10 V. It is fed via a

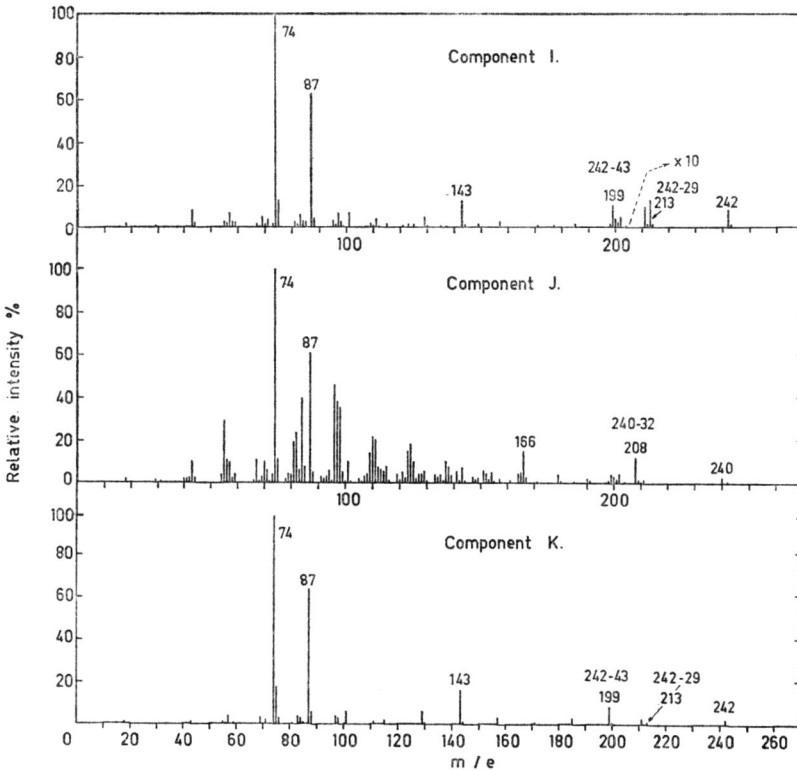

FIG. 2   Mass spectra of incompletely separated components I, J and K from chromatogram of butterfat.

differential amplifier as a reference into the 17-bit D/A-converter. A fraction of this reference is taken to the summing line. This fraction is defined by the BCD-word in the 17-bit up/down converter. This means that $M/M_{const.}$ times U_R is fed to the summing line where $M_{const.}$ is that mass number which makes $U_S = U_R$ ($M_{const.}$ varies with the operational conditions of the system). A fraction of this reference is taken to the summing line. The signal U_S from the transducer is also fed into the summing line. E is the difference between U_S and $U_R \cdot M/M_{const.}$ ($E = U_S - U_R \cdot M/M_{const.}$). The error analyser will detect the error E and deliver count direction and clock pulses

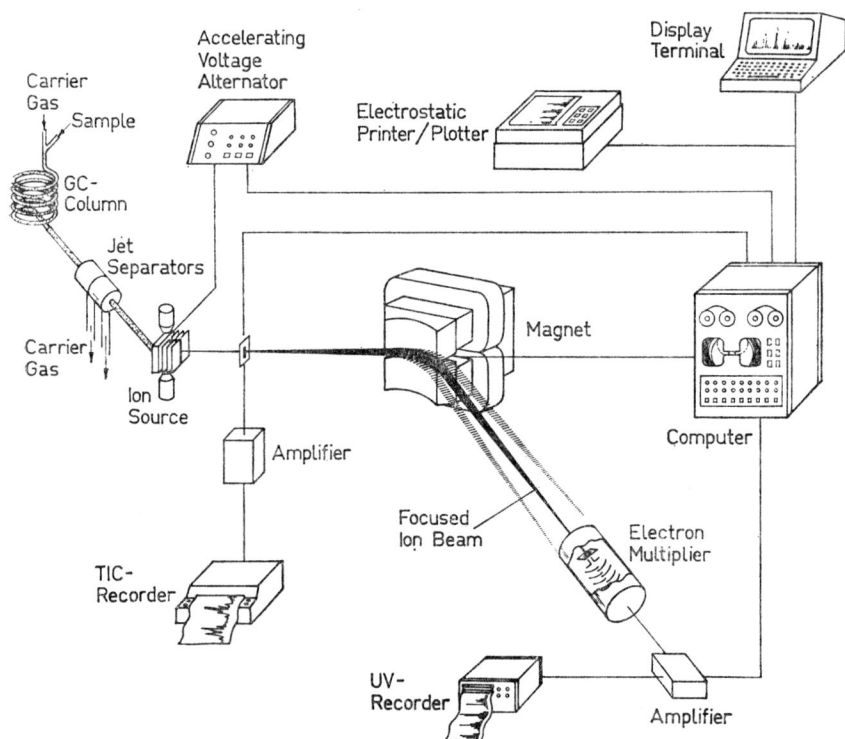

FIG. 3   Block diagram of the GC-MS computer system.

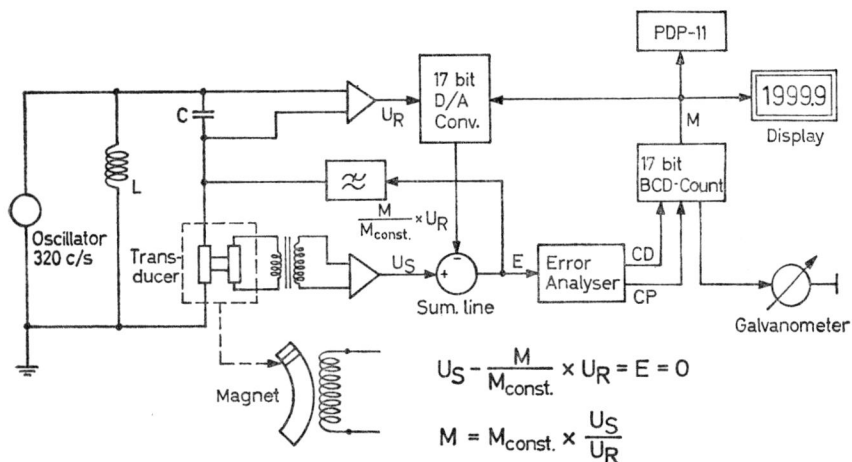

$$U_S - \frac{M}{M_{const.}} \times U_R = E = 0$$

$$M = M_{const.} \times \frac{U_S}{U_R}$$

FIG. 4   Schematic diagram of the mass marker.

TABLE I

Mass Marker Used with the GC-MS Computer System

1. Mass marking on UV paper.
2. Manual setting of the magnetic field is displayed in mass units (as for AVA applications).
3. Scan start and stop directly adjustable and independent of scan speed.
4. Increment sampling (mass number and intensity).
5. Continuous sampling (10 kHz–90 kHz/sec).
6. Computer controlled magnetic field.

to the BCD counter. This changes the fraction of U_R that is fed to the summing line. The error signal is thus reduced and when it is within the resolution of the mass marker the error analyser ceases to deliver clock-pulses. When $E = 0$, $M = M_{const.} U_S/U_R$.

Table I shows different ways to use the mass marker in the GC-MS computer system.

Figure 5 shows the mass marker calibrating diagram used in the increment mode of operation. The diagram shows the mass number of PFK as a function of mass defects. Experimentation has shown that a difference in mass up to 0·5 mass units can be accepted for a given calibration. However, this is dependent upon the resolution. If the mass marker is calibrated for saturated

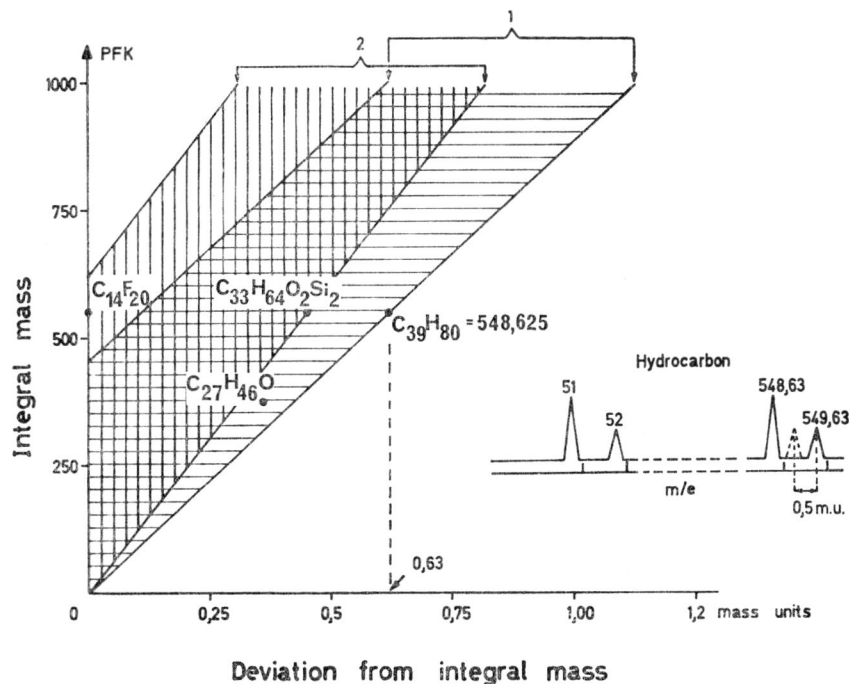

Fig. 5 Mass marker calibration diagram.

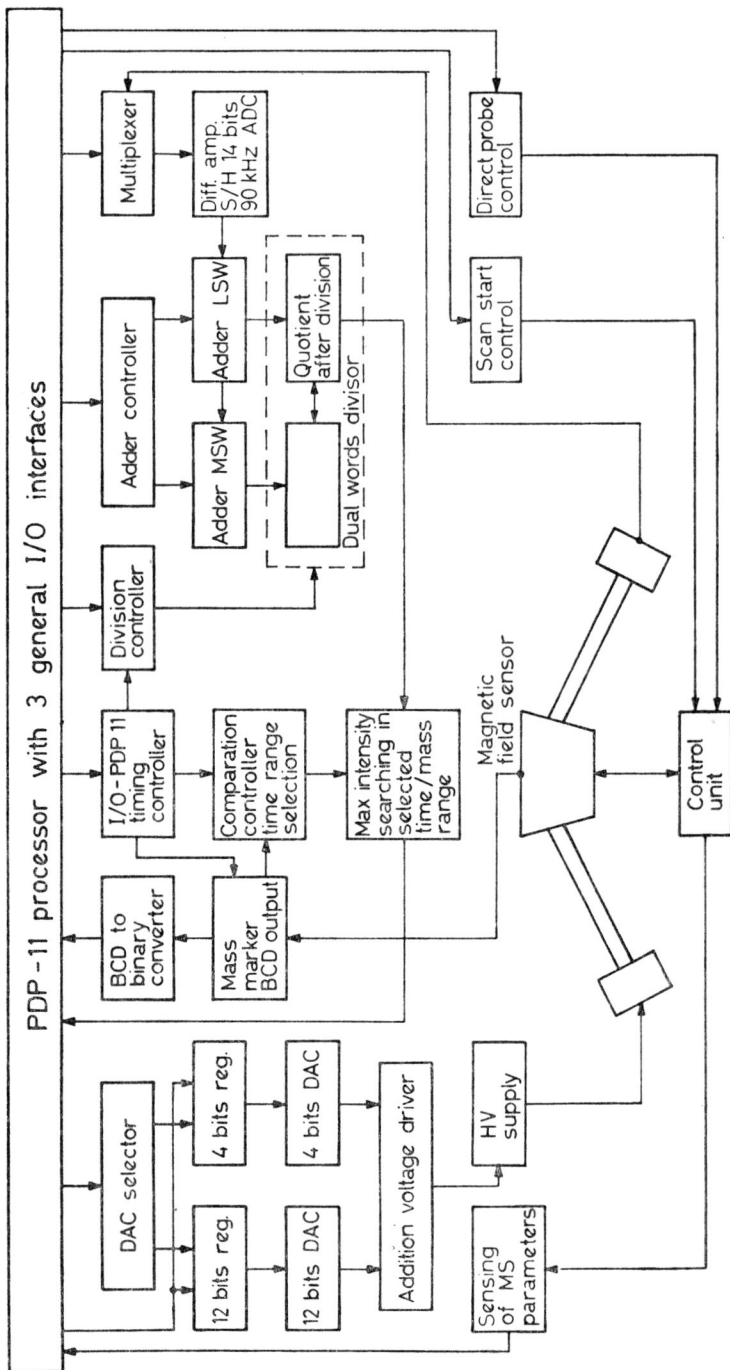

FIG. 6   Block diagram of the GC-MS computer interface.

hydrocarbons; area 1 shows the working range of the incremental registration system. Most organic compounds are to be found within this area. Aromatic compounds with a molecular weight above 500 or compounds which have halogens incorporated require a resetting of the calibration which is a simple procedure.

Figure 6 shows a block diagram of the GC-MS computer interface. A 14-bit 90 kHz AD converter is used for sampling of the analogue signal from

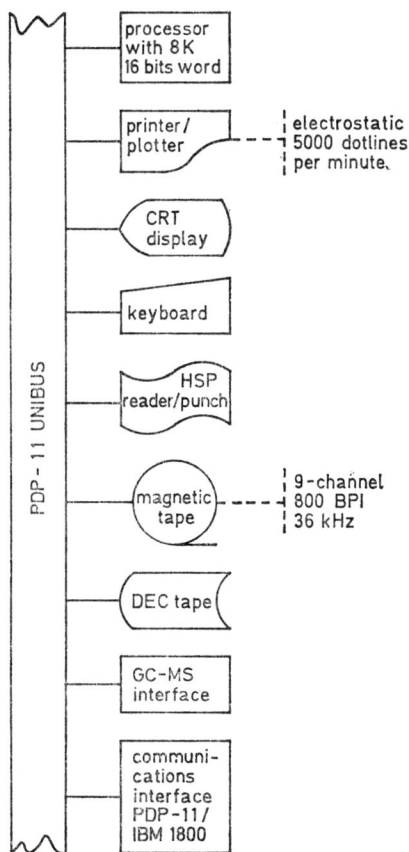

FIG. 7   The PDP-11 computer configuration (only peripheral equipment used for the GC-MS dataprogram shown).

the mass spectrometer. In our work the PDP-11 with an 8 K core memory can accept a sampling rate of about 10 kHz. To make use of the 90 kHz sampling rate special hardware is built into the interface where each ten values from the AD converter are added together and read in two words into the computer or the added sum can be divided so it can be read into the computer as one word and this will increase the accuracy of the ion intensity. By using the mass marker as a trigger for each mass number the scanning rate can be as fast as 0·5 sec for one mass range m/e 10–500.

A 12-bit D/A-converter is used for switching the accelerating voltage from one peak to another and a 4-bit D/A-converter for sweeping the voltage over the peak maximum. The on-line AVA system has some advantages over the off-line in that more peaks can be studied, and when small and highly intense peaks are studied simultaneously it is possible to increase the registration time for the smallest peaks.

The sample from the direct probe is controlled by the computer, and the heater is switched off when an acceptable ion intensity is obtained. The magnetic field will also be controlled by the computer, but the interface is not yet ready. The mass marking will in this case be used as a reference.

Figure 7 shows the PDP-11 computer configuration where only peripheral equipment used for the GC-MS dataprogram are viewed. After the switches of the data register are set, the punched tape read and the start switched on, an asterisk is shown on the screen which means that the program is ready to command. If for instance R, S or T is given several questions are automatically asked. If for instance L is set a list of commands is displayed. A list of the commands is shown in Table II.

TABLE II

A List of the Computer Commands

*	Restart monitor	A	Automatic scan start
B	Background	C	Column
D	Date	E	Each scan out
F	Fresh up display	G	Galvanometer
H	Hard copy output	I	Base peak mass
J	Joy-stick on	L	List commands
M	Mass range	N	Name
O	Output latest scan	P	Position on tape
R	Read from tape	S	Scanning of MS
T	Total or single ions	V	Min. voltage to accept

The question HARD COPY OUTPUT can be answered with C, D, G, I, N, P or R. All of these commands can be given simultaneously except G, C and R.

C represents two mass spectra being shown simultaneously on the screen. D, absolute intensity in decimal units. G, normalized mass spectrum on the screen. I, normalized mass units in absolute intensities. N, nominal mass units with two decimal intensity. P, plotting of normalized spectrum. R, store spectra on DEC-tape. For this command the joy-stick can be set at any position on the GC-curve of the total ion current to select a spectrum to be stored on DEC-tape. The joy-stick can also be used to subtract one spectrum from another. Depending upon how fast and accurate the results from the combined GC-MS will be studied, various parameters can be chosen:

(1) Single scan can be taken whenever a gas chromatographic peak is encountered and from which spectrum a specified background is subtracted. The mass spectrum is normalized, printed and plotted in about 5 seconds for interpretation.

(2) Continuous scanning of the mass spectrometer every other second during the entire gas chromatogram if the spectrum is stored only on magnetic tape and from every other scan a normalized printed or plotted mass spectrum for interpretation can be immediately obtained.

(3) Plot of the intensity of specified ions during the entire gas chromatogram and plot of the sum of all intensities within a spectrum (computer-generated gas chromatogram). An approximate amount of each component is calculated.

(4) Mass spectrum and operating information can be displayed for every other scan for continuous checking of the registered mass spectra.

A comparison of increment and continuous mode of operation using repetitive scanning of GC-components from a sample of urine extract of silylated methaqualone metabolites has been made. A scanning time of one second was used for the mass range m/e 10–500.

GC-MS applications of complex mixtures where packed columns are used often contain unseparated components for which mass spectra are difficult to identify by computer search. The use of capillary columns will aid in increasing the separation and thereby give a higher purity for its mass spectra which should be easier to identify. In order to obtain good results for GC-components from capillary columns it is necessary that a mass range of m/e 10–400 can be scanned in about 0·5 sec.

In registration of isothermal analysis it is possible to set the starting time for the beginning of registration of mass spectra as well as initial interval time between each scan. The GC-peaks are sharper at the beginning of the run, and to reduce the number of scans for the components which have longer retention time the time interval between the scan can be increased by a factor of 0·6 sec/min, 1·2 sec/min or 1·8 sec/min or by other chosen factors.

Maintaining a constant magnetic field and using alternating accelerating voltage (AVA) the mass spectrometer functions as a multiple ion detector

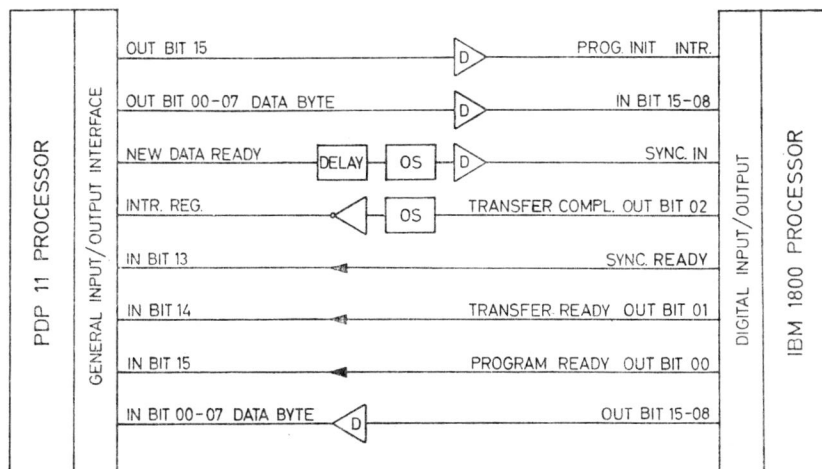

FIG. 8   Communications interface PDP-11/IBM 1800.

for studying characteristic ions in a mass spectrum. This technique is used in the identification and quantitative determination of compounds of small quantity. It is also used for isotope determination.

The identification of a mass spectrum of an unknown compound is normally made by comparison with reference spectra. This is done manually or by computer search. The computer identification procedure can be successful if mass spectra of pure compounds are available. In our laboratory library searching of mass spectra has been done in different ways. Since we have an IBM 1800 computer available and about 12 000 known spectra in the reference library the PDP-11 is used only for searching among about 400 drugs and drug metabolites. A communications interface between PDP-11 and IBM 1800 makes it possible to rapidly transfer the unknown mass spectra to the larger computer for identification, Fig. 8. There is, however, a weakness in the GC-MS analysis in that the retention times of each GC-component cannot from time to time be given with satisfactory accuracy. A library search of mass spectra where also the retention time is given should increase the accuracy of the searching methods.

122

Logos-II: A Large-scale, Real-time Computer System for Multiple-instrument Mass Spectrometry, Including Low and High Resolution, GC-MS, and Spectrum Management Applications†

By A. L. BURLINGAME, R. W. OLSEN
and R. McPHERRON

(*Space Sciences Laboratory, University of California,
Berkeley, California, U.S.A.*)

INTRODUCTION

In recent years, a variety of laboratory computer systems have been developed for applications in mass spectrometry.[1,2] Although the majority of these systems are small and consequently limited in their capabilities, they have been impressively successful. In a few cases, large-scale, time-sharing systems have been applied to the automation of entire laboratories.[3] These time-sharing systems provide a diverse users group with relatively unspecialized facilities for data acquisition and computing. These systems have a potential for on-line mass spectrometry far in excess of smaller, simpler systems. This is due to their ability to support multiple users and to execute many programs concurrently, to their elaborate file structures for the storage and retrieval of data, and to the large amounts of core and disc storage typically associated with such systems. This potential has not been fully realized in any existing system, however, because the commercial time-sharing systems used lack the necessary speed and flexibility.

Logos-II

This paper introduces Logos-II, a time-sharing computer system currently being developed in our laboratory explicitly for on-line mass spectrometry. Logos-II is an on-line, real-time, data acquisition and processing system consisting of a large Xerox Sigma-7 computer, a time-sharing supervisor program, and a collection of processing programs that provide specific control, display, acquisition, and processing services.

† A detailed technical report is available from the authors upon request.

Logos-II differs significantly from existing time-sharing laboratory computer systems in many of its features. The principal difference, however, is in terms of its organizational possibilities. Logos-II is designed so that its acquisition, processing, and control functions can be separated into independent, concurrently executed programs. This kind of organization requires that concurrently executed programs be able to synchronize and communicate, and that a single user be able to execute more than one program at a given time. Unlike Logos-II, most time-sharing systems lack these features and so cannot be organized as above in any simple way.

Because Logos-II is organized in terms of functionally independent, concurrently executed programs, its users can be made independent of the computational processes they initiate. Logos-II's users are never irrevocably locked to a computational process, nor are they constrained to real-time. Users can initiate a real-time process and then monitor or ignore it as they see fit.

In Logos-II, users do not usually interface directly with processing programs (except in the case where the processing program is interactive). Instead, they interface with a control program through which they can select, initialize and activate the processing programs of their choice. Typically, a user will activate a sequence of processing programs to acquire and process mass spectra. Once activated, the processing programs in such a sequence are independent of the control program and the user; hence he is free to do something else without concern for disturbing the on-going, real-time process. Usually, however, a user will elect to monitor the process by having the plots or listings generated by the processing programs displayed on his CRT terminal.

Displaying output data does not compromise a user's independence or constrain him to real-time. This is because processing programs do not send output directly to a user's CRT. Instead, they generate plots and listings in the form of files in disc storage. These files can be displayed as they are generated or at any time thereafter through an interactive processing program specifically intended for display purposes.

In Logos-II, processing programs communicate through one-way message channels. Each such channel is a queue† from which a particular processing program gets its processing instructions. The messages sent to a processing program's queue are referred to as work requests. Each specifies data to be processed and the applicable parameters. Work requests originate either in the control program (when a processing program is activated) or in another processing program. As each processing program in a sequence completes its processing, it sends a work request to the next program and then gets the next work request from its own queue. In the event that its queue is empty, the processing program suspends its execution until it receives another work request.

In addition to their communication function, Logos-II's queues serve to synchronize processing programs and to uncouple them from real-time. A

---

† A queue is a first-in, first-out waiting list that grows or shrinks depending upon the relative rates at which items are inserted or removed. Hence, a queue is a kind of buffer.

processing program is synchronized with its input by the simple expedient of suspending its execution when its queue is empty. Since its queue buffers work requests, a processing program is not required to synchronize with its input in a strict sense; that is, it need not keep up with its input except on the average. Hence, processing programs are effectively uncoupled from strict adherence to real-time.

The synchronization and communication scheme outlined above is Logos-II's basic framework. Logos-II's flexibility derives in large part from this scheme and from its independent control program. Together, these features uncouple users and programs from each other and from real-time and in so doing greatly enhance Logos-II's organizational possibilities.

Goals

In undertaking the development of Logos-II, our goal was to develop a system having a multiple instrument data acquisition and real-time processing capability as well as having extensive facilities for the working scientist to access and process mass spectra. Specifically, our first goal was to achieve complete overlap between the acquisition and processing of mass spectra from at least four high or low resolution mass spectrometers in either batch or GC-MS modes. Our second goal was to develop an on-line data base and programs for processing and displaying selected mass spectra. The data base was to consist of a large collection of individual mass spectra with an internal organization determined by the individual scientist, and a variety of spectrum matching libraries organized for efficient search and retrieval. The processing and display programs were to provide for the generation of the usual variety of plots and listings. Finally, Logos-II was to provide a framework within which we could develop interactive processing programs for assistance in the analysis of mass spectra.

HARDWARE

As might reasonably be expected from the foregoing, Logos-II's hardware requirements are non-trivial. As mentioned earlier, Logos-II is based upon a large Xerox Sigma-7 computer. Figure 1 represents the system's hardware configuration.

Memory and Disc Storage

The large core memory and the large disc pack storage unit shown in Fig. 1 are Logos-II's most important resources. In fact, core memory and disc storage in large amounts are fundamental to Logos-II's capabilities.

Logos-II achieves speed through extensive usage of core memory. Its supervisor program uses substantial amounts for time-sharing tables, queues, and for data base indexes. Processing programs also require substantial amounts. While these are usually small, they are continuously resident in memory when active; that is, Logos-II does not swap active processing programs between memory and disc storage. Hence, all simultaneously active processing programs are in memory at the same time.

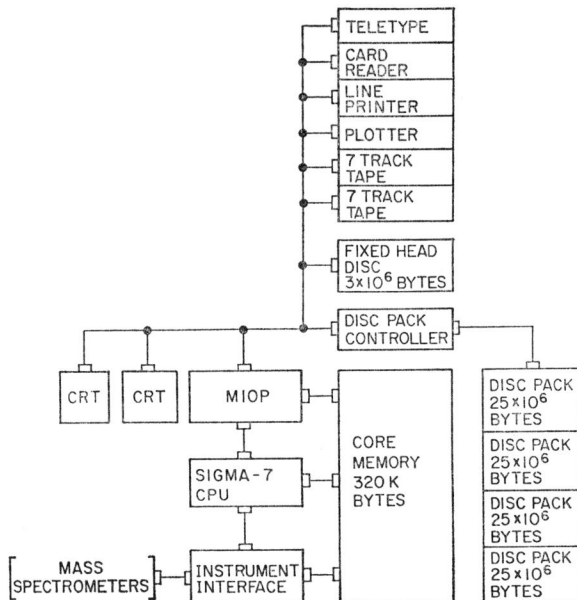

FIG. 1   Logos-II hardware configuration.

The above memory usage accounts for about half of Logos-II's total usage; the remainder is used for buffering data to and from disc storage and for buffering data from the mass spectrometers. Logos-II is a disc-based system: its data normally reside in disc storage and are in memory only when they are being processed. To prevent intolerable delays due to disc access time, Logos-II moves data between memory and disc storage in large blocks (2048 bytes). Each of these blocks must be retained in memory as it is built up or used and so each requires a buffer. Because Logos-II is a multi-user, multi-instrument system, the sum of the memory required for all the buffers in use at any given time can be very large.

CRT Terminals

The CRT terminals (Tektronix type 4010; 1200 characters/sec, 150 vectors/sec) shown in Fig. 1 provide each user with an input keyboard and a high speed alpha-numeric and graphic output display. These are not luxuries; a CRT terminal's speed is a necessity for real-time mass spectrometry and extremely desirable for on-line mass spectrometry in general.

Logos-II enjoys an advantage over most time-sharing systems in that its CRT terminals are connected directly to its input-output processor. Because of this direct connection, Logos-II can output data to its CRT's much faster than would be possible otherwise.

Interrupts

Although not shown in Fig. 1, the Sigma-7 is equipped with an interrupt system and a memory map. The interrupt system is used for synchronizing, timing and signalling purposes; it includes the usual I/O, real-time clock, and external interrupts. While most of the interrupt usage is internal to the supervisor program, one of the external interrupts is connected to the instrument interface, and one to a push button on each CRT terminal. The instrument interface interrupt signals that a data buffer is full, while the terminal interrupt signals that a user requires attention.

Memory Map

The memory map is a programmable, hardware address, translation table that determines the correspondence between program and memory addresses. This correspondence is in terms of a division of program address space and memory into 256 contiguous 2048 byte segments called pages. Through an appropriate setup of the memory map's translation table, any page in address space can be mapped into any page in memory. This means that specific, contiguous regions in address space can be made to correspond to arbitrary, non-contiguous regions in memory. Hence, programs can be segmented into pages and loaded into whatever memory space is available. This relieves Logos-II from dependence upon specific, contiguous memory regions and results in a more efficient management of the available space.

Instrument Interface

The instrument interface shown in Fig. 1 automatically extracts peak profiles from the detector signals of one or more mass spectrometers. This extraction is strictly a hardware function; the interface stores peak profile data directly into a core memory buffer designated for the particular instrument and generates an interrupt when it is full. Hence, once the interface is started, CPU intervention is required only to unload the filled data buffers to disc storage.

The instrument interface is equipped with sixteen input channels and is capable of a combined input sampling rate of 100-kHz samples/sec. The number of mass spectrometers which can be simultaneously on line is determined by the sampling rate required by each instrument. In our laboratory, this is usually 25-kHz samples/sec/instrument.

Instrument Buffering

Although the instrument interface is the key to Logos-II's multiple instrument capability and to its ability to overlap acquisition and processing, there is an extra price paid for these capabilities in terms of the memory required for buffering. This is relatively modest if the peak profiles are converted to centres of gravity before being written to disc storage. This is because the compression which results from the conversion reduces the effective data input rate to the point where disc access time is not a limiting factor. If, on the other hand, peak profiles must be saved, the memory requirements can be substantially greater. This is because the instantaneous buffer filling rate can exceed the average 10 buffers/sec unloading rate determined by the disc access time. This rate difference must, of course, be made up by allowing full buffers to accumulate in memory until the excess input rate is averaged out.

SOFTWARE

Logos-II's software is too extensive to be described within the confines of this paper. What follows is necessarily the briefest possible outline of what is involved.

Supervisor Program

Figure 2 summarizes Logos-II's supervisor program in terms of its component sub-programs. The sub-programs in the control group provide Logos-II's basic control and interface machinery. The scheduler shares CPU time between

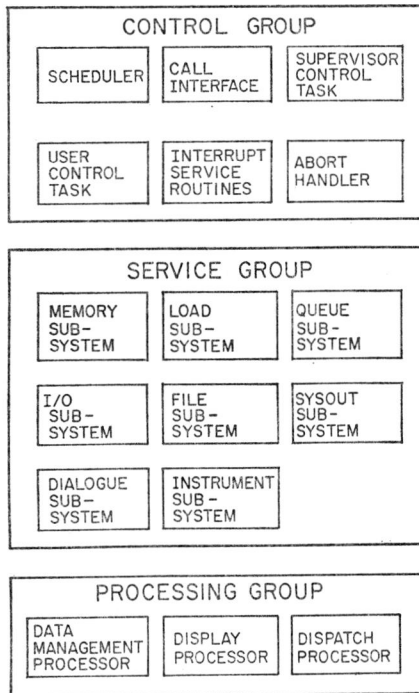

FIG. 2   Logos-II supervisor program.

concurrently executed programs. The user control task is the control program referred to earlier. The call interface is the linkage mechanism between processing programs and the supervisor.

The sub-programs in the service group manage Logos-II's hardware and provide services to the processing programs and to the supervisor itself. The memory and *I/O* sub-systems manage memory and the *I/O* devices and allocate memory and disc storage. The file sub-system organizes disc storage in terms of files and maintains directories through which particular data can be referenced and retrieved. In essence, it is Logos-II's data base.

The queue and sysout sub-systems manage the previously mentioned queues and displayable output files. The load and dialogue sub-systems provide the user control program with means for loading and releasing processing programs from memory and for generating parameter lists from user dialogues. The instrument sub-system manages the instrument interface and buffers mass spectrometer data to Logos-II's acquisition programs.

The sub-programs in the processing group are interactive processing programs built into the supervisor for reasons of efficiency. The data management processor provides users with means for examining Logos-II's data base indexes and for deleting or copying data base files. The display processor displays plot and listing output files generated by other processing programs. The dispatch processor defines and controls sequences of processing programs.

The Basic Real-Time Mass Spectrometry System

The basic real-time mass spectrometry system consists of three processing programs: acquisition, reduction and display. The acquisition program receives peak profile data from the supervisor's instrument sub-system through a buffer queue provided for each on-line instrument. The acquisition program generates a file of time/intensity data for each instrument scan and, optionally, a file of peak profile data. At the end of each scan, the acquisition program generates a display file summarizing scan statistics, and then sends a work request to the reduction program.

The reduction program generates a file of mass/intensity data from the time/intensity data specified in the work request received from the acquisition program. For GC-MS purposes, the reduction program can also build a file (from scan to scan) of total ionization and selected mass chromatogram data. At the end of each scan's processing, the reduction program sends a work request to the display processor.

The display processor generates a displayable low resolution plot output file from the data specified in the work request received from the reduction program and, optionally, a calibration summary listing.

The display processor can also generate total ionization and mass chromatogram plots from the GC-MS file generated by the reduction program. These plots are generated in response to work requests received from the user through the control program.

ACKNOWLEDGMENTS

The authors wish to thank Robert Couse for his contribution to the hardware development and James Holsworth for his contribution to the software development. The financial support of the National Aeronautics and Space Administration (Grants NSG 243, Suppl. 5 and NGR 05-003-435; Contract NAS 9-7889) and the National Institutes of Health (Grant NIH-RR-719-01) is gratefully acknowledged.

REFERENCES

1. Ward, S. D., in 'Mass Spectrometry Vol. 2', Williams, D. H., Senior Reporter, The Chemical Society, Burlington House, London, 1973, chapter 6.
2. Burlingame, A. L. and Johanson, Gary A., *Anal. Chem. Ann. Reviews*, 1972, **44**, 337R.
3. Ziegler, E., Henneberg, D. and Schomberg, G., *Anal. Chem.*, 1970, **42**, 51A.

123

The Detection and Classification of Abnormal Atmospheric Pollutants Using a Quadrupole Mass Spectrometer and Cluster Analysis

By J. HARDY and I. JARDINE

(*Department of Chemistry, The University of Glasgow, Glasgow, U.K.*)

INTRODUCTION

THE atmosphere is normally polluted by aromatic and other hydrocarbons.[1] Such normal pollutants together with the normal constituents of the atmosphere give rise to typical background spectra, one of which is shown in Fig. 1.

The atmospheric pollutants in which we are interested are not normally encountered but are released deliberately or accidentally. As these pollutants are possibly toxic, detection and identification or at least classification must be completed in the shortest possible time.

Any method used must:

(a) detect the pollutants at low concentration
(b) identify or classify accurately
(c) hardware should be compact and portable
(d) cost should be low as a number of detectors may be required to cover an area.

Because of these restrictions we have considered the use of a single-stage membrane-inlet coupled to a Quadrupole Mass Spectrometer, data acquisition system and small computer. In this configuration the system is relatively light and can be made portable.

HARDWARE

The present system is shown in block form in Fig. 2. The level of detection which we hope to attain is $1:10^9$; however present detection sensitivities, with a clean system, are $1:10^7$. It has been demonstrated by Collins and Uteley that a single-stage membrane permits the enrichment of the high mass components by three orders of magnitude relative to nitrogen. One side of the

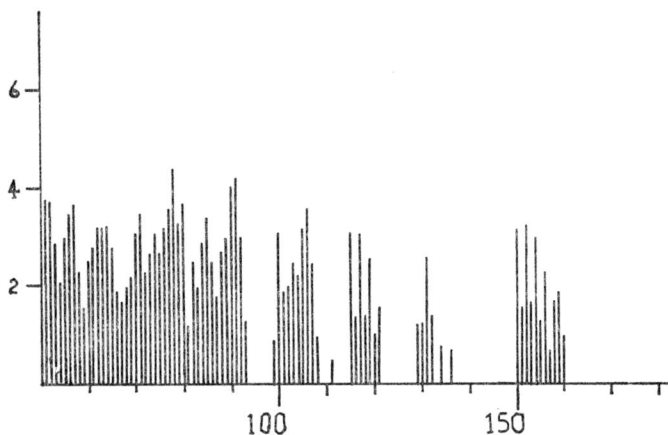

FIG. 1   A typical background spectrum.

membrane is exposed to the atmosphere whilst the other side is exposed to the Quadrupole, coupling taking place through a capillary tube the system responds rapidly to changes in the composition of the atmosphere (even at room temperature). A disadvantage of the single-stage separator is that the low mass region of the spectrum is obscured by the major components of the atmosphere. The use of a two-stage separator will reduce the major atmospheric components but the required additional pumping is complex.

The spectrometer draws in a continuous sample and a spectrum is recorded, digitally, at fixed time intervals. Where an abnormal pollutant is introduced, or a change in one of the background components takes place there will be a

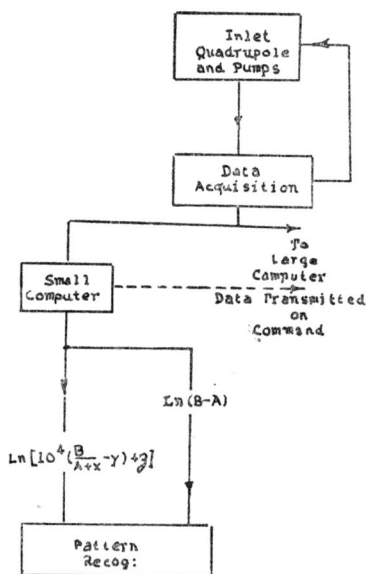

FIG. 2   Block diagram of system.

change in the spectrum. Thus if successive spectra are compared any difference outwith normal statistical variation indicates a change. Initially we subtracted the first spectrum from successive spectra until a change was observed. The difference between the two spectra was then compared to a library. Recently we have used the formula

$$\ln \left[10^4 \left(\frac{B}{A + x} - y \right) + z \right]$$

where B is the value of the peaks appearing in subsequent spectra, A is the same peak in the initial spectrum, x is a small number comparable with the smallest peak to be expected such that $B/(A + x)$ has a value greater than y when $A = 0$; y is used for statistical correction or as in our initial experiments has a value approaching 1. z has a small value so that the expression can be evaluated.

A typical spectrum where the pollutant has a concentration of $5:10^9$ is shown in Fig. 3. The recovered spectra using the two methods outlined are shown in Fig. 4 (subtraction) and Fig. 5 (division). It can be seen that a greater detail is present in Fig. 5. Using the division method the complete positional spectra can be recovered for concentrations one order of magnitude less than using the subtraction method.

Once the spectrum has been recovered identification is carried out in three stages:

(a) binary comparison using a limited library
(b) direct comparison using a larger library with additional abundance information.
(c) cluster theory.

Should a spectrum be identified in step (a) no external action is taken, should it not be identified the full spectrum available at that time is transferred to a large computer to be identified using steps (b) and/or (c).

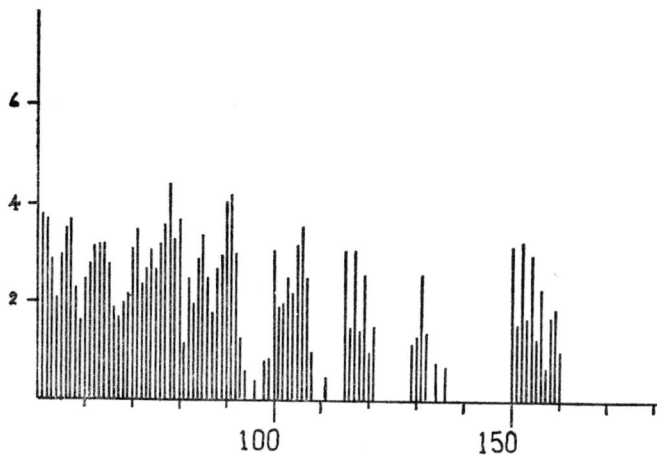

FIG. 3   A spectrum with $5:10^9$ of pollutant present.

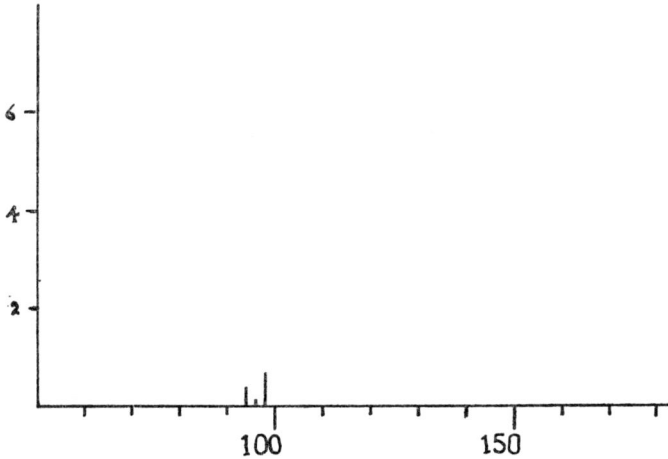

FIG. 4 Recovery of pollutant spectrum by subtraction.

Considering system (a) we have slightly modified Grotch's Criterion for best match.[2,3,4] by making N the number of peaks in the incoming spectrum's classification.

$$c = \mu N + \Sigma [(XOR) - \mu(AND)]$$

where c = the matching criterion

$$\mu = 2 \cdot 3 \text{ (approximately)}$$

N = the number of group/peaks in the recovered spectra. When there are two zeros c is not affected, a disagreement causes c to increment by one whilst an agreement causes c to decrease by 2·3. The spectra being coded are divided into groups of 14 a.m.u. Where no peak appears in the group it is designated by the 4 bit binary number 0000. Otherwise the largest peak is

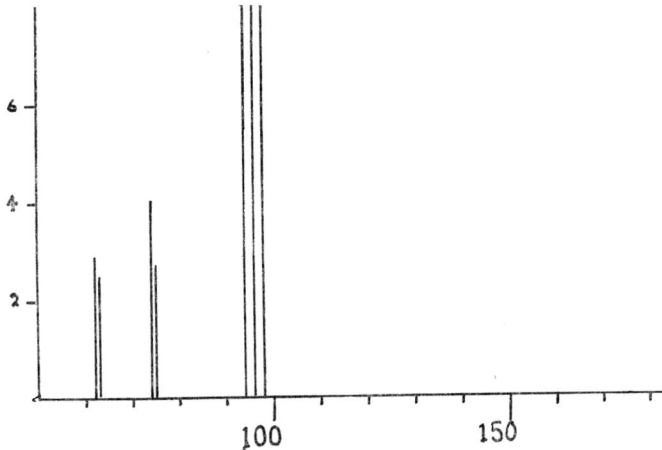

FIG. 5 Recovery of pollutant spectrum by division.

coded by its position within the group: for example if Mass 105 were the largest peak within the group 99 to 112, *i.e.* the 7th peak in the group it would be coded 0111. The group 1111 signifies the end of the spectrum. Examples of the coding are:

1 methyl-4-ethylbenzene 1001 0111 0111 0111 0111 1111
1 methyl-4-n-propylbenzene 1001 0111 0111 0111 0011 1000 1111

Any peaks below mass 56 have been excluded at this stage due to the expected masking by the large normal air components. It is expected that these groups will in the future be able to be considered. Peaks below 1 % of full scale are also disregarded.

Let us take for example the emergence of the spectrum of 1 methyl-4-n-propylbenzene. The peaks will appear in the order 105, 134, 77, etc. as the concentration is increased. The emergence of 105 activates the search programme 'yanking' all library spectra containing 105 for search. The appearance of peaks at 134 and 77 would narrow the search to spectra of the form

xxxx 0111 xxxx 0111 xxxx 1000

Initially when the spectrum is in the form:

0000 0000 0000 0111 0000 0000

if one or more of the library spectra containing this group were toxic, or the group did not exist in the library, an alarm would be activated and the complete spectrum would be transmitted for more detailed analysis. With the emergence of 134 the spectrum will take the form:

0000 0000 0000 0111 0000 1000

If this grouping is contained within one of the toxics the alarm condition remains, should the field be narrowed to the non-toxics the alarm condition and data transmission ceases. The reverse condition cannot occur.

CLUSTER ANALYSIS

Mass spectra are recorded and tabulated as a data matrix of Mass Spectral Characteristics for each spectrum. The choice of characters is very important. The first choice is usually the intensity of the molecular ion for each spectrum. Then follows the intensities of every mass number for each spectrum less the mass numbers where no spectrum makes a significant contribution. The remaining characteristics are the intensities of the ions formed by the loss of fragments from the molecular ion for each spectrum.

By computing the mean and standard deviations of each character a Data Matrix with Standardised Characters is formed from this Matrix, correlation coefficients are calculated for all pairs of spectra. A cluster analysis algorithm now operates where according to the correlation coefficient spectra are grouped according to their similarity.

The results of this analysis are presented as a Dendogram where similar spectra are grouped with a measure of similarity given as a correlation coefficient.

TABLE I
Normal Atmospheric Pollutants

No.	Cat. No.	Mol. Wt.	Name	Semi-structural formula
1	1 612	78	Benzene	(benzene ring)
2	3*	92	Toluene	(ring)—CH₃
3	177	106	Ethylbenzene	(ring)—CH₂CH₃
4	180	106	1,4-Dimethylbenzene	CH₃—(ring)—CH₃
5	179	106	1,3-Dimethylbenzene	CH₃—(ring)—CH₃
6	178	106	1,2-Dimethylbenzene	CH₃—(ring)—CH₃
7	256	120	n-Propylbenzene	(ring)—CH₂CH₂CH₃
8	260	120	1-Ethyl-4-methylbenzene	CH₃—(ring)—CH₂CH₃
9	259	120	1-Ethyl-3-methylbenzene	CH₃—(ring)—CH₂CH₃
10	263	120	1,3,5-Trimethylbenzene	CH₃—(ring)(—CH₃)—CH₃
11	258	120	1-Ethyl-2-methylbenzene	CH₃—(ring)—CH₂CH₃
12	262	120	1,2,4-Trimethylbenzene	CH₃—(ring)(—CH₃)—CH₃
13	462	134	n-Isopropyl-4-methylbenzene	CH₃—(ring)—CH(CH₃)₂
14	494	134	n-Butylbenzene	(ring)—(CH₂)₃CH₃
15	1 431	134	Propylmethylbenzene	CH₃—(ring)(—CH₂CH₂CH₃)

TABLE I—cont'd
Normal Atmospheric Pollutants

No.	Cat. No.	Mol. wt	Name	Semi-structural formula
16	460	134	*sec*-Butylbenzene	—CH(CH₃)CH₂CH₃
17	261	120	1,2,3-Trimethylbenzene	—CH₃, —CH₃, —CH₃
18	264	134	Ethyldimethylbenzene	—CH₃, —CH₃, —CH₂CH₃
19	1 432	134	Propylmethylbenzene	CH₃——CH₃CH₂CH₃
20	853	134	Isopropylmethylbenzene	CH₃——CH(CH₃)₂
21	320	134	1,2,4,5-Tetramethylbenzene	CH₃, CH₃, CH₃, CH₃
22	463	134	1,2,3,5-Tetramethylbenzene	CH₃, CH₃, CH₃, CH₃
23	1 957	134	1,2,3,4-Tetramethylbenzene	CH₃, CH₃, CH₃, CH₃
24	705	282	*n*-Eicosane	CH₃(CH₂)₁₈CH₃
25	*61	166	Fluorene	
26	633	168	Dibenzofuran	
27	1 436	154	Acenaphthene	
28	906	168	Methyldiphenyl	—CH₂
29	1 007	254	*n*-Octadecane	CH₃(CH₂)₁₆CH₃
30	613	154	Diphenyl	

TABLE I—cont'd
Normal Atmospheric Pollutants

No.	Cat. No.	Mol. wt	Name	Semi-structural formula
31	1 006	240	n-Heptadecane	$CH_3(CH_2)_{15}CH_3$
32	1 759	135	Benzothiazole	
33	1 005	226	n-Hexadecane	$CH_3(CH_2)_{14}CH_3$
34	1 004	212	n-Pentadecane	$CH_3(CH_2)_{13}CH_3$
35	410	128	Naphthalene	
36	894	142	2-Methylnaphthalene	
37	881	186	1-Dodecanol	$CH_3(CH_2)_{10}CH_2OH$
38	880	158	1-Decanol	$CH_3(CH_2)_8CH_2OH$
39	87*	120	Acetophenone	$-COCH_3$
40	1 003	198	n-Tetradecane	$CH_3(CH_2)_{12}CH_3$
41	704	198	7-Methyltridecane	$CH_3(CH_2)_5CH(CH_3)(CH_2)_5CH_3$
42	1 103	132	Methylindan	CH_3
43	86*	118	Benzofuran	
44	83	106	Benzaldehyde	$-CHO$
45	**	147	Dichlorobenzene	$Cl-$ $-Cl$
46	404	170	n-Dodecane	$CH_3(CH_2)_{10}CH_3$

TABLE I—cont'd

Normal Atmospheric Pollutants

No.	Cat. No.	Mol. wt	Name	Semi-structural formula
47	523	184	n-Tridecane	$CH_3(CH_2)_{11}CH_3$
48	403	156	n-Undecane	$CH_3(CH_2)_9CH_3$
49	840	156	Isoundecane	$(CH_3)_2CH(CH_2)_7CH_3$
50	109	142	n-Decane	$CH_3(CH_2)_8CH_3$
51	484	140	Isodecene	$(CH_3)_2CHCH=CH(CH_2)_4CH_3$
52	89	99	Dichloroethane	$ClCH_2CH_2Cl$
53	479	142	Isodecane	$(CH_3)_2CH(CH_2)_6CH_3$
54	132	128	n-Nonane	$CH_3(CH_2)_7CH_3$
55	245	128	Isononane	$(CH_3)_2CH(CH_2)_5CH_3$
56	39	114	n-Octane	$CH_3(CH_2)_6CH_3$
57	40	114	2-Methylheptane	$(CH_3)_2CH(CH_2)_4CH_3$
58	14	100	n-Heptane	$CH_3(CH_2)_5CH_3$
59	15	100	2-Methylhexane	$(CH_3)_2CH(CH_2)_3CH_3$
60	990	142	1-Methylnaphthalene	

All Cat. Nos (Catalogue Numbers) refer to the A.P.I. Collection except * which refers to the M.C.A.R.P. Collection and ** which refers to an Uncertified Spectrum.

60 'ZURICH' COMPOUNDS, 37 CHARACTERS,   WARD'S METHOD.

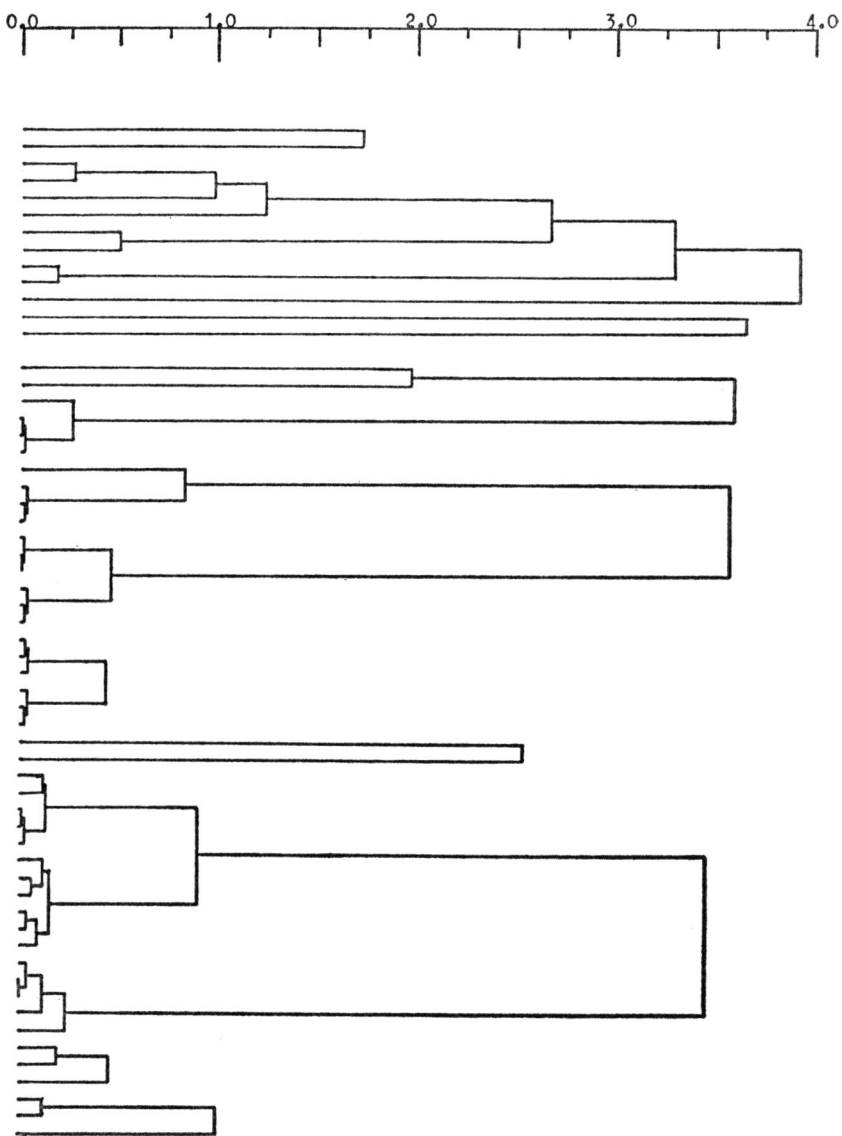

Fig. 6   Correlation of compounds in Table I.

The mass spectra of sixty compounds normally found in the atmosphere were obtained from catalogued data. The compounds are listed in Table I. The compounds were listed according to the square root of the intensities at each m/e value. The characters of the spectra, intensity of the molecular ion and fragment ions M^+-1 to M^+-59 were included. These characters were scanned and an abbreviated list of 37 were retained. Clustering was performed using Ward's method. The results are shown in Fig. 6.

CONCLUSION

A workable system has been demonstrated, further development is now being carried out to improve the sensitivity of the mass spectrometer and the inlet system.

REFERENCES

1. Grob, K. and Grob, G., *J. Chromatogr.*, 1971, **62**, 1–13.
2. Grotch, S. L., *Anal. Chem.*, 1970, **42**, 11.
3. Grotch, S. L., *Anal. Chem.*, 1971, **43**, 11.
4. Grotch, S. L., *Anal. Chem.*, 1973, **45**, 1.

Discussion

R. C. Lao (Air Pollution Control Directorate, Canada): How did you sample the air and what was the air flow-rate? Also what kind of membrane was used in your system?

J. Hardy: The air sampling system consists of a 15 cm capillary tube of 1 mm bore which is connected to the source by a swage-lock fitting. The side exposed to the atmosphere is covered by a melanex membrane mounted on an electron microscope grid. The flow-rate is adjusted by a precision valve such that source pressure is 10^{-6} torr. When the valve is closed the source pressure is $5 \cdot 10^{-9}$ torr.

124
Developments of Data Analysis for Small Computers

By N. A. B. GRAY

(*Organic Geochemistry Unit, University of Bristol, Bristol, U.K.*)

In the Organic Geochemistry Unit, University of Bristol, we have been developing programs for the analysis of low resolution mass spectral data using a small laboratory computer. Our systems are being developed for a 12k PDP8e computer with DEC tapes and a video display; with interface to a mass spectrometer, such a system costs in the region £15 000–£20 000. We have two projects which might interest delegates.

Our conventional data analysis systems are based on file search methods. Instead of using a large file of reference compounds with each reference compound characterized by a set of intense ions, we use a set of files each containing characteristics of reference compounds with related chemical structures. The ions used to characterize a reference spectrum are picked to try to distinguish it from other similar structures and can be ions of relatively minor intensity in the complete spectrum of the reference compound.† We have two methods of automatically selecting the file containing appropriate reference spectra for comparison with a particular unknown. One method is based on the 'ion-series' spectra used to classify compound types by Smith;‡ this method is particularly suitable for compounds containing large alkyl chains or alicyclic rings. The other file selection method uses a 'profile' identifying constraints on the intensities of ions in the spectrum of the unknown compound.†† Thus, our file search methods work by mapping the unknown spectrum onto a specific search file and then checking for the features defined for the reference compound, *e.g.* parent peak, in the complete spectrum of the unknown.

---

† The ions used to characterize a reference compound can be chosen by the user; or there is a program which will read complete reference spectra and pick ten ions from the higher mass region.

‡ The spectra are classified using the 'ion-series' method of Smith (Anal. Chem. 1972). This classification gives a compound class, suggests the appropriate parent ion and provides the key which identifies the file of reference spectra comprising that ion-series class.

†† The profile consists of a set of statements about intensities of specific ions in the spectrum, *e.g.* 61 a.m.u. more intense than 60, or mass 85 exceeds 20%.

The less conventional data analysis method that we have developed for the laboratory computer attempts to take advantage of the availability of the large number of empirical rules for interpreting low resolution mass spectra. The presence of a particular intense key ion can suggest a particular skeletal class to the spectroscopist, *e.g.* 191 a.m.u. frequently characterizes triterpane type skeletons. The experienced spectroscopist usually proceeds then using the presence of related ions to distinguish increasingly specific subclasses and finally derive a suggested structure. We have an interactive programme (INTERACT) that permits the spectroscopist to specify features that characterize particular classes and subclasses of compounds and build up a decision network for identifying structural types using the normal empirical methods. A data table produced by this interactive programme characterizes the methods the spectroscopist has defined; this data table is stored on DEC tape for future use. This data table can then be used by a second programme (INTERPRETER) to suggest structures for spectra acquired in the course of a GC-MS run. Thus, specific mass spectral knowledge is kept strictly separate from the underlying programme. The spectroscopist never has to write any programme code to make the programme handle a new situation, it is given a new table constructed using the conversational interactive programme. We have used this programme successfully for identifying the components in a mixture of lactones, methyl and ethyl esters, dialkyl phthalates, etc., and in another example, to characterize the compounds present in a set of pesticide residues. Some of the advantages of this method are:

speed: the analysis could easily take place simultaneously with continued data acquisition.

scope: the programmes can suggest structures for compounds that are not present in the search files

and the fact that the analysis can be taken to the degree appropriate to the application, *e.g.* a compound may be identified as 'phosphate pesticide', or 'methyl thiophosphate', or most specifically as 'CYGON-267'.

Abbreviations

ADC	Analogue to digital converter	EDD	Electron distribution difference
AES	Auger electron spectroscopy	EDTA	Ethylenediamine tetra-acetic acid
AMS	Anion mass spectrometry	EED	Electron energy
AP	Appearance potential		distribution
API	American Petroleum Institute	EH	Electrohydrodynamic
		EI	Electron impact
ASTM	American Society for Testing and Materials	ESA	Electrostatic energy analyser
AVA	Accelerating voltage alternation	ESR	Electron spin resonance
		FD	Field desorption
BaP	Benzo (a) pyrene	FEM	Field emission
BeP	Benzo (e) pyrene		microscope
BkF	Benzo (k) fluoranthene	FI	Field ionization
CAS	Chemical Abstracts Service	FID	Field impulse desorption
		FIK	Field ionization kinetic
CE	Charge exchange	FIM	Field ion microscope
CI	Chemical ionization	FIMS	Field ion mass
CIMS	Chemical ionization mass spectrometry		spectrometry
		FI–MS	Field ionization mass
CMS	Cation mass spectrometry		spectrometry
		FWHM	Ion beam width at half
CRT	Cathode ray tube		maximum intensity
CSCM	Critical slope curve matching	GC	Gas chromatography
		GC–MS	Gas chromatography-
DAC	Digital to analogue converter		mass spectrometry
		GE	General Electric
DADI	Direct analysis daughter ions	GLC	Gas–liquid chromato-graphy
DCRT	Division of Computer Research and Technology	ICDR	Ion cyclotron double resonance
DCT	Dissociative charge transfer	ICR	Ion cyclotron resonance
		IE	Ionization efficiency
DNA	Deoxyribonucleic acid	IGR	Insect growth regulator
ED	Energy deposition	IKE	Ion kinetic energy

Ikes	Ion Kinetic Energy Spectroscopy	RDA	Retro Diels Alder
IMR	Ion-molecule reaction	RE	Recombination energy
INMS	Ionized neutral mass spectrometry	REE	Rare earth elements
IP	Ionization potential	REGN	Registry numbers
ISS	Ion scattering spectrometer	RPD	Retarding potential difference
JH	Juvenile hormones	RSF	Relative sensitivity factor
LC	Liquid chromatography	SAMS	Secondary atom mass spectrometry
LE	Linear extrapolation	SDIE	Second derivative of total ionization efficiency
MF	Match factor	SIM	Single ion monitoring
MID	Multiple ion detector	SIMS	Secondary ion mass spectrometry
MS	Mass spectrometry		
MSDC	Mass Spectrometry Data Centre	SSMS	Spark source mass spectrometry
MSS	Mass spectral search	STIRS	Self-training interpretive and retrieval system
NHI	National Health Institute		
NHLI	National Heart and Lung Institute	TE	Translational energy
		TEM	Trochoidal electron monochromator
NMR	Nuclear magnetic resonance	TFA	Trifluoroacetyl
PAH	Polyaromatic hydrocarbons	TIC	Total ion current
		TMA	Trimellitic anhydride
PCB	Polychlorinated biphenyls	TMS	Trimethylsilyl
		TOF	Time of flight
PE	Photoelectric spectroscopy	TRMS	Time resolved mass spectrometry
PES	Photoelectron spectroscopy	TVA	Thermal volatilization analysis
PFP	Pentafluoropropionyl	UHV	Ultra-high vacuum
PGA/B/E/F	Prostaglandins	UMPA	Universal microprobe analyser
PI	Photoionization		
PRO MIN	Panel programmable multiple ion monitor	UV	Ultraviolet
		VC	Vanishing current
QET	Quasi-equilibrium theory	VTMS	Vinyl trimethyl silane
RA	Relative abundance	WLN	Wiswesser line notation

Subject Index

Abundance, 26, 55–7, 69–75, 81, 85, 87, 329, 330, 521, 714, 718, 736, 737, 762, 849–53, 865–9, 888
Accelerating voltage alternation, 91, 245, 246
Accuracy, 649–54
Acetaldehyde, 246, 546
Acetamides, 42–4, 400
Acetates, 504
Acetic acid, 306, 307
Acetone, 117, 119, 123, 137, 308, 309
Acetonitrile, 279, 280
Acetophenone, 518, 521, 851
Acetoxyanthraquinones, 285, 286
Acetyl carbromal, 476
Acetylene, 759–65, 776, 777, 797–9, 885
Acetylsalicylic acid, 500–2
Activation energy, 871
Activity coefficient, 608
Adamantanes, 725–33
Adenosine, 518
Adipic acid, 241
Adsorbed atoms, 775
Adsorption, 109
  layers, 117
Afterglow technique, 797–803
Air pollution,
  see Atmospheric pollution
Alanine, 143, 146, 147
Alcoholism, 248
Alcohols, 305–7, 520, 522
Aldehydes, 307, 521, 522
Aldrin, 503
Alicyclic compounds, 47, 48
Aliphatic nitro compounds, 31
Alkadienals, 212

Alkaloids, 137–9
Alkanes, 520
Alkenes, 520, 522, 878
Alkoxide ions, 277–9
Alkoxy
  benzoic acid, 17
  group, 173, 176, 179
Alkyl
  benzoylbenzoates, 17
  dimethyltrimellitates, 17, 18, 20, 23
  esters, 17–23
  formates, 277
  furans, 212
  group migration, 31
  methylisophthalates, 17, 18, 22
  methylterephthalates, 17, 18, 22
  oximes, 173–9
  trimellitates, 17–23
  trimellitic esters, 17–23
  trimellitimidates, 17, 18
  trimethyldodecadienols, 251
Allobarbitone, 181–3
Allothreonine, 223, 224
Alloys, 114, 568, 569, 603–9, 667–9
Allylbarbituric acid, 181
Allylic cleavage, 164
Altosid, 251–6
Aluminium, 630, 631, 653, 666–9
  oxide, 579–85
Amides, 44, 520, 521
Amines, 246, 518, 520
Amino
  acid antagonists, 223–6
  acids, 124, 143–8, 187, 200, 201, 204, 223–6, 400, 499
  butyric acids, 144, 146, 147